# PROGRAMMING and CUSTOMIZING PICmicro® MICROCONTROLLERS

SCHAUMBURG TOWNSHIP DISTRICT LIBRARY
3 1257 01564 0922

1 BOOK
1 DISC

# PROGRAMMING and CUSTOMIZING PICmicro® MICROCONTROLLERS

**MYKE PREDKO**

McGraw-Hill

New York San Francisco Washington, D.C. Auckland Bogotá
Caracas Lisbon London Madrid Mexico City Milan
Montreal New Delhi San Juan Singapore
Sydney Tokyo Toronto

**Library of Congress Cataloging-in-Publication Data**

Predko, Myke.
Programming and customizing PICmicro® MCU microcontrollers / Myke Predko
p. cm.
ISBN 0-07-136172-3
1. Programmable controllers. I. Title.

TJ223.P76 P737 2002
629.8'9—dc21 00-051541

*McGraw-Hill*
*A Division of The **McGraw-Hill** Companies*



5 6 7 8 9 0 DOC/DOC 0 9 8 7 6 5 4 3 2

P/N 0-07-136173-1
PART OF
ISBN 0-07-136172-3

*The sponsoring editor for this book was Scott Grillo, the editing supervisor was John Baker, and the production supervisor was Sherri Souffrance.*

Printed and bound by R.R. Donnelley & Sons Company.

# CONTENTS

# ACKNOWLEDGMENTS

While my name is on the cover of this book, this edition (as well as the first) would not have been possible without the generous help of a multitude of people and companies. This book is immeasurably richer due to their efforts and suggestions. I have over one thousand emails between myself and various individuals consisting of suggestions and ideas for making this second edition better than the first—I hope I have been able to produce something that is truly useful.

The first "Thank you" goes to everyone on MIT's "PICList." The two thousand or so individuals that subscribe to this list server have made the PICmicro® MCU probably the best supported and most interesting chips available on the market today. While I could probably fill several pages of names listing everyone who has answered my questions and made suggestions on how this second edition could be better, I am going to refrain in fear that I will miss someone. Everyone on the list is to be commended for their generosity in helping others and providing support when the chips are (literally) down.

This book wouldn't have been possible except for the patience and enthusiasm of Scott Grillo for trying out new ideas. Scott was exceedingly understanding when I missed the deadline for the book and also was willing to take on the adventure of figuring out what was required to ship a book with both a CD-ROM *and* a PCB. What I appreciated most of all was the time he devoted to me listening to my (mostly hare-brained) ideas and helping me to come up with the book concept presented here.

Ben Wirz has been an invaluable resource on this book, helping me to define the tools (the "El Cheapo," "YAP-II" and "EMU-II") and their constituent parts to making sure that I don't specify components that can only be found between Spadina and University on Queen Street West in Toronto. Ben, I really appreciate your critiques of the table of contents and your suggestions on what the book needed to make it better for everyone. Next time we go to "Wonderland," it will just be the two of us.

Along with Ben, I would like to thank Don McKenzie, Kalle Pihlajasaari, Mick Gulovsen and Philippe Techer for your suggestions and ideas. A lot of the projects in this book wouldn't exist without your help, ideas or the SimmStick.

To Stephanie Lentz and everyone at the PRD Group—thank you for the hours spent on this book (and the pocketbook) despite its extreme size and complexity. I don't know how you managed to decode the chicken-scratches put in the margins of the Galley Proofs. The remaining errors are all mine even though you did your best to prevent me from making them.

Bruce Reynolds (of "Rentron") was an invaluable resource as well for this edition to help me work through the table of contents and make sure that all the information needed for new application developers was in the book. I really appreciate the ability to write articles for your web site as a precursor for much of the information presented here.

I am pleased to say that Microchip has been behind me every step of the way for this book project. Along with (early) part samples, tool references and information, I appreciate the fast response to questions and the help with making sure I had the correct information. A big thank you goes out to Len Chiella and Greg Anderson of the local (Toronto) Microchip offices as well as Al Lovrich, Kris Aman, Elizabeth Hancock and Eric Sells for the time spent on the phone, the many emails, graphics, parts and suggestions. I know that supporting authors is not in any of your job descriptions and I appreciate the time you were able to devote to me.

Along with the Microchip employees listed above, I would also like to thank Kris Van Herk for the time she spent with me and the manuscript making sure that I used the correct terms for Microchip products. Special thanks to Steve Sanghi, Microchip's CEO for taking time to review the manuscript and write the forword.

Along with the efforts of the Microchip employees, I would like to thank Dave Cochran of Pipe-Thompson Technologies who made sure that I always had everything I needed and all my questions were answered. Dave, I also appreciated the lunches at Grazie with you. Len, and Greg, not only did we agree on what should be in the book, but also on what to order.

Jeff Schmoyer of microEngineering Labs, Inc. was an excellent resource for me to understand how "PicBasic" worked and was always enthusiastic and helpful for all the questions that I had. PicBasic and the "EPIC" programmer are outstanding tools that I recommend to both new PICmicro® MCU developers and experienced application designers alike.

I learned more about compiler operation from Walter Banks of Bytecraft Limited in a few hours of telephone conversations than I did in my two senior years at university. While much of this information came after I had finished this book, the time spent allowed me to go back over the experiments and applications presented in this book with a much better eye toward making the code more efficient.

There are five other companies that I have grown to rely on an awful lot for creating books as well as doing my own home projects. I recognized two of these companies in the first edition and I felt I should include three others for their excellent service in the Toronto area.

Since writing the first edition of this book, Digi-Key has continued their excellent customer support and improved upon it with their web pages and overnight home delivery to Canada. AP Circuits are still the best quick turn PCB prototyping house in the business and I recommend that you use them for all your projects.

For each of my books, I have relied upon M & A Cameras and LightLabs here in Toronto for equipment rentals, photofinishing and advice. I realize that M & A also rent equipment to the professional photographers for the movie industry, but they have always taken the time to answer my questions and help me become a better photographer. LightLabs has always done their level best to ensure the poor pictures I have taken come out as clear and scanner ready as possible. I know I can still do a lot better, but both these companies have done a lot to hide my mistakes. Lastly, I want to thank the people at Supremetronic on Queen Street in Toronto for their unbelievably well stocked shelves of all the "little stuff" that I need for developing circuits and applications along with the time spent helping me find (and count) the parts that I have needed.

When I wrote "PC PhD," almost twenty percent of the book was written on airliners and I thanked the flight attendants for their time (and ice water) and villified the airline owners for stacking their business passengers like cord wood. For this book, an astounding forty or more percent of the book was written and proof-read on Air Canada flights and I would like to repeat the thanks to the attendants. I also appreciated the concerns of Air Canada's management for the one flight I had with them that really sets the record for how awful a flight from Denver to Toronto can be. Please keep the excellent staff flying and work at making more space for those of us that are of above average height (or even below average height).

I would like to thank everyone at Celestica (my regular employer) for helping to make the first edition of this book happen as well as recognizing that I needed to work on this edition even though the sky was falling. Celestica is an amazing company that is rich in the resources that really matter—the people that work there. I want to thank Karim Osman again in particular for testing the first versions of the "YAP" and helping me to come up with the RS-232 interface philosophies that I have used since the first edition of this book.

To my children, Joel, Elliot, and Marya; thank you for recognizing that the notes, parts and projects left lying around the house are not to be touched, and when I'm mumbling about strange things, I'm probably not listening to how your day went. The three of you would be absolutely perfect if you would just finish your homework before it was due. This book is something you should be proud of as well.

Marya, thank you for the hugs when it was painfully obvious I needed them.

Finally, the biggest "Thank You" has to go to my aptly named wife, Patience. Thank you for letting me spend all those hours in front of my PC, and then spending a similar number of hours helping me out by keying in the never ending pages of scrawl that was written in airport bars, hotel rooms and cramped airline seats. Thank you for putting up with the incessant FedEx, Purolator, and UPS couriers, and organizing the sale of the old house and designing the kitchen and bathroom of the new one. Writing something like this book is an unbelievably arduous task and it never would have been possible without your love and support.

Let's go and enjoy the new house,

myke predko<br>
Toronto, Canada<br>
April 2000

# FOREWORD

The Computer Revolution is at least 40 years old and shows no signs of slowing down, but with standardization of network protocols and spectacular advances in broadband technology the revolution has taken a new turn: Universal Connectivity. It's now possible to connect almost any device, no matter how small, to almost any other device anywhere in the world, provided suitable pipelines are available.

Of course Universal Connectivity means nothing without Distributed Intelligence, which enables cooperation in the execution of tasks as well as the sharing of information between networked devices. In multipurpose business and personal computers the core of Distributed Intelligence is the microprocessor, but in embedded (usually single-purpose) systems it is the microcontroller unit.

Microcontrollers are found in a multitude of applications in the automotive, consumer, communications, office automation and industrial control markets. For example, a modern car may have 50 or more microcontrollers controlling anti-lock brakes, keyless entry, air bags, burglary alarm system, sun roof, ignition, and other engine functions. Even a technologically clueless consumer now carries a cell phone and/or pager and is likely to have an answering machine along with a dozen home appliances that are controlled by microcontrollers. He may own a laptop computer with one or two microprocessors and several microcontrollers in it. On the other hand, his home is likely to have at least 30 and perhaps as many as 200 microcontrollers embedded in such applications as the washing machine, clothes dryer, furnace, air conditioner, security system, refrigerator, watches, sprinkler system, microwave oven, toys, toaster, hair dryer, radio, TV, VCR, calculator, electronic games, clocks, garage door openers, and smoke detectors.

It shouldn't be a surprise, then, that in 1999 the embedded market used approximately 5.2 billion microcontrollers a year, compared to 346 million microprocessors for PCs (according to industry analyst Dataquest)—or 15 times as many microcontrollers as microprocessors.

But this is just the beginning. At present many of the devices containing microcontrollers are isolated—that is, they are not connected to any other device nor to a central control system. This is particularly true of home appliances. But as Universal Connectivity makes Home Networking a ubiquitous reality, demand for microcontroller-controlled appliances will explode exponentially. The consumer will want to turn on her porch light and check the home security system from her car before she walks to the house in the dark. She will want the smoke detector to send a message to her office computer if it senses smoke, and she will want the security system to alert the police if it detects that a window has just been broken or a lock has been jimmied. She will want to be able to monitor the baby's room from the kitchen.

Other market segments also will experience explosive growth as new applications are developed, more sophisticated protocols are created and standardized, and pipelines are upgraded or installed where they presently do not exist.

Here is a detailed example of what an 8-bit microcontroller can do. In a hotel a single 8-bit microcontroller can control door access, temperature, and lighting, and detect smoke and motion for each room via a RS-485 network link to a host computer. The desk clerk can then assign a key to the room, turn on the courtesy light and set the system temperature when a person checks in. In addition, the microcontroller will allow the host computer to keep track of room access by housekeepers, repairpersons, and so on.

The microcontroller in the room control node has two functions. First, it has to continuously monitor the sensors and controls located in the hotel room; any room status changes are sent to the host computer. And it must monitor incoming control bytes from the network host computer. If the floor and room data match, the microcontroller must process the data bytes and execute the command or request for data. A PIC18C452 8-bit microcontroller has the bandwidth required to simultaneously handle communication on the network and perform the room functions.

The microcontroller has many peripherals that allow simple implementation of the room control node. For example, the universal synchronous/asynchronous receiver/transmitter (USART) on the PIC18C452 has an address-detect feature that makes it well suited for RS-485 applications. The A/D converter can be used for temperature and smoke sensor inputs. The two available pulse-width modulation modules can be used to generate analog voltage outputs for light dimmers. The on-chip master synchronous serial port module provides an easy interface to a serial EEPROM device. Finally, the device has plenty of general-purpose I/O for connecting keypads, displays and logic-activated devices. As onboard Flash memory grows in popularity, many companies are expanding their Flash microcontroller offerings.

The presence of the microcontroller in the room control node allows it to make decisions based on sensor input without intervention from the host computer. For example, the microcontroller can use motion-sensor, temperature-sensor and door-access information to provide automatic energy savings. It can automatically set the HVAC thermostat for lower energy consumption if the room has been unoccupied for a long time.

A control panel located near the entrance door of the hotel room contains the microcontroller, a status display and a keypad, and the RS-485 twisted-pair network connected to the microcontroller through a low-cost transceiver IC.

The status display and keypad can be used to access a variety of functions such as setting the light levels and room temperature. The room control node forwards any access codes received by the card reader to the host computer. In a similar fashion, the microcontroller can store valid access code data received from the host computer in nonvolatile EEPROM memory. This increases security benefits because access privileges to the hotel room can be changed instantly from the host computer.

That's a lot of performance considering that 8-bit microcontrollers are the cheapest class of microcontrollers available.

It is widely recognized that where cost is a significant factor, 8-bit microcontrollers provide the most bang for the buck, and the PICmicro® MCU microcontroller architecture provides one of the broadest product offerings, ranging from 8-pin, 12-bit instruction word to 84-pin, 16-bit instruction word devices. This book is intended to prepare you to take the fullest advantage of one of the most important enabling technologies for embedded devices, whether networked or standalone, yet one of the most economical ones. Original equipment manufacturers (OEMs) place a substantial value on those design engineers who can apply this technology to achieve one of the Holy Grails of electronic design: optimum performance with low design and manufacturing costs.

Note: The Microchip name and logo, PIC, and PICmicro® MCU are registered trademarks of Microchip Technology Inc. in the USA and other countries. All other trademarks are the property of their respective owners.

# INTRODUCTION

In the introduction to the first edition of this book, I explained my fascination with the Intel 8048 microcontroller. I first discovered this device when I was looking at the first IBM PC's schematics. While the PC's schematic itself took up twenty pages, the keyboard's schematic simply consisted of single 8048 chips which provided a "bridge" between the keys on the keyboard and the PC system unit. I got a copy of the 8048's datasheet and was amazed at the features available, making the single chip device very analogous to a complete computer system with a processor, application storage, variable storage, timers, processor interrupt capability and Input/Output ("I/O"). This single chip gave designers the capability of developing highly sophisticated applications while having extremely simple electronic circuits.

**Figure 1** **An assortment of PICmicro® Microcontrollers**

Microcontrollers have become such an integral part of our lives that I daresay the true "Computer Revolution" was the insinuation of the microcontroller into our daily lives. It has come to the point where virtually every electronic product we come in contact with has at least one microcontroller built into it. Electrical appliances often use them for motor and element control as well as operating interfacing. Cellular phones would not be possible without them. Even your car's power seat has a few built into it! The true foundation of our modern society is the more than *five billion* microcontrollers that are built into the products we buy each year.

One of the most popular and easy to use microcontroller families available on the market today is the Microchip "PICmicro®" microcontroller. Originally known as the "PIC" (for "Peripheral Interface Controller"), the PICmicro® microcontroller (*MCU*) consists of over two hundred variations (or "Part Numbers"), each designed to be optimal in different applications. These variations consist of a number of memory configurations, different I/O pin arrangements, amount of support hardware required, packaging and available peripheral functions. This wide range of device options is not unique to the PICmicro® MCU; many other microcontrollers can boast a similar menu of part numbers with different options for the designer.

What has made the PICmicro® MCU successful is:

- The availability of excellent, low-cost (free) development tools;
- The largest and strongest user Internet based community of probably any silicon chip family;
- An outstanding distributor network with a wide variety of parts available in very short notice;
- A wide range of devices with various features that just about guarantees that there is a PICmicro® microcontroller suitable for any application;
- Microchip's efforts at continually improving and enhancing the PICmicro® MCU family based on customer's needs.

These strengths have contributed to the Microchip PICmicro® MCU being the second most popular microcontroller product in the world today (2000). This is up from the twentieth position in 1995. I would not be surprised if the PICmicro® MCU became the largest selling family of architectures in the next few years. Along with the strong line up of current part numbers, the addition of more Flash program memory parts, the PIC18Cxx architecture and new built-in features (like USB) will make the PICmicro® MCU a strong contender for the foreseeable future.

The first edition of this book has been very popular, an award winner (Amazon.com's 1998 "Robotics and Automation" award) and been translated into other languages. Despite this, there are aspects of the book that should be improved upon and changes in the PICmicro® MCU lineup as well as my own personal philosophies on microcontroller application development that have made it appropriate to create this second edition. In the "What's new in this edition" later in this Introduction, I have detailed the changes between this edition and the first one.

One consistent concern about the first edition is the relatively high level it started at. This made the book difficult for the beginners to grasp the concepts of what was being explained or why something was done in a particular way. To help provide a bridge for people new to digital electronics and programming, I have included a number of introductory chapters, appendices and pdf files on the CD-ROM for this material.

Three types of applications have been included in this book. "Experiments" are simple applications that, for the most part, do not do anything useful other than help explain how the PICmicro® MCU is programmed and used in circuits. These applications along with the code and circuitry used with them can be used in your own applications, avoiding the need for you to come up with them on your own.

The "Projects" are complete applications that demonstrate how the PICmicro® MCU can interface with different devices. While some of the applications are quite complex, I have worked at keeping them all as simple as possible and designed them so they can be built and tested in one evening. The applications have also been designed to avoid high speed execution and AC transmission line issues however possible to make prototype builds as robust as possible.

The last type of application presented in this book are the various developer's tools used for application software development. In this book, I have included the design for two different types of PICmicro® MCU programmers and a device emulator that can be used to help with your own PICmicro® MCU application development. I am pleased that I have been able to include a PCB to allow you to build your own programmer quite easily and inexpensively. These application development tools, along with the application development tools included on the CD-ROM gives you the ability to quickly start working with the PICmicro® MCU in very short order.

There is no better feeling in the world than building a circuit, connecting a programmed PICmicro device into it, applying power, and watching it work properly the first time. To help you get to this point as quickly as possible, along with providing and showing how the "MPLAB", "Integrated Development Environment" ("IDE") is used, I focus a great deal of this book on understanding the PICmicro® MCU processor, I/O hardware, and application development process.

I have tried to come up with a "real world" set of applications with a variety of interfaces that you can use as a basis for your own applications. I have tried to make the applications as realistic as possible and I have included explanations for the theory behind the interfaces and the approach taken to connect a PICmicro® MCU to them.

Included on the CD-ROM that comes with this book is the source code for all the PICmicro® MCU applications presented in this book. PICmicro® MCU assembler *Macros* and other source files are available on the CD-ROM as well. Executable and "installation" files are also included for PC application code that are used to run the PICmicro® MCU experiments as well as the source code for selected applications. I have not included source file to the "Tool" code as I want to avoid having to support different people's modifications.

Also on the CD-ROM is a very complete tool base that you can work with for creating and simulating your PICmicro® MCU application code. Microchip's "MPLAB," Virtual Micro Design's "UMPS" IDEs are available for Microsoft "Windows" PCs. These are excellent tools for you to learn about the PICmicro® MCU and develop application for it. I have also included copies of the Linux "GPL" assembler and simulator that are outstanding tools for learning about the PICmicro® MCU (but you do not want to run Microsoft "Windows").

Finally, on the CD-ROM I have included datasheets for all the PICmicro® MCU devices that have been used in this book along with their programming specifications. Further reference information is available on the CD-ROM and everything can be accessed via an "HTML" interface from any web browser. The CD-ROM is as complete as I could make it and I hope hundreds of Mega-Bytes of data on it are useful for you.

In the three years since the publication of the first edition of "Programming and Customizing the PIC Microcontroller," I have received almost a thousand emails from readers with questions, comments, and suggestions. I hope this continues with this edition. Please spend a few moments and let me know what you think of this book, and anything I can do to improve it.

## Conventions Used in this Book

| | |
|---|---|
| Ω | Ohms |
| k kΩ | (Thousands of Ohms) |
| MΩ | Millions of Ohms |
| μF | microFarads (1/1,000,000 Farads) |
| pF | picoFarads (1/1,000,000,000,000 Farads) |
| secs | seconds |
| msecs | milliseconds (1/1,000 of a second) |
| μsecs | microseconds (1/1,000,000 of a second) |
| nsecs | nanoseconds (1/1,000,000,000 of a second) |
| Hz | Hertz (Number of Cycles per Second) |
| kHz | kiloHertz (1,000 cycles per second) |
| MHz | MegaHertz (1,000,000 cycles per second) |
| GHz | GigaHertz (1,000,000,000 cycles per second) |
| #### | Decimal Number |
| –#### | Negative Decimal Number |
| 0×0#### | Hexadecimal Number |
| 0b0#### | Binary Number |
| n.nn × 10**e | The number n.nn times ten to the power "e" |
| *{}* | Optional Parameters within Italicized Braces |
| / | Either or Parameters |
| _Label | Negatively active Signal/Bit |
| Register.Bit | Specific Bit in a Register |
| Monospace Font | Example Code |
| // | Text/Code/Information that Follows is Commented out |
| :or . . . | "And So On." This is put in to avoid having to put in meaningless (and confusing) material to the discussion text |
| & | Two Input Bitwise AND Truth Table: |

| Inputs A | Inputs B | Output |
|---|---|---|
| 0 | 0 | 0 |
| 0 | 1 | 0 |
| 1 | 0 | 0 |
| 1 | 1 | 1 |

| | |
|---|---|
| AND, && | Logical AND |
| \| | Two Input Bitwise OR Truth Table: |

| Inputs A | Inputs B | Output |
|---|---|---|
| 0 | 0 | 0 |
| 0 | 1 | 1 |
| 1 | 0 | 1 |
| 1 | 1 | 1 |

| | |
|---|---|
| OR, \|\| | Logical OR |
| ^ | Two Input Bitwise XOR Truth Table: |

| Inputs A | B | Output |
|---|---|---|
| 0 | 0 | 0 |
| 0 | 1 | 1 |
| 1 | 0 | |
| 1 | 1 | 0 |

| | |
|---|---|
| XOR | Logical XOR |
| ! | Single Input Bitwise Inversion<br>Truth Table: |

| Input A | Output |
|---|---|
| 0 | 1 |
| 1 | 0 |

| | |
|---|---|
| NOT | Logical Inversion |
| + | Addition |
| – | Subtraction or Negation of a Decimal |
| * | Multiplication |
| / | Division |
| % | Modulus of two Numbers. The modulus is the "remainder" of integer division |
| <<# | Shift Value to the Left "#" Times |
| >># | Shift Value to the Right "#" Times |

Along with defining Units there are a few terms and expressions I should define here to make sure you are clear on what I am saying in the text. These terms are often used in electronics and programming, although my use of them is specific to microcontrollers and the PICmicro® MCU.

| | |
|---|---|
| Application | The hardware circuit and programming code used to make up a microcontroller project. Both are required for the microcontroller to work properly. |
| Source Code | The human-readable instructions used in an application that are converted by a compiler or assembler into instructions that the microcontroller's processor can execute directly. |
| Software | I use the generic term "Software" for the application's code. You may have seen the term replaced with "Firmware" in some references. |

# PICmicro® MCU Resources and Tools

Over the past three or four years, the PICmicro® MCU has become a real powerhouse with regards to available support. Along with this book, there are several others, numerous Internet and commercial resources that you can access. In many ways, the PICmicro® MCU has more support available than any other electronic device. These resources are available for the beginner and hobbyist as well as the professional.

In this book, I have provided you with an explanation of the device, how it is programmed (both developing the software and burning it into the chip) and interfaces to other devices. Along with the text material, the CD-ROM includes source code for all the experiments, projects, and tools presented in the book. For creating code, I have included the Microchip "MPLAB" and Virtual Devices "UMPS" IDEs along with the Microchip datasheets for the devices that are presented in the book. The CD-ROM also includes the "www.rentron.com" articles that I have been writing to help people understand the PICmicro® MCU better. These articles demonstrate the PICmicro® MCU from some different perspectives and will give you an idea of what was the genesis for many of the concepts presented in this book.

Microchip has been very generous in making the datasheets, MPLAB, and other materials used in this book available for the CD-ROM. I want to point out that chances are, at least two files will be updated in the time it takes to "master" the CD-ROM for the book to when it is put out on the shelf. Because Microchip frequently updates the information that they publish, I recommend that you check with their website:

**http://www.microchip.com**

before embarking on a complex project of your own to ensure that you don't miss any updates or errata that become available. The datasheets on the CD-ROM are an excellent quick reference for your own projects but shouldn't be relied upon if you are using advanced peripherals.

The embedded PCB that is included with this book is designed to be part of an "El Cheapo" programmer. For this programmer, I have tried to select parts that can be found in most electronics stores (I have included a list of Digi-Key part numbers for reference). MS-DOS and Microsoft "Windows" ("Win32") are available for this programmer and have been placed on the CD-ROM.

Along with the "El Cheapo," I have also made the other PCB designs that are used with applications in this book available in "Gerber" format on the CD-ROM. All the information necessary for you to build the PCBs (or have them built) is included on the CD-ROM.

Some of the projects and tools presented in this book will be available for sale from Wirz Electronics (http://www.wirz.com) and there is more information on how to order them later in the text as well as on the CD-ROM. I should point out that the experiments, projects, and tools in this book are available for your personal use only. They cannot be reproduced for sale, except with my written permission.

If you don't have an Internet connection, then I highly recommend that you sign up with an Internet Service Provider ("ISP") as soon as you can. I will be keeping track of any errors in the book that are discovered along with their corrections on my website:

**http://www.myke.com**

The Internet is also the number one source for any code, tools and help that you may need. The best Internet source you will find is MIT's "PICList." This "List Server" (or "ListServ"), with two thousand subscribers, is an outstanding resource and will help you get on your way. The variety of experience you can draw upon can be overwhelming with questions often answered within minutes of being asked.

The last resource that you should be aware of is Microchip itself. Microchip, along with its network of distributors provide a very strong set of developer's tools, application engineering help and their yearly seminar series. The Microchip development tools (such as "PICStart Plus," "MPLAB 2000," "MPLAB ICD" which are presented later in the book) are all of excellent quality and reasonable cost. The yearly seminars are a great way to buy these development tools at reduced prices as well as meet your Microchip representatives and others in your area that work with the PICmicro® MCU professionally or as hobbyists.

# What's New in this Edition

While much of the material from the first edition has been retained for this one, there have been some significant changes and additions for this edition.

They are:

1. Changes to make the book more accessible for people new to electronics and easier to find specific information.
2. The book has been laid out in a new format. This format was first used in later books of the "Programming and Customizing" series and provides a list of the subchapters on the first page of each chapter.
3. To help you find important information faster I have included the "List of Figures", "List of Tables" and "List of Code Examples" as well as the information located in the book's index. These references are meant to help find specific information and code very quickly.
4. Some of the information was given in the first edition before prerequisite information was presented. Some chapters have been reordered and changed to eliminate this from being a problem in the second edition.
5. All "pseudo-code" examples are written in "C." "C" is the most popular high level language for PICmicro® MCU application development (as well as most technical programming). I have followed "C" conventions in all areas of the book when presenting information or data, wherever possible.
6. A table format for register definitions has been used in this edition to help make finding specific information easier. Bits are defined from the most significant to least significant to make translating the bit numbers to values simpler.
7. A glossary of terms used in the book has been included. This glossary has been optimized for the PICmicro® MCU and the concepts required for it.
8. "Holes" in information and data have been eliminated.
9. Introductory electronics and programming information has been included on the CD-ROM. This information will give you a better background to what is being presented as well as allow me to space out information and introduce one concept at a time.
10. More glossary/appendix reference data has been provided.
11. The "Conventions Used in this Book" section of the introduction has been expanded to include all mathematical operators and symbols used in the text and formulas.
12. The example experiments, projects and tools have been enhanced.
13. The experiments, projects and tools have been relabeled to avoid confusion regarding the order in which information is presented. In the original edition, the applications were labeled according to the order in which they were developed. In this edition, the experiments and projects have been labeled according to what category of application they come under and the order in which they appear.

14. There are a number of new experiments, projects and tools added to this book. These additions are used to demonstrate new functions or clarify areas that were ambiguous.
15. Complete schematics and bills of material are available for all the applications that are presented in this book.
16. The "El Cheapo" programmer PCB is included with the book to allow you to quickly and easily program PICmicro® MCUs and start building your own applications very quickly and inexpensively.
17. All the PICmicro® MCU code written for the book has been tested with "MPLAB" and programmed into PICmicro® MCU devices, where appropriate, on multiple development tools. The MPLAB IDE is an outstanding tool designed for Microsoft "Windows" which is available free of charge from Microchip. Also included with the book is Virtual Micro Design's "UMPS" and the "GPL" Linux assembler and simulator.
18. PC Interface code has been tested on a variety of PCs. While I cannot guarantee that the code will work on all PCs, it should be robust enough to work on most without problems. I have tried to include both "MS-DOS" as well as Microsoft "Windows" code for the projects.
19. All parts specified in this book are available easily from a variety of sources. Where there can be confusion with regards to the parts, I have listed distributor part numbers in the text.
20. The latest PICmicro® MCU devices and features are presented. The eight and fourteen pin PICmicro® MCUs along with the latest EEPROM/Flash and PIC18Cxx parts and their features have been added to this book. I realize that between the time I write this and when the book comes to print, even more parts will be added. Please consult the Microchip web site for the latest list of available PICmicro® MCU part numbers.
21. With the description of each interface, I have included sample code that can be placed directly into your applications. These "snippets" of code are written with constants or variables that are described in the accompanying text.
22. Update on software development. Text has been expanded to include discussions on application development as it relates to the operation of the PICmicro® MCU. The concept of "Event Programming" is introduced as to how it relates to the PICmicro® MCU and microcontroller programming in general.
23. Two new chapters on assembly language and macro programming have been added to help you understand how optimal code is developed and how it is measured. The measurements that I introduce may be considered somewhat unusual, but I believe they are appropriate for real-time microcontroller applications.

# Copyright and Trademarks

# 1

# MICROCONTROLLERS

## CONTENTS AT A GLANCE



In a time when digital electronics is becoming more complex and less accessible to students and low-end circuit developers, microcontrollers have become excellent tools for learning about electronics and programming, as well as providing the capabilities to create sophisticated electronic applications fairly easily and inexpensively. Microcontrollers and, more recently, *Programmable Logic Devices (PLDs)* provide a method to learn about digital interfacing and programming, and to provide the capability to easily create applications that control real-world devices.

This chapter presents how microcontrollers, as computer processor-based devices that are completely self-contained with memory and I/O, can be used as the heart of a limitless variety of different electronics applications. PLDs offer many of the same capabilities that are suitable for similar, but different applications because they do not have the internal processor. Understanding the strengths and weaknesses of the microcontroller and the basic rules for application design is important for being able to successfully develop usable applications in an efficient manner. As you progress through the book, you will gain the knowledge necessary to develop your own applications for the Microchip PICmicro® MCU eight-bit microcontroller.

# Microcontroller Chips

If you were to investigate all the different types of microcontrollers, you would probably be amazed at the number of different ones available. The different types of devices break down as:

- Embedded (self-contained) eight-bit microcontrollers
- 16- to 32-bit microcontrollers
- Digital signal processors

A wide range of embedded (self contained) devices are available. In an embedded microcontroller, all of the necessary resources (memory, I/O, etc.) are available on the chip; in your application circuit, you only have to provide power and clocking. These microcontrollers can be based on an established microprocessor core or use a design specific to microcontroller functions. Thus, there is a great deal of variability of operation—even in devices that are performing the same tasks.

The primary role of these microcontrollers is to provide inexpensive, programmable logic control and interfacing to external devices. Thus, they typically are not required to provide highly complex functions. But, they are capable of providing surprisingly sophisticated control of different applications.

When I say that the devices are inexpensive, I mean that they range in cost from $1 to more than $20 each (with the cost depending on complexity, which is a function of internal features and external pin count, and quantity purchased). Twenty dollars for a chip might seem expensive, but this is largely mitigated by the lack of additional chips required for the application. With the PICmicro® MCU, the external parts needed to support the microcontroller's execution in an application can literally cost as little as five cents.

In Fig. 1-1, you will get an idea of how an embedded microcontroller chip is laid out and the interfaces that are available to the outside world.

Generally, these microcontrollers have the following characteristics that allow them to be simply wired into a circuit with very little support requirements. The following list will give you an idea of what's always available in the embedded microcontroller package:

- Processor reset
- Device clocking
- Central processor
- PROM/EPROM/EEPROM/Flash program memory and programming interface

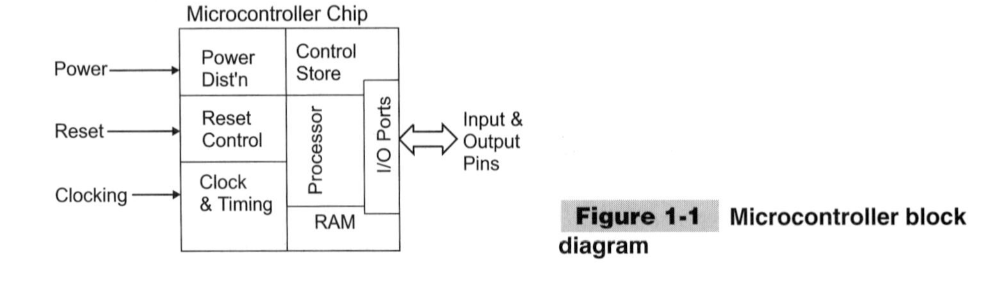


**Figure 1-1 Microcontroller block diagram**

- Variable RAM
- I/O pins
- Instruction cycle timers

Along with these basic requirements for a microcontroller, most modern microcontrollers have many of following features built in:

- Built-in monitor/debugger program
- Built-in program memory programming from a direct host connection
- Interrupt capability (from a variety of sources)
- Analog input and output (I/O) (both PWM and variable DC I/O)
- Serial I/O (synchronous and asynchronous data transfers)
- Bus/external memory interfaces (for RAM and ROM)

All of these features increase the flexibility of the device considerably and not only make developing applications easier, but possible in some cases. Note that most of these options enhance the I/O pins' function and do not affect the basic operation of the processor. These options can usually be disabled (leaving the pins used for simple I/O), bringing the microcontroller's capabilities equal to the basic set listed previously.

Early microcontrollers were manufactured using bipolar or NMOS technologies. Most modern devices are fabricated using CMOS technology, which decreases the current chip's size and the power requirements considerably. For most modern microcontrollers, the current required is anywhere from a few microamperes ($\mu$A) in "Sleep" mode to up to about 10 milliamperes (mA) for a microcontroller running at 20 MHz. A smaller chip size means the chip requires less power and that more chips can be built on a single "wafer." The more chips that are built on a wafer, the lower the unit price is.

Maximum speeds for the different devices are typically in the low tens of megahertz (MHz). The primary limiting factor is the access time of the memory used in the microcontrollers. For the typical applications that are based on microcontrollers, this is generally not an issue. The issue is the ability to provide relatively complex interfaces for applications using a simple microcontroller. The execution cycles required to provide these interfaces can take away from the time needed to process the input and output data. This book covers the advanced PICmicro® MCU hardware features that provide interfacing functions, as well as "bit-banging" methods for simulating the interfaces while still leaving enough processor cycles to work through the application processing.

Some microcontrollers (most notably devices with 16- or 32-bit data paths) rely completely on external memory, instead of built-in program memory and variable memory. Despite the tremendous advantages that a microcontroller has with built-in program storage and internal variable RAM, there are times (and applications) where you will want to add external (both program and variable) memory to your microcontroller.

There are two basic ways of doing this. The first is to add memory devices to the microcontroller as if it were a microprocessor. Many microcontrollers are designed with built-in hardware to allow this technique. For an external memory device, the microcontroller diagram can be changed to Fig. 1-2.

The classic example of this type of microcontroller is the Intel 8051. Some members of the family are shipped without any built-in program memory, which necessitated the ad-

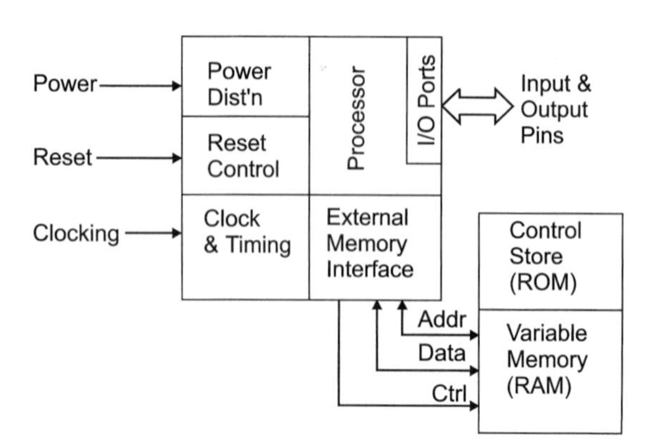


**Figure 1-2** **Microcontroller with external memory**

dition of external ROM. The bus designs of the 8051 added up to 64K as well as 64K variable RAM. Other members of the family (the 8751 and 8752) had built-in EPROM program memory that could be used alone, shared with external ROM and RAM, or even bypassed completely, allowing the microcontroller to run without the contents of the chip's program memory being accessed. This latter feature allowed 8051s that were programmed with incorrect or "downlevel" program memory to be used in other applications and not have to be scrapped.

In the PICmicro® MCU family, some of the PIC17Cxx devices are capable of implementing external memory. In this book, along with explaining how the high-end PICmicro® MCU devices work in the external memory modes, I have included a simple external memory bus access application to show how memory devices are interfaced to the PIC17Cxx.

A typical application for external memory microcontroller is as a hard-disk controller/buffer that provides a simple interface to a system processor and distributes large amounts of data (usually measured in megabytes (MB). The external memory allows much more buffer memory than would be normally possible in an embedded microcontroller.

The second way is to add a bus interface to the microcontroller and control the bus I/O in software or using built-in hardware. The two-wire *Inter-Inter Computer (I2C)* protocol is a very common bus standard for this method. This will allow simple I/O devices without complex bus interfaces although if the external memory device is to store application code, then an interpreter for the data will have to be built into the microcontroller's built in memory to be able to decode the information being passed to it.

*Digital Signal Processors (DSPs)* are a relatively new category of processor. The purpose of a DSP is to take sample data from an analog system and calculate an appropriate response. DSPs and their *Arithmetic Logic Units (ALUs),* the hardware that performs calculations, run at very high speed to allow this control in real time. DSPs are often used in such applications as active noise-canceling microphones in aircraft (a second, ambient noise microphone provides a signal that is subtracted from the primary microphone signal, canceling the ambient noise, leaving just the pilot's voice) or eliminating "ghosting" in broadcast television signals.

Developing DSP algorithms is a science unto itself, a subbranch of control theory. The science of control theory requires very advanced mathematics and is beyond the scope of this book (later, I will discuss "fuzzy logic," which is a nontraditional method to use a computer to interface with analog systems). DSPs come in a variety of designs that use features that are most often found in the embedded microcontrollers and external-memory microcontrollers. Although typically not designed for stand-alone applications, digital signal processors can be used to control external digital hardware, as well as process the input signals and formulate appropriate output signals.

One of the biggest questions often asked about the PICmicro® MCU's capabilities is whether or not it can do audio-frequency (0 to 7,000 Hz) digital signal processing. The short answer to this question is "no" because the PICmicro® MCU lacks the processor features (such as a built-in multiplier) and speed to be able to carry out the functions. If you are planning on doing digital signal processing with the PICmicro® MCU, then I recommend running the PICmicro® MCU at the highest possible speed (usually 20 MHz) and do not process any signals that are faster than 1 kHz.

## APPLICATIONS

In this book, *application* describes all of the different aspects of a microcontroller-based circuit. I think it is important to note that a project is based on a collection of different development efforts (for the hardware and software) and not a development process in a single discipline. This section introduces you to the five different elements to a microcontroller project and explains some of the terms and concepts relating to them.

The five different aspects to every microcontroller project are:

1. Microcontroller and support circuitry
2. Project power
3. Application software
4. User input/output (I/O)
5. Device interface (I/F)

These different elements are shown working together in Fig. 1-3.

This book goes through each of these aspects with respect to the Microchip PICmicro® MCU to give you the knowledge needed to develop your own PICmicro® MCU application. Before doing that, I want to introduce you to the different elements and help you to understand how they interrelate and how applications are built from them.

The microcontroller and support circuitry is simply the microcontroller (PICmicro® MCU) chip and any electronic parts that are required for it to work in the application.

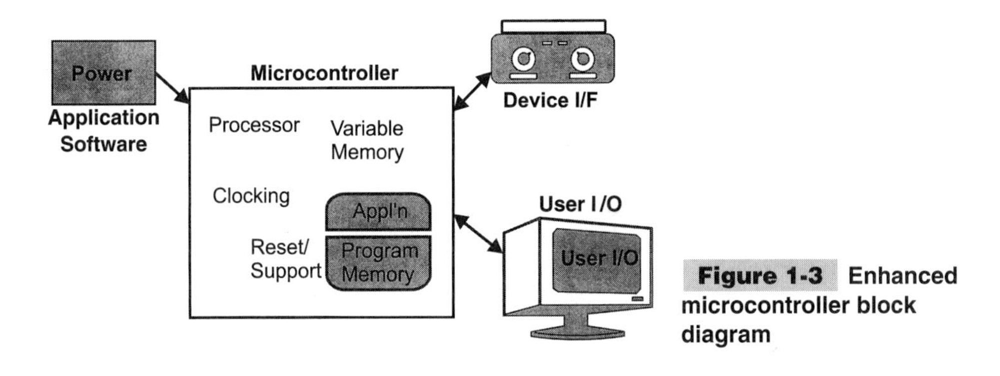


**Figure 1-3** Enhanced microcontroller block diagram

Along with the chip itself, most PICmicro® MCU applications require a decoupling capacitor, a reset circuit, and an oscillator. The design of the PICmicro® MCU (as is true for most microcontrollers) makes the specification of these parts very simple; in fact, in some of the newer devices, these circuits are built into the chip itself.

The decoupling capacitor is used to minimize the effects of rapid changes in power demands by devices in the circuit (including the PICmicro® MCU itself). A decoupling capacitor can be thought of as a filter that smoothes out the rough spots of the power supply and provides additional current for high load situations on the part. As shown later in the book, a decoupling capacitor is crucial for the PICmicro® MCU and should never be left out of an application's circuit.

The reset circuit on the PICmicro® MCU is very simple; often it is just a "pull-up." This simple circuit is often enhanced in many applications to ensure that marginal-voltage situations (such as the case of a battery that is losing its charge) do not cause problems with the operation of the PICmicro® MCU. The book explains how reset works in the PICmicro® MCU and what can be done with the reset to prevent problems with the PICmicro® MCU's operation.

For any computer processor to run, a clock is required to provide timing for each instruction operation. This clock is provided by an oscillator built into the PICmicro® MCU that uses a crystal, ceramic resonator, or an RC network to provide the time base of the PICmicro® MCU's clocks circuitry. One of the nice features of many PICmicro® MCUs is a built-in 4-MHz oscillator that avoids the need to provide the oscillator circuit and frees up two pins for I/O functions.

Power is an often-neglected part of a microcontroller application's design. Proper specification of the power supply design and selection of components are important for an application's reliability and cost effectiveness. This book describes the important aspects of a PICmicro® MCU's power supply, along with a number of circuits to help you specify the best power supply for specific applications.

I feel that the user input/output interface is crucial to the success of a microcontroller application. This book shows a number of different ways to pass data between a user and a PICmicro® MCU. Some of these methods might seem frivolous or trivial, but having an easy-to-use interface between your application and the user is a differentiator in today's marketplace. Along with information on developing user I/O circuitry are some of my thoughts on what is appropriate for users.

Device interfacing is really what microcontroller applications are all about. Looking over Chapter 16, you should get the idea that there is a myriad of different devices that microcontrollers can interface with control or monitor. I have tried to present a good sampling of different devices to show different methods of interfacing to the PICmicro® MCU that can be used in your own applications.

Within the microcontroller is the application software stored in program memory. Although this is one-fifth of the different elements that make up a microcontroller application, it will seem like it will take up six-fifths of the work. Microcontroller software development is really an art form. Information presented in this book should give you an introduction to the knowledge required to develop your own applications. Also, there are a lot of code "snippets" you can add to your own applications.

If you are new to electronics programming, you should look at the appendices "Introduction to Electronics" and "Introduction to Programming" on the CD-ROM. These appendices will help reference the information presented on the PICmicro® MCU and the applications built from the microcontroller presented in this book. I have tried to focus the information with respect to developing electronic circuits for microcontrollers in these appendices.

**Instructions and software** When a processor executes program or code statements, it is reading a set of bits from program memory and using the pattern of bits to carry out a specific function. These bits are known as an *instruction* and each pattern carries out a different function in the processor. A collection of instructions is known as *program*. When the program is executing, the instructions are fetched from program memory using a program counter address. After an instruction is executed, the program counter is incremented to point to the next instruction in program memory.

The four types of instructions are:

- Data movement
- Data processing
- Execution change
- Processor control

The data-movement instructions move data or constants to and from processor registers, variable memory and program memory (which, in some processors, are the same thing), and peripheral I/O ports. The many types of data-movement instructions are based on the processor architecture, number of internal addressing modes, and the organization of the I/O ports.

In the PICmicro® MCU, variable memory and program memory is separate. The I/O registers are part of the variable memory space. Because the PICmicro® MCU only has one processor specific register (the accumulator), only five addressing modes are available and given in the following list. In Chapters 3 and 4, these operations are explained in more detail as how they work with regard to the PICmicro® MCU.

1. Constant value put into the accumulator.
2. Variable/register contents put into the accumulator.
3. Indexed address variable/register put into the accumulator.
4. Accumulator contents put into a variable/register.
5. Accumulator contents put into indexed address variable/register.

I say "only five" addressing modes because, as you investigate other processor architectures, you will find that many devices can have more than a dozen different ways to access data within the memory spaces. The previous list is a good base reference for a processor and can provide virtually any function that is required of an application. The only "missing" function that I feel is necessary in a processor is the ability to access the program counter stack. This feature is available in the PIC18Cxx devices, but not in the low-end, mid-range, and PIC17Cxx PICmicro® MCUs.

Data-processing instructions consist of basic arithmetic and bitwise operations. A typical processor will have the following data-processing instructions:

1. Addition
2. Subtraction

**3** Incrementing
**4** Decrementing
**5** Bitwise AND
**6** Bitwise OR
**7** Bitwise XOR
**8** Bitwise negation

These instructions work the number of bits of the data word size (for the PICmicro® MCU, this is eight bits). Many processors are capable of carrying out multiplication, division comparisons, and data types of varying sizes (and not just the "word" size). For most microcontrollers (the PICmicro® MCU included), the word size is eight bits and advanced data-processing operations are not available.

Execution change instructions include *gotos* (or *jumps*), *calls, interrupts, branches,* and *skips.* For branches and gotos, the new address is specified as part of the instruction. Branches and gotos are very similar, except that branches are used for "short jumps," which cannot access the entire program memory and are used because they require less memory and execute in fewer instruction cycles. These execution changes are called *non-conditional* because they are always executed when encountered by the processor. Skips are instructions that will skip over the following instruction. Skips are normally conditional and based on a specific status condition.

If you have developed applications on other processors, you might interpret the word *status* to mean the bits built into the STATUS register. These bits are set after an instruction to indicate such things as whether or not the result of the instruction was equal to zero or caused an overflow. These status bits are available in the PICmicro® MCU, but are supplemented with all the other bits in the processor, each of which can be accessed and tested individually. This provides a great deal of additional capabilities in the PICmicro® MCU that are not present in many other devices, and it allows some amazing efficiencies in application software that will be discussed later in the book.

An example of using status bits is shown in how a 16-bit variable increment is implemented in an eight-bit processor. If the processor's zero flag is not set after the low eight-bit increment, then the following instruction (which increments the upper eight-bit increment) is skipped. But, if the result of the lower eight-bit increment is equal to zero, then the upper eight-bit increment is executed:

```
Increment           LowEightBits
SkipIfZero
   Increment        HighEightBits
```

The skip is used in the PICmicro® MCU to provide conditional execution. Other processors only have conditional "branches" or "gotos."

Other execution change instructions include the *call and interrupt,* which cause execution to jump to a routine and return back to the instruction after the Call/Interrupt instruction. A call is similar to a branch of goto and has the address of the routine to jump to included in the instruction. The address jumped to is known as a *routine* and includes a Return instruction at its end, which returns execution to the previous address.

The two types of *hardware interrupts* are explained in more detail in the next section. Hardware interrupts can be thought of calls to routines that are initiated by a "hardware event." *Software interrupts* are instructions that make calls to interrupt-handler routines.

Software interrupts are not often used in smaller microcontrollers, but they are used to advantage in the IBM PC.

In most processors, the *program counter* cannot be accessed directly, to jump or call arbitrary addresses in program memory. In some processors, such as the PICmicro® MCUs, the program counter cannot be updated by the contents of a variable. In these cases, the program counter can be directly accessed and updated. Care must be taken when updating the processor's program counter to ensure that the correct address is calculated before it is updated.

Processor control instructions are specific and control the operation of the processor. Common process or control instructions are *Sleep,* which puts the processor (and microcontroller) into a low-power mode. Another processor control instruction is *Interrupt Mask,* which stops hardware interrupt requests from being processed.

## PERIPHERAL FUNCTIONS

All microcontrollers have built-in I/O functions. These functions can range from I/O pins of remarkable simplicity (literally, just a pull up and a transistor) to full Ethernet interfaces or video on-screen display functions. After understanding a microcontroller's processor architecture, the next area to work through is the peripherals, understanding how I/O pins work, as well as how more complex functions are carried out.

Peripherals are normally added to a microcontroller design by the use of design "macros." These macros are specified during the design of the chip. As different functions are specified, a unique microcontroller is created. Microchip has developed a rich feature set for the PICmicro® MCU that has resulted in many different PICmicro® MCU part numbers. Each part number has a different set of features that will give you, as the application designer, a selection of parts that will allow you to choose a device that is well suited to your application.

When I said that an I/O pin could be as simple as a transistor and a pull-up resistor, I wasn't being facetious. The Intel 8051 uses an I/O pin that is this simple (Fig. 1-4).

This pin design is somewhat austere and is designed to be used as an input when the output is set high so that another driver on the pin can change the pin's state to high or low easily against the high-impedance pull up.

A more-typical I/O pin is shown in Fig. 1-5. It provides a tristatable output from the control register. This pin can be used for digital input as well (with the output driver turned off). This is the PICmicro® MCU's I/O pin design and uses the control register to select the pin mode as *input* or *output*. Later in this book I will discuss the operation of the I/O pins in greater detail.

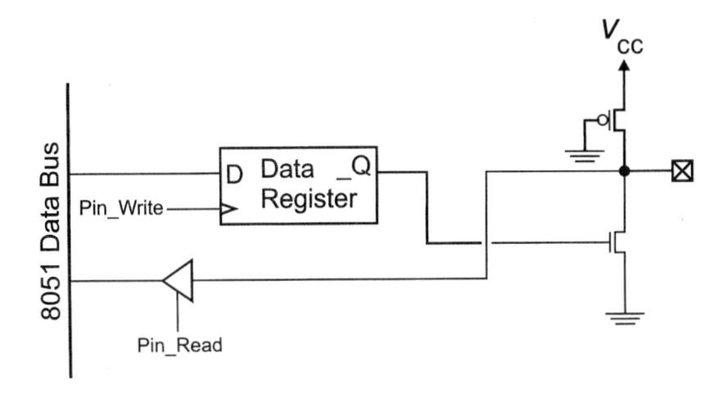


**Figure 1-4** 8051 Parallel I/O pins

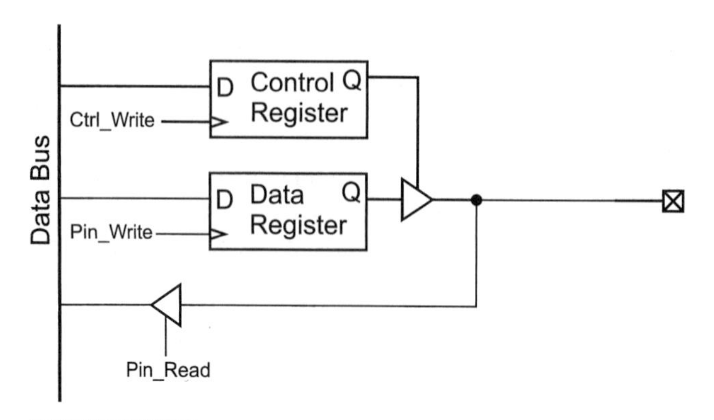


**Figure 1-5** **Standard bi-directional parallel I/O pins**

A microcontroller can also have more advanced peripheral I/O functions, such as serial I/O. These peripheral functions are designed to simplify the operation or interfacing to other devices. I call this a *peripheral function* because it provides a service that would normally be performed by peripherals that are added to a peripheral bus in a microcontroller. The peripheral bus in a microcontroller is located inside the device with all of the peripheral functions.

How functions are programmed in a microcontroller is half the battle in understanding how they are used. Along with changing the function of an I/O pin, they might also require the services of a timer or the microcontroller's interrupt controller. The block diagram in Fig. 1-6 shows an I/O pin with a serial data transmitter.

**Bit-banging I/O** Despite the plethora of features available in different PICmicro® MCUs, you might want to use peripheral functions that are not available or perhaps the

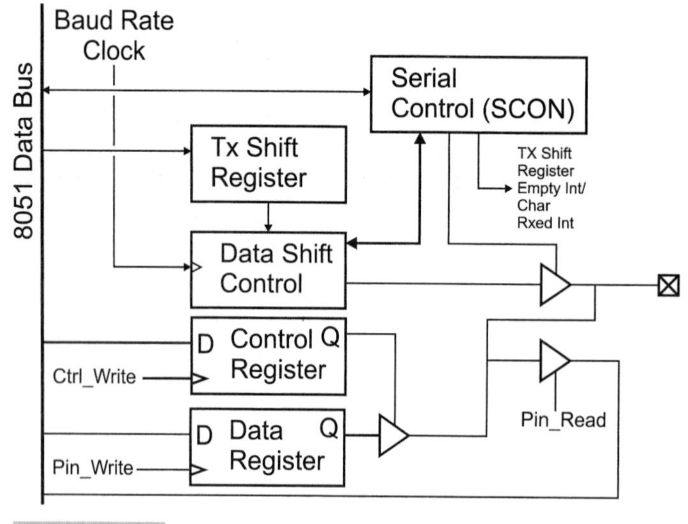


**Figure 1-6** **Serial I/O with parallel pin functions**

built-in features are not designed to work with the specific hardware that you want to use in the application. These functions can be provided by writing code that executes the desired I/O operations using the I/O pins of the microcontroller. This is known as *bit banging,* which is very commonly used in PICmicro® MCU microcontroller application development.

There are two philosophies behind the methods used to provide bit-banging peripheral functions. The first is to carry out the operation directly in the execution line of the code and suspend all other operations. An example of this is the "pseudo-code" for a serial receiver shown below:

```
Int SerRXWait() {                   // Wait for and Return the Next

                                    // Asynchronous Character

int i;
int OutByte;

  while (IP_Bit == High);           // Wait for the "Start" Bit

  HalfBitDlay();                    // Delay to Middle of Bit

  for (i = 0; I < 8; I++) {         // Read the 8 Bits of Data

    BitDlay();                      // Delay one Bit Period

    OutByte = (OutByte >> 1) + (IP_Bit << 7);

  }

  return OutByte;                   // Return the Byte Read In

} // End SerRXWait
```

The advantage of this method is that it is relatively easy to code, but the downside is that it requires all other operations in the PICmicro® MCU to stop. The serial receive function waits literally forever for data to come in. While this function is waiting or receiving a character, nothing else can execute in the microcontroller.

The other method of providing bit-banging functions is to periodically interrupt the mainline execution to provide a periodic input/output function. For the serial receive function, I could sample the input line at three times the incoming bit speed and react to the start bit. The code that executes each time the timer interrupt handler is *involved*, is:

```
Interrupt IntSerRX() {

  if (rSTATUS != startRX)       // Is Something Being Received?
    Is (IP_Bit == Low) {        // No - Check for Start Bit

      rSTATUS = startRX;        // Start Bit Found
      dlayCount = 4;            // Wait four Timer Dlays to middle
                                //   of 2nd Bit
      bitCount = 8;             // Eight Bits are Read

    } else;
else                            // Reading a Bit
```

```
if (--dlayCount == 0) {      // If Bit Dlay is Finished

  OutByte = (OutByte >> 1) + (IP_Bit << 7);
  dlayCount = 3;

  if (--BitCount == 0)       // Read all 8 Bits?
    rSTATUS = byteRX;

  }

} // End Serial RX Interrupt Handler
```

I realize that this function seems quite complex and, at first glance, is probably just about impossible to understand exactly how it works. I will explain how this function works in more detail later in the book. What I wanted to show was a bit-banging function that does not prevent other microcontroller operations from being carried out while it is operating. As well, *Application Program Interfaces (APIs)* can be written for this function to make it "behave," from the application code's perspective, like hardware built into the microcontroller.

While this function is operating as a periodic interrupt, it is taking processor cycles away from the mainline code, but the overall percentage of lost cycles is very low. For this reason, I prefer this type of bit banging to the inline application code. As I discuss PICmicro® MCU peripheral functions later in the book, I will also present methods for implementing them as bit-banging functions using timer interrupts.

## PROCESSOR ARCHITECTURES

Here's a hint when you are inviting computer scientists to dinner; be sure that they all agree on what is the best type of computer architecture. There are a lot of strong points for supporting the different options that are available in computer architectures. Although RISC is in vogue right now, a lot of people feel that CISC has been unfairly maligned. This is also true for proponents of Harvard over Princeton computer architectures and whether or not a processor's instructions should be hard-coded or micro-coded. Trust me when I say that if you don't type your guests properly, you will have a dinner with lots of shouting, name calling, and bun throwing.

The following sections provide some background for the different processor types, explain the advantages/disadvantages of the different features, and show why the designers of the PICmicro® MCU would make specific choices over others. These sections are not meant to provide you with a complete understanding of computer processor architecture design, but should help explain some concepts behind the PICmicro® MCU and reasons why some aspects of the design are implemented the way they are.

**CISC versus RISC** Currently, many processors are called *RISC* (*Reduced Instruction Set Computers,* pronounced "risk") because there is a perception that RISC is faster than *CISC (Complex Instruction Set Computers).* This can be confusing because many processors available are identified as being "RISC-like," but are, in fact, CISC processors. And, in some applications, CISC processors will execute code faster than RISC processors or execute applications that RISC processors cannot.

What is the real difference between RISC and CISC? CISC processors tend to have a large number of instructions, each carrying out a different permutation of the same operation (accessing data directly, through index registers, etc.) with instructions perceived to be useful by the processor's designer.

In a RISC system, the instructions are as minimal as possible to allow the user to design their own operations, rather than use what the processor designer has given them. Later in the book, I show how a stack "push" and "pop" would be done by RISC system in two instructions that allow the two simple constituent instructions to be used for different operations (or compound instructions, such as push and pop).

This ability to write to all the registers in the processor as if they were the same is known as the *orthogonality* or *symmetry* of the processor. This allows some operations to be unexpectedly powerful and flexible. This can be seen in conditional jumping. In a CISC system, a conditional jump is usually based on status register bits. In a RISC system, such as the PICmicro® MCU, a conditional jump can be based on a bit anywhere in memory. This greatly simplifies the operation of flags and executing code based on their state.

For a RISC system to be successful, the designer must do more than just reduce the number of tasks performed in an instruction. By carefully designing the processor's architecture, the flexibility can be increased to the point where a very small instruction set, able to execute in very few instruction cycles, can be used to provide extremely complex functions in the most efficient manner.

The PICmicro® MCU takes advantage of the aspects of register symmetry in the design of the register/RAM addressing in the processor. As I go through the PICmicro® MCU architecture, instructions and applications in the following chapters, you will see that fast data-processing operations within the processor can be very easily implemented with the operations being the same, regardless of the data type being modified.

**Harvard versus Princeton** Many years ago, the United States government asked Harvard and Princeton universities to create a computer architecture to be used in computing tables of Navel artillery shell distances for varying elevations and environmental conditions.

Princeton's response was for a computer that had common memory for storing the control program, as well as variables and other data structures. It was best known by the chief scientist's name, John Von Neumann. Figure 1-7 is a block diagram of the architecture.

The memory interface unit is responsible for arbitrating access to the memory space be-

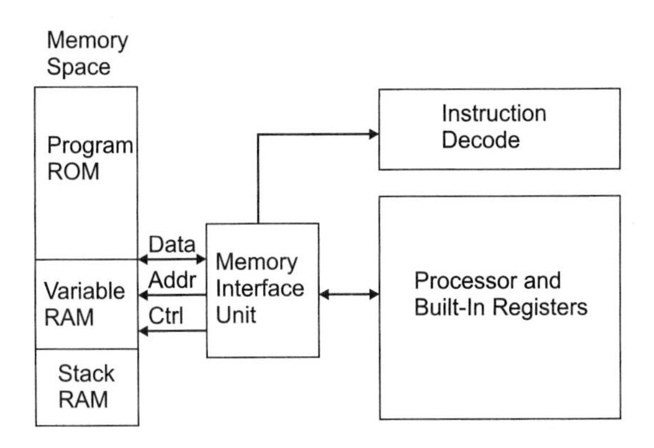


**Figure 1-7** Princeton architecture block diagram

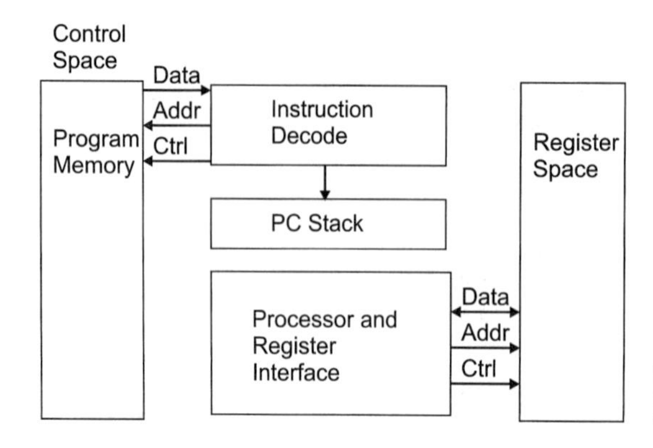


**Figure 1-8 Harvard architecture block diagram**

tween reading instructions (based upon the current program counter) and passing data back and forth among the processor and its internal registers.

It might, at first, seem that the memory interface unit is a bottle-neck between the processor and the variable/RAM space—especially with the requirement for fetching instructions at the same time. In many Princeton architected processors, this is not the case because the time required to execute an instruction is normally used to fetch the next instruction (this is known as *prefetching*). Other processors (most notably, the Pentium processor in your PC) have separate programs and data caches that can be accessed directly while other address accesses are taking place.

Harvard's response was a design that used separate memory banks for program storage, the processor stack, and variable RAM. Figure 1-8 is the block diagram for a typical Harvard processor, such as the PICmicro® MCU.

The Princeton architecture won the competition because it was better suited to the technology of the time. Using one memory was preferable because of the unreliability of then-current electronics (this was before transistors were in widespread general use). A single memory and associated interface would have fewer things that could fail.

The Harvard architecture was largely ignored until the late 1970s when microcontroller manufacturers realized that the architecture had advantages for the devices that they were currently designing.

The Von Neumann architecture's largest advantage is that it simplifies the microcontroller chip design because only one memory is accessed. For microcontrollers, its biggest asset is that the contents of *Random Access Memory (RAM)* can be used for both variable (data) storage, as well as program instruction storage. An advantage for some applications is the program counter stack contents, which are available for access by the program via the single memory area. This allows greater flexibility in developing software, primarily in the area of real-time operating systems (which are covered in greater detail later in the book).

The Harvard architecture tends to execute instructions in fewer instruction cycles than the Von Neumann architecture. This is because a much greater amount of instruction parallelism is possible in the Harvard architecture. *Parallelism* means that instruction fetches can occur during previous instruction execution and not wait for either a "dead" cycle of the instruction's execution or have to stop the processor's operation while the next instruction is being fetched even if variables are accessed by the instruction.

For example, if a Princeton-architected processor was to execute a read byte and store

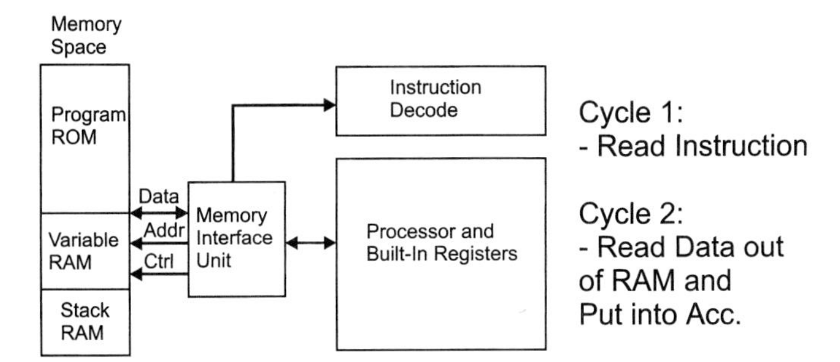


**Figure 1-9** **"move Acc, Reg" in Princeton Arch.**

in the accumulator instruction, it would carry out the instruction sequence shown in Fig. 1-9. In the first cycle of the instruction execution, the instruction is read in from the memory space. In the next cycle, the data to be put in the accumulator is read from the memory space.

The Harvard architecture, because of its increased parallelism, would be able to carry out the instruction while the next instruction is being fetched from memory (the current instruction was fetched during the previous instruction's execution). As is shown in Fig. 1-10, executing this instruction in the Harvard architecture also occurs over two instructions, but the instruction read takes place while the previous instruction is carried out. This allows the instruction to execute in only one instruction cycle (while the next instruction is being read in).

This method of execution (parallelism), like RISC instructions, also helps instructions take the same number of cycles for easier timing of loops and crucial code. This point, although seemingly made in passing, is probably the most important aspect that I would consider when choosing a microcontroller for a timing-sensitive application.

For example, the Microchip PICmicro® MCU executes every instruction, except ones that modify the program counter in four clock cycles (one *instruction cycle*). This makes

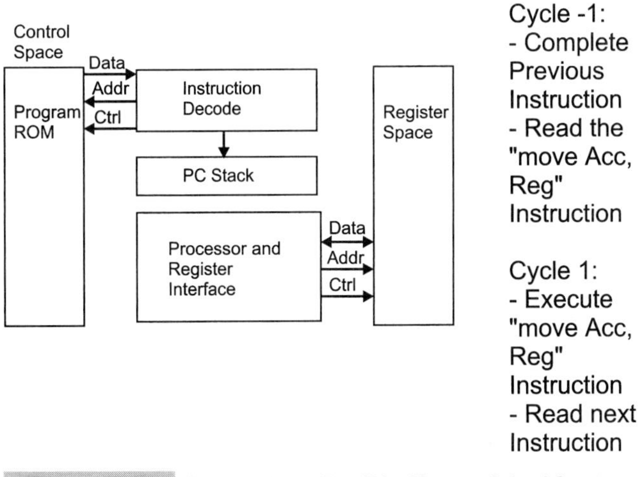


**Figure 1-10** **"move Acc, Reg" in Harvard Architect**

crucial timing operations much easier than, for example, an Intel 8051, which can take anywhere from 12 to 64 clock cycles to complete an instruction. Often, a simulator or a hardware emulator will be required to accurately time a function, rather than trying to figure out manually how many cycles will be used from the code.

I should caution you (I will continue to do so throughout the book) that these types of performance comparisons might not be representative to all the processors using these two types of architectures. The comparison that matters is the actual application and different architectures and devices will offer unique features that might make it easier to do a different application. In some cases, certain applications will not only be more efficiently executed by a specific architecture, but can *only* execute in a specific architecture.

After reading this section, you probably feel that a Harvard-architected microcontroller is the only way to go. But, the Harvard architecture lacks the flexibility of the Princeton in some applications that are typically found in "high-end" systems, such as servers and workstations.

The Harvard architecture is really best suited for processors that do not process large amounts of memory from different sources (where the Von Neumann architecture is best) and applications have to access this small amount of memory very quickly. This feature of the Harvard architecture (which is what is used in the PICmicro® MCU's processor) makes it well suited for microcontroller applications.

**Micro-coded versus hard-coded processors** Once the processor's architecture has been decided upon, then the design of the architecture goes to the engineers who are responsible for implementing the design in silicon. Most of these details are left "under the covers" and do not affect how the application designer interfaces with the application. But one detail that can have a big effect on how applications execute is whether or not the processor is a *hard-coded* or *micro-coded* device.

Each processor instruction is, in fact, a series of instructions that are executed to carry out the instruction. For example, to load the accumulator in a processor, the following steps could be taken:

1. Output address in instruction to the data memory address bus drivers.
2. Configure internal bus for data memory value to be stored in accumulator.
3. Enable bus read.
4. Store the data into the accumulator.
5. Compare data read in to zero or any other important conditions and set bits in the STATUS register.
6. Disable bus read.

Each instruction for a processor has a series of steps that must be executed in order to carry out the instruction's function. To execute these steps, the processor is designed to either fetch this series of instructions from a memory or execute a set of logic functions unique to the instruction.

A micro-coded processor is really a processor within a processor. In a micro-coded processor, a state machine executes each different instruction as the address to a subroutine of instructions. When an instruction is loaded into the instruction-holding register, certain bits of the instruction are used to point to the start of the instruction routine (or *microcode*) and the *uCode Instruction Decode and Processor* logic executes the microcode instructions until an instruction end is encountered (Fig. 1-11).

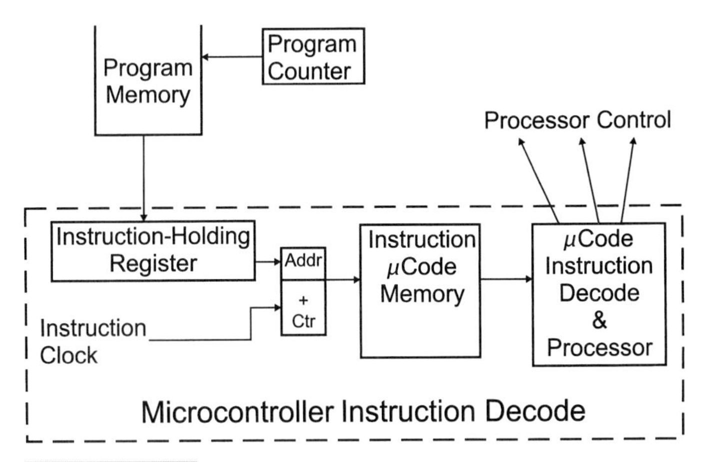


**Figure 1-11** Microcoded processor

Having the instruction-holding register wider than the program memory is not a mistake. In some processors, the program memory is only eight bits wide, although the full instruction might be some multiple of this (for example, in the 8051, most instructions are 16 bits wide). In this case, multiple program memory reads occur to load the instruction holding register before the instruction can be executed.

The width of the program memory, and the speed in which the instruction holding register can be loaded into it, is a factor in the speed of execution of the processor. In Harvard-architected processors, such the PICmicro® MCU, the program memory can be the width of the instruction so the instruction-holding register can be loaded in one cycle. In most Princeton-architected processors, which have an eight-bit data bus, the instruction-holding register is loaded through multiple data reads.

A *hardwired processor* uses the bit pattern of the instruction to access specific logic gates (possibly unique to the instruction), which are executed as a combinatorial circuit to carry out the instruction. Figure 1-12 shows how the instruction loaded into the instruction-holding register is used to initiate a specific portion of the execution logic, which carries out all the functions of the instruction.

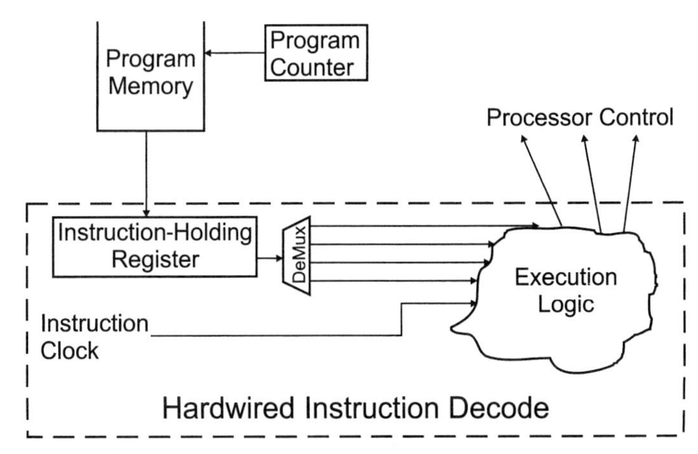


**Figure 1-12** Hardwired instruction processor

Each of the two methods offers advantages over the other. A micro-coded processor is usually simpler than a hardwired one to design and can be implemented faster with less chance of having problems at specific conditions. If problems are found, revised "steppings" of the silicon can be made with a relatively small amount of design effort.

A great example of the quick and easy changes that micro-coded processors allow was a number of years ago when IBM wanted to have a microprocessor that could run 370 assembly language instructions. Before IBM began to design their own microprocessor, they looked around at existing designs and noticed that the Motorola 68000 had the same hardware organization (architecture) as the 370 (although the instructions were completely different). IBM ended up contracting Motorola to rewrite the microcode for the 68000 and came up with a new microprocessor that was able to run 370 instructions, but at a small fraction of the cost of developing a new device.

A hardwired processor is usually a lot more complex because the same functions have to be repeated over and over again in hardware. How many times do you think that a register read or write function has to be repeated for each type of instruction? This means that the processor design will probably be harder to debug and be less flexible than a micro-coded design, but instructions will execute in fewer clock cycles.

In most processors, each instruction executes in a set number of clock cycles. This set number of cycles is known as the *instruction cycle* and is measured as a multiple of clock cycles. Each instruction cycle in the mid-range PICmicro® MCU family of devices takes four clock cycles. This means that a PICmicro® MCU running at 4 MHz is executing the instructions at a rate of one million instructions per second. This is a very popular speed to run the PICmicro® MCU at because each instruction cycle is 1 μsec long and it is very easy to "time" an application or set of instructions. When I present some bit-banging I/O functions, I will cover how the code is timed and I will take advantage of the PICmicro® MCU's consistent four-clock-cycle instruction cycle when I do this.

Using a hard-coded over micro-coded processor can result in some significant performance gains. For example, the original 8051 was designed to execute one instruction in 12 cycles. This large number of cycles requires a 12-MHz clock to execute code at a rate of 1 *Million Instructions Per Second (MIPS),* whereas the PICmicro® MCU must have a clock running at 4 MHz to get the same performance.

Personally, I try to stay away from simplistic comparisons of computer architectures, such as those in the last paragraph. Instead, I try to pick the device that is best suited for an application. Throughout the book, I will work at explaining what I consider to be the most relevant performance measurements, as well as how to select the best PICmicro® MCU device for a given set of requirements.

**Memory types** Memory is probably not something you normally think about when you create applications for a personal computer. The memory in a modern PC available for an application can be up to 4.3 gigabytes (GB) in size and can be "swapped" in and out of memory, as required.

In a PC, when you execute an application, you read the application from disk and store it into an allocated section of memory. In a microcontroller, this is not possible because there is no disk to read from. The application is stored in *nonvolatile* memory and is always the only software that the microcontroller will execute. Having the program always available in memory makes its writing somewhat different than PC or workstation applications.

Understanding how much memory is available in a microcontroller and how it is architected (wired and accessed) is crucial to understand—especially when you are planning on how to implement the application code. In a microcontroller, memory for different purposes is typically segregated and arranged to allow the device to execute most efficiently.

*Program memory* is known by a number of different names, including *control store* and *firmware* (as well as some permutations of the various names). The name really isn't important, instead understanding that this memory space is the maximum size of the application code that can be loaded into the microcontroller and that the application code also includes all the low-level code and device interfaces necessary to execute an application.

A PC program could be as simple as:

```
// Print a simple message

main(){

  printf( "Hello World\n" );

}
```

and the actual application code to carry out this function could be as simple as:

```
MAIN:                         ;  Program Execution Starts Here

  mov     AH, 9               ;  Print the Message
  lds     DX, Msg
  int     021h

  mov     AH, 04Ch            ;  End the Program
  xor     AL, AL              ;  With a Return Code of "0"
  int     021h

Msg       db      "Hello World," 00Dh, 00Ah, 0
```

This code simply calls the operating system routine that will print the message stored at location `Msg` and then returns execution to the operating system.

If you were to execute this program in a microcontroller, this code would be considerably longer because *you* have to:

1 Write the display subroutine (not to mention that you would have to figure out what kind of device to output to and initialize it).
2 Be sure that any other hardware in the device or connected to it is properly initialized.
3 At the end of the application, you will have to determine how the application ends (there is no command line to return to).

This extra code is put into program memory along with the application code. For such devices as an LCD, it's not inconceivable that the support routines require more program memory space than the application.

Rather than scaring you, I hope I've sparked your interest. This aspect of microcontrollers is what really gets me excited about doing applications and projects with them.

With this understanding of how applications execute in a microcontroller, you can look at how it is actually implemented in the device. Earlier in this section, I mentioned that the program memory was nonvolatile. *Nonvolatile memory* is memory that does not lose its contents even when power is removed. Normal or volatile memory circuits lose their contents when power is lost. Volatile memory is most commonly known as *Random Access Memory (RAM)* and can be read from or written to by the microcontroller's processor. The nonvolatile program memory is often known as *Read-Only Memory (ROM)* because during execution, the processor can only read from it, not write new information into it.

The PICmicro® MCU has four different types of program memory available in different devices and applications: none (external ROM), mask ROM, EPROM, and EEPROM/Flash. Although these four types of memory all provide the same function, memory for the processor to read and execute, they each have different characteristics and are best suited for different purposes.

"None" probably seems like a strange option, but in the high-end PICmicro® MCUs running in Microprocessor mode, it is a very legitimate choice. With no internal program memory, the device has to be connected to an external ROM chip (Fig. 1-13)

The external ROM features are primarily used when more application program memory is required or when applications and data are to be loaded into RAM while the application is running.

Microcontrollers are available with read-only memory program memory. When the chips are built, they are completed, except for etching the last metal layer. When an order comes in for a batch of microcontrollers with a ROM with a customer specified application, these chips are pulled from stock and the last metal layer is then etched using a mask made from the customer-supplied software program. This is known as *mask ROM programming.*

With the program put into the chip, the customer will have a device that they can use in their product without having to worry about programming them. Often, ROM contents cannot be read out of the microcontroller to thwart others trying to pirate or reverse engineer the product.

There are some significant downsides to this method of buying microcontrollers. The first two are the cost and "lead time" of having the customized chips built. Although the actual piece price of a ROM program memory chip is less than a device with customer-programmable (field) program memory, the *NonRecurring Expenses (NRE)* costs of getting the mask made only makes this process cost effective in lots of 10,000 (or more) chips. As well, the *lead time* (the time from when the chips are ordered until they are required) for getting mask ROM devices built is typically on the order of six to 10 weeks.

For certain applications, such as for the automotive market, the downsides of mask ROM microcontrollers are not significant. Here, the parts are ordered well in advance of their use and a large guaranteed order is ensured.

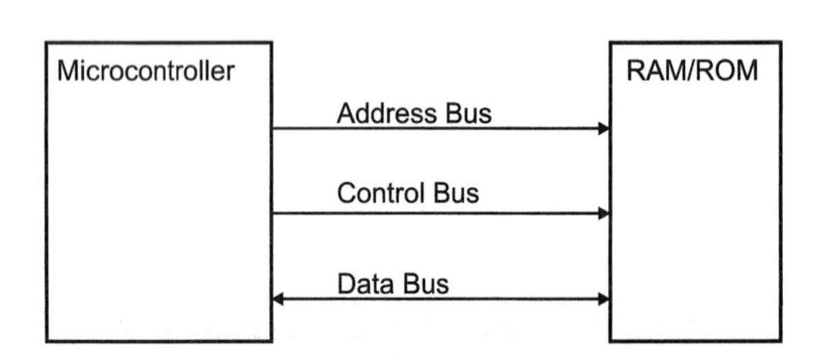


**Figure 1-13** External memory wired to a MCU

The first reprogrammable program memory or *Erasable PROM (EPROM)* program memory microcontrollers were introduced in the late 1970s. These devices used ultraviolet light erasable memory cells that could be loaded with a program and then erased so that another application could be programmed into them. An EPROM memory cell consists of a transistor that can be can be set to be always on or off. Figure 1-14 shows the side view of the EPROM transistor.

The EPROM transistor is a MOSFET-like transistor with a floating gate surrounded by silicon dioxide above the substrate of the device. *Silicon dioxide* is best known as *glass* and is a very good insulator. To program the floating gate, the control gate above the floating gate is raised to a high enough voltage potential to have the silicon dioxide surrounding it to break down and allow a charge to pass into the floating gate. With a charge in the floating gate, the transistor is turned on at all times, until the charge escapes (which will take a very long time that is usually measured in tens of years).

Before programming, all of the floating gates of all the cells are uncharged and the act of programming the program memory will load a charge into some of these cells. By convention, the memory cell acts as a switch to a pulled-up bit. If an unprogrammed memory cell is read, a "1" will be returned because the switch is off. After the cell is programmed and pulls the line to ground, a "0" is returned.

To see if a program memory is ready to be programmed, each byte is read out of it and compared with 0x0FF (all bits set).

To erase a programmed EPROM cell, UV light energizes the trapped electrons in the floating gate to an energy level where they can escape. In some devices, you might discover that some EPROM cells are protected from UV light by a metal layer over them. The purpose of this metal layer is to prevent the cell from being erased. This is often done in memory-protection schemes in which crucial bits, if erased, will allow reading out of the software in the device. By placing the metal shield over the bit, UV light targeted to just the code-protection bit cannot reach the floating gate and the programmed cell cannot be erased.

This might seem like an unreliable method of storing data, but EPROM memories are normally rated as being able to keep their contents without any bits changing state for 30 years or more. This specification is largely a statistical one, based on the probability of the charge in one of the cells to leak away enough in 30 years to change the state of the transistor from "on" to "off." If you are willing to wait long enough, you will find that many EPROM devices will actually store data for much longer than 30 years.

Microcontrollers with EPROM program memory can be placed in two types of packages. If you've worked with EPROM before, you probably have seen the ceramic packages with a small "window" built in for erasing the device (Fig. 1-15). EPROM mi-

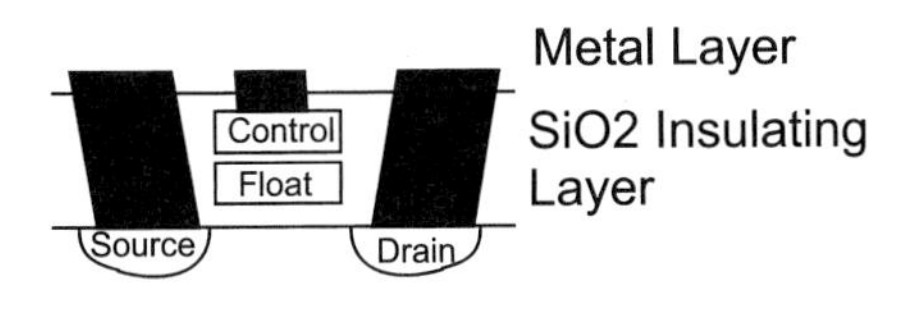


**Figure 1-14** "EPROM" memory cell

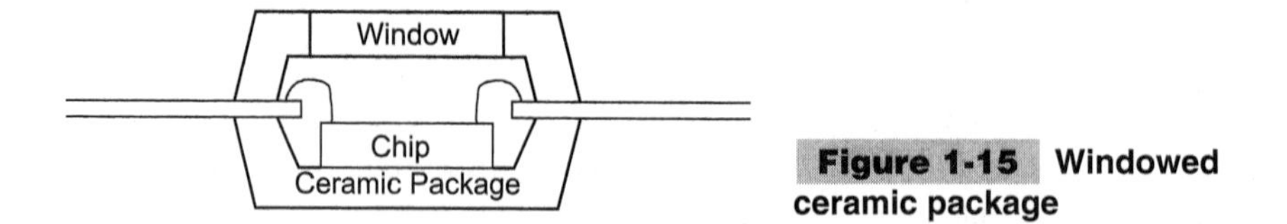


**Figure 1-15** Windowed ceramic package

crocontrollers are also available in packages with no window and are known as *One-Time Programmable (OTP)* which is shown in Fig. 1-16.

OTP devices might seem odd when you consider that the advantage of the EPROM is its ability to be erased and reprogrammed. Taking away the window for UV light to access it, you might as well go with a PROM or mask ROM device.

OTP devices actually fill a large market niche. Windowed ceramic packages can cost 10 times more than cheap plastic packages and, in most microcontroller applications and products, the device will never be reprogrammed. So, by using OTP packaging, the part can still be field programmed, will be electrically identical to the part used to develop the application, and will be very cost effective for quantities less than the break-even point for mask ROM.

Using OTP parts has a significant advantage for the card assembler or manufacturer. If the manufacturer is building several products with the same device (which is not unheard of for microcontrollers), then keeping unprogrammed parts in stock and programming only the number required for the current build run can be a real advantage in terms of inventory-carrying costs. Rather than keeping five or six of the same devices, at a minimum chip-manufacturing build quantity in stock, the manufacturer just needs to keep the minimum number of parts on hand to satisfy their current orders for the various products.

An improvement over UV-erasable EPROM technology is *Electrically Erasable PROM (EEPROM)*. This nonvolatile memory is built with the same technology as EPROM, but the floating gate's charge can be removed by circuits on the chip and no UV light is required.

Two types of EEPROM are used in microcontrollers. The first type is simply known as *EEPROM* and it allows each bit (and byte) in the program memory array to be reprogrammed without affecting any other cells in the array. This type of memory first became available in the early 1980s and found its way into microcontrollers in the early 1990s. EEPROM has been very successful when implemented in small, easy-to-access packages.

In the late 1980s, Intel introduced a modification to EEPROM that was called *Flash*. The difference between Flash and EEPROM is Flash's use of a bussed circuit for erasing multiple cells' floating gates, rather than making each cell independent. This reduced the cost of the EEPROM memory and speeded up the time required to program a device (rather than having to erase each cell in the EEPROM individually, in Flash, the erase cycle, which takes as long for one byte, erases all the memory in the array).

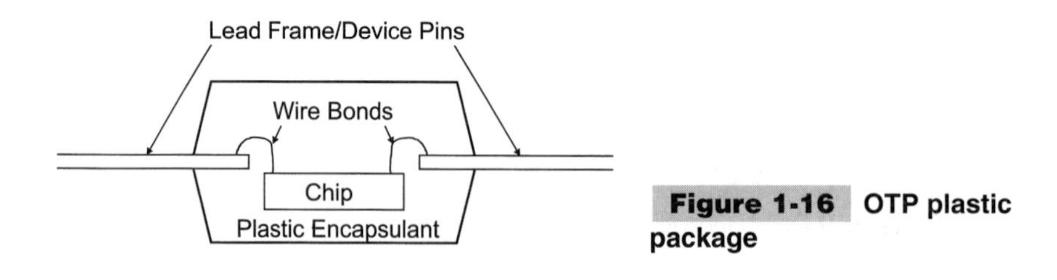


**Figure 1-16** OTP plastic package

Microchip's Flash PICmicro® MCU microcontroller parts are actually EEPROM based. This has caused some confusion in the market because the original EEPROM devices were called *EEPROM devices* and later devices, which use the same technology, were called *Flash. Flash* is the current term used by Microchip to describe their devices with electrically reprogrammable program memory, so I will stick with this convention in this book.

If you've spent some time programming PC applications, you've probably never worried about the space that variables and data structures fill. Most modern PC languages will allow just about unlimited variable storage. If you looked at PICmicro® MCU datasheet before reading this book, you would have seen very little memory in the file registers of the PICmicro® MCU and you probably wondered how complex applications could be written for the device.

Creating complex applications with limited variable RAM in the PICmicro® MCU is not difficult, although large arrays cannot be implemented without external memory. This book presents some very substantial applications without requiring any external memory. These applications also include sophisticated text-based user interfaces that use the ability of the PICmicro® MCU to read program memory for text output data.

A microcontroller has four types of internal variable data storage: bits, registers, variable RAM, and the program counter stack.

All variable storage in the PICmicro® MCU is implemented as *Static Random-Access Memory (SRAM),* which will retain the current contents as long as power is applied to it. This is in contrast to the ROM used by program memory, which does not lose its contents when power is taken away. *SRAM* can be referred to as *volatile memory.*

Each bit in an SRAM memory array is made up of the six-transistor memory cell shown in Fig. 1-17. This memory cell (probably known to you as a *flip flop*) will stay in one state until the write enable transistor is turned on and the write data is "over powers" the state of the SRAM cell.

The P-channel/N-channel transistor pair on the write side of the flip flop will hold this value as a voltage level because it will cause the P-channel/N-channel transistor pair on the read side to output the complemented value. This complemented value will then be fed back to the write side's transistors, which complements the value again, resulting in the actual value that had been set in the flip flop.

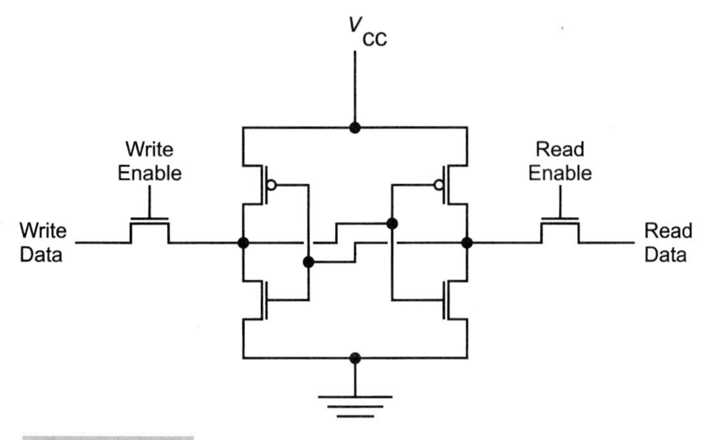


**Figure 1-17** Static RAM ("SRAM") memory cell

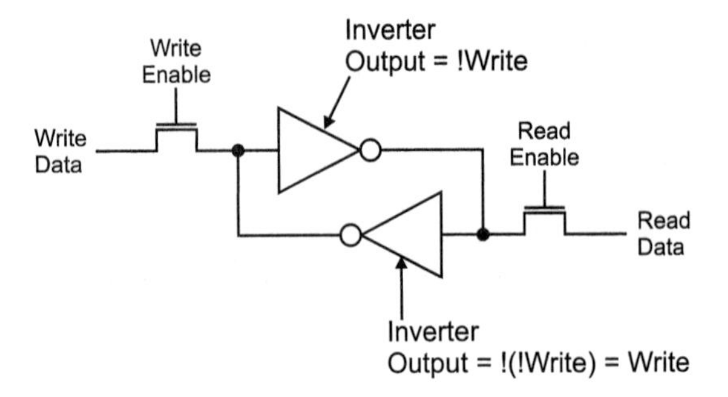


**Figure 1-18** SRAM memory cell block diagram

This circuit is really a pair of inverters feeding back to each other (Fig. 1-18)

Once a value has been set in the inverters' feedback loop, it will stay there until changed.

Reading data is accomplished by asserting the read enable line and inverting the value output (because the read side contains the inverted write side's data). The driver to the SRAM cell must be able to overpower the output of the inverter in order for it to change state.

This method of implementing a SRAM cell is well suited to a microcontroller because it uses very little power (current only flows when the state is changed) and is quite fast. It is not very efficient in terms of silicon space, the six transistors required for each memory cell actually fill quite a bit of silicon surface area ("real estate").

Bits might not seem that useful if you are used to writing PC applications, but in a microcontroller, they can really help make the application much more efficient by allowing fast manipulation of pin states, flags, or state variables. In PCs and workstations, these functions can be carried out in byte-level logical statements.

Working with bits will probably not be something that comes easily to you. If you've been working with "typical" processors and hardware, you will probably be used to thinking in terms of byte-wise instructions. The PICmicro® MCU can manipulate bits very easily and efficiently, although I often find myself thinking in terms of byte operations when I am planning an application.

When eight bits are grouped together, they are known as a *byte.* The byte is really the basic unit of storage in most simple microcontrollers, such as the PICmicro® MCU. Memory bytes are known as *file registers* in the PICmicro® MCU. *Registers* are defined as a limited number of RAM locations that can be accessed very easily by the processor itself. This differs from RAM, which requires some external support and decoding logic and might take a few instruction cycles to access.

Processor stacks are a simple and fast way of saving data during program execution. This is a bit of a dry explanation of stacks and does not really tell you what *you* would use them for. Stacks (Fig. 1-19) save data in a processor the same way that you save papers on your desk. As you are working, the work piles up in front of you and you do the task that is at the top of the pile.

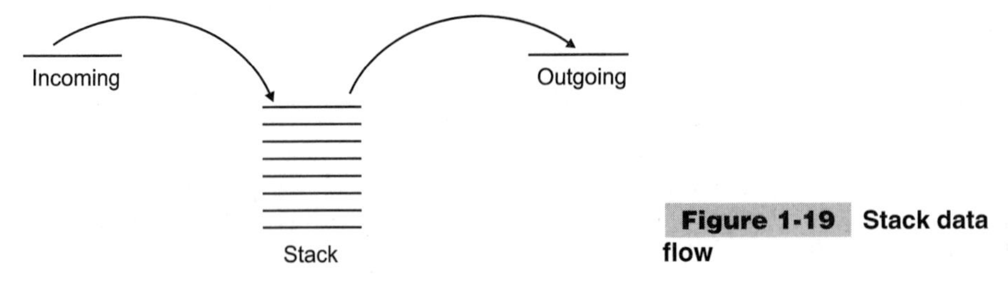


**Figure 1-19** Stack data flow

A stack is known as a *Last-In/First-Out (LIFO) memory.* This should be pretty obvious because the first work item on the stack of paper will be last one you get to.

In a computer processor, a stack works in exactly the same manner as the stacked paper example. Data most recently put onto the stack (known as a *push*) is the first item pulled off the stack (known as a *pop*).

The PICmicro® MCU has only a program counter stack available (which is used to point to the next instruction to be executed) and nothing for data itself. If you want to save data in a stack-like fashion, you will have to simulate the stack functions using code:

```
push( data ) {

  SP++;                          // Point to the Next Address in Memory
  Stack[ SP ] = data;            // Store the Data

} // End push

int pop() {

int i;

  i = Stack[ SP ];               // Get Data Pointed to by the SP
  SP--;                          // Decrement the SP

  return i;

} // End pop
```

The PICmicro® MCU's stack is covered in more detail later in the book. This code is intended introduce you to the concept of the LIFO stack, what is, and what isn't available in the PICmicro® MCU. Not having a stack for data can make some applications awkward in the PICmicro® MCU, but I present a number of methods and rules used to implement the data stack that will simplify the task of adding one to your application. In the higher end PICmicro® MCUs, the data stacks can be very easily implemented using their advanced indexed addressing options.

**Interrupts** Interrupts can improve the efficiency of your applications and simplify your application code. Despite this, they are seldom used and are often avoided as much as possible. For many application developers, interrupts are perceived as being difficult to work with and something that actively complicates your application. This book presents a lot of information on interrupts and applications that use interrupts because of the ease of programming and processing speed improvements that they can offer PICmicro® MCU applications.

Computer interrupts are very analogous to interrupts in your every day life. As the computer processor is executing application code, a *hardware event* might occur, which requests the processor to stop executing and respond or handle the hardware event. The hardware event requesting the interrupt can be a timer overflow, a serial character received (or finished sending), a user pressing a button, and so on. Many different hardware events can cause an interrupt to occur.

As the processor is executing application statements if an interrupt request is received by the processor and the interrupt acknowledge hardware is enabled, the processor saves the current application information before executing the interrupt handler (which is often referred to as the *interrupt service routine*). I refer to the executing application code as the *application mainline* and the interrupt handler as just the *handler.* Interrupt operation is shown in Fig. 1-20.

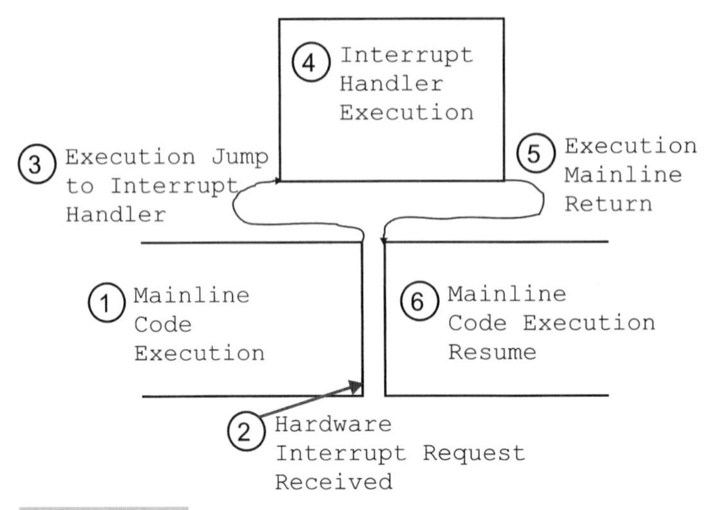


**Figure 1-20** Interrupt execution

By following the process laid out in Fig. 1-20, the operation receiving an interrupt request and processing it is explained. As the mainline code is executing (1), a hardware interrupt request is encountered by the processor (2). When the interrupt is responded to (3), all of the information pertinent to the mainline execution (known as the *context information*) is saved. Once this is done, the interrupt handler code is executed, which saves the hardware interrupt request information and resets the interrupt request hardware (4). With the interrupt "handled," execution can now return to the mainline. The context registers are restored (5) and execution (6) can resume. Normally, interrupts are enabled when execution returns to the mainline, but they can be enabled inside the handler to allow *nested* interrupts.

When an interrupt request is acknowledged and execution jumps to the handler, the address it jumps to is known as the *interrupt vector.* This address or vector can be specified by the application, or a specific address is used by the processor. In the PICmicro® MCU, this address is constant. Some processors (including the higher-end PICmicro® MCUs) might have multiple interrupt vectors for handling multiple interrupt sources with different interrupt handlers.

The context register saving and restoring can occur totally within the interrupt acknowledge (in which execution jumps from the mainline to the interrupt vector), it might have to be done totally manually (the PICmicro® MCU falls into this category), or somewhere in between. If context register saving has been done manually, then I recommend initially saving all appropriate registers and, as you gain more experience, you can reduce the number of registers stored.

In the interrupt handler, I recommend processing interrupts by first saving the interrupt occurrence, resetting the hardware so that another interrupt can occur, and finally processing the data. Depending on the application, this three-step process might be truncated or changed. The important point is to have interrupts execute as quickly as possible to ensure that no interrupt requests are missed.

In some processors and applications, you might want to allow interrupt requests to execute from interrupt handlers. This is known as *nested interrupts* and their execution looks like the operation shown in Fig. 1-21.

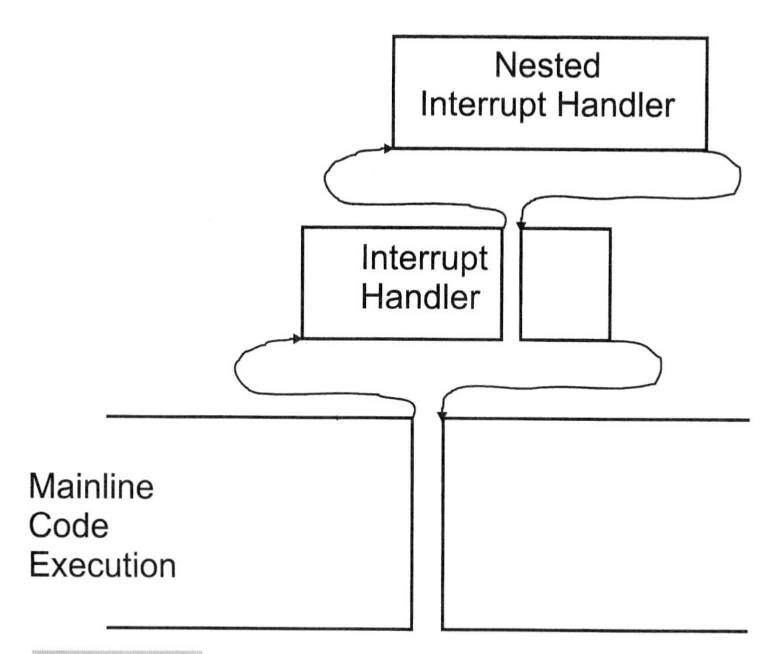


**Figure 1-21** "Nested" interrupt execution

Nested interrupts can be difficult to code properly. Depending on how the application executes, requests might be missed, unless you are very sure of your application and how requests come in and are handled.

Not using interrupts does not mean that in-line event handlers should not follow the three steps (save interrupt information, reset interrupt hardware, and process interrupt) used by interrupt handlers. The three-step interrupt handler process minimizes the opportunity for subsequent interrupts to be missed, which is especially important if they are coming in at an undetermined interval. Ideally, interrupt handlers should be written to execute as quickly as possible so that no subsequent requests occur while the interrupt handler is active and unable to respond to them.

An important concept to understand about interrupts is that they are *requests.* Depending on the application, the request might be refused or postponed. In an application, the code might be executing what is known as *crucial code.* This code is not interruptible because of its priority or that it is executing code that must execute in a crucially timed application. In these cases, the application can ignore the interrupt altogether or postpone it until the crucial code has completed instead of executing as soon as it receives an interrupt request.

As I was writing this, I looked over the first edition of this book, as well as the other books I have written and saw that I have shown how computer interrupts are analogous to the following every-day activities:

1. Watching *The X-Files.*
2. Watching *Star Trek.*
3. Working at your desk.
4. Driving your car.

If I was to include a discussion on how interrupts are analogous to eating, I think I would encompass every normal activity that a typical reader of my books would do during their day. With this perspective, you should realize that interrupts are something that you should consider using in your PICmicro® MCU applications because they are something that you are actually very familiar with. When used appropriately, interrupts will simplify your application code.

Before reading through additional information, I would suggest that you read through the "Interrupts" section of "Introduction to Programming" on the CD-ROM, as well as the following section, "Event-Driven Programming," which will give you a better idea of how they can be used in microcontroller applications.

## SOFTWARE-DEVELOPMENT TOOLS

When the first computers were built more than 50 years ago, there were no automated software-development tools. Applications were written by hand and converted to instruction bits and then manually loaded into the computer ROM. ROM in the first computers consisted of setting switches, which was a pretty onerous task, compared to today's relatively simple chip programming. This process was very human intensive and had to be done very carefully, lest an error occur in the conversion and loading of instructions, which would be difficult to recognize and debug.

Today, a number of development tools are available to simplify the process of developing application software and loading it into the computer, or, in this book's case, the PICmicro® MCU microcontroller.

The most basic software development tool is the assembler, which converts program instructions written by the programmer into the bits required by the processor. Even though I describe the assembler as the "most basic" development tool, it can be very complex. Along with processing straight instructions, data can be brought in from other sources, and macros and defines can be used to simplify the application. This book focuses on assembler and applications written using it.

*Compilers* convert high-level languages into processor instructions and actual bits and bytes in a similar manner to the assembler. The difference between an assembler and a compiler is that an assembler processes device-specific instructions while a compiler converts a statement:

```
A = B + C
```

into device-specific instructions.

Both assemblers and compilers can produce complete application *hex files,* which can be loaded into a computer (or microcontroller) ROM directly and executed. They can also produce *object files,* which hold a subset of the entire application and can be *linked* together to form the application's hex file.

Although PICmicro® MCU linking tools are available, I will recommend not using them, except in the situation where source languages are mixed. For the relatively small applications written for the PICmicro® MCU, having the complete source code available in one application makes reading through the code easier, the assembly faster, and avoids the issue of keeping track of multiple files for an application.

Along with the assemblers and compilers, which convert source code into the bits required to execute on the PICmicro® MCU microcontroller, are *simulators,* which allow you to watch the execution of an application on a PC or workstation. This eliminates the need for *burning* (programming) a PICmicro® MCU and watch it attempt to execute an application. It allows you to watch how the application is executing instead of just programming it and hoping for the best.

An *emulator* is a step up from a simulator in that it is an actual PICmicro® MCU that can be inserted into a circuit. In this case, an application can run in the actual circuit, rather than be simulated in another computer. Emulators tend to be quite expensive, but can be a godsend when you can't figure out what is happening in an application.

Simulators and emulators, although excellent tools for checking an application, are not development tools and do not replace the process of designing an application and writing the code in a structured manner. Elsewhere in this book, both simulators and emulators are described in detail, along with how I believe they are best used in the application-development process.

# Programmable Logic Devices

*Programmable Logic Devices (PLDs)* are chips that have logic gates and flip flops built in, but are not interconnected. The application designer will specify how the gates and flip flops are interconnected in order to create a portion of the application's circuit. Most people feel that programmable logic devices are a relatively new invention, but they have been around for many years. It has only been quite recently (in the last 10 years or so) that reusable chip technology (i.e., EPROM and Flash) PLDs have been available at prices that hobbyists and small companies could afford.

There are two types of PLDs. The first is the simple array of logic gates and devices, known as *PALs* and *GALs* (I generically refer to them as *PALs*). The chips themselves are quite simple and relatively easy to design circuits for. These circuits are normally arranged as a "sum of products," in which signals on the chip can be easily interconnected to form more complex logic functions. The chips are normally blocked out as a series of inputs and outputs (Fig. 1-22).

The vertical lines (busses) in Fig. 1-22 are referenced to the gates and I/O pins to which they are connected.

To form logic functions, the "sum of products" is used. In Fig. 1-22, a simple four-I/O, 12-gate PAL is shown. Every output is driven on a bus in both positive, as well as negative format. Connections are made between the gates and the busses to create logic functions.

For example, the XOR gate, which is characterized by:

```
A XOR B   = A ^ B
          = _A AND B OR A AND _B
          = (_A * B) + (A * _B)
```

is not often available in standard logic. Taking Fig. 1-22 and connecting the busses to the different I/O pins and gates within the PAL, I can implement the XOR gate (Fig. 1-23).

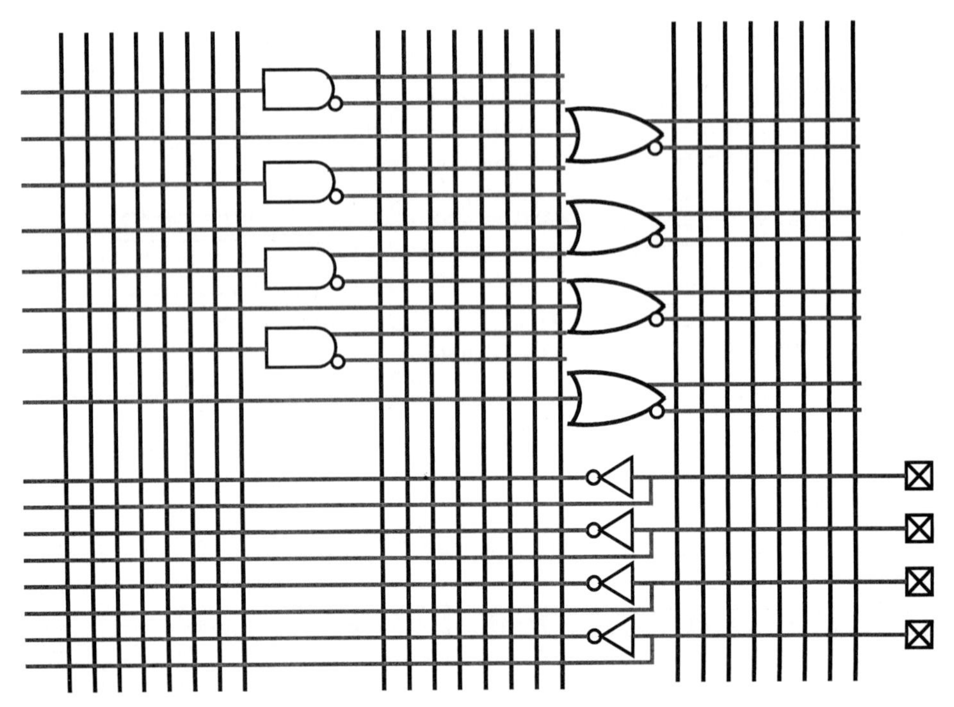

**Figure 1-22** PLD matrix connections

Note in Fig. 1-23 that an I/O pin changes from an input to an output by simply connecting it directly to a gate output. This feature allows the pins to be used as either inputs or outputs.

Options for PALs include varying numbers of inputs to the internal AND and OR gates. For the PLD shown in Fig. 1-22, I have left open the option that any of the pins can be used for any purpose. This is a bit unusual; normally in PALs, the number of inputs to a gate is restricted. Another option is to include built-in flip-flops to store states and turn the PAL from a combinatorial circuit into a sequential one.

PALs might seem simple, but they can result in large decreases in the chip count for an application. In some cases, PALs are more expensive than the chips they replace, but they reduce application power and board-space chip requirements. These savings could result in overall product savings. It is not unusual for 10 TTL chips to be replaced by a single PAL, resulting in huge PC board and power-supply cost savings.

At the high end of the programmable-logic device family range, some devices are virtually *ASICs (Application-Specific Integrated Circuits)* and use the same programming language (VHDL) and development tools as ASICS. These complex parts generally have their functions broken into macros. An ASIC/PLD macro can be an AND, or XOR not, logic gate, flip flops, or collections of functions (such as multiplexers and arithmetic logic units), which simplify the task of circuit development and eliminate the need for wiring individual gates into basic functions.

The high-end programmable logic device's programming information is often directly transferable to the technology. This allows initial production to use programmable logic

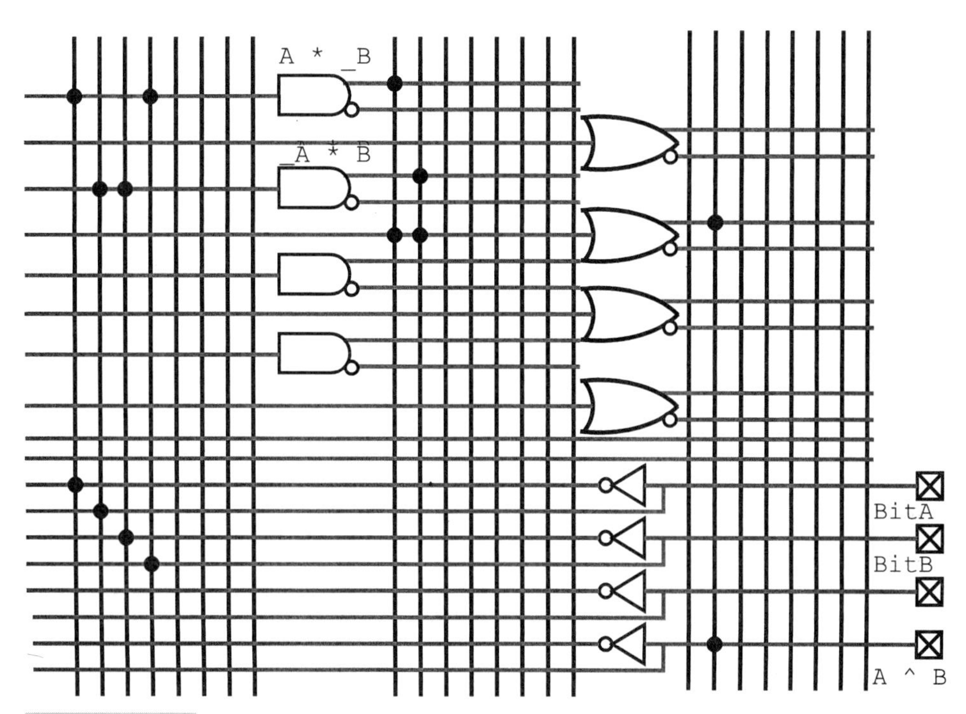


**Figure 1-23** PLD matrix connections for an "XOR" gate

devices that requires little cost to program and when the design is qualified, ASICS can be built at a chip foundry for reduced costs per unit.

Programmable logic devices have the advantage of being able to implement fast (less than 10 ns) logic switching, but they typically do not have the ability to store more than a few bits of data.

Often programmable logic devices are used in proprietary circuits because their functions cannot be easily traced and decoded.

Programmable logic device and ASIC development tools are generally function text based, as opposed to graphically based applications (like a schematic drawing). This means that a text format, like the XOR definition, must be used to define the functions. Most compilers for these statements are intelligent enough to pick the best gates within the device to work with and pick the best paths without your intervention. They are typically much more sophisticated (and expensive) than the compilers used to convert high-level program statements into instructions for a processor.

# Deciding Which Device to Use in Your Application

Deciding what kind of part to use before beginning to do the design work can have significant impact on the cost, performance, and development effort required for an application.

As you gain more experience with electronic devices and application development, you will be better able to decide which device is most appropriate for a given set of requirements.

Notice that I didn't say "best" with regard to the device choice. Part selection is very subjective. What one individual considers best for a set of requirements probably will not be same as what others consider the best. When deciding which device to use, you will probably consider different issues:

- Built-in features
- Device cost
- Tool availability
- Familiarity/expertise availability

Many of these comparisons are very subjective and, as I said, will be different for different individuals.

Personally, I try to list devices in applications that meet the requirements as completely as possible and then work at deciding at what is the "correct" device.

PLDs are excellent for situations where fast and "dumb" responses to input are required. PLDs are very well suited for replacing "glue" chips that provide complex combinatorial functions. Timed responses and sequential circuits can be very difficult to implement with PLDs.

Microcontrollers are best suited for intelligent interfaces with varying timing requirements. These interfaces can include keypads, LCDs, and busses. For these types of applications, the Microchip PICmicro® MCU is very well suited for use as a single-chip solution.

2

# THE MICROCHIP PICmicro® MCU

## CONTENTS AT A GLANCE



Over the years, many different eight-bit embedded microcontrollers have appeared on the market. Many of these designs were created by companies that developed the microcontrollers for use in other products that they sell (often cellular phones and pagers). I call this the *captured market* because no matter what, a microcontroller family will always have a customer that will buy the microcontroller with any product left over when offered on the open market for sale. Third-party product developers can take advantage of left-over stock that the microcontroller manufacturers have made available.

Independent microcontroller manufacturers tend to be much smaller companies than the captive device manufacturers (mostly because they do not manufacture products other than semiconductors) and tend to have to "scramble" to get customers interested in their

…any of these companies and microcontroller designs are niche players who tar… …arkets and opportunities.

…hip, with their PICmicro® MCU families of microcontrollers, has avoided these …es and has established itself as a market leader (going from the number 20 to … two supplier of microcontrollers on the open market in five years). The company …en able to establish this stellar growth by continuously improving and enhancing the …icro® MCU families, according to customer demands. Also, they have provided wo…d-class application development tools for the PICmicro® MCU at competitive prices.

I use the term *families* when describing the PICmicro® MCU because four different distinct processor architectures are described in this book. From a high level, each PICmicro® MCU can be described as a *microcontroller* with an eight-bit RISC Harvard-architected processor that has built-in clocking, reset, memory, timers, interrupts, input/output (I/O) pins, and optional advanced interfacing capabilities. The advanced interfacing options are common hardware macros that are available to the different PICmicro® MCU processor families. The PICmicro® MCU is a remarkably easy to work with microcontroller that is well optimized for many different applications.

Since writing the first edition of this book, Microchip has released more than *100* new versions of the PICmicro® MCU (with each version known as a new *part number*). Each part number has unique processor features, program and register memory, as well as I/O capabilities that help make it easier for the application designer to select a part that is best for their application. Among the features that make the PICmicro® MCU extremely attractive to new developers is the ease in which the devices can be programmed using very simple hardware (as you can see in the El Cheapo programmer that is included with this book).

Along with the new part numbers, Microchip has also been continually improving the development tools that are available for the PICmicro® MCU. The MPLAB IDE is one of the best *Integrated Development Environments (IDE)* that I have ever worked with and it continues to be improved. Microchip also has its own line of programmers and emulators that are quite affordable, of excellent quality, and are integrated to work with the MPLAB IDE.

The PICmicro® MCU is astonishingly well supported by users with well over 1,000 Web sites (as of this writing) that are devoted to it and feature applications, third-party (commercial and GNU license) development tools and *Frequently Asked Questions (FAQs)*. These sites make the PICmicro® MCU one of the most-represented and best-supported devices on the Internet. MIT's PICList is one of the busiest List Servers on the Internet with more than 2,000 subscribers discussing PICmicro® MCU issues and helping out with other's problems. The PICmicro® MCU is probably the second-best Internet-supported electronic device on the Internet, with only the PC having more Web sites and information.

All of these factors make the Microchip PICmicro® MCU an excellent microcontroller to learn about and work with. This is true if you are just starting out in electronics and programming or if you are an expert looking to add some new skills to your repertoire.

# Device and Feature Summary

When you first look at the selection of different part numbers for the PICmicro® MCU, you will probably be overwhelmed by the number of choices. As I write this, 117 different PICmicro® MCU part numbers are available, with as many as thirty new ones available each year. It is a credit to Microchip that the part numbers are reasonably logically ordered and the number of different part numbers are not a hindrance, but actually an advantage for you when selecting what is the best part for you to use for your applications.

I tend to group the part numbers according to the processor architecture and the features on the chips (Table 2-1).

**TABLE 2-1 PICmicro® MCU Part Numbers To Feature Table**

| PART NUMBER | ARCHITECTURE | FEATURES | APPLICATIONS |
|---|---|---|---|
| 12C5xx | Low-End | Internal | Simple Interfacing Osc/Reset |
| 12C6xx | Mid-Range | ADC/Internal | Simple Interfacing Osc/Reset/ Data EERPOM |
| 14C000 | Mid-Range | ADC/Vref | Power Supply Control |
| 16C5x | Low-End | | Basic Applications |
| 16C505 | Low-End | | Basic Applications |
| 16HV540 | Low-End | Voltage Regulator | Basic Applications |
| 16C55x | Mid-Range | | Basic Applications |
| 16C6x | Mid-Range | | Digital Applications |
| 16C62x | Mid-Range | Voltage Comparator | Analog Monitoring |
| 16F62x | Mid-Range | Voltage Comparator/ Flash Program Memory | Analog Monitoring |
| 16C7x | Mid-Range | ADC | Analog Interfacing |
| 16x8x | Mid-Range | Flash Program Memory | Application Development |
| 16F87x | Mid-Range | ADC/Flash Program Memory | Analog Interfacing/ Applications Development |
| 16C9xx | Mid-Range | ADC/I2C | Analog Interfacing |
| 17Cxx | High-End | External Memory | Advanced Applications |
| 18Cxxx | 18Cxx | ADC/I2C | Analog/Digital Interfacing |

This book covers the four different processor architectures and the different hardware features that are available in the different PICmicro® MCU devices. The applications presented use the PICmicro® MCUs that I consider the best suited for them.

For device-specific information, I have included a number of PICmicro® MCU datasheets on the CD-ROM that comes with this book. If you are looking for specific features, you should check the Microchip Web site for the latest part list, as well as the Microchip product line card, which details all the parts that are available, along with their features.

## LOW-END ARCHITECTURES

When the PICmicro® MCU first became available from General Instruments in the early 1980s, the microcontroller consisted of a very simple processor executing 12-bit wide instructions with basic I/O functions. Some devices were able to work with external program memory, and others had built-in programmable ROM on board. Variable RAM in these devices consisted of a few tens of bytes. The chips themselves were built using numerous manufacturing processes and were quite inefficient in terms of power consumption compared to modern devices. Over the years, these microcontrollers have been improved by redesigning them with CMOS technology, providing better program memory that can be easily programmed (or burned) in the field, along with additional features.

Despite these improvements, the original PICmicro® MCU architecture has become the low-end of the PICmicro® MCU microcontroller families. The devices do not have many of the features of the other PICmicro® MCU families that make them less attractive to work with for many applications. This is not to say that the low-end devices, (which have been given the part numbers: 12C5xx, 16C5x, and 16C505), are not useful and should not be considered when planning an application. Instead, they should be considered for specific application niches. The application niches should be chosen with the following application requirements in mind:

- Simple interface functions
- Limited variable memory
- Simple digital interfaces

I make these recommendations primarily because the low-end PICmicro® MCU's do not have a lot of program memory and cannot implement involved application code. Many devices only have 512 instructions available with the maximum for the architecture being only 2048 (2K). This is not to say complex applications and code cannot be implemented, just that code with complex interface functions (and complex text interfaces) should not be attempted with low-end PICmicro® MCUs.

When the low-end PICmicro® MCUs first came out, they were given their own unique register names and conventions that differed from the mid-range devices. These variances caused some confusion with people transitioning between the parts. To help alleviate this problem, over the past few years, Microchip has been changing the low-end device register names and documentation to better match the mid-ranges. This has resulted in some confusion for people who have worked with the low-end devices or who have older documentation.

I recommend that you only use the latest low-end documentation and stick with the register and resource names that match the mid-range devices to simplify the effort in moving (porting) applications between architectures. With these changes to the documentation, the low-end devices have become much more like the mid-range, although with fewer features. For this reason, I tend to call the low-end architecture a *subset* of the mid-range architecture.

When I first started working with PICmicro® MCUs, I didn't feel like the low-end architectures were that useful because of the limited program and variable memory, no interrupts, no advanced peripheral features, and no serial programming. This conclusion was felt even more strongly because of the availability of the low cost mid-range parts that do not have these limitations. Microchip has kept the low-end architecture viable with the release of the 12C5xx and 16C505 devices that are ideally suited for simple, digital interfacing applications. These parts are extremely low cost and can be used to replace common clocking (such as the 555 timer) and logic chips.

I believe that the low-end projects presented later in this book are well suited to the low-end devices. I focus on the 12C5xx and 16C505 parts that have built-in reset and clocking capabilities that can be used to simplify overall applications.

## MID-RANGE APPLICATIONS

When you look at a list of PICmicro® MCU part numbers, you will probably realize that the mid-range processor architecture is used in an overwhelming majority of the microcontrollers that Microchip makes. This is the reason why I focus on the mid-range in this book and show the other differences of PICmicro® MCU architectures as variations on the mid-ranges.

The mid-range PICmicro® MCUs also have the widest range of peripheral enhancements available to any of the other PICmicro® MCU families and more complete than this diversity means that there are over 200 different part numbers from which to choose from to help you decide the optimal solution to your application needs. Depending in your experience with other microcontrollers, you might be taking this statement with a grain of salt; other manufacturers have had problems supplying all the parts they advertise or only make certain part numbers available to low-volume customers. Microchip has gone to great efforts to ensure that all PICmicro® MCUs are available and virtually all package types from distributors. The only exception to this would be bare dies in waffle pack shipping containers that are only available for high-volume customers ordering directly from Microchip.

If you are familiar with the classic Von Neumann architecture, all of the PICmicro® MCU families (not just the mid-range) will seem pretty strange. This book focuses on and presents the mid-range architecture from a block-diagram perspective. Each aspect of it is explained so that you will understand how the instructions execute and what makes them attractive to optimizing the application. This point is probably the most important: when you are familiar and comfortable with the PICmicro® MCU's architecture, you will be amazed at what you can come up with.

The mid-range PICmicro® MCU architecture uses the low-end's and enhances it with the ability to access many more registers. This ability is used to implement advanced I/O peripheral devices, as well as more variable memory. In either device architecture, you will find that the PICmicro® MCU processor architecture will allow you to optimize code in ways that cannot be done in other architectures.

## PIC17Cxx DEVICES

The PIC17Cxx PICmicro® MCU is the most different from the other three PICmicro® MCU processor architectures presented in this book. The PIC17Cxx has the ability to interface with eight- and 16-bit parallel bus devices, as well as having quite good built-in serial (asynchronous and synchronous) interfaces. Besides the built-in parallel bus and serial interfaces, the PIC17Cxx also has a number of timers that have good support for pulse generation and measurement.

The true strength of the PIC17Cxx is in its architecture and computational abilities. The PIC17Cxx processor architecture is quite different from the low-end and mid-range components and has some features that allow faster internal data movement and processing. Along with the addition of a built-in hardware multiplier and faster clock speed, the PIC17Cxx can digitally process audio signals for some applications. This capability is really not available in the low-end and mid-range PICmicro® MCUs.

## PIC18Cxx DEVICES

As I write this, the first production samples of the newest PICmicro® MCU family, the PIC18Cxx, are becoming available. The PIC18Cxx is well positioned to be the architecture family of choice in the PICmicro® MCU lines with the peripheral features of the mid-range and an enhanced processor that has significantly increased capabilities over both the mid-range and PIC17Cxx devices. When I first became interested in the PICmicro® MCU, the PIC16C54 was the device of choice for new users. When I wrote the first edition of this book, the PIC16C84 was the most popular device. For this edition, I am focusing on the PIC16F84 and PIC16F877, but when the next edition comes out, I wouldn't be surprised if a flash-based PIC18Cxx is featured.

The PIC18Cxx processor offers the following advantages:

- Up to 1 MB of instructions can be addressed in the program memory
- Up to 4 KB of file and hardware registers
- A software-accessible stack
- Improved oscillator options

The PIC18Cxx has a 16-bit instruction word and instruction set that is source-code compatible with the mid-range devices. As the PIC18Cxx architecture and applications are presented, mid-range compatible code is used as much as possible to simplify the task of porting mid-range applications to the PIC18Cxx parts.

Although increased memory access and more oscillator options (including the use of the OSC2 pin as I/O) are useful advantages of the PIC18Cxx over the other PICmicro® MCU families, the ability to read and write the stack is very exciting. As shown later in the book, this feature can be used to implement a true real-time operating system, something that cannot be done in the other PICmicro® MCU architecture families.

## ROM/EPROM/FLASH

There are a lot of options for the PICmicro® MCU and one of the most important is the type of program memory that is used for an application's PICmicro® MCU. Choosing program memory has implications on cost, customer response, and security. The choice should not be taken lightly because it can very well be a determining factor in the success of your product.

Some microcontrollers, such as the 8051, have an option for external memory to be used instead of internal memory. This option makes it easier to develop applications in some cases, but it does mean that fewer I/O lines are available for the application. This option is not available for the low-end and mid-range PICmicro® MCUs because the instruction word size is not evenly divisible by eight (the most common size of RAM, ROM, PROM, EPROM, and Flash memory chips). The PIC17Cxx devices, with their 16-bit instruction word do have an external memory capability. The PIC18Cxx currently does not have this capability, but it could be added, like the PIC17Cxx's, in the future.

I tend to avoid using external program memory because of the loss of pins, increased board and wiring requirements, and extra component costs. The different PICmicro® MCU families have varying amounts of program memory built in, which allows you to choose the part number that best suits your application.

The most basic program memory option for the PICmicro® MCU is EPROM. This is probably surprising because, based on experience with other devices, most people would consider ROM to be the most basic option available for microcontrollers. Microchip has made mask-programmable ROM available for most devices, but will ask a customer to consider EPROM first—especially to support initial builds and application debugging. Some low-end, all mid-range, and the high-end chips support serial programming, which can be done on assembled products. This is known as *In-Circuit Serial Programming (ICSP)* and is covered in detail later in this book.

EPROM parts are available in two types of packages. "Windowed" packages are quartz-windowed ceramic packages that allow ultraviolet light to the chip for erasing. The windowed ceramic packages are usually given the package code *JW*. *One-Time Programmable (OTP)* packaging consists of plastic encapsulant used for the chip that is soldered to the board. In this type of package, no ultraviolet erasing light can reach the device to erase it after it has been programmed (which is why it is called *one-time programmable*).

Figure 2-1 shows an OTP PIC12C508 along with a windowed ceramic "JW" PIC12CE673 and a Surface Mount Technology (SMT) PIC12C508.

One-time programmable EPROM parts are designed for production and can only be programmed once. Using these parts has three advantages over ROM:

- There is no part procurement lead time.
- There is no NRE needed for buying product.
- Product can be shared between customers.

*NonRecurring Expenses (NRE)* required for making ROM masks and the stocking costs of unprogrammed parts can be shared for different products as demand warrants. Stocking unprogrammed EPROM parts are an advantage if a code problem is uncovered, compared to stocking a batch of ROM parts that are no longer usable in the product. From this kind of perspective, it is easy to see that EPROM is preferable to ROM for many situations.

Microchip was one of the first microcontroller manufacturers to provide Flash (or EEPROM) program memory. The advantage of this type of program memory is that it can be programmed without an ultraviolet erasure step. Thus, applications can be built into a

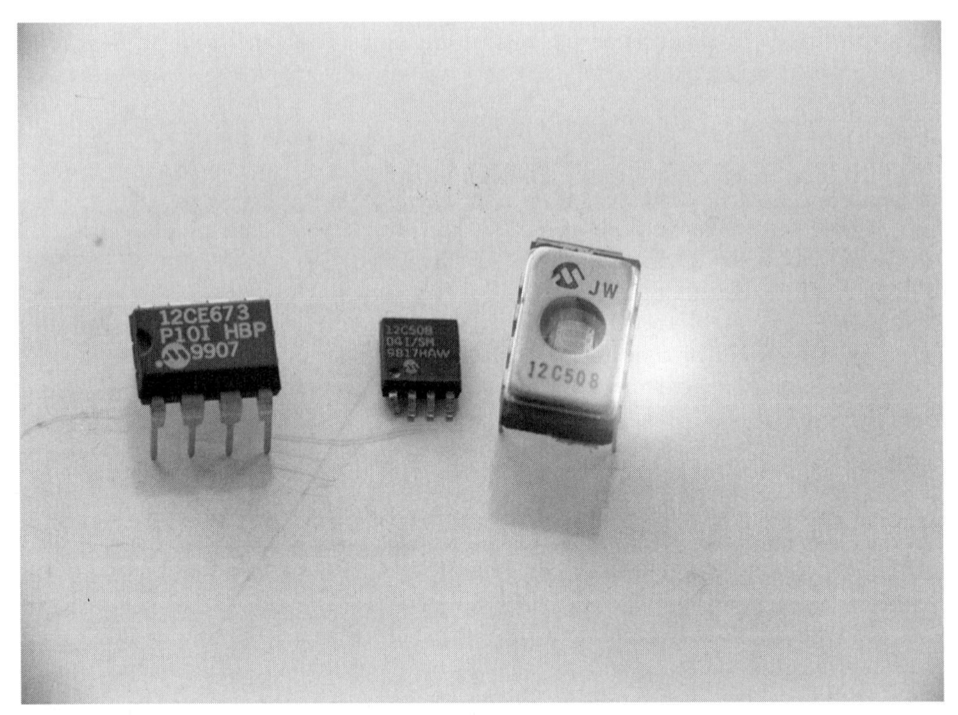


**Figure 2-1** Different 8-pin PICmicro® MCU packages

circuit and be reprogrammed during application development. Microchip refers to its electrically reprogrammable PICmicro® MCUs as *Flash*—even though it does have the electrically erasable program (EEPROM) capability of modifying individual words in the program memory. In some newer devices, program memory can be changed from within the application and this ability is taken advantage of in the PICmicro® MCU emulator (EMU-II) presented later in this book.

One of the biggest complaints that I see about the choices that Microchip has made regarding programming is the unavailability of Flash versions for all devices. As time has gone on, more Flash-based PICmicro® MCUs have become available. Now, representative devices are available for virtually all of the mid-range, with new devices planned to fill the gaps, as well as be provided for the 18Cxx architecture.

Chapter 15 ("Experiments") uses Flash-based PICmicro® MCUs almost exclusively. This is to take advantage of the reprogrammablility of the Flash to allow fast application changes for doing "what if" testing, as well as to keep your costs down when creating the experiments. Chapter 16 ("Projects") uses more EPROM-based parts because these applications are more permanent and less likely to be changed over time.

## PERIPHERALS

A good portion of this book is dedicated to the built-in peripherals of the PICmicro® MCU. These peripherals range from advanced serial I/O capabilities, analog-to-digital conversion, pulse timing, and output. Mastering these features will simplify your applications and make them much more cost effective than if you had to add external chips to per-

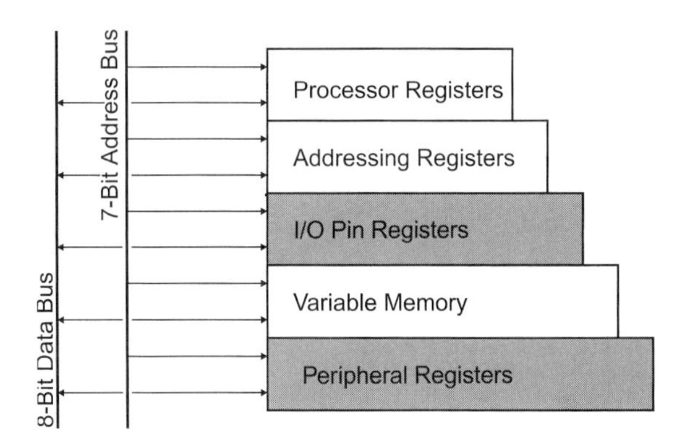


**Figure 2-2** **Block diagram of peripherals in a PICmicro® MCU**

form the functions. Along with covering how the built-in peripherals are used, information is provided on how these functions can be simulated in bit-banging routines.

All of the built-in peripherals are on the PICmicro® MCU's internal data bus and take up addresses in the register space of the processor. This register space is used to locate processor-specific registers (such as STATUS), addressing registers (such as INDF and FSR), I/O pin registers (for example, PORTA and TRISA are really peripherals themselves), variable memory, and the peripheral I/O registers. One way to visualize this is shown in Fig. 2-2.

The peripherals can often also request hardware interrupts. The base interrupt control and acknowledge register is INTCON, but many PICmicro® MCUs have additional registers available for multiple interrupt sources.

In most PICmicro® MCUs, the peripherals use I/O pins that are shared with other I/O functions (usually parallel I/O). Depending on the peripheral, when the peripheral hardware is engaged, the I/O pins automatically are devoted to the task or they might have to be explicitly written to for the peripheral function to be enabled. For some analog I/O digital converter input pins, they are set in ADC mode on power up, and some instructions are required to put them in a mode where they can be used as digital I/O.

This book covers many of the different peripherals available within the PICmicro® MCU and some of the things to watch out for. To make your job of interfacing to them easier, I have included macros that will help you to develop assembler application code, requiring just a few simple parameters to use in your applications.

## DEVICE PACKAGING

When I use the term *device packaging,* I am describing the *encapsulant* that is used to protect the chip and the interconnect technology used to connect the chip electrically to the printed circuit card (which I call the *raw card* or *Printed Circuit Board* or *PCB*). There are quite a few options in this area; selecting the appropriate ones to use can have a significant impact on the final application's cost, size, and robustness.

The two primary types of encapsulation used to protect chips are plastic and ceramic. *Plastic encapsulants* are the most prevalent and use an epoxy potting compound that is injected around a chip after it has been wired to a lead frame. The lead frame becomes the pins used on the package and is wired to the chip via very thin aluminum wires ultrasonically bonded to both the chip and the lead frame. Some chips are attached to the lead frame using C4 technology that is described later.

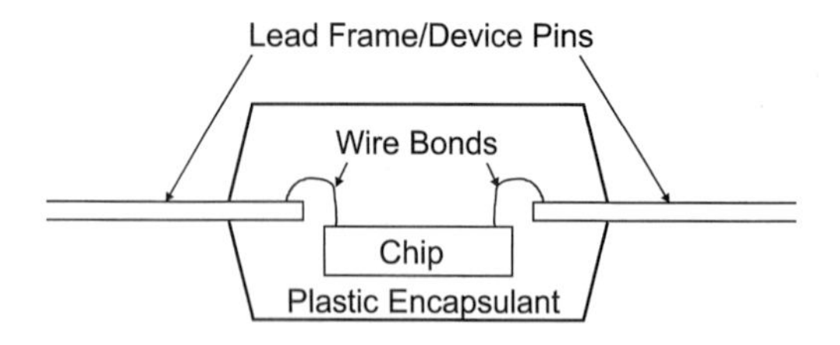


**Figure 2-3** OTP plastic package

Once the encapsulant has hardened, the chip is protected from light, moisture, and physical damage. As I indicated earlier, EPROM microcontrollers in a plastic package are generally referred to as a *One-Time Programmable (OTP)* package (Fig. 2-3). Once the EPROM has been programmed, the device cannot be used for anything else.

The primary purpose of putting a microcontroller into a ceramic package is that a quartz window can be built into the package to allow ultra-violet light to erase the EPROM program memory.

When a ceramic package is used, the chip is glued to the bottom half and is wired to the lead frame. Ceramic packaging is normally only available as a PTH device, where plastic packages can be in a very wide range of different card-attachment technologies.

Ceramic packaging can drive up the cost of a single chip dramatically (as much as ten times more than the price of a plastic OTP packaged device). This makes this type of packaging only suitable for such uses as application debugging, where the advantage of the window for erasing outweighs the extra cost of the package.

The technology used to attach the chip to the board has changed dramatically over the past 10 years. In the 1980s, most devices were only available in *Pin-Through-Hole (PTH) technology* (Fig. 2-5), in which the lead frame pins are soldered into holes in the raw card.

This type of attachment technology is very easy to work with (very little specialized knowledge or equipment is required to manufacture or rework boards built with PTH chips). The primary disadvantage of PTH is the amount of space required to put the hole in the card. As well, the requirements for space around each hole makes the spacing between lead centers quite large, by comparison to *Surface-Mount Technology (SMT)* (shown in Fig. 2-6) in which the pins are soldered to the surface of the card.

The pin through hole is normally built with pins 0.100″ (100 thousandths of an inch) between pin centers. For some pin grid array parts (in which the pins are put on a two-dimensional matrix), lead centers can be as low as 0.071″ (71 thousandths of an inch) between pin centers. The measurement between lead centers is a crucial one for electronics because it is directly related how densely a board can be "populated" with electronic components.

The two primary types of SMT leads are the *gull wing* and the *J lead* (Fig. 2-7).

The two different types of packages offer advantages in certain situations. The gull-wing package allows for hand assembly of parts and easier inspection of the solder joints. The J lead reduces the size of the part's overall footprint. Right now, gull wing parts are

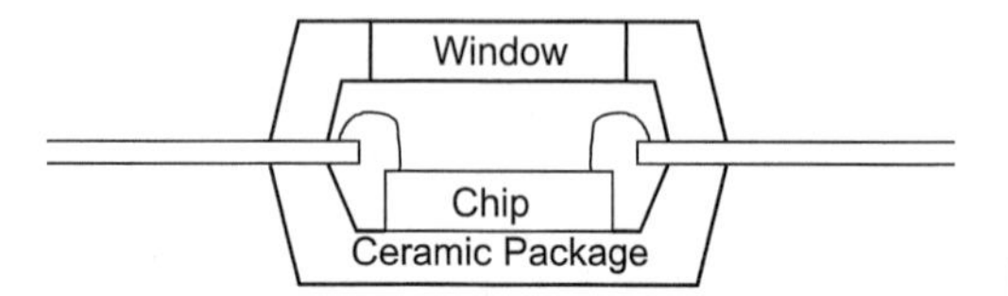


**Figure 2-4** Windowed ceramic package

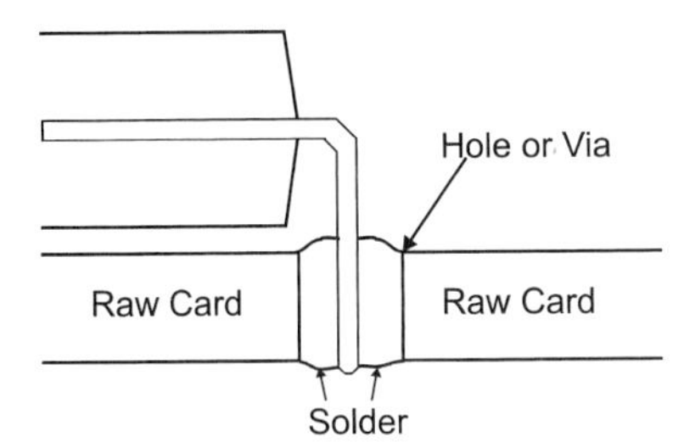


**Figure 2-5** Pin through hole connection

significantly more popular because this style of pin allows easier manufacturing and reworking of very small leads (with lead centers down to 0.016″).

The smaller size and lead centers of the SMT devices has resulted in significantly higher board densities (measured in chips per square inch) than PTH. As noted, typical PTH lead centers are 0.100″ apart, and SMT starts at 0.050″ and can go as low as 0.16″. The SMT parts with small lead centers are known as *fine-pitch parts.*

To give an idea of what this means in terms of board density, consider a PTH package that has pins at 0.100″ lead centers and an SMT package with pins at 0.050″ lead centers. With the smaller lead sizes, the SMT package can be about half the size of the PTH part in each dimension (which means that four SMT parts can be placed in approximately the same space as one PTH part). As well, without holes through the card, components can be put on both sides of the card. Thus, in the raw card space required for one PTH part, up to eight SMT parts can be placed on the card.

Assembling and reworking SMT parts is actually easier in a manufacturing setting than PTH. Raw cards have a solder/flux mixture, called *solder paste,* screened onto the SMT pads of the boards. This screening process consists of a metal stencil with holes cut into in the locations where the solder paste is to be put. A squeegee-like device spreads the paste over the stencil and the paste is deposited on the card where there are holes.

Once the paste has been deposited, the parts are placed onto the paste and then run through an oven to melt the solder paste, soldering the parts to the board. To rework a component, hot air (or nitrogen gas) is flowed over the solder joints to melt the solder, allowing the part to be pulled off. Although SMT is easier to work with in a manufacturing setting, it is much more difficult for the hobbyist or developers to work with (especially if parts have to be pulled off a board to be reprogrammed).

*Chip On Board (COB)* packaging is very descriptive because in this type of packaging, a chip is literally placed on the raw card. Chip on board is useful in microcontroller appli-

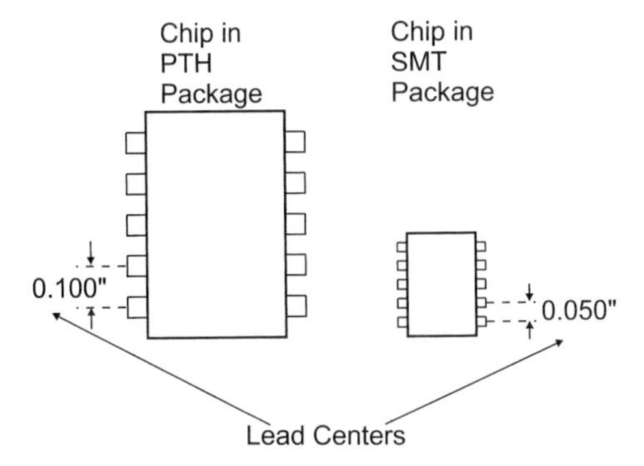


**Figure 2-6** PTH package Vrs SMT

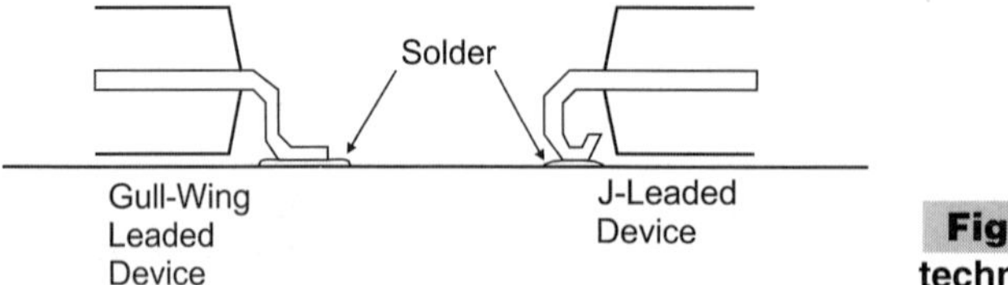


**Figure 2-7** Surface mount technology packages

cations that require a very small form factor for the final product; because the chip is used directly, there is no overhead of a package for the application. Typical applications for COB include telephone smart cards and satellite or cable TV descramblers.

Two methods of COB attachment are currently in use. The first method is to place the chip on the card and wire the pads of the chip to the pads on the card using the same technology as wiring a chip to its "lead frame" inside a package. This is done using small aluminum wires ultrasonically welded to the chip and raw card. (Fig. 2-8)

The chip itself can either be glued or soldered to the raw card. Soldering the chip to the raw card is used in applications where the raw card is to be used as a heatsink for the chip (which reduces the overall cost of the assembly).

The other method of COB, known as *C4* (Fig. 2-9), is actually very similar to the SMT process.

The solder balls used in this process are called *bumps* (because they are so small). This technology was originally developed by IBM for attaching chips to ceramic substrates or backplanes without having to go through a wire-bonding step.

C4 attach requires a very significant investment in tools for placement and a very specialized process (because of the small distance between the chip and the card, water used to wash the card can be trapped with flux residue, causing reliability problems later). C4 attachment is really in the experimental stage at this point for chip attachment to printed circuit boards. This is caused by the difficulty in reliability in putting the chip down onto a raw card, the opportunity for fatigue failure in the bumps caused by the chip, and raw card expanding and contracting at different rates (caused by heating and cooling).

PICmicro® MCUs are available in a variety of packaging, as can be seen in the following diagrams. Figure 2-10 shows the different windowed ceramic packages. Figure 2-11 illustrates the different sizes of the plastic OTP packages. Standard SMT packages are shown in Fig. 2-12 and the fine-pitch lead devices are shown in Fig. 2-13.

All of the SMT packages have leads with 0.050″ lead centers, except for some of the small, fine-pitch parts. Unpackaged chips are available for some parts from Microchip for use in COB applications. In these cases, the chips are available in *waffle* packs for automated chip pickup and placement.

Each package has a one- or two-letter suffix code in the part number to describe the chip's package. For example, the 16F84 is normally sold as a PIC16F84P, which indicates the 16F84 has a plastic, DIP package. Normally, windowed ceramic parts are PTH, except for the CL PLCC package.

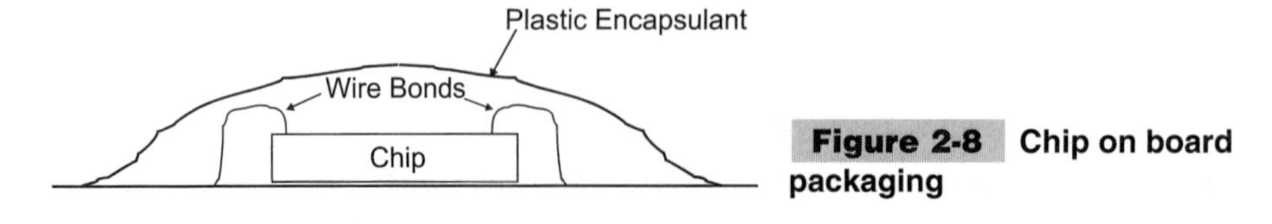


**Figure 2-8** Chip on board packaging

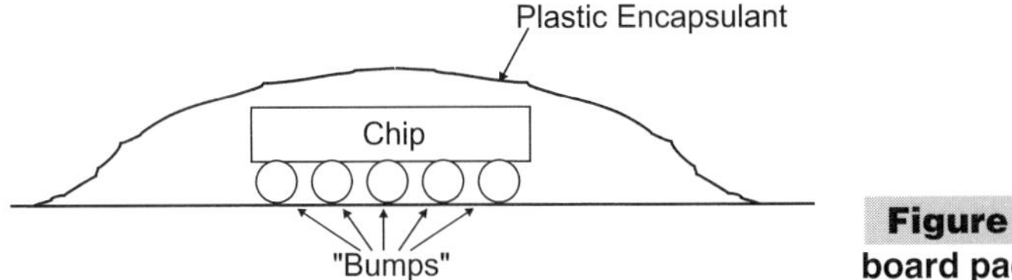


**Figure 2-9** "C4" chip on board packaging

## PART NUMBER CONVENTIONS AND ORDERING

Determining which part number to specify when ordering a PICmicro® MCU for an application or product is very consistent across the entire line. Figure 2-14 shows the conventions for how the part numbers are specified by Microchip.

In Fig. 2-14, please note that not all the options presented are available for each part. The data sheet for the part will have specific information for the part number.

**Letter suffixes** When you look for a specific PICmicro® MCU part number, you might discover that you have more than one part to choose from. As I write this, three versions of the 16C73 are available: the PIC16C73, the PIC16C73A, and the PIC16C73B. The letter suffixes indicate different versions of the part, but documentation on the differences is often very sketchy and will seem incomplete.

Microchip, like many other integrated circuit manufacturers, continually tracks the quality of their products, as well as their conformance to specifications. They are also continually replacing their manufacturing equipment with newer tools that are capable of producing better-quality chips with smaller device dimensions. The quality information and manufacturing process improvements make the updating of parts attractive in different situations. These updates are the new letter suffixes that you will see in PICmicro® MCU catalogs.

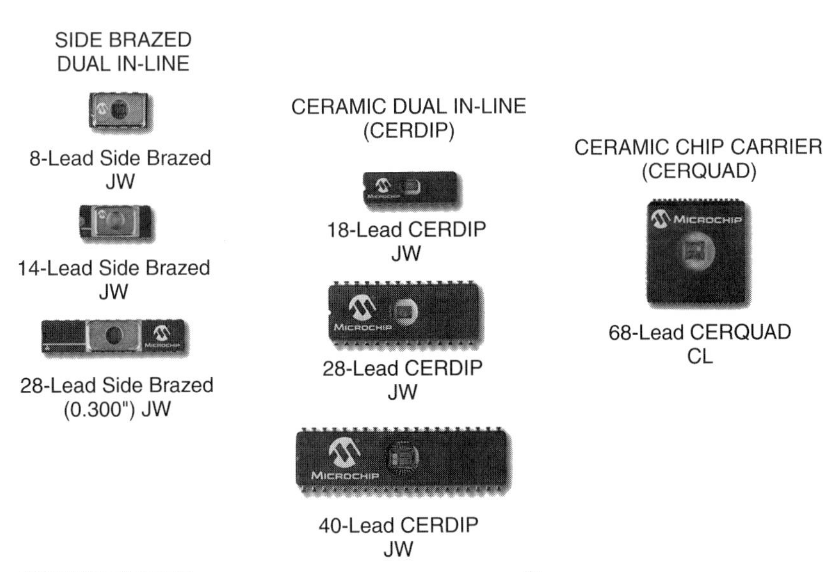


**Figure 2-10** Windowed ceramic PICmicro® MCU packages

PLASTIC DUAL IN-LINE
(PDIP)

8-Lead PDIP
P

28-Lead PDIP
P

14-Lead PDIP
P

28-Lead Skinney PDIP
SP

18-Lead PDIP
P

40-Lead PDIP
P

24-Lead PDIP
P

64-Lead Shrink PDIP
SP

**Figure 2-11** **Plastic DIP "OTP" packages**

PLASTIC LEADED
CHIP CARRIER (PLCC)

32- Lead PLCC
L

44-Lead PLCC
L

68-Lead PLCC
L

84-Lead PLCC
L

PLASTIC SMALL OUTLINE
(SOIC)

8-Lead SOIC (EIAJ)
(0.208") SM

16-Lead SOIC
(0.150") SL

8-Lead SOIC
(0.150") SN

18-Lead SOIC
SO

14-Lead SOIC
(0.150") SL

28-Lead SOIC
SO

**Figure 2-12** **Standard plastic surface mount packages**

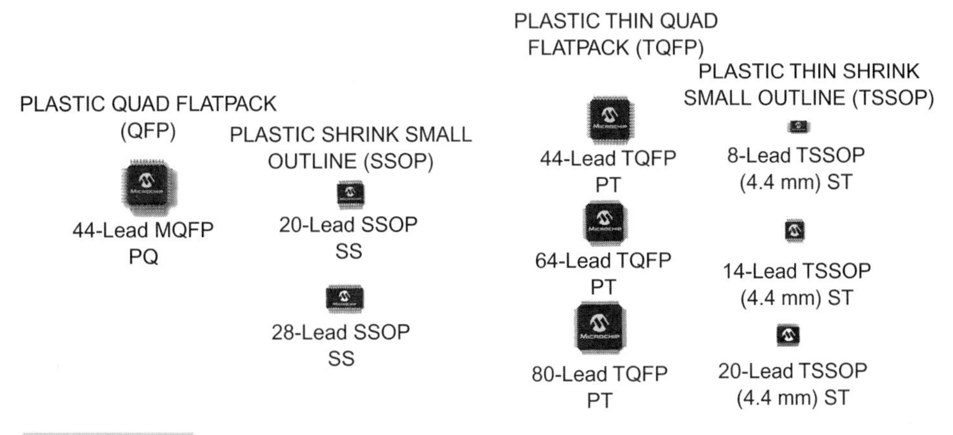


**Figure 2-13** **"Fine Pitch" plastic surface mount packages**

These suffixes represent an entirely new chip design (often referred to as a *respin*). Microchip continually updates their parts to use smaller chips, as well as eliminate the use of circuits that have proven to be unreliable in manufacturing. The function (speed and features) of the part is never changed in these revisions, lest compatibility with previous versions of the device be lost. For the most part, different letter codes of the same PICmicro® MCU part number will work in an application, regardless of its suffix letter.

Even though I've mentioned product quality as a driver in implementing a respin, it is not the main driver. The main reason to carry out a respin is to provide smaller chips that perform the same function. Smaller chips bring two advantages to Microchip: power reduction and part cost reduction.

Power is reduced as the part size is reduced. The longer the path an electron has to travel, the greater the overall resistance it will encounter. By reducing the distances that

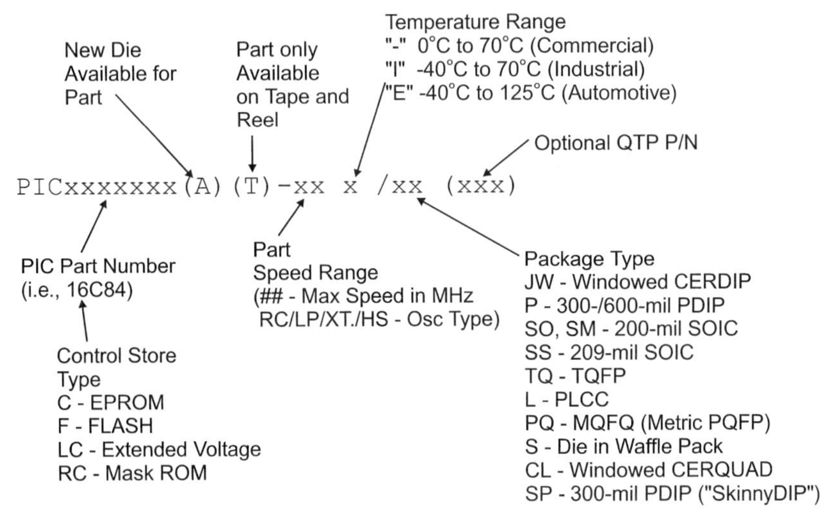


**Figure 2-14** **PICmicro® MCU part number definition**

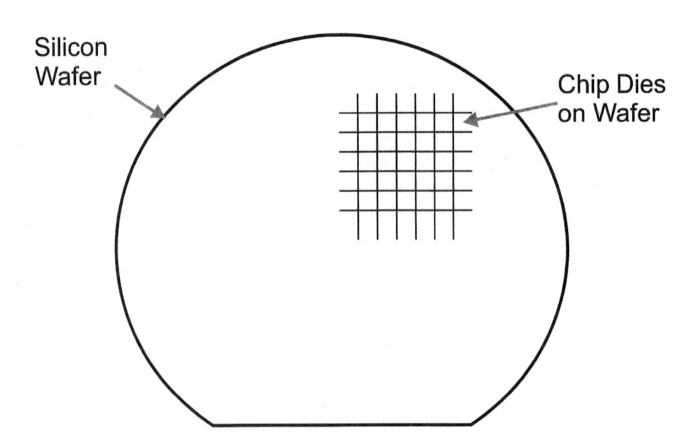


**Figure 2-15** **Wafer and die relationship**

electrical currents have to travel on the chips, the resistances (which are the cause of power dissipation within the chip) are reduced.

It might be surprising that redesigning a part reduces its cost, but even a small reduction in part size can have huge cost advantages for a chip manufacturer. In chip manufacturing, cost is directly related to the number of *wafers* required to build a specific number of chips. By increasing the number of chips on a wafer, the throughput of the manufacturing process will increase without a significant increase in cost.

When chips are manufactured, silicon wafers have the circuits for a number of chip *dies* imprinted on them. During the manufacturing process, circuits are laid out on the dies using a photographic process. Once the circuits are laid out, chemical processes will convert the pictures on the dies into circuits. When the manufacturing process has finished, the dies are cut apart and put into packages for what are normally known as *chips* (Fig. 2-15).

The vast majority of the time that the wafers and dies spend in the manufacturing process are devoted to the chemical processes that "build" the circuits. These chemical processes do not change based on the number of circuits on a wafer. Thus, the manufacturing process cost per die goes down the more dies that can be put through the process (or the more dies that can be put on a single wafer).

For example, if a reduction of 25% can be achieved on a die axis, an area reduction of almost 50% per die can be achieved. This means that by reducing the size of a chip by a quarter on each side, almost twice as many chips can be put on a wafer. The new chip cost will be a little more than one half of what it was originally to Microchip.

# MPLAB

Microchip's MPLAB (Fig. 2-16) *Integrated Development Environment (IDE)* is, by far, the best application development tool offered by any chip manufacturer I know of. The MPLAB IDE is easy to work with, fairly intuitive and, best of all, free. It is capable of integrating all of the software application development tasks. This is provided in a source-code level, rather than instruction-level or even bits'n'bytes, format.

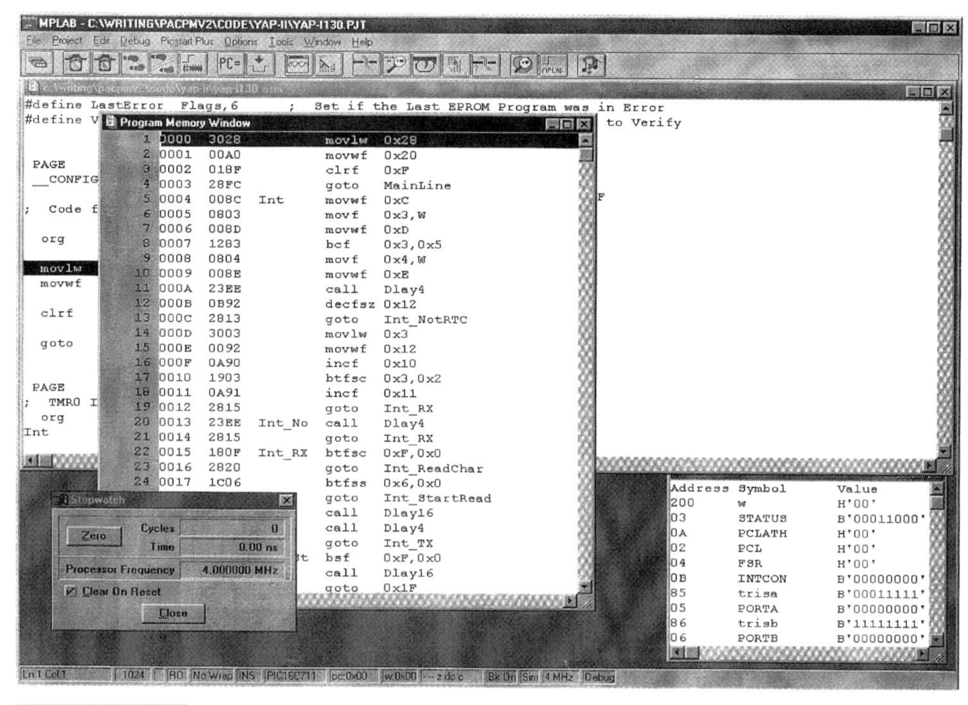


**Figure 2-16 The MPLAB IDE executing**

The MPLAB IDE is a 16-bit Microsoft Windows application that is designed for Win16 (Windows 3.x) and Win32 (Windows 95/98/NT/2000) operating systems. It requires about 10 MB of hard drive space on your PC. I have included version 5.11 (or later, depending on CD-ROM revisions), which can be used to edit, assemble, simulate, emulate, and program the code given in this book. It can also be used to develop applications for any PICmicro® MCU devices announced at the time of release of the used version of the MPLAB IDE.

The MPLAB IDE brings together the following functions:

- Editor
- Assembler
- Compiler
- Linker
- Simulator
- Emulator
- Programmer

together for an application into a package known as a *project*. Projects keep track of all the options used with the MPLAB IDE for creating an application and displaying it on the MPLAB IDE desktop.

When the simulator/emulator functions are used, code can be displayed and monitored either from the source code window or as individual instructions in the processor memory window. This capability allows you to see the application executing, as well as observe and modify variable and hardware register contents.

MPLAB is designed to work with the following Microchip development tools:

- PICStart Plus programmer
- PRO MATE II programmer
- PICMaster emulator
- MPLAB-ICE emulator
- MPLAB-ICD debugger

Along with this hardware, C compilers from High-Tech, CCS, and Bytecraft interface to the MPLAB IDE seamlessly, without requiring any special interfaces.

All of the applications, projects, and tools presented in this book used the MPLAB IDE when they were created for writing the source code, assembling and compiling it, simulating and emulating the code, and programming the parts with the code. The version of the MPLAB IDE that comes with the book is the latest version available from Microchip at the time of writing. Although this version is very comprehensive and well debugged, you should check the Microchip Web page for the latest version before working through the applications presented here.

# FUZZYtech

To help you develop your own fuzzy logic applications; Microchip has created FUZZYtech. This tool will allow you to easily and graphically create fuzzy logic applications. Included in the kit is a simple demonstration board, consisting of a resistive (ohmic) heater and a temperature sensor (thermistor). This demonstration board can be used to set up a first fuzzy logic application for a potentiometer (desired temperature setting), the temperature sensor, and the PWM heater output.

If you go through FUZZYtech, you will discover that the rule generation still must be done manually (although entered into the system through the FUZZYtech Microsoft Windows GUI) and then the system creates the output code. The application programmer is responsible for the interfaces to external hardware, as well as any other functions to be put into the PICmicro® MCU.

FUZZYtech does not turn the PICmicro® MCU into a fuzzy logic device tailored to a specific application. FUZZYtech creates MPASM-compatible output that can be integrated into the application code. It provides source code to be used as a fuzzy logic control for given inputs as an enhancement to the application.

In this book I will demonstrate the operation of MPLAB along with an example application.

# KEELOQ

Microchip's KEELOQ development system is a security algorithm for allowing control of hardware devices over media that can be monitored by third parties. My car's theft alarm uses a KEELOQ-equipped PICmicro® MCU in the locking/unlocking fob and in the car

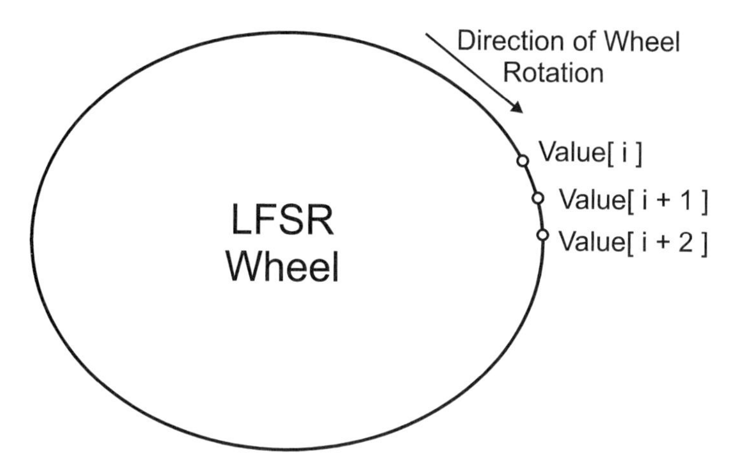


**Figure 2-17** "KEELOQ" psuedo-random wheel

itself. The fob transmits a seemingly random number to the receiver and increments to a new (pseudo) random value for the next time a signal is to be sent.

Each fob/receiver combination has a unique linear feedback shift register to create a pseudo-random number, based on a specific "seed" value. I like to think of the pseudo-random number generator as a wheel with a large number of points on it. When you are at any point on the wheel, you can figure out the next value as it rotates (Fig. 2-17).

When the transmitter sends a signal, the receiver checks it against its wheel and responds if it is equal to the expected value at the next value on the wheel. If it isn't, the receiver marks the place and identifies the next point on the wheel. If the next time the transmitter sends a value and it matches the next expected value, the receiver responds.

The reason the receiver checks the incoming signal and the chance that something is missed is because the transmitter's signal could be lost or garbled in the receiver (or, like me, you might have a small child that likes to push buttons, any buttons).

Now, I have really simplified this and probably missed some very important points of how KEELOQ works. But, this is okay because if you would like to use KEELOQ technology, you will have to sign a nondisclosure agreement with Microchip. KEELOQ is quite complex and is able to handle a number of different cases (such as sending different commands to the receiver) that aren't discussed here.

# The Parallax Basic Stamp

A number of years ago, Parallax came out with an innovative use for the PICmicro® MCU. They packaged it with a voltage regulator and serial EEPROM to create a self-contained unit that could be used by beginners and experts alike to create small applications quickly and very easily (Fig. 2-18). The STAMP1 consists of a PIC16F54 running at 4 MHz, provides eight I/O pins, and can execute PBASIC statements (the programming language of the Basic Stamp) at a rate of about 1,000 per second. The STAMP2, which uses a PIC16F57, has 16 I/O pins, more EEPROM memory, can run at about twice the speed of the STAMP1.

**Figure 2-18** Parallax "Basic Stamp1" (left); "Basic Stamp2" (right)

The basic stamp consists of a small circuit; the original devices were the size of a postage stamp (where the product got its name) that can be blocked out as shown in (Fig. 2-19).

The hardware itself is quite robust, except for the voltage regulator. The devices used in the stamp can be easily burned you if too much current is drawn through them. When this happens, I recommend replacing the built-in voltage regulator with a 78L05, which shuts down if excessive current is drawn.

The basic stamps are programmed from a PC using a parallel port (BS1) or serial port (BS2). Programming normally takes just a few seconds and is done from the Parallax Stamp Integrated Development Environment, which consists of an MS-DOS editor, compiler, and programmer interface.

To make creating the applications easier, Parallax created PBASIC, a BASIC language subset that was designed for use in interfacing applications. To be honest, this version of

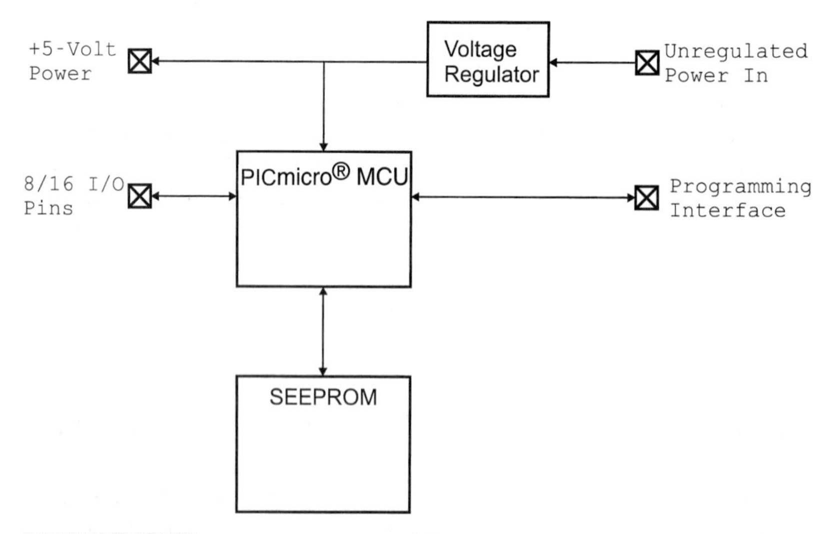


**Figure 2-19** Block diagram of parallax "basic stamp"

BASIC is very limited and can be counter-intuitive to work with in the areas of assignments and conditional execution. But, for many people, the stamps are the fastest way to get into Microcontroller programming and hardware interfacing.

Both BASIC for the stamps and PicBasic for the PICmicro® MCU are compiled languages. In the basic stamp's case, the PBASIC statements are converted into "tokens" that are executed by an interpreter built into the basic stamp. The Microelectronics Lab's PicBasic converts PBASIC statements into actual PICmicro® MCU instructions that are executed from the PICmicro® MCU's program memory.

The appendices include a fairly detailed explanation of the PicBasic statements and I have created quite a complex application using the PicBasic compiler. This compiler is quite straightforward and easy to work with.

The Parallax Basic Stamp PBASIC language also works quite well, but you should be aware of two issues.

First, assignment statements are executed left to right and not in order of operations, as you might be familiar with. For example, the statement:

```
A = B + C * D
```

is executed by first adding *B* to *C* and then multiplying this sum to *D*. I tend to think of this as being similar to the *Reverse Polish Notation (RPN)* data-entry method used in HP calculators and try to write the code this way. So, in PBASIC, the line above should be written as:

```
A = C * D + B
```

to ensure that the product of *D* and *C* is added to *B*.

Another way of doing this is to store intermediate values to make the statement easier to understand. For example, the single line about can be broken up into:

```
temp = D * C
   A = temp + B
```

The second issue about PBASIC is how variable memory is used. Instead of providing variables that can be declared, memory in the PICmicro® MCU is defined by bits, bytes, and words consisting of two bytes. The bytes are B0, B1, etc. It is important that, for a given word number, the byte number times two and the next one are used. Thus, changing a byte variable will also change a word variable. Care must be taken to ensure the variables are declared in such a way that no words or bytes use the same memory space.

# PICmicro® MCU-Compatible Devices

The PICmicro® MCU has spawned a number of successful products that are based on the device and its architecture. These devices have ranged from development tools (such as a number of the programmers and emulators presented in this book) to source code and pin-compatible "clones." These devices have generated a lot of interest. In many cases, such as the Parallax Basic Stamp, they have been truly innovative and interesting products.

Microchip has not licensed any of the PICmicro® MCU circuits or architecture to other companies. Despite this, a number of chips are available that claim to be PICmicro® MCU compatible from an electrical or software perspective. This book only focuses on the Microchip products and the products and applications that are designed to work with them.

# 3

# THE PICmicro® MCU PROCESSOR ARCHITECTURE

## CONTENTS AT A GLANCE



I've always found that when you really understand something, you can describe it in a number of different ways and formats. At the end of this chapter, you will be able to understand the PICmicro® MCU architecture and how the various functional blocks interconnect. By going through each feature of the PICmicro® MCU processor, I hope to give you a better understanding of how each of the functional parts of the PICmicro® MCU interface with each other and how they work.

This chapter covers the various features of the PICmicro® MCU and develops a block diagram of its internal workings.

The PIC16C61 Microchip documentation includes a block diagram (Fig. 3-1). Looking at it for the first time can be confusing and frustrating.

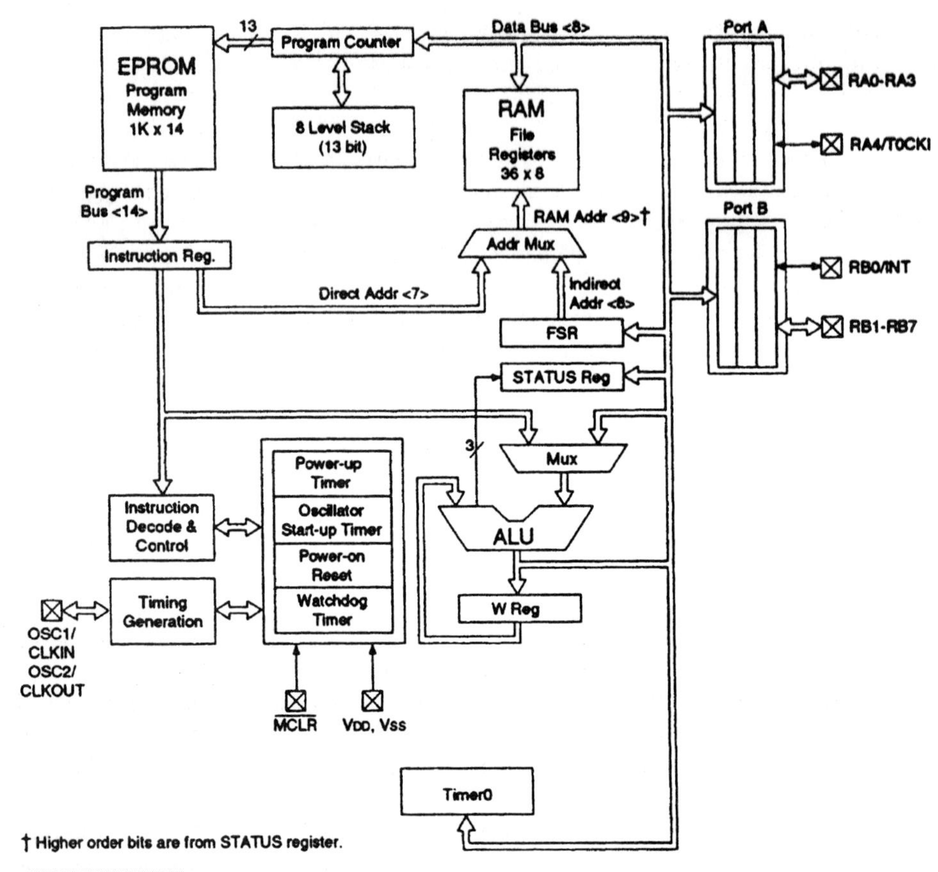


**Figure 3-1** PIC16C61 Block diagram

I have made the material presented in Chapter 3 of the first edition easier to understand. For the second edition of the book, I have removed references and uses of PICmicro® MCU instructions, except where their use is pertinent to the material and helps with the explanation of the architecture and how the PICmicro® MCU works. For these changes, the code is now written in C pseudo-code to illustrate how the architecture features are accessed.

To avoid confusion by trying to explain all the different architecture differences among the PICmicro® MCU families, the mid-range family's architecture is covered first and the other families are presented afterward.

# The CPU

In the documentation, you'll find that the PICmicro® MCU processor is described as a "RISC-like architecture . . . separate instruction and data memory (Harvard architecture)." This chapter explains what this means for people who do not have PhDs in computer architectures. Also, this chapter explains how application code executes in the PICmicro® MCU processor.

The PICmicro® MCU processor can be thought of as an *Arithmetic/Logic Unit (ALU),* receiving, processing, and storing data to and from the various registers. A number of specific-use registers control the operation of the CPU, as well as I/O-control registers and RAM registers, which can be used by the application software for variable storage. In this book, I call the specific use registers *hardware registers* or *I/O registers,* depending on the function that they perform. RAM registers or variable registers are called *file registers* by Microchip.

The registers are completely separate from the program memory and are said to be in their own spaces. This is known as *Harvard architecture.* In Fig. 3-2, notice that the program memory and the hardware it is connected to is completely separate from the register space. This is not quite 100% true (as is explained later in this chapter regarding immediate addressing and table operation).

The PICmicro® MCU has three primary methods of accessing data. *Direct addressing* means that the register address within the register bank is specified in the instruction. If a constant is going to be specified, then it specified *immediately* in the instruction. The last method of addressing is to use an index register that points to the address of the register to be accessed. *Indexed addressing* is used because the address to be accessed can be arithmetically changed. Other processors have additional methods of addressing data.

When accessing registers directly, seven address bits are explicitly defined as part of the instructions. In mid-range PICmicro® MCU direct addressing instructions, these seven bits result in up to 128 addresses that can be accessed (Fig. 3-3).

These 128 register addresses are known as a *bank.* To expand the register space beyond 128 addresses for hardware and variable registers, Microchip has added the capability of accessing multiple banks of registers, each being able to register 128 addresses in the mid-range PICmicro® MCUs.

The low-end PICmicro® MCUs can access 32 registers per bank; they also have the opportunity to have four banks accessible by the processor (up to 128 register addresses total). This is explained later in this chapter, along with how register addressing is implemented for the PIC17Cxx and PIC18Cxx devices.

The "ALU" shown in Fig. 3-3 is the arithmetic/logic unit. This circuit is responsible for doing all of the arithmetic and bitwise operations, as well as initiating conditional execution requested by the PICmicro® MCU's instructions. Every computer processor available today has an ALU that integrates these functions into one block of circuits. The ALU is discussed in detail later in this chapter.

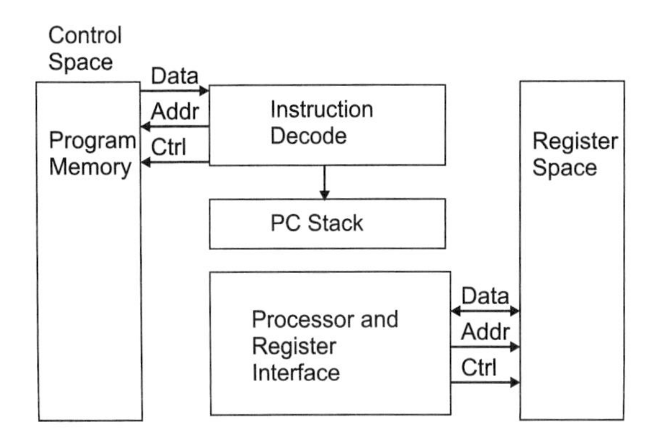


**Figure 3-2** Harvard architecture block diagram

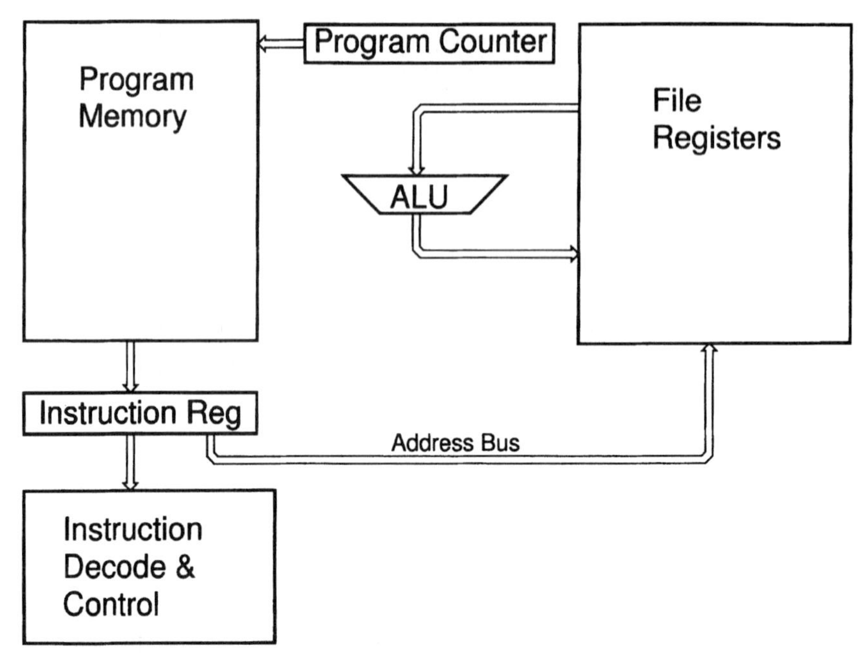


**Figure 3-3** **Basic PICmicro® MCU architecture**

The program counter provides addresses into the program memory (which contains the instructions for the PICmicro® MCU processor), which are then read out and stored in the instruction register and then decoded by the instruction decode and control circuitry. If an instruction is received that provides a direct address into the file registers, then the least-significant seven bits of the instruction are used as the address into the file registers.

This little fact is important to remember because it will give you an idea of whether or not the assembled source code is correct. Sometimes you will write instructions incorrectly and although they will assemble without error, their operation will not be correct. To find this, you can look at the instruction bits in the listing file and check that the correct address and instruction is being used.

The *program memory* contains the code that is executed as the PICmicro® MCU application. The contents of the program memory consists of the full instruction at each address (which is 12 bits for the low end, 14 bits for the mid-range, and 16 bits for both the PIC17Cxx and PIC18Cxx devices). This differs from many other microcontrollers in which the program memory is only eight bits wide and instructions that are larger than eight bits are read in subsequent reads. Providing the full instruction in program memory and reading it at the same time results in the PICmicro® MCU being somewhat faster in instruction fetches than other microcontrollers.

The block diagram (Fig. 3-3), although having 80 or more percent of the circuits needed for the PICmicro® MCU's processor, is not a viable processor design. There is no way to pass data from the program memory and the file registers for immediate addressing, and there is no way to modify the program counter. As this chapter continues, Fig. 3-3 is fleshed out until it becomes a viable processor that can execute PICmicro® MCU instructions.

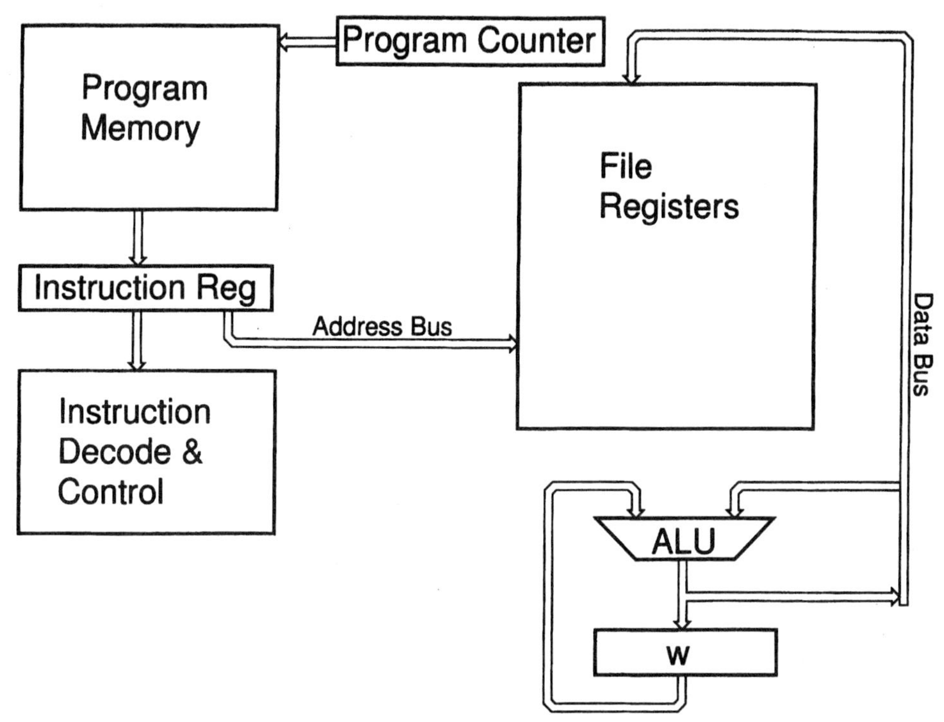


**Figure 3-4** PICmicro® MCU Processor with "w" register as an "accumulator"

To implement two ALU operation parameters, a temporary holding register, often known as an *accumulator,* is required. In the PICmicro® MCU, the accumulator is known as the *w register.* The w register cannot be accessed directly (in the low-end and mid-range PICmicro® MCUs), instead the contents must be moved to other registers that can be accessed directly.

Every arithmetic operation that occurs in the PICmicro® MCU uses the w register. If you want to add the contents of two registers together, you would first move the contents of one register into w and then add the contents of the second to it.

The PICmicro® MCU architecture is very powerful from the perspective that the result of this operation can be stored either in the w register or the source of the data. Storing the result back into the source effectively eliminates the need for an additional instruction for saving the result of the operation. This allows movement of results easily and efficiently.

This changes the processor diagram to Fig. 3-4. Notice that the ALU has changed to a device with two inputs (which is the case in the actual PICmicro® MCU's ALU) and that the contents of the w register are used as one of the inputs. Also notice that when a result is passed from the ALU, it could either be stored into the w register or in the file registers. This is a bit of foreshadowing into one of the most important features of the PICmicro® MCU architecture and how instructions execute.

The diagram shows the PICmicro® MCU at its simplest level. Despite this, well over half of the instructions can be run only using this hardware.

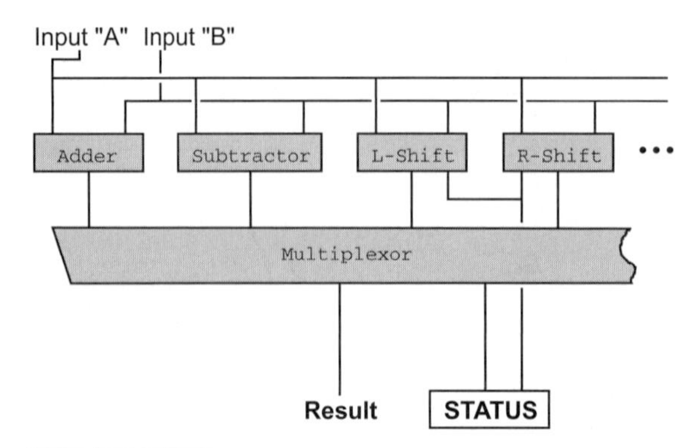


**Figure 3-5** ALU using multiplexor to select operation

# The PICmicro® MCU's ALU

The *arithmetic logic unit (ALU)* of the PICmicro® MCU processor performs arithmetic, bitwise and shifting operations on one or two bytes of data at a time. These three simple functions have been optimized to help maximize the performance of the PICmicro® MCU and minimize the cost of building the chips. An in-depth understanding of the ALU's function is not crucial to developing applications for the PICmicro® MCU. However, having an idea of the trade-offs that were made when designing the ALU will give you a better idea of how PICmicro® MCU instructions execute and what is the best way to create your applications. Within this discussion of how the PICmicro® MCU's ALU operates and is designed, I will discuss the hardware used in 27 of the 37 instructions available in the mid-range PICmicro® MCU processor.

Twenty years ago, when the PICmicro® MCU was first developed, any savings in circuits used in the ALU (or anywhere else in the device) paid huge dividends in the final cost of manufacturing the device. This legacy has stuck with all of the different PICmicro® MCU architecture's ALU design.

I think of the ALU as a number of processor operations that execute in parallel with a single multiplexer, which is used to select the result that is to be used by the application. Graphically, this looks like the block diagram shown in Fig. 3-5.

The STATUS register stores the results of the operations and is described in more detail in the next section. The ALU is the primary modifier of the STATUS bits that are used to record the result of operations, as well as provide input to the data shift instructions.

The circuit shown in Fig. 3-5 would certainly work as drawn, but it would involve a number of redundant circuits. Many of these functions could be combined into a single circuit by looking for opportunities, such as noting that an increment is addition by one and combining the two functions. A list of arithmetic and bitwise functions available within the PICmicro® MCU, along with the combinations, are shown in Table 3-1.

The twelve operations listed in Table 3-1 could be reduced to six basic operations with the constants *1* and *0*x*0FF* provided as extra inputs, along with immediate and register data. Notice that the basic bitwise operations (AND, OR, XOR, and Shift Right) do not have equivalencies; this is not a problem because they are usually simple functions to implement in logic. This is not true for the arithmetic operations.

**TABLE 3-1 Available PICmicro® MCU ALU Operations**

| OPERATION | EQUIVALENT OPERATION |
|---|---|
| Move | AND with 0x0FF |
| Addition | None |
| Subtraction | Addition to a Negative |
| Negationn | XOR with 0x0FF (Bitwise "Invert") and Increment |
| Increment | Addition to One |
| Decrement | Subtraction by One/Addition by 0x0FF |
| AND | None |
| OR | None |
| XOR | None |
| Complement | XOR with 0x0FF |
| Shift Left | Add value to itself plus carry |
| Shift Right | None |

For example, instead of providing a separate subtractor, the ALU's adder could be used with the addition of some simple circuits to provide addition and subtraction capability (Fig. 3-6).

I have used subtract as an example here because it is an instruction you will probably learn to hate as you start working with the PICmicro® MCU. The reason for the problems with subtraction is because the result of the operation probably won't make sense to you unless you look at how the operation is carried out and how the hardware is implemented (Fig. 3-6). To introduce subtraction and help show how the PICmicro® MCU's ALU works, I wanted to show how an adder, with some a few additional circuits, could be used to provide addition and subtraction instructions using only an adder and a selectable negation circuit. The other instructions in the PICmicro® MCU work as you would expect and

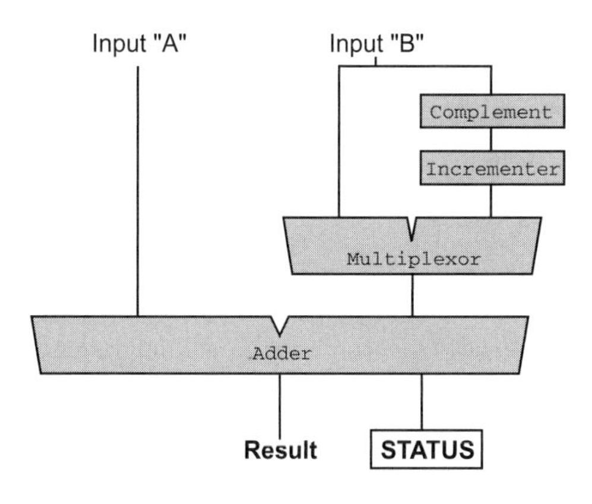


**Figure 3-6** Adder Input using multiplexed straight and inverted to simply implement subtractor

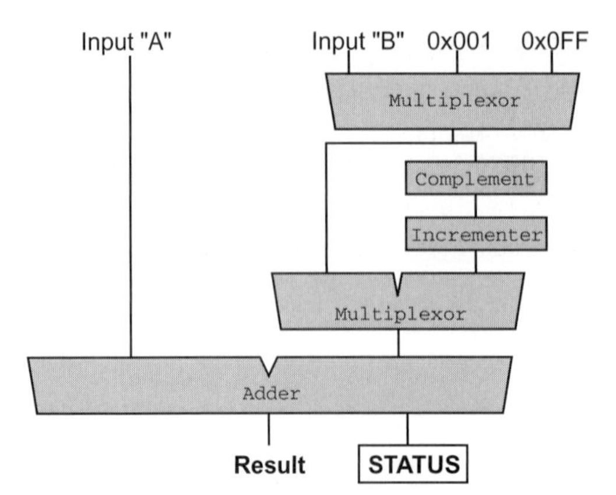


**Figure 3-7** Muxing inputs for multiple functions

the optimization of the ALU does not result in any other nonconventional instruction execution.

The circuit in Fig. 3-6 could be further enhanced by adding 0x001, 0x0FF as part of the ALU input B selection. Figure 3-7 shows a relatively simple circuit that can do addition, subtraction, incrementing, and decrementing. Adding the capability of complementing Input A, would give the ability of negating a number as well. By adding just a few inputs to the ALU's adder, it is able to perform all the arithmetic operations of the PICmicro® MCU.

Notice that I do not include a 0 as an input to the circuit's block diagram. I didn't because 0 is very easy to derive by incrementing 0x0FF. The actual result is 0x0100. Because the eight least-significant bits are used, the value is effectively 0x000. This kind of value generation avoids the need for an additional multiplexer input and decoding circuitry in the ALU.

Like many microcontrollers, the PICmicro® MCU instruction set has the capability of modifying and testing individual bits in registers. These instructions are not as clever as you might think and are, in fact, implemented with the ALU hardware that I've described in this section.

A bit set instruction simply ORs a register with a value that has the appropriate bit set. A bit clear (or reset) instruction ANDs the contents of a register with a byte that has all the bits set, except for the one to be cleared. It is important to realize that entire registers are read in, modified by the AND/OR functions and then written back to the register. As shown later in the book, not being aware of the method used in the PICmicro® MCU for setting and clearing bits can result in some vexing problems with some application's execution.

## THE STATUS REGISTER

The STATUS register is the primary CPU execution control register used to control the execution of the application code and monitor the status of arithmetic and bitwise operations. The STATUS register is common for the different PICmicro® MCU architectures, but some bits relate to the ALU that I want to include in this section.

Each of the different mid-range PICmicro® MCU processor architectures have three bits ("flags"), which are set or reset, depending on the result of an arithmetic or bitwise operation. They are the *carry, digit carry*, and *zero* bits; in the low-end and mid-range devices, they are bits zero, one, and two, respectively, of the STATUS register. These bits are often referred to as the *execution status flags* (Z, DC, and C).

The *zero flag (Z)* is set when the result of an operation is zero. For example, ANDing 0x05A with 0x0A5:

```
0x05A AND 0x0A5 = 0b001011010 & 0b010100101
                = 0b000000000
```

results in zero which will set the zero flag.

Adding zero and zero together will obviously produce a zero result, but so will the addition of two values that add up to 0x0100 (256). In this case:

```
0x080 + 0x080 = 0b010000000 + 0b010000000
              = 0b100000000
```

produces the nine-bit result 0x0100. Because all of the processor-accessible registers in the PICmicro® MCU are only eight bits in size, only the least-significant eight bits will be stored in the destination. These least-significant eight bits are all zeros, so the zero flag will be set as well.

The *carry flag (C)* is set when the result of an operation is greater than 255 (0x0FF) and is meant to indicate that any higher-order bytes should be updated as well. In the previous example (0x080 + 0x080), the result was 0x0100 that stored 0x000 in the destination and set the zero flag. In this case, the ninth bit of the result (the 1) would be stored in the carry flag. If the result was less than 0x0100, then the carry flag would have been reset.

Along with being used for addition, the carry flag is used for subtraction and shift instructions.

Previously, I have noted that subtraction was actually negative addition. For example, one subtracted from two would be:

```
2 - 1 = 2 + (-1)
```

The two's complement equivalent of the negative number can be calculated by complementing it and incrementing the result:

```
-1 = (1^0x0FF) + 1
```

Putting this value back into the previous formula, subtraction becomes:

```
2 - 1 = 2 + (-1)
      = 2 + (1^0x0FF) + 1
      = 2 + 0x0FE + 1
      = 0 x 0101
```

This value stored into the (eight-bit) destination is 0x001, but the ninth bit, which is used as the carry flag, is set. Thus, the actual subtraction result will set the carry flag. This is different from most other processors, in which a positive (or zero) result from a subtraction operation resets the carry flag and sets it if the result is less than zero. In these processors, the carry flag becomes a borrow flag and indicates when a value has to be borrowed from a higher-order byte must occur.

In the PICmicro® MCU, the carry flag is really a positive flag when it comes to subtraction. If the carry flag is set, then the result is zero or positive. If the carry flag is reset, then the result is negative. This difference between the PICmicro® MCU and other processors can make it difficult to port assembly-language applications directly from other processors to the PICmicro® MCU.

In the latest Microchip documentation, the carry flag is referred to as a *negative borrow flag,* with respect to subtraction. This is a reasonable way of looking at the execution of the instruction as it is reset when a borrow from the next significant byte is required.

The *digit carry* is set when the least-significant *nybble (four bits)* of the result is greater than 15 after an arithmetic operation (add or subtract). It behaves identically to the carry flag, except that it is only changed by the result of the least-significant four bits, instead of by the whole byte.

For example, the operation:

```
0x0A + 0x0A = 0x014
```

in the PICmicro® MCU, the digit-carry flag will be set (and the zero and carry flags reset).

The digit-carry flag might seem to be unnecessary, but as you understand the PICmicro® MCU more and more, you will find opportunities where it is very useful. Later, the book has some examples of how the DC flag can be used and the functions that it can provide.

The execution status bits and how different instructions change them are explained in more detail in the next chapter. The STATUS register itself, along with the zero, carry, and digit carry flags are explained in Chapter 5.

To change the three arithmetic STATUS bits, a new value must be explicitly written into them (using the *movwf, bcf,* or *bsf* instructions). If the STATUS register is the destination of an arithmetic or bitwise operation (as is explained in the next section), these bits will contain their bit values of the result of the operation, not the value resulting from the operation. In the first edition of this book, I made the mistake of thinking the result would override the values for these bits in the first experiment.

The STATUS register can be added to the PICmicro® MCU architecture block diagram to show how the results from the ALU are stored in them. Figure 3-8 shows the PICmicro® MCU processor with the STATUS register being written to by the ALU.

# Data Movement

Earlier in the book, I presented the concept that there are five methods of accessing (reading and writing) data within the PICmicro® MCU application. These five methods correspond to the traditional used by other computer processors. The five addressing *modes* built into the PICmicro® MCU can be manipulated to make the built in data addressing modes much richer and more capable. This and the next section cover the different addressing modes and how the PICmicro® MCU architecture has been designed to give much more flexibility to instruction execution than you might first suspect when looking at the architecture or the instruction set.

When a seven-bit register address is specified within an instruction, it is known as *direct addressing* and any register within a 128-address bank can be accessed. Arithmetic and bitwise operations that access a register (instead of provide a parameter explicitly) can store the result in either the w register or back in the source register.

Earlier in the chapter, I introduced this capability as something to note in the architecture block diagrams. Figure 3-8 shows that the result from the ALU can be stored either back into the file registers or into the w register. When storing the result back into the file registers, the same address as the source is used for the destination.

This capability gives you the option of performing an operation without changing the value saved in either w or the source register. The obvious use of this feature is to subtract two values together without saving the result to place the result of the comparison in the STATUS bits instead of changing the source.

To facilitate this, the last parameter of a register arithmetic or bitwise instruction is either a *0* or a *1* and could be the labels *w* or *f*, respectively, as is shown in the *addwf* instruction:

```
addwf register, w|f
```

In this instruction, the contents of the w register are added to the contents of *register*. If *w* (or *0*) is specified as the destination, then the result is stored in the w register. If *f* (or *1*) is specified, then the result of the addition instruction is stored in *register*.

This is one of the most confusing and powerful concepts of the PICmicro® MCU and it can be a problem for many new PICmicro® MCU programmers. The ability to specify an arithmetic operation's result is unusual in eight-bit processors and is not described in most beginner courses in assembly-language programming.

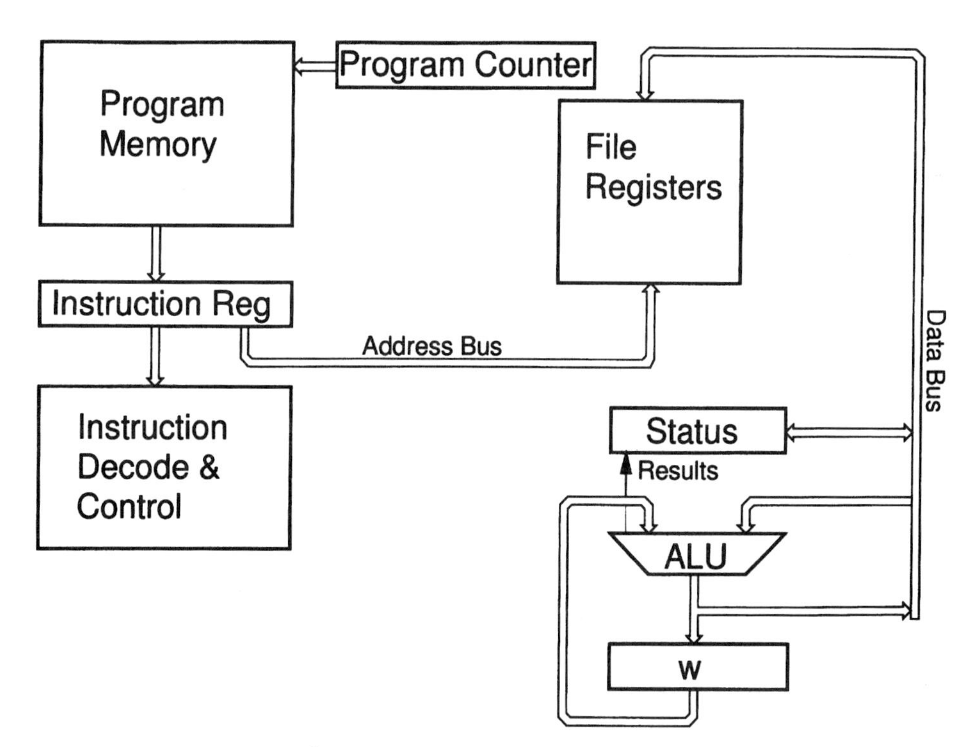


**Figure 3-8** **PICmicro® MCU Processor with "Status" register**

This feature will make applications more efficient and often simpler than what could be written in less radical processor architectures. For example, if you had to implement the statement:

```
A = A + 4
```

in a typical processor, the instructions would be:

```
Accumulator = 4
Accumulator = Accumulator + 4
          A = Accumulator
```

If the register destination option in the PICmicro® MCU is used, then the code could be simplified to:

```
Accumulator = 4
          A = A + Accumulator
```

This example shows one of the ways in which the PICmicro® MCU's processor is more efficient than many other microcontroller's. In this example, by simply storing the addition result back into the accumulator (w in the PICmicro® MCU), I decreased the space and cycles required for implementing the $A = A + 4$ statement in PICmicro® MCU assembler by one third over what would be required by other devices.

When I write PICmicro® MCU assembly language, I continually look for opportunities to save the result in one of the parameters, instead of saving it temporarily in the w register and then providing an explicit store instruction.

In the previous example, I could have written the typical assembler statements as:

```
Accumulator = A
Accumulator = Accumulator + 4
          A = Accumulator
```

which makes the opportunity for this optimization less obvious. When you are first learning to program the PICmicro® MCU, try to look at things as many different ways as possible; you will be amazed at what will fall out of the equations.

As a quick hint that will become easier to understand as you work with the PICmicro® MCU architecture: always order assembler statements so that the statements accessing the same variables are always grouped together. By doing this, the opportunity for saving the result as the destination and eliminating the need to explicitly store it back becomes much more obvious.

Leaving the result in w is useful for instructions where you are comparing and don't want to change the source. It is also used in cases where the result is an intermediate value or the result is to be stored in another register. The immediate data is passed as the least-significant eight bits of the instruction. This addressing mode is not surprisingly known as *immediate addressing.*

To provide immediate addressing, a multiplexer is placed before the ALU to select the data source from either the eight least-significant bits of the instruction of the registers of the PICmicro® MCU. As is shown in Fig. 3-9, if immediate (or "explicit") data is to be processed, then data from the instruction can be selected as an ALU parameter instead of file registers.

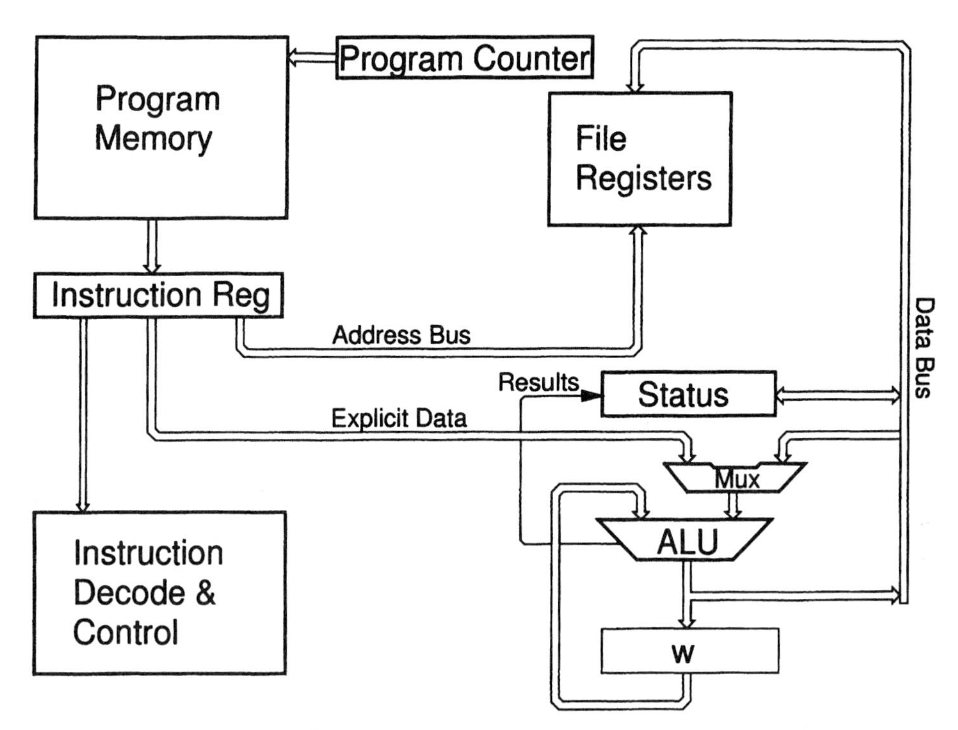


**Figure 3-9** **PICmicro® MCU Processor with immediate ("Explicit") data source**

There will be instances in applications where the ability to directly address a register or explicitly specify a value will not be sufficient and some method of arithmetically specifying an address will be required. In the PICmicro® MCU, *indexed addressing* is carried out by loading the FSR register with the address that you want to access. This eight-bit register has some bank considerations for data movement (which are discussed in the next section).

The contents of the FSR register are multiplexed with the seven immediate address bits (Fig. 3-10). The address source to be used is selected as the least-significant bits of the instruction if the register is not equal to 0x000, the INDF register, which indicates that the register pointed to by the FSR register is to be used.

Indexed addressing is typically described in high-level languages as specifying the index to an array variable. This method of addressing is called *array addressing* because the array variable can simply be known as an *array*. Adding one to an array variable could be written out as:

```
Array[Index] = Array[Index] + 1;
```

In this statement, the start of the array variable is the label *Array* while the byte (element) within it is the *Index*.

When specifying the array variable and element in the PICmicro® MCU, the offset to the start of the array variable must be added to the element number to get the register address within the PICmicro® MCU.

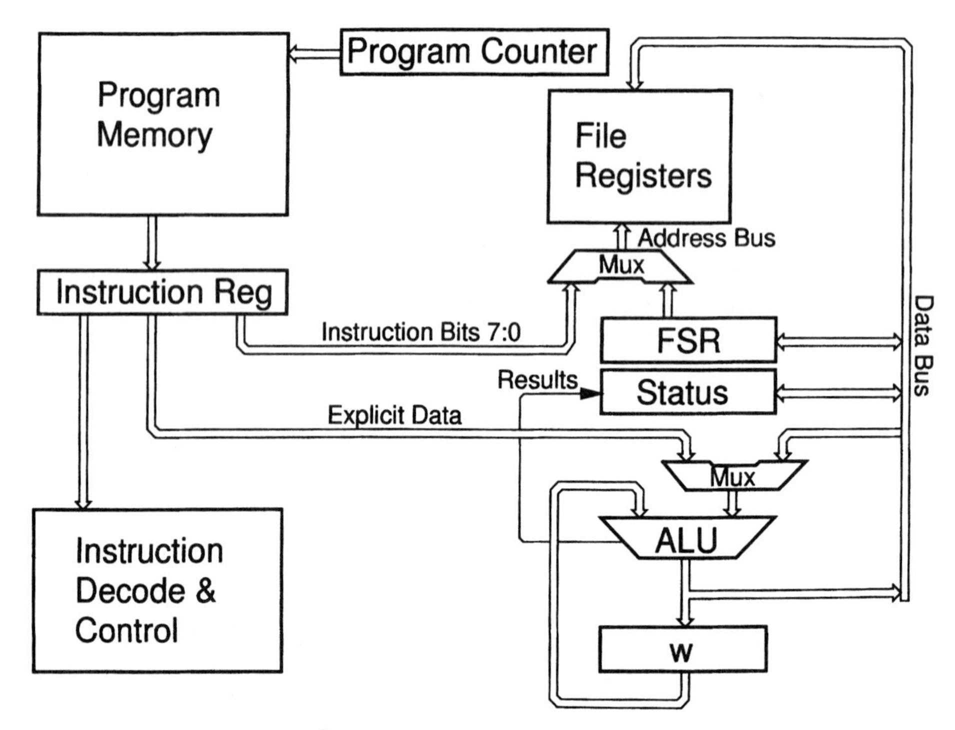


**Figure 3-10** PICmicro® MCU Processor with indexed addressing

So, to carry out this array increment operation in the PICmicro® MCU, the following steps would have to be taken:

```
w = Index;
w = w + Array;              // The Element Address is the Index into the
                            // array variable added to the start of the
                            // array variable
FSR = w;                    // Load the Index register with the Element
                            // Address
[FSR] = [FSR] + 1;          // Increment the Element Address
```

Notice that I take advantage of the ability of the PICmicro® MCU to load, change, and store a register.

One thing is wrong with this list of instructions; the *[FSR]* format used to indicate the register pointed to by the FSR register isn't used. Instead, the INDF register (which has an address of zero in the low-end and mid-range PICmicro® MCUs) is accessed. This register actually doesn't exist; it is the register that is pointed to by the FSR register. This is a bit of a hard concept to understand, but it will become clearer as I work through the instructions and sample code later in the book.

Using the *INDF* register instead of *[FSR],* the instruction operation listed becomes:

```
w = Index;
w = w + Array;              // The Element Address is the Index into the
                            // array variable added to the start of the
                            // array variable
```

```
FSR = w;                    //  Load the Index register with the Element
                            //  Address
INDF = INDF + 1;            //  Increment the Element Address
```

This example is fairly simple. Accessing array variables that have elements larger than one byte or cases where the destination is not the same as the source (and a constant isn't added to them) make the operations of the PICmicro® MCU somewhat more complex.

Single-byte, single-dimensional arrays can be implemented quite easily as can multidimensional arrays. Multidimensional arrays are treated like a single-dimensional arrays, but the index is arithmetically calculated from each parameter (i.e., the index for element 3, 5 in an eight-by-eight array would be $3 * 8 + 5$).

## BANK ADDRESSING

One of the most-difficult concepts for most people to understand when they first start working with the PICmicro® MCU is of the register banks used in the low-end and mid-range PICmicro® MCUs. The number of registers available for direct addressing in the PICmicro® MCU is limited to the number of bits in the instruction that can be devoted to the task. The low-end PICmicro® MCUs have only five bits (for a total of 32 registers per bank), but the mid-range PICmicro® MCUs have seven bits available (for a total of 128 registers per bank). I focus on the mid-range PICmicro® MCUs; the low-end PICmicro® MCUs, the PIC17Cxx, and the PIC18Cxx use different methods that are presented later in the chapter.

To provide additional register addresses, Microchip has introduced the concept of *banks* for the registers. Each bank consists of an address space the maximum size allowable by the number of bits provided for the address. When an application is executing, it is executing out of a specific bank, with the 128 registers devoted to the bank directly accessible.

Each PICmicro® MCU has a number of common hardware registers that are available across all the banks. For the mid-range devices, these registers are INDF and FSR, STATUS, INTCON (presented later), PCL, and PCLATH (also presented later). These registers can be accessed regardless of the bank that has been selected. Other hardware registers can be common across all or some of the banks as well. All mid-range PICmicro® MCUs have file registers that are common across banks to allow data to be transferred across them.

Figure 3-11 shows the PIC16C84's register space for Bank 0 and Bank 1. When execution has selected Bank 0, the PORTA and PORTB registers can be directly addressed. When Bank 1 is selected, the TRISA and TRISB registers are accessed at the same address as PORTA and PORTB when Bank 0 is selected.

To change the current bank where the application is executing, the RPx bits of the STATUS register are changed. To change between Bank 0 and Bank 1 or Bank 2 and Bank 3, RP0 is modified. Another way of looking at RP0 is that it selects between odd or even banks. RP1 selects between the upper (Bank 2 and Bank 3) and lower (Bank 0 and Bank 1) bank pairs. For most of the basic PICmicro® MCU applications presented in this book, you will only be concerned with Bank 0, Bank 1, and RP0.

At the risk of getting ahead of myself, the TRIS registers are used to specify the input or output operation of the I/O port bits. When one of the TRIS register bits are set, the corresponding PORT bit is in *input mode*. When the TRIS bit is reset, then the PORT bit is in

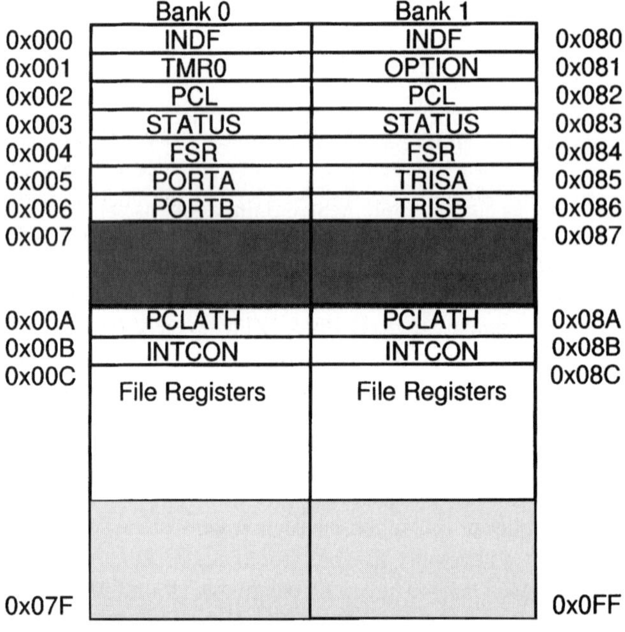


**Figure 3-11** **PIC16C84 Register square**

*output mode.* To access the PORT bits, Bank 0 must be selected; access to the TRIS bits requires Bank 1 to be selected.

For example, to set PORTB bit 0 as an output and loaded with a 1, the PICmicro® MCU code would execute as:

```
PORTB.Bit0 = 1;        // Load PORTB.Bit0 with a "1"
STATUS.RPO = 1;        // Start Executing out of Bank 1
TRISB.Bit0 = 0;        // Make PORTB.Bit0 Output
STATUS.RPO = 0;        // Resume Execution in Bank 0
```

The "RPO" bit is a bank select bit built into the Status register and will be described in more detail below.

To help clarify the bank operations, the Bank 1 registers are defined with bit 7 set in their address specification. For the mid-range PICmicro® MCUs, the Bank 0 register addresses are in the range of 0 to 0x07F; Bank 1 register addresses are in the range of 0x080 to 0x0FF. Once the RP0 bit is set to select the appropriate bank, the least-significant seven bits of the address are used to access a specific register.

As you start working with more complex mid-range PICmicro® MCUs, which use all four banks, you will see registers with address bit eight set, which indicates that the registers are in banks two and three. To access these registers, the appropriate bank select bits of the STATUS register are set and the least-significant seven bits of the Microchip-specified address are used as the address.

The Microchip TRISB label is given the value 0x086, which has bit seven set and is in bank 1. PORTB has an address value of 0x006 and can only be accessed when bank 0 is selected.

Specifying an address with bit seven (or eight) set will result in the message

```
Register in operand not in bank 0. Ensure that bank bits are correct.
```

This indicates that an invalid register address has been specified and to be sure that execution is in the correct bits. Most people clear bits seven and eight of the defined register address to avoid this message. This can be done by simply ANDing the address with 0x07F to clear bit seven, but a much more clever operation is normally done to ensure that the correct registers are accessed from the correct bank.

Instead of ANDing with 0x07F, to clear bit seven for Bank 1, the address is XORed with 0x080. By doing this, if the register is supposed to be in Bank 1 (bit seven of the address is set), then it will be cleared. If the register can only be accessed in Bank 0 (bit seven of the address is reset), then this operation will result in bit seven being set and will cause the message listed above to be given. This is a nice way to ensure that you are not accessing registers not in the currently selected bank.

Using the XOR operation, the example becomes:

```
PORTB.Bit0 = 1;                 // Load PORTB.Bit0 with a "1"
STATUS.RP0 = 1;                 // Start Executing out of Bank 1
(TRISB^0x080).Bit0 = 0;         // Make PORTB.Bit0 Output
STATUS.RP0 = 0;                 // Resume Execution in Bank 0
```

This is also true for banks two and three, which have address bit eight set. Table 3-2 lists the value to XOR registers for specific banks. If the error message comes out of the register access, then you will know you are accessing a register in the wrong bank. Notice that the INDF, PCL, STATUS, FSR, PCLATH, and INTCON registers are common across all the banks and do not have to have their addresses be XORed with a constant value to be accessed correctly from within any bank.

Direct bank addressing is a very confusing concept and, unfortunately, very important to PICmicro® MCU application development. I realize that it will probably be difficult for you to understand exactly what I am saying here, but it will become clearer as you work through the example application code.

The index register (FSR), as indicated, is eight bits in size and its bit seven is used to select between the odd and even banks (Bank 0 and Bank 2 versus Bank 1 and Bank 3). Put another way, if bit seven of the FSR is set, then the register being pointed to is in the odd register bank. This "straddling" of the banks makes it very easy to access different banks without changing the RP0 bit.

**TABLE 3-2 Proper PICmicro® MCU PORT and TRIS Bit Access with Correct Bank Addresses**

| BANK | RP0 | RP1 | ADDRESS RANGE | XOR VALUE |
|---|---|---|---|---|
| 0 | 0 | 0 | 0x0000 to 0x007F | None |
| 1 | 1 | 0 | 0x0080 to 0x00FF | 0x0080 |
| 2 | 0 | 1 | 0x0100 to 0x017F | 0x0100 |
| 3 | 1 | 1 | 0x0180 to 0x01FF | 0x0180 |

For example, if I were to use the FSR register to point to TRISB instead of accessing it directly, I could use the code:

```
PORTB.Bit0 = 1;          //  Load PORTB.Bit0 with a "1"
FSR = TRISB;             //  FSR Points to TRISB
INDF.Bit0 = 0;           //  Make PORTB.Bit0 Output
```

This ability of the mid-range FSR register to access both Banks 0 and 1 is why I recommend that, for many applications, array variables are placed in odd banks while single-element variables are placed in even banks. Of course, this is only possible if the entire file-register range is not "shadowed" across the banks (such as the PIC16F84, PIC16C711, and other simple mid-range PICmicro® MCUs that you are likely to use when you are starting out).

To select between banks two/three and banks zero/one with the FSR, the IRP bit of the STATUS register is used. This bit is analogous to the RP1 bit for direct addressing. Having separate bits for selecting between the high and low bank pairs means that data can be transferred between banks using direct and index addressing without having to change the bank-select bits for either case.

Even though the FSR register can access 256 different register addresses across two banks, it *cannot* be used to access more than 128 file registers contiguously (all in a row). The reason for this is the control registers contained at the first few addresses of each bank. If you try to "wrap" around a 128-byte bank boundary, you will corrupt the PICmicro® MCU's control registers with disastrous results.

# The Program Counter and Stack

The mid-range's program counter can be represented by the block diagram in Fig. 3-12. Looking across the different families of the PICmicro® MCU device, implementing gotos, calls, and table writes (writing to the program counter registers directly) will seem inconsistent and difficult to understand. Actually, these operations work according to a similar philosophy in the different architectures. Once you understand it, they really won't seem all that scary. This section shows how the program counter works in the mid-range devices; later, the chapter explains how the program counter works for the other families.

In all PICmicro® MCU devices, instructions take one word or address. This is part of the RISC philosophy that is used for the design. This might mean that there is not sufficient space in a goto or call instruction for the entire address of the new location of the program counter. A certain number of the address's least-significant bits are put in the instruction. These bits reflect the "page" size of the PICmicro® MCU.

*Tables* are an important feature of the PICmicro® MCU to allow for conditional jumping or data access. Many of the applications presented in this book use tables for user interfaces or conditional execution. Tables are code artifacts in which the program counter is written to force a jump to a specific location in program memory. The least-significant eight bits of the program counter has been given a software interface and is known as the *PCL register*. Writing to these bits will change the program counter to a new value.

When the eight least-significant bits are written to the PCL, the remaining, more significant bits are taken from the PCLATH register and concatenated to the eight bits written to PCL. The value in the PCLATH register is written into the program counter any time

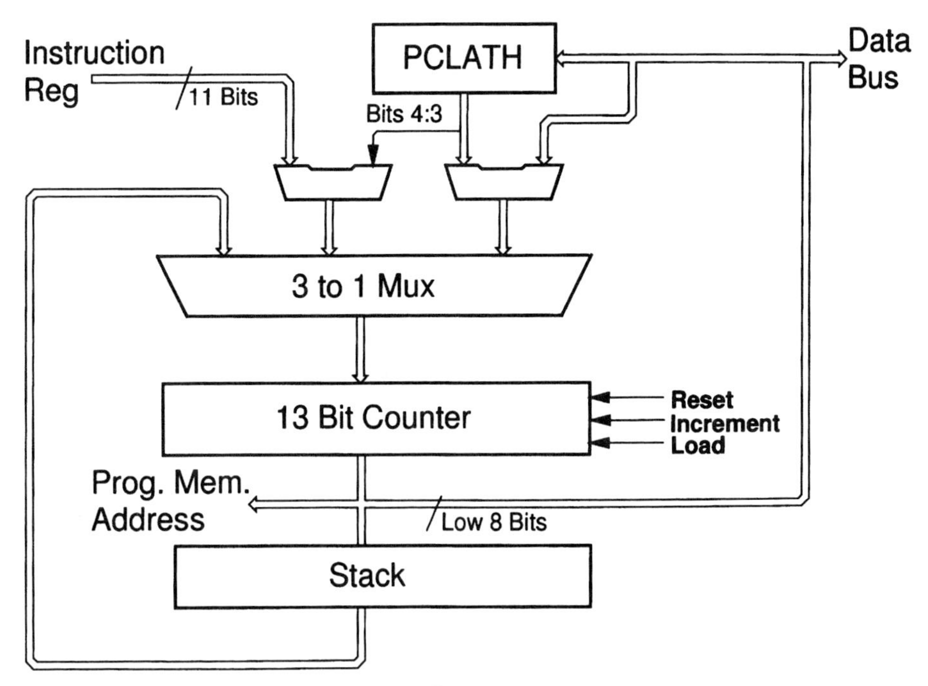


**Figure 3-12** Mid-range PICmicro® MCU program counter and stack block diagram

PCL is changed. This is also true for goto and call instructions, but it works somewhat differently in these cases.

To demonstrate how direct writes to the PICmicro® MCU's program counter works, you could consider the example of wanting to jump to address 0x01234 within a mid-range PICmicro® MCU's program memory using a direct write to the program counter. First, the value *0x012* is written into the PCLATH register. Next the value *0x034* is written into the PCL register. When the write to the PCL register is made, the upper bits of the program counter are loaded from the PCLATH register. This operation could be modeled as:

```
PCLATH = 0x012;          // Set the PCLATH Value
PCL = 0x034;             // Change the Program Counter
                         // Program Counter = (PCLATH << 8) + PCL
                         //                 = (0x012 << 8) + 0x034
                         //                 = 0x01200 + 0x034
                         //                 = 0x01234
```

Another way of approaching how the write to the PICmicro® MCU's program counter is to look at the block diagram of the PICmicro® MCU's program counter hardware and see how the data flows from the processor into the program counter. In Fig. 3-13, the *addwf PCL, f* instruction, which adds the current value in PCL to the contents of w and put the result back into the program counter, is shown. In the diagram, you can see that the PCLATH bits are combined with the data coming out of the ALU after the addition operation and then passed back to the 13-bit counter (the actual PICmicro® MCU's program counter) through the 3-to-1 *mux* (multiplexer).

When the *addwf PCL, f* instruction is executed, eight bits of data are added to the pro-

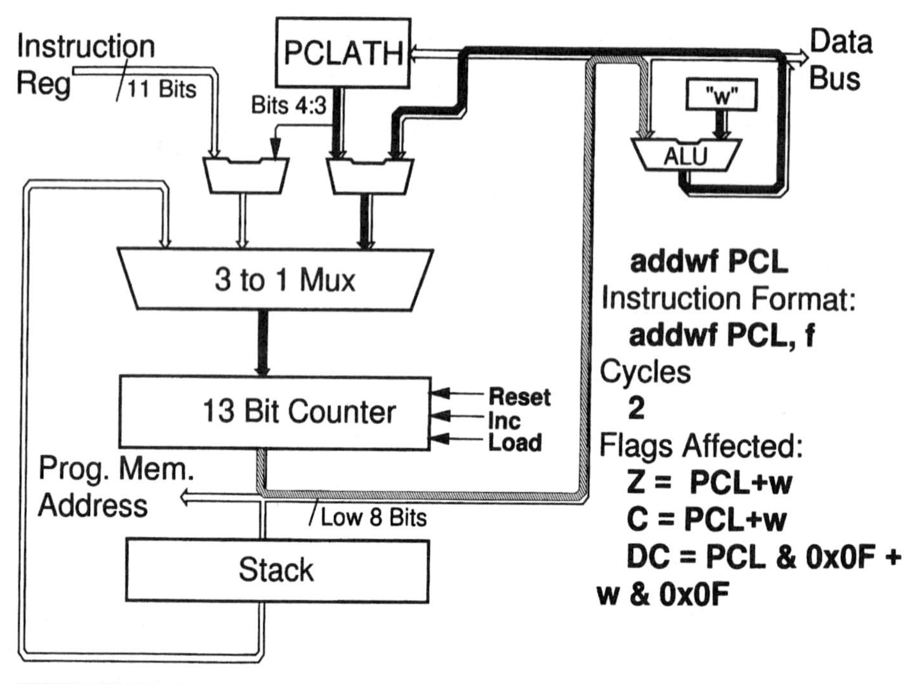


**Figure 3-13** "addwf PCL, f" Instruction operation

gram counter. This means that only 256 unique addresses can be accessed (they can be anywhere in the PICmicro® MCU's program memory because the PCLATH register will provide the upper address bits). Although a table size of 255 seems to be the maximum, there are some tricks to increase the size significantly.

In each PICmicro® MCU, a *page* is the number of instructions that can be conveniently jumped within using the available bits in the instruction. The page size for the low-end PICmicro® MCU is determined by the nine-bit address that is embedded in the 12-bit instruction. These nine bits can address 512 (0x0200) instructions, which is the low-end PICmicro® MCU's page size. In the mid-range devices, 11 bits are used for the address within an instruction, which gives the devices a 2,048 (0x0800) instruction page size. Any address within a page can be accessed directly by a *goto* or *call* instruction.

The addresses specified by *gotos* and *calls* instructions are zero based within the page and are not relative to the location of the *goto* or *call* instruction. This important point can be confusing because, in MPASM assembly-language programming, *goto* and *call* instructions can jump to instructions that are relative to the *goto* and *call* instructions without regard to the start of the page.

If addresses outside the page have to be accessed, then the new page must be selected. In the mid-range devices, the selected page is provided to the program counter by the PCLATH register. In this case, only the bits that are not specified by the *goto* or *call* instruction are added to the address that is loaded into the PICmicro® MCU's program counter. The PCLATH bits that are in conflict with the instruction's address are ignored and the instruction's address bits are used instead. For the mid-range PICmicro® MCU, this means that PCLATH bits zero through two are ignored when a *goto* or *call* instruction is encountered.

Going back to the previous example, if PCLATH was loaded with 0x012 and the instruction *goto 0x0567* was encountered, the PICmicro® MCU's program counter would be loaded with 0x0567 for the 11 least-significant bits and the least-significant three bits of PCLATH (0b0010) are ignored:

```
PCLATH = 0x012;        // Set the Page Value
goto 0x0567            // PC = ((PCLATH & 0x018)<<8) + Address
                       //    = ((0x012 & 0x018)<<8) + 0x0567
                       //    = (0x010<<8) + 0x0567
                       //    = 0x01000 + 0x0567
                       //    = 0x01567
```

For this example, when the *goto* instruction is executed, the PICmicro® MCU's program counter will be loaded with 0x01567.

The previous example's 0x01234 is correct because PCL is updated directly. If a *goto 0x034* instruction was in place, then the address jumped to will be 0x01034 because the most-significant three bits of the address to goto are equal to zero.

So a goto or call typically gets its address from the instruction and the PCLATH register. Figure 3-14 shows that the PCLATH register is accessed to make up the complete address, but that only two bits (four and three) are used when the new address is calculated.

Subroutine calls work very similarly to gotos or writes to the PICmicro® MCU's pro-

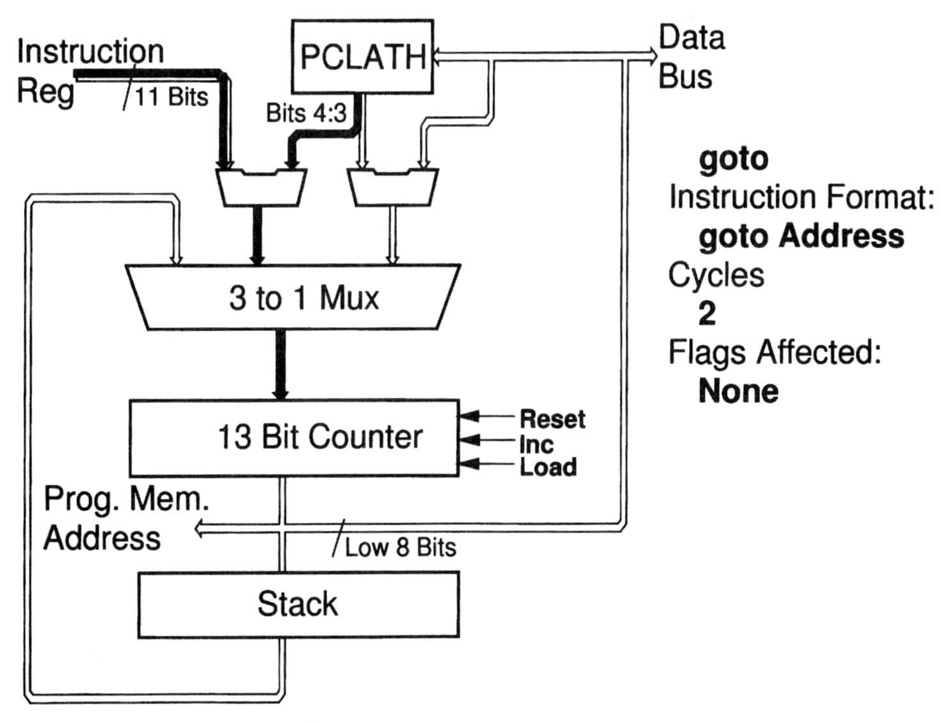


**Figure 3-14** PICmicro® MCU "Goto" instruction operation

gram counter, except that before the program counter is updated, it is "pushed" into the stack shown in Fig. 3-12. The value pushed onto the stack is not the address of the *call* instruction, but the address of the instruction *after* the call, which is the return address for the subroutine. In virtually all processors (the PICmicro® MCU included), as soon as the instruction is fetched from program memory, the program counter is incremented. When a *call* instruction is executed, this incremented value is saved on the stack, not the original value.

The PICmicro® MCU's stack is a bit unusual in that it is devoted to the program counter, cannot be accessed by software and is quite limited. In most other devices, the stack is part of variable memory and can be accessed by the application code. By placing the stack in variable memory, almost infinitely large stacks can be implemented, allowing such programming constructs as recursive subroutines and data pushing and popping onto and off of the stack.

These limitations to the PICmicro® MCU stack means that nested subroutine calls and nested interrupt-request handlers have to be limited in an application. As well, data will have to be stored using the FSR index register into a simulated stack. This is not really a significant problem for your application code and the application code in this book shows you how to implement your own data stack for saving and passing data between subroutines.

Before moving on, I just want to point out that the PIC18Cxx devices do not have all the stack limitations of the other PICmicro® MCU family architectures. The PIC18Cxx's stack can be updated from the application code, which allows it to implement such applications as debuggers and real-time operating systems, which are presented later in the book. Later, this chapter presents the PIC18Cxx's program counter stack and how it can be manipulated.

# Reset

Six different situations cause the PICmicro® MCU's reset (hardware re-initialized and processor stopped) to become active, followed by execution restarting at the reset vector address, and begin to execute the application again. The operation of the PICmicro® MCU is almost exactly the same in the different situations, although applications might use the different reset options or check different indicators.

The six reset options are:

1. Power On Reset (POR)
2. _MCLR reset during operation
3. Brown-Out Detect reset (BOR)
4. Watchdog Timer (WDT) reset
5. _MCLR reset during sleep
6. WDT reset during sleep

*_MCLR* is the PICmicro® MCU's negatively active reset pin. Negatively active means that when the pin is pulled to ground, it makes the reset circuit active, stops the internal PICmicro® MCU oscillator, reinitializes the PICmicro® MCU hardware, and holds the

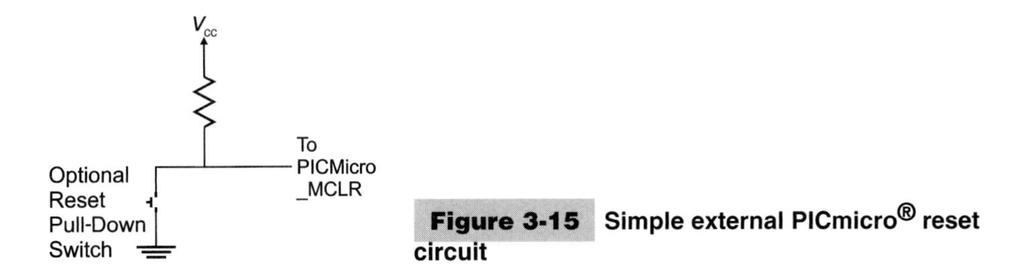


**Figure 3-15** Simple external PICmicro® reset circuit

PICmicro® MCU in an inactive state until the _MCLR line goes high again. A typical PICmicro® MCU reset circuit is shown in Fig. 3-15.

In the circuit shown in Fig. 3-15, when power is applied to the PICmicro® MCU ($V_{CC}$ becomes active), reset will be disabled, and the PICmicro® MCU will be allowed to execute. When _MCLR goes high, the internal operation looks like Fig. 3-16.

In this situation, when _MCLR goes high, the internal oscillator is started. After 1024 cycles (and an optional PWRTE internal 72-ms delay), the application code begins to execute at the reset vector.

The *Brown-Out Detect* is a function that is built into some PICmicro® MCUs in which the reset circuit is activated when the input power drops below 4.0 or 1.8 volts (for low-voltage operations). This feature is typically used with battery-powered applications in which $V_{CC}$ is not regulated.

*Sleep* is a state in which the PICmicro® MCU can be placed when active operation is not required. Sleep can be turned off, and the PICmicro® MCU allowed to execute again by a _MCLR reset, a watchdog timer reset, TMRO interrupt, or an external interrupt request. Sleep is examined in more detail elsewhere in this book.

The last hardware feature that can cause a reset is the *WatchDog Timer (WDT)*. This timer must be reset within a specified interval or the PICmicro® MCU will be reset automatically. The purpose of the watchdog timer is to reset the PICmicro® MCU when it has been upset by an external event and it is unable to continue executing properly.

When the PICmicro® MCU resets, two bits in the status register and two other bits in the optional PCON register will change state. The PCON register, available on later-designed PICmicro® MCUs, makes it much easier to determine the cause of a reset.

The two bits affected by the reset are _TO and _PD. _TO is active (low), when the

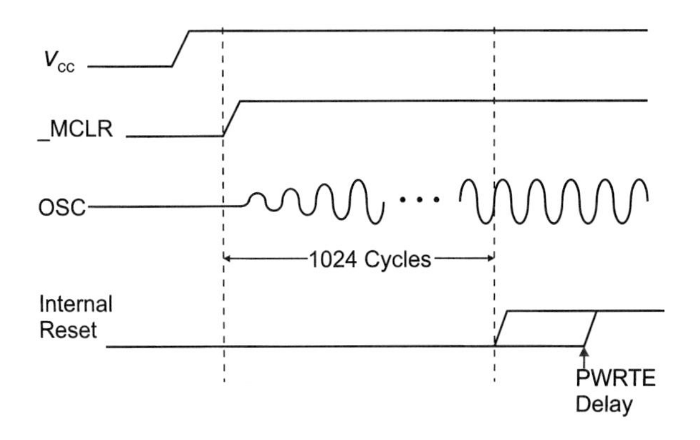


**Figure 3-16** PICmicro® MCU reset waveform

**TABLE 3-3 Reset Event to Status Bit Values**

| EVENT | _TO | _PD | _POR | _BOR |
|---|---|---|---|---|
| Power On | 1 | 1 | 0 | 1 |
| Brown Out | 1 | 1 | 1 | 0 |
| _MCLR Reset | U | U | 1 | 1 |
| Watchdog Timer Reset | 0 | 1 | 1 | 1 |
| _MCLR During Sleep | 1 | 0 | 1 | 1 |
| Watchdog Timer Reset During Sleep | 0 | 0 | 1 | 1 |

watchdog timer has caused a reset. _PD is active (low) when the reset takes place after sleep. The _PCON _BOR register bit is active when a brown-out reset has occurred. And the PCON _POR bit is active when the reset follows the PICmicro® MCU being powered up. Table 3-3 shows how these bits are set for the six different reset situations.

Upon *any* reset, file registers have the same values as they had before reset and the hardware registers are given their power-up settings. Thus, the I/O pins are returned to input and the peripheral functions are disabled. To restore operation after reset, you might have to save the hardware register contents values before the expected reset operation so that they can be restored later.

The w and file register contents on power up are undetermined and can be any value. When you work with MPLAB and other simulators, these values are generally zero, which will lead to problems if they are not initialized. This issue is covered later in the book.

If a _MCLR or WDT reset occurs, the file register contents are the same as before the reset. This allows you to determine the reset type by placing a known value into these registers and checking them immediately following reset. This is shown in Chapter 15.

The reset vector is the program memory address that the application starts executing after reset. For the mid-range and high-end devices, this address is zero (0x0000). For the low-end PICmicro® MCUs, the reset vector is the highest address of the program memory (i.e., for a 512-instruction device, this is address 511 decimal, 0x01FF).

Most people leave the low-end PICmicro® MCU reset vector address unprogrammed (0x0FFF, which translates to the *xorlw 0x0FF* instruction) and let the program counter roll over to zero and start executing the application from there, as if the reset vector was address 0 (like the other PICmicro® MCUs).

# Interrupts

In the various sections so far where I describe interrupts, I have used diagrams like Fig. 3-17 to describe how interrupt requests are passed to the PICmicro® MCU, responded to and then execution returns to the mainline. Along with providing this explanation, I have included philosophies behind use of interrupts and how applications have been written to exploit them, but I have not spent a lot of time covering what happens in the

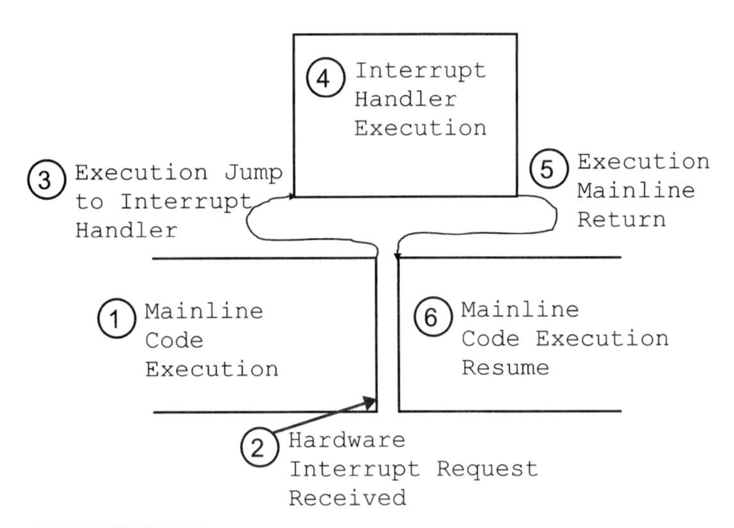


**Figure 3-17** Interrupt execution

PICmicro® MCU when interrupts execute. To understand what happens inside the PICmicro® MCU, I have already gone through many of the important features that make interrupt execution possible. What I want to do in this section is to describe the hardware used by the mid-range PICmicro® MCU to receive, acknowledge, and process interrupts. The interrupt hardware used by the PIC17Cxx and PIC18Cxx is similar and works in much the same way.

Because of the operation of the PICmicro® MCU, some interrupt-requesting events are coming in to the processor all the time. These requests are a result of TMR0 overflowing during normal operation, PORTB input pins changing state, etc. In fact, many of the peripheral hardware events don't have a completion bit or flag; instead, they rely on the interrupt request flag to indicate that the operation has completed or the input event has occurred. Later in the book, these bits are described as *F bits* because their labels always end in the letter *"F."*

To have these interrupt event requests passed to the PICmicro® MCU processor, the interrupt request enable bit (which I call the *E bits*) respective for the interrupt event request has to be set along with the GIE bit of the INTCON register. For the three basic interrupts in the PICmicro® MCU (TMR0 overflow, RBO/INT pin state change, and PORTB pin change), the E and F flags are in the INTCON register. Other interrupt event E and F flags can be located in the PIR and PIE registers or in peripheral control registers, depending on the peripheral requesting the interrupt and the PICmicro® MCU part number.

When the interrupt request comes in the F bit is set. If the GIE bit is set, then on the next instruction, instead of executing the next instruction, the address of the next instruction (or destination address) is saved in the program counter stack and execution jumps to address 0x0004 (for the mid-range PICmicro® MCUs) which is the *interrupt vector*. At this time, the GIE bit is reset, preventing any other interrupts from being acknowledged.

The code starting at address 0x0004 is known as the *interrupt handler* and its purpose is to respond to the incoming event, reset the interrupt-requesting hardware, and prepare it for requesting another interrupt event and reset the interrupt controller hardware. For many interrupt events, all that is required to reset the requesting hardware and the interrupt controller is to simply reset the F-bit requesting the interrupt. During the interrupt handler, GIE is reset, which prevents other interrupt events from interrupting the interrupt handler, which could cause problems with the PICmicro® MCU having to handle a nested interrupt.

Execution continues from here until the *retfie* (*return from interrupt*) instruction, which sets the GIE bit again to allow additional interrupts to execute and returns the PICmicro® MCU's program counter to the address after the interrupt was acknowledged. This entire process is shown in Fig. 3-18.

Figure 3-18 shows the different aspects of the interrupt handler's execution. There are a few things to notice in this diagram. The first is the two instruction cycles required for the jump to the interrupt handler and the two cycles required for the *retfie* instruction to execute. As covered in the earlier sections, when a jump occurs in the PICmicro® MCU, two cycles are required to flush the prefetch buffer and to load in the next instruction before it can be executed.

Along with looking at the two-instruction cycle operation of the execution changes, notice that the jump to the interrupt vector cannot occur until the current instruction has finished executing. This is important because it means that the timing for the interrupt handler is not 100% predictable. The operation of the interrupt handler will lengthen by one cycle if a call, jump, or PCL update is occurring when the interrupt request comes in. In these cases, the jump to the interrupt handler will have to wait for the two-cycle instruction to complete before the jump can occur, which results in a maximum four-instruction-cycle interrupt latency instead of the best-case situation of three-instruction-cycle latency.

I am mentioning this because where you are most likely to see a difference in the response to an interrupt request is in the MPLAB IDE simulator, where the jump to the interrupt

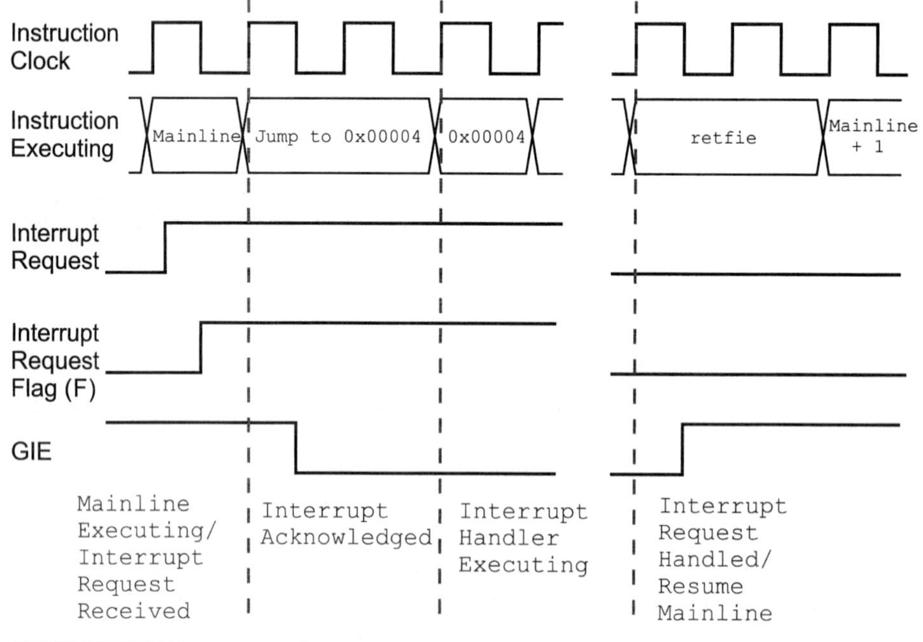


**Figure 3-18** PICmicro® MCU interrupt execution

handler might happen one cycle later than you expect. Not expecting that this can cause you to look through the code, trying to find the reason for this anomaly. In very few cases, the one instruction cycle delay will be a problem, but when you are first working with the PICmicro® MCU, this can cause some confusion.

If you forget to reset the F flag or if another interrupt event requests the interrupt handler before the current request has completed, you will find that execution will not seem to return to the mainline. Instead of immediately following the *retfie* instruction, the first instruction of the interrupt handler (at address 0x0004) will be executed. You can *starve* the mainline code of cycles if interrupt requests come in too quickly or are not reset.

Starving the mainline of instruction cycles because of interrupt operation is something that is very hard to find when you are debugging your application. In fact, I would consider it to be one of the hardest problems to find someone new to the PICmicro® MCU because the simulator will probably not show what happens unless there is a volume of requests (most likely from peripherals).

# Architecture Differences

The four different PICmicro® MCU architectures have a number of similarities and many of the differences are a result of the instruction word sizes than features added or deleted in the processor itself. As pointed out elsewhere in the book, I have focused on the mid-range PICmicro® MCU processor architecture because it has the most commonality with the other architectures.

The two differences you should understand before working with other architectures for applications are:

1. Program counter circuitry
2. Register organization

The following sections examine these differences, how execution changes are accomplished, and how data is accessed and moved within the different PICmicro® MCU architectures.

## LOW-END DEVICES

The low-end PICmicro® MCU devices have a very similar architecture to that of the mid-range devices, although it is missing some of the features of the mid-range. The most obvious omission is the lack of the *addlw* and *sublw* instructions, but there are some other subtler differences that you will have to deal with as well.

One of these differences is a change in the reset vector compared to the mid-range PICmicros® MCU. In the mid-range devices, reset is always 0, but in the low-end devices, this address is always the last address in program memory. Table 3-4 lists the reset vector addresses for different low-end devices' program memory sizes.

As covered in more detail later in the book; I recommend ignoring the reset vector ad-

**TABLE 3-4 Low-end PICmicro® MCU Program Memory Size to Reset Vector**

| PROGRAM MEMORY SIZE | RESET VECTOR |
|---|---|
| 512 | 0x01FF |
| 1024 | 0x03FF |
| 2048 | 0x07FF |

dress and instead use address 0, which will be the next instruction after the instruction at program memory end's executes. When the last instruction in program memory executes, the low-end (and, actually, all) PICmicro® MCU's program counter resets to zero and execution continues from there. If the instruction is left unprogrammed, then it will be read as *(xorlw 0x0FF),* which essentially negates the initial contents of w, which are unknown because the value in w is undefined at power up, as are all the other file registers.

By ignoring the last instruction, you are allowing applications to be written similarly to mid-range applications and not have any differences in regard to reset. It is important to remember that this last address must be left unprogrammed with no instructions placed in it. I tend to remember this by always subtracting one from the specified program memory size.

The following two sections describe the differences in the program counter hardware and the register addressing hardware between the low-end devices and the mid-range devices. These are the major differences between the two architectures (along with the availability of interrupts in the mid-range). Later, I present strategies for writing applications in such a way that moving code and full applications between the two architectures is relatively simple.

**Register access** I consider the register organization of the low-end PICmicro® MCUs to be the largest differentiator between them and the mid-range devices. The use of a 32 bank with no bank-select bits considerably reduces the possible number of file registers and the usability of (relatively) large tables in the low-end PICmicro® MCUs. Although I am disappointed by how few file registers are available and the difficulty in accessing what is available, I do think the low-end PICmicro® MCUs are usable and should be considered when specifying which PICmicro® MCU to use in an application.

The low-end register space is shown in Fig. 3-20.

The low-end PICmicro® MCU's TRIS and OPTION registers can only be written to using the *tris* and *option* instructions. These instructions are explained in detail in the next chapter, but notice that I write them in lower case. This is done to differentiate them from the TRIS and OPTION registers, which, like all registers identified in the text, are denoted in upper case.

Low-end instructions only provide five bits for a register address in a direct addressing instruction and take the form:

```
INSTRTdRRRRR
```

where *INSTRT* is the bit pattern for the instruction. *d* is the destination (1 stores the result back in the register and 0 stores the result in the w register) and *RRRRR* is the register

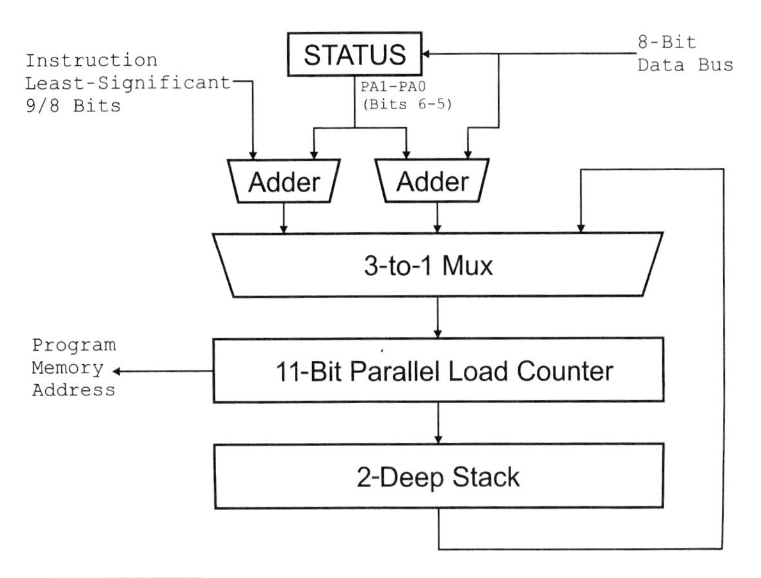


**Figure 3-19** Low-End program counter block diagram

address. In these direct addressing instructions, only the registers in the first bank can be accessed. Accessing registers in other banks require the use of the FSR register.

As can be seen in Fig. 3-20, the first 16 addresses of each bank are common. The 16-bank unique file registers are located in all last 16 addresses of the bank. This limitation of only being able to address data 16 bytes at a time prevents the construction of arrays or other data structures longer than 16 bytes.

Of course, you could work out an algorithm for changing the FSR's high-order bits (bits five and six) to simulate an array of greater than 16 bytes, but, rather than doing this, I would recommend that you go to one of the other PICmicro® MCU architectures for the application instead.

There can be up to four banks in the low-end devices. If 16 file register bytes are available in the last half of each bank and eight or nine file registers are available in the first half (depending on whether or not port C is available), the maximum number of unique file registers in the low-end PICmicro® MCU is 72 or 73.

One quirk is that the low-end PICmicro® MCU's FSR register can never equal zero. Instead of ignoring unused high-order FSR bits, Microchip's designers instead elected to set them. Even if all four bank registers are used (for a total of 128 FSR accessible registers), the FSR register cannot be equal to zero; the FSR register bit seven will be set. Table 3-5

**TABLE 3-5 Low-end PICmicro® MCU Minimum FSR Value to Number of Banks**

| NUMBER OF BANKS | SET FSR BITS | MINUMUM FSR VALUE |
|---|---|---|
| 1 | 7, 6, 5 | 0x0E0 |
| 2 | 7, 6 | 0x0C0 |
| 4 | 7 | 0x080 |

lists which bits will be set in the low-end's FSR, depending on how many bank registers the PICmicro® MCU has.

It can be hard to remember that the low-end's FSR register can never be zero. Chances are, you'll only remember it after you've tested the contents of FSR with an instruction sequence like:

```
movlw 0
xorwf FSR, w
```

and discover that the result is never zeroed.

**Program counter** The low-end PICmicro® MCU's program counter is quite a bit different from that of the mid-range PICmicro® MCU. If you look at the standard register set for the low end, you'll see that there is no PCLATH register; the page-select bits are part of the STATUS register (where the bank select bits are in the mid-range PICmicro® MCUs). As well, because of limitations in the low-end architecture, there are some problems with being able to place and work with tables and subroutines that you should be aware of.

The differences in the low-end PICmicro® MCU's program counter to the mid-range's are partially based on the 512-instruction page size of the low-end PICmicro® MCUs (the mid-range has a 2,048-instruction page size). In the low-end devices, execution stays within these 512 instructions, unless an interpage jump on call is executed or execution simply passes from a lower page to an upper.

The low-end PICmicro® MCU's program counter block diagram is given in Fig. 3-19.

The PA0 and PA1 bits of the STATUS register (bits five and six) perform the same function as the PCLATH register of the mid-range PICmicro® MCUs. Bit PA0 is used to provide bit nine of the destination address to jump to during a *goto* or *call* instruction or when PCL is written to. Bit PA1 is address bit 10. In some low-end PICmicro® MCUs,

| Bank 0 | Bank 1 | Bank 2 | Bank 3 | |
|---|---|---|---|---|
| Addr - Reg | Addr - Reg | Addr - Reg | Addr - Reg | |
| 00 - INDF<br>01 - TMR0<br>02 - PCL<br>03 - STATUS<br>04 - FSR<br>05 - PORTA*<br>06 - PORTB<br>07 - PORTC<br>08-0F Shared File Regs | 20 - INDF<br>21 - TMR0<br>22 - PCL<br>23 - STATUS<br>24 - FSR<br>25 - PORTA*<br>26 - PORTB<br>27 - PORTC<br>28-2F Shared File Regs | 40 - INDF<br>41 - TMR0<br>42 - PCL<br>43 - STATUS<br>44 - FSR<br>45 - PORTA*<br>46 - PORTB<br>47 - PORTC<br>28-2F Shared File Regs | 60 - INDF<br>61 - TMR0<br>62 - PCL<br>63 - STATUS<br>64 - FSR<br>65 - PORTA*<br>66 - PORTB<br>67 - PORTC<br>68-8F Shared File Regs | Shared Registers |
| 10-1F Bank 0 File Regs | 30-3F Bank 1 File Regs | 50-4F Bank 2 File Regs | 70-7F Bank 3 File Regs | Bank Unique Registers |

*OSCCAL can take place of PORTA in PICmicro® MCUs with internal oscillators

**Figure 3-20** Low-End PICmicro® MCU register map

you will see bit seven of the STATUS register being referred to as *PA2*. This bit is not used by any of the current PICmicro® MCUs.

In the mid-range devices, to perform a jump based on changing PCL, the following code is used:

```
PCLATH = HIGH new_address;
   PCL = LOW new_address;
```

In the low-end PICmicro® MCUs, this operation is quite a bit more complex because although PAO-PA2 bits are updated, none of the other bits in the status register should be changed. The equivalent low-end PICmicro® MCU operation to the PCLATH/PCL updating of the program counter is:

```
STATUS = (STATUS & 0x01F) + ((HIGH new_address & 0x0FE) << 4);
PCL    = LOW new_address;
```

In the low-end PICmicro® MCU code, the contents of the status register are ANDed with 0x01F to reset the PAO to PA2 bits. In this formula, which changes the STATUS register, notice that I also delete the least-significant bit of the upper byte of the *new_address.* As discussed in the next paragraph, the bit eight of the destination address can never be specified within the STATUS register. Once the correct bits for PA0 and PA1 have been calculated, they are added to the other bits of the STATUS register. For call instructions, bit eight of the new address is always zero because the instruction word only provides an eight-bit address and PA0 becomes the ninth bit of the new address. This is not a problem for the *goto* instruction, because nine bits, which compass a full low-end page of 512 instruction addresses can be specified within the *goto* instruction itself. For the *call* instruction, in which only eight bits address bits are specified as the destination address, the last 256 instructions of a low-end page cannot be accessed.

Table jumps (direct writes to the PCL register) also have the same restriction as the *call* instruction addresses; they all must be in the first 256 instructions of an instruction page. Larger than 256 entry tables could be created, but they would require a bit of software to calculate the jump across page boundaries to make the table appear contiguous.

Using any of the PA bits for flags in the status register should *never* be done. Incorrect updates of these bits, which are not returned to the correct value before the next table operation, *goto* or *call* will result in the application jumping being invalid. This will be almost impossible for you to debug, so avoid any potential problems and don't modify these bits, except when you are about to change your address location to another page.

## PIC17CXX ARCHITECTURE

When you compare the features of the PIC17Cxx to the low-end and mid-range PICmicro® MCUs, you probably feel like a completely different architecture was created for it. The unique features of the PIC17Cxx compared to the other PICmicro® MCUs include:

**1** The ability to access external, parallel memory
**2** Up to seven I/O ports

3 A built-in 8x8 multiplier
4 Up to 902 file registers in up to 16 banks
5 Up to 64K address space
6 The ability to read and write program memory
7 Multiple interrupt vectors

Along with these enhanced features, block diagrams of the PIC17Cxx (Fig. 3-21) further make you feel like the PIC17Cxx is unique and not that portable between the other PICmicro® MCU architectures.

Despite this apparently different architecture, you can think of the PIC17Cxx's processor as architected like Fig. 3-22, which is not that much different from the low-end and mid-range PICmicro® MCU processors. Data for arithmetic operations are still routed through the registers on a data bus and temporary values are stored in a w register (known as *WREG* in the PIC17Cxx and PIC18Cxx). Instructions can also save arithmetic results in either WREG or the source register. With this basis, you should better understand why I consider the PIC17Cxx to be similar to that of the low-end and mid-range PICmicro® MCUs.

The important differences in the PIC17Cxx architecture are:

1 The accumulator, WREG, can be addressed in the register space
2 The STATUS and OPTION register functions are spread across different register
3 The program counter works slightly differently than for the other architectures
4 The registers are accessed differently and accesses can bypass the WREG
5 An 8x8 multiplier is built in

The next two sections cover the program counter and register differences. This really only leaves two points to consider. The first is the status register changes and the 8x8 multiplier hardware.

Instead of a status register, the PIC17Cxx provides the same functions in the ALUSTA and CPUSTA register. The ALUSTA register is defined in Table 3-6.

As covered later in the book, the FSR index registers of the PIC17Cxx and PIC18Cxx can be incremented or decremented during an access. This feature is useful for implementing high-level languages.

The overflow (OV) bit indicates when the arithmetic operation incorrectly changes the polarity of a two's complement number. For example, when adding numbers like 0x042 and 0x053, the result will be 0x095, which is the two's compliment value -105 (decimal) and not the desired 149 (decimal). After this case, the OV flag will be set to indicate that the result is invalid as a two's complement number. The CPUSTA which is used to monitor the PICmicro® MCU's execution status, is shown in Table 3-7.

Interrupts in the PIC17Cxx are similar in operation to the mid-range PICmicro® MCUs, with an E bit enabling the interrupt-request flag bit (the F bit) to request that the processor execute the appropriate interrupt handler. The PIC17Cxx does not have a GIE bit, which enables interrupts, but does have the GLINTD bit, which must be reset for interrupt requests to be passed to the processor. I like to think of it as the *_GIE (negative GIE) bit.*

Depending on which interrupt is requested and acknowledged, execution will jump to a different interrupt vector address. If multiple interrupts are requested at the same time, the highest priority one will be serviced first. The interrupts, their priorities, and vectors are listed in Table 3-8.

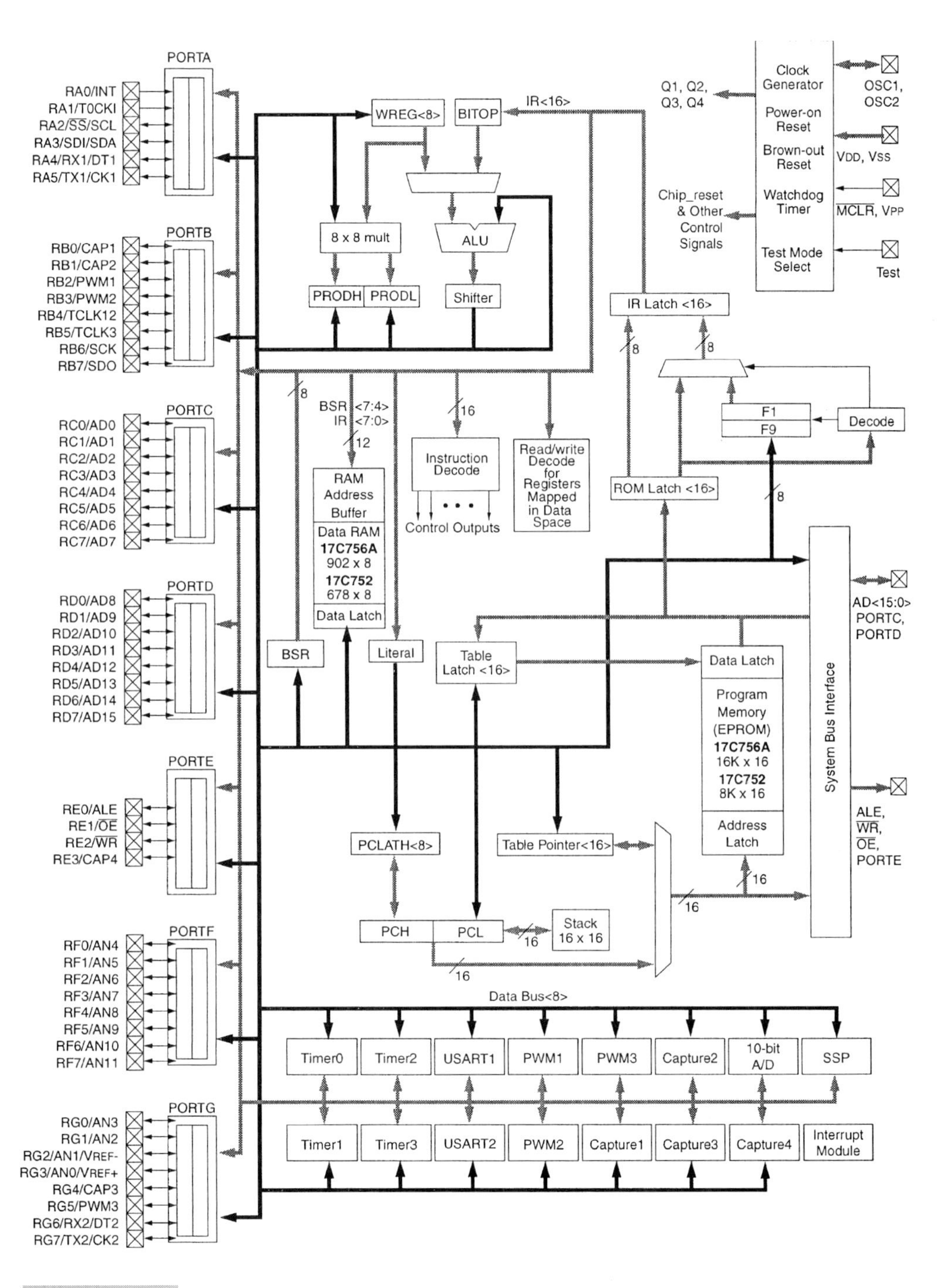


**Figure 3-21** PIC17Cxx processor architecture

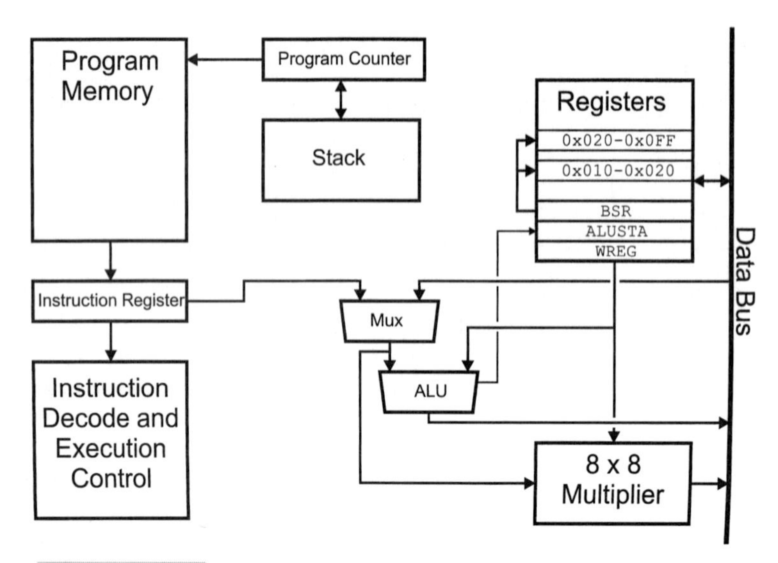


**Figure 3-22** PIC17Cxx processor architecture

The 8x8 multiplier is a very simple piece of hardware that bypasses the PICmicro® MCU's ALU and ALUSTA register to multiply the contents of WREG with another value. The 16-bit result is stored in the PRODL and PRODH registers.

Remembering a trick from high school, you can use the eight-bit multiply capability to multiply two 16-bit numbers together. When you were learning basic algebra and multiplying two two-value expressions together, you were taught to multiply the two "first" val-

**TABLE 3-6 PIC17CXX ALUSTA Bit Definitions**

| BIT | FUNCTION |
|---|---|
| 7-6 | FS3: FS2- FSR1 Mode Select<br>1x - FSR1 does not Change after Access<br>01 - Increment FSR1 after Access<br>00 - Decrement FSR1 after Access |
| 5-4 | FS1: FS0 - FSR0 Mode Select<br>1x - FSR1 does not Change after Access<br>01 - Increment FSR1 after Access<br>00 - Decrement FSR1 after Access |
| 3 | OV - Overflow bit Indicated Overflow in a seven bit, two's complement number |
| 2 | Z - Set when Result is Zero |
| 1 | DC - set when Add or Subtract Operation on the Least Significant four bits Affects the Most Significant four bits |
| 0 | C - Set on Add or Subtract if Operation Affects a More Significant Byte |

**TABLE 3-7 PIC17CXX CPUSTA Bit Definitions**

| BIT | FUNCTION |
|---|---|
| 7-6 | Unused |
| 5 | STKAV - When Set, the Program Counter Stack has Space Available |
| 4 | GLINTD - When Set, all Interrupt Requests are disabled |
| 3 | _TO - Reset after a Watchdog Timer Reset |
| 2 | _PD - Reset after a "sleep" Instruction |
| 1 | _POR - Reset by the PICmicro® MCU if a "Power On Reset" |
| 0 | _POR - Reset by the PICmicro® MCU if a "Brown Out Reset" |

ues together, followed by the "outside," "inside," and "last." This FOIL algorithm could be described as:

```
(A + B) x (C + D) = AC + AD + BC + BD
```

By breaking a 16-bit number into two bytes and recognizing that the high byte is multiplied by 0x0100, *A* and *B* can be written as:

```
A = (AH * 0x0100) + AL
B = (BH * 0x0100) + BL
```

For *A* x *B,* the numbers can be broken up into two parts and then FOILed:

```
A x B = (AH * 0x0100 * AL) + (BH * 0x0100 * BL)
      = (AH * 0x0100 * BH * 0x0100) + (AH * 0x0100 * BL)
        + (AL * BH * 0x0100) + (AL * BL)
```

Knowing that multiplying by 0x0100 is the same as shifting up by one byte (or by eight bits), the two 16-bit variables, *A* and *B* can be multiplied together into the 32-bit "product" using the code:

```
Product  = MUL(AL,BL);
TProduct = MUL(AL,BH)<<8;
Product  = Product + TProduct;
TProduct = MUL(AH, BL)<<8;
Product  = Product + TProduct;
```

**TABLE 3-8 PIC17Cxx Interrupt Vector Address and Priorities for Different Sources**

| PRIORITY | VECTOR ADDRESS | SOURCE |
|---|---|---|
| HIgh | 0x0008 | RAO/INT Pin Interrupt |
| | 0x0010 | TMRO Overflow Interrupt |
| | 0x0018 | TOCKI Pin Interrupt |
| Low | 0x0020 | Peripheral Device Interrupt |

```
TProduct = (MUL(AH, BH)<<8)<<8;
Product  = Product + TProduct;
```

This process can be very easily converted to assembly language.

**Register access** I feel that the most-significant differentiator in the four PICmicro® MCU architectures is how registers are accessed. The low-end and mid-range PICmicro® MCU architectures are similar, but the lack of the same hardware I/O registers in each bank of the mid-range differentiates it from the low-end in which the first 16 addresses of each bank are "shadow copies" of bank zero's registers. The PIC18Cxx register architecture is a completely "flat" twelve-address bit space. The PIC17Cxx has a unique register architecture from the other PICmicro® MCUs.

The PIC17Cxx's register space is designed around a single 8-bit register address built into the instruction set. Like the low-end and mid-range PICmicro® MCUs, the PIC17Cxx uses multiple register banks to allow the access more registers than just this base number. Unlike the low-end and mid-range PICmicro® MCUs, there are two bank areas to access registers and each one has its own set of address bits. I normally think of the PIC17Cxx's registers as being organized like Fig. 3-23.

The first 32-register addresses (0x000 to 0x0IF) is known as the *primary register set*. These registers are the primary processor and PICmicro® MCU hardware features. The hardware

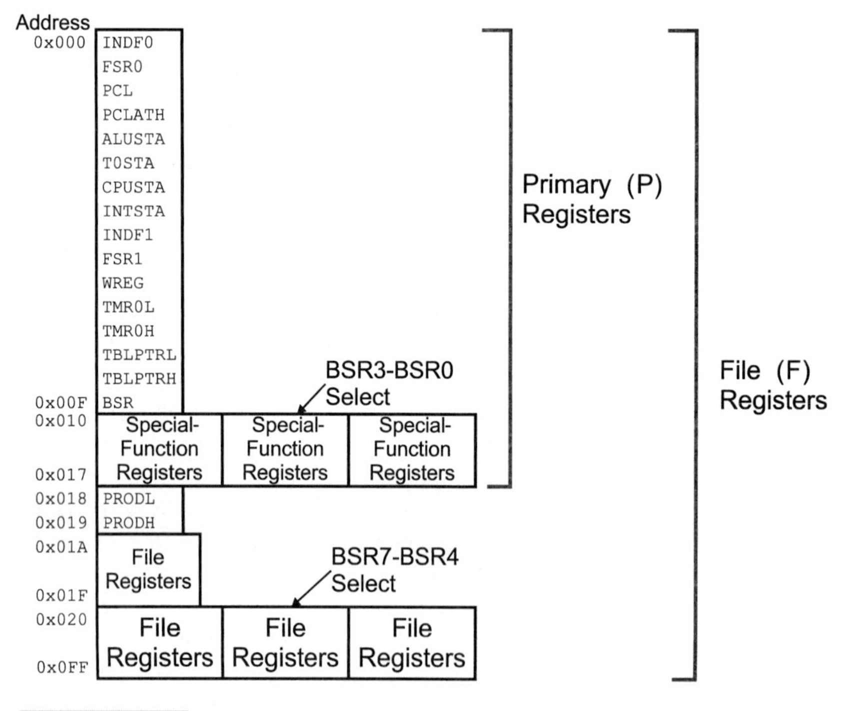


**Figure 3-23** PIC17Cxx register block

interface registers are the *Special-Function Registers (SFR)* located at the register banks in 0x010 to 0x01F, with up to 16 banks. The P register special-function register banks are selected by the least-significant four bits of the *Bank-Select Register (BSR).*

The entire register address range can be accessed as the "F" register set in the full set, banked file (variable) registers are located in the register space above the "P" registers. These file registers are banked, with the selection made by the most-significant four bits of the BSR register.

The five address bits of the primary register set and the eight of the full register set allows data to be moved quickly and easily within the register space without having to go through the WREG. For example, moving the contents of 0x042 and 0x043 to the TMR0L and TMR0H registers could be accomplished by using the *movfp* instructions, which pass data from the F (full) register set to the primary:

```
movfp   0x042, TMR0L
movfp   0x043, TMR0H
```

These two instructions perform the same operations as the mid-range operations:

```
temp = w;
w = contents of 0x042;
TMR0L = w;
w = contents of 0x043;
TMR0H = w;
w = temp;
```

without requiring the *temp* variable to store the current copy of w.

A nice feature of the PIC17Cxx (and PIC18Cxx) architecture is the accessibility of the w register in the register map. Using instructions like *movfp,* which doesn't modify the status (ALUSTA in the PIC17Cxx) register bits, is much easier than using the double *swapf* of the mid-range PICmicro® MCU when saving and restoring context registers during interrupt processing.

**Program counter** If you are familiar with the mid-range PICmicro® MCU's program counter, you should not have any problems working with the PIC17Cxx's. The PIC17Cxx behaves very similarly to the mid-range PICmicro® MCUs, with only a few differences that you should be aware of.

The PIC17Cxx's processor can access 64k 16-bit words of program memory, either internally or externally to the chip. Each instruction word is given a single address, so to address the 64k words (or 128k bytes), 16 bits are required. From the application developer's perspective, these 16 bits can be accessed via the PCL and PCLATH registers in exactly the same way as the low-end and mid-range PICmicro® MCUs. The PIC17Cxx's program-counter block diagram is shown in Fig. 3-24.

The block diagram in Fig. 3-24 differs from the mid-range PICmicro® MCU's program counter block diagram in one important respect, when the *goto* and *call* instructions are executed, the upper five bits of the specified instruction overwrite the lower five bits of the PCLATH register. After executing a *goto* or *call* instruction PCLATH has been changed to the offset of the destination within the current page.

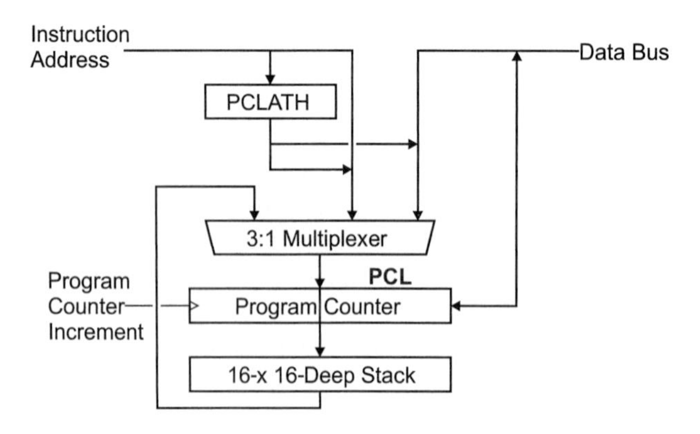


**Figure 3-24** **PIC17Cxx program counter**

For the most part, this is not a problem, but you will have to beware of one case: when the "table" jumps. When you first saw that a *goto* statement updated the PCLATH register, you probably thought that this made table operations easier. For example, a table read could be as simple as:

```
  call TableRd(offset);

  :

int TableRd(int entry)
{

  PCL = Table + entry

Table:
  return "A"
  return "B"
  :

} // End TableRd
```

Looking at this code, there should be no problems as long as the table does not go over a 256-instruction word boundary. Even if it did, the code could be changed to:

```
  Call TableRd(offset)

  :

int TableRd(int entry)
{

int temp;

  temp = (table & 0x0FF) + entry;

  if (temp > 0x0FF)
    PCLATH = PCLATH + 1;
  PCL = temp
```

```
Table:
  Return "A"
  Return "B"
  :

} // End TableRd
```

This is better, but still does not address the real problem if an interrupt occurs sometime between the *call* instruction and the PCL update. If an interrupt occurs and if a *goto* or *call* instruction is executed within the interrupt handler, then the PCLATH register will be changed, resulting in an invalid table jump.

To prevent this from happening, the interrupt handler's entry and exits must always save and restore PCLATH as shown in the code example:

```
Int
  movpf  ALUSTA, _alusta
  movpf  BSR, _bsr
  movpf  WREG, _wreg
  movpf  PCLATH, _pclath

    :                              ; Interrupt Handler Code.

  movfp  _pclath, PCLATH
  movfp  _wreg, WREG
  movfp  _bsr, BSR
  movfp  _alusta, ALUSTA

  retfie                           ; Return to Interrupted Code.
```

This code will make more sense as you read about PIC17Cxx register addressing and how the instructions work, later in this chapter and the book. For now, keep this code as the template for PIC17Cxx interrupt handlers.

## PIC18CXX ARCHITECTURE

The PIC18Cxx architecture (Fig. 3-25) is probably the easiest PICmicro® MCU to develop assembly-language code for. This is because of its large linear register space that can be accessed simply and multiple index registers able to operate like a data stack with pushes and pops. Microchip advertises that the PIC18Cxx is source-code compatible with other PICmicro® MCUs. This point is a bit of a stretch because the PIC18Cxx instruction set has many capabilities and functions not in the other PICmicro® MCUs.

The block diagram I like to use for the PIC18Cxx is shown in Fig. 3-26.

Figure 3-26 shows that the registers are all contained in a 4,096-byte contiguous register space. What is important to realize (and might not be very clear in Fig. 3-26), is that the WREG register provides an input to the ALU, as well as a possible destination to all the arithmetic and bitwise instructions.

When registers are accessed directly, an eight-bit address is specified in the instruction. To access every byte within the register space, a four-bit BSR register has been provided with the ability to select each 256-register bank. As shown in the next section, direct register access has some short cuts that you can take advantage of to avoid using the BSR in your applications.

Like the PIC17Cxx, the PIC18Cxx has a number of FSR index registers with FSR post-

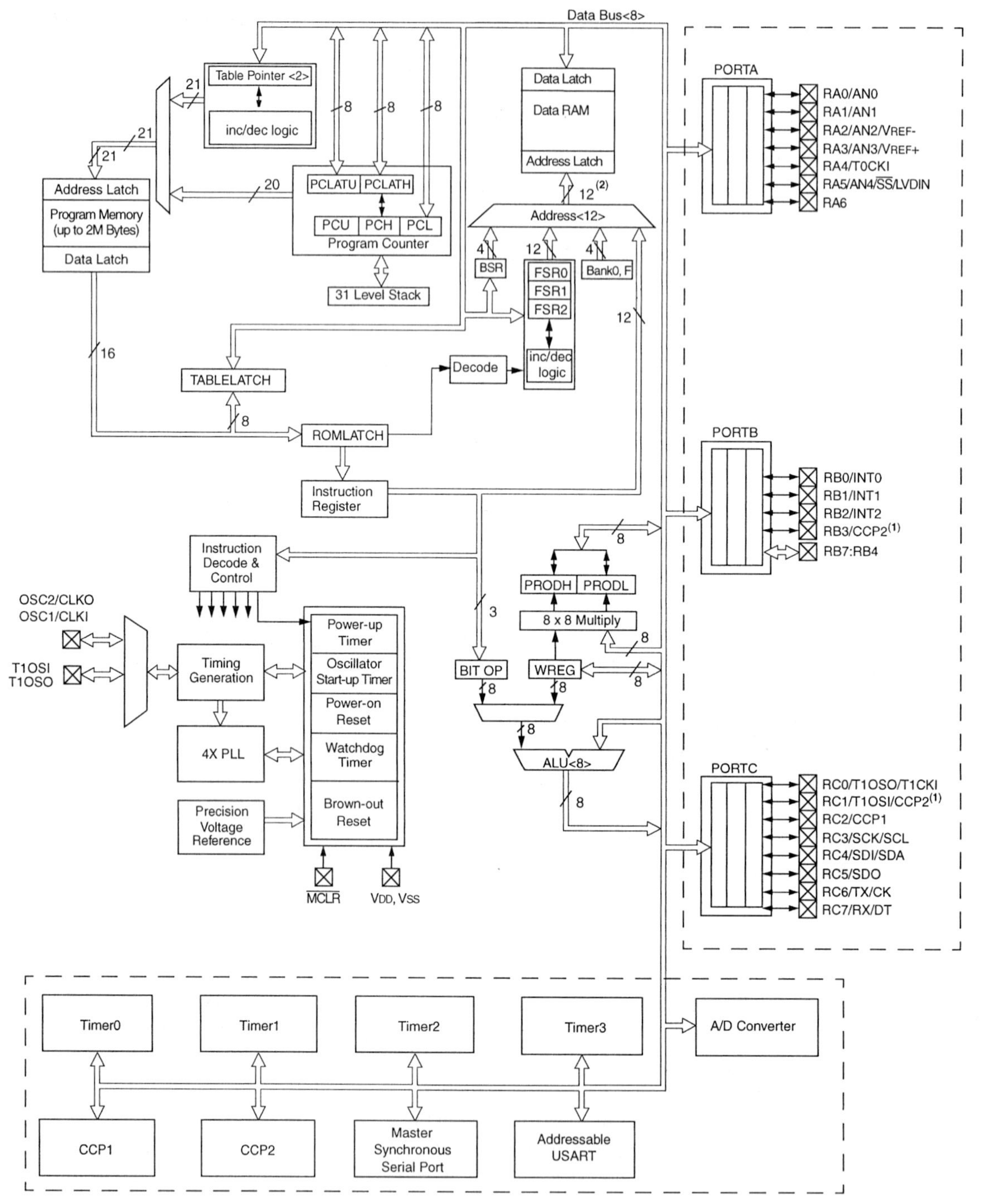


**Note 1:** Optional multiplexing of CCP2 input/output with RB3 is enabled by selection of configuration bit.

**2:** The high order bits of the Direct Address for the RAM are from the BSR register (except for the MOVFF instruction).

**3:** Many of the general purpose I/O pins are multiplexed with one or more peripheral module functions. The multiplexing combinations are device dependent.

**Figure 3-25** **PIC18Cxx processor architecture**

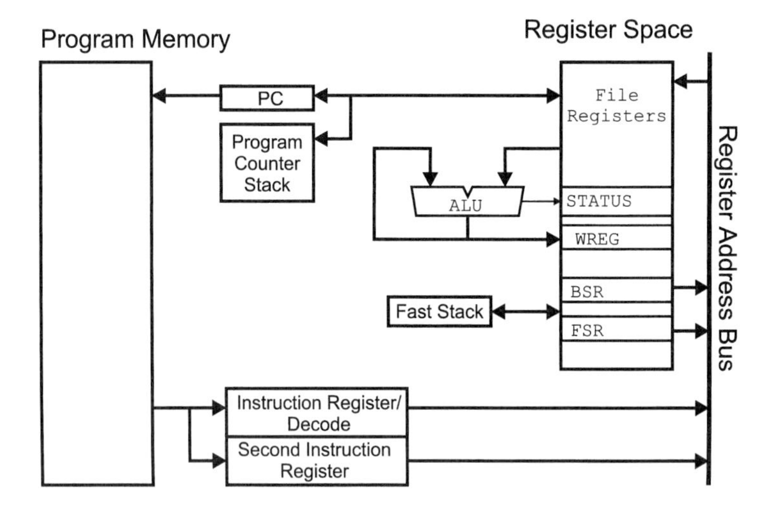


**Figure 3-26** PIC18Cxx architecture block diagram

and pre-increment, post-decrement, and the ability to access data, relative to the FSR. This will make compiler development much simpler for the PIC18CXx than for any of the other PICmicro® MCUs.

The PIC18Cxx's register space is designed to minimize the work required by assembly-language developers and compiler writers.

As I mentioned, instruction, formatting, and execution is similar in the PIC18Cxx with the major difference being the direct register addressing options. The PIC18Cxx has a number of word instructions that allow *goto* and *call* instructions throughout the entire 2.1-MB maximum program memory space, as well as the ability to move register contents directly within the memory space.

New instructions include enhanced subtraction capabilities, including compare operations in which two parameters (one in WREG) are subtracted with the status bits changed and the result discarded. When coupled with the *branch on condition* instructions, the PIC18Cxx has many of the more traditional conditional execution options that are not available in other PICmicro® MCUs. When the total condition execution of the PIC18Cxx are compared to other processors, the PIC18Cxx has many more options than are available in the average device.

The PIC18Cxx's ALU has been enhanced, compared to the other PICmicro® MCUs by the inclusion of *add with carry* and true *subtract with borrow* instructions. The true *subtract with borrow* instruction works as you would expect, instead of the typical reversed PICmicro® MCU instruction:

```
subwfb Reg, Dest                ; Dest = Reg - WREG - B
```

the PIC18Cxx offers the:

```
subfwb Reg, Dest                ; Dest = WREG - Reg - B
```

that works in a more traditional manner than the standard PICmicro® MCU subtraction instructions. This new subtraction instruction make the transition to the new PICmicro® MCUs easier for people familiar with other computer processors.

The program counter and its stack are similar in operation to the other PICmicro® MCU, but have the ability to be modified under application software control. This new capability greatly enhances the PIC18Cxx's ability to run multitasking operating systems or monitor programs compared to the other PICmicro® MCUs. This exciting feature is taken advantage of later in the book.

Lastly, the PIC18Cxx does have some unique interrupt capabilities. Rather than having multiple interrupt vectors, each dedicated to a type of interrupt, such as the PIC17Cxx, the PIC18Cxx has the ability to specify high-priority interrupts, which are given a different address from low-priority interrupts. As well as providing a fast path for the high-priority interrupts, this feature allows splitting up of interrupts to avoid having to check F bits to determine which interrupt is active.

The last architecture feature that can be taken advantage of in PIC18Cxx is the *fast stack,* which consists of three register bytes that save the context registers automatically when an interrupt request is acknowledged. These registers can be restored upon return from the interrupt handler when the *retfie (Return From Interrupt)* instruction is executed. The fast stack can also be taken advantage of within basic subroutine calls, but this is a feature I do not recommend using because it will cause problems if interrupts are used in an application.

**Register access** The PIC18Cxx register architecture is probably the nicest of the four PICmicro® MCU families. Although there is still banking, the variable placement rules covered elsewhere in the book still apply with the ability of directly accessing key variables, as well as the I/O hardware registers (called the *SFR* registers in the PIC18Cxx). In the applications that I have done for the PIC18Cxx, I have found that I have had to think the least about the variable placement and hardware register accessing. The PIC18Cxx can access up to 4,096 eight-bit registers that are available in a contiguous memory space. Twelve address bits are used to access each address within the register map space shown in Fig. 3-27.

Although there are still register banks, the file registers from one bank to the next can be accessed by simply incrementing one of the three FSR registers, instead of redirecting the FSR register into the next bank that is required in the other PICmicro® MCUs. The FSR registers can either be loaded with a full 12-bit address using the *lfsr* instruction or the FSR registers can be accessed directly by the application.

To access a register directly, the PIC18Cxx's *BSR (Bank-Select Register)* register must be set to the bank the register is located in. The BSR register contains the upper four bits of the register's address, with the lower eight bits explicitly specified within the instruction. The direct address is calculated using the formula:

```
Address = (BSR << 8) + Direct Address
```

To simplify directly accessing variables, the first 128 addresses are combined with the second 128 addresses (Fig. 3-27) to make up the *access bank.* This bank allows direct addressing of the special-function registers in the PICmicro® MCU, as well as a collection of variables without having to worry about the BSR register.

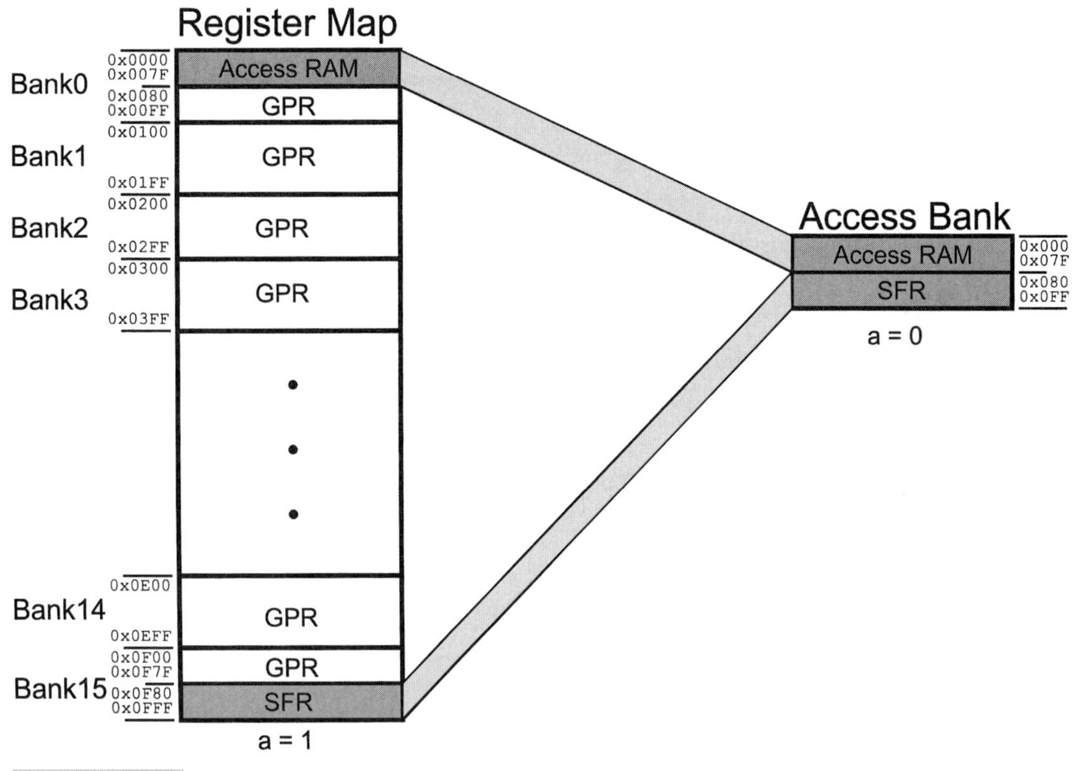


**Figure 3-27** **PIC18Cxx register format**

Practically speaking, the access bank means that, for the first time in the PICmicro® MCU, you will be able to access most, if not all, of the registers required in an application without having to specify a bank or use a special instruction (such as *tris* and *option*). This greatly simplifies the task of developing PIC18Cxx applications and avoids some of the more-difficult aspects of learning how to program the PICmicro® MCU in assembly language: how to access data in different banks.

The hardware I/O registers (SFR) are located in the last 128 addresses of the register space. This might seem limiting, but remember that this is more dedicated hardware register space than is available in any of the other PICmicro® MCUs.

The first 128 file registers in the access bank are more than enough for most applications. As stated, these registers should be used for individual byte and 16-bit variables used in an application. Placing these variables in this address space will avoid the need for you to specify a BSR address for accessing data.

If registers must be specified outside of the access RAM area, they could be accessed directly using the *movff* instruction that moves a source byte from anywhere in the register map to any other location. This two-word instruction is covered in more detail in the next chapter.

The index register operation of the PIC18Cxx is very well organized and will make it much easier for compiler writers to create PIC18Cxx compilers than for other PICmicro®

**TABLE 3-9 PIC18Cxx INDF Register Options**

| REGISTER | FUNCTION |
|---|---|
| INDFx | Register Pointed to by FSRx without modification |
| PLUSWx | Register Pointed to by FSRx + WREG |
| PREINCx | FSRx Incremented before data movement. Register Pointed to by FSRx +1 |
| POSTDECx | Register Pointed to by FSRx. FSRx Decremented After Access |
| POSTINCx | Register Pointed to by FSRx. FSRx Incremented after Access |

MCUs. Along with the three 12-bit-long FSR registers, when data is accessed, the operation can result in the FSR being incremented before or after the data access, decremented after, or access the address of the FSR contents added to the contents of the w register. A specific access option is selected by accessing different INDF register address. Table 3-9 lists the different INDF registers and their options concerning their respective FSR register.

This FSR capability is taken advantage of in the real-time operating system presented later in the book.

When compilers are covered later in the book, included are the operations carried out based on the traditional PICmicro® MCU architectures. The capabilities of the FSR register in the PIC18Cxx allow the FSR registers to simulate stack operations. For example, to simulate a push of the contents of the WREG using FSR0 as a stack pointer, the operation:

```
POSTDEC0 = WREG;
```

could be used. Going the other way, a pop WREG could be implemented as:

```
WREG = PREINC0;
```

In the first case, the stack is decremented after a data value is placed on it. When the data value is to be popped off, the stack pointer (FSR0) is incremented and the data value it is pointing to is returned.

I specified this order of operations to allow access to pushed stack items. Each time that a value is pushed, the FSR register is decremented. To go back and access other items, I can use the PLUSW0 register to read a stack element. For example, to read the element placed three pushes earlier, I would use the code:

```
WREG = 3;
WREG = PLUSW0;
```

This example, while showing how the FSR access with offset works, does not take into account the abilities of the PIC18Cxx instruction set. Using this example as a basis, you would probably assume that writing into the FSR stack at a specific offset is not simple. This is because there is no way to add a constant from the WREG and having a value somewhere that can be accessed and written by the FSR register. The PIC18Cxx's *movff* instruction allows data transfers using an FSR index register and the WREG offset without accessing WREG in any way.

**Program counter** The PIC18Cxx program counter and stack is similar to the hardware used in the other devices, except for three important differences. The first difference is the need for accessing more than 16 address bits for the maximum one million possible instructions of program memory. The second difference is the availability of the fast stack, which allows interrupt context register saves and restores to occur without requiring any special code. The last difference is the ability to read and write from the stack. These differences add a lot of capabilities to the 18Cxx that allow applications that are not possible in the other PICmicro® MCU architectures to be implemented.

In the PIC18Cxx, when handling addresses outside the current program counter, not only is a PCLATH register (or PA bits, as in the low-end devices) update required, but also a high-order register for addresses above the first 64 instruction words. This register is known as *PCLATU*. PCLATU works identically to the PCLATH register and its contents are loaded into the PIC18Cxx PICmicro® MCUs program counter when PCL is updated.

Each instruction in the PIC18Cxx starts on an even address. Thus, the first instruction starts at address zero, the second at address two, the third at address four, etc. Setting the program counter to an odd address will result in the MPLAB simulator halting and the PIC18Cxx working unpredictably. Changing the convention used in the previous PICmicro® MCUs to one where each byte is addressed means that some rules about addressing will have to be relearned for the PIC18Cxx.

As I write this, no PIC18Cxxs have more than 16k instructions, which means that the PCLATU register will never be anything other than zero. The case where PCLATU will be updated is when you have a PIC18Cxx that has more than 32k instructions, which require 64k byte addresses.

The fast stack is an interesting feature that will simplify your subroutine calls (in applications that don't have interrupts enabled), as well as working with interrupt handlers. To use the fast stack in the *call* and *return* instructions, a *1* parameter is put at the end of the instructions. To prevent the fast stack from being used, a *0* parameter is put at the end of the *call* and *return* instructions.

The *fast stack* is a three-byte memory location where the w, STATUS, and BSR registers are stored automatically when an interrupt request is acknowledged and execution jumps to the interrupt vector. If interrupts are not used in an application, then these registers can be saved or restored with a *call* and *return*, such as:

```
  Call sub, 1          ;  Call "sub" after saving "w", "STATUS"
                       ;  and "BSR"
     :
sub                    ;  Execute "Sub", Ignore "w", "STATUS"
    :                  ;  and "BSR
  return 1             ;  Restore "w", "STATUS" and "BSR" before
                       ;  Return to Caller
```

The fast stack option is not recommended in applications in which interrupts are enabled because of the interrupt overwriting the saved data when it executes.

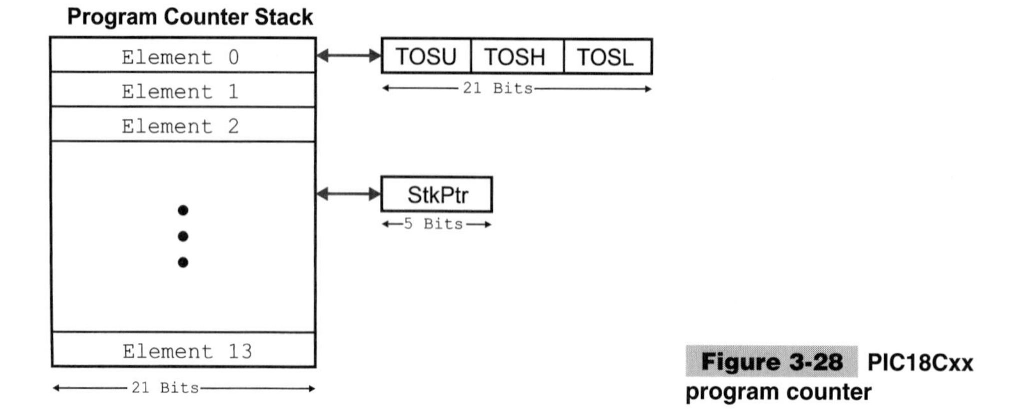


**Figure 3-28** PIC18Cxx program counter

If the fast option is not used with interrupts, then the three context registers can be saved, restored by using the code:

```
Int
  movwf       _w                    ; Save Context Registers
  movff       STATUS, _status
  movff       BSR, _bsr
    :                               ; Interrupt Handler Code
  movff       _bsr, BSR             ; Restore Context Registers
  movf        _w, w
  movff       _status, STATUS
  retfie
```

Notice that in the interrupt handler, the w register is restored before the status register so that the status flags are not changed by the *movf* instruction after they have been restored. As in the example interrupt code given for the PIC17Cxx architecture, this code should be kept in your hip pocket until it is required.

The last difference is also the most significant. The ability to access the stack is quite profound and a deeper understanding of the PIC18Cxx's stack is required than for the other PICmicro® MCU processor architectures.

**TABLE 3-10 PIC18Cxx STKPTR Register Bit Definitions**

| BIT | DESCRIPTION |
|---|---|
| 7 | STKFUL—Stack Full Flag which is set when the Stack is Full or Overflowed |
| 6 | STKUNF—Stack Underflow Flag which is set when more Stack Elements have been Popped than Pushed. |
| 5 | Unused |
| 4-0 | SP4:SPO—Stack Pointer |

The stack itself, at 31 entries is deeper than the other PICmicro® MCU stacks, and the hardware monitoring the stack is available as the STKPTR register. A block diagram of the stack is shown in Fig. 3-28.

The STKPTR register is shown in Table 3-10. The STKUNF and STKFUL bits will be set if their respective conditions are met. If the STVREN bit of the configuration fuses is set, when the STKUNF and STKFUL conditions are true, the PICmicro® MCU will be reset.

I'm of a mixed mind as to the appropriateness of resetting the PICmicro® MCU after an invalid stack operation. Although a reset will definitely indicate an error has occurred (just like a watchdog timer timeout), there would be a problem with decoding what has happened.

The value at the top of the stack can be read (or written) using the top or stack (TOSU, TOSH, and TOSL) registers. These registers are pseudo-registers, like INDF. These registers access the top of the stack directly.

When an address is pushed onto the stack, the SP bits of the STKPTR register are incremented and then the TOSU, TOSH, and TOSL registers are updated. Address pops occur in the reverse order; data is taken out of the TOSU, TOSH, and TOSL registers and then the stack pointer bits are decremented. The 18Cxx has push and pop instructions, which increment and decrement the stack pointer and SP bits. These instructions should be used to change the stack pointer; the SP bits should never be written to directly to avoid any possible damage to the stack and being unable to return to the caller.

Using the TOS registers, the stack can be recorded or changed, for example, if you wanted to implement a computed return statement "into" a table which starts at "Table Start." The following instructions could be used:

```
TOSU = ((TableStart & 0x0FF0000) >> 8) + ((offset & 0x0FF0000) >> 8);
if ((((TableStart & 0x0FF00) >> 4) + ((offset & 0x0FF00) >> 4)) >
   0x0FF)
```

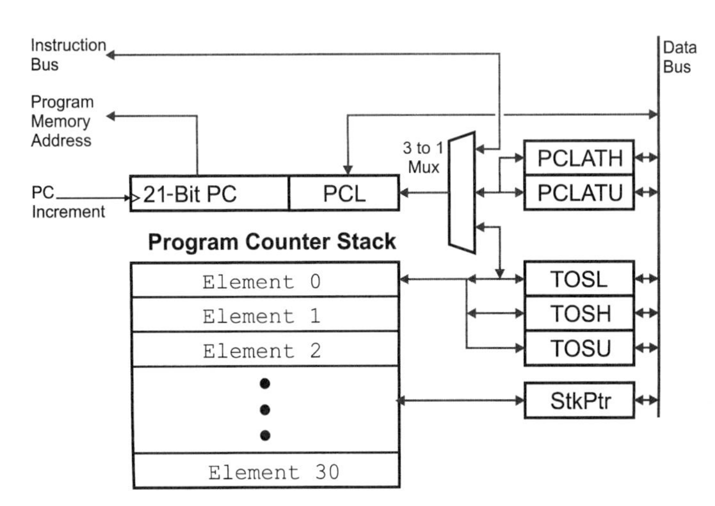


**Figure 3-29** PIC18Cxx program counter stack

```
  TOSU = TOSU + 1;
TOSH = ((TableStart & 0x0FF00) >> 4) + ((offset & 0x0FF00) >> 4);
if (((TableStart & 0x0FF) + (offset & 0x0FF)) > 0x0FF)
  if ((TOSH + 1) > 0x0FF) {
    TOSH = 0;
    TOSU = TOSU + 1;
  } else
    TOSH = TOSH + 1;
TOSL = (TableStart + offset) & 0x0FF;
Return
```

The block diagram for the 18Cxx's program counter and stack is similar to that of the other PICmicro® MCUs, but it incorporates the differences that have been discussed (Fig. 3-29).

In Fig. 3-29, you can see that the 21-bit program counter can be updated either from the stack, the PCL registers of the processor, a new address from the instruction or incremented after normal instruction execution. The output value is used to access program memory during execution and/or the configuration fuses during programming.

# 4

# THE PICmicro® MCU INSTRUCTION SET

CONTENTS AT A GLANCE



This book emphasizes understanding the PICmicro® MCU's processor architecture and visualizing how application code and instructions move data through the PICmicro® MCU. I do this because the PICmicro® MCU's instruction set is somewhat unusual. Most people first learn assembly-language programming on a conventional Von Neumann processor, such as the Motorola 6800. When presented with the PICmicro® MCU, they feel like they are starting all over again.

By developing a good understanding of the PICmicro® MCU device, you will be able to code it quite easily. Along with being able to develop software for the PICmicro® MCU quite easily, you will also be able to look for opportunities to optimize your application and simplify it using the PICmicro® MCU's architectural features.

To characterize a processor's instruction set, I find that it is best to break the instructions into functional groups. The instruction sets used by the four different PICmicro® MCU architectures can be broken up into four such groups.

The first group contains the data-movement instructions that are used to move data in and out of the processor. As indicated earlier in the previous book, data movement within the PICmicro® MCU generally passes through the w register, although register arithmetic instructions have the option of storing the result into the w register or back into the source register.

Data-processing instructions includes adding and subtracting from registers, along with incrementing, decrementing, and doing bitwise operations. The arithmetic instruction group can be broken up into two subgroups, the register arithmetic (where only the contents of registers are used) and the immediate arithmetic (where an explicitly stated constant value is used for the operation).

Execution change instructions make up the next functional group. These are the *gotos, calls, returns,* and conditional instruction skips. The PICmicro® MCU instruction sets differ from other traditional processor instruction sets in that a *jump on condition* requires two instructions instead of a single explicit one. To carry out conditional jumps or other conditional operations, a *skip next* instruction is executed before the actual operation.

Along with traditional *gotos* and *calls* instructions is the opportunity to write to the PICmicro® MCU's program counter directly. This ability gives you the opportunity to create conditional jumps, based on arithmetic values or implement data tables, which return constants for different values. This very powerful function in the PICmicro® MCU is in the applications presented in this book.

Finally, a number of processor control instructions are used to control the operation of the PICmicro® MCU's processor. These instructions are quite typical for most microprocessors (not only microcontrollers) and there shouldn't be any surprises in this set.

As in the previous chapter, I initially focus on the mid-range devices to explain the instruction's operation. Later in the chapter, the differences between the instruction sets used by PICmicro® MCU families are reviewed. The appendices include instruction set summaries for your use.

You should always remember that PICmicro® MCU instructions execute in one instruction cycle unless the program counter is changed so that execution does not continue to the following instruction. This makes programming timing crucial applications much easier than in other microcontrollers; later in the book, I'll present examples of time-critical code and methodologies used to write it.

# The Mid-Range Instruction Set

It would be wrong to say that I choose the mid-range to concentrate on in this book because it is a bridge between the "high-end" devices (PIC17Cxx and PIC18Cxx) and the low-end. Yes, the mid-range's instruction set has more instructions than the low-end, and the high-end has more, but I choose the mid-range device because it will most likely be the family for which you will develop your applications. As shown in the previous chapters, the mid-range has a substantial number of different part numbers with a myriad of different built-in features.

As I work through the instructions, I have included a diagram with the data flow for each instruction as well as the number of cycles required for the instruction to execute and the STATUS register flags that are affected. This format will be optionally used in the explanation of the instruction sets for the other PICmicro® MCU processor architectures because the instruction operation can be identical to mid-range's instructions explained previously.

In the appendices, I have included a reference for each instruction.

## DATA-MOVEMENT INSTRUCTIONS

If you are familiar with the Intel 8086 (which is the base processor used in the IBM PC), you will probably appreciate the ability of some instructions to execute without having to store temporary results in the accumulator registers. This feature can significantly simplify applications and avoid the need to temporarily modify the contents of the accumulators.

Unfortunately, the PICmicro® MCU does not have this capability and data must pass through the w register (which is the PICmicro® MCU's accumulator) before it can be put in a destination register. The typical data flow for information in the PICmicro® MCU is:

```
w = Source;                   // Load "w" with the Source Data
Destination = w;              // Store the contents of "w"
```

To load w, two primary instructions are used. The *movlw* instruction (Fig. 4-1) loads w with a constant value. This constant can be any eight-bit value. The format of the instruction is:

```
movlw Constant         ;  Load "w" with "Constant"
```

This is the basic method of loading w with a value, but it depends on what has been programmed into the application code.

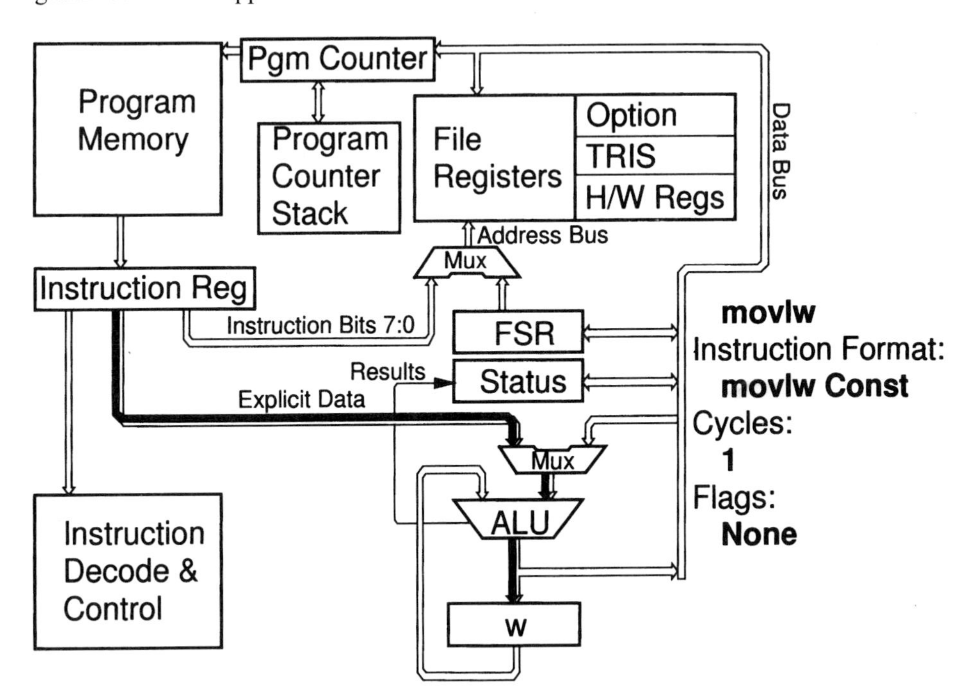


**Figure 4-1** "movlw" Instruction operation

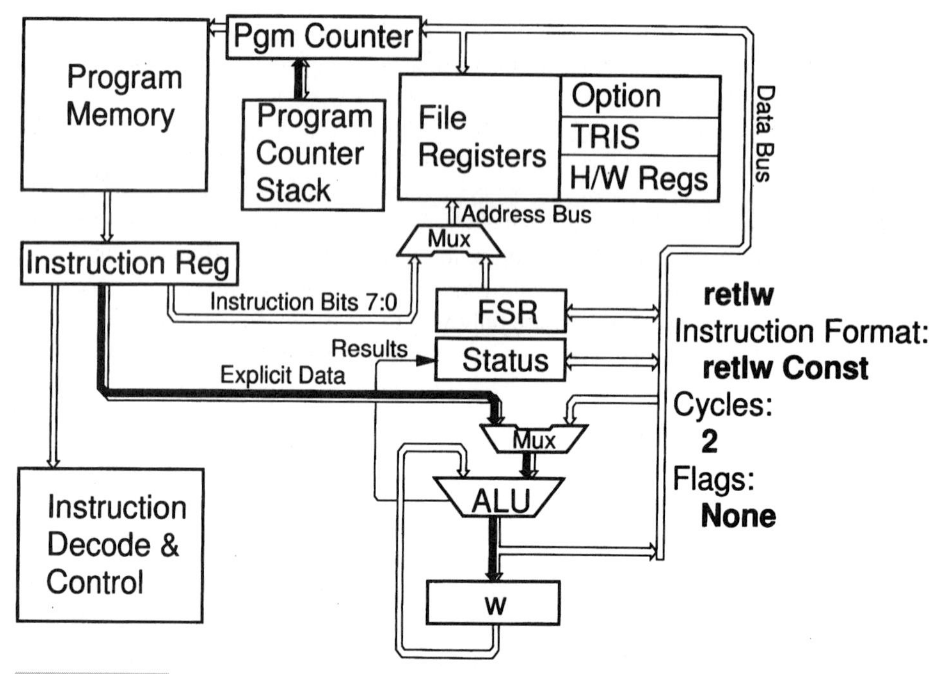


**Figure 4-2** "retlw" Instruction operation

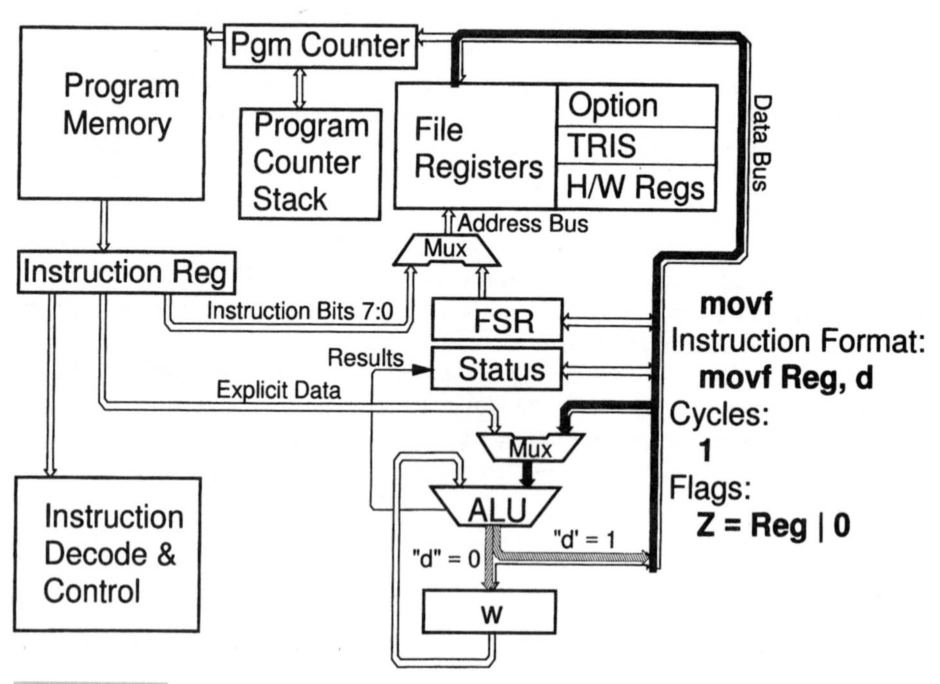


**Figure 4-3** "movf" Instruction operation

None of the STATUS flags are changed by the *movlw* instruction.

In the "movlw constant" instruction on page 105, the text that follows the semi-colon (;) is ignored by the assembler. This text is known as a *comment* and is used to indicate to a person reading the source code what the code is doing. In "C," double *slashes* ("//") perform the same function. Comments are described in more detail later in the book.

Later, this chapter introduces how subroutines are implemented in the PICmicro® MCU. One of the subroutine *return* instructions is the *retlw* instruction, which returns from a subroutine and loads w with a constant value (the operation of this instruction in shown in Fig. 4-2).

This instruction is useful for implementing tables that return constant values (which are explained later in the chapter) and are the only way of returning from a subroutine in the low-end devices. The operation of the *retlw* instruction, which is defined as:

```
retlw Constant
```

could be simulated by the use of the two instructions:

```
movlw Constant
return
```

To load w with the contents of a register, the *movf* instruction (Fig. 4-3) is used. The format of the *movf* instruction is:

```
movf  Variable, d    ;  Move the contents of "Variable" through the
                     ;  ALU and set the "Zero" flag based on its
                     ;  value. Store "Variable" according to "d"
```

where *d* is the destination of the contents of the variable. As written earlier in the book, *d* can be either 0 or 1. When you use the MPLAB assembler, the values *w* or *f* can be used instead of 0 or 1, respectively, for *d* to indicate where the byte read in is going to be stored. If *d* is *w* or 0, then the value of *Variable* read by *movf* will be stored in the w register.

If *d* is *f* or 1, then the value of *Variable* read by *movf* will pass through the PICmicro® MCU's ALU, change the zero flag (according to its value), and then be stored back into *Variable* without changing the contents of the w register.

I like to define the *movf* instruction as being used to set the zero flag according to the contents of the register and optionally load w with the contents of the register. This might seem like a backward way to describe the instruction, but it is actually quite accurate in terms of how the instruction executes.

To test the value, the ALU ORs the value read from the register with 0x000, which sets the zero flag if the result of this operation is zero.

Another way to set a register value is to use the clear instructions, *clrw* and *clrf* (Fig. 4-4). *Clrw* clears the w register and sets the zero flag. *Clrf* clears the specified register and also sets the zero flag.

The *clrw* instruction does not have any parameters and is invoked by simply entering:

```
clrw
```

The *clrf* instruction only has one parameter and that is the register that is specified:

```
clrf Register
```

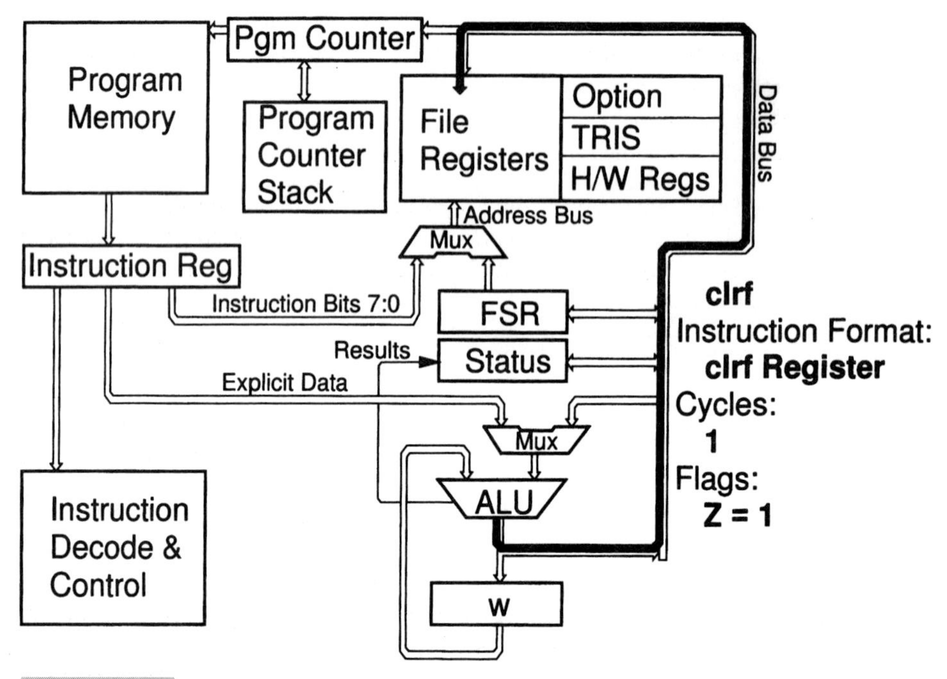


**Figure 4-4** "clrf" Instruction operation

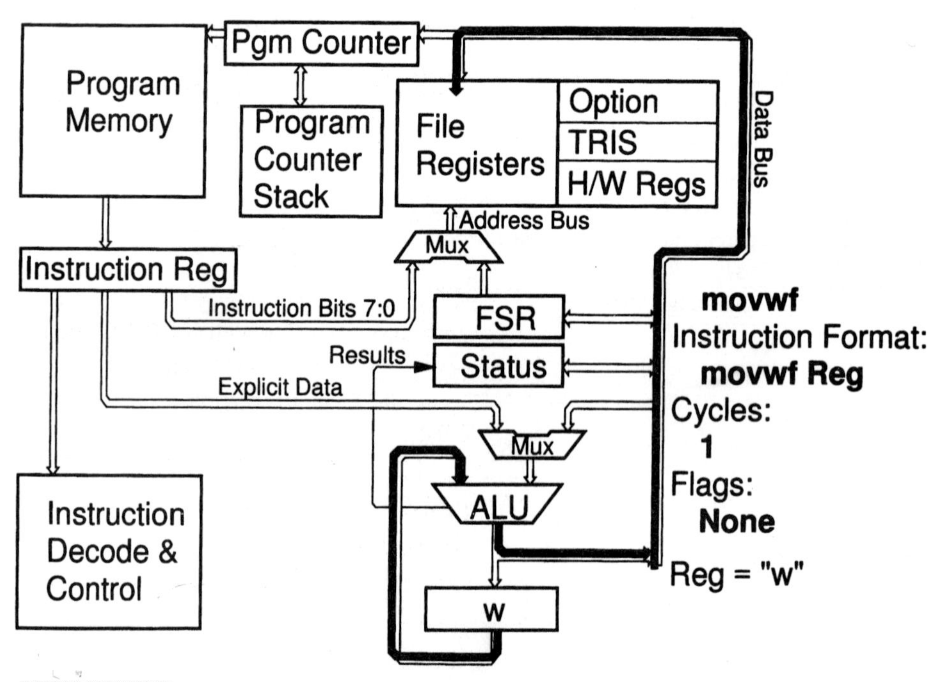


**Figure 4-5** "movwf" Instruction operation

Storing the contents of w is accomplished using the *movwf* instruction (Fig. 4-5). This instruction simply loads the specified register with the contents of the w register. No STATUS flags are affected by the operation.

The format of *movwf* is:

```
movwf Register
```

where *Register* is the destination register for the contents of the w register.

Along with the *movwf* instruction, the *option* and *tris* instructions store the contents of w into specific registers. None of the STATUS flags are changed (or are required to be changed) for these instructions.

The *option* instruction (which doesn't have any parameters (Fig. 4-6) is specified as:

```
option
```

This instruction saves the contents of w into the OPTION register (at address 0x081), bypassing the need (in the mid-range PICmicro® MCU) to set the RP0 bit of the STATUS register to set the contents of OPTION. Notice that in the Microchip PICmicro® MCU manuals, you will most often see the OPTION register referred to as *OPTION_REG*. The reason for this is because of the same label being given to both the OPTION register and *option* instruction. This book does not use the OPTION_REG convention to specify the register instead of the instruction, but the register is capitalized (as has been done with all register names to identify them). The instruction is in lowercase.

*Tris* (Fig. 4-7) is used to load an I/O port driver-enable (TRIS) register with the contents of w. The TRISA, TRISB, and TRISC registers can be accessed using these instructions, which have the format:

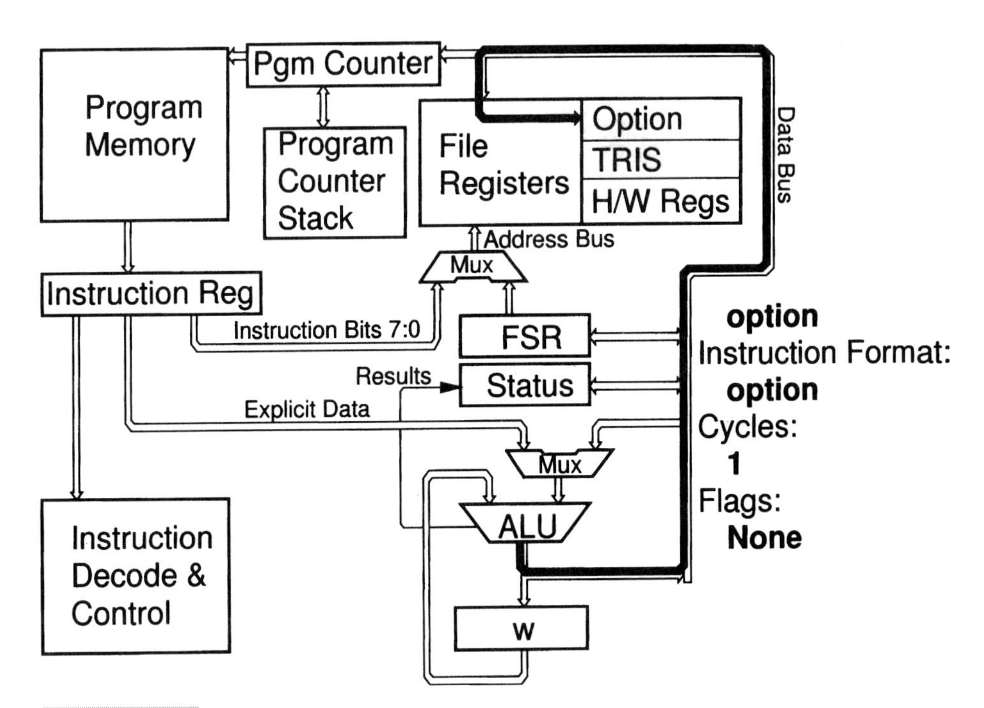


**Figure 4-6** "option" Instruction operation

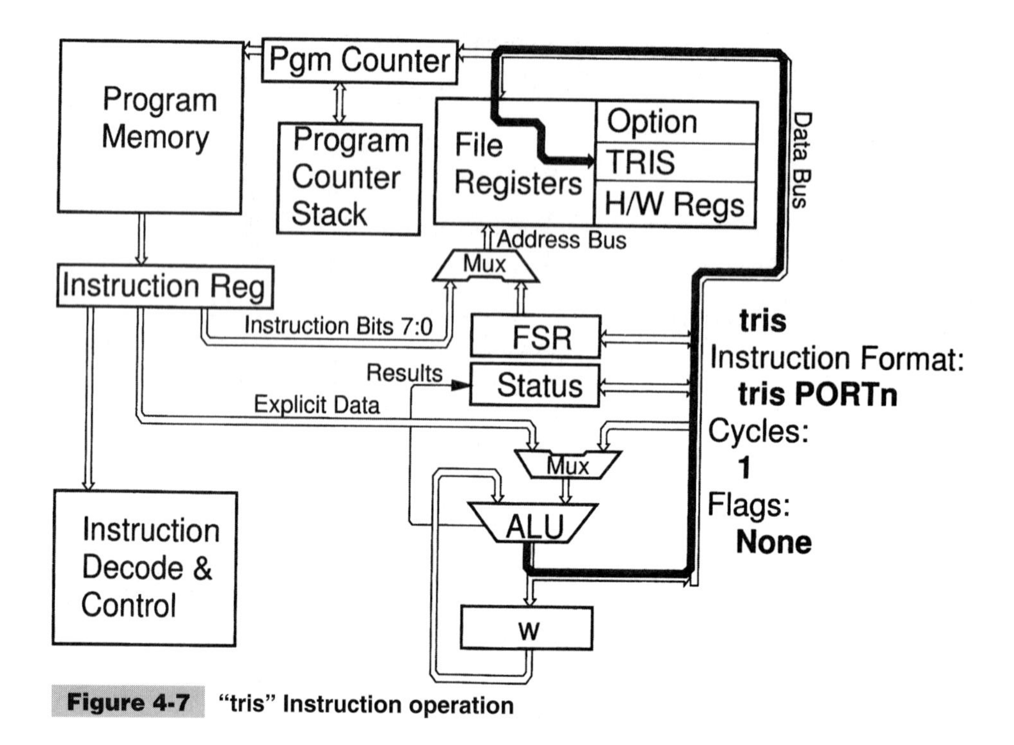


**Figure 4-7** "tris" Instruction operation

```
tris    PORTx
```

or by setting the RP0 bit of the STATUS register, which makes execution use the Bank 1 set of registers and access the TRIS registers directly.

Microchip does not recommend the use of the *option* and *tris* instructions in the mid-range PICmicro® MCUs. These instructions were originally created for the low-end PICmicro® MCUs, which do not have the OPTION and TRIS registers in the different banks (as do the mid-range). Microchip, while continuing their use in the current mid-range PICmicro® MCUs, might not continue them in future devices, which is why their use is not recommended.

As indicated elsewhere in the book, I personally don't recommend their use because they do not access all the PORT registers in all PICmicro® MCUs. If you look at the bit pattern for the instruction:

```
0b0 00 0000 0110 0fff
```

where fff is the PORT register written to by the instruction, you will see that the TRIS registers for PORTA (address 5), PORTB (address 6), and PORTC (address 7) are the only ones that can be written to. With PICmicros® MCU that have a PORTD (address 0x008) and PORTE (address 0x009), the *tris* instruction cannot be used because these TRIS registers cannot be accessed by the *tris* instruction.

Despite this, you still might want to use the *tris* and *option* instructions—especially when debugging an application on a mid-range PICmicro® MCU that is designed to be used with a low-end one.

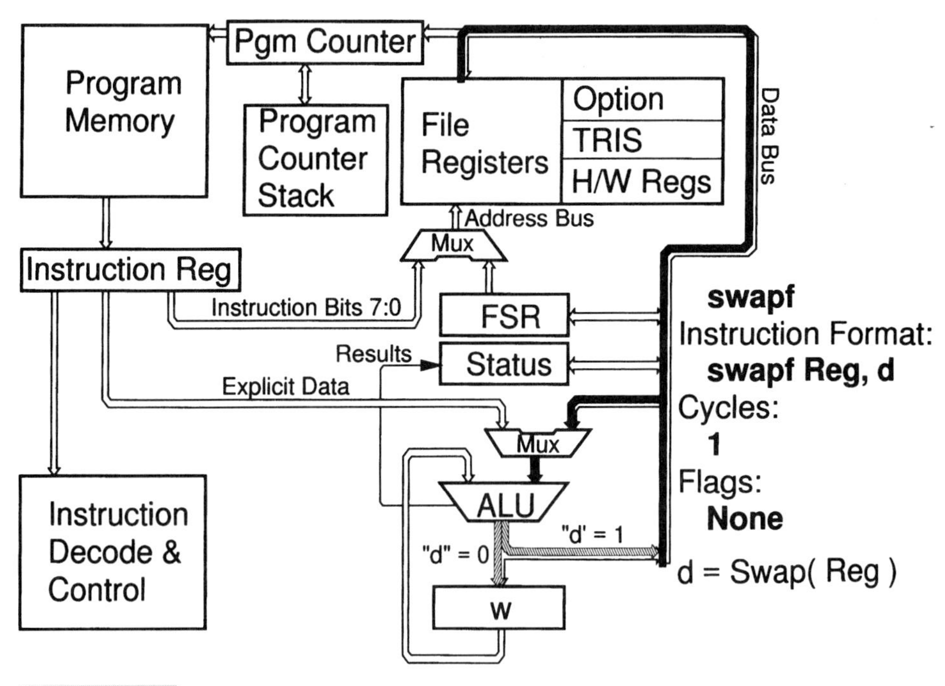


**Figure 4-8** "swapf" Instruction operation

One of the interesting instructions in the PICmicro® MCU is the *swapf* instruction (Fig. 4-8). This instruction exchanges (swaps) the contents of the high and low nybbles of source register and stores the value in w or back in the source register, depending on the value of *d* in its invocation:

```
swapf Register, d
```

The most obvious use for *swapf* is to use it to display a byte as two ASCII nybbles.

```
swapf   Register, w
call    NybbleDisplay     ;  Output the Most Significant four bits of
                          ;    Register as an ASCII Character

movf    Register, w       ;  Output the Least Significant four bits of
                          ;    Register as an ASCII Character

call    NybbleDisplay     ;  Register as an ASCII Character
```

The code loads the least-significant four bits of w with the digit to display before calling *NybbleDisplay*, which converts these four bits into an ASCII hex code representation. This example code will first output the most-significant four bits of the contents of *Register* followed by the least-significant four bits.

*Swapf* does not modify any of the STATUS flags, which makes it useful when loading w without changing any of the status flags.

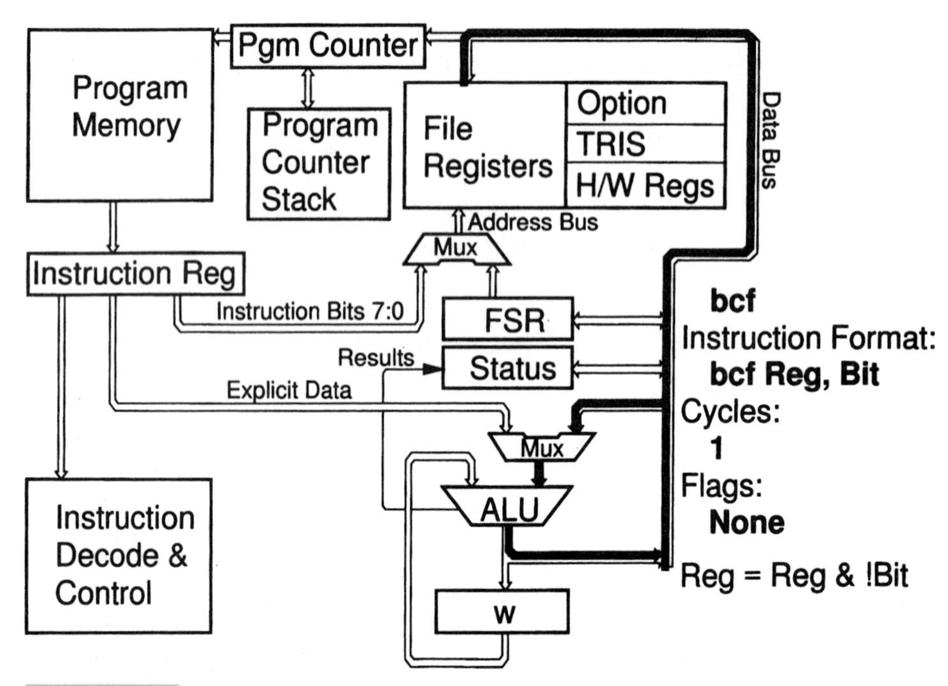


**Figure 4-9** **"bcf" Instruction operation**

```
swapf Register, f
swapf Register, w
```

The code snippet exchanges the high and low nybbles of *Register* and stores the result back into *Register* before exchanging them again and loading the contents into w. This double exchange returns the contents of *Register* to the original value for loading into the w register without modifying any of the STATUS register bits.

The ability of loading w with a register value without affecting the contents of the STATUS bits (specifically the zero flag, which is modified by the *movf* instruction) is something that is taken advantage of in PICmicro® MCU interrupt handlers. In the PIC17Cxx and PIC18Cxx, data-movement instructions have been included that do not modify any of the STATUS bits, so this use of *swapf* is not necessary in these PICmicro® MCU processors.

The last two instructions used for data movement are the *bcf* and *bsf* instructions that reset or set a specific bit in a register, respectively. The operation of the *bcf* instruction is shown in Fig. 4-9 and is put into the source code as:

```
bcf Register, Bit
```

In the *bcf* instruction, the *Bit* of *Register* is reset. The operation of this instruction could be characterized as:

```
Register = Register & (0x0FF ^ (1 << Bit))
```

In this operation, the contents of *Register* are ANDed with a value that has all of the bits set, except for the one that you want to be reset. In this equation, the « operation shifts the value 1 over *Bit* times to the left. When 1 has been shifted over *Bit* times to the left, it is XORed with 0x0FF, resulting in the specified bit reset. When this is ANDed with the contents of the register, that bit will be reset.

The *bsf* instruction sets the register bit specified by the instruction as:

```
bsf   Register, Bit
```

Like *bcf*, *bsf* can be characterized by an equation:

```
Register = Register | (1 « Bit)
```

In *bsf*, the value 1 is shifted to the left *Bit* number of times and ORed with the contents of *Register*.

Both *bcf* and *bsf* are useful instructions when you just want to change the state of a single bit. They are paired with the *btfsc* and *btfss* instructions that test the state of a register bit and skip the next instruction accordingly. When I wrote the first edition of this book, I put these four instructions in a section called *Bit Operators*. I changed this for the second edition because this confused the section on execution change and made it more difficult to relate how the instructions operate.

The *bcf* and *bsf* really belong in this section because they move specific bit values into registers. The *btfsc* and *btfss*' instructions in the "Execution Change Instructions" section because they are the primary method that is used to conditionally change how a PICmicro® MCU application is executing.

## DATA-PROCESSING INSTRUCTIONS

The PICmicro® MCU does not have a very wide range of instructions that algorithmically or logically change data values. The seven unique operations (implemented over 15 instructions) available in the PICmicro® MCU might not seem to be that comprehensive, but they are an excellent base to work from and provide a set of operations from which more complex operations can be implemented. As I work through the different operations, I will show simple optimizations or tricks that will help you with your applications as well as explain exactly how the instructions work.

The arithmetic operation that probably first comes to mind is addition. In the PICmicro® MCU, addition is carried out in a very straightforward manner, with the contents of the register specified by the *addwf* instruction added to the contents of w with the result stored in either the specified register of w. The operation of *addwf* is shown in Fig. 4-10.

The format used for the *addwf* instruction is:

```
addwf Register, d
```

Where *Register* contains the value to be added to the contents of w. If *d* is 1 or *f*, then the result will be stored back into *Register*. If *d* is 0 or *w*, the result is stored in the w register.

*Addlw* is used to add an immediate (constant) value to the contents of the w register with the result being stored back into the w register. The operation of the instruction is shown in Fig. 4-11 and its source code format is:

```
addlw Constant
```

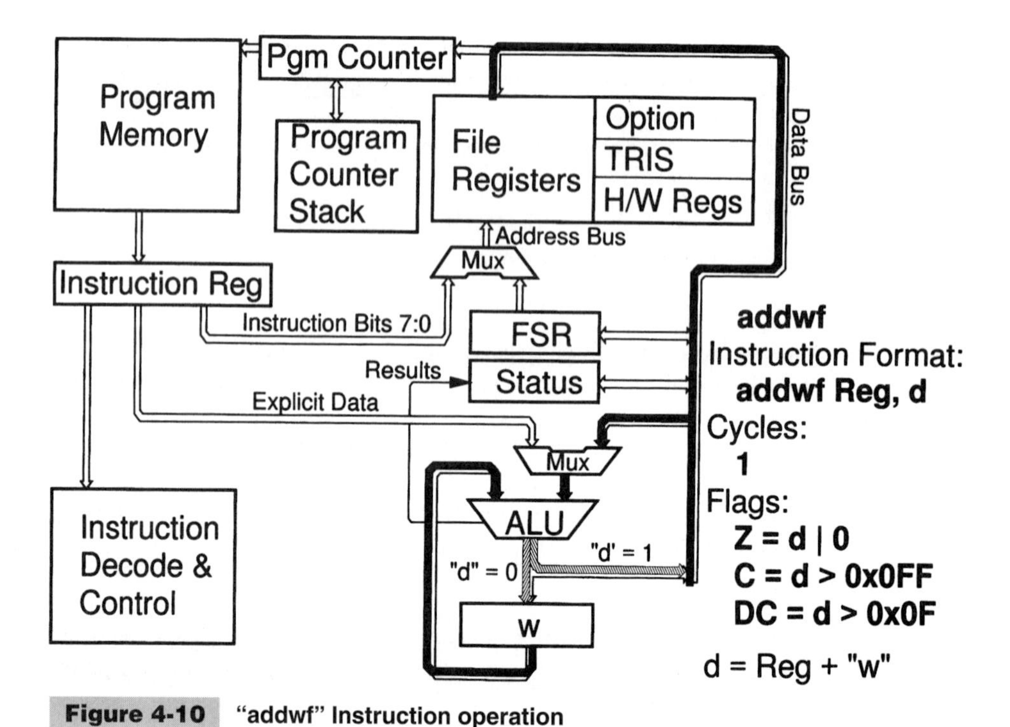


**Figure 4-10** "addwf" Instruction operation

Pgm Counter
Program Memory
Program Counter Stack
File Registers
Option
TRIS
H/W Regs
Data Bus
Address Bus
Mux
Instruction Reg
Instruction Bits 7:0
FSR
Results
Status
Explicit Data
Mux
ALU
w
Instruction Decode & Control

**addlw**
Instruction Format:
**addlw Const**
Cycles:
**1**
Flags:
**Z = Result**
**C = Result**
**DC = Result**

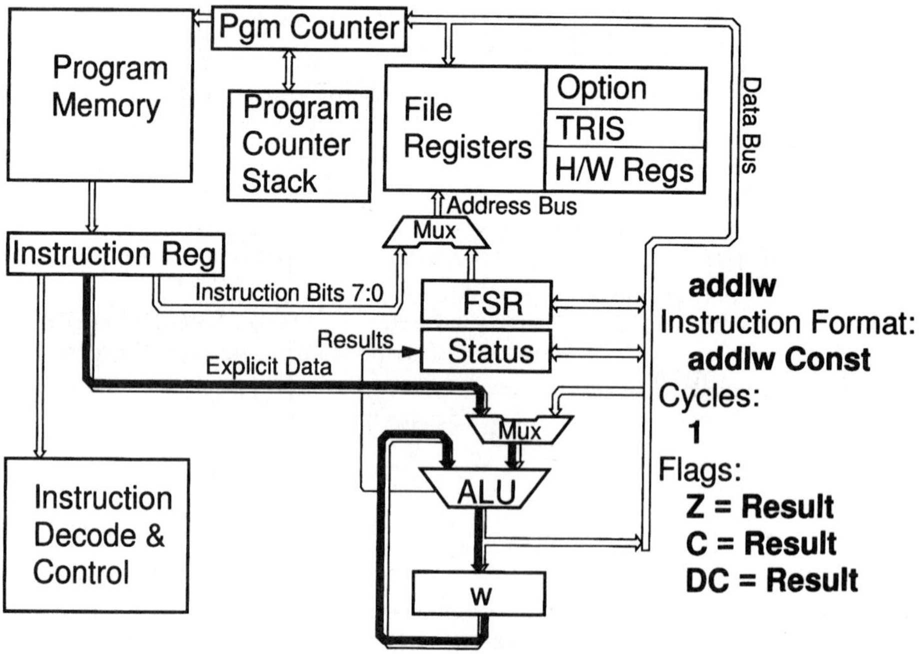

**Figure 4-11** "addlw" Instruction operation

As indicated in Chapter 3, all of the operation bits are affected by the addition and subtraction instructions. The zero flag is set if the result ANDed with 0x0FF is equal to zero. The carry flag is set if the result is greater than 0x0FF (255 decimal).

The digit carry flag is set when the sum of the least-significant four bits (also called a *nybble*) is greater than 0x0F (15).

For example, if you had the code:

```
movlw    10        ;  Add 0x00A to 0x00A
movwf    Reg
addwf    Reg, w    ;  Put the Result in w
```

At the end of execution, the w register would contain 20 (or 0x014), *Reg* would have 10 (0x00A), the zero and carry flags would be reset (equal to zero), while the digit carry flag would be set.

Subtraction in the PIC is something that you should look over and understand thoroughly before you use it. The *subwf* instruction (Fig. 4-12) probably works backwards as to how you would expect it to work (especially if you have experience with other processors).

The instruction invocation is:

```
subwf    Register, d
```

in which the contents of *Register* have the contents of w subtracted from it and the result placed either in w or *Register*, based on the destination (*d*). The subtraction is the negative addition described in the previous chapter.

This description probably doesn't make a lot of sense; the best way to explain subtrac-

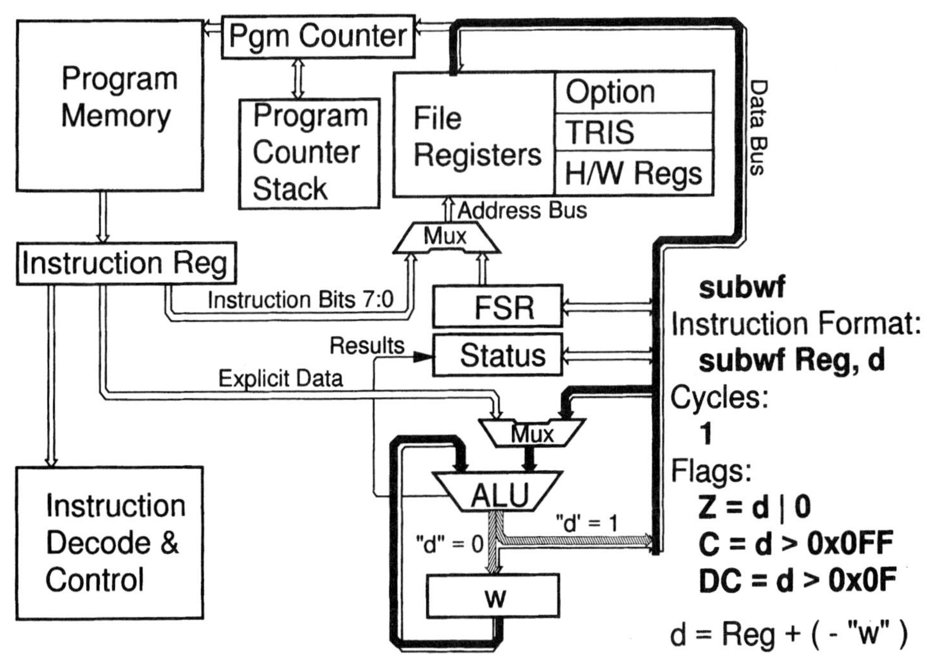


**Figure 4-12** "subwf" Instruction operation

tion in the PICmicro® MCU is to note that it is not subtraction at all. Instead, it is adding a negative value to the source.

Thus, instead of *subwf* operating as:

```
Destination = Source - w
```

which is executed in other processors, subtraction in the PICmicro® MCU is actually:

```
Destination = Source + (-w)
```

The negative w term of this equation is found by substituting in the two's complement negation formula:

```
Negative = (Positive ^ 0x0FF) + 1

for "-Positive".
```

This means that the *subwf* instruction operation formula presented is really:

```
Destination = Source + (w ^ 0x0FF) + 1
```

I find that when I am using the instruction, it helps to remember this formula because I can easily understand what *subwf* is doing and predict how it will behave.

Remembering this formula also helps me to understand how the carry flags work. Looking at the instruction, the carry and digit carry flags probably run nonintuitively to what you expect (and might have experienced with other processors).

For example, to show what happens when you subtract 2 from 1 in the PIC:

```
Source      = 1
w           = 2
Instruction = subwf Source, w
```

The formula:

```
Destination = Source + (w ^ 0x0FF) + 1
```

is used, which yields (for the *subwf Source, w* instruction):

```
w = Source + (w ^ 0x0FF) + 1
  = 1 + (2 ^ 0x0FF) + 1
  = 1 + 0x0FD + 1
  = 0x0FF
  = -1
```

which is what is expected. Notice that, in this case, the carry flag is reset, which is not expected in a typical subtraction operation where carry is the "borrow" flag. If a negative result was produced in a traditional processor, we would expect the carry (or an explicit borrow) flag to be set.

Working through the same instruction (*subwf Source, w*) and the registers loaded with:

```
Source = 2
w      = 1
```

The *subwf* formula can be applied as:

```
w = Source + (w ^ 0x0FF) + 1
  = 2 + (1 ^ 0x0FF) + 1
  = 2 + 0x0FE + 1
  = 0x0101
```

In this example, the value 0x001 (0x0101 and 0x0FF) is actually stored in w. But, note that, in this case of subtracting a lower value from a higher value, the carry flag (and, possibly, the digit carry flag) are set! If you work through an example for the contents of w equal to *Source,* you will find that the result is 0x0100 and the carry flag will also be set in this situation.

As indicated, the carry flag is often used as a borrow flag in computer processors. Thus, the bit is set when the result is less than zero and indicates to the application code that a borrow from the next-significant byte must be made for the result to be a valid result.

The behavior of the *subwf, sublw,* or adding a negative number works somewhat differently from the typical situation. For this reason, I like to refer to the carry flag as the *positive flag* after subtraction.

Microchip refers to the carry flag as the *negative borrow flag* for reasons that will become apparent later in this chapter. If you were to invert the carry flag in the PICmicro® MCU after subtraction operations, you would discover that it behaves exactly as you would expect a borrow flag to behave.

The *sublw* subtracts the value in w from the literal value of the operation of the instruction (Fig. 4-13). This probably sounds confusing and it is. The invocation of the *sublw* instruction is quite straightforward:

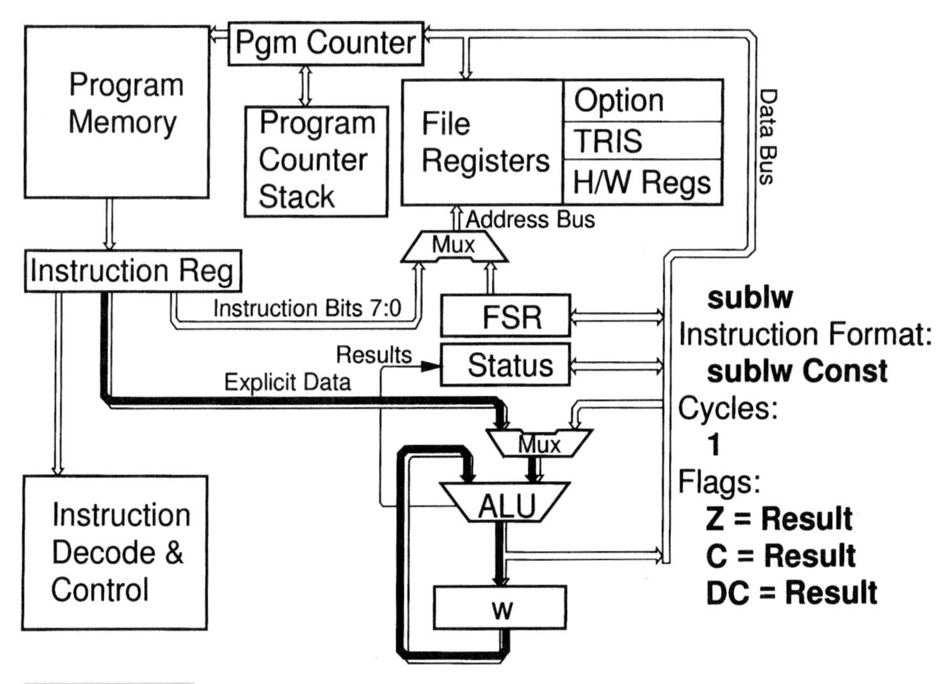


**Figure 4-13** "sublw" Instruction operation

```
sublw Constant
```

*Sublw,* like *subwf,* subtracts the contents of the w register from the passed parameter. In this case, the contents of w are taken away from the constant. Writing this down as a formula, the operation is:

```
w = Constant - w
```

Using the same subtraction operation as described for *subwf,* the *sublw* operation is actually:

```
w = Constant + ( w ^ 0x0FF ) + 1
```

*sublw* changes the flags in a similar manner to that of *subwf.*

When I wrote the first edition of the book, I commented that I avoid this instruction except for the case:

```
sublw   0       ; Negate Value in w
                ;   w = 0 - w
```

which negates the value in w because I found it to be nonintuitive. Over the three years since I finished the first edition, I would probably disagree with this statement and note that I now use *sublw* quite a bit in my complex operations because it can simplify applications.

When you first start working with the PICmicro® MCU, if you want to subtract a constant value from the value in w, I recommend adding the negative instead of attempting to subtract the negative. For example, if you had to subtract the constant 47 from the contents of w, then the instruction:

```
addlw  -47
```

should be used. If you wanted to carry out this same operation using the subtraction instructions, the required code would be:

```
movwf  Temp
movlw  47
subwf  Temp, w
```

which takes up three times the number of instructions and requires one file register as well.

Notice that the *addlw* and *sublw* instructions are *not* available in the 16C5X devices.

Registers can have one added or taken away from themselves using the *incf* and *decf* instructions. Figure 4-14 shows how the *decf* instruction works. To add one to a register, it is "incremented" and taking one away from a register is "decrementing."

To invoke the instructions, the format:

```
incf Register, d
```

is used for incrementing a register and:

```
decf Register, d
```

is used for decrementing the register.

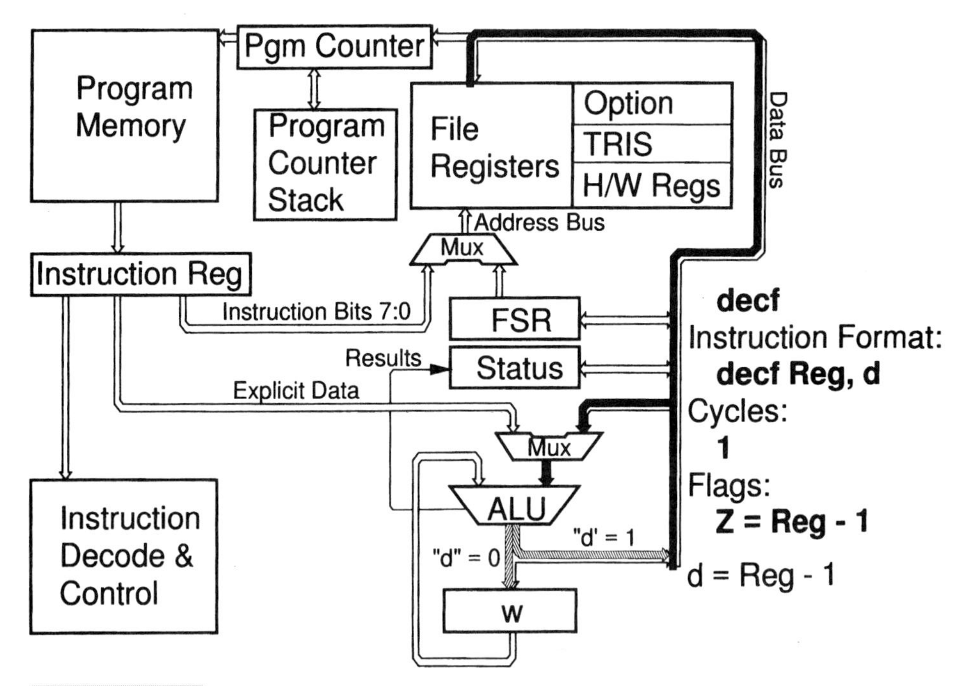


**Figure 4-14** "decf" Instruction operation

*Decf* is somewhat less obvious with the operation of the PICmicro® MCU's ALU, as described in the previous chapter. To decrement the value, instead of subtracting one from the register, 0x0FF (−1) is added to it. This causes the correct result (one less than the original value) to be saved in the destination.

The result of the increment and decrement can be stored back in the original register (*d* set to *f* or 1) or into the w register (*d* set to *w* or 0). Upon completion of the instruction, the zero flag of the STATUS register is either set (the result is equal to zero) or reset (the result was not equal to zero). The carry and digit carry flags are not affected by the operation of these two instructions.

*Comf* inverts the contents of a register and its operation is shown in Fig. 4-15. This operation is the same as XORing the contents of a register with 0x0FF to invert or complement each bit.

To invoke *comf,* the instruction has the format:

```
comf    Register, d
```

where *d* is the destination and is 0 or *w* to store the result in w and is 1 or *f* to store the result back into *Register.*

It is important to remember that complementing a register is not the same as negating it. To two's complement negate a register, the register has to be complemented and then incremented. To negate a *Register* in the PICmicro® MCU, the following two instructions should be used:

```
comf   Register, f
incf   Register, f
```

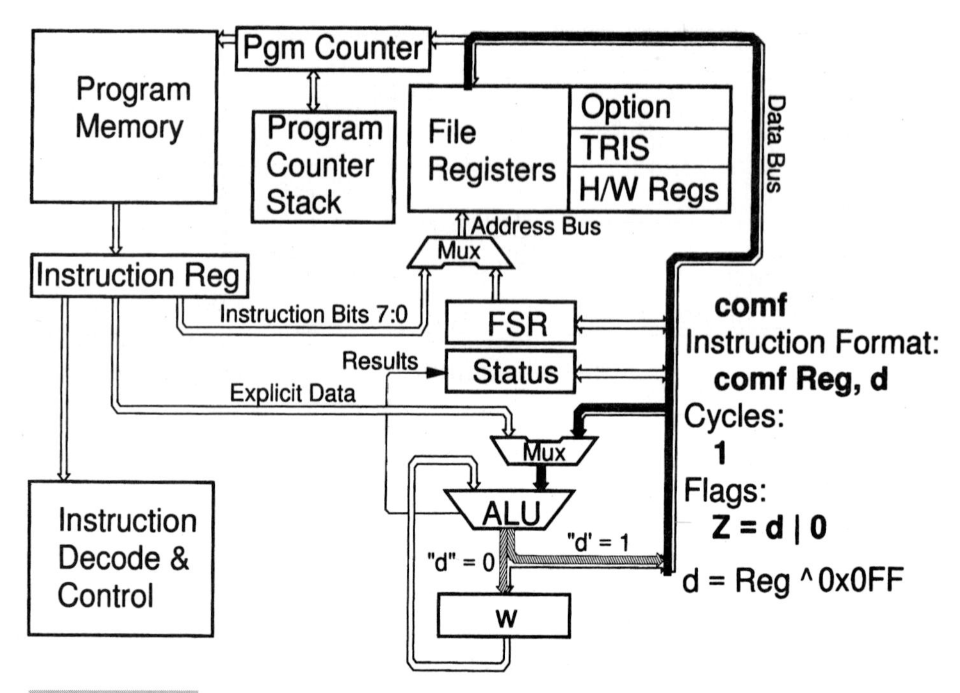


**Figure 4-15** "comf" Instruction operation

Along with arithmetic operations, the PICmicro® MCU can also perform the bitwise logical functions AND, OR, and XOR. These instructions are available for combining the contents of a register, along with the contents of w or combining the contents of w with a constant value.

When two bits are ANDed together, the result will be 1 if both inputs are 1. If either input is not 1, then the result will be 0. The PICmicro® MCU has two AND instructions, *andwf* and *andlw,* which perform these functions.

To invoke *andwf,* the instruction format:

```
andwf Register, d
```

where *d* can specify that the result of ANDing the contents of w with *Register* is placed into w or back into *Register.*

*Andlw* is invoked using the instruction format:

```
andlw Constant
```

In both AND instructions, the zero flag is set when the result of the AND operation is equal to zero.

You might find it a bit confusing to discover that Microchip refers to what I call *ORing* bits together as *inclusive ORing*. In this operation, when either bit of an OR input is set (1), the result will be set (1). The two instructions which execute this function, *iorwf* (Fig. 4-16) and *iorlw.*

*Iorwf* is invoked using the instruction format:

```
iorwf   Register, d
```

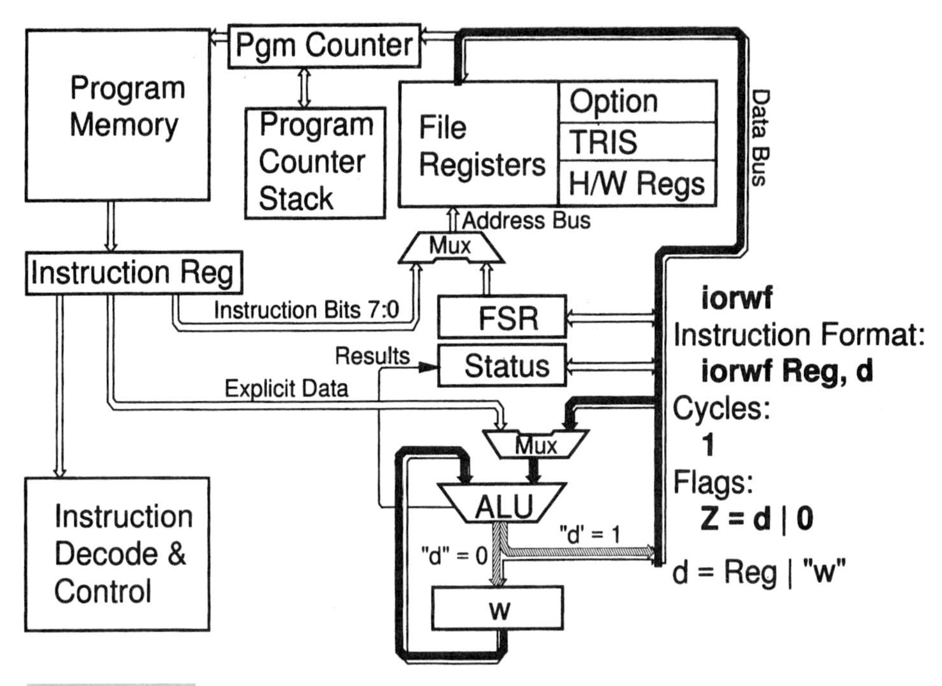


**Figure 4-16** "iorwf" Instruction operation

and ORs together the contents of *Register,* along with the contents of w and stores the result in either w or *Register,* depending on the value of *d.*

ORing a constant with the contents of w is accomplished by the *iorlw* instruction that has the format:

```
iorlw   Constant
```

In both PICmicro® MCU *Inclusive OR* instructions, the zero flag set to the result of the operation is zero.

The last bitwise logical operation is the *Exclusive OR (XOR)*. In this operation, if one of two inputs is set (1) and the other reset (0), the result will be set (1). If both inputs are at the same state (set or reset), then the result will be reset (0).

Like ANDing and ORing, XORing in the PICmicro® MCU can be done with either the contents of a register being XORed with the contents of w or the contents of w are XORed with a constant.

The *xorwf* instruction XORs the contents of a register with w and stores the result according to the value of destination *d* in the format:

```
xorwf Register, d
```

To XOR the contents of w with a constant and place the result back into w, the *xorlw* instruction (Fig. 4-17) is used:

```
xorlw Constant
```

Like ANDing and ORing, XORing will set the zero register if the result of the operation is equal to 0x000.

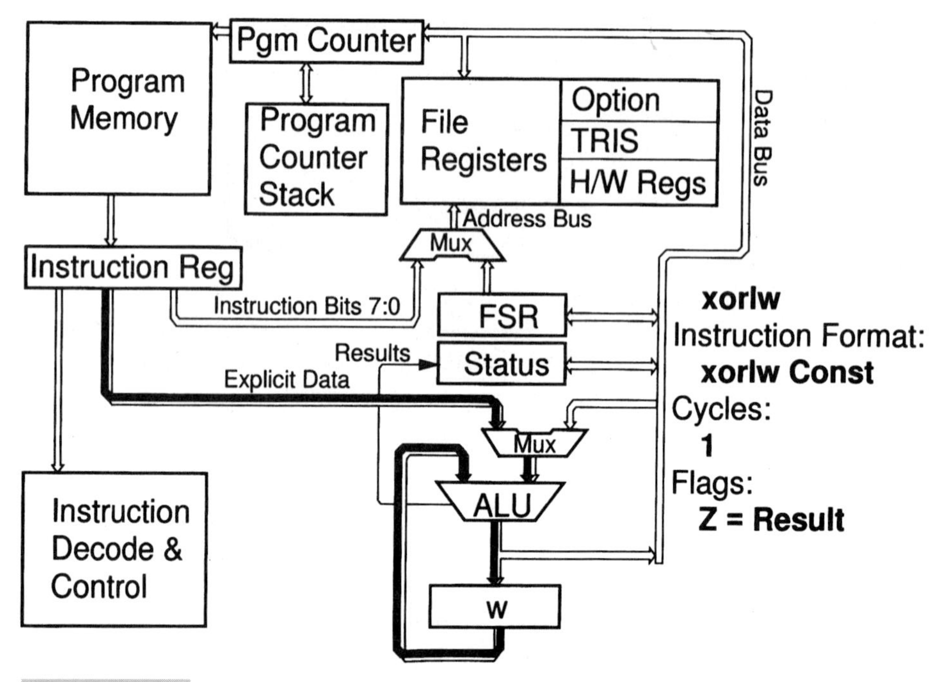


**Figure 4-17** "xorlw" Instruction operation

The rotate left (*rlf*, see Fig. 4-18) and rotate right (*rrf*) instructions are useful for a number of reasons. Their basic function of rotate is to move a register one bit to the left (upward) or to the right (downward), with the least-significant value being loaded from the carry flag, and the most-significant value put into the carry flag.

To rotate the contents of a register to the left (up), the *rlf* instruction is used and has the format:

```
rlf    Register, d
```

When this instruction executes, the contents of the STATUS carry flag are stored in the least-significant bit of the destination (either w or Register) while the contents of Register are shifted up by one bit. To shift a register up by one bit, the contents of bit zero are stored in bit one, the contents of bit one are stored in bit two, etc. When the register is shifted up by one, bit zero is left open and is given the contents of the carry flag. Bit seven of Register is stored in the STATUS carry flag to complete the rotation.

To rotate a register to the right by one bit, the *rrf* instruction is used:

```
rrf    Register, d
```

Instead of moving the registers "up," in a right rotate, the registers are moved "down;" the contents of bit 7 are stored in bit 6, the contents of bit 6 are stored in bit 5, etc. The contents of the carry flag before the invocation of the instruction is stored in bit 7 and upon completion of the *rrf* instruction, the original contents of bit 0 of Register are stored in the carry flag.

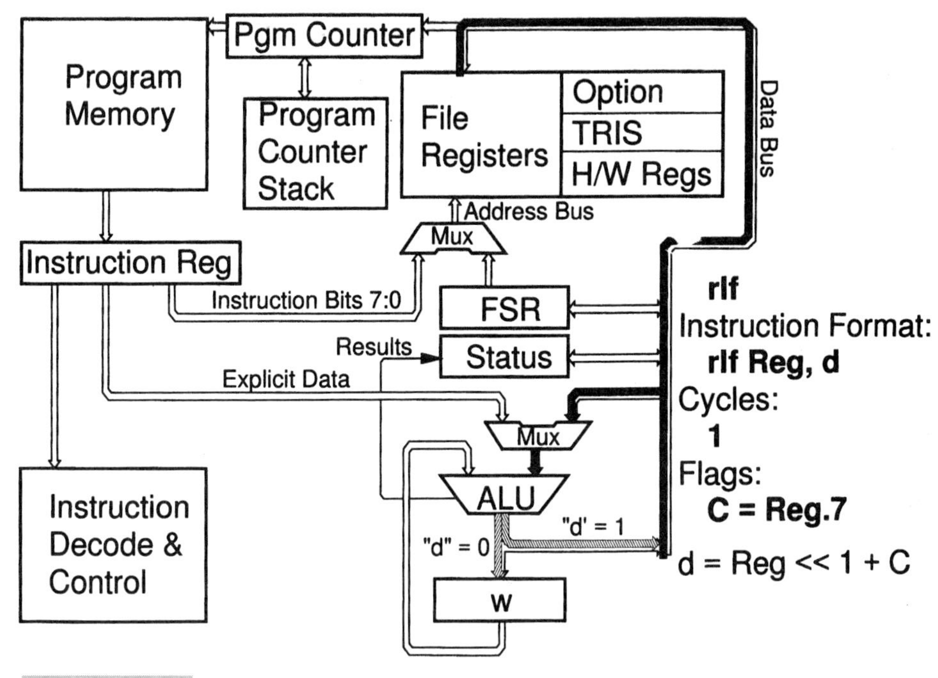


**Figure 4-18** **"rlf" Instruction operation**

A little trick that can be used to rotate a register and not lose a bit is to execute the snippet:

```
rrf    Register, w
rrf    Register, f
```

In the first of these two instructions, the carry flag is loaded with the contents of bit zero and the shifted value is placed into w, where it can be ignored. The second instruction then does the *rotate* instruction with carry loaded with its least-significant bit, which places it in the most-significant bit of the destination. This trick also works for the *rlf* instruction.

The *rotate* instructions can be used for carrying out multiplication and division on a value with powers of two. This can also be done on 16-bit values. The following example shows how to multiply a 16-bit number by four:

```
bcf  STATUS, C          ;  Clear the Carry flag before Rotating
rlf  Reg, f             ;  Shift the Value up (Multiply by 2)
rlf  Reg + 1, f
bcf  STATUS, C          ;  Now, Repeat to Multiply by 2 again
rlf  Reg, f
rlf  Reg, f
```

Another use is to shift data in/out of a register serially (this is shown later when data is shifted in/out of various registers from external devices).

## EXECUTION CHANGE OPERATORS

Before attempting to use a *goto* or *call* instruction, it is imperative that you understand how they work. If you haven't done so, go back a read "The Program Counter and Stack" section of Chapter 3. The *goto* and *call* instructions can behave strangely in some circumstances and the PICmicro® MCU's program counter will not have the correct destination address.

The reason for the unusual actions taken during the *goto* or *call* instructions is because in the PICmicro® MCU instruction set; all instructions are the same length. This means that the size of the destination address could be larger than what the instruction has space to pass along to the PICmicro® MCU's program counter.

Both *goto* and *call* have to be explicitly specify an absolute address with a PICmicro® MCU "page" (2048 instruction addresses in the mid-range chip). If the destination address is outside the page, the PCLATH register (or appropriate STATUS bits, for the low-end chip) must be set with the correct page information.

For example, jumping between pages in the mid-range PICmicro® MCU can be accomplished by the code:

```
movlw   HIGH Label    ;  Interpage goto
movwf   PCLATH
goto    Label
```

In this snippet of code, the PCLATH register is updated with the new page before the *goto* instruction (Fig. 4-19) is executed. This forces the program counter to be loaded with the correct and full label address when the *goto* instruction is executed.

In all of the instructions that change the program counter presented in this section, execution will take two instruction cycles instead of the customary one of the data-movement, data-processing, or processor-control instructions. This is caused by the PICmicro® MCU's instruction register being already loaded with the instruction at the next address, having to be loaded with a new address, and then prefetching the instruction at that address before it can be executed. The actual timing for the operation is two cycles (Fig. 4-20) when the *goto* (or any other program counter changing) instruction is executed, the "prefetched" instruction (*Goto + 1* in Fig. 4-20) is no longer valid and it must be changed with the *Destination* instruction. Before the *Destination* instruction can be executed, it must be loaded into the PICmicro® MCU's prefetch register so that it can be executed in the next instruction cycle. This is the second instruction cycle after the *goto* instruction instead of the first, which would be the case for any of the other instructions that don't change the PICmicro® MCU's program counter.

The *call* instruction (Fig. 4-21) works almost exactly the same way as *goto*, except that the pointer to the next instruction is stored on the program counter stack.

Three different types of return statements are used in the mid-range and PIC17Cxx and 18Cxx PICmicro® MCUs. Each one of these instructions "pops" the current value from the top of the hardware stack and stores it in the program counter. These addresses are the saved next instructions that were saved when the subroutine was called or an interrupt request was acknowledged.

The simple *return* instruction (Fig. 4-22) returns the stack pointer to the address pointed after the call to subroutine. No registers or control bits are changed.

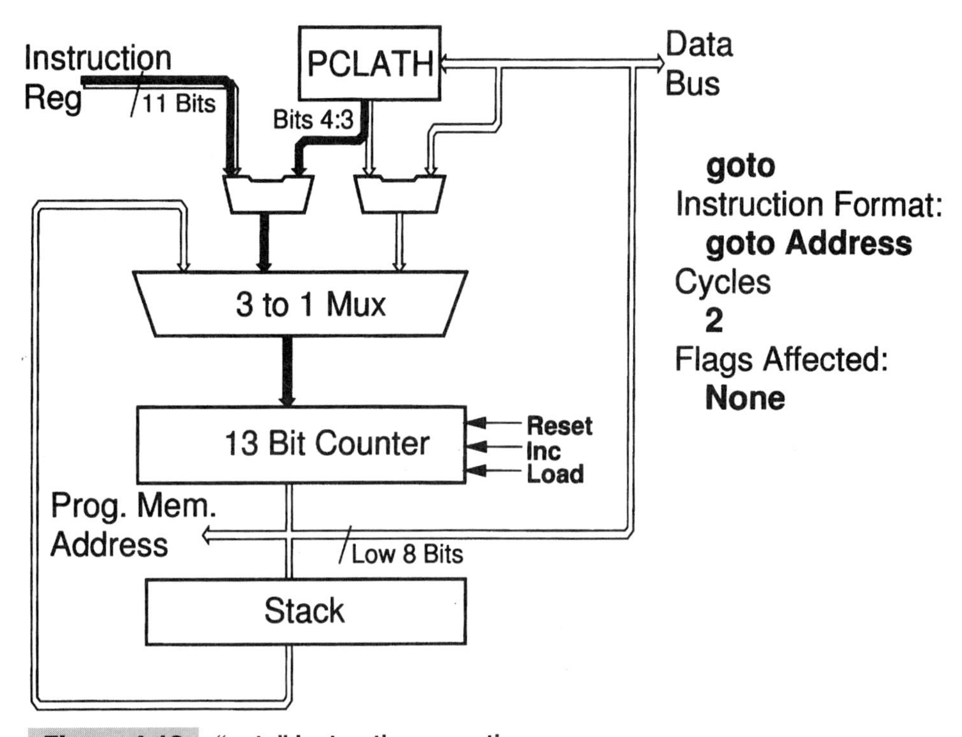


**Figure 4-19** "goto" Instruction operation

As noted earlier in the chapter, the *retlw* instruction is also available for you to return from a subroutine with a new value stored in the w register. This instruction is the only *return* available in the low-end devices (and it is described later in this chapter). The instruction is useful for returning table information (which is explained in the next section) or returning condition information in w that can be tested by the calling program. The instruction loads w with an immediate value before executing a return from interrupt.

The *retfie* instruction is similar to the *return* instruction, except that it is used to return from an interrupt. The only difference between the two instructions is that the GIE bit in the interrupt control register (INTCON) is set during the instruction. This allows interrupt requests to be acknowledged immediately following the execution of the instruction and the interrupt handler. This operation in the *retfie* instruction simplifies the interrupt handler to acknowledge different interrupts sequentially, rather than having to provide a check before ending the handler to ensure that nothing is pending; if there is, it can handle them.

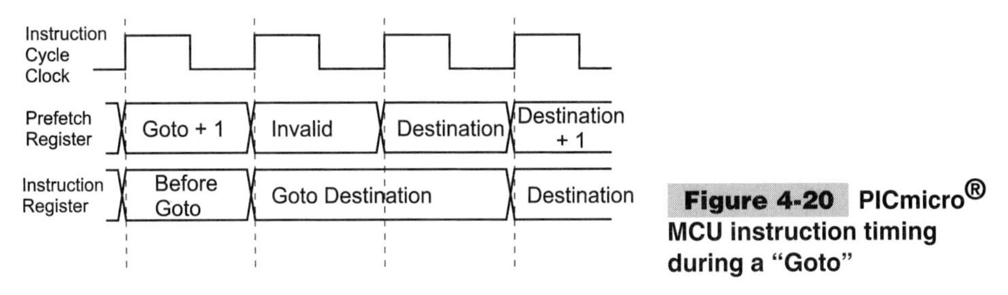


**Figure 4-20** PICmicro® MCU instruction timing during a "Goto"

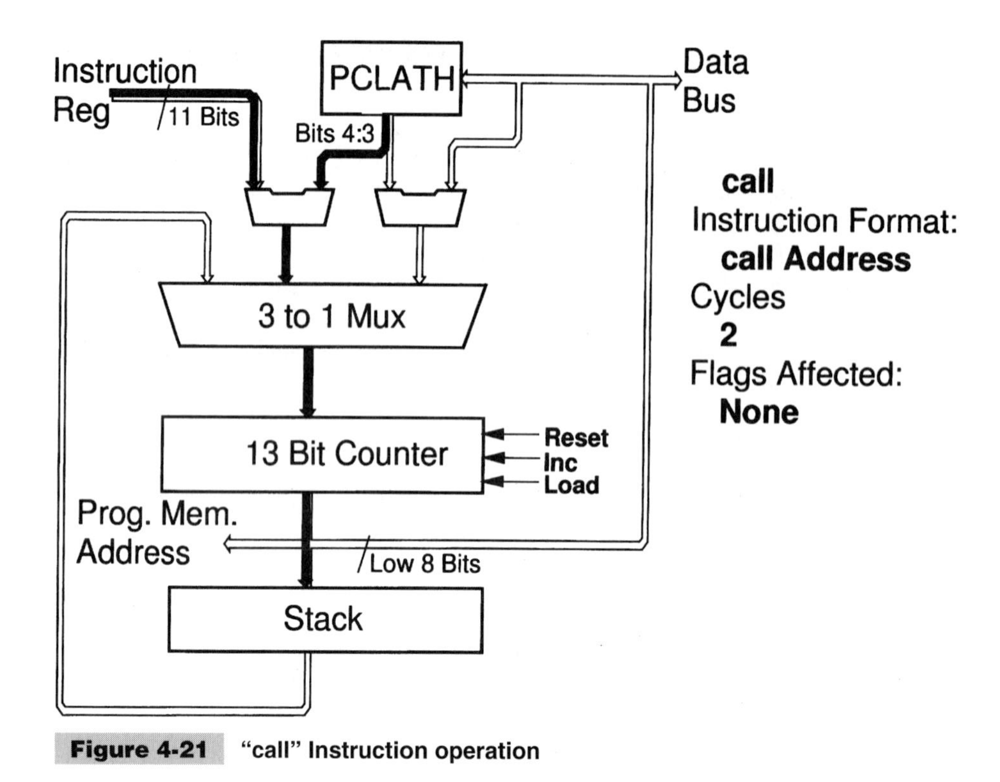


**Figure 4-21** "call" Instruction operation

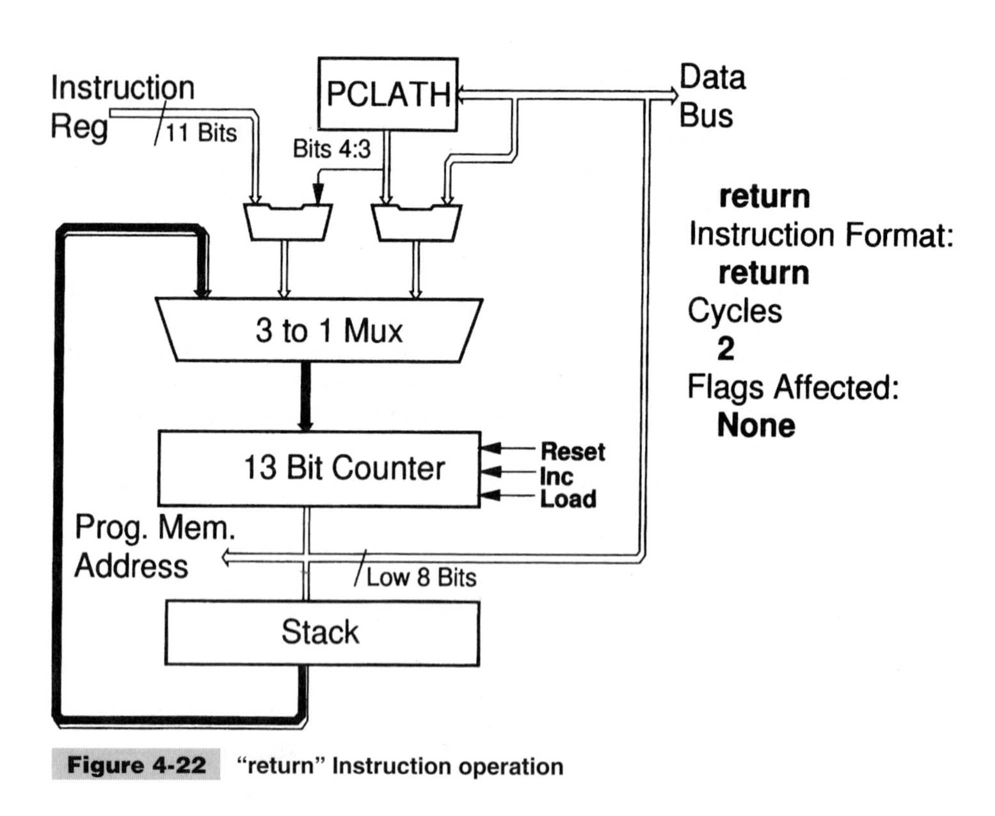


**Figure 4-22** "return" Instruction operation

As I have said elsewhere in this book, the basic PICmicro® MCU architecture doesn't use *jump on condition* instructions. Instead, a number of instructions allow skipping the next instruction in line based on specific conditions. The basic instructions that carry out this function are the *skip on bit condition* instructions, *btfsc* and *btfss*. These two instructions use the same format as the bit set and reset instructions (*bcf* and *bsf*), although they test the condition of bits, rather than change them.

The *btfsc* instruction skips over the next instruction if the bit condition is reset (0). The format of the instruction is:

```
btfsc   Register, Bit
```

*Btfss* skips over the next instruction if the bit condition is set (1). The format for the instruction is:

```
btfss   Register, Bit
```

These two instructions are used as the basic method of conditionally changing the execution of an application. For example, to jump to an address if the zero flag is set, the two instructions:

```
btfsc STATUS, Z;  Test Zero, Skip over Next if Zero Flag is Reset
 goto Label    ;  Zero Flag Set – Jump to "Label"
```

Notice that in this snippet, I test for the negative condition and skip over the next instruction if it is true. This is something to beware of when you first start working with the PICmicro® MCU. Conditional jumps based on the state of the carry flag after a comparison (subtraction) operation can be very confusing. I'm sure that you will have problems implementing it the first time you attempt it. To make things easier on yourself, think of the positive test first and then simply swap the instructions from *btfsc* to *btfss* and vice versa.

If the bit condition is not true and the skip is not executed, then the *btfsc* and *btfss* instructions execute in one instruction cycle. If the condition is true, then the instruction executes in two cycles, essentially treating the following instruction like a *nop*.

In a traditional processor, these two instructions would be the single *jz* instruction. By combining the *btfsc* and *btfss* instructions with the zero and carry flags, basic instructions can be implemented (Table 4-1). The complement jump in Table 4-1 is the jump that is used to jump if the opposite condition is true.

Built into the MPLAB assembler are a number of pseudo-instructions that provide similar functions. Personally, I would recommend that you avoid using these conditional jump instructions because they mask the true usefulness of the *btfsc* and *btfss* instructions.

This usefulness comes about because the instructions can operate on *any* bit in *any* register. Flag bits can be very easily changed in the PICmicro® MCU using the *bcf* and *bsf* instructions and the *btfsc* and *btfsc* instructions to test them. In a traditional processor, to set a bit, such an instruction would have to be used:

```
movf    Register, w
iorlw   1 << Bit
movwf   Register
```

**TABLE 4-1 Different Conditional Jumps and Equivalent PICmicro® MCU Code**

| CONDITIONAL JUMP | PICmicro® MCU INSTRUCTIONS | COMPLEMENT JUMP |
|---|---|---|
| jz Label | btfsc STATUS, Z<br>goto Label | jnz Label |
| jnz Label | btfss STATUS, Z<br>goto Label | jz Label |
| jc Label | btfsc STATUS, C<br>goto Label | jnc Label |
| jnc Label | btfss STATUS, C<br>goto Label | jc Label |

In the PICmicro® MCU, the single instruction:

```
bsf  Register, Bit
```

is just required.

Similarly, if a bit is to be tested in a register and execution jumps to a label if it is set, a traditional processor would require the code:

```
movf    Register, w
andlw   1 << Bit
jnz     Label
```

In the PICmicro® MCU, the code would be:

```
btfsc   Register, Bit
 goto   Label
```

In both cases, the PICmicro® MCU is able to change a bit and test its condition in fewer instructions and without affecting the contents of the accumulator or the STATUS register. Keep this very powerful feature in mind when you are developing your applications. By spending a few minutes planning how the registers are going to operate, you can improve the efficiency of your application by some remarkable margins.

*Bit skip* instructions are useful in a variety of cases, from checking interrupt active flag bits to seeing if a number is negative (checking the most-significant bit). Throughout the book, a number of different ways are presented in which the bit skip commands can be used to simplify the software development.

The bit commands are quite unique to the PIC compared to other microcontrollers. I believe that they give the application developer a significant advantage in developing applications, allowing far easier methods of bit manipulation that would be otherwise possible.

Along with the bit test (*btfsc* and *btfss*) instructions, two other instructions skip on a given instruction. They increment or decrement a register and skip the next instruction if the result is equal to zero. Figure 4-23 shows the operation of the *incfsz* instruction while the instruction format is:

```
incfsz Register, d
```

In the *incfsz* instruction, the contents of *Register* are incremented and stored according to the value of *d*. If *d* was *w* or 0, then the w register would be stored with the result of the operation. If *d* is *f* or 1, then the register would be updated with the result of the operation. No STATUS flags are modified by the operation of *incfsz* or *decfsz* (which is a difference between them and the *incf* and *decf* instructions).

*Decfsz* is similar in operation to *incfsz,* except that the register is decremented and the skip occurs if the result is equal to zero. Its format is:

```
decfsz Register, d
```

These two instructions work exactly the same as the *incf* and *decf* instructions in terms of data processing. One is added or subtracted from *Register*. The result is then stored either in w or back in the source register. The important difference to these instructions are, if the result is equal to zero following the increment/decrement, the next instruction is skipped.

If the result is not equal to zero (and the next instruction is not skipped), *incfsz* and *decfsz* execute in one instruction cycle. *Incfsz* and *decfsz* execute in two instruction cycles if the result is equal to zero. If the result is zero then the next instruction is skipped over and treated like a *nop*. Often, these instructions are used in crucially timed loops, so understanding the exact timing of the two instructions is important.

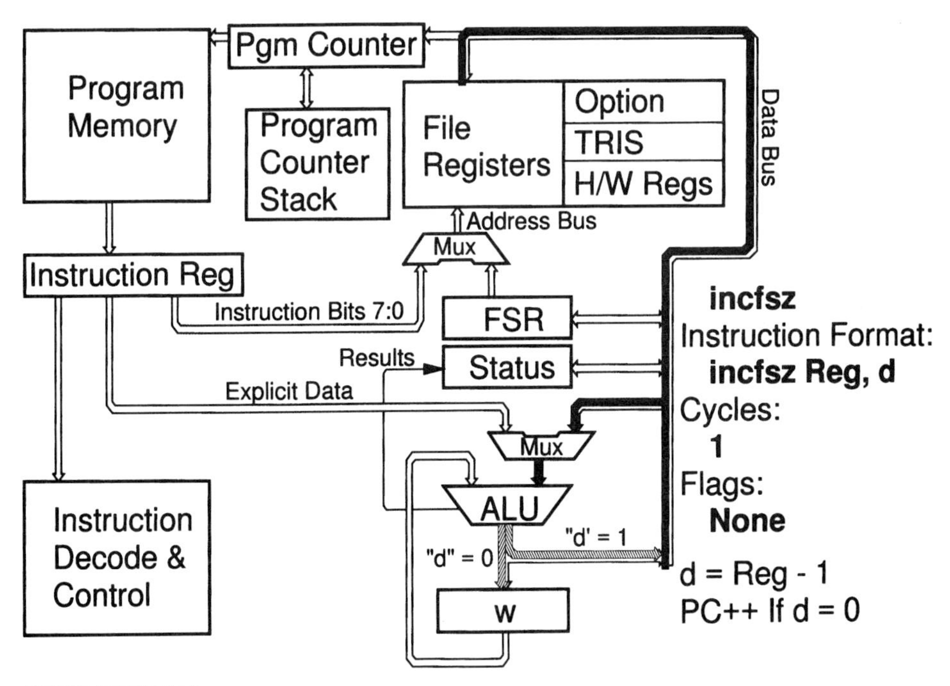


**Figure 4-23** "incfsz" Instruction operation

I should say that *decfsz* is normally used for loop control. The following code example shows how a loop can be repeated 37 times with very little software overhead:

```
  movlw    37                  ;    Load the Count Register
  movwf    LoopCounter
Loop                           ;    Repeat for each iteration of the loop
  :
  decfsz LoopCounter, f        ;    Decrement the Count Register
   goto   Loop                 ;      If not == zero, loop again
                               ;    Continue on with the Program
```

This code can be used anywhere a loop is required. As you can see, the overhead code is only four instructions long and it only requires two or three instruction cycles each loop.

If you are using these two instructions on processor registers, take care to ensure that the hardware registers are capable of reaching zero. In the low-end PICmicro® MCUs, the FSR can never be equal to zero, which makes the *incfsz* and *decfsz* instructions useless with this register.

As well, these instructions do not affect any status flags (zero would probably be expected). Thus, you might want to put a *bsf STATUS, Z* after the instruction following the *incfsz/decfsz* instruction.

For example, in a loop:

```
  decfsz   Count           ; Decrement the Count Value
   goto    Loop            ; Jump back to Loop if Count  != 0

  bsf  STATUS, Z           ; Set Zero Flag to Indicate Loop End
```

**Tables** So far, this book has touched upon explicitly changing the contents of the PICmicro® MCU's program counter to provide explicit jumps within an application. As I work through some of the more advanced programming techniques that can be used with the PICmicro® MCU, the need for being able to explicitly jump to a location will become more obvious. Before getting into those techniques, I want to introduce you to a programming construct in the PICmicro® MCU that I am sure you will use a lot of in your application programming.

When implementing a PICmicro® MCU application that can communicate with humans, the ability to send text messages will be required. Tables of text messages can be implemented in the PICmicro® MCU quite simply with the advantage that they will execute quickly and with a consistent number of cycles, no matter where the data in the table to be retrieved is.

The most traditional method of implementing a table is to provide a subroutine that adds a constant to a known point in the application and stores this value in the PICmicro® MCU's program counter. At the new address, a *retlw* instruction is used to store the table value in the w register and return to the caller's code.

The most basic way of doing this is to use the *addwf PCL, f* instruction to update the PICmicro® MCU's program counter with the table immediately following it as the table's destination addresses. A simple version of this subroutine could be:

```
Table            ;  Return Table Value for Contents of w
  addwf  PCL, f  ;  Add the Table Index to the Program Counter
```

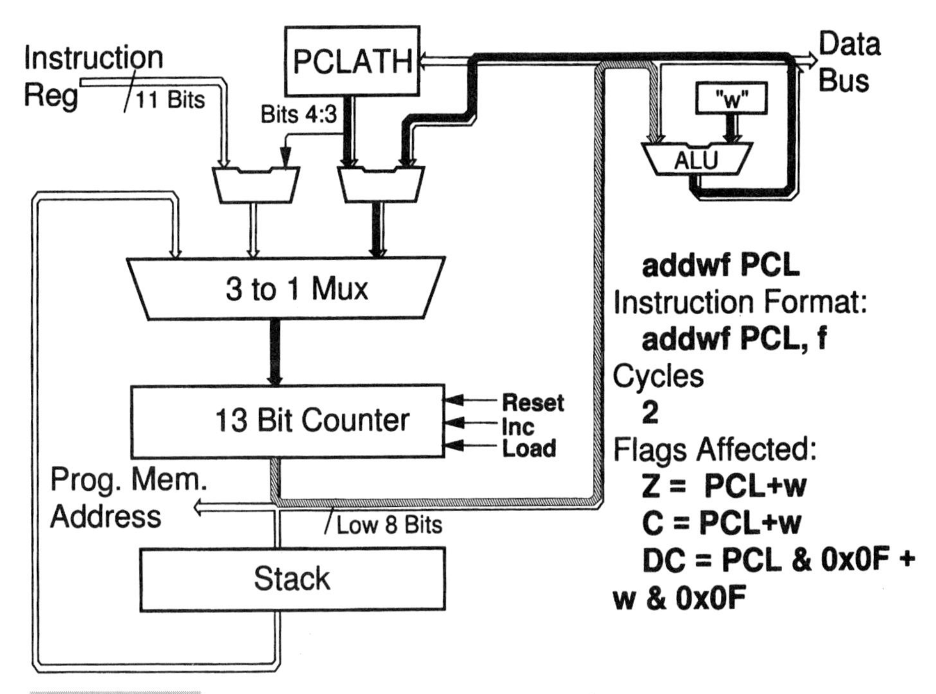


**Figure 4-24** "addwf PCL, f" Instruction operation

```
TableEntries
  retlw "T"       ;  ASCII "Table" to be returned
  retlw "a"
  retlw "b"
  retlw "l"
  retlw "e"
  retlw 0
```

The *addwf PCL, f* instruction (Fig. 4-24) adds the contents of w (which is the table value to return that has been passed to the *Table* subroutine) to the program counter via PCL. When the *addwf PCL, f* instruction executes, the program counter is already incremented to the next instruction. To return the *T* in the table, a value of zero must be passed in w. To return *a*, a value of 1 is passed in w, etc.

The zero value at the end of *Table* is used to indicate that the table value has ended. Normally, when I am using a text table like this one, I want to have some way to determine when I am at the end of the table, a NULL character (ASCII 0x000) is the choice that I normally use. I like ending a table with 0x000 because when it is ORed with 0x000 or ANDed with 0x0FF, the zero flag will be set without changing the value of the contents of the w register.

The table subroutine can be enhanced by using the *dt* assembler directive (command), which combines the table's *retlw* instructions. Using *dt,* the subroutine becomes:

```
Table2                   ; Return Table Value for Contents of w

  addwf PCL, f           ; Add the Table Index to the Program Counter
```

```
TableEntries
  dt    "Table", 0
```

If you were to compare the instructions produced by the Table and Table2 subroutines, you would find that they are identical, including the number of instructions used by each subroutine. Later, the book describes directives and list the entire set available to you for the MPLAB PICmicro® MCU assembler, which will be used later in the book for the experiments and applications.

This simple table is useful for many applications, but only under one condition; the table itself has to be located in the first 256 instructions of the PICmicro® MCU's program memory and PCLATH cannot have been changed since reset. If the table is not located in the first 256 instructions or straddles, first 256-instruction boundary, then execution will jump to an invalid address because the PCLATH register is not correctly set up for the table.

To rectify this, I tend to use the generic table code:

```
Table3                          ; Return Table Value for Contents of w
                                ;   Anywhere in PICmicro Memory
  movwf  Temp                   ; Save the Table Index
  movlw  HIGH TableEntries      ; Get the Current 256 Instruction Block
  movwf  PCLATH                 ; Store it so the Next Jump is Correct
  movwf  Temp, w                ; Compute the Offset within the 256
  addlw  LOW TableEntries       ;   Instruction Block
  btfsc  STATUS, C
   incf  PCLATH, f              ; If in next, increment PCLATH
  movwf  PCL                    ; Write the correct address to the
                                ; Program Counter

TableEntries

   dt    "Table", 0
```

In this example, the PCLATH register is updated according to the starting location of the instructions at *TableEntries*. When I calculate the address of the actual table element to access, I increment PCLATH if the destination is outside the initial 256-instruction address block.

This chapter and the book have given you examples of typical processor code and how the PICmicro® MCU's instruction set and features can be used to improve the operation of the code. This case shows you a very general set of code and I recommend that you always use it for your tables instead of the single instruction program counter update.

The code at *Table3* is not highly complex, executes in just a few extra cycles than the smallest possible method and, most importantly, will work anywhere within the PICmicro® MCU's program memory. This last point is crucial in making sure that you do not create a subroutine that might have problems executing properly if it is moved between applications or if code is added before it and it moves over the initial 256-instruction boundary.

Tables do not have to only be used for ASCII information, as shown in the table examples. They can also be used for conditional execution and for the execution state machines, as shown later in the book. Tables are not difficult to implement, but care must be taken to ensure that incorrect updates to the program counter are not made. Generally, problems

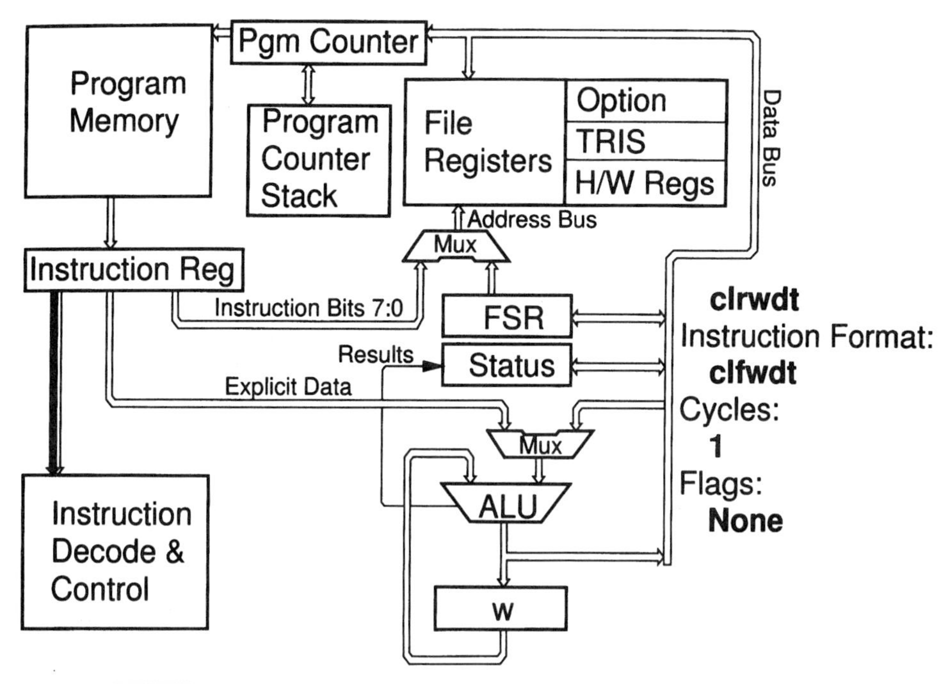


**Figure 4-25** "clrwdt" Instruction operation

will occur in applications that are executing after a while and will be extremely difficult to characterize, find, and correct.

## PROCESSOR-CONTROL INSTRUCTIONS

Only three instructions are used to explicitly control the operation of the PICmicro® MCU's processor. The first, *clrwdt* (Fig. 4-25) is used to reset the watchdog counter. The second, *sleep,* is used to hold the PIC in the current state until some condition changes and allows the PIC to continue execution. The last processor-control instruction is the *nop* (no operation), which simply delays one instruction cycle of time.

*Clrwdt* clears the watchdog timer (and the TMR0/WDT prescaler, if it is used with the watchdog timer), resetting the interval in which a timeout can occur. The purpose of the watchdog timer is to reset the PICmicro® MCU if execution is running improperly (i.e., caused by an external EMI "upset" or if a problem with the application code causes execution to run amok). To ensure that a watchdog timer timeout (and reset) is not executed at an inappropriate time, a *clrwdt* instruction is inserted in the code to reset the timer before the watchdog timer timeout, if the application is running properly.

Ideally, the application code should only have one *clrwdt* instruction written into it and this should only be executed through one path (i.e., every time an input event has processed and the queue for the next input event is about to be checked).

There are two purposes of the *sleep* instruction that executes as shown in Fig. 4-26). The first is to shut down the PICmicro® MCU once it has finished processing the program. This prevents the PICmicro® MCU from continuing to run and potentially affecting any other

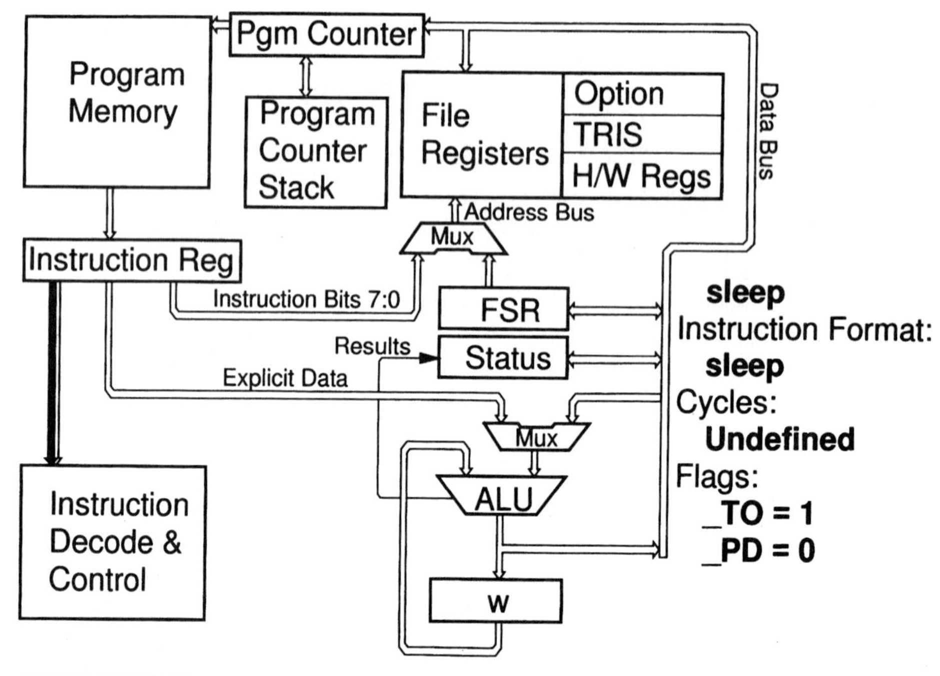


**Figure 4-26** "sleep" Instruction operation

hardware in the application while it is executing. By using the PICmicro® MCU in this manner, you are presuming that the PICmicro® MCU is only required for a certain aspect of the application (i.e., initialization of the hardware) and that it will not be required after this function has completed.

The second purpose of the *sleep* instruction is to provide a method to allow the PICmicro® MCU to wait for a certain event to happen. A "sleeping" PICmicro® MCU can be flagged of this event in one of three ways. The first is a reset on the _MCLR pin (which will cause the PICmicro® MCU to begin executing again at Address 0), the second is if the watchdog timer wakes up the PICmicro® MCU, and the third method is to cause wakeup by some external event (i.e., interrupt).

Using *sleep* for any of these methods will allow you to eliminate the need for wait loops, could simplify your software or allow a "quiet" PICmicro® MCU to do an ADC conversion. The Parallax Basic Stamp and the PicBasic compiler uses *sleep* for its *nap* instruction to simplify the operation of the code and to minimize the current requirements of the PICmicro® MCU while it is stopped.

Figure 4-27 shows how a sleeping PICmicro® MCU can be awakened by an interrupt. During sleep, the built-in oscillator is turned off; when the PICmicro® MCU is awakened, it restarts in a similar manner to the initial power up of the microcontroller. This wake up takes a relatively long time (1024 clock cycles) to wait for the built-in oscillator to stabilize before it resets the PICmicro® MCU and resumes executing the application code.

*Nop* means *no operation*. When this instruction (Fig. 4-28) is executed, the processor will just skip through it, with nothing (registers or STATUS register bits) changed. If you study a number of different processors, you will find that they all have an *nop* instruction.

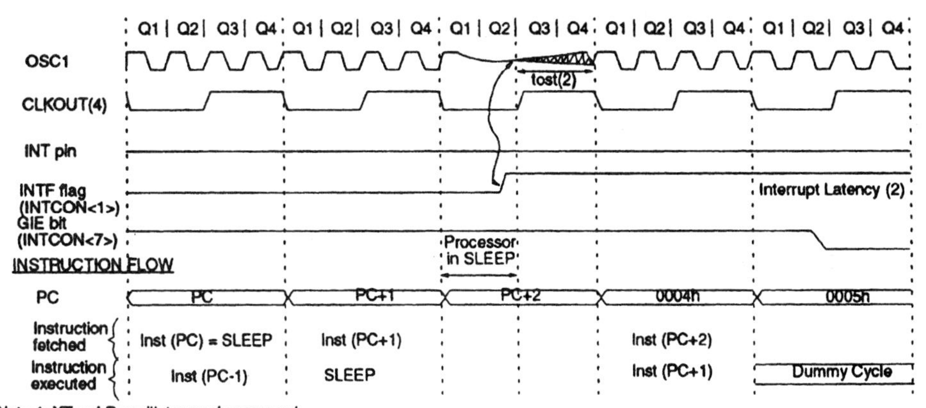


Note 1: XT or LP oscillator mode assumed.
2: tost = 1024tosc (drawing not to scale). This delay will not be there for RC osc mode.
3: When GIE = 0 processor jumps to interrupt routine after wake-up. If GIE = 1, execution will continue in line.
4: CLKOUT is not available in these osc modes, but shown here for timing reference.

**Figure 4-27** PICmicro® MCU "Awakened" by an Interrupt

*Nops* are traditionally used for two purposes. The first is to provide time synchronizing code for an application; if you look at the bit-banging routines in this book, you will see that I use them to synchronize the output of data out on the serial line. When I say to *synchronize* the output, I am ensuring that the 1s and 0s occur at exactly the same time within the data output loops.

Elsewhere, the book features constant delays. Still, you should always remember that

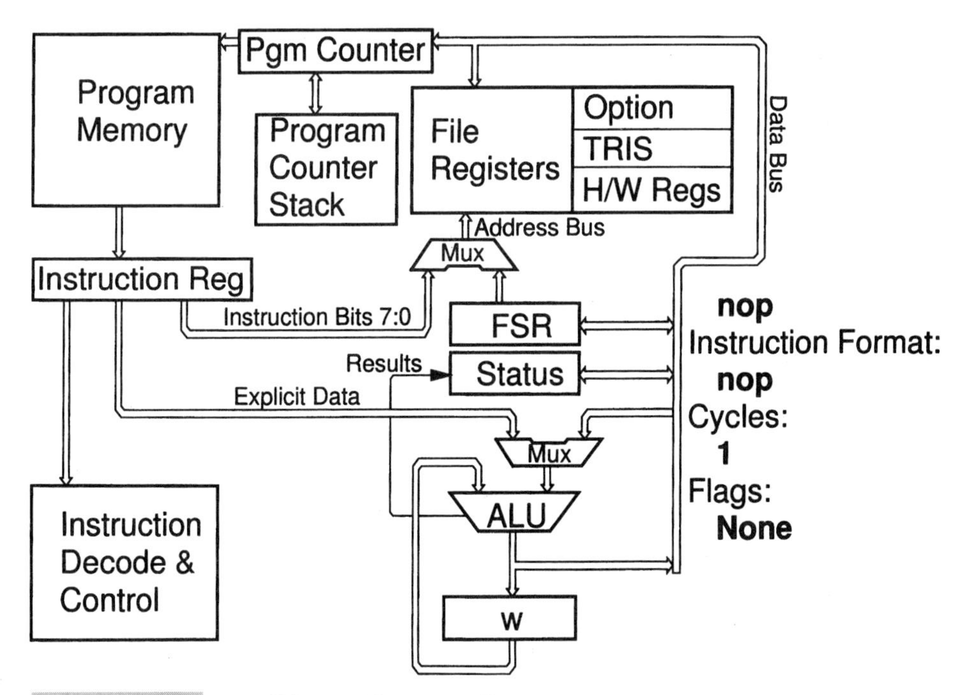


**Figure 4-28** "nop" Instruction operation

when you need to delay two cycles, use a *goto the next instruction* instead of two *nops*. The format for this *goto* instruction is:

```
goto    $ + 1
```

The *goto* instruction, as described, always executes in two cycles. If the destination is the next instruction, it will behave identically to the *nop*, but will execute in two instruction words, rather than one. This and a few other tricks described later in the book will allow you to implement long timing delays in the PICmicro® MCU using just a few instructions.

The second traditional use of *nops* is to provide space for "patch" code.

*Patching code* in processors is usually done by replacing instruction locations that have all of their bits set with instructions that were placed "in line" to see how the operation of the application is affected. In the PICmicro® MCU, it is inconvenient to use *nops* in this manner because of the programmable memory used in the PICmicro® MCU. As explained elsewhere in the book, in EPROM, EEPROM, and Flash memory technology, when the memory is ready to be programmed (i.e., cleared) all of the cells are set (equal to 1). During programming, the zeros are added to make the various instructions.

This is a problem for all the PICmicro® MCUs except for the PIC18Cxx because the *nop* instruction is all zeros. Thus, an instruction cannot be burned out of a *nop* because there are no ones to change to zeros. The PIC18Cxx has a nop instruction in which all bits are set and allows patch code to be implemented in a traditional manner.

Despite this, one method provides space in the code for patches in the PICmicro® MCU. To do this, unprogrammed code must be left in the application that can be changed by a programmer. To do this, the reverse of making instructions from *nops* is used.

For example, you might put the following code in your mid-range PICmicro® MCU application to provide patch code space:

```
goto   $ + 6          ;  Skip Over five patch addresses
dw     0x03FFF        ;  Instruction Word with all bits set
dw     0x03FFF
dw     0x03FFF
dw     0x03FFF
dw     0x03FFF
```

To enter some patch code, all of the 1s in the *goto* statement must be programmed to 0s, changing the instruction into a *nop*. The *dw 0x03FFF* assembler directives are used because they keep all of the bits set at the instruction word where the *dw* directive is located. This code snippet will allow you up to add up to five instructions without having to reassemble your code.

To add patch code, convert the *goto $ + 6* instruction to a *nop* and then write over the *dw* statements with the instructions needed for the patch. For example, adding code to invert the contents of w by using *xorlw 0x0FF* could be accomplished by changing the six previous instructions to:

```
nop                   ;  FORMERLY: goto    $ + 6
xorlw  0x0FF          ;  FORMERLY: dw      0x03FFF
goto   $ + 4          ;  FORMERLY: dw      0x03FFF
dw     0x03FFF
dw     0x03FFF
dw     0x03FFF
```

# Other PICmicro® MCU Instruction Sets

Although the different PICmicro® MCU processor families share many similarities to each other, they are all unique processors with their own architectures and instruction sets. Microchip, in an effort to make the transition between devices simpler, has enforced commonality of instructions and register names.

Most of this book concentrates on the mid-range PICmicro® MCU's architecture. This chapter has presented a detailed description of the mid-range device's instruction set and how the instructions operate. The mid-range PICmicro® MCUs are currently the most popular and "feature-rich" family, which makes them the obvious choice for most applications.

However, this does not mean that the low-end, PIC17Cxx, and 18Cxx products are better suited for many other applications than the mid-range PICmicro® MCU. The rest of this chapter presents the other architecture's instruction sets, as well as their differences to the mid-range to help you understand how to port your applications between families.

## LOW-END PICmicro® MCU INSTRUCTION SET

The mid-range PICmicro® MCU's instruction set is based on the low-end PICmicro® MCUs. All of the mid-range PICmicro® MCU's instructions (except for *addlw, sublw, retfie,* and *return*) are available in the low-end PICmicro® MCUs and the same techniques are used for application development. Rather than go through each instruction, I just want to cover the instruction differences and issues that you will have with directly addressing registers in the low-end PICmicro® MCUs.

The direct addition and subtraction instructions are not included in the low-end PICmicro® MCU's instruction set. Instead of executing a simple *addlw Constant* instruction, you will have to execute:

```
movwf  TempReg             ;  Save the Contents of w
movlw  Constant
addwf  TempReg, w          ;  Load Accumulator with the Original w Value
                           ;  added to the Constant
```

The loss of the immediate subtraction (*sublw Constant*) operation is a bit more complex because there is a definite order of operations, with the contents of w subtracted from the constant so that a constant value will have to be put into a temporary register. The code for doing this could be:

```
movwf   TempReg         ; Save the Contents of w
movlw   Constant
xorwf   TempReg, f      ; Swap w and "TempReg" Constants
                        ;   w = Constant, TempReg = wOrig ^ Constant
xorwf   TempReg, w      ; w = Constant ^ (wOrig ^ Constant),
                        ;   TempReg = wOrig ^ Constant
                        ; w = wOrig, TempReg = wOrig ^ Constant
xorwf   TempReg, f      ; w = wOrig,
                        ;   TempReg = wOrig ^ Constant ^ (wOrig)
                        ; w = wOrig, TempReg = Constant
```

```
subwf   TempReg, w      ; Load Accumulator with the Original w
                        ;   subtracted from the Constant Value
```

I realize that these operations add quite a few instructions (and require a file register), but they will simulate the *addlw* and *sublw* instructions and can be placed in a macro for your use.

The lack of a *retfie* instruction should not be surprising because there are no interrupts in the low-end PICmicro® MCUs.

You should be surprised about the lack of the *return* instruction. This is further confused by different versions of the Microchip assembler accepting the *return* instruction for the low-end devices and substituting a *retlw 0* instruction for it. This has caused problems for a number of people and is the reason why I tend to not return subroutine parameters in w because I might slip up in my low-end programming or port code from a mid-range application into a low-end PICmicro® MCU and find that it doesn't work properly.

The latest versions of the MPLAB MPASM assembler will return a warning if a *return* instruction has been inserted into low-end PICmicro® MCU application code.

In the previous sections, I noted that *tris* and *option* were optional instructions and not recommended because they could not access all the TRIS registers available in some PICmicro® MCUs. These instructions are mandatory in the low-end PICmicro® MCUs because there is no way to directly access registers in anything other than in a 32-register bank.

To access registers in banks other than 0, the FSR index register will have to be used. This means that you have to be very careful when designing your PICmicro® MCU applications to ensure that the single- and double-byte variables are located in Bank 0 and that all of the array variables are located in the other banks. Ideally, you want to avoid having to set up FSR in order to access variables in your application code.

For the *call* and *goto* instructions, as well as those in which the contents of PCL are modified, the PA0 to PA2 bits of the STATUS register are used to create the actual jump address within a specified bank. This operation, which is similar to the operation of the mid-range devices and its PCLATH register, is explained in Chapter 3.

## PIC17Cxx INSTRUCTION SET

When I first wrote this chapter, I simply created a single section for the PIC17Cxx and PIC18Cxx and just highlighted the differences between the two architectures and the mid-range. As I worked through them, I discovered that substantial differences exist between the architectures. The thinking process required to develop applications for them differed, so the instructions would be approached differently than would be used for the mid-range architecture.

The PIC17Cxx processor architecture is well suited for what I would consider to be external memory applications. In these applications, external devices, such as EPROMs, are used to store the executable code, rather than the internal program memory. These applications normally rely on the use of standard EPROM parts and development tools, rather than specialized microcontroller-development tools.

The PIC17Cxx's ability to access program memory both inside and outside the device makes it ideal for applications that are first developed in a traditional microprocessor environment and then transferred to internal microcontroller program memory. Keep this in mind when you are developing applications using external EPROMs because you will probably want to avoid *table read/write* instructions (which work differently between the internal EPROM and external memory).

When debugging PIC17Cxx applications using external EPROM, rather than relying on an emulator, you can use a logic (state) analyzer and monitor the application's execution each time the EPROM is accessed. From the state analyzer results, you can convert the bits into instructions and see how the application executed.

**Data-movement instructions** The largest difference between the PIC17Cxx and the other PICmicro® MCUs is how data-movement instructions are implemented. Instead of having the instructions centered on the w or WREG register, the PIC17Cxx data-movement instructions focus on movement between the primary and the full (or file) register address spaces.

The PIC17Cxx does not have a *movf* instruction; instead it has *movfp* (Fig. 4-29) and *movpf* (Fig. 4-30) instructions, which move data from the full register (addresses 0x000 to 0x0FF) set to the primary (addresses 0x000 to 0x01F) register set and visa versa, respectively. To remember which instruction to use, I always think of the instruction's data flow as:

*Source → Destination*

With the source on the left and the first character after *mov* and the destination being on the right and the second character after *mov*.

The *movfp* and *movpf* instructions allow bypassing the WREG, just as the PIC18Cxx's *movff* instruction does and allows you to use the unused hardware I/O registers in the primary register space for variables. I recommend just using the TBLPTR registers as your temporary storage in the Primary register set because they are the registers that are least likely to be accessed during normal execution.

Even with the *movfp* and *movpf* instructions, the PIC17Cxx still has the *"movlw"* and *movwf* instructions of the other PICmicro processors.

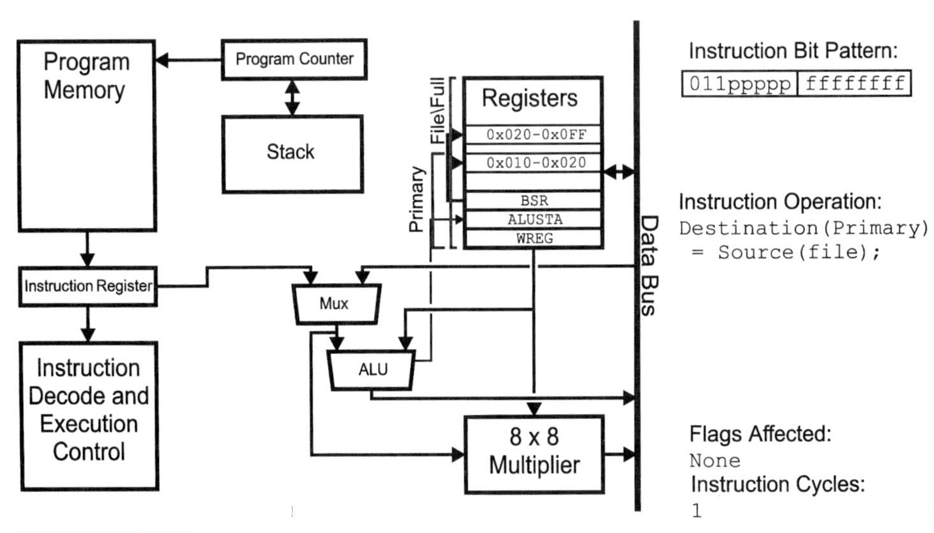


**Figure 4-29** PIC17Cxx "movfp f, p" instruction

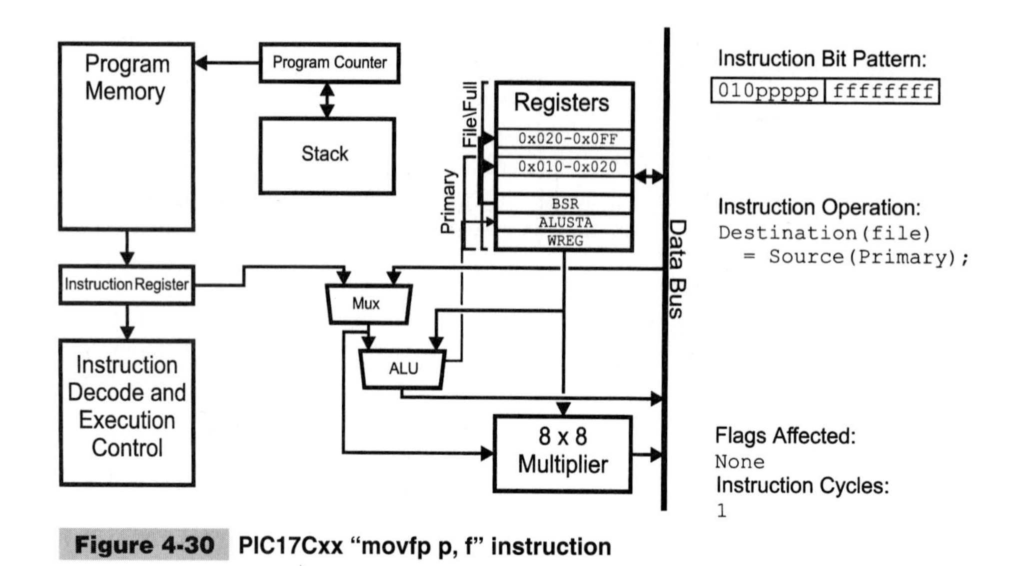


**Figure 4-30** PIC17Cxx "movfp p, f" instruction

When I first started using the PIC17Cxx, I thought of the *movwf* instruction as the *movpf* instruction and could substitute *movwf* using the instruction:

```
movpf  WREG, Reg
```

*Movwf* in the PIC17Cxx is a unique instruction and has a different bit pattern than the *movpf WREG, Reg* instruction.

Along with the *movfp* and *movpf* instructions, the PIC17Cxx also has the *movlb* (Fig. 4-31) and the *movlr* (Fig. 4-32) instructions, which are used to move explicit values into the lower and upper nybbles of the BSR register, respectively. As covered elsewhere in the book, the lower nybble of the BSR selects the bank at address range 0x010 to 0x017 while the upper nybble of the BSR selects the bank at address range 0x020 to 0x0FF.

The PIC17Cxx (and PIC18Cxx) have the ability to read and write program memory. In the PIC17Cxx, this is done using the *tablrd* (Fig. 4-33) and *tablwt* (Fig. 4-34) instructions. These instructions only pass program memory data eight bits at a time (although 16 bits are transferred in each instruction). To access the "missing" eight bits, the *tlrd* (Fig. 4-35) and *tlwt* (Fig. 4-36) have to be used.

For example, to do a 16-bit program memory read, the following instruction sequence is used:

```
movlw    LOW PMAddr            ;  Setup the Read Address
movwf    TBLPTRL
movlw    HIGH PMAddr
movwf    TBLPTRH

tablrd   0, 0, DestReg         ;  Read Lower 8 Bits of Program
                               ;    Memory Address
tlrd     1, DestReg + 1        ;  Read the Upper 8 Bits of Program
                               ;    Memory
```

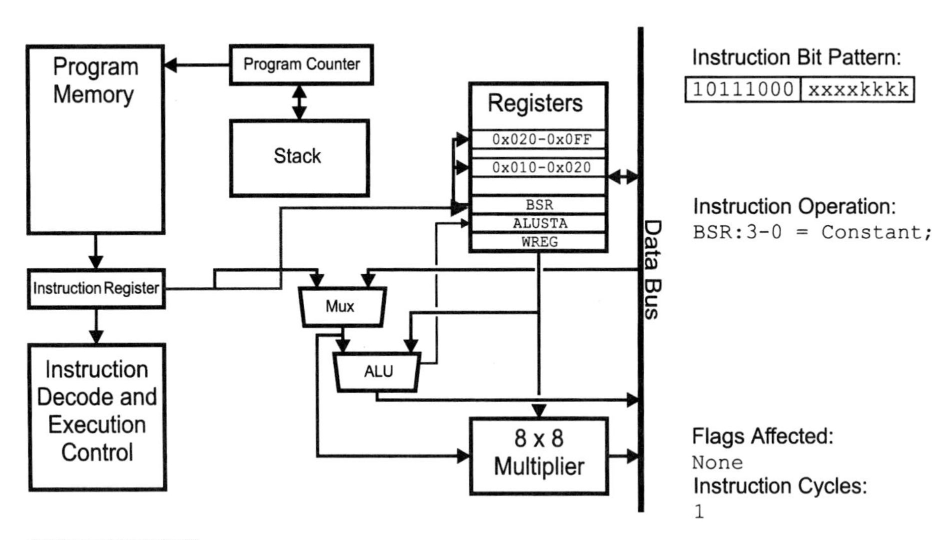


**Figure 4-31** PIC17Cxx "movlb Constant" instruction

Writing to external program memory would be accomplished using the instruction sequence:

```
movlw   LOW PMAddr          ; Setup the Write Address
movwf   TBLPTRL
movlw   HIGH PMAddr
movwf   TBLPTRH
```

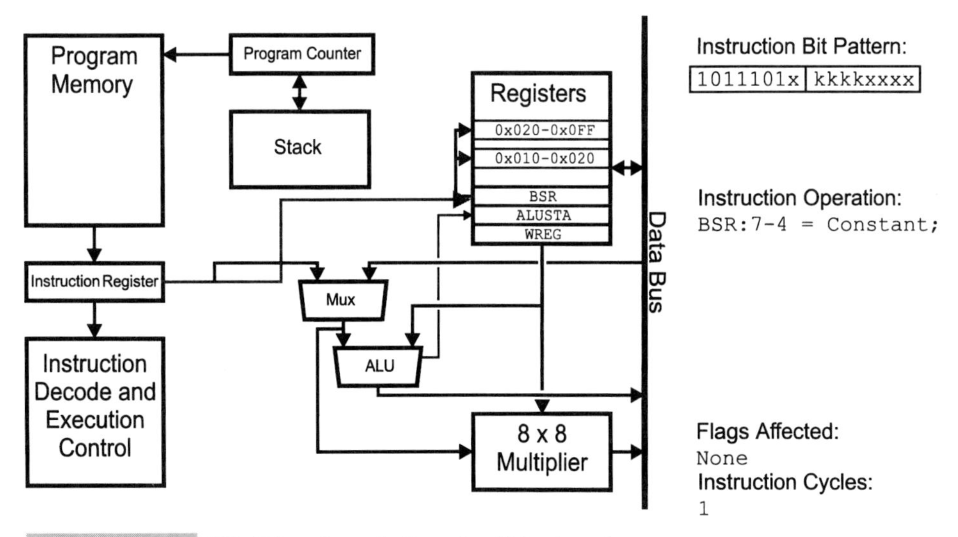


**Figure 4-32** PIC17Cxx "movlr Constant" instruction

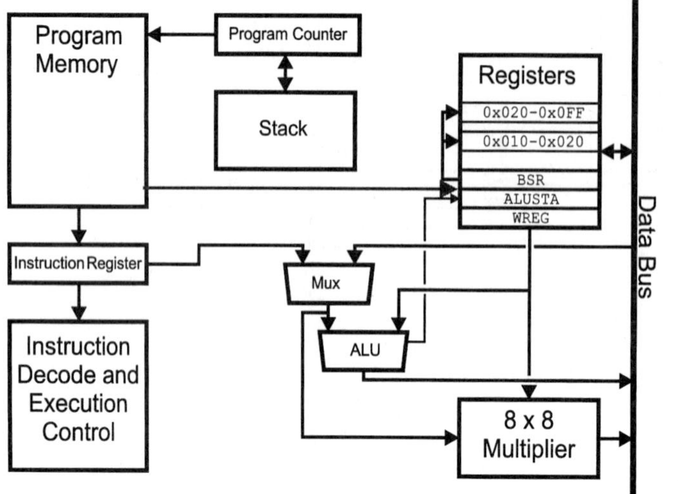


Instruction Bit Pattern:

| 101010ti | ffffffff |
|---|---|

Instruction Operation:

```
if (t == 1)
  Reg = TBLATH;
else
  TBLAT =
    PM(TBLPTR);
  Reg = TBLATL;
if (i == 1)
  TBLPTR = TBLPTR
    + 1;
```

Flags Affected:
None

Instruction Cycles:
2 (3 if Reg= PCL)

**Figure 4-33** PIC17Cxx "tablrd t, i, Reg" instruction

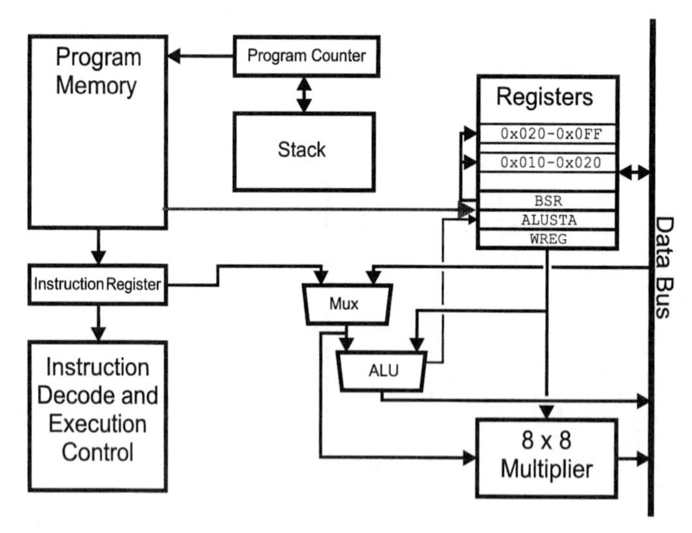


Instruction Bit Pattern:

| 101011ti | ffffffff |
|---|---|

Instruction Operation:

```
if (t == 1)
  TBLATH = Reg;
else
  TBLATL = Reg;
  PM(TBLPTR) =
    TBLAT;
if (i == 1)
  TBLPTR = TBLPTR
    + 1;
```

Flags Affected:
None

Instruction Cycles:
2 (Many if EPROM Write)

Notes: Writes to EPROM are terminated by reset or interrupts.

**Figure 4-34** PIC17Cxx "tablwt t, i, Reg" instruction

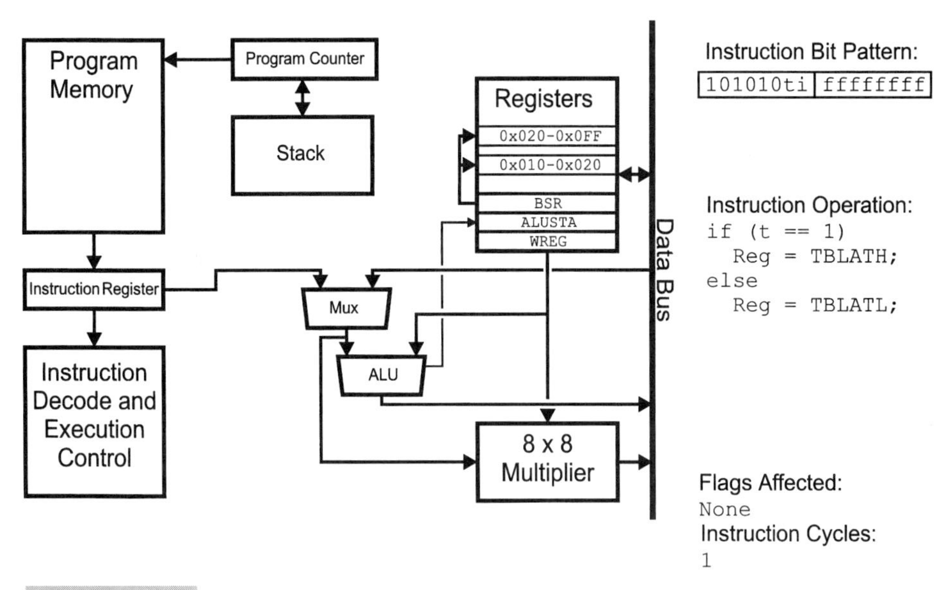


**Figure 4-35** PIC17Cxx "tlrd t, Reg" instruction

```
tlwt     0, SourceReg          ; Load TABLATH with the Lower 8
                               ;   Bits
tablwt   1, 0, SourceReg + 1   ; Carry out the Write with the
                               ;   Upper 8 Bits Specified
```

If the destination of the program memory write (PMAddr) is not EPROM, then the *tablrd* and *tablwt* instructions execute in only two cycles. If the PIC17Cxx's internal pro-

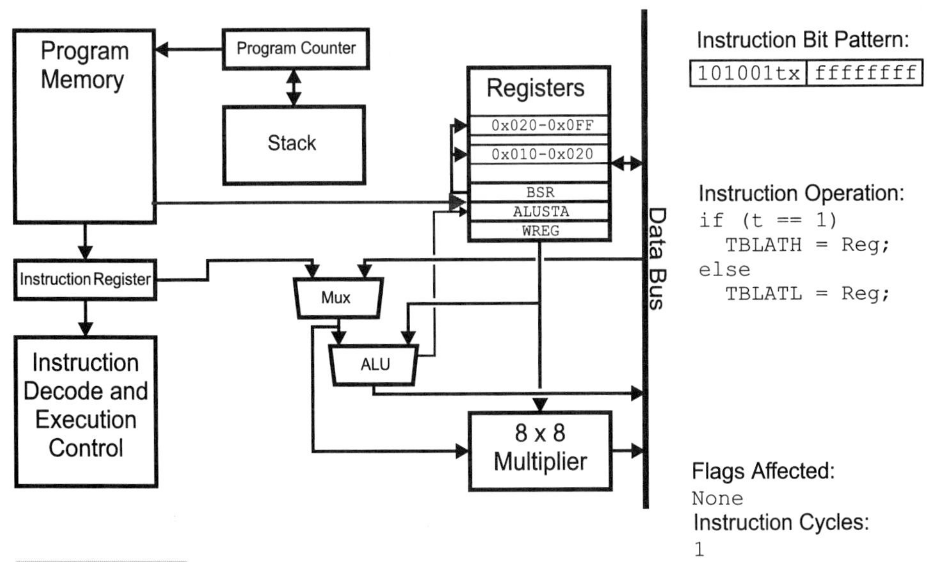


**Figure 4-36** PIC17Cxx "tlwt t, Reg" instruction

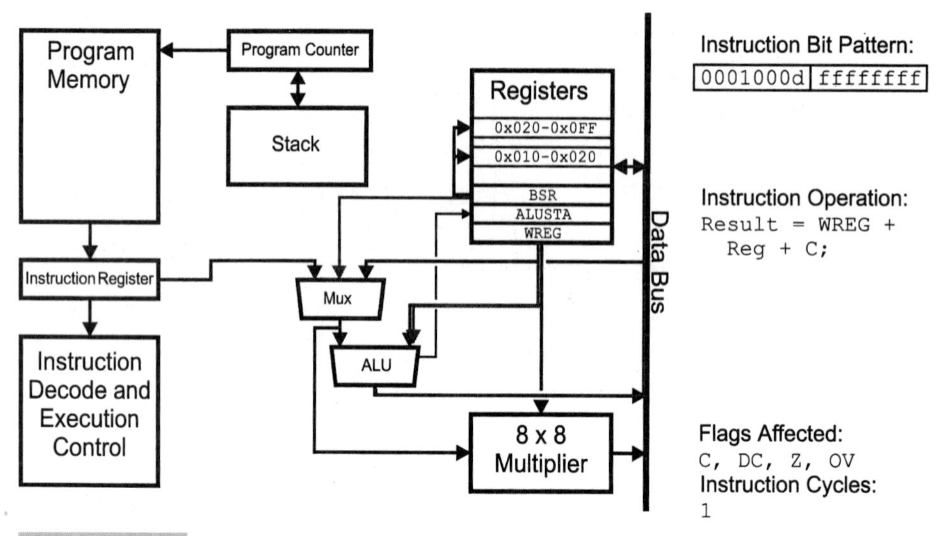


**Figure 4-37** PIC17Cxx "addwfc f, d" instruction

gram memory is being accessed, then the write operation has to be terminated by an interrupt or a reset. The *tablwt* instruction is used to load the PIC17Cxx's program memory with instructions and data. The PIC17Cxx programming operation using these instructions is explained in detail later in this book.

**Data-processing instructions** Along with the "base" data-processing instructions of the low-end and mid-range PICmicro® MCU architectures, the PIC17Cxx (and PIC18Cxx) have some enhanced data-processing instructions that make processing application data faster and more efficient. This improvement in efficiency will also result in your applications being smaller and easier to go through and understand.

Addition and subtraction has been enhanced by the inclusion of the *addwfc* (Fig. 4-37) and the *subwfb* (Fig. 4-38) instructions. These instructions use the carry flag to pass the carry/borrow result from the previous less-significant byte result to the current byte result. For example, the 16-bit subtraction operation:

```
A = B - C
```

is written in low-end and mid-range PICmicro® MCU assembler as:

```
movf    C + 1, w        ; Process High Byte First
subwf   B + 1, w
movwf   A + 1
movf    C, w            ; Get Low Byte Result
subwf   B, w
movwf   A
btfss   STATUS, C       ; Take Away from High Byte Result
 decf   A + 1, f
```

can be simplified to:

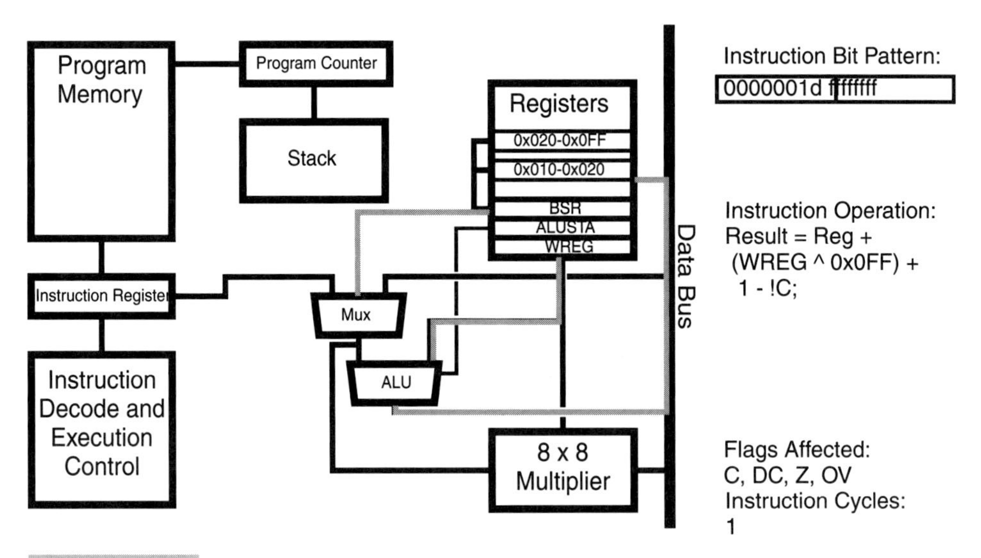


**Figure 4-38** PIC17Cxx "subwfb f, d" instruction

```
movfp    C, WREG          ; Do Low Byte First
subwf    B, w
movwf    A
movfp    C + 1, WREG      ; Process the High Byte
subwfb   B + 1, w
movwf    A
```

The *subwfb* instruction takes advantage of the operation of the carry flag as a negative borrow flag and subtracts it from the upper byte result, eliminating the need for the carry test and high byte decrement of the low-end/mid-range 16-bit subtraction code.

The *addwfc* instruction adds the carry flag from the previous, lower-order byte result to the current byte sum. Using *addwfc,* the 32-bit addition operation:

```
A = B + C
```

can be quite simply implemented as:

```
movfp    B, WREG
addwf    C, w
movwf    A
movfp    B + 1, WREG
addwfc   C + 1, w
movwf    A + 1
movfp    B + 2, WREG
addwfc   C + 2, w
movwf    A + 2
movfp    B + 3, WREG
addwfc   C + 3, w
movwf    A + 3
```

This code is very simple compared to a 32-bit addition not having this instruction. The most efficient way that I can figure out how to code a 32-bit addition for the low-end and mid-range PICmicro® MCUs is:

```
clrf     A + 1               ; Clear Upper Bytes for Carry Addition
clrf     A + 2
clrf     A + 3
movf     B, w                ; Do First Byte Addition
addwf    C, w
movwf    A
btfsc    STATUS, C           ; If Carry, Increment
 incf    A + 1, f
movf     B + 1, w            ; Second Byte Addition
addwf    C + 1, w
btfsc    STATUS, C           ; If Carry, Increment the Third Byte
 incf    A + 2, f
addwf    A + 1, f            ; Put it into the Second Byte Result
btfsc    STATUS, C
 incf    A + 2, f
movf     B + 2, w            ; Third Byte Addition
addwf    C + 2, w
btfsc    STATUS, C           ; If Carry, Increment the Forth Byte
 incf    A + 3, f
addwf    A + 2, f            ; Put it into the Third Byte Result
btfsc    STATUS, C
 incf    A + 3, f
movf     B + 3, w            ; Perform the Forth Byte Addition
movwf    C + 3, w
addwf    A + 3, f
```

The second example's code is more than twice the size of the PIC17Cxx's and the carry flag is not correct at the end for all conditions, whereas it is for the PIC17Cxx example code. The PIC17Cxx example code is a lot easier to understand—especially compared to the low-end and mid-range example code.

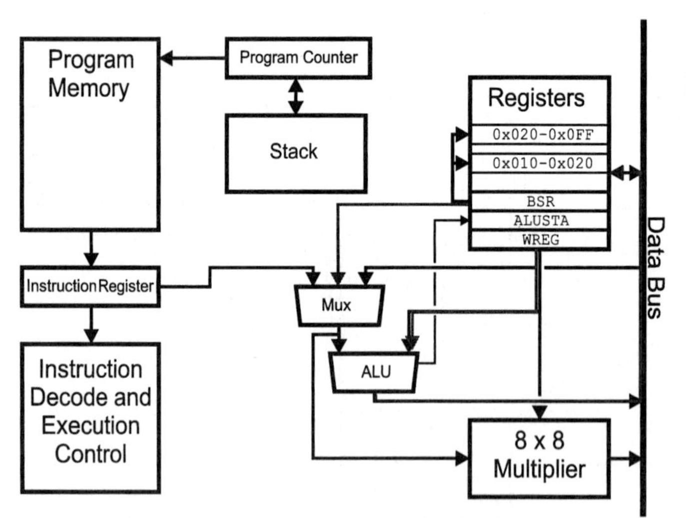


Instruction Bit Pattern:

| 0010111s | ffffffff |
|---|---|

Instruction Operation:

```
if (((WREG & 15)
    > 9) |
    (DC == 1))
  WREG = Reg + 6;
if (((WREG & 250)
    > 144) |
    (C == 1))
  WREG = Reg + 96;
if (s ==1)
  Reg = WREG;
```

Flags Affected:
C

Instruction Cycles:
1

**Figure 4-39** PIC17Cxx "daw f, s" instruction

Another new instruction for the PIC17Cxx is the *Decimal Adjust for Addition* (*DAW*, Fig. 4-39). This instruction is executed after the addition of two *Binary Coded Decimal (BCD)* numbers and converts the result into a valid BCD number.

BCD numbers are two-digit decimal numbers in which each digit is stored as a nybble in a byte. For example, *21 decimal* would be stored as 0x021 in a BCD byte instead of 0x015 in a hexadecimal byte.

*Daw* is used to ensure that the result of a BCD addition operation is valid. This instruction is needed if the lower digit is greater than nine; the value must be put into the range of 0 to 9 and the upper digit incremented.

If you wanted to add 15 and 26 in BCD, the result should be 0x041, but using the straight *addwf* instruction the result will be 0x03B. After executing the *addwf* instruction, if *daw* is executed, the result will be converted to 0x041 using the formulas shown in Fig. 4-39, which is the correct BCD result of adding 0x015 and 0x026 BCD together.

*Daw* is unusual in that it always places the operation's result back into WREG. It can also store the result in a file register, but, at the same time, the result will be stored in WREG. *Daw* also produces a correct carry result if the upper nybble's result was greater than nine. For example, adding two four-digit BCD numbers could be accomplished by the code:

```
movfp    B, w            ;  Get Low Byte Addition Result
addwf    C, w
daw      A,f             ;  Convert the Result and Store in "A"
movfp    B + 1, w        ;  Get the High Byte Addition Result
addwfc   C + 1, w        ;  Add with Carry from Low Byte Addition
daw      A + 1, f        ;  Convert the Result back to BCD and
                         ;  Store
```

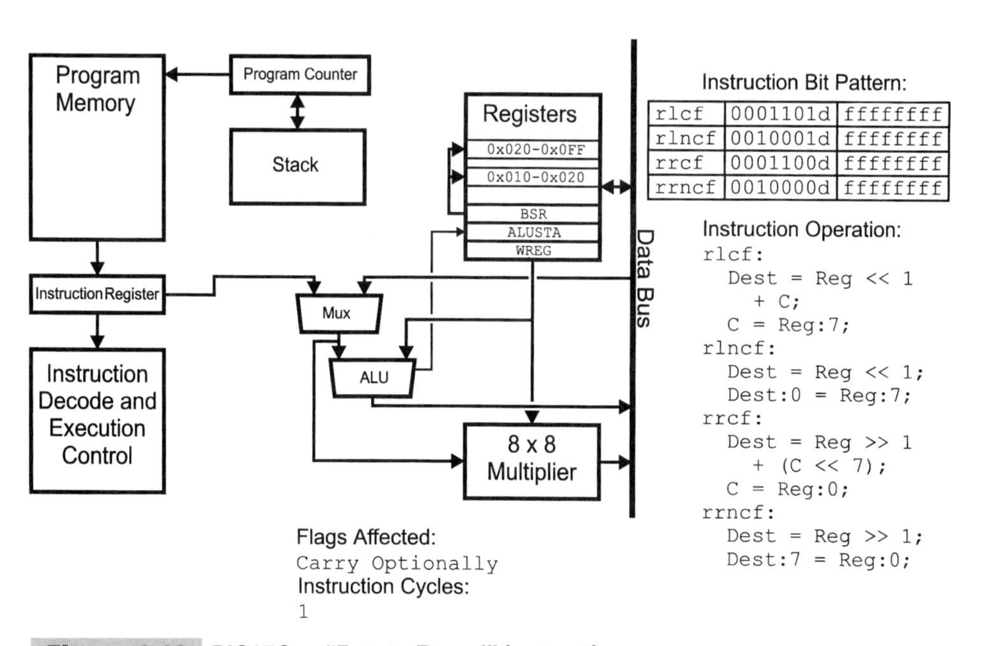


Instruction Bit Pattern:

| | | |
|---|---|---|
| rlcf | 0001101d | ffffffff |
| rlncf | 0010001d | ffffffff |
| rrcf | 0001100d | ffffffff |
| rrncf | 0010000d | ffffffff |

Instruction Operation:

```
rlcf:
  Dest = Reg << 1
    + C;
  C = Reg:7;
rlncf:
  Dest = Reg << 1;
  Dest:0 = Reg:7;
rrcf:
  Dest = Reg >> 1
    + (C << 7);
  C = Reg:0;
rrncf:
  Dest = Reg >> 1;
  Dest:7 = Reg:0;
```

**Figure 4-40** PIC17Cxx "Rotate Reg, d" instructions

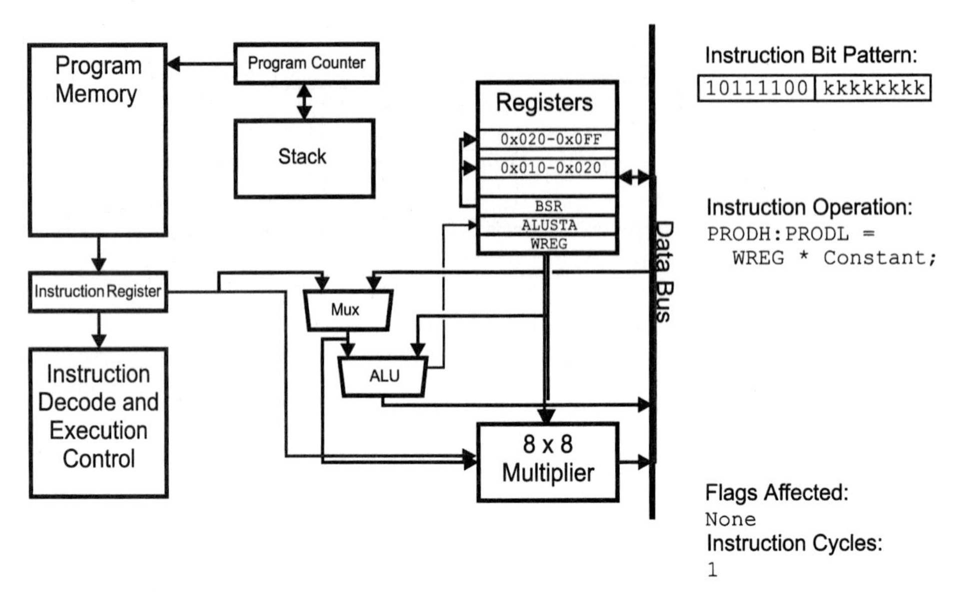


**Figure 4-41** PIC17Cxx "Mullw Constant" instruction

When you look at the rotate instructions in the PIC17Cxx (and PIC18Cxx), it will seem like they have been completely changed. This isn't quite true, the ability to select whether or not the instructions rotate through carry have just been added. The *RxCF* instructions (where *x* is "l" or "r" for "left" and "right," respectively) rotates the value in the register through the carry flag, which makes its operations identical to the low-end and mid-range PICmicro®MCUs. The *RxNCF* instructions rotate the contents of the register without involving the carry flag. The instructions are described in Fig. 4-40.

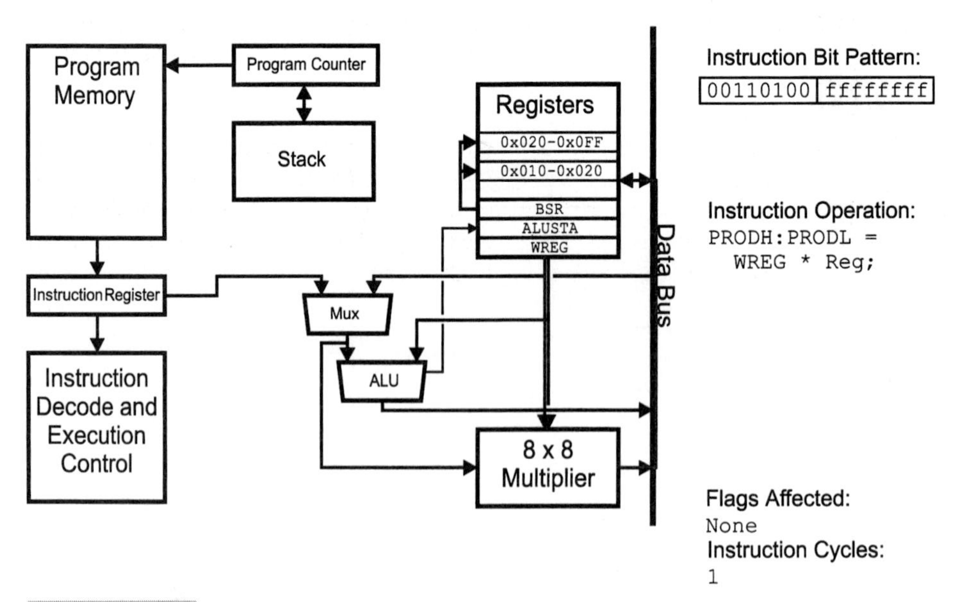


**Figure 4-42** PIC17Cxx "Mullwf Reg" instruction

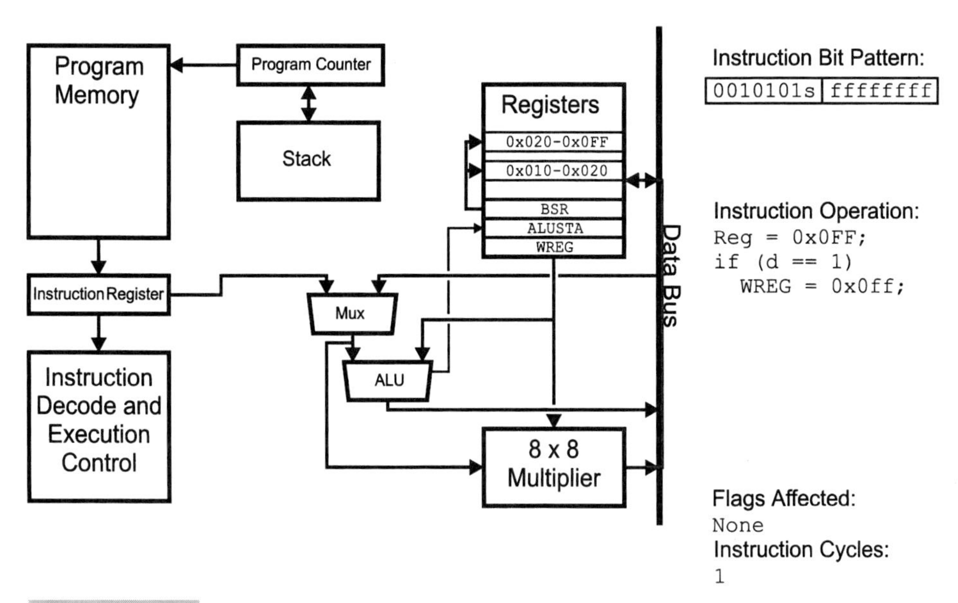


**Figure 4-43** PIC17Cxx "setf Reg, d" instruction

The most-significant enhancement of the PIC17Cxx and PIC18Cxx over the low-end and mid-range PICmicro® MCU processors is the inclusion of the *single instruction cycle multiply* instructions. Both instructions multiply two eight-bit numbers and places the 16-bit result in the PRODH and PRODL registers. *Mullw* (Fig. 4-41) multiplies the contents of WREG with a constant and *mullwf* (Fig. 4-42) multiplies the contents of the specified register with the contents of WREG. Both instructions are very well behaved and give you the capability to implement low-frequency DSP functions.

The remaining two instructions are quite basic. The *setf* instruction (Fig. 4-43) sets all the bits in a register and, optionally, WREG. The numeric result of *setf* can be thought of as 0x0FF, 255 or -1, depending on the application. *Setf* is the complementary instruction to *clrf* (which clears all the bits of a register).

*Btg* (Fig. 4-44) simply complements (toggles) the state of a bit. Whereas the *bcf* instruction has been shown to be an AND with the appropriate bit reset and *bsf* to be an OR with the appropriate bit set, *btg* can be thought of an XOR with the appropriate bit set.

*Btg* is best used in situations where a bit value is dynamically set and has to be changed for some reason using constantly timed application code. Instead of the low-end or mid-range bit toggle code:

```
btfsc    Register, Bit
  goto   $ + 4
nop
bsf      Register, Bit
goto     $ + 3
bcf      Register, Bit
goto     $ + 1
```

or

```
movlw    1 << Bit
xovwf    Register,f
```

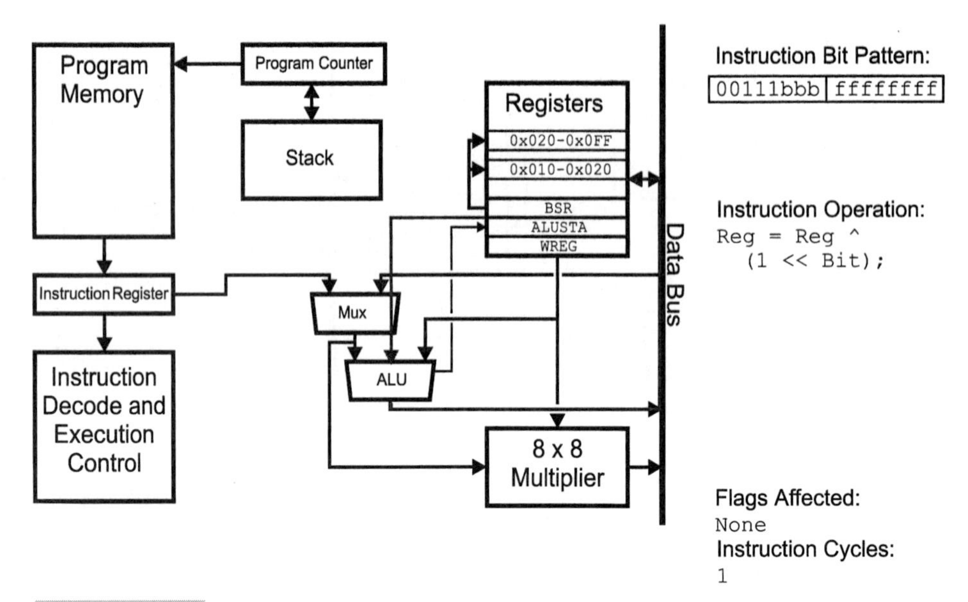


**Figure 4-44** PIC17Cxx "btg Reg, Bit" instruction

In which *Register*'s *Bit* is complemented with constant (two or six instruction cycle) timing. In the PIC17Cxx and PIC18Cxx, *btg* can be simply used to replace these seven instructions:

```
btg  Register, Bit
```

and the specific bit will be toggled in one instruction cycle and not change the contents of WREG or the ALUSTA bits.

**Execution change instructions** The PIC17Cxx does not have significantly different execute change operations compared to the low-end and mid-range PICmicro® MCUs. For the most part, the PIC17Cxx has variations on themes, rather than totally new functions. Despite this statement, some new "compare and skip" instructions have been added to the PIC17Cxx (and PIC18Cxx) that are very useful.

The only additions to the low-end and mid-range PICmicro® MCU standard execution instructions are the inclusion of the *tstfsz* (Fig. 4-45) and *lcall* instructions. *Tstfsz* will cause a skip if the contents of a register are equal to zero. The most immediate use that I can see for this instruction is in a 16-bit decrement. Instead of the four-instruction decrement shown elsewhere in the book, the code can be simplified to:

```
tstfsz   Reg
  decf   Reg + 1, f
decf     Reg, f
```

The *lcall* instruction (Fig. 4-46) uses the full eight bits of the PCLATH register to specify the destination address of the *call* instruction. Conceivably, this instruction could be used for table-based calls, each call table entry was placed on 256 instruction offsets. This is probably not practical for most applications, but I'm sure that some time over the years, I'll come up with an application that will implement tables using this instruction.

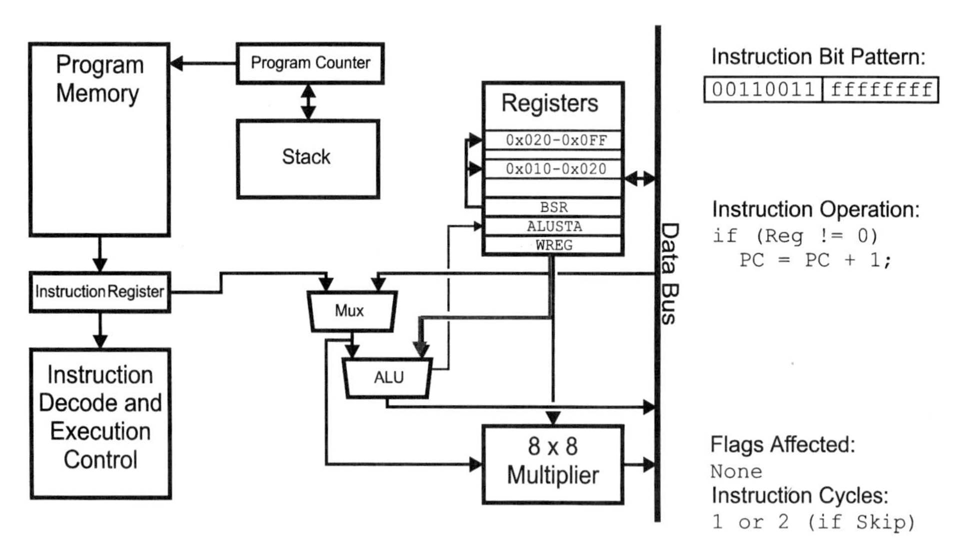


**Figure 4-45** PIC17Cxx "Tstfsz Reg" instruction

The most important enhancement of the PIC17Cxx's processor over the low-end and mid-range PICmicro® MCU architecture conditional execution instructions is the addition of the compare and skip instructions (Fig. 4-47). These instructions compare the contents of the specified register to the contents of WREG and skip the next instruction if the result is equal to zero (*cpfseq*), less than zero (*cpfslt*), or greater than zero (*cpfsgt*).

These instructions follow the same subtraction process as the other instructions and,

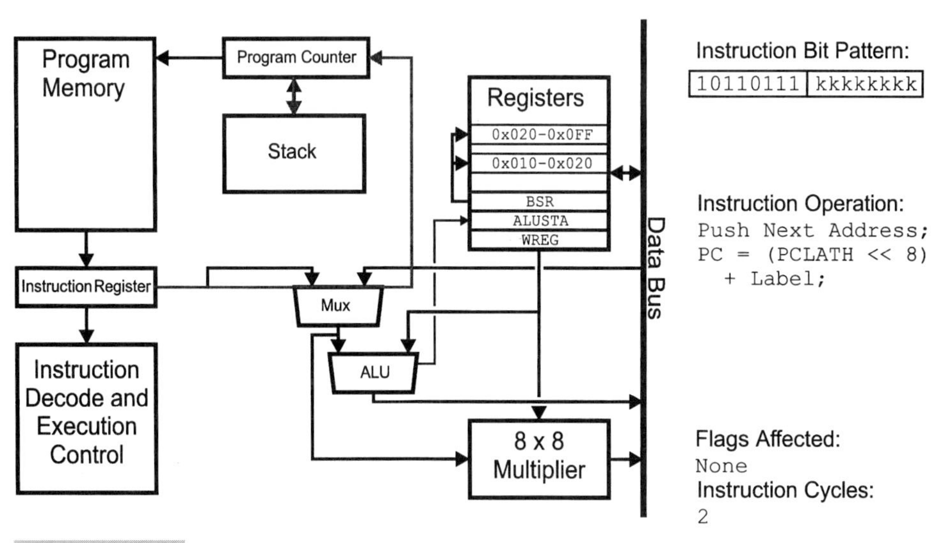


**Figure 4-46** PIC17Cxx "Lcall Label" instruction

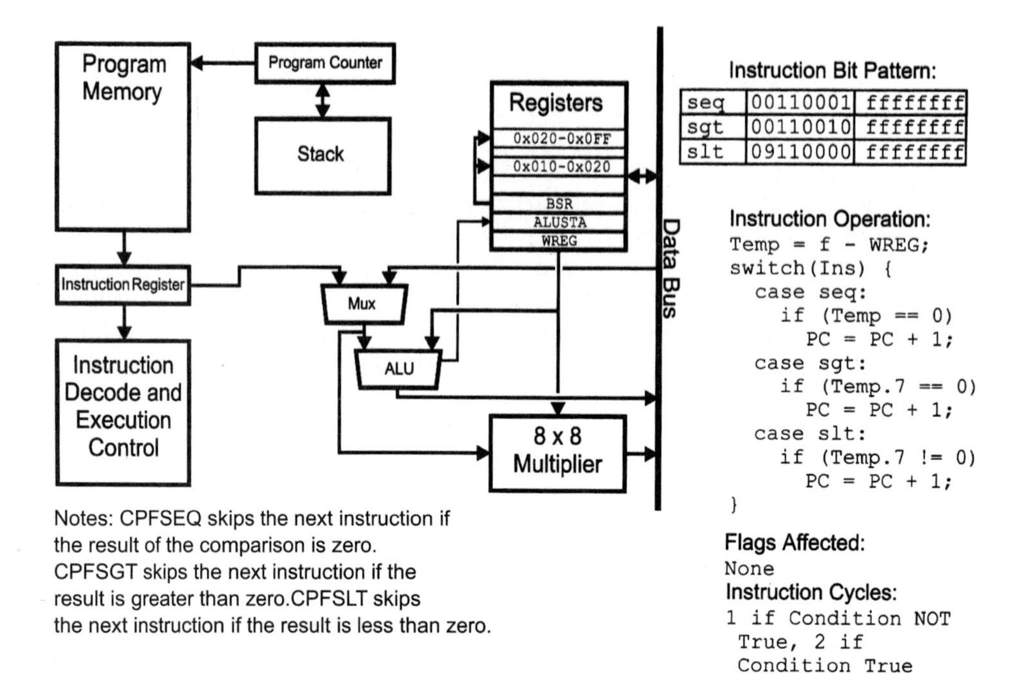
Program Memory
Program Counter
Stack
Registers
0x020-0x0FF
0x010-0x020
BSR
ALUSTA
WREG
Instruction Register
Mux
ALU
Instruction Decode and Execution Control
8 x 8 Multiplier
Data Bus
Instruction Bit Pattern:
seq 00110001 ffffffff
sgt 00110010 ffffffff
slt 09110000 ffffffff
Instruction Operation:
Temp = f - WREG;
switch(Ins) {
case seq:
if (Temp == 0)
PC = PC + 1;
case sgt:
if (Temp.7 == 0)
PC = PC + 1;
case slt:
if (Temp.7 != 0)
PC = PC + 1;
}
Flags Affected:
None
Instruction Cycles:
1 if Condition NOT True, 2 if Condition True
Notes: CPFSEQ skips the next instruction if the result of the comparison is zero. CPFSGT skips the next instruction if the result is greater than zero.CPFSLT skips the next instruction if the result is less than zero.

**Figure 4-47** PIC17Cxx "Compare" instructions

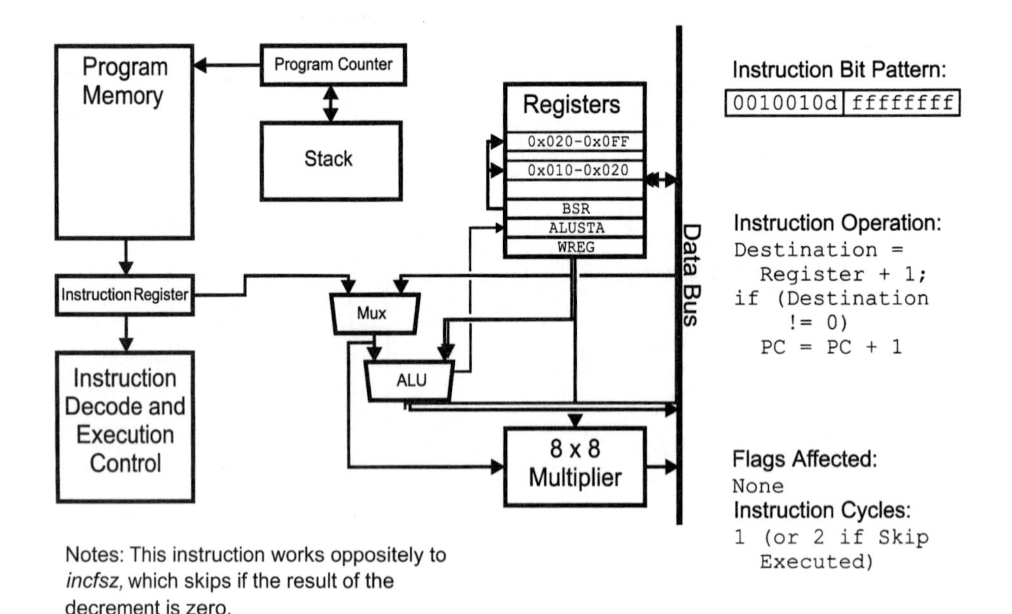
Program Memory
Program Counter
Stack
Registers
0x020-0x0FF
0x010-0x020
BSR
ALUSTA
WREG
Instruction Register
Mux
ALU
Instruction Decode and Execution Control
8 x 8 Multiplier
Data Bus
Instruction Bit Pattern:
0010010d ffffffff
Instruction Operation:
Destination = Register + 1;
if (Destination != 0)
PC = PC + 1
Flags Affected:
None
Instruction Cycles:
1 (or 2 if Skip Executed)
Notes: This instruction works oppositely to incfsz, which skips if the result of the decrement is zero.

**Figure 4-48** PIC17Cxx "infsnz Reg, D" instruction

as such, are somewhat difficult to follow if you look at the Microchip datasheets. For this reason, I tend to ignore what is going on in the instructions and just focus on the results. Are the contents of the register equal to, less than, or greater than the contents of the WREG?

Another enhancement to the PIC17Cxx and PIC18Cxx processor architectures is the increment and skip if the result is *not* equal to zero (*infsnz*, Fig. 4-48) and the decrement and skip if the result is not equal to zero (*dcfsnz*, Fig. 4-49).

Along with all of these instructions, you have to remember all the different capabilities of the enhanced instructions. This can be somewhat unnatural if you have worked with the low-end and mid-range PICmicro® MCU devices quite a bit before working with the PIC17Cxx. For example, a 16-bit increment can be simplified from the three-instruction sequence:

```
incf    Reg, f
btfsc   STATUS, C
  incf  Reg + 1, f
```

to:

```
infsnz  Reg, f
  incf  Reg + 1, f
```

I must confess that I don't always look for the most efficient way to execute instructions in the PIC17Cxx and PIC18Cxx because of my experience and familiarity with the low-end and mid-range PICmicro® MCU architectures.

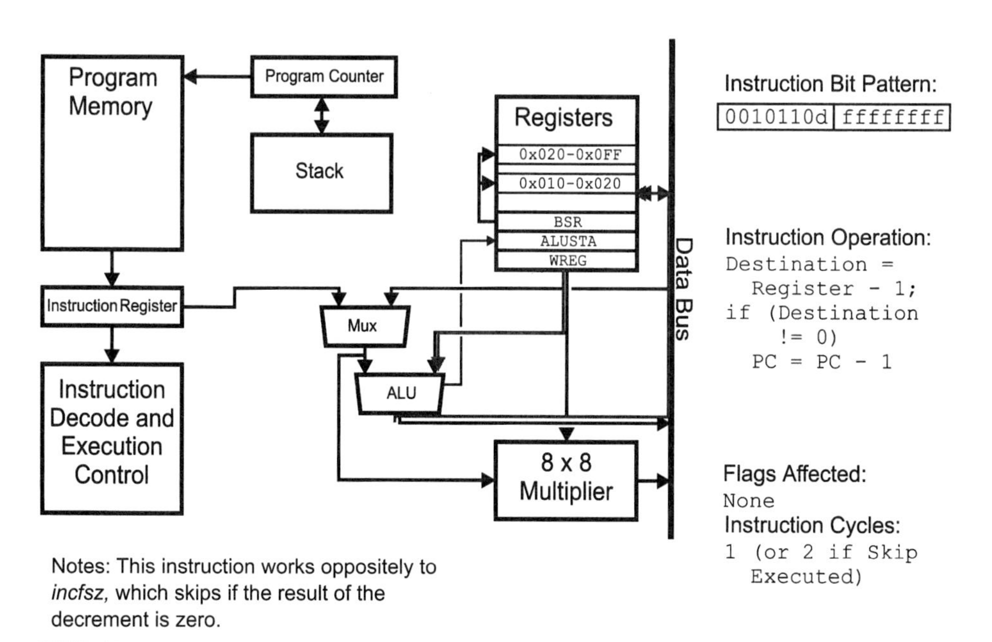


**Figure 4-49** PIC17Cxx "dcfsnz Reg, D" instruction

**Processor-control instructions** The PIC17Cxx does not have any processor-control instructions that are different from the low-end or mid-range PICmicro® MCUs. The PIC17Cxx *sleep* and *nop* instructions behave identically to their low-end and mid-range counterparts.

The only aspect of processor control that you will have to be aware of in the PIC17Cxx is how configuration fuses are programmed. This is covered elsewhere in the book, including applications showing how the fuses are programmed are available on the CD-ROM.

## PIC18Cxx INSTRUCTION SET

One of the first points made about the PIC18Cxx architecture in the Microchip PIC18Cxx datasheets is that it is "source-code compatible with the PIC16Cxx ('mid-range') instruction set." Although many of the instructions are common between the two architectures, I feel that this statement is somewhat optimistic because of two points. As I write this, I have successfully completed a real-time operating system for the PIC18Cxx, have done a fuzzy-logic fan motor governor for the architecture, and rewritten many of my 16-bit code snippets to reflect the new instructions in the PIC18Cxx instruction set. Although I certainly don't qualify as an expert programmer on the PIC18Cxx architecture, I hope you can respect my opinions when I say that the PIC18Cxx architecture is *better* than the mid-range's. I also think that once you become proficient with programming the PIC18Cxx, you will be able to create applications more easily and more efficiently for the PIC18Cxx than for the mid-range PICmicro® MCU architecture.

The first difference I would like to point out about the PIC18Cxx compared to the mid-range PICmicro® MCU architecture is the different addressing capabilities of the two families. As you work with the mid-range PICmicro® MCUs, you will become very familiar with the operation of the bank registers and how to negotiate them quickly in application code. This need (and skill) is unneeded in the PIC18Cxx and you will find that you will order your PIC18Cxx applications differently. When you re-order the PIC18Cxx applications, you will also find that they are more efficient than the equivalent mid-range PICmicro® MCU applications by 30 to 50 percent.

I found that I very quickly organized my applications very similarly to the way I would organize 8051 applications. Accesses to the *Special-Function Registers* (*SFRs*) can be executed directly along with individual registers without worrying about banking addresses or loading index registers. The multiple full-memory wide index registers allow very fast access to all the file registers in the PIC18Cxx and really eliminate the need to access the BSR register in your applications.

The second difference I found in the PIC18Cxx over the mid-range PICmicro® MCU architecture is that it is a superset of the original PICmicro® MCU architecture. As a superset, there are operations that can be carried out much more efficiently and much easier in the PIC18Cxx than in the mid-range devices. Although I have noted that the standard PICmicro® MCU mid-range to be at least 30 percent more efficient than other eight-bit processor architectures, the improved instruction set of the PIC18Cxx easily doubles that value or more.

This makes the PIC18Cxx up two twice as efficient in terms of application instruction size and execution cycles than other microcontroller's processors in the marketplace. Coupled with the very fast (four times) instruction clock, the PIC18Cxx is capable of remarkably fast and powerful applications without excessive heat dissipation or current requirements.

**Data-movement instructions** In all PICmicro® MCU applications, I have found the first thing to be done when planning an application is deciding where variables are to be placed. This is probably the single-most important aspect of how successful the application is going to be. In the mid-range PICmicro® MCU devices, placing single- and double-byte variables in Bank 0 cuts down on the number of execution bank changes that have to be made within the application. To access array variables, I use a different bank, which can be accessed separately from the directly addressed variables.

In the PIC18Cxx, I follow a similar philosophy for single- and double-byte variable placement. These variables are always located in the access RAM, and array variables are placed in the upper banks of memory, where they can be accessed by the index (FSR) registers, which do not require special bank access. An excellent example of this technique is in the PIC18Cxx RTOS presented in Chapter 17. Task information blocks (the memory used to store the task-operating information and variables) is located outside of the access RAM and when it is time to load the information for an active task. This block of memory is passed down into the access RAM so that to the executing task its variables appear to always be in the access RAM.

Placing the single- and double-byte variables in the access RAM means that the access Bank does not have to be specified for most applications. This results in application code appearing to be almost identical to mid-range application code. I still stand by my statements that mid-range and PIC18Cxx code will look different as the enhanced PIC18Cxx instructions allow much more efficient code and require a different thought process from the mid-range architecture.

Data can be loaded and stored in the WREG register using the *movf*, *movlw*, and *movwf* instructions. The important difference between these instructions and the mid-range's analogs is the addition of the access bit in the instruction bit patterns. When a 1 is specified as the last parameter in the instruction, like:

```
movf i, w, 1
```

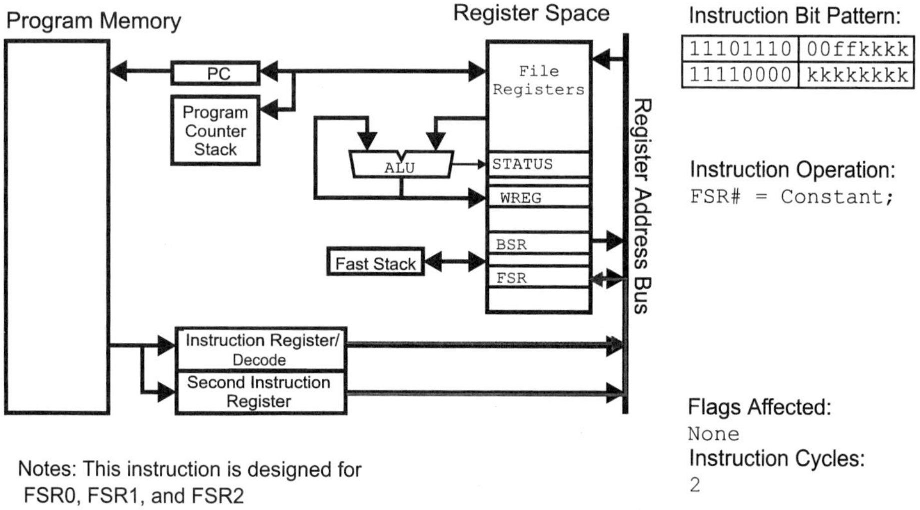

Instruction Bit Pattern:

| 11101110 | 00ffkkkk |
|---|---|
| 11110000 | kkkkkkkk |

Instruction Operation:
FSR# = Constant;

Flags Affected:
None

Instruction Cycles:
2

Notes: This instruction is designed for FSR0, FSR1, and FSR2

**Figure 4-50** PIC18Cxx "lfsr Value" instruction

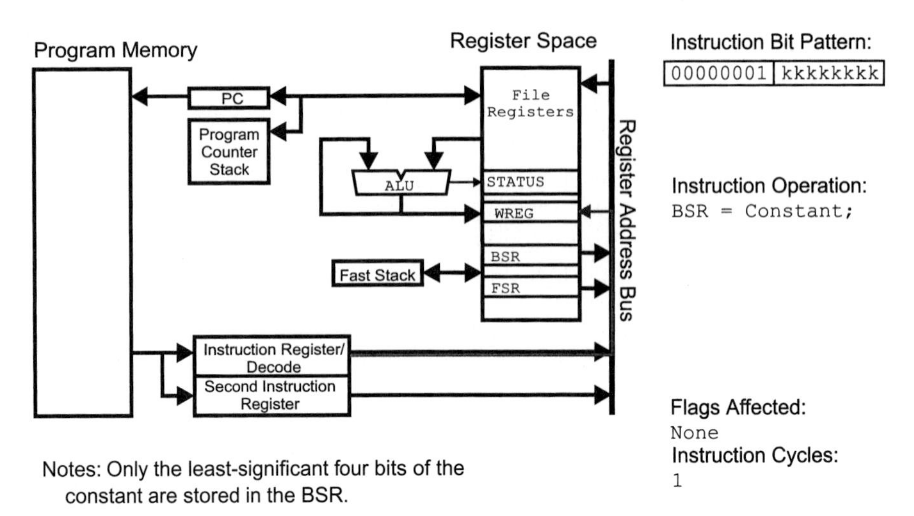


**Figure 4-51** PIC18Cxx "movlb Value" instruction

the BSR register is used to select the bank, *i* is addressed within. If a 0 or nothing is specified as the instruction's last parameter, i.e.:

```
movf i, w
```

the value for *i* is taken out of the access bank or PIC18Cxx addresses 0x0000 to 0x07F for file registers or 0x0F80 to 0x0FFF for the *special-function registers (SFRs)*.

The new instructions added to the PIC18Cxx architecture are the *lfsr* (Fig. 4-50), *movlb* (Fig. 4-51), and *movff* (Fig. 4-52) instructions. These instructions are used to specify addresses anywhere in the PIC18Cxx's register space.

The first two instructions are used to load constant full-register addresses into the FSR index register and the *Bank Select Register (BSR),* respectively. *Lfsr* loads a 12-bit con-

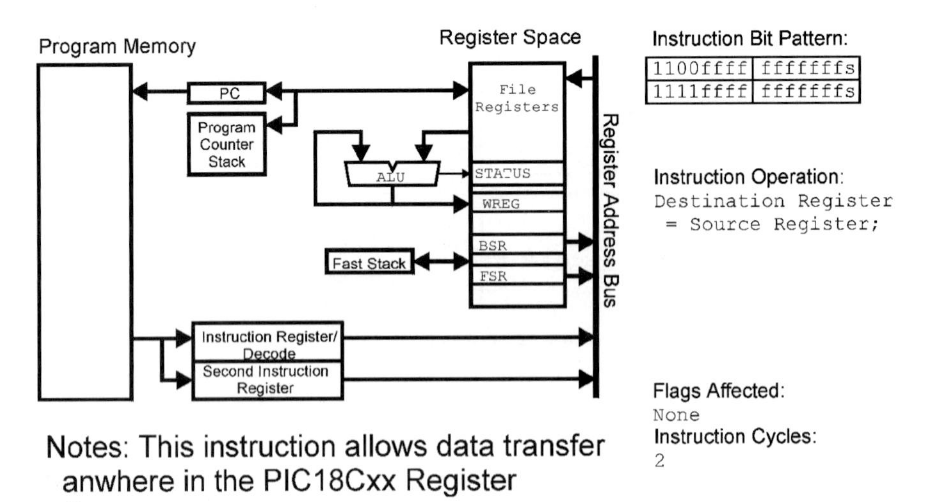


**Figure 4-52** PIC18Cxx "movff s, d" instruction

stant into the specified FSR register as the start to a table. *Movlb* loads a four-bit constant into the BSR register that specifies which bank variables are to be taken from. Although I find *lfsr* very useful, I doubt I will use *movlb* very often because I won't directly access file registers outside of the access bank unless I absolutely have to.

Even if I had to access file registers outside of the Access RAM bank, I would probably still avoid using the BSR simply because of the availability of the *movff* instruction (Fig. 4-52). This instruction allows movement of byte data between any two addresses in the PIC18Cxx's register space. There are not a lot of points to be noted about this instruction, other than that it is a two-word instruction. In the second word of the instruction, like all the two-word instructions, the second word has the four most-significant bits set. These four bits indicate to the PIC18Cxx processor instruction-decode circuitry that the second word is valid. If it is not present, the processor jumps over the instruction and then executes the next one in sequence. If these bits are set when the processor isn't expecting them, it just treats the instruction as an *nop*.

The *movff* instruction is particularly useful because it does not change the STATUS register bits and does not affect the contents of WREG. In the RTOS, I found this instruction to allow very simple string data (single-dimensional array) copies. For example, to copy five bytes of data from one string to another, the code is simply:

```
 lfsr     FSR0, SourceString    ;  Point to the Start of the Strings
 lfsr     FSR1, DestString
 movlw    5                     ;  Load WREG with 5
Loop

 movff    POSTINC0, POSTINC1    ;  Copy String and Increment the FSR
 decfsz   WREG, f               ;   Registers
  bra     Loop
```

in the mid-range architecture, the same function would be accomplished by the following code:

```
 clrf     Count                 ;  Reset the Offset Within the String
Loop
 movlw    SourceString          ;  Get the Current Source Element
 addwf    Count, w
 movwf    FSR
 movf     INDF, w
 movwf    Temp
 movlw    DestString            ;  Save the Source in the Destination
 addwf    Count, w
 movwf    FSR
 movf     Temp, w
 movwf    INDF
 incf     Count, f              ;  Increment the Current Element
 movlw    5                     ;  Loop Until "Count" == 5
 xorwf    Count, w
 btfss    STATUS, Z
  goto    Loop
```

Most PIC18Cxx code will not result in as dramatic improvements as this, but you can see where the *movff* instruction, along with the ability to post increment FSR registers can improve an application's code efficiency (no matter how you measure it) significantly.

In the mid-range PICmicro® MCU string move example code, notice that the source

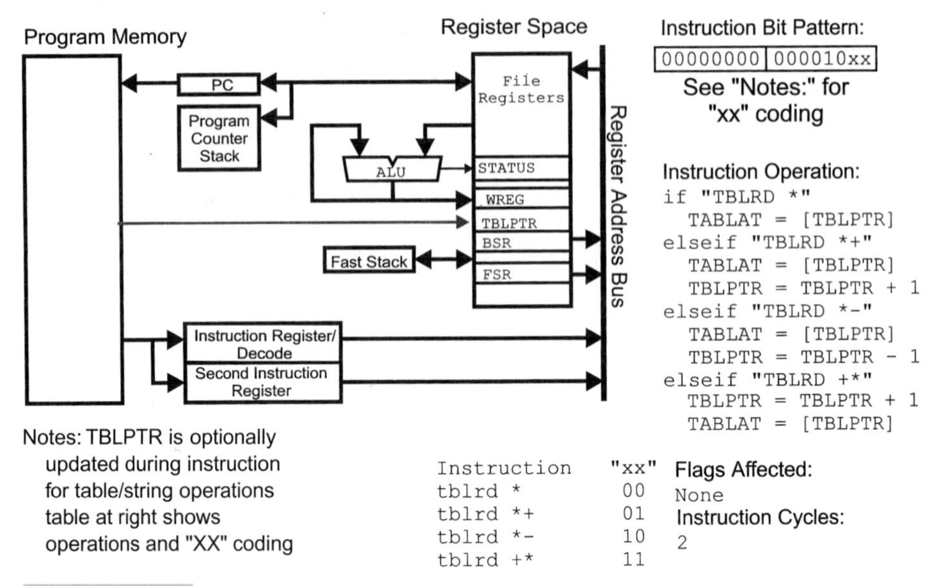


**Figure 4-53** PIC18Cxx "tblrd*" instruction

and destination are located in the same bank pairs. If different bank pairs used for the two strings (such as Bank 0 and Bank 3), the code would become more cumbersome. This is not an issue with the PIC18Cxx and its 12-bit FSR registers.

Along with accessing data in the register space, the PIC18Cxx can also access its own program memory, like the PIC17Cxx. The *tblrd* (Fig. 4-53) instruction will place the 16-bit contents of the program memory, at the TBLPTR-specified address, into the TABLAT registers. TBLPTR is a 21-bit-long address; an instruction must have its least-significant bit reset (clear) so that the 16-bit address does not go over (straddle) a word boundary.

To carry out a program memory read, the following instruction sequence could be used:

```
movlw     UPPER ReadAddr     ;  Load the Top 5 Address Bits
movwf     TBLPTRU
movlw     HIGH ReadAddr      ;  Load the "Middle" 8 Address Bits
movwf     TBLPTRH
movlw     LOW ReadAddr       ;  Load the Bottom 8 Address Bits
movwf     TBLPTRL

tblrd     *                  ;  Read the Program Memory

movf      TABLATL, w         ;  Process the Low Byte of the Program
  :                          ;    Memory Instruction

movf      TABLATH, w         ;  Process the High Byte of the Program
  :                          ;    Memory Instruction
```

In tblrd (and tblwt, which follows), a TBLPTR increment or decrement specification can be optionally put in. Figure 4-53 includes the four options and how the bit pattern is changed.

*Table write* (*tblwt*, Fig. 4-54) instructions are only available to write to the EPROM program memory at the current time. For this instruction to successfully execute, the LWRT

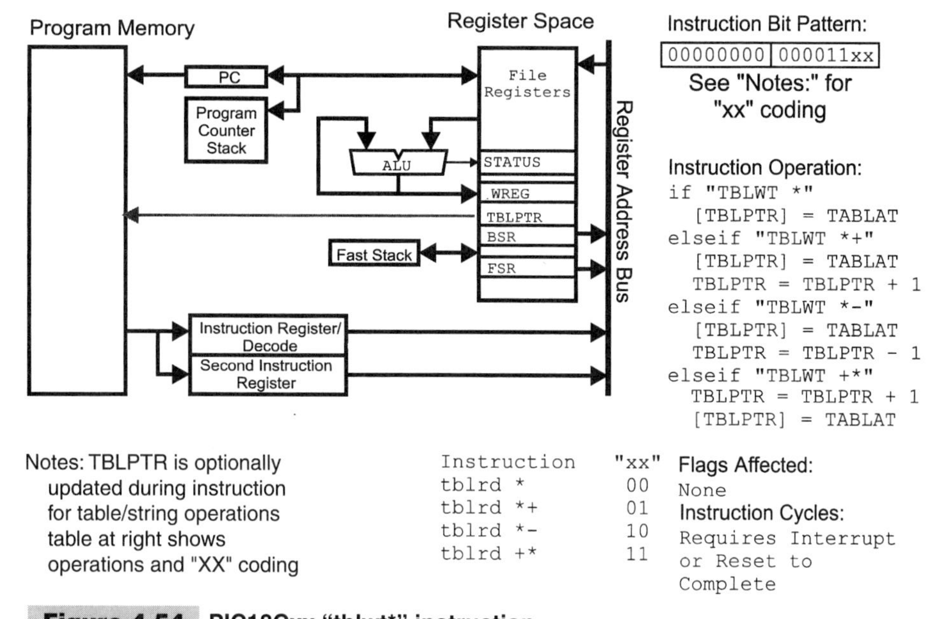
Program Memory
Register Space
PC
Program Counter Stack
ALU
File Registers
STATUS
WREG
TBLPTR
BSR
FSR
Fast Stack
Instruction Register/ Decode
Second Instruction Register
Register Address Bus
Instruction Bit Pattern:
00000000 000011xx
See "Notes:" for "xx" coding
Instruction Operation:
if "TBLWT *"
[TBLPTR] = TABLAT
elseif "TBLWT *+"
[TBLPTR] = TABLAT
TBLPTR = TBLPTR + 1
elseif "TBLWT *-"
[TBLPTR] = TABLAT
TBLPTR = TBLPTR - 1
elseif "TBLWT +*"
TBLPTR = TBLPTR + 1
[TBLPTR] = TABLAT
Notes: TBLPTR is optionally updated during instruction for table/string operations table at right shows operations and "XX" coding
Instruction "xx"
tblrd * 00
tblrd *+ 01
tblrd *- 10
tblrd +* 11
Flags Affected:
None
Instruction Cycles:
Requires Interrupt or Reset to Complete

**Figure 4-54** PIC18Cxx "tblwt*" instruction

bit of the RCON register must be set and _MCLR driven to $V_{pp}$ (13 to 14 volts), rather than just five volts. Like the PIC17Cxx's internal *write* instruction, the PIC18Cxx *tblwt* instruction must be terminated by a reset or interrupt.

As currently implemented, the *tblrd* and *tblwt* instructions can only access internal program memory. I would expect that sometime in the future, the PIC18Cxx will be available with external program memory interfaces, such as the PIC17Cxx.

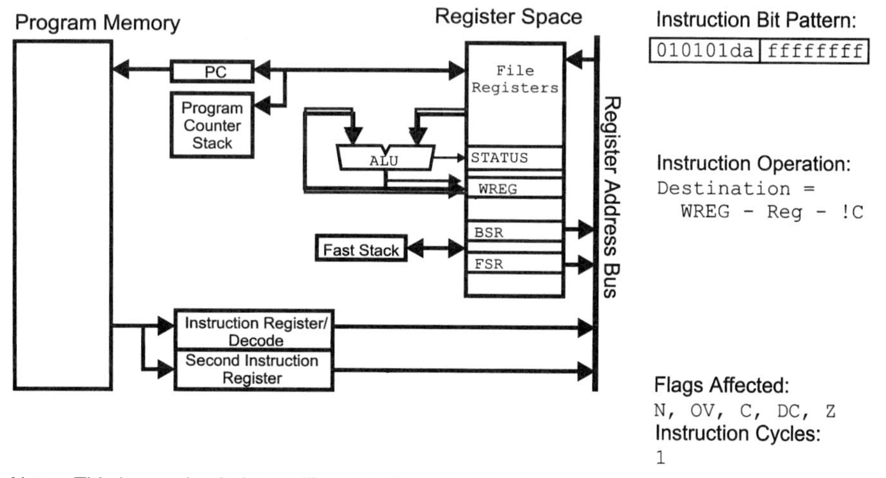
Program Memory
Register Space
PC
Program Counter Stack
ALU
File Registers
STATUS
WREG
BSR
FSR
Fast Stack
Instruction Register/ Decode
Second Instruction Register
Register Address Bus
Instruction Bit Pattern:
010101da ffffffff
Instruction Operation:
Destination =
WREG - Reg - !C
Flags Affected:
N, OV, C, DC, Z
Instruction Cycles:
1

Notes: This instruction behaves like a traditional subtract and is different from the standard subtraction instructions available in the other PICmicro® MCU architectures.

**Figure 4-55** PIC18Cxx "subfwb Reg, f, a" instruction

**Data-processing instructions** The PIC18Cxx has essentially the same data-processing capabilities as the PIC17Cxx for two cases that add some flexibility and conventional capabilities, compared to the other PICmicro® MCU processors. As you look through the PIC18Cxx instruction set, you will see that the additions and modifications to the PIC18Cxx instruction set make it more similar to that of other processors while retaining the PIC18Cxx's ability to create very efficient code.

The most-significant addition to the PIC18Cxx's data-processing instructions is the *subfwb* (Fig. 4-55) instruction. This instruction carries out a *subtract with borrow* instruction in the order that most people are familiar with if they have worked with other processors. Instead of the typical PICmicro® MCU subtraction instruction:

```
Result = (Source Value) - WREG [-  !C]
```

the *subfwb* instruction executes as:

```
Result = WREG - (Source Value) -  !C
```

This instruction frees you from the need of thinking backwards when subtraction instructions are used in an application. To use the *subfwb* instruction, the WREG is loaded with the value to be subtracted from (the *subtend*) and the value to take away (the *subtractor*) is specified in the instruction. This means that if you have the statement:

```
A = B - C
```

the values of the expression can be loaded in the same left to right order as the PICmicro® MCU instructions and use the sequence:

```
bcf       STATUS, C
movf      B, w
subfwb    C, w
movwf     A
```

This is the same order as would be used in most other processors. Notice that I reset the carry flag before the instruction sequence to avoid any possibilities of the carry being reset unexpectedly and taking away an extra 1, which would be very hard to find in application code.

A PIC18Cxx 16-bit subtraction operation could be:

```
bcf       STATUS, C
movf      B, w
subfwb    C, w
movwf     A
movf      B + 1, w
subfwb    C + 1, w
movwf     A + 1
```

Or, if you want to save on the instruction used to clear the carry flag at the start of the sequence:

```
movf      C, w
subwf     B, w
movwf     A
```

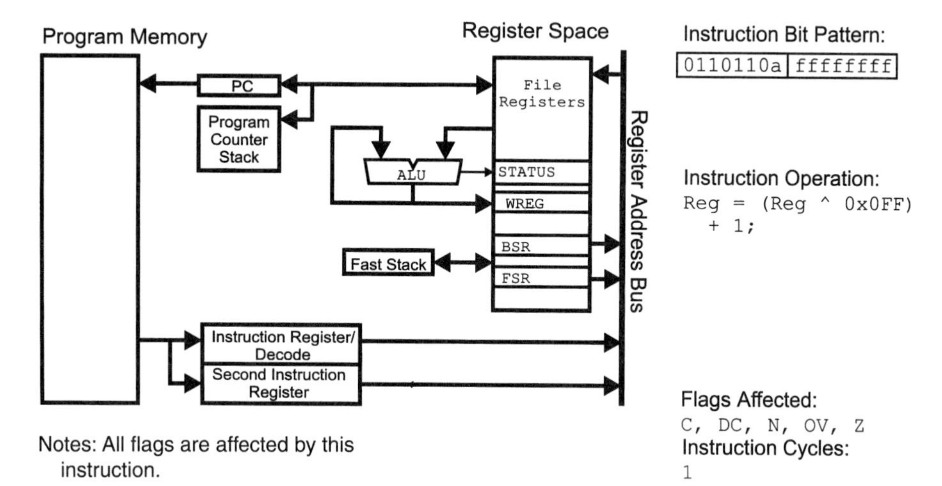


**Figure 4-56** PIC18Cxx "negf Reg, a" instruction

```
movf     B + 1, w
subfwb   C + 1, w
movwf    A + 1
```

The other difference between the 18Cxx and the other PICmicro® MCU processors is the inclusion of the *negf* (Fig. 4-56) instruction. This instruction differs from the PIC17Cxx *negw* instruction, which can only negate the contents of WREG in that *negf* can negate any register in the PIC18Cxx's register space.

**Execution change instructions** The PIC18Cxx's execution-change instructions, upon first glance, should be very familiar to you if you are familiar with the other PICmicro® MCU families. The PIC18Cxx has the *btfsc, btfss, goto,* and *call* of the low-end, mid-range, and PIC17Cxx PICmicro® MCUs, along with the *compare and skip on equals (cpfseq), greater than (cpfsgt),* and *less than (cpfslt)*. The PIC18Cxx also has the *enhanced increment* and *skip on result not equal to zero* (*infsnz* and *dcfsnz*). Along with these similarities, the PIC18Cxx has four new features that you should be aware of (and remember their availability) when you are developing applications for it.

The first feature that you should be aware of is the presence of the multi-word instructions. These instructions take up two 16-bit words instead of the single word of all the instructions in the other PICmicro® MCU architecture families. These instructions are:

- movff
- goto
- call

For your first applications, I would recommend that you use the instruction set's single-word instructions only. The *branch* and *relative call* set of instructions provide relative address jumps, rather than the *goto* and *call* instructions, respectively (Fig. 4-57), which can access any program memory location available to the PIC18Cxx processor.

I recommend using the *branch* and *rcall* instructions as much as possible. This is not only to save the amount of program memory space used by the application, but also to

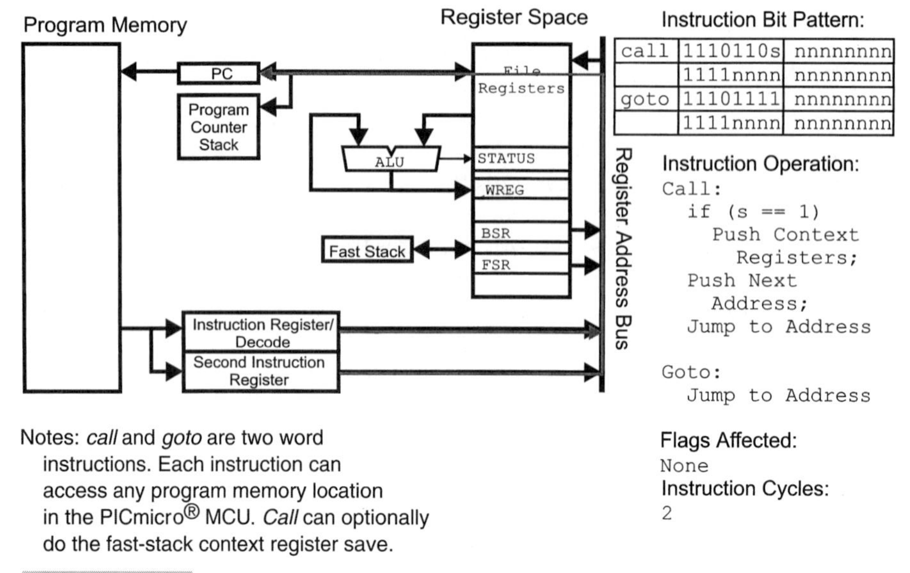


**Figure 4-57** PIC18Cxx "Goto/Call Label" instructions

avoid putting you in the situation where you are used to single-word instructions and are unfamiliar with multi-word instructions. This can be a particular problem when counting instructions for a relative jump. The only time you should be using the *goto* or *call* instructions is if you have to access a memory location outside the range of the relative branches. This range is −512 to +511 instruction addresses for the *bra* (branch always) and *rcall* (relative call) instructions and -64 to +63 instruction addresses for the conditional branch instructions. The *rcall* instruction information is shown in Fig. 4-58.

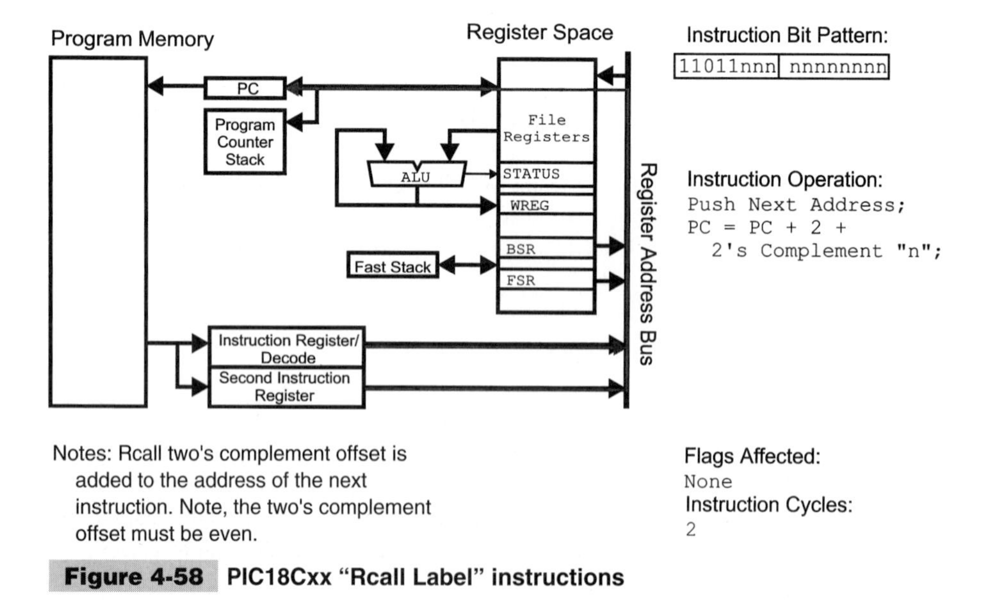


**Figure 4-58** PIC18Cxx "Rcall Label" instructions

Along with using the single-word execution change instructions, I also recommend that you avoid using the *$* directive and branching relative to it. The reason for this is that the PIC18Cxx breaks with the tradition established by the other PICmicro® MCU architectures and places an address location on each byte address boundary and not each instruction ("word") boundary. As well, the PIC18Cxx cannot execute on odd-byte instructions; this will cause your application to lock up or behave strangely, as if the instructions burned into program memory weren't actually there.

The *$ directive* (assembler command) simply returns the address of the current instruction. It is often used to simplify PICmicro® MCU assembly code and avoid the need for multiple labels. The *$* directive is particularly useful for avoiding having to come up with multiple labels for short and numerous loops.

Please bear with me as I explain what the problems are with having instructions stored at byte addresses, rather than word. If you've worked with other architectures, you probably don't understand what the issue is because other processors have instructions start on odd-byte boundaries and no problems arise from this procedure.

Throughout the book, when I am working with the low-end, mid-range, and PIC17Cxx architectures, you are going to see instructions like:

```
decfsz    Count, f
 goto     $ - 1
```

which runs without any problems.

If this code was placed directly into a PIC18Cxx application, you would first find that the simulator would stop executing when the *goto $ − 1* instruction was encountered. This is because of the destination starting at the second byte of the *decfsz* instruction, rather than the first. The second byte picked up would be the first byte of the following *goto* instruction. By stopping, the simulator is indicating that there is a problem.

If you were to burn the code into a PIC18Cxx and attempt to execute it, you would discover that it would work correctly up to this point and then behave strangely afterward (if you could observe the application working correctly at all). This is because the PIC18Cxx's "instruction decode and control" circuitry is trying to correctly execute the instruction made up of the second half of the *decfsz* instruction *concatenated* (added to) the first half of the *goto* instruction. I'm sure you can see that the resulting instruction in no way was desired for your application.

Looking at it from the perspective of the code, when the processor executes the first two instructions, it is reading in the data:

```
Addr       2E Count                 ;  decfsz Count, f
Addr + 2   EF (Addr + 1) & 0x0FF    ;  goto $ - 1
Addr + 4   FF (Addr + 1) >> 8
```

After the *goto $ − 1*, the processor will be executing:

```
Addr + 1   Count EF  ;  ????
```

As you can see, the instruction that would be executed after the *goto $ − 1* instruction would be *Count EF* instead of *2E Count*. What the processor does when it encounters *Count EF* is totally dependent on the address of *Count*. Trying to figure out what the instruction was going to execute beforehand and understand what the failure actually is caused by will be just about impossible.

To avoid these types of headaches, I recommend that you *always* use labels in your PIC18Cxx application code. After putting a label in the snippet, the code becomes:

```
Loop
  decfsz  Count, f
   goto   Loop
```

and the opportunity for the problem is eliminated.

I have experimented with different ways to create computed addresses with the *$* directive and I've never really come up with one that I am comfortable with. In the PIC18Cxx application code presented in this book, you'll see me using the format:

```
  goto    $ +/- (2 * n)
```

where the number multiplied by two (*n*) is the number of instructions to jump.

Another situation where the byte addressing of the PIC18Cxx will cause problems is in the table jumps. Elsewhere in the book, I showed that a byte could be used to jump to any address within a 256-element table in the mid-range PICmicro® MCUs using the code:

```
  movlw     HIGH Table              ; Set PCLATH to Table Start
  movwf     PCLATH
  movlw     LOW Table               ; Compute the Low Byte Address of the
  addwf     Index, w                ; Table Element
  btfsc     STATUS, C               ; If > 256, Increment PCLATH
   incf     PCLATH
  movwf     PCL                     ; Jump to the Address

  Table
  :
```

This code will have problems because of the byte-wide addressing of the PIC18Cxx. To avoid this problem and ensure that the correct element is jumped to, I double the offset before adding it to the three register (PCLATU, PCLATH, and PCL) program counter registers:

```
  clrf      Offset + 1              ;  Use "Offset" for Index * 2
  movff     Index, Offset
  bcf       STATUS, C
  rlcf      Offset, f
  rlcf      Offset + 1, f
  movlw     UPPER Table             ;  Setup the High 13 Bits of the Table
  movwf     PCLATU                  ;    Address
  movlw     HIGH Table
  movwf     PCLATH
  movlw     LOW Table               ;  Get Correct Offset in the First 256
  addwf     Offset, f               ;    Addresses of the Table
  movlw     0
  addwfc    Offset + 1, w           ;  Add the Carry to the High Offset Byte
  addwf     PCLATH, f
  movlw     0                       ;  Add the Resulting Carry to the Upper
  addwf     PCLATU, f               ;    Offset Byte
  movf      Offset, w               ;  Execute Jump by Writing to PCL
  movwf     PCL

  Table
  :
```

This is (obviously) one of the cases where the PIC18Cxx actually requires more complex instructions to carry out an operation. This code will always execute in the same number of instruction cycles (20), just like in the mid-range example, although the mid-range general case executes in only eight cycles.

Notice that the instruction:

```
movff  Offset, PCL
```

cannot be used as PCL cannot be the destination of a *movff* instruction.

I indicated that there was a one word *goto* instruction called *bra* for *branch*. This instruction type (Fig. 4-59) changes the program counter according to the two's complement offset provided in the instruction, according to the formula:

```
PCnew = PCcurrent + 2 + Offset
```

where *PCcurrent* is the current address of the executing branch instruction. The *2* added to *PCcurrent* results in the address after the current one. *Offset* is the two's complement value which is added or subtracted (if the offset is negative) from the sum of *PCcurrent* and two.

The MPASM assembler computes the correct offset for you when the destination of a branch instruction is a label. The two's complement offset is computed using the formula:

```
Offset = Destination - (Current Address)
```

If the destination is outside the range of the instruction, it is flagged as an error by the MPASM assembler.

Along with the nonconditional branch, eight conditional branch instructions are available in the PIC18Cxx (Fig. 4-59). They are *Branch on Zero Flag Set (bz), Branch on Zero Flag Reset (bnz), Branch on Carry Flag Set (bc), Branch on Carry Flag Reset (bnc), Branch on Negative Flag Set (bn), Branch on Negative Flag Reset (bnn), Branch on Over-*

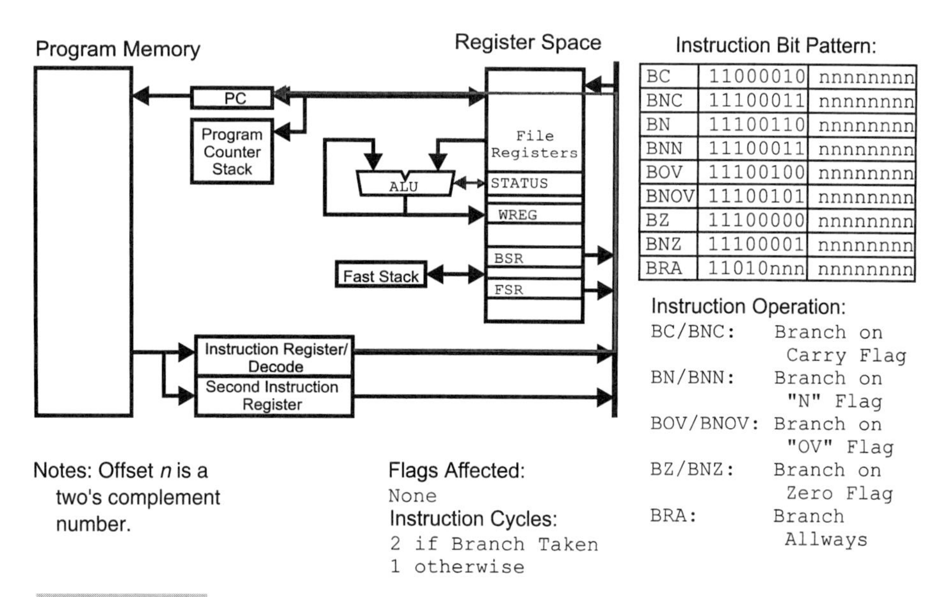


**Figure 4-59** PIC18Cxx "Branch Offset" instructions

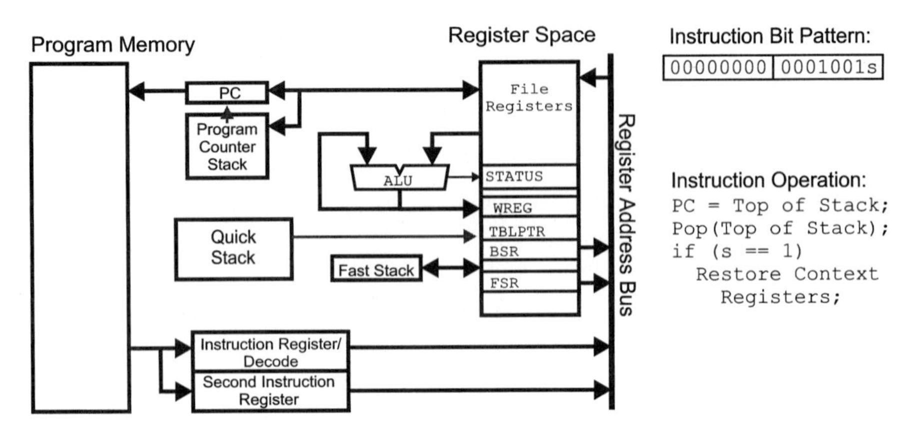


**Figure 4-60** PIC18Cxx "Return s" instruction

*flow Flag Set (bov)*, and *Branch on Overflow Flag Reset (bnov)*. These instructions are equivalent to the *branch-on condition* instructions found in other processors.

These instructions behave similarly to the *bra* instruction, except that they have eight bits for the offset address (to the *bra* instruction's 11). This gives the instructions the ability to change the program counter by −64 to +63 instructions.

The last new feature of the PIC18Cxx architecture, which is different from the other architectures, is the "fast stack" in which the WREG, STATUS, and BSR registers are saved nonconditionally upon the interrupt acknowledge and vector jump and conditionally during a subroutine *call* instruction. These registers can be optionally restored after a *return* (Fig. 4-60) or *retfie* (Fig. 4-61) instruction.

This feature must be used sparingly as the fast stack and it is really just a single set of memory locations. An interrupt acknowledgement followed by a subroutine call that saves WREG, STATUS, and BSR or the fast stack will be overwritten.

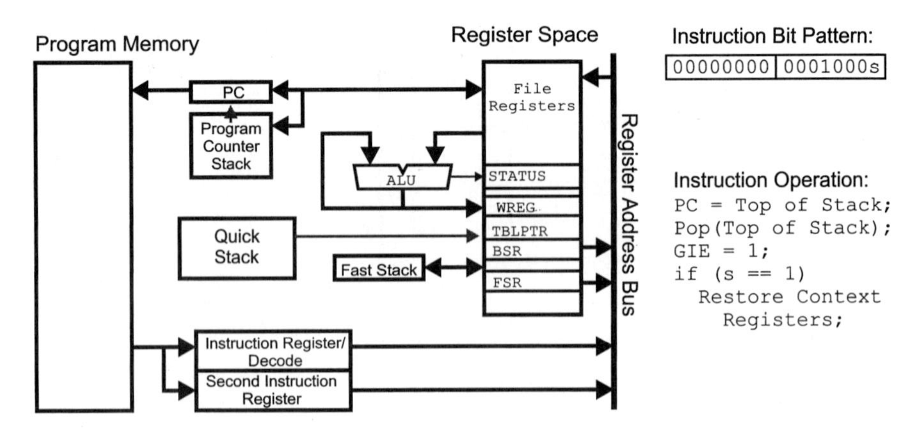


**Figure 4-61** PIC18Cxx "Retfie s" instruction

When I show a basic interrupt handler for the mid-range PICmicro® MCUs, along with the w and STATUS registers, I also include saving the contents of the FSR and the PCLATH registers. This is not required in the PIC18Cxx because of the multiple FSR registers available and the ability to jump anywhere within the application. If an FSR register is required within an interrupt handler, then chances are that one can be reserved for this use within the application when resources are allocated. Later, this book presents the need for listing how the built-in hardware functions (resources) are specified or allocated by an application.

The PCLATH (and PCLATU) registers should not have to be saved in the interrupt handler, unless a table is accessed in the application. The *goto* and *branch* instructions update the program counter and not the PCLATH and PCLATU registers.

Thus, a PIC18Cxx interrupt handler can be as simple as:

```
org 8

Int

; #### - Execute Interrupt Handler Code

retfie 1
```

so long as nested interrupts are not allowed and subroutine calls do not use the fast stack.

**Processor-control instructions** The PIC18Cxx has the same processor instructions as the other PICmicro® MCUs, but instruction enhancement is worth bringing to your attention. When designing the PIC18Cxx, the Microchip designers did something I've wanted for years. They created a *nop* instruction (Fig. 4-62) that has two bit patterns: all bits set and all bits reset.

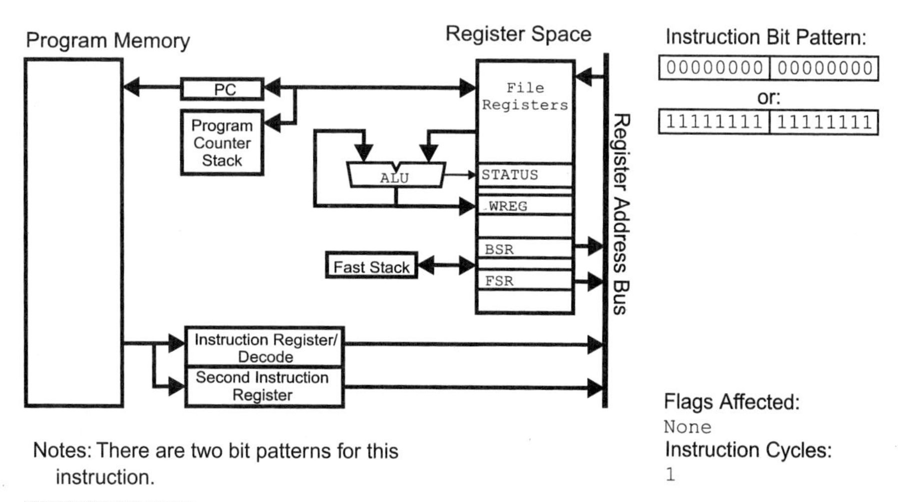


Instruction Bit Pattern:

00000000 00000000

or:

11111111 11111111

Flags Affected:
None

Instruction Cycles:
1

**Figure 4-62** PIC18Cxx "nop" instruction bit patterns

The profoundness of this instruction and what can be done with it will probably not be immediately obvious to you. But, it will be a significant advantage to people who put patch space into their applications. Earlier in this chapter, I showed how PICmicro® MCU patch space is implemented out of a jump over a series of instructions with all the bits set. For the mid-range PICmicro® MCU, five instructions of patch space actually looks like:

```
goto   $ + 6
dw     0x03FFF
dw     0x03FFF
dw     0x03FFF
dw     0x03FFF
dw     0x03FFF
```

To add instructions here, the *goto* instruction is converted to a *nop* by programming (clearing) all bits of the instruction and then putting in the required instruction, along with a jump over any remaining *dw 0×03FFF* instructions. For example, putting in a general case eight-bit addition will change the patch space to:

```
nop                              ;  Formerly "goto $ + 6"
movf    B, w                     ;  Formerly "dw 0×03FFF"
addwf   C, w                     ;  Formerly "dw 0×03FFF"
movwf   A                        ;  Formerly "dw 0×03FFF"
goto    $ + 2                    ;  Formerly "dw 0×03FFF"
dw      0x03FFF
```

Notice that even though only three instructions were added, five instruction addresses were affected. This makes this method of implementing patch space quite inefficient. In the PIC18Cxx, just the patch space instructions that are to be modified are changed and no space is required for jumping around instructions. For the same example in the PIC18Cxx, the patch space would be:

```
dw      0x0FFFF                  ;  nop
dw      0x0FFFF                  ;  nop
dw      0x0FFFF                  ;  nop
dw      0x0FFFF                  ;  nop
dw      0x0FFFF                  ;  nop
dw      0x0FFFF                  ;  nop
```

To add the three instructions to the patch space, the changes:

```
movf    B, w                     ;  Formerly "dw 0x0FFFF"
addwf   C, w                     ;  Formerly "dw 0x0FFFF"
movwf   A                        ;  Formerly "dw 0x0FFFF"
dw      0x0FFFF                  ;  nop
dw      0x0FFFF                  ;  nop
dw      0x0FFFF                  ;  nop
```

are made. In the PIC18Cxx, to add three instructions, only three instructions of the patch space are modified.

# 5

# PICmicro® MCU HARDWARE FEATURES

## CONTENTS AT A GLANCE

To get a good idea of what a remarkable device the PICmicro® MCU is, take a look at how the 8 pin PICmicro® MCUs are packaged in Fig. 5-1. The PIC12C508, although being the device with the fewest built-in features, contains 11 of the PICmicro® MCU hardware features that are presented in this chapter. The PIC12C673 shown in Figure 5-1 has everything except serial I/O and bus interfaces. As time has passed, the PICmicro® MCU families have been continually improved and updated with new features and expanded memory.

The next chapter shows how the basic digital I/O functions of the PICmicro® MCU can be used to simulate basic logic and chip functions, but many PICmicro® MCU part numbers are available with advanced interface functions to make application development easier.

As time has gone on, more PICmicro® MCU part numbers with advanced features and lower costs have been released. This process of providing enhanced features has continued to the point where PICmicro® MCUs with advanced interfaces have become cheaper than PICmicro® MCUs without these features with external hardware to provide required function.

When I wrote the first edition of this book, I tended to focus on the PIC16F84, which is a Flash/EEPROM-based program memory part with no I/O features because it was cheap, widely available, and tended to be the device most new users of the PICmicro® MCU turned to. With this edition, I have expanded the part numbers that I have used considerably and have tried to take advantage of as many built-in features as possible.

When presenting the different hardware features, I also include sample code for accessing the features. The code is written to allow easy "cut and pasting." To help facilitate this, I have written the code without constants and specific variable names. When you copy these code snippets, notice that labels should be replaced with application specific values.

I have also included the snippet source on the CD-ROM in the "PICmicro® MCU Interfaces" page. This will allow you to more easily cut and paste the snippets into your application.

# Power Input and Decoupling

The three basic aspects of application hardware design for the PICmicro® MCU power, reset and clocking, are very simple and extremely robust. The PICmicro® MCU is extremely tolerant of wide variances of power input and is quite tolerant of noise or sags in the supplied power. This tolerance makes the PICmicro® MCU ideal for learning about digital electronics, microcontroller application development, and applications where the quality of power cannot be guaranteed.

Connecting a PICmicro® MCU is very simple. It only requires a 0.0.1-$\mu$F to 0.1-$\mu$F decoupling capacitor across the $V_{dd}$ and $V_{ss}$ pins. A typical power connection is shown in Fig. 5-2.

A few points about Fig. 5-2; the first is, my bias towards my TTL education will show through as I talk about power. In TTL, +5-V power is known as $V_{cc}$ and the common negative connection, known as *ground* or *Gnd,* is $V_{ss}$. This book uses the TTL conventions of $V_{cc}$ and *Gnd* because most people are familiar with these terms. I believe that *Gnd* makes the most sense in digital circuits.

The decoupling capacitor is used to filter the voltage both within and without the PICmicro® MCU. During the transition of the circuits from one state to another (and from low-to-high current requirements), the internal (and external) voltages inside the chip will fluctuate, which could cause the chip to lock up, reset, or behave unpredictably in other ways.

**Figure 5-1** 8 Pin PICmicro® MCU's in PTH OTP, SOLC, and windowed PTH packages

Figure 5-2 shows the decoupling capacitor to be polarized. The most typically used values for the decoupling capacitors are 0.01 to 0.1 $\mu$F. Personally, I prefer using a 0.1-$\mu$F tantalum capacitor because it gives some extra capacity to guard against electrical upsets and can respond to upsets very quickly due to its *low equivalent series resistance* (ESR).

The capacitor is shown as being polarized because tantalum capacitors have to be installed correctly although non-polarized capacitors such as ceramic disk or polyester capacitors can be used for this purpose. Electrolytic capacitors should not be used for this purpose because they cannot react quickly enough to quickly fluctuating voltage levels. Tantalum capacitors are the best, but watch for a few things.

First, watch the polarity of the capacitor. Tantalum capacitors inserted backwards, like electrolytic capacitors, can catch fire or explode. This is not an issue with ceramic disk or polyester caps.

Second, be sure to derate the specified voltage of the part to 40% or less for your application. *Derating capacitors* means that instead of using them at their rated voltage, you should choose capacitors that are rated at about three times (or more) of the specified volt-

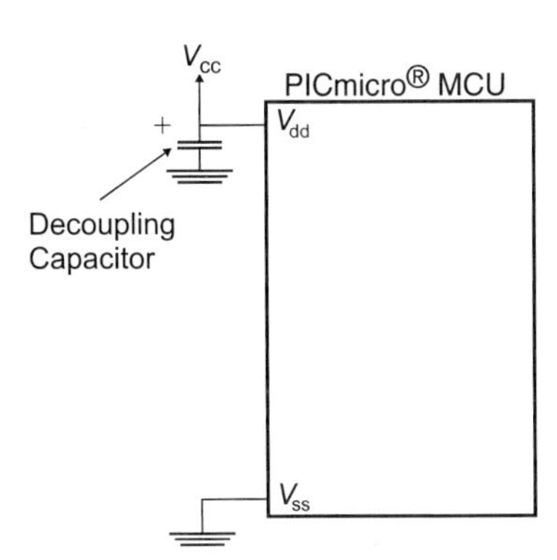


**Figure 5-2** PICmicro® MCU power connections

age for the application. For my applications, which use tantalum decoupling capacitors at 5 volts, I use parts that are rated at 16 volts (which is a derated value less than one third of the rated value).

When tantalum capacitors are first powered up in an application, voltage spikes (which are caused by power supply start up and current transients with in chips) can cause them to develop "pin-hole" breakthroughs in the dielectric, which can cause the plates to touch. When the plates touch, heat is generated by the short circuit, which boils away more dielectric. This process snowballs until the part explodes or catches fire. By derating the capacitor values, the dielectric layer is thicker, minimizing the opportunity for pinholes to develop.

Decoupling capacitors should be placed across $V_{cc}$ and Gnd of all chips, which can develop significant current transients. This does complicate the wiring somewhat, but does make for much more reliable applications. Notice that many chips with $V_{cc}$ and Gnd at opposite corners can use sockets with capacitors built in. Chapter 15 shows how decoupling caps are required for PICmicro® MCU applications.

The actual power circuit required for the PICmicro® MCU is very simple; in the McGraw-Hill *Handbook of Microcontrollers,* I stated that if you had anything more complex, you should find another microcontroller. I still believe this to be true and unless there are very concrete reasons why you want to use a microcontroller (or any other device) that requires more than just five volts of input power and a decoupling capacitor; you should find another device to use. This device should use the same single power voltage input as the rest of the chips in the application and only require one standard decoupling capacitor.

Standard PICmicro® MCUs are designed for power anywhere from 4.0 to 6.0 volts. Some PICmicro® MCUs have been "qualified" to run from 2.0 to 6.0 volts and are identified as having this capability as being "low-voltage" devices. These low-voltage parts are identical to the high-voltage supply parts except that they have been tested at the factory to run with input voltages down to 2.0 volts.

Thus, most high-voltage devices will have the capability of running a low voltage. You should refrain from taking advantage of this capability because you will find that some parts do not work with lower voltages applied to $V_{dd}$. Microchip will not support problems with components that have problems at low voltages, unless the components are qualified for this type of application.

Notice that the brown-out reset built into many PICmicro® MCUs is designed to become active at 4.5 volts. This makes the brown-out reset incompatible with most low-voltage applications, although some PICmicro® MCUs have a programmable brown-out reset voltage level to allow different power voltage inputs.

If you are working with a PICmicro® MCU in a low-voltage application that does not have the programmable brown-out reset capability, you will have to develop your own brown-out detect circuits, such as the one shown in Fig. 5-3.

In Fig. 5-3, if $V_{dd}$ drops below the characteristic voltage of the Zener diode, then _MCLR will be pulled low and the PICmicro® MCU will reset.

Low-voltage PICmicro® MCU parts are identified by the addition of the letter *L* before the *C* or *F* in its part number.

It is probably funny to hear this, but the biggest problem I have had (and have heard from other people) with PICmicro® MCU power is when the chip is plugged into the application backwards. This might sound like a strange problem, but only if you've never worked on a failing application that is driving you crazy and it is 3:00 A.M.

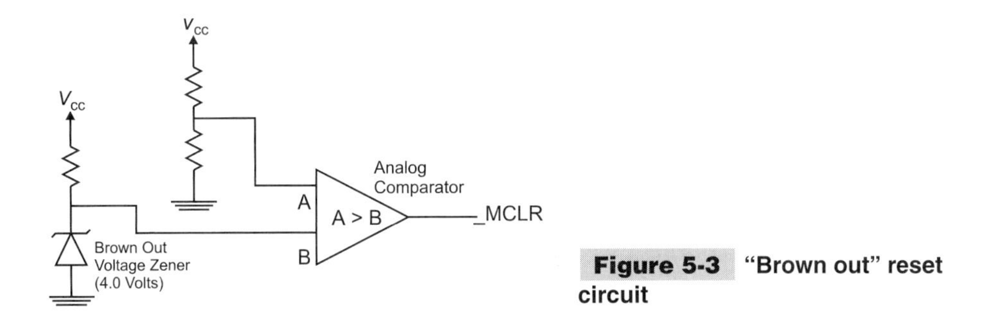


**Figure 5-3** "Brown out" reset circuit

As you get more and more frustrated and tired, the notch that indicates the end of the part, which has pin 1, seems to jump to the other end of the chip.

As I said, I have heard of many people; including myself that have done this and found that after the PICmicro® MCU cools off (it can get very hot when it's plugged in backwards) and is placed correctly in the circuit, it will continue to run perfectly and can be reprogrammed. It seems that unless the chip is allowed to get so hot that the plastic top pops off (which is possible), the PICmicro® MCU will work fine after this type of abuse.

This leads to a couple of points. The first point is the unconventional pinout on all PICmicro® MCU packages. The 18-pin devices (such as the 16F84) are very susceptible to the problem of applying power incorrectly and care must be taken to ensure that the part isn't plugged in backwards. I have not (nor do I know of anybody that has) done an experiment on how many times a PICmicro® MCU can be plugged in backwards, but I know doing this will decrease the reliability of the chip. If you have a part that was plugged in backwards, do not ship it out in a product. Any part that goes through this situation should not be considered "as good as new," despite what your experiences might lead you to believe.

The second point about this is that your power supply should be capable of *crowbarring* (shutting down) if the current draw increases dramatically (such as in the case of a reversed PICmicro® MCU). This is one of the reasons why I like the 78(L)xx voltage regulators; they might be more expensive than some other parts, but they won't burn out when they experience overcurrent conditions or cause reversed parts that they are driving to burn out.

## HIGH-VOLTAGE DEVICES

As I go through the practical issues of the PICmicro® MCU, one aspect always seems annoying: creating a power-supply circuit that has a voltage regulator with sufficient current rating to drive the circuit. Although this is not terribly difficult to produce, it can take up valuable real estate and drive up the cost of your application. The appendices describe different power supplies and how they are used with the PICmicro® MCU.

As I write this, the PIC16HV540 has just become available with a built-in voltage regulator that allows the PICmicro® MCU to be driven without any external regulators for battery application or poorly regulated power input. The PICmicro® MCU itself is pin and program compatible with the PIC16F54, which will allow the use of the PICmicro® MCU in applications that you probably haven't considered before. To support the voltage regulator, there are a few tricks that you should be aware of, as well as some enhanced features that affect the operation of the part. Other than these, the PIC16HV540 works identically to the PIC16C54.

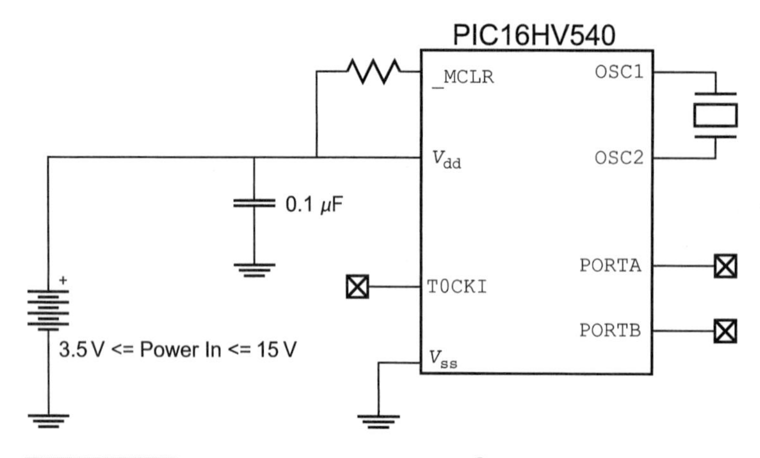


**Figure 5-4** **High-voltage PICmicro® MCU connections**

To connect a PIC16HV540 to a battery, the circuit can be as simple as Fig. 5-4.

Figure 5-4 shows the PIC16HV540 connected directly to a battery with only a 0.1-$\mu$F decoupling capacitor. I don't show a switch because *sleep* can be used to turn off the device and put it into a low current (no more than 14 $\mu$A required) mode. Wake up from sleep can be accomplished either by watchdog timer time out, _MCLR reset, or a PORTB pin change.

The way I've drawn the PIC16HV540, you might think that the device is similar to a PIC16C54 with a voltage regulator in front of it (Fig. 5-5).

This is not quite true because the I/O port pins can use the device's regulated voltage for the input voltage. The PORTA pins can provide up to the regulated voltage, but PORTB provides swings from ground to the input voltage. The actual device's block diagram looks like Fig. 5-6.

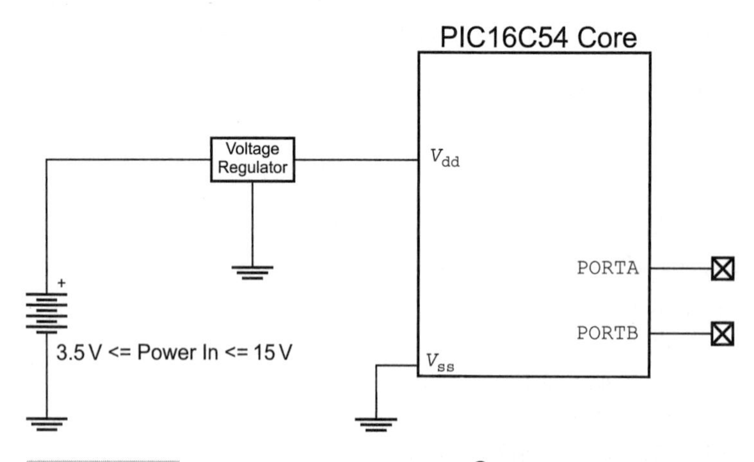


**Figure 5-5** **High-voltage PICmicro® MCU analogous circuit**

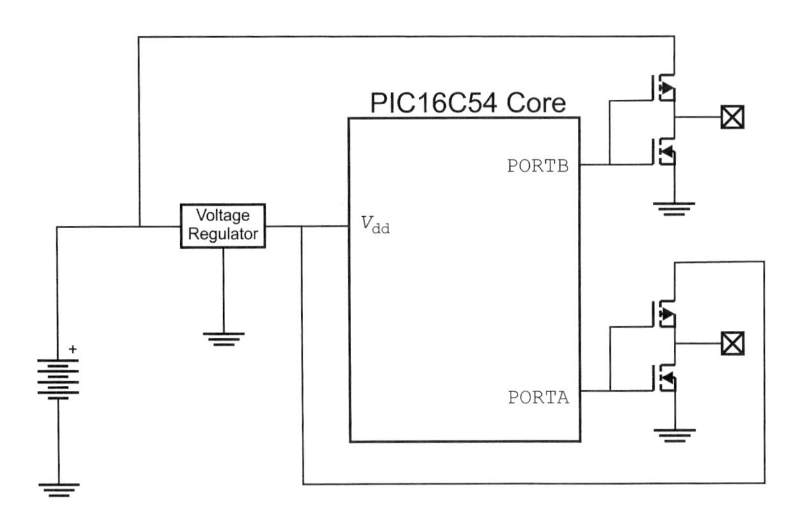


**Figure 5-6 Actual high-voltage PICmicro® MCU circuit**

In the actual circuit, PORTA can be used to power +5-/+3-volt TTL CMOS devices, and PORTB is well suited for high-voltage I/O. PORTB's threshold voltage is similar to the PIC16C54's (i.e., anything greater than 2.5 volts is a 1) and it can be used for buttons and LED I/O.

The voltage regulator can work as either a 5- or 3-volt regulator by setting or resetting, respectively, the RL bit of the OPTION2 register, which is in the OPTION/TRIS address space of the low-end PICmicro® MCU processor. This register is an auxiliary-configuration fuses register, which can be modified within an application. The bits of the OPTION2 register are defined in Table 5-1.

**TABLE 5-1 PIC16HV540 "OPTION2" Register Definition**

| BIT | DESCRIPTION |
|---|---|
| 7-6 | Unused |
| 5 | WPC—When set, device will Wake Up On RBO-RB3 changing |
| 4 | SWE—Software Watchdog Timer. If the WDT is not Enabled in the Configuration Fuses, setting this bit will enable it in software |
| 3 | RL—Regulated voltage select bit (Set for 5 Volts, Reset for 3 Volts) |
| 2 | SL—Sleep Voltage Level Setting (if Set, use "RL" Voltage, when Reset, use 3 Volts) |
| 1 | BL—Brown Out Voltage Select. When Set—3.1 volts for 5 Volt Operation and when Reset—2.2 Volts for 3 Volt Operation |
| 0 | BE—Brown Out Checking Enabled when Set. |

OPTION2 is written to using the TRIS instruction as:

```
TRIS   7
```

Or

```
TRIS   OPTION2
```

When the PIC16HV540 is powered on, the RL bit is set, which selects 5-volt output from the voltage regulator. A voltage less than 5 volts can be input, but it must be greater than 3.1 volts to avoid brown-out reset (which is enabled on power up) from holding the device reset. If you are powering the PIC16HV540 form a source that is less than +5 volts, you should first set the PICmicro® MCU for 3-volt operation.

The OPTION2 register description (and throughout this section) has introduced a number of concepts that haven't been covered yet. If you go through this chapter, you'll find introductions to the watchdog timer, sleep, and the I/O pins that will make the information provided in this section easier to understand.

# Reset

*Reset* in the PICmicro® MCU is very simple and easy to implement. As well, over the past few years, the task of implementing reset has been simplified even more with the addition of internal reset and a built-in *Brown-Out Detect (BOD)* function in some of the new parts. These functions, along with a 72-ms power-up delay function (the PWRTE bit of the CONFIGURATION register) make it easy to come up with an application with good reset characteristics in a variety of different situations, with little or no external hardware.

The internal reset frees the _MCLR pin from being used as the reset control source and allows it to be used as an input pin. When this feature is enabled, the device reset is active as long as more than 4.0 volts is available at $V_{dd}$ for regular parts; 2.0 volts is available in extended-voltage parts. The freed _MCLR pin can only be used as an input (no output drivers are built in) and does not have the clamping diodes of the other PICmicro® MCU I/O pins.

The brown-out reset function, if it is on the PICmicro® MCU that you are working with and is enabled, will cause the internal reset circuitry to become active when the $V_{dd}$ voltage becomes less than 4.0 volts in 5-volt applications. This can be useful in battery-powered applications, where a drop in battery power can cause intermittent application execution. This function is becoming available in most new PICmicro® MCU devices and is enabled through the configuration register. It cannot be accessed within the application code, except for checking the PCON_BOD bit.

The PWRTE function will delay start of PICmicro® MCU application for 72 ms. This feature is designed to allow the PICmicro® MCU internal clock to stabilize before the application starts executing. The PWRTE bit of the configuration register should always be active, unless the application has a clock external to the PICmicro® MCU, that is stable when reset becomes disabled and the PICmicro® MCU starts executing.

# Watchdog Timer

Environments with "noisy" power or large electrical fields can cause the circuits within the PICmicro® MCU to become "upset" and stop executing properly. These events can cause the program counter to change to an invalid address or the instruction decoder to process an instruction improperly. Often, when this happens, the PICmicro® MCU locks up and stops executing the application. To help counter this problem, Microchip has designed the PICmicro® MCU with a *WatchDog Timer (WDT),* which will reset the PICmicro® MCU if normal application execution is lost and the PICmicro® MCU starts executing incorrectly or locks up.

The watchdog timer is an 18-ms delay that will reset the PICmicro® MCU if it times out. Normally, in an application, it is reset before timing out by executing a *clrwdt* instruction. The block diagram of the WDT is shown in Fig. 5-7.

The WDT oscillator in Fig. 5-7 is an RC oscillator, which drives a counter. The *O/F (overflow)* output of the counter is optionally passed to a prescaler before going to the PICmicro® MCU's reset circuit. The prescaler is covered in more detail later in this chapter, but its basic purpose is to count overflow events, either from the PICmicro® MCU's TMR0 or the watchdog timer. The overflow is passed to the PICmicro® MCU's reset circuit when the specified number of watchdog timer overflows have occurred. The prescaler allows watchdog-timer reset delays from 18 ms to 2.3 s.

When the watchdog timer causes a reset in the PICmicro® MCU, the _TO bit of the status register is reset. In the application, if watchdog-timer resets are used, then the _TO bit should be checked to determine if the current file register settings are properly set from previous application execution.

I recommend that the watchdog timer should be reset by the *clrwdt* instruction after half of the reset period has passed. The nominal error in the RC oscillator used for the watchdog-timer function is 20%, which means that watchdog-timer timeouts can occur anywhere from 14 to 22 ms, (when no prescaler is used). To be on the safe side, executing *clrwdt* every 9 ms in this situation will avoid any potential invalid watchdog-timer resets.

The watchdog timer is enabled from within the configuration word and cannot be disabled within the application. You must be very careful to avoid enabling the watchdog timer unless you have provided support for it in the application code. *Providing support*

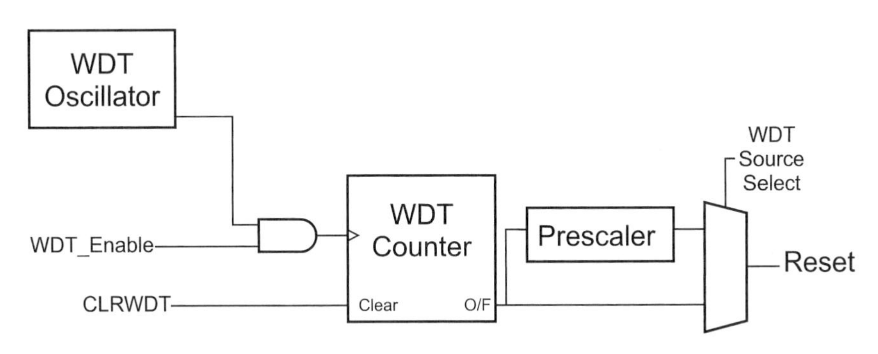


**Figure 5-7 PICmicro® MCU watchdog timer block diagram**

for the watchdog timer means that the *clrwdt* instruction is executed repeatedly to avoid the watchdog timer from resetting the PICmicro® MCU unexpectedly.

In many PICmicro® MCUs, the watchdog-timer enable bit of the configuration word is positive active and enabled when the bit is set (i.e., unprogrammed). This can cause some problems with new application developers that forget to disable the watchdog timer explicitly. Later, the book covers configuration register programming, but I wanted to recommend that you always have a _WDT_OFF parameter in your __CONFIG statement, unless you are developing applications that use the watchdog timer. This will avoid any potential problems with the WDT from being inadvertently set and result in the application begin continually reset.

If you are going to create an application that uses the watchdog timer, I recommend that it should only be enabled when you are about to release the application. It can be a problem during debug and can cause the PICmicro® MCU to reset itself when you least expect it, making it difficult to debug the application.

# System Clock/Oscillators

There are five different methods of clocking the PICmicro® MCU. The different options are designed to fill different application requirements for cost, speed, and accuracy. A few applications that require extreme accuracy allow the use of cheaper clock designs.

The clock options are:

- Internal clocking
- R/C networks
- Crystals
- Ceramic resonators
- External oscillators

The built-in oscillator is becoming more prevalent in low-cost PICmicro® MCUs. This option consists of a capacitor and variable resistor for the oscillator.

The OSCAL register (Fig. 5-8) is loaded with a calibration value that is provided by Microchip. This type of oscillator will have an accuracy of 1.5% from 4 MHz or better while running at 4 MHz.

This single oscillator speed of 4 MHz might seem like an unreasonable restriction, but it is actually quite a useful speed for many applications. The internal oscillator is enabled within the configuration register and cannot be changed during application execution.

The internal oscillator does simplify application wiring. If it is selected, the pins that are normally used by it can be used as I/O pins. For such devices as the eight-pin 12C5xx parts, this drastically improves the efficiency of the device in the application's circuit.

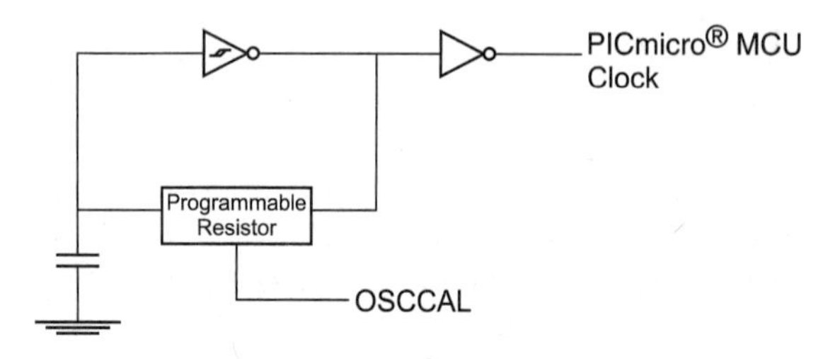


**Figure 5-8** PICmicro® MCU built in oscillator

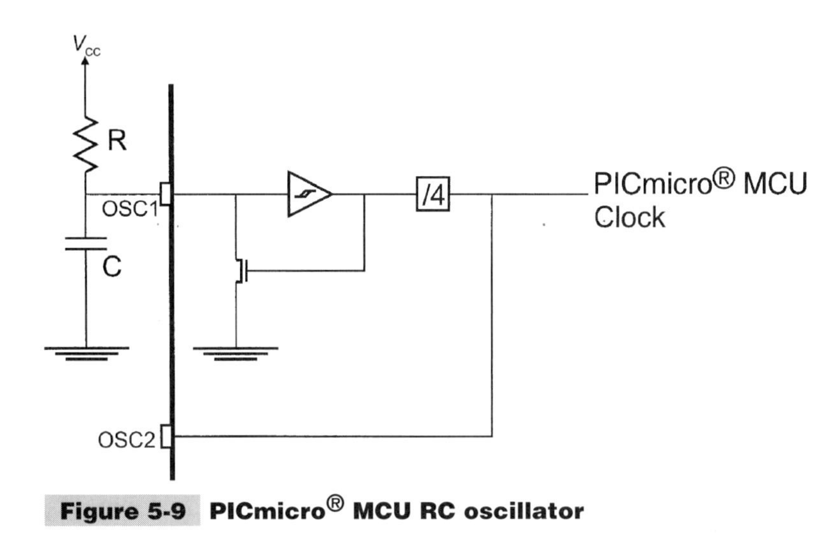


**Figure 5-9 PICmicro® MCU RC oscillator**

The second type of oscillator is the external RC oscillator in which a resistor capacitor network provides the clocking for the PICmicro® MCU (Fig. 5-9).

The resistor-capacitor charging/discharging voltage is buffered through a Schmidt trigger noninverting buffer, which is used to enable or disable an N-channel MOSFET transistor pull-down switch.

The advantages of the RC oscillator is the cheapness for which it can be added to an application (the cost of the RC components can be just a few cents) and the instruction clock is output for use by other devices in the circuit. The disadvantages of this method are the relatively low speed that this oscillator runs at (generally to a maximum of 1 MHz) and the inaccuracy of this method (an error of 30% to the target speed are not unheard of). This method is best used for low-cost, timing-insensitive applications.

When using this method, it is important to only use the RC values specified on the data sheet of the PICmicro® MCU that you want to use. Selecting the wrong RC values could result in the oscillator to run erratically, at a speed that is orders of a magnitude off from the target speed, or to not run at all.

Crystals and ceramic resonators use a similar connection scheme for operation. The crystal or ceramic resonator is wired into the circuit as shown in Fig. 5-10.

Crystals and ceramic resonators delay the propagation of a signal a set amount of time. This set amount is dependent on how the crystal is cut. As well, for best results, a parallel circuit crystal should be used.

The two capacitors shown in Fig. 5-10 are connected to one side of the crystal or ceramic resonator and the other to ground. The values of the two capacitors are specified in the PICmicro® MCU device's data sheet and usually match one another. You might find that for some speeds and devices, different values are specified for the two capacitors.

Three speed ranges are defined for each device, with the speed specification defining the current output in the PICmicro® MCU's oscillator circuit. Microchip refers to these speed ranges as *oscillator types,* but I prefer thinking of them as *speed ranges* because this seems to be a more accurate description of how they work.

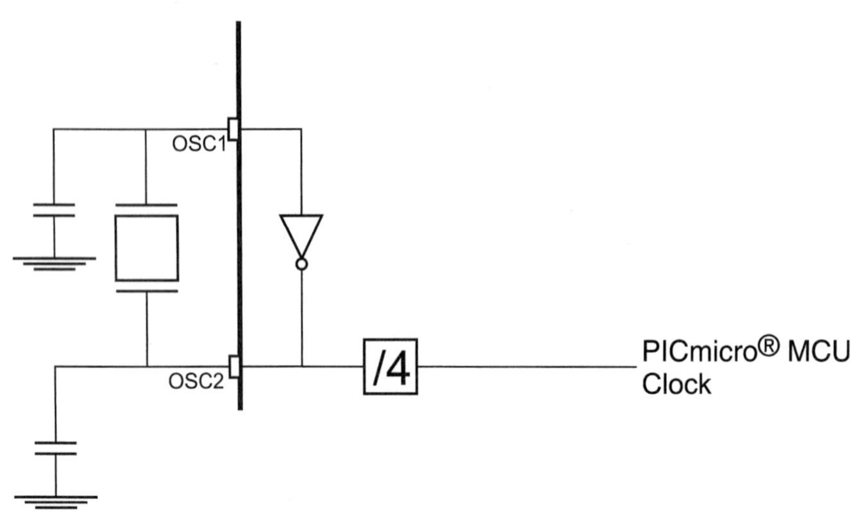


**Figure 5-10 PICmicro® MCU crystal oscillator**

The speed ranges are shown in Table 5-2. These speed ranges are selected in the configuration register.

Using the crystal or ceramic resonator, the OSC2 pin can be used to drive one CMOS input (Fig. 5-11).

Personally, I wouldn't recommend implementing this in an application because the capacitance of the CMOS buffer could affect the operation of the PICmicro® MCU's oscillator (i.e., causing it to not run at all). I recommend that if you want to drive multiple devices from a single clock that you use an external oscillator and *fan out* the clock to the PICmicro® MCU and other devices.

A ceramic resonator behaves similarly to a crystal, except that they are much more robust (i.e., can withstand more severe physical shocks) than crystals and, in large quantities, they can be much cheaper. Many ceramic resonators are available as three-pin devices, which have the external capacitors built in to them, meaning that along with just wiring them to the PICmicro® MCU's OSC1 and OSC2 pins, just a ground connection is required.

The downsides of ceramic resonators is that their accuracy is not as good as crystals (usually accurate to 0.5 percent versus a crystal accuracy of 0.02 to 0.1 percent) and some devices with built-in capacitors might not be suitable for use with the PICmicro® MCU.

**TABLE 5-2 PICmicro® MCU Oscillator Frequency Ranges**

| RANGE | FREQUENCY |
|---|---|
| LP | 0–200 kHz |
| XT | 200 kHz–4 MHz |
| HS | 4 MHz–20 MHz (or the Device Maximum) |

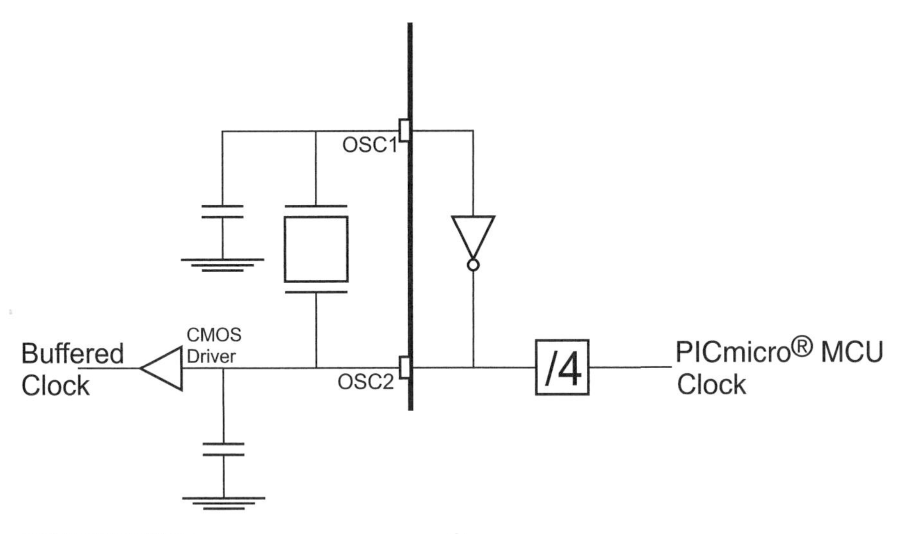


**Figure 5-11** **Buffered PICmicro® MCU crystal oscillator**

The last type of oscillator is the external oscillator, which is driven directly into the OSC1 pin (Fig. 5-12).

In Fig. 5-12, the OSC2 pin can be used to redrive the clock (although inverted) to other devices. The LP, XT, or HS speed selections are made to correspond to the input oscillator speed.

If you subscribe to the PICList list, you'll periodically see e-mails from people who drive their PICmicro® MCUs much beyond the rated speed. This is possible for most devices because of the Microchip's design qualification, which ensures that all PICmicro® MCUs of a given part number run at the rated speeds. It is not recommended that you count on your PICmicro® MCUs to run at faster-than-rated speeds because this capability is not guaranteed for all PICmicro® MCU's; you will find devices that will not run at the speeds required by your application.

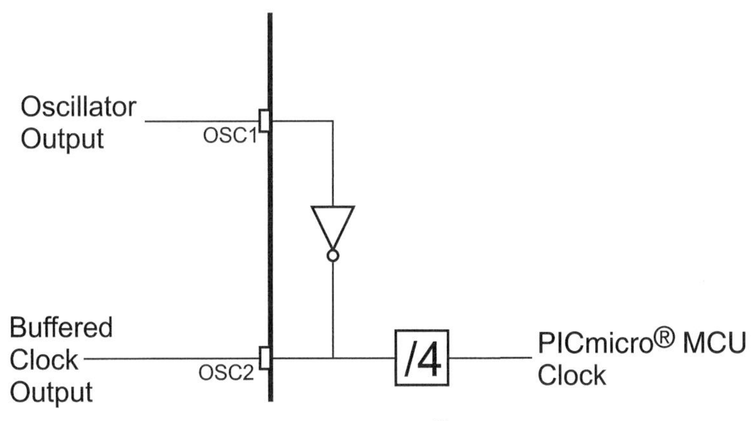


**Figure 5-12** **External PICmicro® MCU oscillator**

When designing and laying your application circuit, always remember to keep the parts of the oscillator as close as possible to the PICmicro and circuit grounds. When the specified parts are used for the oscillators and they are kept close to the PICmicro® MCU, you'll be impressed with the stability and robustness of the PICmicro® MCU's oscillator and instruction clocking.

## BUILT-IN OSCILLATORS

One of the features I like that Microchip is adding to more new PICmicro® MCUs is a built-in RC oscillator that can be selected when the device is programmed. This 4-MHz oscillator avoids the need for adding external parts and frees up pins for use in interfacing to hardware. The inclusion of the device is one of the reasons that the eight-pin PICmicro® MCUs (the 12Cxxx family) can use up to six I/O pins in an application.

The built-in oscillator consists of an RC network that provides a delay for a ring oscillator (Fig. 5-13).

In this circuit, the programmable resistor and capacitor delay the voltage output from the inverter from reaching its input for 250 ns.

The OSCAL register holds the value setting for the programmable resistor. The calibration value to be put into this register is determined by Microchip when the part is tested at the factory, the correct value is calculated and burned into the chip at its reset vector address. The mechanics of working with the OSCAL value is covered in a later section of this chapter.

The 4-MHz oscillator speed was determined by Microchip to be a good clock speed for many applications. With the calibration value and OSCAL register, the actual clock value can be expected to be within 1.5 percent of 4 MHz.

## 18Cxx OSCILLATOR AND INSTRUCTION CLOCK

In the year and a half before this book's release (mid 2000), I found that the PIC18Cxx's clock was described in a number of ways, some of which were contradictory and some that made the PIC18Cxx oscillator to seem better than possible. I wanted to use this section to explain how the oscillator of the 18Cxx works and what new modes you can take advantage of in your applications.

The typical PICmicro® MCU has four oscillator modes, which are selected from the configuration fuses when the PICmicro® MCU starts up. The PIC18Cxx has one more oscillator type, along with the ability to change an unused OSC2 pin into an additional I/O pin. A PLL clock four-times multiplier circuit is available, which allows the PICmicro® MCU to run with a one-instruction-cycle-per-clock cycle. As well as the PIC clock multi-

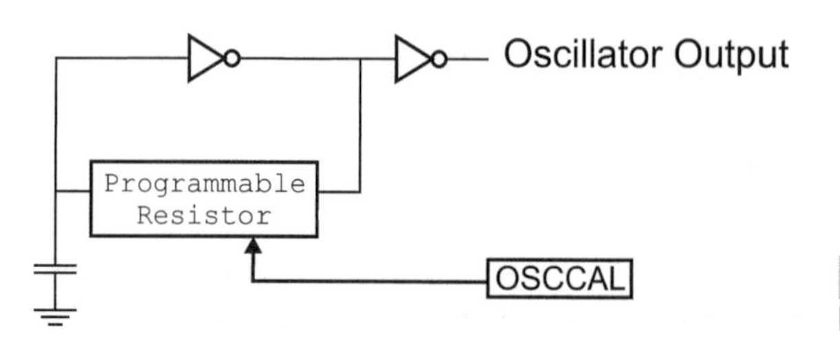


**Figure 5-13** **PICmicro® MCU built-in oscillator**

plier, the PICmicro® MCU can be driven by the TMR1 oscillator for reduced power or faster application execution. This might seem like a dizzying array of options:

- RC oscillator
- LP oscillator
- HS oscillator
- XT oscillator
- External oscillator
- TMR1 clock

LP and XT modes execute exactly the same way on the PIC18Cxx as the other PICmicro® MCUs. RC can work exactly the same with a one-quarter-speed clock driven out of the OS2 pin or can have the OSC2 pin changed to the RA6 I/O pin and is known as *RCIO mode*. The external oscillator option will take in an external clock signal and output a one-quarter-speed clock on OSC2, unless the OSC2 pin is to be used as RA6 (like the RC oscillator mode and known as *ECIO*). The external oscillator will work for all data speeds from DC to 40 MHz, which the 18Cxx can run at.

One of the most interesting features of the PIC18Cxx is the HS mode, which can run like a typical PICmicro® MCU, from 4 MHz to 40 MHz. An optional *Phased-Locked Loop (PLL)* clock multiplier can be added to the external clock (with the external clock limited to 10 MHz). This feature (known as *HSPLL*) allows a 10-MHz PIC18Cxx to run as if it were being driven with a 40-MHz clock. The advantages of this feature include reduced EMI emissions from the PICmicro® MCU, along with reduced the oscillator's power consumption.

The phase-locked loop cannot be used to multiply a 40-MHz clock by four to get the equivalent of a PICmicro® MCU running at 160 MHz (but, it would be nice).

The RC, RCIO, XT, LP, ES, ESIO, HS, and HSPLL oscillator modes are selected at programming time and selected within the PIC18Cxx's configuration fuses.

The last oscillator mode consists of software that enables the TMR1 oscillator as the PICmicro® MCU's processor clock source. This feature allows application execution with I/O polling and interrupts to continue, except at a vastly reduced speed (and power consumption). This mode is enabled and disabled by setting and resetting, respectively, the SCS bit of the PIC18Cxx's OSCON register. For this mode to work, the OSCON bit of the configuration fuses must be reset.

When the TMR1 oscillator is enabled (by setting the SCS bit), execution moves over immediately to the TMR1 clock and the standard oscillator is shut down. This transition is very fast, with only eight TMR1 clock cycles lost before execution resumes with TMR1 as the clock source.

When transitioning from TMR1 to the standard oscillator, the oscillator is restarted with a 1024-cycle delay for the clock to stabilize before resuming execution.

The oscillator circuit in the PIC18Cxx appears in block diagram form (Fig. 5-14).

# Configuration Registers

One feature of the PICmicro® MCU that is somewhat unusual compared to other microcontrollers is its configuration register. This instruction-sized word, which is "outside" of

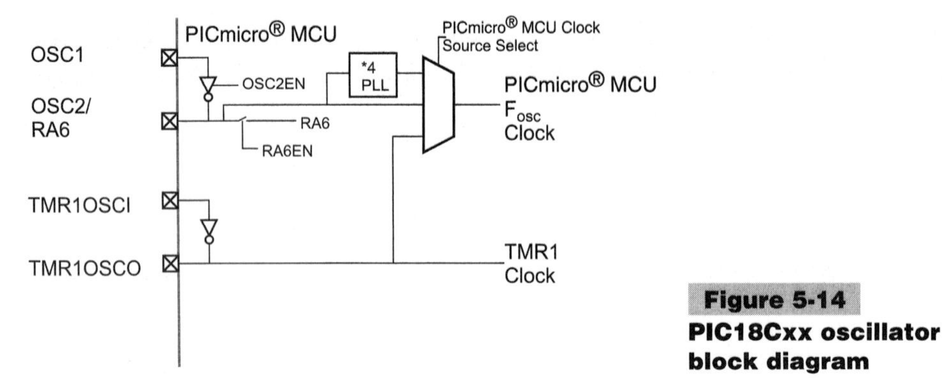


**Figure 5-14**
**PIC18Cxx oscillator block diagram**

the program memory area in the low-end and mid-range PICmicro® MCUs, is responsible for specifying:

- Oscillator mode used
- Program memory protection
- Reset parameters
- Watchdog timer
- 16F87x debug mode

The value for the configuration register is written when the device is being programmed, but it cannot be accessed by the application in the low-end and mid-range parts. During programming, the PICmicro® MCU's program counter is used to access the configuration register's address. The configuration register is at address 0x0FF for the low-end devices and in the mid-range PICmicro® MCUs; the configuration register is always at 0x02007. The register bits are used to enable hardware or set hardware states, but cannot be read by the program and their states can only be directly determined within software by the operation of the application.

In the PIC17Cxx, the configuration register is located at addresses 0x0FE00 to 0x0FE07 for the low byte, or the register and addresses 0x0FE08 to 0x0FE0F for the high byte. These registers can be accessed during application execution by a *tablrd* instruction. At the end of the *tablrd* instruction, the address specified for the configuration byte will be assessed in the TABLATL register with TABATH always being 0x0FE.

In each PICmicro® MCU's MPLAB device .INC files is a list of parameters for the different options. These parameters are used with the __CONFIG statement of an assembler file. The recommendations I present here are repeated throughout the book. Note that there are *two* underscore characters before the "CONFIG" characters.

The __CONFIG statement is used to AND together all the constants in the .INC file. This is because some bits will be left unprogrammed (set) and others will be programmed (reset). To simplify putting them all together, when creating the configuration constants, they are ANDed together.

For example, if you had a simple mid-range PICmicro® MCU with only two parameters (which doesn't exist), the watchdog timer and code protection, the .INC file values are shown in Table 5-3.

To set up a PICmicro® MCU with the watchdog timer enabled (_WDT_ON) and code protection on (_LP_ON), the configuration statement would be:

```
__CONFIG _WDT_ON & _LP_ON
```

**TABLE 5-3 Sample PICmicro® MCU "Configuration Fuse" Values**

| | |
|---|---|
| _WDT_ON | 0x03FFF |
| _WDT_OFF | 0x03FFD |
| _LP_ON | 0x03FFE |
| _LP_OFF | 0x03FFF |

Notice that one value for each option has all the bits set. By ANDing the different parameter values with each other, the programmed (reset) bits should be specified correctly.

When creating an application, always use the __CONFIG statement because this will store the configuration values in the .HEX file for you at the correct address. When doing this, only use the Microchip .INC file values and don't try to calculate your own values. Chances are that you will make a mistake in coming up with your own and, in any case, using the Microchip values is easier than coming up with than on your own.

Personally, I prefer a programmer that picks up the __CONFIG values and loads them into the PICmicro® MCU automatically. I recommend that you should use a programmer that has feature as well. If you have to manually set the configuration fuses, you will forget to do it or select the wrong values. This is especially true when you are new to the PICmicro® MCU or are frustrated by problems. The Microchip PICSTART Plus and the programmer designs given in this book use the configuration values created from the __CONFIG statement.

The last recommendation is to always specify a state for all the configuration fuse options for the PICmicro® MCU using the __CONFIG statement. In many cases, the unprogrammed (set) state will not be at a default value that will be useful. I can't tell you how often I have been caught forgetting to specify _WDT_OFF (disable the watchdog timer), only to discover that the feature is enabled with the configuration register bit set.

# Sleep

In some applications (notably battery-powered circuits), it might be appropriate to shut down the PICmicro® MCU until it is required again. The PICmicro® MCU's *sleep* function and instruction provides this capability by turning off the oscillator and holding the PICmicro® MCU waiting for reset (_MCLR or WDT) an external interrupt, or some timer interrupts. Most internal timer interrupt requests are not able to become active because the PICmicro® MCU instruction clock driving the internal hardware clocks is shut off.

Turning off the instruction clock results in a current consumption measured in microamperes (as opposed to the milliamperes required by a normally running PICmicro® MCU).

Entering sleep is accomplished by simply executing the *sleep* instruction. Sleep can be terminated by the events listed in Table 5-4.

**TABLE 5-4 Sleep Termination Events and Execution Resume Addresses**

| EVENT | EXECUTION RESUME |
|---|---|
| _MCLR Reset | Reset Vector |
| WDT Reset | Reset Vector |
| External Interrupt | Next/Instructions or Interrupt Vector |
| TMR1 Interrupt | Next Instructions or Interrupt Vector |

The interrupt requests can only wake the device if the appropriate E bits are set. After the *sleep* instruction, the next instruction is always executed—even if the GIE bit is set. For this reason, a *nop* should always be placed after the sleep instruction to ensure that no invalid instruction is executed before the interrupt handler:

```
sleep
nop
```

For one of the different reset types to restart the PICmicro® MCU from sleep, execution restarts at the reset vector (address 0 in the mid-range). If an interrupt source is used, then if the GIE bit of INTCON is set, then after restarting the clock, the nop instruction will execute and execution will jump to the interrupt vector. If the GIE bit is reset, execution will continue after the nop instruction.

The clock restart from sleep will be similar to that of a power-on reset, with the clock executing for 1024 cycles before the *nop* instruction is executed, *Inst(PC + 1)* as shown in Fig. 5-15.

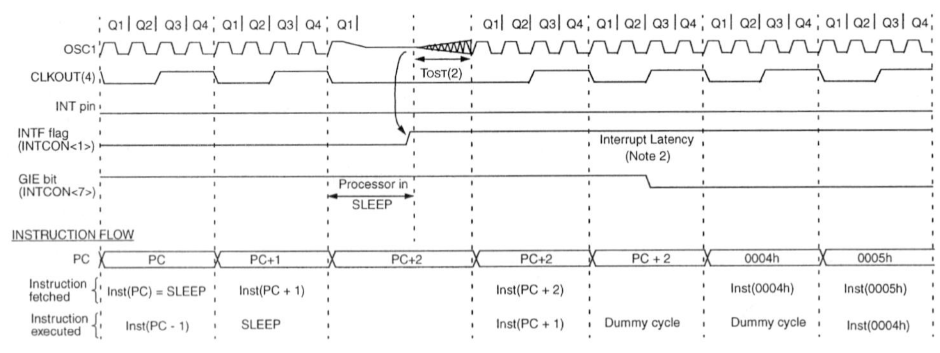


Note 1: XT, HS or LP oscillator mode assumed.
2: TOST = 1024TOSC (drawing not to scale) This delay will not be there for RC osc mode.
3: GIE = '1' assumed. In this case after wake- up, the processor jumps to the interrupt routine. If GIE = '0', execution will continue in-line.
4: CLKOUT is not available in these osc modes, but shown here for timing reference.

**Figure 5-15**

# Hardware and File Registers

If you have worked with other processors and computer systems, you are probably surprised by the close coupling and shared memory space of the processor's registers, hardware I/O registers, and variable RAM. This is a result of the small (five-bit addressing in low-end and seven bit for the mid range, eight bit for the PIC17Cxx and twelve bits for the PIC18Cxx) register space accessible to the processors. This aspect means that applications have to be written to take advantage of the characteristics of the register spaces.

In the mid-range PICmicro® MCU, each instruction that accesses a register contains the addresses within the given bank with a maximum bank size of seven bits, which allows as many as 128 different addresses. In each bank, the registers fall within four distinct groups:

- Processor registers
- I/O hardware registers
- Variable memory
- Shared (shadowed) variable memory

The processor registers consist of STATUS, PCL, PCLATH, (from the mid range), FSR, INDIF, and WREG (for the high end). These registers are always at the same addresses within the different PICmicro® MCU families. These addresses are shown in Table 5-5. These registers can be accessed from within any of the register banks. In the two previous chapters, I explained how these registers work.

The I/O hardware registers consist of the OPTION, TMR0, PORT, I/O pins and enable registers, INTCON, other interrupt control and flag registers, and any other hardware features built into the particular PICmicro® MCU. The important difference between these registers and processor registers is that, except for INTCON, these registers are bank specific. Although some conventions are used for the placement of these functions, for part numbers and specific functions, the registers are located in different addresses. The registers with conventions are shown in Table 5-6.

**TABLE 5-5 Base Register Addresses by PICmicro® MCU Architecture Family**

| | LOW-END | MID-RANGE | PIC17CXY | 18CXX |
|---|---|---|---|---|
| w | N/A | N/A | 0x00A | 0x0FE8 |
| STATUS | 0x003 | 0x003 | 0x004/0x006 | 0x0FD8 |
| PCL | 0x002 | 0x002 | 0x002 | 0x0FF9 |
| PCLATH | in "STATUS" | 0x00A | 0x003 | 0x0FFA |
| FSR | 0x004 | 0x004 | 0x001, 0x009 | 0x0FEA/0x0FE9 |
| INDF | 0x000 | 0x000 | 0x000, 0x008 | 0x0FEF |

**TABLE 5-6 I/O Register Addresses by PICmicro® MCU Architecture Family**

| | LOW-END | MID-RANGE | HIGH-END | 18CXX |
|---|---|---|---|---|
| OPTION | "Option" ins | 0x081 | 0x005 | 0x0FD0 |
| TMRO | 0x001 | 0x001 | 0x00B/0x00C | 0x0FD7/0x0FD6 |
| PORTA-PORTC | 0x005–0x007 | 0x005–0x007 | Varies | 0x0F82/0x0F80 |
| TRISA-TRISC | 'tris" ins | 0x085-0x087 | Varies | 0x0FD4/0x0FD2 |
| PORTD/TRISD | N/A | 0x008/0x088 | Varies | 0x0F83/0x0FD5 |
| PORTE/TRISE | N/A | 0x009/0x089 | Varies | 0x0F84/0x0FD6 |
| INTCON | N/A | 0x0xB | 0x007 | 0x0FF2 |
| OSCAL | 0x005 | N/A | N/A | N/A |

As time goes on and more features become standard, you'll probably see the mid-range PICmicro® MCU's standardize on a 32-byte processor and I/O hardware register block (also known as the *Special-Function Registers, SFRs*) at the start of each bank.

Above the processor and I/O hardware registers are the file registers or variable memory. This memory can be bank specific or shared between banks. In all PICmicro® MCUs are a number of bytes that are always available (shared, or what I call *shadowed*) across all the register banks. This memory is used to pass data between the banks, or, as I prefer to use them, to provide a common variable for sharing context register data during interrupts without having to change the bank specification in the status register. The shared memory is PICmicro® MCU part number specific and can be common across all banks or pairs of banks.

In the low-end PICmicro® MCUs, many devices have multiple banks, but these multiple banks are strictly used to provide additional file registers. Normally in these PICmicro® MCUs, the first 16 addresses of each bank (address 0 to 0x00F) are common with the lower 16 addresses of the other banks, which has file registers that are specific to them.

## ZERO REGISTERS

I don't really know if this qualifies as a "feature," but unused registers in a PICmicro® MCU's register map will return zero (0x000) when they are read. This capability can be useful in some applications.

*Zero registers* (undefined registers that return zero when read), are normally defined in the Microchip documentation as *shaded addresses* in the device register map documentation. Figure 5-16 shows the 16F84's register map with addresses 7 (PORTC in other PICmicro® MCU devices) in each bank shaded, indicating that they return zero when read.

Of course, when these registers are written to, their values are lost and not stored in the register. One might say that the information has "gone to the great bit bucket in the sky." I am hesitant to recommend using the zero registers when programming. It is important to note that, in different PICmicro® MCU part numbers, the zero registers are at different locations. Because of this, if code is transferred directly from one application to another and the zero register chosen is not available in the PICmicro® MCU device destination

| Bank 0 | Bank 1 |
|---|---|
| Addr - Reg | Addr - Reg |
| 00 - INDF<br>01 - TMR0<br>02 - PCL<br>03 - STATUS<br>04 - FSR<br>05 - PORTA<br>06 - PORTB | 80 - INDF<br>81 - OPTION<br>82 - PCL<br>83 - STATUS<br>84 - FSR<br>85 - TRISA<br>86 - TRISB |
| 07 - | 87 - |
| 08 - EEDATA<br>09 - EEADR<br>0A - PCLATH<br>0B - INTCON | 88 - EECON1<br>89 - EECON2<br>8A - PCLATH<br>8B - INTCON |
| 0C-4F Shared - File Regs | 8C-CF Shared - File Regs |
| 50-7F - Unused | D0-FF - Unused |

Shaded areas indicate unused registers - 0x000 Returned when these registers are read

**Figure 5-16** **PIC16F84 register map**

(for example, a valid file or hardware register is at this location), then the code will not work correctly. Instead of using a hardware zero register, I would recommend that a file register is defined and initialized for the purpose of always returning zero.

# Parallel Input/Output

The most basic way to get data in and out of the PICmicro® MCU is via the parallel I/O bits that are located in the ports. In many PICmicro® MCUs, these pins have peripherals "behind" them to provide advanced I/O capabilities. Despite this capability, in virtually every PICmicro® MCU application that you create, the straight I/O port functions will be required.

The PICmicro® MCU's typical I/O pin is capable of being either an input pin or an output pin. When in Output mode, the pins are able to source or sink roughly 20 mA of current.

The block diagram of a PICmicro® MCU I/O pin is shown in Fig. 5-17. Each register port consists of a number of these circuits, one for each I/O pin.

Depending on your previous experiences, this I/O pin can look as if it is very complex, needlessly complex, or pretty basic. When you compare this to other microcontrollers, you'll find that the PICmicro® MCU's I/O pins are very typical of some devices and more complex than others. Regardless of your own feelings about the I/O pins, you should be aware of a few aspects of the pins.

I/O pins are associated to the bit number of the port where they belong. The maximum size for an I/O port is eight bits (or pins) for one byte. The convention used by Microchip is to label the pins according to their bit number and port where they're associated.

The convention is:

```
R%#
```

Where % is the port letter (port A, port B, etc.) and # is the bit number.

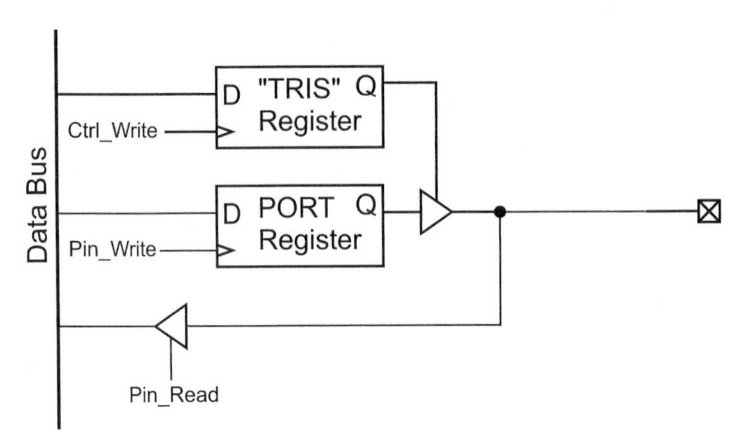


**Figure 5-17** Standard PICmicro® MCU I/O pin block diagram

Using this convention, *RB3* is port B, pin 3. I will use it throughout the book to label individual pins. In some places, you will see the convention:

```
PORT%.#
```

which uses the same values for % and # as the *R%#* format.

The *TRIS (TRI-State buffer enable)* register is used to control the operating mode of the I/O pin. When the register is loaded with a 1 (which is the power-up default), the pin is input only (or in *Input mode*), with the tristate buffer disabled and not driving the pin. When a 0 is loaded into a pin's TRIS bit, the tristate buffer is enabled *Output mode* and the value that is in the Data Out register is driven onto the pin.

The TRIS register can be confusing in regard to where it is located in the low-end and mid-range PICmicro® MCU's register maps. In low-end devices, the TRIS register can only be written to using the *tris* instruction. In mid-range (and higher) PICmicro® MCUs, the TRIS register is often in a different bank from the port data register or use the *tris* instruction. These two methods are somewhat awkward to use and both have their own quirks.

The *tris* instruction has the format:

```
tris    PORT #
```

where # is A, B, or C. When this instruction executes, the contents of w are loaded into the TRIS register for the specified port. If the TRIS value is going to be updated in the application, then I recommend saving the value loaded into the TRIS register into a file register and updating that (and saving it again) before executing. The *tris* instruction code to do this could be:

```
movlw    0x0FF ^ (1 << Bit)      ;  Find the Bit to Reset
andwf    TRIS#SaveReg, f         ;  Update the Saved TRIS Value
movf     TRIS#SaveReg, w         ;  Use as the TRIS Value
tris     PORT#
```

The mid-range parts can execute the *tris* instruction, but it is not recommended by Microchip because support for it is not guaranteed for all future products. Personally, I would not recommend it for use with mid-range devices simply because not all of the possible registers can be controlled with this instruction. Mid-range PICmicro® MCUs can have five I/O ports (PORTA through PORTE); the TRIS instruction can only access ports A, B, and C, so access to D and E are not possible.

The recommended way to access the mid-range PICmicro® MCU's TRIS registers is to change the RPO bit of the STATUS register and read or write the register directly as:

```
bsf     STATUS, RPO
movlw   NewTRISA
movwf   TRISA ^ 0x080
bcf     STATUS, RPO
```

Notice that I XORed the TRISA register address with 0x080 to avoid any messages telling you to check the page that is being accessed. TRISA is a bank 1 register and has bit seven of its set; its address/value is 0x085. XORing 0x085 with 0x080 will return a value of 0x005, which is a valid address within the 7 bit mid-range PICmicro® MCU bank. Bank addressing is explained in greater detail in Chapter 3.

A number of hints that will make working with TRIS registers easier in the mid-range PICmicro® MCUs can be translated to the PIC17Cxx and PIC18Cxx. The first is, notice that putting a port bit into Output mode is loading it with a 0; putting it in Input mode is loading it with a 1. I always remember which is which by: *0 = Output and 1 = Input.* The digit approximates the first letter of the corresponding word.

The second hint (and this can be used for other registers as well) is to use the assembler's arithmetic operations to figure out the TRIS bit pattern for you. The *left shift* can be used to shift a 1 to a specific bit position.

For example, setting bit 2 could be accomplished by shifting 1 to the left twice:

```
Bit value = 1 << 2
          = 4              ; "4" is Equal to 2 ** 2
```

The << and >> operators shift a value by a specified number of bits, as shown. If an application had to make all of port B's I/O pins output, except for bits 2 and 5, the TRIS value could be calculated and loaded into the TRIS register using the code:

```
bsf    STATUS, RPO
movlw  (1 << 2) + (1 << 5)
movwf  TRISB ^ 0x080
bcf    STATUS, RPO
```

If a register is predominantly input, with just a few outputs, then an initial value of 0x0FF (all input) could be XORed with the bits that are to be used for output. For example, if all of the port C bits, except for bit 3 were in Input mode, the code:

```
bsf     STATUS, RPO
movlw   0x0FF ^ (1 << 3)
movwf   TRISC
bcf     STATUS, RPO
```

could be used. This method takes all the thinking and converting out of determining bit values, which eliminates a potential for errors.

Going back to Fig. 5-17, notice that the pin input is read from the I/O pin and not from the DATA OUT register. This is important to remember because sometimes this arrangement will cause problems. This issue seems to be unique to the PICmicro® MCU; other microcontrollers (including the PIC18Cxx) either have a separate input address or have a method of selecting which (the data out register or the pin) to read, based on whether or not the pin is input or Output mode.

The problem lies in that many instructions read and write the register contents in ways in that you would not expect. Probably the biggest culprits of this are the *bcf/bsf* instructions, which perform a register read, bit OR or AND, and then write the new value back to the register. There is the opportunity that a port bit will be in Input mode with a specific value set into it. If the pin is at a different value, then this value will be read and written back by the PICmicro® MCU. This problem is covered elsewhere in the book and I have included an experiment showing how it can be a problem for you. The best way to avoid this issue is to always write the desired output values to the port register before changing the TRIS register.

I wish I kept track of problems that people report on the PICList; I would bet that the number one or two problem encountered by people is with RA4 or PORTA.4 pin (Fig. 5-18). This pin is an *open drain,* which can be used on a pulled-up dotted AND bus. When RA4 is used by most people for the first time (I'm guilty of this as well), they forget or don't realize that the pin does not have the ability to drive a high voltage and can't understand why this pin seems to be broken.

This pin cannot drive a positive voltage out unless it is pulled up (I normally use a 1K to 10-K resistor, depending on the input capacitance of what is being driven).

Also notice that this pin has a Schmidt trigger input, which has a different threshold voltage for signals going low to high than the threshold for going high to low (Fig. 5-19).

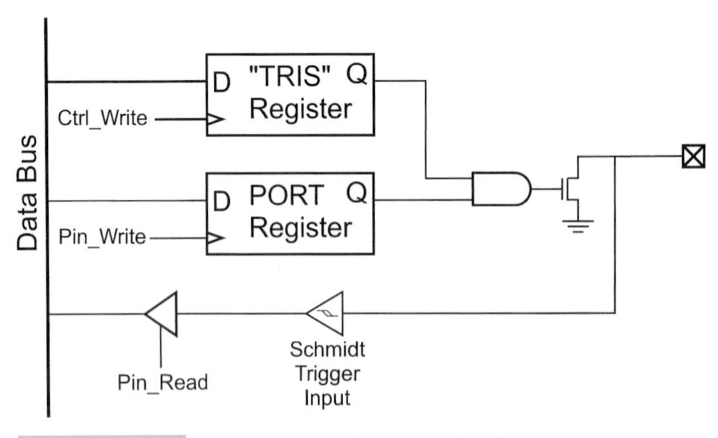


**Figure 5-18** **PORTA bit 4 I/O pin block diagram**

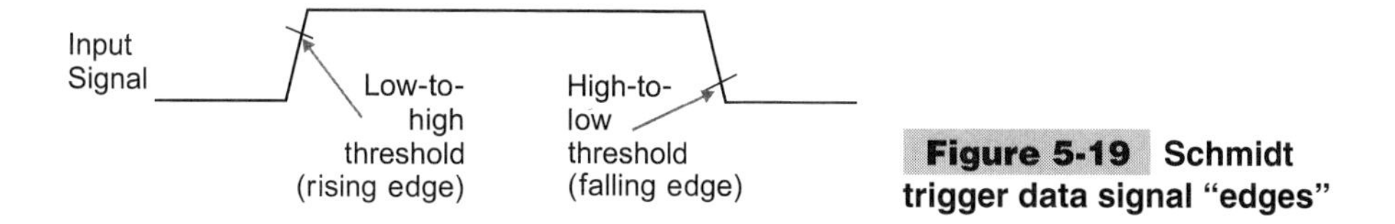


**Figure 5-19** **Schmidt trigger data signal "edges"**

The purpose of the Schmidt trigger is to provide *hysteresis* for an input signal and try to eliminate some "bouncing" errors. This input causes RA4 to behave differently than other I/O pins in some circumstances; most notably, this pin should never be used for an RC circuit to read a potentiometer's position. In this case, the Schmidt trigger input will allow some current flow from the pin into it, which affects the result of the potentiometer read operation.

A feature you should be aware of in the mid-range parts is the availability of a controllable pull-up on the PORTB pins. This pull-up is controlled by the _RPBU bit of the OPTION register and is enabled when this bit is reset and the bit itself is set for output. The port B pin block diagram is shown in Fig. 5-20.

The "weak" pull-up is approximately 50 K and can simplify button inputs, eliminating the need for an external pull-up resistor.

Changing pin inputs can initiate interrupt requests to the processor. The function that is normally used is the RBO/INT pin, which can request an interrupt (if the INTE bit is set in the INTCON register). An interrupt request can be made on rising or falling edges, as selected by the INTEDG bit of the OPTION register. If INTEDG is set, interrupts can be requested on the rising edge of signal into RBO/INT. INTEDG reset will cause an interrupt on a falling edge. I like to think of the RBO/INT interrupt request hardware as the block diagram shown in Fig. 5-21.

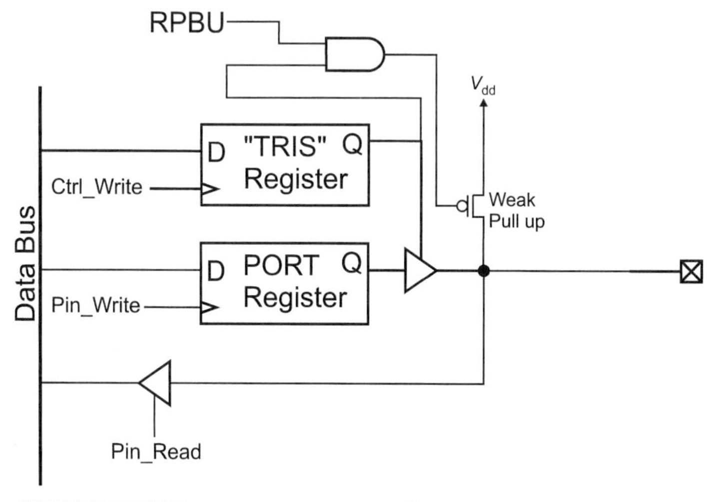


**Figure 5-20** **Standard PICmicro® MCU I/O pin block diagram**

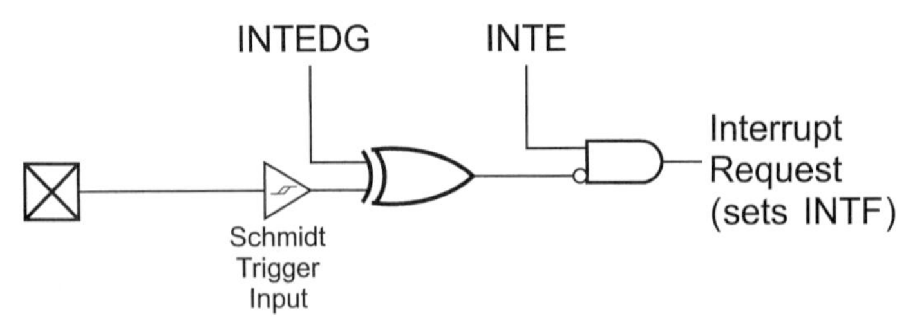


**Figure 5-21** "RBO/INT" input interrupt circuitry

Once the interrupt is acknowledged by the processor, the INTF bit of the INTCON register has to be reset to enable another interruption RBO/INT pin transition. The interrupt request goes through a Schmidt trigger buffer, but the STD interface pin does not. This will affect the loading of the pin somewhat.

The other type of interrupt that can be requested from the I/O pins is the port B change on interrupt. If the RBIE bit of the INTCON register is set, then any changes to the RB4 to RB7 pins while they are in Input mode will request an interrupt on port change and set the RBIF flag in the INTCON register. To clear this interrupt request, PORTB must be first read to set the current value that is followed by resetting the RBIF flag.

The port change on interrupt is only available on RB4 to RB7 when they are in Input mode. Changing the state of any of these pins while they are in Output mode will not cause a port change interrupt request.

This interrupt is a bit tricky to use, but remember that PORTB should never be polled to eliminate any possibilities of an unexpected interrupt request reset occurring before it can be acknowledged. To avoid this problem, when I use the port change interrupt feature, I do not make any other PORTB pins inputs except for the Interrupt Sources.

## 12C5XX AND 16C505 I/O PIN ACCESSING

I really like 12C5xx parts and the 16C505, but the first time you use them, you will probably find that you cannot access all the I/O pins by default. Some of the pins are seemingly lost in a morass of configuration fuses and features. I found that the first time I used the 12C508, it took me two hours (with a lot of erasures) to figure out how to do it. This section is devoted to giving you the information to avoid this and go straight ahead and use the 12C5xx or 16C505 with the built-in oscillator and reset without experiencing any nasty surprises or "missing pins."

It is unfortunate that these functions are not the default values for the device. In the 12C5xx's case, the addition of an external reset and clock drastically reduces the usability of the parts. In fact, these built-in features are the only reason why I select these parts in the first place.

The built-in oscillator is selected by the _INTRC_OSC parameter of the __CONFIG statement in your source file. When the PICmicro® MCU is programmed, a value for the calibration register (OSCCAL) has to be inserted. By convention, a:

```
movlw OSCCAL-value
```

is put in at the reset address and then at address zero (when the program counter overflows and resets to zero), this value is saved into the OSCCAL register using a:

```
movwf OSCCAL
```

instruction.

When choosing the programmer that you are going to use for this part, be sure that it can read the calibration value of the PICmicro® MCUs with the built-in oscillator feature and that it can program the calibration value into the application. This calibration value should be specified by the user separately from the application's .HEX file.

This type of programmer, although providing the basic functions required, is going to be difficult to work with. You should select a programmer that will read the factory-set calibration value and allow it to be programmed into the device separately from the application. This will reduce the opportunities for errors when working with the internal oscillator.

It almost shouldn't have to be said, but when you get a new windowed (JW) PICmicro® MCU with the internal oscillator feature, you should read it out and write it on the part itself. I put a label on the underside (Fig. 5-22), but other people scratch the value in or write it on the back in a light ink or paint.

Notice that combination values cannot be predicted. Figure 5-22 shows the calibration values from a number of PIC12C508 JWs (Windowed) that I bought at the same time, all marked with the same lot code. The values of the various parts' calibration values were:

- 0x0D0
- 0x090
- 0x030

**Figure 5-22** **8 PIN PICmicro® MCU with the calibration valve marked on the backside**

- 0x080
- 0x0A0
- 0x090

Notice that there is no consistency.

Developing your own calibration value could be done by trying values out against a calibrated PICmicro® MCU and seeing which value has the minimum difference in timing with the calibrated device. But it is much easier to record the factory calibration value when you first buy the part before you program it.

The internal reset is enabled by the _MCLRE_ OFF parameter of __CONFIG. This parameter disables the external (the *E* in _MCLRE_OFF) _MCLR pin and ties the PICmicro® MCU's internal reset to $V_{dd}$. The _MCLR pin now becomes an input pin for the application.

Once the _IntRC_OSC and _MCLRE_OFF parameters are put in to the __CONFIG statement and the OSCCAL value in w is saved, you will find that the pin that provides the clock input (GP2 in the 12C5xx) cannot be used for I/O. This is because of the reset value of the T0CS bit of the OPTION register being set, which causes the pin to be selected for TMR0 input. This overrides the I/O functions of the pin. Simply resetting the bit in the OPTION register will allow the pin to be used for normal I/O.

When I create a PIC12C5xx or PIC16C505 application, the initial code that I use is:

```
 __CONFIG _MCLRE_OFF & _IntRC_OSC    ; Add Application Specific
                                     ; "CP" and "WDT" parameters
 org 0
   movf OSCCAL
   movlw 0x0FF ^ (1 << T0CS)
   option
;     All I/O pins are NOW Available and Internal 4 MHz Clock is Running
;     - Start Application
```

There are a few points to remember. First, notice that the pin that can be used for _MCLR is only available for input. Secondly, this pin is not clamped with diodes inside the PICmicro® MCU and is used for the $V_{pp}$ pin, which means that high-voltage inputs could reset the PICmicro® MCU and put it into Programming mode. Lastly, remember that any and all writes to the OPTION register must keep the T0CS bit at reset or the bit that can be used for TMR0 input might stop being I/O capable.

# Interrupts

I recommend that you use interrupts in your applications because they can simplify the effort required to write the application code and allows code that can respond to inputs much faster. The book also covers how interrupt-handler software is written and what to look for. This section covers the mid-range PICmicro® MCU's interrupt hardware and how the PICmicro® MCU processor responds to an interrupt request. The PIC17Cxx and PIC18Cxx parts behave similarly and most of the information provided here is applicable for the interrupt hardware built into them.

**TABLE 5-7 Mid-Range PICmicro® MCU "INTCON" Register Definition**

| BIT | DESCRIPTION |
|---|---|
| 7 | GIE—Global Interrupt Enable |
| 6 | Device Specific |
| 5 | TOIE—TMR0 Overflow Interrupt Enable |
| 4 | INTE—RBO/INT Pin Interrupt Enable |
| 3 | RBIE—PORTB Input Change Interrrupt Enable |
| 2 | TOIF—TMR0 Overflow Interrupt Request Active |
| 1 | INTF—RBO/INT Pin Interrupt Request Active |
| 0 | RBIF—PORTB Input Change Interrupt Request Active |

In the mid-range PICmicro® MCU, the INTCON register is the central focus point for interrupts. This register is used to globally enable interrupts and control the response to different interrupt inputs. The register consists of four different bit types and is located at address 0x0B, in all active register banks. The bit usage is similar for all mid-range devices. All the bits are positive (set) active.

The GIE bit must be set for interrupt requests to be passed to the processor. This bit can globally prevent (mask) or allow (unmask) interrupt requests to go to the processor. If a crucial section of code is being entered, by resetting this bit, the interrupt request will be ignored. This bit is reset upon acceptance of the interrupt request and then set upon exit from the interrupt handler.

The INTCON bit names that ground in "E" are the interrupt enable flags. When these bits are set, any incoming interrupt requests will set the corresponding interrupt request active flags, which has a bit name that ends in "F." The request active flag must be reset in hardware and is not reset automatically by the operation of the interrupt acceptance (which I also call *acknowledgment* elsewhere in the book). As well, the requesting hardware might have to be reset before the F (request active flag) can be reset.

In Fig. 5-21, if the E bit for the INT interrupt source is set, then interrupt requests, will request an interrupt of the PICmicro® MCU processor if the GIE bit is set.

When the processor receives the interrupt request, it completes the current instruction before jumping to the interrupt vector. Instruction execution in the PICmicro® MCU can be one or two cycles long and, when added to the two-instruction delay for calling the interrupt handler, the total delay (which is known as *interrupt latency*) is three or four instruction cycles. In the mid-range devices, the interrupt vector's address is 0x00004 for all interrupt sources. What happens during execution is shown in Fig. 5-23.

The PIC17Cxx's interrupt execution varies from this in one respect: different interrupt sources result in jumps to different interrupt address vectors, according to their priority. Table 5-8 lists the interrupt sources, the vector addresses, and their priorities.

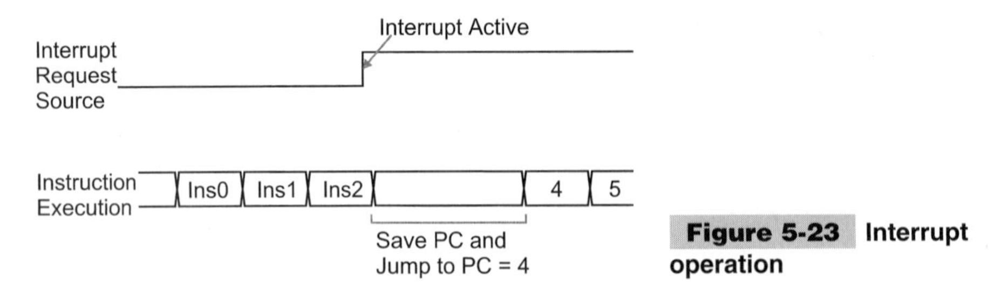


**Figure 5-23** Interrupt operation

As noted, when the processor jumps to the interrupt vector, the GIE bit is reset. The return address (the address of the instruction after the interrupted one) is saved on the program counter stack. With GIE reset, no subsequent interrupts can stop the interrupt handler's execution.

Elsewhere, the book covers setting GIE to allow *nested interrupts* (interrupts that are acknowledged from within an executing interrupt handler). Because of the lack of a data stack in the PICmicro® MCU's processor, I do not recommend setting the GIE bit within the interrupt handler. Instead, it should be set by the *retfie* instruction after the one that the interrupt execution jumped from.

Interrupts will execute and return properly from any instruction or instruction combination. Some reports stated that mid-range interrupts would have problems if they were acknowledged after a PCL update, but this is not true; there are no PICmicro® MCU hardware deficiencies when interrupts can execute.

In the mid-range parts, three interrupt sources are handled from within the INTCON register. The TOIF and TOIE bits allow an interrupt request when TMR0 overflows (equal to 255 or 0x0100). The second responds to an input on the RBO pin, (usually marked *RBO/INT* by Microchip), when the input on the pin goes high or low (depending on the OPTION register state).

The last interrupt source, PORTB change interrupt requires some comments because there is some confusion about the operation of this interrupt.

The PORTB change interrupt will request an interrupt if any of the RB7 to RB4 pins, which are in Input mode, change state. The pins will not request an interrupt if they are in Output mode. To reset the interrupt, PORTB must be read before resetting the RBIF flag. If this is not done, the RBIF flag will remain set and requesting an interrupt no matter if you try to reset it.

Along with these three interrupt sources, many PICmicro® MCUs have a number of built-in peripherals that can request interrupts to allow these additional sources. Bit six of

**TABLE 5-8 PIC17Cxx Interrupt Vector Definition**

| INTERRUPT SOURCE | VECTOR ADDRESS | PRIORITY |
|---|---|---|
| RAO/INT PIN | 0x00008 | 1 (highest) |
| TMR0 Overflow | 0x00010 | 2 |
| External Timer1 Interrupt | 0x00018 | 3 |
| Peripheral Interrupts | 0x00020 | 4 (longest) |

INTCON can be used as another E bit (with the F bit in another register) or an enable bit for the "PIE" and "PIR" registers which contains the mask and control bits for the other sources.

There can be one or two sets of PIE and PIR registers, according to the features built into the PICmicro® MCU. Unfortunately, there doesn't seem to be any convention used for how the bits are set. But by looking at the device data sheet, you will get a specification for the interrupt bits.

## TMR0

TMR0 is the basic eight-bit timer available to all PICmicro® MCUs. Although the low-end PICmicro® MCUs do not support interrupt requests from TMR0 (the other PICmicro® MCU architectures do), TMR0 can still provide many useful functions for applications. TMR0 has a few peculiarities with regard to its operation that you should be aware of, but for the most part, it is quite straight forward to use. Some example applications using it are presented later in the book.

TMR0 is an eight-bit incrementing counter that can be preset (loaded) by application code with a specific value. The counter can either be clocked by an external source or by the instruction clock. Each TMR0 input is matched to two instruction clocks for synchronization. This feature limits the maximum speed of the timer to one half of the instruction clock speed. The TMR0 block diagram is shown in Fig. 5-24.

The TOCS and TOCE bits are used to select the clock source and the clock edge that increments TMR0 (rising or falling edge). These bits are located in the OPTION register that is covered later in the chapter.

The synchronizer is a glitch-elimination feature of the PICmicro® MCU. This circuit only updates TMR0 when two instruction cycles have passed without a change to the input. For most applications, this circuit means that TMR0 is updated after two instruction cycles have passed.

TMR0 can be driven by external devices through the T0CKI pin. The T0CKI pin is dedicated to this function in the low-end devices (although, in the 12C5xx and 16C505 PICmicro® MCUs, the pin can be used for digital I/O). In the other PICmicro® MCU architectures, the pin can also be used to provide digital I/O. When a clock is driven into the TMR0 input, the input is buffered by an internal a Schmidt trigger to help minimize noise-related problems with the input.

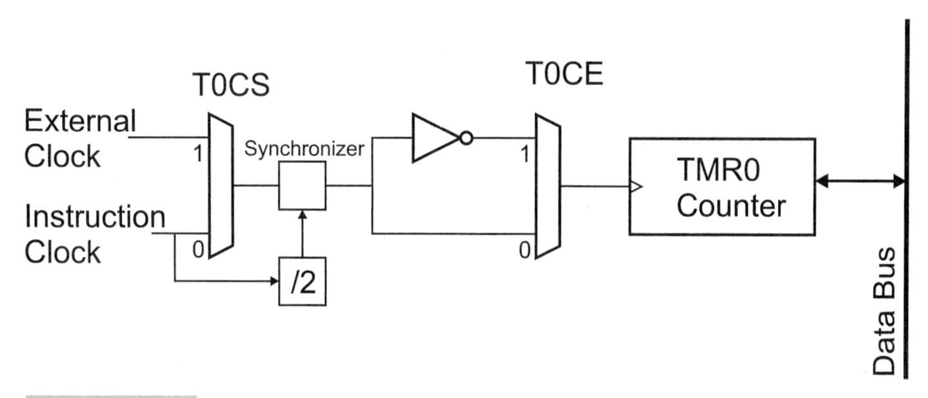


**Figure 5-24** TMR0 block diagram

The input to TMR0 can be made with and without the prescaler, which provides a divide-by feature to the TMR0 input. As covered later in the chapter, the prescaler can count anywhere from one to 128 cycles before passing along an increment signal to the PICmicro® MCU.

For the low-end and mid-range PICmicro® MCUs, TMR0 is located at register address 0x001. The contents of TMR0 can be read from and written to directly. Always remember that when TMR0 is being updated, the synchronizer (and prescaler if connected) will be reset.

As shown in the experiments chapter, if TMR0 is polled for being equal to zero using the code:

```
movf     TMR0, f
btfss    STATUS,Z
  goto   $ - 2
```

and a 4 : 1 or greater prescaler is used with TMR0, it will loop forever. Even if the prescaler is not used, you will find that the delay will be twice as long as you would expect because the *movf TMR0, f* instruction will reload TMR0 and reset the synchronizer. This mistake is very common and I know of many people that have not quite believed this until they simulated it and saw it actually happening.

When polling TMR0, the code:

```
movf     TMR0, w                 ; Load "w" with the contents of TMR0
btfss    STATUS, Z               ; Skip if the Result is Equal to Zero
  goto   $ - 2
```

should be used.

It might seem unusual to poll TMR0 for reaching zero because the timer is incremented with each synchronized clock tick, instead of checking it against a specific value, but this is the conventional way to check the result. In the mid-range, PICmicro® MCU's TMR0 interrupt is requested when TMR0 rolls over or overflows from 0x0FF to 0x000, so this method fits in well with the interrupt request.

Remember that the *bcf* and *bsf* instructions (as well as any others that read and store data into TMR0) will reset the synchronizer (and prescaler if selected) and can affect the operation of your application in unexpected ways.

In the mid-range (and higher) PICmicro® MCU architectures, when TMR0 transitions from 0x0FF to 0x000, an interrupt request can be generated. This is first accomplished by loading TMR0 to the specified delay value and then setting the T0IE bit and resetting to 0x000. T0IF will be set and, if GIE is set, the interrupt request will be acknowledged.

Any time that TMR0 transitions to 0x000 from 0x0FF, the T0IF flag will be set. This flag must be reset explicitly before enabling TMR0 interrupts to prevent the chance for invalid interrupt request when the TMR0 interrupt is enabled and TMR0 has already overflowed. The *T0IF (TMR0 Interrupt Request Flag)* must be reset in the interrupt handler. It is not reset automatically when the interrupt request is acknowledged.

Creating specific delays with TMR0 is a bit unusual because of the function of counting up to 0x000 from 0x0FF. As covered in the next section, calculating the initial TMR0 value to provide a specific delay requires essentially negative logic. Although it isn't hard to understand and do, it is not completely intuitive.

## CALCULATING DELAYS

One of the basic operations of a timer is to provide a set time delay. In many of the experiments and applications, I often use TMR0 to request an interrupt after a specific time delay. It isn't difficult to use a timer to calculate a delay, although you should be aware of a few things before you attempt it.

First, the interrupt is requested when TMR0 overflows or equals 256. 255 or 0x0FF is the largest value that can be saved in the bit timer register. 0x0FF does not cause an interrupt (although an interrupt will be requested when the timer increments or overflows to 256 or 0x0100). Therefore, the delay is calculated for the time when the timer equals 256 and not 255.

The initial timer value must be the number of clock increments (or "ticks") required to get to 256 within a specific period of time because the timer can only count up. To calculate the initial value (with no prescaler, described later), the following formula is used:

$$TMR0\ initial = 256 - delay\ cycles$$

The delay cycles value is found by taking the delay time, dividing by the frequency, and multiplying by 4. The result is divided by two because the input to TMR0 divides the cycles by two.

$$\begin{aligned} delay\ cycles &= (delay\ time * frequency/4)/2 \\ &= delay\ time * frequency/8 \end{aligned}$$

These formulas can be used to calculate a 160-$\mu$s delay in a 4-MHz PICmicro® MCU application. Starting with the second formula, the number of delay cycles timed is calculated as:

$$\begin{aligned} delay\ cycles &= delay\ time * frequency/8 \\ &= 160\ \mu s * 4MHz/8 \\ &= 160\ (10^{-6})\ secs * 4(10^{-6})/secs/8 \\ &= 160 * 4/8 \\ &= 80 \end{aligned}$$

This value is then used to calculate the initial timer value, using the first formula:

$$\begin{aligned} initial &= 256 - delay\ cycles \\ &= 256 - 80 \\ &= 176 \end{aligned}$$

So, to delay 160 $\mu$s before a TMR0 overflow interrupt in a PICmicro® MCU running at 4 MHz, an initial value of 176 must be loaded into TMR0.

For longer delays than 512 cycles, the prescaler can be used to divide the number of cycles input into the timer. The prescaler, as covered in the next section, divides the incoming data by powers of two from 1 to 128.

When calculating the delay, the delay cycles are continually halved until the value is less than 256. For example, if a delay of 5 ms was required; before TMR0 in the 4-MHz PICmicro® MCU could be initialized, the delay cycles would be calculated as:

$$
\begin{aligned}
\textit{delay cycles} &= \textit{delay time} * \textit{frequency}/8 \\
&= 5 \text{ ms} * 4 \text{ MHz}/8 \\
&= 5(10^{-3}) \text{ secs} * 4(10^{6})/\text{secs}/8 \\
&= 5(10^{3})/2 \\
&= 2.5(10^{3}) \\
&= 2{,}500
\end{aligned}
$$

Because the calculated delay cycles are greater than 256, it is continually divided by 2 to get an appropriate prescaler divisor. Dividing by 2, the delay count would be 1,250, which is still greater than 256. Divided by 2 again, delay count would be 625, which is also greater than 256. Dividing the original value by eight yields a delay count of 312.5. Finally, dividing the original by 16 yields a delay count of 156.25. Rounding it off, the value can be put into the initial value formula.

$$
\begin{aligned}
\textit{initial} &= 256 - 156 \\
&= 100
\end{aligned}
$$

Now, the rounding that I did (taking off 0.25) will result in a difference of four cycles between the actual delay of 5 ms and the delay calculated within TMR0. If this is a crucial loss, then I suggest you do one of two things. First, you can add four delay cycles to your code, which can be four *nops,* a dummy call, or even delaying loading the timer by four instruction cycles.

If the rounding of four cycles is that crucial to your application (a loss of 0.32 percent in delay accuracy), then I recommend that you dispense with this method altogether and use an inline delay method described elsewhere in this book. I'm not trying to be facetious; a few cycles over a long period of time can cause significant problems in some applications. I am also making this suggestion because of the unpredictable interrupt response latency that can happen if a two-instruction-cycle instruction is executing when the interrupt is received. Another case for the unpredictable interrupt latency is if the interrupt request is received *after* the bit is polled in the PICmicro® MCU execution state machine.

# Prescaler

The *prescaler* is a power-of-two counter that can be selected for use with either the watchdog timer or TMR0. Its purpose is to divide the incoming clock signals by a software-selectable power of two value to allow the eight-bit TMR0 to time longer events or increase the watchdog delay from 18 ms to 2.3 s.

The prescaler's operation is controlled by four bits contained within the OPTION register. "PSA" selects whether the watchdog timer uses the prescaler (when PSA is set) or TMR uses the prescaler (when PSA is reset). Notice that the prescaler has to be assigned to either the watchdog timer or TMR0. Both functions are able to execute with no prescaler or with the prescaler's delay count set to one, then the prescaler is effectively "turned" off.

The prescaler itself is a power-of-two counter, with the trigger point selected by the PSO, PS1, and PS2 bits of the OPTION register. I call prescaler operation as power of two

**TABLE 5-9 Prescaler Values to Delays**

| PS2—PS0 | PRESCALER DELAY |
|---|---|
| 000 | 1 cycle |
| 001 | 2 cycles |
| 010 | 4 cycles |
| 011 | 8 cycles |
| 100 | 16 cycles |
| 101 | 32 cycles |
| 110 | 64 cycles |
| 111 | 128 cycles |

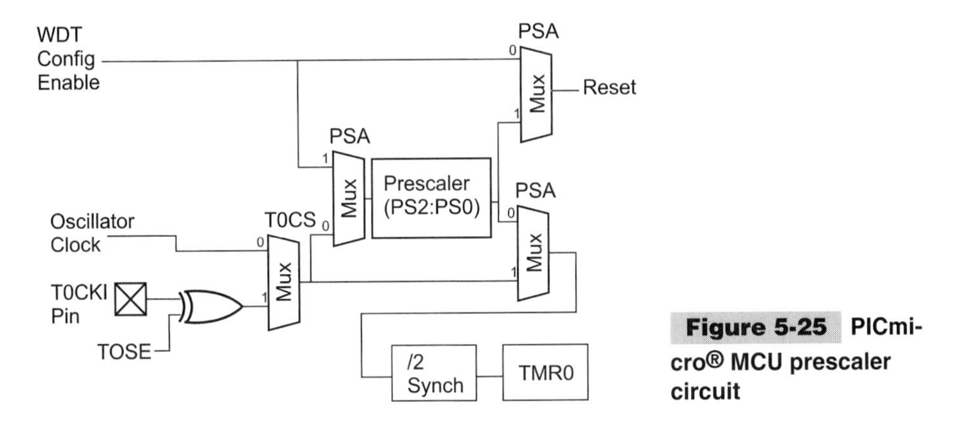


**Figure 5-25** PICmicro® MCU prescaler circuit

because the number of cycles delay is a function of the PSA bit value. Table 5-9 shows the prescaler cycle delay for varying values of the PS# bits.

For PS2 to PS0 equal to 011 one of every eight clock cycles input into the prescaler will be passed to the device that the prescaler is driving.

The watchdog timer and TMR0's block diagrams can be combined with the prescaler to show how the functions work (Fig. 5-25).

# The OPTION Register

The OPTION register is a cornerstone register for the operation of the PICmicro® MCU. The register controls the operation and delay of the prescaler, selects the clock source and specifies the operation of the interrupt source pin. It is a very useful register. You should always keep it in the back of your mind and be sure that it is set up properly for your application.

In the low-end devices, the OPTION register is defined in Table 5-10. Updating the OPTION register in the low end, requires the *option* instruction which moves the contents of w into the OPTION_REG (which is the MPLAB label for the OPTION register).

The mid-range device's option register is similar, but does not have any device-specific bits (Table 5-11). The mid-range device's OPTION register can be written to using the *option* instruction, like the low end, or by setting the RPO bit or the STATUS register and accessing it at address one. The address of the OPTION_REG is 0x081 or 0x001 in bank 1 of the address space.

Using the *option* instruction is not recommended in the mid-range devices and might not be supported in them in the future. Instead, writing to it directly from bank 1 should be

**TABLE 5-10 Low-End PICmicro® MCU "OPTION" Register Definition**

| BIT | LABEL/FUNCTION |
|---|---|
| 7 | _GPWU—Enable wakeup on pin change ] Device Specific |
| 6 | _GPPU—Enable I/O PortB Weak Pull-ups ] Device Specific |
| 5 | TOCS—TMR0 clock source select |
| | 1—Tockl pin |
| | 0—Instruction clock |
| 4 | TOSE—TMR0 Increment Source Edge Select |
| | 1—High to Low on Tockl Pin |
| | 0—Low to High on Tockl Pin |
| 3 | PSA—Prescaler Assignment Bit |
| | 1—Prescaler Assigned to Watchdog Timer |
| | 0—Prescaler Assigned to TMR0 |
| 2-0 | PS2-PSO - Prescaler Rate Select |
| | 000—1:1 |
| | 001—1:2 |
| | 010—1:4 |
| | 011—1:8 |
| | 100—1:16 |
| | 101—1:32 |
| | 110—1:64 |
| | 111—1:128 |

**TABLE 5-11 Mid-range Picmicro® MCU "Option" Register Definition**

| BIT | LABEL/FUNCTION |
|---|---|
| 7 | _RBPU—Enable PortB Weak Pull-ups |
| | 1—Pull-ups Disabled |
| | 0—Pull-ups Enabled |
| 6 | INTEDG—Interrupt Request On: |
| | 1—low to high on RBO/INT |
| | 0—high to low on RBO/INT |
| 5 | TOCS—TMRO clock source select |
| | 1—Tockl Pin |
| | 0—Instruction Clock |
| 4 | TOSE—TMR0 Update Edge Select |
| | 1—Increment on High to Low |
| | 0—Increment on Low to High |
| 3 | PSA—Prescaler Assignment Bit |
| | 1—Prescaler Assigned to Watchdog Timer |
| | 0—Prescaler Assigned to TMR0 |
| 2-0 | PS2-PS0—prescaler rate select |
| | 000—1:1 |
| | 001—1:2 |
| | 010—1:4 |
| | 011—1:8 |
| | 100—1:16 |
| | 101—1:32 |
| | 110—1:64 |
| | 111—1:128 |

used always and, except for a few special cases (explained later), the *option* instruction should never be used.

The PIC17Cxx PICmicro® MCU's do not have an OPTION register because many of the functions continued by option are either not present (such as the prescaler and PORTB weak pull-ups) or are provided in other registers. The PIC18Cxx provides a mid-range "compatible" OPTION register, but it is not at the same address as the mid-range devices.

# Mid-Range Built-In EEPROM/Flash Access

An increasingly popular feature in PICmicro® MCU devices is the availability of built-in EEPROM memory that can be used to store configuration, calibration, or software data. In one of the example applications, I use it to store application source code. In the mid-range devices, this feature can be accessed using the registers. In some low-end devices, EEPROM is accessed as if it were an I2C device attached to the PICmicro® MCU. The next section covers how these devices work.

Along with accessing data, some Flash devices (the 16F62x and 16F87x) can also read and write Flash program memory from within the application. This is useful for storing larger amounts of data in a nonvolatile memory and changing applications, as I do in the EMU-II emulator presented later in this book. The differences between data EEPROM I/O and program memory Flash I/O are covered at the end of this section.

For data EEPROM I/O, you should be aware of four registers: EECON1, EECON2, EEADR, and EEDATA. These registers are used to control access to the EEPROM. As you would expect, EEADR and EEDATA are used to provide the address and data interface into the up to 256-byte data EEPROM memory. EECON and EECON2 are used to initiate the type of access, as well as indicate that the operation has completed. EECON2 is a pseudo-register that cannot be read from, but is written to with the data, 0x055/0x0AA to indicate the write is valid.

EECON1 contains the bits for controlling the access that are shown in Table 5-12. These bits can be in different bit positions in different devices, which is why I have not specified the bit values in Table 5-12.

Using these bits, a read can be initiated as:

```
movf / movlw          address/ADDR, w
```

**TABLE 5-12 Critical EECON1 Bits**

| BIT | FUNCTION |
|---|---|
| EEPCD | Set to Access Program Memory. Reset to Access Data EEPROM only in 16F62x and 16F87x. |
| WRERR | Set if a write Error is Terminated early to indicate Data Write may not have been Successful. |
| WREN | When set, a write to EEPROM begins. |
| WR | Set to indicate an upcoming Write Operation. Cleared when the Write Operation is complete. |
| RD | Set to indicate Read Operation. Cleared by next Instruction Automatically. |

```
bcf                     STATUS, RP0
movwf                   EEADR
bsf                     STATUS, RP0
bsf                     EECON1, ^ 0x08, RD
bcf                     STATUS, RP0
movf                    EEDATA, w           ; w = EEPROM [address/ADDR]
```

In this code example, it is assumed that these registers are in banks 0 and 1, which is true for the 16F84. But for devices like the 16F87x, where the ADC registers use these addresses, the EEPROM register addresses are actually in banks 2 and 3. The read operation must have a *bsf STATUS, RP1* at the start and *bcf STATUS, RP1* at the end:

```
movf /movlw             address/ADDR, w
bsf                     STATUS, RP1
bcf                     STATUS, RP0
movwf                   EEADR ^ 0x0100
bsf                     STATUS, RP0
bsf                     EECON 1 ^ 0x0180, RD
bcf                     STATUS, RP0
movf                    EEDATA ^ 0x0100, w     ; w = EEPROM [address/ADDR]
bcf                     STATUS, RP1
```

Write operations are similar, but have two important differences. The first is that the operation can take up to 10 ms to complete, which means the WR bit of EECON1 has to be polled for completion or in the EPROM interrupt request hardware enabled. The second difference, as mentioned, is that a timed write must be implemented to carry out the operation.

The code to carry out an EPROM write could be:

```
movlw /movf             constant/DATA, w
bcf                     STATUS, RP0
movwf                   EEDATA
movlw /movf             address/ADDR, w
movwf                   EEADR
bsf                     STATUS,RP0
bsf                     EECON1 ^ 0x080, WREN
bcf                     INTCON,GIE
movlw                   0x055               ]       CRITICAL SECTION
movwf                   EECON2 ^ 0x080      ]
movlw                   0x0AA               ]
movwf                   EECON2 ^ 0x080      ]
bsf                     EECON1 ^ 0x080, WR  ]
bsf                     INTCON, GIE
btfsc                   EECON1 ^ 0x080, WR  ] Poll for Operation Ended
  goto                  $ - 1               ]
bcf                     EECON1 ^ 0x080, WREN
bcf                     STATUS, RP0
bsf                     INTCON, GIE
```

For the devices with the EE access registers in banks 2 and 3, this code is modified in the same manner as the EEPROM read code.

Notice that EEPROM cannot be accessed in any way until WR is reset; otherwise, there will be a WRERR.

The critically timed code is used to indicate to the EEPROM access-control hardware that the application is under control and that a write is desired. Any deviation in these instructions (including interrupts during the sequence) will cause the write request to be ignored by the EEPROM access-control hardware.

Instead of polling, after the WR bit is set, the EEIE interrupt-request bit can be set. Once the EEPROM write has completed, the EEIF file is set and the hardware interrupt is requested.

For the 16F62x and 16F87x PICmicro® MCUs, program memory can be read or written to in a similar way to EEPROM data memory. The difference is the inclusion of the EEPGD bit in EECON1 that is not present in the devices with just EEPROM data memory. In the devices that do have programmable data and program memory, this bit should always be set (program memory) or reset (data memory), according to the memory access.

Along with the inclusion of the EEPGD bit, two additional registers are used to address and access the greater than eight-bit data and number of address bits. These bits are known as (not too surprisingly) *EEADRH* and *EEDATH.* Notice that the maximum data value for EEDATH is 3F because 14 bits per instruction is used for program memory.

To read to program memory, the following code is used for the 16F87x. Notice the two *nops* to allow the operation to complete before the instruction is available for reading:

```
bsf                    STATUS, RP1
movlw /movwf           LOW address/ADDR, w
movwf                  EEADR ^ 0x0100
movlw /movwf           HIGH address/ADDR, w
movwf                  EEADRH ^ 0x0100
bsf                    STATUS, RP0
bsf                    EECON1 ^ 0x0180, EEPGD
bsf                    EECON1 ^ 0x0180, RD
nop
nop
bcf                    STATUS, RP0
movf                   EEDATA, w
movwf                  ----                  ; Store Lo Byte of Program Memory
movwf                  EEDATH, w
movwf                  ----                  ; Store Hi Byte of Program Memory
bcf                    STATUS, RP1
```

Writing to program memory is similar to writing to data, but also has the two *nops* in which the operation occurs. No polling or interrupts are available for this operation. Instead, the processor halts during this operation. Even though the processor has stopped for a program memory write, peripheral function (ADCs, serial I/O, etc.) are still active.

```
bsf                    STATUS, RP1
movlw /movf            LOW address/ADDR, w
movwf                  EEADR
movlw /movwf           HIGH address/ADDR, w
movwf                  EEADRH
movlw /movwf           LOW Constant/DATA, w
movwf                  EEDATA
movlw /movwf           HIGH Constant/DATA, w  ; Maximum 0x03F
movwf                  EEDATH
bsf                    STATUS, RP0
bsf                    EECON1 ^ 0x0180, EEPGO
bsf                    EECON1 ^ 0x0180, WREN
bcf                    INTCON, GIE           ]      Critically
movlw                  0x055                 ]      timed
movwf                  EECON2 ^ 0x0180       ]      code.
movlw                  0x0AA                 ]
```

```
movwf          EECON2 ^ 0x0180, OR    ]
nop                                   ] operation
nop                                   ] executes
bcf            EECON1 ^ 0x0180, WREN
bsf            INTCON, GIE
```

## LOW-END BUILT-IN DATA EEPROM

The low-end architecture, with its 32-address register page does not have many registers or space that can be devoted to advanced peripheral I/O functions. For this reason, when the PIC12C5xx parts were given built-in EEPROM (and called the *PIC16CE5xx*), a fairly clever interface had to be developed. This interface consists of connecting the EEPROM within the PICmicro® MCU as if it were an external I2C device. Reading and writing data is more complex than in the mid-range devices (where there are registers for I/O operations). But the read/write operations are relatively simple to code.

The EEPROM-included PIC12CE5xx parts use the most significant bits of the *GPIO (General-Purpose I/O)* register and its corresponding TRIS register. The PIC12CE5xx's EEPROM interface is shown in Fig. 5-26.

As shown in Fig. 5-26, GPIO bits 6 and 7 do not have TRIS control bits. As well, bit 6 (the 12CEEPROM bit, SDA) has an open-drain driver. This driver circuit is designed to let both the PICmicro® MCU and the EEPROM drive the data line at different intervals without having to disable the output bit writing to the EEPROM.

Information is written to the EEPROM device using the waveform shown in Fig. 5-27. Note that for timing I use instruction cycles for a PICmicro® MCU running at 4 MHz.

The start and stop bits are used to indicate the beginning and end of an operation and can be used halfway through to halt an operation. The start and stop bits are actually invalid cases (data cannot change while one clock is active or high).

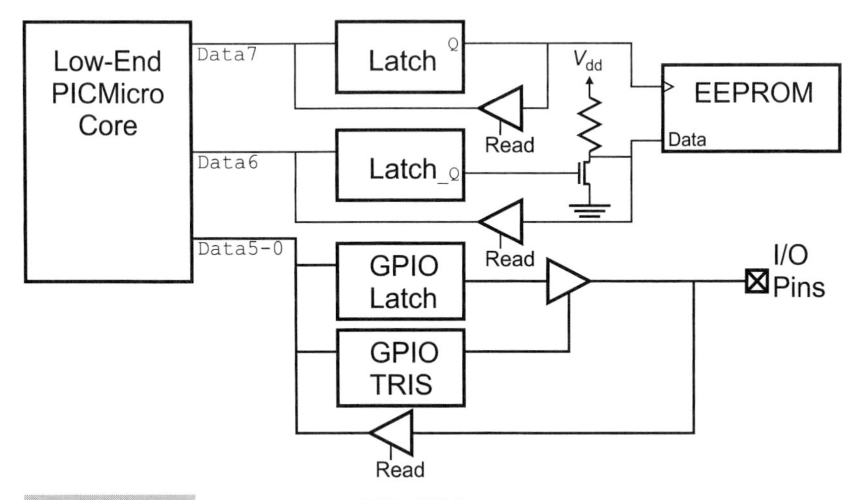


**Figure 5-26** PIC12CE5xx EEPROM interface

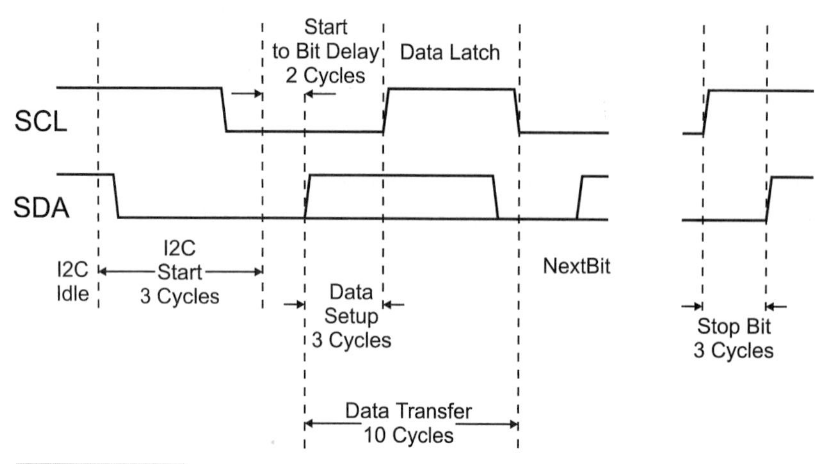


**Figure 5-27** **PIC12CE5xx EEPROM interface waveform**

This operation means that the GPIO port must be accessed carefully; always be sure that the SDA and SCL GPIO bits have a "1" in them or else the built-in EEPROM might be accessed incorrectly, causing problems with subsequent reads. You should never use the instruction:

```
clrf    GPIO
```

Data is written most-significant bit first, which is probably backwards to most applications. Before any transfer, a control byte must be written. The control byte's data is in the format:

```
0b01010000R
```

where *R* is the Read/_Write byte (indicating what is coming next). If the read/write bit is set, then a read of the EEPROM at the current address pointer will occur. If a write is to take place, the read/write bit is reset.

After a byte is sent, the SDA line is pulled low to indicate an acknowledgment (ACK or just A, in the bit-stream representations). This bit is set low (as an acknowledgment) when the operation has completed successfully. If the acknowledgment bit is high (NACK), it does not necessarily mean there was a failure; if it is issued by the EEPROM, then it indicates that a previous write has not completed. The PICmicro® MCU will issue it to stop the EEPROM from preparing to send additional bytes out of its memory in a multibyte read.

Five operations can be carried out with the EEPROM that is built into the PIC12CE5xx. They are:

**1** Current address set
**2** Current address set/data byte write
**3** Data byte read at current address
**4** Sequential (multibyte) read at current address
**5** Write completion poll

The EEPROM in the PIC12CE5xx is only 16 bytes in size. Each byte is accessed using a four-bit address. This address is set using a control byte with the R bit reset, followed by the address. The bit stream looks like:

```
idle - Start - 1010000A - 0000addrA - DataByteA - Stop - idle
```

In the second byte sent, the 0b00000addr pattern indicates that the four *addr* address bits become the address to set the EEPROM's internal address pointer to for subsequent operations.

After the two bytes have been sent, the SCL and SDA lines are returned to IDLE for three cycles, using the instruction:

```
movlw           0x0C0
iorwf           GPIO,f              ; set SDA /SCL
```

before another operation can complete.

The address data write is similar to the address write, but does not force the two lines into IDLE mode and it passes along a data byte before stopping the transfer:

```
Idle - Start - 10100000A - 0000addrA - DataByteA - Stop - idle
```

Data bytes can be read singly or sequentially, depending on the state of ACK from the PICmicro® MCU to the EEPROM after reading a byte. To halt a read, when the last byte to be read has been received, the PICmicro® MCU issues a NACK ("N" in the bitstream listing) to indicate that the operation has completed.

A single-byte read looks like:

```
idle - Start - 10100001A - DataByteN - Stop - idle
```

A two-byte read looks like:

```
idle - Start - 10100001A - DataByteA - DataByteN - Stop - idle
```

The last operation is sending dummy write control bytes to poll the EEPROM to see whether or not a byte write has completed (10 ms are required). If the write has completed, then an ACK will be returned or else a NACK will be returned.

One point on the Flash15x-ASM code is referenced in the 12CE5xx data sheet and found on the Microchip Web page. This file is designed to be linked into your application and provide the necessary I2C routines to access the EEPROM memory. Unfortunately, this file is quite difficult to set up correctly and there are no instructions for using it.

If you want to use the Flash15x.ASM file, here are some hints that should make it easier:

1 Install the Flash15×.ASM code so that it occupies memory in the first 256 bytes of the PICmicro® MCU. The file should not be put at the start of program memory because this will interfere with the PICmicro® MCU's reset.
2 Declare EEADDR and EEDATA in your file-register variable declarations.
3 Be sure that the *#define emulated* line is commented out or changed. If this line is left in, code will be generated that will attempt to write to the SDA and SCL bits (which don't exist); in the process, it will set all of the GPIO bits to output.

# TMR1 and TMR2

Along with TMR0, many PICmicro® MCUs have an additional 16-bit (TMR1) and 8-bit (TMR2) timer built into them. These timers are designed to work with the compare/capture program hardware feature. Along with enhancing this module, they can also be used as "straight' timers within the application. This section presents their basic operation and the next section shows how they work with the CCP module hardware.

TMR1 (Fig. 5-28) is a 16-bit timer that has four possible inputs. What is most interesting about TMR1 is that it can use its own crystal to clock it. This allows TMR1 to run while the PICmicro® MCU's processor is asleep.

To access TMR1 data, the TMR1L and TMR1H registers are read and written. Just like in TMR0, if the TMR1 value registers are written, the TMR1 prescaler is reset. A TMR1 interrupt request (TMR1IF) is made when TMR1 overflows. TMR1 interrupt requests are passed to the PICmicro® MCU's processor when the TMR1IE bit is set.

TMR1IF and TMR1IE are normally located in the PIR and PIE registers. To request an interrupt, along with TMR1IE and GIE being set, the INTCON PIE bit must also be set.

To control the operation of TMR1, the T1CON register is accessed with its bits defined in Table 5-13.

The external oscillator is designed for fairly low-speed real-time clock applications. Normally, a 32.768-kHz watch crystal is used, along with two 33-pF capacitors. 100-kHz or 200-kHz crystals could be used with TMR1, but the capacitance required for the circuit changes to 15 pF. The TMR1 oscillator circuit is shown in Fig. 5-29.

When TMR1 is running at the same time as the processor, the T1SYNCH bit should be reset. This bit will cause TMR1 to be synchronized with the instruction clock. If the TMR1 registers are to be accessed during processor execution, resetting T1SYNCH will ensure that no clock transitions occur during TMR1 access. T1SYNCH must be set (no synchro-

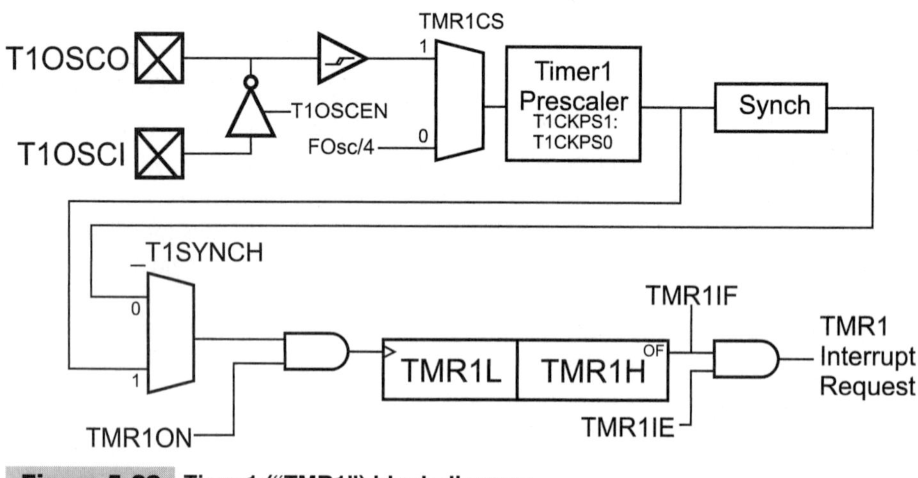


**Figure 5-28** Timer1 ("TMR1") block diagram

**TABLE 5-13 TICON Bit Definition**

| BIT | DESCRIPTION |
|---|---|
| 7-6 | Unused |
| 5-4 | T1CPS1:T1CPS0—Select TMR1 Prescaler Value<br>11—1:8 prescaler<br>10—1:4 prescaler<br>01—1:2 prescaler<br>00—1:1 prescaler |
| 3 | T1OSLEN—Set to Enable TMR1's built in Oscillator |
| 2 | T1SYNCH—when TMR1CS reset the TMR1 clock is synchronized to the Instruction Clock |
| 1 | TMR1CS—When Set, External Clock is Used |
| 0 | TMR1ON—When Set, TMR1 is Enabled |

nized input) when the PICmicro® MCU is in Sleep mode. In Sleep mode, the main oscillator is stopped, stopping the synchronization clock to TMR1.

In the PIC18Cxx devices, TMR1 can be specified as the processor clock. I know I will take advantage of this feature as I work more with the PIC18Cxx to create an application to run in a low-power mode (less than a 1-mA intrinsic current) without disabling the entire device and its built-in functions. Notice that returning to the normal program oscillator will require the 1024-instruction cycle and optional 72-ms power-up reset delay that occurs when the PICmicro® MCU powers up normally.

The TMR1 prescaler allows 24-bit instruction cycle delay values to be used with TMR1. These delays can either be a constant value or an overflow, similar to TMR0. To calculate a delay, the formula:

Delay = (65,536 − *TMR1Init*) * *prescaler/T1frequency*

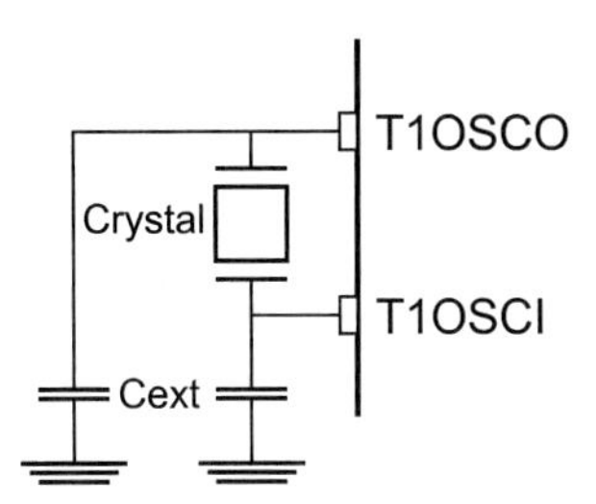


**Figure 5-29** Timer1 ("TMR1") oscillator circuit

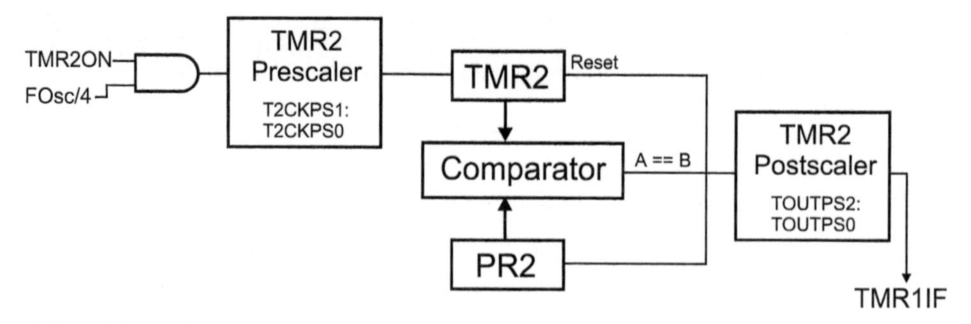


**Figure 5-30** Timer2 ("TMR2") block diagram

where the *T1frequency* can be the instruction clock, TMR1 oscillator, or an external clock driving TMR1. Rearranging the formula, the *TMR1init* initial value can be calculated as:

$$\text{TMR1Init} = 65{,}536 - (\textit{Delay} \times \textit{T1Frequency/prescaler})$$

When calculating delays, the prescaler will have to be increased until the calculated *TMR1Init* is positive. This is similar as to how the TMR0 prescaler and initial value is calculated for TMR0.

TMR2 (Fig. 5-30) is used as a recurring event timer. When it is used with the CCP module, it is used to provide a pulse-width-modulated timebase frequency. In normal operations, it can be used to create a 16-bit instruction cycle delay.

TMR2 is continually compared against the value in PR2. When the contents of TMR2 and PR2 match, TMR2 is reset, and the event is passed to the CCP as *TMR2 Reset.* If the TMR2 is to be used to produce a delay within the application, a postscaler is incremented when TMR2 overflows and eventually passes an interrupt request to the processor.

TMR2 is controlled by the T2CON register, which is defined in Table 5-14.

The TMR2 register can be read or written at any time with the usual note that writes cause the prescaler to be zeroed. Updates to TMR2 do not reset the TMR2 prescaler.

The timer itself is not synchronized with the instruction clock because it can only be used with the instruction clock. Thus, TMR2 can be incremented on a 1-to-1 instruction clock ratio.

PR2 contains the reset or count up to value. The delay before reset is defined as:

$$\textit{Delay} = \textit{prescaler} \times (\textit{PR2} + 1)/(F_{osc}/4)$$

If PR2 is equal to zero, the delay is:

$$\textit{Delay} = (\textit{prescaler} \times 256)/(F_{osc}/4)$$

I do not usually calculate TMR2 delays with an initial *TMR2INIT* value. Instead, I take advantage of the PR2 register to provide a repeating delay and just reset TMR2 before starting the delay.

To calculate the delay between TMR2 overflows (and interrupt requests):

$$\textit{Delay} = (\textit{prescaler} \times [\textit{PR2} + 1|256])/((F_{osc}/4) * \textit{postscaler})$$

**TABLE 5-14 T2CON Bit Definition**

| BIT | DESCRIPTION |
|---|---|
| 7 | Unused |
| 6-5 | TOUTPS3 : TOUTPS0—TMR2 Postscaler Select |
| | 1111—16 : 1 Postscaler |
| | 1110—15 : 1 Postscaler |
| | : |
| | 0000—1 : 1 Postscaler |
| 2 | TMR2ON—When Set, TMR2 Prescaler is Enabled |
| 1-0 | T2CKPS1 : T2CKPS0—TMR2 Prescaler Select |
| | 1x—16 : 1 prescaler |
| | 01— 4 : 1 prescaler |
| | 00— 1 : 1 prescaler |

Interrupts use the TMR2IE and TMR2IF bits that are similar to the corresponding bits in TMR1. These bits are located in the PIR and PIE registers. Because of the exact interrupt frequency, TMR2 is well suited for applications that provide bit-banging functions, such as asynchronous serial communications or pulse-width-modulated signal outputs.

## COMPARE/CAPTURE/PWM (CCP) MODULE

Included with TMR1 and TMR2 is a control register and a set of logic functions (known as the *CCP*), which enhances the operation of the timers and can simplify your applications. This hardware can be provided singly or in pairs, which allows multiple functions to execute at the same time. If two CCP modules are built into the PICmicro® MCU, then one is known as *CCP1* and the other as *CCP2*. In the case where two CCP modules are built in, then all the registers are identified with the CCP1 or CCP2 prefix.

The CCP hardware is controlled by the CCP1CON (or CCP2CON) register, which is defined in Table 5-15.

The most basic CCP mode is capture, which loads the CCPR registers (CCPR1H, CCPR1L, CCPR2H, and CCPR2L), according to the mode that the CCP register is set in. This function is shown in Fig. 5-31 and shows that the current TMR1 value is saved when the specified compare condition is met.

Before enabling the Capture mode, TMR1 must be enabled (usually running with the PICmicro® MCU clock). The edge-detect circuit in Fig. 5-31 is a four-to-one multiplexer, which chooses between the prescaled rising edge input or a falling-edge input. It passes the selected edge to latch the current TMR1 value and optionally request an interrupt.

In Capture mode, TMR1 is running continuously and is loaded when the condition on the CCPx pin matches the condition specified by the CCPxMS:CCPxM0 bits. When a cap-

**TABLE 5-15 CCPxCON Bit Definition**

| BIT | FUNCTION |
|---|---|
| 7-6 | Unused |
| 5-4 | DC1B1 : DC1B0—CEPST significant 2 bits of the PWM compare value. |
| 3-0 | CCP1M3 : CCP1M0—CCP module operating mode. |
| | 11xx—PWM Mode |
| | 1011—Compare Mode—Trigger Special Event |
| | 1010—Compare Mode—Generate Software Interrupt |
| | 1001—Compare Mode—on Match CCP pin low |
| | 1000—Compare Mode—on Match CCP pin high |
| | 0111—Capture on every 16th rising edge |
| | 0110—Capture on every 4th rising edge |
| | 0101—Capture on every rising edge |
| | 0100—Capture on every falling edge |
| | 00xx—CCP off |

ture occurs, then an interrupt request is made. This interrupt request should be acknowledged and the contents of CCPRxH and CCPRxL saved to avoid having them written over and the value lost.

Capture mode is used to time repeating functions or determine the length of a PWM pulse. If a PWM pulse is to be timed, then when the start value is loaded, the polarity is reversed to get to the end of the pulse. When timing a PWM pulse, the TMR1 clock must be fast enough to get a meaningful value with a high enough resolution so that there will be an accurate representation of the timing.

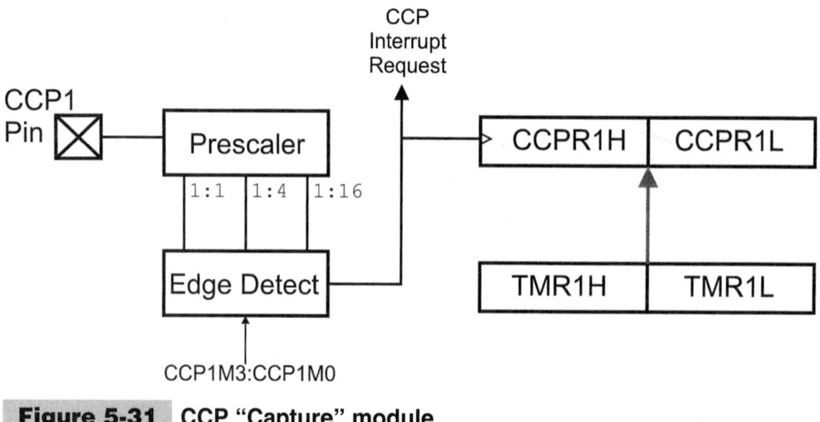


**Figure 5-31** CCP "Capture" module

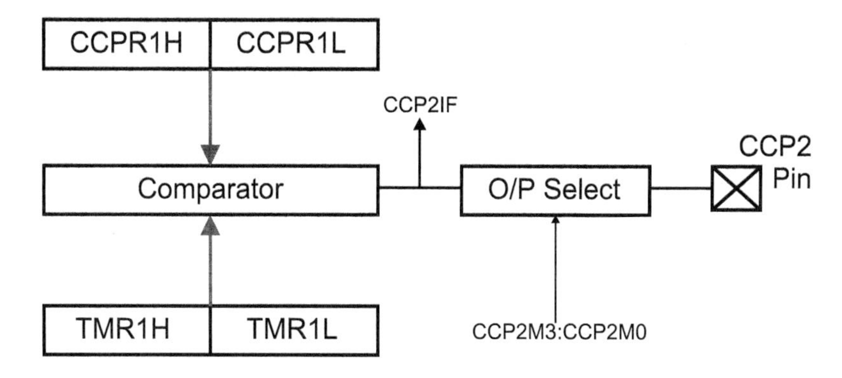


**Figure 5-32** **CCP "Compare" module**

Compare mode changes the state of the CCPx pin of the PICmicro® MCU when the contents of TMR1 match the value in the CCPRxM and CCPRxL registers (Fig. 5-32). This mode is used to trigger or control application code or external hardware after a specific delay.

The most interesting use I've seen for the Compare mode of the CCP is to turn the PICmicro® MCU into a watchdog for a complex system. As shown in Fig. 5-33, the PICmicro® MCU controls reset to the system processor. Upon power up, the PICmicro® MCU holds the processor reset until $V_{cc}$ has stabilized and then TMR1 is reset each time the system writes to the PICmicro® MCU. System reset is enabled if $V_{cc}$ falls below a specific level.

Using event-driven code, the PICmicro® MCU application would look like:

```
PowerUpEvent()                              // PICmicro® MCU Power Up
{
  TMR1 = 0; TMR1 = on;                      // Start TMR1
  CCPRx = PowerUpDelay;                     // Put in Watchdog Delay
  CCPxCON = 0b000001000;                    // Drive Pin Low and /then High
                                            //   on Compare Match
```

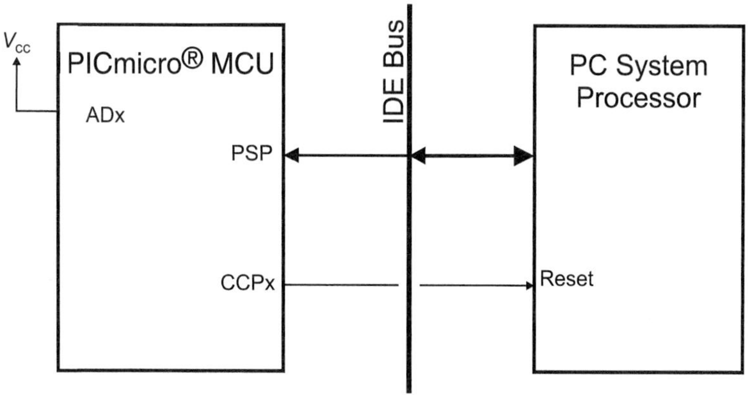

$V_{cc}$
PICmicro® MCU
ADx
PSP
CCPx
IDE Bus
PC System Processor
Reset

**Figure 5-33** **CCP Compare module PC reset controller**

```
  ADCIE = on;                            // Start ADC Check of Vcc
} // End PowerUpEvent
CompareMatchEvent( )                     // TMR1 = Compare / WDT T/O.
{
  CCPxCON = 0;                           // turn off compare.
  CCPx = 1;                              // reset system
} // End CompareMatchEvent
PSPWriteEvent( )                         // PSP Written to Reset WDT
{                                        // Count
 TMR1 = 0;
} // End PSPWriteEvent
ADCIFEvent()                             // ADC Finished Vcc check
{
 if (ADC < OperatingMinimum) {
   CCPxCON = 0;                          // Turn Off ADC
   CCPx = high;                          // Reset system program
   }
   ADCIF = 0;                            // Reset Interrupt Request
} // End ADCIFEvent
```

Of the three CCP modes, I find the PWM signal generator to be the most useful. This mode outputs a PWM signal using the TMR2 reset at a specific value capability. The block diagram of PWM mode is shown in Fig. 5-34. The mode is a combination of the normal execution of TMR2 and capture mode; the standard TMR2 provides the PWM period and the compare control provides the on-time specification.

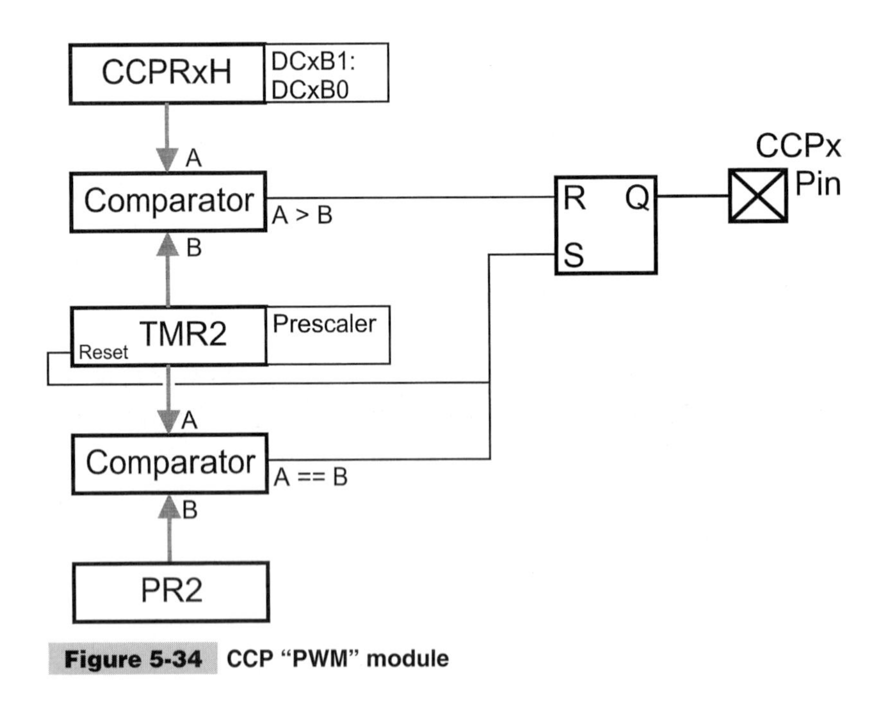


**Figure 5-34** CCP "PWM" module

When the PWM circuit executes, TMR1 counts until its most-significant eight bits are equal to the contents of PR2. When TMR2 equals PR2, TMR2 is reset to zero, and the CCPx pin is set high. TMR2 is run in a 10-bit mode (the 4:1 prescaler is enabled before PWM operation). This 10-bit value is then compared to a program value in CCPRxM (along with the two DCxBx bits in CCPxCON); when they match, the CCPx output pin is reset low.

To set up a 65% duty cycle in a 20-kHz PWM executing in a PICmicro® MCU clocked at 4 MHz, the following steps are taken.

First, the CCPRxM and PR2 values are calculated for TMR2, the 4:1 prescaler must be enabled, resulting in a delay of:

```
Delay = (PR2 + 1) frequency/4)

  PR2 = delay  frequency - 1
      = 50 ms 4 MHz - 1
      = 200 - 1
      = 199
 65% of 200 is 130, which is then loaded into CCPRxM.
```

The code for creating the 65%, 20-kHz PWM is:

```
 movlw                199
 movwf                PR2                      ; Set up TMR2 Operation
 movlw                (1 << TMR2ON) + 11
 movwf                T2CON                    ; Start it Running with a 50
                                               ; Period
 movlw                130                      ; 65% of the Period
 movwf                CCPRxH
 movlw                (1<<DCxB1) + 0x00F
 movwf                CCPxCON                  ; Start PWM
; PWM is operating
```

Notice that in this code, I don't enable interrupts or have to monitor the signal output. As well, notice that I don't use the fractional bits. To use the two least-significant bits, I assume they are fractional values. For this example, if I wanted to fine tune the PWM frequency to 65.875%, I would recalculate the value as a fraction of the total period.

For a period of 200 TMR2 counts with a prescaler of 4, the CCPRxH value becomes 131.75. To operate the PWM, I would load 130 into CCPRxh (subtracting 1 to match TMR2's zero start) and then the fractional value 0.75 into DCxB1 and DCxB0 bits. I as-

**TABLE 5-16 CCP DCxBX Bit Definition**

| FRACTION | DCxB1:DCxB0 |
|---|---|
| 0.00 | 00 |
| 0.25 | 01 |
| 0.50 | 10 |
| 0.75 | 11 |

sume that DCxB1 has a value of 0.50 and DCxB0 has a fractional value of 0.25. So, to get a PWM in this case, CCPRxH is loaded with 130, and DCxB1 and DCxB0 are both set. Table 5-16 gives the fractional DCxBX bit values.

The least-significant two bits of the PWM are obviously not that important unless a very small PWM on period is used in an application. A good example of this is using the PWM module for an RC servo. In this case, the PWM period of 20 ms with an on time of 1 to 2 ms. This gives a PWM on range of 5% to 10%, which makes the DCxB1 and DCxB0 bits important in accurately positioning the servo.

# Serial I/O

Like many microcontrollers, the PICmicro® MCU has optional built-in serial I/O interfacing hardware. These interfaces, which are available on certain PICmicro® MCU part numbers, allows a PICmicro® MCU to interface with external memory and enhanced I/O functions (such as ADCs) or communicate with a PC using RS-232. Like other enhanced peripheral features, the serial I/O hardware is available on different PICmicro® MCUs and the hardware might be available dissimilarly in different devices.

You will find that the serial I/O hardware is not as flexible as you might want. The lack of flexibility is a concern with regards to interfacing RS-232 cheaply to PCs or using the synchronous serial interfaces in different situations. The following sections and Chapters 15 and 16 cover how these interfaces are used practically and how they can be modified for more flexible and efficient serial interfaces.

## SYNCHRONOUS SERIAL PORT (SSP) COMMUNICATIONS MODULE

The basic operations of the *Synchronous Serial Port (SSP)* are covered in this section, followed by the I2C operations in the next section. I break the operation of the SSP into two parts because I2C is quite a complex operation that I felt would be best served by covering it in its own section.

The second reason for splitting out the I2C function is the inability of two SSP versions to provide the full range of I2C operations. The SSP and BSSP modules, which are available in many PICmicro® MCUs, do not have I2C Master mode capabilities. This limits their usefulness in working with I2C (where, typically, the PICmicro® MCU is a master), as compared to PICmicro® MCUs equipped with the *MSSP (Master SSP)* module that does have I2C multimaster capabilities.

This section covers the SSP module and how it works in SPI mode. SPI is an 8-bit synchronous serial protocol that uses three data bits to interface to external devices. Data is clocked out, with the most-significant bit first, on rising or falling edges of the clock. The clock itself is generated within the PICmicro® MCU (Master mode), or it is provided by an external device and used by the PICmicro® MCU (slave mode) to clock out the data. The SPI data stream looks like Fig. 5-35.

The SSP clock can be positive with a 0 idle or negative (high line idle) with a 1 idle and the clock pulsing to 0 and back again. The data-receive latch is generally on the return-to-idle state transition.

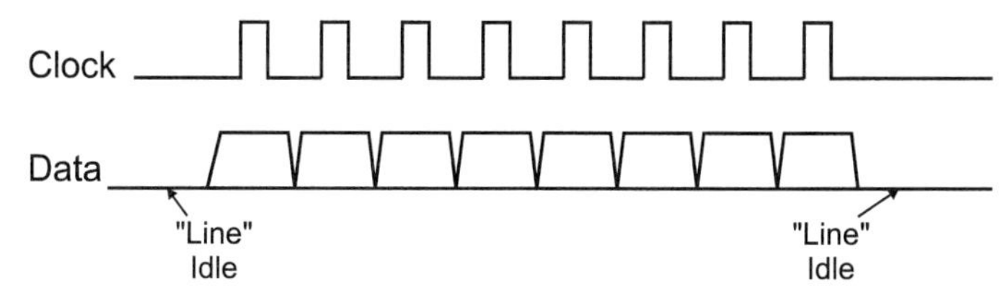


**Figure 5-35** SPI synchronous serial data waveform

The BSSP module is the basic SSP module, which provides data polling on the return-to-idle clock edge. The original SSP module provides the ability to vary when data is output and read.

Controlling the operation of the different SSP modules is the SSPCON register (Table 5-17). In describing the operational bits, notice that I only describe the SPI-specific operations. The block diagram for the SSP module is shown in Fig. 5-36.

In Master mode, when a byte is written to SSPBUF (an 8-bit, most-significant bit first), the data-transfer process is initiated. The status of the transfer can be checked by the SSPSTAT register BF flag; the SSPSTAT register is defined in Table 5-18.

**TABLE 5-17 SSP/BSSP SSPCON Bit Definition**

| BIT | FUNCTION |
|---|---|
| 7 | WCOL—Write collision, set when new byte written to SSPBUF while transfer is taking place |
| 6 | SSPOV—Receive Overflow, indicates that the unread byte is SSPBUF Over written while in SPI slave mode |
| 5 | SSPEN—Set to enable the SSP module |
| 4 | CKP—Clock polarity select, set to have a high idle |
| 3-0 | SSPM3:SSPMO SPI mode select |
| | 1xxx—I2C and reserved modes |
| | 011x—I2C slave modes |
| | 0101—SPI slave mode, clock = SCK pin, _SS not used |
| | 0100—SPI slave mode, clock = SCK pin, _SS enabled |
| | 0011—SPI master mode, TMR2 clock used |
| | 0010—SPI master mode, INSCK/16 |
| | 0001—SPI master mode, INSCK/4 |
| | 0000—SPI master mode, INSCK |

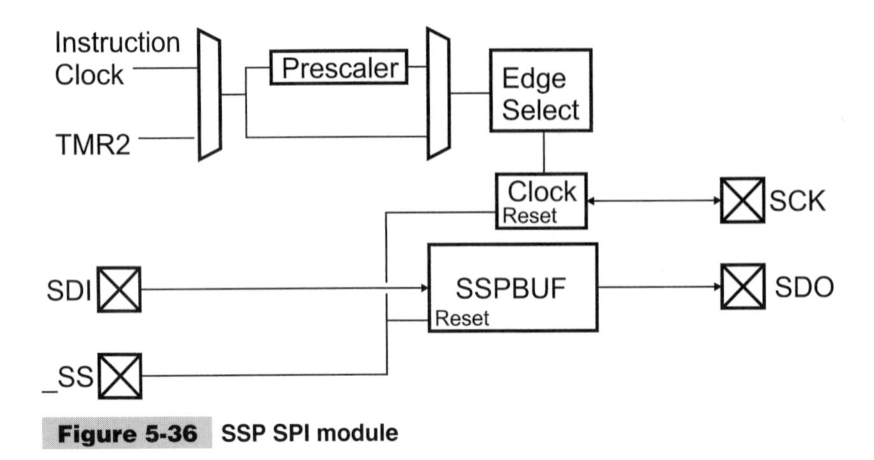


**Figure 5-36** SSP SPI module

The connection of a PICmicro® MCU to an SPI bus is quite straight-forward. Figure 5-37 shows two PICmicro® MCUs with the SD0's and SDI's sides connected. To initiate a byte transfer, a byte is written to the SSPBUF of the master. Writing to the SSPBUF of the slave will not initiate a transfer. When SPI mode is enabled, the SDI, SDO, and SCK bits' TRIS functions are set appropriately (Fig. 5-37).

The SSP SPI transfers can be used for single-byte synchronous serial transmits of receivers with serial devices. Figure 5-38 shows the circuit to transmit a byte to a 74LS374 wired as a serial in, parallel out shift register. Figure 5-39 shows a 74LS374 being used with a 74LS244 as a synchronous parallel-in/ serial-out register. Both of these operations are initiated by a write to SSPBUF.

**TABLE 5-18 SSP/BSSP SSPSTAT Bit Definition**

| BIT | FUNCTION |
|---|---|
| 7 | SMP—Set to have data sampled after active to idle transition, reset to sample at active to idle transition, not available in BSSP |
| 6 | CKE—Set to TX data on idle to active transition, else TX data on active to idle transition, not available in BSSP |
| 5 | D/_A—Used by I2C |
| 4 | P—Used by I2C |
| 3 | S—Used by I2C |
| 2 | R/_W—Used by I2C |
| 1 | UA—Used by I2C |
| 0 | BF—Busy flag, reset while SPI operation active |

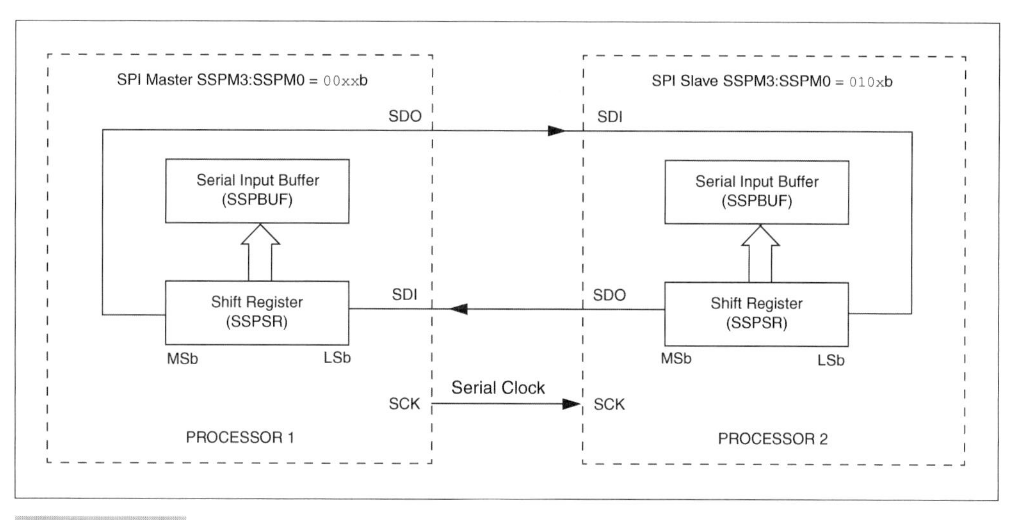


**Figure 5-37** SPI Master/Slave Connection

The SPI data receive operation might not be immediately obvious. To latch data into the '374, the I/O pin is driven high, which disables the '374's drivers allowing the parallel data to be latched in. When the I/O pin is low, the '244's drivers are disabled and the '374 behaves like a shift register.

To show this, I have listed the code to read the input state of Fig. 5-39. Notice that I disable the SSPEN bit when a transfer is not occurring, to allow the I/O pin and SCK to strobe in the data.

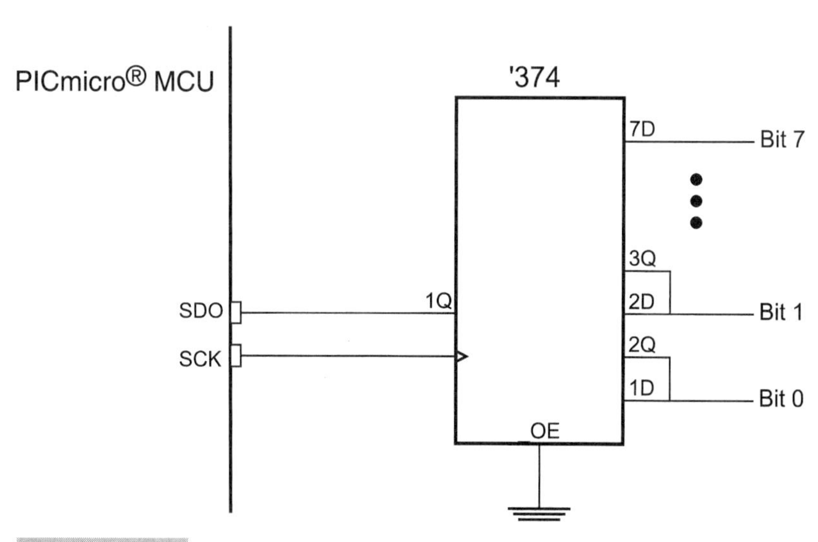


**Figure 5-38** SSP SPI module used to shift data out

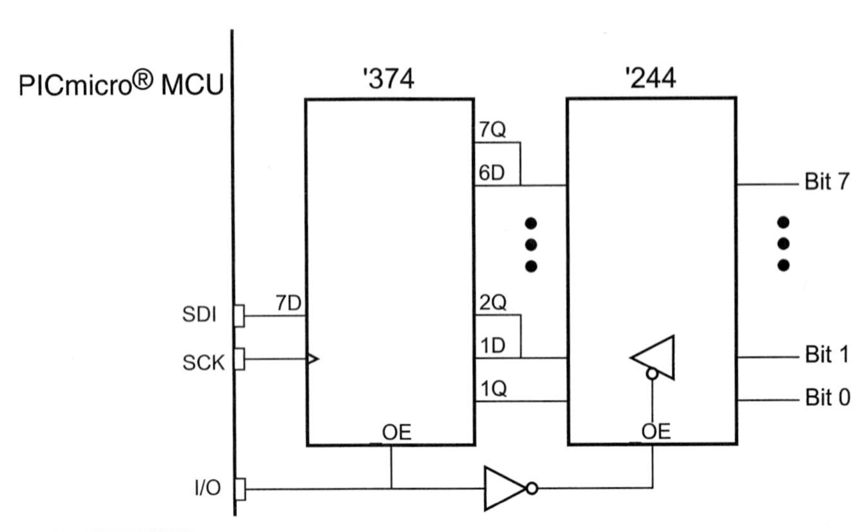


**Figure 5-39** SSP SPI module used to shift data in

```
  bsf          IOPin                       ; Want to Latch Data into the '374
  bcf          SCK
  bsf          STATUS,                     RPO
  bcf          IOpin
  bcf          SCK
  bcf          STATUS, RPO
  bsf          SCK                         ; Latch the Data into the '374
  bcf          SCK
  bcf          IOpin                       ; Disable '244 output, Enable '374
  movlw        (I << SMP) + (I << CKE)
  movwf        SSPSTAT                     ; Set up the SSP Shift In
  movlw        (I << SSPEN) + (I << CKP) +0x000
  movwf        SSPCON
  movf         TXData, f                   ; Load the Byte to Send
  movwf        SSPBUF                      ; Start Data Transfer
  btfss        SSPTAT, BF
   goto        $ - 1                       ; Wait for Data Receive to Complete
;  Data Ready in SSPBUF when Execution
;  Here
  bcf SSPCON,$SPEN                         ; Turn off SSP
```

When using the SSP, the data rate can either be selected as a multiple of the executing clock or use the TMR2 overflow output. The actual timing depends on the hardware to which the PICmicro® MCU SSP master is communicating.

When in Slave mode, along with an external clock being provided, there is a transmit reset pin, known as *_SS*. When this pin is asserted high, the SSP output is stopped (the SDO TRIS bit is changed to Input mode) and the SSP is reset with a count of zero. When the bit is reset, the clock will start up again, the original most-significant bit is reset, followed by the remaining seven bits.

**Master SSP and I2C operation** When I wrote the first edition of this book, one of the most-significant concerns that people have had with the PICmicro® MCU's built-in hardware was the lack of master and multimastering I2C capability. This concern has been resolved with the availability of the *MSSP (Master SSP)* module that is included in new PICmicro® MCU devices. The original SSP and BSSP will continue to be available in devices that were originally designed with them, but the enhanced MSSP will be designed into all new devices that have the SSP module.

The MSSP data sheets contain 33 pages documenting how the function works. When you actually work with the MSSP, you will find that very few instructions are actually required to implement a basic master mode I2C interface and that their use is quite easy to understand. This section concentrates on a single master I2C interface and points out the issues that you will have to be aware of when working in a multimaster system.

Five registers are accessed for MSSP I2C operation: the SSP control registers (SSPCON and SSPCON2), the SSP status register (SSPSTAT), the SSP receive/transmit register (SSPBUF), and the SSP address register (SSPADD). These registers are available in the SSP and BSSP, but are slightly different for the MSSP. The MSSP control registers are defined in Tables 5-19 through 5-21.

The status of the transfer can be checked by the SSPSTAT register BF flag; the SSPSTAT register is defined as:

I2C connections between the PICmicro® MCU's I2C SDA (data) and SCL (clock) pins is very simple, with just a pull up on each line (Fig. 5-40). I typically use a 1-K resistor for 400-kHz data transfers and a 10 K for 100-kHz data rates.

Notice that before any of the I2C modes are to be used, the TRIS bits of the respective SDA and SCL pins *must* be in Input mode. Unlike many of the other built-in advanced I/O functions, MSSP does not control the TRIS bits. Not having the TRIS bits in Input mode will not allow the I2C functions to operate.

In Master mode, the PICmicro® MCU is responsible for driving the clock (SCL) line for the I2C network. This is done by selecting one of the SPI Master modes and loading the SSPADD register with a value to provide a data rate that is defined by the formula:

$$I2C\ Data\ Rate = F_{osc}/[4 * (SSPADD + 1)]$$

This can be rearranged to:

$$SSPADD = [F_{osc}/(4 * I2C\ Data\ Rate)] - 1$$

So, in a 4-MHz PICmicro® MCU, to define a 100-kHz I2C data rate, the previous formula would be used to calculate the value loaded into SSPADD:

$$\begin{aligned} SSPADD &= [F_{osc}/(4 * (I2C\ Data\ Rate)] - 1 \\ &= [4\ \text{MHz}/(4 * (100\ \text{kHz})] - 1 \\ &= [4 * 10^6 / (4 * (100 * 10^3)] - 1 \\ &= 10 - 1 \\ &= 9 \end{aligned}$$

In a PICmicro® MCU running at 4 MHz, a value of 9 must be loaded into the SSPADD to have a 100-kHz I2C data transfer.

**TABLE 5-19 MSSP SSPCON Bit Definition**

| BIT | FUNCTION |
|---|---|
| 7 | WCOL—Write collision, set when new byte written to SSPBUF while transfer is taking place |
| 6 | SSPOV—Receive Overflow, indicates that the unread byte is SSPBUF over written |
| 5 | SSPEN—Set to enable the SSP module |
| 4 | CKP—In I2C Modes, if bit is reset, the I2C "SCL" Clock Line is Low. Keep this bit set. |
| 3-0 | SSPM3:SSPM0 SPI mode select |
| | 1111—I2C 10 Bit Master Mode/Start and Stop Bit Interrupts |
| | 1110—I2C 7 Bit Master Mode/Start and Stop Bit Interrupts |
| | 1101—Reserved |
| | 1100—Reserved |
| | 1011—I2C Master Mode with Slave Idle |
| | 1010—Reserved |
| | 1001—Reserved |
| | 1000—I2C Master Mode with SSPADD Clock Definition |
| | 0111—I2C Slave Mode, 10 Bit Address |
| | 0110—I2C Slave Mode, 7 Bit Address |
| | 0101—SPI slave mode, clock = SCK pin, _SS not used |
| | 0100—SPI slave mode, clock = SCK pin, _SS enabled |
| | 0011—SPI master mode, TMR2 clock used |
| | 0010—SPI master mode, INSCK/16 |
| | 0001—SPI master mode, INSCK/4 |
| | 0000—SPI master mode, INSCK |

To send data from the PICmicro® MCU to an I2C device using the MSSP, the following steps must be taken:

1. The SDA/SCL lines must be put into Input mode (i.e., their respective TRIS bits must be set)
2. I2C Master mode is enabled. This is accomplished by setting the SSPEN bit of SSPCON and writing 0b01000 to the SSPM3:SSPM0 bits of the SSPCON register.

**TABLE 5-20 MSSP SSPCON2 Bit Definition**

| BIT | FUNCTION |
|---|---|
| 7 | GCEN—Enable Interrupt when "General Call Address" (0x000) is Received |
| 6 | ACKSTAT—Received Acknowledge Status. Set when Acknowledge was Received |
| 5 | ACKDT—Acknowledge Value Driven out on Data Write |
| 4 | ACKEN—Acknowledge Sequence Enable Bit which when Set will Initiate an Acknowledge sequence on SDA/SCL. Cleared by Hardware |
| 3 | RCEN—I2C Receive Enable Bit |
| 2 | PEN—Stop Condition Initiate Bit. When Set, Stop Condition on SDA/SCL. Cleared by Hardware |
| 1 | RSEN—Set to Initiate the Repeated Start Condition on SDA/SCL. Cleared by Hardware |
| 0 | SEN—When Set, a Start Condition is Initiated on the SDA/SCL. Cleared by hardware. |

3 A start condition is initiated by setting the SEN bit of SSPCON2. This bit is then polled until it is reset.
4 SSPBUF is loaded with the address of the device to access. Notice that for many I2C devices, the least-significant bit transmitted is the read/write bit. The R/_W bit of SSPSTAT is polled until it is reset (which indicates the transmit has been completed).

**TABLE 5-21 MSSP SSPSTAT Bit Definition**

| BIT | FUNCTION |
|---|---|
| 7 | SMP—Set to have data sampled after active to idle transition, reset to sample at active to idle transition, not available in BSSP |
| 6 | CKE—Set to TX data on idle to active transition, else TX data on active to idle transition, not available in BSSP |
| 5 | D/_A—Used by I2C |
| 4 | P—Used by I2C |
| 3 | S—Used by I2C |
| 2 | R/_W—Used by I2C |
| 1 | UA—Used by I2C |
| 0 | BF—Busy flag, reset while SPI operation active |

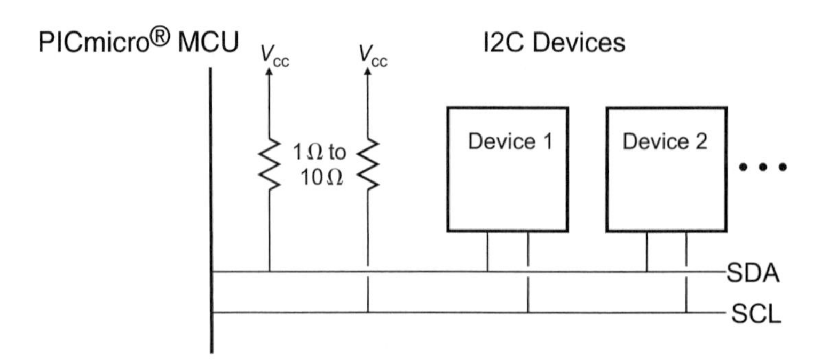


**Figure 5-40** 12C connection to PICmicro® MCU

5. The ACK bit from the receiving device is checked by reading the "ACKDT" bit of the SSPCON2 register
6. SSPBUF is loaded with the first eight bits of data or a secondary address that is within the device being accessed. The R/_W bit of SSPSTAT is polled until it is reset.
7. The ACK bit from the receiving device is checked by reading the ACKDT bit of the SSPCON2 register.
8. A new start condition might have to be initiated between the first and subsequent data bytes. This is initiated by setting the SEN bit of SSPCON2. This bit is then polled until it is reset.
9. Operations 6 through 8 are repeated until all data is sent or a *NACK (Negative ACKnowledge)* is received from the receiving device.
10. A stop condition is initiated by setting the PEN bit of SSPCON2. This bit is then polled until it is reset

This sequence of operations is shown in Fig. 5-41. Notice that in Fig. 5-41, the SSPIF interrupt request flag operation is shown. In this sequence, I avoid interrupts, but the SSPIF bit can be used to either request an interrupt or to avoid the need to poll different bits to wait for the various operations to complete.

To receive data from a device employs a similar set of operations with the only difference being that after the address byte(s) have been sent, the MSSP is configured to receive data when the transfer is initiated:

1. The SDA/SCL lines must be put into Input mode (i.e., their respective TRIS bits must be set).
2. I2C Master mode is enabled. This is accomplished by setting the SSPEN bit of SSPCON and writing 0b01000 to the SSPM3:SSPM0 bits of the SSPCON register
3. A start condition is initiated by setting the SEN bit of SSPCON2. This bit is then polled until it is reset.
4. SSPBUF is loaded with the address of the device to access. Notice that for many I2C devices, the least-significant bit transmitted is the read/write bit. The R/_W bit of SSPSTAT is polled until it is reset (which indicates that the transmit has been completed).
5. The ACK bit from the receiving device is checked by reading the ACKDT bit of the SSPCON2 register.
6. SSPBUF is optionally loaded with the secondary address within the device being read from. The R/_W bit of SSPSTAT is polled until it is reset.

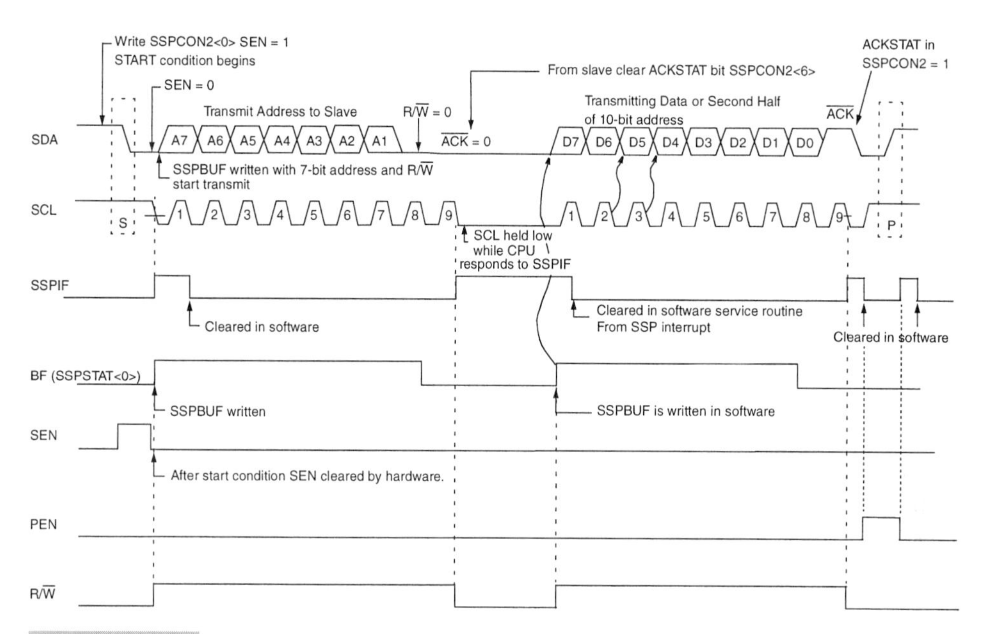


**Figure 5-41** MSSP I2C data address/transmission

**7** If a secondary address was written to the device being read from, reading the ACKDT bit of the SSPCON2 register checks the ACK bit from the receiving device.

**8** A new start condition might have to be initiated between the first and subsequent data bytes. This is initiated by setting the SEN bit of SSPCON2. This bit is then polled until it is reset.

**9** If the secondary address byte was sent, then a second device address byte (with the Read indicated) might have to be sent to the device being read. The R/_W bit of SSPSTAT is polled until it is reset.

**10** The ACKDT will be set (NACK) or reset (ACK) to indicate whether or not the data byte transfer is to be acknowledged in the device being read.

**11** The RCEN bit in the SSPCON2 register is set to start a data byte receive. The BF bit of the SSPSTAT register is polled until the data byte has been received.

**12** Operations 10 through 11 are repeated until all data is received and a *NACK (Negative ACKnowledge)* is sent to the device being read

**13** A stop condition is initiated by setting the PEN bit of SSPCON2. This bit is then polled until it is reset

Figure 5-42 shows the data-receive operation waveform.

Along with the single Master mode, the MSSP is also capable of driving data in Multi-master mode. In this mode, if a data write collision is detected, it stops transmitting data and requests an interrupt to indicate that there is a problem. An I2C *collision* is where the current device is transmitting a high data value, but a low data value is on the SDA line. This condition is shown in Fig. 5-43. The WCOL bit of the SSPCON register indicates that the collision has occurred.

When the collision occurs, the I2C software must wait some period of time (I use the time required to transmit three bytes) before polling the SDA and SCL lines to ensure that

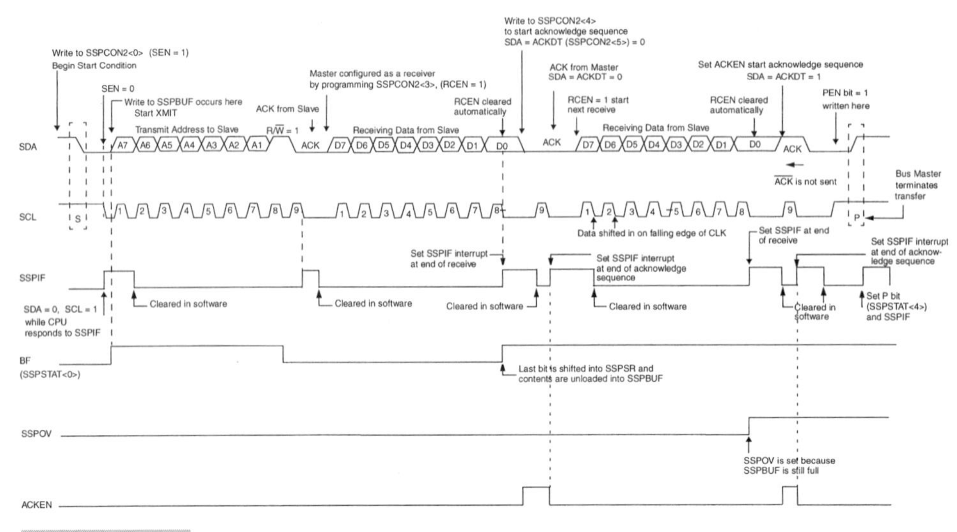


**Figure 5-42** MSSP I2C data address/read

they are high and then initiating a repeated start condition operation. A *repeated start condition* is the process of restarting the I2C data transfer right from the beginning (even if it was halfway through when the collision occurred).

## USART ASYNCHRONOUS SERIAL COMMUNICATIONS

The PICmicro® MCU's *USART (Universal Asynchronous Synchronous Receiver Transmitter)* hardware allows you to interface with serial devices like a PC using RS-232 or for synchronous serial devices with the PICmicro® MCU providing the clock or having an external clock drive the data rate. The USART module is best suited for asynchronous serial data transmission and this section concentrates on its capabilities.

Asynchronous data has been covered elsewhere in more detail in this book. The PICmicro® MCU transmits and receives *NRZ (No Return to Zero)* asynchronous data in the format shown in Fig. 5-44. Figure 5-44 shows five bits of serial data. The PICmicro® MCU can transfer eight or nine bits, although, by setting the high-order bits of the output word, smaller data packets can be sent.

Synchronous data is sent with a clock and is in the format shown in Fig. 5-45.

Synchronous data is latched into the destination on the failing edge of the clock. In both these cases, a byte is sent within a packet. Although packet decoding is covered in detail

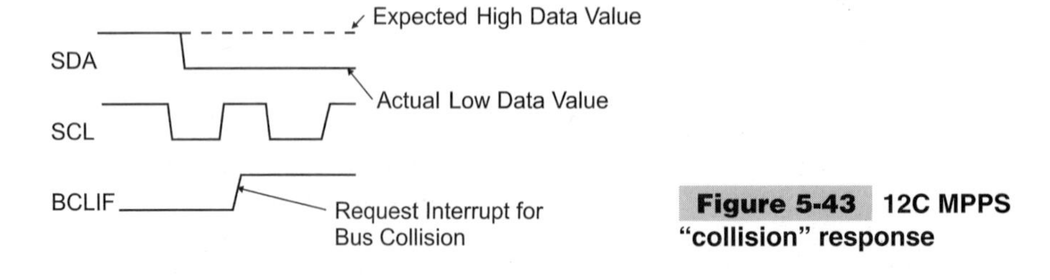


**Figure 5-43** 12C MPPS "collision" response

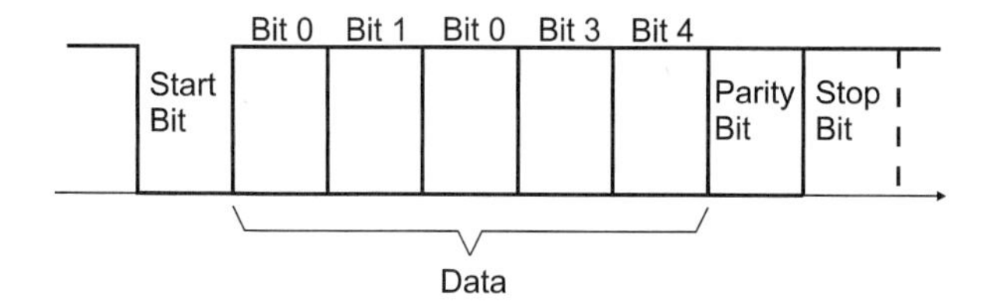


**Figure 5-44** Baudot asynchronous serial data

elsewhere in this book, this section treats the packet encoding and decoding a "black box" part of the USART and deal with how the data bytes are transmitted and received.

The three modules to the USART are the clock generator, the serial data transmission unit, and the serial data reception unit. The two serial I/O units require the clock generator for shifting data out at the write interval. The clock generator's block diagram is shown in Fig. 5-46.

In the clock generator circuit, the SPBRG register is used as a comparison value for the counter. When the counter is equal to the SPBRG register's value, a clock tick output is made and the counter is reset. The counter operation is gated and controlled by the *SPEN (Serial Port ENable)* bit, along with the synch (which selects whether the port is in synchronous or asynchronous mode) and BRGH, which selects the data rate.

In the PICmicro® MCU USART, the bits used to control the operation of the clock generator, transmit unit, and receive unit are spread between the TXSTA and RCSTA registers, along with the interrupt enable and acknowledge registers. The individual bits are defined at the end of this section, after the three functions of the USART are explained.

For asynchronous operation, the data speed is specified by the formula:

$$Data\ Rate = F_{osc}/\{16 * [4 ** (1 - BRGH)] * (SPBRG + 1)\}$$

This formula can be rearranged so that the SPBRG value can be derived from the desired data rate:

$$SPBRG = F_{osc}/\{Data\ Rate * 16 * [4 ** (1 - BRGH)]\} - 1$$

So, for a PICmicro® MCU running at 4-MHz, the SPBRG value for a 1200-bps data rate with BRGH reset is calculated as:

$$\begin{aligned} SPBRG &= F_{osc}/\{Data\ Rate * 16 * [4 ** (1 - BRGH)]\} - 1 \\ &= 4\ \text{MHz}/\{1200/\text{sec} * 16\ [4 ** (1 - 0)]\} - 1 \\ &= 4\ (10 ** 6)/(1200 * 16 * 4) - 1 \\ &= 52.0833 - 1 \\ &= 51.0833 \end{aligned}$$

With 51 stored in SPBRG, the PICmicro® MCU will communicate at an actual data rate of 1,201.9 bps, which has an error of 0.16% to the target data rate of 1200 bps. This error is well within limits to prevent any bits being read in error.

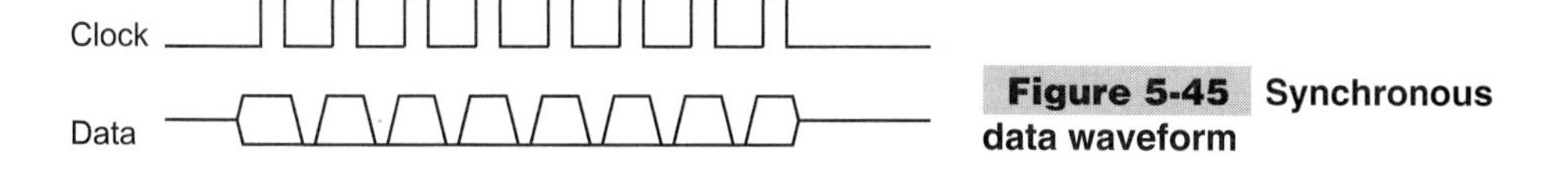


**Figure 5-45** Synchronous data waveform

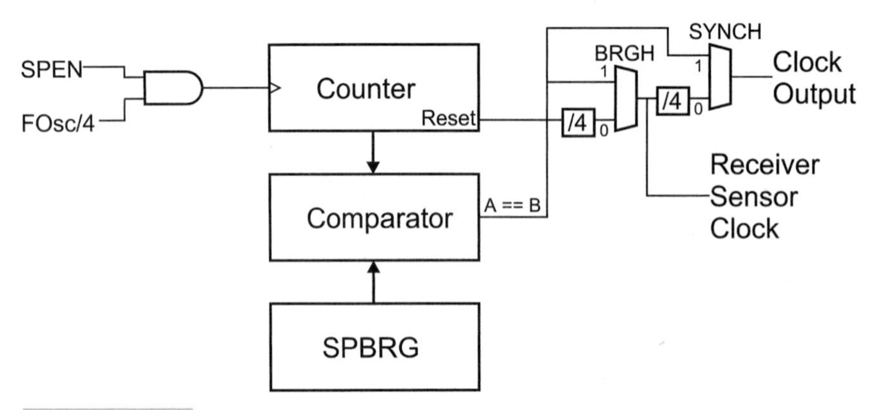


**Figure 5-46** USART clock block diagram

It's worth noting that, for many early PICmicro® MCU part numbers, the BRGH bit does not work properly when it is set. Most later letter revisions (i.e., PIC16C73B) and recently released part numbers have the problem fixed. It can be hard to determine whether or not you have a corrected PICmicro® MCU. To be on the safe side, I recommend that you always develop your applications with the BRGH bit reset. If you need data rates faster than what is possible for the PICmicro® MCU clock (2400 bps is the maximum for a 4-MHz clock), I recommend that you increase the PICmicro® MCU's clock speed, rather than risk setting BRGH in a device in which it doesn't work properly.

The transmission unit of the USART can send eight or nine bits in a clocked (synchronous) or unclocked (synchronous) manner. The block diagram of the hardware is shown in Fig. 5-47.

If the synch bit is set, then data is driven out on the PICmicro® MCU's RX pin with the data clock being either driven into, or out of the TX pin. When data is loaded into the TXREG, if CSRC is reset, then an external device will clock it out. If CSRC can be shifted

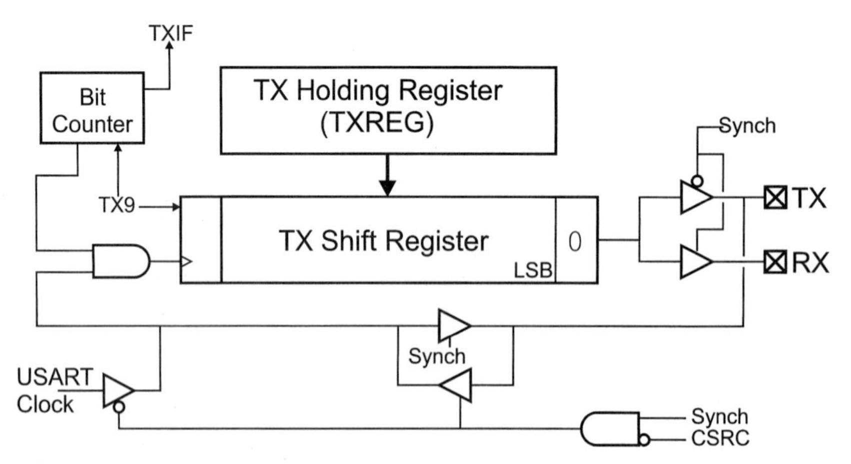


**Figure 5-47** USART transmit hardware block diagram

eight or nine bits at a time with the operation stopping when the data has been shifted out. An interrupt can be requested when the operation is complete.

In Asynchronous mode, once data is loaded into the TXREG, it is shifted out with a 0 leading start bit in NRZ format.

The transmit hold register can be loaded with a new value to be sent immediately following the passing of the byte in the transmit shift register. This single buffering of the data allows data to be sent continuously without the software polling the TXREG to find out when is the correct time to send out another byte. USART transmit interrupt requests are made when the TX holding register is empty. This feature is available for both synchronous and asynchronous transmission modes.

The USART receive unit is the most complex of the USART's three parts. This complexity comes from the need for it to determine whether or not the incoming asynchronous data is valid or not using the pin buffer and control unit built into the USART receive pin. The block diagram for the USART's receiver is shown in Fig. 5-48.

If the port is in Synchronous mode, data is shifted in, either according to the USART's clock or using an external device's clock.

For asynchronous data, the receiver sensor clock is used to provide a polling clock for the incoming data. This sixteen-time data rate clock's input into the pin buffer and control unit provides a polling clock for the hardware. When the input data line is low for three receive sensor clock periods, data is then read in from the middle of the next bit (Fig. 5-49). When data is being received, the line is polled three times and the majority states read is determined to be the correct data value. This repeats for the eight or nine bits of data with the stop bit being the final check.

Like the TX unit, the RX unit has a holding register. If data is not immediately processed and an incoming byte is received, the data will not be lost. But, if the data is not picked up by the time the next byte has been received, then an overrun error will occur. Another type of error is the framing error, which is set if the stop bit of the incoming NRZ

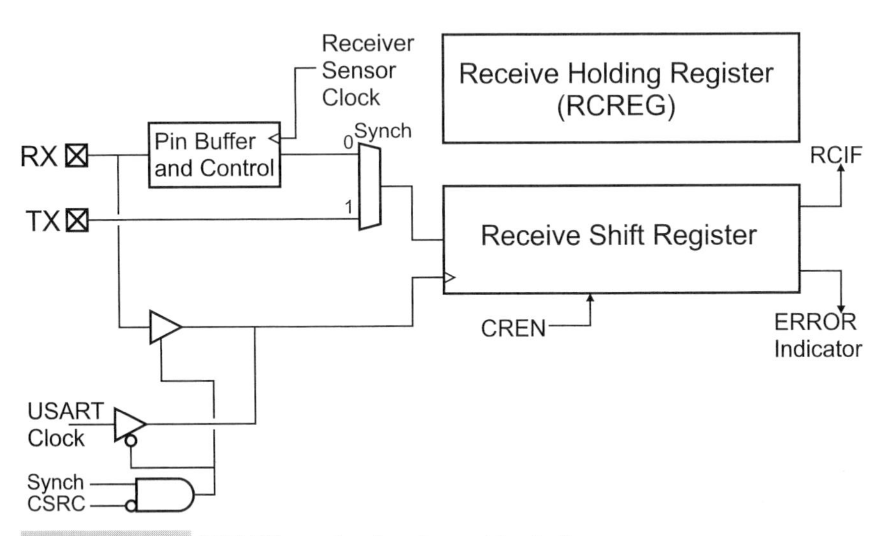


**Figure 5-48** USART receive hardware block diagram

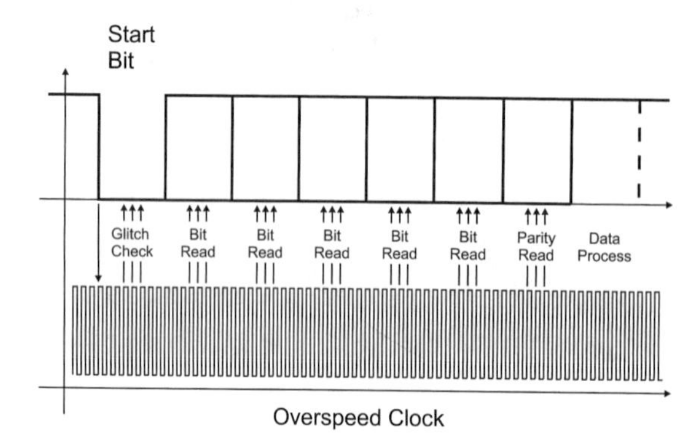


**Figure 5-49** Reading an asynch data packet

packet is not zero. These errors are recorded in the *RCSTA (ReCeiver STAtus)* register and have to be reset by software.

In some PICmicro® MCUs, the USART receive unit can also be used to receive two synchronous bytes in the format *data:address,* where *address* is a byte destined for a specific device on a bus. When the ADDEN bit of the RCSTA register is set, no interrupts will be requested until both the address and data bytes have been received. To distinguish between the bytes, the ninth address bit is set (while the ninth bit of data bytes are reset). When this interrupt request is received, the interrupt handler checks the device address for its value before responding to the data byte.

To control the USART, two registers are used explicitly. The *TXSTA (Transmitter STAtus)* register is located at address 0x098 and has the bit definitions shown in Table 5-22.

The SPBRG register is at address 0x099. The RCSTA register is at address 0x018 and is defined in Table 5-23.

**TABLE 5-22 USART TXSTA Bit Definition**

| BIT | DEFINITION |
|---|---|
| 7 | CSRC—Clock Source Select used in Synchronous Mode. When Set, the US ART Clock Generator is Used |
| 6 | TX9—Set to Enable nine bit Serial I/O |
| 5 | TXEN—Set to Enable Data Transmission |
| 4 | SYNC—Set to Enable Synchronous Transmission |
| 3 | Unused |
| 2 | BRGH—Used in Asynchronous Mode to Enable Fast Data Transmission. It is Recommended to keep this bit Reset |
| 1 | TRMT—Set if the Transmission Shift Register is Empty |
| 0 | TXD—Nine bit of Transmitted Data |

**TABLE 5-23 USART RCSTA Bit Definition**

| BIT | DEFINITION |
|---|---|
| 7 | SPEN—Set to Enable the USART |
| 6 | RX9—Set to Enable nine bit USART Receive |
| 5 | SREN—Set to Enable Single Byte Synchronous Data Receive. Reset when data has been received |
| 4 | CREN—Set to Enable Continuous Receive |
| 3 | ADDEN—Set to Receive Data:Address Information. May be unused in many PICmicro® MCU Part Numbers |
| 2 | FERR—"Framing Error" bit |
| 1 | OERR—"Overrun Error" bit |
| 0 | RX9D—Received ninth bit |

The TXREG is normally at address 0x019 and RCREG is normally at address 0x01A. The TXIF, TXIE, RCIE, and RCIF bits are in different interrupt enable request registers and bit numbers specific to the part being used.

To set up asynchronous serial communication transmit, the following code is used:

```
bsf        STATUS, RP0
bcf        TXSTA, SYNCH              ; Not in Synchronous mode
bcf        TXSTA, BRGH               ; BRGH = 0
movlw      DataRate                  ; Set USART Data Rate
movwf      SPBRG
bcf        STATUS, RP0               ; Enable serial port
bsf        RCSTA ^ 0x080, SPEN
bsf        STATUS, RP0
bcf        TXSTA, TX9                ; Only 8 bits to send
bsf        TXSTA, TXEN               ; Enable Data Transmit
bcf        STATUS, RP0
```

To send the data byte in w, use the code:

```
btfss TXSTA, TRMT
 goto $ - 1                          ; Wait for Holding Register to
                                     ; become Free/Empty
movwf TXREG                          ; Load Holding Register
                                     ; If Transmit Shift Register is
                                     ; Empty, byte will be sent
```

In the data send code, the TRMT bit, which indicates when the TX holding register is empty is polled. When the register is empty, the next byte to send is put into the transmit shift register. This polling loop can be eliminated by setting the TXIE bit in the interrupt control register and then in your interrupt handler, checking to see if the TXIF flag is set before saving a byte in TXREG.

To set up an asynchronous read, the following code is used:

```
bsf          STATUS, RPO
bcf          TXSTA, SYNCH              ; Want Asynch Communications
bcf          TXSTA, BRGH               ; Low Speed Clock
movlw        DataRate                  ; Set Data Rate
movwf        SPBRG
bsf          RCSTA ^ 0x080, SPEN       ; Enable Serial Port
bcf          TCSTA ^ 0x080, RX9        ; Eight Bits to Receive
```

To receive data, use the code:

```
btfss        PIR1, RXIF                ; Wait for a Character to be
 goto        $ - 1                     ; Received
movf         RCREG, w                  ; Get the byte Received
bcf          PIR1, RXIF                ; Reset the RX byte Interrupt
                                       ; Request Flag
```

# Analog I/O

Depending on your experience level, you might feel that the PICmicro® MCU has quite limited analog I/O capabilities. This is especially true if you are looking for high-speed analog operation from the PICmicro® MCU. The Microchip PICmicro® MCU mid-range devices actually have relatively good analog I/O capabilities, although, for high-speed analog I/O, you might want to look at external ADCs and DACs (or even other microcontrollers or circuits), which provide high-speed capabilities.

The ADC built in the PIC16C7x PICmicro® MCUs can sample and process signals as fast as 25 kHz (or so) accurately. Looking at the ADC's specifications, you might feel that the best analog signal frequency that can be processed is 50 kHz (as the examples in the data sheet show a 19-ms acquisition/processing time). I specify 25 kHz because of Nyquist's sampling theorem, which says that to properly sample an analog signal, you must sample at twice the highest data frequency expected in the signal.

25 kHz might seem like a reasonably fast signal to sample. After all, speech only requires 2.5 kHz and full-spectrum audio only reaches about 18 kHz. But for most electronic signals, 25 kHz is actually quite a low speed and not very useful (for example, the AM radio band starts at 66 kHz or the NTSC *colorbust* clock runs at 3.5 MHz.

Along with the slow ADC sampling and processing speeds, *Digital Signal Processing (DSP)* algorithms are difficult to implement on the PICmicro® MCU because of the processor's ability to interface with only eight bits of data and lack of multiply or divide instructions. Limited DSP functions can be implemented, but they will be challenging for data input waveforms that are faster than 1 kHz or so.

For these reasons, I don't recommend that the PICmicro® MCU's built-in ADC's be used for anything other than measuring DC voltages. With up to 12 bits available with built-in ADCs, the PICmicro® MCU is very well suited for accurately measuring *DC (Direct Current)* analog voltages.

Personally, I find the Microchip documentation to be quite complex and difficult to figure out concerning how to use the built-in ADC hardware for applications. The following sections cover how the analog input and processing works on the PICmicro® MCU. Some hints are included for using the features without having to wade through all the documentation.

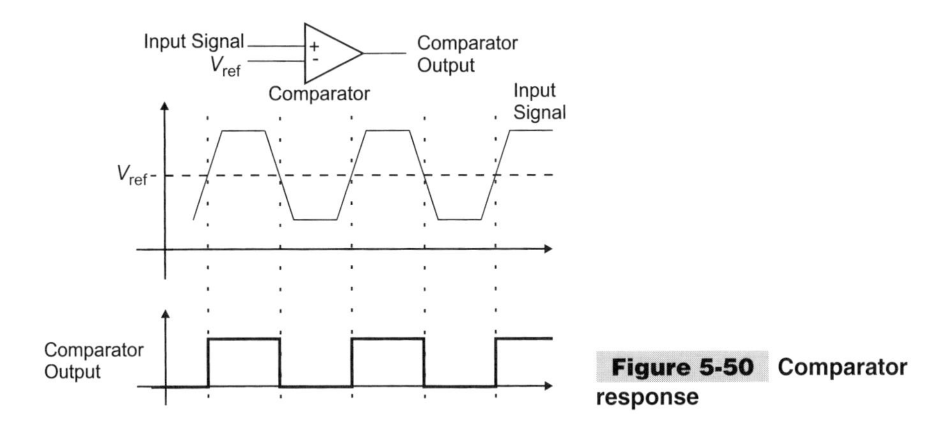


**Figure 5-50** Comparator response

## PIC16C62x: VOLTAGE COMPARISON

Along with the ADCs of the PIC16C2x, analog voltages can be processed by the use of comparators that indicate when a voltage is greater than another voltage. The inputs compared can be switched between different I/O pins as well as ground or a reference voltage that can be generated inside of the PICmicro® MCU chip.

Enabling comparators is a very straightforward operation with the only prerequisite being that the pins used for the analog compare must be in Input mode. Comparator response is virtually instantaneous, which allows alarm or other fast responses from changes in the comparator inputs.

The comparator works very conventionally (Fig. 5-50). If the value of *+input* is greater than the *−input,* the output is high. The two comparators in the PIC16C82X are controlled by the CMCON register, which is defined in Table 5-24.

The CIS and CM2:CM0 bits work together to select the operation of the comparators. The following numbered notes are the details to the notes in Table 5-25.

1. For CM2:CM0 equal to 000, RA3 through RA0 cannot be used for digital I/O.
2. For CM2:CM0 equal to 000, RA2 and RA1 cannot be used for digital I/O.
3. RA3 can be used for digital I/O.

**TABLE 5-24 CMCON Bit Definition**

| BIT | DESCRIPTION |
|---|---|
| 7 | C2OUT—Comparator 2 Output (High if + > −) |
| 6 | C1OUT—Comparator 1 Output (High if + > −) |
| 5-4 | Unused |
| 3 | CIS—Comparator Input switch |
| 2-0 | CM2:CM0—Comparator Mode |

**TABLE 5-25 Comparator Module I/E Specification**

| CM | CIS | COMP 1 | | COMP 2 | |
|---|---|---|---|---|---|
| | | + input | – input | + input | – input |
| 000 | X | RA0 | RA3 (1) | RA2 | RA1 (4) |
| 001 | 0 | RA2 | RA0 | RA2 | RA1 |
| 001 | 1 | RA2 | RA3 | RA2 | RA1 |
| 010 | 0 | Vref | RA3 | Vref | RA1 |
| 010 | 1 | Vref | RA3 | Vref | RA2 |
| 011 | X | RA2 | RA0 (3) | RA2 | RA1 |
| 100 | X | RA3 | RA0 (4) | RA2 | RA1 |
| 101 | X | DON'T | CARE | RA2 | RA1 |
| 110 | X | RA2 | RA0 (5) | RA2 | RA1 (6) |
| 111 | X | RA3 | RA0 (7) | RA2 | RA1 (8) |

**4** RAO and RA3 can be used for digital I/O.
**5** RA3 is a digital output, same as comparator 1 output.
**6** RA4 us the open drain output of comparator 2.
**7** RA0 and RA3 can be used for digital I/O.
**8** RA1 and RA2 can be used for digital I/O.

Upon power up, the comparator CM bits are all reset, which means that RA0 to RA3 are in analog Input mode. If you want to disable analog input, the CM bits must be set (write 0x007 to CMCOM).

Interrupts can be enabled, which will interrupt the processor when one of the comparator output changes. This is enabled differently for each PICmicro® MCU with built-in comparators. Like the PORTB change on interrupt, after a comparator change interrupt request has been received, the CMCOM register must be read to reset the interrupt handler.

Along with comparing to external values, the PIC16C62x can also generate a reference voltage ($V_{ref}$ in Table 5-25) using its own built-in four-bit digital-to-analog converter (Fig. 5-51).

The $V_{ref}$ control bits are found in the VRCON register and are defined in Table 5-26.

The $V_{ref}$ output is dependent on the state of the VRR bit. The $V_{ref}$ voltage output can be expressed mathematically if VRR is set as:

$$V_{ref} = V_{dd} * (Vfcon \,\&\, 0\text{x}00F)/24$$

or, if it is reset as:

$$V_{ref} = V_{dd} * [8 + (Vrcon \,\&\, 0\text{x}00F)]/32$$

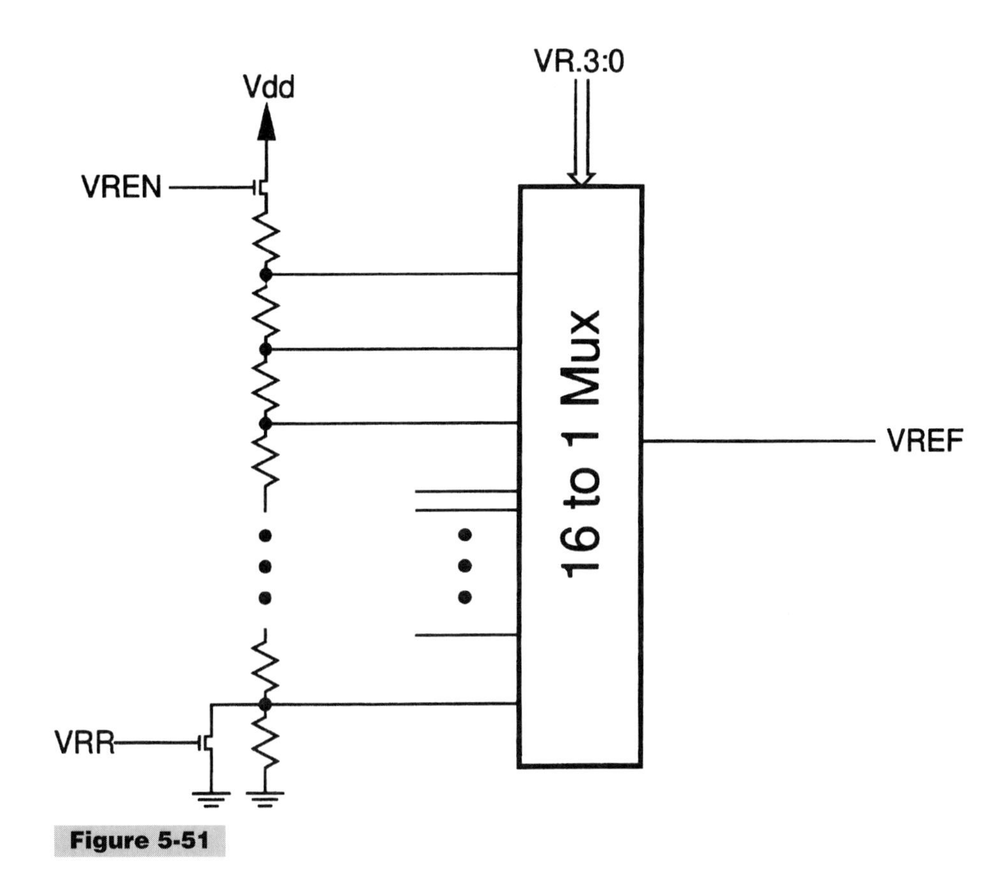


Figure 5-51

Notice that when VRR is set, the maximum voltage of $V_{ref}$ is 15/24 of $V_{dd}$, or just less than two-thirds of $V_{dd}$. When VRR is reset, $V_{ref}$ can be almost three-quarters of $V_{dd}$.

$V_{ref}$ can be output on RA2, but you will probably want it to be used internally only with the comparators because using it to drive RA2 requires extra current from the application. If an analog voltage output were required, I would recommend implementing a PWM output (as discussed elsewhere in the book).

## PIC16C7x: ANALOG INPUT

When I first started using the *Analog-to-Digital Converter (ADC)* built into the PIC16C7x devices, I felt like the feature was very complex and difficult to work with. When you read through the Microchip data sheets on the ADC that is built into the PIC16C7x, PIC16F87x, and other devices that have it built in, you will find that there are multiple, 20-page descriptions of the ADC. Each of these descriptions is slightly different, depending on the part number and its features (such as the number of I/O pins that can provide ADC input). Instead of repeating all of this information, I want to give you a brief overview of the basics of the ADCs, along with the important concepts that you will have to know. At the end of this section is some sample code to help guide you in successfully using the ADC in your own applications.

**TABLE 5-26 Comparator VRCON Bit Definitions**

| BIT | DESCRIPTION |
|---|---|
| 7 | VREN—Vref Enable when Set |
| 6 | VROE—Vref output enable when set<br>RA2—Vref |
| 5 | VRR—Vref Range Select<br>1 = Low Range<br>0 = High Range |
| 4 | Unused |
| 3–0 | VR3:VR0—Voltage Selection Bits |

All PICmicro® MCU devices that have a *7* as the character after the *C* or *F* of the part number have a built-in analog-to-digital converter, which will indicate an analog voltage level from 0 to $V_{dd}$, with 8- to 12-bit accuracy. The PORTA pins can be used as either digital I/O or analog inputs. The actual bit accuracy, utilization of pins, and operating speed is a function of the PICmicro® MCU part number and the clock speed at which the PICmicro® MCU runs.

When a pin is configured for analog input, it follows the models shown in Fig. 5-52.

$R_s$ in the $V_{source}$ circuit is the in-line resistance of the power supply. In order to get reasonable times for charging the ADC's holding capacitor, this value should be less than 10 K.

If you look through the ADC documentation, you will find that the time required for the holding capacitor to load the analog voltage and to stabilize is:

$$T_{ack} = 5 \text{ ms} + [(Temp - 25\text{C}) \times 0.05 \text{ ms/C}] \\ + (3.19\text{C} * 10^7) \times (8 \text{ K} + R_s)$$

Which works out to anywhere from 7.6 to 10.7 $\mu$s at room temperature. I usually avoid this calculation altogether and assume that 15 to 20 $\mu$s is required for the holding capacitor voltage to stabilize to the input voltage.

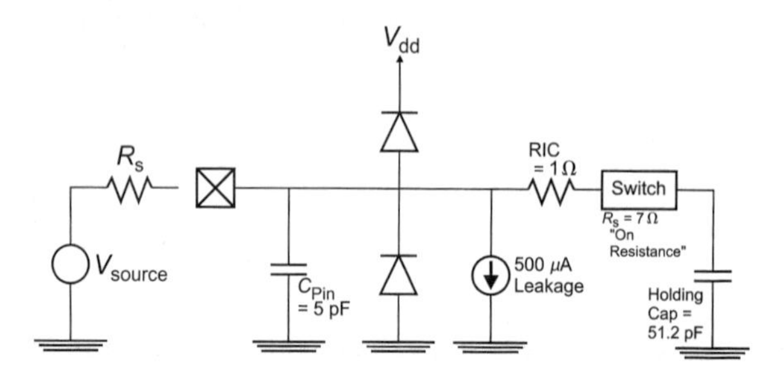


**Figure 5-52** PICmicro® MCU internal ADC equivalent input

Once the voltage is stabilized at the capacitor, a test for each bit is made. 9.5 cycles are required to do an eight-bit conversion. The bit conversion cycle time (known as *TAD*) can be anywhere from 1.6 to 6.4 $\mu$s and can either use the PICmicro® MCU's instruction clock or a built-in 250-kHz RC oscillator. To get a valid TAD time using the PICmicro® MCU's instruction clock, a two, eight, or 32 prescaler is built into the ADC.

For example, a 4-MHz clock using the divide-by-eight prescaler will have a 2-$\mu$s TAD time, which is acceptable for the ADC. If the divide-by-two counter would be used, the TAD would be 500 ns, which is much too fast for the ADC to work correctly.

The built-in 250-kHz oscillator is used to carry out the ADC conversion when the PICmicro® MCU is asleep. For maximum ADC accuracy, Microchip recommends that the PICmicro® MCU be put to sleep during the ADC conversion for minimum internal voltage or current upsets. If the PICmicro® MCU is put to sleep, then the minimum conversion time is much longer than what is possible using the built-in clock because the PICmicro® MCU has to restart when the ADC completion interrupt has been requested and wakes the PICmicro® MCU from sleep.

The minimum conversion time is defined as the total time required for the holding capacitor to stabilize at the input voltage and for the ADC operation to complete. Assuming that an 8-$\mu$s holding time could be implemented along with a 15-$\mu$s ADC conversion time, the maximum time is about 24 $\mu$s (a rate of 41,000 ADC samples per second can be made).

This is not fast enough for most electronics operations and probably not fast enough for audio decoding (especially with the slow digital processing capabilities of the PICmicro® MCU's processor). I'm pointing this out to indicate that the PICmicro® MCU's ADC is best used for very slowly changing inputs.

To measure analog voltages, the analog input pins or the PICmicro® MCU, which are usually in port A, have to be set to analog input on power up; the analog input pins are normally set to analog input and not digital I/O. To specify the modes, the ADCON1 register is written to. Table 5-27 shows the two least-significant bits (known as *PCFG1:PCFG0*) of the ADCON1 register with the types of I/O pin operation selected in a PIC16C71.

When I use an ADC-equipped PICmicro® MCU in an application where all the PORTA pins have to be digital I/O, I normally do the conversion right at the start of the application before writing to the TRISA register or initializing the state of the pins. Until the pins are changed to digital I/O, they will always return zero and cannot be set to an output value of one.

Normally, when the ADC is used in a PICmicro® MCU, the voltage reference is from ground to $V_{dd}$. If this range is not acceptable or if the power supply is unreliable, a new reference voltage can be specified. In some devices that are equipped with ADCs, the lower

**TABLE 5-27 Sample ADCON1 Bit Definitions for the PIC16C71**

| ADCON1 BITS | AN3 | AN2 | AN1 | AN0 |
|---|---|---|---|---|
| 11 | D | D | D | D |
| 10 | D | D | A | A |
| 01 | Vref | A | A | A |
| 00 | A | A | A | A |

**TABLE 5-28 ADCON0 Bit Definitions**

| BIT | FUNCTION |
|---|---|
| 7-6 | ADCS1: ADCS0 bits used to select the TAD clock.<br>11—Internal 250 kHz Oscillator<br>10—FOSC/32<br>01—FOSC/8<br>00—FOSC/2 |
| 5-3 | CHS2:CHS0—Bits used to Select which Analog Input is to be Measured. These bits and their operation is Part Number Specific |
| 2 | GO/_DONE—Set Bit to Start ADC Conversion, Reset by Hardware when ADC Conversion is Complete. |
| 1 | ADIF—Set upon Completion of ADC Conversion and Requests an Interrupt. |
| 0 | ADON—Set to Enable the ADC |

voltage reference can be externally specified as well. The bit definition of ADCON1 is part-number specific and changes based on the device part number, number of PORTA pins, and the number of bit resolution provided by the ADC.

The ADCON0 register is used to control the operation of the ADC. The bits of the register are typically defined as shown in Table 5-28.

The ADC consumes power even when it is not being used and for this reason, if the ADC is not being used, ADON should be reset.

If the PICmicro® MCU's ADC is capable of returning a 10-bit result, the data is stored in the two ADRESH and ADRES registers.

When 10-bit ADC results are available, the data can be stored in ADRESH/ADRESL in two different formats. The first is to store the data right justified with the most-significant six bits of ADRESH loaded with zero and the least two significant bits loaded with the two most-significant bits of the result. This format is useful if the result is going to be used as a 16-bit number, with all the bits used to calculate an average.

The second 10-bit ADC result format is left justified, in which the eight most-significant bits are stored in ADRESH. This format is used when only an eight-bit value is required in the application and the two least-significant bits can be lopped off or ignored.

To do an analog-to-digital conversion, the following steps should be taken:

1. Write to ADCON1 indicating which are the digital I/O pins and which are the analog I/O pins. At this time, if a 10-bit conversion is going to be done, then set the format flag in ADCON 1 appropriately.
2. Write to ADCON0, setting ADON, resetting ADIF and GO/_DONE, and specifying the ADC TAD clock and the pin to be used.
3. Wait for the input signal to stabilize.
4. Set the GO/_DONE bit. If this is a high-accuracy measurement, ADIE should be enabled for interrupts and then the PICmicro® MCU put to sleep.
5. Poll GO/_DONE until it is reset (conversion done).

**6** Read the result from ADRES and, optionally, ADRESH.

To read an analog voltage from the RAO pin of a PIC167C1 running a 4-MHz PICmicro® MCU, the code would be:

```
bsf            STATUS, RPO
movlw          0x002
movwf          ADCON1 ^ 0x080        ; AN1/AN0 are Analog Inputs
bcf            STATUS, RPO
movlw          0x041                 ; Start up the ADC
movwf          ADCON0
movlw          5
addlw          0x0FF                 ; Delay 20 usec for Holding
btfss          STATUS, Z             ; Capacitor to Stabilize
 goto          $ - 2
bsf            ADCON0, GO            ; start the ADC conversion
btfsc          ADCON0, GO            ; Wait for the ADC Conversion
 goto          $ - 1                 ; to End
movf           ADRES, w              ; Read the ADC result
```

As you read the Microchip data sheets on the ADC, you will see that there are methods of implementing shorter, less-accurate conversions. I do not recommend implementing these conversions because they decrease the accuracy of the ADC conversion, but do not affect the biggest delay to doing the ADC conversion; the delay for the holding capacitor. Thus, while the ADC can operate with a modest increase in speed, the total number of samples per second that can be made with the ADC cannot be substantially increased.

# Parallel Slave Port (PSP)

One of the most interesting features of the 40-pin mid-range and PIC18Cxx PICmicro® MCUs is the *Parallel Slave Port (PSP),* which is built into the PORTD and PORTE I/O pins. This feature allows the PICmicro® MCU to act like an intelligent peripheral to any eight-bit data bus device.

The PSP is very easy to wire up with separate chip select and read/write pins for enabling the data transfer. The block diagram of the PSP is shown in Fig. 5-53.

The actual read/write I/O operations occur as you would expect for a typical I/O device connected to a microprocessor's address/data control bus. A read and write operation waveform is shown in Fig. 5-54.

The minimum access time is one clock (not instruction clock) cycle. For a PICmicro® MCU running at 20 MHz, the minimum access time is 50 ns.

To enable the parallel slave port, the PSP mode bit of the TRISE register must be set. When this bit is set, port D becomes driven from the _CS, _RD, and _WR bits, which are RE2, RE1, and RE0, respectively. When the PSP mode bit is set, the values in PORTD, PORTE, TRISD, and TRISE are ignored.

PSP mode should be enabled the whole time that the PICmicro® MCU is active. Changing the pins between modes could cause data contention problems with the device driving the bus connected to PORTD and PORTE. As well, the contents of PORTD and PORTE are unknown upon return from PSP mode.

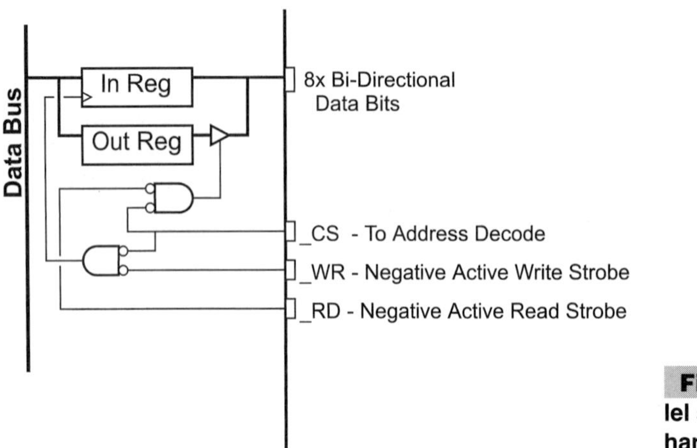


**Figure 5-53** **Parallel slave port ("PSP") hardware**

When PSP mode is enabled and _CS and _RD are active, PORTD drives out the contents of OUTREG. When OUTREG (which is at PORTD's address) is written to, the *OBF (Output Buffer Full)* bit of TRISE is set. This feature and the input data flags in TRISE are not available in all devices. The PBF bit will become reset automatically when the byte in the OUTREG is read by the device driving the external parallel bus.

When a byte is written into the parallel slave port (_CS and _WR are active), the value is saved in INREG until it is overwritten by a new value. If the optional status registers are available, the IBF bit is set when the INREG is written to and cleared when the byte in INREG read. If the byte is not read before the next byte is written into INREG, the IBOV bit, which indicates the overwrite condition, is set.

In older PICmicro® MCUs that have PSP port, the IBF, OBF, and IBOV bits are not available in TRISE. Although I recommend only using parallel slave port devices that have the IBF, OBF, and IBOV flags, there will be times when this is not possible. If you have use a part that doesn't have these bits, be sure to create a method of protocol of sending data to ensure that no data byte transfers are missed. This can be done by sending the complement of the previous byte to the PICmicro® MCU before the next byte is

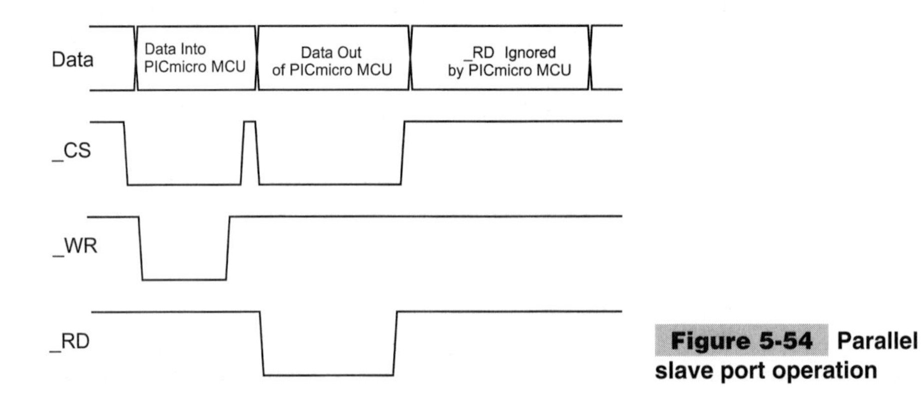


**Figure 5-54** **Parallel slave port operation**

sent and responding to reading the byte in OUTREG by writing its complement back into INREG.

With the parallel slave port working, all the other PICmicro® MCU resources are available. This means you can use ADCs (ensuring that the PORTE bits are not set for analog input, which will cause problems with the parallel slave port), serial I/O and other features that allow advanced I/O to and from the PICmicro® MCU. In Chapter 16, the PSP is used to implement a custom serial interface to a PC's ISA bus.

# 17Cxx External Memory Connections

Parallel memory devices can be connected to the 17Cxx PICmicro® MCU devices to enhance the PICmicro® MCUs program memory space. The interface provided is up to 64K of 16-bit data words via a multiplexed address/data bus. The multiplexed bus might seem somewhat difficult to use, but it actually isn't; memory devices can be added quite easily and quickly. The four memory modes available to the 17Cxx PICmicro® MCUs are shown in Table 5-29 and Fig. 5-55.

An unprogrammed PC17Cxx's configuration fuses sets the PICmicro® MCU into Microprocessor mode, which cannot access any internal program memory. This allows output devices to be placed into applications, with external program memory providing the application code. This feature allows a way of debugging an application before it is burned into the PICmicro® MCU.

External memory can be read from or written to, using the *tablrd* and *tablwt* instructions. In extended microcontrollers and microprocessor modes, the internal program memory can be read using the *tablrd* instruction in the microcontroller modes. These *table* instructions use the table pointer register (TBLPTRH for the high eight bits and TBLPTRL for the low eight bits) to address the operation. During table reads and writes, the table

**TABLE 5-29 PIC17Cxx Memory Modes**

| MODE | PROGRAM MEMORY CHARACTERISTICS |
|---|---|
| Microcontroller | Internal to the PICmicro® MCU, able to read Configuration Fuses and Read and Write Program Memory |
| Protected Microcontroller | Internal to the PICmicro® MCU, able to read Configuration fuses, Program Memory can be read but not Written |
| Extended Microcontroller | Program Memory Internal to PICmicro® MCU Accessible. External Memory in Address Space Above Read and Writeable as well. Unable to read Configuration Fuses. |
| Microprocessor | No internal Program Memory or Configuration Fuses Accessible. Whole 64 K program memory space Accessible outside PICmicro® MCU |

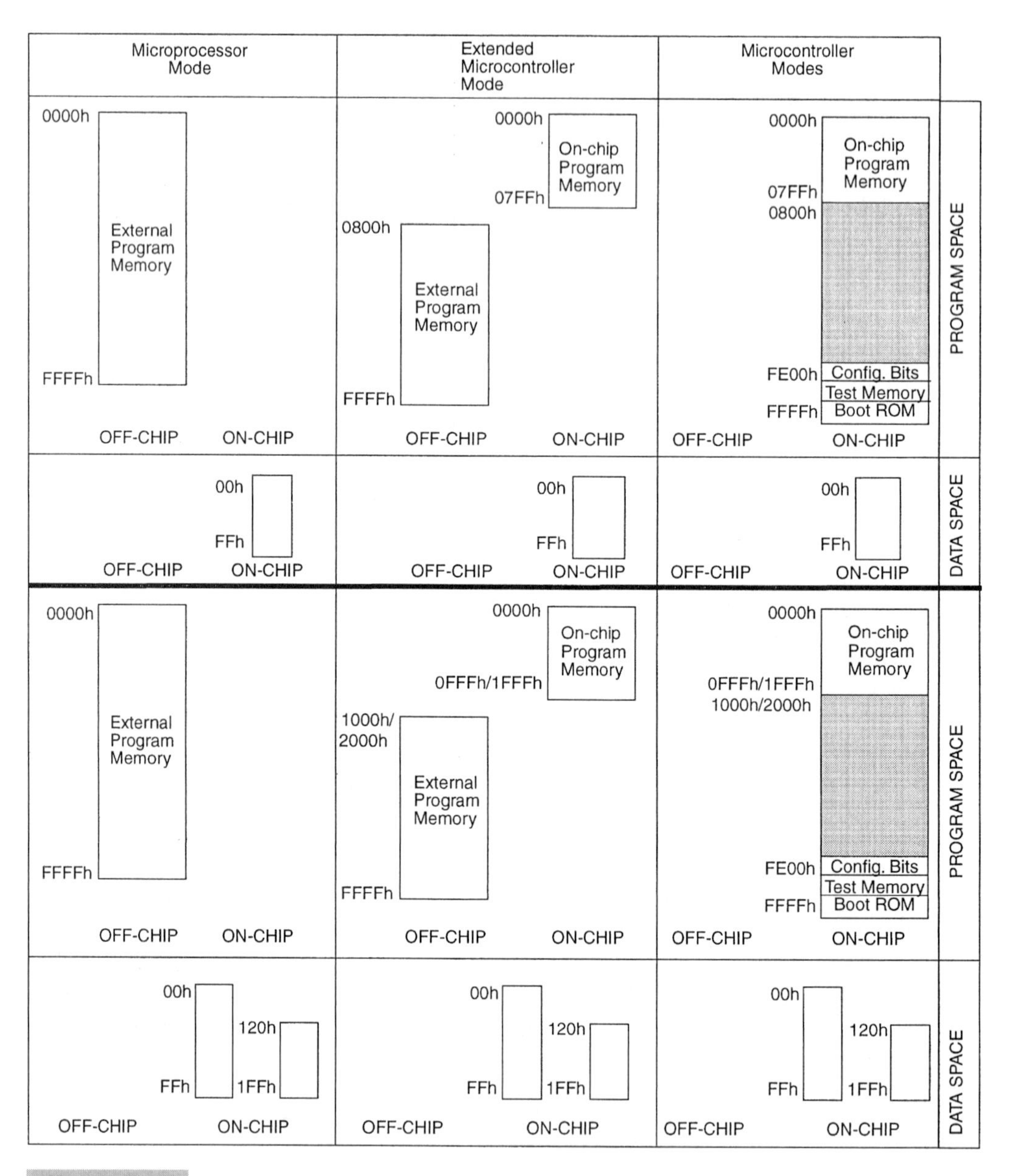


**Figure 5-55**

latch register (TABLATH for the high byte and TABLATL for the low byte) is used to buffer the 16 bits during the transfer because the 17Cxx PICmicro® MCU's processor can only access data eight bits at a time.

The block diagram for accessing program memory in the 17Cxx family of PICmicro® MCUs is shown in Fig. 5-56.

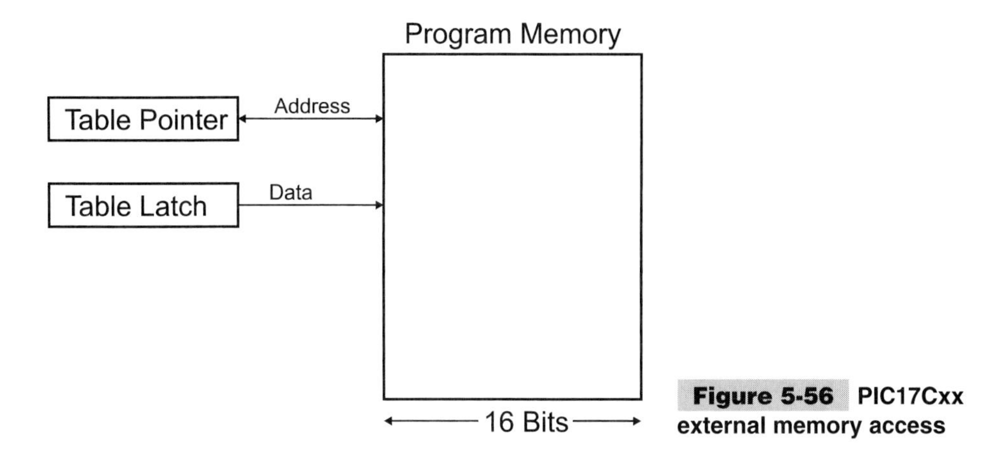


**Figure 5-56** PIC17Cxx external memory access

To execute a read or write to program memory, the address in the table pointer has to be first set up. Writing to each of the two 8-bit registers does this. Next, if the operation is a read, the *tablrd* instruction is executed with a dummy destination to update the table latch register. Once this is done, two *read* instructions are carried out to read the 16 bits at the specified address. This instruction sequence is:

```
movlw     HIGH PM_address          ; Set up Table Pointer
movwf     TBLPTRH
movlw     LOW PM_address
movwf     TABLPTRL
tablrd    0, 0, WREG               ; Update Latch Register
tlrd      1, WREG                  ; Read High Byte
movwf     HIGH Destination
tablrd    0, 0, WREG               ; Read Low Byte
movwf     LOW Destination
```

Even if you are going to only read eight bits, the dummy read set to the opposite byte is required.

Writing to program memory is similar, but with one important difference; in PIC17Cxx devices, the PICmicro® MCU's internal EPROM program memory can be written to from within the application. This feature is used by the PIC17Cxx's ICSP programming, which is covered later in the book.

The instruction sequence for the write is:

```
movlw     HIGH PM_address
movwf     TBLPTRH
movlw     LOW PM_address
movwf     TBLPTRL
movlw     HIGH Data
tlwt      1, WREG
movlw     LOW data
tablwt    0,0,WREG
```

In this sequence, after the address is set up, the high byte of the data to be written is stored in the table latch register, then the low byte is written into the low byte of the table latch. When this has been done, the whole 16-bit word is written into program memory. This feature can be useful in an application to load constants into program memory for calibration values of device serial numbers that are unknown when the PICmicro® MCU is initially programmed.

To write a new value into the internal EPROM word, the following sequence is required:

1. Disable all interrupts, except for the EPROM write terminate interrupt.
2. Raise _MCLR to a $V_{pp}$ voltage of 13 volts.
3. Clear the watchdog timer (WDT) if it is enabled.
4. Perform write.
5. Wait for terminate interrupt.
6. Verify the memory location before continuing.

The *tablwt* instruction executes as is shown in Fig. 5-57.

Notice that when the actual program memory write occurs, the write goes over into the clock cycle of the next instruction. The *tablrd* instruction executes similarly, but with the program memory access reversed (Fig. 5-58).

Connecting external devices to the 17Cxx PICmicro® MCU is relatively simple, with two 74LS373 latches used for buffering the address before the I/O operation occurs. The circuit for adding external memory to the PIC17Cxx is shown in Fig. 5-59.

The address bits can be decoded to provide access for multiple devices. A 74LS138 can be used to decode three lines into eight negative active outputs.

When performing a read, the AD bus, and ALE and _OE lines look like Fig. 5-60.

A write will look like Fig. 5-61.

These waveforms are actually quite traditional and match up with many microprocessor's (such as the Intel 8080), but should be reconciled with the waveforms for the *tablwt* and *tablrd* instructions presented. Notice that only one of the two data transfers will be visible on the PICmicro® MCU's external bus.

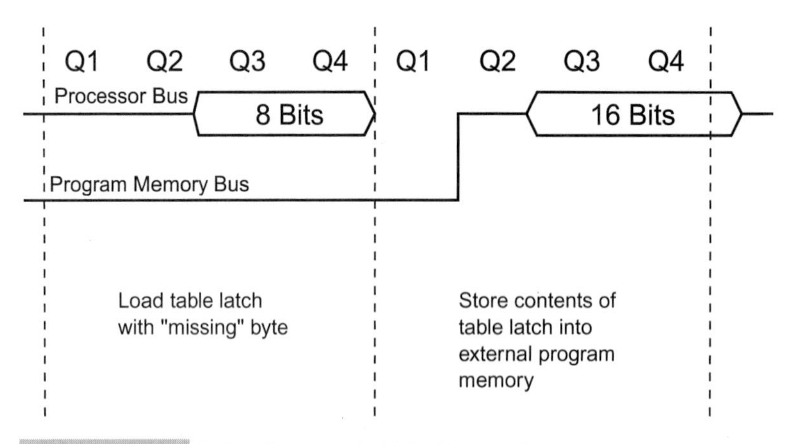


**Figure 5-57** PIC17Cxx "TABLWT" instruction execution

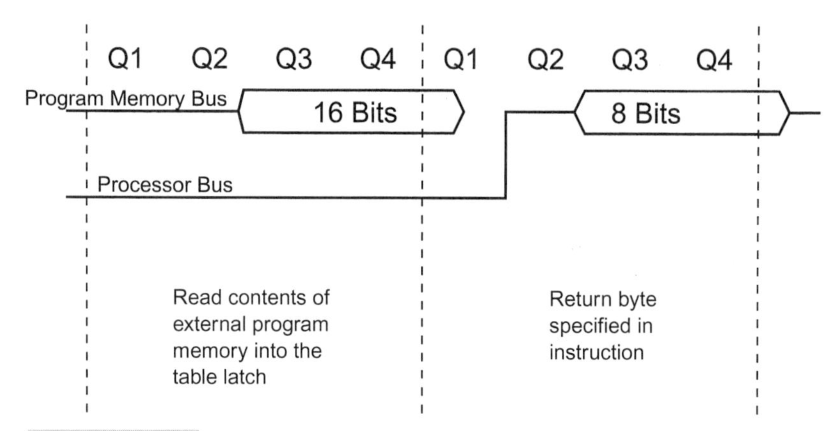


**Figure 5-58** PIC17Cxx "TABLRD" instruction execution

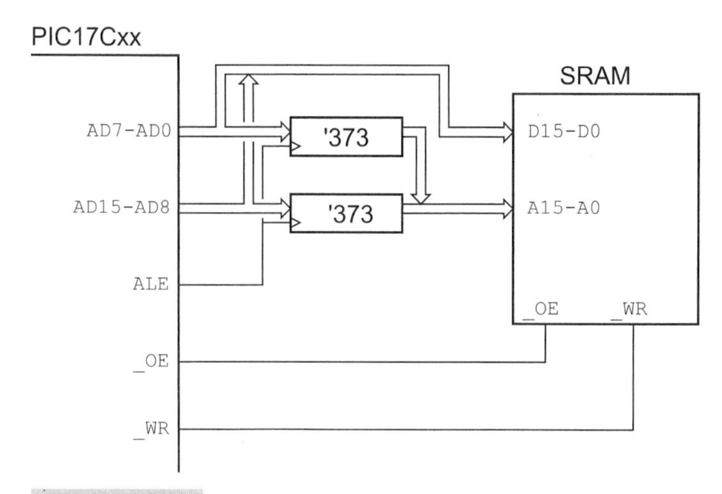


**Figure 5-59** PIC17Cxx external memory connections

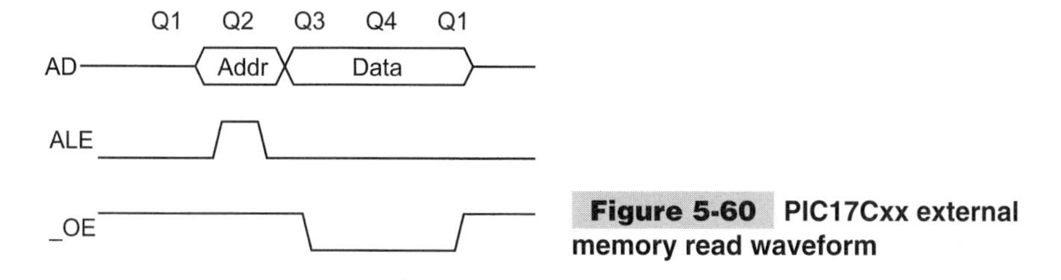


**Figure 5-60** PIC17Cxx external memory read waveform

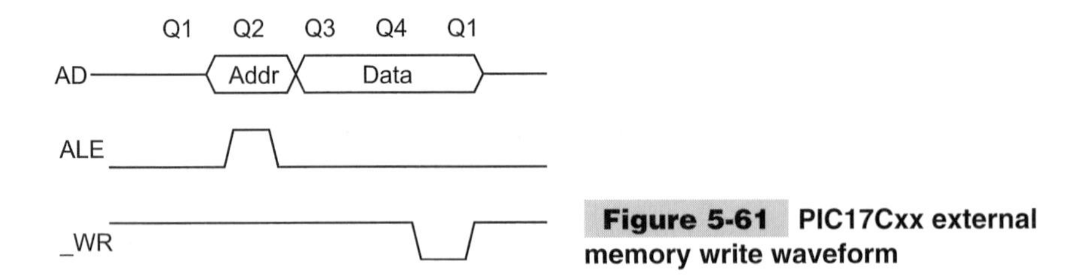


**Figure 5-61** PIC17Cxx external memory write waveform

Also notice that two clock cycles are used for the data transfer. Thus, the data-access speed of the external device must be less than twice the period of the PICmicro® MCU's clock. For example, if the PICmicro® MCU was running at 10 MHz, the clock period would be 100 ns and any external devices connected to the bus would have to have an access time of 200 ns or less.

# In-Circuit Serial Programming (ICSP)

Microchip was one of the first manufacturers to produce microcontrollers that could be programmed after being wired into an application. This capability was first provided in mid-range PICmicro® MCUs, but has since become a feature in all new PICmicro® microcontrollers. *ICSP* can also be used for parts that have not yet been soldered into a circuit, which minimizes the cost of creating a PICmicro® MCU device programmer.

The ICSP connector for the mid-range PICmicro® MCUs is a 5-pin, 0.100 spacing IDC connector with the pin out shown in Table 5-30.

As is shown in Chapter 13, this capability can be exploited to create simple programmers. It can also be used for *Surface-Mount Technology (SMT)* parts, which cannot be easily fixtured, to be "burned" with application code after it has been assembled into a circuit.

This function is very useful and, for me, it is an important advantage of the PICmicro® MCU over other microcontrollers. In this book, I only use ICSP programmable parts, as well as provide designs for ICSP-compatible programmers and driver software for you to easily program your own PICmicro® MCU parts and build the applications presented in this book.

**TABLE 5-30 ICSP Pin Connections**

| PIN | FUNCTION |
|---|---|
| 1 | Vpp (Programming Voltage) |
| 2 | Vcc (+5 Volts) |
| 3 | Gnd |
| 4 | Data |
| 5 | Clock |

# Future Devices

In writing this book, I have received a lot of support and information from Microchip on current and future parts. I have done everything possible to try and ensure that all of the information regarding hardware devices is current and looking ahead into the future as I write this. But, I know that over time, new interfacing features, new parts, and even new PICmicro® MCU architectures will become available.

I have tried to count the number of new PICmicro® MCU part numbers released by Microchip since I submitted the first edition manuscript. The number I have come up with is 132.

In the space of three years, there has been more Flash program memory parts, different-sized program and variable memory devices with similar I/O capabilities, new I/O interfaces, and a new PICmicro® MCU device architecture (the 18Cxx). I have no doubt that the next three years will see just as significant changes. As I do the final proofreading of this book, the USB-equipped PICmicro® MCUs are becoming available as "engineering samples." Over the next three years, I hope to see CAN-enabled PICmicro® MCUs become available, and the 18Cxx architecture start to have interfacing features and Flash program memory (previously only available in the mid-range parts).

# 6

# PICmicro® MCU APPLICATION DESIGN AND HARDWARE INTERFACING

CONTENTS AT A GLANCE



After reading the previous two chapters and "Introduction to Electronics" on the CD-ROM, you must feel like there is nothing left to understand about designing PICmicro® MCU applications. In these chapters on the CD-ROM I have provided background on how microcontroller interfacing is carried out and some of the theories and pitfalls that you should be aware of. This chapter covers some specifics about how the PICmicro® MCU works when wired to different interfaces.

I am not trying to say that the PICmicro® MCU is a difficult microcontroller to interface to. Actually, it is one of the easiest eight-bit microcontrollers to develop interface applications for. In different versions, with built-in oscillators and reset circuits, interfaces can be unbelievably easy to implement. Despite this, you should be aware of a few things that will help you make your applications more efficient and keep you from having difficult to debug problems.

# Estimating Application Power Requirements

Accurate power estimating for your applications is important because the expected power consumption affects how much heat is produced by the circuit and what kind of power supply is required by it. Although the PICmicro® MCU power can usually only be estimated to an order of a magnitude, this is usually good enough to ensure that the circuit won't overheat, the power supply won't burn out or the batteries will not be exhausted before the required execution time has passed.

For the PICmicro® MCU itself, the "intrinsic" current, what I call the current consumed by the microcontroller with nothing connected to it, is available from Microchip in the data sheets. For the PIC16F87x, rated at 4 MHz, Table 6-1 lists the IDD (the supply current or intrinsic current), according to the oscillator type.

The current rating for the oscillator type selected in the configuration fuses should be the basis for your estimate. Notice that as the clock frequency changes, the intrinsic current consumption will go up at a rate of about one mA per MHz. For estimating purposes, I recommend that you use the worst case for the oscillator selected.

Next, select the current requirements for the devices that connect directly to the PICmicro® MCU. Depending on their operation, the current requirements can change drastically. For example, an LED that is off consumes just about no current, and one that is on can consume from five to 20 mA. Again, for these devices, the worst cases should be used with the estimate.

Also, note that different devices will respond differently, depending on how they are used. A very good example of this is a LCD display. Many modern LCDs have built-in pull-ups to make interfacing easier for electronic devices that have open collector outputs (such as the 8051). Typically, these devices current requirements will be quoted with the

**TABLE 6-1 PICmicro® MCU Oscillator Current Consumption Comparison**

| OSCILLATOR | IDD | FREQUENCY |
|---|---|---|
| LP | 52.5 μA | 32 kHz |
| RC | 5 mA | 4 MHz |
| XT | 5 mA | 4 MHz |
| HS | 13.5 mA | 4 MHz |

minimum value, rather than the maximum. To ensure you have an accurate estimate, you will have to check the current drain with the LCD connected and operating with the PICmicro® MCU.

Lastly, the power consumption of other devices connected to the circuit (but not the PICmicro® MCU) will have to be determined through the device's data sheets. Again, the worst-case situation should be taken into account.

Once these three current values have been found, they can be summed together to get the total application power and then multiplied by the voltage applied to get the power consumed by the application. Once I have this value, I normally multiply it by a 25- to 50-percent "derater" to ensure that I have the absolute worst case.

In the applications in this book where I have specified the current, I have continually sought out the worst case and then derated the power to make it seem even worse. This is to ensure that you will not have any problems with your application power supply. Power can really make or break an application. Incorrectly specifying a supply can lead to problems with the application not powering up properly, failing intermittently, or not running as long on batteries as expected.

Marginal power supply problems can be extremely difficult to find as well. By going with a derated worst case for my application power requirements, I have eliminated one possible point in the application from going bad.

# Reset

Reset in many new PICmicro® MCU part numbers can be simply implemented, eliminating the need for a separate circuit or having a built-in brown-out reset sensor. Even putting your own reset circuit into an application is simple; only a couple of rules must be followed.

Adding external reset circuit to the PICmicro® MCU consists of a pull-up connected to the _MCLR pin of the PICmicro® MCU. As shown in Fig. 6-1, a switch pulling _MCLR to ground (to reset the device) can be implemented with a momentary on switch.

A resistor of 1 to 10 K is probably appropriate; the input is CMOS and does not draw any current through the resistor. The resistor is primarily used as a current-limiting device for the momentary-on switch.

In the configuration registers of the mid-range parts there is a bit known as *PWRTE*. This bit will insert a 72-ms delay during PICmicro® MCU power up before the first instruction is fetched and executed. The purpose of this function is to allow the PICmicro®

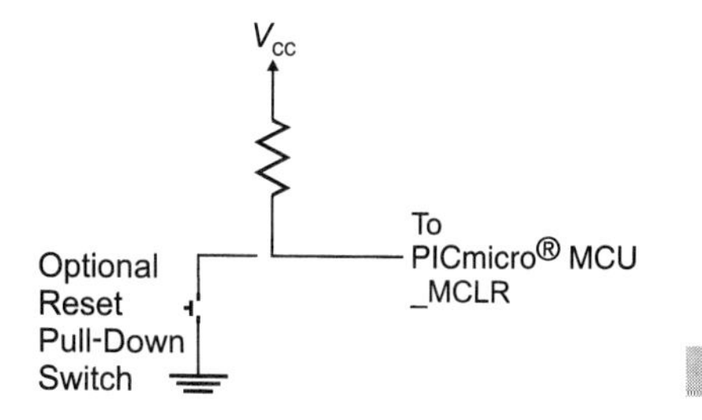


**Figure 6-1**

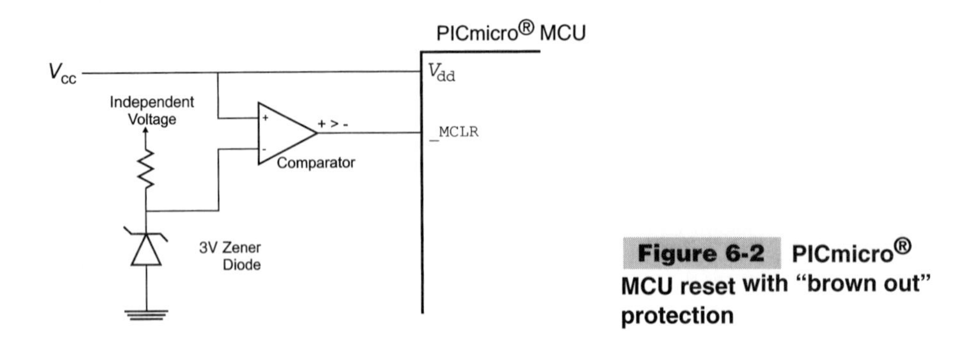


**Figure 6-2** PICmicro® MCU reset with "brown out" protection

MCU's clock to stabilize before the application starts. In the low-end and high-end PICmicro® MCU's, this function is not always available.

PWRTE does not have to be enabled if a stable clock is being input to the PICmicro® MCU, such as in the case where a "canned oscillator" is used as the PICmicro® MCU's clock source instead of a crystal, ceramic resonator, or RC network.

When the _MCLR pin is made active (pulled low), the oscillator stops until the pin is brought back high. As well, the oscillator is also stopped during sleep mode to minimize power consumption. The PWRTE 72 ms delay is required in these cases, as well to ensure the clock is stable before the application's code starts executing.

If the PICmicro® MCU is run at low voltage (less than 4.0 volts), do not enable the built-in *Brown-Out Reset (BOR),* unless it is available with a low-voltage selector option. Once power drops below 4.0 volts, then circuit will become active and will hold the PICmicro® MCU reset—even though it is receiving a valid voltage for the application. The *low-voltage option* usually means that the brown-out reset will reset the PICmicro® MCU when the input voltage is below 4.0 volts or 1.8 volts.

If you are going to use low voltage and want a brown-out detect function, this can be added with a Zener diode and a comparator as is shown in Fig. 6-2.

In this circuit, voltage is reduced by the Zener diode voltage regulator to 3 volts. If $V_{cc}$ goes below three volts, this circuit will put the PICmicro® MCU into reset. The voltage-divider values can be changed for different ratios and $R$ can be quite high (100 K+) to minimize current drain in a battery-driven application.

The PIC12C5xx and 16C505 PICmicro® MCU's can provide an internal reset function, which uses $V_{dd}$ as the _MCLR pin input. This frees the _MCLR pin for use as an input. The freed _MCLR pins generally cannot be used as an output.

A common use for this pin is RS-232 input using a resistor as a current limiter and providing bit-banging software to read the incoming values. If you use the _MCLR/I/O pin in this fashion, be sure that you "clamp" the pin shown in Fig. 6-3.

If the incoming negative voltage is not clamped within the PICmicro® MCU, the negative voltage could cause the PICmicro® MCU to be forced into reset mode. If the positive voltage is not clamped, then the PICmicro® MCU could go into the programming mode. The general-purpose pins are designed with clamping diodes built in and will not allow inputs to be driven outside $V_{dd}$ or ground. Not clamping the input pins can cause some confusing problems when you first work with the PICmicro® MCU in this type of application. The use of clamping diodes for RS-232 interfacing is shown in the "Serial LCD Interface" in Chapter 16.

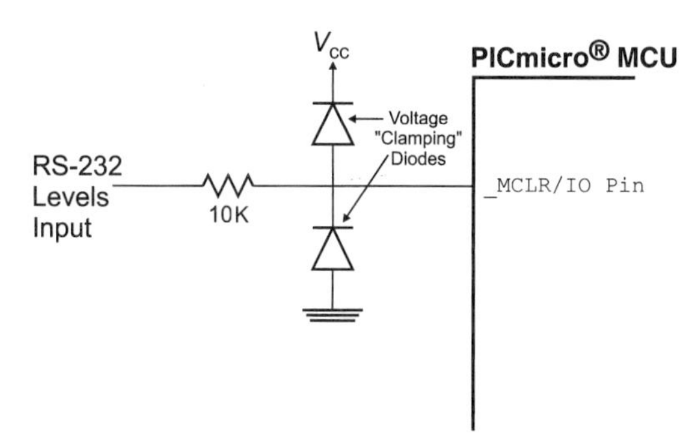


**Figure 6-3** Internal reset allowing "_MCLR" pin to be used for RS-232 input

# Interfacing to External Devices

The previous chapters have provided a lot of information about the peripheral hardware built into the PICmicro® MCU that will help making applications easier. Coupled with the information contained in the appendices, you would have thought I had it all covered.

The following sections cover some of the hints and tips I've learned over the years for interfacing PICmicro® MCUs to other devices. With this information, I have also included source code for the different interface hardware. This code is available on the CD-ROM as snippets that can be cut and pasted into your applications, as well as into macros that can be added to the applications.

Much of this information and code is used later in the book when I go through the experiments and projects. Some of the interfaces will seem very complex or difficult to create, but I have tried to work through many of the difficulties and provide you with sample circuits that are simple and cheap to implement.

## DIGITAL LOGIC

It should not be surprising that the PICmicro® MCU can interface directly to TTL and CMOS digital logic devices. The PICmicro® MCU's parallel I/O pins provide digital output levels that can be detected properly by both logic technologies and inputs that can detect logic levels output from these logic families.

If you check the PICmicro® MCU data sheets, you will see that the output characteristics are:

$V_{ol}$ (output low voltage) = 0.6 V (max.)
$V_{oh}$ (output high voltage) = $V_{dd}$ − 0.7 V (min.)

This specification is given to allow for different $V_{dd}$ power inputs. For a $V_{dd}$ of 5 volts, you can expect a "high" output of 4.3 volts or greater (normally, I see 4.7 volts when a PICmicro® MCU pin is not under load). If the power voltage input ($V_{dd}$) was reduced to 2 volts, low output would still be 0.6 volts and high output becomes 1.3 volts ($V_{dd}$ − 0.7) or greater.

The PICmicro® MCU pins are specified to drive (source) up to 20 mA and sink (pull the output to ground) 25 mA. These current capabilities easily allow the PICmicro® MCU to drive LEDs. The total current sourced or sunk by the PICmicro® MCU should not exceed 150 mA (which is six I/O pins sinking the maximum current).

The input "threshold" voltage, the point at which the input changes from an *I* to an *O* and visa versa, is also dependent on the input power ($V_{dd}$) voltage level. The threshold is different for different devices and the data sheet should be consulted for precise values. In general, for a number of different PICmicro® MCU part numbers, this value is specified as being in the range:

$$0.25\ V_{dd} + 0.8\ V >= V_{threshold} >= 0.48\ V_{dd}$$

As a rule, you should use the higher value. For higher $V_{dd}$s, this is approximately one half $V_{dd}$. At lower $V_{dd}$ voltages, (2 volts), the threshold becomes approximately two-thirds of $V_{dd}$.

For the most part, interfacing to conventional logic devices is very straightforward in the PICmicro® MCU. The following sections feature some special interfacing cases and how the PICmicro® MCU can be wired to other devices to take advantage of the PICmicro® MCU's electrical properties. Using these properties, applications can use standard I/O pins to simulate different interface types and, in some cases, reduce the numbers of pins required for an application.

**Parallel bus devices** Although the PICmicro® MCU is very well suited for stand-alone applications, many applications have to connect to external devices. There are built-in PICmicro® MCU interfaces for *Non-Return to Zero (NRZ)* asynchronous I/O and two-wire serial I/O, but sometimes the best interface is a simulated parallel I/O bus. A simulated parallel bus is useful for increasing the I/O capabilities of the PICmicro® MCU using standard I/O chips. These devices can be accessed fairly easily using an eight-bit I/O port and a few extra control pins from the PICmicro® MCU.

I realize that the PIC17Cxx has the ability to drive a parallel bus, but I tend to shy away from using these devices for this purpose because the PIC17Cxx I/O data bus is 16 bits wide and can only access the full word of data at any one time. Most parallel buses require an eight-bit bus and the PIC17Cxx devices tend to be more expensive than using the mid-range parts. Looking at the extra costs of the PIC17Cxx devices, coupled with the complexity of adding the address buffers, you're probably better off using the mid-range devices and simulating the bus as shown here.

When I create a parallel bus, I normally use PORTB for eight data bits and use other PORT pins for the _RD and _WR lines. To avoid the extra costs and complexity of decode circuitry, it is probably best to devote one I/O line to each device. Before writing from the PICmicro® MCU to the device, TRISB is set to output mode and the value to be written is output on PORTB. Next, the _CS and _WR lines are pulled low and remain active until the device's minimum access times are met. _RD is similar with TRISB being put in input mode, the _CS and _RD pins are held active until the devices minimum read access time is met, at which point the data is strobed into w, _CS, and _RD are driven high.

The circuit in Fig. 6-4 requires two parallel output bytes and one parallel input byte. This could be implemented with a 40-pin PICmicro® MCU, using the I/O pins directly, but it is much more cost effective to use an 18-pin PICmicro® MCU (such as the PIC16F84) and a tristate output buffer and two eight bit registers. In Fig. 6-4, it is assumed that data is clocked in or output with negative active signal pulses.

With this circuit, RA0 to RA2 would be set for output and initialized to 1 (high voltage). To read the eight data bits on the tristate buffer, the following code could be used:

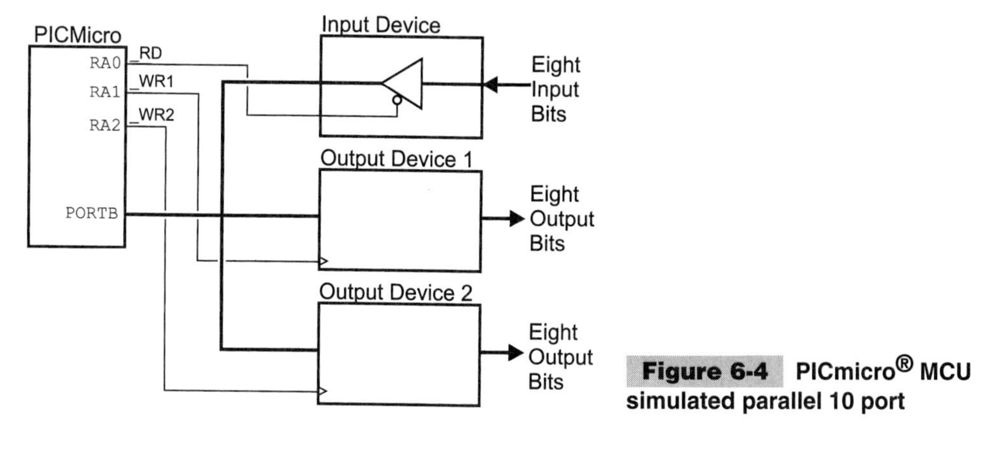


**Figure 6-4** PICmicro® MCU simulated parallel 10 port

```
bsf      STATUS, RPO          ; Put PORTB into Input Mode
movlw    0x0FF
movwf    TRISB ^ 0X080
bcf      STATUS, RPO
bcf      PORTA, 0             ; Drop the "_RD" line
call     Dlay                 ; Delay until Data Output Valid
movf     PORT B, w            ; Read Data from the Port
bsf      PORT A, 0            ; "_RD" = 1 (disable "_RD" Line)
```

Writing to one of the output registers is similar:

```
bsf      STATUS, RPO
clrf     TRIS B ^ 0X080       ; PORTB Output
bcf      STATUS, RPO
bcf      PORTA, 1             ; Enable the "_WR1" Line
movwf    PORTB                ; output the Data
call     Dlay                 ; Wait Data Receive Valid
bsf      PORTA, 1             ; "_WR1" = 1.
```

**Combining input and output** Often when working on applications, you will find some situations where peripheral devices will use more than one pin for I/O. Another case would be when you are connecting two devices, one input and one output and would like to combine them somehow so that you can reduce the number of PICmicro® MCU pins required. Fewer PICmicro® MCU pins means that you can use cheaper PICmicro® MCUs and avoid complex application wiring. This section presents two techniques for doing this and the rules governing their use. This might at first appear problematic and asking for issues with "bus contention," but they really do work and can greatly simplify your application.

When interfacing the PICmicro® MCU to a driver and receiver (such as a memory with a separate output and input), a resistor can be used to avoid bus contention at any of the pins (Fig. 6-5).

In this situation, when the PICmicro® MCU's I/O pin is driving out, it will be driving the Data In pin register and regardless of the output of the Data Out pin. If the PICmicro® MCU and Data Out pins are driving different logic levels, the resister will limit the current flowing between the PICmicro® MCU and the memory Data Out pin. The value received on the Data In pin will always be the PICmicro® MCU's output due to the voltage drop within the registor.

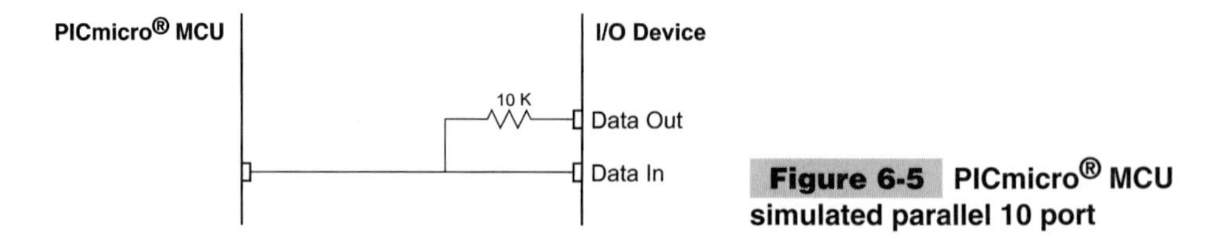


**Figure 6-5** PICmicro® MCU simulated parallel 10 port

When the PICmicro® MCU is receiving data from the memory, its I/O pin will be put in input mode and the Data Out pin will drive its value to not only the PICmicro® MCU's I/O pin, but the Data In pin as well. In this situation, the Data In pin should not be latching any Data In. To avoid this, in most cases where this circuit combines input and output, the two input and output pins are on the same device and the data mode is controlled by the PICmicro® MCU to prevent invalid data from being input into the device. This is an important point because it defines how this trick should be used. The I/O pins that the PICmicro® MCU are connected to must be mutually exclusive and can never be transmitting data at the same time. A common use for this method of connection data in and data out pins is used in SPI memories, which have separate data input and output pins.

The second trick is to have button input, along with an external device receiver. As is shown in Fig. 6-6, a button can be put on the same "net" as an input device and the PICmicro® MCU pin that drives it.

When the button is open or closed, the PICmicro® MCU can drive data to the input device, the 100-K and 10-K resistors will limit the current flow between $V_{cc}$ and ground. If the PICmicro® MCU is going to read the button high (switch open) or low (switch closed) will be driven on the bus at low currents when the pin is in input mode. If the button switch is open, then the 100-K resistor acts like a pull up and a *1* is returned. When the button switch is closed, then there will be approximately 0.5 volt across the 10-K resistor, which will be read as a *0*.

The button with the two resistors pulling up and down are like a low-current driver and the voltage produced by them is easily "overpowered" by active drivers. Like the first method, the external input device cannot receive data except when the PICmicro® MCU is driving the circuit. A separate clock or enable should be used to ensure that input data is received when the PICmicro® MCU is driving the line.

Two points about this method; this can be extrapolated to work with a switch matrix keyboard. The circuit can become very complex, but it will work. Secondly, a

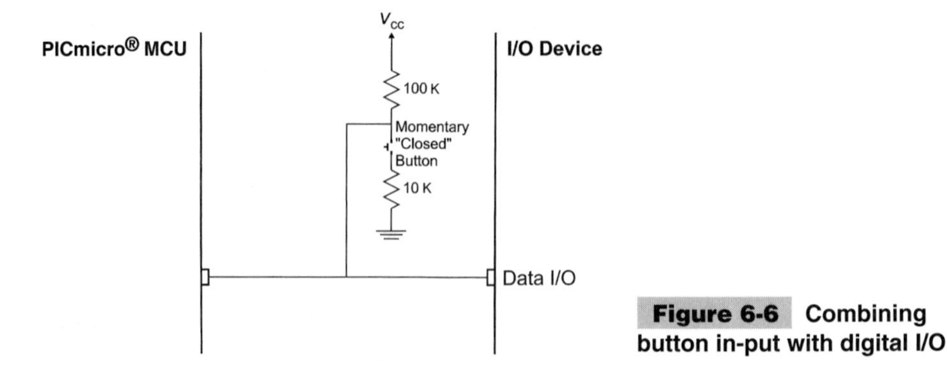


**Figure 6-6** Combining button in-put with digital I/O

resistor/capacitor network for debouncing the button cannot be used with this circuit because a resistor-capacitor network will "slow down" the response of the PICmicro® MCU driving the data input pin and will cause problems with the correct value being accepted. When a button is shared with an input device; such as is shown in Fig. 6-6, software button debouncing will have been done inside the PICmicro® MCU.

**Simulated open collector/open drain I/O** Open collector/open drain outputs are useful in several different interfacing situations. Along with providing a node in a dotted AND bus, they are also useful to interface with I2C and other networks. I find that the single open drain pin available in the different PICmicro® MCU devices to be insufficient for many applications, which is why I find it useful to simulate an open drain driver with a standard I/O pin.

An open drain pin (shown in Fig. 6-7) consists of a N-channel MOSFET transistor with its source connected to the I/O pin. Because there is no P-channel high driver in the pin circuit, when a *1* is being output the only transistor will be turned off and the pin is allowed to float. *Floating,* in most applications, means having a pull up and multiple open-collector/open-drain drivers on the net.

When the data out bit is low (and TRIS is enabled for output), the pin is pulled low. Otherwise, it is not electrically connected to anything ("tristated").

This action can be simulated by using the following code, which enables the I/O pin output to be low if the carry flag is reset. If the carry flag is set, then the pin is put into input mode.

```
bcf         PORT#, pin              ; Make Sure PORT# Pin Bit is "0"
bsf         STATUS, RP0
btfss       STATUS, C               ; If Carry Set, Disable Open Collector
 goto       $ + 4                   ; Carry Reset, Enable Open Collector
nop
bsf         TRIS# ^ 0x080, pin
goto        $ + 3
bcf         TRIS# ^ 0x080, pin
goto        $ + 1
bcf         STATUS, RP0
```

This code, which is designed for mid-range PICmicro® MCUs, will either set the pin to input (tristate) mode or pulled low on the sixth cycle after it has been invoked, I normally put this code into a macro, with the port and pin specified as parameters.

It will seem like I went to some lengths to ensure that the timing was the same for making the bit tristate (input mode) or pulled low to 0, as well as the state specified by the carry flag. Regardless of the execution path, this code will take eight instruction cycles and the

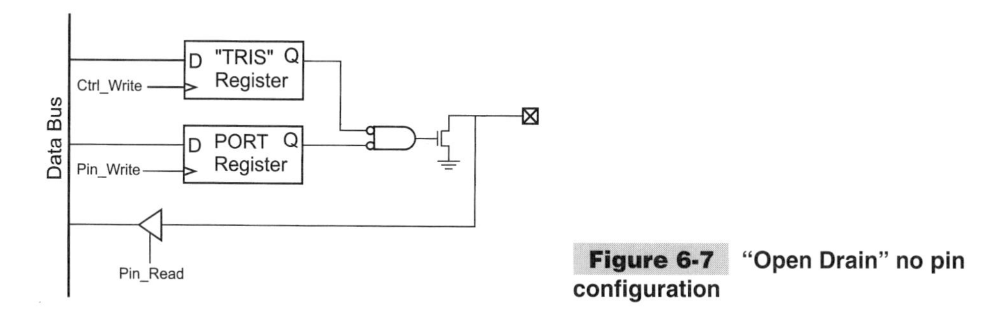


**Figure 6-7** **"Open Drain" no pin configuration**

I/O pin value will be changed at five cycles. I did this because this function is often used with I2C or other network protocols and using the carry flag allows bits to be shifted through this code easily.

In the sample open-drain simulation code, I reset the specified pin before potentially changing its TRIS value. This is to prevent it from being the wrong value, based on reads and writes of other I/O pins. This is the "inadvertent pin changes" that I've written about elsewhere in the book.

## DIFFERENT LOGIC LEVELS WITH ECL AND LEVEL SHIFTING

Often, when working with PICmicro® MCUs and other microcontrollers, you will have to interface devices of different logic families together. For standard positive voltage logic families (i.e., TTL to CMOS), this is not a problem; the devices can be connected directly. But, interfacing a negative voltage logic to a positive voltage logic family (i.e., ECL to CMOS) can cause some problems.

Although chips usually are available to provide this interface function (for both input and output), they typically only work in only one direction (which precludes bi-directional busses—even if the logic family allows it) and the chips can add a significant cost to the application.

The most typical method of providing level conversion is to match the switching threshold voltage levels of the two logic families.

As shown in Fig. 6-8, the ground level for the COMS microcontroller has been shifted below ground (the microcontroller's "ground" is actually the CMOS 0 level). The purpose of this is to place the point where the microcontroller's input logic switches between a *0* and a *1* (known as the *input logic threshold voltage*) is the same as the ECL Logic. The resistor (which is between 1 K and 10 K) is used to limit the current flow because of the different logic swings of the two different families.

Looking at the circuit block diagram, you're probably thinking that the cost of shifting the microcontroller power supply is much greater than just a few interface chips.

Actually, this isn't a big concern because of the low power requirements of modern CMOS microcontrollers. In Fig. 6-8, the microcontroller's ground reference can be produced by placing a silicon diode (which has a 0.7 voltage drop across it) between it and the ECL's 2-volt power supply negative output. The 5-volt and 2-volt supplies' positive output have a common "threshold" voltage for this circuit and the registor limits the CMOS

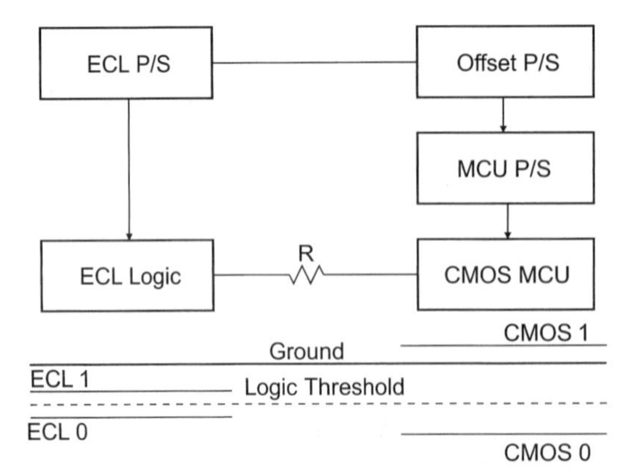


**Figure 6-8** ECL to CMOS logic level conversion

logic swing to the 1 volt ECL swing. This example might seem simplistic, but it would provide the ability to connect a CMOS 0- to +5-volt microcontroller to ECL logic (and allow signals to be sent in either direction, between the PICmicro® MCU and the ECL Logic) at a very low cost.

# LEDs

The most common form of output from a microcontroller is the *Light-Emitting Diode (LED).* As an output device, it is cheap and easy to wire to a microcontroller. Generally, LEDs require anywhere from 5 mA of current to light (which is within the output sink/source specification for most microcontrollers). But, what has to be remembered is, LEDs are diodes, which means that current flows in one direction only. The typical circuit that I use to control an LED from a PICmicro® MCU I/O pin is shown in Fig. 6-9.

With this circuit, the LED will light when the microcontroller's output pin is set to *0* (ground potential). When the pin is set to input or outputs a *1,* the LED will be turned off.

The 220 Ohm resistor is used for current limiting and will prevent excessive current that can damage the microcontroller, LED and the power supply. Elsewhere in the book, I have shown how this resistor value is calculated. Some microcontrollers (such as the PICmicro® MCU) already have current-limiting output pins, which lessens the need for the current-limiting resistor. But, I prefer to always put in the resistor to guarantee that a short (either to ground or $V_{cc}$) cannot ever damage the microcontroller of the circuit it's connected to (including the power supply).

Probably the easiest way to output numeric (both decimal and hex) data is via seven-segment LED displays. These displays were very popular in the 1970s (if you're old enough, your first digital watch probably had seven-segment LED displays), but have been largely replaced by LCDs.

Seven-segment LED displays (Fig. 6-10) are still useful devices that can be added to a circuit without a lot of software effort. By turning on specific LEDs (each of which light up a segment in the display), the display can be used to output decimal numbers.

Each one of the LEDs in the display is given an identifier and a single pin of the LED is brought out of the package. The other LED pins are connected together and wired to a common pin. This common LED pin is used to identify the type of seven-segment display (as either common cathode or common anode).

Wiring one display to a microcontroller is quite easy—it is typically wired as seven (or eight, if the decimal point, DP, is used) LEDs wired to individual pins.

The most important piece of work you'll do when setting up seven-segment LED displays is matching and documenting the microcontroller bits to the LEDs. Spending a few moments at the start of a project will simplify wiring and debugging the display later.

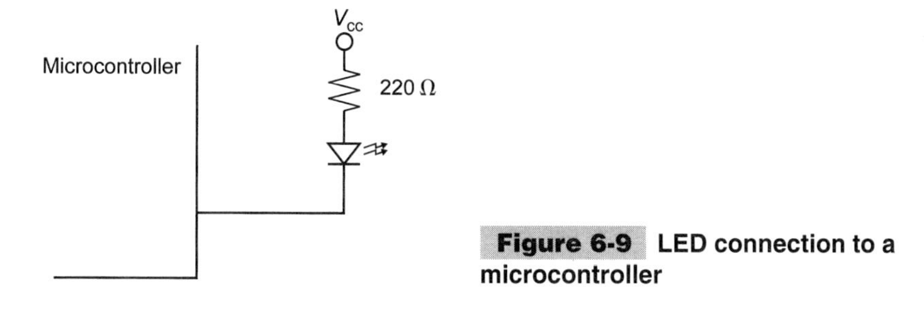


**Figure 6-9** LED connection to a microcontroller

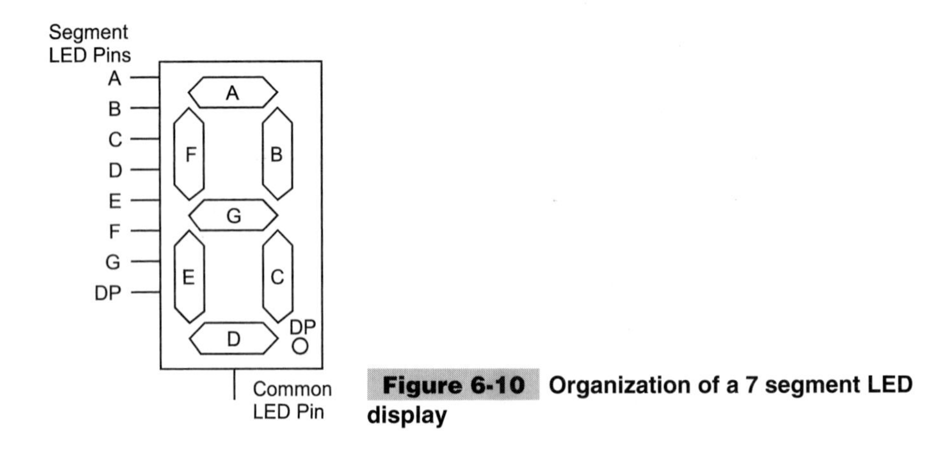


**Figure 6-10** Organization of a 7 segment LED display

The typical method of wiring multiple seven-segment LED displays together is to wire them all in parallel and then control the current flow through the common pin. Because the current is generally too high for a single microcontroller pin, a transistor is used to pass the current to the common power signal. This transistor selects which display is active.

Figure 6-11 shows common-cathode seven-segment displays connected to a microcontroller.

In this circuit, the microcontroller will shift between the displays showing each digit in a very short time slice. This is usually done in a timer interrupt handler. The basis for the interrupt handler's code is:

```
Int
  - Save Context Registers
  - Reset Timer and Interrupt
  - LED_Display = 0                ;  Turn Off all the LEDs
  - LED_Output = Display[ Cur ]
  - Cur = (Cur + 1) mod #Displays  ;  Point to Next Sequence Display
  - LED_Display = 1 << Cur         ;  Display LED for Current Display
  - Restore Context Registers
  - Return from Interrupt
```

This code will cycle through each of the digits (and displays), having current go through the transistors for each one. To avoid flicker, I generally run the code so that each digit is turned on/off at least 50 times per second. The more digits you have, the faster you have

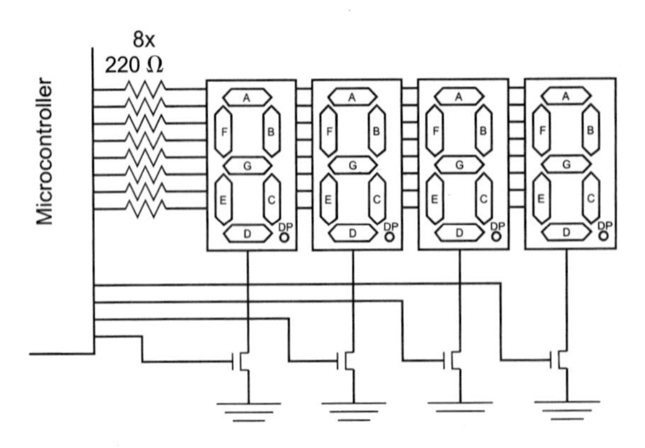


**Figure 6-11** Wiring four 7 segment LED displays

to cycle the interrupt handler (i.e., eight seven-segment displays must cycle at least 400 digits per second, which is eight times as fast as a single display).

You might feel that assigning a microcontroller bit to select each display LED to be somewhat wasteful (at least I do). I have used high-current TTL demultiplexer (i.e., 74S138) outputs as the cathode path to ground (instead of discrete transistors). When the output is selected from the demultiplexer, it goes low, allowing current to flow through the LEDs of that display (and turning it on). This actually simplifies the wiring of the final application as well. The only issue is to ensure that the demultiplexer output can sink the maximum of 140 mA of current that will come through the common cathode connection.

Along with seven-segment displays, 14- and 16-segment LED displays are available, which can be used to display alphanumeric characters (A to Z and 0 to 9). By following the same rules as used when wiring up a seven-segment display, you shouldn't have any problems wiring the display to a PICmicro® MCU. Chapter 16 shows how seven- and 16-segment LEDs can be used to display letters and numbers.

# Switch Bounce

When a button is opened or closed, we perceive that it is a clean operation that really looks like a step function. In reality, the contacts of a switch bounce when they make contact, resulting in a jagged signal (Fig. 6-12).

When this signal is passed to a PICmicro® MCU, the microcontroller can recognize this as multiple button presses, which will cause the application software to act as if multiple, very fast button presses have occurred. To avoid this problem the "noisy" switch press is "debounced" into an idealized "press," or the step function (Fig. 6-13). Two common methods are used to debounce button inputs.

The first is to poll the switch line at short intervals until the switch line stays at the same level for an extended period of time. A button is normally considered to be debounced if it

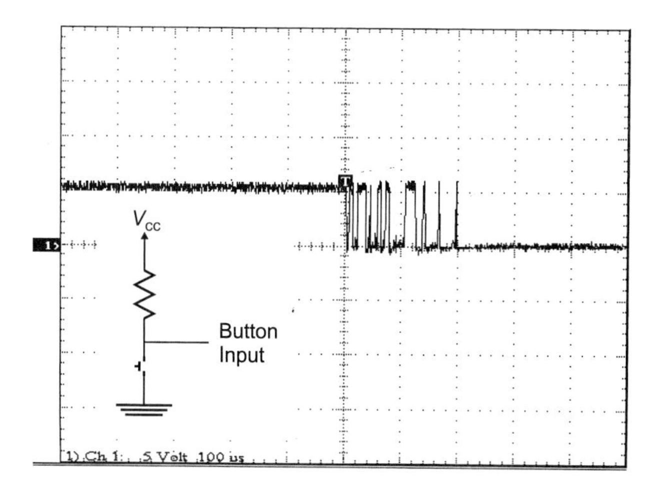


**Figure 6-12** Oscilloscope picture of a switch "bounce"

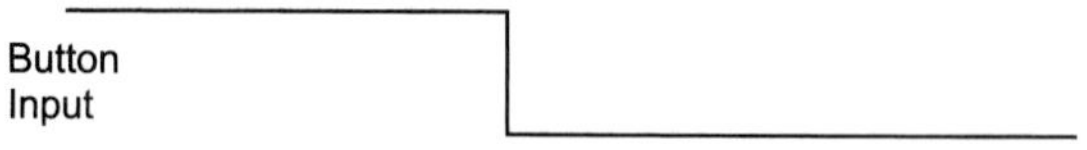

Button
Input

**Figure 6-13** Idealized switch operation

does not change state for 20 ms or longer. By polling the line every 5 ms, this debouncing method can be conceptualized quite easily (Fig. 6-14).

The advantage of this method is that it can be done in an interrupt handler and the line can be scanned periodically with a flag set if the line is high and another flag in the line is low. For the "indeterminate" stage, neither bit would be set. This method of debouncing is good for debouncing keyboard inputs.

The second method is to continually poll the line and wait for 20 ms to go by without the line changing state. The algorithm that I use for this function is:

```
ButLoop:
  while (Button == High);     // Poll Until Button is Pressed
  for (Dlay = 0; (Dlay < 20msec) and (Button == Low); Dlay++);
  if (Dlay != 20 msec)        // Repeat Process if 20 msecs have
    goto ButLoop;             // Not gone by with Button Pressed
```

This code will wait for the button to be pressed and then poll it continuously until either 20 ms has passed or the switch has gone high again. If the switch goes high, the process is repeated until it is held low for 20 ms.

This method is well suited to applications that don't have interrupts, only have one button input, and have no need for processing while polling the button. As restrictive as it sounds, many applications fit these criteria.

This method can also be used with interrupt inputs along with TMR0 in the PICmicro® MCU, which eliminates these restrictions. The interrupt handler behaves like the following pseudo-code when one of the port changes on interrupt bits is used for the button input:

```
interrupt ButtonDebounce()            // Set Flags According to the
{                                     //   Debounced State of the Button

 if (T0IF == 1) {                     // TMR0 Overflow, Button Debounced
    T0IF = 0; T0IE = 0;               // Reset and Turn off TMR0 Interrupts
    if (Button == High) {
      Pressed = 0; NotPressed = 0;    // Set the State of the Button
    } else {
      Pressed = 1; NotPressed = 0;
```

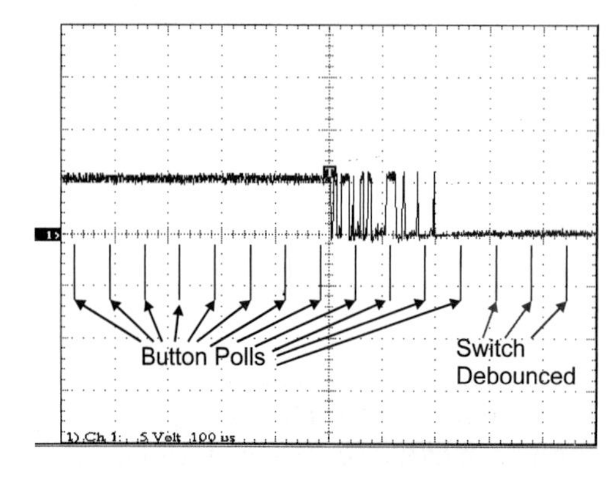


**Figure 6-14** Polling to eliminate "bounce"

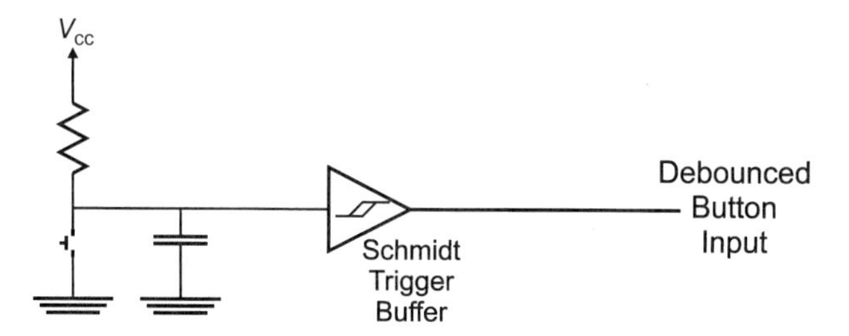


**Figure 6-15** **Debounced switch using a Schmidt Trigger**

```
    }
  } else {                          // Port Change Interrupt
    NotPressed = 1;                 // Nothing True
    RBIF = 0;                       // Reset the Interrupt
    TMR0 = 20msecDlay;              // Reset Timer 0 for 20 ms
    T0IF = 0; T0IE = 1;             // Enable the Timer Interrupt
  }
} // End ButtonDebounce
```

This code waits for the input pin to change state and then resets the two flags that indicate the button state and starts TRM0 to request an interrupt after 20 ms. After a port change interrupt, notice that I reset the button state flags to indicate to the mainline that the button is in a transition state and is not yet debounced. If TMR0 overflows, then the button is polled for its state and the appropriate button state flags are set and reset.

The mainline code should poll the Pressed and NotPressed flags when it is waiting for a specific state. Chapter 15 shows this method of using TMR0 and how interrupts can be implemented with or without interrupts.

If you don't want to use the software approaches, you can use a capacitor to filter the bouncing signal and pass it into a Schmidt trigger input. Schmidt trigger inputs have different thresholds, depending on whether the signal is rising or falling. For rising edges, the trigger point is higher than falling. Schmidt trigger inputs have the "hysteresis" symbol put in the buffer (Fig. 6-15).

This method is fairly reliable, but requires an available Schmidt trigger gate in your circuit. A Schmidt trigger input might be available in your PICmicro® MCU, but check the data sheet to find out which states and peripheral hardware functions can take advantage of it.

Lastly, choose buttons with a positive "click" when they are pressed and released. These have reduced bouncing, often have a "self cleaning" feature to reduce poor contacts, and are a lot easier to work with than other switches that don't have this feature. I have used a number of switches over the years that don't have this click and they can be a real problem in circuits with intermittent connections and unexpected "bouncing" that occurs while the button is pressed and held down.

# Matrix Keypads

Switch matrix keyboards and keypads are really just an extension of the button concepts, with many of the same concerns and issues to watch out for. The big advantage that the matrix keyboards gives you is that they provide a large number of button inputs for a rel-

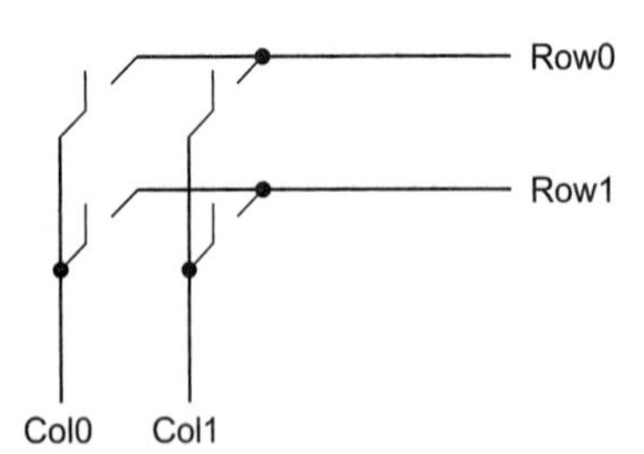


**Figure 6-16** 2x2 switch matrix

atively small number of PICmicro® MCU pins. The PICmicro® MCU is well designed for simply implementing switch matrix keypads, which, like LCD displays that are explained in the next section, can add a lot to your application with a very small investment in hardware and software.

A switch matrix is simply a two-dimensional matrix of wires, with switches at each vertex. The switch is used to interconnect rows and columns in the matrix (Fig. 6-16).

This diagram might not look like the simple button, but it will become more familiar when I add switchable ground connections on the columns (Fig. 6-17).

In this case, by connecting one of the columns to ground, if a switch is closed, the pull down on the row will connect the line to ground. When the row is polled by an I/O pin, a *0* or low voltage will be returned instead of a *1* (which is what will be returned if the switch in the row that is connected to the ground is open).

As stated, the PICmicro® MCU is well suited to implement switch matrix keyboards with PORTB's internal pull-ups and the ability of the I/O ports to simulate the open-drain pull-downs of the columns (Fig. 6-18). Normally, the pins connected to the columns are left in tristate (input) mode. When a column is being scanned, the column pin is output enabled driving a *0* and the four input bits are scanned to see if any are pulled low.

In this case, the keyboard can be scanned for any closed switches (buttons pressed) using the code:

```
int KeyScan()                    // Scan the Keyboard and Return when a
{                                //   key is pressed

int  i = 0;
int  key = -1;

    while (key == -1) {

      for (i = 0; (i < 4) & ((PORTB & 0x00F) == 0x0F0); i++);
```

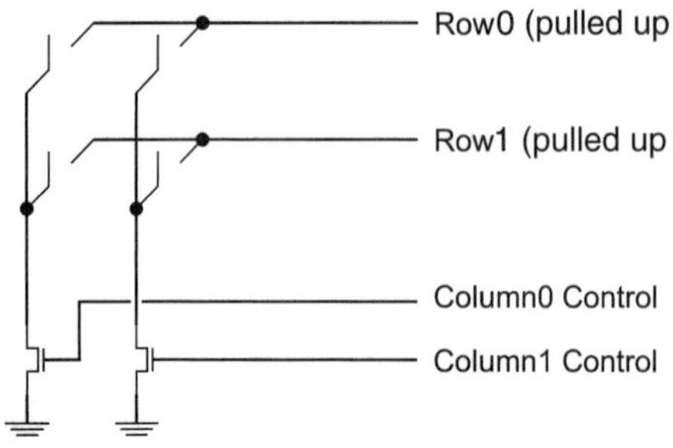

Row0 (pulled up internally)
Row1 (pulled up internally)
Column0 Control
Column1 Control

**Figure 6-17** Switch matrix with pull down transistors

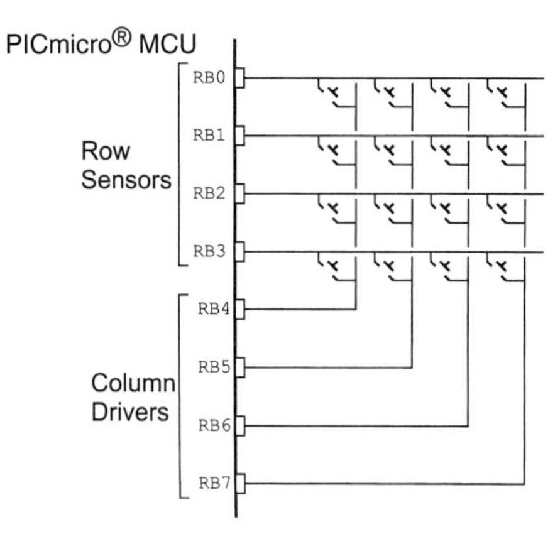


**Figure 6-18** 4x4 switch matrix connected to PORTB

```
      switch (PORTB & 0x00F) { // Find Key that is Pressed
        case 0x00E:            // Row 0
          key = i;
          break;
        case 0x00D:            // Row1
        case 0x00C:
          key = 0x04 + i;
          break;
        case 0x00B:            // Row2
        case 0x00A:
        case 0x009:
        case 0x008:
          key = 0x08 + i;
          break;
        else                   // Row3
          key = 0x0C + i;
          break;
      }//end switch
    }// end while

    return key;

} // End KeyScan
```

The KeyScan function will only return when a key has been pressed. This routine will not allow keys to be debounced or for other code to execute while it is running.

These issues can be resolved by putting the key scan into an interrupt handler, which executes every 5 ms:

```
Interrupt KeyScan( )          // 5 msec Interval Keyboard Scan
{

int i =0;
int key = -1

  for (i = 0; (i <4) & ((PORTB & 0x00F) == 0x00F)); i++);
  if (PORTB & 0x00F) ! = 0x00F) {      // Key Pressed
    switch (PORTB & 0x00F) {  // Find Key that is Pressed
      case 0x00E:             // Row 0
        key = i;
```

```
        break;
      case 0x00D:                 // Row1
      case 0x00C:
        key = 0x04 + i;
        break;
      case 0x00B:                 // Row2
      case 0x00A:
      case 0x009:
      case 0x008:
        key = 0x08 + i;
        break;
      else                        // Row3
        key = 0x0C + i;
        break;
    }//end switch
    if (key == KeySave) {
      keycount = keycount + 1; // Increment Count
                                // <-- Put in Auto Repeat Code Here
      if (keycount == 4)
        keyvalid = key;         // Debounced Key
    } else
      keycount = 0;             // No match - Start Again
    KeySave = key;              // Save Current key for next 5 msec
  }                             // Interval
}// End KeySave
```

This interrupt handler will set the *keyvalid* variable to the row/column combination of the key button (which is known as a *scan code*) when the same value comes up four times in a row. This for time scan is the debounce routine for the keypad. If the value doesn't change for four intervals (20 ms in total), the key is determined to be debounced.

There are two things to notice about this code. First, in both routines, I handle the row with the highest priority. If multiple buttons are pressed, then the one with the highest bit number will be the one that is returned to the user.

The second point is, this code can have an auto repeat function added to it very easily. To do this, a secondary counter has to be first cleared and then incremented each time the *keycount* variable is four or greater. To add an auto repeat key every second (200 intervals), the following code is added in the interrupt handler at the comment.

```
if (keycount == 4) {
  keyrepeat = keyrepeat - 1; // Decrement the Key Auto Repeat Value
  if (keyrepeat == 0) {
    keyrepeat = 200;     // Restart the 1 second Auto Repeat Count
    keycount = 3;        // Reset the counter
    keyvalid = key;      // Return the key
  }
} else                   // Reset the Auto Repeat Counter
keyrepeat = 1;           // End Outputting the Value with Auto
                         //   Repeat
```

The code and methodology for handling switch matrix keypad scans I've outlined here probably seems pretty simple. I'm sure you'll be surprised that, with a scanned keyboard, it is most difficult to figure out the scan codes for specific keys and how to wire the keyboard. Chapter 15 demonstrates how this can be done.

# LCDs

LCDs can add a lot to your application in terms of providing a useful interface for the user, debugging an application, or just giving it a professional look. The most common type of LCD controller is the Hitachi 44780, which provides a relatively simple interface between a processor and an LCD. Using this interface is often not attempted by new designers and programmers because it is difficult to find good documentation on the interface, initializing the interface can be a problem, and the displays themselves are expensive.

I have worked with Hitachi 44780-based LCDs for a while now and I don't believe any of these perceptions. LCDs can be added quite easily to an application and use as few as two digital output pins for control. As for cost, LCDs can be often pulled out of old devices or found in surplus stores for less than a dollar.

The purpose of this section is to give a brief tutorial on how to interface with Hitachi 44780-based LCDs. I have tried to provide all of the data necessary for successfully adding LCDs to your application. In the book, I use Hitachi 44780-based LCDs for a number of different projects.

The most common connector used for the 44780-based LCDs is 14 pins in a row, with pin centers 0.100" apart. The pins are wired as in Table 6-2.

As you would probably guess from this description, the interface is a parallel bus, allowing simple and fast reading/writing of data to and from the LCD.

The waveform shown in Fig. 6-19 will write an ASCII byte out to the LCD's screen. The ASCII code to be displayed is eight bits long and is sent to the LCD either four or eight bits at a time. If four-bit mode is used, two nybbles of data (sent high four bits and then low four bits with an E clock pulse with each nybble) are sent to make up a full eight-bit transfer. The E clock is used to initiate the data transfer within the LCD.

Sending parallel data as either four or eight bits are the two primary modes of operation. Although there are secondary considerations and modes, deciding how to send the data to the LCD is the most crucial decision to be made for an LCD interface application.

**TABLE 6-2 Hitachi 44780 Based LCD Pinout**

| PIN | DESCRIPTION |
|---|---|
| 1 | Ground |
| 2 | Vcc |
| 3 | Contrast Voltage |
| 4 | "R/S" - Instruction/Register Select |
| 5 | "R/W" - Read/Write LCD Registers |
| 6 | "E" - Clock |
| 7-14 | D0-D7 Data Pins |

Eight-bit mode is best used when speed is required in an application and 10 I/O pins are available. Four-bit mode requires six bits. To wire a microcontroller to an LCD in four-bit mode, just the top four bits (DB4-7) are written to.

The R/S bit is used to select whether data or an instruction is being transferred between the microcontroller and the LCD. If the bit is set, then the byte at the current LCD cursor position can be read or written. When the bit is reset, either an instruction is being sent to the LCD or the execution status of the last instruction is read back (whether or not it has completed).

The different instructions available for use with the 44780 are shown in Table 6-3.

The bit descriptions for the different commands are:

```
*Not used/ignored. This bit can be either 1 or 0
Set cursor move direction:
   ID   Increment the cursor after each byte written to display if set
    S   Shift display when byte written to display
Enable display/cursor
    D   Turn display on(1)/off(0)
    C   Turn cursor on(1)/off(0)
    B   Cursor blink on(1)/off(0)
Move cursor/shift display
   SC   Display shift on(1)/off(0)
   RL   Direction of shift right(1)/left(0)
Set interface length
   DL   Set data interface length 8(1)/4(0)
    N   Number of display lines 1(0)/2(1)
    F   Character font 5x10(1)/5x7(0)
Poll the busy flag
   BF   This bit is set while the LCD is processing
Move cursor to CGRAM/display
    A   Address
Read/write ASCII to the display
    H   Data
```

Reading data back is best used in applications that require data to be moved back and forth on the LCD (such as in applications that scroll data between lines). The busy flag can be polled to determine when the last instruction that has been sent has completed processing.

For most applications, there really is no reason to read from the LCD. I usually tie R/W to ground and just wait the maximum amount of time for each instruction (4.1 ms for clearing the display or moving the cursor/display to the home position, 160 $\mu$s for all other commands). As well as making my application software simpler, it also frees up a microcontroller pin for other uses. Different LCDs execute instructions at different rates and to

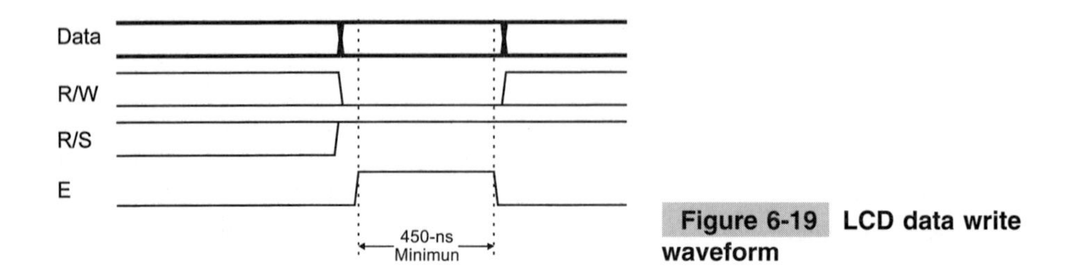


**Figure 6-19 LCD data write waveform**

**TABLE 6-3 Hitachi 44780 Based LCD Commands**

| R/S | R/W | D7 | D6 | D5 | D4 | D3 | D2 | D1 | D0 | INSTRUCTION/DESCRIPTION |
|---|---|---|---|---|---|---|---|---|---|---|
| 4 | 5 | 14 | 13 | 12 | 11 | 10 | 9 | 8 | 7 | Pins |
| 0 | 0 | 0 | 0 | 0 | 0 | 0 | 0 | 0 | 1 | Clear Display |
| 0 | 0 | 0 | 0 | 0 | 0 | 0 | 0 | 1 | * | Return Cursor and LCD to Home Position |
| 0 | 0 | 0 | 0 | 0 | 0 | 0 | 1 | ID | S | Set Cursor Move Direction |
| 0 | 0 | 0 | 0 | 0 | 0 | 1 | D | C | B | Enable Display/Cursor |
| 0 | 0 | 0 | 0 | 0 | 1 | SC | RL | * | * | Move Cursor/Shift Display |
| 0 | 0 | 0 | 0 | 1 | DL | N | F | * | * | Reset/Set Interface Length |
| 0 | 0 | 0 | 1 | A | A | A | A | A | A | Move Cursor to CGRAM |
| 0 | 0 | 1 | A | A | A | A | A | A | A | Move Cursor to Display |
| 0 | 1 | BF | * | * | * | * | * | * | * | Poll the "Busy Flag" |
| 1 | 0 | H | H | H | H | H | H | H | H | Write Hex Character to the Display at the Current Cursor Position |
| 1 | 1 | H | H | H | H | H | H | H | H | Read Hex Character at the Current Cursor Positon on the Display |

avoid problems later on (such as if the LCD is changed to a slower unit), I recommend just using the maximum delays listed here.

In terms of options, I have never seen a 5x10 pixel character LCD display. This means that the F bit in the *set interface* instruction should always be reset (equal to 0).

Before you can send commands or data to the LCD module, the module must be initialized. For eight-bit mode, this is done using the following series of operations:

```
1. Wait more than 15 ms after power is applied.
2. Write 0x030 to LCD and wait 5 ms for the instruction to complete.
3. Write 0x030 to LCD and wait 160 usecs for instruction to complete.
4. Write 0x030 AGAIN to LCD and wait 160 usecs or Poll the Busy Flag.
5. Set the Operating Characteristics of the LCD.
   - Write "Set Interface Length"
   - Write 0x010 to prevent shifting after character write.
   - Write 0x001 to Clear the Display
   - Write "Set Cursor Move Direction" Setting Cursor Behavior Bits
   - Write "Enable Display/Cursor" & enable Display and Optional Cursor
```

In describing how the LCD should be initialized in four-bit mode, I specify writing to the LCD in terms of nybbles. This is because initially, just single nybbles are sent (and not two nybbles, which make up a byte and a full instruction). As mentioned, when a byte is sent, the high nybble is sent before the low nybble and the E pin is toggled each time that four bits are sent to the LCD.

To initialize in four-bit mode the following nybbles are first sent to the LCD:

```
1. Wait more than 15 ms after power is applied.
2. Write 0x03 to LCD and wait 5 ms for the instruction to complete.
3. Write 0x03 to LCD and wait 160 usecs for instruction to complete.
4. Write 0x03 AGAIN to LCD and wait 160 usecs (or poll the Busy Flag).
5. Set the Operating Characteristics of the LCD.
   - Write 0x02 to the LCD to Enable Four Bit Mode
```

All following instruction/data writes require two nybble writes:

```
- Write "Set Interface Length"
- Write 0x01/0x00 to prevent shifting of the display
- Write 0x00/0x01 to Clear the Display
- Write "Set Cursor Move Direction" Setting Cursor Behavior Bits
- Write "Enable Display/Cursor" & enable Display and Optional Cursor
```

Once the initialization is complete, the LCD can be written to with data or instructions as required. Each character to display is written like the control bytes, except that the R/S line is set. During initialization, by setting the S/C bit during the *Move Cursor/Shift Display* command, after each character is sent to the LCD, the cursor built into the LCD will increment to the next position (either right or left). Normally, the S/C bit is set (equal to 1), along with the R/L bit in the *Move Cursor/Shift Display* command for characters to be written from left to right (as with a "TeleType" video display).

One area of confusion is how to move to different locations on the display and, as a follow on, how to move to different lines on an LCD display. Table 6-4 shows how different LCD displays that use a single 44780 can be set up with the addresses for specific character locations. The LCDs listed are the most popular arrangements available and the layout is given as number of columns by number of lines.

The ninth character is the position of the ninth character on the first line. Most LCD displays have a 44780 and support chip to control the operation of the LCD. The 44780 is responsible for the external interface and provides sufficient control lines for 16 characters on the LCD. The support chip enhances the I/O of the 44780 to support up to 128 characters on an LCD in two lines of eight. From Table 6-4, it should be noted that the first two entries (8x1, 16x1) only have the 44780 and not the support chip. This is why the ninth character in the 16x1 does not appear at address 8 and shows up at the address that is common for a two-line LCD.

I've included the 40 character by 4 line (40x4) LCD because it is quite common. Normally, the LCD is wired as two 40x2 displays. The actual connector is normally 16 bits wide with all the 14 connections of the 44780 in common, except for the E (strobe) pins. The E strobes are used to address between the areas of the display used by the two devices. The actual pinouts and character addresses for this type of display can vary between manufacturers and display part numbers.

When using any kind of multiple 44780 LCD display, you should probably only display one 44780's cursor at a time to avoid confusing the user.

Cursors for the 44780 can be turned on as a simple underscore at any time using the *Enable Display/Cursor* LCD instruction and setting the C bit. I don't recommend using the *B (Block mode) bit* because this causes a flashing full-character square to be displayed and it is very annoying.

The LCD can be thought of as a TeleType display because in normal operation, after a character has been sent to the LCD, the internal cursor is moved one character to the right. The *clear display* and *return cursor and LCD to home position* instructions are used to reset the cursor's position to the top right character on the display.

**TABLE 6-4 Hitachi 44780 Based LCD Types and Character Locations**

| LCD | TOP LEFT | NINTH | SECOND LINE | THIRD LINE | FOURTH LINE | COMMENTS |
|---|---|---|---|---|---|---|
| 8x1 | 0 | N/A | N/A | N/A | N/A | Note 1. |
| 16x1 | 0 | 0x040 | N/A | N/A | N/A | Note 1. |
| 16x1 | 0 | 8 | N/A | N/A | N/A | Note 3. |
| 8x2 | 0 | N/A | 0x040 | N/A | N/A | Note 1. |
| 10x2 | 0 | 0x008 | 0x040 | N/A | N/A | Note 2. |
| 16x2 | 0 | 0x008 | 0x040 | N/A | N/A | Note 2. |
| 20x2 | 0 | 0x008 | 0x040 | N/A | N/A | Note 2. |
| 24x2 | 0 | 0x008 | 0x040 | N/A | N/A | Note 2. |
| 30x2 | 0 | 0x008 | 0x040 | N/A | N/A | Note 2. |
| 32x2 | 0 | 0x008 | 0x040 | N/A | N/A | Note 2. |
| 40x2 | 0 | 0x008 | 0x040 | N/A | N/A | Note 2. |
| 16x4 | 0 | 0x008 | 0x040 | 0x010 | 0x050 | Note 2. |
| 20x4 | 0 | 0x008 | 0x040 | 0x014 | 0x054 | Note 2. |
| 40x4 | 0 | N/A | N/A | N/A | N/A | Note 4. |

Note 1: Single 44780/No Support Chip.
Note 2: 44780 with Support Chip.
Note 3: 44780 with Support Chip. This is quite rare.
Note 4: Two 44780s with Support Chips. Addressing is device specific.

An example of moving the cursor is shown in Fig. 6-20.

To move the cursor, the *move cursor to display* instruction is used. For this instruction, bit 7 of the instruction byte is set with the remaining seven bits used as the address of the character on the LCD the cursor is to move to. These seven bits provide 128 addresses, which matches the maximum number of LCD character addresses available. Table 6-4 should be used to determine the address of a character offset on a particular line of an LCD display.

The character set available in the 44780 is basically ASCII. I say "basically" because some characters do not follow the ASCII convention fully (probably the most significant

Initial LCD Condition

After string is written, LCD curser after "u"

Moving LCD Cursor

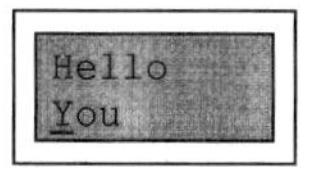


LCD curser is moved to start of second line using `0x0C0` instructior

Final LCD Condition

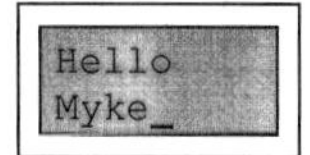


New string is written and overwrites "`You`"

**Figure 6-20** Moving an LCD cursor

Char. code

| | 0000 | 0010 | 0011 | 0100 | 0101 | 0110 | 0111 | 1010 | 1011 | 1100 | 1101 | 1110 | 1111 |
|---|---|---|---|---|---|---|---|---|---|---|---|---|---|
| xxxx0000 | | | 0 | @ | P | ` | p | | ー | タ | ミ | α | p |
| xxxx0001 | | ! | 1 | A | Q | a | q | 。 | ア | チ | ム | ä | q |
| xxxx0010 | | " | 2 | B | R | b | r | 「 | イ | ツ | メ | β | θ |
| xxxx0011 | | # | 3 | C | S | c | s | 」 | ウ | テ | モ | ε | ∞ |
| xxxx0100 | | $ | 4 | D | T | d | t | 、 | エ | ト | ヤ | μ | Ω |
| xxxx0101 | | % | 5 | E | U | e | u | ・ | オ | ナ | ユ | σ | ü |
| xxxx0110 | | & | 6 | F | V | f | v | ヲ | カ | ニ | ヨ | ρ | Σ |
| xxxx0111 | | ' | 7 | G | W | g | w | ァ | キ | ヌ | ラ | g | π |
| xxxx1000 | | ( | 8 | H | X | h | x | ィ | ク | ネ | リ | √ | x̄ |
| xxxx1001 | | ) | 9 | I | Y | i | y | ゥ | ケ | ノ | ル | ⁻¹ | y |
| xxxx1010 | | * | : | J | Z | j | z | ェ | コ | ハ | レ | j | 千 |
| xxxx1011 | | + | ; | K | [ | k | { | ォ | サ | ヒ | ロ | ˣ | 万 |
| xxxx1100 | | , | < | L | ¥ | l | \| | ャ | シ | フ | ワ | ¢ | 円 |
| xxxx1101 | | - | = | M | ] | m | } | ュ | ス | ヘ | ン | Ł | ÷ |
| xxxx1110 | | . | > | N | ^ | n | → | ョ | セ | ホ | ゛ | ñ | |
| xxxx1111 | | / | ? | O | _ | o | ← | ッ | ソ | マ | ゜ | ö | █ |

**Figure 6-21** LCD character set

difference is 0x05B or is not available). The ASCII control characters (0x008 to 0x01F) do not respond as control characters and might display "funny" (Japanese) characters. The LCD character set is shown in Fig. 6-21.

Eight programmable characters are available and use codes 0x000 to 0x007. They are programmed by pointing the LCD's "cursor" to the character generator RAM (CGRAM) area at eight times the character address. The next eight bytes written to the RAM are the line information of the programmable character, starting from the top (Fig. 6-22).

I like to represent this as eight squares by five. Most displays were seven pixels by five for each character, so the extra row may be confusing. Each LCD character is actually eight pixels high, with the bottom row normally used for the underscore cursor. The bottom row can be used for graphic characters, although if you are going to use a visible underscore cursor and have it at the character, I recommend that you don't use this line at all (i.e., set the line to 0x000).

Using this box, you can draw in the pixels that define your special character and then use the bits to determine what the actual data codes are. When I do this, I normally use a

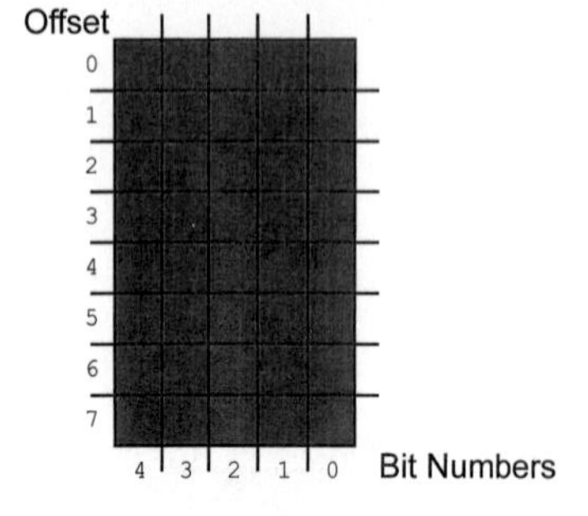


**Figure 6-22** LCD character "box"

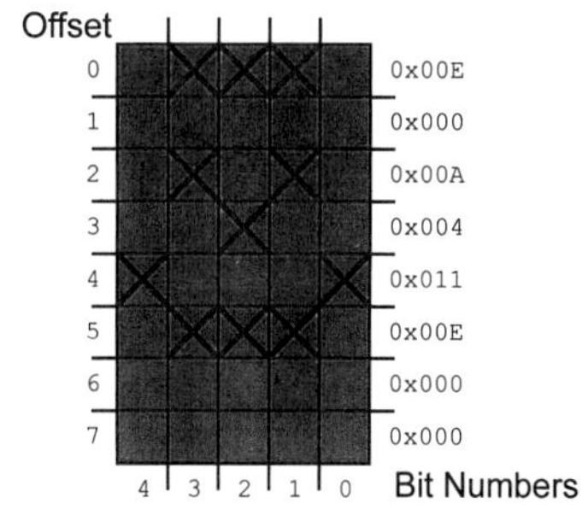


**Figure 6-23** Example LCD custom character

piece of graph paper and then write hex codes for each line, as shown in the lower right diagram of a "Smiley Face" (Fig. 6-23).

For some the "animate" applications, I use character rotation for the animations. This means that, instead of changing the character each time the character moves, I simply display a different character. Doing this means that only two bytes (moving the cursor to the character and the new character to display) have to be sent to the LCD. If animation were accomplished by redefining the characters, then 10 characters would have to be sent to the LCD (one to move into the CGRAM space, the eight defining characters and an instruction returning to display RAM). If multiple characters are going to be used or more than eight pictures for the animation, then you will have to rewrite the character each time.

The user-defined character line information is saved in the LCD's CGRAM area. This 64 bytes of memory is accessed using the *move cursor into CGRAM* instruction in a similar manner to that of moving the cursor to a specific address in the memory with one important difference.

This difference is that each character starts at eight times its character value. This means that user-definable character 0 has its data starting at address 0 of the CGRAM, character 1 starts at address 8, character 2 starts at address 0x010 (16), etc. To get a specific line within the user-definable character, its offset from the top (the top line has an offset of 0) is added to the starting address. In most applications, characters are written to all at one time with character 0 first. In this case, the instruction 0x040 is written to the LCD, followed by all of the user-defined characters.

The last aspect of the LCD to discuss is how to specify a contrast voltage to the display. I typically use a potentiometer wired as a voltage divider. This will provide an easily variable voltage between ground and $V_{CC}$, which will be used to specify the contrast (or darkness) of the characters on the LCD screen. You might find that different LCDs work differently with lower voltages providing darker characters in some and higher voltages do the same thing in others.

There are a variety of different ways of wiring up an LCD. I noted that the 44780 could interface with four or eight bits. To simplify the demands in microcontrollers, a shift register is often used to reduce the number of I/O pins to three.

This can be further reduced by using the circuit shown in which the serial data is combined with the contents of the shift register to produce the E strobe at the appropriate in-

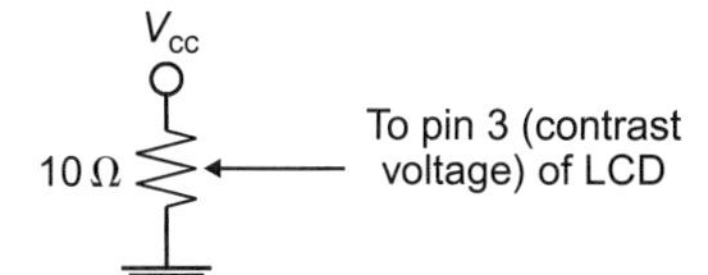


**Figure 6-24** LCD contrast voltage circuit

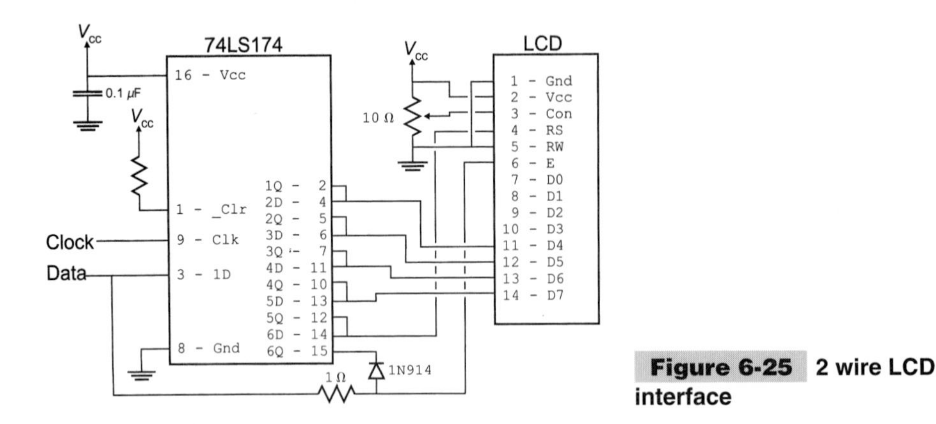


**Figure 6-25** 2 wire LCD interface

terval. This circuit ANDs (using the 1-K resistor and IN914 diode) the output of the sixth D-flip-flop of the 74LS174 and the data bit from the device writing to the LCD to form the E strobe. This method requires one less pin than the three-wire interface and a few more instructions of code. The two-wire LCD interface circuit is shown in Fig. 6-25.

I normally use a 74LS174 wired as a shift register (as shown in the schematic diagram) instead of a serial-in/parallel-out shift register. This circuit should work without any problems with a dedicated serial-in/parallel-out shift register chip, but the timing/clock polarities might be different. When the 74LS174 is used, notice that the data is latched on the rising (from logic low to high) edge of the clock signal. Figure 6-26 is a timing diagram for the two-wire interface and it shows the 74LS174 being cleared, loaded, and then the E strobe when the data is valid and 6Q and incoming data is high.

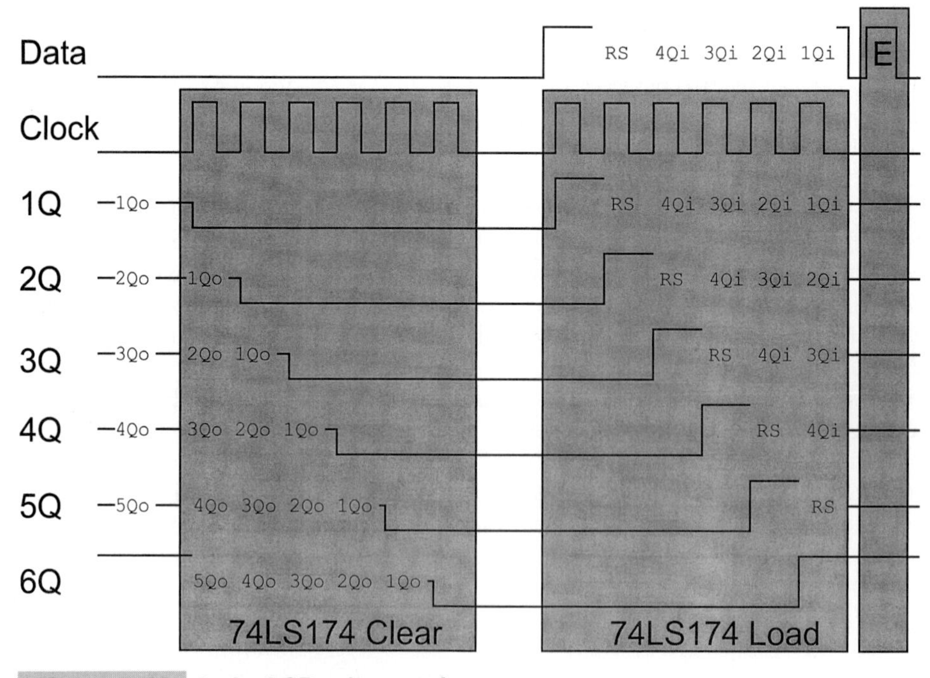


**Figure 6-26** 2 wire LCD write waveform

The right side of this diagram shows how the shift register is written to for this circuit to work. Before data can be written to it, loading every latch with zeros clears the shift register. Next, a *1* (to provide the E gate) is written followed by the R/S bit and the four data bits. Once the latch is loaded in correctly, the data line is pulsed to strobe the E bit. The biggest difference between the three-wire and two-wire interface is that the shift register has to be cleared before it can be loaded and the two-wire operation requires more than six times the number of clock cycles to load four bits into the LCD.

I've used this circuit with the PICmicro® MCU, Basic Stamp, 8051, and AVR, and it really makes the wiring of an LCD to a microcontroller very simple. The biggest issue to watch for is to ensure the E strobe's timing is within specification (i.e., greater than 450 ns); the shift register loads can be interrupted without affecting the actual write. This circuit will not work with open-drain only outputs (something that catches up many people).

One note about the LCD's E strobe is that in some documentation it is specified as *high-level active;* in others, it is specified as *falling-edge active.* It *seems* to be falling-edge active, which is why the two-wire LCD interface works even if the line ends up being high at the end of data being shifted in. If the falling edge is used (like in the two-wire interface) then ensure that before the E line is output on 0, there is at least a 450-ns delay with no lines changing state.

# Analog I/O

Before reading through the following sections, I suggest that you familiarize yourself with *Analog-to-Digital Converters (ADCs)* and *Digital-to-Analog Converters (DACs)* in "Introduction to Electronics" on the CD-ROM. These sections will introduce you to the theory of converting analog voltages and digital values between each other and the problems that can arise with them.

The following sections introduce you to some of the practical aspects of working with analog data with the PICmicro® MCU. This includes position sensing using potentiometers. At first glance, you might think that an ADC-equipped microcontroller is required for this operation, but there are a number of ways of doing this with strictly digital inputs. The IBM PC carries out the same function (it doesn't use an ADC either) for reading joystick positions.

For analog output, I focus on the theory and operation behind *Pulse-Width Modulated (PWM)* analog control signals. This method of control is very popular and is a relatively simple way of providing analog control of a device. It can also be used to communicate analog values between devices without needing any type of communication protocol. The PICmicro® MCU has some built-in hardware that makes the implementation of pulse-width-modulated input and output quite easy to work with.

I want to make it clear that audio input and output capabilities cannot be provided in the PICmicro® MCU without significant front-end signal processing and filtering. Output from the PICmicro® MCU can be simple "beeps" and "boops" without special hardware, but anything more complex will require specialized hardware and software.

## POTENTIOMETERS

One of the more useful human input devices is the *dial*. Rather than relying on some kind of digital data, like a button or character string, the dial allows users a freer range of inputs,

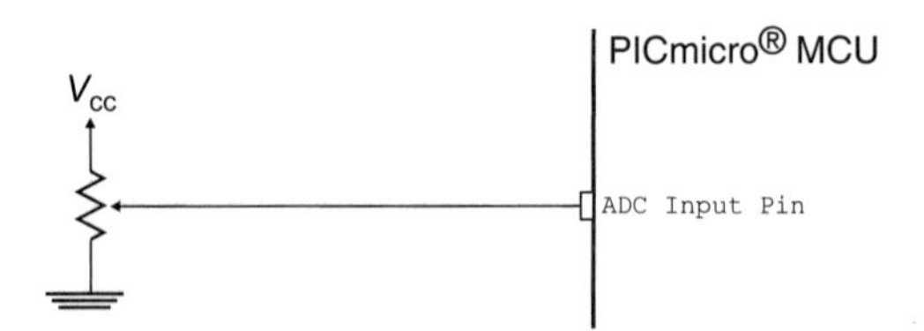


**Figure 6-27** Potentiometer input to PICmicro® MCU ADC

as well as positional feedback information in a mechanical device. For most people, reading a potentiometer value requires setting the potentiometer as a voltage divider and reading the voltage between the two extremes at the "wiper" (Fig. 6-27). A very elegant way of reading a potentiometer's position using the digital input of a PICmicro® MCU is shown in this section.

Notice that I consider the measurement to be of the potentiometer's position and not its resistance. This is an important semantic point; as far as using a potentiometer as an input device, I do not care about the actual resistance of its position, just what its position is. The method of reading a potentiometer's position using a digital I/O pin that I am going to show you is very dependent on the parts used and will vary significantly between implementations.

The method of reading a potentiometer uses the characteristics of charged capacitor discharging through a resistor. If the charge is constant in the capacitor, then the time to discharge varies according to the exponential curve shown in Fig. 6-28.

The charge in a capacitor is proportional with its voltage. If a constant voltage (i.e., from a PICmicro® MCU I/O pin) can be applied to a capacitor, then its charge will be constant. This means that in the voltage discharge curve shown in Fig. 6-28, if the initial voltage is known along with the capacitance and resistance, then the voltage at any point in time can be predicted.

The equation in Fig. 6-28:

$$V_{(t)} = V_{Start}(1 - e ** -t/RC)$$

can be reworked to find $R$, if $V$, $V_{Start}$, $t$, and $C$ are known:

$$R = -t/C * ln[(V_{Start} - V)/V_{Start}]$$

Rather than calculate the value through, you can make the approximation of 2 ms for a resistance of 10 k and a capacitance of 0.1 $\mu$F with a PICmicro® MCU, which has a high-to-low threshold of 1.5 volts. To measure the resistance in a PICmicro® MCU, I use the circuit shown in Fig. 6-29.

For this circuit, the PICmicro® MCU's I/O pin outputs a high, which charges the capacitor (which is limited by the potentiometer).

After the capacitor is charged, the pin is changed to input, the charge in the capacitor draws through the resistor with a voltage determined by the $V_{(t)}$ formula. When the pin first changes state, the voltage across the resistor will be greater than the threshold for some period of time. When the voltage across the potentiometer falls below the voltage threshold, the input pin value returned to the software will be zero. If the time required for

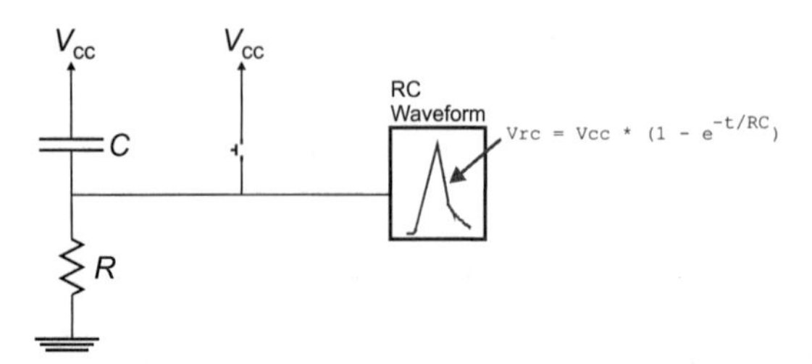


**Figure 6-28** Measure RC time delay

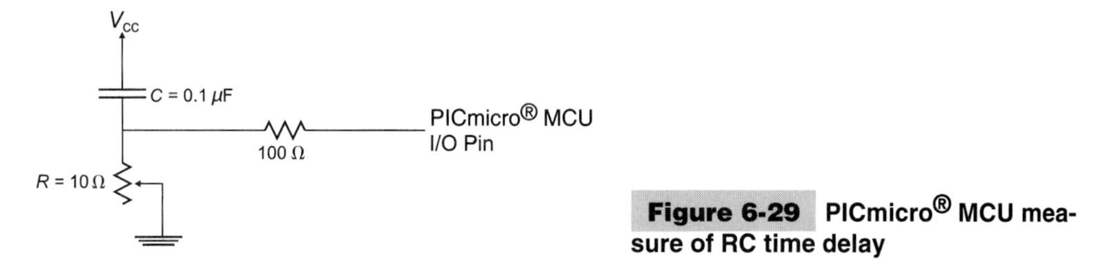


**Figure 6-29** PICmicro® MCU measure of RC time delay

voltage across the pin to change from reading of 1 to a 0 is recorded, it will be proportional to the resistance between the potentiometer's wiper and the capacitor.

The pseudo code for carrying out the potentiometer read is:

```
int ReadPot()                // Return the Potentiometer's Position
{
int i;

   pin = output; pin = 1;          // Charge the Capacitor
   for (i = 0; i < charge; i++);
   pin = input;                    // Let the Capacitor Discharge
   for ( i = 0; pin == 1; i++);

   return I ;

} // End ReadPot
```

The PICmicro® MCU assembly code for implementing this potentiometer read is not much more complex than this pseudo code. Later, the book provides some examples of how potentiometer reads are actually accomplished.

The 100-Ohm resistor between the PICmicro® MCU pin and the RC network is used to prevent any short circuits to ground if the potentiometer is set so that no resistance is in the circuit when the capacitor is changed.

This method of reading a potentiometer's position is very reliable, but not very accurate, nor is it particularly fast. When setting this up for the first time, in a specific circuit, you will have to experiment to find the actual range that it will display. This is because of part variances (including the PICmicro® MCU) and the power supply characteristics. For this reason, I do not recommend using the potentiometer/capacitor circuit in any products. Tuning the values returned will be much more expensive than the costs of a PICmicro® MCU with a built-in ADC.

## PULSE-WIDTH MODULATION (PWM) I/O

The PICmicro® MCU, like most other digital devices, does not handle analog voltages very well. This is especially true for situations where high-current voltages are involved. The best way to handle analog voltages is by using a string of varying wide pulses to indicate the actual voltage level. This string of pulses is known as a *Pulse-Width-Modulated (PWM) analog signal* and it can be used to pass analog data from a digital device, control DC devices, or even output an analog voltage.

This section covers PWM signals and how they can be used with the PICmicro® MCU. In the discussion of TMR1 and TMR2, earlier in the book, I presented how PWM signals are implemented and read using the CCP built-in hardware of the PICmicro® MCU. This

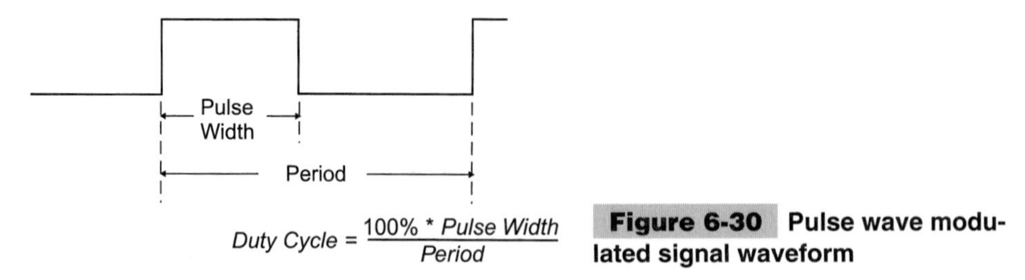


**Figure 6-30** Pulse wave modulated signal waveform

section shows how PWM signals can be used for I/O in PICmicro® MCUs that do not have the CCP module built in.

A PWM signal (Fig. 6-30) is a repeating signal that is on for a set period of time that is proportional to the voltage that is being output. I call the *on time* the *pulse width* in Fig. 6-30. The *duty cycle* is the percentage of time that the on time is relative to the PWM signal's *period.*

To output a PWM signal, the following code is used:

```
Period = PWMPeriod;                     // Initialize the Output
On = PWMperiod - PWMoff;                // Parameters

while (1 == 1) {

  PWM = ON;                             // Start the Pulse
  for (i = 0; i < On; i++ );            // Output ON for "On" Period of
                                        //   Time
  PWM = off;                            // Turn off the Pulse
  For ( ; i < PWMPeriod; i++ );         // Output off for the rest of the
                                        //   PWM Period
 } // end while
```

This code can be implemented remarkably simply in the PICmicro® MCU, but rather than showing it, there is one aspect I want to change. This method is not recommended because it uses all the processor resources of the PICmicro® MCU and does not allow for any other processing.

To avoid this problem, I would recommend using the TMR0 interrupt as:

```
Interrupt PWMOutput()         // When Timer Overflows, Toggle "On" and "Off"
{                             //   and Reset Timer to the correct delay for
                              // the Value

  if (PWM == ON) {            // If PWM is ON, Turn it off and Set Timer
    PWM = off;                //   Value
    TMR0 = PWMPeriod - PWMOn;
  } else {                    // If PWM is off, Turn it ON and Set Timer
    PWM = ON;                 //   Value
    TMR0 = PWMOn;
  } // end if

  INTCON.T0IF = 0;            // Reset Interrupts

} // End PWMOutput TMR0 Interrupt Handler
```

This code is quite easy to port to PICmicro® MCU assembly language. For example, if the PWM period was 1 ms (executing in a 4-MHz PICmicro® MCU), a divide-by-four prescaler value could be used with the timer and the interrupt-handler assembly-language code would be:

```
org 4
Int                         ; Interrupt Handler

  movwf _w                  ; Save Context Registers
  movf STATUS, w            ;  - Assume TMR0 is the only enabled Interrupt
  movwf _status

  btfsc PWM                 ; Is PWM O/P Currently High or Low?
   goto PWM_ON
  nop                       ; Low - Nop to Match Cycles with High

  bsf PWM                   ; Output the Start of the Pulse

  movlw 6 + 6               ; Get the PWM On Period
  subwf PWMOn, w            ; Add to PWM to Get Correct Period for
                            ; Interrupt Handler Delay and Missed cycles
                            ; in maximum 1024 usec Cycles
  goto PWM_Done

PWM_ON                      ; PWM is On - Turn it Off

  bcf PWM                   ; Output the "Low" of the PWM Cycle

  movf PWMOn, w             ; Calculate the "Off" Period
  sublw 6 + 6               ; Subtract from the Period for the Interrupt
                            ; Handler Delay and Missed cycles in maximum
                            ; 1024 usec Cycles
  goto PWM_Done

PWM_Done                    ; Have Finished Changing the PWM Value

  sublw 0                   ; Get the Value to Load into the Timer
  movwf TMR0

  bcf INTCON, T0IF          ; Reset the Interrupt Handler

  movf _status, w           ; Restore the Context Registers
  movwf STATUS
  swapf _w, f
  swapf _w, w

  retfie
```

In this code, TMR0 is loaded in such a way that the PWM's period is always 1 ms (a count of 250 "ticks" with the prescaler value of four). To get the value added and subtracted from the total, I first took the difference between the number of ticks to get 1 ms (250) and the full timer range (256). Next, I counted the total number of instruction cycles of the interrupt handler (which is 23), divided it by four, and added the result to the 1-ms difference. The operation probably seems confusing because I was able to optimize the time for the PWM signal off:

```
Time Off = Period - ON
```

to:

```
movf  ON, w
sublw 6 + 6
sublw 0
movwf TMR0
```

I realize that this doesn't make any sense the first time you look at it, so I will go through it to show how it works.

Using the original equation, you should note that this calculates the number of cycles to be delayed by TMR0, but the actual value to be loaded into TMR0 is calculated as:

```
TMRO Delay Value   = 0x0100 - (Time Off)
                   = 0x0100 - (Period - ON)
                   = 0x0100 - (256 - 250 + Interrupt Execution- ON)
                   = 0x0100 - (6 + 6 - ON)
                   = 0x0100 - (12 - ON)
                   = 0x0100 - 12 + ON
                   = 0x00F4 + ON
```

Going back to the three instructions that load TMR0, you can show that they execute as:

```
movf ON, w              ; w = ON
sublw 6 + 6             ; w = 6 + 6 - w
                        ;   = 12 - ON
                        ;   = 12 + 0x0FF ^ ON + 1
                        ;   = 13 + 0x0FF ^ ON
sublw 0                 ; w = 0 - w
                        ;   = 0 - (13 + 0x0FF ^ ON)
                        ;   = 0 + 0x0FF ^ (13 + 0x0FF ^ ON) + 1
                        ;   = 0x0FF ^ 13 + 0x0FF ^ 0x0FF ^ ON + 1
                        ;   = 0x0FF ^ 13 + ON + 1
                        ;   = 0x0F4 + ON
```

which is (surprisingly enough) the same result as what was found with the "TMR0 delay value" equation. The formula in itself is not that impressive, except that it "dovetails" very well with the *PWM on* half of the code. The process of coming up with this code probably belongs in another chapter on optimization, but to be honest with you, I came up with it using nothing but trial and error along with the feeling that this kind of optimization was possible.

This is an example of what I mean when I say that you should look for opportunities when processing data in the PICmicro® MCU. More often than not, you will come up with something like these few instructions, which are very efficient, and integrate different cases startlingly well.

Notice that, in this code, the PWM signal will never fully be on (a high DC voltage) or fully off (a ground-level DC voltage). This is because when the routine enters the subroutine handler, it changes the output, regardless of whether or not it is required for the length of the interrupt handler. In actuality, if you time it out, you will see that the 23 instruction cycles that the interrupt handler takes between changing the value works out to a 2.4-percent loss of full on and full off. This should not be significant in most applications and will serve as a "heartbeat" to let the receiver "know" the PICmicro® MCU is still functioning—even though the output is stuck at an extreme.

In this example, I have expended quite a bit of energy to ensure that the period remains the same, regardless of the on time. This was done to ensure that the changes in the duty cycle remained proportional to the changes in the on period. This is important if the PWM output is going to be passed through a low-pass filter (Fig. 6-31) to output an analog voltage.

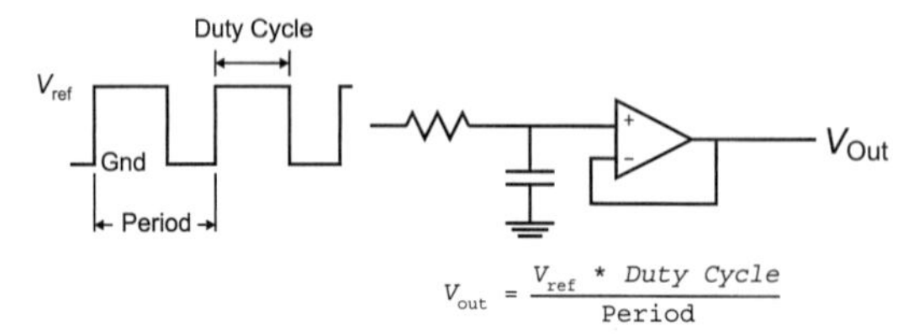


**Figure 6-31** "Pulse width modulated" analog voltage

In many applications where a PWM signal is communicating with another digital device, this effort to ensure that the period is constant and is not required. In these cases, a timer is used to time the *on* period. This can be shown as the pseudo-code:

```
Int TimeOn()                          // Time the Width of an incoming Pulse
{
int i = 0;
  while (PWMIP == off);               // Wait for the Pulse to Start
  for ( ; PWMIP == ON; i++ );         // Time the Pulse Width
  return i;                           // Return the Pulse Width
} // end TimeOn
```

With the actual PICmicro® MCU assembly-language code being quite simple, but dependent on the maximum pulse width value being timed, very long pulses will require large counters or delays in between the PWM input (*PWMIP* in *TimeOn*) poll.

Passing analog data back and forth between digital devices in any format is not going to be accurate because of the errors in digitizing the value and restoring it. This is especially true for PWM signals, which can have very large errors because of the sender and receiver not being properly synched and the receiver not starting to poll at the correct time interval. In fact, the measured value could have an error of upwards of 10 percent from the actual value. This loss of data accuracy means that the analog signals should not be used for data transfers. But, as is shown in Chapters 15 and 16, PWM signals are an excellent way to control analog devices, such as lights and motors.

When using a PWM to drive an analog device, it is important to be sure that the frequency is faster than what a human can perceive. As noted in the "LED" section, this frequency is 30 Hz or more. But for motors and other devices that might have an audible "whine," the PWM signal should have a frequency of 20 kHz or more to ensure that the signal does not bother the user. In Chapter 16, I discuss this in more detail and demonstrate how an audible 10 KHz "whine" can be produced.

The problem with the higher frequencies is that the granularity of the PWM signal decreases. This is because of the inability of the PICmicro® MCU (or what ever digital device is driving the PWM output) to change the output in relatively small time increments from on to off, relative to the size of the PWM signal's period. In the previous example code, four instruction cycles (of 1 $\mu$s each) are the lowest level of granularity for the PWM signal that results in about 250 unique output values. If the PWM signal's period was decreased to 100 $\mu$s, from 1 ms, for a 10-kHz frequency, the same code would only have 25 or so unique output values that could be output. In this case, to retain the original code's granularity, the PICmicro® MCU would have to be sped up 10 times (not possible for most applications) or another way of implementing the PWM will have to be found.

## AUDIO OUTPUT

When I was originally blocking out this book, I wanted to include PICmicro® MCU audio output and input. After some experimentation and a search on the Internet to see what other people have done (or not done, as in the case of audio input to the PICmicro® MCU), I have come to the conclusion that audio input is not appropriate for the PICmicro® MCU. The PICmicro® MCU could have some kind of filtering input, but it simply does not have the processing capability to do more than respond to a specific frequency at a threshold volume. For this reason, I would discourage you from passing audio input to the PICmicro® MCU and just use the PICmicro® MCU to provide audio output using the techniques outlined in this section.

When I discuss the PICmicro® MCU's processing capabilities with regard to audio, I tend to be quite disparaging. The reason for this is the lack of hardware multipliers in the low-end and mid-range PICmicro® MCUs and the inability of all the devices to "natively" handle greater than eight bits of in a floating-point format. The PICmicro® MCU processor has been optimized to respond to digital inputs and cannot implement the real-time processing routines needed for complex analog I/O.

The circuit in Fig. 6-32 passes DC waveforms through the capacitor (which filters out the kickback spikes) to the speaker or piezo buzzer. When a tone is output, your ear will hear a reasonably good tone, but if you were to look at the actual signal on an oscilloscope, you would see the waveform shown in Fig. 6-33, both from the PICmicro® MCU's I/O pin and the piezo buzzer itself.

The PICmicro® MCU pin, capacitor, and speaker are actually a quite complex analog circuit. Notice that the voltage output on the I/O pin is changed from a straight waveform. This is because of the inductive effects of the piezo buffer. The important thing to note in Fig. 6-33 is that the upward spikes appear at the correct period for the output signal.

Timing the output signal is generally accomplished by toggling an output pin at a set period within the TMRO interrupt handler. To generate a 1-kHz signal shown in a PICmicro® MCU running a 4 MHz, you can use the code (which does not use the prescaler) for TMR0 and the PICmicro® MCU's interrupt providing the speaker charges are in the background.

```
   org 4
int
  movwf _w                ; Save Context Registers
  bcf INTCON, TOIF        ; Reset the Interrupt
  movlw 256 - (250 - 4)
  movwf TMRO              ; Reset TMRO for another 500 usecs
  btfsc SPKR              ; Toggle the Speaker
   goto $ + 2
  bsf SPKER               ; Speaker Output High
   goto $ + 2
  bcf SPKER               ; Speaker Output Low
  swapf _w, f             ; Restore Context Registers
  swapf _w, w
  retfie
```

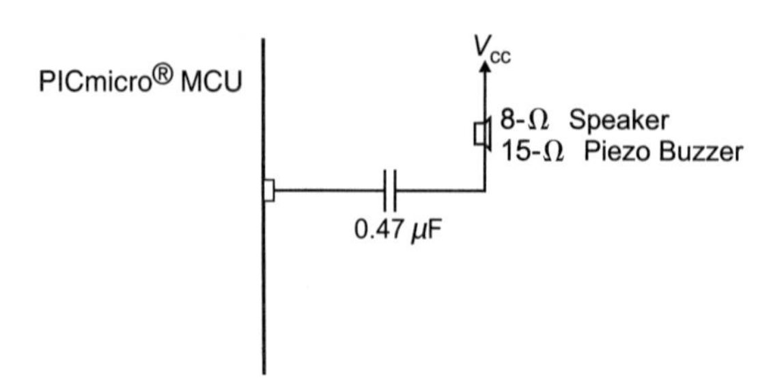


**Figure 6-32** Circuit for driving PICmicro® MCU audio

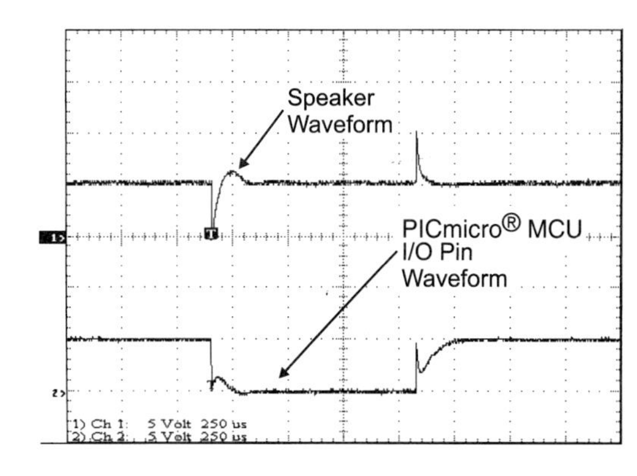


**Figure 6-33** PICmicro® MCU driving a speaker output

There are two points to notice about this interrupt handler. First, I don't bother saving the STATUS register's contents because neither the zero, neither carry nor digit carry flags are changed by any of the instructions used in the handler.

The second point to notice is the reload value of TMRO to generate a 1-kHz output in a 4-MHz PICmicro® MCU (an instruction clock period of 1 $\mu$s), I have to delay 500 cycles for the wave's high and low. Because TMRO has a divide-by-two counts on its input, I have to wait a total of 250 ticks. When I record TMRO, notice that I also take into account the cycles taken to get to the reload (which is seven or eight), divide them by two and take them away from the reload value.

For this handler, the reload value might be off by one cycle, depending on how the mainline executes, for a worst-case error of 0.2% (2,000 ppm). This level of accuracy is approximately the same as what you would get for a ceramic resonator; so, the actual frequency should not be off an appreciable amount from the expected. This level of accuracy will not cause noticeable warbling (changes in the frequency) caused by the changing interrupt latency as you run the application. The actual changes are very small and not detectable by the human ear.

When developing applications that output audio signals, I try to keep the tone within the range of 500 Hz to 2 kHz. This is well within the range of human hearing and is quite easy to implement in a PICmicro® MCU. When you look at the Christmas tree project in Chapter 16, you can see how this is done to create simple tunes on the PICmicro® MCU.

I wanted to finish off this section by describing how the PICmicro® MCU can create more complex sounds, such as telephone TouchTone audio signals. The problem with TouchTones is that they are a combination of two frequencies (that are listed in the appendices). When they're added together, you get a signal like the bottom one shown in Fig. 6-34.

This signal is very hard to replicate with a digital circuit. If you want to create a complex waveform like this, I recommend mixing two PICmicro® MCU-controlled frequency outputs together or using a chip that is designed for the specific purpose that you have in mind. The Parallax Basic Stamp II and PicBasic Pro do have instructions that allow you to drive telephone TouchTones, but they require a filter circuit to work properly. I have found that even with the filter, the phone system does not recognize the data being sent very reliably.

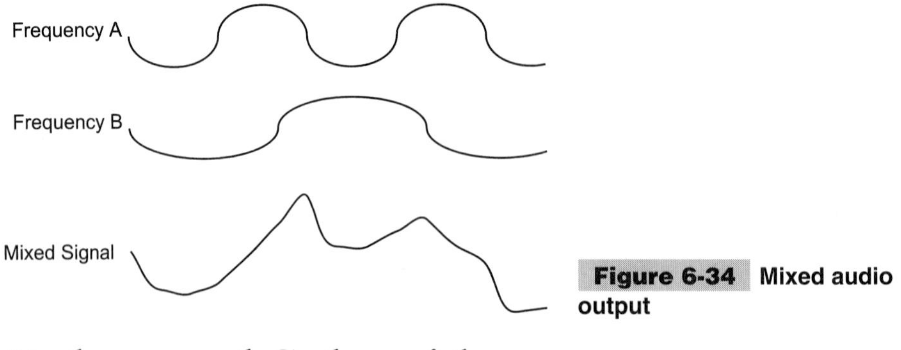


**Figure 6-34** Mixed audio output

# Relays and Solenoids

Some real-life devices that you might have to control by a microcontroller are electromagnetic, such as relays, solenoids, and motors. These devices cannot be driven directly by a microcontroller because of the current required and the noise generated by them. This means that special interfaces must be used to control electromagnetic devices.

The simplest method to control these devices is to just switch them on and off and by supplying power to the coil in the device. The circuit shown in Fig. 6-35 is true for relays (as shown), solenoids (which are coils that draw an iron bar into them when they are energized), or a DC motor (which will only turn in one direction).

In this circuit, the microcontroller turns on the Darlington transistor pair, causing current to pass through the relay coils, closing the contacts. To open the relay, the output is turned off (or a 0 is output). The shunt diode across the coil is used as a kick-back suppressor. When the current is turned off, the magnetic flux in the coil will induce a large back EMF (voltage), which must be absorbed by the circuit or a voltage spike will occur, which can damage the relay power supply and even the microcontroller. This diode must *never* be forgotten in a circuit that controls an electromagnetic device. The kick-back voltage is usually on the order of several hundred volts for a few nanoseconds. This voltage causes the diode to breakdown and allows current to flow, attenuating the induced voltage which can damage the PICmicro® MCU and other electronic devices in the application circuit.

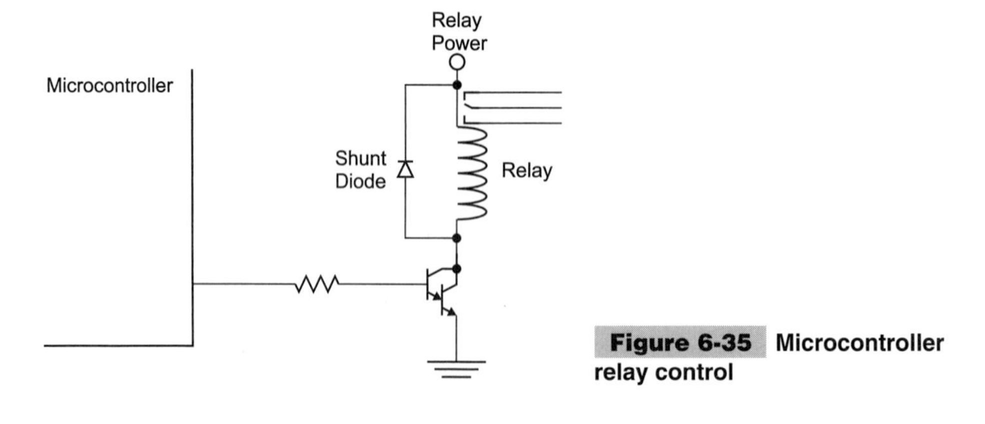


**Figure 6-35** Microcontroller relay control

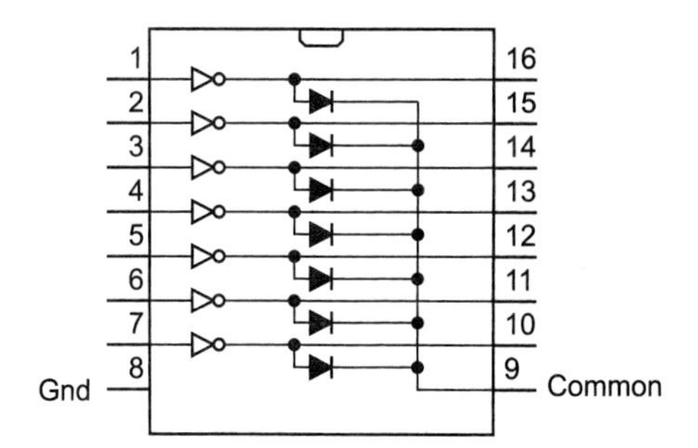


**Figure 6-36** ULN2003A driver array

Rather than designing discrete circuits to carry out this function, I like to use integrated chips for the task. One of the most useful devices is the ULN2003A (Fig. 6-36) or the ULN2803 series of chips, which have Darlington transistor pairs and shunt diodes built in for multiple drivers.

# DC and Stepper Motors

Motors can be controlled by exactly the same hardware as shown in the previous section, but as I noted, they will only run in one direction. A network of switches (transistors) can be used to control turning a motor in either direction; this is known as an *H-bridge* (Fig. 6-37).

In this circuit, if all the switches are open, no current will flow and the motor won't turn. If switches 1 and 4 are closed, the motor will turn in one direction. If switches 2 and 3 are closed, the motor will turn in the other direction. Both switches on one side of the bridge should *never* be closed at the same time because this will cause the motor power supply to burn out or a fuse will blow because a short circuit is directly between the motor power and ground.

Controlling a motor's speed is normally done by "pulsing" the control signals in the form of a PWM signal, as shown previously in this chapter (Fig. 6-30). This will control the average power delivered to the motors. The higher the ratio of the pulse width to the period (the duty cycle), the more power delivered to the motor.

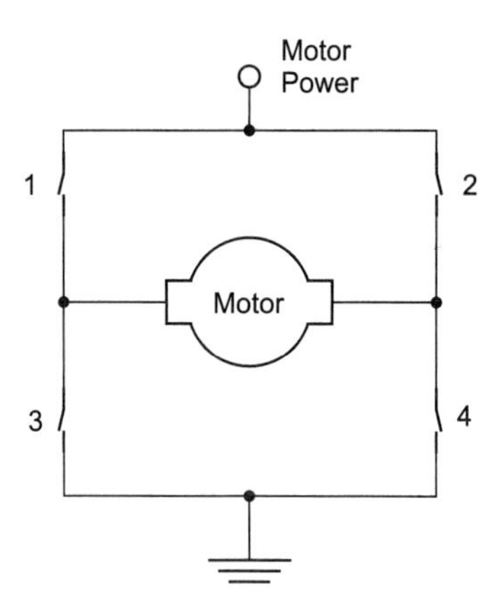


**Figure 6-37** "H" bridge motor driver

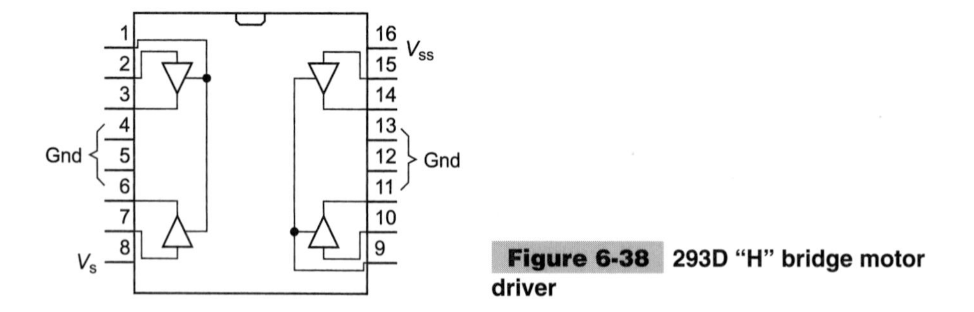


**Figure 6-38** 293D "H" bridge motor driver

The frequency of the PWM signal should be greater than 20 kHz to prevent the PWM from producing an audible signal in the motors as the field is turned on and off.

Like the ULN2003A simplified the wiring of a relay control, the 293D (Fig. 6-38) or 298 chips can be used to control a motor.

The 293D chip can control two motors (one on each side) connected to the buffer outputs (pins 3, 6, 11, and 14). Pins 2, 7, 10, and 15 are used to control the voltage level (the switches in the H-bridge diagram) of the buffer outputs. Pins 1 and 9 are used to control whether or not the buffers are enabled. The buffer controls can be PWM inputs, which make control of the motor speed very easy to implement.

$V_S$ is +5 V used to power the logic in the chip and $V_{SS}$ is the power supplied to the motors (anywhere from 4.5 to 36 volts). A maximum of 500 mA can be supplied to the motors. Like the ULN2003A, the 293D contains integral shunt diodes. This means that to attach a motor to the 293D, no external shunt diodes are required.

In this example circuit, you'll notice that I've included an optional snubber resistor and capacitor. These two components, wired across the brush contacts of the motor, will help reduce electromagnetic emissions and noise spikes from the motor. In the motor-control circuits that I have built, I have never found them to be necessary. But if you find erratic operation from the microcontroller when the motors are running, you might want to put in the 0.1-$\mu$F capacitor and 5 Ohm (2 watt) resistor snubber across the motor's brushes (as shown in the circuit).

An issue with using the 293D and 298 motor controller chips and is that they are bipolar devices with a 0.7-volt drop across each driver (1.4 to 1.5 volts for a dual driver circuit, Fig. 6-39). This drop, with the significant amount of current required for a motor, results

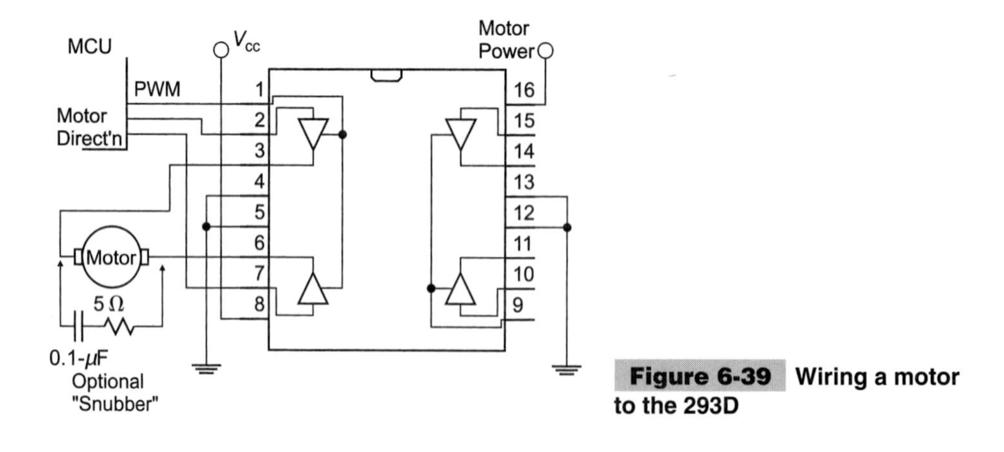


**Figure 6-39** Wiring a motor to the 293D

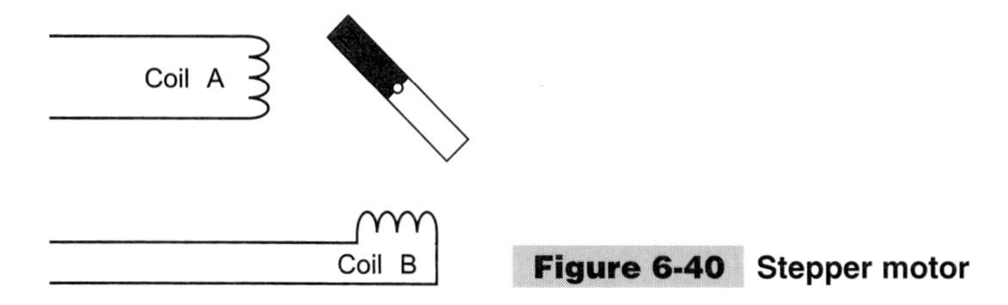


**Figure 6-40** Stepper motor

in a fairly significant amount of power dissipation within the driver. The 293D is limited to 1 amp total output and the 298 is limited to 3 amps. For these circuits to work best, a large heatsink is required.

To minimize the problem of heating and power loss, I have more recently been looking at using power MOSFETS to control motors. Later, the book shows how to wire these transistors in a circuit to control a motor.

Stepper motors are much simpler to develop control software for than a regular DC motor. This is because the motor is turned one step at a time or can turn at a specific rate (specified by the speed in which the steps are executed). In terms of the hardware interface, stepper motors are a bit more complex to wire and require more current (meaning that they are less efficient), but these are offset by the advantages in software control.

A bipolar stepper motor consists of a permanent magnet on the motor's shaft that has its position specified by a pair of coils (Fig. 6-40).

To move the magnet and the shafts, the coils are energized in different patterns to attract the magnet. For the motor shown in Fig. 6-40, the following sequence would be used to turn the magnet (and the shaft) clockwise (Table 6-5).

In this sequence, coil A attracts the north pole of the magnet to put the magnet in an initial position. Then, coil B attracts the south pole, turning the magnet 90 degrees. This continues on to turn the motor 90 degrees for each step.

The output shaft of a stepper motor is often geared down so that each step causes a very small angular deflection (a couple of degrees at most, rather than the 90 degrees in the previous example). This provides more torque output from the motor and greater positional control of the output shaft.

A stepper motor can be controlled by something like a 293D (each side driving one coil). But there are also stepper-motor controller chips, such as the UC1517.

**TABLE 6-5 Commands to Move a Stepper Motor**

| STEP | ANGLE | COIL "A" | COIL "B" |
|---|---|---|---|
| 1 | 0 | S | |
| 2 | 90 | | N |
| 3 | 180 | N | |
| 4 | 270 | | S |
| 5 | 360/0 | S | |

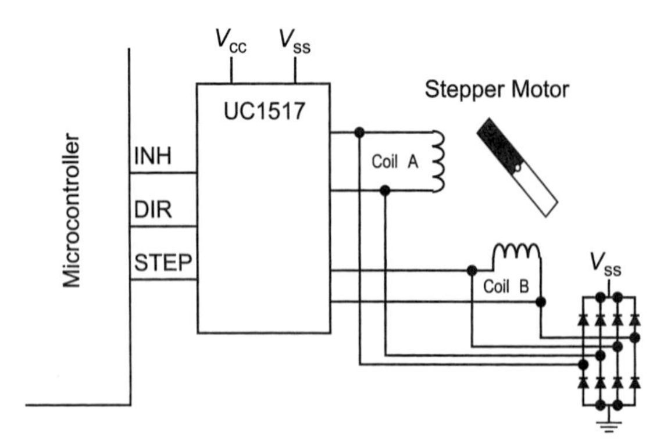


**Figure 6-41** UC1517 control of a stepper motor

In this chip, a step pulse is sent from the microcontroller along with a specified direction. The INH pin will turn off the output drivers and allow the stepper shaft to be moved manually. The UC1517 is capable of outputting bilevel coil levels (which improves efficiency and reduces induced noise), as well as half stepping the motor (which involves energizing both coils to move the magnet/shaft by 45 degrees and not just 90 degrees). These options are specific to the motor/controller used (a bipolar stepper motor can have four to eight wires coming out of it). Before deciding on features to be used, a thorough understanding of the motor and its operation is required.

# R/C Servo Control

Servos designed for use in radio-controlled airplanes, cars, and boats can be easily interfaced to a PICmicro® MCU. They are often used for robots and applications where simple mechanical movement is required. This might be surprising because a positional servo is considered to be an analog device.

The output of an R/C servo is usually a wheel that can be rotated from 0 to 90 degrees. (some servos can turn from 0 to 180 and others have very high torque outputs for special applications). Typically, they only require +5 V, ground, and an input signal.

An R/C servo is an analog device, the input is a PWM signal at digital voltage levels. This pulse is between 1.0 and 2.0 ms long and repeats every 20 ms (Fig. 6-42).

The length of the PWM pulse determines the position of the servo's wheel. A 1.0-ms pulse will cause the wheel to go to 0 degrees and a 2.0-ms pulse will cause the wheel to go to 90 degrees.

With the PICmicro® MCU's TMR2 capable of outputting a PWM signal, controlling a servo could be considered very easy, although the TMR2 output will probably not give you the positional accuracy that you will want.

To produce a PWM signal using a PICmicro® MCU, I normally use a timer interrupt (set every 18 ms) that outputs a 1.0- to 2.0-ms PWM signal using pseudo-code:

```
Interrupt() {                    // Interrupt Handler Code
int i = 0;
 BitOutput( Servo, 1);           // Output the Signal
```

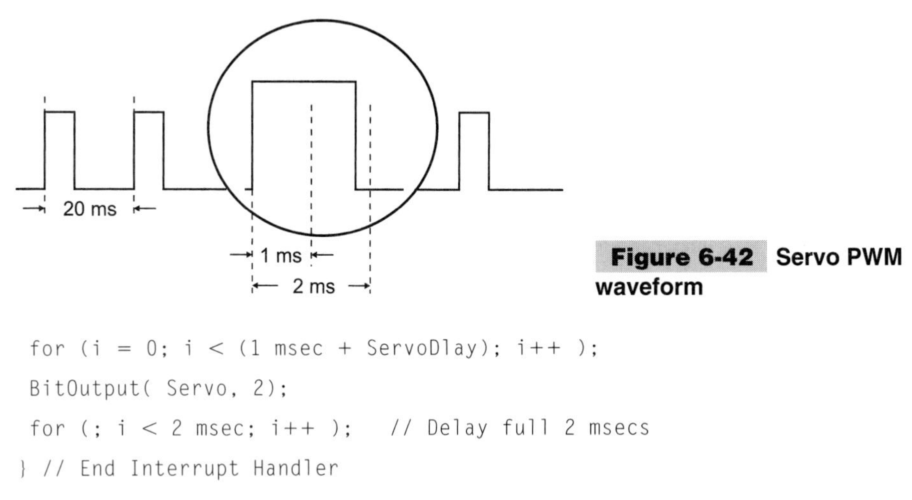


**Figure 6-42** Servo PWM waveform

```
for (i = 0; i < (1 msec + ServoDlay); i++ );
BitOutput( Servo, 2);
for (; i < 2 msec; i++ );     // Delay full 2 msecs
} // End Interrupt Handler
```

This code can be easily expanded to control more than one servo (by adding more output lines and *ServoDlay* variables). This method of controlling servos is also nice because the *ServoDlay* variables can be updated without affecting the operation of the interrupt handler.

The interrupt handler takes two ms out of every 20. Thus, a 10-percent cycle overhead provides the PWM function (and this doesn't change even if more servo outputs are added to the device).

# Serial Interfaces

Most intersystem (or intercomputer) communications are done serially. Thus, a byte of data is sent over a single wire, one bit at a time, with the timing coordinated between the sender and the receiver. The obvious advantage of transmitting data serially is that fewer connections are required.

A number of common serial communication protocols are used by microcontrollers. In some devices, these protocols are built into the chip itself, to simplify the effort required developing software for the application.

## SYNCHRONOUS

For synchronous data communications in a microcontroller, a clock signal is sent along with serial data (Fig. 6-43).

The clock signal strobes the data into the receiver and the transfer can occur on the rising or falling edge of the clock. A typical circuit, using discrete devices, could be like that shown in Fig. 6-44.

This circuit converts serial data into eight digital outputs, which all are available at the same time (When the O/P clock is strobed). For most applications, the second '374 (which

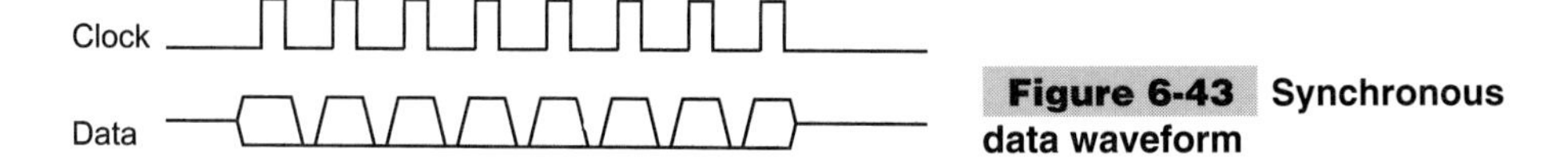


**Figure 6-43** Synchronous data waveform

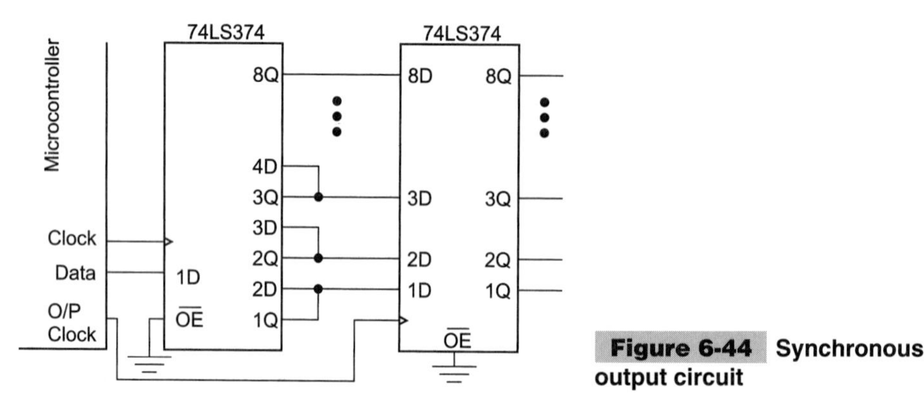


**Figure 6-44** Synchronous output circuit

provides the parallel data) is not required. This serial-to-parallel conversion can also be accomplished using serial-to-parallel chips, but I prefer using eight-bit registers because they are generally easier to find than other TTL parts.

The two very common synchronous data protocols are *Microwire* and *SPI.* These methods of interfacing are used in a number of chips (such as the serial EEPROMs used in the BASIC Stamps). Although the Microwire and SPI standards are quite similar, there are a number of differences.

I consider these protocols to be methods of transferring synchronous serial data, rather than microcontroller network protocols because each device is individually addressed (even though the clock/data lines can be common between multiple devices). If the chip select for the device is not asserted, the device ignores the clock and data lines. With these protocols, only a single master can be on the bus. A possible connection of two devices is shown in Fig. 6-45.

If a synchronous serial port is built into the microcontroller, the data transmit circuitry might look like that shown in Fig. 6-46. This circuit will shift out eight bits of data. For protocols, such as Microwire, where a start bit is initially sent, the start bit is sent using direct reads and writes to the I/O pins. To receive data, a similar circuit would be used, but data would be shifted into the shift register and then read by the microcontroller.

The Microwire protocol is capable of transferring data at up to 1 Mbps. Sixteen bits are transferred at a time. To read 16 bits of data, the waveform looks like that shown in Fig. 6-47.

After selecting a chip and sending a start bit, the clock strobes out an eight-bit command byte (labeled *OP1, OP2, A5* to *A0* in the previous diagram), followed by (optionally) a

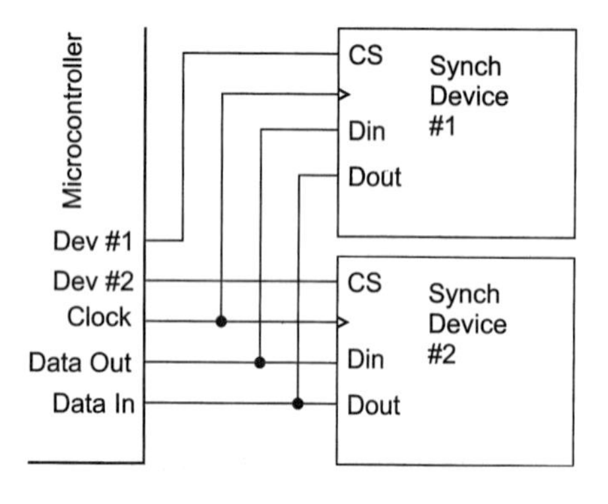


**Figure 6-45** Synchronous device bus

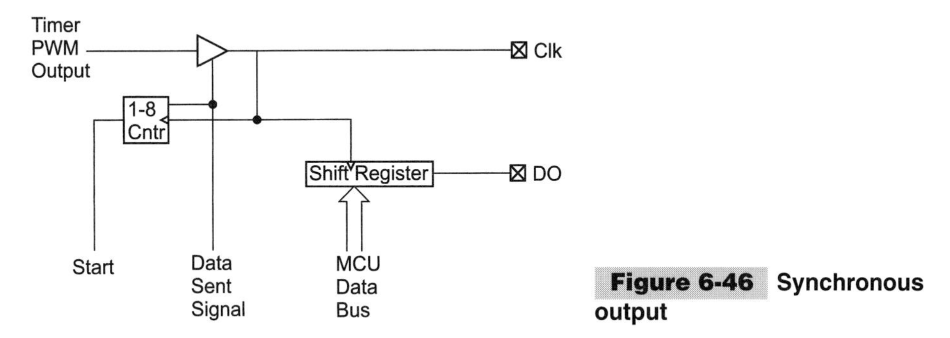


**Figure 6-46** Synchronous output

16-bit address word transmitted and then another 16-bit word, either written or read by the microcontroller.

With a 1-Mbps maximum speed, the clock is both high and low for 500 ns. Transmitted bits should be sent 100 ns before the rising edge of the clock. When reading a bit, it should be checked 100 ns before the falling edge of the clock. Although these timings will work for most devices, you should understand the requirements of the device being interfaced to.

The SPI protocol is similar to Microwire, but with a few differences.

1 SPI is capable of up to 3-Mbps data-transfer rate.
2 The SPI data word size is eight bits.
3 SPI has a hold that allows transmitter to suspend data transfer.
4 Data in SPI can be transferred as multiple bytes, known as *blocks* or *pages*.

Like Microwire, SPI first sends a byte instruction to the receiving device. After the byte is sent, a 16-bit address is optionally sent, followed by eight bits of I/O. As noted, SPI does allow for multiple byte transfers. An SPI data transfer is shown in Fig. 6-48.

The SPI clock is symmetrical (an equal low and high time). Output data should be available at least 30 ns before the clock line goes high and read 30 ns before the falling edge of the clock.

When wiring a Microwire or SPI device, you can do a trick to simplify the microcontroller connection; combine the DI and DO lines into one pin. Figure 6-49 is identical to what was shown earlier in this chapter when interfacing the PICmicro® MCU into a circuit where there is another driver.

In this method of connecting the two devices, when the data pin on the microcontroller has completed sending the serial data, the output driver can be turned off and the microcontroller can read the data coming from the device. The current-limiting resistor between

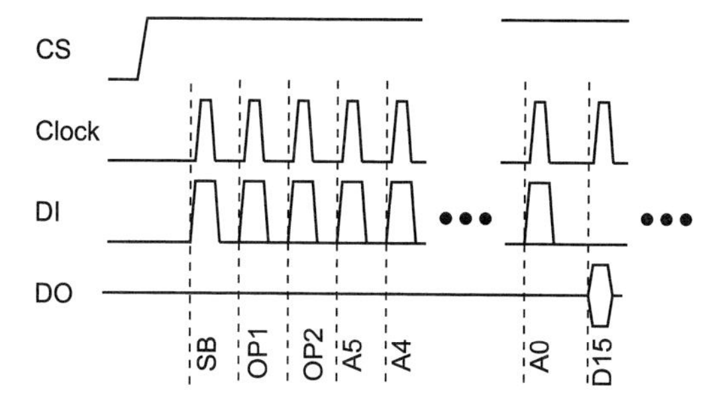


**Figure 6-47** Microwire data read

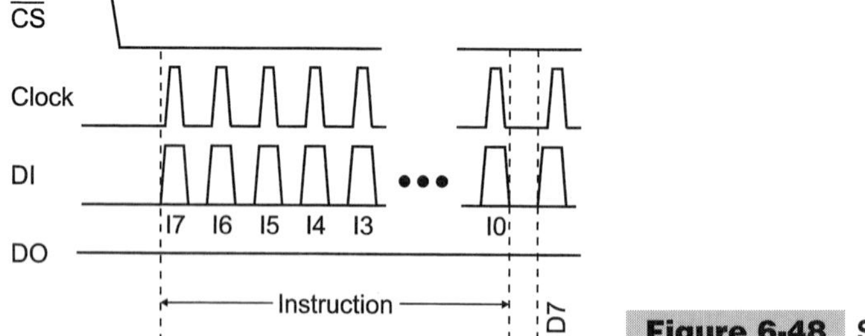


**Figure 6-48** SPI data write

the data pin and DI/DO limits any current flows when both the microcontroller and device are driving the line.

**I2C** The most popular form of microcontroller network is *I2C (Inter-Intercomputer Communications)*. This standard was originally developed by Philips in the late '70s as a method to provide an interface between microprocessors and peripheral devices without wiring full address, data, and control busses between devices. I2C also allows sharing of network resources between processors (which is known as *multi-mastering*).

The I2C bus consists of two lines, a clock line (SCL), which is used to strobe data (from the SDA line) from or to the master that currently has control over the bus. Both of these bus lines are pulled up (to allow multiple devices to drive them). An I2C-controlled stereo system might be wired as in Fig. 6-50.

The two bus lines are used to indicate that a data transmission is about to begin, as well as pass the data on the bus.

To begin a data transfer, a master puts a start condition on the bus. Normally, (when the bus is in the idle state, both the clock and data lines are not being driven (and are pulled high). To initiate a data transfer, the master requesting the bus pulls down the SDA bus line, followed by the SCL bus line. During data transmission, this is an invalid condition (because the data line is changing while the clock line is active/high).

Each bit is then transmitted to or from the slave (the device the message is being communicated with by the master) with the negative clock edge being used to latch in the data (Fig. 6-51). To end data transmission, the reverse is executed, the clock line is allowed to go high, which is followed by the data line.

Data is transmitted in a synchronous (clocked) fashion. The most-significant bit sent first and, after eight bits are sent, the master allows the data line to float (it doesn't drive it low) while strobing the clock to allow the receiving device to pull the data line low as an

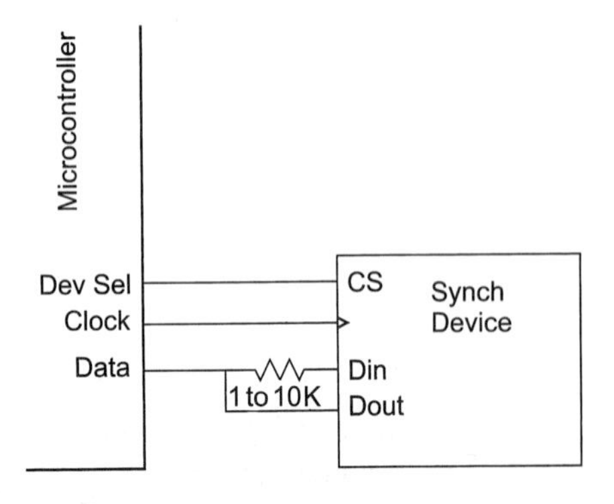


**Figure 6-49** Combining "DO" & "DI"

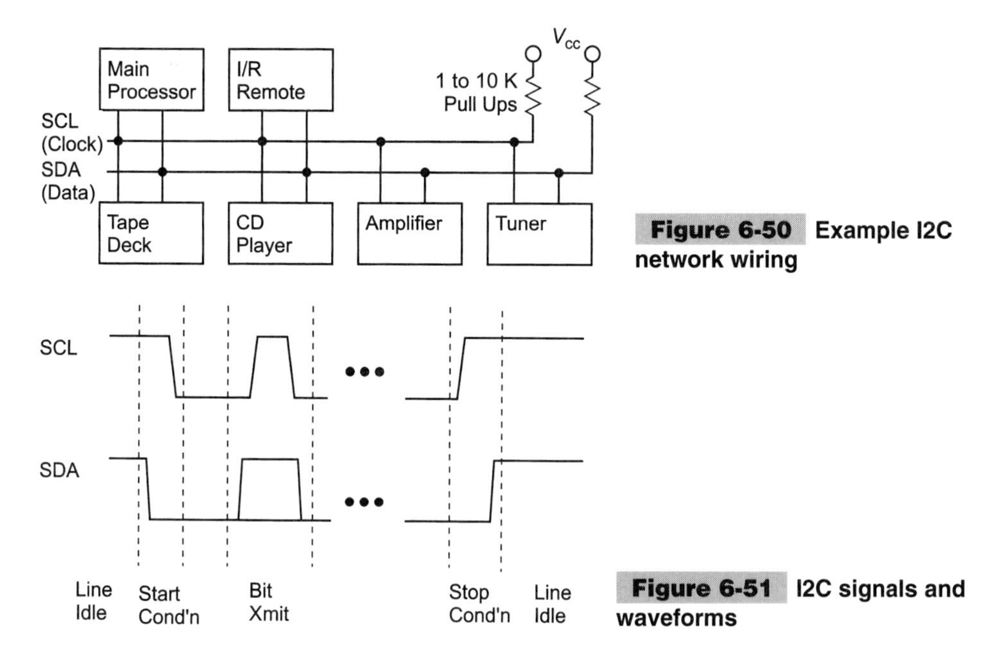


**Figure 6-50** Example I2C network wiring

**Figure 6-51** I2C signals and waveforms

acknowledgment that the data was received. After the acknowledge bit, both the clock and data lines are pulled low in preparation for the next byte to be transmitted or a stop/start condition is put on the bus. Figure 6-52 shows the data waveform.

Sometimes, the acknowledge bit will be allowed to float high—even though the data transfer has completed successfully. This is done to indicate that the data transfer has completed and the receiver (usually a slave device or a master, which is unable to initiate data transfer) can prepare for the next data request.

The two maximum speeds for I2C (because the clock is produced by a master, there really is no minimum speed) are *standard mode* and *fast mode*. Standard mode runs at up to 100 kbps and fast mode can transfer data at up to 400 kbps. Figure 6-53 shows the timing specifications for both the standard (*Std.,* 100-kHz data rate) and fast (400-kHz data rate).

A command is sent from the master to the receiver in the format shown in Fig. 6-54. The receiver address is seven bits long and is the bus address of the receiver. There is a loose standard to use the most significant four bits are used to identify the type of device; the next three bits are used to specify one-of-eight devices of this type (or further specify the device type).

As stated, this is a loose standard. Some devices require certain patterns for the second three bits and others (such as some large serial EEPROMS) use these bits to specify an ad-

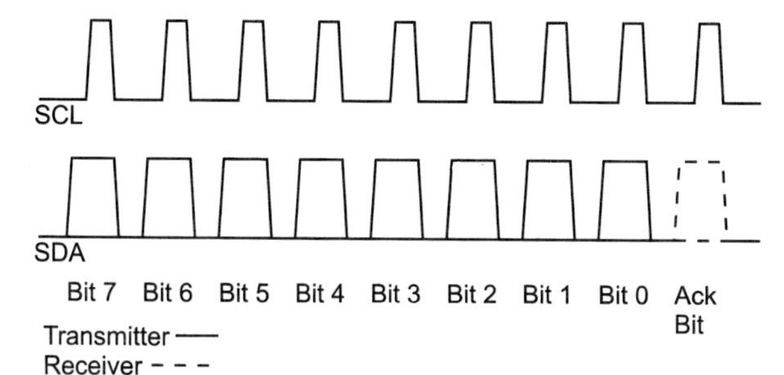


**Figure 6-52** I2C data byte transmission

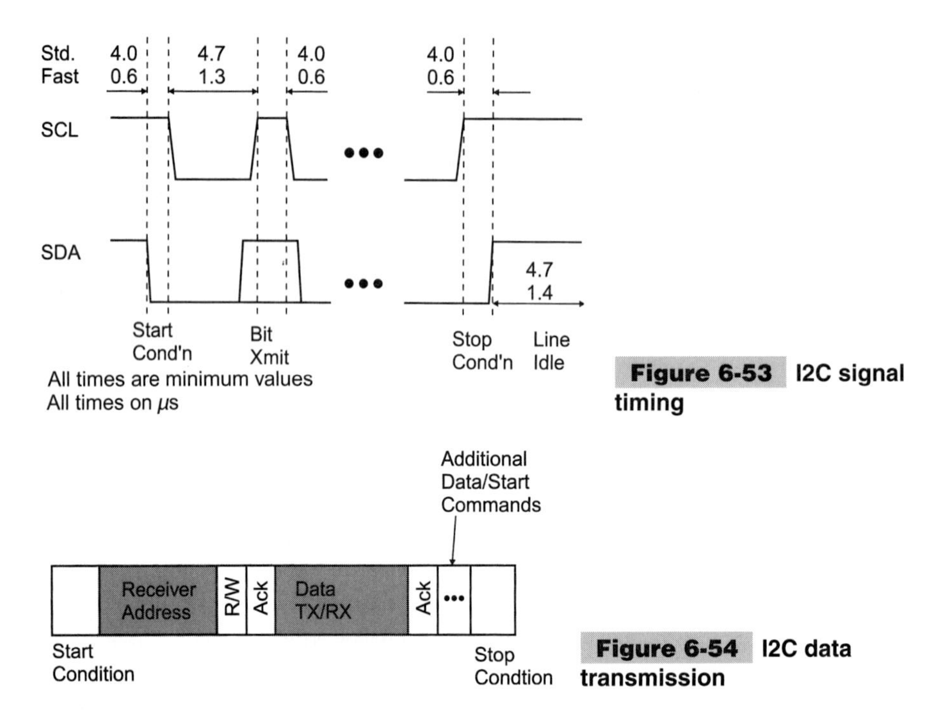


**Figure 6-53** I2C signal timing

**Figure 6-54** I2C data transmission

dress inside of the device. As well, there is a ten-bit address standard in which the first four bits are all set and the next bit reset. The last two are the most-significant two bits of the address, with the final eight bits being sent in a following byte. All this means is that it is very important to map out the devices to be put on the bus and all of their addresses.The first four bit patterns in Table 6-6 generally follow this convention for different devices.

This is really all there is to I2C communication, except for a few points. In some de-

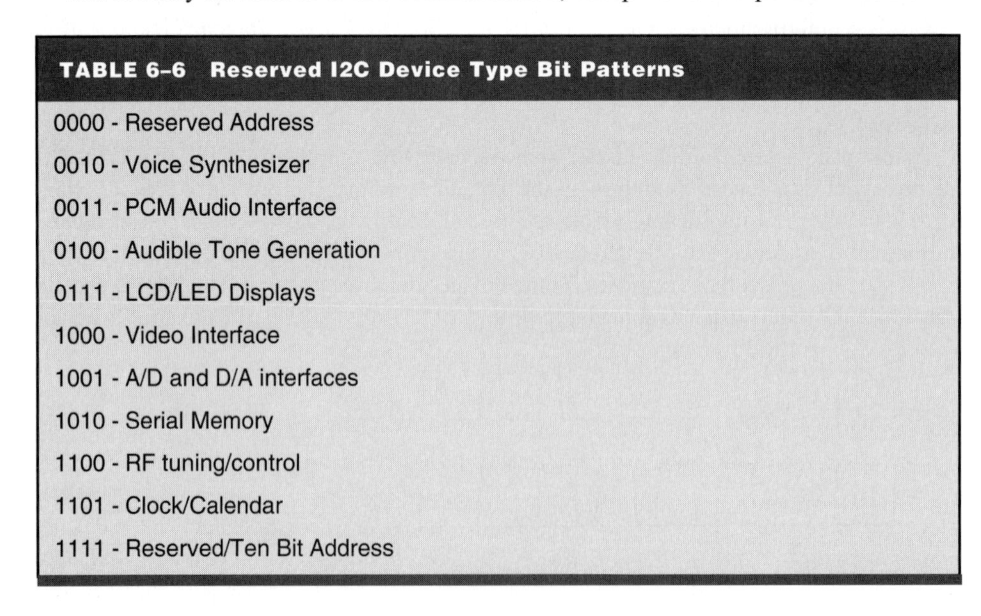

| TABLE 6-6 Reserved I2C Device Type Bit Patterns |
|---|
| 0000 - Reserved Address |
| 0010 - Voice Synthesizer |
| 0011 - PCM Audio Interface |
| 0100 - Audible Tone Generation |
| 0111 - LCD/LED Displays |
| 1000 - Video Interface |
| 1001 - A/D and D/A interfaces |
| 1010 - Serial Memory |
| 1100 - RF tuning/control |
| 1101 - Clock/Calendar |
| 1111 - Reserved/Ten Bit Address |

vices, a start bit has to be resent to reset the receiving device for the next command (i.e., in a serial EEPROM read, the first command sends the address to read from and the second reads the data at that address).

Note that I2C is *multi-mastering,* which is to say that multiple microcontrollers can initiate data transfers on a single I2C bus. This obviously results in possible collisions (which is when two devices attempt to drive the bus at the same time). Obviously, if one microcontroller takes the bus (sends a start condition) before another one attempts to do so, there is no problem. The problem arises when multiple devices initiate the start condition at the same time.

Actually, arbitration in this case is really quite simple. During the data transmission, hardware (or software) in both transmitters synchronize the clock pulses so that they match each other exactly. During the address transmission, if a bit that is expected to be a *1* by a master is actually a *0,* then it drops off the bus because another master is on the bus. The master that drops off will wait until the stop condition and then re-initiate the message. I realize that this is hard to understand with just a written description. The "CAN" section shows how this is done with an asynchronous bus, which is very analogous to this situation.

A bit-banging I2C interface can be implemented in software of the PICmicro® MCU quite easily. But, because of software overhead, the fast mode probably cannot be implemented—even the standard mode's 100 kbps will be a stretch for most devices. I find that implementing I2C in software to be best when the PICmicro® MCU is the single master in a network. That way it doesn't have to be synchronized to any other devices or accept messages from any other devices that are masters and are running a hardware implementation of I2C that might be too fast for the software slave.

## ASYNCHRONOUS (NRZ) SERIAL

Asynchronous long-distance communications came about as a result of the Baudot teletype. This device mechanically (and, later, electronically) sent a string of electrical signals (now called *bits*) to a receiving printer.

This data packet format is still used today for the electrical asynchronous transmission protocols described in the following sections. With the invention of the teletype, data could be sent and retrieved automatically without having to an operator to be sitting by the teletype all night, unless an urgent message was expected. Normally, the nightly messages could be read in the morning.

Before going on, some people get unreasonably angry about the definition and usage of the terms *data rate* and *baud rate. Baud rate* is the maximum number of possible data-bit transitions per second. This includes the start, parity, and stop bits at the ends of the data packet shown in Fig. 6-55, as well as the five data bits in the middle. I use the term *packet* because we are including more than just data (some additional information is in there as well), so *character* or *byte* (if there were eight bits of data) are not appropriate terms.

This means that for every five data bits transmitted, eight bits in total are transmitted (thus, nearly 40% of the data-transmission bandwidth is lost in teletype asynchronous serial communications).

The *data rate* is the number of data bits that are transmitted per second. For this example, if you were transmitting at 110 baud (which is a common teletype data speed), the actual data rate is 68.75 bits per second (or, assuming five bits per character, 13.75 characters per second).

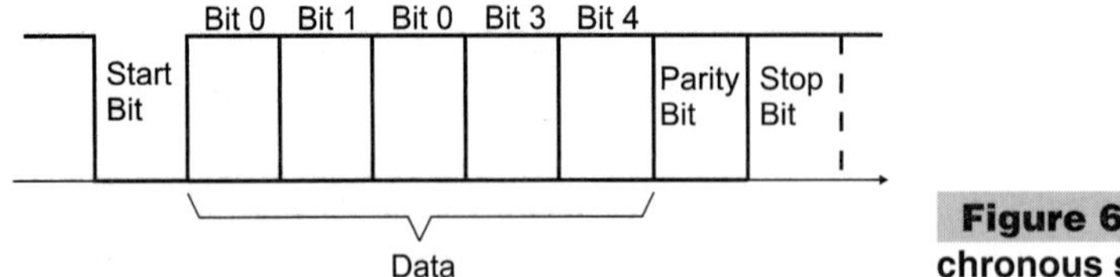


**Figure 6-55** Baudot asynchronous serial data

I use the term *data rate* to describe the *baud rate*. This means that when I say *data rate*, I am specifying the number of bits of *all* types that can be transmitted in a given period of time (usually one second). I realize that this is not absolutely correct, but it makes sense to me to use it in this form and this book is consistent throughout (and I will not use the term *baud rate*).

With only five data bits, the Baudot code could only transmit up to 32 distinct characters. To handle a complete character set, a specific five-digit code was used to notify the receiving teletype that the next five-bit character would be an extended character. With the alphabet and most common punctuation characters in the primary 32, this second data packet wasn't required very often.

The data packet diagram shows three control bits. The start bit is used to synchronize the receiver to the incoming data. The PICmicro® MCU *USART (Universal Synchronous/Asynchronous Receiver/Transmitter)* has an overspeed clock (running at 16 times the incoming bit speed), which samples the incoming data and verifies whether or not the data is valid (Fig. 6-56).

When waiting for a character, the receiver hardware polls the line repeatedly at 1/16-bit period intervals until a 0 (space) is detected. The receiver then waits half a cycle before polling the line again to see if a glitch was detected and not a start bit. Notice that polling occurs in the middle of each bit to avoid problems with bit transitions (if the transmitter's clock is slightly different from the receiver's, the chance of misreading a bit will be minimized).

Once the start bit is validated, the receiver hardware polls the incoming data once every bit period multiple times (again, to ensure that glitches are not read as incorrect data).

The stop bit was originally provided to give both the receiver and the transmitter some time before the next packet is transferred (in early computers, the serial data stream was created and processed by the computers and not by custom hardware, as in modern computers).

The parity bit is a crude method of error detection that was first brought in with teletypes. The purpose of the parity bit is to indicate whether the data was received correctly. An *odd*

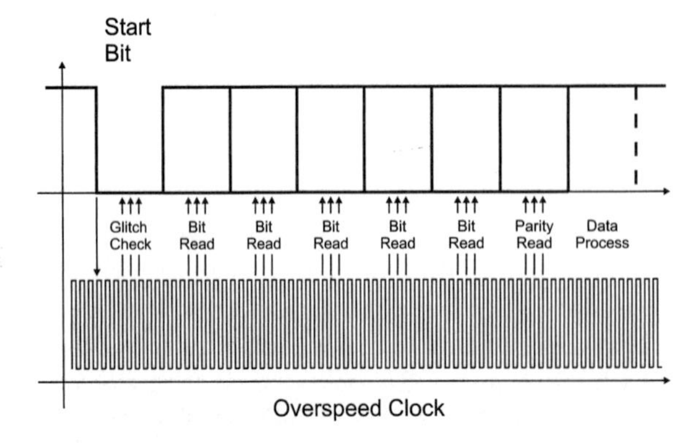


**Figure 6-56** Reading an asynch data packet

*parity* meant that if all the data bits and parity bits set to a mark were counted, then the result would be an odd number. *Even parity* is checking all the data and parity bits and seeing if the number of mark bits is an odd number. Along with even and odd parity are mark, space, and no parity. *Mark parity* means that the parity bit is always set to a 1, *space parity* is always having a 0 for the parity bit, and *no parity* is eliminating the parity bit altogether.

The most common form of asynchronous serial data packet is *8-N-1,* which means eight data bits, no parity, and one stop bit. This reflects the capabilities of modern computers to handle the maximum amount of data with the minimum amount of overhead and with a very high degree of confidence that the data will be correct.

I stated that parity bits are a "crude" form of error detection. I said that because they can only detect one bit error (e.g., if two bits are in error, the parity check will not detect the problem). If you are working in a high-induced-noise environment, you might want to consider using a data protocol that can detect (and, ideally, correct) multiple-bit errors.

**RS-232** In the early days of computing (the 1950s), although data could be transmitted at high speed, it couldn't be read and processed and continuously. So, a set of handshaking lines and protocols were developed for what became known as *RS-232 serial communications.*

With RS-232, the typical packet contains seven bits, (which is the number of bits that each ASCII character contained). This simplified the transmission of man-readable text, but made sending object code and data (which were arranged as bytes) more complex because each byte would have to be split up into two *nybbles* (which are four bits long). Further complicating this is that the first 32 characters of the ASCII character set are defined as "special" characters (e.g., carriage return, back space, etc.). This meant the data nybbles would have to be converted or (shifted up) into valid characters (this is why, if you ever see binary data transmitted from a modem or embedded in an e-mail message, data is either sent as hex codes or the letters *A* to *Q*). With this protocol, to send a single byte of data, two bytes (with the overhead bits resulting in 20 bits in total) would have to be sent.

As pointed out, modern asynchronous serial data transmission normally occur eight bits at a time, which will avoid this problem and allow transmission of full bytes without breaking them up or converting them.

The actual RS-232 communications model is shown in Fig. 6-57. In RS-232, different equipment is wired according to the functions that they perform.

*DTE (Data Terminal Equipment)* and is meant to be the connector used for computers (the PC uses this type of connection). *DCE (Data Communications Equipment)* was meant for modems and terminals that transfer the data.

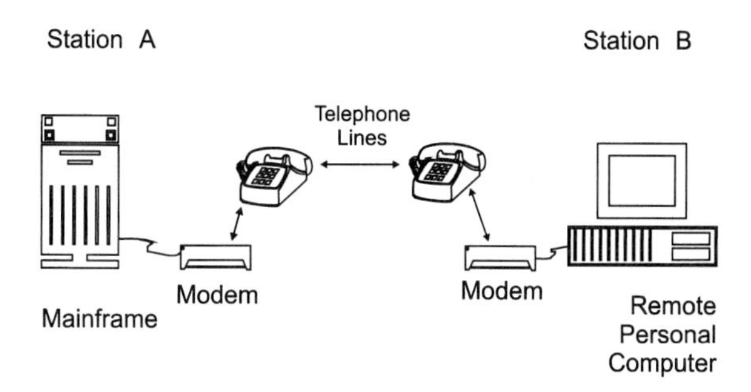


**Figure 6-57** **2 computer communication via modem**

Understanding what the RS-232 model different equipment fits under is crucial to successfully connecting two devices by RS-232. With a pretty good understanding of the serial data, you can now look at the actual voltage signals.

As mentioned, when RS-232 was first developed into a standard, computers and the electronics that drive them were still very primitive and unreliable. Because of that, we've got a couple of legacies to deal with.

The first is the voltage levels of the data. A mark (1) is actually −12 Volts and a space (0) is +12 V (Fig. 6-58).

From Fig. 6-58, you should see that the hardware interface is not simply a TTL or CMOS-level buffer. Later, this section introduces you to some methods of generating and detecting these interface voltages. Voltages in the switching region (+/− 3 V) might not be read as a 0 or 1, depending on the device. You should always be sure that the voltages going into or out of a PICmicro® MCU RS-232 circuit are in the valid regions.

Of more concern are the handshaking signals. These six additional lines (which are at the same logic levels as the transmit/receive lines and shown in Fig. 6-58) are used to interface between devices and control the flow of information between computers.

The *Request To Send (RTS)* and *Clear To Send (CTS)* lines are used to control data flow between the computer (DCE) and the modem (DTE device). When the PC is ready to send data, it asserts (outputs a mark) on RTS. If the DTE device is capable of receiving data, it will assert the CTS line. If the PC is unable to receiver data (i.e., the buffer is full or it is processing what it already has), it will de-assert the RTS line to notify the DTE device that it cannot receive any additional information.

The *Data Transmitter Ready (DTR)* and *Data Set Ready (DSR)* lines are used to establish communications. When the PC is ready to communicate with the DTE device, it asserts DTR. If the DTE device is available and ready to accept data, it will assert DSR to notify the computer than the link is up and ready for data transmission. If a hardware error is in the link, then the DTE device will de-assert the DSR line to notify the computer of the problem. Modems, if the carrier between the receiver is lost, will de-assert the DSR line.

Two more handshaking lines are available in the RS-232 standard that you should be aware of. Even so, chances are that you will never connect anything to them. The first is the *Data Carrier Detect (DCD),* which is asserted when the modem has connected with another device (i.e., the other device has "picked up the phone"). The *Ring Indicator (RI)* is used to indicate to a PC whether or not the phone on the other end of the line is ringing or if it is busy. This line is very rarely used in PICmicro® MCU applications.

A common ground connection exists between the DCE and DTE devices. This connection is crucial for the RS-232-level converters to determine the actual incoming voltages. The ground pin should never be connected to a chassis or shield ground (to avoid large cur-

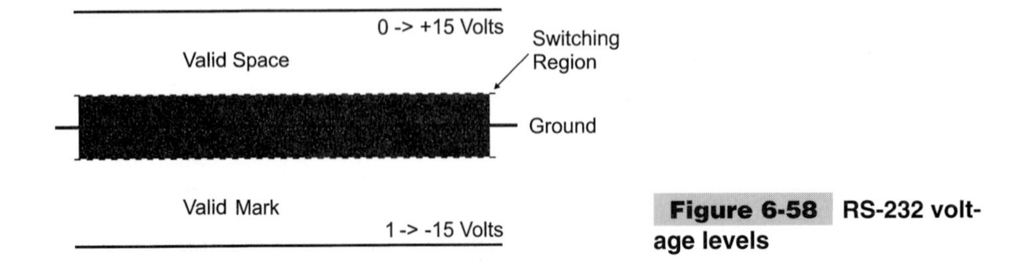


**Figure 6-58** RS-232 voltage levels

rent flows or be shifted and prevent accurate reading of incoming voltage signals). Incorrect grounding of an application can result in the computer or the device it is interfacing to reset or have they power supplies blow a fuse or burn out. The latter consequences are unlikely, but I have seen it happen in a few cases.

To avoid these problems, be sure that the chassis and signal grounds are separate or connected by a high-value (hundreds of k) resistor.

Before going too much further, I should expose you to an ugly truth; the handshaking lines are almost never used in RS-232 (and not just PICmicro® MCU RS-232) communications. The handshaking protocols were added to the RS-232 standard when computers were very slow and unreliable. In this environment, data transmission had to be stopped periodically to allow the receiving equipment to catch up.

Today, this is much less of a concern. Normally, three-wire RS-232 connections are implemented (Fig. 6-59). I normally accomplish this by shorting the DTR/DSR and RTS/CTS lines together at the PICmicro® MCU end. The DCD and RI lines are left unconnected.

With the handshaking lines shorted together, data can be sent and received without having to develop software to handle the different handshaking protocols.

A couple of points on three-wire RS-232 are worth noting. First, it cannot be implemented blindly. In about 20% of the RS-232 applications that I have had to do over the years, I have had to implement some subset of the total seven-wire (transmit, receive, ground, and four handshaking lines) protocol lines. Interestingly enough, I have never had to implement the full hardware protocol. This still means that four out of five times, if you wire the connection (Fig. 6-59), the application would have worked, but you have to be prepared for the other twenty percent of cases where more work is required.

With the three-wire RS-232 protocol, there might be applications where you don't want to implement the hardware handshaking (the DTR, DSR, RTS, and CTS lines), but you might want software handshaking. Two primary standards are in place. The first is known as the *XON/XOFF protocol,* in which the receiver sends an *XOFF* (DC3, character 0x013) when it can't accept any more data. When it is able to receive data, it sends an *XON* (DC1, character 0x011) to notify the transmitter that it can receive more data.

The final aspect of the RS-232 to cover here is the speeds in which data is transferred. When you first see the speeds (such as 300, 2,400, and 9,600 bits per second), they seem rather arbitrary. The original serial data speeds were chosen for teletypes because they gave the mechanical device enough time to print the current character and reset before the next one came in. Over time, these speeds have become standards and as faster devices have become available, they've just been doubled (e.g, 9,600 bps is 300 bps doubled five times).

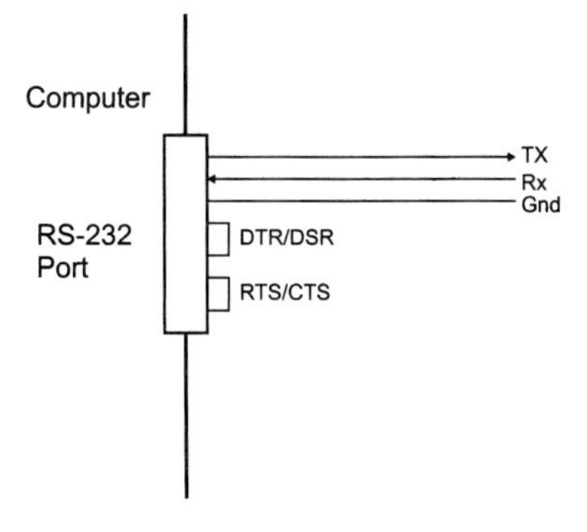


**Figure 6-59** Typical RS-232 wiring

In producing these data rates, the PICmicro® MCU's USART uses clock divider to produce a clock 16 times the data rate. The PICmicro® MCU's operating clock is divided by integers to get the nominal RS-232 speeds. This might seem like it won't work out well, but because of the RS-232's strange relationship with the number 13, the situation isn't as bad as it might seem.

If you invert (to get the period of a bit) the data speeds and convert the units to microseconds, you will discover that the periods are almost exactly divisible by 13. Thus, you can use an even-MHz oscillator in the hardware to communicate over RS-232 using standard frequencies.

For example, if you had a PICmicro® MCU running with a 20-MHz instruction clock and you wanted to communicate with a PC at 9,600 bps, you would determine the number of cycles to delay by:

1. Find the bit period in microseconds. For 9,600 bps, this is 104 $\mu$s.
2. Divide this bit period by 13 to get a multiple number. For 104 $\mu$s, this is 8.

Now, if the external device is running at 20 MHz (which means a 200-ns cycle time), you can figure out the number of cycles as multiples of 8 $\times$ 13 times in the number of cycles in 1 $\mu$s. For 20 MHz, five cycles execute per microsecond. To get the total number of cycles for the 104-$\mu$s bit period, you simply evaluate:

$$5 \text{ cycles}/\mu\text{s} * 13 * 8\ \mu\text{s/bit} = 520 \text{ instruction cycles/bit}$$

The device you are most likely to interface to is the PC. Its serial ports consist of basically the same hardware and BIOS interfaces that were first introduced with the first PC in 1981. Since that time, a nine-pin connector has been specified for the port (in the PC/AT) and one significant hardware upgrade has been introduced when the PS/2 was announced. For the most part, the serial port has changed the least of any component in the PC for the past 20 or so years.

Either a male 25-pin or male 9-pin connector is available on the back of the PC for each serial port. These connectors are shown in Fig. 6-60 and Table 6-7.

The 9-pin standard was originally developed for the PC/AT because the serial port was put on the same adapter card as the printer port. There wasn't enough room for the serial port *and* parallel port to both use 25-pin D-shell connectors. I prefer the smaller form-factor connector.

RS-232 serial communications have a reputation for being difficult to use and I have to disagree. Implementing RS-232 isn't very hard when a few rules are followed and you have a good idea of what's possible.

When implementing an RS-232 interface, you can make your life easier by doing a few simple things. The first is the connection. Whenever I do an application, I standardize by

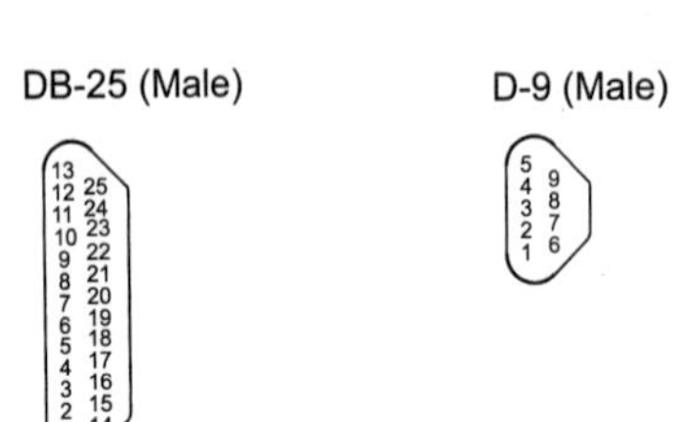


**Figure 6-60** IBM PC DB-25 and D-9 pin RS-232 connectors

**TABLE 6-7 Standard RS-232 Connector Pinouts**

| PIN NAME | 25 PIN | 9 PIN | I/O DIRECTION |
|---|---|---|---|
| TxD | 2 | 3 | Output ("O") |
| RxD | 3 | 2 | Input ("I") |
| Gnd | 7 | 5 | |
| RTS | 4 | 7 | O |
| CTS | 5 | 8 | I |
| DTR | 20 | 4 | O |
| DSR | 6 | 6 | I |
| RI | 22 | 9 | I |
| DCD | 8 | 1 | I |

using a nine-pin D-shell with the DTE interface (the one that comes out of the PC) and use standard "straight-through" cables. In doing so, I always know what my pinout is at the end of the cable when I'm about to hookup a PICmicro® MCU to a PC.

By making the external device DCE always and using a standard pinout, I don't have to fool around with null modems or by making my own cables.

When I create the external device, I also loop back the DTR/DSR and CTS/RTS data pairs inside the external device, rather than at the PC or in the cable. This allows me to use a standard PC and cable without having to do any wiring on my own or any modifications. It actually looks a lot more professional as well.

Tying DTR/DSR and CTS/RTS also means that I can take advantage of built-in terminal emulators. All operating systems have a "dumb" terminal emulator that can be used to debug the external device without requiring the PC code to run. Getting the external device working before debugging the PC application code should simplify your work.

As I went through the RS-232 electrical standard earlier in this section, you were probably concerned about interfacing standard, modern technology (i.e., TTL and CMOS) devices to other RS-232 devices. This is a legitimate concern because, without proper voltage-level conversion, you will not be able to read from or write to external TTL or CMOS devices. Fortunately, this conversion isn't all that difficult because there are methods to make it quite easy.

If you look at the original IBM PC RS-232 port specification, you'll see that 1488/1489 RS-232 level-converter circuits were used for the RS-232 serial port interfaces. The pinout and wiring for these devices in a PC are shown in Fig. 6-61.

There are a few points about the 1488/1489 components that should be discussed. When transmitting data, each transceiver (except for 1) is actually a NAND gate (with the inputs being #A and #B outputting on #Y). When I wire in a 1488, I ensure that the second input to a driver is always pulled high (as I've done with 2B in Fig. 6-61).

The second comment has to do with the 1489 receiver. The #C input is a flow control for the gates (normally RS-232 comes in the #A pin and is driven as TTL out of #Y). This pin is normally left floating (unconnected).

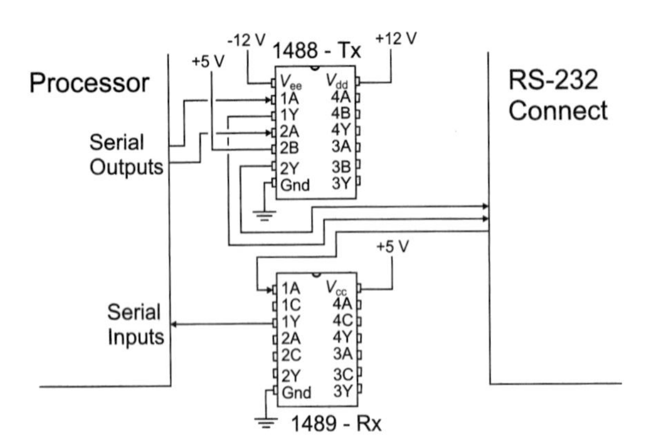


**Figure 6-61** 1488/1489 RS-232 connections

These chips are still available and work very well (up to 115,200 bps maximum RS-232 data rate), only I'd never use them in my own projects because the 1488 (transmitter) requires +/−12-V sources in order to produce valid RS-232 signal voltages.

This section presents three methods that you can choose from to convert RS-232 signal levels to TTL/CMOS (and back again) when you are creating PICmicro® MCU RS-232 projects. These three methods do not require +/−12 V. In fact, they just require the +5-V supply that is used for logic power.

The first method is using an RS-232 converter that has a built-in charge pump to create the +/−12 V required for the RS-232 signal levels. Probably the most well-known chip that is used for this function is the Maxim MAX232 (Fig. 6-62).

This chip is ideal for implementing three-wire RS-232 interfaces (or adding a simple DTR/DSR or RTS/CTS handshaking interface). The ground for the incoming signal is connected to the processor ground (which is not the case's ground).

Along with the MAX232, Maxim and some other chip vendors have a number of other RS-232 charge-pump-equipped devices that will allow you to handle more RS-232 lines (to include the handshaking lines). Some available charge-pump devices do not require the external capacitors that the MAX232 chip does, which will simplify the layout of your circuit (although these chips do cost quite a bit more).

The next method of translating RS-232 and TTL/CMOS voltage levels is to use the transmitter's negative voltage. The circuit in Fig. 6-63 shows how this can be done and is demonstrated in Chapter 15.

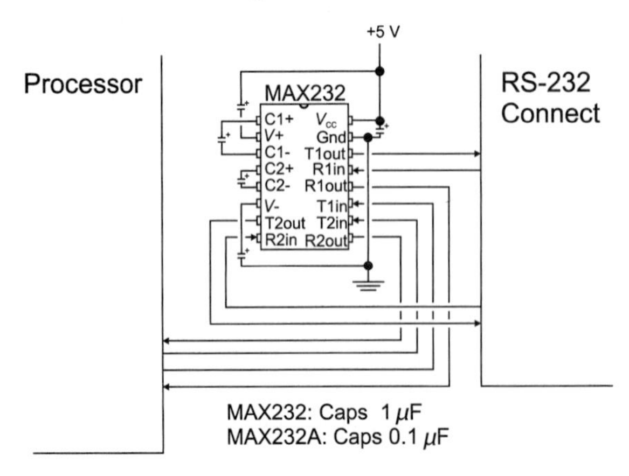


**Figure 6-62** MAXIM MAX232 RS-232 connections

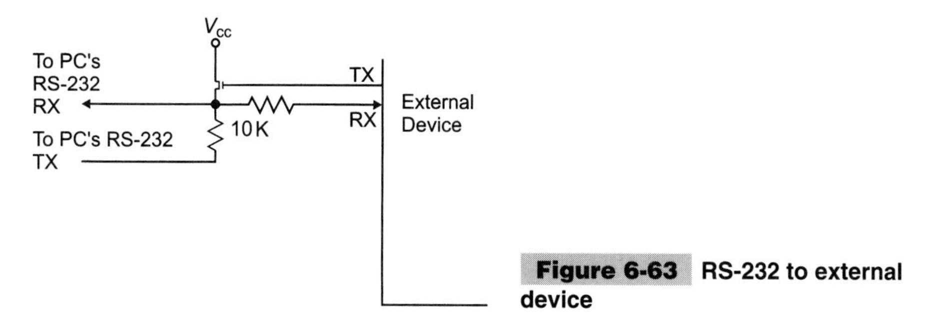


**Figure 6-63** RS-232 to external device

This circuit relies on the RS-232 communications only running in half-duplex mode (i.e., only one device can transmit at a given time). When the external device "wants" to transmit to the PC, it sends the data either as a mark (leaving the voltage being returned to the PC as a negative value) or as a space (by turning on the transistor and enabling the positive voltage output to the PC's receivers). If you go back to the RS-232 voltage-specification drawing, you'll see that +5 V is within the valid voltage range for RS-232 spaces.

This method works very well (consuming just about no power) and is obviously a very cheap way to implement a three-wire RS-232 bidirectional interface. Figure 6-63 shows a N-Channel MOSFET transistor simply because it does not require a base current-limiting transistor the same way that a bipolar transistor would.

When the PICmicro® MCU transmits a byte to the external device through this circuit, it will receive the packet it's sent because this circuit connects the PICmicro® MCU's receiving pin (more or less) directly to its transmitting pin. The software running in the PICmicro® MCU (as well as the external device) will have to handle this.

You also have to be absolutely sure that you are only transmitting in half-duplex mode. If both the PICmicro® MCU and the external device attempt to transmit at the same time, then both messages will be garbled. Instead, the transmission protocol that you use should wait for requested responses, rather than sending them asynchronously (or, you could have one device, either the PC or external devices, wait for a request for data from the other).

Another issue to notice is that data out of the external device will have to be inverted to get the correct transmission voltage levels (e.g., a *0* will output a *1*) to ensure that the transistor turns on at the right time (e.g., a positive voltage for a space).

Unfortunately, this means that the built-in serial port for many microcontrollers cannot be used because they cannot invert the data output as is required by the circuit. An inverter could be put between the serial port and the RS-232 conversion circuit to avoid this problem.

The Dallas Semiconductor DS275 incorporates this circuit (with built-in inverters) into the single package shown in Fig. 6-64.

The DS1275 and the DS275 have the same pinout. Both work exactly the same way, but the DS275 is a later version of the part.

As an aside, before going on to the last interface circuit, this circuit has a big advantage. The PICmicro® MCU's RS-232 transmitter is connected to the PC's RS-232 receiver if the external device (with this circuit) is connected to a PC. Thus, the PICmicro® MCU can *ping* (send a command that is ignored by the external device) via the RS-232 port and see if the external device is connected to it. But, this technique has a hitch. Even though the circuit is connected, how do you know that the external PICmicro® MCU device is working from the PC?

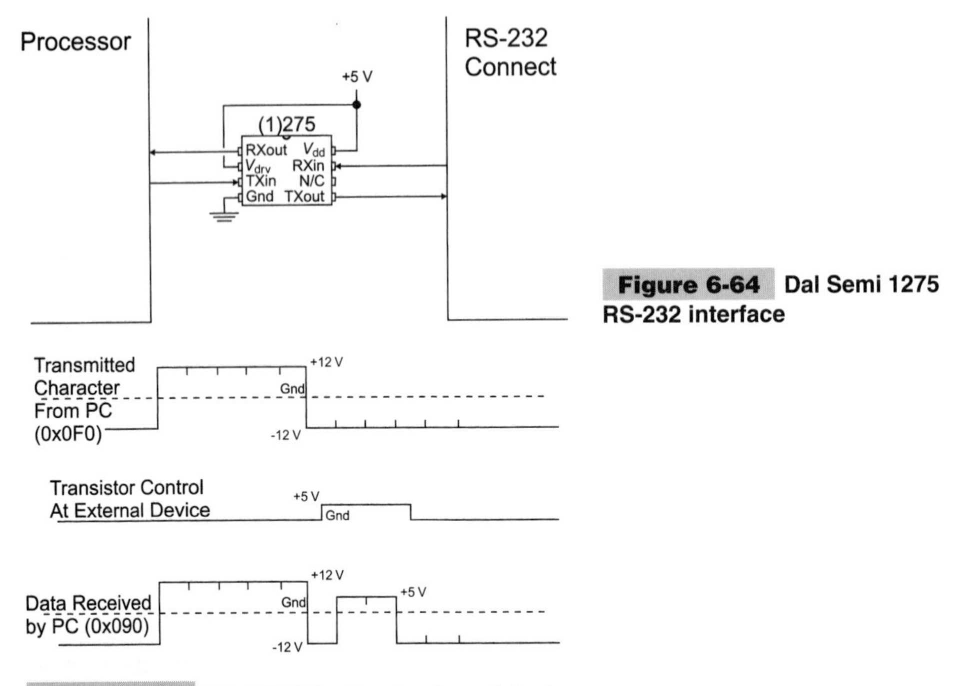


**Figure 6-64** Dal Semi 1275 RS-232 interface

**Figure 6-65** RS-232 "Ping" using transistor/resistor TTL/CMOS voltage conversion circuit

Actually, this isn't that difficult. Simply specify the ping character as something that the external device can recognize and have it modify it so that the PC's software can recognize that the interface is working.

In the past, I have used a microcontroller with bit-banging software (described in the next section) to change some mark bits when it recognizes that a ping character is being received.

Figure 6-65 shows a ping character of 0x0F0 that is modified by the external device (by turning on the transistor) to change some bits into spaces. If the PC receives nothing or 0x0F0, then the external device is not working.

The last interface circuit presented here is simply a resistor (Fig. 6-66). This method of receiving data from an RS-232 device to a logic input probably seems absurdly simple and as if could not work. But it does and very well.

This interface works by relying on clamping diodes in the receiver holding the voltage at the maximum allowable for the receiver. The 10-K resistor limits any current flows and provides a voltage drop (Fig. 6-67).

Figure 6-68 is the actual circuit of the 10-K current-limiting resistor and PICmicro® MCU I/O pin with internal clamping diodes.

There are some things to watch out for when using this receive circuit in the PICmicro® MCU. Although most I/O pins are clamped internally, in some cases, you will have to add your own clamping diodes externally. These cases are when the open drain (RA4) I/O pin is used or the optional _MCLR I/O pin is used for input.

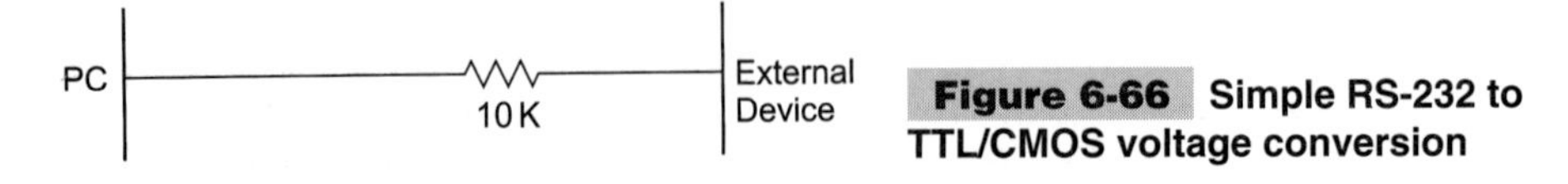


**Figure 6-66** Simple RS-232 to TTL/CMOS voltage conversion

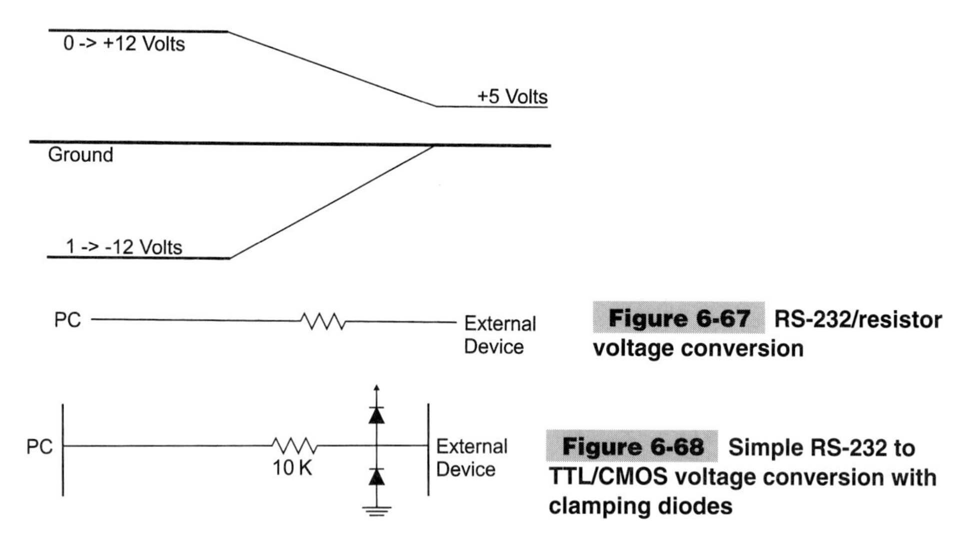


**Figure 6-67** RS-232/resistor voltage conversion

**Figure 6-68** Simple RS-232 to TTL/CMOS voltage conversion with clamping diodes

There are a few rules for this implementation. Some people like to use this configuration with a 100-K resistor (or higher value) instead of the 10-K shown. Personally, I don't like to use anything higher than 10 K because of the possibilities of induced noise with a CMOS input (this is less likely with TTL) causing an incorrect bit in the stream to be read. 10 K will maximize the induced current required to change the state of the I/O pin.

With the availability of many CMOS devices requiring very minimal amounts of current to operate, you might be wondering about different options for powering your circuit. One of the most innovative that I have come across is using the PC's RS-232 ports itself as powering devices that are attached to it using the circuit shown in Fig. 6-69.

When the DTR and RTS lines are outputting a space, a positive voltage (relative to ground) is available. This voltage can be regulated and the output used to power the devices attached to the serial port (up to about 5 mA). For extra current, the Tx line can also be added into the circuit, as well with a break being sent from the PC to output a positive voltage.

The 5 mA is enough current to power the transistor/resistor type of RS-232 transmitter and a PICmicro® MCU running at 4 MHz, along with some additional hardware (such as an LCD). You will not be able to drive an LED with this circuit and you might find that some circuits that you normally use for such things as pull-ups and pull-downs will consume too much power.

Now, with this method of powering the external device, you do not have use of the handshaking lines, but the savings of not having to provide an external power supply (or battery) will outweigh the disadvantages of having to come up with a software pinging and handshaking protocol. Externally powering a device attached to the RS-232 port is ideal for input devices, such as serial mice, which do not require a lot of power.

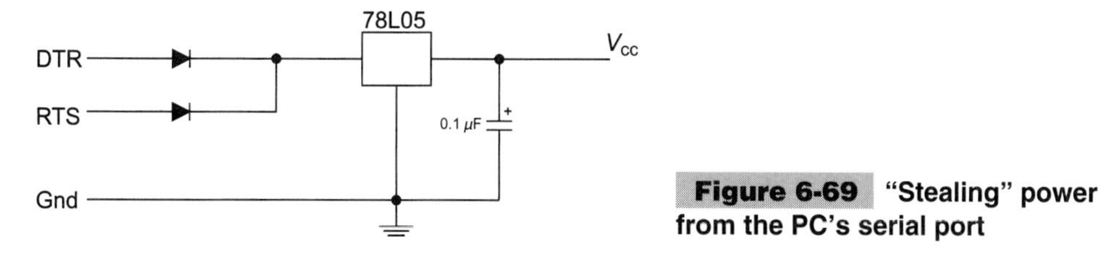


**Figure 6-69** "Stealing" power from the PC's serial port

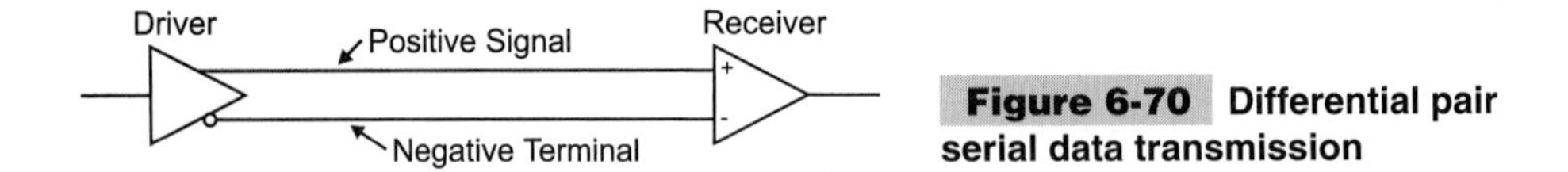


**Figure 6-70** Differential pair serial data transmission

**RS-485/RS-422** So far, this book has covered single-ended asynchronous serial communications methods, such as RS-232 and direct NRZ device interfaces. These interfaces work well in home and office environments, but can be unreliable in environments where power surges and electrical noise can be significant. In these environments, a double-ended or differential pair connection is optimal to ensure the most accurate communications.

A differential-pair serial communications electrical standard consists of a balanced driver with positive and negative outputs that are fed into a comparator, which outputs a *1* or a *0,* depending on whether or not the positive line is at a higher voltage than the negative. Figure 6-70 shows the normal symbols used to describe a differential pair connection.

This data-connection method has several advantages. The most obvious one is that the differential pair doubles the voltage swing sent to the receiver which increases its noise immunity. This is shown in Fig. 6-71, when the positive signal goes high and the negative voltage goes low. The change in the two receiver inputs is 10 volts, rather than the 5 volts of a single line. This is assuming that the voltage swing is 5 volts for the positive and negative terminals of the receiver. This effective doubling of the signal voltage reduces the impact that electrical interface has on the transmitted signal.

Another benefit of differential pair wiring is that if one connection breaks and there is common ground, the circuit will continue to operate (although at reduced noise reduction efficiency). This feature makes differential pairs very attractive in aircraft, and spacecraft, where loss of a connection could be catastrophic.

To minimize AC transmission-line effects, the two wires should be twisted around each other. Twisted-pair wiring can either be bought commercially or made by simply twisting two wires together. Twisted wires have a typically characteristic impedance of 300 Ohms or greater.

A common standard for differential pair communications is RS-422. This standard, which uses many commercially available chips provides:

**1** Multiple receiver operation.
**2** Maximum data rate of 10MBps.
**3** Maximum cable length of 4000 meters (with a 100-kHz signal).

Multiple receiver operation (Fig. 6-72) allows signals to be "broadcasted" to multiple devices.

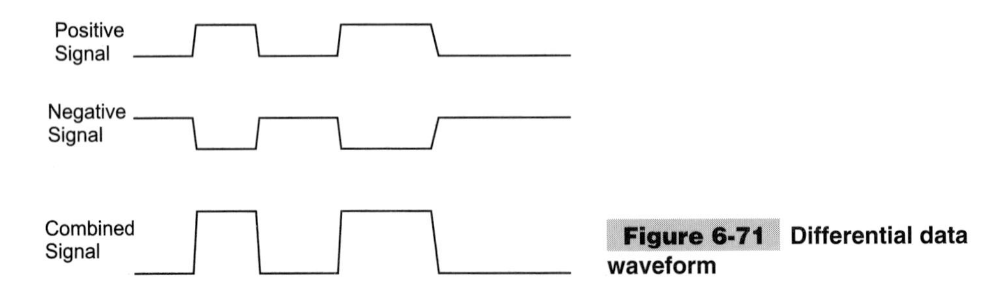


**Figure 6-71** Differential data waveform

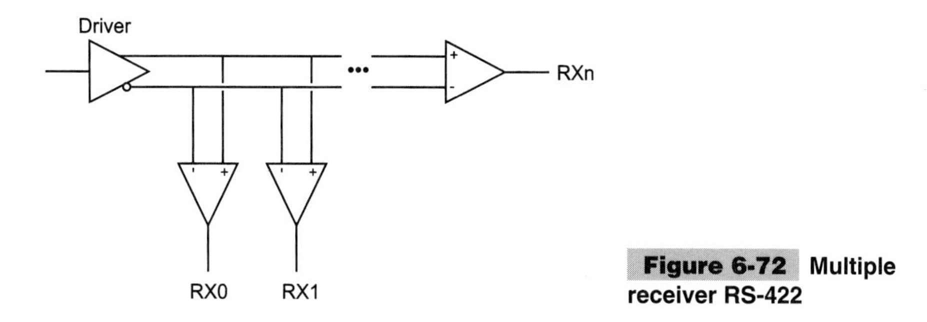


**Figure 6-72** Multiple receiver RS-422

The best distance and speed changes with the number of receivers of the differential pair along with its length. The 4000 m at 100 kHz or 40 m at 10 MHz are examples of this balancing between line length and data rate. For long data lengths, a few hundred ohm terminating resistor might be required between the positive terminal and negative terminal at the end of the lines to minimize reflections coming from the receiver and affecting other receivers.

RS-422 is not as widely used as you might expect, instead, RS-485 is much more popular. RS-485 is very similar to RS-422, except that it allows multiple drivers on the same network. The common chip is the 75176, which has the ability to drive and receive on the lines (Fig. 6-73).

I have drawn the right 75176 of Fig. 6-73 with the Rx and TX and two enables tied together. This results in a two-wire differential I/O device. Normally, the 75176 s are left in RX mode (pin 2 reset), unless they are driving a signal onto the bus. When the unused 75176 s on the lines are all in receive mode, any one can "take over" the lines and transmit data.

Like RS-422, multiple 75176s (up to 32) can be on the RS-485 lines with the capability of driving or receiving. When all the devices are receiving, a high (1) is output from the 75176. This means that the behavior of the 75176 in the RS-485 (because these are multiple drivers) is similar to that of a dotted AND bus. When one driver pulls down the line, all receivers are pulled low. For the RS-485 network to be high, all unused drivers must be off or all active drivers must be transmitting a *1*. This feature of RS-485 is taken advantage in small system networks, such as CAN, which is covered later in the chapter.

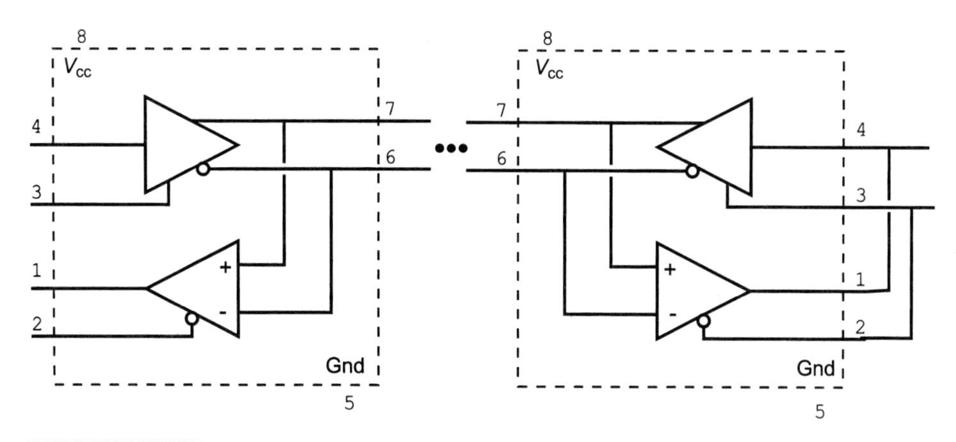


**Figure 6-73** RS-485 connection using a 75176

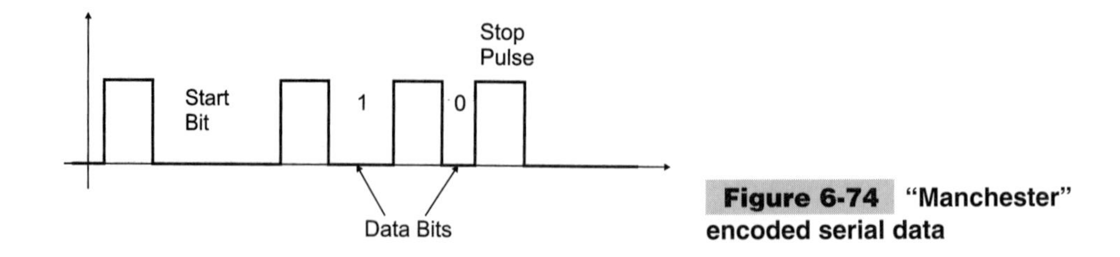


**Figure 6-74** "Manchester" encoded serial data

The only issue to be on the lookout for when creating RS-485/RS-422 connections is to keep the cable polarities correct (positive to positive and negative to negative). Reversing the connectors will result in lost signals and misread transmission values.

**Manchester serial data transfer** Another common method of serially transmitting data asynchronously is to use the Manchester encoding format. In this type of data transfer, each bit is synchronized to a start bit and the following data bits are read with the space dependent on the value of the bit (Fig. 6-74).

In this type of data transmission, the data size is known, so when the start pulse is recognized and the space afterward is not treated as incoming data. Manchester encoding is unique in that the start bit of a packet is quantitatively different from a 1 or a 0. This allows a receiver to determine whether or not the data packet being received is actually at the start of the packet or somewhere in the middle (and should be ignored until a start bit is encountered).

Manchester encoding is well suited for situations where data is not interrupted or restarted anywhere except at the beginning. Because of this, it is the primary method of data transmission for infrared control (such as used in your TV's remote control). Chapter 16 presents two algorithms for reading a TV remote control's IR Manchester encoded data packets.

**Can** The *CAN (Controller Area Network)* protocol was originally developed by Bosch a number of years ago as a networking scheme that could be used to interconnect the computing systems used within automobiles. At the time, there was no single standard for linking digital devices in automobiles. Before the advent of CAN (and J1850, which is the similar North American standard) cars could contain as much as three miles of wiring, weighing 200 pounds, interconnecting the various parts and systems within the car.

CAN was designed to be:

1 Fast (1 Mbps).
2 Insensitive to electromagnetic interference.
3 Simple with few pins in connectors for mechanical reliability.
4 Devices could be added or deleted from the network easily (and during manufacturing).

Although CAN is similar to J1850 and does rely on the same first two layers of the OSI seven-layer communications model, the two standards are electrically incompatible. CAN was the first standard and is thoroughly entrenched in European and Japanese cars and is rapidly making inroads (please excuse the pun) with North American automotive manufacturers.

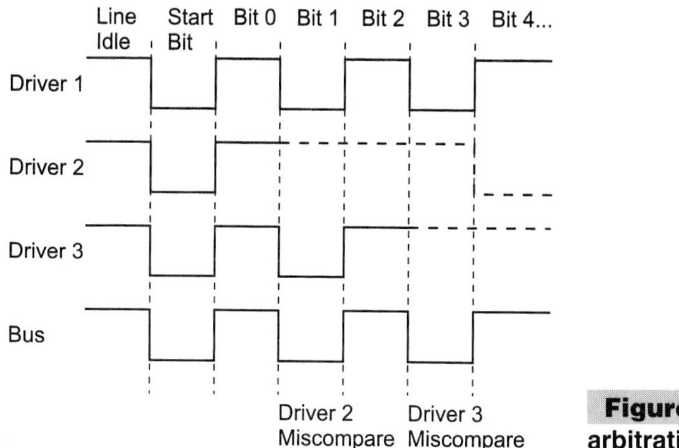


**Figure 6-75** CAN transmission arbitration

CAN is built from a dotted AND bus that is similar to that used in I2C. Electrically, RS-485 drivers are used to provide a differential voltage network that will work even if one of the two conductors is shorted or disconnected (giving the network high reliability inside of the very extreme environment of the automobile). This dotted AND bus allows arbitration between different devices (when the device's "drivers" are active, the bus is pulled down, like in I2C signal). An example of how this method of arbitration works is shown in Fig. 6-75.

In this example, when a driver has a miscompare with what it is transmitting (e.g., when it is sending a *1* and a *0* shows up on the bus), then that driver stops sending data until the current message (which is known as a *frame*) has completed. This is a very simple and effective way to arbitrate multiple signals without having to retransmit all of the colliding messages over again.

The frame (Fig. 6-76) is transmitted as an asynchronous serial stream (which means that no clocking data is transmitted). Thus, both the transmitter and receiver must be working at the same speed (typically, data rates are in the range of 200 kbps to 1 Mbps).

In CAN, a *0* is known as a *dominant bit* and a *1* is known as a *recessive bit*. The different fields of the frame are defined in Table 6-8.

The last important note about CAN is that devices are not given specific names or addresses. Instead, the message is identified (using the 11- or 19-bit message identifier). This method of addressing can provide you with very flexible messaging (which is what CAN is all about).

The CAN frame is very complex, as is the checking that has to be done, both for receiving a message and transmitting a message. Although it can be done using a microcontroller and programming the functions in software, I would recommend only imple-

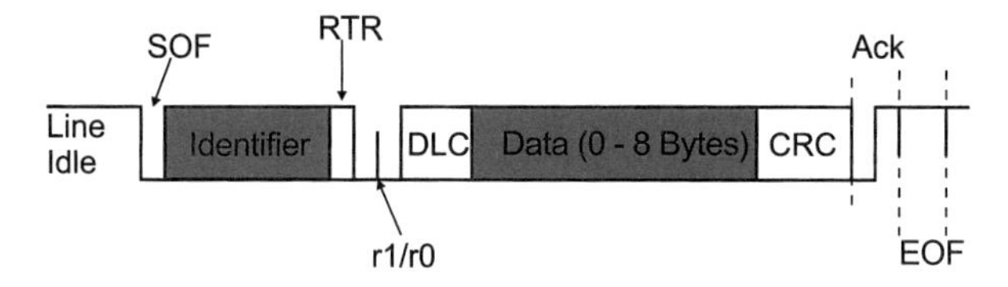


**Figure 6-76** CAN 11 bit identifier frame

**TABLE 6-8 CAN Frame Field Definitions**

| | |
|---|---|
| SOF | Start of Frame, A single Dominant Bit |
| Identifier | 11 or 19 Bit Message Identifier |
| RTR | This Bit is set if the transmitter is also TX'ing Data |
| r1/r0 | Reserved Bits, Should always be Dominant |
| DLC | Four Bits indicating the number of bytes that follow |
| Data | Zero to 8 Bytes of Data, Sent MSB First |
| CRC | 15 bits pf CRC data followed by a recessive bit |
| Ack | Two Bit field, Dominant/ Recessive Bits |
| EOF | End of Frame, at least 7 Recessive Bits |

menting CAN using hardware interfaces. Several microcontroller vendors provide CAN interfaces as part of the devices and quite a few different "standard" chips (the Microchip MCP2510 is a very popular device) are available to carry out CAN interface functions effectively and cheaply.

Currently, no PICmicro® MCU devices support CAN. Microchip has announced plans to create a PICmicro® MCU with CAN built in, but, as I write this, no such device is available nor are there any preliminary datasheets.

## DALLAS SEMICONDUCTOR 1-WIRE INTERFACE

Dallas Semiconductor has created a line of peripherals that are very attractive for use with microcontrollers because they only require one line for transferring data. This single-wire protocol is available in a variety of devices, but the most popular are the DS1820 and DS1821 digital thermometers. These devices can be networked together on the same bus (they have a built-in serial number to allow multiple devices to operate on the same bus) and are accurate to within one degree Fahrenheit. This book demonstrates a few applications using these parts, but I wanted to first introduce you to the protocol and how the devices are used.

Probably the most substantive demonstration of what the DS1820 can do is in an application where it is connected to a PICmicro® MCU (Fig. 6-77). The DS1820 is available in

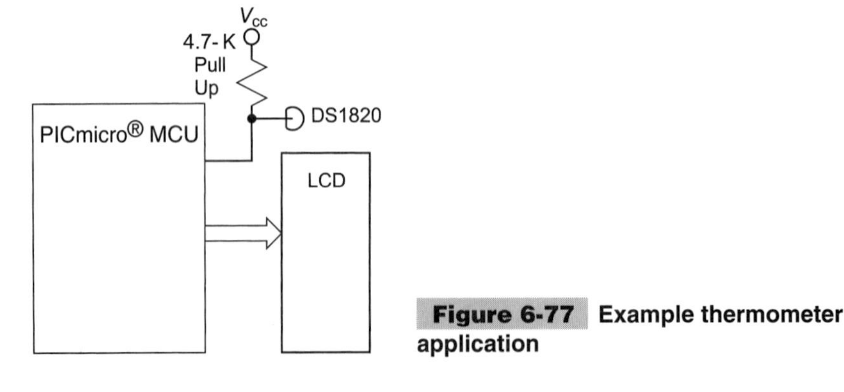


**Figure 6-77** Example thermometer application

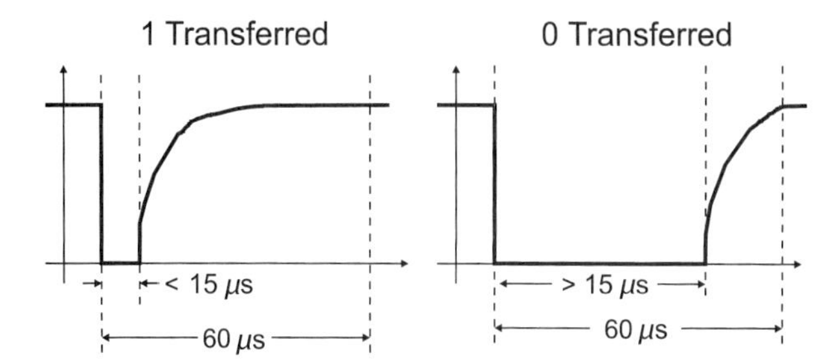


**Figure 6-78** **Dallas Semi. "1-Wire" data transfer**

a variety of packages. The one that I use the most is a three-pin TO-92 that looks like a "tall" plastic transistor package.

The DS1820 has many features that would be useful in a variety of different applications. These include the ability of sharing the single-wire bus with other devices using a unique serial number burned into the device that allows it to be written to individually. The DS1820 has the ability to be powered by the host device. In the applications that use it in the book, I power DS1820 from $V_{CC}$ available to the microcontroller and only have one device on the bus at a given time. I refrained from covering the single-wire bus interface in the networking section earlier in the chapter because it is specific to Dallas Semiconductor products.

Data transfers over the single-wire bus are initiated by the host system (in the application cases, this is the PICmicro® MCU) and are carried out eight bits at a time (with the least significant bit transmitted first). Each bit transfer requires at least 60 $\mu$s. The single-wire bus is pulled up externally (Fig. 6-77) and is pulled down by either the host or the peripheral device to transfer data. If the bus is pulled down for a very short interval, a *1* is being transmitted. If the bus is pulled down for more than 15 $\mu$s, then a *0* is being transmitted. The differences in the *1* and *0* bits are shown in Fig. 6-78.

All data transfers are initiated by the host system. If it is transmitting data, then it holds down the line for the specified period. If it is receiving data from the DS1820, then the host pulls down the line, releases it, and polls the line to see how long it takes for it to go back up. During data transfers, I have ensured that interrupts cannot occur (because this would affect how the data is sent or read if the interrupt takes place during the data transfer).

Before each command is set to the DS1820, a reset and presence pulse is transferred. A reset pulse consists of the host pulling down the line for 480 to 960 $\mu$s. The DS1820 replies by pulling down the line for roughly 100 $\mu$s (60 to 240 $\mu$s is the specified value). To simplify the software, I typically did not check for the presence pulse (because I knew in the application that I had the thermometer connected to the bus). In another application (where the thermometer can be disconnected), putting a check in for the presence pulse might be required.

To carry out a temperature read of a single DS1820 connected to a microcontroller, you can use the following instruction sequence:

**1** Send a reset pulse and delay while the presence pulse is returned.
**2** Send 0x0CC, which is the skip ROM command, which tells the DS1820 to assume that the next command is directed toward it.
**3** Send 0x044, which is the *temperature conversion initiate* instruction. The current temperature will be sampled and stored for later read back.

4. Wait 500+ μs for the temperature conversion to complete.
5. Send a reset pulse and delay while the Presence pulse is returned.
6. Send 0x0CC, skip ROM command again.
7. Send 0x0BE to read the Scratchpad RAM, which contains the current temperature (in degrees Celsius times two).
8. Read the nine bytes of scratchpad RAM.
9. Display the temperature (if bit 0 of the second byte is returned from the Scratchpad RAM is set, the first byte is negated and a – is put on the LCD Display) by dividing the first byte by 2 and sending the converted value to the LCD.

The total procedure for doing a temperature measurement takes about 5 ms. A good (and simple) test of whether or not the thermometer is working is to pinch it between your fingers and see if the returned temperature value goes upward as the application is executing.

When I have implemented PICmicro® MCU code for accessing the DS1820, I have had problems ensuring that my timing delays are correct. When sending data to the DS1820, be sure that a *1* has a delay of about 10 μs and a *0* has a delay of 45 μs. This will ensure that the data the DS1820 reads is not ambiguous and doesn't have any problems with differentiating the data sent to it.

# 7

# PC INTERFACING

## CONTENTS AT A GLANCE



Chances are good that the device that you are most interested in interfacing the PICmicro® MCU to is an IBM PC compatible. Unfortunately, this is also the most difficult device that you will probably try to interface to because of the hardware complexities of modern systems and understanding the different paths that data takes to get from one part of the PC to another. Further complicating the task of interfacing the PICmicro® MCU to the PC is the wide range of devices and features available in the marketplace. Basic can access the ports directly via the MSComm object.

When designing PC port interfaces, I recommend developing the interface code to pass data in a "man-readable" format and use standard "text" data transfers, rather than compressed data. If this is serial data, it will allow a terminal emulator, such as HyperTerminal, to be used when you are debugging the PICmicro® MCU interface. This allows the PC application to be debugged separately, after the PICmicro® MCU interface, which eliminates the two variables when something doesn't work. The time penalty is actually quite small for the text transfers and is more than offset by the ease in which applications can be debugged.

Along with several serial port applications, this book includes a parallel port interface application (the El Cheapo programmer), and an *Industry Standard Architecture (ISA)* adapter card interface. These projects will introduce you to the different PC interfaces and give me a chance to explain their shortcomings and how to get around them.

The *Universal Serial Bus (USB)* is the PC (and Macintosh) interface that I am most looking forward to using with the PICmicro® MCU. This interface is dynamically reconfigurable and provides 100 mA at 5 volts for peripheral devices. As this book was going to press, Microchip was making the first USB-interface-equipped PICmicro® MCUs available as engineering samples. This interface offers the ability to create PC to PICmicro® MCU applications much easier than what I show in this chapter.

There are many ways in which each PC interface can be wired to the PICmicro® MCU. For this reason, this chapter really is more of a theory chapter, providing only information specific to the PC.

# PC Interface Software

This book includes many examples of PC-to-PICmicro® MCU interface software for you to look at. Almost all the code is written in Microsoft's Visual Basic to take advantage of that development environment's ease of use for creating Microsoft Windows applications. Many add-ons can enhance applications and allow you to interface with the PC's hardware I/O ports and device drivers. In addition to working with an easy-to-use development tool (Visual Basic), I include a number of rules for creating hardware interfacing applications to maximize the probability that your application will be successful.

First, in the PICmicro® MCU, instruction timing is predictable; in a modern PC (i.e., anything later than an 80386 running MS-DOS 3.x, Microsoft Windows 95, or any later operating systems), software execution is anything but predictable. If you were to write an application that wrote to two ports and then read back from a third, such as:

```
int exampleIO()
{

int Result;
   IOwrite(regA, valueA);
   IOwrite(regB, valueB);

   Result = IOread(regC);

   return Result;

} // end exampleIO
```

You cannot predict when any of these operations will occur, only that they will be carried out in the order that you've specified. The reasons why the timing cannot be guaranteed include:

- Code location in operating-system "swapper" files, main memory, or caches
- Process code execution optimizations.

- Bus utilization
- Operating system scheduling

As well, the actual timing of operations can range from very fast to slow, depending on how the hardware is implemented. A particularly big culprit of this is the parallel port.

All of this leads to the requirement that the hardware that interfaces to the PC should either provide its own timing through handshaking or it should be as timing insensitive as possible. Making a device timing insensitive can be somewhat difficult because very extreme changes in timing can occur. As I write this, a modern PC has a Pentium III or Itanium processor running at 800 MHz (or more). At the same time, many people are still running with 80386-based PCs or even 80286 based PCs. This is especially true in manufacturing companies, where custom-shielded equipment was installed years ago and still have time scheduled in production to fully recoup the investment in it.

For this reason, I suggest that timing is provided by external hardware as much as possible and internal timings are 10 ms or more apart. This will mean that your applications' data rates will be limited to a maximum of 50 kbps on less, but data transfer will be reliable.

If this hardware is going to interface with a bus with strong timing protocol in place, then these restrictions are not required. The PICmicro® MCU can interface to the ISA bus or the PC's serial ports directly without any handshaking or timing restrictions. Some hardware handshaking will have to be built in for other applications.

For the software itself, much of the current PC interface applications are written in C for the MS-DOS command line. This method is okay, although there will be problems with trying to access specific hardware registers and resources in Windows NT or Windows 2000 without administrator access. As well, MS-DOS applications become very complex very quickly from the user perspective as more features are added. For example, the list of command line options for Pkzip are shown in Table 7-1.

```
Usage: PKZIP [command] [options] zipfile [@list] [files. . .]
```

The best way to avoid this confusion is to use Microsoft Windows *Graphical Interface User (GUI)* for the application. For example, the El Cheapo programmer, included with this book has the Windows interface (Fig. 7-1).

This application handles hex files from anywhere in the PC, can program a variety of different devices (both EPROM and FLASH, as well as built-in oscillator combination values), can verify blank check devices and has a programmer hardware interface built in. The operator of the code is quite intuitive, with a message box giving instructions to the user. As the mode changes, the three operations buttons change for the different operations, keeping the user interface very simple.

**TABLE 7-1 PKZIP Command Summary**

| | | | |
|---|---|---|---|
| Add | Default | Header | Sfx |
| Comment | Delete | Help | Test |
| Configuration | Extract | License | Version |
| Console | Fix | Print | View |

**TABLE 7-2 PKZIP Option Summary**

| | | | |
|---|---|---|---|
| 204 | Header | NoExtended | Shortname |
| After | Include | NoFix | Silent |
| Attributes | Level | Normal | Sort |
| Authenticity | ListChar | NoZipExtension | Span |
| Before | ListSfxTypes | OptionChar | Speed |
| Comment | Locale | Overwrite | Store |
| Decode | Mask | Password | Temp |
| Directories | Maximum | Path | Times |
| Encode | More | Preview | Volume |
| Exclude | Move | Recurse | Warning |

One feature I really like is the graphic of the card. During programming, this graphic shows the user how to install the PICmicro® MCUs and helps to load them through the application. This graphic also leads me to the last point, which is the graphics must be pertinent and clear. I have several applications where the developer has created a gorgeous dialog box, but the interfaces are not clear (including where they are) and how they work. Like the El Cheapo interface shown in Fig. 7-1, keep your user interfaces simple and conventional and help "lead" the user through using the application.

Now, I know there will be the argument from people saying that Visual Basic does not produce the most efficient and smallest code possible. To this, I reply "who cares?"

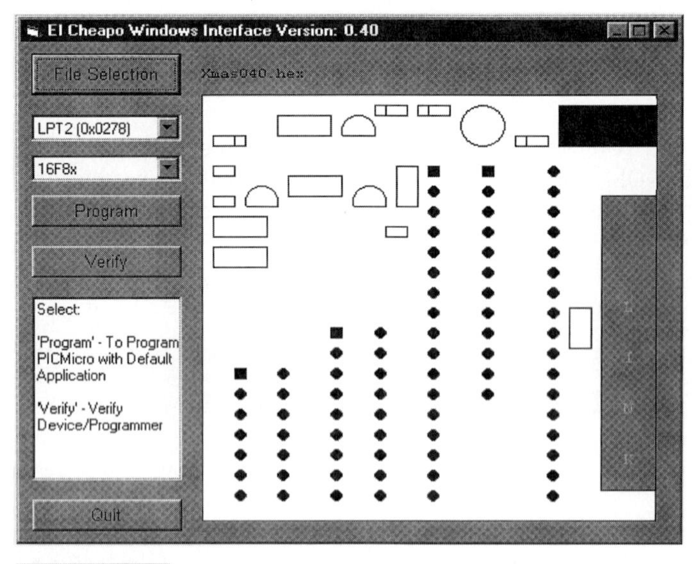


**Figure 7-1** "El cheapo" windows interface

With the performance improvement made in the basic PC over the years, any inefficiency will become more and more insignificant. As I was writing this book, I heard an interesting statistic that a 200-MHz Pentium II has a similar amount of power to a Cray supercomputer of 1980. Coupled this to the gigabytes of storage available (and growing), application "bloat" is really not an issue.

These increases in execution speed will make an application's "efficiency" less of a critical measurement if the device is not continually polling. If you have an application that runs extended periods of time without giving up the processor to other applications, you will have problems with your PC while you are executing applications.

In Visual Basic, to poll hardware, you have to use a timer as a state machine. When an operation is about to start, a mainline event handler has to initialize state variable and start a timer. For example, in the El Cheapo programmer, to carry out a blank check, I use the code:

```
   BlankState = 0;
   BlankAddress = 0;
   Timer1.Interval = 10
   Timer1.Enabled = true
   :

private sub Timer1_Ttimer( )
'timer 1 overflow

   Timer1.Enabled = False      ' Stop Flag = True don't interrupt

   select case (BlankState)
          case 0:       ' Ready Values
                 if (ProgRead = 0x07FFE)        ' Read Program Memory
                        BlankAddress = BlankAddress + 1
                        if (BlankAddress = DeviceEnd)  then
                               State = 1
                        else
                               'Display Messages
                               StopFlag = 0
                        endif
          case 1:       ' Jump to Config Memory
                 ProgWrite(Load_CONFIG, 0x07FFF)
                 State = 2:     ' Blank Address = &H2007
          case 2:               ' Check High 7 Bytes
                 if (ProgRead = 0x07FFE)
                        BlankAddress = BlankAddress + 1
                        if (BlankAddress = &H2008) then State = 3
                 else
                        ' Display Messages
                        StopFlag = 0
                 endif
          case 3;       ' All Done
   ' Display Messages/Set Up Next Step
                 StopFlag = 0
   end select
   if (StopFlag <> 0) then Timer1.Enabled = true
end sub
```

Creating a timer-driven state machine will be a bit different than what you might be used to. But they aren't that hard to do. In the previous example, I could have started with the code:

```
retvalue =0
for BlankAddress = 0 to DeviceEnd
      if (ProgRead <> &H07FFE)
               ' Put in Error Message
               retvalue = -1
      endif
end for

if (retvalue = 0)
      ProgWrite(Load_CONFIG, 0x07FFE)

      for BlankAddress = &H2000 to &H2007
               if (ProgR <> &H07FFE)
                    ' Put in Error Message
                    retvalue = -1
               endif
      end for
endif

if (retvalue = 0)
       ' Display Messages/Put in Next Step
endif
```

If this code has encountered in the application, all the other processes in the PC would be stopped because the code is active and doesn't allow any other applications to execute.

For breaking up the interface code, I try to use the general rule that each operation should not execute for more than 1 ms before pausing or executing a timer wait. Following this rule, you will have a Windows application that is PC independent while only giving up a marginal amount of execution time that for most applications.

The last point about PC interfaces to remember is to give the user some feedback on what is happening. This could be a bar graph, a spinning wheel, or a text percentage display. As long as the user can see what is happening, they will not be that anxious about the time required. You can probably relate to this yourself if you've ever had a PC application that just sat there.

# Serial Ports

The PCs serial ports consist of basically the same hardware and BIOS interfaces that were first introduced with the first PC in 1981. Since that time, a nine-pin connector has been specified for the port (in the PC/AT) and one significant hardware upgrade was introduced when the PS/2 was announced. For the most part, the serial port has changed the least of any component in the PC for the past twenty years.

I feel that the serial ports are currently (this will change when USB PICmicro® MCUs become available) the best way to interface a PICmicro® MCU to the PC. This is because the serial port's timing is standard across virtually all PCs (something that isn't available in the parallel port) and signal current is very low, which minimizes the chances that a problem with the circuit connected to it will damage the PC or PICmicro® MCU.

Notice that I said that it *minimizes* the chances that the PC can be damaged. Very high voltage inputs can damage the PC's RS-232 interface circuitry or the PC itself. This can be avoided by only using proper RS-232 interfaces on your PICmicro® MCU application.

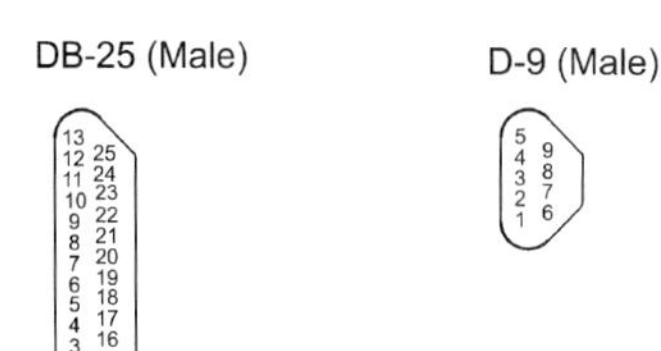


**Figure 7-2** IBM PC DB-25 and D-9 pin RS-232 connectors

Either a male 25-pin or male nine-pin connector is available on the back of the PC for each serial port (Fig. 7-2).

These connectors are wired as in Table 7-3.

The nine-pin standard was originally developed for the PC/AT because the serial port was put on the same adapter card as the printer port and there wasn't enough room for the serial port *and* parallel port to both use 25-pin D-Shell connectors. Actually, I prefer the smaller form-factor connector.

Up to four serial ports can be addressed by the PC. Of these, probably only two will be usable for connecting external devices to the PC. The serial port base addresses are shown in Table 7-4.

Each base address is used as an initial offset to eight registers that are used by the serial port controller (the 8250). The *interrupt number* is the interrupt vector requested when an interrupt condition is encountered.

When IBM was specifying port addresses for the PS/2 and developed the 8514/A graphics adapter (which has become known as *Super VGA*), the designers found that they didn't have enough available I/O port addresses left to support the functions that they wanted to put in. So, they used the addresses reserved for COM4, assuming that very few people connect four serial ports to their PC. This means that virtually all modern PCs cannot work with COM4 and display data at the same time.

**TABLE 7-3 PC RS-232 Pinout**

| PIN NAME | 25 PIN | 9 PIN | I/O DIRECTION OUTPUT ("O") INPUT ("I") |
|---|---|---|---|
| TxD | 2 | 3 | |
| RxD | 3 | 2 | |
| Gnd | 7 | 5 | |
| RTS | 4 | 7 | O |
| CTS | 5 | 8 | I |
| DTR | 20 | 4 | O |
| DSR | 6 | 6 | I |
| RI | 22 | 9 | I |
| DCD | 8 | 1 | I |

**TABLE 7-4 PC Serial Port Base Addresses**

| PORT | BASE ADDRESS | INTERRUPT NUMBER |
|---|---|---|
| COM1 | 0x03F8 | 0x00C |
| COM2 | 0x02F8 | 0x00B |
| COM3 | 0x03E8 | 0x00C |
| COM4 | 0x02E8 | 0x00B |

When I was interfacing a device to a PC with four serial ports (it was a bad assumption), I found that writing to the COM4 port addresses caused my PC's screen to go black. The application was changed to only work with three serial ports and an ISA adapter. To be on the safe side, don't attempt to put a serial port at COM4.

Chances are, you'll have a modem in your PC. As I've noted in the PC's architecture, although the PC has four sets of I/O addresses set aside for serial devices, for practical purposes, only two interrupt channels are available because your modem takes up one address. This is reduced to one if a serial mouse is used.

For serial (COM) ports 1 and 3, interrupt level 4 (which is actually interrupt 12, or 0x00C, in the PC's interrupt table) is used. COM ports 2 and 4 use level 3 (interrupt 11, or 0x00B). By setting your modem to COM2, you are leaving COM1 and COM3 free for interfacing to external devices without worrying about sharing interrupts with the software used by COM2. Although interrupts can be shared between applications and many try to work together, you should avoid sharing whenever possible.

The block diagram of the 8250 *Universal Asynchronous Receiver/Transmitter (UART)* is quite simple and if you were to design your own device, its block diagram would probably look like the 8250's.

In Fig. 7-3, I have shown the data paths for the serial communications. You might want to refer back to this diagram as I explain how the various registers work.

What is not shown in Fig. 7-3 are the interrupt registers and data paths. Interrupts are generated by changes in the status of the different hardware functions. Interrupts are covered in greater detail later in this chapter.

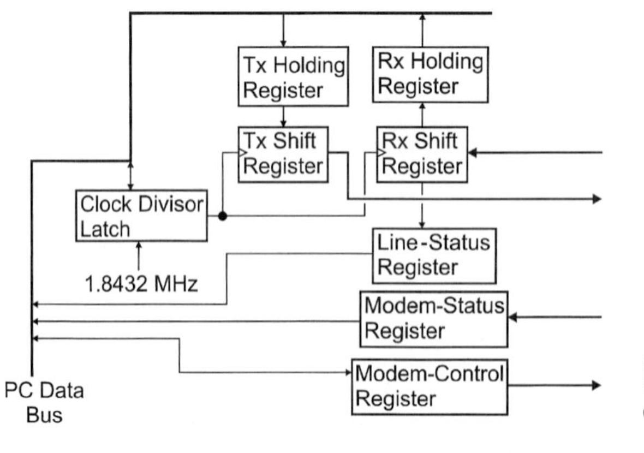


**Figure 7-3** 8250 block diagram

The 8250 serial interface chip is the primary chip used for serial communications. Along with the 8250, the 16450 and the 16550 with the PS/2 as 8250 replacements. These chips can execute the same basic instructions as the 8250, but with additional circuits put into the chips to allow buffering of data (using *First In/First Out, FIFO,* memories) coming in and out.

I'll focus in on the 8250's operation and bypass the FIFO-equipped chips for two reasons. The first is, the FIFO is really not needed, except for very high data speeds (57,600 bps and above). Most software programs written for the RS-232 ports don't attempt to use the buffering FIFO hardware.

This might be surprising, but the reason why most programs ignore the FIFOs is the same reason why I will: many of the FIFO-equipped chips simply don't work properly or according to specifications. It is much safer to use the standard 8250 functions that are guaranteed to work and are compatible for most PC devices.

There are methods to detect whether or not you have a working FIFO-equipped chip, but for virtually all PICmicro® MCU serial communication applications, the FIFO is not required simply because the PICmicro® MCU works best at speeds of 19,200 bps or slower.

The 8250 consists of eight registers offset from the base address (Table 7-5). Each of these registers controls or monitors hardware actions in the 8250. Before I go through each register and explain their operations, a few things must be explained.

The first is the operation of setting the data speed. This is done by setting a 16-bit divisor value into the RX/TX holding register and interrupt-enable register addresses.

To change the function of these two registers, bit 7 of the LINE-CONTROL register is set. The value loaded into the register is multiplied by 16 and divided into 1.8432 MHz to get the actual data rate.

As a formula, this is:

*Data rate* = 1.8432 MHz/(16 * Divisor)

**TABLE 7-5 PC Serial Port Register Offsets**

| BASE ADDRESS OFFSET | REGISTER NAME |
|---|---|
| 0 | Transmitter Holding Register/Receiver Character Buffer/LSB Divisor Latch |
| 1 | Interrupt Enable Register/MSB Divisor Latch |
| 2 | Interrupt Identification Register |
| 3 | Line Control Register |
| 4 | Modem Control Register |
| 5 | Line Status Register |
| 6 | Modem Status Register |
| 7 | Scratchpad Register |

**TABLE 7-6 PC Serial Port Speed Divisor Table**

| DATA RATE | DIVISOR |
|---|---|
| 110 bps | 0x0417 |
| 300 bps | 0x0180 |
| 600 bps | 0x00C0 |
| 1200 bps | 0x0060 |
| 2400 bps | 0x0030 |
| 9600 bps | 0x000C |
| 19200 bps | 0x0006 |
| 115200 bps | 0x0001 |

The divisors for different standard data rates are shown in Table 7-6. The divisors can either be loaded or by using the BIOS-interrupt interface.

When you have finished loading in the data rate, resetting bit 7 of the LINE-CONTROL register returns operation to its "Normal" mode.

After a character is received, it will set a number of conditions (including error conditions) that can only be reset by reading the character in the RECEIVE HOLDING register. If you are waiting for a character (but don't care what it is) and if you don't read it, you will end up with overrun errors as "live" data comes in because a new interrupt won't be generated for it.

For this reason, it's always a good idea to read the serial port at the start of an application. By reading the port, you are clearing out any status and leftover characters.

By reading a character from the serial port at the start of the application, you are also "claiming" the serial port resource for the application in Win32 operating systems if you are executing from the MS-DOS prompt. By reading the port, you are in no way affecting the serial port hardware.

When you write to the base address (with no offset added), you are loading a character into the TRANSMIT HOLDING register, which will be loaded as soon as the SHIFT OUT register has completed sending the previous character. Often, when starting transmission, nothing will be in the SHIFT register, so the character is loaded immediately into the SHIFT register, freeing up the HOLDING register for the next character.

This means that if the HOLDING register is kept full at all times (reloading when the value is moved into the SHIFT register), you can send a continuous stream of data, the start bit of one packet immediately following the stop bit of the packet before it.

By reading the base address, you are reading the contents of the SERIAL RECEIVER HOLDING register. If you're going to read the register, be sure that a new character is read by checking the LINE STATUS register for data ready.

Interrupts can be generated upon the completion of the current packet transmission and the contents of the HOLDING register stored into the SHIFT register can be shifted out. If the modem-status bits change state or if data has been received and is either in error or

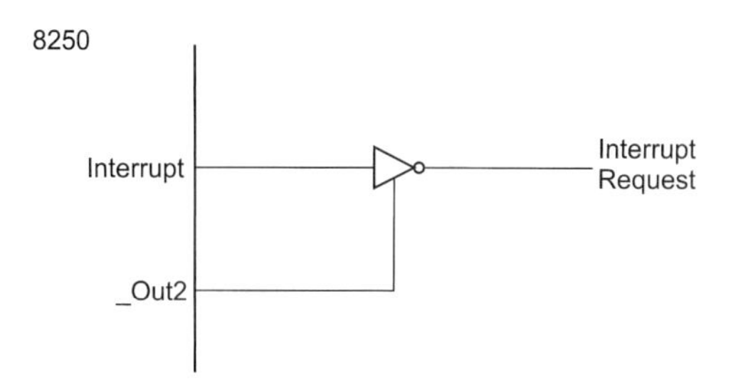


**Figure 7-4** IBM PC serial interrupt enable hardware

ready to be processed. If any of these events occur with the appropriate INTERRUPT-ENABLE register, which is located at the base address plus one, then an interrupt request will be generated by the 8250.

When any interrupts are enabled in the 8250, they will output an interrupt request. This might not be desirable, so, in the PC, some hardware was added to globally mask the interrupt (Fig. 7-4).

_Out2 is controlled within the MODEM CONTROL register. The INTERRUPT-ENABLE register (at the base address plus one) is used to enable the interrupt requests out of the 8250, based on different events (Table 7-7).

The modem input lines are DSR, CTS, RI, and DCD and can cause an interrupt request in the PC when they change state. The event that caused the interrupt request is stored in the MODEM STATUS register. Just as a note, in the 8250, the modem bits are also known as *handshaking lines.*

The receiver line-status bits will request an interrupt when something unusual is happening on the receiver line. These conditions include a data overrun, incorrect parity in the data packet received, a framing error in the received data packet, or if a break in the line is detected.

This interrupt request is given the highest priority so that errors can be responded to first.

The TRANSMIT HOLDING register empty interrupt indicates that the HOLDING register is able to accept another character and is shifting out the one that used to be in the HOLDING register.

It might seem surprising to use a HOLDING register rather than the actual SHIFT register when transmitting data. Using the TRANSMITTER HOLDING register is actually an

**TABLE 7-7 PC Serial Port Interrupt Enable Register**

| BIT | DESCRIPTION |
|---|---|
| 7-4 | Unused, normally set to zero. |
| 3 | When set an interrupt request on change of state for modem interface lines. |
| 2 | Request interrupt for change in receiver holding register status |
| 1 | Request interrupt if the holding register is empty |
| 0 | Request interrupt for received character |

advantage in applications where data has to be transmitted as quickly as possible with a minimum of overhead or "dead" air (in which nothing is being sent).

When sending data at full speed, the start bit of the next packet has to follow the stop bit of the previous packet. One option is to continuously poll the serial port to wait for the stop bit to be sent and load the SHIFT register with the next character to send when it is empty. You could also let the TRANSMIT HOLDING register load the SHIFT register when the SHIFT register has finished sending the previous packet and then reload the HOLDING register when it's convenient to the application (i.e., when an interrupt request is received).

The last event that can generate an interrupt request is when a character is received. This interrupt request is at the second highest priority (after the receive line status interrupt request) to ensure that the received characters and status are serviced as quickly as possible (and avoid framing errors).

When the serial port interrupt request is being serviced, the INTERRUPT-IDENTIFICATION register (base address plus two) is used to determine the cause of the interrupt (Table 7-8).

These bits correspond to the interrupt mask bits used in the INTERRUPT-ENABLE register and should be checked before responding to what you think is the interrupt. I put this caution in because, after reading from the received data register (at the base address), the contents of the INTERRUPT-IDENTIFICATION register are lost.

The LINE-CONTROL register (at base address plus three) is used to set the operating conditions of the data transmitter/receiver (Table 7-9).

Most modern serial communications use 8-N-1 (eight data bits, no parity, one stop bit) for data transmission. This is the simplest and most flexible (fewest overhead bits and can transfer a full byte) method to transfer data. To set the LINE-CONTROL register to 8-N-1, load it with 0b000000011 (or 0x003).

The MODEM-CONTROL register (at the base address plus four) is used to control the modem-control pins, as well as two other functions (Table 7-10).

The loop bit (or feature) is useful when debugging an application or seeing how the serial port works. By simply setting this bit, transmitted data is passed directly to the receiver without going outside the chip.

**TABLE 7-8 PC Serial Port Interrupt Identification Register**

| BIT | DESCRIPTION |
|---|---|
| 7-3 | Unused, Normally set to zero. |
| 2-1 | Interrupt ID Bits<br>B2 B1 Priority Request Type<br>0 0 Lowest Change in Modem Status Lines<br>0 1 Third Transmitter Holding Register Empty<br>1 0 Second Data Received<br>1 1 Highest Receive Line Status Change |

**TABLE 7-9 PC Serial Port Line Control Register**

| BIT | DESCRIPTION |
|---|---|
| 7 | When set, the Transmitter Holding and Interrupt Enable Registers are used for loading the data speed divisor |
| 6 | When set, the 8250 outputs a "Break Conditions (sending a space) until this bit is reset |
| 5-3 | Parity Type Specification<br>B5 B4 B3<br>0 0 0 - No Parity<br>0 0 1 - Odd Parity<br>0 1 0 - No Parity<br>0 1 1 - Even Parity<br>1 0 0 - No Parity<br>1 1 1 - "Space" Parity |
| 2 | When set, two stop bits are sent in the Packet, otherwise one |
| 1-0 | Number of Data Bits sent in a Packet<br>B1 B1<br>0 0 - 5 Bits<br>0 1 - 6 Bits<br>1 0 - 7 Bits<br>1 1 - 8 Bits |

**TABLE 7-10 PC Serial Port Modem Control Register**

| BIT | PIN | DESCRIPTION |
|---|---|---|
| 7-5 | | Unused, normally set to zero |
| 4 | Loop | When Set, Data from the transmitter is looped internally to the receiver |
| 3 | Out2 | When Set, Interrupt Requests from the 8250 are unmasked |
| 2 | Out1 | This bit/pin is not controlling any hardware features in the serial port |
| 1 | _RTS | When this bit is Reset, the RTS line is at "Mark" State |
| 0 | _DTR | When this bit is Reset, the DTR line is at "Mark State" |

The LINE STATUS register (base address plus five) is a read-only register with the current status of the 8250 (Table 7-11).

When you are debugging a serial port application, always remember that the contents of the LINE STATUS register should normally be 0x060 or 0x061. *0x060* indicates that no character is being sent and no character in the TRANSMITTER HOLDING register waiting to be sent. *0x061* indicates that no character is being sent, no character in the

**TABLE 7-11 PC Serial Port Line Status Register**

| BIT | DESCRIPTION |
|---|---|
| 7 | Unused, Normally set to zero |
| 6 | Set when the transmitter shift register is empty |
| 5 | Set when the transmitter holding register is empty |
| 4 | Set when the receive line is held at a space value for longer than the current packet size |
| 3 | This bit is Set when the last character had a framing error (ie stop bit set to "Space") |
| 2 | Set when the last character had a parity error |
| 1 | Set when the latest character has overrun the receiver holding register |
| 0 | Set when a character has been received but not read |

TRANSMITTER HOLDING register waiting to be sent and a character has been received, but not read in. If you see any other value in this register, then an error has occurred with transmitting or receiving the data.

The MODEM-STATUS register (at the base address plus six) is a read-only register dedicated to returning the current status of the modem (DCE) device connected to the PC's serial port (Table 7-12).

**TABLE 7-12 PC Serial Modem Status Register**

| BIT | PIN | DESCRIPTION |
|---|---|---|
| 7 | DCD | When Set, an asserted DCD signal is being received |
| 6 | RI | When Set, the modem is detecting a ring on the device it is connected to |
| 5 | DSR | When Set, a DSR "Mark" is being received |
| 4 | CTS | When Set, a CTS "Mark" is being received |
| 3 | DCD | When this bit is set, the DCD line has changed state since the last check |
| 2 | RI | When set, this bit indicates that the Ring Indicator line has changed from a Mark to a Space |
| 1 | DSR | When this bit is set, the DSR line has changed state since the last check |
| 0 | CTS | When this bit is set, the CTS line has changed state since the last check |

When this register is read, the least-significant four (known as the *delta bits*) are reset. Therefore, two reads of this register, one immediately after the other, could result in different values being returned.

The delta bits are used for the modem status change interrupt request. If interrupts are not used in your application, these bits can be polled for changes in the modem-control lines.

At base address plus seven, in some serial hardware, is the SCRATCHPAD register, which you can read from and write to without affecting any other hardware. This register should not be counted upon being present in all serial interfaces.

Like the SCRATCHPAD register, the *First In/First Out (FIFO)* transmitter and receiver buffers might not be present in the serial interface (or worse, they might not work correctly). As well, very few applications really require this hardware because it really isn't very hard for the PC's processor to keep up with most incoming data speeds. The register definitions present the "lowest common denominator," the 8250.

The only application where I could see the FIFO being required is if the PC was communicating with more than one other device continuously at high speed. If this were the case, instead of using the standard serial ports, I would recommend using some of the available multiple serial port adapters with built-in buffers to reduce the processor's workload.

Like the other hardware devices in the PC, the serial ports have a BIOS interrupt (Int 0x014) for generic operations. These BIOS functions allow you to set up the interface, as well as send and receive characters. When I am creating RS-232 applications, I very rarely use the BIOS functions (or when I do, I just use the port setup) because the BIOS values are quite limited in their capabilities and force you to work in a specific manner.

The most useful serial BIOS requests are port initialize (AH = 0) and extended port initialize (AH = 4). Extended port initialize is available in PS/2 and later PCs (which really means all PCs built after 1987). These functions will set up the serial port with regard to speed. The following data speeds are supported:

- 110 bps
- 150 bps
- 300 bps
- 600 bps
- 1,200 bps
- 2,400 bps
- 4,800 bps
- 9,600 bps
- 19,200 bps

as well as parity type (none, odd, even, mark, and space), number of stop bits (1 or 2), and the number of data bits (7 or 8). The extended port initialize will also allow you to transmit a break character (which is a space longer than a data packet). Typically, data communications works at 8-N-1 (eight data bits, no parity, and one stop bit), which can be written into the LINE-CONTROL register directly. But, you might want to use the BIOS port-initialize functions to set up the speed of the port without having to set the divisor manually.

Actually, the only reason why I ever use the port initialize function is to set the data speed (although if I'm creating an application that runs faster than 19,200 bps, then I will have to write to the divisor registers manually).

Causing an interrupt on the TX HOLDING register empty or on a character received is quite easy. Now, you can also have an interrupt on the changing modem-status lines or on a change of state in the received line (such as an Overrun or Break condition), but for three-wire operations with interrupts being used, I really haven't found a need for this.

To enable interrupts for COM1/COM3 (at Interrupt 0x00C), the following code can be used:

```
SetInt( 0x0C, SerIntHndlr );              // Point the Interrupt Handler to
                                          //   the Correct Handler
Dummy = inp( RxHoldingRegister );         // Turn Off any Pending Interrupts

outp( IntMaskRegister, inp( IntMaskRegister ) & 0x0FB );
                                          // Enable COM1/COM3
                                          // Interrupts in Controller
outp( InterruptEnableRegister, 0x003 );
                                          // Request Interrupts on TxHolding
                                          // Register Empty and RxHolding
                                          // Register Full
outp( ModemControlRegister, inp( ModemControlRegister ) | Out2) ;
                                          // Unmask Interrupt Requests from
                                          // 8250
```

Notice that before I enable any interrupts, I read from the serial port to clear out anything that is pending. This is very important to do because it both "claims" the serial port in the operating system and clears out any inadvertent problems that you can't expect (such as an earlier executing program leaving the serial port in an unknown state).

The best way to "claim" a port is to use an operating system API. This is beyond the scope of this book, but I recommend using system API's and the Registry when accessing the serial ports in a Microsoft Windows PC.

Once an interrupt request is made by the hardware, control is passed to the service routine:

```
SerIntHndlr:                    // Serial Interrupt Handler

// #### - Assume that the Interrupting COM port is identified

  switch ( InterruptIDRegister ) {     // Handle the Interrupt Request
    case 4:                     // Received Character
      InString[i++] = RxHoldingRegister;
      break;
    case 2:                     // TxHolding Register Empty
      TxHoldingRegister = OutString [ j++ ]; // Send the Next
                                // Character
      break;
    default:                    // Some other kind of Interrupt

      Dummy = RxHoldingRegister; // Clear the Receiving Data
  } // endswitch

InterruptControlRegister = EOI; // Reset the Interrupt Controller

returnFromInterrupt;            // Return from the Interrupt.
```

This code will pass data to and from the strings, incrementing their pointers each time. In a real application, the interrupt handler would check that it was executing on the correct port (this code assumes that only one port can cause an interrupt request). The application would be notified when data was available (or, if a table was being saved. If an end character, such as a carriage return, was encountered, then the handler would notify the mainline).

When implementing an RS-232 interface, you can make your life easier by doing a few simple things.

The first step involves the connection. Whenever I do an application, I standardize it by using a nine-pin D-shell with the DTE interface (the one that comes out of the PC) and use standard cables wherever possible. In doing this, I can always identify the pinout at the end of the cable when I'm about to hook up another device to a PC.

When I am creating the external device, I also loop back the DTR/DSR and CTS/RTS data pairs inside the external device, rather than at the PC or in the cable. This allows me to use a standard PC and cable without having to do any wiring or make any modifications. It actually looks a lot more professional as well.

Tying DTR/DSR and CTS/RTS also means that I can take advantage of built-in terminal emulators. Virtually all operating systems have a built-in dumb terminal emulator that can be used to debug the external device without requiring the PC code to run. Getting the external device working before debugging the PC application code should simplify your work.

The last point is, although I can develop applications that run up to 115,200 bps (by writing a *1* into the two data-divisor registers), I typically run at 9,600 bps or less. By keeping the data rates reasonable, then I can run the applications to reasonable lengths (up to about 1,000 feet with shielded cabling) without requiring special protocols because of bit errors.

Using a modem and the local phone lines instead of cabling between the PC and device yourself (which will probably be cheaper and easier for you) can accommodate longer cable lengths. I have not covered hooking a modem to your PC, but this information is typically found in a modem's documentation. If you are running with a Hayes-compatible or AT command set modem, you can send the AT commands directly to the modem as character strings.

One of the biggest problem causes for applications that use RS-232 and are difficult to get running is how RS-232 is used. In these situations, the PC's RS-232 port is modified to work in a manner in which it was not originally designed, or some feature has been exploited to provide a specific function. Often, these types of applications do not work well, or if they work on one type of PC, they won't work on others. I feel that these types of applications are not appropriate for the serial port; other interfaces should be considered first.

A PICmicro® MCU application that is appropriate for PC serial communications has the following characteristics:

1. A standard PC serial port is to be used.
2. Only two computing devices are connected together.
3. These two devices can be an arbitrary distance apart (from inches to miles to astronomical distances).
4. Relatively small amounts of data need to be transferred in real time (on the order of hundreds of kilobytes per hour).

**5** Man-readable commands are transmitted between the devices and data is transmitted using standard protocols.

If an application does not fit all of these criteria, then you should probably be looking for another method of communicating between the PC and the PICmicro® MCU.

## HYPERTERMINAL

Depending on when you bought your PC and the operation system that is running on it, you might not have Higraeve's HyperTerminal ASCII terminal emulator loaded on to your PC under the "Communications" pull-down menu. This section introduces the application and shows you the different features of it as an example of what you should be looking for in a terminal emulator to which you will interface a PICmicro® MCU. As I said in the previous section, I feel that the serial port is the best way to interface to a PICmicro® MCU. As part of the interface, I recommended that the communications are to be as "human readable" as possible. This allows the PICmicro® MCU application to be tested and debugged separate from the PC application to eliminate as many variables as possible.

In demonstrating PC terminal emulators, I focus on Hilgraeve, Inc's *HyperTerminal.* This program is bundled with many copies of Microsoft Windows 95/98/NT/2000 and provides all of the basic services required in a terminal emulator. If you do not have a copy of HyperTerminal or want to upgrade to the latest level (as I write this, version 5.0 is available). You can download and install HyperTerminal for free for personal use from the Hilgraeve website: http://www.hilgraeve.com/.

Once HyperTerminal is installed, you will have to configure it. To do this, start up HyperTerminal and you should get a dialog box that pops up (Fig. 7-5).

To configure the terminal emulator, first disconnect it (the program connects itself automatically) by clicking on the telephone with the receiver off it or, click on Call -> Disconnect. Now select File -> Properties to display the Properties dialog box (which looks

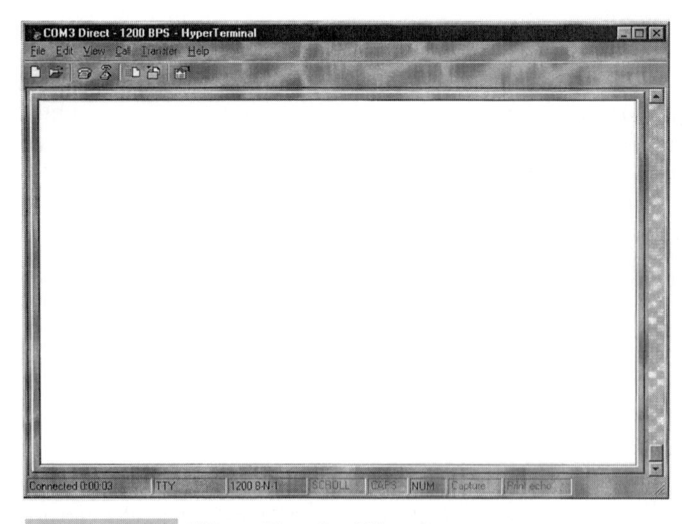


**Figure 7-5** "HyperTerminal" boot screen

like Fig. 7-6). This dialog box specifies how you would like HyperTerminal to work for you.

Click on the Settings tab and be sure that Terminal Keys, ANSI, Emulator, and 500 backspace buffer is selected (Fig. 7-7).

The Terminal Setup menu will allow you to define the cursor settings used in the HyperTerminal dialog box.

ASCII setup will allow you to tailor the data to the application. These parameters are really user and application specific. For the most part, I leave these at the default values.

With the Settings defined, you can click on the Phone Number tab and select the appropriate direct Connect Using using the tab (Fig. 7-8).

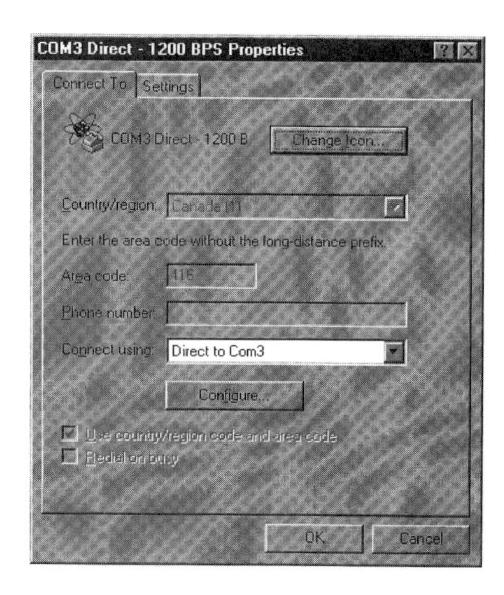


**Figure 7-6** "HyperTerminal" properties select

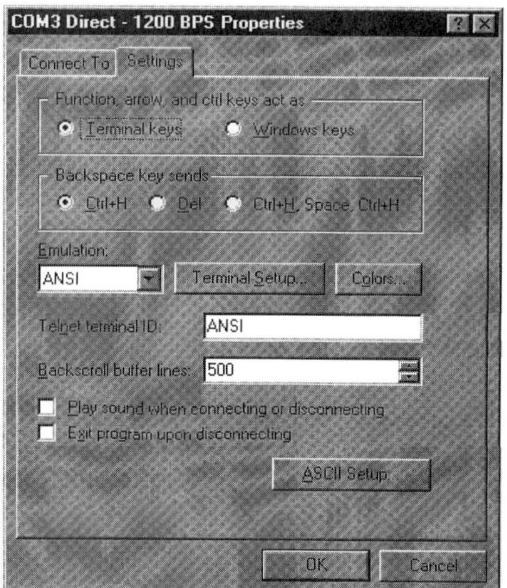


**Figure 7-7** "HyperTerminal" setting select

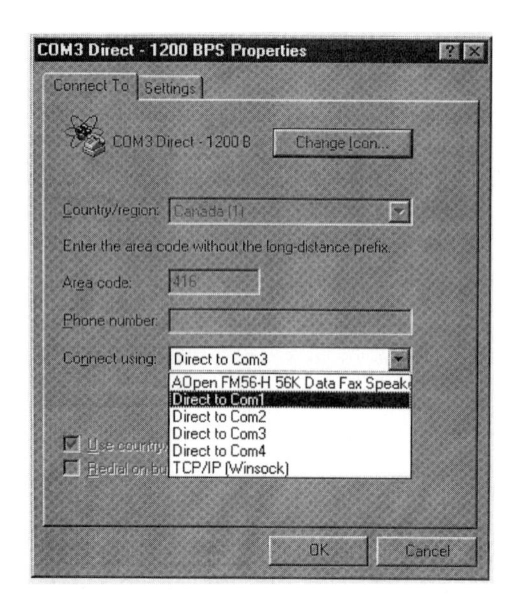


**Figure 7-8** "HyperTerminal" connection select

The serial port can be any serial port within the PC to which you want to connect. The modem that comes up is, not surprisingly, your Internet access server, which can be used for your terminal emulator if you are going to be dialing outside.

Next, click on Configure and look at the parameters to set up. For the PICmicro® MCU applications that connect serially in this book, select Eight bits, None for parity, 1 stop bit, and None for flow control. The data rate (bits per second) is the value you want to use (usually *1200 bps* for the projects presented in this book). Selecting None for flow control means a three-wire RS-232 can be used (Fig. 7-9).

Don't worry about Advanced; it is normally used to specify the serial port's FIFO data buffers. To be on the safe side, select No FIFO Operation.

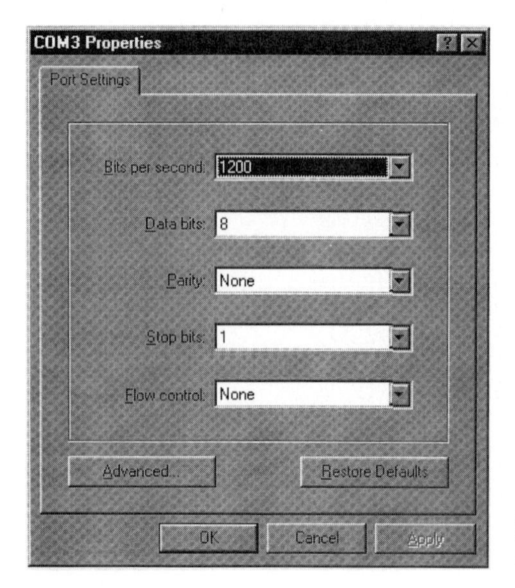


**Figure 7-9** "HyperTerminal" connection select

You can now click on OK for both the Com Properties and the Properties dialog boxes to save the information.

Next, click on View → Font and select a Monospace font like Courier New (Fig. 7-10).

After you click on OK to select the font you want, click on File → Save As. . . so that you cannot lose these parameters. The default directory should be Desktop. For the file name, enter something descriptive like:

```
DIRECT COM 1 – 1200 bps
```

and click on Save. If you minimize all the active windows now, you'll see that a new tab is on your desktop that will bring up the HyperTerminal with the parameters that you just entered.

When HyperTerminal has the desktop's focus and is connected, when you press a key, the ASCII code for it will be transmitted out of the PC's serial port. Data coming in on the serial port will be directed to HyperTerminal's display window.

Files can also be transferred; this is an excellent way to create test cases, instead of repeatedly typing them in manually. Doing a Send File will send the file exactly as it is saved on disk. Send Text File is preferable because it transfers the file's ASCII data in the same format as if it were typed in, including the carriage return/line feed line-end delimiter.

Elsewhere, the book gives some more practical information on using HyperTerminal with different applications. But what I wanted to leave you with is a list of the features I think that are needed for a terminal emulator that connects to a PICmicro® MCU application:

1 TTY and ANSI terminal emulation.
2 Varying data rates.
3 8-N-1 data format.

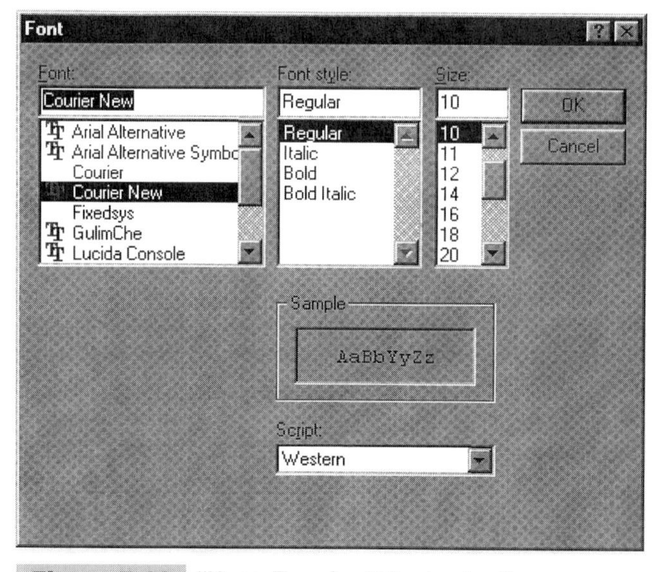


**Figure 7-10** "HyperTerminal" font selection

4 Monospace fonts.
5 User-selectable com port access.
6 User-selectable handshaking.
7 Configuration save.
8 Text file transfer.

This is a pretty basic list of features, but if your terminal emulator provides these capabilities, you will be able to use it to interface to the PICmicro® MCU applications presented here, as well as any that you create on your own.

## VISUAL BASIC MSCOMM SERIAL CONTROL

I really like Microsoft's Visual Basic as a quick-and-dirty Microsoft Windows application-development tool. I find that I can create Visual Basic applications very quickly and can update them as the application gets more complex. It is also an excellent tool for experimenting with (although my "experiments" usually turn into applications). If you are willing to prowl around and look through what kind of extra controls are available, you will find the MSComm serial communications control. This control allows you direct access to the serial ports within the PC and allows you to interface directly with the serial port hardware without having to load in device drivers.

MSComm is an excellent way to provide an RS-232 interface from your PC to your PICmicro® MCU applications. Projects in this book shows how Visual Basic, with MSComm, can be used to interface with PICmicro® MCU applications very quickly. I used Visual Basic with MSComm for my YAPII programmer user interface.

The MSComm control itself is very easy to use; the biggest problem is trying to figure out how to enable it. When you first load up Visual Basic, you are given a basic number of controls in the Toolbox down the left side of the development screen. These controls are the basic ones needed to execute most initial (beginner's) Visual Basic applications. The basic controls can be expanded with not only the MSComm serial port controls; but also Microsoft file objects, ActiveX, and OLE controls; Kodak Image and Macrovision Shockwave controls, and a lot of other controls and objects that you can use.

To add the MSComm serial port control to the available selection, you can click on Project, followed by Component, and then Apply Microsoft Comm Control (Fig. 7-11).

With the control added to the toolbox, you can now use MSComm with your applications. The YAP programmer is a fairly complex programmer that was designed to interface only with PC and workstation serial ports to program PICmicro® MCUs. When I originally designed the YAP, I designed it for use with a generic terminal emulator. By using MSComm, I was able to come up with a reasonably attractive Windows front end that runs quickly and easily for your applications. With the programmer working, I wanted to provide a simple Windows front end to demonstrate and simplify how the programmer works. The initial dialog box I came up with is shown in Fig. 7-12.

To work with the MSComm control, after loading the control onto the toolbox, I placed MSComm's Telephone icon on the dialog box, similarly as I would with the timer. When the application is executing, the icon is invisible to the user.

To initialize the MSComm control, I used the recommended sequence that consists of:

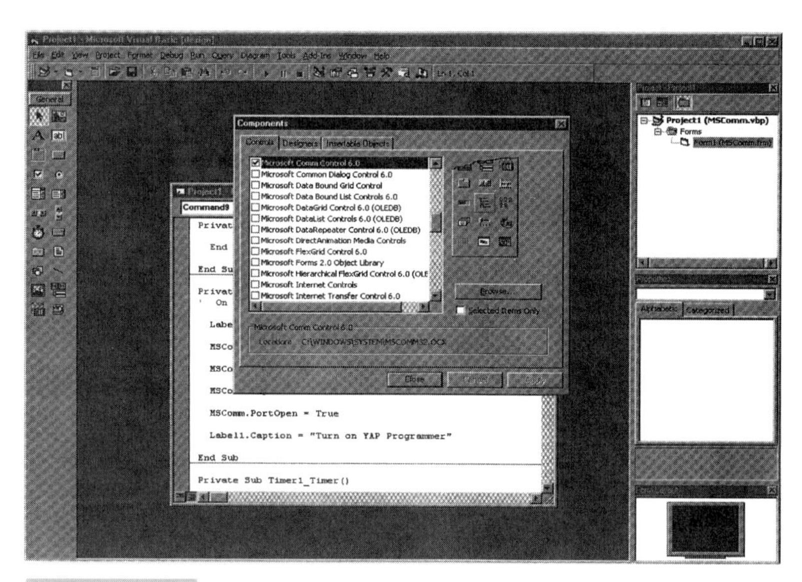


Figure 7-11 Setting up the "MSComm" control

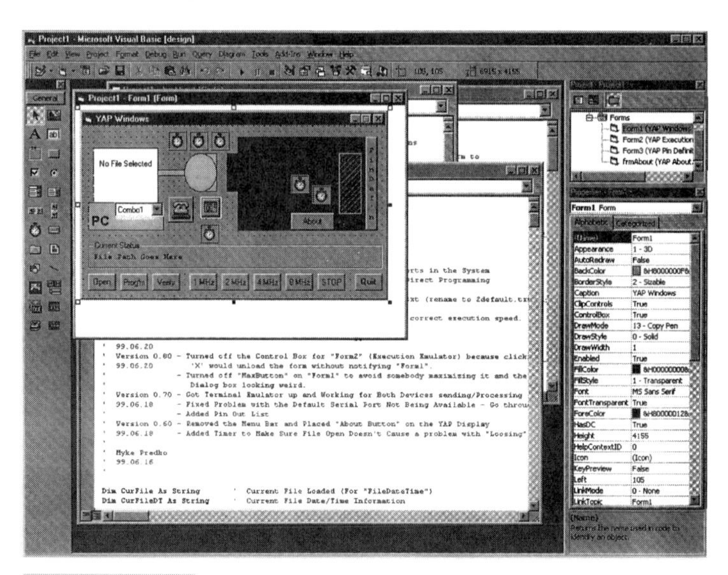


Figure 7-12 "YAP" "MSComm" control

**1** Specify the hardware serial port to be used.
**2** Set the speed and data format to be used.
**3** Define the buffer size.
**4** Open the port and begin to use it.

The code used to perform these functions is placed in the Form_Load subroutine, which means that the port is enabled before the primary dialog box is executing. Although the control values should be self-explanatory, I will go into greater detail about them later in the chapter.

```
Private Sub Form_Load()
' On Form Load, Setup Serial Port 3 for YAP Programmer
  MSComm3.CommPort = 3
  MSComm3.Settings = "1200,N,8,1"
  MSComm3.InputLen = 0
  MSComm3.PortOpen = True
  Text1.Text = "Turn on YAP Programmer"
End Sub
```

With the port initialized and executing, I use a 50-ms timer to continually poll the serial port and display data in the Text box when it is received.

```
Private Sub Timer1_Timer()
' Interrupt every 50 msecs and Read in the Buffer
Dim InputString

  InputString = MSComm3.Input

  If (InputString <> "") Then
    If (Text1.Text = "Turn on YAP Programmer") Then
    Text1.Text = "" ' Clear the Display Buffer
  End If
  Text1.Text = Text1.Text + InputString
End If

End Sub
```

This application code first prompts the user to turn on the programmer and if it is when data is received, the text display is cleared and data is placed in sequence on the display.

For specialized operations (such as selecting flash versus EPROM control store types), I used CommandButton controls, which send data to the YAP via the serial port using the code:

```
Private Sub Command1_Click()
' Put the Programmer into "Flash" Mode
  Text1.Text = ""
  MSComm3.Output = "f" + Chr$(13)
```

With these controls, the YAP can be controlled using the buttons and the mouse with the dialog much more quickly and efficiently (i.e., little chance for error) than if the commands were entered manually by the user. One nice feature of this application is the text box that is continually updated by the timer interrupt routine, showing what is actually happening with the YAP and allowing the user to debug problems very quickly.

Once the MSComm control is placed on the display, the properties listed in Table 7-13 are used to control it.

**TABLE 7-13 Visual Basic MSComm Control Summary**

| PROPERTY | SETTING | DESCRIPTION |
|---|---|---|
| Break | True/False | When set to "True", Break Sends a "0" break signal until the property is changed to "False". |
| CDHolding | True/False | Read only property that indicates if the "Carrier Detect" line is active. This is an important line to poll in applications which use modems. |
| CommEvent | Integer | Read only property that is only available while the application is running. If the application is running without any problems, this property returns zero. This property is read by the "OnComm" event handler code to process the reason why the "event" was caused. |
| CommID | Object | Read only property that returns an identifier for the serial port assigned to the MSComm control. |
| CommPort | Integer | Specify the "COMx" (1–3) serial port that is used by the MSComm control. |
| CTSHolding | True/False | Read only property that returns the current state of the serial port's "Clear To Send" line. |
| DSRHolding | True/False | Read only property that returns the current state of the serial port's "Data Set Ready" line. |
| DTREnable | True/False | Property used to specify the state of the "Data Terminal Ready" line. |
| EOFEnable | True/False | Specify whether or not an "OnComm" event will be generated if an "End Of File" character (0x01A) is encountered. |
| Handshaking | 0, 1, 2, or 3 | Sets the current handshaking protocol for the serial port:<br>0—No handshaking (default)<br>1—XON/XOFF Handshaking<br>2—RTS/CTS (Hardware) Handshaking<br>3—Both XON/XOFF and RTS/CTS Handshaking |
| InBufferCount | Integer | Read only property indicating how many characters have been received by the serial port. |
| InBufferSize | Integer | Property used to specify the number of bytes available for the Input Data Buffer. The default size is 1024 bytes. |
| Input | String | Return a String of Characters from the Input Buffer. |

(continued)

**TABLE 7-13 Visual Basic MSComm Control Summary *(Continued)***

| PROPERTY | SETTING | DESCRIPTION |
|---|---|---|
| InputMode | Integer | Specify how data is to be retrieved using the "Input" property. Zero specifies data will be received as Text (Default). One will specify that data will be passed without editing ("Binary" format). |
| InputLen | Integer | Sets the Maximum Number of characters that will be returned when the "Input" property is accessed. Setting this value to zero will return the entire buffer. |
| NullDiscard | True/False | Specify whether or not Null Characters are transferred from the port to the receiver buffer. |
| OutBufferCount | Integer | Read only property that returns the Number of Characters waiting in the Output Buffer. |
| OutBufferSize | Integer | Specify the size of the Output Buffer. The default is 512 Bytes. |
| Output | Integer | Output a string of characters through the serial port. |
| ParityReplace | Integer | Specify the character that will replace characters which have a "Parity" Error. The default character is "?" and the ASCII code for the replacement character must be specified. |
| PortOpen | True/False | Specify whether or not the data port is to be transmitting and receiving data. Normally a port is closed ("False"). |
| Rthreshold | Integer | Specify the number of characters before there is an "OnComm" event. The default value of zero disables event generation. Setting the "Rthreshold" to one will cause an "OnComm" event each time a character is received. |
| RTSEnable | True/False | Specify the value output on the "Request To Send" line. |
| Settings | String | Send a String to the Serial Port to specify its operating characteristics. The String is in the format "Speed, Parity, Length, Stop" with the following valid parameter values:<br>Speed: Data Rate of the Communication<br>110<br>300<br>600<br>1200<br>2400 |

| TABLE 7-13 Visual Basic MSComm Control Summary *(Continued)* | | |
|---|---|---|
| PROPERTY | SETTING | DESCRIPTION |
| | | 9600 (Default)<br>14400<br>19200<br>28800<br>38400<br>56000<br>128000<br>256000<br>Parity: The type of error checking sent with the byte<br>E—Even Parity<br>M—Mark Parity<br>N—No Parity (Default)<br>O—Odd Parity<br>S—Space Parity<br>Length: The number of bits transmitted at a time<br>4—4 Bits<br>5—5 Bits<br>6—6 Bits<br>7—7 Bits<br>8—8 Bits (Default)<br>Stop: The number of stop bits transmitted with the byte<br>1—1 Stop Bit (Default)<br>1.5—1.5 Stop Bits<br>2—2 Stop Bits |
| Sthreshold | Integer | Specify the number of bytes to be transmitted before an "OnComm" event is generated. The default is zero (which means no "OnComm" event is generated for transmission). Setting this value to one will cause an "OnComm" event after each character is transmitted. |

These properties are quite simple to use and will allow you to quickly develop and debug serial Visual Basic applications.

Further enhancing the usefulness of the MSComm control is the OnComm event. This routine is similar to an interrupt, as it is requested after specified events in the serial port. The CommEvent property contains the reason code for the event (Table 7-14).

These values can be processed in the OnComm event handler like:

```
Private Sub Object_OnComm()
' Handle Serial Port Events

   Select Case Object.CommEvent
     Case comEventBreak ' Handle a "Break" Received
        Beep
```

**TABLE 7-14 Visual Basic MSComm "OnComm" Event Summary**

| COMMEVENT IDENTIFIER | COMMEVENT CODE | DESCRIPTION |
|---|---|---|
| comEvSend | 1 | Specified Number of Characters Sent |
| comEvReceive | 2 | Specified Number of Characters Received |
| comEvCTS | 3 | Change in the "Clear To Send" line |
| comEvDSR | 4 | Change in the "Data Set Ready" line |
| comEvCD | 5 | Change in the "Carrier Detect" line |
| comEvRing | 6 | Ring Detect is Active |
| comEvEOF | 7 | "End Of File" Character Detected |
| comEventBreak | 1001 | Break Signal Received |
| comEventFrame | 1004 | Framing Error in incoming data |
| comEventOverrun | 1006 | Receive Port Overrun |
| comEventRxOver | 1008 | Receive Buffer Overflow |
| comEventRxParity | 1009 | Parity Error in Received Data |
| comEventTxFull | 1010 | Transmit Buffer Full |
| comEventDCB | 1011 | Unexpected Device Control Block Error |

```
    : ' Handle other events
  End Select
End Sub
```

To identify the serial port object, I have italicized the word *Object* in the OnComm event handler to make the label used for the serial port more noticeable.

For basic OnComm events to be responded to, the OnComm handler code has to be added to the application. If the handler is not present and the error or event takes place, then it will be simply ignored by Visual Basic.

# Parallel Port

Several years before the IBM PC first became available, the Centronics Corporation built and sold printers using a simple parallel bus interface. This bus was used to pass data from a computer to a printer and poll the printer status, waiting until additional characters could be sent. As part of the format, a special connector was also used.

This connector format became very popular, was adopted by a number of printer manufacturers, and quickly became an industry standard. The Centronics printer port's advantages were that its hardware could be replicated by using a few simple components, it was relatively fast compared to RS-232 ports, and software could be easily written for it.

Today, the parallel port is the first device most people look to when simple I/O expansion must be implemented in the PC. I consider this unfortunate because this port is actually poorly designed for the purpose. If you are looking for efficient digital input/digital output in the PC, ISA or USB busses should be considered first.

The parallel port itself is very simple; the design used in the PC/AT consists of just seven TTL chips and provides a simple, byte-wide parallel bi-directional interface into the PC. Over the last 20 years, as PCs have gotten more complex, so have their printers. When the PC was first introduced, the standard printer was a relabeled Epson (with an IBM logo) dot-matrix Centronics-compatible graphics printer, which used the parallel port's data and handshaking lines to control the data transfer, a byte at a time from the PC.

The early printer interfaces in the PC, after sending a byte, would wait for a handshaking line to indicate that the printer was ready for the next character before sending the next one. As you could imagine, this method was very slow and took up all of the PC's cycles in printing a file. As printers have improved, data buffers have been built into them to allow faster data transfers, as well as byte-wide checking of data information. Specialized devices, such as scanners, have been added to the PC's parallel port because of the reasonably high bandwidth that can be obtained using this port. This method of passing data to a printer works reasonably well, but can be inefficient for large volumes of data. It can be difficult to create drivers that work under Windows to share the printer port with another device.

To model the parallel port, I usually go right to the base circuit shown in Fig. 7-13. This diagram shows the parallel-port connector pin out, along with the registers involved with passing data and the appropriate bits for the different functions. The CONTROL register is used to enable the data output latch drivers and enable interrupt requests from the parallel-port hardware.

In Fig. 7-13, it is assumed that the parallel port is a 25-pin DB-25 male connector. The true Centronics printer connector is a 36-pin shell, but this shell is connected to the PC's DB-25M parallel port connector via a female connector and several feet of cable. When developing hardware that interfaces to the PC, I normally use straight-through DB-25M to

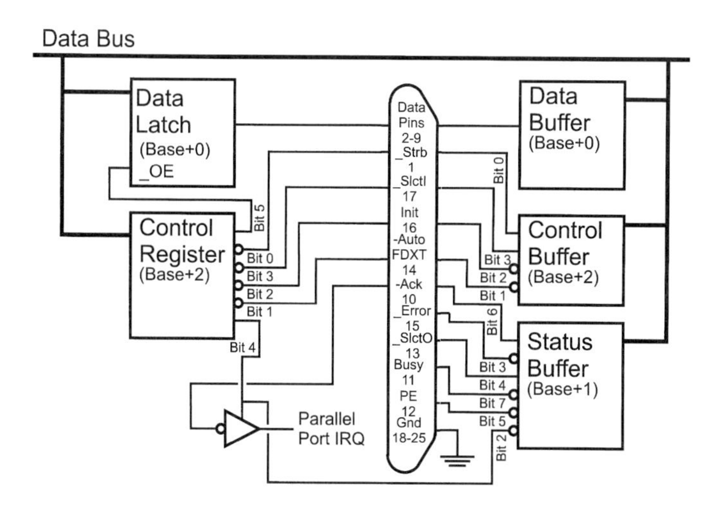


**Figure 7-13** Parallel port block diagram

DB-25F (Female) cable or a "printer extension" cable (which is a DB-25F to DB-25F). The straight-through cable is normally used as a straight-through serial cable and extension to the Centronics connector cables. The printer extension cable is used for connecting PCs to a selector box to inexpensively share PCs. The advantage of using these types of cables is that the output can be brought from your PC to your bench and not be translated in any way. This is an advantage in applications where the hardware interface will be connected directly to the parallel port on the PC.

If you look at the PC/AT technical reference manual, you will see that the parallel port is designed with 74LS TTL logic, which is capable of driving 20 mA (or greater) loads. It is a very dangerous assumption to make that all PCs have this kind of parallel-port drive capability. Most modern PCs have the parallel port function embedded in the SuperIO chip. This chip is an ASIC, which has most of the user I/O functions, as well as timers and the interrupt controllers for the PC. In this case, the parallel port, at best, will only be able to source a couple of mA or so and sink 20 mA. The port bits are designed to behave like the 8051 I/O pins discussed elsewhere in this book. To be on the safe side, I recommend that interfaces requiring no more than 1 mA to be sourced by a parallel-port pin. The registers that provide the functions are shown in Table 7-15.

The base addresses are 0x0378 and 0x0278 for LPT1 and LPT2, respectively. Hardware interrupt requests can be initiated from the hardware, interrupt 0x00F for LPT1, and 0x00D from LPT2. Although some standards provide for an LPT3 and LPT4, these are somewhat loose and might not be possible to implement in some PCs, adapter types, and configurations.

When the parallel port passes data to a printer, the I/O pins create the basic waveform shown in Fig. 7-14.

It is important to notice that the Printer BIOS routines will not send a new character to the printer until it is no longer busy. When Busy is no longer active, the Ack line is pulsed active, which can be used to request an interrupt to indicate to data output code that the printer is ready to accept another character.

The timing of the circuit for printer applications is quite simple, with 0.5-$\mu$s minimum delays needed between edges on the waveforms in Fig. 7-14.

When interfacing to the parallel port, because the different port pins are seemingly inverted at random, I use a set of functions that I created a number of years ago to eliminate the confusion. These routines change all the input and output lines to being positively active (to simplify trying to figure out what is happening).

```
PPortOut( int BaseAddr )        // Enable Data Bit Drivers
{
 outp( BaseAddr + 2, inp( BaseAddr + 2 ) & 0x0DF );
```

**TABLE 7-15 PC Parallel Port Register Addresses**

| REGISTER | ADDRESS | FUNCTION |
|---|---|---|
| Data | Base + 0 | Pass 8 Bits of Data to and From the PC |
| Control | Base + 2 | Pass Control Signals to External device |
| Status | Base + 1 | Return the Printer Status |

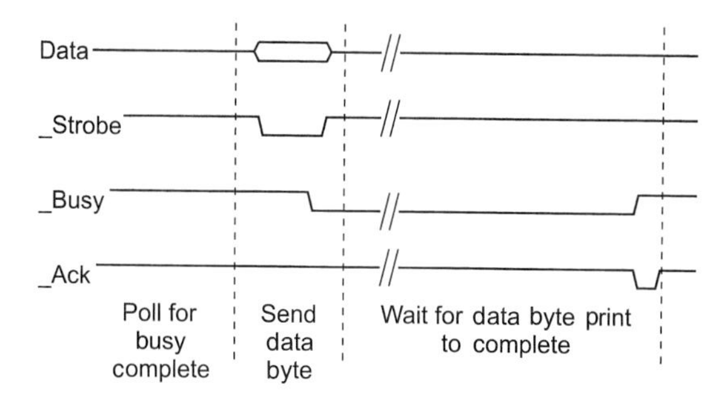


**Figure 7-14** **Parallel port printer byte write waveform**

```
} // End PPortOut

PPortIn( int BaseAddr )        // Disable Data Bit Drivers
{

 outp( BaseAddr + 2, inp( BaseAddr + 2 ) | 0x020 );

} // End PPortIn

PPortIRQEn( int BaseAddr )     // Enable the Parallel Ports Interrupt
{                              //   Requesting Hardware

 outp( BaseAddr + 2, inp( BaseAddr + 2 ) | 0x010 );

} // End PPortIRQEn

PPortIRQDis( int BaseAddr )    // Disable the Parallel Ports Interrupt
{                              //   Requesting Hardware

 outp( BaseAddr + 2, inp( BaseAddr + 2 ) & 0x0EF );

} // End PPortIRQDis

PPortHiSLCT( Int BaseAddr )    // Set "SLCT In" (Pin 17) to an
{                              //   Electrical "High"

 outp( BaseAddr + 2, inp( BaseAddr + 2 ) | 0x008 );

} // End PPortHiSLCT

PPortLoSLCT( Int BaseAddr )    // Set "SLCT In" (Pin 17) to an
{                              //   Electrical "low"

 outp( BaseAddr + 2, inp( BaseAddr + 2 ) & 0x0F7 );

} // End PPortLoSLCT

PPortHiInit( Int BaseAddr )    // Set "Init" (Pin 16) to an
{                              //   Electrical "High"

 outp( BaseAddr + 2, inp( BaseAddr + 2 ) & 0x0FB );

} // End PPortHiInit

PPortLoInit( Int BaseAddr )    // Set "Init" (Pin 16) to an
{                              //   Electrical "low"

 outp( BaseAddr + 2, inp( BaseAddr + 2 ) | 0x004 );

} // End PPortLoInit

PPortHiAuto( Int BaseAddr )    // Set "Auto FDXT" (Pin 14) to an
{                              //   Electrical "High"

 outp( BaseAddr + 2, inp( BaseAddr + 2 ) & 0x0FD );
```

```
} // End PPortHiAuto

PPortLoAuto( Int BaseAddr )     // Set "Auto FDXT" (Pin 14) to an
{                               //   Electrical "low"

 outp( BaseAddr + 2, inp( BaseAddr + 2 ) | 0x002 );

} // End PPortLoAuto

PPortHiStrobe( Int BaseAddr )   // Set "Strobe" (Pin 1) to an
{                               //   Electrical "High"

 outp( BaseAddr + 2, inp( BaseAddr + 2 ) & 0x0FE );

} // End PPortHiStrobe

PPortLoStrobe( Int BaseAddr )   // Set "Strobe" (Pin 1) to an
{                               //   Electrical "low"

 outp( BaseAddr + 2, inp( BaseAddr + 2 ) | 0x001 );

} // End PPortLoStrobe
```

In the Status bit read routines, notice that I have not included reads for bits that are driven in from the port. I assume that the CONTROL register latches are good and that the device connected to the port is not holding it high or low. Also, for the status bit read routines, the result returned is either zero or one and inverted, if appropriate.

```
PPortRdBusy( Int BaseAddr )     // Read the "Busy" (Pin 11) handshaking
{                               //   line

 return 1 ^ (( inp( BaseAddr + 1 ) & 0x080 ) >> 7 );

} // End PPortRdBusy

PPortRdError( Int BaseAddr )    // Read the "Printer Error" (Pin 12)
{                               //   handshaking line

 return 1 ^ (( inp( BaseAddr + 1 ) & 0x020 ) >> 5 );

} // End PPortRdBusy

PPortRdSLCTO( Int BaseAddr )    // Read the "SLCT Out" (Pin 13)
{                               //   handshaking line

 return 1 ^ (( inp( BaseAddr + 1 ) & 0x010 ) >> 4 );

} // End PPortRdSLCTO

PPortRdAck( Int BaseAddr )      // Read the "Ack" (Pin 10) handshaking
{                               //   line

 return ( inp( BaseAddr + 1 ) & 0x008 ) >> 3;

} // End PPortRdAck
```

With these routines, external digital hardware can be controlled by the four output bits driven to the parallel port and data is either read from or written to and an additional four input pins are available.

I give one very good example in how to interface to the PC's parallel port in the El Cheapo programmer, which has a PC board that is included with this book. This circuit is the result of multiple design passes, trying to find a circuit that would work with a number of different PCs.

The original design was published on the web and took advantage of the bi-directional pin capability of the printer port's data bits. Almost immediately after publishing the design, I received e-mails stating that the programmer did not work with specific PCs. This was surprising to me because I thought the initial design was fairly "bullet proof" with the inclusion of the RC network for timing the application to meet the minimum requirements of the PICmicro® MCU programming specification.

To make a long story short, I learned that there is no such thing as a *standard* PC printer port. I found many examples of ports with fairly hefty pull-ups, some ports with reasonable pull-ups, and some with no bi-directional capabilities. All of these PCs had Pentium II or later processors in them.

The bottom line is that you have to design your application to be as universal as possible. Thus, you must assume that the control lines are weakly pulled up and the data lines are not bi-directional. As well, put in an external timing interface (like the RC delay circuit in the El Cheapo) to ensure that PCs with high-speed printer-port interfaces will not send data/pulses faster than your application can accept them.

# Keyboard and Mouse Ports

The PC's keyboard and mouse ports operate with a synchronous serial data protocol that was first introduced with the original IBM PC. This protocol allows data to be sent from the keyboard in such a way that multiple pressed keys can be recognized within the PC without any key presses being lost. The standard was enhanced with the PC/AT as a bi-directional communication method. Three years later, when the PS/2 was introduced, the mouse interface also used the keyboard's protocol, freeing up a serial port or ISA slot, which, up to this point, was needed for the mouse interface. The keyboard protocol used in the PC was so successful that IBM used it for all of its PC, terminal, and workstation product lines that have been developed since 1981 and it is also used by many other PC vendors.

A measure of how good this protocol is the industry that has sprung up for creating compatible keyboards, mice (and other pointing devices), and peripherals that enhance the operation of the PC. These peripherals are usually such devices as barcode scanners and magnetic-stripe readers that pass data to the PC as if it were coming from the keyboard. The operation of these peripherals brings up an important point: the keyboard and mouse interface ports have a specific function to provide keyboard and mouse data into the PC. They should not be used for any other functions. I have seen some projects in the past that use the keyboard as a way to interface with outside devices. Using the keyboard port in this manner should be discouraged.

It is possible to piggy back hardware onto the keyboard port and send instructions between the PC and keyboard, but it is quite impossible to get this same function in Windows unless you would be willing to rewrite the keyboard interface.

The bi-directional keyboard protocol should not present any surprises if you've worked with I2C, Microwire, or other synchronous data protocols. The PC's keyboard communication protocol is set up well to pass data between the PC and keyboard with auxiliary devices in parallel that can monitor and pass data with the other devices.

The female keyboard connector facing out of the PC is shown in Fig. 7-15. With data passed via the clock and data lines, the port can usually supply up to 100 mA over and above the keyboard requirements. The power (+5 VDC) might not be fused, so any hardware put on the port must not draw excessive current to prevent damage to the motherboard or PC's power supply.

Data from the keyboard looks like the waveform shown in Fig. 7-16. The parity bit is odd, which is to say that the eight data bits plus the parity is an odd number.

The data line should not change for at least 5 $\mu$s from the change of the clock line. The clock line should be high or low for at least 30 $\mu$s (with 40 $\mu$s being typical).

Data that is sent from the system unit to the keyboard is similar, but with the clock inverted. The data changes while the clock is low and is latched in when the clock goes high (Fig. 7-17).

When data is sent from the keyboard, the clock is pulled low, then data is sent with the keyboard accepting data when the clock is pulsed high. The bit timings are the same as data from the keyboard. These two protocols are used to allow a device wired in parallel to monitor the communication to and from the PC.

In MS-DOS, the keyboard codes are normally a combination of the keyboard scan code and appropriate ASCII code. Table 7-16 lists the different codes returned for keystrokes by themselves, and with a *Shift, Ctrl,* or *Alt* modifier. The table shows the codes in scan/ASCII configuration for the extended function keyboard characters. The standard function codes are the same, except that *F11, F12,* and the keypad *Center* key do not return any codes and for the explicit arrow and explicit *Insert, Home, Page Up, Delete, End,* and *Page Down* keys, the 0x0E0 ASCII code is actually 0x000.

All of the values in Table 7-16 are in hex and I have put in the scan codes as they appear on my PC. I have not made allowances for upper and lower case in this table because this is differentiated by the PC itself. *KP* indicates the keypad and it, or a single *A* (which indicates alternate arrow and other keys), followed by *UA, DA, LA,* or *RA* indicates an arrow. *I, D, H, PU, PD,* or *E* with *KP* or *A* indicates the Insert, Delete, Home, Page up, Page Down, or End on the keypad, respectively.

The keypad numbers, when *Alt* is pressed is used to enter in specific ASCII codes in decimal. For example, *Alt, 6, 5* will enter in an ASCII *A* character. I have marked these keys in Table 7-16 with #.

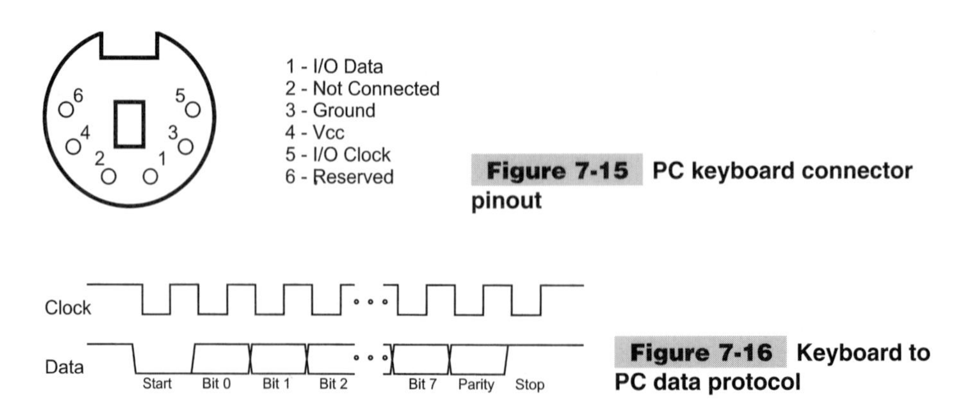


**Figure 7-15** PC keyboard connector pinout

**Figure 7-16** Keyboard to PC data protocol

**TABLE 7-16 PC Keyboard Scan Codes**

| KEY | STANDARD CODES | "SHIFT" CODES | "CTRL" CODES | "ALT" CODES |
|---|---|---|---|---|
| Esc | 01/1B | 01/1B | 01/1B | 01/00 |
| 1 | 02/31 | 02/21 | — | 78/00 |
| 2 | 03/32 | 03/40 | 03/00 | 79/00 |
| 3 | 04/33 | 04/23 | — | 7A/00 |
| 4 | 05/34 | 05/24 | — | 7B/00 |
| 5 | 06/35 | 06/25 | — | 7C/00 |
| 6 | 07/36 | 07/5E | 07/1E | 7D/00 |
| 7 | 08/37 | 08/26 | — | 7E/00 |
| 8 | 09/38 | 09/2A | — | 7F/00 |
| 9 | 0A/39 | 0A/28 | — | 80/00 |
| 0 | 0B/30 | 0B/29 | — | 81/00 |
| - | 0C/2D | 0C/5F | 0C/1F | 82/00 |
| = | 0D/3D | 9C/2B | — | 83/00 |
| BS | 0E/08 | 0E/08 | 0E/7F | 0E/00 |
| Tab | 0F/09 | 0F/00 | 94/00 | A5/00 |
| Q | 10/71 | 10/51 | 10/11 | 10/00 |
| W | 11/77 | 11/57 | 11/17 | 11/00 |
| E | 12/65 | 12/45 | 12/05 | 12/00 |
| R | 13/72 | 13/52 | 13/12 | 13/00 |
| T | 14/74 | 14/54 | 14/14 | 14/00 |
| Y | 15/79 | 15/59 | 15/19 | 15/00 |
| U | 16/75 | 16/55 | 16/15 | 16/00 |
| I | 17/69 | 17/49 | 17/09 | 17/00 |
| O | 18/6F | 18/4F | 18/0F | 18/00 |
| P | 19/70 | 19/50 | 19/10 | 19/00 |
| [ | 1A/5B | 1A/7B | 1A/1B | 1A/00 |
| ] | 1B/5D | 1B/7D | 1B/1D | 1B/00 |
| Enter | 1C/0D | 1C/0D | 1C/0A | 1C/00 |
| A | 1D/61 | 1E/41 | 1E/01 | 1E/00 |
| S | 1F/73 | 1F/53 | 1F/13 | 1F/00 |
| D | 20/64 | 20/44 | 20/04 | 20/00 |

*(continued)*

**TABLE 7-16 PC Keyboard Scan Codes *(Continued)***

| KEY | STANDARD CODES | "SHIFT" CODES | "CTRL" CODES | "ALT" CODES |
|---|---|---|---|---|
| F | 21/66 | 21/46 | 21/06 | 21/00 |
| G | 22/67 | 22/47 | 22/07 | 22/00 |
| H | 23/68 | 23/48 | 23/08 | 23/00 |
| J | 24/6A | 24/4A | 24/0A | 24/00 |
| K | 25/6B | 25/4B | 25/0B | 25/00 |
| L | 26/6C | 26/4C | 26/0C | 26/00 |
| ; | 27/3B | 27/3A | — | 27/00 |
| ‘ | 28/27 | 28/22 | — | 28/00 |
| ‘ | 29/60 | 29/7E | — | 29/00 |
| \ | 2B/5C | 2B/7C | 2B/1C | 2B/00 |
| Z | 2C/7A | 2C/5A | 2C/1A | 2C/00 |
| X | 2D/78 | 2D/58 | 2D/18 | 2D/00 |
| C | 2E/63 | 2E/43 | 2E/03 | 2E/00 |
| V | 2F/76 | 2F/56 | 2F/18 | 2F/00 |
| B | 30/62 | 30/42 | 30/02 | 30/00 |
| N | 31/6E | 31/4E | 31/0E | 31/00 |
| M | 32/6D | 32/4D | 32/0D | 32/00 |
| ‘ | 33/2C | 33/3C | — | 33/00 |
| . | 34/2E | 34/3E | — | 34/00 |
| / | 35/2F | 35/3F | — | 35/00 |
| KP * | 37/2A | 37/2A | 96/00 | 37/00 |
| SPACE | 39/20 | 39/20 | 39/20 | 39/20 |
| F1 | 3B/00 | 54/00 | 5E/00 | 68/00 |
| F2 | 3C/00 | 55/00 | 5F/00 | 69/00 |
| F3 | 3D/00 | 56/00 | 60/00 | 6A/00 |
| F4 | 3E/00 | 57/00 | 61/00 | 6B/00 |
| F5 | 3F/00 | 58/00 | 62/00 | 6C/00 |
| F6 | 40/00 | 59/00 | 63/00 | 6D/00 |
| F7 | 41/00 | 5A/00 | 64/00 | 6E/00 |
| F8 | 42/00 | 5B/00 | 65/00 | 6F/00 |
| F9 | 43/00 | 5C/00 | 66/00 | 70/00 |

**TABLE 7-16 PC Keyboard Scan Codes *(Continued)***

| KEY | STANDARD CODES | "SHIFT" CODES | "CTRL" CODES | "ALT" CODES |
|---|---|---|---|---|
| F10 | 44/00 | 5D/00 | 67/00 | 71/00 |
| F11 | 85/00 | 87/00 | 89/00 | 8B/00 |
| F12 | 86/00 | 88/00 | 8A/00 | 8C/00 |
| KP H | 47/00 | 47/37 | 77/00 | # |
| KP UA | 48/00 | 48/38 | 8D/00 | # |
| KP PU | 49/00 | 49/39 | 84/00 | # |
| KP - | 4A/2D | 4A/2D | 8E/00 | 4A/00 |
| KP LA | 4B/00 | 4B/34 | 73/00 | # |
| KP C | 4C/00 | 4C/35 | 8F/00 | # |
| KP RA | 4D/00 | 4D/36 | 74/00 | # |
| KP + | 4E/2B | 4E/2B | 90/00 | 4E/00 |
| KP E | 4F/00 | 4F/31 | 75/00 | # |
| KP DA | 50/00 | 50/32 | 91/00 | # |
| KP PD | 51/00 | 51/33 | 76/00 | # |
| KP I | 52/00 | 52/30 | 92/00 | — |
| KP D | 53/00 | 53/2E | 93/00 | — |
| KP Enter | E0/0D | E0/0D | E0/0A | — |
| KP / | E0/2F | E0/2F | 95/00 | — |
| PAUSE | — | — | 72/00 | — |
| BREAK | — | — | 00/00 | — |
| A H | 47/E0 | 47/E0 | 77/E0 | 97/00 |
| A UA | 48/E0 | 48/E0 | 8D/E0 | 98/00 |
| A PU | 49/E0 | 49/E0 | 84/E0 | 99/00 |
| A LA | 4B/E0 | 4B/E0 | 73/E0 | 9B/00 |
| A RA | 4D/E0 | 4D/E0 | 74/E0 | 9D/00 |
| A E | 4F/E0 | 4F/E0 | 75/E0 | 9F/00 |
| A DA | 50/E0 | 50/E0 | 91/E0 | A0/00 |
| A PD | 51/E0 | 51/E0 | 76/E0 | A1/00 |
| A I | 52/E0 | 52/E0 | 92/E0 | A2/00 |
| A D | 53/E0 | 53/E0 | 93/E0 | A3/00 |

**TABLE 7-17 PC to Keyboard Commands**

| CODE | FUNCTION |
|---|---|
| 0x0ED | Set Indicator LED's. The next Character out is the LED status |
| 0x0EE | Echo — Keyboard Returns 0x0EE |
| 0x0EF-0x0F2 | Ignored by the Keyboard |
| 0x0F3 | Set Typematic rate, next character is the rate |
| 0x0F4 | Enable Key Scanning |
| 0x0F5 | Set to Default (no LEDs on, default Typematic rate) and disable Key Scanning |
| 0x0F6 | Set to Default (no LEDs on, default Typematic rate) and enable Key Scanning |
| 0x0F7-0x0FD | Ignored by the Keyboard |
| 0x0FE | Request Keyboard to resend the last character |
| 0x0FF | Reset the Keyboard's Microcontroller |

The PC itself has a number of commands that it can send to the keyboard that are shown in Table 7-17. In all of these cases (except for the *ignore* and *echo* commands), the keyboard sends back the acknowledge character, 0x0FA.

For each byte sent, a 0x0F0 character is sent first to notify the PC that the character is coming, followed by the actual keyboard scan code (Fig. 7-17). I have shown an expanded oscilloscope picture of the actual character in Fig. 7-18.

The key I pressed (I can't say that I sent a character) is labeled with *E* and sends a scan code of 0x012. If you look at the set bits in Fig. 7-18, you will see that if you check the bits starting after a *0* start bit, that *0x012* is indeed sent to the PC. The last bit sent is a parity bit.

Looking at the actual oscilloscope pictures in Fig. 7-18 and 7-19, you'll see that the actual timing can be seen. For the keyboard that I have scoped, which is a Cirque Wave Keyboard 2, the clock high and low are each roughly 30 $\mu$s in duration.

The long low final clock value is used to give the PC time to process the incoming data byte before another keyboard device is allowed to transmit on the bus.

To access keyboard data in an application, the most important thing to remember is to use the extended BIOS APIs instead of the basic APIs. The extended APIs will give you

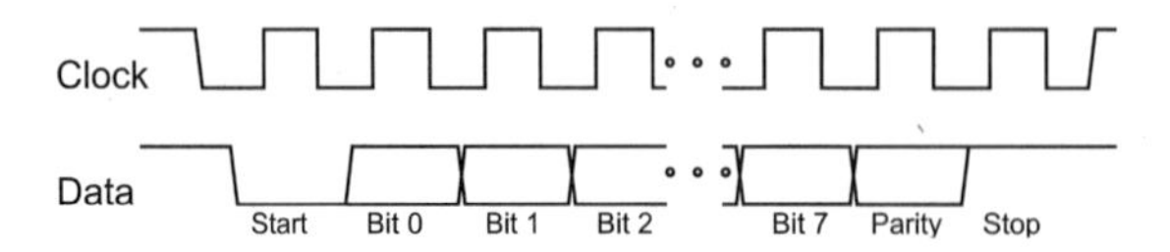


**Figure 7-17** PC to keyboard data protocol

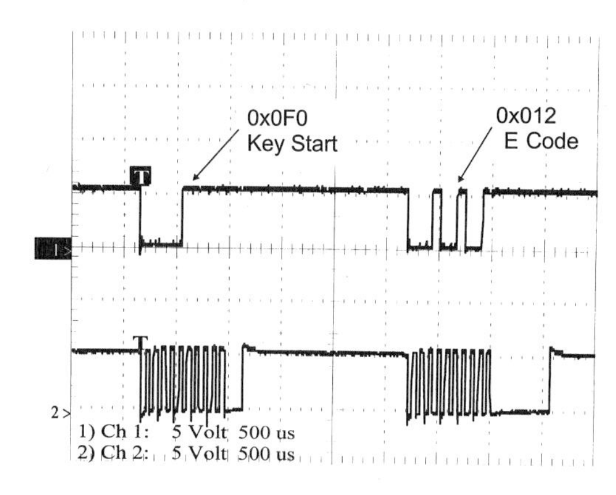


**Figure 7-18** Keyboard sending "E" to the PC

full access to the PC's keyboard, including *F11, F12,* the numeric keypad, and cursor-control keys of the advanced PC/AT keyboard.

I do not recommend accessing any APIs, other than the Read Keyboard Buffer, Check Keyboard Buffer Status, and Read Keyboard Shift Status that are provided in BIOS interrupt 0x016. Setting the Typematic rate or sending characters to the keyboard, especially in Windows, could adversely affect how the PC operates when the application has finished. Changes to the keyboard's functionality should only be made through the operating system interfaces so that it does not receive keyboard data unexpectedly and run erratically.

When keystrokes are returned from the APIs, two bytes are returned. The most-significant byte is the scan code with the least-significant byte being the ASCII code of the key pressed. For most applications, just the least-significant byte can be used and the scan code discarded.

In the PC, either a 0x000 or 0x0E0 character in the ASCII portion of the two-byte scan and ASCII code return represents non-ASCII keys. If you receive one of these characters after issuing a keyboard read, then the actual key read can be decoded by looking at Table 7-16. This feature makes it very easy to process keystrokes in an application.

For example, to simply process a keystroke in C, the following switch code could be used:

```
switch(( KeySave = KEYREAD()) & 0x0FF ) {   // Process the Key
  case 0x000:                              // Special Function Keys
```

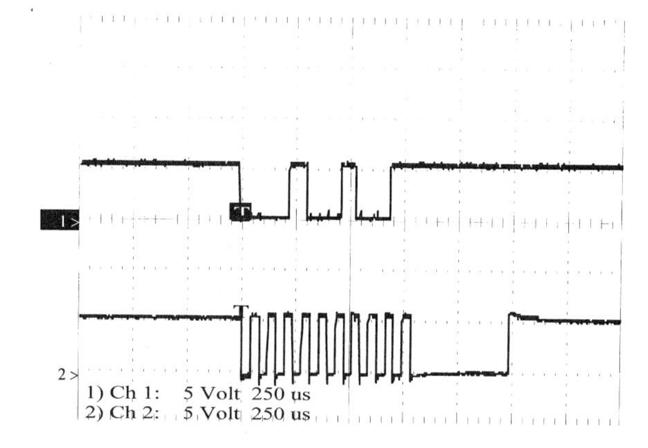


**Figure 7-19** Detail of keyboard sending "E"

```
case 0x0E0:
  KeySave = ( KeySave >> 8 ) & 0x0FF;// process the Scan Code
  :
break;
case 0x00D:                               // Handle "Enter"
  :
break;
  :                                       // Handle Other Special Keys
default:                                  // Other, Unneeded Keys
  :
} // endswitch
```

The PC keyboard is a marvelous invention that is useful for much more than just operating as a keyboard. Each key press sends at least two different packets of information. The first is the *make code,* which is repeated at the Typematic rate while the key is pressed. The second is the *break code,* which is sent when the key is released.

In the true IBM design, when two keys are pressed and one is released, the keyboard will continue to send make codes from the last pressed key until one is released. This can be of particular advantage in games or applications where the keyboard is used for device control. But, to take advantage of it, your application will have to take over the keyboard BIOS functions and redirect the BIOS interrupt-handler vector. This is not very difficult to do, but you will have to save the original vector to the keyboard interrupt handler and restore it when you are returning to MS-DOS.

The best way to add a device to the PC's keyboard port is by connecting it in parallel to the keyboard. Providing the parallel interface will allow you to pass data to the PC without having to buffer keystrokes of the PC instructions (Fig. 7-20).

When connecting a device to the keyboard port, it is important to remember that the keyboard port's bus is open collector, which means that your device can only pull down the clock and data lines. A "high" voltage cannot be driven onto the keyboard connector.

To send data to the PC, the device simply has to send keyboard scan codes in the same format as the keyboard. While the device is operating, it must monitor the traffic between the PC and keyboard and wait to send data to the PC until the currently transmitting packet has completed. This is relatively easy to do. Problems can happen when your device begins to transmit a scan code and either the keyboard or PC transmitting a byte at the same time. The actual data will become garbled and both of the messages could be lost. The chances of this happening are quite remote, but if data got garbled, the PC might stop working correctly or wrong data could be processed. To avoid this problem, your device should stop transmitting when unexpected data shows up on the line. This is similar to I2C and CAN multi-mastering, which is covered in Chapter 6.

The mouse port itself uses an identical data protocol as that of the keyboard. The big difference between the two is that the PC does not drive data back to the mouse. Mouse (also known as the *pointing device* within Windows) information can be passed to the PC one of two ways. When the mouse was first provided to the PC, it used a serial port and typically transmitted three data bytes at 9,600 bps.

The PS/2-style mouse uses the same connectors and protocol as the keyboard port. When power is applied to the mouse's microcontroller, an ASCII *M* is sent to the PC to indicate that the device on this port is a mouse/pointing device. The PS/2-style mouse was first introduced by IBM with the, not surprisingly, PS/2 computer in 1987. Most modern PCs and motherboards use the PS/2-style mouse.

The PS/2-style mouse sends three bytes every 50 ms, when the mouse has been moved, or if a button has been pressed (Table 7-18). The three bytes can be seen in Fig. 7-21.

Note that a delay of roughly 200 $\mu$s occurs between each byte transmission and at the end of each byte, the clock "dead zone" is active to allow the PC to process the incoming position bytes.

The Delta X and Delta Y values are the position differences of the mouse/pointing device since the last time the three bytes were sent (just less than 50 ms before). These "deltas" are used by the mouse BIOS software to calculate the current position of the mouse on the screen, based on the previous position.

The 20 times per second position update (from the 50-ms interval between bytes) is a reasonable update rate and allows the user to move the mouse on the display very smoothly without any perceived discontinuities or jumps in its motion.

**TABLE 7-18 Mouse Data Byte Format**

| BYTE NUMBER | FUNCTION/VALUE |
|---|---|
| 1 | Start of Data |
| | Bit 7 — 0 |
| | Bit 6 — 1 |
| | Bit 5 — Set if the "left button" is pressed |
| | Bit 4 — Set if the "right button" is pressed |
| | Bit 3 — Bit 6 of the "Delta Y" |
| | Bit 2 — Bit 7 of the "Delta Y" |
| | Bit 1 — Bit 6 of the "Delta X" |
| | Bit 0 — Bit 7 of the "Delta X" |
| 2 | "Delta X" Position |
| | Bit 7 — 0 |
| | Bit 6 — 0 |
| | Bit 5 — 0 — Bits 5 — 0 of "Delta X" |
| 3 | "Delta Y" Position |
| | Bit 7 — 0 |
| | Bit 6 — 0 |
| | Bit 5 — 0 — Bits 5 — 0 of "Delta Y" |

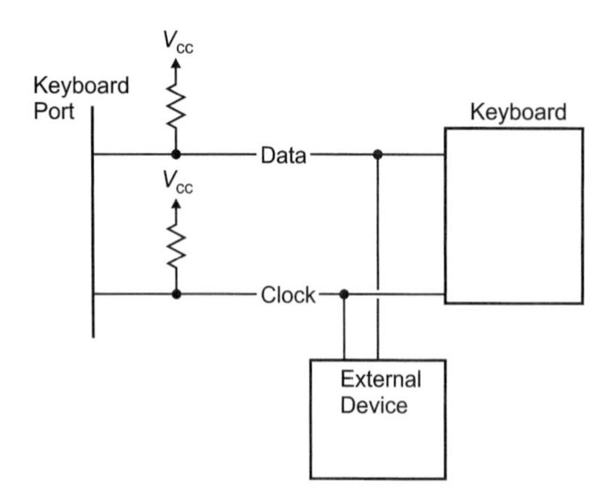


**Figure 7-20** Sharing a keyboard with another device

How the mouse/pointing device actually sends data is handled within the operating system/mouse device drivers. The Int 33h mouse BIOS and operating-system APIs do not make any differentiation, as far as the application is concerned.

# ISA Bus

The ISA bus is still widely available and the recommended interface to use for your first applications—even with the advent of the PC/99 specification that excludes its use. The ISA bus was originally designed as a method to convert the 8088's multiplexed bus into a 20-bit address, 8-data-bit bus for peripherals. Along with the processor bus, ISA also provides interrupt and DMA interface to adapters. I believe that this complete, easy-to-interface-to bus is the primary reason why the PC became so popular after it was first introduced.

ISA is ideally suited for simple, low-cost, and relatively slow hardware interfaces, including the PICmicro® MCU's PSP. This section introduces the PC's ISA bus and covers some of the issues for interfacing the PICmicro® MCU to it. I do not include *Direct Memory Access (DMA)* operations because this interface is difficult and inappropriate to use with the PICmicro® MCU. Chapter 16 presents the design for an ISA adapter that uses a PICmicro® MCU as a peripheral interface for the PC.

If you look at the Intel 8088's datasheet, you'll see that the multiplex address/data bus with DMA and interrupts is actually quite complex. When the PC was designed, IBM designed the motherboard and specified the ISA slots in such a way that the complexity of the bus was hidden from the user (Fig. 7-22).

Looking at the PC's motherboard and ISA slots from this perspective, it should be no surprise that the read/write cycle on the ISA bus looks like that in Fig. 7-23. This waveform is identical for the I/O address space reads and writes.

A little surprise is the ALE bit, which becomes active when the address is valid. The original purpose of this bit was to initiate the RAS/CAS address multiplexing for reading and writing to memory on the ISA bus. Today, with modern memory speeds, this bit is of little use.

To transfer data, the time between ALE active and the data available is normally 760 ns for eight-bit transfers and 125 ns for 16-bit transfers. The faster access for 16-bit transfers

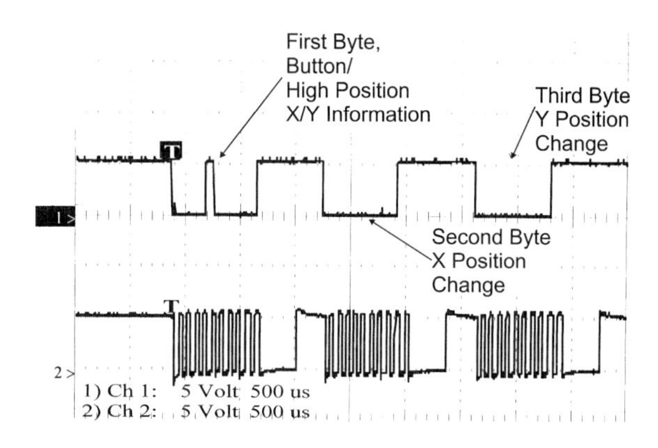


**Figure 7-21** Mouse position waveforms

was a function of the PC/AT and the need for moving memory data faster in the 80286 system.

The eight-bit ISA bus consists of a two-sided 31-pin card edge connector with the pins defined in Table 7-19.

There is a 16-bit extension connector that is a separate two-sided 18-pin connector on the end of the eight-bit interface. I have not included this connector definition because the PICmicro® MCU cannot access data 16 bits at a time.

The data and address busses are buffered to the processor, this should not be surprising, but it is important to remember. Addresses from 0x00000 to 0x0FFFFF (zero to one megabyte) can be accessed with the eight-bit connector. Memory devices can be located 0x0C0000 to 0x0DFFF, but care must be taken to avoid contention with other devices located within this memory space. For the PICmicro® MCU, memory-mapped devices are not recommended, which will avoid the problem with trying to find addresses that can be used in different PCs.

*BALE (Buffered ALE)* was the term used in the original PC because the ALE line was produced by the 8088's instruction sequence clock. This pin was buffered to avoid having the ISA bus directly processor driven. Today, this bit is more commonly known as *ALE* and it provides essentially the same operation and timing as BALE.

*-I/O CH CHK* was designed for use with parity-checked memory. If a byte was read that did not match the saved parity, a NMI interrupt request was made of the processor.

*I/O CH RDY* is a line driven low by an adapter if it needs more time to complete an operation. This is yet another pin that was really specific to the times when DRAM was put on the ISA bus. During each processor cycle (of 760 ns in the 8088), while the pin was high, the ISA read/write cycle would wait full processor cycles until the line was

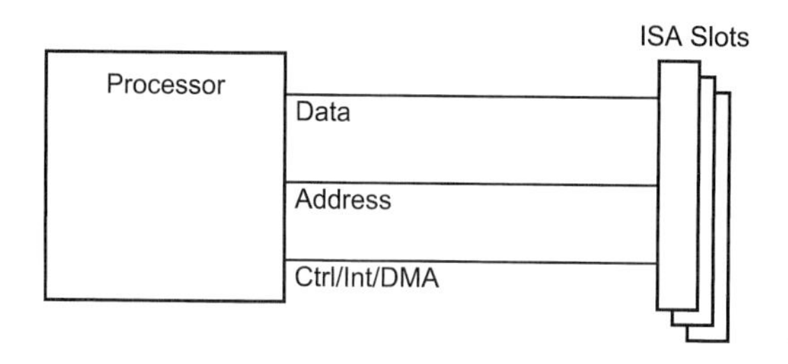


**Figure 7-22** Processor/ISA block diagram

**TABLE 7-19 ISA Bus Pinout**

| PIN | "A" (CONNECTOR) | "B" (SOLDER) |
|---|---|---|
| 1 | I/O CH CHK | Ground |
| 2 | D7 | Reset |
| 3 | D6 | +5 V |
| 4 | D5 | IRQ2 |
| 5 | D4 | +5 V |
| 6 | D3 | DRQ2 |
| 7 | D2 | −12 V |
| 8 | D1 | −CARD SLCTD |
| 9 | D0 | +12 V |
| 10 | +IO CH RDY | Ground |
| 11 | AEN | −MEMW |
| 12 | A19 | −MEMR |
| 13 | A18 | −IOW |
| 14 | A17 | −IOR |
| 15 | A16 | −DACK3 |
| 16 | A15 | DRQ3 |
| 17 | A14 | −DACK1 |
| 18 | A13 | DRQ1 |
| 19 | A12 | −DACK0 (−REFRESH) |
| 20 | A11 | OSC |
| 21 | A10 | IRQ7 |
| 22 | A9 | IRQ6 |
| 23 | A8 | IRQ5 |
| 24 | A7 | IRQ4 |
| 25 | A6 | IRQ3 |
| 26 | A5 | −DACK2 |
| 27 | A4 | T/C |
| 28 | A3 | BALE |
| 29 | A2 | +5 V |
| 30 | A1 | CLOCK — 14.31818 MHz |
| 31 | A0 | Gnd |

dropped low. In the PC/AT (and later) PCs, this delay has a granularity of 210 ns. When using PICmicro® MCUs with PSP, no delays are necessary.

*-IOR* and *-IOW* are used to read and write data from the ISA bus into the processor's I/O space. Only I/O addresses 0x0000 to 0x03FF are used for the ISA bus hardware of these 1K addresses; there are only a few available for use with adapters. For PICmicro® MCU applications, I recommend 0x0300 to 0x031F (that is the prototype card adapter) because no commercial devices are designed for this address range.

*-MEMR* and *-MEMW* pins indicate that the processor is reading and writing to ISA bus memory.

Pins IRQ3 through IRQ7, IRQ9 through IRQ12, IRQ14, and IRQ15 request hardware interrupts. When these lines are driven high, the 8259As on the motherboard (which are known as *Programmable Interrupt Controllers, PICs*) will process the request in a descending order of priority. These lines are driven high to request an interrupt. A PICmicro® MCU can drive these lines, but it should only be active when a high is driven onto the interrupt line to allow other devices to share the interrupt pins.

The *CLOCK line* is a four-times color burst clock that runs at 14.31818 MHz and was the system clock of the original PC. Divided by three, this clock was used for the 4.77-MHz 8088 clock in the original PC. This clock is not running at the 200+ MHz of your Pentium processor. The 14.31818-MHz clock was distributed to the system to provide clocking for the MDA and CGA video display cards. This clock can be useful when providing a simple clock for microcontroller and other clocked devices on adapter cards.

The *OSC pin* is driven at up to 8 MHz. The pin was originally added to the 16-bit ISA specification as the actual clock speed for memory adapters connected to the bus. As PC clock speeds have gone up, this pin's speed hasn't. The signal runs at 8 MHz to provide a clock for system operations and not provide significant radiated noise problems.

The *DRQ# pins* request a DMA transfer to take place. When the corresponding DACK# pin is driven high, the DMA controller is reading or writing an I/O address of an adapter card. When the DMA controllers have control of the bus over the processor, the AEN pin is active to indicate to other adapters that a DMA operation is in process. When all of the DMA data has been transferred, the T/C bit is pulsed high to indicate that the operation has completed. When the T/C bit becomes active, the adapter should request a hardware interrupt to indicate to the software that the operation is complete.

DRQ1 to DRQ4 are used to request eight-bit DMA transfers (DMA channel 0 is used for the DRAM refresh circuitry) and DRQ4 through DRQ7 are used to request 16-bit DMA transfers. DMA is not covered in detail in this book because it would be very difficult to implement with the PICmicro® MCU. It is really only designed for applications that pass large amounts of data between a peripheral and the PC.

The *-REFRESH* or *-DACK0* pin is active when the DMA controller's channel 0 is active and doing a RAS-only refresh of the system memory. The purpose of this pin is to be distributed to ISA memory cards and use the address on the lower eight to 10 bits of the bus for refreshing the memory card's DRAM. In modern PCs, this line might not be active, depending on how memory is implemented.

The last pin is the *-MASTER,* which is driven by an adapter when it wants to take over the bus and drive its own signals. This is a way of providing DMA to the system without using the 8237s or allowing another processor to access the system resources. There are a few potential problems with this pin and its usage. One of the most important things to watch for is holding the line active for longer than 15 $\mu$s could result in a missed refresh interval. The -MASTER pin is also not all that useful because it cannot be used to drive data to the motherboard's hardware.

Despite the plethora of pins in this interface, you can create ISA adapters quite easily. For example, to provide a PICmicro® MCU ISA peripheral interface, the circuit shown in Fig. 7-24 is all that is required to pass data to the LCD, and read status and data back from the device.

Notice that I do not use ALE and just use the address data and I/O read and write pins. Eight-bit transfers have the -IOR and -IOW pins active for roughly 500 ns that allows the PICmicro® MCU to pass data to and from the PSP data port.

It is important to know that the original PC and PC/AT only decode the least-significant 10 bits of the 16-bit I/O address bus. Modern PCs often have registers above this limit or might simulate the PC and PC/AT and not decode any of the bits above this limit. The PC was originally released with 1,024 device addresses that only decoded ISA address pins zero to nine. If an I/O access was made to an address above address 1,023 or 0x03FF in the PC or PC/AT, the most-significant addresses are ignored when the device selections are used. You might have an idea for a device that you want to put at I/O space address 0x0420, but, in the PC, the motherboard will interpret this as address 0x020, which is the primary interrupt controller.

In the current Pentium III "Northbridge" and "Southbridge" chips, registers are built in at addresses above 0x03FF (the 10-bit limit). These addresses have been chosen to work with the standard I/O addresses within the PC and not cause any conflicts with them. As well, in these chips, many of the standard functions are decoded within the chips; the full 16 bits are used for the decoding, which simplifies the operation.

To be on the safe side, just use the I/O addresses from 0x0300 to 0x031F for PICmicro® MCU addressing.

Interrupts in the ISA bus are positive-active, edge-triggered, TTL/CMOS inputs that are passed to two 8259As on the PC's motherboard. The interrupts work reasonably well, although you should be aware of a few issues. The biggest issue for me is the inability of the pins to safely work with multiple sources under all conditions. If multiple devices are on the interrupt line and one attempts to drive it high, "bus contention" will result, which might cause an interrupt request to be passed to the processor.

Thus, when planning which interrupt to use, understanding the current usage in the PC is important before a new adapter is configured to a specific request line. Using the serial- or parallel-port interrupts (IRQ3, IRQ4, or IRQ7) is usually a safe bet because most mod-

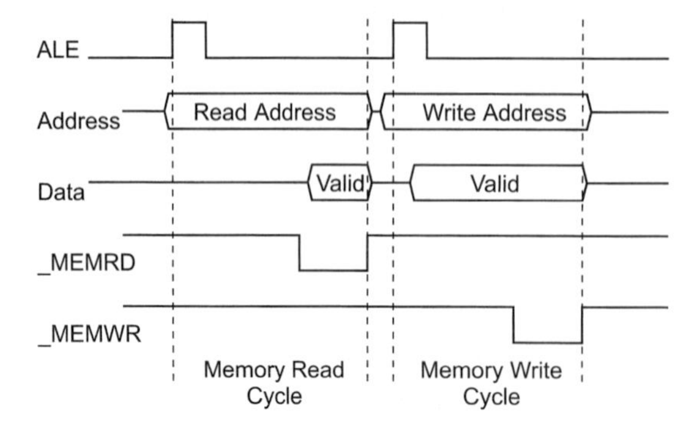


**Figure 7-23** ISA bus timing

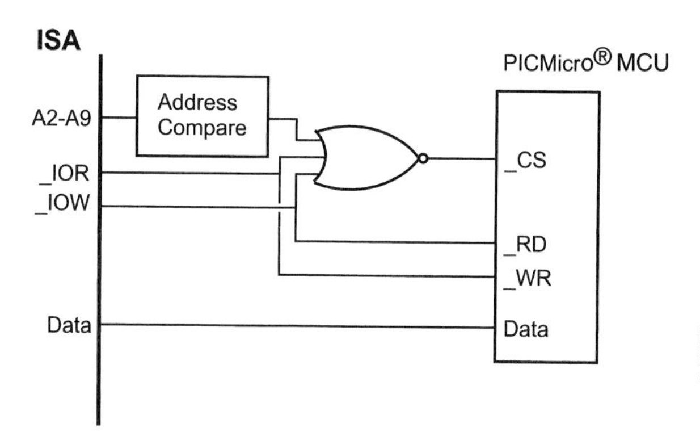


**Figure 7-24** ISA connection to a PSP equipped PICmicro® MCU

ern PC motherboards have this hardware built in and there is no danger of a problem from having multiple busses drive the same line.

In terms of which interrupts to select, I have suggested IRQs 3, 4, and 7. If they are not available for whatever reason, then you should consider IRQs 10, 11, 12, 14, or 15.

If an interrupt appears to be used, check to see if anything is driving the line on the motherboard's ISA slots. This can be done with a common logic probe, which can indicate high, low, or open conditions. If nothing is normally driving the line, then you can use the circuit shown in Fig. 7-25 to enable a tri-state driver output as a high logic level only when the interrupt is being requested. This should drastically reduce the opportunity for bus contention or missed interrupt requests.

Once you have specified the interrupt number in your PC application code, you can set the interrupt vector to the handler using the following steps:

1. Save the original vector using MS-DOS interrupt *021h AH = 035h API.*
2. Set the new vector using MS-DOS interrupt *021h AH = 025h API.*
3. Enable the interrupt request mask bit in the 8259.

To enable the interrupt-request mask bit in the 8259, the appropriate interrupt mask register bit has to be reset. This register is at the 8259's base address plus one. This can be done with the following statement:

```
outp( IntBase + 1, inp ( IntBase + 1 ) & (( 0x0FF ^ ( 1 << Bit )));
```

In this statement, the interrupt mask register is read in, the appropriate is bit cleared, and then written back. Nothing more needs to be done with the 8259.

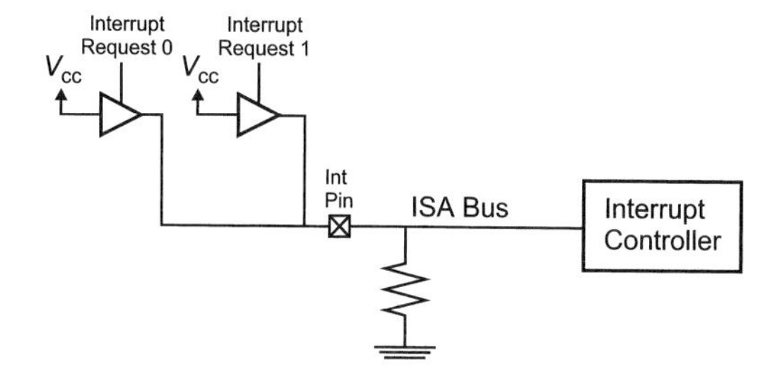


**Figure 7-25** Multiple interrupt request circuit

To "release" the interrupt vector and the interrupt source at the end of the application, the following steps must be taken:

**1** Disable the interrupt request mask bit in the 8259.
**2** Restore the original vector using MS-DOS interrupt *021h AH = 025h API.*

I restore the interrupt vector to the original value because it is normally pointing to the D11 routine, which causes the requesting interrupt to be masked off and never responded to again.

Unless you are designing a "Plug and Play" adapter card or an adapter for only one application in a PC, you are probably going to have to set up a switch block with multiple interrupt-request numbers and addresses. I do not recommend hard wiring the interrupt-request number because this might make your application unworkable in some situations. I realize that this solution is suboptimal, especially considering that most users are like me and will lose their documentation on how to set the application's software for different device addresses and interrupts. The only suggestion I have to this problem is to create a web page with the documentation available for download.

# 8

# PICmicro® MCU APPLICATION SOFTWARE-DEVELOPMENT TOOLS

CONTENTS AT A GLANCE



Taking a cue from the "chick and egg" question, a common question about the PICmicro® MCU is, which came first, the PICmicro® MCU's popularity or the excellent software-development tools that are available for it? As the popularity of the PICmicro® MCU increases, the availability of excellent software-development tools increases, which, in turn, makes the PICmicro® MCU more attractive to new users.

When I first started working with the PICmicro® MCU, Microchip provided MS-DOS command-line software development tools, which did the job quite competently. The MS-DOS command-line tools were, however, quite clunky and difficult to see the full picture of what was happening during application execution. Figure 8-1 shows a screen shot of *MPSIM* (the MS-DOS command-line simulator) executing. This tool is a PICmicro® MCU simulator that requires a user to provide an MPSIM-INI configuration file, which declares which registers are to be monitored by the user as the application executes. *Stimulus* files can be added to the application to avoid the need for the user to input specific inputs manually.

As indicated, MPSIM was competent, but because the interface displayed only one line at a time, it was very hard to see the "full picture" with this tool. When I was finishing the first edition of this book, Microchip released version 3.00 of MPLAB (Fig. 8-2). This Microsoft Windows application eliminated a lot of the difficulty in developing and simulating a PICmicro® MCU application that was inherent in the MS-DOS command-line environment. MPLAB also gave an *Integrated Development Environment (IDE)* that ran under Microsoft Windows and gives the user a tool in which they can edit, assemble, or compile the source code, simulate the application, run it on an emulated device, and "burn" the application code into a PICmicro® MCU chip. This all in one tool is excellent for new PICmicro® MCU developers and is a great way to see your applications executing and responding to different input conditions. I focus on MPLAB in this book and use it exclusively for PICmicro® MCU application development because of the ease in which applications can be created with it.

This chapter introduces the different tools that are available for PICmicro® MCU application development from editors to assemblers, compilers, simulators, and emulators.

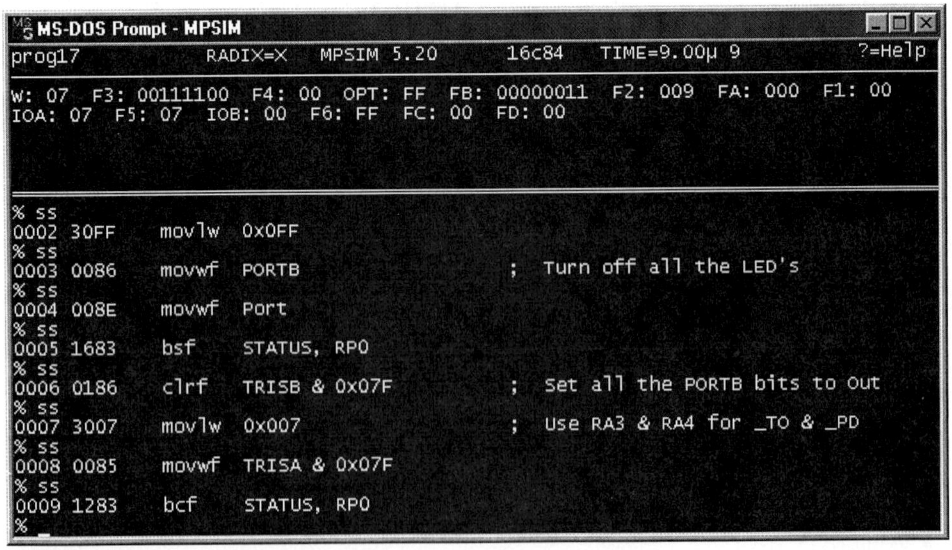

```
MS-DOS Prompt - MPSIM
prog17            RADIX=X    MPSIM 5.20       16c84    TIME=9.00µ 9           ?=Help

W: 07  F3: 00111100  F4: 00  OPT: FF  FB: 00000011  F2: 009  FA: 000  F1: 00
IOA: 07  F5: 07  IOB: 00  F6: FF  FC: 00  FD: 00

% ss
0002 30FF     movlw   0x0FF
% ss
0003 0086     movwf   PORTB                  ;  Turn off all the LED's
% ss
0004 008E     movwf   Port
% ss
0005 1683     bsf     STATUS, RP0
% ss
0006 0186     clrf    TRISB & 0x07F          ;  Set all the PORTB bits to Out
% ss
0007 3007     movlw   0x007                  ;  Use RA3 & RA4 for _TO & _PD
% ss
0008 0085     movwf   TRISA & 0x07F
% ss
0009 1283     bcf     STATUS, RP0
% _
```

**Figure 8-1** "MPSIM" operation

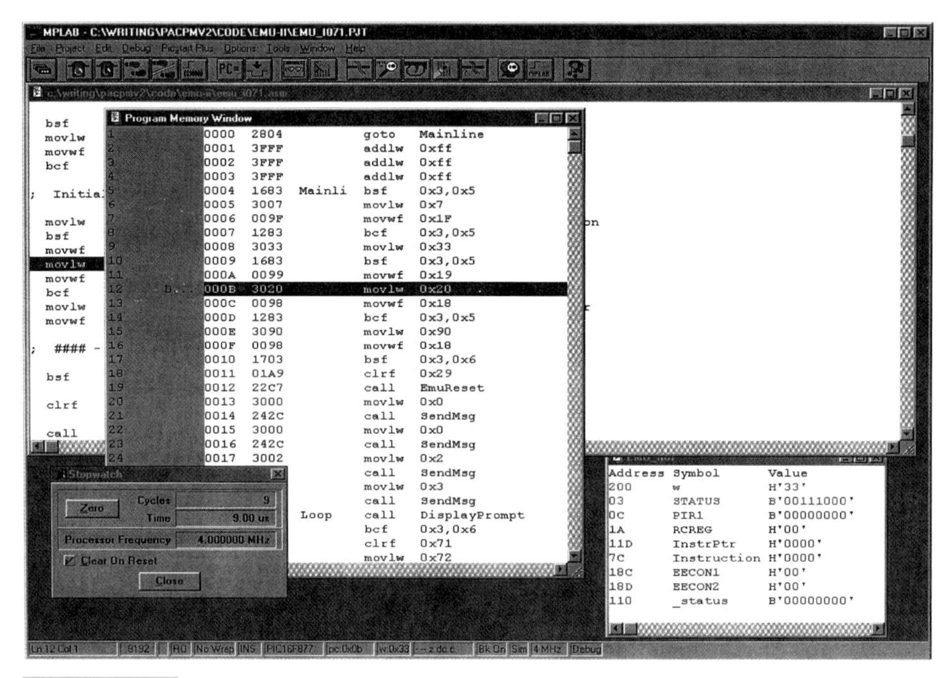


**Figure 8-2** "MPLAB" operation

Although I focus on the MPLAB IDE and assembly-language programming in this book, this chapter also presents some other tools that you can use for PICmicro® MCU development. The tools available for creating PICmicro® MCU application run the gamut from freeware, shareware, GNU Tools, and commercial products.

The next two chapters present hardware emulators and programmers to round out the development tools required for the PICmicro® MCU. The emulators and programmers that are presented use the features and functions described in this chapter.

# Software-Development Tools

What a mouthful this chapter's title is! When I first wrote the book, I called the chapter *Programming Tools*, but changed it when I realized just how broad a topic it is and give you a better idea of what was encompassed in it. As you will see, I'm pretty laid back about how I think people should develop code. The reason for this (and I'll re-emphasize it over and over) is that I believe that everybody should be allowed to develop code and applications in a manner in which they are most comfortable. This doesn't mean that you should stick with what you know because you like it, but find what works for you and then exploit and improve upon it.

In terms of the language to be used, this is really up to the application software developer. For most of my PICmicro® MCU application development, I use Microchip's MPASM assembler. The primary reason for this is because of the price (free) and the fact the Microchip assembler uses the instructions documented in the Microchip databooks. Some shortcuts are provided in the assembler and, although I touch on some of them, I have tried to avoid them where ever possible to create code and examples that are as easily understandable as possible. I've avoided these shortcuts because they tend to make the code harder to read for the beginner, and I like having less than forty instructions to remember.

This does not mean that I advocate the writing all applications in assembler; nothing could be further from the truth. There are a lot of good development languages out there and when you go out on your own and develop code, you should use the language that you are most comfortable with.

The high-level language you choose should be able to produce efficient code; efficiency should be measured in number of assembler instructions produced for each line of high-level-language source code. The high-level language should also be able to access all the features and registers of the microcontroller. Lastly, the high-level language should be something you are comfortable with; learning a new language for a new processor/system should be a choice, not a requirement. The next section of this chapter covers languages and their requirements.

Some vendors supply tools that do not use the instruction format specified in the Microchip data books. The best example of this is Parallax's Programming Tools, which create instructions that simulate Intel 8051 instructions. This might make you feel more comfortable (and using tools that make you comfortable is what this section is all about), but before trying out the tools, be sure that you understand what you are getting and what the implications are. PICLIST, STAMPLIST, and sci.electronics.design are excellent Internet resources for beginning your research on what is the best language for you.

One tool I have found to be invaluable for developing code on a microcontroller is the *simulator*. A simulator is a tool that allows the application software developer to try out his code and monitor it during execution with a variety of "stimulus." The stimulus can be in the form of the application developer manually changing simulated registers or input pins. Stimulus input can also be a file of I/O pin data values to eliminate the need for manual specifications. This chapter (and elsewhere in the book) covers the different methods of stimulus available to the MPLAB IDE and how they are used to debug applications.

I always try to get as close to 100% confidence in my application code before I program it into a microcontroller and try it out. The only way I can do this is by running the code on a simulator. The MPLAB IDE simulator is presented as an important tool in testing out your applications before they are programmed into a PICmicro® MCU and you are left trying to figure out why they don't work.

## EDITORS

An editor is an application program that runs on a PC or workstation to allow a "man-readable" file to be created or changed. The editor can also be used to look at files, (which is known as *browsing*). As I review editors in this section, you'll see that I haven't been very adventurous in what I use.

Over the years, I have probably tried out 100 or more different editors to develop and modify application code, browse and change hex files, create text (like this book or web page HTML), and send e-mails. Most of these trials were taken at the suggestion of someone else because they had found a "wonderful new editor."

Out of all the different editors I've tried, I've only found two that I've liked and used for any length of time. For standard editing requirements, I just use the standard Microsoft Windows WordPad, NotePad, and Word editors. I find that I'm very picky in what I consider to be an outstanding editor. For the most part, I rely on the Microsoft data-entry standards with basic text-editing features. The MPLAB IDE's editor using the standard Microsoft editing conventions has reinforced the validity of this decision.

A standard Microsoft editor, such as WordPad (Fig. 8-3), is a Word-based editor. The cursor, which is a vertical bar, indicates where characters will be placed. It is moved onto the window by either the arrow keys or by using the mouse and setting it's position by a left click of the mouse. When I'm editing a file, I very rarely use the mouse, instead I use the arrow keys, *Home, End, Page Up,* and *Page Down* almost exclusively. Table 8-1 lists the standard Microsoft editor operations.

To delete and move text, I use the cut and paste functions. To select text to cut and paste, I mark the text first pressing a *Shift* key while moving the cursor to highlight the text to be relocated. If you have marked text incorrectly, then simply move the cursor without a *Shift* key pressed to delete the text marking. Pressing your mouse's left button and moving the mouse across the desired text will also mark it. Left clicking on another part of the screen will move the cursor there and eliminate the text marking.

Next, the keystrokes *Ctrl-X* remove the marked text from the file and place it into the Windows clipboard. *Ctrl-C* copies the marked text into the clipboard and doesn't delete it. To put the text at a specific location within a file after the current cursor location, use *Ctrl-V*.

Notice that I do not use *Delete* or *Insert*. Deleting marked text destroys it completely whereas *Ctrl-X* saves it in the clipboard so that it can be restored if you made a mistake.

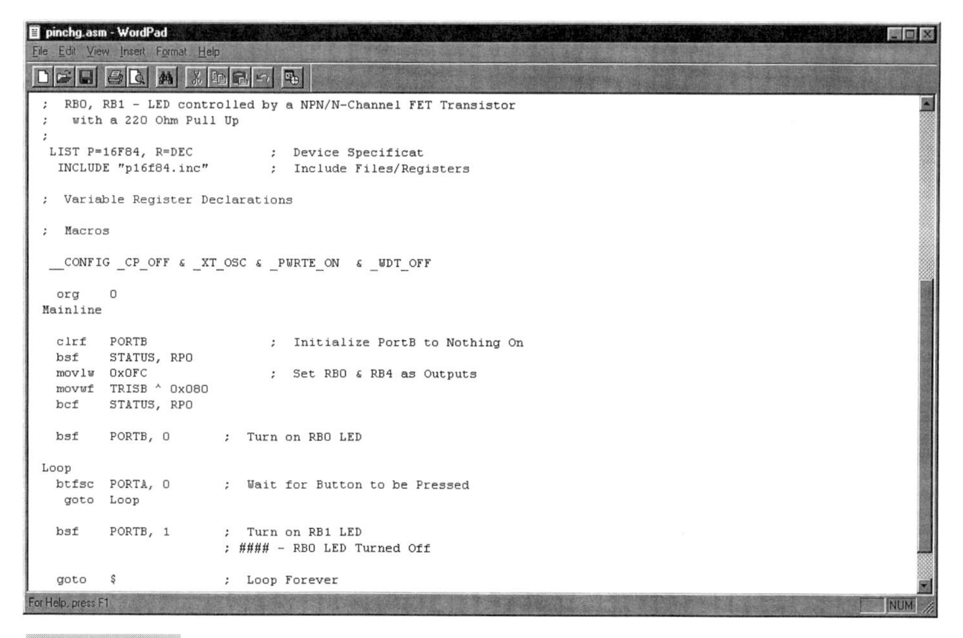


**Figure 8-3** "WordPad" MicroSoft windows editor

**TABLE 8-1 Microsoft Editor Key Stroke Operations**

| KEYSTROKES | OPERATION |
|---|---|
| Up Arrow | Move Cursor up one Line |
| Down Arrow | Move Cursor down one Line |
| Left Arrow | Move Cursor left on Character |
| Right Arrow | Move Cursor right on Arrow |
| Page Up | Move viewed Window Up |
| Page Down | Move viewed Window Down |
| Ctrl—Left Arrow | Jump to Start of Word |
| Ctrl—Right Arrow | Jump to Start of next Word |
| Ctrl—Page Up | Move Cursor to Top of viewed Window |
| Ctrl—Page Down | Move Cursor to Bottom of viewed Window |
| Home | Move Cursor to Start of Line |
| End | Move Cursor to End of Line |
| Ctrl—Home | Jump to Start of File |
| Ctrl—End | Jump to End of File |
| Shift—Left Arrow | Increase the Marked Block by one character to the left |
| Shift—Right Arrow | Increase the Marked Block by one character to the right |
| Shift—Up Arrow | Increase the Marked Block by one line up |
| Shift—Down Arrow | Increase the Marked Block by one line down |
| Ctrl Shift—Left Arrow | Increase the Marked Block by one word to the left |
| Ctrl Shift—Right Arrow | Increase the Marked Block by one word to the Right |

The *Insert* key toggles between data Insert or Replace mode for Microsoft keystroke-compatible editors. Normally, when an editor boots up, it is in *Insert mode,* which means that any keystrokes are inserted into the text rather than replace what is already there. This is the preferable mode to be in.

If you make a mistake, you can restore what you had before the last operation using "ctrl-Z." MPLAB uses a Microsoft-compatible editor, as does UMPS, which is why I have changed from a line-based editor. In a line-based editor, a CR/LF character combination terminates each line displayed on the screen. In a Microsoft-compatible editor, the CR/LF is used to separate paragraphs and lines are broken up on the display in order to show all the data on them. If you were to look at a paragraph produced by a Microsoft-compatible editor on a line editor, you would find that the paragraph would be one line and most of it would not be displayed because it was past the right edge of the editor's text window.

The advantage of a line-based editor is the ease in which blocks of data can be moved. As well, the editor that I use is programmable to respond to specific keystroke sequences. For example, typing in the word

```
 if
```

The editor would automatically put in:

```
if ( ) {
} else {
} // endif
```

This feature made it very fast for me to create C programs, as well as PICmicro® MCU assembly-language programs before the advent of MPLAB (and Microsoft's Visual Studio, which has its own built-in editors that follow the conventions presented here).

I also keep a *bit editor* handy, in case I want to look at a binary file. A bit editor is a tool that shows each byte as two nybbles. This tool can be useful for patching code (and defeating copy-protection in MS-DOS games), but I do not recommend patching applications like the PICmicro® MCU applications. I recommend that code patches should never be implemented in a bit editor, instead, the source code should be copied into a new file and the application rebuilt.

With older editors, you must watch out that they might put a 0x01A character at the end of a file. In the MPLAB IDE and other Microsoft-compatible editors, this character displayed as a small square box at the end of the file. The 0x01A character was required by MS-DOS 1.x to indicate a file's end. For some reason, even though MS-DOS and Windows no longer require this file-end delimiter and as MS-DOS and Windows have become more sophisticated in their file handling, this file-end indicator has not completely disappeared. Although MPLAB and most other tools do not have a problem with this character placed at the end of the file, some other development tools do. If you are using a tool where this is a problem, you can simply delete the character (which shows up as a small square).

## ASSEMBLERS

The most popular way to program the PICmicro® MCU is to use the Microchip MPASM assembler, which can be downloaded free of charge from Microchip's website and is provided on this book's CD-ROM. This tool converts a set of human-readable symbols into the bit patterns that will be programmed into the PICmicro® MCU and executed by the processor. This section introduces assembly-language programming and how an assembler works.

So far, this book has introduced the PICmicro® MCU's hardware, processor architecture, and the instructions that are executed as the application code. With this background, I want to go deeper into the application software to help explain how an assembler works and how you can take advantage of its operation when you are creating your own applications.

Earlier in the book, when I discussed the PICmicro® MCU's instruction sets, I was really referring to the set of symbols that are used by a PICmicro® MCU assembler to convert application code into the data used by the program. This description implies that an instruction is a symbol that is converted into a hex number by the assembler. This would be correct and helps explain that an assembler is really an application that compares strings to tables; when it finds a match, it replaces the string with a numeric value.

For the example code:

```
 movlw  7
Loop
 addlw    0 - 1
 btfsc    STATUS, Z
  goto    Loop
```

**TABLE 8-2 Assembler First Pass Instruction Information**

| CODE | | ADDRESS | NUMERIC | EXPECTED PARAMETERS | LABEL VALUE |
|---|---|---|---|---|---|
| movlw | 7 | 123 | 3000 | 8 bits | |
| Loop | | | | | loop = 124 |
| addlw | 0 -1 | 124 | 3E00 | 8 bits | |
| btfsc | STATUS, Z | 125 | 1C00 | reg, bit | |
| goto | Loop | 126 | 2800 | 11 bits | |

The MPASM assembler performs the following operations: First, it determines the current program counter for the code. Next, it reads through the first parameter. If it is an "instruction" label, it converts the instruction label into a numeric value. Along with the instruction's code, it also identifies the expected parameters. If the label is not in the instruction table, then the assembler assumes that the string is an address label and stores the value in an address label and stores the value in an address table. This operation is known as a *pass* and it processes the entire source file before going on.

If this example code started at address 0x0123, the first pass of the assembler would produce the information shown in Table 8-2. Once this pass is complete, a second pass is executed in which the parameter values for each instruction are evaluated and added to the instruction's numeric. When the parameters are evaluated, any labels that are checked against a file register/address label table. The second pass produces complete instruction values (Table 8-3).

In the parameter *evaluation,* MPASM produces a 32-bit value that is used from the least-significant bits upwards, according to the instruction format. This is done by what I call the *assembler calculator* (covered in detail in Chapter 10). In the third line (*addlw 0-1*), the 32-bit value for negative one (–1) is:

```
0x0FFFFFFFF
```

even though the least-significant eight bits are used with the instruction. Notice that when the instruction is first decoded, the parameter values are left zeroed. When the parameters

**TABLE 8-3 Assembler Second Pass Instruction Information**

| CODE | | ADDRESS | NUMERIC |
|---|---|---|---|
| movlw | 7 | 123 | 3007 |
| Loop | | | |
| addlw | 0-1 | 124 | 3EFF |
| btfsc | STATUS, Z | 125 | 1D03 |
| goto | Loop | 126 | 2924 |

are evaluated, the correct values can be simply added, in the second pass, to the instruction numeric to create a correct instruction for the PICmicro® MCU's processor.

This total operation is "two-pass assembly," the basis for most assemblers on the market today.

The MPASM assembler works almost identically to this, but with a few differences. The first is the inclusion of a macro processor, which inserts macros into the source code. The MPASM assembler also accesses one of four different instruction types to numeric- and parameter-type tables, based on the type of PICmicro® MCU for which the source is written. MPASM also imbeds "included" files into the source code before the passes begin. Once these operations are complete, the two-pass assembler can be invoked to convert the resulting source into an instruction bit pattern for the PICmicro® MCU's processor. These initial steps and operations are also very common for many assemblers.

Errors in the MPASM assembler can be flagged in one of three categories: error, warning, and message. *Errors* indicate that significant problems have occurred with the source code that will have to be corrected before the application can be programmed into a PICmicro® MCU. *Warnings* indicate that the code does not follow the Microchip recommended format and might not work as you expect. *Messages* are used to indicate that something looks funny. Although the code will probably run, you should be sure that you understand the problem.

As I will say throughout this book, you should never attempt to program a PICmicro® MCU and expect it to work properly if there are any errors, warnings, or messages. In some of the examples here, I show what kind of messages you can get and then show what happens when the assembler messages are ignored. The errors, warnings, and messages are well thought out. If something comes up, you should heed them. Chances are, they will save you from having to debug a problem when the application is running in a PICmicro® MCU.

This also goes for suppressing messages. In the first edition, I noted that leaving off the *,f* when an operation's result is put back into the source register could be safely used with only a "message" being displayed. Over the three years since writing that book, I've discovered cases where I have had problems with an application because I left off*, w* from the instruction. Now, I very rarely suppress any messages or warnings and only program a PICmicro® MCU if the application assembles cleanly (without any errors, warnings, or messages).

The only times that I suppress a message using the *error level* directive or warning is when I expect it. In these cases, I suppress the message or warning generation, place the instruction, and then re-enable the message or warning generation. This way, I acknowledge the expected message or warning and am notified if it comes up again elsewhere (and unexpectedly) in the application source code.

In the assembler description, I touched on the idea that the assembler keeps track of the current program counter and can perform mathematical calculations for you. Both these features should be taken advantage of as much as possible to minimize the effort you have to make in having to come up with repetitive (yet unique) labels and constants. The following chapters present how to do this in more detail.

## COMPILERS

When you first see the results of a compiler converting high-level source code into assembly language, you will probably feel like the process of changing high-level language source code into processor assembly code is more a result of magic than a series of mathematical operations executed by a computer. As shown in this section, compilers work through a series of reasonably simple rules to convert high-level language statements into assembly language. There is no magic involved—even though the results sometimes are pretty amazing.

Modern compilers also look for opportunities to simplify assembly code, further resulting in smaller and more efficient applications. If you are a beginner to PICmicro® MCU assembly-language development, you should not be surprised to discover that modern compilers can produce more efficient assembly code than *you* can.

This section, introduces some of the strategies that I used and issues that I encountered when I developed the various PICmicro® MCU compilers over the years for myself. The low-end and mid-range PICmicro® MCU's might seem like they are poorly designed for compilers to develop efficient code for them, but efficient compilers can be created for them reasonably easily if the compiler is well thought out beforehand. Later, this chapter presents my PICLite language for the PICmicro® MCU.

Throughout this book, you will see statements like:

```
A = B + (C * D);
```

A number of steps must be carried out to convert this high-level statement into a PICmicro® MCU assembly-language instruction. Most compilers take advantage of a processor data stack to work through these types of instructions and keeping track of temporary values. When processing the statements, the values are pushed onto the stack in the reverse order that they are required. When the statement is executed, this data is popped off the stack and processed. Because most of the PICmicro® MCU architectures do not have a built-in data stack, the compiler developer will have to decide how to push and pop the data. To explain how compilers work, I introduce their operation using a stack and then discuss the options available in the PICmicro® MCU .

First, a compiler determines what the type of statement it has to work next. Most high-level languages have five different types of statements (Table 8-4).

In the first types (assignment statement, conditional execution, and subroutine/function call), the part of the statement shown as "..." can be considered as the statement. This is the data that will be put on the data stack and then executed. Putting the statement in a heap stored in a post-fix order (the least-significant operations given the highest priority) does this.

**TABLE 8-4 High Level Language Statement Types**

| STATEMENT | TYPE |
|---|---|
| variable = ... | Assignment Statement |
| if (...) | Conditional Execution Statement |
| label (...) | Subroutine/Function Call |
| type variable [= constant] | Variable Declaration |
| [type] label (variable, ...) | Subroutine/Function Declaration |

For the operation in the assignment statement:

```
A = B + (C * D);
```

The post-fix heap is shown in Fig. 8-4.

Next, the order of operations is determined by pulling the data from the head in a "prefix," left to right order. In this operation, the lowest operation on the left is pushed onto the stack, followed by the lowest on the right and the operator is pushed onto the stack. When the operator executes, it pops the previous two (or one) stack elements and then pushes the result onto the stack. The stack operations for:

```
A = B + (C * D);
```

are:

```
Push        C
Push        D
Execute     *
Push        B
Execute     +
Pop         A
```

In this sequence of stack operations, C, followed by D, is pushed onto the stack. Next, they are popped off the stack and multiplied together. The result of $C * D$ is pushed back onto the stack. The results of $C * D$ are popped off the stack, along with $B$ and added together with the result pushed onto the stack. To finish off the instruction, the final result $(B+ (C*D)$ is popped from the stack and stored in $A$.

Depending on your age, you might remember Hewlett-Packard calculators that worked this way. Data entry took the form of the stack instructions listed and were known as *Reverse Polish Notation (RPN)*. It took a bit of getting used to, but once you were able to think in RPN, it actually was easier working through complex problems because you didn't have to remember how many parenthesis were active. In universities and colleges all over the world, the true "propeller heads" people could think in RPN and not use a pencil and paper to plan out how they were going to enter statements into their calculators.

Leaving a result on the stack is important for the other two types of statements. In the "if" statement, if the value on the top of the stack is not equal to zero, then the condition is determined to be true. For the *subroutine/function call* statement, the parameters passed to the *subroutine/function* are accessed from the stack by the *subroutine/function* code, according to their position, relative to the top of the stack. Array elements are also stack values that are popped off when an element is to be accessed.

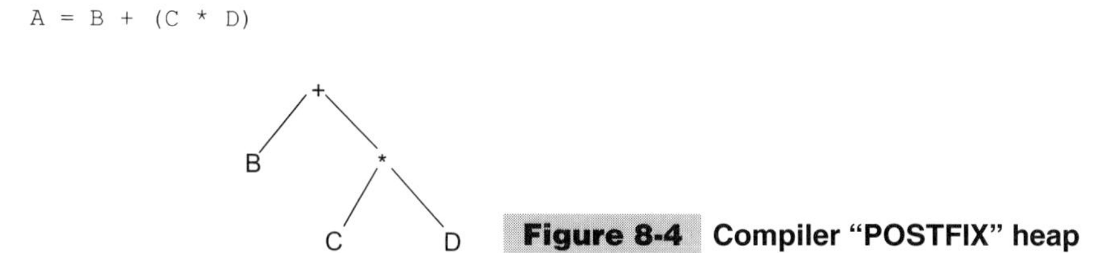


**Figure 8-4** **Compiler "POSTFIX" heap**

In the PICmicro® MCU, a data stack can be implemented using the FSR register. To push an element onto the stack, I recommend that the code:

```
incf    FSR, f
movwf   INDF
```

is used and to pop a value, the code:

```
movf    INDF
decf    FSR, f
```

is used. The push snippet increments the FSR before writing to the stack to ensure there is no way the stack values can be corrupted if the pop operation is interrupted halfway through.

The biggest problem with this method is that the FSR index register is dedicated to the compiled code and is not available to any linked assembly code. The contents of FSR can be stored in a temporary variable before it is changing assembly-language code and restored upon leaving the assembly language code. This seems a bit clumsy, however.

Another way the PICmicro® MCU can implement a data stack is to use temporary variables and access them directly. For the previous assignment statement example, the operations would be:

```
Temp1  = C;
Temp2  = D;
Temp3  = B;

Temp1  = Temp1 * Temp2;
Temp1  = Temp1 + Temp3;

A = temp 1
```

This method gets very complex when statements that have more than one stack entry left on the stack during execution. The two types of statements that best come to mind are ones that use array elements and those that call subroutines or functions that require more than one input parameter. In these cases, the data is evaluated and left on the stack for later operation.

Adding an array element read to the example statement:

```
A = B + (C[4] * D);
```

would result in the post-fix heap shown in Fig. 8-5 and the stack operations:

```
Push          D
Push          4
Push          C[
Execute       *
Push          B
Execute       +
```

In this example, the order of operations would be to push D onto the stack, followed by *4* and when the *Push C[* operation was encountered, the previous element (the *4*) would be popped off the stack and used as the index into array *C*. Once *C[4]* was evaluated, it would be pushed onto the stack and the operation would continue as before with the result left on the stack.

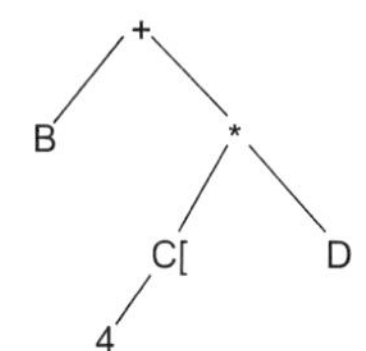


**Figure 8-5** Compiler POSTFIX heap for array statement

Many subroutines and functions have multiple parameters passed to them. For the function:

```
int Func(int varA, int varB)
```

A calling assignment statement could be:

```
A = B + Func(C[4], D);
```

which would require the post fix heap shown in Fig. 8-6.
And the stack loading:

```
Push        D
Push        4
Push        C [
Call        Func
Push        B
Execute     +
```

The stack is executed through similarly to the previous example, but when the *Func* call is encountered, the two previous values are left on the stack and then *Func* is called with them as local variable parameters.

In *Func,* the two parameters are referenced according to their position, relative to the top of the stack. The first parameter will be one position below the stack top and the second parameter was the stack top.

If *Func,* was:

```
Int Func(int varA, int varB)
{

  varA = varA + 1;

  return varA * varB;

} // end Func
```

A = B + Func(C[4], D)

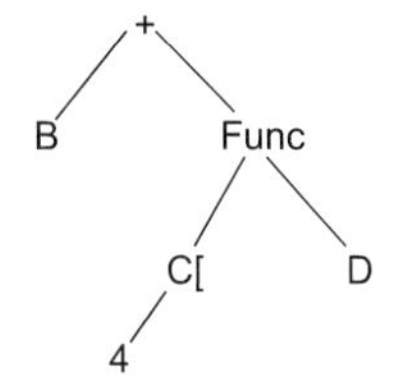


**Figure 8-6** Compiler POSTFIX heap for function call

The actual compiled code for the function could be:

```
Func:
; varA = varA + 1
 Push        StackTop - 1          ; Push "varA" as the New Stack Top
 Push        1                     ; Push 1 onto the Stack
 Execute     +                     ; Pop "varA" & "1", Add, Push Result
 Pop         StackTop - 1          ; Pop Stack Top and Store in "varA"
; return varA * varB
 Push        StackTop - 1          ; Push "varA" as the New Stack Top
 Push        StackTop              ; Push "varB" as the New Stack Top
 Execute     *                     ; Pop "varA" & "varB", Multiply, Push
                                   ;  Result
 return
```

in the calling code after the initial call statement, the first stack item would be popped off and saved and the next two would be popped off and discarded. The saved value would then be pushed onto the stack and execution would continue.

Another way of doing this would be to pop the new top value off the stack and place it as the first stack element before the call. Once this is done, any other values could be popped off the stack and discarded. The code making this call and popping off the unneeded values would look like:

```
Push         varA
Push         varB
Call         Func
Pop          StackTop - 2       ; Put Result in the First Parameter
                                ;  Position
Pop          BitBucket          ; Get Rid of Second Parameter ("varB")
; Result of "Func" is the Top Stack Element
```

When you look at actual code produced by a compiler, you will probably see deviations from the strict stack operations outlined in this section. Optimizing code in the compiler causes these deviations. For example, statements like:

```
A = B + (C * (4 * 2));
```

should be executed as:

```
A = B + (C * 8);
```

With the stack operations looking like:

```
Push          8
Push          C
Execute       *
Push          B
Execute       +
Pop           A
```

In this case, the data stack might not be required by the compiler. Instead, the operations could be carried out using temporary registers. The only time a stack would be needed is if the assignment statement executed in a subroutine was called by the main line code and the interrupt handler or if it was inside a recursive subroutine.

Remember, when using compiled code with the PICmicro® MCU, to be sure that subtraction is handled correctly. Addition (and bitwise) operators can be handled in the order in which they were received. The value being subtracted has to be stored in w before the value is subtracted from is operated upon. This issue complicates the operation of the post-fix heap and stack operations for both subtraction, as well as comparison operations.

## LINKERS

Object files (which normally end in .OBJ) can be produced by a compiler or assembler instead of .HEX files (which are complete applications). Object files are portions of an application that are linked together to create a complete application. Linker programs are very complex and not only concatenate object files together, but also provide address references between the object files. They can also be used in conjunction with "make" files, which are used to specify how the final application is put together.

Object files are very similar to hex files, except labels that are dependent on references outside the current source file are flagged and not replaced by constant values. Along with this, references to addresses outside the current object file are also flagged. Some of the purposes of a linker are to resolve the references to external labels with the public labels in the object files and correlate the addresses to the instructions that use them.

Another function of the linker is to provide addresses for the different object files. To do this, the code must be written so that it is relocatable. *Relocatable* means that no specific addresses can be used within the source code. For high-level languages, this is usually not a problem, although it can make assembly-language development more difficult.

To show what I mean, consider the case where three object files must be linked together (Table 8-5).

In this example, ObjectB or ObjectC can be put in any order after ObjectA. Thus, ObjectB can start at address 100 or address 250, depending on whether or not ObjectC is between it and ObjectA.

Once the linker has calculated the addresses for each object file, it can calculate the addresses for the labels that are accessed outside of each object file and put the addresses into the code. The resulting hex file can then be "burned" into a PICmicro® MCU.

The Microchip linker currently does not have a make capability, but many other linkers do. A *make* file is a simple program that specifies how an application is to be built. The make utility executes the make program and selects files for assembly, compiling and linking with different options. The make file can be very complex, with different options specified for conditional debugging or partial linking. Because the make utility can also assemble and compile source files, parameters can be passed to the assembler and linker from it.

**TABLE 8-5 Linker File Lengths and Positioning**

| FILE | LENGTH | COMMENTS |
|---|---|---|
| ObjectA | 100 Instructions | Contains Application Header — must always be first |
| ObjectB | 200 Instructions | |
| ObjectC | 150 Instructions | |

This book does very little with the MPLAB linker and object files other than show how this capability can be used. For simple assembly-language applications written by one person, linking object files does not make a lot of sense because the process of creating a hex file is more complicated and slower (which means more opportunities for errors). To provide a simulated link capability, I will create "include" files with common functions that are loaded into the application code when it is assembled.

Linking applications does make a lot of sense when multiple languages are involved (each object file is created from one language type), multiple people are involved with the application development or the source application is very long (longer than 10,000 lines). In these cases, linking object files together will make the application easier to understand and probably faster to create and debug.

## SIMULATORS

Throughout this book, I have tried to stress the importance of simulators when developing PICmicro® MCU applications. A simulator is a software tool in which the operation of an application can be observed working in a software-created processor, allowing you to find and fix problems before a PICmicro® MCU is programmed and you are stuck with a circuit that just sits there and does nothing. As presented later in the chapter and the book, Microchip's MPLAB IDE with its built-in simulator is used for testing out applications.

A simulator consists of a software model of the PICmicro® MCU processor, which can be controlled along with the ability to pass test I/O signals back and forth to the software model processor. It is important to remember that the processor and I/O in a simulator are software models. Some situations that can arise in simulator hardware cannot be properly simulated. The MPLAB IDE is an excellent tool for testing out application code, but it is somewhat limited in its ability to model advanced peripheral I/O ports.

I tend to think of a simulator as a collection of "black boxes" that are controlled by the simulator host software (Fig. 8-7). I drew the simulator this way because it allows different boxes to be swapped in and out to make up different part numbers without significant effort to create simulators for different functions. Probably more actual simulator modules are used in the MPLAB IDE, but for the purposes of this chapter, the five presented in Fig. 8-7 are adequate.

The program memory block is loaded with the hex file that will be programmed into the PICmicro® MCU . The processor model pulls data from this simulated program memory as required. The file registers are similar, but the processor model can read and write to the file registers.

The I/O and hardware registers block provides some kind of model of the I/O pins. For basic digital I/O, MPLAB does provide a good model of the I/O hardware, but for advanced peripherals (the USART, MSSP, and other peripherals), the MPLAB IDE model does little more than just accept input from the processor and put it into the registers.

To help debug your applications, the I/O pins have the capability of being driven with simulated external inputs. This stimulus box in Fig. 8-7 can either be directly used in changing I/O register bits or it can run a stimulus file. A *stimulus file* is a file that is created with I/O information that can be processed by the simulator processor without much intervention by the user. I focus on stimulus files as a tool to be used with the MPLAB IDE simulator, which is used to debug application code in this book.

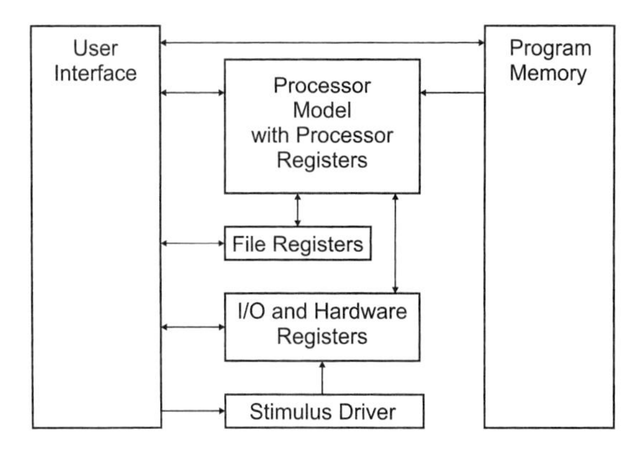


**Figure 8-7** Simulator software architecture

The topic of stimulus files brings up an important point. When debugging an application, the simulator must be set up in such a way that it will always run the same way. This is an important philosophical point that I want to make sure you follow. If the simulated application runs differently each time, you will have problems trying to identify problems and fixing them. When you debug your application using a simulator, you should focus on fixing the problem, not learning how the simulator is set up for this application's debugging.

The processor model block is obviously the heart of the simulator, with the user interface commanding it to execute, single step, or step. The processor model is an extremely complex piece of software, which not only has to fetch and execute instructions and access registers, but also has to manage such peripheral functions as TMRO interrupts and the watch-dog timer. Making the creation of this module even more complex, the execution of the simulated instructions must be as fast as possible.

The user interface is the primary window into the application code. As such, it should be as configurable as possible to allow for user preferences, as well as display different variables and I/O registers, based on the users preferences. Along with displaying customizable registers, the simulator should work through the code source, rather than the simple instructions.

In the first edition of this book, I provided Microchip's MPSIM MS-DOS command-line simulator. This simulator provided all the features covered in this section, except for a full-source file display. The ability of seeing how code executes from a source-code view is crucial for me and made working with the MPSIM IDE difficult. Having this capability in the MPLAB IDE makes debugging application software much easier.

The MPLAB IDE does have all of the simulator features discussed here and it integrates with the assembler. This book concentrates on the MPLAB IDE and its simulator to demonstrate how application code runs in the PICmicro® MCU. It provides you with the ability to observe the application execution and, hopefully, debug any potential problems before you program any hardware and try out the applications.

## EMULATORS

The next step up from a simulator is an *emulator,* a device that is connected to your development PC or work station and allows you to monitor application execution in hardware. The emulator block diagram looks like that shown in Fig. 8-8.

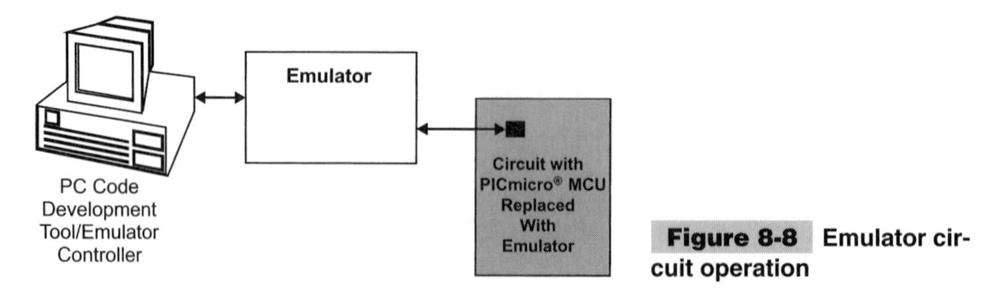


**Figure 8-8** Emulator circuit operation

An advantage of an emulator over the simulator is that actual pin I/O signals can be observed, both from the processor's perspective, as well as from the circuit's. As well, the emulator often has the same hardware as the actual device, so there are no missing peripheral interface functions.

The best method of providing an emulator is to use a *bond-out chip*. This is an actual PICmicro® MCU with specific access provided to the processor so that it can be started or stopped and provide emulator-unique instructions. The Microchip emulators all use actual PICmicro® MCU chips, so the hardware interfaces are accurate to actual devices.

There are two disadvantages to working with an emulator. First is cost. Emulators generally cost $2,000 (or more) and require separate "pods" for each device being emulated. This cost is not significant for many companies, but, for small companies and individuals, it can be.

To help offset the costs, there are a number of simplified emulators, such as Microchip's MPLAB-ICD or the EMU-II, presented later in the book. These devices often provide many of the same features as a full emulator, but are restricted in different ways.

Another problem with emulators is the difficulty in connecting them to a circuit and the unreliability of the connection once it is in the circuit. The MPLAB-2000 emulator, as shown later in the book, utilizes a small tripod to hold the emulator close to the target circuit without placing any strain on the PICmicro® MCU connector. This is the best method that I have seen to minimize connection problems.

Like the simulator, having a direct connection to a source-code browser is crucial for efficient operation of the emulator. Without this capability, you will be forced to cross reference to a listing file for absolute address. I feel that a simulator and an emulator user interface should do this for me.

As reiterated elsewhere in this book, an emulator is not a development tool; it is a debugging or failure-analysis tool. When developing applications, try to avoid getting into the "code a little, debug a little" cycle because having an emulator makes it easy to do. When developing applications, it is important to focus on the big picture and not lose sight of it by playing around with different functions.

## IDEs

When I'm developing software, I always find that I'm the happiest (most productive, debugging fastest) using an *Integrated Development Environment (IDE)*. This extends to PC programming, where I always liked the original Borland Turbo Pascal and the modern Microsoft Visual Basic and Visual C++ development tools (part of Microsoft's Visual Development Studio IDE). I don't believe that I am alone in feeling this way. The Microchip PICmicro® MCU MPLAB Integrated Development Environment (Fig. 8-9) has done a lot

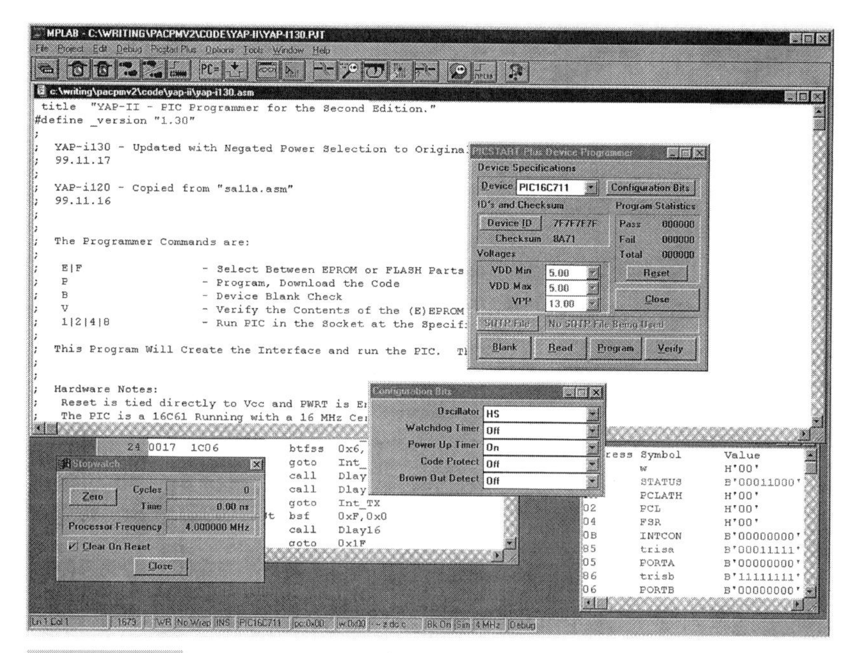


**Figure 8-9** "MPLAB" IDE with PICstart plus interface

to make the PICmicro® MCU the most popular microcontroller for people looking for a place to learn about microcontrollers.

An integrated development environment integrates all of the software-development tools I've described in the previous sections. A microcontroller IDE brings the following tools together:

- Editor
- Assembler
- Compiler
- Linker
- Simulator
- Emulator
- Programmer

The purpose of an integrated development environment is to provide the data in a system that allows it to be shared seamlessly, as far as the user is concerned and does not require any special input on their part. When I first started working with the PICmicro® MCU with MS-DOS command-line tools, to change some source code, assemble and program a part, I had to go through the following steps:

1. Invoke a source-code editor.
2. Change/enter application code.
3. Save and exit editor.
4. Invoke the assembler.

5 Select the source file.
6 Invoke the programmer.
7 Select the correct configuration bits.
8 Program the PICmicro® MCU.

In the MPLAB IDE, the steps for changing source code and programming a part are reduced to:

1 Invoke the MPLAB IDE.
2 Change/enter application code.
3 Click on *Project -> Build.*
4 Click on *Enable PICStart Plus.*
5 Click on *Start.*

Not only is this process faster and less prone to errors, problems with the source code can be fixed immediately in the editor and the application can be rebuilt without having to go back to the editor and assembler.

The MPLAB IDE's projects are used to record the user's preferences for an application, the source code files currently edited, the PICmicro® MCU part number (along with its execution parameters, such as the clock speed used with it), and the user's preferences in terms of window placement on the MPLAB IDE desktop. Later sections of this chapter describe the different features of the MPLAB IDE and how to work with it when developing your own applications.

When learning a new programming language or microcontroller, you should only look at devices that have a well-supported IDE available. This will make the process of learning the new device easier for you and it will minimize the possible mistakes that you will experience when transferring data among the editor, assembler, simulator, and programmer. This book works with the MPLAB IDE exclusively because of its ease in integrating the PICmicro® MCU application-development steps with application code development, simulation, and programming.

# High-Level Languages

The PICmicro® MCU has been around long enough for there to be a number of different languages (and versions of each language) to choose from. The appendices list a number of sources of languages that you can choose from. This book includes the examples in assembler, with C pseudo-code to explain what the code is doing. Again, staying with the theme of this chapter, the choice of language is up to you; I feel it is important that you program in the language in which you are most comfortable.

Having said this, I still have some comments. I feel that a few features of a language implementation are important.

The first has to do with memory. Few microcontrollers (and the PICmicro® MCUs in particular) are not blessed with unlimited memory (or disk-caching schemes), either in terms of control store or variable RAM. The language and implementation that you use should be very frugal with its use of the memory resources.

I have found that, when developing PICmicro® MCU applications, a well-designed application does not require a lot of memory. Although in several of the applications presented in this book do use up most of the available resources, in none of them am I totally hamstrung with meeting the application's requirements. In many cases, I am trying to use the PICmicro® MCU with the smallest amount of memory required for the application to minimize the total cost.

But, you are probably thinking, everything is written in assembler and, almost by definition, assembler only provides exactly what you require. Languages can use a lot of memory—especially if they aren't optimized. Before investing in a compiler, be sure that you understand what type of code is produced and the typical amount of code you can use. Efficiency is measured in terms of response time and program space. Poor compilers can be many *hundreds* of times less efficient than good compilers.

Another aspect of how much assembler code a compiler produces affects the efficiency of how you debug the code. When developing code for the IBM PC, I often find it useful to look at the produced assembler code to see why the program doesn't work. If the compiler doesn't produce efficient code, you might have a lot of problems understanding what the program is doing at a given point.

The next important aspect of the language is the data types used. Many times I've found that I would like to use more than eight bits for counters and such. Native PICmicro® MCUs only run 8-bit code, you should be sure that the compiler you use provides a variety of data types (16 bit can be extremely useful). Many of the programs presented in here use 16-bit variables and the appendices include a number of 16-bit mathematical algorithms for your use.

Hardware support and initializations are another important aspect to look at when evaluating PICmicro® MCU compilers. The questions center around what does the compiler's initial code do before starting the application code (i.e., does it set certain features, such as timers and I/O ports in specific states that might cause problems later?). If the compiler uses resources that you will want to use (i.e., the FSR register), you might have to change the way you were planning to develop an application because of the way the compiler uses and sets up the resources that are available within the PICmicro® MCU.

Applicability across the whole PICmicro® MCU line is another consideration when looking at different languages. The language that you choose should produce code for all of the devices in the PICmicro® MCU lineup. This is very important because you might be putting your application on a 16F84 for development and debugging, but your ultimate application might use a 16C54 (which is much cheaper). Using a compiler capable of producing code for all members of the PICmicro® MCU line means that new code doesn't have to be written when porting functions (or even whole applications) to different members of the PICmicro® MCU family.

Modifying the compiler-produced assembly code is less than optimal from the perspective of code support. Errors can creep in when changing code to support two different PICmicro® MCUs. These errors include having to change a number of locations, with some ending up being missed or having to change code for different hardware implementations in different devices. Some programming techniques (such as only accessing specific hardware in subroutines) can minimize the opportunity for these errors in the future. If at all possible, the compiler you choose should be able to create code for different PICmicro® MCU devices.

Optimization of the produced code is an important aspect of the compiler. The compiler should be able to understand what is going on and use the most efficient code possible. The other important aspect of optimization is making sure the compiler only uses the code that is required for the application. I personally like to have all my subroutines available in one file or library. Only the subroutines that a particular application calls should be included in the final object file. In the MPASM assembly language (and many other languages), this is known as *conditional assembly*.

In much of the code in this book, I have used conditional assembly to allow easy/fast debug of code. Chapter 10 covers conditional assembly in greater detail, along with how it can be used to enhance your application development. Often, long delays are necessary in when code is running in real time. These delays can make the operation of the simulator/debugger very tedious and difficult to understand the flow of the program. By eliminating these blocks of code for simulator debugging, the time required to execute the code is much faster and it is much easier for you to follow the code operation.

Even simple optimization routines can improve the size and speed of the compiled code by orders of magnitude and make a high-level language approximate the performance (speed and code size) that of assembled code.

The last (and probably most important) piece of advice that I have about languages is: don't pick one that hides the features of the microcontroller. When I say this, you should have direct access to all the PICmicro® MCU registers.

The reason for this, as paradoxically as it seems, is simplicity. If the language controls the interface to the hardware, you must learn how to use the language controls. This means, in addition to understanding the PICmicro® MCU, you also have to learn the language and its hardware interface. To make matters worse, if the interface is designed by somebody else who doesn't have your application in mind, they've just written the most universally designed interface they could.

For example, to print "Hello World" and start a new line would be:

```
printf( "Hello World\n" );     // In "C"
```

The *printf* routine depends on specific hardware to function properly. If this function changes in different PICmicro® MCU part numbers, it is important that you are able to access the required resources so that the necessary functions can be coded easily.

After reading this section, you probably think assembler is the best way to go for efficient PICmicro® MCU application development. I think that high-level code development has significant advantages in terms of development effort and speed. Writing the code in a high-level language does prevent a lot of the typos and confusion endemic in assembler coding. Take care to ensure that the compiler and language chosen does not limit the type of applications that you can create. Instead, it should simplify them and enhance the development process.

## GLOBAL AND LOCAL VARIABLES

When working with assembler and high-level languages, two types of variables can be used in applications: global and local variables. *Global* variables can be accessed anywhere in an application, but *local* variables can only be accessed within the routine in which they are declared. Deciding which type of variable to use will affect the operation of your PICmicro® MCU application and could make it easier to avoid problems with your application code. You might assume that local variables can only be used in such high-level languages as C, but they can also be used in assembler subroutines as well.

Global variables are normally declared and local variables are declared within a subroutine, as shown in the following C code snippet:

```
int row, col;

int videoPos(int i int j) {

int temp;

 i = i * 80;
 temp = I + j;

 Return temp;

} // End videoPos Function

main() {                        // Example Mainline

 for (row = 0; row < 25; row++ )
  for (col = 0; col < 80; col++ )
   VideoMemory[videoPos(row , col)] = ' ';

} // End Example Mainline
```

The variables *row* and *col* are global and can be accessed by any function in the application. The variables *i, j,* and *temp,* are local to the *videoPos* function. In C, global variables are defined outside of any functions, as shown in the example. Local variables are defined as being inside functions and include the parameters to the functions (*i* and *j* in the example).

These variable names can be reused in different functions and there will not be any confusion by the compiler. For example, in the mainline (*main*), *row* and *col,* could be replaced by the local variables *I* and *j* and execution would not be any different.

If *row* and *col* were changed to *i* and *j,* the input parameters to *videoPos* changes *i* to a new temporary value within the function. If *i* was a global variable, *i* would be an invalid value for the rest of the application.

This is why I always stress that local variables (or at least variables that are specific to a function and subroutine) should be used instead of global variables as much as possible. If the variables are local to the subroutine, then there is no opportunity for data to be destroyed by other routines and no opportunity for the routine to destroy data required for other routines.

In most processors, local variables are stored in and referenced from the data stack. This allows the data to be allocated dynamically by the active subroutines, as required. The input parameters of a subroutine (*i* and *j* in *videoPos*) are usually explicitly pushed onto the stack when the subroutine is called. Temporary variables declared inside of the subroutine are actually data pushes to make space on the stack. Local variables inside a function are relative to the current stack position. In the previous example, the call to *videoPos* and its operation can be shown using a pseudo-assembler.

```
Main:

   :

; VideoMemory[videoPos(row , col)] = ' ';
 push row                       ; Parameters Pushed in Same Order as in
 push col                       ; Subroutines
```

```
 call videoPos
 pop _____                ; Restore the stack to the precall value
 pop _____

; videoPos(int i, int j)

; int temp;
 push #0               ; Define "temp"

; i = i * 80

 move accumulator, (sp - 3)    ; Get first Parameter ("i") from stack
 mul accumulator, #80
 move (sp - 3), accumulator    ; Store the Result back onto the stack
                               ;  (in "i")

; temp = i + j

 move accumulator, (sp - 3)    ; Load Accumulator with "I"
 add accumulator, (sp - 2)     ; Add "j" to Accumulator
 move (sp - 1), accumulator    ; Save the Result in "temp"

; return temp

 pop accumulator              ; Restore the Stack to its Original Value and
                              ;  load the Accumulator with "temp"

 return                       ; End of "videoPos"
```

In this example, the parameter and local variables are pushed onto the stack before the subroutine's code is executed. By doing this, unique memory blocks are allocated for local variables each time the subroutine is called.

Most of the PICmicro® MCU architectures do not have a data stack with push, pop, and reference instructions, which makes the operation more complex. In the PICmicro® MCU, unique names could be chosen for each local variable in each subroutine. This isn't a bad method, except when subroutines are called from interrupt handlers or from recursive subroutines.

In the case where a subroutine can be called from mainline code or on interrupt handler, you probably would be best off making two copies of the subroutine, each with their own local variables.

If you are going to use recursive subroutines with local variables (which I don't recommend because of the small program counter stack in the PICmicro® MCU), you will have to simulate a stack using the FSR index register.

## BASIC AND PICBASIC

When I look at the progression of the *Beginner's All-purpose Symbolic Instruction Code (BASIC)* language for the PICmicro® MCU, I see some amazing parallels to the progression of BASIC for personal computers. Twenty five years ago, when the first PCs were coming into existence, if a user did not want to program in assembler, the only option available was BASIC. Today, some very sophisticated BASICs are available for the PC. However, the BASICs available for the PICmicro® MCU are more similar to the PC BASICs of a quarter century ago than what is available today, like *Visual Basic*.

When I first started programming, I started with BASIC on an Apple ][. This basic was written by Steve Wozniak and assembled by hand (I remember seeing an interview in which he boasted that the original Apple BASIC source code was never entered into a computer; it was written in a notebook, where it was continually updated and modified). The basic itself was very primitive. It just had the ability to display and edit source code, save it onto a cassette player (seriously), and execute the applications. The original Apple Basic was able to access all hardware addresses within the computer directly and optionally execute some assembly-language source.

This is in stark contrast to the latest versions of Microsoft's Visual Basic. This book presents a number of PC to PICmicro® MCU interface applications that are Visual Basic based. Visual Basic can be used to access hardware and hardware functions within the PC (as I do with the applications presented in this book). It can be used to provide front-end interfaces to database programs and Visual Basic can provide access to networked information, including information on the Internet. Visual Basic provides a very user-friendly development system and the ability to develop code that runs under code-protected operating systems. Visual Basic does have one drawback for hardware interfacing applications; it cannot access system resources, except through DLLs and system files, such as the package provided with the El Cheapo programmer.

The first BASIC that was generally available for the PICmicro® MCU was Parallax's PBASIC. This language is a very simple BASIC that provides all the basic features, but with some things to watch out for. "Introduction to Programming" on the CD-ROM provides a detailed description of meLab's PicBasic (which is a derivative of Parallax's original PBASIC), but I want to make you aware of a few points.

The first point is how variables are handled. In "true" BASICs, variables are *implicitly* declared. This means that they are not declared and no is space reserved for them until they are referenced in the source code. The type of variable is defined by the suffix to the label (with no suffix, the variable is a floating value, a # suffix means that the variable is an integer, and a *$* suffix means that the variable is a text string). Specific types of variables (including arrays) are defined using the *DIM* statement.

In PBASIC, eight-bit variables are predefined and are given the labels *B0* to *B15*. These variables can be concatenated together to form 16-bit integers or broken up into eight individual bit variables. Arbitrary label names can be assigned to these variables as a *define*. There are no arrays in PBASIC.

Assignment statements are very usual in PBASIC. Instead of working through an order of operations, assignments are executed left to right. For example, the statement:

```
A = B + (C * D)
```

would be expected to execute in the order:

1. Multiply the contents of *C* with the contents of *D*
2. Add the contents of *B* with the product of *C* and *D*
3. Store the result in *A*

In PBASIC, the addition would execute first, followed by the multiplication. Written out, this would be:

1. Add the contents of *B* to the contents of *C*
2. Multiply *D* with the sum of *B* and *C*
3. Store the result in *A*

This can be a real problem for people who have experience with other languages. To avoid these problems when I work with PBASIC, I normally avoid compound statements (such as the example). Instead, I break it up into its constituent parts and save any intermediate results in temporary variables.

This would mean that the preceding statement would be written in PBASIC as:

```
Temp = C * D
A = Temp + B
```

to execute correctly.

When writing compound statements like this, be sure that the intermediate values are not stored in the final destination. Storing intermediate values in the final destination could cause them to be used by peripheral hardware in the PICmicro® MCU, causing invalid operation of the application.

PicBasic executes instructions in the conventional order that can make things just as confusing as PBASIC's execution orders. To avoid any problems moving code between the PICmicro® MCU and the Basic Stamp, I avoid writing complex statements and instead use a *Temp* variable (as shown) any time I am developing BASIC code for the PICmicro® MCU.

Along with the lack of arrays, there is no way to jump to an arithmetically calculated address in PBASIC, but there are some useful branch and table instructions available in the language that can perform similar functions. In PBASIC, all jumps and calls are to labels; no absolute addresses are allowed in the source code.

PBASIC cannot access hardware resources directly. Like the variables, eight (Basic Stamp1) or 16 (Basic Stamp2) I/O port bits can be controlled and accessed within the application, but the registers are never accessed directly.

meLab's PicBasic compiler has faithfully reproduced the PBASIC language, and has allowed the addition of assembly-language code and hardware access to the application source code.

## C FOR THE PICmicro® MCU

Currently, the most popular high-level language application development tool is C. This high-level language is used in application development for virtually every computer available on the market for the past 20 years or so. The reasons for this popularity is the logical design of the language's statements and advanced features, such as data structures and the efficient processor assembly-language code that can be generated.

C's popularity, like many things, has fed on itself. As the language has become more popular, more courses have been taught in it, resulting in more C programmers. The Internet has allowed people to share source code (which can be used in a variety of different systems), further increasing its popularity and usefulness. This popularity makes C an important language to be comfortable with for anyone interested in developing systems applications.

The low-end and mid-range PICmicro® MCU architectures are not well suited for procedural-based languages, like C. However, the higher-end PICmicro® MCUs are better suited and can handle the high-level language statements reasonably efficiently for most cases. This section covers the different PICmicro® MCU devices and the features that make them poorly or well suited for high-level languages, as well as the features you should look for when choosing a PICmicro® MCU C compiler.

The largest point of concern for implementing a high-level language, such as C, is the need for a data stack that can save intermediate values in the high-level language statements. This is shown in the traditional methods used to program complex statements:

```
A = B + (C * D);
```

where the statement parameters are pushed onto a stack and processed from here. For this statement, the stack operations would be:

```
push B
push C
push D
mul
add
pop  A
```

Where the *mul* and *add* statements pop the two top elements off the stack, perform the operations, and then push the result back onto the stack.

The problem with the low-end and mid-range PICmicro® MCUs is the lack of a data stack. This stack can be simulated using the macros:

```
Push Macro
 incf  FSR, f
 movwf INDF
 endm
```

and

```
Pop Macro
 movf  INDF, w
 decf  FSR, f
 endm
```

The first macros push the contents of the accumulator (w) onto the stack after incrementing the data stack pointer (FSR). I increment first because, if there is an interrupt, which also uses the stack and the data was written to before the increment, there is the opportunity that the value put onto the stack is overwritten during the interrupt handler.

The PIC17Cxx and PIC18Cxx do not have this issue because indirect addressing operations have the capability of incrementing and decrementing the FSR register during data transfers. Ideally, in a C language, the pre-increment and post-decrement should be used to provide stack functions that are compatible with these statements.

Even though the PIC17Cxx has the FSR pre- and post-increment and decrement operations, they do not have an offset add/subtract capability, which would make accessing stack data within subroutines easier. For example, the call and subroutine code:

```
Subr(int YAxis, int ZAxis)
{

  :

 Address = 4 + (YAxis * Width) + (ZAxis * Length * Width);

  :
```

```
} // End Subr

  :

 A = Subr(j, k);
```

requires accessing data on the stack at arbitrary locations. Before *Subr* is called, both the *j* and *k* variables are pushed onto the stack and accessed by the statements within *Subr*.

For the *Address* assignment statement, these stack values have to be pulled off the stack and push temporary values onto the stack. Assuming the pre-increment and post-decrement push and pop operations described, this statement would become:

```
; Address = 4 + (YAxis * Width) + (ZAxis * Length * Width);
 push Stack - 1                 ; Push "ZAxis" Stack Parameter
 push Length
 mul
 push Width
 mul
 push Stack - 3                 ; Push "YAxis" Stack Parameter
 push Width
 mul
 add                            ; Add the Two Products Together
 push 4
 add
 pop Address
```

The problem is the *Stack - 1* and *Stack - 3* operations. In the PIC18Cxx and many other processors, an offset can be added to the index to carry out an access. This is not possible in the PICmicro® MCU, except in the case of the PIC18Cxx (which has an indirect access with adding the offset to the FSR register) and the *movff* (move from file register to file register) instruction.

In the low-end, mid-range, and PIC17Cxx PICmicro® MCUs, the *push Stack - n* (which I call an *offset push*) operation would be executed as:

```
bcf       INTCON, GIE        ; Disable Interrupts in the Mid-Range
                             ;  PICmicro® MCUs
movlw     n                  ; Decrement the Stack to the Correct
subwf     FSR, f             ;  Position
movf      INDF, w
movwf     PushPopTemp
movlw     n + 1              ; Increment to the Current Position + 1
addwf     FSR, f             ;  for Push
bsf       INTCON, GIE        ; Enable Interrupts in the Mid-Range
                             ;  PICmicro® MCUs
movf      PushPopTemp, w     ; Save the Data on the Stack
movwf     INDF
```

This is obviously quite a complex operation and it uses up a lot of execution cycles and space in the PICmicro® MCU's program memory.

In the PIC18Cxx, the data stack pushes and pops should use post-decrement and pre-increments to make the previous stack values be "above" the current. By doing this, the code required to execute the *Address* assignment is:

```
movlw   1                  ; Push "ZAxis" onto the Stack
movff   PLUSW0, POSTDEC0   ; Push the Previous Stack Element Again
                             onto Stack
movff   Length, POSTDEC0   ; Push "Length" onto the Stack
```

```
call    mul                  ; Multiply the Two Top Stack Elements
                               Together
movff   Width, POSTDEC0      ; Push "Width" onto the Stack
call    mul
movlw   3                    ; Push "YAxis" onto the Stack
movff   PLUSW0, POSTDEC0
movff   Width
call    mul
call    add                  ; Add the Two Parameters together and
                               put the Result back onto the Stack
movlw   4
movwf   POSTDEC0             ; Push 4 Onto the Stack
call    add
movff   PREINC0, Address
```

Compared to the initial case (shown above) of 12 instructions, the PIC18Cxx can execute the same statement in 15 instructions. The increase of 25% over optimum is actually very good for any processor despite the seemingly high inefficiency. The mid-range PICmicro® MCU architecture (using the push, pop, and offset push code described), carrying out these operations would require 36 instructions (200% more).

The other issue with high-level languages that the low-end and mid-range PICmicro® MCUs are not well suited for is *pointers*. As I've indicated elsewhere in the book, pointers are difficult to implement. For most people, this is not a bad thing.

The PIC17Cxx and PIC18Cxx have multiple FSR registers, which means that the stack pointer FSR can be left alone while another FSR register is used to provide the pointer function. In the PIC17Cxx, I would still not recommend that pointers be used because of the banking scheme built into the PICmicro® MCU. The PIC18Cxx, with its large (up to 4,096 byte) contiguous data memory space, allows for pointer use quite easily and efficiently.

Structures and unions in C also require an additional FSR to access specific elements of the data types. Again, I would recommend avoiding the low-end and mid-range PICmicro® MCUs for C applications that have these programming constructs.

In typical C development environments, a large amount of heap space is made available. This is typically used for the statement data stack (described previously), but it is also used to provide space for local variables, including structures and unions. In all the PICmicro® MCU architectures, this space is very limited. To avoid problems with the data stack running out of space, local structures should not be used in an application.

If you have worked with C and are familiar with how it works, you should be comfortable with the concept that it is almost impossible to use without pointers. Simple statements, like *printf*,:

```
printf("Failure, Expect 0x0%04X Actual 0x0%40X", Expected, Actual);
```

use pointers even though they are not explicitly noted. In the *printf* example, everything within the double quotes (") is stored in memory and the pointer to this string is passed to the *printf* library routine. In a typical C implementation for a Harvard processor, the string in quotes is copied from program memory and stored into register memory and a pointer to this address is returned. Implementing this type of statement in a PICmicro® MCU causes a double whammy to the application developer. The string takes up quite a bit of space in both the program memory and register memory, of which there isn't a lot of in either case.

In PICmicro® MCU compilers, this code should never use a pointer, although it might be reasonable to do it in the PIC18Cxx because of its large, "flat" register space. Instead, the string in quotes should be kept in program memory and a pointer to the string passed to the *printf* function should indicate that the string is not in register memory.

When selecting a C compiler for the PICmicro® MCU, be sure that it provides a basic interrupt-handler procedure header. In this book, I use the "interrupt" data type that indicates that the interrupt-handler code is within the procedure:

```
interrupt InterruptHandler()
{
  :                           // Put in Interrupt Handler Code Here

} // End InterruptHandler
```

The start of the interrupt-handler procedure should save all of the possible context registers by pushing them onto the stack (and popping them off). Pushing the context registers onto the stack will allow "nested" interrupt handlers or, in the case of the PIC17Cxx and PIC18Cxx, multiple handlers to execute.

Notice that when implementing an interrupt handler in the mid-range and PIC17Cxx devices, the w and STATUS register contents will have to be stored in a temporary register first to avoid problems with the stack. A sample mid-range PICmicro® MCU C interrupt-handler starting code that pushes the context registers onto the stack could be:

```
InterruptHandler
 movwf     _w             ; Save "w" and STATUS before placing on stack
 movf      STATUS, w
 bcf       STATUS, RP0    ; Make Sure Execution in Bank 0
 bcf       STATUS, RP1
 movwf     _status
 movf      _w             ; Now, Save "w" onto the stack
 incf      FSR, f
 movwf     INDF
 movf      _status        ; Save the Status Register onto the stack
 incf      FSR, f
 movwf     INDF
 movf      PCLATH, w      ; Save the PCLATH Register
 incf      FSR, f
 movwf     INDF
 clrf      PCLATH         ; Reset PCLATH to Page 0 for the Interrupt
                          ;  Handler
```

In this sample code, notice that I reset the RP1 and RP0 flags (because, in the mid-range devices, I recommended that all arrays and other data structures that are accessed by the FSR register be placed in bank 1 and single variables in bank 0.) and that I reset PCLATH. This code is the truly "general case." Although it takes a few more cycles than what could be considered "best," it should always be used to ensure proper operation in all cases.

After the header is executed, interrupts can be enabled again, allowing for nested interrupts because the saved w and STATUS register values are saved on the data stack. The interrupt entry and exit code shown can be used exactly as shown in assembly-language applications to provide nested interrupt handlers. This is the only case where I feel it is acceptable to allow nested interrupts in the mid-range PICmicro® MCU.

To return from an interrupt in the mid-range PICmicro® MCU, the following code should be used:

```
movf      INDF, w        ; Restore PCLATH
decf      FSR, f
```

```
movwf     PCLATH
movf      INDF, w
decf      FSR, f
bcf       INTCON, GIE          ; Disable Interrupts until operation is
                               ;  complete
movwf     _status               ; Save the Status Register Values
movf      INDF, w
movwf     _w                   ; Save the "w" Register
movf      _status, w           ; Restore STATUS
movwf     STATUS
swapf     _w, f                 ; Restore "w"
swapf     _w, w
retfie
```

A "true" C (ANSI standard) has a number of library routines associated with it. These routines provide the capability of different types of data types (such as floating-point numbers), finding trigonometric and logarithmic values, and providing standard I/O to consoles (*getf* and *printf*). Many standard routines and often a lot of custom ones are available for the environment (processor and operating system) and hardware. Just as an example, for the C that I use for the IBM PC, the library routines take up 286 KB of space on the hard drive. I would expect that a full-featured PICmicro® MCU C compiler would have a similarly sized library.

This makes the importance of a sophisticated application linker, which only includes needed functions from the application code and library, very important. This capability is provided in Microchip's MPLINK, but you should check for it in other tool sets before buying.

When using functions, understanding how they work is even more crucial in the PICmicro® MCU than in a PC or workstation. As I mentioned, the PICmicro® MCU in general only has limited heap space and different PICmicro® MCUs have different hardware features. Standard I/O functions (*getf* and *printf* ) can be designed to work with PICmicro® MCUs that have built-in USARTs, but will not work in lower-end PICmicro® MCUs, which do not have this capability, so their use will either have to be flagged or part number specific routines used.

## THE PICLite LANGUAGE

To demonstrate my thoughts on how PICmicro® MCU compilers should operate, I have included a copy of my PICLite compiler. PICLite is a very simple language designed for the mid-range (14-bit core with Bank 0 and Bank 1 registers) family of PICmicro® MCUs.

Currently, the output is assembler and is executed from a PC-DOS command line. Once the compiler has been run cleanly, the output code can be run through MPASM (version 1.40 or greater). The reason why I specified assembler as the output is because the application developer can insert assembler code in line without any special headings and the compiler can identify the instructions and hold them until the end of the compilation step for processing by the PICLite assembler.

This language is simpler than BASIC, but I feel that it is well organized and optimized. Source programs can be of any size and are not limited in any way (other than your PC's hard drive space). The PICLite language has a very comprehensive macro language capability. Structured programming constructs (e.g., *while*) are available using macros.

Eight-bit (byte data type) and 16-bit (word data type) variables and arrays are supported, but registers (defined as specific addresses) can only be an eight-bit variable. For assignment statement formatting, I have kept to *C* conventions as much as possible.

Creating a PICmicro® MCU interrupt handler is possible and quite simple to implement. At the end of the mainline code, placing the label *int:* will cause an interrupt header to be inserted in the code (at address 0x004) and the code following will be placed starting at the interrupt vector address. At the end of the program, the interrupt return is placed. PICLite takes care of providing the interrupt-handler context save and restore information and sets up any necessary information for executing the PICLite statements.

The format of the language is defined in Tables 8-6 through 8-12.

**TABLE 8-6 PICLite Command Line Invocation**

| PICLITE INVOCATION |
|---|
| PICLite [D:] [Path] FileName [.EXT] [.] [!] |

**TABLE 8-7 PICLite Command Line Options**

| COMMAND LINE OPTIONS: | |
|---|---|
| [D:] [Path] FileName [.EXT] | — Disk/Path Source File is In |
| . | — Debug Enable (Don't Delete .~tm File) |
| ! | — No Response (No Printing out of Results) |
| /W | — Produce an MPLAB IDE "Watch" File |

**TABLE 8-8 PICLite Output Files**

| OUTPUT | |
|---|---|
| FileName.1st | — Listing File |
| FileName.asm | — MPASM Format Assembler File |
| FileName.~tm | — Debug/Temporary File (for Language Debug) |
| FileName.wat | — "Watch" file for MPLAB |

**TABLE 8-9 PICLite Address Label Specification**

| ADDRESS LABELS (CASE SENSITIVE): | |
|---|---|
| Label {:} | — The Label Starts in the first column of the line (all other Instructions have to start in a column other than the first one). |
| | — The colon (":") at the end of the label is optional for program labels, but mandatory if a program statement (ie "A = B + C") is on the same line. |

**TABLE 8-10 PICLite Variable/Constant/Register Specification**

| VARIABLE/CONSTANT/REGISTER DECLARATION | |
|---|---|
| Label {[Array]} DataType {= InitValue} | |
| | — Arrays only on "byte" or "word" |
| | — InitValue ONLY on byte/word and NO ARRAYS |
| | — Must be in the first column |

**TABLE 8-11 PICLite Data Types**

| DATA TYPES (CASE INSENSITIVE) | |
|---|---|
| = value | — Constant Value |
| byte | — file Register |
| word | — two file Registers in Intel Format (16 Bits) |
| @ addr | — Hardware Register/Defined |
| | — Addr 2x for Shadowed Registers |
| | — Registers @ addresses >= 0x02000 will be 14 bits long |
| | — Registers @ addresses >= 0x02000 will be calculated during compilation cannot useRegister or Variable Values. |

**TABLE 8-12 PICLite PICmicro Hardware Specifications**

| PICMICRO® MCU HARDWARE SPECIFICATIONS | |
|---|---|
| type PIC_Type | —PIC Specification (ie "PIC_Type" = "16F84") |
| {s}memory Start:End | —"s" means "shadow"/Bank 0 in Bank 1 |
| size n | —"n" Instructions in Control Store |

Although these statements can be put in the source, I have created some .DEF files that contain this information for specific PICmicro® MCUs (Tables 8-13 through 8-22).

Currently supported devices are:

- 16F84
- 16C71
- 16C73A
- 16C76

PICLite is the fourth PICmicro® MCU compiler that I have written. The first compiler was very C like, the second removed the braces and just used the indentation on the line to

**TABLE 8-13 PICLite Constant Specifications**

**CONSTANT FORMATS**

0x0#### —Hex Number

0b0######## —Binary Number

### —Decimal Number

'c' —Characters

'\r' —Carriage Return (Standard "C" Backslash Data types supported)

"St","ring" —String Data, Values concatenated with comma (",")

**TABLE 8-14 PICLite Assignment Statement Specifications**

**ASSIGNMENT STATEMENTS**

Label = Label | Constant { Operator Label | Constant . . . }
= Expression

**TABLE 8-15 PICLite Assignment Statement Operators**

**OPERATORS (IN ASCENDING ORDER OF PRIORITY)**

"+" —Addition

"–" —Subtraction (Negation if Unary)

"*" —Multiplication

"/" —Division

"//" —Modulus

"<<" —Shift Left

">>" —Shift Right

"&" —Bitwise AND

"|" —Bitwise OR

"^" —Bitwise XOR

"==" —Compare Values, Return 1 if Equals

"!=" —Compare Values, Return 1 if Not Equals

">" —Compare Values, Return 1 if Left Greater

">=" —Compare, 1 if Left Greater or Equal

"<" —Compare Values, Return 1 if Left Less

**TABLE 8-15 PICLite Assignment Statement Operators *(Continued)***

<table>
<tr><th colspan="2">OPERATORS (IN ASCENDING ORDER OF PRIORITY)</th></tr>
<tr><td>“<=”</td><td>—Compare, 1 if Left Less or Equal</td></tr>
<tr><td>“&&”</td><td>—Logical AND</td></tr>
<tr><td>“||”</td><td>—Logical OR</td></tr>
<tr><td></td><td>—Division and Modulus Code taken from Microchip AN617</td></tr>
</table>

**TABLE 8-16 PICLite Bracket Specifications**

| BRACKETS | |
|---|---|
| “(”, “)” | —Used to Establish Order of Operations |
| “[”, “]” | —Array Index |

**TABLE 8-17 PICLite Language Keywords**

| LANGUAGE KEYWORDS | |
|---|---|
| if condition then Label | —If Expression != 0 Jump to Label |
| ; | —Comment, Everything to Right ignored |
| file [D:] [path]FileName.ext | —Load File and Insert Source |
| int: | —Start of Interrupt Code |
| | —Everything After “int:” will be put into the Interrupt Handler |
| | —Interrupt Code *can* call goto/call into Mainline and visa-versa |

determine what the nesting level of the conditional statements, and the third used explicit ends for the conditional statements. When you look at PICLite, you'll see that there has been a definite trend to simplify thc language in each new version.

PICLite is the most successful compiler that I have written, although there is still a lot of room for improvement. The next version of the compiler will include the following features:

1 Better macro processor
2 The ability to use Microchip PICmicro® MCU .INC files instead of having to create them myself
3 The FSR as a stack register for complex operations
4 The addition of functions to the language
5 Produce an MPLAB IDE COD file

**TABLE 8-18 PICLite Macro Processor Keywords**

| MACRO PROCESSOR KEYWORDS | |
|---|---|
| Label({Parm {, . . . }}) macro | |
| : | —Define the Multi-Line Macro |
| | —Ends with the Definition of the Next Macro |
| Label({Parm {, . . .}}) macro . . . code . . . | |
| | —Single line Macro |
| | —End Multi-line Macros with single line Macro |
| %int %Variable{[ ArraySize ]}{ = Initial} | |
| | —Define a Macro Variable |
| %if condition | —if "condition" is true, compile following |
| : | —code to execute if "%if" is true |
| {%else | —"%if" condition is not true, execute following |
| :} | —code to execute if "%if" is not true |
| %end | —Finish "%if" conditional code |
| %error "String" | —Force Error during Compile |

**TABLE 8-19 PICLite Structured Programming Keywords**

**STRUCTURED PROGRAMMING CONSTRUCTS**

The Structured Language Constructs were created using the Macro Processor. The file "structur.mac" must be included in the source file using the "file" command.

```
mif ( condition )          ; If "condition" is true, execute
                           ;  following
;—if "true" code here
{melse()                   ; If "condition" is false, execute
                           ;  following
;—if "false" code here}
mend()                     ; End the "if" statement
mwhile ( condition )       ; Loop While "condition" is true
;—Code to Execute while the "condition" is true
mend()
mfor( Variable, Initial, Stop, Increment )
                           ; "for ( Variable = Initial;
```

**TABLE 8-19 PICLite Structured Programming Keywords *(Continued)***

| STRUCTURED PROGRAMMING CONSTRUCTS | |
|---|---|
| | ; Variable ! = Stop; |
| | ; Variable += Increment )" |
| ;—Code to Execute in the "for" loop | |
| mend() | |

**TABLE 8-20 PICLite Assembler Keywords and Identifiers Keywords**

| ASSEMBLER KEYWORDS AND IDENTIFIERS | |
|---|---|
| "$" | —Current Address in Program |
| "dt" | —Define Table |

**TABLE 8-21 PICLite Configuration Fuse Specification**

| CONFIGURATION FUSES (AT 0X02007) DEFINITIONS | |
|---|---|
| "CONFIG_BASE' | — Variable located at 0x02007 (and defined in the ".DEF" files zipped with PICLite .exe) that contains the Configuration Fuse Values |
| CONFIG_BASE = Parm1 & Parm2 ... | — CONFIG_BASE can be assigned using the predefined operating parameters or constants |
| | — Standard Configuration Values NOTE: These values are different for different PICmicros |
| CP_OFF<br>CP_EN | — Code Protect On/Off |
| PWRTE_DIS<br>PWRTE_EN | — Power Up Write Timer Enable/Disable |
| WDT_EN<br>WDT_DIS | — Watchdog Timer Enable/Disable |
| RC_OSC<br>HS_OSC | — Oscillator Type Definitions |
| XT_OSC<br>LP_OSC | |

| TABLE 8-22 Asynchronous Stimulus Operation Types |
|---|
| Pulse |
| Low |
| High |
| Toggle |

Probably the biggest improvement I would make to the language is to write it under Win32 instead of the MS-DOS command line. Much of the work required to develop this (and the previous compilers) was to try to work through MS-DOS's 640-KB maximum application size, which was compounded by the desire to run a debugger (CodeView) while working on the compiler.

Using Win32 will provide up to 4.3 GB of "flat" memory for the application so that I can keep all iterations of the code within memory and not have to "spool" it out onto disk. This will allow the PICLite compiler application to run much faster than it currently does.

# MPSIM.INI

Although this version of the book does not use the MS-DOS MPASM and MPSIM command-line tools, I do want to say a few words about using MPSIM and its MS-DOS display. The simulator itself requires the specification of different registers for it to display as the application is executing. This section details how this is done. If you find that you have to use the MS-DOS command-line MPSIM program, you can get a complete set of operating instructions by entering *h* followed by pressing the *Enter* key.

The MPSIM.INI file is used to control how the MPSIM simulator starts when you are debugging your program. Such things as the target PICmicro® MCU, the default radix for values, the registers to monitor, and the clock period you are using can be set into the MPSIM.INI.

The simulator (MPSIM) does not have a great method of displaying data; you will probably have to work with it for a while to understand what it is trying to show you. Some patience and perseverance is required to use MPSIM and I find it much more difficult to keep track of application execution than the MPLAB IDE. You might wish to use the MPLAB IDE or GPSIM and run it on your computer, rather than use MPSIM. Please don't think that the method I use for displaying registers is the only one available to you; you can develop your own conventions that make the most sense to you.

Here is a typical MPSIM.INI File:

```
; MPSIM File for PROG2 - Turning on an LED
;
; Myke Predko - 96.05.20
;
P 84            ; Use a 16C84
SR X            ; Hex Numbers in the Simulator
ZR             ; Zero the Registers
RE             ; Reset Elapsed Time and Step Count
```

```
DW D                ; Disable the WDT
V W,X,2              ; Display: the "W" Register
AD F3,B,8             ;   Status Register
AD F4,X,2             ;   FSR Register
AD OPT,X,2            ;   Option Register
AD FB,B,8             ;   INTCON Register
AD F2,X,3             ;   PCL Register
AD FA,X,3             ;   PCLATH Register
AD F1,X,2             ;   TMRO Register
AD IOA,X,2            ;   Port "A" Tris Register
AD F5,X,2             ;   Port "A" Register
AD IOB,X,2            ;   Port "B" Tris Register
AD F6,X,2             ;   Port "B" Register
AD FC,X,2             ;   "Test" Register
rs
sc 4                           ; Set the Clock to 1MHz
lo prog2
di 0,0                         ; Display the First Instruction
```

An important point in this file is the use of the comments and the absence of white space to indicate breaks in the different areas of the file. Blank lines cannot be put into the MPSIM.INI file because at each blank line, the previous instruction will be repeated. The commands and their various parameters is explained in Microchip's *MPSIM User's Guide*.

The first part of the MPSIM.INI file is used to set the simulator up for the processor and the basic defaults. Notice that *84* is used for the 16C84 and not the 16F84. The 16C84 is a similar part to the 16F84, with the most important difference being that it has fewer file registers than the 16F84. The 16F84 code provided in this book can be converted very easily to 16C84 (basically, by changing the PICmicro® MCU device specification) before assembling it.

Register data makes sense to me in hex (because we are talking about 8-bit registers). This might conflict with what I wrote elsewhere about only working with decimal, but I do prefer looking at data in registers in a manner in which I can convert to binary (to understand the bit pattern). This is not to say that I don't use binary for some registers (in which each bit is a separate entity) or decimal for constants. I use the data-display format that displays the important aspects of the data most efficiently and intuitively to me.

I have tried to come up with a standard format for how I work with the simulator. Thus, I can always be pretty sure of where to look for various registers. The best analogy I can come up with how I do the registers is in terms of an airplane cockpit. Every aircraft has had six basic instruments since World War II. They are all clustered in the middle of the cockpit for the pilot to find easily, without much visual scanning. I try to do the same thing with the registers. I put the primary execution registers first. They include:

- The w register
- The STATUS register (this is put in binary format)
- The program counter
- The FSR register
- Port *A* I/O control (TRISA) and data registers
- Port *B* I/O control (TRISB) and data registers

Looking at the MPSIM.INI file listing, you can see these registers are displayed right at the top (at the top of the MPSIM screen), which will make them easy to find.

After these basic registers, I then include any specific hardware control registers that are used by the program (e.g., INTCON or TMR0).

Following the hardware registers, I put in the specified variable registers. To debug a program, using MPSIM, I usually print out a copy of the listing file and use it to follow the action. For this reason, I display the variable registers in the same order as what's in the program.

As can be seen, the labeling on the registers leaves much to be desired. The conventions used here to define where the registers are placed does help make understanding the data displayed easier.

Once all of the registers are defined, then basic setup of the program, including specifying the clock period of the program, is put in the MPSIM.INI file. Finally, a command to load in the program is put into the MPSIM.INI file.

You can put any valid MPSIM command into the MPSIM.INI file. When I'm involved in heavy debugging, I'll often update the file to include breakpoint definitions. Although MPSIM is not great at showing you what's going on, a well-thought-out MPSIM.INI will make your application debugging possible and somewhat easier than just using a haphazard MPSIM.INI.

# Stimulus (.STI) Files

For programs that require input from the hardware around it, you'll need a stimulus file to debug it in the simulator before burning the chip. As noted later in this chapter, the most significant PICmicro® MCU debugging happens at the PC, not the workbench. PICmicro® MCU input can be injected asynchronously into the simulator as the program is executing, but a stimulus file offers the advantage of being able to offer the PICmicro® MCU a consistent data input for debugging.

Stimulus files allow you to define what the program is going to encounter in terms of input. Stimulus files generally allow you to define waveforms and conditions that the PICmicro® MCU will encounter in the application.

The format of the stimulus file is actually quite simple. Input value changes are made at specific step counts. Thus, the pins to be input have to be identified along with the step count. The same stimulus file format works for both MPSIM and MPLAB.

All input bits of the PICmicro® MCU can be referenced using the standard pin identifier listed in the Microchip datasheets and used in this book. The Reset pin uses the label *MCLR*. Port input bits start with an *R*, followed by the port identifier letter, and finally the bit. Port B bit 4 would have the identifier *RB4*.

This is a sample stimulus file (which, by convention, always ends in .STI):

```
!
! Sample Stimulus File
!
Step     MCLR     RB4       ! Define the Bits to be Controlled
  1        1        1       ! Initialize the Bit Values
! Wait for the Program and Hardware to be Initialized
 500       0        1       ! Reset the PICmicro® MCU
1000       1        1
1500       1        0       ! Change the State of the Port Bit
2000       1        1       ! Restore it for rest of program
```

The lines beginning with *!* are comments. The *!* character is used to indicate that comments follow to the end of the line. The first actual line is the "Step. . ." that is used to declare the bits that are controlled by this stimulus file and the order in which they are presented.

The following, noncommented lines are the input data to the simulated PICmicro® MCU. As indicated, the signal value is asserted at the cycle count value. Thus, some interrupt latency issues cannot be simulated because interrupt requests at particular clock cycles within an instruction cycle cannot be specified.

The MPSIM stimulus files are easy to develop, but they are not time-based. Thus, you have to calculate the instruction count ("Step" in the stimulus file) using the formula:

Instruction count = Time delay * Frequency/4

So, to get the instruction count for a 15-ms delay in a 3.58-MHz PICmicro® MCU, the formula would return:

Instruction count = Time delay ** Frequency/4
= 15 ms * 3.58 MHz/4
= 15(10123) seconds 3 3.58 (10 1 6) cycles per second/4
= 13,425 cycles

So, the cycle step count at 15 ms in this example is 13,462. In the stimulus files, the step counts are absolute, so the cycle count should be added to the step values after the data pattern has been determined.

You'll feel as if you are going blind as you create complex stimulus files, but the effort is well worth it when it comes to the final product and the program you burn into your PICmicro® MCU works first time.

# FUZZYtech

One of the more powerful development tools available to you for developing digital control applications with the PICmicro® MCU is Microchip's FUZZYtech (Fig. 8-10). This tool is very useful when developing motor-control applications or other applications that require control of simple parameters, when you don't want to develop a *PID (Proportional Integral Differential)* controller for the application. Although I am really a beginner in developing fuzzy logic applications, I found that FUZZYtech is quite an easy-to-use tool that is well suited for basic control applications. However, there are a few things to watch out for (covered later in this section).

I bought a copy of FUZZYtech Explorer at a Microchip seminar some time ago. The kit consists of a set of diskettes with a GUI-based fuzzy-logic development system and with an example hardware application consisting of a digital heater and thermometer (Fig. 8-13) that can be used to demonstrate fuzzy-logic applications. In about two hours, I was able to create my own set of fuzzy-logic rules to control the temperature of the heater to within one degree Celsius. I have used this kit to develop a fan-speed controller project for the PIC18Cxx.

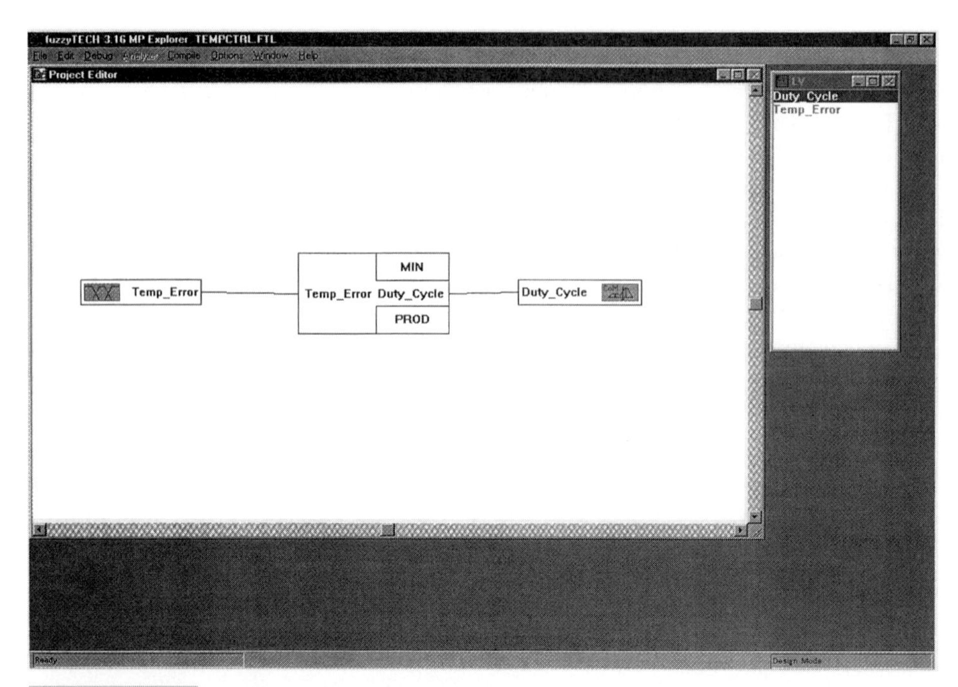


**Figure 8-10** **"FUZZYtech" GUI main display**

FUZZYtech Explorer provides the capability of developing a simple fuzzy-logic system, which is needed to develop a fuzzy-language application with one or two inputs and one output. This capability is more than sufficient for most applications that you will use while developing your initial PICmicro® MCU control applications. As you get more familiar with fuzzy logic, you can upgrade to FUZZYtech Edition, which has multiple inputs and multiple outputs that can be controlled with fuzzy rules.

To create an application, you will first have to define what you want out of it. For a simple motor control, you will have to follow the list of work items that follow. Chapter 16 contains a fuzzy logic fan-speed controller that I developed using this list of work items.

1. Define how the input parameters are to be measured.
2. Define how often the input parameters are polled.
3. Define how the output control function is to be designed.
4. Specify the control hardware within the PICmicro® MCU that is going to be used.
5. Specify the polling interval.
6. Design the hardware to be used along with the user interfaces and an RS-232 interface.
7. Develop sensor, control, and interface code.
8. Develop a simple sensor transmit and control receive application that can be used by the PC to control the system before the final code is developed for the application.

Of probably the greatest consequence to the entire operation is point 5: *Specify the polling interval*. The code produced by FUZZYtech is PICmicro® MCU assembler. As such, it will take some time to execute—especially if the debug function is to be used. I recommend that initially the inputs are polled once per second and the outputs updated once per second as well. For the temperature-control application that comes with FUZZYtech, this isn't an issue because the temperature-control PWM runs at a 10-Hz rate. 10 Hz is not suitable for other PWM applications, such as controlling a motor, so care must be taken to ensure that enough time is left in the application to communicate with the PC host system in Debug mode.

Once this work is done, you are ready to begin to define the rules that are required to control the system. This is done using the debugging capabilities of FUZZYtech. These capabilities consist of having the application send the sensor information, have FUZZYtech process them for a new control value, and send them back to the application. This process does take some time, which is why I recommend that you keep your fuzzy-logic polling interval to one second (or longer).

A typical motor-control system will consist of a sensor input, a processing module, and a PWM-control output. When FUZZYtech is first initialized, this system set up can be defined as shown in Fig. 8-11. The motor input can either be an absolute value or a "delta" from a desired value. Figure 8-11 was taken from the FUZZYtech tutorial in which the device's "delta" temperature to the specified temperature is displayed.

Next, the rules for the input and how they relate to the output are defined. For the FUZZYtech example, Fig. 8-11 shows the four different rules that were designed for the temperature input. If the device's temperature is greater than the target temperature, it is simply described as *Hot* and the heater PWM is turned off. If the temperature is too cool, then the PWM is increased at different rates, depending on the difference between the actual value and the desired value. *Cold* will maximize the PWM output and *Cool* will

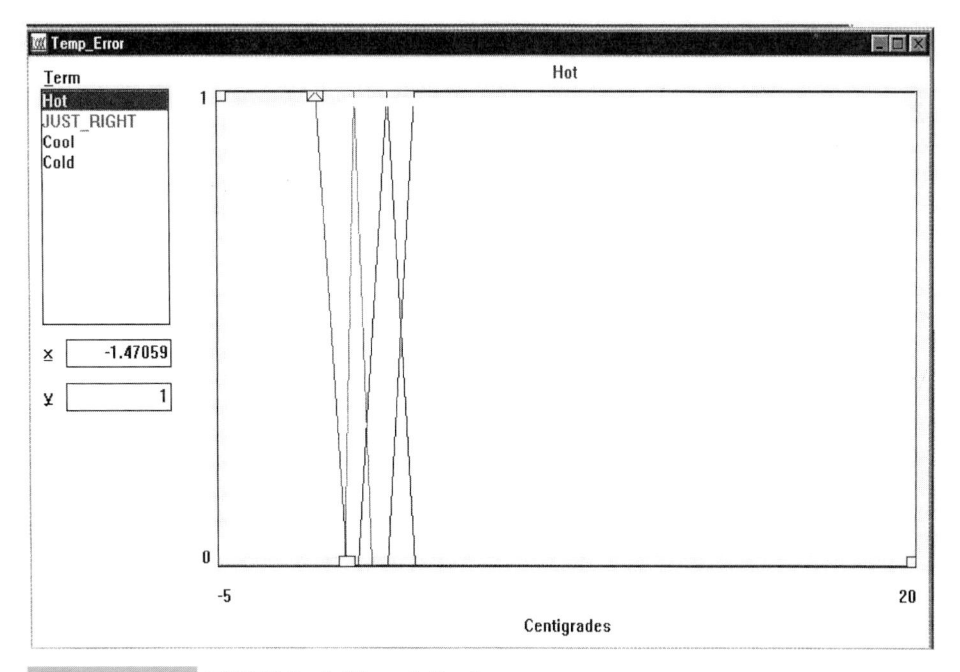


**Figure 8-11** FUZZY "rule" input display

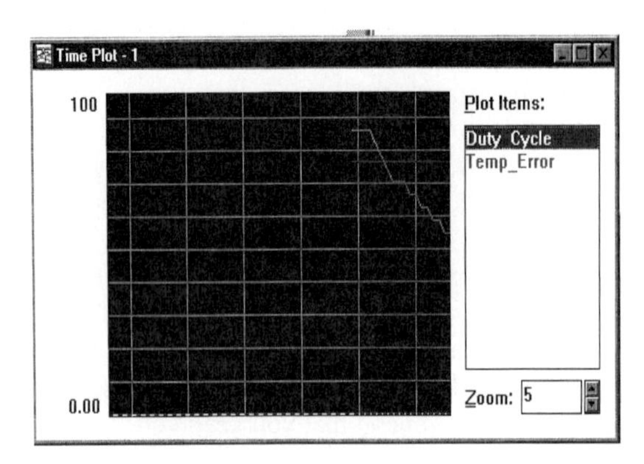


**Figure 8-12** **FUZZY operation time display**

slowly increase the PWM until the *JUST_RIGHT* rule is reach. *JUST_RIGHT* will keep the PWM at the value that set the temperature there.

In Fig. 8-11, it appears that the bands are very close together (which they are). When I created the application, I used the rules defined by Microchip in the FUZZYtech tutorial. If I were to do this application again, I would probably change the delta temperature range from −5 to 20 degrees Celsius to −5 to +5 degrees Celsius to allow the different rules to be more easily seen.

The example application that comes with FUZZYtech allows the PC running FUZZYtech to interface with and control the application. The example application transmits the delta temperature to the PC running FUZZYtech and waits for FUZZYtech to process the data and then transmit back the appropriate PWM value (from 0 to 100).

When I first ran the example application, I tried moving the points on the rule chart in Fig. 8-11 at random to see what would happen. I found that without thinking about what you are doing, you could very easily create an *unstable system* or one that oscillated, rather than settled at a specific point.

This oscillation could be observed with the Time Plot function of FUZZYtech. Figure 8-12 shows the Time Plot function for FUZZYtech when it was first starting to control the application. The Time Plot is an excellent tool to monitor what is happening with your application.

With some thought, I placed *JUST_RIGHT* at plus and minus one degree of the target temperature (delta temperature equals zero), *Hot* at one degree above, and the *Cool* and *Cold* rules at less than one and less than three degrees Celsius. After making these changes to the random parameters that I tried before, I found that the oscillations stopped almost immediately and the thermostat was accurate to within one degree Celsius as long as I ran the application.

With the controls seeming to work correctly, you can add the fuzzy logic code that FUZZYtech will produce for you to your application. The code produced by FUZZYtech's compiler is in assembly-language format and, although it seems reasonably efficient, it is going to require quite a bit of "massaging" in order to work properly with modern application code. I refer to "modern application code" because the assembly-language code produced by FUZZYtech uses older versions of register labels. Probably the worst example I saw was that TMR0 was referred to as *rtcc*. When I created my own application using FUZZYtech, I changed the compiled source code to meet the modern register names and my own programming requirements. This is not a particularly hard task, but it will take you quite a while.

**Figure 8-13** FUZZYtech tutorial board

When the fuzzy-logic application code is "compiled" for the application, it is stored in a PROJECT.ASM file and PROJECT.VAR. The PROJECT.ASM consists of the fuzzy-logic programming code and has the references to the older labels that were covered previously. This file is to be included into your application and consists of two labels. *Initproject* is the label to the fuzzy-logic software initialization subroutine and *project* is the label to the fuzzy-logic processing software. This code will probably also call a number of other subroutines that are used to find the area and center of mass of the rules to determine the correct course of action. These routines will have to be included in the application or, at the very least, included in the subdirectory in which the source is located.

PROJECT.VAR contains a list of the variables required by the compiled fuzzy logic subroutines. These variables consist of the *LV0_MYIN1* and *LV1_MYIN2* input parameter variables and the *LV2_MYOUT1* output parameter variable, as well as the variables required by the *include* mathematical routines.

The mainline of a fuzzy-logic application that will be in the following format:

```
title "FileName - One Line Description"
#define _version "x.xx"
;
; Sample Template for FUZZYtech Application
;
; Update History:
;
; Application Description/Comments
;
; Author
;
; Hardware Notes:
;
LIST R=DEC                       ; Device Specification
 INCLUDE "p16cxx.inc"            ; Include Files/Registers
 INCLUDE "myprojct.inc"          ; FUZZYtech Compiled Source Code

; Variable Register Declarations
 INCLUDE "myprojct.var"          ; FUZZYtech Variable Declarations
```

```
; Macros
 -CONFIG _CP_OFF & _XT_OSC & _PWRTE_ON & _WDT_OFF & _BODEN_OFF

 org  0

Mainline

 goto      Mainline_Code

 org       4                   ; Interrupt Handler at Address 4
Int

MainLine_Code

 call      initmyproject       ; Initialize the FUZZYtech Variables

; Carryout other application initializations

Loop                           ; Return here for Each Control Loop

; Sample the inputs and create the FUZZYtech inputs

 movf      parameter_1, w      ; Initialize the FUZZYtech inputs
 movwf     LV0_MYIN1
 movf      parameter_2, w
 movwf     LV1_MYIN2
 call      myproject           ; Run the Inputs through FUZZYtech

 movf      INVALIDFLAGS, w     ; Were there Errors in FUZZYtech code?
 btfss     STATUS, Z
  goto     FUZZYError          ; If "INVALIDFLAGS" not Zero then yes

; Process the output

; Put in Delay for application execution

 goto  Loop                     ; Return here for next loop

FUZZYERROR                                   ; Error in FUZZYtech Processing

; Put in error Handler Here

 goto      Loop                ; Try to Restore Operation

; Subroutines
 end
```

If you look in the FUZZYtech documentation, you will see that this is not the same as what Microchip specifies. The differences between my template and FUZZYtech's are really in the format that I use to write code. These are largely cosmetic differences. You might choose to ignore my format or just use FUZZYtech's without modification.

# Version Support Tools

*Version support* is a $10 phrase for keeping track of your software and making sure that the correct level is used and released for your applications. I find that when I develop PICmicro® MCU applications (and any software in general) that I tend to go through a number of different versions (starting with a hardware diagnostic going up to a functioning system) and experiment with different ways of doing things. Sometimes, they're not always successful.

What I find works for me is putting each application in a separate subdirectory on my PC. Then, I often start with a program named with version information and work my way up. In the comments, I note when this version is to be used and whether or not it is to be released. If I am outputting any kind of initial messages, I always include the version so that I can see from the software what version I am working with.

To do this simply, I put a *define* at the start of my application with a string defining the version. This usually looks like:

```
#define _version "1.00"
```

In the code itself, I can then insert this string anywhere that the version information is required. For example, if I want to have the version number displayed from a table read, I can insert the define simply into the *dt* text string statement like:

```
 dt "Version: ", _version, 0
```

which, if *_version* is the string *"1.00"*, as shown, it will be processed as if it were:

```
 dt "Version: 1.00", 0
```

If an experimental version is not successful, I make sure that it is clearly marked in the title bar and opening comments of the program.

The MPLAB IDE helps facilitate version control through its use of projects. Chapter 15 explains how to set up projects in MPLAB and how they can simplify your application development.

# The MPLAB IDE

In this book, I have gone on at length about what an excellent tool the MPLAB IDE is if you're working with microcontrollers for the first time and you are trying to learn the PICmicro® MCU architecture. How to program, how to program an assembler, and how to create circuits with the PICmicro® MCU is enough to learn. Learning the MPLAB IDE is just one more thing and it's not all that much of a joy at that. The MPLAB IDE is unique because Microchip is probably the only vendor out there that has a tool of this sophistication available for free. Not having this type of development tool would make learning the PICmicro® MCU much more difficult.

The MPLAB IDE is a very complete *Integrated Development Environment (IDE)* for all the different PICmicro® MCU families that run under Microsoft's Windows version 3.1x (or later) operating systems. The MPLAB IDE integrates the different operations of developing a PICmicro® MCU application. This is done from a user-configurable desktop with different capabilities built into the program.

The MPLAB IDE can integrate the following different functions:

- editor
- assemblers
- compilers
- linkers
- programmers
- emulators

For the rest of the chapter introduces you to the various aspects and features of MPLAB. This information is reintroduced in Chapter 15, which has more concrete examples of how the different features work.

## CREATING PROJECTS AND PROJECT FILES (.PJT)

When creating an application in the MPLAB IDE, the application code itself has to be loaded into a project. The project brings together all of the files necessary to create the hex file that is burned into a PICmicro® MCU. Along with this feature, the MPLAB IDE project records the open windows in the MPLAB IDE desktop, along with their positions and any user preferences. This section introduces the MPLAB IDE projects and the information that is recorded within them.

When you first start up the MPLAB IDE, you will be faced with a blank desktop. Figure 8-14 shows the MPLAB IDE desktop with a project executing within it. The desktop is used to load and process applications as they are being developed. It is the primary interface used within this book for application code.

All of the MPLAB IDE commands are available from the *pulldowns* at the top of the desktop. From this line, source files can be specified and edited; the operating environment selected; and the source code assembled, compiled, linked, simulated, and finally run in a PICmicro® MCU. When you first get the MPLAB IDE up and running on your PC, you might want to browse the pull downs to see what is available.

The next line is the *toolbar*, a set of icons used to access specific functions in the MPLAB IDE. Normally, this line is set to a simulator debug line (Fig. 8-14), which allows

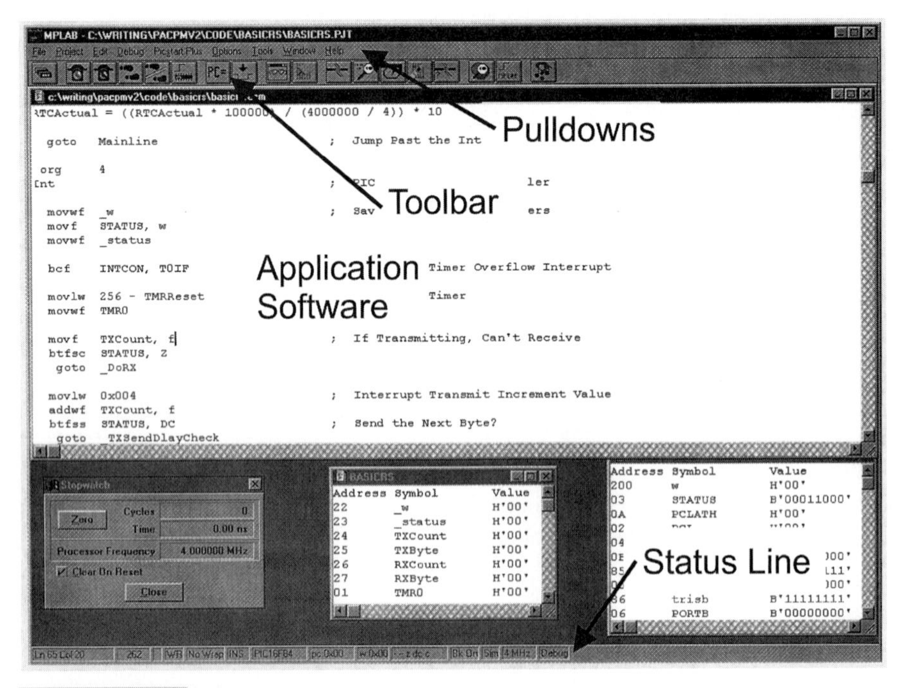


**Figure 8-14** MPLAB labeled desktop

simple selection of go, stop, step, step over, modify the program counter, etc. Other toolbars can be selected by clicking on the left-most toolbar icon. You can select toolbars designed to simplify application editing, assembling, and compiling.

When you are ready to create your application, you will have to click on the Option pull down followed by Development Mode. From Development Mode, you will get a window requesting the type of PICmicro® MCU that you are going to work with, along with the MPLAB IDE operating mode. For the projects presented in this book, the MPLAB-SIM Simulator selection should always be selected.

After choosing the PICmicro® MCU and the operating mode, the project can be created. This is done by selecting the Project pull down, followed by clicking on New. As shown later in the book, the source file used for the project should be placed in its own subdirectory and the project file should be in that subdirectory as well. When you have specified the project, a window like Fig. 8-15 will come up. Click on the hex file name and then on Node Properties.

As an aside, the MPLAB IDE considers source files to be *nodes*. For the most part, I ignore this term in this book, but you will probably come across it as you spend more time with the MPLAB IDE and sometimes wonder exactly what it means.

In previous versions of the MPLAB IDE, some work had to be done to allow the project editor to accept assembly files. In version 5.00 (which is used for the examples) and later versions, this extra work was eliminated. Now, you just have to select the assembly-language source files that make up the hex file.

Once the assembly source file has been selected, click on OK in the Edit Project window and you are ready to load in the source file (using the Files pull down, followed by selecting Open). Once the source file is selected, you can also add other features to the MPLAB IDE desktop features, such as the Stopwatch and Watch windows.

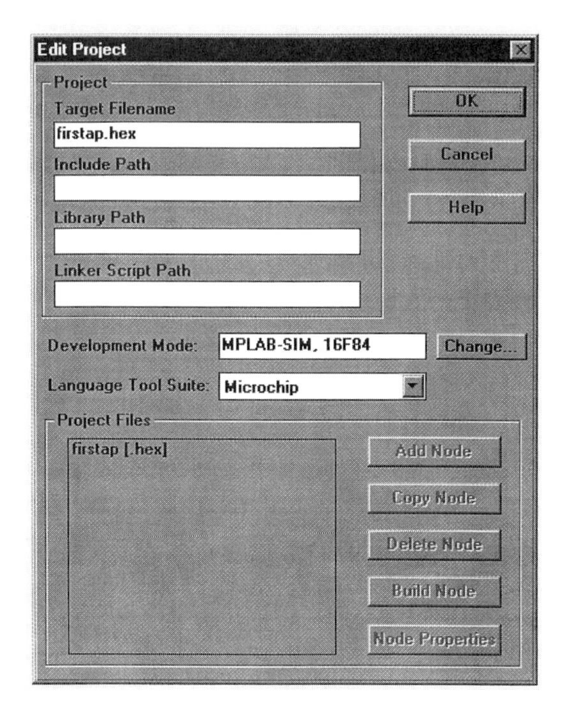


**Figure 8-15** Selecting the MPLAB IDE source files

When the project has been setup and laid out, the project is saved into a .PJT file. This file contains all the relevant information regarding a project. A sample .PJT file is shown for the Cylon experiment that is presented later in the book.

```
[PROJECT]
Target = CYLON.HEX
Development_Mode = 2
Processor = 0x684a
ToolSuite = Microchip

[Fuses]
Frequency = 4000000.000000
ClockSource = 1
Mode = 0
WDTEnable = 0
WDTBreak = 0
WDTPrescale = 142
StackBreak = 0
Freeze = 1
EmulatorMemory = 1
ShortWrites = 0
PwrSetting = 0
OSC_Settings = 0
ProgBankRegEnable=0
BankRegStartAddress=0
BankRegEndAddress=0
DisableIoPins=0

[Directories]
Include =
Library =
LinkerScript =

[Target Data]
FileList = CYLON.ASM;
BuildTool = MPASM
OptionString =
AdditionalOptionString =
BuildRequired = 0

[CYLON.ASM]
Type = 4
Path =
FileList = P16F84.INC;
BuildTool =
OptionString =
AdditionalOptionString =

[P16F84.INC]
Type = 5
Path =
FileList =
BuildTool =
OptionString =
AdditionalOptionString =

[Trace-Config]
GetAddress = 1
AddressWidth = 5.000000
GetOpcode = 1
OpcodeWidth = 5.000000
GetLabel = 1
```

```
LabelWidth = 8.000000
GetInstruction = 1
InstructionWidth = 10.000000
InstructionRaw = 1
GetSrcAddr = 1
SrcAddrWidth = 4.000000
SrcAddrRaw = 1
GetSrcValue = 1
SrcValueWidth = 4.000000
GetDstAddr = 1
DstAddrWidth = 4.000000
DstAddrRaw = 1
GetDstValue = 1
DstValueWidth = 4.000000
GetExtInputs = 1
ExtInputsWidth = 9.000000
ExtInputsHexFmt = 0
GetTimeStamp = 1
TimeStampWidth = 11.000000
TimeStampCycles = 1
TraceLines = 100
TraceFull = 0

[Windows]
2 = 0000 c:\writing\pacpmv2\code\cylon\cylon.asm 0 0 1016 451 9 0
1 = 0010 19 82 559 510 9 4
0 = 0002 C:\WRITING\PACPMV2\CODE\CYLON\16F84.wat 698 435 1016 668 9 0
Stopwatch = 16 541 284 700 9 0

[mru-list]
1 = c:\writing\pacpmv2\code\cylon\cylon.asm
2 = c:\writing\pacpmv2\code\sleep\sleep.asm
3 = c:\writing\pacpmv2\code\intdeb\intdeb.asm
4 = c:\writing\pacpmv2\code\eeprom\eeprom.asm
5 = c:\writing\pacpmv2\code\vladder\vladder.asm
```

In this file, I am sure that you will be able to pick out the file names of the source and hex files, as well as where the windows are placed on the desktop. Many of the headings will be immediately familiar to you, but others will be somewhat cryptic. I recommend that the .PJT files never be edited because changes to this file could result in unexpected changes in the operation of the MPLAB IDE.

## PROVIDING INPUT TO THE SIMULATED APPLICATION

I will discuss the practical aspects simulating applications in greater detail later in the book. This section presents the five different methodologies of passing input to the application code as it executes in the MPLAB IDE simulator. The range of choices is a bit overwhelming (especially because most other simulators only have one or two input methodologies). The five different simulator input methodologies allow you to tailor the input to your actual requirements to find and fix your problems. Chapter 15 presents some examples on how some of these simulation features are used to check the operation of an application before it is burned into a PICmicro® MCU.

The most basic input method is the Asynchronous Stimulus window (Fig. 8-16), which consists of a set of buttons that can be programmed to drive any of the simulated PICmicro® MCU pins. The button can be set to change the pin by:

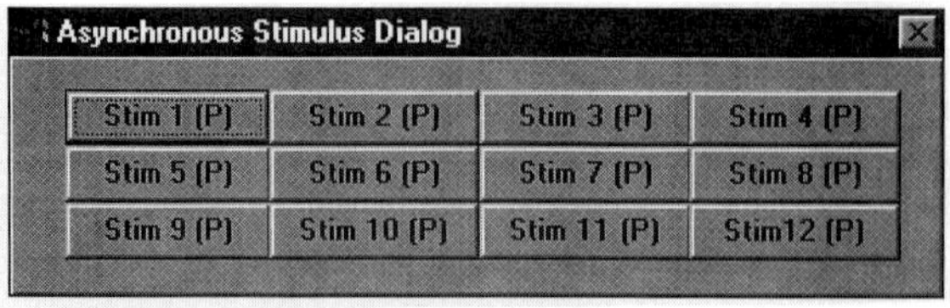

Asynchronous Stimulus Dialog

Stim 1 (P) Stim 2 (P) Stim 3 (P) Stim 4 (P)
Stim 5 (P) Stim 6 (P) Stim 7 (P) Stim 8 (P)
Stim 9 (P) Stim 10 (P) Stim 11 (P) Stim12 (P)

**Figure 8-16** **"Asynchronous stimulus" window**

- Pulse
- Low
- High
- Toggle

The *Pulse option* pulses the input pin to the complemented state and then back to the original state within one instruction cycle. This mode is useful for clocking TMR0 or requesting an external interrupt. Setting the pin High or Low will drive the set value onto the pin. To change the value of the pin between the two states, you can program two buttons in parallel with each other and each button changes the state. This can also be done with a single toggle button, which changes the input state each time that the button is pressed.

I do not like to use the asynchronous inputs a lot because they are completely user generated. There have been many times when simulating an application that I have forgotten to press the appropriate buttons or have pressed them at the wrong point in time.

To eliminate the problem of not pressing an asynchronous input button at the correct time, I normally develop a stimulus file for the application. This file provides an instruction counter value (*step*) and a pin state value at this point. An example stimulus file showing RS-232 data being transferred to a simulated PICmicro® MCU bit-banging application is shown:

```
! RS232CHK
!
! Look at RS-232 Operation at 1200 BPS for a 16 MHz 16C711
!
Step          RB0
     1 1
!
!   Data Sent to the PICmicro® MCU
!
 1000  0       ! - Send 0x065/0x00D to See Response
 3333  1       ! Bit 0
 6667  0
10000  1
13333  0
16667  0       ! Bit 4
20000  1
23333  1
26667  0
30000  1       ! Stop Bit
!
40000  0       ! Send 0x00D
43333  1       ! Bit 0
46667  0
50000  1
53333  1
56667  0       ! Bit 4
60000  0
```

```
63333  0
66667  0
70000  1       ! Stop Bit
```

The *!* character in the stimulus file is a comment (everything to the right of it is ignored). The first actual line of a stimulus file contains the "step" directive, followed by the pins to be driven. In this example, the stimulus data is passed to the RA1 pin.

The input pins can be any of the pins in the simulated device using the Microchip convention of *R$#*, where *$* is the port identifier (*A*, *B*, *C*, etc.) and # is the number of the pin in the port. Reset can be driven from a stimulus file as the MCLR pin.

Multiple pins can be specified within a file. For example, a stimulus file simulating the inputs from a mouse's angular motion sensor could be:

```
Step     RA0      RA1
  1       0        0
100       0        1
200       1        1
300       1        0
400       0        0
500       0        1
600       1        1
700       1        0
800       0        0
```

Notice that in the stimulus file, two movement cycles were included to work the code through without having to reset the application. Stimulus data will not repeat once the counter has gone beyond 800 and the RA0 and RA1 inputs will stay reset until the simulated PICmicro® MCU is reset.

To count the instruction cycles, the StopWatch window (Fig. 8-17) is invoked and put onto the MPLAB IDE desktop. Clicking on the Window pull down and then Stopwatch. . . initiate this window. The Stopwatch function allows you to time applications. I generally keep the Clear on Reset button checked and never click on the Zero button unless I know exactly what I am doing.

The reason for this is that resetting the stopwatch will cause any stimulus input to become reset. Start from the beginning and not where the application code is. The only time I click on Zero is for timing loops (such as for composite video output), where the timing is crucial and I want to avoid having to subtract a previous value, which is an opportunity to make an error.

The speed of the stopwatch (and the simulated PICmicro® MCU) is set by clicking on the Option pull down, then selecting Development Mode, and clicking on the Clock tab. The processor speed can be specified at any time when you are working with the MPLAB IDE.

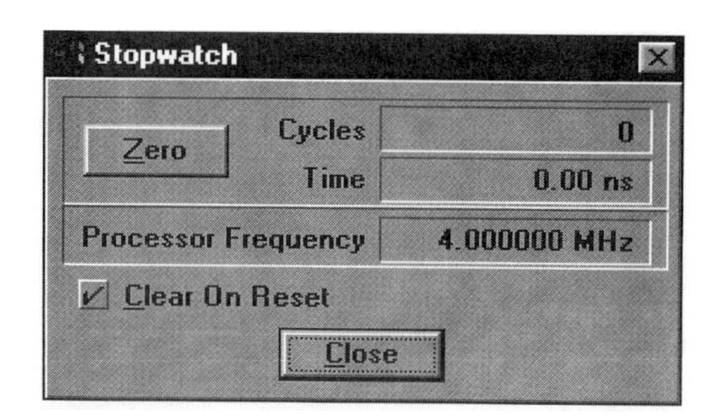


**Figure 8-17** **The MPLAB IDE "stop watch"**

When calculating the steps for the stimulus file, remember that they occur at four times the PICmicro® MCU clock period. I run most of my applications at 4 MHz because the instruction clock period is 1 $\mu$s.

In the example stimulus file, if the clock was changed to 3.58 MHz and assuming that the original values were specified for a 4-MHz clock, the step values would have to change by the ratio of 3.58 MHz/4 MHz.

The example stimulus file becomes:

```
Step     RA0      RA1
  1       0        0
 90       0        1
179       1        1
269       1        0
358       0        0
448       0        1
537       1        1
627       1        0
716       0        0
```

The task of changing the timing or specifying it in the first place is not difficult. The formula that I use is:

Step = Time delay * Clock frequency/4

Using this formula, a 250-$\mu$s step point would be calculated as:

Step = Time delay * Clock frequency/4
= 250 $\mu$s * 3.58MHz/4
= 223.75
= 224 (after rounding)

Clocks can be input into the simulated PICmicro® MCU by clicking on the *Debug* pull down, *Simulator Stimulus,* and *Clock Stimulus* selections. The clock stimulus dialog box (Fig. 8-18) can input regular clocks into a PICmicro® MCU by selecting the pin and then the high and low time of the clock, along with whether or not the clock is inverted (which means at reset, the clock will be low rather than high). The clock counts in Fig. 8-18 are in instruction cycles, which are found exactly the same way as shown previously.

Clock stimulus can be used for simple I/O tests, but it is really best suited for putting in repeating inputs that drive clocks or interrupts.

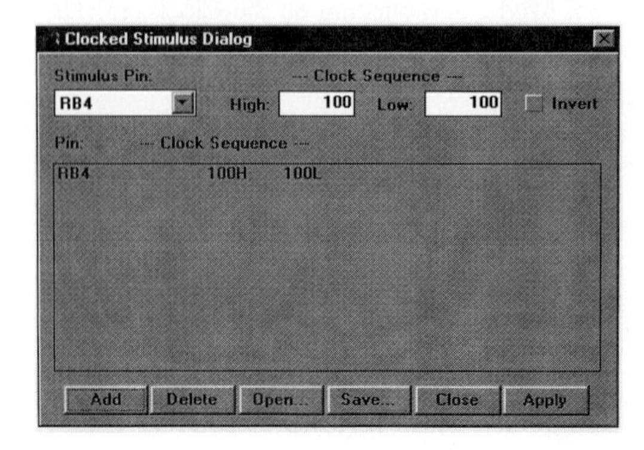


**Figure 8-18** The MPLAB IDE "clocked stimulus"

As indicated elsewhere, the MPLAB simulator does not handle advanced peripheral operations well. This deficiency can be alleviated somewhat by using the register stimulus feature of the MPLAB IDE. This feature will store a two digit hex value in a specified register every time that a specific address is encountered in the simulated application execution. To load the operating parameters of the register stimulus method, click the *Debug* pull down, followed by *Simulator Stimulus,* and then *Register Stimulus* is enabled. This brings up the small window (Fig. 8-19) in which you will select the address of the register to change, as well as the address where this happens. Once the addresses have been specified, the register stimulus file is selected by clicking on Browse.

The register stimulus file is a simple text file consisting of a column of two digit hex numbers. The following example file lists all the values from 0x000 to 0x0FF:

```
00
01
02
 :
40
41
 :
C0
C1
 :
FF
```

No comments or multi-byte values are allowed in this file. The colons (:) are used to indicate that the file continues with the data.

I do not use register stimulus that much, largely because it does not work in a way that is very useful for me. Generally, when I want to change a hardware register's value to see the application code processing it, I am going to have to change both the register's value, as well as a flag indicating that the data is available. The register stimulus feature does not allow more than one register to be modified at any given time.

To modify more than one register or modify a register asynchronously, I use the register Modify window (Fig. 8-20). This window is available from clicking on the *Window* pull down and then selecting *Modify*. This window can access any register in the simulated device, including w, which cannot be directly addressed in the low-end and mid-range devices.

All five of the simulator pin/register input methodologies presented in this section can work together. Thus, any two (or more) inputs can be enabled simultaneously to interface with the application and help you to debug it and find any potential problems. Doing this

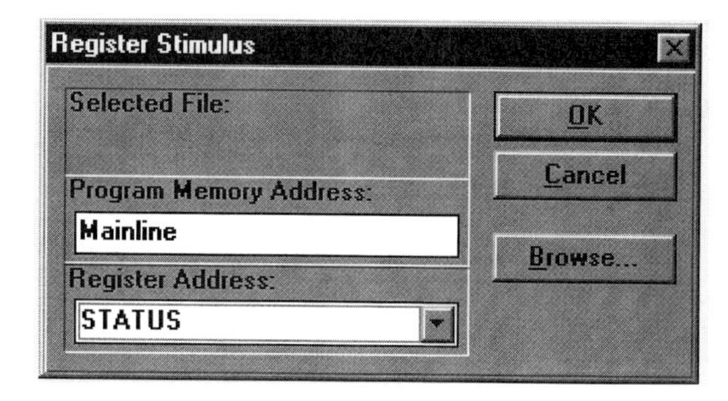


**Figure 8-19** The MPLAB IDE "register stimulus" specification

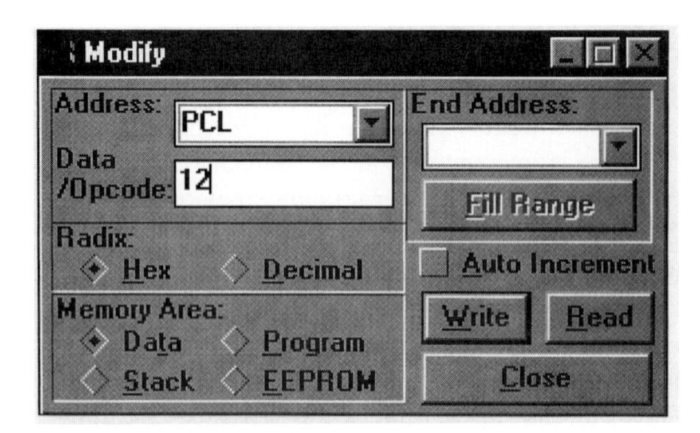


**Figure 8-20** The MPLAB IDE "register modify" window

is not always as easy as it sounds because of the difficulty in synchronizing the inputs so that they work together.

When you work with the different methods of specifying the inputs to the PICmicro® MCU or the registers, you should notice that the MPLAB IDE allows you to use simplified pin names (such as RA0), defined register names, and program memory address labels inside the source code. This makes the input features much easier to work with and avoids your having to read through the listing file to find absolute addresses or values.

## MPASM (.ASM) SOURCE

If you have looked at the Chapter 15 application code, you would probably be surprised to find that only two types of statements are required for a PICmicro® MCU application. This will be hard to reconcile if you have looked at the applications in the book because they seem to be just *full* of different types of statements, each one seeming to provide a different feature to the PICmicro® MCU. Actually, all of these different statement types are meaningless to the assembler. Instead, it just looks through the application code for instructions and an indication of the end of the code.

The most basic application source I could come up with is called MINIMUM.ASM and can be found in the Code\Minimum subdirectory of your PC's PICmicro® MCU directory. This code clears PORTB and then clears the TRISB register, which enables all eight bits for output. Once this has completed, the application goes into an endless loop. The code that does this is simply:

```
 clrf   6
 bsf    3, 5
 clrf   6
 bcf    3, 5
 goto   4
end
```

Comparing this source file to what I have produced in Chapter 15, you will feel like something is missing. I can say that nothing is missing from the perspective of what the assembler needs to convert the source code to a hex file that can be programmed into the application.

The source code in Chapter 15 looks so different because different statements have been added to the MPASM assembler to make applications easier for you to write and follow what is happening. This section goes through the different aspects of the source file and explains what the different statements are and why you would like to use them.

The two statement types that are required for an application are the PICmicro® MCU instructions and the *end* directive. The instructions are the application themselves and the *end* directive is a flag to the assembler to stop. The only requirement on these two statements is that they cannot start in the first column of the file.

*Directives* are instructions to an assembler. The next section lists all the directives that are recognized by the MPASM assembler and what they do. Later chapters cover different types of directives (such as macros) in more detail and how they can be used to simplify application development. This section just introduces the basic directives needed to develop a readable PICmicro® MCU application.

Just using these two statements will certainly make your application efficient, but just about impossible for other people (and probably yourself) to read. By adding different types of statements, the readability of the MPASM source is improved considerably and the ease by which you develop applications will be improved as well.

When you look at MINIMUM.ASM, the first problem you will have is that you don't have any idea what the different instructions are pointing to. *Labels* and *defines* are added to applications that allow you to reference addresses and certain constants with text strings that should make understanding the code somewhat easier. By taking MINIMUM.ASM and adding the register name labels (from the documentation), I can improve the readability of the application considerably:

```
 clrf  PORTB
 bsf  STATUS, 5
 clrf  TRISB ^ 0x080
 bcf  STATUS, 5
 goto  4
end
```

The bit labels given in the documentation can also be used to further enhance the readability of the application source code:

```
  clrf    PORTB
  bsf     STATUS, RP0
  clrf    TRISB ^ 80
  bcf     STATUS, RP0
  goto    4
end
```

The XORing *TRISB* with 80 clears the most significant bit of the address. This is explained in more detail in Chapter 3. When MPASM starts executing, the default numbering system (radix) is hexadecimal. This means that the *80* that is XORed with the address of *TRISB* is actually 128 decimal. I show how radices are specified later in this section.

These register and bit labels are not available automatically to the assembler. They must be loaded in from the Microchip "include files" (.INC). The include files have all the labels in the documentation, as well as other information required by the application. The *INCLUDE* directive is used to copy a text file (like the .INC file) into the source file.

```
  include  "p16F84.inc"
   clrf    PORTB
   bsf     STATUS, RP0
   clrf    TRISB ^ 80
```

```
   bcf     STATUS, RP0
   goto    4
 end
```

For this application, I have assumed that the PIC16F84 is the PICmicro® MCU used in the application and loaded its .INC file using the *include* directive.

Labels can also be used as addresses within the application and are located in the first column of the application. This avoids having to keep track of absolute or relative addresses. In *minimum*, I can add the *forever* label to eliminate the need for counting the number of instructions and explicitly put in the address to jump to.

```
include  "p16F84.inc"
  clrf   PORTB
  bsf    STATUS, RP0
  clrf   TRISB ^ 80
  bcf    STATUS, RP0
forever
  goto   forever
 end
```

In the PICmicro® MCU assembler, a colon character (:) is not needed to identify a label. The label should be in the first column to indicate that it is not an instruction or directive. When a label definition, like the *forever* line is encountered, the label (*forever*, in this case) is assigned the value of the current address.

Another way of doing the same thing in this case is to use the *$* directive as the destination of the *goto* instruction:

```
include  "p16F84.inc"
  clrf   PORTB
  bsf    STATUS, RP0
  clrf   TRISB ^ 80
  bcf    STATUS, RP0
  goto   $
 end
```

The *$* directive returns the address of the current instruction. In this case, the *goto $* instruction statement puts the PICmicro® MCU processor into an endless loop. The *$* can be used with arithmetic operations to jump to an address, which is relative to the current one. For example, *$ - 1* will place the address of the previous instruction into the source code.

Labels can be used for variables that are defined as file registers. I usually use the *CBLOCK* directive, which has the single parameter as the start of the register block. Following the CBLOCK and starting address statement, the variables are listed. If more than one byte is required for a variable, then a colon (:) followed by the number of bytes is specified. Once all the variables have been included, the *ENDC* directive is used. The variable declaration looks like:

```
 CBLOCK 0x020
i                    ; 8 Bit Variable
j:2                  ; 16 Bit Variable
k:4                  ; 32 Bit Variable
 ENDC
```

After each variable declaration, a counter initialized to the starting address (the parameter of the *CBLOCK* statement) is incremented by the number of bytes of the variables. For the previous example, *i* is at address 0x020, *j* is at address 0x021, and *k* is at address 0x023.

Accessing multi-byte constants is accomplished either by mathematically making the values "palatable" for the assembler or by specifying the *UPPER*, *HIGH*, or *LOW* directives. Using the variables defined previously, they can be loaded with constants as:

```
movlw       0x012               ; Load "i" with Decimal 18
movwf       i
movlw       HIGH 1234           ; Load "j" with Decimal 4660
movwf       j + 1               ; High Byte Loaded with 0x012
movlw       LOW 1234
movwf       j                   ; Low Byte Loaded with 0x034
```

*LOW* always returns the least-significant byte of the constant, *HIGH* returns the second least significant byte of the constant, and *UPPER* returns the most-significant byte.

I tend to avoid using *UPPER* because its use can be ambiguous to the assembler and unexpected results will be returned. Instead, I use the assembler calculator, as in the following example to load *k* with a 32-bit constant:

```
movlw       LOW 0x012345678     ; Load "k" with the 32 Bit Constant
movwf       k                   ; Load Byte 0 of "k" with 0x078
movlw       HIGH 0x012345678
movwf       k + 1               ; Load Byte 1 of "k" with 0x056
movlw       (0x012345678 & 0x0FF0000) >> 8
movwf       k + 2               ; Load Byte 2 of "k" with 0x034
movlw       (0x012345678 & 0x0FF000000) >> 12
movwf       k + 3               ; Load Byte 3 of "k" with 0x012
```

The second way of defining variables is to define their addresses as constants. *Constants* are text labels that have been assigned a numeric value using the *EQU* directive and can be referred to as *equates*. For example, the statement:

```
PORTB_REG EQU 6         ; Define a different value
```

is used to assign the value *6* to the string *PORTB_REG*. Each time that *PORTB_REG* is encountered in the application code, the MPASM assembler substitutes the string for the constant *6*.

Constants can be set to immediate values, as shown, or they can be set to an arithmetic value that is calculated when the assembler encounters the statement. An example of a constant declare using an arithmetic statement is:

```
TRISB_REG EQU PORTB_REG + 0x080
```

In the second EQU statement, the TRISB register is assigned the offset of the PORTB register plus 0x080 to indicate that it is in Bank 1.

I do not recommend using equates for variable definitions. The *CBLOCK* directive is somewhat simpler (and requires fewer keystrokes) and keeps track of variable addresses if you add or delete variables.

The address of code can be set explicitly with the *org* directive. This directive sets the starting location of the assembly programming. Normally, the start of a PICmicro® MCU application is given the *org 0* statement to ensure that code is placed at the beginning of the application, such as:

```
include "p16F84.inc"
 org      0
  clrf    PORTB
  bsf     STATUS, RP0
  clrf    TRISB ^ 80
  bcf     STATUS, RP0
  goto    $
 end
```

This is not absolutely required for this application because the assembler is reset to zero before it starts executing. It is a good idea to do it, however, to ensure that somebody reading the code understands where it begins.

For your initial PICmicro® MCU applications, the only time you will not use the *org 0* statement is when you are specifying the address of the PICmicro® MCU's interrupt vector (which is at address 0x0004). A typical application that uses interrupts will have initial statements like:

```
 org    0
 goto   Mainline

Int
 org    4
  :            ; Interrupt Handler
Mainline       ; Mainline Code
  :
```

One of the biggest differences between the PICmicro® MCU and other microcontrollers is the CONFIGURATION FUSE register. This register is defined differently for each PICmicro® MCU part number and contains operating information including:

- Program memory (code) protection
- Oscillator type
- Watchdog timer enable
- Power-up wait timer enable
- Emulator mode enable

These fuses are specified in the source file using the *__CONFIG* directive. This directive takes the bit value of its single parameter and stores it at the CONFIGURATION FUSE register address. For the mid-range devices, this is address 0x02007. So, the statement:

```
 __CONFIG 0x01234
```

stores the value *0x01234* at address 0x02007. This statement is equivalent to:

```
 org   0x02007
 dw    0x01234
```

The fuse values and states are defined in the PICmicro® MCU include (.INC) file. As indicated elsewhere, when you begin working with a PICmicro® MCU device, you should understand the different configuration options and ensure that you include all of these options in your *__CONFIG* statement.

When specifying configuration fuse values from the include file, each parameter should be ANDed together. This way any reset bits will be combined together to produce the value that is loaded into the configuration fuse register.

In the minimum application, which uses the PIC16F84, you should be aware of four different configuration fuses :

- Oscillator
- Watchdog timer
- Power-up timer
- Program memory code protection.

In this application, I would want to use some pretty typical settings, which are the crystal oscillator (_XT_OSC), no watchdog timer enabled (_WDT_OFF), the power up timer enabled (_PWRTE_ON), and no program memory code protection (_CP_OFF). To combine these settings into a single value for the configuration fuses, I add the statement:

```
_CONFIG _XT_OSC & _WDT_OFF & _PWRTE_ON & _CP_OFF
```

to the application code, which changes it to:

```
 include "p16F84.inc"
_CONFIG _XT_OSC & _WDT_OFF & _PWRTE_ON & _CP_OFF
 org     0
  clrf   PORTB
  bsf    STATUS, RP0
  clrf   TRISB ^ 80
  bcf    STATUS, RP0
  goto   $
 end
```

I recommend that PICmicro® MCU programmers that extract the configuration fuse information from the produced hex file should only be used. This eliminates the need for you to manually set the configuration fuses during programming (which saves you some work and avoids an opportunity for the configuration fuses to be set incorrectly).

When MPASM executes, the default numbering system is hexadecimal (base 16). Personally, I prefer working in a base 10 (decimal) numbering system, so I change the *radix* (which specifies the default numbering system base). This is done using the *LIST* directive. The *LIST* directive is used to enable or disable listing of the source file or specify operating parameters for the assembler.

In all applications, I add the *LIST R=DEC* statement, which changes the default number base to base 10, rather than base 16. After adding it to MINIMUM.ASM, all values have to be checked to be in the correct base. The immediate value XORed with the address of TRISB will have to be changed to be explicitly specified as hex (using the *0x* prefix):

```
LIST R=DEC
include "p16F84.inc"
_CONFIG _XT_OSC & _WDT_OFF & _PWRTE_ON & _CP_OFF
```

```
org      0
  clrf     PORTB
  bsf      STATUS, RP0
  clrf     TRISB ^ 0x080
  bcf      STATUS, RP0
  goto     $
 end
```

All of this was done in the interests of making the source code easier to read and understand. But when you look over what I've done to the source, I sure haven't made it that much easier to figure out what it is done by just looking. Adding comments to the source will make the application much easier to understand because they will explain what the application does, who is the author, what changes have been made to it, and what the code is doing. The semicolon (;) is used to indicate that the text to the right is to be ignored and is just for the application developer's use.

After adding comments to the application, it looks like:

```
; minimum.asm - A simple Application that turns on the LEDs that are
;  connected to all the PORTB Pins.
; Author: Myke Predko
; 00.01.06
 LIST R=DEC
 include "p16F84.inc"
 _CONFIG _XT_OSC & _WDT_OFF & _PWRTE_ON & _CP_OFF
 org        0
 clrf       PORTB                ; LED is ON when Port Bit is Low
 bsf        STATUS, RP0
 clrf       TRISB ^ 0x080        ; Enable all of PORTB for Output
 bcf        STATUS, RP0
 goto       $                    ; When done, Loop forever
 end
```

This adds a lot to help understanding what is happening in the application. Notice that not every line has a comment; I have tried to only comment the "functional blocks" to allow the programmer who will be working with this code to try and better understand what is happening.

It probably seems to be a bit tight. To help alleviate this, blank lines are added to break up functional blocks of code and try to make the code easier to understand.

```
; minimum.asm - A simple Application that turns on the LEDs that are
;  connected to all the PORTB Pins.
; Author: Myke Predko
; 00.01.06

 LIST R=DEC

 include "p16F84.inc"
 _CONFIG _XT_OSC & _WDT_OFF & _PWRTE_ON & _CP_OFF

 org        0

 clrf       PORTB                ; LED is ON when Port Bit is Low

 bsf        STATUS, RP0
 clrf       TRISB ^ 0x080        ; Enable all of PORTB for Output
 bcf        STATUS, RP0

 goto       $                    ; When done, Loop forever

 end
```

Now, the application is in a format that should be reasonably easy to understand. The header comments are in a bit different format than what I will use in the book, but you should get an idea of what each line is responsible for. Using labels, comments, and white space, you will greatly enhance the readability of your application so that you can figure out what the applications are doing.

To successfully program the PICmicro® MCU, there are quite a few conventions and directives that you should follow when developing an application. To help you familiarize yourself with these requirements for PICmicro® MCU applications, I suggest that you work through Chapter 15 to see how they all work together. After you have worked through a few applications, creating your own PICmicro® MCU application will seem like second nature to you.

The next chapter introduces a PICmicro® MCU template that you can use to develop your own applications without having to memorize everything that has been presented in this section.

**Assembler directives** The MPASM assembler is rich with directives or assembler instructions that can be used to make your application programming easier. The directives are executed by the assembler to modify, enhance, or format the source file before assembling the source code into an object or hex file.

Directives are not placed at the first column of a source file. To help me differentiate them, I place them on the second column of the source file, with only labels starting in the first column. I place source code at the third column.

Table 8-23 lists all the assembler directives used in MPLAB, along with examples for their use and any comments that I have on using them. For directives that can only be used with another directive, I have provided a notation to the prerequisite directive.

## STANDARD DECLARATION AND INCLUDE (.INC) FILES

One of the first things I often see when new PICmicro® MCU programmers have problems with their initial applications is that they have defined the hardware registers that they are going to use in their applications. The programmer is mystified why their code doesn't work or, in some cases, assembly correctly. Almost invariably, the problems with the application are a result of a typo or transposition of a register address or label.

To fix the problems, I tell them to delete their hardware register declarations and use the *include* directive to load the Microchip-written .INC register definition files into their applications. These files were written by Microchip to provide the application developer with the addresses of the PICmicro® MCU hardware registers, along with some other parameters, in the same format as the documentation. Usually, when the programmer-defined hardware register declares are deleted and the .INC file is added to the source, the application problems disappear.

An .INC file is available for every PICmicro® MCU part number in the format:

*pPICmicro .inc*

where *PICmicro* is the PICmicro® MCU part number. For example, the include file for the PIC16F84 is P16F84.INC and the include file for the PIC12C508 is P12C508.INC. This is true for all the PICmicro® MCU devices, except for the original, low-end (PIC16C5x) parts. For these devices, the include file is P16C5X.INC.

**TABLE 8-23 MPLAB Assembler Directives**

| DIRECTIVE | USAGE EXAMPLE | COMMENTS |
|---|---|---|
| _BADRAM | _BADRAM Start, End | Flag a range of file registers that are unimplemented |
| BANKISEL | BANKISEL <label> | Update the IRP bit of the STATUS register before the FSR register is used to access a register indirectly. This directive is normally used with linked source files. |
| BANKSEL | BANKSEL Label | Update the RPx bits of the STATUS register before accessing a file register directly. This directive is not available for the low-end devices (for these devices, the FSR register should be used to access the specific address indirectly). This directive is also not available for the high-end PICmicro® MCUs, which should use the *movlb* instruction. |
| CBLOCK | CBLOCK Address<br>Var1, Var2<br>VarA:2<br>ENDC | Used to define a starting address for variables or constants that require increasing values. To declare multiple-byte variables or constants that increment by more than one, a colon (:) is placed after the label and before the number to increment by. This is shown for *VarA* in the usage example. The *ENDC* directive is required to "turn off" CBLOCK operation. |
| CODE | CODE [Address] | Used with an object file to define the start of application code in the source file. A *Label* can be specified before the directive to give a specific label to the object file block of code. If no *Address* is specified, then MPLINK will calculate the appropriate address for the *CODE* state- |

**TABLE 8-23 MPLAB Assembler Directives (*Continued*)**

| DIRECTIVE | USAGE EXAMPLE | COMMENTS |
|---|---|---|
| | | ment and the instructions that follow it. |
| _CONFIG | _CONFIG Value | This directive is used to set the PICmicro® MCU's configuration bits to a specific value. *_CONFIG* automatically sets the correct address for the specific PICmicro® MCU. The *Value* consists of constants declared in the PICmicro® MCU's .INC file. |
| CONSTANT/ =/EQU | CONSTANT Label = Value<br>or<br>Label = Value<br>or<br>Label EQU Value | Define a constant using one of the three formatting methods shown in usage example. The constant *Value* references to the *Label* and is evaluated when the *Label* is defined. To replace a label with a string, use # *DEFINE*. |
| DA/DATA/DB | DA Value\|"string"<br>or<br>DATA Value\|"string"<br>or<br>DB Value\|"string" | Set program memory words with the specified data values. If a "*string*" is defined, then each byte is put into its own word. The *DW* directive is recommended to be used instead of *DATA* or *DB* because its operation is less ambiguous when it comes to how the data is stored. Notice that *DATA/DB/DW* do not store the data according as part of a *retlw* instruction. For the *retlw* instruction to be included with the data, the *DT* directive must be used. These directives are best suited for use in serial EEPROM source files. |
| DE | ORG 0x02100<br>DE Value\|"string" | This instruction is used to save initialization data for the PICmicro® MCU's built-in data EEPROM. Notice that an *org 0x02100* statement has to precede the *de* |

*(continued)*

TABLE 8-23 MPLAB Assembler Directives *(Continued)*

| DIRECTIVE | USAGE EXAMPLE | COMMENTS |
|---|---|---|
| | | directive to ensure that the PICmicro® MCU's program counter will be at the correct address for programming. |
| #DEFINE | #DEFINE Label [string] | Specify that any time *Label* is encountered, it is replaced by the string. Notice that *string* is optional and the defined *Label* can be used for conditional assembly. If *Label* is to be replaced by a constant, then one of the *CONSTANT* declarations should be used. This directive is placed in the first column of the source file. |
| DT | DT Value[,Value. . .]\| "string" | Place the *Value* in a *retlw* statement. If *DT*'s parameter is part of a *"string"*, then each byte of the string is given its own *retlw* statement. This directive is used to implement read-only tables in the PICmicro® MCU. |
| DW | DW Value [,Value. . .] | Reserve program memory for the specified *Value*. This value will be placed in a full-program-memory word. |
| Else | | Used in conjunction with *IF*, *IFDEF*, or *IFNDEF* to provide an alternative path for conditional assembly. Look at these directives for examples of how *ELSE* is used. |
| END | END | End the program block. This directive is required at the end of all application source files. |
| ENDC | | Used to end the *CBLOCK* label constant value saving and updating. See *CBLOCK* for an example of how this directive is used. |

**TABLE 8-23 MPLAB Assembler Directives (*Continued*)**

<table>
<tr><th>DIRECTIVE</th><th>USAGE EXAMPLE</th><th>COMMENTS</th></tr>
<tr><td>ENDIF</td><td></td><td>Used to end an IF statement conditional code block. See IF, IFDEF, or IFNDEF for an example of how this directive is used.</td></tr>
<tr><td>ENDM</td><td></td><td>Used to end the MACRO definition. See CBLOCK for an example of how this directive is used.</td></tr>
<tr><td>ENDW</td><td></td><td>Used to end the block of code repeated by the WHILE conditional loop instruction. See WHILE for an example of how this directive is used.</td></tr>
<tr><td>ERROR</td><td>ERROR “string”</td><td>Force an ERROR into the code with the “string” message inserted into the listing/error files.</td></tr>
<tr><td>ERRORLEVEL</td><td>ERRORLEVEL 0|1|2, +#|−#</td><td>Change the assembler’s response to the specific Error (2), Warning (1) or Message (0) Number (#). Specifying − before the number will cause any occurrences of the error, warning, or message to be ignored by the assembler and not be reported. Specifying + before the number will cause any occurrences of the error, warning, or message to be output by the assembler.</td></tr>
<tr><td>EXITM</td><td></td><td>For use within a MACRO to force the stopping of the MACRO expansion. Using this directive is not recommended, except in the case where the MACRO’s execution is in error and should not continue until the error has been fixed. Using EXITM in the body of the MACRO could result in phase errors, which can be very hard to find.</td></tr>
</table>

*(continued)*

**TABLE 8-23 MPLAB Assembler Directives (*Continued*)**

| DIRECTIVE | USAGE EXAMPLE | COMMENTS |
|---|---|---|
| EXPAND | EXPAND | Enable printing *MACRO* expansions in the listing file after they have been disabled by the *NOEXPAND* directive. Printing of *MACRO* expansions is the default in the MPLAB IDE. |
| EXTERN | EXTERN Label | Make a program memory label in an object file available to other object files. |
| FILL | FILL Value, Count | Put in *Value* for Count words. If *Value* is surrounded by parentheses, then an instruction can be put in, i.e., (*goto 0*). In earlier versions of MPLAB, *Fill* did not have a *Count* parameter and replaced any program memory address that did not have an instruction assigned to it or areas that were not reserved (using *RES*) with the *Value*. |
| GLOBAL | GLOBAL Label | Specify a label within an object file that can be accessed by other object files. *GLOBAL* is different from *EXTERN* because it can only be put into the source after the label is defined. |
| IDATA | IDATA [Address] | Used to specify a data area within an object file. If no *Address* is specified, then the assembler calculates the address. A label can be used with *IDATA* to reference it. |
| _IDLOCS | _IDLOCS Value | Set the four ID locations of the PICmicro® MCU with the four nybbles of *Value*. This directive is not available for the 17Cxx devices. |
| IF | IF Parm1 COND Parm2<br>; "True" Code | If *Parm1 COND Parm2* is *"true"*, then insert and |

**TABLE 8-23 MPLAB Assembler Directives (*Continued*)**

| DIRECTIVE | USAGE EXAMPLE | COMMENTS |
|---|---|---|
| | ELSE<br>; "False" Code<br>ENDIF | assemble the *"True"* code. Otherwise, insert and assemble the *False* code. The *Else* directive and *False* codes are optional. |
| IFDEF | IFDEF Label<br>; "True" Code<br>ELSE<br>; "False" Code | If the label has been defined (using *#DEFINE*), then insert and assemble the *"True"* code. Otherwise insert and assemble the *"False"* code. The *Else* directive and *"False"* codes are optional. |
| IFNDEF | IFNDEF Label<br>; "True" Code<br>ELSE<br>; "False" Code | If the label has not been defined (using *#DEFINE*), then insert and assemble the *"True"* code. Otherwise insert and assemble the *"False"* code. The *Else* directive and *"False"* codes are optional. |
| INCLUDE | INCLUDE "FileName.Ext" | Load *"FileName.Ext"* at the current location within the source code. |
| LIST | LIST option[, . . .] | Define the assembler options for the source file. The available options are:<br>Option / Default / Description<br>b=nnn / 8 / Set tab spaces<br>c=nnn / 132 / Set column width.<br>f=format / INHX8M / Set the hex file output.<br>free / FIXED / Use free-format parser.<br>fixed / FIXED / Use fixed-format parser.<br>mm=ON\|OFF / ON / Print memory map in list file.<br>n=nnn / 60 / Set lines per page. |

*(continued)*

<table>
<tr><th colspan="5">TABLE 8-23 MPLAB Assembler Directives (Continued)</th></tr>
<tr><th>DIRECTIVE</th><th>USAGE EXAMPLE</th><th colspan="3">COMMENTS</th></tr>
<tr><td></td><td></td><td>p=type</td><td>None</td><td>Set PICmicro® MCU type.</td></tr>
<tr><td></td><td></td><td>r=radix</td><td>HEX</td><td>Set default radix (HEX, DEC, or OCT available)</td></tr>
<tr><td></td><td></td><td>st=ON|OFF</td><td>ON</td><td>Print symbol table in list file.</td></tr>
<tr><td></td><td></td><td>t=ON|OFF</td><td>OFF</td><td>Truncate lines of listing.</td></tr>
<tr><td></td><td></td><td>w=0|1|2</td><td>0</td><td>Set the message level.</td></tr>
<tr><td></td><td></td><td>x=ON|OFF</td><td>ON</td><td>Turn macro expansion on or off.</td></tr>
<tr><td>LOCAL</td><td>Fillup MACRO Size<br>Local i<br>i = 0<br>WHILE (i < Size)<br>DW 0x015AA<br>i = i + 1<br>ENDW<br>ENDM</td><td colspan="3">Define a variable that is local to a MACRO and cannot be accessed outside of the MACRO.</td></tr>
<tr><td>MACRO</td><td>Label MACRO [Parm[, . . .]]<br>bsf Parm, 0<br>ENDM</td><td colspan="3">Define a block of code that will replace the Label every time it is encountered. The optional parameters will replace the parameters in the MACRO itself.</td></tr>
<tr><td>_MAXRAM</td><td>_MAXRAM End</td><td colspan="3">Define the last file register address in a PICmicro® MCU that can be used.</td></tr>
<tr><td>MESSG</td><td>MESSG “string”</td><td colspan="3">Cause “string” to be inserted into the source file at the MESSG statement. No errors or warnings are generated for this instruction.</td></tr>
</table>

**TABLE 8-23 MPLAB Assembler Directives (*Continued*)**

| DIRECTIVE | USAGE EXAMPLE | COMMENTS |
|---|---|---|
| NOEXPAND | NOEXPAND | Turn off Macro expansion in the listing file. |
| NOLIST | NOLIST | Turn off source-code listing output in listing file. |
| ORG | ORG Address | Set the starting address for the following code to be placed. |
| PAGE | PAGE | Insert a page break before the *PAGE* directive. |
| PAGESEL | PAGESEL Label<br>goto Label | Insert the instruction page of a label before jumping to that label or calling the subroutine at it. |
| PROCESSOR | PROCESSOR type | This directive is available for commonality with earlier Microchip PICmicro® MCU assemblers. The *processor* option of the *LIST* directive should be used instead. |
| RADIX | RADIX Radix | This directive is available for commonality with earlier Microchip PICmicro® MCU assemblers. Available options are *HEX*, *DEC*, and *OCT*. The default radix should be selected in the *LIST* directive instead. |
| RES | RES MemorySize | Reserve a block of program memory in an object file for use by another. A label can be placed before the *RES* directive to save what the value is. |
| SET | Label SET Value | *SET* is similar to the *CONSTANT*, *EQU*, and = directives, except that the *Label* can be changed later in the code with another *SET* directive statement. |
| SPACE | SPACE Value | Insert a Set number of blank lines into a listing file. |

*(continued)*

**TABLE 8-23 MPLAB Assembler Directives *(Continued)***

| DIRECTIVE | USAGE EXAMPLE | COMMENTS |
|---|---|---|
| SUBTITLE | SUBTITLE "string" | Insert *"string"* on the line following the *TITLE* string on each page of a listing file. |
| TITLE | TITLE "string" | Insert *"string"* on the top line on each page of a listing file. |
| UPDATA | UDATA [Address]<br>Label1 RES 1<br>Label2 RES 2 | Declare the beginning of an uninitialized data section. *RES* labels should follow to mark variables in the uninitialized data space. This command is designed for serial EEPROMS. |
| UPDATA_ACS | UDATA_ACS [Address]<br>Label1 RES 1<br>Label2 RES 2 | Declare the beginning of an uninitialized data section in a 18Cxx PICmicro® MCU. *RES* labels should follow to mark variables in the uninitialized data space. |
| UPDATA_OVR | UDATA_OVR [Address]<br>Label1 RES 1<br>Label2 RES 2 | Declare the beginning of an uninitialized data section that can be overwritten by other files (as an *Overlay*). *RES* labels should follow to mark variables in the uninitialized data space. This command is designed for serial EEPROMs. |
| UDATA_SHR | UATA_SHR [Address]<br>Label1 RES 1 | Declare the beginning of data memory that is shared across all of the register banks. |
| #UNDEFINE | #UNDEFINE Label | Delete a label that was *#DEFINED.* |
| VARIABLE | VARIABLE Label<br>[= Value] | Declare an assembly-time variable that can be updated within the code using a simple assignment statement. |
| WHILE | WHILE Parm1 COND Parm2<br>; while "True"<br>ENDW | Execute code within the *WHILE/ENDW* directives while the *Parm1 COND Parm2* test is true. Notice |

**TABLE 8-23 MPLAB Assembler Directives *(Continued)***

| DIRECTIVE | USAGE EXAMPLE | COMMENTS |
|---|---|---|
| | | that in the listing file, the code will appear as if the code within the *WHILE/WEND* directives was repeated a number of times. |

The following file is the P12C508.INC, which is relatively small, but has all the elements that you should look for in the include files.

```
        LIST
; P12C508.INC Standard Header File, Version 1.02  Microchip Technology,
Inc.
        NOLIST

; This header file defines configurations, registers, and other useful
bits of
; information for the PIC12C508 microcontroller. These names are taken to
match
; the data sheets as closely as possible.

; Note that the processor must be selected before this file is
; included. The processor may be selected the following ways:

;     1. Command line switch:
;         C:\ MPASM MYFILE.ASM /P12C508
;     2. LIST directive in the source file
;         LIST  P=12C508
;     3. Processor Type entry in the MPASM full-screen interface

;==========================================================
;
;         Revision History
;
;==========================================================

;Rev:  Date:  Reason:

;1.02  05/12/97 Correct STATUS and OPTION register bits
;1.01  08/21/96 Removed VCLMP fuse, corrected oscillators
;1.00  04/10/96 Initial Release

;==========================================================
;
;          Verify Processor
;
;==========================================================

          IFNDEF __12C508
          MESSG "Processor-header file mismatch. Verify selected proces-
sor."
    ENDIF

;==========================================================
;
;          Register Definitions
;
;==========================================================
```

```
W               EQU   H'0000'
F               EQU   H'0001'

;—- Register Files ----------------------------------------------------
---

INDF              EQU   H'0000'
TMR0              EQU   H'0001'
PCL             EQU   H'0002'
STATUS            EQU   H'0003'
FSR             EQU   H'0004'
OSCCAL            EQU   H'0005'
GPIO              EQU   H'0006'

;—- STATUS Bits ------------------------------------------------------

GPWUF             EQU   H'0007'
PA0             EQU   H'0005'
NOT_TO            EQU   H'0004'
NOT_PD            EQU   H'0003'
Z               EQU   H'0002'
DC              EQU   H'0001'
C               EQU   H'0000'

;—- OPTION Bits ------------------------------------------------------

NOT_GPWU            EQU   H'0007'
NOT_GPPU            EQU   H'0006'
T0CS              EQU   H'0005'
T0SE              EQU   H'0004'
PSA             EQU   H'0003'
PS2             EQU   H'0002'
PS1             EQU   H'0001'
PS0             EQU   H'0000'
;==========================================================================
;
;     RAM Definition
;
;==========================================================================

      __MAXRAM H'1F'

;==========================================================================
;
;     Configuration Bits
;
;==========================================================================

_MCLRE_ON           EQU   H'0FFF'
_MCLRE_OFF            EQU   H'0FEF'
_CP_ON            EQU   H'0FF7'
_CP_OFF           EQU   H'0FFF'
_WDT_ON           EQU   H'0FFF'
_WDT_OFF            EQU   H'0FFB'
_LP_OSC           EQU   H'0FFC'
_XT_OSC           EQU   H'0FFD'
_IntRC_OSC            EQU   H'0FFE'
_ExtRC_OSC            EQU   H'0FFF'

      LIST
```

At the start of the file, the PICmicro® MCU specified within the MPLAB IDE is checked against the file to be sure that they match. When the MPLAB IDE has a PICmicro® MCU selected, the part number label with two underscore characters (_) is defined when the assembler is invoked. For the PIC12C508, this label is *__12C508*; for the PIC16F84, it is *__16F84* and so on for other part numbers. This label is used in the experiments to provide the correct include file and access hardware appropriately, instead of having to define multiple source files for users of the PIC16F84 and the PIC16F87x devices.

Once the PICmicro® MCU type is verified, then the hardware register addresses (under Register Files) are defined. The registers are given the same labels as the Microchip documentation and have their addresses specified with them. Following the hardware register address definitions, the bit definitions for hardware registers that have unique, accessible bits are defined.

After the hardware register files are defined, then the file registers are defined. The *__MAXRAM* and *__BADRAM* directives are used to indicate what are the valid addresses for variables. One thing lacking with these directives is that the addresses are not given labels (a label indicating the start of the file registers would be useful) and the registers "shadowed" across banks are not defined. This information could make application development somewhat easier and avoid having to look up the file register address ranges from the data books.

Lastly, the configuration fuse bits are defined. When I start working with a new PICmicro® MCU, one of the first things I always do is to open up the .INC file and look at the configuration fuses. As covered in other areas of the book, a very common mistake is to forget one of the configuration fuse options, which causes your PICmicro® MCU application to not work as expected. I always want to be sure to access each configuration fuse option, either enabling or disabling it to avoid any unexpected problems.

The .INC values are defined with the *NOLIST* parameter specified. Thus, the actual definitions will not be seen in the listing file, but will show up in the symbol table at the end of the listing file.

## OUTPUT FILES

When you first start a MPLAB IDE project, you will generally only have one file (the .ASM source file). After creating the MPLAB IDE project for the source file and assembling it, you will be amazed to find that five or six additional files have been created by MPLAB, all with the same filename as the source file and project. This section goes through the different files and explains how to read the listing (.LST) file to get information out of it.

After executing the MPLAB IDE, you can expect to see a number of files with the filename extensions shown in Table 8-24.

The previous sections introduced you to the assembly source file and the project file. This section introduces the other files that are created by the MPLAB IDE.

When the MPLAB IDE is invoked and a project is loaded with a source file in the editor, a .$$$ file is created. This file contains the source code before any changes by the MPLAB editor. Typically, this file is not required unless the source is corrupted in some way (which often means that you have done something that you didn't mean to). This is the same for the .BKX file, which is a backup of the hex file created by MPLAB for the project when it was invoked previously.

The hex file (explained in more detail in Chapter 13, is the result of the MPLAB *build* operation and is the code (ones and zeros) that are to be programmed into the PICmicro® MCU.

**TABLE 8-24 The MPLAB IDE Output Files**

| FILE EXTENSION | FUNCTION |
|---|---|
| .asm | Application Source File |
| .$$$ | Backup of the Application Source File |
| .cod | "Label Reference" for the MPLAB Simulator/Emulator |
| .err | Error Summary File |
| .lst | Listing File |
| .hex | Hex File to be loaded into the PICmicro® MCU |
| .bkx | Backup of the Hex File |
| .pjt | Project file |

This file is in human-readable format, although the first time you look at it, it will be somewhat confusing.

The .COD file is the label reference table for MPLAB to be able to run the simulator and put pointers on the correct line of the source. The .COD file is not in human-readable format and is actually quite complex to understand and work through. You should not attempt to edit or change this file in any way because changes to it will affect how the simulator displays the running application.

Any errors in the assembly/link operations will be recorded in the .ERR file. This file, which is in human-readable format should be of size zero when the source code builds correctly. If any errors are encountered while processing the source, then the file lists the line where the error was found and the error type.

It is important to remember that the .ERR file returns the syntax errors of the application source code. *Syntax errors* are errors in spelling instructions or labels, or using statements in an incorrect format. The .ERR file does not know what your application is supposed to do and will not tell you when you make an error. This is an important difference that many people do not seem to understand.

I can't tell you how many e-mails I have gotten from people saying that MPASM doesn't work properly because it reports that there aren't any errors and yet the application doesn't work correctly. This might sound funny and something worthy of a web page detailing mistakes made by new users, but it is something that many new developers have trouble understanding (especially if they are new to programming). If starting out developing applications, you have to understand that the errors only list problems with how the application code is put into the source code file.

I don't like a few aspects of how errors are reported in the MPLAB IDE. The first is that the error descriptions can be somewhat terse and vague and not fully understood by new PICmicro® MCU application developers. This means that if you get an error and you don't understand what it means, don't feel bad about it. Instead, jump to the line that is referenced (by double clicking on the line in the error window displayed in the MPLAB IDE) and see if you can figure out what the problem is.

The second thing I don't like is how errors with macros are reported. Macro errors are referenced back to the invoking line, not the line in the macro. This can make debugging macros a challenge—especially if they are very complex.

As stated elsewhere, you should only attempt to program a PICmicro® MCU if the .ERR file is zero bytes long. Syntax errors can result in misinterpreted data by MPASM and the result could be an application that doesn't behave as expected.

The last type of file is the *listing file*, which is a complete set of information that is summarized in the different files. The following listing was taken from one of the experiments and shows the different elements of the listing file. I go through them to show you what is being displayed. To make the file easier to read, I have truncated the lines to the end of the page and deleted anything that would be wrapped around to the next line. As well, I have taken away the page breaks (except for the one at the start of the application) to save space in the book.

```
MPASM 02.30 Released        ADC.ASM  12-27-1999 14:26:06     PAGE 1

LOC       OBJECT CODE     LINE SOURCE TEXT
  VALUE
                          00001 title "ADC - Reading a Resistor Value with
                          00002 ;
                          00003 ; This Program Uses the ADC built into a
                          00004 ; Reads an ADC Value and displays it on
                          00005 ;
                          00006 ; Hardware Notes:
                          00007 ;  PIC16C711 running at 4 MHz
                          00008 ;  Reset is tied directly to Vcc and PWRT is
                          00009 ;  A 10K Pot Wired as a Voltage Divider on
                          00010 ;  A 220 Ohm Resistor and LED is attached to
                          00011 ;
                          00012 ; Myke Predko
                          00013 ; 99.12.27
                          00014 ;
                          00015  LIST P=16C711, R=DEC
                          00016  INCLUDE "p16c711.inc"
                          00001     LIST
                          00002 ; P16C711.INC Standard Header File, Version
                          00151     LIST
                          00017
                          00018 ; Registers
                          00019
2007      3FF1            00020 __CONFIG _CP_OFF&_WDT_OFF&_XT_OSC&_PWRTE_ON
                          00021
                          00022  PAGE
                          00023 ; Mainline of ADC
                          00024
00000                     00025 org    0
                          00026
0000 30FF                 00027  movlw    0x0FF
0001 0086                 00028  movwf    PORTB          ; Turn off
0002 0185                 00029  clrf     PORTA          ; Use PORTA
                          00030
0003 1683                 00031  bsf      STATUS, RP0   ; Have to go
0004 0186                 00032  clrf     TRISB & 0x07F ; Set all
0005 0188                 00033  clrf     ADCON1 ^ 0x080; Make RA0
0006 1283                 00034  bcf      STATUS, RP0    ; Go back to
                          00035
```

```
0007 3081           00036  movlw    0x081        ; Setup
0008 0088           00037  movwf    ADCON0       ;
                    00038                        ;
                    00039                        ;  CHS1:CHS0
                    00040                        ;  Go/_Done
                    00041                        ;  ADIF - 0
                    00042                        ;  ADON - 1
                    00043
0009                00044 Loop
                    00045
0009 3003           00046  movlw    3            ; Wait 12
000A 3EFF           00047  addlw    0x0FF        ; Take One
000B 1D03           00048  btfss    STATUS, Z
000C 280A           00049  goto     $ - 2
                    00050
000D 1508           00051  bsf      ADCON0, GO   ; Turn on
000E 1908           00052  btfsc    ADCON0, GO   ; Wait for
000F 280E           00053  goto     $ - 1
                    00054
0010 1683           00055  bsf      STATUS, RP0
0011 0909           00056  comf     ADRES, w     ; Get the
0012 1283           00057  bcf      STATUS, RP0
0013 0086           00058  movwf    PORTB
                    00059
0014 2809           00060  goto     Loop         ; Get
                    00061
                    00062
                    00063  end
SYMBOL TABLE
 LABEL                              VALUE

ADCON0                              00000008
ADCON1                              00000088
ADCS0                               00000006
ADCS1                               00000007
ADIE                                00000006
ADIF                                00000001
ADON                                00000000
ADRES                               00000009
C                                   00000000
CHS0                                00000003
CHS1                                00000004
DC                                  00000001
F                                   00000001
FSR                                 00000004
GIE                                 00000007
GO                                  00000002
GO_DONE                             00000002
INDF                                00000000
INTCON                              0000000B
INTE                                00000004
INTEDG                              00000006
INTF                                00000001
IRP                                 00000007
Loop                                00000009
NOT_BO                              00000000
NOT_BOR                             00000000
NOT_DONE                            00000002
NOT_PD                              00000003
NOT_POR                             00000001
NOT_RBPU                            00000007
```

```
NOT_TO                                    00000004
OPTION_REG                                00000081
PCFG0                                     00000000
PCFG1                                     00000001
PCL                                       00000002
PCLATH                                    0000000A
PCON                                      00000087
PORTA                                     00000005
PORTB                                     00000006
PS0                                       00000000
PS1                                       00000001
PS2                                       00000002
PSA                                       00000003
RBIE                                      00000003
RBIF                                      00000000
RP0                                       00000005
RP1                                       00000006
STATUS                                    00000003
T0CS                                      00000005
T0IE                                      00000005
T0IF                                      00000002
T0SE                                      00000004
TMR0                                      00000001
TRISA                                     00000085
TRISB                                     00000086
W                                         00000000
Z                                         00000002
_BODEN_OFF                                00003FBF
_BODEN_ON                                 00003FFF
_CP_OFF                                   00003FFF
_CP_ON                                    0000004F
_HS_OSC                                   00003FFE
_LP_OSC                                   00003FFC
_PWRTE_OFF                                00003FFF
_PWRTE_ON                                 00003FF7
_RC_OSC                                   00003FFF
_WDT_OFF                                  00003FFB
_WDT_ON                                   00003FFF
_XT_OSC                                   00003FFD
__16C711                                  00000001

MEMORY USAGE MAP ('X' = Used, '-' = Unused)

0000 : XXXXXXXXXXXXXXXX XXXXX----------- ---------------- ------------
2000 : -------X-------- ---------------- ---------------- ------------

All other memory blocks unused.

Program Memory Words Used:  21
Program Memory Words Free: 1003


Errors   :    0
Warnings :    0 reported,    0 suppressed
Messages :    0 reported,    0 suppressed
```

You should be aware of three separate areas in the listing file. The first is the source code area in which the object code (the hex instruction value) is given to the left of the source file line, along with the address where it is located. Each line is repeated with its line number in the source file listed. With this information, instructions can be found either by their address within the PICmicro® MCU's program memory space or by the line where they are found on in the source code.

With the MPLAB IDE's ability to work as a source code editor, looking at the bits and bytes is largely unnecessary. Having said this, as you first start working through your first problem application, you might want to take a look at the code, which is produced in order to ensure that what you think is being programmed in actually is. As I was writing this, I received a question from a "newbie" who wanted to know what was wrong with his application that had the following instruction:

```
bsf 0, PORTB
```

in it. I must confess that I spent quite a bit of time looking at the code until I noticed what the problem actually was.

To find the problem, I first simulated the application and found that PORTB bit zero was not reset as I would have expected by the instruction. Because it looked correct, I looked at what the assembler produced for the instruction. I found that bit pattern 0x01680 was printed in the listing file instead of the expected 0x01405. To produce 0x01405, the instruction should have been:

```
bsf PORTB, 0
```

You are probably thinking that this wasn't a very good example because few people have the bit patterns for every instruction memorized. I don't expect anyone to have the bit patterns memorized. Instead, they should remember that the last seven bits of an instruction that accesses a register is the register address. When I saw the listing file for the instruction, I was expecting to see *0x05* as the last seven bits. When I didn't see that, I looked again at the instruction and immediately saw the problem.

The next section in the listing file is a list of the values of all of the labels in the application. Notice that hardware register addresses, bit numbers, labels, and variable file register addresses are included in this section, listed in alphabetical order. If you are familiar with other assemblers, you might expect that the label types and references to them are also included. In the MPLAB IDE, just the label and its value at the end of the application are listed.

The last section is a summary of the addresses used by the application, along with a total of any errors, warnings, or messages. The program memory address summary can be very useful when you are using a large fraction of the address space in the PICmicro® MCU and you want an idea of what is available.

## LINKING AND LINKED APPLICATIONS

A relatively new feature of MPLAB is the ability to create object files and link them together as applications. I do not use this feature a lot because most of my applications are just written by myself. Linking code is probably best served when applications are developed in groups. The other advantage of object code is that the source language used to develop application software can be mixed (such as C and assembler) and then linked together for the final application.

When one developer has created a part of an application, the object file can be distributed to other developers as a method of code control. Nobody else can change the source code, but they can use it with their own application code.

Personally, I do not use link files for three reasons. First, I do not tend to work with others when I am developing PICmicro® MCU application code. Second, it tends to complicate the application development process. Third, the assembly and linkage process takes a lot longer than simply using one file.

This does not mean that I do not want to combine different code sources together. In this case, I normally make up my own include files, which are assembled all at one time. I find this to be the most efficient for my requirements.

Before creating the linked application object, you will have to create the source files that are to be linked together. This is done by writing out your application source files and then providing links to the addresses that have to be accessed between the different files. It is important to remember that variables and instruction addresses have to be common.

In the *code\linking* subdirectory of the *PICmicro* directory, you will find two source files that are meant to be linked together. The first file, TEST3.ASM is:

```
TITLE - Test3 - Jump to Test3A
;
; Example Application using the MPLAB Linker
;
; Myke Predko
; 2000.02.02
;
; Hardware Notes:
; PIC16F84 running in a Simulator
;
 LIST      R=DEC                ; list directive to define processor
 #include  "p16F84.inc"         ; processor specific variable definitions

 __CONFIG _CP_OFF & _WDT_ON & _PWRTE_ON & _XT_OSC

 EXTERN TEST3A                  ; Specify Mainline Location
 GLOBAL TEST3AStart
 GLOBAL flag                    ; Variable passed to Linked File


;***** VARIABLE DEFINITIONS (examples)

; example of using Uninitialized Data Section
INT_VAR      UDATA   0x0C
w_temp       RES     1          ; variable used for context saving
status_temp  RES     1          ; variable used for context saving
flag         RES     2          ; temporary variable (shared locations -
                                  G_DATA)

;**********************************************************************

RESET_VECTOR CODE    0x000      ; processor reset vector
               goto    start    ; go to beginning of program

INT_VECTOR   CODE    0x004      ; interrupt vector location
               movwf    w_temp
PROG CODE 0x005
               movf   STATUS, w
               movwf  status_temp

; isr code can go here or be located as a call subroutine elsewhere

             movf    status_temp, w  ; Restore Context Registers
             movwf   STATUS
             swapf   w_temp, f
```

```
             swapf   w_temp, w
             retfie
start
             goto    TEST3A
TEST3AStart
             END                     ;        directive 'end of program'
```

This file has a slightly different format than the others presented in this book. That is because I have copied template files from the Microchip website for this section. This code declares a common (global) variable (flag) and then, once execution is past, the interrupt handler jumps to *TEST3A*. Notice that the external (shared between the source files) addresses are declared at the start of the application. I like to do this to keep track of where everything should be.

The *GLOBAL* directive indicates that the label is local to the current source file and *EXTERN* indicates the label is declared externally to the source file.

The second file initializes the global variable *flag* to 77 decimal and then decrements it before ending up in an infinite loop.

```
TITLE "Test3A - Actually Execute the Code"

;

; Example Application using the MPLAB Linker

;

; Myke Predko
; 2000.02.02

;

; Hardware Notes:
; PIC16F84 running in a Simulator

;

 EXTERN flag                         ; External Values Linked into Code
 EXTERN TEST3AStart

 list r=dec
 #include "p16f84.inc"

TEST3ACODE CODE
TEST3A
 GLOBAL TEST3A                       ; Address to Pass to Linked File
    banksel flag                     ; example
    clrf  flag                       ; example
; remaining code goes here
 movlw 77
 movwf flag
 movlw 0x001
 subwf flag, f
 btfss STATUS, Z
  goto $ - 2

 goto $                              ; Loop Here Forever

              END                    ; directive 'end of program'
```

These two files are then combined together when the project is built by setting up the application project specially to link the two files together.

Creating the linking project is accomplished by creating a New project (clicking on the *MPLAB Project* pull down, followed by *New*) and selecting the hex file name. Next, the hex file *Node* is selected and *Node Properties* clicked on. Instead of using MPASM, change the *Language Tool* to *MPLINK* (Fig. 8-21).

When this is done, you can now add the two nodes (TEST3.ASM and TEST3A.ASM), as well as a link script file. This is shown in Fig. 8-22. The link script file is provided for your use by Microchip; you can copy the (PICmicro® MCU part number) appropriate file from the *\Program Files\MPLAB* subdirectory of your PC. For this application, I used 16F84.LKR, which looks like:

```
// File: 16f84.lkr
// Sample linker command file for 16F84, 16CR84A
// 12/05/97

LIBPATH .

CODEPAGE   NAME=vectors   START=0x0      END=0x4      PROTECTED
CODEPAGE   NAME=page      START=0x5      END=0x3FF
CODEPAGE   NAME=.idlocs   START=0x2000   END=0x2003
CODEPAGE   NAME=.config   START=0x2007   END=0x2007

DATABANK   NAME=gprs      START=0xC      END=0x4F

DATABANK   NAME=sfr0      START=0x0      END=0xB      PROTECTED
DATABANK   NAME=sfr1      START=0x80     END=0x8B     PROTECTED
```

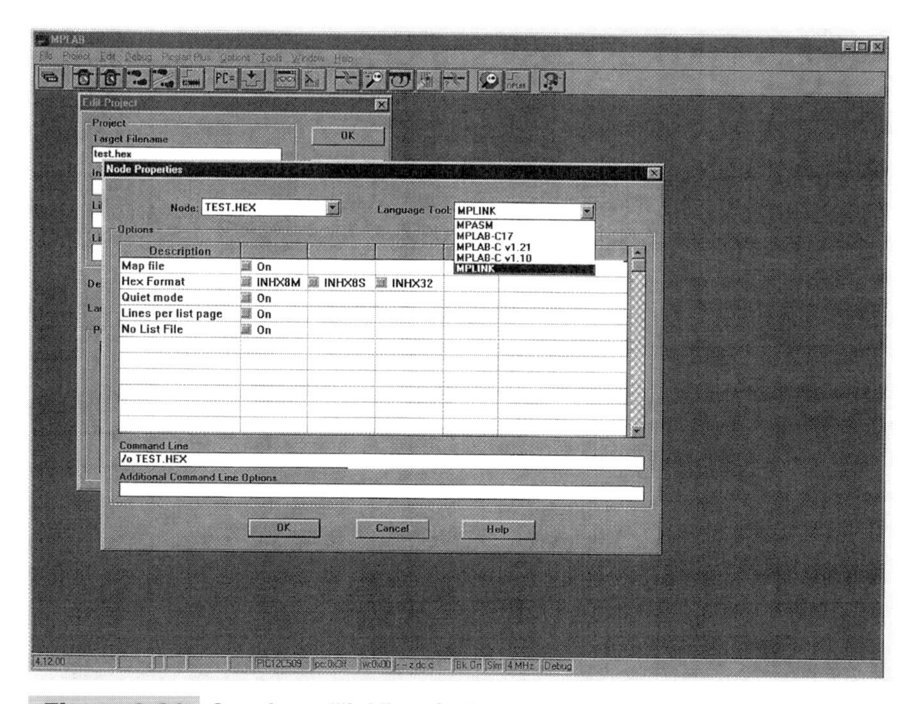


**Figure 8-21** Creating a "link" project

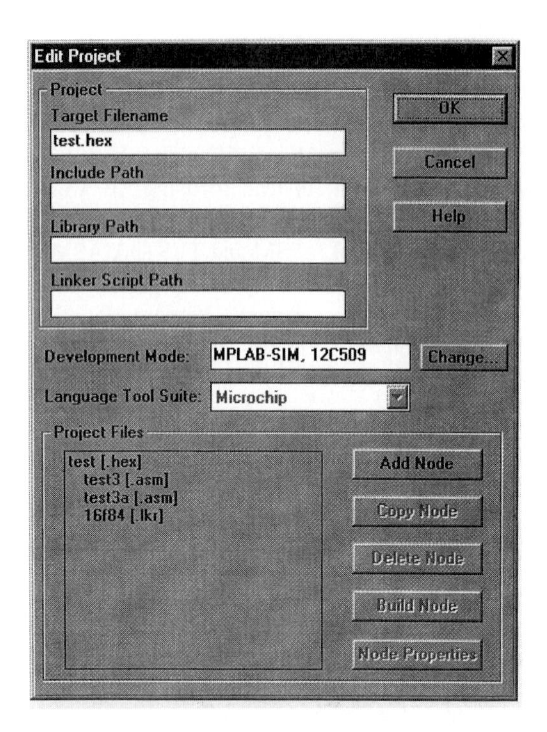


**Figure 8-22** Specifying link files

```
SECTION     NAME=STARTUP   ROM=vectors     // Reset and interrupt vectors
SECTION     NAME=PROG      ROM=page        // ROM code space
SECTION     NAME=IDLOCS    ROM=.idlocs     // ID locations
SECTION     NAME=CONFIG    ROM=.config     // Configuration bits location
```

This file can be modified, but I would recommend against it for your first few applications.

Once the files are specified, you can click on *OK* in the *Edit Project* window and test your application. Figure 8-22 shows how I have put in TEST3.ASM and TEST3A.ASM, one on top of the other with a watch window (described later in this chapter) to check to see how the context registers (w and STATUS) and the *flag* variable as the application executes.

To create the hex file, the standard *Ctrl-F10* key sequence or clicking on the *Project* pull down, followed by clicking on *Build All*, is used. The build process for linked files consists of assembling all of the source files and then linking them together. This operation will take three or four times longer than just a straight assemble to a hex file and produce the *Build Results* window that contains information like the following listing.

```
Building TEST.HEX...

Compiling TEST3.ASM:
Command line: "C:\PROGRA~1\MPLAB\MPASMWIN.EXE /e+ /l+ /x- /c+ /p16F84
/o+ /q C:\WRITING\PACPMV2\CODE\LINKING\TEST3.ASM"

Compiling TEST3A.ASM:

Command line: "C:\PROGRA~1\MPLAB\MPASMWIN.EXE /e+ /l+ /x- /c+ /p16F84
/o+ /q C:\WRITING\PACPMV2\CODE\LINKING\TEST3A.ASM"
```

```
Linking:
Command line: "C:\PROGRA~1\MPLAB\MPLINK.EXE /o TEST.HEX TEST3.O TEST3A.O
16F84.LKR "

MPLINK v1.30.01, Linker
Copyright (c) 1999 Microchip Technology Inc.

Errors   : 0
Warnings : 0

MP2COD v1.30.01, COFF to COD File Converter
Copyright (c) 1999 Microchip Technology Inc.

Errors   : 0
Warnings : 0

MP2HEX v1.30.01, COFF to HEX File Converter
Copyright (c) 1999 Microchip Technology Inc.

Errors   : 0
Warnings : 0
```

Once the results window has the message "Build completed successfully" at the end, you are ready to simulate the application or burn it into a PICmicro® MCU. If you look at Fig. 8-23, you'll see that the MPLAB simulator will follow the execution from one file to another and allow you to step through the code in each source file.

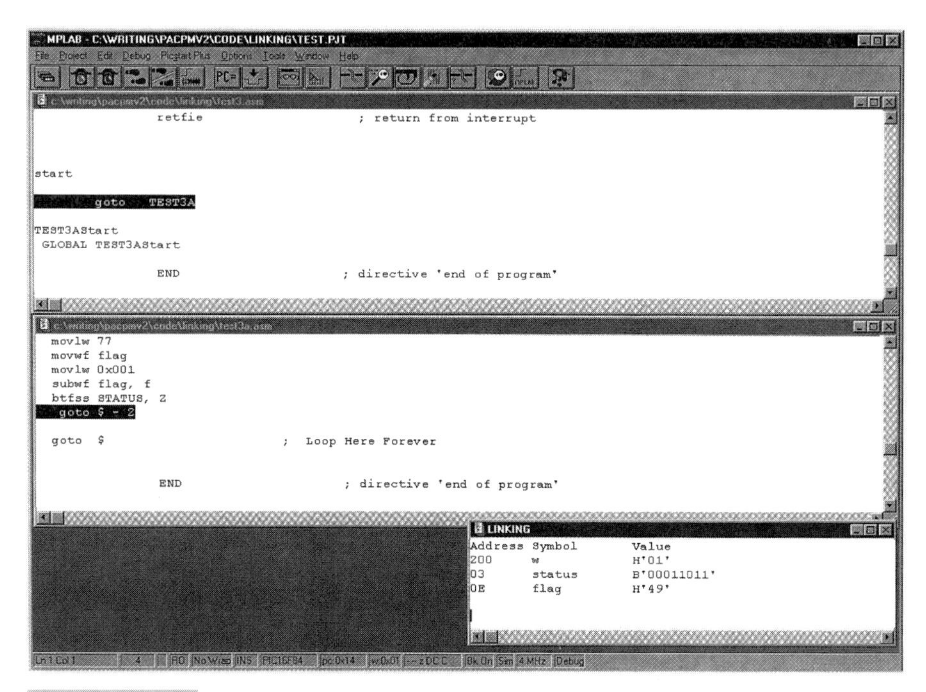


**Figure 8-23** Link files executing in MPLAB

When you are linking files together, I cannot emphasize enough the importance of ensuring that there are no errors or warnings before you start simulating or programming a PICmicro® MCU with the application. It might be possible to get a hex file from MPLINK, but chances are that the application will not execute correctly.

## WATCH WINDOW FILES

The MPLAB IDE has the capability to display specific register and bit contents in the PICmicro® MCU. For all of the experiments presented in this book, I have included specifications for at least one watch window. These windows (such as the one shown in Fig. 8-24) allow you to select the registers to monitor. Chapter 15 covers creating and using watch windows, but in this section, I want to introduce you to the concept and explain the information that is used with them.

To define a watch window or add more registers to it, the *Register Selection* window is brought up for you to select the registers that you would like to monitor. The *Properties* window is selected from the *Register Selection* window (Fig. 8-25) to specify the characteristics of the register that is displayed.

Watch windows should be started after the application has assembled without any errors, warnings, or messages. If there are errors when the watch window is created, the file-register information is not available to the MPLAB IDE and the list of registers available is restricted to the basic set available to the device.

Registers can be displayed in decimal, hex, binary, and ASCII formats and can be one, two, three, or four bytes in size. Multiple byte data can be selected as being the least-significant byte first (my normal default) or most-significant byte first. Along with registers, individual bits within them can also be displayed. This gives you a pretty complete range of options for displaying data.

When the watch window has been customized, the display data is saved in a text file like the one for the watch windows shown in Fig. 8-24:

```
[Watch_Name]
Watch_Name=3RS


[Watch_Offset]
0=HIGH
1=HIGH
2=HIGH
3=HIGH
```

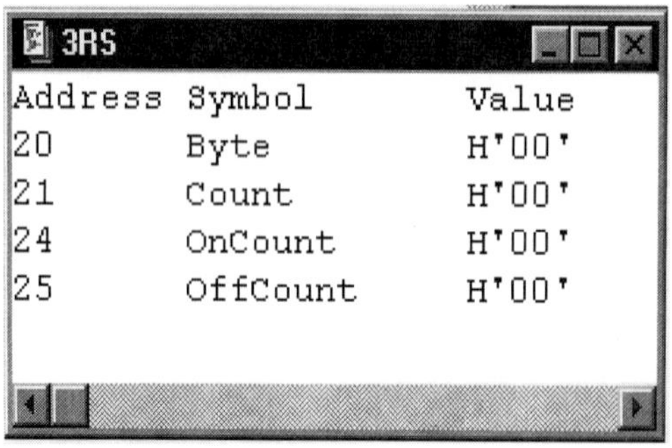

3RS

| Address | Symbol | Value |
|---|---|---|
| 20 | Byte | H'00' |
| 21 | Count | H'00' |
| 24 | OnCount | H'00' |
| 25 | OffCount | H'00' |

**Figure 8-24** Sample "watch window"

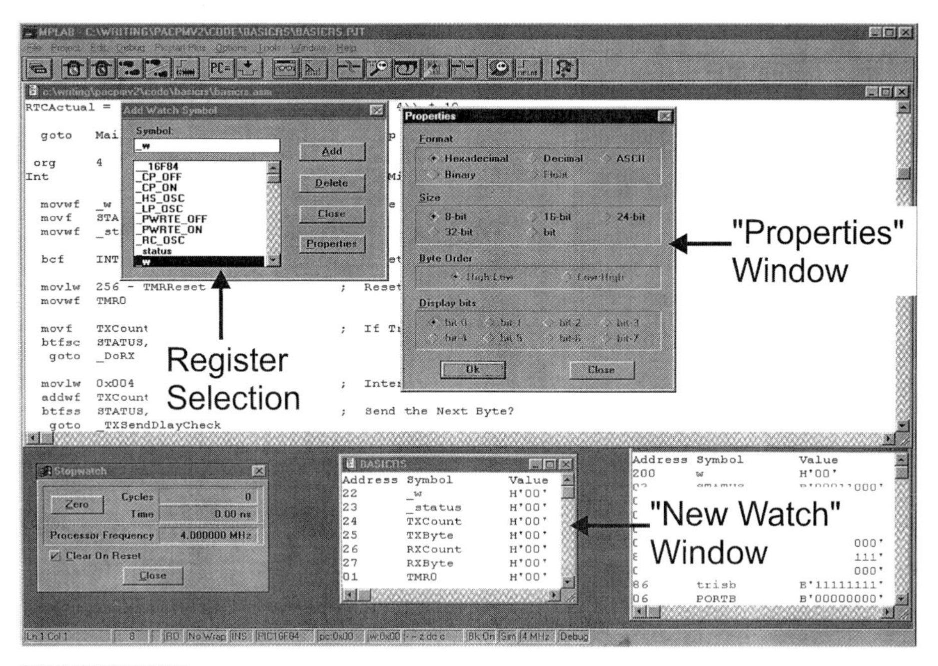


**Figure 8-25** The MPLAB watch window create

```
[Watch_Registers]
0=Byte
1=Count
2=OnCount
3=OffCount

[Watch_Format]
0=HEX
1=HEX
2=HEX
3=HEX

[Watch_Size]
0=BIT8
1=BIT8
2=BIT8
3=BIT8

[Watch_Bit]
0=
1=
2=
3=
```

The watch window file (having the filename extension .WAT) consists of six different parts. The first part is the file name, used by the MPLAB IDE to ensure that the file is not corrupted or invalidly copied from another watch file.

Next, the registers to monitor are listed in the order in which they are listed in the watch window. Notice that for *3RS*, the four registers are passed to each of the following sections. The second section of the file lists the order in which the data is to be displayed (most-significant or high byte is the default), followed by the register name. Next comes the data format, data size, and bit definitions. The information is quite logical and easy to understand.

I am going through this amount of detail about the watch file because sometimes you are going to feel the need to edit it. The watch window interface is not that easy to work with. To change the format of a displayed register, you have to delete all the registers after the one you want. Then, you will have to reinsert it with the correct parameters. Once this is done, you can put back the other registers that were deleted to gain access to the register to be modified.

This process, although not particularly arduous, can be difficult to remember the set up of all the registers that are deleted and then have to be reinstalled. Rather than going through this process, you can edit the specific watch register using an editor (such as NotePad) and avoid having to remember the list and format of the other registers. If you forget a register, you might have to go back and change the watch window *again*.

I do not recommend creating your own watch windows (although I do it automatically in PICLite) simply because the effort in doing this will be much greater than just creating a new watch window and adding valid register values that can be edited later.

# 9

# PICmicro® MCU ASSEMBLY-LANGUAGE SOFTWARE-DEVELOPMENT TECHNIQUES

## CONTENTS AT A GLANCE



The PICmicro® MCU is an interesting device to write application software for. If you have experience with other processors, you will probably consider the PICmicro® MCU to be quite a bit different and maybe even "low end," compared to other microcontrollers and microprocessors. Despite this first impression, very sophisticated software applications can be written for the PICmicro® MCU. By following the tricks and suggestions presented in this chapter, they will be surprisingly efficient as well.

At the risk of sounding Zen, the PICmicro® MCU is best programmed when you are in the right "headspace." As you become more familiar with the architecture, you will begin to see how to exploit the architecture and instruction set features to best implement your applications. The PICmicro® MCU has been designed to pass and manipulate bits and bytes very quickly between locations in the chip. Being able to plan your applications with an understanding of the data paths will allow you to write applications that can require as little as one third of the clock cycles and instructions as would be required in other microcontrollers.

This level of optimization is not a function of learning the instruction set and some rules. Instead, it is a result of thoroughly understanding how the PICmicro® MCU works and being able to visualize the best path for data within the processor.

# Creating and Supporting Your Code

When you've been in the business long enough, you'll discover that you just cannot escape some applications. For me, it's a bunch of 8086 assembly language token-ring test routines that I wrote in the mid 1980s. For some reason, this code comes back to haunt me every three of four years as people bring in new applications that use the base hardware and need some way to test it. I've often joked that I won't be allowed to die until this code has been finally "obsoleted."

One of the reasons why I can support this code is that I have used the same format for my coding for almost 20 years now. This format works very well for me and allows me to follow through with (much) later fixes and updates. The next section presents the template that I use for the PICmicro® MCU which is based on my 8086 assembly language program template.

When I present the template, I am doing this as an example and show what I consider to be the important information on it. If you have something that works better for you, then by all means use it. The next section lists what I consider the important things to capture in the source code file and what have been the important changes. How this information is actually formatted is not a big deal as long as you can find and understand it easily.

Another source for this type of template is on the Microchip web site. Microchip has created a number of template files for assembler (which can be used as a stand alone or linked with other files) for people's use. Either Microchip's or my templates are a good way to start working with the PICmicro® MCU without having to learn the MPLAB assembler basics.

Along with the source-code format, you should also think about how you are going to keep track of different versions. The ideal way of doing this is to take advantage of Microsoft Windows' long file names and name your application something like:

```
PICmicro MCU NTSC Composite Video Output Version 0-20.ASM
```

Unfortunately, this is not allowed in the MPLAB IDE, so I find myself using an acronym like:

```
PNLV0020.ASM
```

which requires you to open up the file to see what the code is actually used for.

A good technique to use to track your code is to place it in a separate subdirectory for each application. I used this technique for all the files in this book, although it probably looks best for something like the El Cheapo's PC source code (Fig. 9-1).

With the long file-name capabilities of Visual Basic, the code can be easily identified in the El Cheapo subdirectory with versions easy to identify.

Before starting any new version of code, I always am sure that I have a concrete plan of what I want to do or add to existing files. Sometimes, I'll make a list of to do's and check them off with each version as I do them. The important point here is to keep the code and what you want it to do straight and manageable. Creating a large application in one version is just about impossible and very hard to debug.

Supporting your applications really comes under two headings. The first is *fixing problems* and the second is *updating the code for new requirements.* These actions really encompass many of the points covered in later chapters on application design and debugging. By organizing your code properly, the task of updating your applications will be easier and much less painful.

Regardless of what you do, you will never escape your own code. I know of a CEO of a company that has several billion dollars in sales, but is still supporting the core technology of the products they sell. I know for a fact that he hates it, but he has never found the time to adequately train somebody else on his technology.

Accept the fact that you will have to maintain your code for as long as you live. Write it in such a way that it makes sense to you and keep the files in a format that you can easily work with when you are called from your deathbed to make that one last "little change."

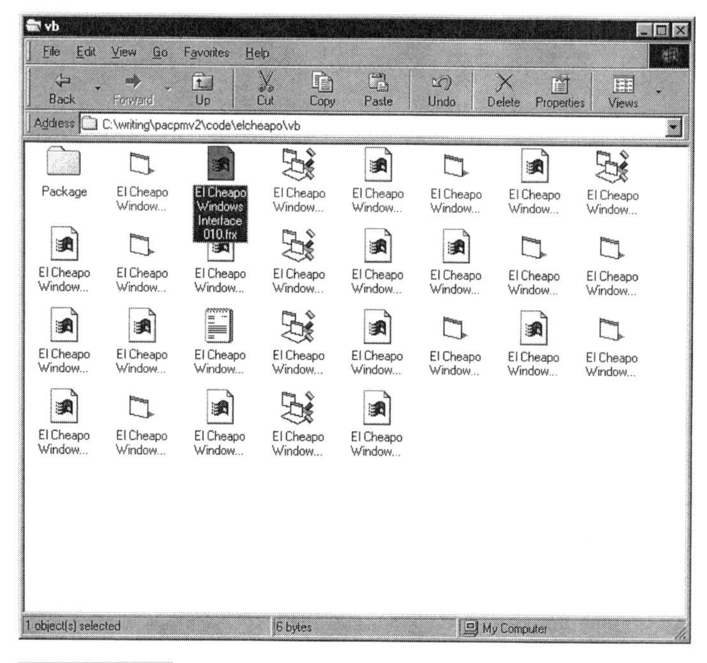


**Figure 9-1** Saving files in the PC for easy retrieval

## SAMPLE TEMPLATE

When I create my own mid-range PICmicro® MCU applications, I always start with the following template:

```
 title  "FileName - One Line Description"
#define _version "x.xx"

;
;  Update History:
;
;  Application Description/Comments
;
;  Author
;
;  Hardware Notes:
;
 LIST R=DEC                    ;  Device Specification
  INCLUDE "p16cxx.inc"         ;  Include Files/Registers

;  Variable Register Declarations

;  Macros

 __CONFIG _CP_OFF & _XT_OSC & _PWRTE_ON  & _WDT_OFF & _BODEN_OFF

  org   0
Mainline

  goto  Mainline_Code

  org   4                        ;  Interrupt Handler at Address 4

Int

MainLine_Code

;  Subroutines

  end
```

This template structures my applications and ensures that I don't forget anything that I consider crucial. The file TEMPLATE.ASM can be found in the *code/template* subdirectory of the hardfile *PICmicro* directory. Before starting any application, this file should be copied from the subdirectory and the specifics for the application added to it.

The *title* and *_version* at the top of the file shows what the application does so that I can scan the files very quickly, instead of going by what the file name (which can only be eight characters long for the MPLAB IDE) indicates. The *title* line will show up on the top of each new listing file page. The *_version* define statement will then show the code revision level and can be inserted in any text displayed by the application.

A *Debug* define directive might appear after the *_version* define directive if the "Debug" label is going to be tested for in the application. This directive is used with conditionally assembling code to take out long delays and hardware polling that won't be available within the MPLAB simulator. Before building the application for burning into a PICmicro® MCU, the *Debug* define will be changed so that the "proper" code will be used with the application. Later, the book covers *Debug* defines in more detail.

Next, I put in the update history (with specific changes), along with the description of what the application does. In other books, I've joked about how dumb it can be to sign your name to the application (because you will never lose the responsibility of maintaining it).

One thing that I do that seems to be a bit unusual is that I list the hardware operating parameters of the application. I started doing this so that I could create stimulus files easily for applications. This seems to have grown into a much more comprehensive list that provides a cross-reference between an application's hardware circuit and the PICmicro® MCU software.

Before declaring anything myself, I load in the required include files and specify that the default radix is decimal. There is no real reason to put these two types of statements together, other than if you change the device you are working with and you have all the places that need to be changed in the same place. As commented on elsewhere, I only use the Microchip PICmicro® MCU include files because these have all the documented registers and bits of the data sheets, which means that I don't have to create my own.

Instead of specifying the PICmicro® MCU part number in the source code, I let the definition in the MPLAB IDE do it for me. This allows multiple part numbers to be supported by an application. This may sound a bit unusual, but I often design an application to use a PIC16F84 for development and then use a PIC16C711 or PIC16C621 for the actual "product."

With the device declarations completed, I then do my variable defines and macro declarations. When doing this, remember to always specify prerequisites before they are used. The MPASM assembler will not be able to resolve any labels that are defined below them in the source code when doing macros or defines. This is not true for labels that are referenced before their use in the application instruction code that follows the declarations and operating parameters.

Finally, I declare the device operating parameters that are to be programmed into the CONFIGURATION register or configuration fuses (which are explained in more detail later in this chapter), followed by the application code. I put subroutines at the end of the application code simply because the reset and interrupt-handler vectors are at the "beginning" of the data space. Putting the subroutines after the mainline and interrupt handler seems to be most appropriate under this condition.

This template is used for single source-file applications, which make up the vast majority of my PICmicro® MCU applications. If multi-source file applications are created, then the *__CONFIG* line is left out of anything, other than the first (or header) file, which are explained in the previous chapter. *Publics* and *Externals* are added in its place with the code following, as it does in the single source-file template. Variables should be declared in the header file and then passed to the linked in files as publics.

This template can be modified and used with the other PICmicro® MCU device architectures.

# PICmicro® MCU Programming Tips

As I go over the manuscript of this book, I feel like I could create a 300-page book, based simply on "hints and kinks" for programming the PICmicro® MCU. In the more than seven years I have been programming the device, I am amazed at the different ideas people have come up with for programming the chip in different situations. Coupled with the ability to tailor applications to take advantage of the hardware, you could create some remarkably efficient applications.

Much of the information provided in this book will leave you scratching your head and asking "how did somebody come up with that?" The answer often lies in necessity: the application developer had to implement some features in fewer instructions, in fewer cycles, or in less variable memory ("file registers" in the PICmicro® MCU). For most of these programming tips, the person who came up with them not only had the need to do them, but also understood the PICmicro® MCU architecture and instruction set well enough to look for better ways to implement the functions, rather than the most obvious methods.

The following sections provide some tricks and better ideas on how to implement different functions in the PICmicro® MCU. Along with the information in this chapter, I have included code hints in other chapters, in the appendices, and on the CD-ROM. I hope that you will look through these ideas as inspiration to help with your own applications. If you create something unique on your own, please let me know so that I can distribute it to other people.

## LABELS, ADDRESSES, AND FLAGS

If you have skipped ahead in the book and taken a look at some of the code examples in Chapters 15 and 16, you will probably be a bit concerned because addresses are used in a number of different ways that probably won't be familiar to you. The PICmicro® MCU's architecture and instruction require a more careful watch of absolute addresses than you are probably used to. This section covers the different memory spaces in the PICmicro® MCU and how they are accessed.

The most familiar memory space to you is probably the instruction program memory. As I said earlier in the book, the program memory is the width of the instruction word. The maximum depth of the memory is based on the word size, which can be the instruction word size minus one for addressing for the low-end, mid-range, and PIC18Cxx PICmicro® MCUs.

To figure out the maximum depth of program memory in the low-end and mid-range PICmicro® MCUs, the formula:

$$\textit{Maximum program memory} = 2^{(\text{Word size}-1)}$$

is used. The PIC17Cxx's program memory size is two to the power of the word size (16 bits, for a total of 65,536 16-bit words). The PIC18Cxx has a program memory size that is not based upon the Instruction Size.

It is important that although the low-end, mid-range, and PIC18Cxx's program counter is the size of the instruction word (12 and 14 bits, respectively), the upper half of the addressable space is not available to the application. This upper half to the program memory is used to store the configuration fuses, IDLOC nybbles (which are covered in the following section), and provide test addresses used during PICmicro® MCU manufacturing.

The PIC17Cxx does not have the separate address space and the configuration fuses can be accessed from within the application and potentially used as instructions for the processor. Actually, the PIC18Cxx configuration fuses can be accessed from the application as well, but the *Table* instructions are required.

When an application is executing, it can only jump with a set number of instructions, which is known as the *page*. The *page* concept was covered in more detail earlier in the book. The size of the page in the various PICmicro® MCU architecture families is based on the maximum number of bits that could be put into an instruction for the address. In the low-end PICmicro® MCU, a maximum of nine bits are available in the *goto* instruction that is used to address a page of 512 instructions. In the mid-range PICmicro® MCU, the number of bits specified in *goto* is 11, for a page size of 2,048 instructions. The PIC17Cxx has 13 address bits in its *goto* instruction, for a page size of 8,192 instructions. The PIC18Cxx can jump anywhere within the application, without regard to the page size.

The point of covering this is to note that in these three families, addresses are always absolute values within the current page. For example, in the code:

```
 org      0
  goto    Mainline

   :                                      ;  Code to Skip Over

 org      0x0123
Mainline
```

the address value loaded into the *goto* instruction would be the destination address (Mainline), which is 0x0123. In other processors that you might be familiar with, an offset would be added to the program counter, making the value in the *goto* instruction *0x0122* because the program counter had been incremented to point to the next instruction.

This can be further confused by an instruction sequence like:

```
  btfsc  Button, Down          ;  Wait for Button to be Pressed
   goto  $ - 1
```

The *$* returns a constant integer value, which is the address of the current instruction. In the *goto $ - 1* instruction, the address that is loaded is the address of the *goto $ - 1* instruction minus one.

The PIC18Cxx behaves more like a traditional processor and has absolute address jumps and relative address branches. It does not have a "page," per se. The *goto* and *call* instructions in the PIC18Cxx can change application execution anywhere within the PICmicro® MCU's 1-MB program memory address range.

As well, the PIC18Cxx has the ability to "branch" with an eight-bit or an 11-bit two's complement number. The branch instructions do not use an absolute address. Instead, they add the two's complement to the current address plus two. In the PIC18Cxx, the instruction sequence:

```
  btfsc  Button, Down, 1       ;  Wait for Button to be Pressed
   bra   $ - (2 * 1)
```

would perform the same operation as the previous example, but I replaced the *goto $ - 1* instruction with a *branch always*.

In the PIC18Cxx example, if an odd address is specified, the MPLAB simulator will halt without a message. If the code is burned into the PIC18Cxx and a jump to an odd address occurs, execution can branch to an unexpected address. As noted earlier, each byte is addressed in the PIC18Cxx and not the 16 bit word. Further complicating the use of relative jumps is the instructions that take up more than one instruction word. These complexities lead me to recommend that you do not use relative jumps with the *$* directive with the PIC18Cxx.

You will probably be comfortable with how the destination values for either the *goto* and *bra* instructions are calculated depending on your previous experience. If this is your first assembly-language experience, the absolute addresses of the low-end, mid-range, and PIC17Cxx will probably make a lot of sense to you. If you have worked with other processors before, the PIC18Cxx will make more sense to you.

Regardless of which method is most comfortable for you, I recommend writing your applications in such a way that the actual addresses not used are a concern. Thus, labels should be used in your source code at all times and the *org* directive statement should only be used for the reset vector and interrupt vectors. For all other addresses, the assembler should be allowed to calculate the absolute addresses for you. By allowing the assembler to generate addresses, you will simplify your application coding and make it much more "portable" to multiple locations within the same source file or others.

For example, the mid-range code:

```
 org      0x012
  btfsc   Button, Down                ; Address 0x012
   goto   0x012                       ; Address 0x013
```

will do everything that the previous example code will do, but it is specific to one address in the PICmicro® MCU. By using the *goto $ - 1* instruction, the code can be cut and pasted anywhere within the application or used in other applications.

Letting the assembler generate the addresses for you is accomplished one of two ways. The first is the label, which is placed in the first column of the source code and should not be one of the reserved instructions or directives used by the assembler. In the MPLAB assembler, a *label* is defined as any string of characters starting with a letter or "_" character and can be up to 33 characters long. In the body of the label, numeric digits can be used. When one of these strings is encountered, it is loaded into a table, along with the current program counter value for when it is referenced elsewhere in the application.

Labels in MPLAB's assembler can have a colon (*:*) optionally put on the end of the string. This feature was put into MPLAB to retain compatibility with other assemblers that users might be familiar with. Instead of flagging a following colon with an error (it really isn't needed in the MPLAB assembler), it is simply ignored.

Using this example, a *Loop* label can be added to make the code completely portable:

```
Loop
  btfsc   Button, Down                ; Address = "Loop"
   goto   Loop                        ; Address = "Loop" + 1
```

The disadvantage of this method is that only one *Loop* (and *Skip*) can be put into an application. Program memory labels are really best suited for cases where they can be more global in scope. For example, an application should really only have one main *Loop,* which is where this label should be used.

Personally, I always like to use the labels *Loop, Skip,* and *End* in my applications. To allow their use, I will usually preface them with something like the acronym of the current subroutine's name. For example, if the code was in the subroutine *GetButton,* I would change it to:

```
GB_Loop
  btfsc   Button, Down                 ; Address = "Loop"
   goto   GB_Loop                      ; Address = "Loop" + 1
```

Instead of using labels in program memory, I prefer using the *$* directive, which returns the current address of the program counter as an integer constant and can be manipulated to point to the correct address. Going back to the original code for the button poll snippet, the *$* directive eliminates the need for a label altogether:

```
  btfsc  Button, Down         ;  Wait for Button to be Pressed
   goto  $ - 1
```

This is the optimal way to write this code for the low-end, mid-range, and PIC17Cxx PICmicro® MCUs (as I noted previously). For the PIC18Cxx, you should be sure that you understand exactly how the offset is calculated, or find the number of instruction words (as in the earlier example) and then multiply it by two (as shown earlier in this section).

You do not have to expend the effort trying to come up with a unique label (which can become hard in a complex application). As you get more comfortable with the *$* directive, you will see that it's happening faster than if a label was used.

The problem with using the *$* directive is that it can be difficult to count out the offset to the current instruction (either positive or negative). To avoid making mistakes in counting, the *$* should only be done in short sections of code. Also, beware of using the *$* directive in large sections of code that have instructions added or deleted between the destination and the *goto* instruction. The best way to avoid this is to only use the *$* in situations like the previous one, where code will not be added between the *goto* and the destination.

If you have worked with assemblers for other (Von Neumann) processors, chances are you have had to request memory where variables were going to be placed. This operation was a result of variable memory being in the same memory space as program memory. This is not a requirement of the PICmicro® MCU in which the register space (where variables are located) is separate from the program memory space.

To allocate a variable in the PICmicro® MCU, you have to create an equate, which references a label to a file register (or multiple file registers if a 16-bit, or greater, variable is required). In the first edition, I specified that this was done by finding the first file register in the processor and then starting a list of equates from there.

As covered elsewhere, an *equate* is when a label is given a specific constant value. Every time that the label is encountered, the constant that is associated with it is used. Program memory labels can be thought of as equates that have been given the current value of the program counter.

For the PIC16F84, variable equate declarations look like:

```
i EQU 0x00C
j EQU 0x00D                    ;  Note, "j" is Sixteen Bits in Size
k EQU 0x00F
    :
```

The problems with this method are that adding and deleting variables are a problem—especially if not many free file registers are available.

To eliminate this problem, Microchip has come up with the CBLOCK directive, which has a single parameter that indicates the start of a label equate. Each label is given an ascending address. If any labels need more than one byte, a colon (:) and the number of bytes is specified after the label. When the definitions are finished, the *ENDC* directive is specified.

Using the *CBLOCK* and *ENDC* directives, the previous variable declarations could be implemented as:

```
 CBLOCK 0x00C           ;  Define the PIC16F84 File Register Start
i, j:2, k
 ENDC
```

This is obviously much simpler than the previous method (i.e., it requires less thinking) and easier to read.

What I don't like about *CBLOCK* is that specific addresses cannot be specified within it. For most variables, this is not a problem, but as presented elsewhere in the book, I tend to put variable arrays on power of two byte boundaries to take advantage of the PICmicro® MCU's bit set/reset instructions to keep the index within the correct range. To ensure that I don't have a problem, I specify an equate for the variable array specifically and ensure that it does not conflict with the variables defined in the *CBLOCK.*

The last type of data to define in the PICmicro® MCU is the bit. If you look through the Microchip MPLAB assembler documentation, you will discover that no bit data types are built in. This is not a significant problem if you are willing to use the *#define* directive to create a *define* label that includes the register and bit together.

For example, you could define the STATUS register's zero flag as:

```
#DEFINE zeroflag STATUS, Z
```

A *define* is like an equate, except where the equate associates a constant to the label, a define associates a string to the label. For the *zeroflag* define, if it were used in the code:

```
  movf   TMR0, w                  ;  Wait for TMR0 to Overflow
  btfss  zeroflag
   goto  $ - 2
```

as part of the *btfss* instruction, the string *STATUS, Z* would replace *zeroflag,* as shown when the application was assembled:

```
  movf   TMR0, w                  ;  Wait for TMR0 to Overflow
  btfss  STATUS, Z
   goto  $ - 2
```

Defining bits like this is a very effective method of putting labels to bits. Using this method, you no longer have to remember the register and bit number of a flag. This can be particularly useful when a number of bits are defined in a register (or multiple registers). Instead of remembering the register and bit numbers for a specific flag, all you have to remember is the *define* label. Using the bit "define" with the bit instructions of the PICmicro® MCU allows you to work with single bit variables in your application.

## CONFIGURATION FUSES AND ID LOCATIONS

When I discuss programming PICmicro® MCUs, I recommend using only programmers that can pick up the value specified in the *__CONFIG* statement and program it into the PICmicro® MCU automatically, without any intervention from the user. This is important because it eliminates one step that is often missed or done incorrectly by the user. This section presents the methods to specify the configuration fuse values and a serial number/version code for your applications in a way that will be picked up by your PICmicro® MCU programmer.

The *__CONFIG* (note the two underscores) directive is used to specify the configuration fuse options for the PICmicro® MCU. Along with the *__CONFIG* directive, a number of constants are added together to provide the configuration fuse values.

A typical set of configuration fuse values (taken from the PIC16F877) is shown in Table 9-1. These files can be found in the "P————.INC." file that Microchip includes with MPLAB, where "———" is the PICmicro® MCU's part number.

These values are meant to be ANDed together to get the actual bit pattern to be loaded into the configuration register. For example, for the watchdog timer to be disabled, bit two

**TABLE 9-1 PIC18F877 Configuration Fused Labels and Values**

| LABEL | VALUE | OPERATION |
|---|---|---|
| _BODEN_ON | 0x03FFF | Brown Out Reset Detect Enabled |
| _BODEN_OFF | 0x03FBF | Brown Out Reset Detect Disabled |
| _CP_ALL | 0x00FCF | All Program Memory Protected |
| _CP_HALF | 0x01FDF | Upper Half of Program Memory Protected |
| _CP_UPPER 256 | 0x02FEF | Upper 256 Bytes of Program Memory Protected |
| _CP_OFF | 0x03FFF | No Code Protection |
| _WRT_ENABLE_OFF | 0x03DFF | Program Memory can be Written to from Application |
| _PWRTE_OFF | 0x03FFF | 72 msec TIme Up Delay Off |
| _PWRTE_ON | 0x03FF7 | 72 msec Time Up Delay On |
| _WDT_ON | 0x03FFF | Watchdog TImer On |
| _WDT_OFF | 0x03FFB | Watchdog TImer Off |
| _LP_OSC | 0x03FFC | LP Oscillator Type |
| _XT_OSC | 0x03FFD | XT Oscillator Type |
| _HS_OSC | 0x03FFE | HS Oscillator Type |
| _RC_OSC | 0x03FFF | RC Oscillator |
| _DEBUG_ON | 0x037FF | Internal Emulator Enabled |
| _DEBUG_OFF | 0x03FFF | Internal Emulator Disabled |
| _CPD_ON | 0x03EFF | Data EEPROM Write Protected |
| _CPD_OFF | 0x03FFF | Data EEPROM Not Write Protected |
| _LVP_ON | 0x03FFF | Low Voltage Program Enabled |
| _LVP_OFF | 0x03F7F | Low Voltage Program Disabled |

of the configuration word has to be reset. If you look through the label values in Table 9-1, you'll see that this is the only parameter that has bit two reset.

Because the different options controlled by the configuration register depend on bits being reset, I recommend that you select one of every option in this list. This is done by matching up each parameter with the same character string inside both. For example, *_CPD_OFF* and *_CPD_ON* control to the same option, as does *_LP_ OSC, _HS_OSC, _XT_OSC,* and *_RC _OSC.* The different option for the PIC16F877 are shown in Table 9-2.

After listing the different options, the *__CONFIG* statement for the application can be created in the application like:

```
__CONFIG _BODEN_ON & _CP_OFF & _WRTE_ENABLE_ON & _PWRTE_ON & _WDT_OFF &
_XT_OSC & _DEBUG_OFF & _CPD_ON & _LVP_OFF
```

When this statement is evaluated, the constant values are replaced with the constants above and evaluated together into a constant to be loaded into the configuration fuses. For the labels selected the *__CONFIG* statement could be reduced to:

```
__CONFIG 0x03F71
```

which is the value that will be loaded into the configuration fuse register.

One feature of the *__CONFIG* directive is that it works differently for the different PICmicro® MCU families. For the low-end devices, the configuration fuses are at address 0x0FFF. For the mid-range, they are at 0x02007. The PIC17Cxx's configuration fuses are in the address range 0x0FE00 to 0x0FE07.

For the low-end and mid-range parts, I like to think of the *__CONFIG* directive being the equivalent to an *ORG* statement with a *dw* statement. Using the previous example, this would make the mid-range PICmicro® MCU *__CONFIG* statement equivalent to:

```
org 0x02007
dw  0x03F71
```

Along with the configuration fuses, the four nybbles that are available for identifications or version information can be accessed by the *__IDLOCS* (two underscores) direc-

**TABLE 9-2 PIC18F877 Configuration Fuse Categories**

| LABEL | OPTION |
|---|---|
| _BODEN_## | Brown Out Detect |
| _CP_## | Code Protect |
| _WRT_ENABLE_## | Program Memory Application Write Enable |
| _PWRTE_## | 72 msec Power Up Wait Timer |
| _##_OSC | Oscillator Type |
| _DEBUG_## | Built in Emulator |
| _CPD_## | Data EEPROM Write Enable |
| _LVP_## | Low Voltage Programming Enable |

TABLE 9-3 IDLOCS Data Placement

| ADDRESS | DATA |
|---|---|
| 0x0200 | 0x00001 |
| 0x02001 | 0x00002 |
| 0x02002 | 0x00003 |
| 0x02003 | 0x00004 |

tive. Each nybble of the 16-bit *__IDLOCS* statement parameter are placed in one of the IDLOCS program memory location. For example, in a mid-range PICmicro® MCU, the statement:

```
__IDLOCS 0x01234
```

will load the program memory with the data shown in Table 9-3.

Taking a clue from my description of how *__CONFIG* works, you could code the data into the four *__IDLOCS* location as:

```
org           0x02000
 dw           IDLOCS_VALUE1
 dw           IDLOCS_VALUE2
 dw           IDLOCS_VALUE3
 dw           IDLOCS_VALUE4
```

As a final reminder, the *__IDLOCS* and *__CONFIG* data are stored outside of the application-accessible program memory in the low-end and mid-range PICmicro® MCUs. This makes IDLOCS best suited for identifying parts for manufacturing rather than storing application specific information in them. The values put into the respective memory areas can be accessed directly by the 17Cxx. The self-program capability of the 17Cxx devices could result in some interesting options in how these devices are customized after being soldered into a circuit.

## SUBROUTINES WITH PARAMETER PASSING

For subroutines to work effectively, there must be the ability to pass data (known as *parameters*) from the "caller" to the subroutine. The three ways to pass parameters in the PICmicro® MCU each have their own advantages and potential problems. This section introduces each of the three methods and shows how they can be implemented in the PICmicro® MCU.

In most modern-structured high-level languages, parameters are passed to subroutines as if they were parameters to a mathematical function. One value (or parameter) is returned. An example subroutine (function) that has data passed to it would look like:

```
A = subroutine(parm1, parm2);
```

in C source code.

The subroutine's input parameters (*parm1* and *parm2* ) and output parameter (which is stored in A in this example) can be shared and are common to the caller and subroutine by the following methods:

1. Global variables.
2. Unique shared variables
3. Data stack.

Passing parameters using global variables really isn't passing anything to a subroutine and back. Instead, the variables, which can be accessed anywhere in the code, are used by both the main line and the subroutine to call a subroutine that uses global variables. Just a "call" statement is used.

```
call subroutine
```

The advantage of this method is that it requires the least amount of amount of code and executes in the least number of cycles. The problems with this method are that it can be very confusing to work through the source code and changes to the global variables can affect other subroutines. As well, this method cannot be used for subroutines called from the mainline and interrupts, as well; recursive subroutines cannot use this method. This method of parameter passing is the most basic and often used by new programmers.

The second method is to use a set of unique variables for each subroutine. Before the call, the unique variables have to be set up and after the call, the returned parameter is taken from the shared variable. In this case, the statement:

```
A = subroutine(parml, parm2);
```

becomes:

```
  movf   parml, w             ;  Save Parameters
  movwf  subroutineparml      ;   passed to Subroutine
  movf   parm2, w
  movwf  subroutineparm2
  call   subroutine
  movf   subroutinereturn, w ;  Get Returned
  movwf  A                    ;   Parameter
```

This method has a number of advantages over using global variables. The difficulty in following the code is minimized and the opportunity for changing global variables is eliminated. Like using global variables, recursive subroutines, or subroutines called from both the mainline and interrupt handlers is not possible. This method can also be a problem in PICmicro® MCUs with very limited variable (file register) space.

The method normally used by most processors and high-level languages is to save parameters on a stack and then access the parameters from the stack. The PICmicro® MCU cannot access stack data directly, but the offsets can be easily calculated. Before any subroutine calls can take place, the FSR has to be offset with the start of a buffer.

```
movlwbufferstart - 1
movwf  FSR
```

When the parameters are pushed onto the simulated stack, the operation is:

```
incf   FSR, f
movwf  INDF
```

The increment of FSR is done first so that if an interrupt request is acknowledged during this operation, any pushes in the interrupt handler will not affect the data in the mainline.

Popping data from the stack uses the format:

```
movf   INDF , w
decf   FSR, f
```

With the simulated stack, the example call to subroutine used the code:

```
movf   parm1 , w              ; Save Parameters
incf   FSR, t
movwf  INDF
movf   parm 2, w
incf   FSR, f                 ; Make Space for Return
call   subroutine
movf   INDF, w                ; Get Returned Value
decf   FSR, f
movwf  A
decf   FSR, f                 ; Reset the STACK
decf   FSR, f
```

This method is very good because it does not require global variables of any type and allows for subroutines that are called from both the execution main line and from the interrupt handler or recursively. As well, data on the stack can be changed (this operation has created local variables).

The disadvantage of this method is the complexity required to access data within the subroutine and adding additional variables. When accessing the variables and changing FSR, you will have to disable interrupts. For the example above, to read "parm1," the code below would have to be used:

```
movlw  0 - 3
bcf    INTCON, GIE
addwf  FSR, f
movf   INDF, w
movwf  SUBRTN_TEMP            ; Read "parm1"
movlw  3
bcf    INTCON, GIE
addwf  FSR, f
movf   SUBRTN + TEMP, w
```

The *SUBRTN_TEMP* variable is used to save the value read from the stack while the FSR is updated. For most changes in the FSR, simple increment and decrement instructions could be used instead. This would require fewer instructions and render the temporary variable unnecessary. The previous code could be rewritten as:

```
bcf    INTCON, GIE
decf   FSR, f
decf   FSR, f
decf   FSR, f
movf   INDF, w
incf   FSR, f
incf   FSR, f
incf   FSR, f
bsf    INTCON, GIE
```

Although this code seems reasonably easy to work with, it will become a lot more complex as you add 16-bit variables and arrays.

One method of passing parameters that I haven't touched upon is that of using the processor's registers. For the PICmicro® MCU, there is only the w register, which is eight bits wide. As well, in the low-end devices, the availability of just the *retlw* instruction precludes passing data back. This limits the usability of passing parameters by only using the w register, but it can be used for very simple data read and write cases.

## WORKING WITH THE ARCHITECTURE

This section was originally entitled *Working With the Architecture's Quirks*. The architecture has some unusual features that make pristine assembly-language applications directly from another microcontroller to the PIC microcontroller difficult. However, as I started listing what I wanted to do in this and the following sections, I realized that there were many advantages to the PICmicro® MCU's architecture. Many of the "quirks" actually allow very efficient code to be written for different situations. The following sections present how the PICmicro® MCU architecture can be used to produce some code that is best described as "funky."

Another reason why I took out the word "quirks" in this section's title is because as I looked at how the PICmicro® MCU is coded, I realized that it wasn't all that unconventional. The actual code from other, more traditional microcontrollers can be passed directly to the PICmicro® MCU with only small changes.

In addition, the basic operation sequence of adding two numbers together is:

1. Load the accumulator with the first additional RAM.
2. Add the second additional RAM to the contents of the accumulator.
3. Store the contents of the accumulator into the destination.

In PICmicro® MCU assembly-language code, this is:

```
movf      Parm1, w
addwf     Parm2, w
movwf     Destination
```

where the *movf* and *addwf* instructions can be changed to *movlw* or *addlw,* respectively, if either parameter is a constant.

Subtraction in the PICmicro® MCU follows a similar set of instructions, but because of the way the subtraction operation works, the number that is being subtracted must always be loaded first. For example, for the high-level language statement:

```
Destination = Parm1 - Parm2
```

The sequence of operations is:

1. Load w with the second parameter (which is the value to be taken away from the first).
2. Subtract the contents of w from the first parameter and store the result in w.
3. Store the contents of w in the destination.

In PICmicro® MCU assembly code, this is:

```
movf      Parm2, w
subwf     Parm1, w
movwf     Destination
```

Like the addition operation, the *movf* and *subwf* instructions can be replaced with *movlw* or *sublw,* respectively if either *Parm1* or *Parm2* are constants.

The PICmicro® MCU's instructions contrasts to the 8051 and other microcontroller architectures in which the subtract instruction takes away the its parameter from the contents of the accumulator.

**Subtraction, comparing, and negation** The previous section reviewed how subtraction worked from a numerical point of view, but one of the most confusing aspects of subtraction is how carry works and how conditional status code is created. As well, negation of numbers is always an issue. A few tricks can be done with the subtraction instructions to quickly negate numbers. These negation tricks also give you some better ideas of how the subtract instructions work.

As indicated elsewhere, the subtract instructions work as:

```
PICmicro® MCU subtract – parameter - w
                       = parameter + (w ^ 0x0FF) +1
```

This operation affects the Zero, Carry, and Digit Carry STATUS register flags. In most applications, it is how the carry flag is affected that is of the most importance. This flag will be set if the result is equal to or greater than zero. This is in contrast to how the carry or borrow flags work in most processors. Previously, I described the carry flag after a subtract operation as a "positive flag." If the carry flag is set after a subtract operation, then a borrow of the next significant byte is not required. It also means that the result is negative if the carry flag is reset.

This can be seen in more detail by evaluating the subtract instruction sequence for:

```
Result A - B
```

which is:

```
movlw   B                          ;  Assume A and B are Constants
sublw   A
movwf   Result
```

By starting with *A* equals to one, different values of *B* can be used with this sequence to show how the carry flag is set after subtract instructions.

Table 9-4 shows the result, carry, and zero flags after the previous snippet.

I did not include the Digit Carry (DC) flag in Table 9-4 because it would be the same as Carry. In subtraction of more complex numbers (i.e., two-digit hex), the DC file becomes difficult to work with and specific examples for its use (like the ASCII/nybble-conversion routines) have to be designed.

When you are first learning how to program in assembly language, you might want to convert high-level language statements into assembly language using formulas or basic

**TABLE 9-4 Subtraction Carry and Zero Flag Results Table**

| "A" | "B" | RESULT | CARRY | ZERO |
|---|---|---|---|---|
| 1 | 0 | 1 | 1 | 0 |
| 1 | 1 | 0 | 1 | 1 |
| 1 | 2 | 0xOFF(−1) | 0 | 0 |

guides. When you look at subtraction for comparing, the code seems very complex. In actuality, using the PICmicro® MCU subtract isn't that complex and the instruction sequence:

```
movf    Parm1, w/movlw   Parm1
subwf   Parm2, w/sublw   Parm2
btfsc   status, C
 goto   label
```

can be used each time the statement

```
if (A Cond B) then go to label
```

is encountered.

Where *Cond* is defined by the definitions in Table 9-5.

Selecting the status flag (carry or zero) is used to skip over the *goto* instruction and which order the two parameters are used. The actual condition operation can be implemented. Table 9-6 shows how the different cases can be coded. This table is useful to remember when you are working on PIC applications—even if you aren't simply converting high-level language source code by hand into PICmicro® MCU assembly.

Negation of the contents of a file register is accomplished by performing the two's complement operation. This is to invert the contents of a register and increment it.

```
comf    reg, f
incf    reg, f
```

**TABLE 9-5 "if" Condition Definitions**

| COND | OPERATION |
|---|---|
| == | jump if equal |
| != | jump if not equal |
| > | jump if FIRST is Greater than the second |
| >= | jump if FIRST is Greater than or Equal to the second |
| < | jump if FIRST is Less than second |
| <= | jump if FIRST is Less than or Equal to second |

**TABLE 9-6 Condition to Subtraction Instruction Cross Reference**

| JUMP "IF" | CONDITION TO CHECK | CODE |
|---|---|---|
| A == B | A − B = 0 | movf A, w/movlw A<br>subwf B, w/sublwB<br>btfsc STATUS, Z<br>goto Label ; Jump if (Z=5=1) |
| A != B | A − B = 0 | movf A, w/movlw A<br>subwf B, w/sublwB<br>btfss STATUS, Z<br>goto Label ; Jump if (Z=5=0) |
| A > B | B − A < 0 | movf A, w/movlw A<br>subwf B, w/sublwB<br>btfss STATUS, C<br>goto Label ; Jump if (C=5=0) |
| A >= B | A − B >= 0 | movf B, w/movlw B<br>subwf A, w/sublwA<br>btfsc STATUS, C<br>goto Label ; Jump if (C=5=1) |
| A < B | A − B < 0 | movf B, w/movlw B<br>subwf A, w/sublwA<br>btfss STATUS, C<br>goto Label ; Jump if (C=5=0) |
| A <= B | B − A > 0 | movf A, w/movlw A<br>subwf B, w/movlwB<br>btfsc STATUS, C<br>goto Label ; Jump if (C=5=1) |

If the contents to be negated are in w, there are a couple of tricks that can be used to carry this out. For mid-range devices, the *sublw 0* instruction can be used:

```
sublw  0      ; w = 0 - w
              ;   = 0 + (w ^ 0x0FF) +1
              ;   = (w ^ 0x0ff) + 1
              ;   = -w
```

But, in low-end PICmicro® MCU devices, you can add and subtract the w register contents with a register as shown:

```
addwf  Reg, w              ; w =  w + Reg
subwf  Reg, w              ; w = Reg - (w + Reg)
                           ;   = -w
```

*Reg* should be chosen from the file registers and not any of the hardware registers that might change between execution of the instructions.

**Bit "AND" and "OR"** One of the most frustrating things to do is to respond based on the status of two bits. In the past, I found that I had to come up with some pretty clever code, only to feel like it was not good enough. In the past year or so, I've discovered the usefulness of combining two *skip on condition* instructions to get a result for two parameters. The two skip parameters are used in such a way as the first one jumps to an instruction if a case is true and the second jumps over the instruction if the second case is not true.

For example, if you wanted to set a bit if two other bits were true (the result is the AND of two arbitrary bits), you could use the code:

```
  bcf     Result              ;  Assume A and C = 0
  btfss   A
   goto   Skip                ;  A = 0, don't set Result

  btfsc   B                   ;  B = 0, don't set Result
   bsf    Result              ;  A = B = 1, set result
Skip
```

This code is quite complex and somewhat difficult to understand. A further problem is that it can return after a different number of cycles, depending on the state of *A*. If *A* is reset, the code will return after four instruction cycles. If it is set, six instruction cycles will pass before execution gets to *Skip*.

By combining the two tests, the following code could be used to provide the same function:

```
  bsf     Result              ;  Assume A = B = 1
  btfsc   A                   ;  A == 0, Result = 0
   btfss  B                   ;  B == 1, Result = 1
    bcf   Result              ;  A == 0 or B == 0, Result = 0
```

This code is smaller, always executes in the same number of cycles, and is easier to work through and see what is happening.

An *OR* function can be implemented similarly:

```
  bcf     Result              ;  Assume A = B = 0
  btfss   A                   ;  A == 1, Result = 1
   btfsc  B                   ;  A == B == 0, Result = 0
    bsf   Result              ;  A == 1 or B == 1, Result = 1
```

This trick of using two conditions to either skip to or skip over an instruction is useful in many cases. As shown later in this chapter, this capability is used to implement constant-loop timing for 16-bit delay loops.

**16-bit operations** As you start creating your own PICmicro® MCU applications, you'll discover that eight bits for data is often insufficient for the task at hand. Instead, larger base values have to be used for the saving and operating on data. The appendices present a number of snippets for accessing 16-bit data values, but this section introduces the concepts of declaring and accessing 16-bit (and greater) variables and constants.

The first edition of this book presented variable declaration as using either *EQU* or the *CBLOCK* directives. This edition focuses exclusively on just using the *CBLOCK* directive. To define larger than eight-bit data structures using *CBLOCK*, a colon (*:*), followed by the number of bytes, is specified.

For example, eight-, 16-, and 32-bit variables are declared in the MPLAB assembler as:

```
 CBLOCK 0x00C
i
j:2
k:4
 ENDC
```

To access data a byte at a time, the address with an offset is used. The *low, high,* and *upper* directives cannot be used as they return different parts of the register address and do not modify them. Using these methods, the instructions:

```
movf    j + 1, w          ; Retrieves the high byte of "j"
movf    HIGH j, w         ; Retrieves the contents of the register at
                            (j and 0xff00)>>8
movf    UPPER j, w        ; Retrieves the contents of the register at
                            address 0 of the bank.
```

One confusing aspect of MPLAB for me is the default of high/low data storage in the MPLAB simulator and MPASM. The low/high format works better for using application code and makes more sense to me. As well, you will notice that all 16-bit registers in the PICmicro® MCU are defined in low (byte/address), followed by high (byte/address), data format.

The preceding paragraph might be confusing for you, but let me explain exactly what I mean. If 16-bit data is saved in the high/low (what I think of as "Motorola format," which is where I first saw it), when 16-bit information is displayed in memory, it looks "correct." For example, if 0x01234 was stored in high/low format starting at address 0x010, the file register display would show:

```
0010    1234
```

which appears "natural."

If the data is stored in low/high ("Intel") format, 0x01234 at 0x010 would appear as:

```
0010    3412
```

which is somewhat confusing.

I recommend storing data in low/high (Intel) format for two reasons. First, it makes logical sense saving the low-value byte at the low address. The second reason is that 16 bit hardware registers in the PICmicro® MCUs are all in this format and putting variables in this format will keep things consistent. The act of mentally reversing the two bytes becomes second nature very quickly and I dare say that you will become very familiar and comfortable with it after working through just a few applications.

When multi-byte data is displayed in MPLAB watch windows, the default is in the high/low format. Be sure that when you add a multi-byte variable to the window, click on the *low/high* selection.

Working with multi-byte variables is not as simple as working with single-byte variables because the entire variable must be taken into account.

For example, when incrementing a byte, the only considerations are the value of the result and the zero flag. This can be implemented quite easily for a 16-bit variable.

```
incf    LOW variable, f
btfsc   STATUS, Z
 incf   HIGH variable, f
```

Addition with two 16-bit variables becomes much more complex because, along with the result, the zero, carry, and digit carry flags must be involved as well. A "reasonably" correct 16-bit addition could be:

```
movf     HIGH A, w
addwf    HIGH B, w
movwf    HIGH C
movf     LOW A, w
addwf    LOW B, w
movwf    LOW C
btfss    STATUS, C
 goto    $ + 5
movlw    1
addwf    HIGH C, f
movf     LOW A, w
addwf    LOW B, w
xorwf    HIGH C, w
```

This code correctly computes the 16-bit result, as well as correctly sets the zero and digit carry flags. Unfortunately, it requires five more instructions than a simple case and does not set carry correctly. To set carry correctly, a temporary variable and seven additional instructions are required:

```
clrf     Temporary
movf     HIGH A, w
addwf    HIGH B, w
movwf    HIGH C, w
btfsc    STATUS, C
bsf      Temporary, 0
movf     LOW A, w
addwf    LOW B, w
movwf    LOW C
btfsc    STATUS, C
 goto    $ + 6
incf     HIGH C, f
btfsc    STATUS, Z
bsf      Temporary, 0
movf     LOW A, w
addwf    LOW B, w
xorwf    LOW C, w
bcf      STATUS, C
btfsc    Temporary, 0
 bcf     STATUS, C
```

This level of fidelity is not often required. Instead, you should pick the multi-byte operation that provides the result that you need. The appendices present routines that provide the correct 16-bit result, but the status flags will not be correct for the result. For correct flags, other snippets are used or the result is compared to an expected value.

**MulDiv: Constant multiplication and division** When you get into advanced mathematics (especially if you continue your academic career into electrical engineering), you will learn to appreciate the power of arithmetic series. With a modest amount of computing power, quite impressive results can be produced in terms of calculating data values. A good example of this is using arithmetic series to calculate a sine, cosine, or logarithmic function value for a given parameter. An arithmetic series can be used in analog electronics to prove that summing a number of simple sine waves can result in square wave, sawtooth, or other arbitrary repeating waveforms.

An arithmetic series has the form:

Result = + P1X1 + P2X2 + P3X3 + P4X4 + · · ·

Where *P#* is the prefix value, which is calculated to provide the function value. *X#* is the parameter value, which is modified for each value in the series. The parameter change can be a variety of different operations, including squaring, square rooting, multiplying by the power of a negative number, etc. For the multiplication and division operations shown here, I will be shifting the parameter by one bit for each series element.

The theory and mathematics of calculating the prefix and parameter for an arithmetic series can be quite complex, but it can be used in cases like producing the prefix values for simple multiplication or division operations (as shown in this section). To demonstrate the operations, I have created the multiply and divide macros that can be found in the *code\MulDiv* subdirectory of the *PICmicro* directory.

The two macros providing the multiplication and division functions use MPLAB assembler capabilities that I haven't explained yet (although they are in the next chapter). To try and avoid confusion, I will explain how the macros work from the perspective of a high-level language before presenting the actual code. To further help explain how the macros work, I will present them from the perspective of implementing the function in straight PICmicro® MCU assembler.

Multiplication (and division) can be represented by a number of different methods. When you were taught basic arithmetic, multiplication was repeated addition. If you had to program it, you would use the high-level code:

```
Product = 0;
for (i = 0; i < Multiplier; i++ )
  Product = Product + Multiplicand;
```

This method works very well, but requires a differing amount of time based on the multiplier (i.e., eight times something takes four times longer than two times something). This is not a problem for single-digit multiplication, but when multiplication gets more complex, the operations become significantly longer, which can negatively impact the operation of the application code.

Ideally, a multiplication method (algorithm) that does not have such extreme ranges should be used. As you would expect, this is where the arithmetic series is involved.

As you know, every number consists of constants multiplied by exponents of the number's base. For example, 123 decimal is actually:

123 = 1 * Hundreds + 2 * tens + 3 * ones

This also works for binary numbers and is used to convert constants between numbering systems. 123 decimal is 1111011 binary (0x07B). This can be represented like 123 decimal:

$$123 = 1 * \text{sixty-four} + 1 * \text{thirty-two} + 1 * \text{sixteen} + 1 * \text{eight} + 0 * \text{four} + 1 * \text{two} + 1 * \text{one}$$

In this binary sequence, I have also included any digits that are zero (bit 2 or "4" in the case of 123) because they will be used when multiplying two numbers together. This binary sequence can be used as the "prefix" of a multiplication arithmetic series, if each value is used to add the multiplicand that has been shifted up by the number of the bit. The shifted-up multiplicand can be thought of as the "parameter" of the series. This series can be written out as:

$$\begin{aligned} A * B = &((A \,\&\, (1 << 0)) \mathrel{!=} 0)*(B << 0) \\ &+ ((A \,\&\, (1 << 1)) \mathrel{!=} 0)*(B << 1) \\ &+ ((A \,\&\, (1 << 2)) \mathrel{!=} 0)*(B << 2) \\ &+ \cdots + ((A \,\&\, (1 << 7)) \mathrel{!=} 0)*(B << 7) \end{aligned}$$

This series can be converted to high-level code very easily:

```
int Multiply(int A, int B)           //  Multiply two eight bit values
//   together and return the result
{

int Product = 0;
int i;

  for (i = 0; i < 8; i++) {          //  Repeat for each bit
    if ((A & 1) != 0)
      Product = Product + B;         //  Bit of Multiplier is Set, Add
                                     //   Multiplicand to Product
    A = A >> 1;                      //  Shift down the Multiplier
    B = B << 1;                      //  Shift up the Multiplicand
  }
  return Product;                    //  Finished, Return the Result

}  //  End Multiply
```

This function will only loop eight times. Each time, it will shift up the multiplicand, which is the *B << 2* term in this series, and shift down the multiplier, which is the equivalent of the *(A & (1 << 0)) != 0* term of this series. This term is 100% mathematically correct. If the prefix result is not equal to zero, the shifted term will be added to *Product.*

For example, if you were multiplying together thirteen (0b01101) to ten (0b01010), the terms would be:

```
A * B = ((A & (1 << 0)) != 0)*(B << 0) +
        ((A & (1 << 1)) != 0)*(B << 1) +
        ((A & (1 << 2)) != 0)*(B << 2) +
        ((A & (1 << 3)) != 0)*(B << 3)
      = ((13 & (1 << 0)) != 0)*(10 << 0) +
        ((13 & (1 << 1)) != 0)*(10 << 1) +
        ((13 & (1 << 2)) != 0)*(10 << 2) +
        ((13 & (1 << 3)) != 0)*(10 << 3)
```

```
= ((13 & 1) != 0)* 10 +
  ((13 & 2) != 0)* 20 +
  ((13 & 4) != 0)* 40 +
  ((13 & 8) != 0)* 80
= (1 != 0)* 10 +
  (0 != 0)* 20 +
  (4 != 0)* 40 +
  (8 != 0)* 80
= 1 * 10 +
  0 * 10 +
  1 * 40 +
  1 * 80
= 10 +
  40 +
  80
= 130
```

For humans, this probably seems like a very slow way to implement multiplication, but for the PICmicro® MCU, it is actually very fast and consistent. Doing an eight bit by eight bit multiply, the following PICmicro® MCU code is used:

```
  clrf     Product
  clrf     Product + 1
  clrf     TempMultiplicand + 1
  movlw    8
  movwf    Count
Loop
  btfss    Multiplier, 0            ;  If Bit 0 Set, then Add
   goto    Skip                     ;   "Multiplicand" to the Product
  movf     TempMultiplicand + 1, w  ;  Add the High Eight Bits First
  addwf    Product + 1, f
  movf     Multiplicand, w          ;  Add Low Eight Bits Next
  addwf    Product, f
  btfsc    STATUS, C
   incf    Product + 1, f
Skip
  bcf      STATUS, C
  rlf      Multiplicand, f          ;  Shift the Multiplicand Up
  rlf      TempMultiplicand, f
  rrf      Multiplier, f            ;  Shift the Multiplier Down for
  decfsz   Count, f                 ;   Bit Check
   goto    Loop
```

When this code is exited, *Product* will contain a 16-bit result. Notice that I added a *TempMultiplicand* variable for the high eight bits of the shifted multiplicand. The repeated addition case could be written as:

```
  clrf     Product
  clrf     Product + 1
Loop
  movf     Multiplicand, w          ;  Add Multiplicand to the Product
  addwf    Product, f
  btfsc    STATUS, C
   incf    Product + 1
  decfsz   Multiplier, f            ;  Repeat Multiplier times
   goto    Loop
```

Although the repeated addition code requires less than half the number of instructions as what I consider to be the "better" shifting multiplication and does not require an extra file register, it can execute in anywhere from eight instruction cycles to 1,773 instruction cycles and produces an incorrect result if *multiplier* is zero.

The "better" case executes in anywhere from 84 to 124 instruction cycles and has a variability of no more than 47%. The simple version has a variability of as much as 22,000%. Also notice that the "better" example runs in fewer cycles for all cases, except when the multiplier is less than 17.

This algorithm can be used to multiply a variable value by a constant and to allow this operation to be put into MPLAB assembly code easily. I have created the multiply macro:

```
multiply macro Register, Value          ;  Multiply 8 bit value by a
 variable i = 0, TValue                 ;   constant
TValue = Value                          ;  Save the Constant Multiplier
  movf    Register, w
  movwf   Temporary                     ;  Use "Temporary" as Shifted Value
  clrf    Temporary + 1
  clrf    Product
  clrf    Product + 1
 while (i < 8)
 if ((TValue & 1) != 0)                 ;  If LSB Set, Add the Value
  movf    Temporary + 1, w
  addwf   Product + 1, f
  movf    Temporary, w
  addwf   Product, f
  btfsc   STATUS, C
   incf   Product + 1, f
 endif
  bcf     STATUS, C                     ;  Shift Up Temporary multiplicand
  rlf     Temporary, f
  rlf     Temporary + 1, f
TValue = TValue >> 1                    ;  Shift down to check the Next Bit
i = i + 1
 endw
 endm
```

To multiply the contents of an eight bit file register by a constant, the multiply macro is invoked as:

```
multiply Register, Constant
```

where *Register* is the file register to multiply by *Constant*. The result will be stored in the 16-bit variable *Product*. The macro itself will insert the code needed to perform the operation, but without the looping functions. This means that the code will take anywhere from 29 to 77 instructions. Thus, its worst case is slightly better than the loop code's best case.

This method of shifting data is how I perform all multiplication in the PICmicro® MCU. Appendix G shows that this is the method used to multiply two 16-bit numbers together.

Division is always much more difficult to perform than multiplication. Multiplication is repeated addition, and division could be thought of as repeated subtraction. The basic code version of division could be thought of as:

```
Remainder = Dividend;
for (Quotient = 0; (Dividend - Divisor) > 0; Quotient++)
  Dividend = Dividend - Divisor;
Remainder = Remainder - (Quotient * Divisor);
```

This code also includes returning a remainder from the operation.

To simplify the division operation, I can use an arithmetic series like I did for multiplication, but this one will work differently than multiplication. I would call the series produced for multiplication a *closed* series because the series is defined for a set range of numbers. In division, this is possible for some numbers, but not for all; numbers do not always divide "evenly" (i.e., have a remainder equal to zero) into others. In these cases, a decision has to be made about what to do about them.

To come up with a division arithmetic series, I would want to use the property of division that a number can be divided by a second one by multiplying by the *reciprocal* of the second number. This can be shown as:

$$A/B = A * (1/B)$$

This will probably be unexpected because fractions (what reciprocals really are) are not possible in the PICmicro® MCU, which can only handle positive integers. To get around this problem, you just have to look back in the history of mathematics to see how this problem has been encountered and solved before.

Four hundred years ago, when scientists were taking the results from plotting the path of the planets about the sun and trying to come up with a general mathematical theory about the motion of the planets, they had to work with trigonometric tables. The problem with using these tables was that the decimal point had not yet been invented. Instead of having values based as fractions of 1, these tables returned the numerical fraction over 6,000.

For instance, if you look up the sine of 45 degrees, you would get the value 0.707107. In the 6,000-based table, the sine of 45 degrees would be: 4,243.

This same principle can be applied to the finding of fractions in the PICmicro® MCU. Instead of coming up with a result that is less than 1, the division method presented here calculates the fraction as a result less than 65,536 (0x010000).

By doing this, 1/3 is not processed as 0.33333, but as 21845 (0x05555 or 0b00101010101010101), 1/5 is 13,107 (0x03333 or 0b00011001100110011), 1/7 is 9,362 (0x02492 or 0b00010010010010010), and so on.

It is important to note that these fractions as binary strings are repeating or "open" series, which complicates the division operations somewhat. Powers of two will result in a closed series, but, for the most part, the fractional values will not be closed. To ensure that the result is as correct as possible, the fractional bits should be taken as far as possible; which is why I divided by 65,536 ($2^{16}$) and not 256 ($2^8$).

Now that I have the fraction, I can develop the arithmetic series. For eight-bit division, this series is:

```
A / B = (((65,536 / B) & (1 << 15)) != 0) * (A >> 0) +
        (((65,536 / B) & (1 << 14)) != 0) * (A >> 1) +
        (((65,536 / B) & (1 << 13)) != 0) * (A >> 2) + . . . +
        (((65,536 / B) & (1 << 0)) != 0) * (A >> 15)
```

I shift *A* down for each element in the series because each test of the shifting-down bit in the fraction requires that the dividend be shifted down as well. This operation could be written in a high-level language as:

```
int Divide(int A, int B)              //  Carry out Eight Bit Division
{

int  Quotient = 0;
int  Divisor;
int  TempDividend;
int  i;
  Divisor = 65536 / B;                //  Get the Fractional Divisor
  TempDividend = Dividend << 8;       //  Get the Dividend to be used.
  for (i = 0; i < 8; i++ ) {          //  Repeat for sixteen cycles
    if ((Divisor & 0x08000) != 0)     //  Have to add Quotient Fraction
      Quotient = Quotient + TempDividend;
    TempDividend = TempDividend >> 1;
    Divisor = (Divisor & 0x07FFF) << 1;
  }

  if ((Divisor & 0x080) != 0)         //  Calculate Rounded Divisor
    Divisor = Divisor + 0x0100;
  Divisor = Divisor >> 8;

  return Divisor;

}  //  End Divide
```

As you work through this function, two things should be unexpected in the code. First, I use 16-bit values for an eight-bit result. This was done because I wanted to get a "rounded" (to the nearest one) result. If the result has a fraction of 0.5 or greater (in which bit seven of the result is set), then I increment the returned divisor.

The second is that I shift up the dividend and divisor by eight bits. This is done so that as I shift down the dividend, I do not lose the fractional bits of the result (which would not produce an accurate "rounding" of the result).

When I implemented this function, I did not write it out as straight PICmicro® MCU assembler for use in an application. The multiplication operation, because it is closed, can be carried out within straight code. The division operation laid out previously does not have this advantage because the result is most likely open.

This open result means that the PICmicro® MCU's internal functions cannot be used to calculate the fraction of the divisor. To calculate the divisor fraction, I have written the following macro with the divisor calculated by the macro calculator (explained later in the book).

```
divide macro Register, Value   ;  Divide 8 bit value by a constant
 variable i = 0, TValue
TValue = 0x010000 / Value      ;  Get the Constant Divider
  movf   Register, w
  movwf  Temporary + 1         ;  Use "Temporary" as the Shifted Value
  clrf   Temporary
  clrf   Quotient
  clrf   Quotient + 1
 while (i < 8)
  bcf    STATUS, C             ;  Shift Down the Temporary
  rrf    Temporary + 1, f
  rrf    Temporary, f
 if ((TValue & 0x08000) != 0) ;  If LSB Set, Add the Value
  movf   Temporary + 1, w
  addwf  Quotient + 1, f
```

```
  movf   Temporary, w
  addwf  Quotient, f
  btfsc  STATUS, C
   incf  Quotient + 1, f
 endif
TValue = TValue << 1          ;  Shift up to check the Next Bit
i = i + 1
 endw
  movf   Quotient + 1, w      ;  Provide Result Rounding
  btfsc  Quotient, 7
   incf  Quotient + 1, w
  movwf  Quotient
 endm
```

The divisor fraction is calculated at assembly time and used at assembly time to create the series of instructions to divide a variable value by a constant.

The macro will produce code that ranges from 39 instructions to 81 and takes the same number of instruction cycles to execute. This is only marginally larger than the analogous multiplication code.

There are two concerns with this code. First, if one is selected as the divisor, the code will return a quotient of zero. This is because 0x010000 divided by one is 0x010000 and will not cause any of the loops to add the current value. A divisor *Value* of one could be checked in the macro and an error returned if this is a potential problem.

The second problem is a bit more insidious and is reflective of how division algorithms work. The quotient returned is rounded to the nearest one. In many applications that require a division operation, this would not be acceptable. Instead, the quotient and remainder would have to be returned.

This macro was written to round the value so that indicator operations (like RPM in a tachometer) could be implemented quickly and efficiently. The value returned from the divide macro should not be passed onto any other arithmetic functions to prevent the error in the result from being passed down the line. If the quotient were required for subsequent operations, I would suggest that you use the 16-bit division routine presented in the appendices. If this macro is to be used, then the entire 16-bit quotient calculated by this macro (the lower eight bits being the fractional value less than one) is passed along with the final result divided by 256 (by "lopping off" the least-significant byte).

**Delays** Delays are often crucial aspects of an application. In PICmicro® MCU applications, it is not unusual to have microsecond, millisecond, or even full-second delay routines built in. In the first edition of the book, I didn't do a very good job of explaining how to create useful delays and how they are used in applications. This section clears up the errors I made and helps you to understand how adding delays to an application can make your life simpler, as well as help you to understand how crucially timed application code works.

The basic unit of timing in an application is the *instruction cycle*. The instruction *clock rate* is one quarter of the external clock frequency (as explained earlier in this book). The reciprocal is the instruction cycle *period*. The instruction cycle period is found using the formula:

*Instruction cycle = 4/Clock frequency*

So, for a clock frequency of 3.58 MHz, the instruction cycle period is found as:

*Instruction cycle* = *4/Clock frequency*
= 4/3.58 MHz
= 1.12 ms

Actual time delays should be converted into instruction cycle delays as quickly as possible. The formula that I use for doing this is:

*Instruction delay* = *Time delay* * *Clock frequency/4*

For example, if you had a PICmicro® MCU running at 10 MHz and wanted a 5-ms delay, the formula would be:

*Instruction delay* = *Time delay* * *Clock frequency/4*
= 5 ms * 10 MHz/4
= 50 * $(10^3)/4$
= 1.25 $(10^4)$
= 12,500

So, for a delay of 5 ms in a PICmicro® MCU running at 10 MHz, 12,500 instruction cycles would have to execute.

For a one-instruction delay, an *nop* instruction is used. For two cycles, use the *goto $ + 1* instruction. Four cycles can be implemented by calling a subroutine that simply returns. The two instructions take four instruction cycles to execute:

```
  :
  call     Dlay4                      ;  Delay 4 instruction cycles
  :

Dlay4
  return
```

This won't seem that special until you realize what can be done with it. By adding another layer to the subroutine, which calls *Dlay 4,* you can double the delay very simply. For example, to delay 16 instruction cycles, you could use the code:

```
  :
  call     Dlay16                     ;  Delay 16 instruction cycles
  :

Dlay16
  call     Dlay8
Dlay8
  call     Dlay4
Dlay4
  return
```

In this code, when *Dlay16* is called, the *call* instruction requires two instruction cycles to reach *Dlay16.* The call to *Dlay8* requires an additional two instruction cycles; it calls *Dlay4* for a total of six instruction cycles. When *Dlay4* returns, eight instruction cycles have executed. When the code returns to the call at *Dlay8,* it returns to the call at *Dlay16*, which continues executing to *Dlay8,* and the process continues. As you work through these four instructions, you will find that a total of 16 instruction cycles are delayed.

For longer delays, I recommend a loop for two reasons. The first because of the stack, which the method above requires. The low-end PICmicro® MCUs have a two-entry-deep stack, which means the 16-cycle delay uses more stack entries than is available in the microcontroller. Even if an eight-cycle delay were implemented as shown above, no stack space would be available for subroutine calls. In the mid-range PICmicro® MCUs, the three subroutine calls of the 16-instruction cycle delay code is probably the practical maximum (with interrupts in the application).

The second reason to avoid this method is because it starts to be suboptimal for numbers that are not divisible by two. For example, to get a 31 instruction cycle delay, the following calls and instructions are required:

```
call    Dlay16
call    Dlay 8
call    Dlay 4
goto    $ + 1
nop
```

Using loop code, the same 31-cycle delay could be implemented in one less instruction without changing w and STATUS.

```
movwf   _w
movlw   9
movwf   Dlay
decfsz  Dlay, f
 goto   $ - 1
swapf   _w, f
swapf   _w, w
```

The loop code's only real disadvantage is that it requires two variables (one for saving w and one for counting down the value).

Single variable loops, like this example, can work in three-cycle increments (as in this case) or in four-cycle increments by using the w register as the temporary register:

```
movwf   _w
movlw   7
addlw   0 - 1
btfss   STATUS, Z
 goto   $ - 2
swapf   _w, f
swapf   _w, w
```

This code does not change w, but does change STATUS.

Using these two methods, single-variable loops can delay up to 768 or 1024 instruction cycles.

For longer delays, TMRO (using the instruction clock and the prescaler) could be used, but I don't recommend it, except if the timer overflow interrupt request is going to be used. In this case, the prescaler value would be set and an initial value placed in TMR0 at the start of the delay, followed by the TMR0 interrupt request enable.

Using the prescaler, delays from 512 cycles to 65,536 instruction cycles can be implemented (based on the TMR0 synchronizer requiring a minimum of two instruction cycles for every time it is to be incremented). To determine the delay value of the prescaler, I use the formula:

$$Prescaler\ value = Int[Log2(Instruction\ delay/512)]$$

Where *Log2* is the Base 2 logarithm of the number of 512 blocks that have to be executed in the delay. The *Int* function returns the integer value of its parameter. This is used to make the result an integer that can be loaded directly into the OPTION register.

When working through this formula, the fraction *Instruction Delay/512* should be rounded up to an integer that is a power of 2 (1, 2, 4, 8, 16, 32, 64, or 128). Next, this value's power-of-two equivalent should be returned as the result of the *Int[Log2(. . .)]* operation. If the fraction rounds up to 256 or greater, then TMR0 and the prescaler cannot be used to produce this delay.

Going back to the example of 12,500 instruction cycles required for a 5-ms delay in a PICmicro® MCU running at 10 MHz, the prescaler value can be calculated as:

$$\begin{aligned} Prescaler\ value &= Int[Log2(Instruction\ delay/512)] \\ &= Int[Log2(12{,}500/512)] \\ &= Int[Log2(24.4)] \\ &= Int[Log2(32)] \leftarrow 24.4 \text{ rounded up} \\ &= 5 \end{aligned}$$

With the prescaler value calculated, the initial TMR0 value can be calculated using the formula:

$$TMRO\ initial = 256 - (Delay * Frequency/4)/(2^{Prescaler + 1})$$

One is added to the prescaler before making it the power of two that is divided into the number of instruction cycles to take into account the action of TMR0 to divide each cycle by two because of the synchronizer.

Using the example of a 5-ms wait in a 10-MHz PICmicro® MCU, the formula can be used to calculate the initial TMR0 value:

$$\begin{aligned} TMRO\ initial &= 256 - (Delay * Frequency/4)/(2^{Prescaler + 1}) \\ &= 256 - (5 \text{ ms} * 10 \text{ MHz}/4)/(2^6) \\ &= 256 - (12{,}500/64) \\ &= 256 - 195 \\ &= 61 \end{aligned}$$

The prescaler and TMR0 initial values could be used to generate the 5-ms delay using the code:

```
movlw    0x0D5              ;  Set the PICmicro® MCU's Prescaler to TMR0
option                      ;   and Instruction Cycle Divide by 64
movlw    61                 ;  Set up the Timer Delay
movwf    TMR0
```

```
Loop                          ;  Wait for the Timer to Overflow to 0x000
  movf     TMR0, w
  btfss    STATUS, Z
   goto     Loop
```

This code should only be used when the prescaler is set to divide TMR0's instruction clock input by two or more (for a TMR0 increment of four cycles or more). If the prescaler is not used or is set to 1:1, then TMR0 will increment twice in the loop, which means that there is a 50% chance that the 0x000 will not be observed. Ideally, TMR0 should request an interrupt that would eliminate the restrictions on the prescaler.

Looking at the calculations and this code, you might feel that this method has an error. When I calculated the *TMR0 Initial* value, I rounded the result of 12,500/64 to the nearest integer. 12,500/64 is actually 195.313.

The desired delay is 12,500 cycles. In actuality, the PICmicro® MCU will execute:

$$\begin{aligned} Delay &= (256 - TMR0\ initial) * (2^{Prescaler\ +\ 1}) \\ &= (256 - 61) * (2^{5\ +\ 1}) \\ &= 195 * 64 \\ &= 12{,}480 \end{aligned}$$

Even taking into account the four instructions used to set up the timer, the actual number of cycles is off from the desired value by 16 instruction cycles. This works out to an error rate of:

$$\begin{aligned} Timing\ error &= (Desired\ delay - Actual\ delay)/Actual\ delay * 100\% \\ &= (12{,}500 = 12{,}484)/12{,}500 * 100\% \\ &= 16/12{,}500 * 100\% \\ &= 0.128\% \end{aligned}$$

Using TMR0 and the prescaler to provide the 5-ms delay will result in a 0.128% error, which is not significant for most applications. But it will become significant when compounded over time—especially if the PICmicro® MCU is communicating with another device. If this delay is used for a bit communications protocol, after 390 bits will be a 50% error between the "perfect device" and this one.

I described the other device as being *perfect* to point out that there really are no perfect timers. When I cover oscillators elsewhere in the book, notice that I consider the 50 ppm (0.005%) tolerance for crystals to be outstanding, the 0.5% error for ceramic resonators to be quite good, and the 1% (or so) error for built-in oscillators to be tolerable for most applications. There is no such thing as a perfect clock—even the cesium atomic clocks used as time references have to be reset periodically.

Avoid thinking of your applications or other devices as having to be absolutely perfect in terms of timing. Instead, be sure that you understand the timing requirements for an application and that you provide a mechanism to continually "synch up" between devices. Failing to do this will result in an application that works occasionally; you will seem to be dealing more with errors than with actual data.

For long delays, I recommend using two bytes for the delay and decrementing them. When I wrote the first edition, I suggested the code:

```
  movlw    LOW Value
  movwf    LOW Dlay
  movlw    HIGH Value
  movwf    HIGH Dlay
Loop
  decf     LOW Dlay, f
  btfsc    STATUS, Z
   decfsz  HIGH Dlay, f
    goto   Loop
```

In this code, each time through the loop will take five instruction cycles. To calculate the *Value* to be loaded into the *Dlay* variables, I use the simple formula:

*Value* = [(*Delay* * *Frequency*/4)/5] + 256

The *+256* in the formula increments the high byte so that the value stored into the low byte will execute along with the number of times the high byte has to be decremented with the low byte set to zero (which causes the code to loop 256 times).

For the example of a 5-ms delay in a 10-MHz PICmicro® MCU, *Value* is calculated to be:

*Value* = [(*Delay* * *Frequency*/4)/5] + 256
= [(5 ms * 10 MHz/4)/5] + 256
= (12,500/5) + 256
= 2,500 + 256
= 2,756
= 0x00AC4

For this method, a maximum value of 65,505 (0x0FFFF) can be calculated for *Value,* but I prefer stopping at 50,256 (0x0C450), which gives a 250,000-instruction cycle delay that can be built upon by an outside loop. For example, in the 10-MHz example PICmicro® MCU, to get a three-second delay, I would first calculate the delay of 250,000 instruction cycles:

*Delay* = *Instruction cycles* * 4/*Frequency*
= 250,000 * 4/10 MHz
= 0.1 seconds

To get a three-second delay, this 250,000-instruction cycle (0.1 second) delay would have to execute 30 times. Loading a register with the value of 30 and using the PICmicro® MCU's *decfsz* instruction would accomplish this:

```
  movlw    30          ;  Load the Outside Loop
  movwf    Outside
OuterLoop
  movlw    LOW 0x0C450 ;  Inside 250,000 instruction Delay
  movwf    Dlay
  movlw    HIGH 0x0C450
  movwf    Dlay + 1
Loop
  decf     Dlay,+ 1, f
  btfsc    STATUS, Z
   decfsz  Dlay +1, f
    goto   LOOP
   decfsz  Outside, f
    goto   Outer Loop
```

For a "true" three-second delay, a total of 7,500,000 instruction cycles is required. In the previous code, an extra 212 instruction cycles are added to the code. If you are thinking of changing the *Loop* delay value to better match the desired number of instructions, consider the magnitude of this error over three seconds:

$$\begin{aligned} \textit{Total error} &= 212/7{,}500{,}000 * 100\% \\ &= 0.002827\% \\ &= 28.27 \text{ ppm} \end{aligned}$$

An error of 28 parts per million is at most half of the expected error rate of any clocking scheme that you will use for your PICmicro® MCU applications. With this low error level, I would recommend that any extra cycles in large delays like this be ignored. Even in the case where the timer was used, the error was quite slight and approached the tolerance of the PICmicro® MCU's clock.

**Random numbers** Probably the most difficult thing to do with a computer is to generate a random number. A random number is useful on many kinds of applications (most notably for games) where the random data is used to spice up a display or provide test data for hardware. As you look at most of the different algorithms out there, you'll see that not only are they not random, but they also require input from a user for a "seed" value.

I would like to discourage you from requesting "seed" data from a user. This is annoying for the user and they tend to enter the same data over and over again, to see if they can figure out how to "beat" the application.

Instead, I recommend setting up a high-speed timer and saving one digit of its value as the random number (seed).

For example, if the application was waiting for the user to press a button, it could increment a counter and use bits two to five as a hexadecimal "seed."

```
;  "Button" is debounced open.

Random:
  incf    RandomCounter, f    ;  Increment random number
  btfsc   Button              ;  Check button
   goto   Random

;  Do Button debounce, return to "Random", if not debounced

  rrf     RandomCounter, f    ;  Shift down by 2
  rrf     RandomCounter, w
  andlw   0x00F               ;  w contains a 4 bit random number
```

If the PICmicro® MCU executes at 4 MHz, the "random" loop would take four instruction cycles (4 μs) to run through. This means that the number changes 250,000 times per second (rolls over almost 1,000 per second). With the four bits taken from the value, the changes occur just under 4,000 times per second.

This rate of change makes it impossible for a user to press the button at a certain interval or time to get the same value. But, there can be a few holes if you're not careful. First, be sure that the button is not pressed before you start the random counter. If you don't, the user will learn to hold down the button to get the same value all the time. Second, watch how your debounce routine works and be sure that the random counter is not affected by it.

Another way to come up with a random number is to use the power-up values of the PICmicro® MCU's file registers. As shown in Chapter 15, the power-up values of the file registers change each time the PICmicro® MCU does a power-on reset. The problem with this method is that the contents do not change after a _MCLR or watchdog timer reset. To avoid these problems, the PICmicro® MCU should never be reset, except by cycling its power and restarting the application.

Other methods for generating random values use hardware (such as counting the number of voltage spikes on a reverse-biased diode). The first method presented here is probably the most efficient for the PICmicro® MCU applications because it is based on the reactions of a human, much more random than what you would expect to get from a computer processor.

**Patch space** *Patch code* is instruction space left in the program memory of a computer system to allow changes to be added to an application without having to rebuild it. Using patch code, changes would be made locally to a processor's program memory to allow the developer to try different things before they would change the source, rebuild it, and load it into the system to try again (in the early days of PCs, this was an arduous process).

When I was first taught assembly-language programming, I was always told to leave patch space in my applications. This space would be used to add code to help debug applications. Using patch space was something I never did very well. I always found that I never accurately documented the changes I had made and the entire debugging process was slower than if I simply tried to figure out what was wrong and change the source code directly.

The reasons for providing and using patch space were:

1 Application rebuild and reprogramming was a long and tedious process.
2 Application memory was usually RAM and could be easily changed in a debugger.

Both these arguments are largely false for the PICmicro® MCU and modern systems in general. Using tools like MPLAB, changes to the application source can be done in seconds. Programming, using an ICSP application, can be accomplished within a minute or so. When I first learned assembly-language programming (for the Motorola 6800), a source code change was made and re-assembled on a main frame, downloaded to a mini computer, downloaded onto a cassette tape, and then downloaded into the 6800 computer system. This was very complex and created a lot of opportunities for problems, compared to using MPLAB with a Microchip programmer for developing software and programming PICmicro® MCU devices.

Many PICmicro® MCU applications can use Flash-based parts for application debugging or for an emulator, which can allow changes in source code to be transferred to the application's PICmicro® MCU almost immediately. Both of these capabilities basically invalidate the arguments for providing patch code space in an application.

But, there are some cases where only an EPROM PICmicro® MCU part can be used and no emulator is available for it. In these cases, some program memory can be devoted to providing patch code in the application.

The typical way of doing this is to define a block of code in-line to the application with a jump around it. The assembly code "prototype" to do this is:

```
  goto   $ + size + 1
variable  i = 0
 while (I <  size)
  dw      0x03FFF
i = i + 1
 endw
```

Later, the book covers the format of the *variable, while,* and *wend* directives. The previous code defines a block of program memory, *size + 1* instructions long, which can be loaded with any instructions. The *dw 0x03FFF* instructions leave each memory location unprogrammed and able to accept patch code.

To put instructions into the patch code space, the first *goto* is overwritten with 0x00000, which is the *NOP* instruction. Following this, the instructions can be programmed into the PICmicro® MCU's program memory.

To show what I mean, let's go through an example. An application has the statement:

```
Result = 47 - B
```

which the developer has coded:

```
movlw   47
subwf   B, w
movf    Result
```

When this was entered into the source code, the developer was not sure that it was correct. In case it wasn't, the developer provided three instructions of patch code space to rewrite the code in case the original code was wrong. The actual source code used for the application was:

```
movlw   47
subwf   B, w
movwf   Result
goto    $ + 4                 ;  Jump Over Patch Code
dw      0x03FFF
dw      0x03FFF
dw      0x03FFF
```

In the course of the application's debug, the developer discovered that the original subtraction method was incorrect and the use of *47* and *B* had to be reversed. To try out the code, the original subtraction operation is overwritten with zeros (to make the three instructions *nops*), along with the *goto* instruction. Next, the three correct subtraction instructions are programmed in. These seven instructions in program memory become:

```
; current instructions                  previous instructions
  nop                                   ;  movlw   47
  nop                                   ;  subwf   B, w
  nop                                   ;  movwf   Restart
  nop                                   ;  goto    $ + 4
  movf    B, w                          ;  dw      0x3FFF
  sublw   47                            ;  dw      0x3FFF
  movf    Restart                       ;  dw      0x3FFF
```

Of course, to be able to do this, you will need a programmer that can write to the specific address with specific values. The programmer designs provided in this book do not support this.

If you look at the PIC18Cxx's instruction set, you will see that *nop* instructions can either be 0x0FFFF or 0x00000. This allows patch space to be put inline without the need for jumps around it. For the reasons recommended, I still discourage the use of patch space in the PIC18Cxx and changes to the code should be made in the source code. Then, the device should be reprogrammed with the new application code.

**Structures, pointers, and arrays** If you have been trained in a classical programming environment, you are probably very comfortable with working with structures and pointers. Complicating matters, you might want to use these programming constructs with arrays. Developing applications with these features is possible for the PICmicro® MCU using MPLAB's assembler, although they are rarely used because of the lack of large contiguous file register space in most of the PICmicro® MCU device families and the difficulty in working with the PICmicro® MCU processor. Despite these difficulties, structures, pointers, and arrays can be created in PICmicro® MCU assembly language, although using them will be somewhat more complex than what you are used to with other processors.

The three programming constructs presented here all require the use of the FSR register and the bank-select features of the PICmicro® MCU. In actuality, this will be the most challenging and sophisticated use of the FSR that you will see in this book. You might want to try out the code snippets provided here to see exactly how they work.

MPLAB assembler does not have a *structure* directive that can be used to define a structure within the application. In other assemblers, structure directives are used to set up a single data type, consisting of a number of other data types.

In C, a structure is defined like:

```
struct VarStruct {              //  Define a Structure
  int  varA;                    //  16 Bit Variable
  char varB;                    //  8 Bit Variable
  char * NextVarStruct;         //  Pointer to the next structure in the
                                //     list
 };
```

When a variable is declared, the structure is used as the data type. In C, a variable defined from this structure would be declared as:

```
struct VarStruct VarValue;
```

To access the structure elements in *VarValue,* the element name is added to the variable name with a period or dot (.):

```
 VarValue.varA = VarValue.varB * 4;
```

In this statement, the structure element *varA* is loaded with the contents of the structure element *varB* after it has been multiplied by four. This capability allows quick and easy structure variable element access and modification.

This is not a high-level language-specific capability; many assemblers allow structures to be defined and used with structure variables. The MPLAB assembler does not, but you can still specify a structure and use it in MPLAB using the *CBLOCK* directive. The *VarStruct* structure shown previously can be defined in MPLAB as:

```
;  VarStruct - Define an MPLAB Structure
 CBLOCK 0                     ;  Start with Offset equal to Zero
varA:2                        ;  16 Bit Variable
varB                          ;  8 Bit Variable
NextVarStruct:2               ;  Pointer to the next structure in the list
SizeOfVarStruct               ;  Set to Number of Bytes of "VarStruct"
 ENDC
```

The last entry in the *CBLOCK* statement is assigned the number of bytes in the structure and will not be used, except to keep track of the number of bytes required by the structure. The *SizeOf* built-in C function returns the size of a structure, which is why I chose this identifier for the structure function.

To declare a structure variable in MPLAB, the CBLOCK directive is used again like:

```
 CBLOCK 0x0??               ;  Define Variables at File Register Start
  :
VarValue:SizeOfVarStruct   ;  Define the Structure Variable
  :
 ENDC
```

In these CBLOCK statements, *VarValue* is defined along with the other variables the size of the structure (as returned by *SizeOfVarStruct*). For this declaration to work properly, the structure must be defined before the variable to ensure that *SizeOfVarStruct* is valid and does not change between the assembly passes.

To address the structure elements in the structure variable, the structure element offset (defined by the first CBLOCK) is added to the address of the structure variable. To show how this is done, the PICmicro® MCU assembly code for the example C statement:

```
 VarValue.varA = VarValue.varB * 4;
```

could use the code:

```
  clrf     VarValue + varA + 1      ;  VarValue.varA = VarValue.varB
  movf     VarValue + varB, w
  movwf    VarValue + varA
  bcf      STATUS, C                ;  VarValue.varA = VarValue.varB*2
  rlf      VarValue + varA, f
  rlf      VarValue + varA + 1, f
  rlf      VarValue + varA, f       ;  VarValue.varA = VarValue.varB*4
  rlf      VarValue + varA + 1, f
```

The PIC18Cxx allows for offsets to be used with FSR registers. This capability is demonstrated in the Real-Time Operating System presented later in the book.

Pointers are variable types that can point to other variables anywhere within the application variable memory space. In the entire PICmicro® MCU family, no devices have more than 4,096 file registers, so the data size that I normally use for PICmicro® MCU pointers is a 16-bit variable.

Depending on the PICmicro® MCU architecture, pointers can be somewhat confusing and difficult to work with, although after a few seconds of thought, you could probably come up with a method to access registers in the different devices. For the low-end devices, pointers are quite simple to implement with a single eight-bit register that can be stored in the FSR register. The general case for mid-range PICmicro® MCUs requires nine bits to fully access registers in the four banks, which will require a 16-bit variable. For the PIC17Cxx and PIC18Cxx, I would recommend that the four-bit bank address of the selected register be included with the eight-bit register address in a 16-bit variable.

Having a pointer address in an eight-bit register is quite easy to implement, but pointers become much more complex if they are pointing to a structure. The last structure element in the *VarStruct* examples is designed as a pointer to the next structure variable in a list of *VarStructs*. To implement a list of *VarStructs,* three structure variables can be defined along with a pointer to *VarStructs* that will have a structure variable "strung along" in a list. In C, the variables would be declared using the code:

```
struct VarStruct StructureA;
struct VarStruct StructureB;
struct VarStruct StructureC;
struct * VarStruct StructurePtr;
```

Notice that *StructureA, StructureB,* and *StructureC* are each used to define the five bytes of *VarStruct*. *StructurePtr* simply points to a structure and initially has no value behind it.

In C, returning a pointer address is accomplished by using the asterisk (*) and the ampersand (&) is used to return the address of a value.

To set up the list of *StructureA* pointing to *StructureB* pointing to *StructureC* and *StructurePtr* pointing to the start of the list (*StructureA*), the following C code is used:

```
StructureA.NextVarStruct = &StructureB;
StructureB.NextVarStruct = &StructureC;
StructurePtr = &StructureA;
```

To access a structure pointer in the list, the *StructurePtr* pointer is used with the -> operator, which indicates that it is referring to the pointed to value. To set *StructureB.varB* to 0x012, the following code would be used:

```
StructurePtr -> NextVarStruct -> varB = 0x012;
```

As you would imagine, this is quite complex to do in the PICmicro® MCU. To give you an idea of how this would be done, the code below shows the assembler code for the previous C statement.

This code is written for the mid-range PICmicro® MCU. To simplify the operation of the pointer load, I make my pointer structure nine bits in size with the value being the register address used in the Microchip device .INC file. This address will be put into the FSR register and *IRP* bit of the STATUS register using the macro:

```
SetPointer macro Address
 if ((Address & 0x0100) == 0)
  bcf      STATUS, IRP
```

```
else
 bsf      STATUS, IRP
endif
 movlw    Address & 0x0FF
 movwf    FSR
endm
```

I used this macro in the PICmicro® MCU code to simplify the actual operations of the C statement in assembler:

```
SetPointer  StructurePtr          ;  Point to the Start of the List
                                  ;   (StructureA)
 movlw      NextVarStruct         ;  Point to the next List Element
                                  ;   (StructureB)
 addwf      FSR, f
 movf       INDF, w               ;  Get the Low Byte of the Next List
                                  ;   Element
 movwf      PointerTemp
 incf       FSR, f
 movf       INDF, w               ;  Get the High Byte of the Next List
                                  ;   Element
 movwf      PointerTemp + 1
 bcf        STATUS, IRP           ;  Load the Next List Element into FSR
 btfsc      PointerTemp + 1, 0
  bsf       STATUS, IRP           ;   AND the IRP
 movf       PointerTemp, w
 movwf      FSR
 movlw      varB                  ;  Point to the Element within the
                                  ;   Structure
 addwf      FSR, f
 movlw      0x012                 ;  Finally, do the Assignment
 movwf      INDF
```

Despite the simplification of the macro, the PICmicro® MCU code to implement the pointer statement is still quite complex and, more importantly, very confusing. Trying to follow this code will be a bit difficult, although this is a characteristic of all assembly-language pointer programming.

Looking at this code, one consideration about creating pointers and structures in the PICmicro® MCU that is not readily apparent is the importance of never having a structure go over a bank boundary. This is probably obvious, but I just point it out as something to watch for if your application grows and the latest structure you added to the application code seems to cause the PICmicro® MCU to seem to lock up or fail in strange ways. When a structure goes over a bank boundary, accesses to the structure will modify the PICmicro® MCU's context registers (like the STATUS, PCL, and OPTION registers) at the start of the bank, causing all kinds of chaos.

If you are new to programming, you will find working with pointers confusing. I'm not new to programming and I find pointers confusing. I realize that there are times when programming constructs like the linked list example above is required for an application, but they should be avoided unless absolutely necessary in the PICmicro® MCU. With a bit of thought about the application design, they should not be required at all.

"Not being required" is especially true for modern programming philosophies. With the advent of Java, pointers are currently unfashionable. Right now, the only case where I would consider pointers absolutely necessary is working with allocated variable memory in PCs or workstations. This is not a concern with the PICmicro® MCU.

The last programming construct I want to introduce you to in this section is arrays. PICmicro® MCU array programming is covered elsewhere in the book, but I want to include some advanced aspects of it and how arrays can be used with structures. Arrays with pointers are simply too complex to implement in PICmicro® MCU assembler to even think about it.

Multi-dimensional arrays have to be defined according to the amount of space that is required. If you are going to implement a five-by-five array of bytes, the *CBLOCK* directive can be used like:

```
 CBLOCK 0x0??
   :
Array:3*5                       ;  Define a 5 by 5 array
  :
 ENDC
```

Although the declaration of this array is simple, working with it is not. For example, accessing the byte at array element 2, 3 would require multiplying the first-dimension specification by 5 before the second dimension specification is added to it. For the C code:

```
 Array[2][3] = 0x012;                   ; Load Element 2, 3 with 0x012
```

the PICmicro® MCU assembler code would be:

```
  movf      Parm1, w            ;  Multiply the first dimension specified
  movwf     FSR                 ;   by 5
  bcf       STATUS, C           ;  Multiply the first dimension by 4
  rlf       FSR, f              ;   first
  bcf       STATUS, C
  rlf       FSR, f
  addwf     FSR, f              ;  Add first dimension to 4x first
                                ;   dimension to get 5x
  movf      Parm2, w            ;  Add the second dimension to 5x first
  addwf     FSR, f
  movlw     0x012               ;  Do assignment at array element 2,3
  movwf     INDF
```

Notice that the *multiply by five* uses the function of multiplying by a power of two and then adding the value again to get the odd multiplier. Without using this trick, this array access code would be much more complex and probably require the use of a temporary variable and a subroutine.

To simplify this operation, I suggest two improvements. First, change the way that the array is declared to five by four from five by three. Four is a power of two and is very easy to multiply by. By doing this, five bytes are added to the array. Hopefully, this is not a significant amount of memory in the application (it could be for something like a low-end PICmicro® MCU, where only 16 unique file registers are available in each bank).

Second, I would reverse the order in which data is stored in the array. Instead of the first parameter being multiplied by five I want the second parameter to be multiplied by four. Making these changes, the code becomes:

```
 bcf      STATUS, C         ;  Multiply the second dimension
 rlf      Parm2, w          ;   specified by four
 movwf    FSR
 bcf      STATUS, C
 rlf      FSR, f
 movf     Parm1, w          ;  Add first dimension to second*4
 addwf    FSR, f
 movlw    0x012             ;  Do assignment at array element 2,3
 movwf    INDF
```

Making these two changes results in an over 20% decrease in application code size. It is debatable whether or not it is easier to read and understand than the first example (it is for me).

Arrays can be used with structures to allow quite complex data tracking operations. For example, a five-by-three array of *VarStruct* could be created. This would be done in C using the code:

```
struct VarStruct VarStructArray[3][5];
```

In the MPLAB assembler, this operation is also quite simple:

```
 CBLOCK 0x0??
   :
VarStructArray:3*5*SizeOfVarStruct   ;  Define a 5 by 5 VarStruct array
  :
 ENDC
```

To access elements in the two-dimensional array, the same code is used, but once the result is calculated, it will have to be multiplied by the size of *VarStruct* (which is five). For example, the C code statement:

```
 VarStructArray[2][3].varB = 'A';
```

would be implemented in PICmicro® MCU assembler as:

```
 movf     Parm1, w          ;  Multiply the first dimension specified
 movwf    FSR               ;   by 5
 bcf      STATUS, C
 rlf      FSR, f
 bcf      STATUS, C
 rlf      FSR, f
 addwf    FSR, f
 movf     Parm2, w          ;  Add the second dimension to 5x first
 addwf    FSR, w            ;  Save first * 5 + second and multiply
 movwf    FSR, w            ;   it by the "SizeOfVarStruct" which is
 bcf      STATUS, C         ;   5
 rlf      FSR, f
 bcf      STATUS, C
 rlf      FSR, f
 addwf    FSR, f
 movlw    "A"               ;  Do assignment at structure element in
 movwf    INDF              ;   array element 2,3
```

When making arrays of structures, remember to keep track of the size of the final array. This is why I used the original five-by-three array code for this example. The size of the *VarStruct* structure is five bytes. With a five-by-three array, the total array size would be

75 bytes. If I were to use a five-by-four array for this example, the ultimate array size would be 100 bytes, which would be a problem for most PICmicro® MCUs, except for the PIC18Cxx.

Even when working with single-byte multi-dimensional arrays, always be aware of the ultimate array size. This example produces the same result as if a three-dimensional array of three by five by five was created. If the array was increased in size to five by five by five, 125 file registers are required, which is outside of the capabilities of all of the PICmicro® MCUs except for the PIC18Cxx unless playing around with bank registers (and the base address) is desired. I highly recommend that you avoid implementing single arrays across multiple register banks. If there is no other way to implement the application, you should look at other microcontrollers or use the PIC18Cxx.

**Sorting** The PICmicro® MCU is deficient in sorting data. The reason for this comment is the relatively small file register banks for storing array data, as well as the single FSR index address register in the low-end and mid-range PICmicro® MCUs. These two features make sorting data quite difficult and inefficient.

Having said this, there are always cases where code for sorting data is required. I have come up with my own bubble sort code for the PICmicro® MCU. The input is a single-dimensional array of values and the output is a single-dimensional array of the values sorted in ascending order. The variables used by the code are shown in Table 9-7.

The code, as follows, is designed to sort four values. The size of the array to be sorted can be increased easily by changing the *lend* value before the start of the routine:

```
Sort
;  Now, In the Sorting Routine
  movlw       rega          ;  Setup Where you are Storing the Result
  movwf       next
  movlw       reg4          ;  For Shrinking List, Get the Last Addr
  movwf       lend          ;  Watch for the Ending Value

Loop                        ;  Loop Around Here Until List is Empty
  movlw       regl          ;  Load FSR for Searching for the lowest
  movwf       FSR
  movwf       addr          ;  At Start, Assume the First is Lowest
```

**TABLE 9-7 Sorting Subroutine Required Variables**

| VARIABLE | FUNCTION |
|---|---|
| regl | Start of Array of Values to be Sorted |
| rega | Array of Sorted Values |
| next | Location to put the Next Sorted Value |
| llow | Value of the Last Lowest Number |
| addr | Location of the Last Lowest Number |
| lend | Location of the List End |

```
  movf        INDF, w       ;  Get the Current and Use As the Lowest
  movwf       llow          ;  Save it as the Current Lowest

Loop2                       ;  Loop Here Until FSR = lend
  movf        FSR, w        ;  Are we at the end?
  subwf       lend, w
  btfsc       STATUS, Z     ;  If Zero Flag is Set, We're At the End
   goto       Save          ;  Save the Currently Lowest Value

  incf        FSR, f        ;  Now, Look at the Next Value

  movf        llow, w
  subwf       INDF, w       ;  Do we Have Something that's Lower?
  btfsc       STATUS, C     ;  If Carry Set, then current is Lowest
   goto       Loop2

  movf        INDF, w       ;  Current is the Lowest - Save It
  movwf       llow
  movf        FSR, w        ;  And, Save the Address It's at
  movwf       addr
  goto        Loop2         ;  Loop Around and Look at the Next

;  The List has been Checked and "low" and "addr" have Lowest Current
;   Value and it's Address, Respectively.

Save                        ;  Now, Save the Currently Lowest Value
  movf        next, w       ;  Store it in the FSR
  movwf       FSR

  movf        llow, w       ;  Get the Lowest
  movwf       INDF          ;  Store it in the Sorted List

  movf        next, w       ;  Are we at the End of the List
  sublw       regd
  btfsc       STATUS, Z     ;  If NOT Zero, then Loop Around
   goto       PEnd          ;  Else, They Match, End the Program
  incf        next, f       ;  Increment Pointer to the Next Value

;  The Lowest Current Value has been Put in the "Sorted" List,
;   Now, Shorten the List at the Value we took out
  movf        addr, w       ;  Get Address the Value was Taken Out Of
  movwf       FSR           ;  Put in the FSR for Later

Loop3                       ;  Now, Loop Around Storing the New List
  movf        FSR, w        ;  Are we at the End of the List?
  subwf       lend, w
  btfsc       STATUS, Z     ;  Is the Zero Flag Set?
   goto       Skip          ;  Yes, List Has been Copied

  incf        FSR, f        ;  Get Next Value and Store in Current
  movf        INDF, w
  decf        FSR, f
  movwf       INDF
  incf        FSR           ;  Increment the Index and Loop Around
  goto        Loop3

Skip
  decf        lend, f       ;  Decrement the Ending Address
  goto        Loop

;  Sort is All Finished
PEnd                        ;  Program End
  return
```

There's a bit of a story to this routine. On the PICLIST, somebody asked for a routine that sorted four numbers in one list and put them in another. After working for about three hours, I came up with the solution listed on the previous pages.

It was quite an eye opener when I saw what other people came up with; whereas I improved the baseline code by about three times (three times shorter execution time and about a third of the original code size), the best solutions were almost 100 times better!

The solutions presented were designed to do exactly what was required: sort four numbers from one list and put them in another. Developing a macro for comparing two values and putting the lowest first, the solution was:

```
least  regc, regd              ; Move the Lowest Value to Front of
least  regb, regc              ; beginning of List
least  regb, rega
least  regc, regd              ; Move 2nd Lowest Value to 2nd from
least  regb, regc              ; the Front
least  regc, regd              ; Put the two highest in order
```

The macro used to accomplish this was:

```
least   macro    reg1, reg2
  movf           reg2, w
  subwf          reg1, w
  btfsc STATUS, C               ;  If no Carry, Swap the Values
   goto $+6                     ;  Else, Skip over the Rest
  movf  reg1, w                 ;  Now, Swap the Values
  xorwf reg2, w
  xorwf reg2
  xorwf reg2, w
  movwf reg1
        endm
```

The lesson I learned in all this is to understand what the customer wants. Although the routine that I created is very clever and took a bit of work, it is too general and a more specific solution was perfect for the customer's requirements.

Because of the hardware restrictions in the PICmicro® MCU, I limit the sorting routines to basically what you see here. In other microcontrollers, with processors that can access large amounts of memory indirectly, I would recommend using the *QuickSort* routine. This routine divides an array into two halves, one above and one below the mean. This process is repeated (i.e., the routine is called recursively) until the data is sorted.

The pseudo-code for QuickSort is:

```
QuickSort( int Bottom, int Top )       //  Sort the Array from
{                                      //   Bottom to Top

int i, j;
int mean;
int temp;

  if ( Top == ( Bottom + 1 )){         //  Do A Sort on Last Two Elements

    if ( Array[ Bottom ] > Array[ Top ] ) {
      temp = Array[ Bottom ];          //  Swap the Two Elements
      Array[ Bottom ] = Array[ Top ];
      Array[ Top ] = temp;
    }
```

```
  } else {                              //  Sort > 2 Elements in to 2
                                        //   Halves
    for ( i = Bottom; i < ( Top + 1 ); i++ )       //  Get Array Total
      mean += Array[ i ];

    mean = mean / ( Top + 1 - Bottom )        //  Get Array mean

    i = Bottom;  j = Top;
    while ( i != j ) {                  //  Split the Data into two halves

      while (( Array[ i ] < mean ) && ( i != j ))
        i++;                            //  Find Array Element Above mean

      while (( Array[ j ] < mean ) && ( i != j ))
        j++;                            //  Find Array Element Below Mean

      if ( i != j ) {                   //  Swap the Two Value Positions
        temp = Array[ i ];              //   - Lower Half is Less than mean
        Array[ i ] = Array[ j ];
        Array[ j ] = temp;              //   - Upper Half is >= mean
      }
    }                                   //  Finished Splitting the Data

    QuickSort( j, Top );                //  Sort the Top Half of the Data
    if ( i > Bottom )                   //  Sort the Bottom Half of Data
      QuickSort( Bottom, i );

  }  //  Finished Sorting
}  //  End QuickSort
```

The advantage of QuickSort is that it is an *Order NlogN sort*. The time required for the sort is proportional to the product of number of elements times the base two logarithm of the number of elements. BubbleSort is an *Order N-Squared sort,* which means that the time required for the sort is proportional to the square of the number of elements.

For small arrays of data, BubbleSort is more efficient than QuickSort, but as the array size grows, QuickSort becomes the preferred sorting method.

I do not recommend trying to implement QuickSort in the PICmicro® MCU for a number of reasons. First, QuickSort is a recursive algorithm. You will find that the PICmicro® MCU's program counter stack will be used up very quickly if QuickSort is used. I have implemented QuickSort in BASIC without requiring recursive calls, but it does require a lot of memory to store intermediate array starts, ends, and means. QuickSort might be an option for the 17Cxx or 18Cxx parts, but it is definitely inappropriate for the low-end and mid-range PICmicro® MCUs.

# Interrupts

I've written a lot about interrupts and interrupt handlers in this book. Although they are not terribly hard to create, you should follow some rules and conventions when writing the software handlers:

1. Use the standard header information provided in the next section.
2. Keep them as short as possible with the interrupt-controller hardware reset occurring as early in the interrupt handler as possible.

**3** Avoid nested interrupts.
**4** Do not call subroutines from an interrupt handler. If this is not possible, avoid reentrant subroutines.

By following these four rules, when you first start writing interrupt handlers on your own, you should minimize the problems with the interrupt handler code that will make debugging your application easier.

**STANDARD CONTEXT-SAVING ROUTINES** When working with interrupts, the standard interrupt handler saves the w, STATUS, PCLATH, and any other context registers that are used by both an interrupt handler and a mainline application. The FSR register often fits into this definition and should be saved along with the other three registers.

The standard mid-range PICmicro® MCU context saving and restoring code is:

```
Int
  movwf  _w                  ;  Save "w" Contents
  movf   STATUS, w
  bcf    STATUS, RPO         ;  Save "STATUS" in Bank 0
  bcf    STATUS, RPO
  movwf  _status
  movf   PCLATH, w
  movwf  _pclath
  clrf   PCLATH              ;  Make Sure Execution out of Page 0

   :

  movf   _pclath, w
  movwf  PCLATH
  movf   _status, w
  movwf  STATUS
  swapf  _w, f
  swapf  _w, w
  retfie
```

The first time you see this code, it will seem somewhat strange and difficult to understand what is happening.

The first line, *movwf w,* seems straightforward enough as saving the contents of w into a temporary register. Beware that the bank bits in the STATUS register (*RPO* and *RP1*) can be any value. To ensure that the contents of w can be saved safely, the variable *_w* should either be placed at the same address in each page or in a common register that is shadowed across all pages. Personally, I prefer the latter method because it means that only one declaration is registered for the variable, whereas the other method requires one declaration per page.

Next, the STATUS register is placed in w and the STATUS register's *RP* bits are loaded with the bank where the context-saving registers are. In the example code, I used bank 0, but any bank can be used. The important point to remember is that the RP bits are set after the STATUS register's contents are saved in w.

Finally, the PCLATH register is saved and then reset to zero (the interrupt handler starts executing in page 0 of the PICmicro® MCU). These three lines can be avoided if the PICmicro® MCU you are using has less than one page (2048 instructions for the mid-range) of PICmicro® MCU and the interrupt handler doesn't change PCLATH.

Other context registers (such as the FSR register) can be saved using the process of loading them in w and saving them in a variable.

With the context registers stored, the interrupt handler can now load them with any values required as it responds to the interrupt request. Before completing, the requesting F flag will have to be reset.

The process of restoring the registers is the reverse of saving them, except no instructions are used to set the registers to specific values.

The only surprising aspect of the context registration restore is the two *swapf* instructions before *retfie* instruction. These instructions address the issue of how to load w without changing the STATUS register's Zero flag. If you look on to the instruction definitions, you'll see that the *swapf* instruction does not change any STATUS register flags.

Using the *movf* instruction will modify the Zero Flag, but by first swapping the two nybbles in place and then swapping them again as the value is loaded into w. This operation will load w with the correct mainline value without changing the STATUS register's Zero flag.

This method is quite clever and efficient and avoids having to do something like:

```
movf   _status, w
movwf  STATUS                 ;  Restore the Status Register
movlw  0
btfss  STATUS, Z
 movlw 1
movwf  _status                ;  Save Value According to Zero
movf   _w, w
movf   _status, f             ;  Set Zero According to the Original
retfie                        ;   Value
```

This code, which was the method I came up with originally to return from interrupts seven years ago, uses the *_status* variable and makes it zero or non-zero, based on the Zero flag's original state. Once w is restored, the new *_status* value is compared to zero. This method requires twice the number of instructions of the method using the two *swap f* instructions to restore w from *_w* without changing the Zero flag.

## NO CONTEXT-SAVING INTERRUPT HANDLERS

Sometimes in the PICmicro® MCU, the interrupt handler just resets an interrupt active (F) flag along with a program flag to indicate that the interrupt happened. Another case would be if the interrupt handler, after resetting the F flag incremented or decremented a counter. In these cases, there is no reason to save the context information, which reduces the number of active variables and allows the interrupt handler to execute in as few as seven instruction cycles.

Looking through the mid-range PICmicro® MCU instruction set, eight instructions are appropriate to be used in this type of interrupt handler because they don't change the w or STATUS registers (Table 9-8).

I realize that *goto* and *call* do not change either w or STATUS, but they do use PCLATH, which could be incorrect, depending on whether the application is executing outside of the first page and for this reason they are not included in Table 9-8.

**TABLE 9-8 PICmicro® MCU Instructions That do not Affect "w" or STATUS**

| VARIABLE | FUNCTION |
|---|---|
| bcf | |
| bsf | |
| btfsc | |
| btfss | |
| decfsz | |
| incfsz | |
| swapf | |

Using these instructions, an interrupt handler that sets a flag when TMR0 overflows could be:

```
int
  bcf     INTCON, T0IF
  bsf     T0Flag
  retfie
```

and an interrupt handler that increments a counter on TMR0's overflow could be:

```
int
  bcf     INTCON, T0IF
  incfsz  TMR0Count, f
   nop
  retfie
```

The two interrupt handlers can be combined to set a flag when TMR0 has overflowed a set number of times:

```
int
  bcf     INTCON, T0IF
  bsf     T0Flag
  decfsz  TMR0Count, f
   bcf    T0Flag
  retfie
```

In the last example, nine to 10 instruction cycles are required for the interrupt handler.

In these examples, it is assumed that the register bank is always zero. If you cannot guarantee that the register bank is zero, place the flags in registers that are common across the different banks in the PICmicro® MCU processor.

One style of writing programs is to execute the application functions from entirely within interrupt handlers. After initializing hardware, the mainline just executes an endless loop (*goto $*). When an interrupt occurs, the handler not only handles the interrupt request, but also provides all the responses to the interrupt. In this case, saving the context registers is not required because the mainline code is only executed intermittently and does not perform any logical functions.

I discourage this style of application coding because it has the very definite possibility that interrupts will be missed if an interrupt request is received while another is being processed. This is especially true if multiple interrupt sources are used with an application.

## REENTRANT SUBROUTINES

In many applications, subroutines are required to be used both by interrupt handlers and the mainline code. In these cases, the subroutines can be interrupted and called again from the interrupt handler. Subroutines that can support being called multiple times, from the mainline and interrupt handlers are known as *reentrant*.

For most processors, sharing a subroutine between the mainline and interrupt handler code can be carried out quite safely and efficiently. In the PICmicro® MCU, there can be problems with doing this. I recommend that you duplicate the subroutine and make the two copies specified to the interrupt handler and the mainline.

The reason for this recommendation is because of the PICmicro® MCU's lack of a data stack, which can be used to store "temporary" or "local" variables. Some high-level programming languages provide this capability (and you can provide it as well in assembly-language programming), but doing this will require more complex subroutines and will take *FSR* away from the mainline use in the code. Often, a PICmicro® MCU subroutine that supports interrupt handler and mainline access will be larger than two copies of the routine.

If "nested" interrupts are also allowed, care should be taken to avoid calling subroutines that may have been called and were executing by the previous handlers when the nested interrupt took place. Not properly handling the subroutines can result in data variables that are overwritten or previous interrupt's data being lost.

To avoid these potential problems, simply do not call subroutines from your interrupt-handler code and avoid allowing nested interrupts from executing.

## SIMULATING LOGIC

One of Microchip's early application notes was on how to simulate logic functions using a PICmicro® MCU. This was an interesting example of how logic functions could be processed by the PICmicro® MCU in a very slow-speed application. Included in the application note were details on how the PICmicro® MCU could be made to simulate logic functions.

For example, simulating the eight-bit address compare circuit shown in Fig. 9-2 could be implemented using the hardware shown in Fig. 9-3.

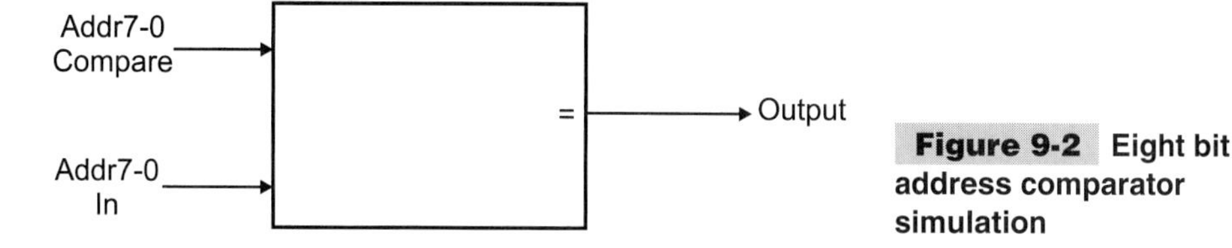


**Figure 9-2** Eight bit address comparator simulation

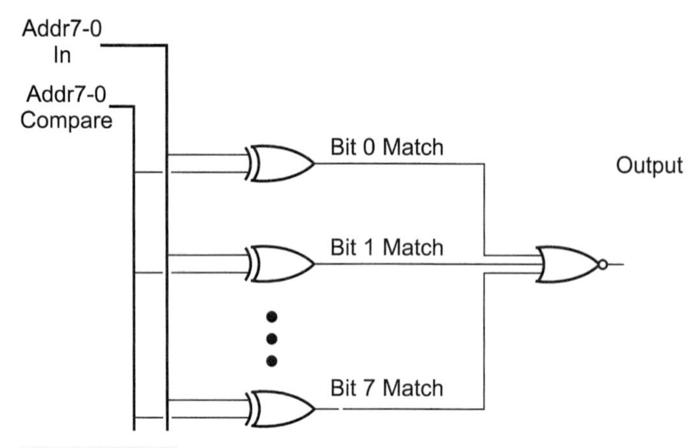


**Figure 9-3** Eight bit address comparator circuit

This circuit could be simulated within a PICmicro® MCU using the model shown in Fig. 9-4, where the comparison to the *set addr,* which is stored in the PICmicro® MCU already, would be:

```
Loop
  movf    PORTB, w          ;  Get Current
  xorwf   SetAddr,w
  btfsc   STATUS, Z         ;  Zero if Match
   goto   matchHI
  nop
  bcf     RA0               ;  No Match
  goto    Loop
matchHI
  bsf     RA0
  goto    Loop
```

This sequence of instructions takes eight instruction cycles under a match or mismatch. Herein lies the problem with using the PICmicro® MCU for simulating logic; this sequence is relatively very slow. For this example, if the PICmicro® MCU was running at 20 MHz, the worst-case switching time you could expect is 1.6 $\mu$s. The best case is 1.2 $\mu$s (if the change is just before the *movf* instruction).

Comparatively, most CMOS and TTL logic would execute this in much less than 100 ns, more than one full order of magnitude speed increase (with two or three orders of magnitude increase being more likely). This improvement becomes even more profound

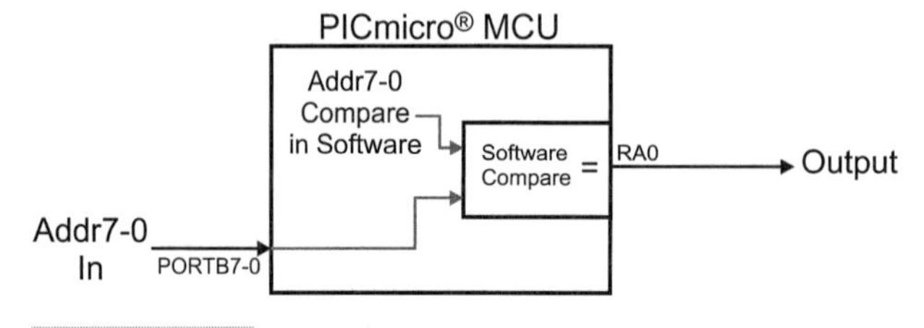


**Figure 9-4** Eight bit address comparator simulation

when more complex operations are required and the PICmicro® MCU code becomes lengthier.

Even a simple case, such as devoting a PICmicro® MCU to ANDing two inputs together, is very slow. For example, if you had two inputs and wanted to implement an AND output, you could use the code:

```
Loop
  btfsc    Bit1
   btfss   Bit2
    goto   reset                 ;  If Bit1 or Bit2 is Low, Turn Off Output
  nop
  bsf      Output
  goto     Loop
reset
  bcf      Output
  goto     Loop
```

This requires seven instruction cycles.

You might want to simplify the code by assuming that the output is high before the comparison:

```
Loop
  bsf      Output
  btfsc    Bit1
   btfss   Bit2
    bcf    Output                ;  If Either Input is Low, Output is Low
  goto     Loop
```

which reduces the number of cycles in the loop by 1, but will either have a constant *1* output or an alternating *1* and *0* (three cycles each) if either one of the inputs is low (Fig. 9-5).

Even at six cycles per bit, this cannot really compete against any kind of logic in terms of speed and the PICmicro® MCU will always be more expensive in terms of cost. In the final analysis, I never recommend using a PICmicro® MCU for a logic replacement. Instead, TTL/CMOS chips should be used.

This is not to say that the PICmicro® MCU cannot be used for implementing simple logic functions in software that will avoid the need for external gates. Logic functions can be implemented in software and let you avoid adding logic externally to the PICmicro® MCU. Providing the logic functions within the PICmicro® MCU will reduce the number of I/O pins required for the PICmicro® MCU and will reduce the cost of the application.

I realize that in some cases, you will consider a larger PICmicro® MCU because one gate is required, and power, space, and cost requirements do not allow an additional chip to be put into the application. In these cases, I recommend that you look at the logic analogs provided in "Introduction to Electronics," provided on the CD-ROM, if you are in a situation where you don't want to add a logic chip to your application for just one gate.

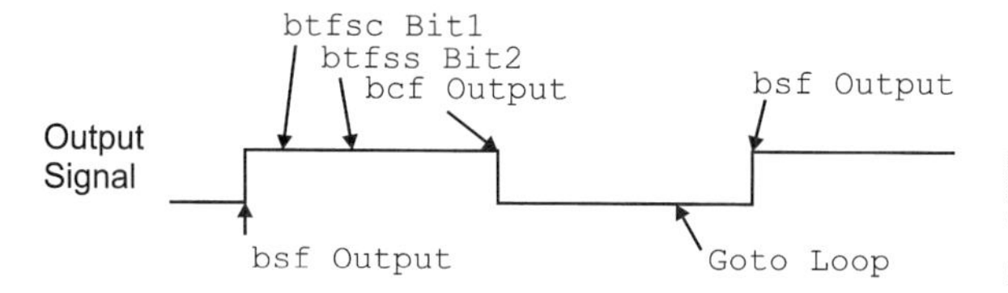


**Figure 9-5** Two input bit "AND" function waveform to software analysis

# Event-Driven Programming

The PICmicro® MCU is particularly well suited to event-driven programming applications because of how its interrupt-handler code works. Many of the experiments and applications presented in this book use event-driven programming to respond to numerous inputs without affecting the processing of other inputs or tasks within the application.

As presented in "Introduction to Programming" on the CD-ROM, the typical high-level format, for event-driven programming application is:

```
main()                  //  Event Driven Updated Initial Application
{

int  i = 0;

  TimerDelay = 1sec;

  interrupts = TimerHandler | ButtonHandler;

  if (Button == Up)   //  Load in the Initial Button State
    LED = Off;
  else
    LED = On;

  while(1 == 1);

}  // end main

interrupt TimerHandler()//  Display "i" and Increment

  TimerInterrupt = Reset;

  output( i &  0x00F);

  i = i + 1;

}  //  End TimerHandler

interrupt ButtonUp( )

{

  ButtonInterrupt = Reset;

  LED = Off;

}  //  end ButtonUp

interrupt ButtonDown( )

{

  button interrupt = reset

  LED = on;

}  //  end ButtonDown
```

This style of programming can be easily copied into the PICmicro® MCU by testing the *F (Interrupt Request Flag)* bits and responding to the first one that is set. The interrupt handler for this application would look like this:

```
 Org          4
Int
  movwf   _w
  movf    STATUS, w
  movwf   _status

  btfsc   INTCON, TOIF
   goto   INTTMRO
  btfsc   INTCON, INTF
   goto   BUTTONINT

INTERRUPT REQUEST               ;  clear all other Interrupt Requests
  clrf    INTCON
  goto    INTEND

TMROINT                         ;  Respond to the TMRO Interrupt

  bcf     INTCON, TOIF

  movf    i, w                  ;  Output (i and 0x00F)
  xorlw   0x00F
  movwf   OUTP

  incf    i, f

  goto    INTEND

BUTTON INT                      ;  Respond to the Button Pressed

  bcf     INTCON, INTF
  btfss   BUTTON, UP
   goto   BUTTONDOWN

BUTTONUP                        ;  Button Released, LED Off

  bsf      STATUS, RPO          ;  Change Button Interrupt Request Direction
  bcf      OPTION_REG & 0x080, INTEDG
  bcf      STATUS, RPO
  bsf      LED                  ;  LED Off

  goto     INTEND

BUTTON DOWN                     ;  Button Pressed, LED On

  bsf      STATUS, RPO          ;  Change Button Interrupt Request Direction
  bcf      OPTION_REG & 0x080, INTEDG
  bcf      STATUS, RPO

  bcf      LED                  ; LED On

  goto     INTEND

INTEND                          ;  Interrupt Handler Finished - Return to

                                ;   Mainline (the "while (1 == 1)" Statement)
```

```
movf     _status, w
movwf    STATUS
swapf    _w, f
swapf    _w, w
retfie
```

This code should be a straightforward conversion of the high-level code, except for the button. Before jumping to button up or button down, the polarity has to be determined. When the polarity is determined, the interrupt edge bit is set to interrupt when the button input changes state. This state determination is used to help differentiate which part of the event handler should execute.

Notice that the event to be responded to can be given a priority with the first check being the interrupt source responded before any of the others. In the example code, TMRO is responded to first—even if a button interrupt has been received. The order (and priority) of event checks and responses can be selected in such a way as to ensure that no events are missed and the most important ones are responded to first.

# State Machine Programming

Two different methods are used to implement a state machine in PICmicro® MCU assembly language. The first method is to use a table like:

```
movf    State, w            ;  Jump to the Appropriate State Table Entry
addwf   PCL, f
goto    State0              ;  Routine for State == 0
goto    State1              ;  Routine for State == 1
goto    State2              ;  Routine for State == 2
  :
```

This method is quite efficient for implementing a state machine. As with implementing any table, execution takes a constant amount of time, regardless of the value of *State*. The only requirement for this method is to have the state values as linear numbers that start with zero.

The second method is to repeatedly test the *State* variable for specific values and then execute the appropriate code. The typical PICmicro® MCU code for this is:

```
  movf    State, w            ;  Load the "State" Variable
  xorlw   0                   ;  Test for State == 0
  btfss    STATUS, Z
   goto    NotState0
    :                         ;  Routine for State == 0
  goto     StateEnd
NotState0
  xorlw   0 ^ 1               ;  Test for State == 1
  btfss    STATUS, Z
   goto    NotState1
    :                         ;  Routine for State == 1
  goto     StateEnd
```

```
NotState1
  xorlw   1 ^ 2               ;  Test for State == 2
  btfss    STATUS, Z
   goto    NotState2
    :                         ;  Routine for State == 2
  goto     StateEnd
NotState2
    :
StateEnd                      ;  Finished with State Machine Execution
```

In this code, *State* never has to be reloaded because the previous value XORed with the contents of STATE is XORed again in the next instruction when it is tested for a different value. The advantages of this method are that nonsequential values of *State* can be implemented. The *State* variable can also be done away with if this method is used and w can be used for the new state value. The obvious drawbacks to this method over the previous one is that many more statements are required and the time to execute statements is different for each state's routine.

I use this code to implement character data processing to state machines when specific values are required. This method is superior to the previous method when a potentially large number of inputs could be received and only a few will be processed.

Either method can be useful when implementing complex applications as state machines in the low-end PICmicro® MCUs, where single states have multiple sources and you do not want to use the limited stack in a subroutine.

# Porting Code Between PICmicro® MCU Device Architectures

I generally design my code in terms of the mid-range PICmicro® MCU. If the application is destined for another PICmicro® MCU family, I then change the methods used for the code by the rules presented in the next two sections. When it comes right down to it, the changes are quite subtle and application code can be created that can be ported between PICmicro® MCU architectures very easily.

## PORTING MID-RANGE APPLICATIONS TO THE LOW-END

It was probably surprising for you to see that I consider the low-end PICmicro® MCU architecture to be a subset of the mid-range PICmicro® MCUs. All of the low-end PICmicro® MCU's instructions are available in the mid-range and work exactly the same way. The lack of some instructions requires some work-around, but I find that by writing mid-range code with the low-end in mind, the code can be ported remarkably easily with few changes.

There are seven primary differences between the mid-range and low-end PICmicro® MCU instruction sets. Some of these differences are a result of the 12-bit instruction word of the low-end devices and others are caused by the low-end's architecture.

1 Smaller instruction page size
2 Four fewer instructions
3 Subroutine *call* instruction differences.
4 Reduced program counter stack
5 Subroutine *return* instruction differences
6 No interrupt capability
7 No register bank select bits

The low-end PICmicro® MCU's page size is 512 instructions and it requires nine address bits. This will seem like a lot smaller than the mid-range, which has a page size of 2048 instructions (11 address bits) and will seem to be difficult to create applications for. With this restriction, along with others in this section, I tend to look at the low-end devices being better suited for applications that require no more than 512 instructions with limited text/data I/O.

Five hundred and twelve instructions might not seem like a lot. To put this in perspective, I always thought the 512 bytes used for booting in the PC was frightfully restrictive. When the PC boots, a single 512-byte sector is loaded into memory from disk, execution jumps to it, and this code loads and starts up an operating system in the PC (such as Windows 2000 or Linux).

In *PC Ph.D.*, I finally wrote my own code for running operating systemless, booting from a diskette in a PC, and found that I only required 107 bytes. To get some perspective on this, note that each 8086 instruction averages about three bytes, you can see that a lot can be done in the 512 bytes of a low-end PICmicro® MCU page.

The mid-range instructions that are not available in the low-end are:

The lack of the two return instructions are addressed in the section on the subroutine *call* and *return* instruction differences.

The lack of *addlw* and *sublw* can make some applications awkward where these instructions are made on immediate values stored in w. To counter the lack of the *addlw* and *sublw* instructions, I keep all values being processed in a temporary variable, rather than in w. For example, instead of:

```
movf    A, w
andlw   0x03F
addlw   3
movf    PORTB
```

I would write the code function as:

```
movlw   0x03f
andlw   A, w
movwf   temp
movlw   3
addwf   temp, w
movwf   PORTB
```

which can execute on both the mid-range and low-end PICmicro® MCU processors.

The *temp* register contains the intermediate value of the calculation. The intermediate value is saved from the source and retrieved for processing and is saved in the destination. This method does require one additional instruction and a *temp* variable, but it can be used by both the mid-range and low-end PICmicro® MCUs without modification.

Subroutine calls can be quite awkward in the low-end because the label must be located in the first 256 bytes of the PICmicro® MCU's instruction page. This is because of how the *goto* and *call* instructions are defined, which is a result of how bits are allocated in the low-end PICmicro® MCUs.

If you look at the instruction bit patterns for the two instructions:

```
goto   0b0 101k kkkk kkkk
call   0b0 1001 kkkk kkkk
```

You'll see that *goto* has nine bits available to the jump to address, and *call* only has eight. The nine address bits in the *goto* instruction means that the *goto* address can be anywhere in the 512 instructions of a page.

The eight bits of the *call* instruction means that only 256 addresses in the instruction page can be accessed. In the low-end PICmicro® MCU, these are defined as the first 256 instructions of a page. This restriction can be a problem in applications if it is not planned for.

To counter this problem, most people make up a list of gotos, each one pointing to a subroutine, in the first 256 instructions of a bank. This looks like:

```
sub1
  goto    subroutine1
sub2
  goto    subroutine2
```

This allows the subroutines to be placed anywhere in the instruction page without worrying whether or not they are in the first 256 instruction addresses of the page.

For table operations, the:

```
addwf PCL, f
```

and

```
movwf PCL
```

have a similar restriction to *call* and can also only be located in the first 256 instructions of a page. This is caused by the inability of accessing bit eight of the address. *PA0* in the STATUS register affects address bit nine and *PCL* affects address bits zero through seven. Bit eight of the address cannot be specified and the PICmicro® MCU defaults this address bit to zero. This restriction means that no tables can be longer than 256 instructions in the low-end PICmicro® MCUs.

You should also be aware of two other restrictions for the low-end devices. The first is the availability of only two positions on the program counter stack. Thus, there can only be one nesting level in low-end applications. Subroutines that have been called by other subroutines cannot call subroutines themselves. This is not a particularly onerous restriction because the number of nesting levels in all PICmicro® MCUs is quite limited. Even in the mid-range and PIC17Cxx devices (which have eight and 16, respectively, position stacks), I don't recommend using many nested subroutines because of the possibility of interrupts causing program counter to overflow the stack and return lost addresses.

The second restriction is the availability of only a *retlw* (return from subroutine after loading w with a constant) instruction for returning from subroutines. No *return* statement that leaves w unchanged is available in the low-end PICmicro® MCUs.

Some versions of MPASM, when it encounters a *return* instruction in low-end code, will convert the instruction into:

```
retlw 0        ; Return with "0" in w.
```

This caused me several hours of confusion one night trying to figure out why code that ran fine in a mid-range PICmicro® MCU wouldn't work in a low-end chip. To avoid this problem, you might wish to always return values in a subroutine in a common variable, rather than in w.

No interrupts in the low-end PICmicro® MCUs change the way that inputs and hardware events are monitored. Instead of waiting for an event, your code will have to poll for the event. If a signal waited for is very short, you might miss it. Appendix H includes a small snippet of code that will set up TMR0 of the PICmicro® MCU to overflow when a signal is received. This will allow you to use the TMR0 hardware in the PICmicro® MCU to monitor short-duration pulses.

The last difference, the lack of bank-select bits is the biggest issue when porting the mid-range applications to the low end. The most obvious problem will be the need to convert all of the bank switch writes to OPTION and the TRIS registers to the single-instruction equivalents. But be aware of a larger problem.

In the low-end devices, a register bank uses only five bits for addressing (in the high-end, seven address bits for each address is used). This only gives a maximum of 32 registers per bank. There can be up to four banks, each with a shared INDF, TMRO, PCL, STATUS, FSR, PORTA, PORTB, PORTC, and file registers. The top-16 registers are usually left as being unique to each bank. This layout leaves a maximum of 25 bytes for an array, with 16 being the maximum in any bank other than bank 0.

The low-end register organization looks like that shown in Fig. 9-6.

| Bank 0 | Bank 1 | Bank 2 | Bank 3 | |
|---|---|---|---|---|
| Addr - Reg | Addr - Reg | Addr - Reg | Addr - Reg | |
| 00 - INDF<br>01 - TMRO<br>02 - PCL<br>03 - STATUS<br>04 - FSR<br>05 - PORTA*<br>06 - PORTB<br>07 - PORTC<br>08-0F Shared File Regs | 20 - INDF<br>21 - TMRO<br>22 - PCL<br>23 - STATUS<br>24 - FSR<br>25 - PORTA*<br>26 - PORTB<br>27 - PORTC<br>28-2F Shared File Regs | 40 - INDF<br>41 - TMRO<br>42 - PCL<br>43 - STATUS<br>44 - FSR<br>45 - PORTA*<br>46 - PORTB<br>47 - PORTC<br>28-2F Shared File Regs | 60 - INDF<br>61 - TMRO<br>62 - PCL<br>63 - STATUS<br>64 - FSR<br>65 - PORTA*<br>66 - PORTB<br>67 - PORTC<br>68-8F Shared File Regs | Shared Registers |
| 10-1F Bank 0 File Regs | 30-3F Bank 1 File Regs | 50-4F Bank 2 File Regs | 70-7F Bank 3 File Regs | Bank Unique Registers |

* - "OSCCAL" may take place of "PORTA" in PICmicro® MCUs with Internal Oscillators

OPTION - Accessed via "option" Instruction

TRIS# - Accessed via "TRIS PORT#" Instruction

**Figure 9-6** **Low-end PICmicro® MCU register map**

In the low-end PICmicro® MCU devices, even though a total of seven address bits are used when all four banks are taken into account, only a maximum of 73 unique addresses are implemented, with 48 of them accessible only by the FSR. In most mid-range devices, the registers in each bank are unique to that bank, although there are "shadowed" registers to allow data movement between banks. In the low end, the first 16 registers of each bank are shared (or shadowed) between the maximum four banks.

Another issue to be aware of with the low end and its restricted register banking operation is that the FSR index register can never be equal to zero in the low-end PICmicro® MCU. Because of this, be sure that you never write any application code like:

```
  movlw   10
  movwf   FSR
Loop
  decfsz  FSR, f
   goto   Loop
```

This loop will never end because FSR can never equal zero.

The FSR can be a useful temporary register in the mid-range PICmicro® MCU because you don't have to define it and if indexed addressing is not used in the application, it really becomes a free file register for your use. Because the FSR can never be zero in the low end, I recommend that it never be used, except in its primary function as an array index register.

## PORTING TO THE PIC17Cxx AND PIC18Cxx

When I was writing this book, this section was one of the last that I wrote because I felt like there were a lot of issues to be aware of when porting applications to the PIC17Cxx and PIC18Cxx architectures from the low-end and mid-range PICmicro® MCUs. After planning to write this section, I have discovered that there are not that many issues with regards to porting between all of the PICmicro® MCU architectures. In fact, using the rules outlined in the previous section will help you to write code that can be passed back and forth between any of the different PICmicro® MCU devices quite simply and with very little device-specific modification.

The significant differences of the PIC17Cxx and the PIC18Cxx devices to the lower-end architectures are:

1. Different PCLATH and *goto/call* instruction operation.
2. Sixteen-bit instruction words with the ability to address them directly.
3. Compare and branch instructions.
4. Different register bank organization.
5. An eight-bit hardware multiplier.
6. Additional arithmetic and bitwise operation instructions.

Other than these differences, application code written for the low-end and mid-range PICmicro® MCUs should execute without any problems in the higher-end devices.

When you develop your first applications for the PIC17Cxx and PIC18Cxx devices, I recommend that you do not access the *PCLATH* (and *PCLATU* in the PIC18Cxx) registers for your first applications. This should not be a significant hardship for you because the page size is quite large (8,192 instructions in the PIC17Cxx and the PIC18Cxx can jump to any address in the PICmicro® MCU).

When you do have to modify the PCLATH register for a jump, then load it with the high byte of the destination address before the *goto* or *call* instruction (the same as would be done for a computed *goto*):

```
movlw   HIGH Destination
movwf   PCLATH
goto    Destination & 0x01FF
```

By using this code always, the destination will always be correct. This code could also be used in the mid-range PICmicro® MCUs, but not the low end. For the PIC18Cxx, I would recommend just using the long *goto* instruction and avoid modifying the PCLATU register all together.

The question will probably come up regarding computed *gotos* and table jumps. The identical code, as used in the mid range, could be used for the PIC17Cxx. As I have shown elsewhere in the book, different code is required for the PIC18Cxx Table jumps.

```
  addlw    LOW Table            ;  Calculate the Actual Address
  movwf    TableOff             ;  Save the Computed Table Offset
  movlw    HIGH Table           ;  Setup PCLATH
  btfsc    STATUS, C            ;  Increment Past the Current?
   addlw   1
  movwf    PCLATH
  movf     TableOff             ;  Get the Computed Table Offset
  movwf    PCL                  ;  make the Jump
Table                           ;  Table Code Follows. . .
```

Notice that in the table jump code example, the operation of the *movf TableOff* instruction would be *movfp TableOff, WREG* for the PIC17Cxx.

If the purpose of the computed *goto* is to return a byte value (using *retlw*), then I would suggest taking advantage of the 16-bit instruction word, store two bytes in an instruction word, and use the table read instructions to read back two values. This is somewhat more efficient in terms of coding and requires approximately the same number of instructions and instruction cycles.

The addition of the compare a register to w and skip the next instruction on a condition is a unique capability to the PIC17Cxx and PIC18Cxx. If these instructions are going to be used, then I recommend that their functions be implemented in the low-end and mid-range PICmicro® MCUs using macros.

For example, the *cpfseq* instruction, which skips the next instruction if the contents of the register are equal to *w*, could be implemented in the low-end and mid-range PICmicro® MCUs as the macro:

```
cpfseq macro Register
  subwf    Register, w                 ;  Subtract "w" from Register
  btfss    STATUS, Z
 endm
```

Compare and skip if less than *(cpfslt)* can also be implemented easily with the macro code:

```
cpfslt macro Register
  subwf    Register, w                 ;  Subtract "w" from Register
  btfsc    STATUS, C
 endm
```

Compare and skip if greater is a bit more complex and requires the use of a double-bit skip on condition as the carry flag will be set if the contents of Register are equal to or greater than the contents of w:

```
cpfsgt macro Register
  subwf   Register, w                 ;  Subtract "w" from Register
  btfss   STATUS, Z                   ;  Don't Skip if Zero Set
   btfss  STATUS, C                   ;  Skip if Zero Reset and Carry Set
 endm
```

Notice that these instructions will change the contents of w and the STATUS register flags. For most applications, this should not be an issue. But if previous flag values are used in the application, when porting to the low-end and mid-range PICmicro® MCUs, you might want to save the STATUS registers before executing the macros.

The PIC18Cxx's branch on condition is somewhat unusual because it uses a relative offset, rather than a page address, as does all other PICmicro® MCU execution change instructions. This should not be an issue when creating macros for porting the function to the low-end and mid-range PICmicro® MCUs, unless a page boundary is going to be crossed. For this reason, I recommend being sure that the appropriate *PAx* bits in the low end and the appropriate PCLATH bits in the mid range are changed before the instruction.

The PIC18Cxx *BC* (branch if carry) instruction can be simulated in the low-end PICmicro® MCUs using the macro:

```
bc macro Address
 local EndAddress
 if ((Address & 0x0400) != 0)        ;  Set up Page Bits for Jump
  bsf   STATUS, PA1
 else
  bcf   STATUS, PA1
 endif
 if ((Address & 0x0200) != 0)
  bsf   STATUS, PA0
 else
  bcf   STATUS, PA0
 endif
  btfsc  STATUS, C
   goto  Address & 0x01FF            ;  If Carry Set, Jump
 if ((EndAddress & 0x0400) != 0)     ;  Restore the Page Bits
  bsf   STATUS, PA1
 else
  bcf   STATUS, PA1
 endif
if ((EndAddress & 0x0200) != 0)
  bsf   STATUS, PA0
 else
  bcf   STATUS, PA0
 endif
EndAddress                           ;  End Address for the macro
 endm
```

The use of the *EndAddress* is to ensure that there is no opportunity for the situation where the page boundary occurs within the macro to cause a problem. The mid-range device's *bc* macro is similar, except that the PCLATH bits are changed and then reset. It is ironic that these macros, which simulate instructions in the PIC18Cxx, actually have a greater address range capability in the low-end and mid-range devices. The simulated macros can jump anywhere in program memory, whereas the actual PIC18Cxx instructions can only jump -128 to +127 addresses from the next instruction.

Along with the *bc* PIC18Cxx instruction, the *bnc, bnz, bra,* and *bz* instructions can all be implemented in the low-end and mid-range PICmicro® MCUs. The *bov, bnov, bn,* and *bnn* instructions cannot be simulated using a macro because the OV and N flags are not implemented in the low-end and mid-range PICmicro® MCUs.

Porting variable access code, despite the different file register bank organization, is not really an issue for either the PIC17Cxx and PIC18Cxxs. For analogous translation of file-register access application code, the BSR register can be updated with new addresses, but when I move applications between architectures, I simply place all my variables starting at address 0x020 and access them directly, rather than keep track of things with BSR.

The PIC17Cxx and PIC18Cxx hardware I/O register organization is quite a bit different than how the hardware I/O registers are implemented in the low end and mid range. When porting applications, changes will have to be made to the source code, according to the destination device. This is actually the most significant amount of code development that will have to be done for porting applications.

The eight-bit hardware multiplier is an obvious omission when going from the PIC17Cxx and PIC18Cxx devices to the low-end and mid-range PICmicro® MCUs. This hardware multiplier capability will have to be simulated in software using the code presented elsewhere in this book.

A number of enhanced instructions available to the PIC17Cxx and PIC18Cxx PICmicro® MCUs can cause some problems when porting devices. If you are expecting to port applications between the low-end and mid-range PICmicro® MCUs, I would suggest that working with the low-end PICmicro® MCU's arithmetic instructions (*addwf* and *subwf*) be the limit of the instructions used. The other instructions, although they will really enhance your application code, will be very difficult to simulate in the lower-end PICmicro® MCU architectures.

Limiting the use of some of the most useful features in the PIC18Cxx (and, to a lesser extent, the PIC17Cxx) really begs the question of whether or not porting the application to a lower-end device is appropriate. The PIC18Cxx has many new instructions and capabilities that are not available in the low-end and mid-range devices. Over time, I expect to see a great number of new PIC18Cxx part numbers that can replace mid-range PICmicro® MCU devices.

When this becomes the case, the reason for considering porting applications becomes much less important. For this reason, I consider this section to be more of a guide to port snippets of code between PICmicro® MCU architectures, rather than full applications. Porting full applications will require a fair amount of work and take quite a bit of time ensuring that all of the necessary changes between the architectures is done correctly. The application will work with all of the differences between the source and destination PICmicro® MCUs properly compensated for.

# Optimizing PICmicro® MCU Applications

As you start developing your own PICmicro® MCU applications, you are going to discover that the device you choose does not have enough program memory, file registers, or I/O pins. One solution might be to try another device that has more of these resources or built-in hardware. The problem with this solution is that it usually costs more and some of the basic problems with the application are not addressed.

Instead of jumping to a new device, you can work through some strategies to allow you to use the device that you have already chosen. This optimization of the application is not terribly hard to do. I find it fun to see how much more function I can cram into a PICmicro® MCU. This section covers some strategies that you can use.

One of the most frustrating things you can experience in an application is running out of file registers. No PICmicro® MCU has a lot to begin with and you will probably find that you will run out of them when you plan a complex application in small (low end) devices. You can look at a few things to try and help alleviate the need for file registers.

The first and most obvious action to take is to look for file registers, which are used as flags. Flags should be implemented as individual bits. This can reduce the requirements from eight registers down to one. One of the reasons why people use a file register for a flag is to avoid having to remember the bit number for a specific flag. This concern can be eliminated by using the *#DEFINE* directive to specify individual bits.

For example, eight bits of a FLAGS register could be declared like:

```
#DEFINE RUNFLAG FLAGS,0
#DEFINE STOPFLAG FLAGS,1
#DEFINE REQUESTFLAG FLAGS,2
```

and to set or reset a flag, the *bcf* or *bsf* instructions are used like:

```
bsf RUNFLAG
```

The *#DEFINE* actually puts two parameters to the label. When *RUNFLAG* and the other *#DEFINE* labels are encountered, the instruction is loaded with the register and the bit number instead of having to put in both parameters in the instruction.

This optimization is interesting because it reduces register requirements and the number of instructions required to test the state of a flag to one from the multiple instructions used in other processors.

For example, to jump if *STOPFLAG* is set, the instructions:

```
movf    FLAGS, w     ;  Clear everything but the stop bit
andlw   1 << 1
btfss   STATUS, Z    ;  Jump to label if the bit is set
 goto   LABEL
```

could be used (and would be typically used for other processors). But in the PICmicro® MCU, the code is just:

```
btfsc   STOPFLAG
 goto   Label
```

which is much easier to read and understand than the previous code.

Some people will try to use hardware registers for temporary storage of data. I would like to dismiss this as much as possible; writing random values to a register can result in unexpected and unwanted hardware operations. The obvious exception to this is using the FSR for saving data. This is not to say that shadow values should be kept off the port I/O registers; they can use the actual registers themselves.

To reduce code space, you can look at a number of things that will not require large changes to the application. First off, check to see that all arrays are in Bank 1, rather than in Bank 0. The FSR register can access data in either bank. Code space will be saved if the single-byte variables are stored in the bank where the application mostly executes from.

If you have to switch banks in the application and access registers, use the multiple shared file-register addresses available in the PICmicro® MCU instead of w. This will allow the sharing of multiple bytes between the banks without having to toggle the *RP0* or *RP1* bits.

Spending some time thinking about how variables are placed in the application will result in huge code savings. Ideally, you should try to never change the bank register when accessing file registers.

Inefficient array and stack accesses can be particularly wasteful in terms of instructions. When implementing arrays, look for built-in features that can help you simplify the accesses. One trick I like to use is to put array elements starting at a specific offset, instead of relying on *CBLOCK* to allocate the address for me.

For example, if I were to put a 16-byte array at 0x040, to access an element, all I would have to do is load the index to the element and set bit six to create a correct address. This avoids the complications of adding the starting offset to the index, which might not seem like a major inconvenience, but it can add to the number of instructions required.

This trick becomes very useful when working with circular buffers. In this example, the 16-byte array is located in file register addresses 0x040 to 0x04F. To increment to the next address within it, the two instructions:

```
incf   ArrayIndex, f
bcf    ArrayIndex, 4
```

are required. These two instructions will increment the address and keep it within the correct range (which has bit four of its address always reset).

If an arbitrary starting and ending point for the circular buffer is used, then the following code would be required:

```
incf    ArrayIndex, w
xorlw   ArrayEnd +1
btfsc   STATUS, Z
 movlw  ArrayStart ^ (ArrayEnd +1)
xorlw   ArrayEnd +1
movwf   ArrayIndex
```

These six instructions test for the array pointer to be past the end of the array after incrementing it and reset to the start of the array, if required. This code also requires six instruction cycles and modifies w, which might require saving and loading the value in it (the two-instruction solution does not).

Jumping between pages can eat up a lot of instructions. To avoid this, try to keep functional blocks of code together in one page as much as possible. You might find that if you copy subroutines into multiple pages, the actual program memory requirements are decreased.

To optimize file register and program memory usage, try to come up with as many possible solutions as you can. This also includes looking at other people's code for examples. Chances are that you will be able to reduce the requirements above and beyond the immediate needs of the application.

Not having enough I/O pins can be a particularly troubling problem, but additional pins can be added to the circuit—either via a parallel or synchronous serial bus (Fig. 9-7).

The obvious drawbacks to the synchronous serial bus are the time required to shift bits and the bit ripple as the bits get shifted in or out. This bit ripple can be avoided by placing a latch between the shift register and the I/O pins. The latch output is updated after the data has been shifted from the PICmicro® MCU.

To shift data out, the following code could be used:

```
bcf     DataBit
btfsc   SourceRegister, 0
 bsf    DataBit
bsf     ClockBit
bcf     ClockBit
```

These six instructions for shifting out a bit can be reduced to four instructions by keeping a file register loaded with a shifted value and by placing the data bit in bit zero or bit seven of the I/O port. The code for a data pin at RBO, with a clock, also in port B (and rising edge active) to shift out a bit first has the port B value saved in shifted format and the clock bit low:

```
rrf     PORTB, w
andwf   0x0FF ^ (1 << (clock -1))
movwf   PORTBSave
```

When the least-significant bit of the data is to be shifted out, the four instructions:

```
rrf     SourceRegister, w
rlf     PORTBSave, w
movwf   PORTB
bsf     ClockBit
```

could be used. These four instructions can be put into a loop to shift out a byte:

```
  movlw    8
  movwf    Count
Loop:
  rrf      SourceRegister, f
  rlf      PORTBSave, w
  movwf    PORTBB
  bsf      ClockBit
  decfsz   Count, f
   goto    Loop
```

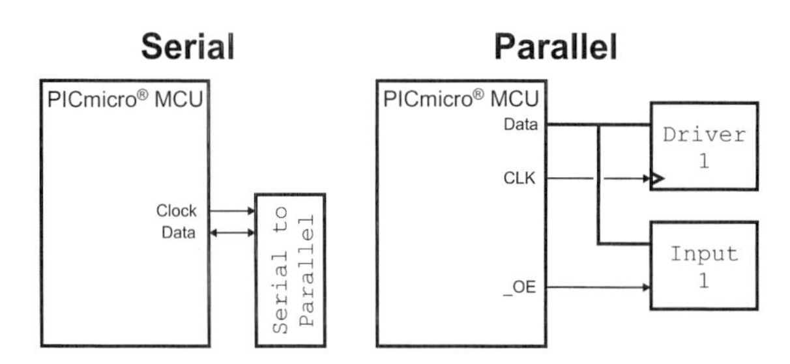


**Figure 9-7** Different types of simulated busses

This code requires 16 instruction cycles less than if the original six instructions were used in the loop:

```
  movlw     8
  movwf     Count
Loop:
  Bcf       DataBit
  btfsc     SourceRegister, 0
  bsf       DataBit
  bsf       ClockBit
  bcf       ClockBit
  rrf       SourceRegister, f
  decfsz    Count, f
   goto     Loop
```

The parallel port can be challenging to wire and can be difficult to wire a directional interface. Despite this, data can be passed quickly and does not have the ripple of the serial method.

Notice that both these methods require additional chips, which can drive the product's cost up to the point where a PICmicro® MCU with more pins is more economical.

The last issue for optimization is often the most difficult to overcome: meeting minimum timing specifications. Often the only solution to this problem is to use a PICmicro® MCU (or another microcontroller) with built-in hardware that provides the basic function of running the PICmicro® MCU at a faster speed.

This is not to say that the problem is insolvable; with a bit of work, you work through your code to find solutions to the problem. Remember to look at as many different solutions as possible and that rearranging the hardware can have some advantages. As was shown in the data-shifting example, a substantial improvement in instruction cycles resulted in a nine-cycle loop for a 22% speed improvement of the application function.

# A Baker's Dozen Rules to Help Avoid Application Software Problems

In PICmicro® MCU programming, I find that a number of rules always apply to prevent the opportunity for basic problems later. I follow these 13 rules when I develop a PICmicro® MCU application in assembly language:

1. *Always initialize your variables*. I have been caught more than a few times reading the contents of a variable before I have written to it. In the PICmicro® MCU, file registers are not initialized to any specific value. They can be any value from 0x000 to 0x0FF. When initializing variables, be sure to initialize them to zero, which matches the initial values that they are given by the MPLAB simulator to at least guarantee that they will work the same way in the application as they do in the simulator.
2. *Indent conditionally executing code after a* skip *instruction.* This is naturally taught for high-level languages, but it does have its place in PICmicro® MCU assembly language to make conditionally executing instruction stand out visually.

3. *Let the compiler/assembler do the calculations for you.* This will help make your code more portable to other applications and save you the hassle of working with a calculator, trying to figure out what the values should be (and potentially making a mistake).
4. *Use Microchip's register definition files without modification in all your applications.* I can't tell you how many times I've helped people with broken applications where the only problem was that they copied in a register or bit address incorrectly. Microchip has spent a lot of time developing the include files that are shipped with MPLAB and ensuring that they are correct. There's no reason for you to duplicate this effort.
   Don't change register/bit labels or use different ones. Even if you are more familiar with different terms, don't use them. By changing the labels to what you are comfortable with, you are making the code more difficult for others to follow, as well as increasing the opportunity for errors to be introduced into your application's source code.
5. *Keep your code as simple as possible.* When the PICmicro® MCU application is working, everybody will be impressed with how it operates, not the complexity or cleverness of your code.
6. *Develop your application in terms of functional blocks and interfaces.* Instead of creating one massive application, develop it as a series of steps. Each step should be simulated or tested on hardware before proceeding.
7. *Establish a plan to test and confirm that your code is correct.* Test your code each step of the way and do not move on to the next step until you are 100% satisfied with the performance of the code up to that point.
8. *Use CBLOCK or other built in tools to allocate variables instead of defining them manually.* Using a variable-define tool will avoid problems later when variables have to be added or deleted, at which time you will have to manually work through all the potentially affected addresses. This also goes for *goto* or *call* instruction destinations. Let the assembler generate the absolute addresses unless a specific address is absolutely required.
9. *Avoid changing the register bank unless it is absolutely necessary.* Ideally, an application should be designed so that all of the bank 1 registers and hardware are initialized after reset and then execute in bank 0 for the rest of the application. Going along with this, design your application so that single-byte/word variables are in bank 0 and array variables in bank 1 (which can be accessed by the FSR register without changing the *RP0* bit in the STATUS register). This will help to avoid having to keep straight what bank the application is currently running in.
10. *Don't allow code to go over page boundaries.* If your code is larger than 2048 instructions, place subroutines in the upper page. Code that is allowed to "drift" over page boundaries can have problems with the correct PCLATH register contents.
11. *Use the __CONFIG statement always in your source code and use a programmer that programs the configuration information automatically (not one that requires manual intervention).* When developing your application, keep the watchdog timer disabled and only enable it in the *__CONFIG* statement when all other functions have been debugged.
12. *Simulate as much of your application as possible.* The time used to develop a stimulus file and single step through it will be saved several times over in the time needed to debug the application if simulation has not been carried out.

**13** *Keep subroutine calling to a minimum.* All the PICmicro® MCU families have limited stacks and are not designed for recursive functions in applications. Always be sure that your maximum calling depth is less than two for the low-end, eight for the mid-range, and 16 for the PIC17Cxx, and 31 for the PIC18Cxx PICmicro® MCUs. Notice that interrupt handlers use the same stack. The maximum depth of their execution must be summed with the maximum depth of the mainline to avoid any potential problems in the application's execution.

# 10

# MACRO DEVELOPMENT

CONTENTS AT A GLANCE



Since writing the first edition, I have discovered that one of the biggest questions people have about developing PICmicro® MCU software applications is understanding how macros work and how they are implemented. This was surprising to me because I have always felt that macros are very simple to understand and use. They are a feature of most assemblers (including MPASM) and they can be used to make application development much easier.

A *macro* is a function that is invoked from within an application's code. This invocation is replaced by the code within the macro. Macros can be thought of as being similar to subroutines, but instead of calling a central routine, the routine is embedded into the source (Fig. 10-1).

When macros are invoked, they are normally invoked with parameters that are replaced in the macro using the referencing labels. These referencing labels can be thought of as variables (or defines) that are set with the parameter value and are replaced by the parameter value every time they are encountered in the macro.

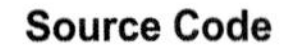


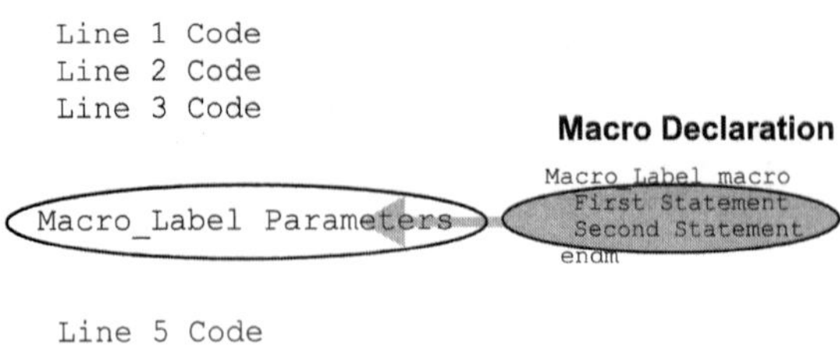


**Figure 10.1** Macro invocation operation

For example, instead of repeatedly writing the code:

```
incf     A, f
btfsc    STATUS, Z
  incf   A + 1, f
```

every time you required a 16-bit increment, the macro:

```
inc16 macro Variable
  incf     Variable, f
  btfsc    STATUS, Z
    incf   Variable + 1, f
endm
```

could be invoked. In this macro, the parameter *Variable* will be replaced by whatever string you specified for it. If the statement:

```
inc16 A
```

was placed in the application when the macroprocessor executed, it would be replaced with:

```
incf   A, f
btfsc  STATUS, Z
 incf  A + 1, f
```

This ability to use parameters goes much beyond customizing file register addresses and constants.

The advantages of using macros are that the overhead of subroutines can be avoided. The immediate overhead that comes to mind is the *call* and *return* instructions, but parameters that are passed to the routine do not have to be saved. This leads to some interesting situations where macros can create much more efficient application code than calling a subroutine.

For example, if a subroutine were used to add two 16-bit numbers together, the parameters have to be saved in temporary variables. They would be processed by the subroutine, the result returned in a temporary variable, and then passed to the final destination.

In the PICmicro® MCU, the code to do this could be:

```
;  C = A + B
```

```
  movf      A, w             ;    Save "A" in a Temporary Register
  movwf     Temp1
  movf      A + 1, w
  movwf     Temp1 + 1
  movf      B, w             ;    Save "B" in a Temporary Register
  movwf     Temp2
  movf      B + 1, w
  movwf     Temp2 + 1

  call      Add16Bits        ;    Call the 16 Bit Addition Subroutine

  movf      Temp1, w         ;    Put the Result in "C"
  movwf     C
  movf      Temp1 + 1, w
  movwf     C + 1

  :

Add16Bits                    ;    16 Bit Addition Subroutine
  movf      Temp2 + 1, w     ;    Add the High Byte First
  addwf     Temp1 + 1, f
  movf      Temp2, w         ;    Add the Low Byte
  addwf     Temp1, f
  btfsc     STATUS, C
   incf     Temp1 + 1, f     ;    If Low Byte Carry, Increment High Byte
  return                     ;    Return to Caller
```

In this example, the actual number of instructions required to support the subroutine is almost twice what the subroutine requires!

In comparison, a macro for the same function:

```
 Add16BitsMacro macro VarA, VarB, VarC
  movf     VarA + 1, w         ; Add the High Byte First
  addwf    VarB + 1, w
  movwf    VarC + 1            ; Save the Result
  movf     VarA, w             ; Add the Low Byte
  addwf    VarB, w
  movwf    VarC
  btfsc    STATUS, C           ; If low byte carry, Increment High Byte
   incf    VarC + 1, f
 endm
```

requires just one more instruction for the general case (and one less if the destination is the same as one of the source variables) and requires none of the instructions used to pass data between the temporary registers and the parameters. In this example, an application that uses this macro instead of the general subroutine will always be more efficient.

An added advantage of a macro over a subroutine is that if the macro is made available and not used, it will not take up any program memory.

In the year before I started writing this second edition, I have undergone a change in my personal philosophy about macros. I have repeated the hoary old tale about how I had to support a *Real-Time Operating System (RTOS)* that was written entirely from macros and how awful that was. The application was a totally wrong method to use macros. It was virtually impossible for me (or anyone else) to support. It colored my perspective on macros for almost 15 years. After this experience, I avoided creating complex macros because I worried about falling into the trap that the RTOS creators did.

This was shortsighted of me and I can see where macros can be used to implement sophisticated interfacing functions that are re-usable between applications. My personal rules for macro development are to avoid application specific logic functions and try to write them for the widest possible set of applications and devices.

This chapter covers the features of macros and how they can be enhanced using the directives built into the PICmicro® MCU assembler to allow you to create complex functions. These functions allow you to write code once and let you use the macros across many different applications.

One point that may not be terribly obvious is that macros execute during "build time" and not at "run time." This means that macros execute when the source code is assembled and not when it is executing in the PICmicro® MCU. If you place variables as macro parameters, the variables' addresses will be used, *not* the contents of the variables themselves. This important point trips many people up when they first start working with macros.

# The Difference Between Defines and Macros

I consider *defines* to be simple macros in their own right because of the way they operate. This view is slightly unusual, but it is a good way to look at how defines work and find good uses for them that might not be readily apparent. Defines are not as flexible as macros, but they can significantly add to the ease of application programming and application readability.

Defines in MPASM are declared using the format:

```
#Define Label String
```

*Label* is a standard alphanumeric text label. What makes defines special is that when the label is encountered, it is replaced by a string. This powerful concept can make PICmicro® MCU assembly-language programming a lot easier both to write and read.

Defines can be used to simplify the operation setting, resetting hardware bits and flags in your application. For example, if bit 3 of the *Flag* variable is used to indicate the ready state of the serial port, it could be declared as:

```
#Define Serflag Flag, 3
```

To reset the bit in the serial initializing routine (which indicates that it is available), instead of using the statement:

```
bsf    Flag,3
```

which is not very helpful or easy to understand, the instruction:

```
bsf Serflag
```

can be used. The single define eliminates remembering (or looking up) where the bit is located and which bit it is. As well, when reading the code, using the *define* directive enhances the readability of the purpose of the instruction.

Defines work because they directly copy in the string and let the assembler evaluate them. This is different from *equates,* which evaluates a string and stores it as a constant referenced to the label. This can cause some subtle differences that can cause problems if you don't know what you are doing.

For example, if you had the code:

```
variable A = 37              ;  "Variable" is a run time variable

Test1 EQU A * 5              ;  Test1 = 185

#define Test2 A * 5          ;  Test2 is Evaluated when it is used

    :

A = A + 5                    ;  A = 42

    :

  movlw   Test1              ;  w = 185

    :

  movlw   Test2              ;  w = A * 5
                             ;    = 42 * 5
                             ;    = 210
```

In this example, even though *Test1* and *Test2* are declared at the same point in the code identically, they are evaluated differently and will be different values in different locations of the application. This is a very useful capability, but it can be a problem elsewhere.

In the sample code, I declare the *variable A*. *Variable* is a specific assembler directive to create a temporary storage value during the application. Variables are discussed in the later sections of this chapter. When it is used in *Test1,* the values of *A* are used when the assembler processor encounters the statement, multiplied by five and assigned to *Test1*. After *Test1* and *Test2* are defined in the code, *A* is modified, which results in a different value being calculated for *Test2* when it is used later in the application.

A useful function that defines can provide is that they can provide constant strings throughout an application. I use this ability to keep track of my application version number. In many of the newer application codes presented in this book, you'll see that the second or third line of an application is:

```
#define _version "1.00"
```

Because it is at the top of the source, I can see the version number as soon as the source code is brought up in an editor. This define can be used throughout the code to provide the version information without me having to update when I come up with a new version. Often, I will output the version information to indicate what is the source code burned into the PICmicro® MCU. This *#define* statement can be put into a *DT* statement and read out of a table using conventional table-read code:

```
dt "Version: ", _version, 0
```

In this code, each byte of the *_version* define is added to the string of characters used with the *dt* statement, as if the code was entered as:

```
dt "Version: 1.00", 0
```

Define labels do not have to have strings associated with them. This might seem unusual (and seem to defeat the purpose of defines), but this allows them to be used as assembly-time execution-control flags that are shown later in this chapter. This function allows fast customization of an application modification for debug.

Defines, although simpler than macros, can be used for many different purposes. Although macros can only be used to replace lines of text, defines can simplify instruction programming and provide common information or text strings. Neither macros nor defines can replace the other, but their functions complement each other.

# The Assembler Calculator

Although not really part of the macroprocessor, the ability of the MPASM assembler (and most compilers) to do constant calculations can make applications easier to develop so that you don't have to use a hand-held calculator to find the values of the constants. Over the past year or so, I've been trying to get out the habit of pulling out my calculator to come up with values and instead let the calculations be done in the assembler. The assembler calculator adds a lot of capabilities that allow your macros to be used in different situations that require different parameters.

The assembler calculator works on algebraic expressions, similar to how they're used in high-level languages. It is important to remember that the calculator works as part of the last pass of the assembler. It inserts the address of an application variable, instead of its contents. This can be confusing because variables can be declared within the source that are only available for use by the assembler calculator.

So far in the book, you have seen the assembler calculator in operation for such instructions as:

```
movlw (1 << GIE) | (1 << TOIE)
```

which loads the w register with a byte, destined for the INTCON register, which has the GIE (7) and T0IF (5) bits set. In this case, the assembler calculator is used to change bit numbers to actual values to be loaded into a byte. Shifting one by the bit number converts the bit number into a constant value that will set the bit when loaded into a register.

This trick is useful and avoids having to figure out what value is used for specific bits being set or reset. In this example, if this trick has not been used, I would have to first find the "bit values" of "GIE" and "TO1E," then remember that bit seven being set is the same as adding 128 and if bit five is set, 32 is added. The result of these two values is 160 decimal or 0x0A0. Using the assembler calculator, I didn't have to worry about any of this.

To reset specific bits, the same trick can be used, but the bits have to be reset, which is done by a bitwise inversion of the bits and then ANDing the result with the current value. XORing the set bit value with 0x0FF accomplishes the bitwise inversion.

For example, to clear bits 4, 2, and 1 in the w register, the following instruction could be used:

```
andlw 0x0FF ^ ((1 << 4) | (1 << 2) | (1 << 1))
```

If you were to do this manually, you would have to follow these steps:

**1** Calculate the values for bits 4, 2, and 1 being set.

$(1 << 4) = 16$
$(1 << 2) = 4$
$(1 << 1) = 2$

which translates to:

$$\begin{aligned}(1 << 4)|(1 << 2)|(1 << 1) &= 16 \mid 4 \mid 2 \\ &= 22 \\ &= 0x016\end{aligned}$$

**2** Calculate the inverse (XOR with 0x0FF):

0x0FF ^ 0x016 = 0x0E9

**3** Put the value into the *andlw* instruction:

```
andlw 0x0E9
```

If you go through the manual process, you can see that there are seven opportunities for you to calculate constant values incorrectly or copy down the wrong value. This is what I meant when I said that the assembly calculator is easier and less prone to mistakes. Table 10-1 lists the arithmetic calculator's arithmetic operators. All the operators have two parameters, except for when - negates a value or the complement (~) operator, which only have one parameter.

In the clear bits example, I could have used the equation format:

```
andlw  ~((1 << 4) | (1 << 2) | (1 << 1))
```

instead of adding the *0x0FF ^* characters in this instruction. Personally, I find XORing with 0x0FF to be more obvious as a bitwise inversion than using the ~ (tilde) character.

Along with the arithmetic operations, parenthesis (the *(* and *)* characters) can be used in the expressions to ensure that the operation is executed in the correct order. In these examples, I have used parenthesis to ensure that the correct order of operations occurs for these instructions.

For 16-bit values, you can use the *low* and *high* assembler directives. For example, if you wanted to jump to a specific address in another page, you could use the code:

**TABLE 10-1 Operators Available to the Assembler Calculator**

| OPERATOR | DESCRIPTION | COMMENTS |
|---|---|---|
| + | Addition | |
| – | Subtraction/Negation | If no First Parameter then Negation |
| * | Multiply | |
| / | Divide | |
| % | Modulus | Return Remainder from Divide Operation |
| << | Shift Left | Shift the First Parameter to the Left by the Second Parameter's Number of Bits. |
| >> | Shift Right | Shift the First Parameter to the Right by the Second Parameter's Number of Bits. |
| & | Bitwise AND | AND Together the Parameter's Bits |
| \| | Bitwise OR | OR Together the Parameter's Bits |
| ^ | Bitwise XOR | XOR Together the Parameter's Bits |
| ~ | Complement | XOR the Parameter with 0x0FF to get the Complemented or "Inverted" value |

```
movlw   HIGH Label          ; "Label" is the
movwf   PCLATH              ;   Destination
movlw   LOW Label
movwf   PCL
```

which is the same as:

```
movlw   ((label && 0x0FF00) >> 8)
movwf   PCLATH
movlw   LABEL && 0x0FF
movwf   PCL
```

In this example, the first example (which use *high* and *low*) are much easier to see what is happening.

The *low* directive will always return the least significant byte of a value. If a value is declared as greater than 16 bits in size, the *upper* directive will return the most-significant eight bits for the 32- or 24-bit number. I'm mentioning this because using these directives can be tricky in some situations.

As covered earlier in the book, the *$* operator returns the current program counter, which is a 16-bit value that can be manipulated using the assembler calculator's operators as if it were a constant.

*Variables* that are only used in assembly, can be declared using the format:

```
variable label [ = constant][, . . .]
```

These "variables" are 32 bits in size and can be set to any value with the operators listed using the = operator to make an assignment statement like:

```
LABEL1 = LABEL1 * 2
```

It is important to remember that the label is not an application variable (i.e., it cannot be modified by the PICmicro® MCU as it is running) and when it is assigned a new value, it must be in the first column of the assembly-language source. When it is being read in another statement, it can appear in any column (except for the first) in the line.

Taking a cue from C, assembler variable assignment statements can be simplified if the destination is one of the source parameters. These operations can be confusing to use and read, unless you are familiar with C. Table 10-2 shows what is available and how they work.

The assembler calculator can also do comparisons between two parameters. If the comparison is true, a *1* is returned, otherwise a *0* is returned. The comparison operators (listed in Table 10-3) are required for the conditional assembly operations presented in the next section.

These comparisons can be compounded with the // and *&&,* which are the logical OR and logical AND operators, respectively. // returns one if either of its two parameters are not equal to zero. *&&* will return a 1 if both parameters are equal to zero. This operation brings up an important point: in the assembler calculator, a *true* condition is *any* nonzero value. The variable *A,* after executing:

```
A = 7 && 5
```

will be loaded with *1,* because *7* and *5* are not zero and both are assumed to be true.

This operation of logical values is not unique to the MPASM assembler calculator; most languages use this convention for *true* and *false.*

**TABLE 10-2 Combined Assignment Operators Available to the Assembler Calculator**

| OPERATOR | EQUIVALENT OPERATION |
|---|---|
| += | Parm1 = Parm1 + Parm2 |
| −= | Parm1 = Parm1 − Parm2 |
| *= | Parm1 = Parm1 * Parm2 |
| /= | Parm1 = Parm1 / Parm2 |
| %= | Parm1 = Parm1 % Parm2 |
| <<= | Parm1 = Parm1 << Parm2 |
| >>= | Parm1 = Parm1 >> Parm2 |
| &= | Parm1 = Parm1 & Parm2 |
| \|= | Parm1 = Parm1 \| Parm2 |
| ^= | Parm1 = Parm1 ^ Parm2 |

**TABLE 10-3 Comparison Operators Available to the Assembler Calculator**

| OPERATOR | OPERATION |
|---|---|
| == | Return 1 if Two Parameters are Equal |
| != | Return 1 if Two Parameters are Different |
| > | Return 1 if the First Parameter is Greater than the Second Parameter |
| >= | Return 1 if the First Parameter is Greater than or Equal to the Second Parameter |
| < | Return 1 if the First Parameter is Less than the Second Parameter |
| <= | Return 1 if the First Parameter is Less than or Equal to the Second Parameter |

The last operator is *!*, which toggles the logical state of a value. For example:

```
A = !4          ;  4 != 0 and is "true"
                ;  A = not true
                ;    = false
                ;    = 0

C = !0          ;  0 == 0 and is "false"
                ;  A = not false
                ;    = true
                ;    = 1
```

The comparison and logical operators might seem unnecessary for arithmetic calculations, but there are cases where they can be useful.

# Conditional Assembly

If you have taken a look at some of the more complex macros presented in this book, you will probably be surprised to see that there are structured language statements (*if, else, endif, while,* and *endw*). At first glance, these statements provide high-level language capabilities to the PICmicro® MCU assembly code. If you look at the listing file after these statements have executed, the code produced will be somewhat strange and not at all what you were expecting.

These statements specify which code and what values are going to be used to assemble the application code. Instead of being executed at run time, they are actually executed when the source code is being assembled and can be used to conditionally change constant values or to add or delete sections of code.

Conditional assembly statements are not part of the macro processor; they can be used outside of macros and (as shown in the next section) they can be used in straight source-level code. The purpose of conditional assembly statements is to selectively change the source code of an application before they are assembled. Conditional assembly statements are actually *directives* that are processed along with other directives (such as *EQU, dt,* etc.).

For example, if an application was to be run on two different PICmicro® MCUs, each with different built-in hardware, conditional assembly could be used in the following manner:

```
#define USARTPres
   :
 ifdef USARTPres
   :                     ;  Put in USART Handler Code
 else
   :                     ;  Put in Non-USART Serial Handler
 endif
```

In this case, the *#define* statement creates a define in the application code. Later in the code, when the *ifdef* statement is encountered, if the *USARTPres* define is present, the first block of code is put into the assembler source code and the second is ignored. If *USARTPres* is not defined, then the first block of code is ignored and the second block is assembled as source code.

For all *if* conditional assembly directives, *endif* is required and *else* is an optional conditional assembly directive that will include the code if the original condition was not true. I realize that some people will put in a null statement after an *if* to have the *else* execute like:

```
ifdef USARTPres        ;  If USART Present, don't add any code
else
  :                    ;  Put in Non-USART Code
endif
```

This can be somewhat difficult for people reading the source to understand what is happening. Instead of using *else* like this, the negative condition should be checked for using the *ifndef* directive.

The convention that I use is to place all conditional assembly and directives statements in the second column of the source code (the instructions start in the third column). This allows me to visually see the difference between the conditional assembly statements and actual instructions. This convention was started with my C programming to be sure that I could see the difference between conditional statements and the actual source code.

Along with the *ifdef* statement is the *ifndef* statement. These instructions test for the presence of defines and respond accordingly. When the define checked for by *ifdef* is present, then the code following will be included. When the define checked for by *ifndef* is present, then the code following will be ignored. The *ifndef* statement can be used in the previous example to simplify it and make it easier to understand:

```
ifndef USARTPres       ;  If USART Present, don't add any code
  :                    ;  Put in Non-USART Code
endif
```

You can do a number of tricks with *ifdef* and *ifndef* conditional assembly statements that can make your code development easier and more flexible.

The first is conditionally deleting code. As you work through an application, you will often want to remove some code to test out different aspects of it. Elsewhere, the book presents the concept of "commenting out the code:"

```
; addlw 0x0FF          ; #### - Instruction not needed, but kept
```

For single instructions, this is not too hard to do and easy to keep track of. For many instructions, it can be difficult to track everything that has to be removed (but kept). An easy way of doing this is to put an *ifdef* and *endif* statement before and after the code:

```
ifdef Do_Not_Assemble                ; #### - Ignore following
  :                                  ; Block of Code NOT Assembled
endif                                ; #### - Ignore above
```

This takes literally just a few seconds to remove the code and it can be disabled just as quickly (by changing the *ifdef* to *ifndef* ).

The second is when multiple PICmicro® MCUs can be used by an application. The previous chapter showed the sample template as having a specific PICmicro® MCU part number put into the *LIST* statement to define the PICmicro® MCU for the application. In MPASM, this is actually not required because the PICmicro® MCU part number is specified within MPLAB (by selecting *Options* and *Development Mode* in the PICmicro® MCU desktop).

As well, when the PICmicro® MCU is specified and assembled, MPASM automatically creates a label for the part number. This number is the device part number with two underscore (_) characters in front of it. For example, the PIC16F84, when assembled, will have the *__16F84* label available within the application during assembly.

I take advantage of this in Chapter 15 as each experiment can be implemented on more than one device, based on the development mode. In Chapter 15, a PIC16F84A or PIC16F877 can be used as the base processor. To make it easier for you to work through the experiments, I have added the *LIST* statement and conditional execution statements:

```
list R=DEC
ifdef __16F84A
  include "p16f84a.inc"
 else
 ifdef __16F877
  include "p16f877.inc"
 else
  error "Code is Not Designed for Specified Processor"
 endif
 endif
```

In these statements, the default radix is specified (always decimal or 10 for my applications). Depending on which processor is specified within MPLAB, the appropriate include file is added to the code or an error is forced.

In this code, notice that I have put in nested *if* statements. Up to eight levels of nesting are possible in MPASM, although this can be very confusing to read. I recommend that no more than two be used (as in the example).

Along with conditional assembly based on the presence or absence of labels, constant and variable condition testing can also be done with conditional assembly statements. For example, tests against addresses could be performed for interpage jumping:

```
if ((($ & 0x01800) ^ (Label & 0x01800)) != 0)
 movlw  HIGH Label          ;  Different Pages - Update PCLATH
```

```
  movwf  PCLATH
 endif
  goto   Label & 0x07FF       ;  Jump to Label
```

In this example, if the destination is in a different page from the current location (which is returned by the *$* directive in MPLAB), then PCLATH is updated before the *goto* statement.

The previous example is less than optimal for four reasons. First is that whether or not PCLATH has to be updated is variable based on the address of the *goto* statement. A more accurate way to do this is:

```
 if ((($ + 2) & 0x01800) ^ (Label & 0x01800)) != 0)
  movlw  HIGH Label           ;  Different Pages - Update PCLATH
  movwf  PCLATH
 endif
  goto   Label & 0x07FF       ;  Jump to Label
```

In this case, the possible address of the *goto* is checked, rather than the current address. There is the opportunity that the current address will be in a different page than the *goto* and PCLATH might not be updated correctly.

The second problem is that this code requires a different amount of space, depending on which path is taken. Doing this can result in an address *phase error,* which indicates that during the different passes in the assembler, required addresses change in a way that makes correct assembly impossible. These different addresses are caused when the conditional code executes for a second time and addresses come out differently. Phase errors are very hard to find. Chances are, if you have one in one location, there will be a number of them.

The best way to avoid phase errors is to always be sure that the same number of instructions is used no matter what path is taken in the conditional assembly. For this code, I can add two *nops* as part of *else* (assembled if the condition is not true) to ensure that no addresses will change.

```
 if ((($ + 2) & 0x01800) ^ (Label & 0x01800)) != 0)
  movlw  HIGH Label           ;  Different Pages - Update PCLATH
  movwf  PCLATH
 else
  nop                         ;  Add Two instructions to prevent
  nop                         ;   "Phase" Errors
 endif
  goto   Label & 0x07FF       ;  Jump to Label
```

The third problem with this code is that a message can be produced, indicating that the jump is to a different page. To avoid this, the *goto* address should have the current page bits added to it. This changes the code to:

```
 if ((($ + 2) & 0x01800) ^ (Label & 0x01800)) != 0)
  movlw  HIGH Label           ;  Different Pages - Update PCLATH
  movwf  PCLATH
 else
  nop                         ;  Add Two instructions to prevent
  nop                         ;   "Phase" Errors
 endif
  goto   (Label & 0x07FF) | ($ & 0x01800)    ;  Jump to Label
```

The next problem with this code is that it changes *w*. Thus, this code cannot be used if the contents of w are going to be passed to the destination label. Instead of explicitly loading PCLATH with the destination, the bits can be changed individually using the code:

```
if ((($ + 2) & 0x01000) ^ (Label & 0x01000)) != 0)
if ((($ + 2) & 0x01000) == 0)
 bsf  PCLATH, 5                 ;  Label in Pages 2 or 3
else
 bcf  PCLATH, 5                 ;  Label in Pages 0 or 1
endif
else
 nop                            ;  No Difference in High Pages
endif
if ((($ + 2) & 0x00800) ^ (Label & 0x00800)) != 0)
if ((($ + 2) & 0x00800) == 0)
 bsf  PCLATH, 4                 ;  Label in Pages 1 or 3
else
 bcf  PCLATH, 4                 ;  Label in Pages 0 or 2
endif
else
 nop                            ;  No Difference in Low Pages
endif
 goto   (Label & 0x07FF) | ($ & 0x01800)   ;  Jump to Label
```

Looking at this mess of conditional assembly statements, it is starting to look a lot like a macro. That is why I have included conditional assembly statements in this chapter. Conditional assembly statements, although simplifying your applications in some ways, will result in fairly complex applications in others. The previous code has really become the *lgoto* macro:

```
lgoto Macro Label
 if ((($ + 2) & 0x01000) ^ (Label & 0x01000)) != 0)
 if ((($ + 2) & 0x01000) == 0)
  bsf  PCLATH, 5                 ;  Label in Pages 2 or 3
 else
  bcf  PCLATH, 5                 ;  Label in Pages 0 or 1
 endif
 else
  nop                            ;  No Difference in High Pages
 endif
 if ((($ + 2) & 0x00800) ^ (Label & 0x00800)) != 0)
 if ((($ + 2) & 0x00800) == 0)
  bsf  PCLATH, 4                 ;  Label in Pages 1 or 3
 else
  bcf  PCLATH, 4                 ;  Label in Pages 0 or 2
 endif
 else
  nop                            ;  No Difference in Low Pages
 endif
  goto   (Label & 0x07FF) | ($ & 0x01800)  ;  Jump to Label
 endm
```

which can be placed anywhere in your application with three instructions specifying a jump to anywhere in the PICmicro® MCU device's program memory each time the macro is encountered.

Along with program constants, you can also declare integer variables, which can be updated during the assembly of the application. The variable directive, appropriately enough, is used to declare the variables with optional initial values:

```
variable i, j=7
```

The variables can be used constants in the application or as parts of labels. As covered in the previous section, variables can be used to avoid having to calculate your own constant values.

I often use variables as counters for use with the *while* conditional assembly statement. For example, if I wanted to loop six times, I could use the code:

```
 variable i=0                  ;  Declare the Counter
 while (i < 6)
   :                           ;  Put in Statements to be repeated six times
i = i + 1                      ;  Increment the Counter
 endw
```

Note that when the variable *i* is updated, the statement starts at the first column of the line. The MPLAB assembler requires this.

In this code, the statements within the *while* and *endw* statements are inserted into the assembly-language source file each time that the condition for *while* is true. Looking at how *while* has executed in the list file can be a bit confusing. For the code:

```
  goto    $ + 7                ;  Put in Patch Space
 variable i = 0
 while (i < 6)
  dw      0x03FFF              ;  Add Dummy Ins
i = i + 1
 endw
```

which puts in a jump over six instructions of "patch" space, the listing file looks like:

```
0000 2807           00036  goto    $ + 7   ;  Put in Patch Space
  0000              00037  variable i = 0
                    00038  while (i < 6)
0001 3FFF           00039    dw      0x03FFF ;  Add Dummy Ins
  00000001          00040 i = i + 1
0002 3FFF           00039    dw      0x03FFF ;  Add Dummy Ins
  00000002          00040 i = i + 1
0003 3FFF           00039    dw      0x03FFF ;  Add Dummy Ins
  00000003          00040 i = i + 1
0004 3FFF           00039    dw      0x03FFF ;  Add Dummy Ins
  00000004          00040 i = i + 1
0005 3FFF           00039    dw      0x03FFF ;  Add Dummy Ins
  00000005          00040 i = i + 1
0006 3FFF           00039    dw      0x03FFF ;  Add Dummy Ins
  00000006          00040 i = i + 1
                    00041 endw
```

This listing looks more like six *dw 0x03FFF* instructions and $i = i + 1$ statements, rather than the two of them being repeated six times.

The conditional assembly instructions, *if* and *while* use the same condition test format as the *if* and *while* statements of the C language. The condition tests can only occur on constant values of up to 32 bits in size. Like C's *if* and *while*, the MPASM

**TABLE 10-4 Comparison Operators and Return Values**

<table>
<tr><th>CONDITION</th><th>FUNCTION</th></tr>
<tr><td>==</td><td>Return "True" if Both Parameters are the Same</td></tr>
<tr><td>!=</td><td>Return "True" if Parameters are not the Same</td></tr>
<tr><td>></td><td>Return "True" if the First Parameter is Greater than the Second</td></tr>
<tr><td>>=</td><td>Return "True" if the First Parameter is Greater than or Equal to the Second</td></tr>
<tr><td><</td><td>Return "True" if the First Parameter is Less than the Second</td></tr>
<tr><td><=</td><td>Return "True" if the First Parameter is Less than or Equal to the Second</td></tr>
<tr><td>&</td><td>Return "True" if Both Parameters are "True"</td></tr>
<tr><td>|</td><td>Return "True" if either one of the Parameters is "True"</td></tr>
</table>

assemblers conditional assembly statements use the two parameter conditions shown in Table 10-4.

When the statements are true, a nonzero value is returned. If the statements are false, then zero is returned.

## DEFINES FOR APPLICATION DEBUG

This book harps on a lot about the need to simulate applications before they are burned into actual PICmicro® MCU hardware. For many applications this isn't an issue, but for applications that have very long delays built into them, this can be a very significant issue. The time required for external hardware initializations can actually take many minutes in MPLAB because the simulation has to go through long delay loops. Another situation could be for hardware that is nonessential to the task at hand, but requires a relatively complex interaction with simulated hardware or use built-in PICmicro® MCU interfaces that are not simulated in MPLAB. An example of the latter situation is a PICmicro® MCU application that uses the USART port to send and receive configuration information (which cannot be easily simulated by the MPLAB and a stimulus file).

Dealing with this problem is relatively simple with the use of the conditional assembly directives built into the MPASM assembler. These instructions will allow execution to skip over problem code areas simply by specifying a flag using the *#DEFINE* directive. The state of the *#DEFINE* flags is generally defined by their presence or absence. The most common one in my code is the *Debug* label, which is defined for simulation by using the statement:

```
#DEFINE Debug
```

This statement usually appears on the second or third line of my applications, although when I am ready to burn an application into a PICmicro® MCU, I change the statement to:

```
#DEFINE nDebug
```

to disable any changes to the source code and ensure that the object file is created with the correct code. The use of the *Debug* label to replace problem areas of code like:

```
ifndef Debug
 call  UnreasonablyLongDlay         ;  Wait for External Hardware
else                                ;  "Debug" #DEFINEd
 nop                                ;  Put in Instruction Anywise
endif
```

In this code, if *Debug* has not been set in a *#DEFINE* statement, then the instruction that is a problem in the simulator is used for the source file. If *Debug* has been defined, then the *nop* is executed in the place of the *call* instruction. It is important to match all of the instructions taken out by the *ifndef Debug* by an equal number of *nops*. If the number of instructions is not matched with *nops,* you could run into the situation where the code assembles to different sizes that can cause problems if you put in fewer *nops* to actual instructions.

Along with being used for taking out lengthy delays or hardware inconsistencies in the MPLAB simulator, *Debug* conditional assembly code can be used to put the application into a specific state. This can save quite a bit of time during simulation and can also be used to initialize the state of the application before simulation begins.

For example, when I was creating the EMU-II emulator application code, I used the *Debug #DEFINE* label to enter in the following commands:

```
ifdef    Debug      ;  Clear simulated application program Flash
 movlw   'P'        ;  Start a program memory Program/Clear Flash
 call    BufferAdd
 movlw   0x00D
 call    BufferAdd
 movlw   0x003      ;  Stop the transfer after the data is cleared
 call    BufferAdd
endif
```

This code initiates the command to load a new application into the Flash (and, in preparation for the code, the Flash is cleared), followed by a *Ctrl-C* to stop the expected transfer. The reason for requiring this code is that any writes to the simulated Flash would not be reset when I rebuilt the application. These commands clear the simulated Flash so that each time that the application code is reset, I would be working with cleared Flash instead of something that I would have to simulate the operation of downloading code into it. This simulation isn't possible because of my use of the USART, which is not simulated within MPLAB.

When the EMU-II application was being debugged, I placed a breakpoint at a location where I wanted to start debugging the EMU-II, knowing that the program memory was cleared and I was ready to start looking at how the application executed.

Multiple *Debug #DEFINE* statements could be put into the application code, but I would not recommend that you do that. Multiple statements get confusing very quickly and often become incompatible. Instead of using multiple *Debug #DEFINE* labels, I would recommend that you just use one. When you have fixed the current problem that you are working on, then you can change the *ifdef Debug* and *ifndef Debug* statements to target the next problem.

# Debugging Macros

The preceding pages provide a lot of information about how to create complex macros. This section covers a few pages on how to debug these monsters. As a macro executes, it is probably going to do some things that will be surprising to you. The result of this execution is some unexpected data placed into your source code. You will have to understand exactly what you should expect and how it should be debugged.

The key to debugging macros is being able to read what is produced. When macros are expanded into the source code, the actual data that is produced can be difficult to see and work through. This section goes through how macros are "expanded" and what to look for in the application listing file in order to determine whether or not there is a problem. This section also covers simulating the application and macros to try and make the problems easier to spot.

When macros execute, the conditional statements (*if, ifdef, ifndef,* and *while*) might not be displayed, depending on the operation. Remember that regular statements are print statements in the macroprocessor. To illustrate this, I have created conditional assembly statements for a macro:

```
variable i = 0
if (i == 0)
 addlw    0 - i
else
 sublw    i
endif
```

When the macro is expanded (or executed), the following information will be displayed in the listing file:

```
00000000          M      variable i = 0
                  M      if (i == 0)
Addr  3E00        M       addlw    0 - i
                  M      else
                  M       sublw    i
                  M      endif
```

As this example illustrates, the setting of the macro variable (*i*) results in a constant value being placed somewhere between the start of the line and the middle. The instructions that are actually added to the listing file are identified by the address of the instruction and its bit pattern is broken up (as shown for the *addlw 0 – i* instruction). In this example, only *addlw 0 – i* is inserted into the source code. The directives *variable, if, else,* and *endif,* and the *sublw i* instruction are all ignored.

*While* directives are somewhat unusual to follow because the code is repeated within them and no start and end reference information can be easily seen. For the example:

```
variable i = 0
while (i < 2)
 movlw    i
i = i + 1
endw
```

the following listing file information is generated:

```
  00000000          M      variable i = 0
                    M      while (i < 2)
 Addr  3000         M       movlw     i
 00000001           M      i = i + 1
 Ad+1  3001         M       movlw     i
 0000002            M      i = i + 1
                    M      endw
```

In this case, the *while* and *endw* directives are not repeated as you would expect. Again, to really understand what is happening, you have to go back and look at the instructions entered into the application code. These instructions are displayed to the left of the column of *Ms,* along with the variable updates.

If the code within the *while* and *endw* directive statements is very complex, you might want to put some comments in to help make the operation easier to read.

Changing the code in the macro to:

```
 variable i = 0
 while (i < 2)
;  while (i < 2)
  movlw i
 i = i + 1
;  endw
 endw
```

The result might be easier for you to follow because the comments *; while (i < 2)* and *; endw* will be inserted within your listing file. This might make it easier for you to follow where the expansion code starts and ends for each interaction of *while.*

Probably the best way to debug a macro is a single step in the simulator through the listing file. If you are doing this in the source file, you will find that the execution jumps to where the macro is defined. This can be quite disconcerting and confusing because the parameters are displayed, not what the parameters actually are.

Breakpoints cannot be placed in MPLAB at a macro invocation for a source file. When I want to set a breakpoint at a macro invocation, I add an *nop* instruction before it to give the MPLAB simulator somewhere to hang the breakpoint. This also works with a listing file, but, in this case, you do not have to add the breakpoint (as I indicated above, breakpoints can be placed at instruction statements within the macro-generated code).

Debugging macros is really the same as debugging a standard application, except that there is no simulator or debugger. This is an important point to remember when you are creating the macro. You might want to flag where you are executing with messages (using the *messg* directive). For example, if you have the code:

```
 if (A == B)
 messg "A == B"
  ;    put in code for A == B
 else
 messg "A != B"
  ;    put in code for A != B
 endif
```

the *messg* directives will flag the path that execution takes through the macro's conditional assembly. This (and the previous) tricks can be used with conditional code external to macros to help you follow their execution and find any problems within them.

# Structured Programming Macros

To finish off this chapter, I wanted to leave you with a few macros that should give you some ideas on how powerful the macro concept is. Also, I wanted to give you a tool to make PICmicro® MCU assembly-language coding easier. The *code\structre* subdirectory of the *PICmicro* directory contains the STRUCTRE.INC file, which contains nine macros that you can use for your own applications to add structured programming conditional execution.

The macros are in the format: _Test*Parm2,* where *Test* can be: *do, until, while, if, else,* and *end*. *Parm2* specifies whether or not the second parameter is a constant or a variable.

When the macros are invoked, the parameters include the two conditions for testing and the condition to test for. The standard C test conditions are shown in Table 10-5.

Only one condition can be accessed within a macro at any time. It has no ability to AND or OR conditions together. These conditions are actually the same ones that are available within the macro processor. Table 10-6 lists the nine different macros and how they are invoked.

Using these macros in your application is quite straightforward. If you are new to PICmicro® MCU programming, you might want to take a look at these macros for your applications.

To execute some code conditionally if two variables were equal would use the macro invocations:

```
_ifv VariableA, ==, VariableB
  :                    ;  Code to Execute if VariableA == VariableB
_end
```

**TABLE 10-5 Comparison Operators Used In Structured Programming Macros**

| CONDITION | DESCRIPTION |
|---|---|
| == | Equals |
| != | Not Equals |
| > | Greater Than |
| >= | Greater Than or Equals to |
| < | Less Than |
| <= | Less Than or Equals to |

**TABLE 10-6 Structured Programming Macros Invocations**

| INVOCATION | DESCRIPTION |
|---|---|
| _ifv Parm1, Condition, Parm2 | Execute the following code if the two variables (*Parm1* and *Parm2*) operated on with the condition is true. *_else* following is optional and the code following it will execute if the condition is false. Must have *_end* following. |
| _ifc Parm1, Condition, Parm2 | Execute the following code if the variable (*Parm1*) and constant (*Parm2*) operated are on with the condition is true. *_else* following is optional and the code following it will execute if the condition is false. Must have *_end* following. |
| _else | Execute the following code if the result of the previous *_if* macro was false. Execution before the *_else* jumps to the next *_end* macro. |
| _end | End the *_if* or *_while* macro invocation. If previous operation was *_while,* then jump back to the *_while* macro. |
| _whilev Parm1, Condition, Parm2 | Execute the following code while the two variables (*Parm1* and *Parm2*) operated on with the condition is true. Must have *_end* following, which will cause execution to jump back to the *_while* macro. |
| _whilec Parm1, Condition, Parm2 | Execute the following code while the variable (*Parm1*) and constant (*Parm2*) operated on with the condition is true. Must have *_end* following, which will cause execution to jump back to the *_while* macro. |
| _do | Start of *do/until* loop. |
| _untilv Parm1, Condition, Parm2 | Jump to previous *_do* if the variables (*Parm1* and *Parm2*) operated on with condition is false. |
| _untilc Parm1, Condition, Parm2 | Jump to previous *_do* if the variable (*Parm1*) and the constant (*Parm2*) operated on with condition is false. |

This can be expanded to include an *else* condition for code that executes if the condition is not true:

```
_ifv VariableA, ==, VariableB
  :                     ;  Code to Execute if VariableA == VariableB
_else
  :                     ;  Code to Execute if VariableA != VariableB
_end
```

For these macros, I decided to use as close a label to actual programming structures as possible, which is why I used the standard names with the underscore (_) character before them.

Five aspects and features of the macro processor have influenced how these structured conditional execution macros were created. First, you cannot distinguish between constants and variable addresses in a macro. When the macro is invoked, a variable address is passed to the macro, instead of the variable label. This was probably done to simplify the effort in writing the macro processor. For this application, it means that either what the value types must be explicitly specified or there must be a way to declare variables so that they can be differentiated from constants.

I looked at a number of different ways to attempt to distinguish the two types of values and found that I could not do it without changing how variables were declared, which would make the application code more complex. Because I could not tell the two different types of data apart, I decided to always make the first parameter a variable and the second one a variable (*v*) or constant (*c*), depending on the character at the end of the macro.

Interestingly enough, this conversion of data does not extend to nonalphanumeric characters. In the macros, the condition test is specified and this is passed into the macro. I take advantage of this to avoid having to create multiple macros, each with a different condition test. Inside the macro, I use this test condition parameter to determine what it actually is (there is no way to compare nonnumeric values in the conditional assembly functions of the macro processor). For example, to find out the condition of the *_ifv* macro, I test the specified condition against different cases:

```
if (1 test 1)                   ;  Check for "=="
 movf   a, w
 subwf  b, w
 btfss  STATUS, Z               ;  Zero Flag Set if True
  else
   if (1 test 0)                ;  Check for "!="/">"/">="
    if (0 test 1)               ;  Check for "!="
     movf   a, w
     subwf  b, w
     btfsc  STATUS, Z           ;  Zero Flag Reset if True
    else                        ;  Else ">"/">="
     if (1 test 1)              ;  Check for ">="
      movf   b, w
      subwf  a, w
      btfss  STATUS, C          ;  Carry Set, ">="
     else
      movf   a, w
      subwf  b, w
      btfsc  STATUS, C          ;  Carry Reset, ">"
     endif
    endif
   else
    if (0 test 1)               ;  Check for "<"/"<="
     if (1 test 1 )
      movf   a, w
      subwf  b, w
      btfss  STATUS, C
     else
      movf   b, w
      subwf  a, w
      btfsc  STATUS, C
     endif
```

```
      else
       error Unknown "if" Condition
      endif
     endif
    endif
```

This method of testing what is being passed is actually very efficient and easy to code. Notice in this condition test code that if none of the tests are satisfied, an error is forced, indicating that the condition is unknown.

The macros themselves use conditional code to produce simple code for the actual functions. For example, in the code:

```
 _ifv Parm1, >, Parm2
   :                            ;  Code to Execute if Parm1 > Parm2
 _else
   :                            ;  Code to Execute if Parm1 <= Parm2
 _end
```

The best-case assembler would be:

```
; _ifv Parm1, >, Parm2
  movf   Parm1, w
  subwf  Parm2, w
  btfss  STATUS, C
   goto  _ifelse1               ;  Not True, Jump to "else" code
   :                            ;  Code to Execute if Parm1 > Parm2
; _else
  goto   _ifend1                ;  Finished with "true" code, Jump to "_end"
_ifelse1
   :                            ;  Code to Execute if Parm1 <= Parm2
; _end
_ifend1
```

If you try out the macros in STRUCTRE.INC, you will find that this exact code is created for this example. To do this, I had to create three stacks to keep track of where I was. The first stack records the nesting level of the structured conditional statements. For these macros, I only allow four nesting levels deep. The next stack records the previous operation. This is important for *_else, _end,* and *_until* to ensure that they are responding correctly. The last stack records the label number for the previous operation. These stacks are combined with a label number to track the correct label to use and jump to.

The label number is appended to the end of the label using the *#v(Number)* feature of the MPASM macro assembler. When a label is encountered with this string at the end, the *Number* is evaluated and concatenated to the end of the string. For the label:

```
_test#v(123)
```

the actual label recognized by MPASM would be:

```
_test123
```

I use this feature to track which label should be jumped to. After every statement, I increment this counter for the next statement to use for its labels. Expanding the previous example to:

```
_ifv Parm1, >, Parm2
_whilec ParmA == ParmB
  :                          ;  Code to Execute if Parm1 > Parm2
  :                          ;   while ParmA == Constant ParmB
_end                         ;  End the "_while"
_else
  :                          ;  Code to Execute if Parm1 <= Parm2
_end
```

The structured programming macros will push the appropriate label number onto the stack and retrieve it when necessary. So, for the expanded example, the actual PICmicro® MCU code would be:

```
; _ifv Parm1, >, Parm2
  movf   Parm2, w
  subwf  Parm1, w
  btfss  STATUS, C
   goto  _ifelse1            ;  Not True, Jump to "else" code
; _whilec ParmA == ParmB
_ifwhile2
  movf   ParmA, w
  subwf  ParmB, w
  btfss  STATUS, Z
   goto  _ifend2
  :                          ;  Code to Execute if Parm1 > Parm2
  :                          ;   while ParmA == Constant ParmB
; _end                       ;  End the "_while"
  goto _ifwhile2
_ifend2
; _else
  goto   _ifend1             ;  Finished with "true" code, Jump to "_end"
_ifelse1
  :                          ;  Code to Execute if Parm1 <= Parm2
; _end
_ifend1
```

I realize that this code is somewhat hard to follow, but if you work through it, you will discover that the *_while* loop is separate from the *_if* code by virtue of the label number, which is *2* for *_while* and *1* for the *_if*. These values are tracked by the three stacks mentioned previously.

The stacks used to store the label number and the other values are not stacks per se, but actual variables that are shifted over by four and then loaded with the value. This limits the total label numbers to 16 different labels, but for most small PICmicro® MCU applications, this should be sufficient. If you feel that more are needed, then you could modify the macros to use a separate stack for *_if*, *_while*, and *_do*, as well as come up with a way of having multiple stack values for each one. With a bit of work, you could have up to 64 label numbers for each type of structured programming macro by expanding the type of macro saved to four different types.

The macros described in this section come under the heading of "out there." These macros were quite a bit of work to get them to the point where they are now. After reading this chapter, you have the knowledge to produce macros like this, but I want to caution you to think through what you are trying to accomplish. Macros, almost by definition, do not produce functions that are easy to debug or even understand. Remember the RTOS totally written in macro, and the effect it had on me for 15 years.

11

# DESIGNING YOUR OWN PICmicro® MCU APPLICATION

CONTENTS AT A GLANCE



As I was creating this book, I received an e-mail with an excellent suggestion: write a chapter on how a reader could develop their own applications. I haven't done this in any of my previous books and I am sure that it would be useful for many people who are new to PICmicro® MCU programming and electronics. This chapter introduces the aspects of developing PICmicro® MCU applications and what steps and roles I find to be useful when I'm designing my own PICmicro® MCU applications and other engineering projects.

Before going through the steps that I use to develop PICmicro® MCU applications, here are a few words of advice. First, document everything. Get into the habit of carrying a notebook or a PDA around with you so that if you get an idea on how to do something, you can record it. Human memory is a pretty fallible storage device and I've spent many hours trying to remember that great idea I had the day before at lunch.

Second, start small. I get a lot of e-mail from people who want to create a substantial project like the Lego "MindStorms" and ask me where they should start. My reply is to get a book and figure out how to program the PICmicro® MCU and learn how to be effective with it before starting to plan a "killer" application.

A large project for any beginner will start off strong and bog down as it seems to drag on and on. I know of some people who rise to the occasion to develop large applications and become experts through them, but these people are few and far between.

Next, don't settle on the first method that you develop. There's always more than one way to do something. For example, the following methods could be used to create a delay in an application which Flashes an LED with the PICmicro® MCU:

1 Delay by code.
2 Delay by timer and polling the timer.
3 TMR0 interrupt delay with interrupt handler.
4 TM2 pulse-width-modulation output.

Which method is best in your application? Spend a few minutes thinking of options that can make the application much easier later on.

It is unfortunate, but I usually discover better methods to carry out something *after* I finish an application. If I figure out a better way, I usually keep it in my notebook for the next time, rather than go back and recreate the application. I might go back and change the application, but only if the change offers a substantial improvement to what was already there (i.e., the circuit costs less to build or can use a PICmicro® MCU with a smaller program memory). The basic rule here is "if it ain't broke, don't fix it."

Lastly, steal other people's ideas and methods. This does not mean stealing their code and circuits, but instead understand how their code, interfaces and applications work and recreating them as it works best for you. This will not only help you avoid getting "blocked," but chances are that you will learn from others and be able to use their ideas in ways that aren't readily apparent.

An excellent resource for this is Rudolf Graf's *Encyclopedia of Electronic Circuits* (more information about the books is available in the appendices). This series of books provides a large selection of circuits from virtually every facet of electronics. Other resources include the various monthly electronics magazines (also listed in the appendices) because these will generally have at least one PICmicro® MCU application in each issue.

The Internet also has a plethora of circuits at various web sites. I just want to caution you not to believe everything that you read on the Internet because the information can be inaccurate (as it is for any information). I've also found a number of web pages that offer a few free circuits as an inducement for you to buy design information from them or using them as consultants.

# Requirements Definition

The most important thing you can do before starting to design and code a PICmicro® MCU application is to create a set of requirements for your application. These requirements will help you to understand what is required to do the project and allow you to check off whether or not you are meeting the original design requirements. When you first start developing PICmicro® MCU applications, you should very rigorously define the requirements of your application because this will allow you to track what you are doing and not get overwhelmed with the task in front of you.

The first thing that I like to do is to create a simple, one-sentence statement that specifies what the project is supposed to do. Some examples, taken from the experiments, projects, and tools in this book are:

"I want to create a PC-controlled PICmicro® MCU programmer that is easy to build and will work with virtually all available PCs."

"I want to demonstrate how the operation of the PICmicro® MCU's I/O pins when an output is programmed high, but held low can lead to inadvertent changing of output states."

"This project will convert an NRZ serial data stream to LCD character data at 1200 and 9600 bps."

Many of these application-description statements are found in the first paragraph of the application write-ups presented in this book. I always feel that you must be able to describe what you want to do, simply and concisely. This statement becomes a guide for you to follow when creating your other requirements definitions.

Next, the physical requirements for the application must be stated. These requirements specify the external requirements that will drive the circuit and the software requirements that follow. A good list to have is:

1. What is the desired size of the project?
2. How much power does the project consume?
3. What is the power source?
4. What type of circuit carrier/board is to be used?
5. Who will be using the project?
6. What are the user interfaces?
7. What are the safety concerns?
8. How much should the project cost?

For many of your answers to these questions, the answer will be "none" or "no restrictions." Even though some answers might be vague, you are still narrowing down what you want to do.

Next, I decide upon the hardware that is appropriate for the project. This list includes:

1. What PICmicro® MCU will be used?
2. What type of clocking will be used?
3. What kind of power supply?
4. What input devices will be used?
5. What output devices will be used?
6. What storage devices will be used?
7. Are the parts available?

These questions probably seem very straight forward, except for the last one. As I write the books, I am discovering more and more what I think is easy to find, often isn't. When I first designed the El Cheapo programmer, I used a P-channel MOSFET that I could find easily from a variety of sources. Unfortunately, this wasn't true for many other people. I changed the transistor to a 2N3906, a very common bipolar PNP transistor. The complaints stopped and acceptance of the programmer increased.

Next, I prepare a list of requirements for software. These requirements include:

1. Language application written in.
2. Simulator availability.
3. Programmer availability.
4. Interrupts to be used.
5. Built-in application interfaces to be used.
6. Data structures to be used.

I should point out that all the lists of requirements are interconnected. You might initially assume that a specific PICmicro® MCU device is to be used only to discover that it is not appropriate for other requirements, or it might have capabilities that you don't require.

Notice that I don't include an expected application size, which would help select the PICmicro® MCU that will be used. When you are first starting out, it can be very difficult to predict the final size of an application. I've been doing it for years and I still can be off by 50% (or more) from what I guess. Often, I specify a device with double the program memory that I think I will need and change it when I better understand what is required.

As you develop your requirements list, you might feel that the different requirements are mutually exclusive; you might want to use a certain PICmicro® MCU part number, but it doesn't support the features required in software. You might find that your suppliers cannot reliably provide you with the parts that are required. If this is the case, you might have to go back and change your original defining statement. This is really not a serious problem. As you work through the requirements if you are willing to work objectively with your requirements and assumptions they're based on, you will find that you can get a solid set of requirements that will lead you to a successful application.

## DEVELOPING A QUALIFICATION PLAN

Part of the requirements that you define for your application should be an application qualification plan. This plan consists of a list of tests that the application must pass before you consider it ready for use. The list of qualification tests presented in this section include some items that might seem obvious, but part of the purpose of the qualification plan is to give you a checklist for when you build other units or instruct others to build the application.

A typical qualification plan would consist of:

- Test for the power supply to supply 4.75 to 5.25 volts of regulated power at the required current load.
- *_MCLR* pulled to an indeterminant logic level.
- The built-in oscillator running when *both* OSC1 and OSC2 are probed.
- User interface output functions are at the correct initial state on power up.
- User interface output functions respond correctly to user inputs.
- Application outputs respond to user inputs as expected.
- Application outputs respond to the timer or application input events as expected.
- Application test can detect all functional failures.
- Reliability calculations have been performed to ensure that the application will not fail before its expected life is finished.

- Environmental qualification.
- FCC, BABT, and other emissions and regulatory testing.
- User documentation is complete and correct.
- Manufacturer documentation is complete and correct.

The first three points of this list (checking power, reset, and clocking) might seem to be obvious and unnecessary, but I would argue that they are needed to ensure that the application will work reliably. If the application isn't executing, then they should be looked at first.

The power range specified in this list probably seems quite restricted—especially with the wide input voltage ranges available to the PICmicro® MCU. You should understand the operating condition of other chips in the circuit and find out their input voltage tolerances. Many active integrated circuits only work reliably within the ±5% window that I have specified in the requirements list.

For the PICmicro® MCU I/O functions, notice that I have tried to order the qualification tests, based upon the order in which the application would be debugged. Initially, just see how anything is coded and qualified at the PICmicro® MCU outputs, followed by responses to user (your) inputs. When this is done, checking to see how the application responds to user inputs, as well as application inputs will give you an idea of how well the application works. If a problem occurs at any one of these steps, you will at least have an idea of where the application failed.

Notice that for many different kinds of inputs, a simple digital multimeter or logic probe will not be sufficient. Ideally, an oscilloscope or logic analyzer should be used to look at the actual I/O and confirm that it is correct. If you do not have access to these tools, you should rigorously simulate the application to ensure that all timings are correct.

In most cases, I would not consider it to be acceptable to "qualify" an application by it working when it is connected to one sample of another piece of hardware. Some variances in the hardware include those that work in one application, but don't work in another.

For example, you might have a problem with a bit-banging RS-232 interface, depending on the tolerance of the clock that is used. If a three-times sample algorithm, as covered elsewhere in the book, is used, then the maximum a bit can be "out" is 30% at the end of the second to last bit. In 8-N-1 transmission, this is bit seven of the application. In this case, the error after nine bits can only be 30% or an error of 3.33 percent in the application clocking.

Depending on the oscillator, 3.33% accuracy is pretty easy to get, but what if the device you are communicating with uses the same bit-banging serial algorithm? Depending on the hardware and the situation, one device can be out by as much as 6.67 percent and the two devices will not properly communicate with each other.

To avoid this type of problem, I would specify a clocking accuracy that does not exceed one third of the worst-case error. In this example, this is 1.11%, again not very hard to get with most clocking schemes available for the PICmicro® MCU, but something to be aware of.

This section on clocking brings up the point of guardbanding and what is the appropriate amount of spacing, slack, or slop you can have in your application's timing and still have it work reliably. In this example, going with the standard rule, there is a 1.11% timing margin or a 100% error bit timing margin allowed for a device's specified clock to be unacceptable.

This might seem like a lot (especially when you phrase it as a margin of 1% out of specification), but in actual terms, it isn't. In the example application, the margin is really only 1% of the total clock's accuracy. To properly qualify an application, you might have to understand the error distribution of the clock circuits to ensure that nothing is shipped that is this far out of tolerance.

Another solution that would be part of the qualification plan is to ensure that a margin check is included in the application test.

Application testing is something that is near and dear to my heart (mostly because I have spent almost all of my professional life ensuring that products work correctly). In this list of qualification items, I noted that the test should be for all functional aspects of an application. If you've worked around electronic circuits for any length of time, you will know that this is almost always impossible because there are so many different ways in which a circuit can fail.

It is important to ensure the application will respond correctly to the expected inputs. In some cases, unexpected inputs will be received (such as when the application connection is "jostled" or when the circuit is "zapped" by static electricity) and these are hard to plan for. You should have a good idea of how the application works under normal operating conditions and be able to test these conditions.

It probably seems surprising that I have included a "reliability" required in the list above—especially because you are probably planning that your first applications are going to turn on an LED, based on a button's input. I will not go into reliability calculations in this book, but if you are going to develop an application for a commercial product, then you will probably want to do the calculations and even test out the application in an accelerated life test to ensure that your calculations are correct.

Reliability qualification is not something that many companies worry about and, with the ruggedness of modern components, it is not perceived as being a crucial factor to a product's success. Personally, I believe that testing the reliability of a product is the most important indicator of the quality of the design and the quality of the components used. Admittedly, reliability testing can add significant costs to a product, but it can prevent unhappy customers and lawsuits because of products that fail prematurely.

*Environmental qualification* is simply ensuring that the product will run reliably in the location where it is expected to operate. Obviously, for most of the applications presented here, this is not a concern because the circuits will be run on a bench or in a home or office environment. Part of the environmental testing is ensuring the application does not radiate excessive noise. Products sold commercially have to be FCC or EU emissions certified.

Many applications work in extremes of voltage, electrical noise, vibration, temperature, or humidity. Along with this, the applications can be sealed into a package and end up being heated by the circuits used within them. If any environmental extremes are expected, the product should be tested and qualified for these environments. Reliability calculations should be performed to ensure that the product will operate for its required lifetime.

Many of the environments where circuits are used will require testing to industry or governmental specifications. For example, if your application was going to be placed in an aircraft, then you should ensure that all specifications are met for equipment operating in this environment. Many of these specifications (especially in the aerospace arena) will seem difficult and expensive to meet, but remember that, in many cases, these specifications were a result of other, earlier products failing (sometimes injuring people).

Lastly, be sure that you have properly documented your application. I consider this aspect of project development to be part of the qualification plan because documentation for users, as well as manufacturers, is crucial to having your product integrated. User interfaces should be designed to make the product easy and intuitive to use (more on this). Being sure that the extra hardware required by the user, as well as how to connect and power the application is always required if anybody other than yourself is going to use the application or built it.

Documentation should be written using an electronic tool that is widely available to any potential users. If you don't feel comfortable with a single format (for example, your product can be used with PCs, Macintoshes, or Unix workstations), then you should document how to use the product in straight ASCII text files.

If straight ASCII text files don't seem appropriate because formatted text is required with graphics, then I would suggest either Adobe PDF format or using *HTML* (HyperText Markup Language) for your documentation. Both data formats are widely available on different computer systems, along with development tools that can be found fairly cheaply and easily.

For the information on the CD-ROM that comes with this book, I choose HTML because I have the tools for working on my web page source code. PDF is often preferred by companies because the content cannot be modified and redistributed from the files.

Documentation format and content should be discussed with the target audience to ensure that it is written appropriately and will be usable. Providing documentation electronically is usually preferred because it can be easily distributed and updated without having to send out multiple text copies.

All these requirements can be summarized in an ISO9000 registration. ISO9000 (the generic term) is a certification required by the European Union and is used to document development and manufacturing processes. Even if you are a one-person company, ISO9000 certification can be achieved for a modest cost and will add enforced rigor to your development process.

ISO9000 vendor certification is a requirement for many companies. For example, when I quoted the El Cheapo PC board that is included in this book, only ISO9000-certified companies were given the opportunity to do the work.

# PICmicro® MCU Resource Allocation

Once you have a requirements plan, you must then decide how the resources in the PICmicro® MCU are to be used by the application. Often, you will find that the application interfaces require more resources than what the chosen PICmicro® MCU can provide or they will conflict in such a way that the two resources cannot be used independently. In this case, you will have to go back and look at the application's requirements definition to see what can be changed so that the resources don't use more than what's available in the PICmicro® MCU. Resource allocation is really a part of the requirements definition; no circuits have been designed and no code has been written, so there should not be any significant lost effort or time in the development process if you have to go back and change the requirements at this point.

Resources in the PICmicro® MCU include:

1 Program memory
2 File registers
3 I/O pins
4 Interruptions
5 TMRO
6 Prescaler
7 TMR1 and TMR2
8 USART
9 SPI
10 I2C
11 ADC
12 Data EEPROM

Looking at the list, you probably are surprised at some of the things I consider to be "resources." Program memory and file register usage was addressed in the requirements definition, but these resources must be allocated for specific uses. For example, to list the file register resources, you might want to list the arrays and variables with the purpose of trying to arrange them so there is a minimum amount of space that cannot be used because of interaction among the variables and arrays.

Estimating actual program memory requirements is much more difficult. Probably the best allocation checking that can be done is to try to break out each block of code and estimate how large it will be. This isn't anywhere near 100% accurate, but it should alert you to any significant problems with not having enough space. As you work through applications, you will gain a better "eye" at estimating how many instructions some codes will require.

Timer allocation can be tricky. You might find yourself with more timed functions than you have timers. In this case, you might have to run the timer with short common-denominator delays and processing, according to the various counters required by the application. For example, to implement two routines that must be run at 50 ms and 60 ms intervals, you could define TMRO to reset once every 10 ms and invoke the two routines using the code:

```
 org            4
int
 movwf          _w
 movf           STATUS, w
 movwf          _status

 bcf            INTCON, TOIF        ;  Reset TMRO Interrupt Request
 movlw          10msecDelay
 movwf          TMRO                ;  Reset 10 msec TMRO delay

 decfsz         50msCounter, f      ;  Check for 50 msec Since Last
  goto          60msCheck           ;  Not Zero, Check 60 msec Routine
 movlw          5
 movwf          50msCounter
 :                                  ;  Execute 50 msec Routine Code

60msCheck                           ;  Check 60 msec Since Last
 decfsz         60msCounter, f      ;    to see if it should execute
  goto          intend
```

```
  movlw       6
  movwf       60msCounter
  :                                   ;  Execute 60 msec Routine code

Int end
  movf        _status, w
  movwf       STATUS, w
  swapf       _w, f
  swapf       _w, w

  retfie
```

The only restriction to this routine is that neither the 50- or 60-ms routine can run for more than 10 ms. Combined with the total execution time, it cannot be more than 10 ms. Ideally, no more than nine ms should be the longest delay to ensure that there will be no problems when the code executes.

Pins can seem like the devil in some applications—especially if built-in functions are used. My recommendations for allocating pins are to keep PORTA free for use as an analog input port or for simple digital I/Os, and try to keep byte-wide functions on a port with no built-in I/O functions that are likely to be used.

Allocating resources is really a part of the application requirements definition process. Like the requirements definition, you might want to repeat the resource allocations planning and work through any conflicts or short falls. This can result in changing the application's requirements, but, in the long run, will simplify the amount of work that has to be done to get the application working.

# Effective User Interfacing

At Celestica, I have to work with an *Environmental Stress Screening (ESS)* chamber that has one of the difficult-to-use interfaces I have ever been exposed to. The (8051) microcontroller-driven LCD/pushbutton interface is poor because the user is not prompted through the process of controlling or programming the ESS chamber. The interface does not give feedback with messages indicating the current operating mode and no prompting on how to jump to other operating modes. One of the biggest challenges is to actually figure out how to start the chamber operating (either in Manual or Automatic modes). Nobody approaches any of the chambers that have these controllers without an inch-thick manual. This is a real mystery to me because the interface has a 16 character ASCII LCD display and six buttons that could be used for feedback and control functions. Creating a well-liked user interface generally isn't that hard to do or expensive (especially when compared against the costs of thick paper manuals, websites, or returned product).

By following a few simple rules, you can come up with a user interface for your application that will make operation of the application intuitive and easy to use effectively by somebody with very little cost.

The first thing you should do is to understand what kind of feedback is appropriate for the application. For example, an oxygen sensor in an automobile does not require a manned user interface, but an RS-232 interface might be appropriate for service center maintenance. A burglar alarm circuit should probably have a light and a siren. A programmable microwave oven should have a keypad and LCD character display for setting the oven strength and cooking time.

When you are deciding the best way to interface to the user, remember that the user probably won't have a manual with them. If an operating mode has to be selected or data entered, the user will have to be guided through the process. This could be done in a manual, but manuals get lost. Keeping the operating instructions in the circuit will keep the difficulty of using the application to a minimum.

In many other circuits, you will see that LEDs are used for this function (Fig. 11-1).

There are three problems with this method. First is that extra money is required to have the panel(s) laid out and manufactured. If the functions change, you will have to change the panel, requiring that the design and layout costs be paid repeatedly. The second concern is that this panel is only in English. If other languages have to be supported, then additional panels must be designed and procured. This can be a headache when the products are configured and shipped.

By far, the biggest concern with using LEDs in a panel like this is that the user will have trouble figuring out how to get to another mode or what to do next in the current mode. Unless a printed manual is included with the application, you will find users that have trouble figuring out how to work the application. Along with these problems are the issues of labeling buttons and ensuring that their labels are appropriate.

A much better method is to use a LCD or other alphanumeric display to not only guide the user, but also give feedback on the device's operation. In the various protects presented in this book, I have a couple of examples using LCDs for data display and user interfacing.

With the alphanumeric display, the software should support a menuing system with prompts. When I wrote *PC Ph.D.,* I created a simple programmable ISA PICmicro® MCU interface. This interface used a two-line LCD and two buttons to select whether or not a device was active and what address it works at (Fig. 11-2). One button was used to select the parameter value and the other was used to select the next menu.

Solutions like using a LCD probably seem expensive, but they should be compared against the cost of writing and printing manuals and designing and ordering custom panels.

When the PICmicro® MCU is connected to another device (or if any two devices are wired together), there should be some kind of constant check (ping) to ensure that the link is active. If you look ahead to Chapter 13, you will see that I go to considerable lengths to ping programmer hardware and to confirm whether or not the connection is active and dis-

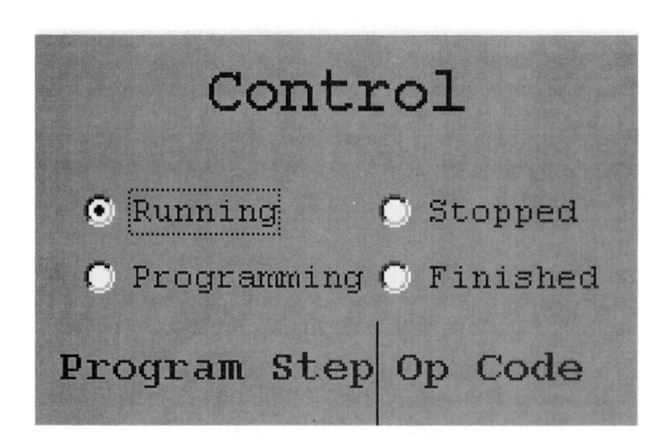


**Figure 11-1** Sample "LED Panel"

**Figure 11-2**

play the current status to the user. It might be a good idea to flag a broken link with a Flashing indicator to bring it to the user's attention.

Data transfer must also be done in a way that the user can see that the link is operational and working. Again, looking at the programmer interfaces provided in this book, you can see that I put in *gauges,* which show the progress of the programming operation.

If you have the ability to store execution parameters (e.g., dataports, file names, etc.), you might want to take advantage of this to allow the user to resume exactly where they left off. This feature makes the application easier to use and avoids reconfiguring between operations.

For input, obviously all inputs must be debounced–either by hardware or within software. Although this section about menus might lead you to the conclusion that the best way to do an interface is with a keypad, you should take a look through the book to see the different interfaces that were created with just a few buttons (and the occasional potentiometer).

# Project Management

Looking over this chapter, I have not done a lot to explain formal project management. Instead, I have just looked at the issues surrounding a hobbyist or new PICmicro® MCU application designer encounter. Although many excellent books cover project-management techniques and issues, I wanted to introduce you to some of the concepts and issues that you will face when developing a PICmicro® MCU application and some of the tools that I have found that are helpful.

To summarize, you should have a plan for the application that includes:

- Tools to be used to develop the application
- Interfaces
- PICmicro® MCU resource specification
- Parts required
- User interface
- Application qualification plan

What is left is to estimate the cost of the work, schedule the work, and create and manage a team of people to do the work. If you are going to build the application for a company, then this information will be required before anybody will give you the money and people needed to do the project.

Costing a project is never simple. Along with planning on the costs of parts that will be used, you also have to take into account the costs of:

- Buying development tools and PCs and Workstations to run them on
- Printed circuit card (PC board) NRE costs
- Shipping costs with customs and duty
- Team members costs

Each of these items will require some research and an understanding of how the costs work for different sources. For example, if you were going to buy an emulator, would it be best to buy it retail or through Microchip? Which emulator is best for this project and why? Can you lease the equipment for the length of time that the project is required? The answers to these questions mean that the emulator costs for a project can be anywhere from 100 dollars to 10,000.

Also in these decisions are people costs. For example, are people available within your company to work on the project? Can you hire temporary employees or use a consultant? Again, the range of options can result in two orders of magnitude differences in out-of-pocket costs (just like the emulator example).

When deciding on the people costs, you will have to first calculate the amount of time and effort required in developing the product and creating a schedule of the important milestones. Calculating the amount of time to develop a project is best done by first breaking the project down into every possible individual work item. When you have a list of work items, you can begin to list the time and effort required for each one. Scheduling this way will help you defend the final number to your management and enable you to create a schedule to monitor how the application's development is progressing.

When all of the work items are listed, you should then be able to produce a Gandt chart. This tool allows you to list all the work items for a project, as well as draw in the crucial path, as done in Fig. 11-3.

Figure 11-3 shows "Tool procurement," "Application circuit design," "PC board layout," and "FCC testing" as the *Crucial Path (CP)* items. These are the work items that must be completed on schedule or the project will not meet its product release date.

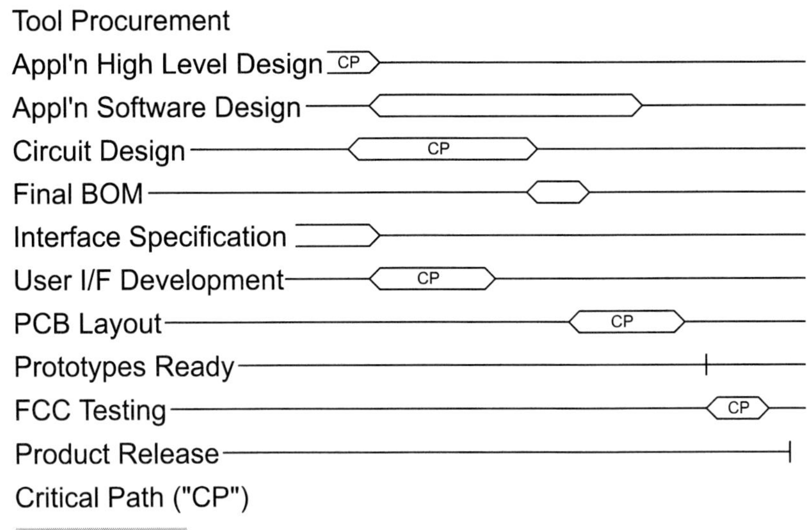


**Figure 11-3** **Sample project "Gandt" chart**

A few things in Fig. 11-3 will probably stick out at you, like the scheduling of the application software development into the FCC testing time. When testing operations are being done, there is often the need to go back and fix some aspect of the application. The project might become behind schedule, so making up time will be crucial. By extending noncrucial path work items in the schedule, then contingency time and resources are available to prevent schedule "slippage" or "movements to the right."

The *Bill Of Material (BOM)* is the list of parts required to building a product. This list will also contain packaging and extra materials (like manuals and cables) and diskettes or CD-ROMs that the product is shipped with. In the projects and experiments listed in this book, I have created simple bills of material for you to ensure that you have the parts on hand before starting the applications.

In any project, you should build in a contingency of money, people, and time for unforeseen problems. I like to have a minimum of 25% for each of these three parameters. You can pride yourself in creating accurate schedules and costs, but unforeseen events can cause problems with making your final date at cost. Some problems I have had to endure include an earthquake at a supplier's factory, a strike at the contract manufacturer building the prototypes, and a national government slapping a duty on one of the parts that I wanted to use.

Always remember that contingency funds can be returned or a release date moved up. Generally, neither thing is bad and can often make you look like a hero.

The hardest aspect of managing a project is to manage the people working on it. Ideally, you would like to have friends working with you, but this is rarely possible. You end up with people you've never worked with before and who you aren't sure you are going to like. I've been involved in many team projects. Here are some things I have found helps make them work. Having a team that doesn't work well together will be the death of a project (and I have been there more times than I care to admit).

There is only one leader with a very clear vision of what the application should be and how it should work. Decisions regarding allowing deviations to the original product concept should only be made by the team leader. Deviations should be made to ensure that the original product vision is not lost. The team leader should also be the only person responsible for presenting status information to management with their interpretation of the progress or problems.

The team should consist of individuals with complementary and reinforcing strengths. It is not always possible to select the best people for the different work items, but the leader should be able to distribute the work to the team members in such a way that individual strengths are capitalized upon.

It is the responsibility of each team member to keep the others apprised to their status and offer suggestions when other members are having problems. Some of the best solutions to hardware problems that I have seen have come from software developers and vice versa. The best way to monitor progress is with a daily meeting. This meeting doesn't have to be very long, but recognizing everyone's contributions and ideas is important to make everyone feel like they are contributing toward the final goal.

Little things like buying T-shirts for each members to wear while working on the project or having somebody bring in donuts and coffee to the daily meetings might sound corny, but they do help to foster a team attitude.

Care must be taken to avoid creating a "poisoned work environment." The obvious example is calling another person a racist name or telling obscene jokes in front of people who could be offended. Not only will this hurt the team's overall spirit, but many actions like these are considered *harassment* and are illegal in many jurisdictions.

Examples of more subtle behaviors that will lead to problems are singling one person out to do secretarial chores, such as getting coffee, office supplies, or taking notes. These activities must be shared amongst the group and the leader should ensure that any tasks that haven't been assigned to a specific individual are shared by the group at large.

Awards are excellent tools to help motivate team members. They can also hurt moral if they are given on a basis of friendship or other reasons that are not based on merit and contribution. Be sure that somebody isn't going to feel badly if someone else receives an award.

People will have personal problems and tragedies. The leader should be aware of any issues and be sure the other team members are aware there is a problem (not necessarily what it is) and ensure latitude is given. Often, this type of action will result in a stronger team because the message that is being given out is that the members of the team can rely on each other.

Many issues relate to being a part of a team, but they can be summarized in four points.

1. Be sensitive to others and do not engage in or allow upsetting or offensive behaviors.
2. Keep everyone apprised to your status and any problems that you are having.
3. Work toward the team's goals and objectives.
4. The leader is the only person who can make decisions that will affect the outcome of the project.

# 12

# DEBUGGING YOUR APPLICATIONS

## CONTENTS AT A GLANCE



In the first edition of this book, I rather facetiously suggested that to debug your application, you should follow the three-step process:

1. Simulate
2. Simulate
3. Simulate

To be fair, after saying this, I did go through some information on how to successfully develop stimulus files and what to look for. I think the information was pretty sketchy and that I could have done better explaining how to find and fix application problems.

Efficiently characterizing and identifying problems is the first topic of this chapter. At work, I try to discourage people from using the term *debug* because it seems to denote a quick fix. Instead, I prefer that the term *failure analysis* be used. It's interesting to see how changing a simple word to a "$10 phrase" can change an operation that is "quick and dirty" into a structured process that allows you to understand exactly what is happening. From this knowledge, the most expedient method of fixing the problem can be found.

The following steps and procedures in this chapter will allow you to fix problems faster than if you just looked for the first problem that appeared and tried to "nail it." This will seem like a paradox because I push for you to "characterize" and understand exactly what is happening and what the symptoms are telling you. With this information and hypothesizing on what the defect is, you will understand exactly what is happening and why your fix should eliminate the problem.

Instead of blindly going ahead and simulating the application in a three-step process, I am pushing for debugging an application by:

1. Characterizing the problem.
2. Hypothesizing the cause of the problem.
3. Testing your hypothesis.

One attitude that I feel is crucial for failure analysis is being suspicious of anything that doesn't "feel right." I have worked with a number of people over the years who have let problems escape out the door simply because they didn't react to something that did not work as expected once or twice during application development. If you have a suspicion that something isn't right, then you should go out and fix it before the problem escapes and you end up with a reputation for shipping shoddy products.

# Characterizing Problems

Even if your PICmicro® MCU just seems to be sitting there and doesn't seem to be working at all, you can still get information out of it to help you figure out the problem. On the other end of the spectrum, if the application works "mostly okay," you should still be able to isolate where the errors are based on understanding what it is actually doing. This process of looking at a failing application and trying to detail its environment and response is known as *error characterization,* which is the crucial aspect of performing failure analysis. This section introduces you to the concept of error characterization and shows how it can be used to help you understand exactly what the problem is.

When I have a PICmicro® MCU application that isn't running properly, I first carry out a scan around the PICmicro® MCU to look at all of the input and output voltage levels. I normally use a logic probe and I record all of my results on a sheet of paper. Digital logic levels, but not voltage levels, can be checked with a logic probe. When you first work with your applications, you might want to check the signal pins with both a logic probe and a voltmeter until you get a good idea of what is happening and what you expect to see.

When checking the PICmicro® MCU's built-in oscillator, check both pins with a logic probe to ensure that the capacitance of the probe doesn't inadvertently set the application running.

When this is done, compare your notes with the expected values. This is an important point because I can't tell you how often I've thought I've seen a problem only to discover that I miscounted and was looking at the wrong pin.

Any discrepancies that you find can be put in one of two "buckets:" those that are caused by an external device driving the pin to the invalid state or the application code driving the pin incorrectly.

To discover which is the case, an ohmmeter check is performed before and after isolating from the circuit. This is known as *floating* the pin. It can be done by removing the PICmicro® MCU and bending the pin out or breaking the connection from the pin to the circuit. Personally, I prefer breaking the connection to bending the pin; the metal used in chip pins tends to be brittle and easily broken if it is bent back and forth even once. If the circuit turns out to be okay, you can go to the application code and look at why the pin could be in an invalid state.

When you are looking around the circuit, it is very important to not make any assumptions about the problem. I call the correct state of mind *totally naive.* This is the ability to see exactly what is happening and not putting your own "spin" on it.

If you don't try to look at everything objectively, you will probably miss the shorted pins, incorrect wiring, or wrong instructions (e.g., *addwf register, w* instead of *addwf register, f*.)

Being totally naive is a skill and I've only met one or two people over the course of my life that has truly mastered it. As humans, we tend to remember what we wanted to do and tend to see that first and miss what is actually in front of us.

Once I was presenting a course on failure analysis with some technicians at work and presented the concept of being "totally naive" when debugging circuits. The class and I were amused when one of the technicians shared with us the perspective he used when debugging defective boards. The technician said that he worked like that all of the time, however, he imagined that he was checking his *boss's* work and hoped to find something he could hold over him. The more obvious a problem, the better it made him feel.

Thinking about it, I have never met anyone who was so happy to be looking for problems.

From this check of the PICmicro® MCU, you will have either found a problem with the circuit or you are going to have to look at the PICmicro® MCU itself. The first thing I do is to pull the PICmicro® MCU out of the circuit and check it on a programmer against the latest application code. The problems that are typically found here are programming errors (often not programmed at all; somehow a step was skipped) or the configuration fuses are set wrong (wrong oscillator type, watchdog timer incorrectly enabled, etc.).

If the circuit and PICmicro® MCU programming are good, you then have to check your application code, but you have the knowledge that the circuit is good and the problem must lie in the software. Later, I present how to use the simulator to find problems.

The important aspect of the characterization process is to document your problems. When I'm setting up a failure-analysis process for a new product at work, I issue notebooks to all the technicians and instruct them in how to fill them out. This notebook is to be kept forever because it is a database of problems and what was done to find them and fix them. This is true even if the failures are recorded on a computer database.

In cases where the application starts up and then fails, the importance of keeping notes detailing what has happened is crucial. For example, if you had an RS-232 interface that locked up, you would have to figure out if the problem was time based, or a data value or data volume (the number of incoming characters). Once this determination was made, then you could look for the problem by simulating the data input and taking advantage of this parameter.

In trying to find a problem, don't be afraid to create a small application to force the problem to the surface. If you can create an automated way to cause the error, then you have correctly characterized the situation and can translate this knowledge to a simulator to observe from the software's "perspective" exactly what is happening within the application code.

The goal of characterizing the problem is to document exactly how the circuit is operating and what its inputs are so that you can reproduce these conditions (to be sure that the problem is fixed once you have made changes) and work at hypothesizing about what is actually happening.

# Hypothesizing and Testing Your Hypothesis

Once you have your problem characterized, you should be able to make a hypothesis on the cause of the problem. This is the most important step because going through the characterization data you collected should point you toward the problem and the fix. Before going on and trying to fix the application using the hypothesis, you should try to test it—either as a "thought experiment" or on actual hardware to see if you can reproduce the problem and have it behave exactly as on the failing board.

Part of making your hypothesis is identifying the problem. This is done using the observations you had made of the problem. I am saying this because it is not unusual to have multiple problems with an application and consider yourself an expert. In the YAP-II, as I was finishing off the code, I identified five concerns that I had about the programmer's execution. After splitting them out, I starting breaking up my observations into what I thought the problem actually was.

The appendices include a table of failures and potential causes. When you have a problem with the PICmicro® MCU, I recommend that you should look here first because it describes what you should see with each problem. As you gain experience debugging your applications, you can rely on this appendix less and less.

When making your hypothesis, you should first list all the observations that seem to be pertinent to the problem at hand. One of the problems with the YAP-II was that ASCII control characters were being accepted from the command prompt instead of being ignored. To build my hypothesis on the problem, I sifted through my notes of the problems (the problem characteristics) and made the following observations:

**1** When the Prompt is active, sometimes characters would be displayed on the HyperTerminal screen as small boxes.
**2** These boxes would cause an <== *Invalid* message when *Enter* was pressed.
**2** I noticed that when the boxes were displayed, I tended to be pressing down on both the *Shift* and *Ctrl* keys at the same time.
**4** When I press the *Ctrl* key with a character key, the small box would appear.

From these observations, I made the hypothesis that ASCII control characters could be produced using the *Ctrl* key and character keys. I further made the hypothesis that the YAP-II's instruction-processing code was not properly filtering out these characters.

To test out the first hypothesis, I disconnected the YAP-II from the PC's serial port and shorted together the receive and transmit lines (pins 2 and 3). I discovered that when I pressed *Ctrl* along with a character key, the box would be displayed on the HyperTerminal window, as it did with the YAP-II. This led me to assume that ASCII control characters were being recognized and echoed back to the PC.

I then made up a stimulus file for the application and tried passing ASCII control characters (hex values less than 0x020) to the code to see what happened. When I did this, I found that the filter I had put in place didn't seem to be working correctly. Using the stimulus file, I confirmed and then fixed the problem that I discovered in the code.

At this point, I felt I was ready to test the fix in the application, so I burned a PICmicro® MCU with the updated code and confirmed that the problem was fixed in the user interface.

This example probably seems trite and pretty obvious, but this is the basic point to following a structured approach to finding and eliminating problems. When you go through the process outlined here, fixing virtually all problems is very simple and forehead-slapping obvious when you work through it.

This is not the case for all problems, but if you go through this process of characterizing and hypothesizing about the problem, you can now talk to an expert about the problem without feeling like you are potentially wasting their time. I know of very few people who will not spend a few minutes helping out with a problem if the up-front work is done to characterize and try to understand what is actually happening.

With this background, an expert can often identify the problem or make suggestions to help understand what exactly is happening and help you find the root cause of the problem.

# Simulating Applications

Before you make any changes to your application code to fix a problem, you should be sure that you can simulate and observe the error. Once you think you have fixed the problem, this process is repeated to confirm that the problem was fixed. This simulation operation helps you close the failure-analysis circle and avoids the problem of running PICmicro® MCUs with unconfirmed "fixes."

This process also has another benefit: it will help force you to learn to use a simulator efficiently and be able to recognize when application code "goes off the rails." This might seem like an inconsequential benefit, but it actually is a major reason to follow the complete failure-analysis process outlined in this book. You cannot expect to efficiently use simulators (and emulators) right from the start. You will have to learn the features and approaches that work best for you when identifying and fixing problems.

When I detailed the operation of MPLAB earlier in the book and covered how simulated application input was provided, I gave quite short shrift to the simulated register input. I personally don't use it because it does not have the capabilities that I like when I simulate application code. I'm sure that many people use this feature of MPLAB and find it useful for debugging problems.

An aspect of simulating an application is deciding where to put breakpoints. By efficiently selecting where a breakpoint belongs, errors can be observed quickly and with a minimum of overhead. I find that correctly placing breakpoints is something of an art; the goal is placing breakpoints in such a way that problems are quickly identified. Once the problem has been found, the same breakpoints can be used to confirm that the fix has been made before another PICmicro® MCU is burned with the corrected code.

Problems that can affect the placement of breakpoints are interrupt handlers, subroutines, and macros. Subroutines and interrupts are a problem because they generally execute multiple times and their execution can be part of the failure mechanism of the application.

A good example of this is a simulated USART serial receiver that polls the line three times per bit. In the example interrupt handler that follows, when a character is received, data is passed to the mainline without any problem, but it is always followed by the byte 0x0FF (or if a byte immediately follows it, it is in error).

```
interrupt serin( )              // 3x Sample USART Serial Input
{                               // Processor

  TMR0 = 1/3 bit;               // Reset the Interrupt
  INTCON = 1 << T0IE;
  switch (Rx state) {
    case 0:                     // Nothing Active
      if (SerialInput == 0)     // Check for Start Bit
        RXState = 1;            // Set up Byte Read
    case 1:                     // Confirm Start Bit
      RXState = 2;
      BitCount = 8;             // Read in 8 bits
      BitDelay = 3;             // Delay 3 Interrupts per bit
      break;
    case 2:                     // Check a Bit For Valid
      BitDelay = BitDelay - 1;
      if (BitDelay == 0) {      // Save the bit?
        RXData = (RXData << 1) + SerialInput;
        BitDelay = 3;
        BitCount = BitCount - 1;
        if (BitCount == 0) {    // Eight Bits Read in?
          DataAvailable = 1;    // Yes
          RXState = 0;          // Restart the polling
        }
    }
    break;
  } // end switch
} // end serin
```

When you walk through the code manually, it seems to be good and should not have any problems. To understand the problem, you can create a stimulus file with incoming data and then try to look for the problem as the code exits.

The first place to look is in the main line code to see if you can repeat the problem. Chances are that you have a mainline poll of the bit-banging RS-232 interface that looks like:

```
while (1 == 1) {                // Loop For Each Incoming Character

  while (DataAvailable == 0);   // Wait for flag to indicate a new
                                //  byte o/p
  DataByte = RXData;            // Save Input Byte
  DataAvailable = 0;            // Reset Flag
    :                           // Process Data
}  // end while forever
```

A breakpoint could be put at the:

```
DataByte = RXData;
```

statement to confirm the problem.

Next you will put in a breakpoint after the:

```
if (BitCount == 0) {
```

statement in the interrupt handler to confirm that it is working twice. Chances are you will be stuck in finding the next problem.

In a case like this, I would suggest that you put a breakpoint at the start of the *serin* interrupt handler and start single stepping through the code to see what happens after the good data save. The first TMRO interrupt after the data will be available when execution tests for the serial input are low (indicating the start of a new data byte's start bit). It will find it because bit 7 is low. After moving to state one in the next TMRO overflow, it will move on and receive all ones for the next eight bits. This can be shown graphically as Fig. 12-1.

The three times sample bit-banging receive works to place the sample as close to the middle of a bit as possible. This is done by waiting one 1/3-bit interrupt period after a low start bit is found. Unfortunately, in this case, if bit 7 is a zero (which is the case for ASCII characters), then the next three times bit sample will detect the still-low bit and start reading what it "thinks" is data.

There are a number of different fixes, depending on the applications requirements. Probably the simplest way is to add a "glitch detect" in state 1 of the *serin* interrupt handler:

```
case 1:
  if (SerialInput == 0)        // If Line is High, Assume a
    RXState = 0;               //   "Glitch" and Reset State
  else {
    RXState = 2                // Else, Line is Low for two
    BitCount = 8;              //   Polling Periods, Read in
    BitDelay = 2;              //   the Character being Sent
  }
  break;
```

With this change in place, you should then simulate the application again to be sure that it works properly before you burn code into a PICmicro® MCU.

If this were real life, chances are that this fix would mostly correct the problem. But some characters would still be followed by a 0x0FF character. This is because of the chance that a pull right on the leading edge of the start bit would allow four samples in bit

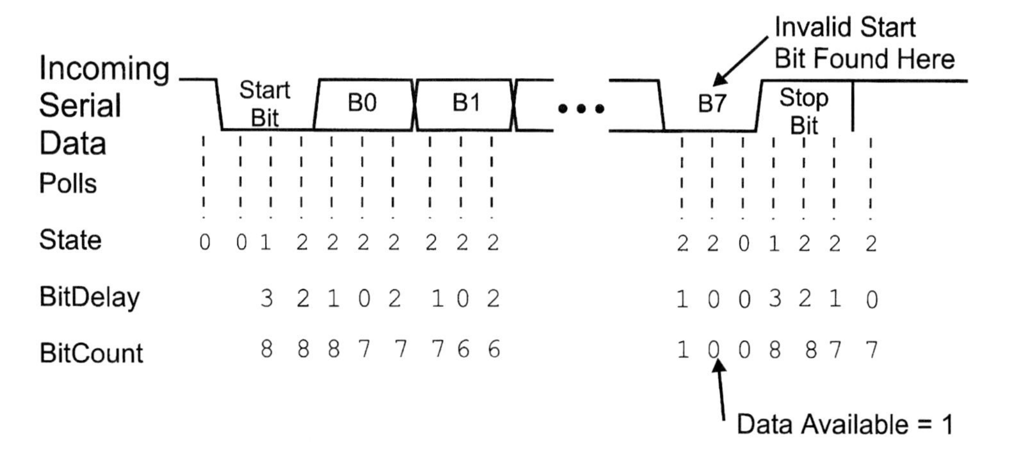


**Figure 12-1** Looking at serial input polling

seven, each of which is low. Because the last two of these would be picked up by *serin,* the new byte read would start again and return 0x0FF.

This is where your hypothesis will have to be used followed by changing the stimulus file so that the start bit fails on the first state 0 pull and the bit period is lengthened slightly (1% is all that is required). This should allow the two low reads at the end of the valid byte, but it will be difficult to time perfectly. To properly fix the code, a check for stop bit is put in after bit 7.

The previous "fix" (to "state 1") actually changes the application so that a problem caused by a different error mechanism would appear. This second error mechanism is also difficult to simulate.

## DESIGNING STIMULUS FILES

Before you burn a PICmicro® MCU with an application, you should have simulated with different inputs to eliminate as many problems as possible. In doing this, you should have created a number of stimulus files. These files are important while debugging an application because they will be used (and modified) to try and simulate the problem and help you see exactly what is happening in the application. This section covers developing stimulus files both before a PICmicro® MCU is programmed with an application, as well as when there are problems.

Stimulus files should be programmed to be as accurate input to the application as possible. When I say "accurate," I mean that it should produce signals that are representative of what the PICmicro® MCU is going to be responding to. For example, if simulating incoming asynchronous serial data at 9600 bps (104.167 μs/bit), 104 μs per bit should be used. To send ASCII A (0x041) to the simulated PICmicro® MCU, the stimulus would file:

```
Step        Rx#        !        Send ASCII "A" - 0x041
1           1          !        @ 9600bps for a PICmicro® MCU running
                       !        at 4 MHz
                       !
10000       0          !        start bit
10104       1          !        bit 0
10268       0          !        bit 1
10313       0          !        bit 2 -note round up.
10417       0          !        bit 3
10521       0          !        bit 4
10625       0          !        bit 5
10729       1          !        bit 6
10833       0          !        bit 7
10938       1          !        stop bit
```

Stimulus files are actually very simple programs that are used to test the operation of an application.

Notice that I round up at bits two and seven. This is because the actual fraction of the data sent is 0.5 of a cycle (or greater). Also notice that I hold off sending the byte until clock cycle 10,000 and not 1. This allows the PICmicro® MCU to initialize and be ready for the incoming data before anything comes in and potentially affects the operation of the PICmicro® MCU. For complex sequences like this, start at a clock cycle that will allow you to easily count offsets without having to add the offsets to a strange base.

Once this file is created, it can be copied and tested for different cases. In your circuit, you might have found that the asynchronous serial source is actually 5.6% off and data is coming at 110-ms bit periods instead of 104.167. The stimulus file to test this out would be:

```
Step    Rx#     !       Send ASCII "A" - 0x041
1       1       !      @ 9600bps - 5.6% for a PICmicro® MCU running
                !      at 4 mhz
                !
10000   0       !      start bit
10110   1       !      bit 0
10220   0       !      bit 1
10330   0       !      bit 2 -note round up.
10440   0       !      bit 3
10550   0       !      bit 4
10660   0       !      bit 5
10770   1       !      bit 6
10880   0       !      bit 7
10990   1       !      stop bit
```

This operation is known as "margining" or "guardbanding." The purpose of using it during application development is to see how tolerant the application is of invalid inputs. Chances are, you will find that a 5.6% error will not be handled properly by a bit-banging serial application.

It is a good idea to carry out this margining and guardbanding for commercial products. This operation will help "qualify" the product and help you to understand whether there is a possible software dependency. In this example, you might want to change the incoming data rate until the read becomes unreliable; combining this data with your hardware clocking specifications will give you an idea of how robust the application will be in the field.

When characterizing what is happening in your application to better understand what is actually being input into the PICmicro® MCU, you will have to use an oscilloscope or a logic analyzer to measure the exact timings of the inputs. With this data, an accurate stimulus file, you can simulate the inputs exactly as they are behaving in hardware and try to determine where the problem is in the application code.

Once you think you have fixed the problem in the application, the stimulus file that was developed to simulate the problem can be used to confirm that the fix was correct.

# 13

# PROGRAMMING PICmicro® MCUs

CONTENTS AT A GLANCE



In my mind, one of the biggest reasons why the PICmicro® MCU has become so popular is because of the ease of which mid-range devices can be programmed. The synchronous serial process allows very simple programming hardware to be created for programming the chips. Microchip was unusual because it published the programming specifications without requiring a *NonDisclosure Agreement (NDA)* with their corporate lawyers. Microchip was also one of the first microcontroller manufacturers to incorporate EEPROM (Flash) for program memory, which allows devices to be packaged cheaply (windowed ceramic packages are not required). These three actions have made Microchip the choice of many people getting into microcontrollers for the first time and why I can offer this book with a PC board with which you can build your own programmer for very little cost.

This chapter covers, in detail, how to program the different PICmicro® MCU families. On the CD-ROM, I have included the Microchip data sheets that specify how to program PICmicro® MCUs. Along with the theory, I have also included the circuit designs and software for two programmers. For the El Cheapo programmer, included with this book is a printed circuit board with which you can build your own programmer. Some third-party programmers (including Microchip's tools) are presented, which you can buy instead of building your own.

# Hex File Format

The purpose of MPLAB and other assemblers along with compilers is to convert PICmicro® MCU application source code into a data format that can be used by a programmer to load the application into a PICmicro® MCU. The most popular format (used by Microchip and most other programmers, including the two presented in this chapter) is the Intel eight-bit hex file format.

This file format can be shown using the simple application:

```
Address Code Source
      title "CPExampl - Code Protect Example"
      ;
      ;  This Example Application Simply Increments 8 LEDs on
      ;   PortB.
      ;
      ;  Hardware Notes:
      ;   A 4 MHz clock is used with this Example Application
      ;   Reset is tied through a 4.7K Resistor to Vcc
      ;   8x 220 Ohm Rs and LEDs is attached to PortB and Vcc
      ;
      ;  Myke Predko
      ;  99.10.10
      ;
      ;  Registers
        CBLOCK 0x020
      Dlay:2
        ENDC

       LIST R=DEC
       INCLUDE "P16F84.inc"

2007     3FF1    __CONFIG _CP_OFF & _WDT_OFF & _XT_OSC & _PWRTE_ON
                 PAGE
         ;  Mainline of CPExampl

                 org      0
0000     30FF     movlw    0x0FF               ;  Start with LEDs off
0001     0086     movwf    PORTB
0002     1683     bsf      STATUS, RP0
0003     0186     clrf     TRISB ^ 0x080       ;  PortB is ALL Output
0004     1283     bcf      STATUS, RP0

0005     01A0     clrf     Dlay                ;  Clear the Delay
```

```
0006      01A1    clrf    Dlay + 1

0007            Loop                    ;  Endlessly Loop
0007      0BA0    decfsz  Dlay, f       ;  Delay 1/5 Second
0008      2807     goto   Loop          ;   before...
0009      0BA1    decfsz  Dlay + 1, f
000A      2807     goto   Loop

000B      0386    decf    PORTB, f      ;  Incrementing the LEDs

000C      2807    goto    Loop

                 end
```

This application is designed to use eight LEDs, connected to PortB, to count up to 0x0FF and then restart at 0x000. Along with the source code instruction statements, in this listing, I have included the address where each one goes and the bit pattern generated by MPLAB's assembler. This source file is named CPEXAMPLE.ASM and is on the accompanying CD-ROM.

When the hex file is assembled, a hex file (CPEXAMPLE.HEX) is generated. It may seem unneeded to explain this, but the file is referred to as a *hex file* because that is the filename extension given to it. This file contains:

```
:10000000FF308600831686018312A001A101A00B98
:0A0010000728A10B07288603072824
:02400E00F13F80
:00000001FF
```

Each line consists of a starting address and data to be placed starting at this address. The different positions of each line are defined in Table 13-1.

Each line starts with a colon (*:* or ASCII Code 0x03A), along with twice the number of instructions that are on the line. This value will often be *0x010* (because it is in the first line of the example hex file), which is 16 decimals and indicates that there are eight instructions on the line.

**TABLE 13-1 Hex File: Line Definition**

| BYTE | FUNCTION |
|---|---|
| First (1) | Always ":" to indicate the Start of the Line. |
| 2–3 | Two Times the Number of Instructions on the Line. |
| 4–7 | Two Times the Starting Address for the Instructions on the Line. This is in "Motorola" Format. |
| 8–9 | The Line Type (00 - Data, 01 - End). |
| 10–13 | The First Instruction to be loaded into the PICmicro® MCU at the Specified Address. This data is in "Intel" Format. |
| : | Additional Instructions to be loaded at Subsequent Addresses. These instructions are also in "Intel" Format. |
| Last 2 | The Checksum of the Line. |

Each pair of characters makes up an ASCII byte, with the most-significant nybble coming first, followed by the least-significant nybble. Some of the data is represented by four bytes (which will translate to two bytes (16 bits) of actual data), with each pair of bytes used to make up a byte of data or address.

The next four bytes indicate the starting address of the data on the line. If there was a break in the code, for something like:

```
Address  Code  Source
0000             org     0

0000     2807     goto    Mainline              ;  Skip Over Interrupt
                                                ;   Handler
                  PAGE
                  org      4                    ;  Start of Interrupt 0004
                int_handler                     ;   Handler
0004     1529     bsf     TMR0Flag
0005     110B     bcf     INTCON, T0IF
0006     0009     retfie
                 PAGE
0007            Mainline                        ;  Application Mainline
```

The hex file would have a separate line for the reset code starting at Address 0 and the interrupt handler code at address 0:

```
:020000000728CF
:0800080029150B1109008316F4
```

Notice that the second line ends at the address 8 boundary (the next line of data will start at address 0x008, the following one 0x010, etc.). This is not necessary, but is a convention used by the MPASM assembler.

One important point about the address file (along with it being twice the actual value) is that it is in what I call *Motorola format.* The high byte of the address is placed before the low byte, so the address can be read straight out from the file without any translation. In the third line of the CPExampl hex file, address 400E is actually 0x02007 (which is the address of the configuration fuses).

After each instruction is loaded into the PICmicro® MCU's program memory, an internal counter is incremented. When a line is finished, this counter is usually at the correct value for the next line, but if it is not, then it is incremented until it is the same as the line's address. This means that if gaps are in the application, the addresses will be left unprogrammed.

The next two bytes specify the line type. Normally, this is *00,* indicating that the line is data, but when it is *01,* it indicates that the line is the end of the file.

The actual data follows the data type bytes. Each of the following four bytes represents the 14 bits that are to be loaded into the PICmicro® MCU's program memory. Unlike the address bytes, the instruction bytes are saved in *Intel format,* which means the first two bytes are the least-significant bytes of the instruction. The instruction bytes are also not multiplied by two.

Breaking the first line up into parts and comparing it to the instructions and generated bytes for loading into the PICmicro® MCU, the Intel format for the instructions can be seen quite clearly when compared against the listing file.

```
:10 0000 00 FF30 8600 8316 8601 8312 A001 A101 A00B 98
```

The actual data is:

```
30FF    movlw   0x0FF              ;  Start with LEDs off
0086    movwf   PORTB
1683    bsf     STATUS, RP0
0186    clrf    TRISB ^ 0x080      ;  PortB is ALL Output
1283    bcf     STATUS, RP0
01A0    clrf    Dlay               ;  Clear the Delay
01A1    clrf    Dlay + 1

0BA0    decfsz  Dlay, f            ;  Delay 1/5 Second
```

When you first work with the Intel format code, it can be somewhat difficult to initially pick out the full instruction word. As you get more familiar with it (especially if you design your own programmer), you will be able to very quickly decode it.

The last two bytes of each line of the hex file are the checksum of the line. This value is used to confirm the contents of the line and ensure that when all bytes of the line are summed, the least-significant eight bits are equal to 0x000. This value is calculated by taking the least-significant eight bits of the sum of the line and subtracting it from 0x0100.

Using the second line of CPExampl,

```
:0A0010000728A10B07288603072824
```

The sum of all the bytes (except for the checksum bytes is):

```
   0A
   00
   10
   00
   07
   28
   A1
   0B
   07
   28
   86
   03
   07
 + 28
------
  1DC
```

The least-significant eight bits (0x0DC) are taken away from 0x0100 to get the checksum:

```
  0x0100
- 0x00DC
----------
  0x0024
```

This calculated checksum value of *0x024* is the same as the last two bytes of the original line.

While I've called the two checksum bytes the end of each line in the hex file, each line in the file is actually terminated by an ASCII carriage return (0x00D) and line feed (0x00A) combination. It is important to understand the line ending for "home-grown" programmers because the different way files can be read, the line-feed character may or may not be present. This caused me quite a few problems with the YAP programmer, as is detailed later in the chapter.

# Code-Protect Features

In all PICmicro® MCU devices, one or more code-protect bits are included in the configuration fuses word. These bits are used to hinder unauthorized copying or downloading of your application code once you have completed and released an application. This section covers how the PICmicro® MCU's code-protection mechanism works and three issues associated with it.

Once the code-protection bit is set for a section (or all) of program memory, data can still be read out, but it is XORed with the adjacent words. For example, using some of the application code presented in the previous section:

```
0086    movwf   PORTB
1683    bsf     STATUS, RP0
0186    clrf    TRISB ^ 0x080           ;  PortB is ALL Output
```

The code-protect function was enabled for this area of program memory and when reading back the middle word (*bsf STATUS, RP0*), the word returned to the programmer would be:

```
Returned Word = 0x00086 ^ 0x01683 ^ 0x00186
              = 0x01783
```

This method of protecting program memory is somewhat unique in the microcontroller world (other methods of providing the code-protect function include preventing the microcontroller from returning the contents of program memory when the code-protect feature is enabled or scrambling the contents with an encryption array). The PICmicro® MCU method allows the contents of program memory to be verified—even if code protection is enabled, which the other two methods make difficult.

In the EPROM program memory PICmicro® MCU part numbers, the EPROM cell(s) of the configuration word are often covered by an opaque layer of aluminum (Fig. 13-1).

The reason for doing this is to prevent the code-protect bits from being selectively erased (normally in the PICmicro® MCU, when code protection is disabled, these cells are left unprogrammed). Allowing the rest of the program memory to be read straight back. The metal layer prevents ultraviolet (erasing) light from reaching the EPROM cell.

For this reason, I recommend that you never enable code protection in the PICmicro® MCU unless you are absolutely sure of what you are doing. Although some people have

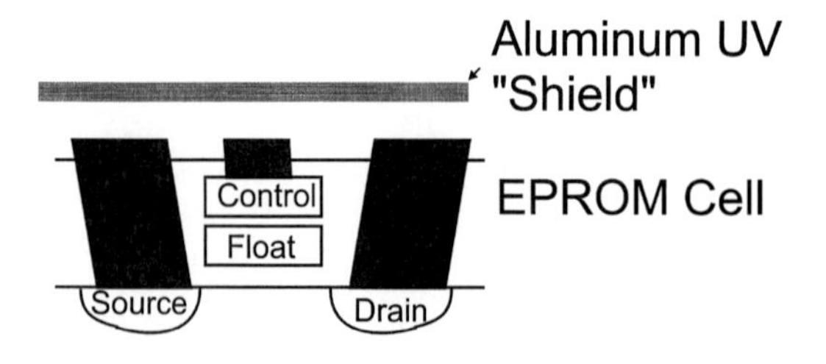


**Figure 13-1** "EPROM" cell with code protection layer

**TABLE 13-2 PIC16F877 Code Protect Configuration Bits Function**

| CODE PROTECTION BITS | FUNCTION |
|---|---|
| 00 | All Program Memory Code Protected |
| 01 | The Last 4K (Half) of Program Memory Protected |
| 10 | The last 256 Instructions Code Protected |
| 11 | No Program Memory is Code Protected |

reported that a "deep" erase cycle of several hours to several days will clear code-protect bits with the layer of aluminum over them, many people have ended up with an interesting (and expensive) piece of abstract art or jewelry.

The EEPROM (Flash) part's code protection is designed so that if it is set, a "complete" erase of the part is required before it can be reused. This will erase all of the contents of the PICmicro® MCU before resetting the code-protection bit. Later, this chapter describes the procedure.

A lot of code-protection options are available in many of the PICmicro® MCUs. For example, the 16F877 has four different ways to specify the code protection (Table 13-2). This allows you some interesting options and protection for your application.

Although the PICmicro® MCU's code-protection hardware is well designed to protect the contents of the PICmicro® MCU's program memory, it is not infallible. Many companies advertise the capabilities of reading code-protected memory (ostensibly for legitimate companies that have lost the source code to a part). The techniques used are somewhat esoteric, but can be accomplished on lab equipment (such as Scanning Electron Microscopes) which are available in many chip-making facilities around the world. I am mentioning this so that you do not develop a false sense of security if you released a product that has a code-protected PICmicro® MCU inside of it. You will still have to be vigilant and watch your competition to be sure that your product hasn't been "pirated."

# Low-End Programming

The following sections present the *In-Circuit Serial Programming (ICSP)* algorithm for programming some low-end and the mid-range PICmicro® MCUs. This programming method can be implemented simply and using a wide variety of different hardware devices. The basic low-end and high-end PICmicro® MCUs do not have this capability; instead, they must be programmed using a parallel algorithm with several additional inputs required to carry out the programming operation. This parallel programming method also restricts the number of pins available in the PICmicro® MCU. The low-end PICmicro® MCU requires access to 17 pins for programming (Table 13-3). A programmer used for the low-end PICmicro® MCUs will have the features shown in Fig. 13-2.

I have drawn in the multiple single shots to ensure that the specified timing is achieved to program the PICmicro® MCU for "normal" programming. A 100-ms pulse is required, but for the configuration word, the timing is 10 ms, which is why I show the separate single shot.

**TABLE 13-3 Low-End PICmicro® MCU Programming Pins**

| PINS | FUNCTION |
|---|---|
| RA0–RA3 | D0–D3 of the Instruction Word |
| RB0–RB7 | D4–D11 of the Instruction Word |
| TOCK1 | Program/Verify Clock |
| OSC1 | Program Counter Input\ |
| _MCLR/Vpp | Programming Power |
| Vdd | PICmicro® MCU Power |
| Vss | Ground ("Gnd") |

When a PICmicro® MCU is to be programmed, the MCLR line is pulled up to 13 volts, although TOCK 1 is held high and OSC 1 is pulled low. The PICmicro® MCU's internal program counter (which is used to keep track of the address) is initialized to 0x0FFF, which is the configuration fuse address.

To program a memory location, the following procedure is used:

1 The new word is driven onto RA0 through RA3 and RB0 through RB7.
2 The *prog* single shot sends a 100-$\mu$s programming pulse to the PICmicro® MCU.
3 The data word drivers (driver enable) is turned off.

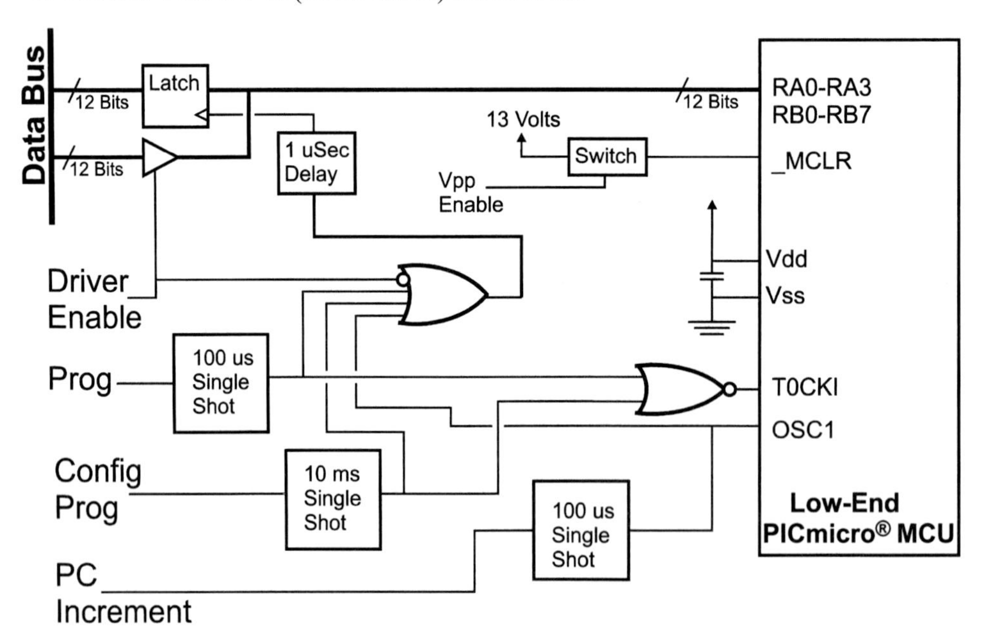


**Figure 13-2** Low-end PICmicro® MCU programmer

4 A programming pulse is driven, which reads back the word address to confirm the programming was correct. In Fig. 13-2, the read back latch is loaded on the falling edge of the on gate to get the data driven by the PICmicro® MCU.
5 Steps two through four are repeated a maximum of 25 times or until the data stored in the latch is correct.
6 Steps one through four are repeated three times that were required to get the correct data out from the PICmicro® MCU. This is used to ensure that the data is programmed reliably.
7 *OSC1* is pulsed to increment to the next address. This operation also causes the PICmicro® MCU to drive out the data at the current address before incrementing the PICmicro® MCU's program counter (which occurs on the falling edge of OSC1).

Looking at the circuit in Fig. 13-2, you are probably thinking that it is needlessly complex. I would disagree if you were thinking of programming a low-end device using a dedicated intelligent programmer, where timing-pulse durations can be algorithmically produced. If you were going to use a simple programmer (like a PC's serial port), the programmable single-shot chips are definitely required.

Programming steps one to four are shown in Fig. 13-3, along with the latch clock signal.

When programming, there must always be two T0CK1 pulses, the first being the programming pulse (10 ms or 100 $\mu$s) and then read back. The program data word must be valid for 1 $\mu$s before the T0CK1 programming pulse is driven into the PICmicro® MCU and data out is available 250 ms after the falling edge of T0CK1. Notice that using the circuit shown in Fig. 13-2 will result in data being driven into the read-back latch because T0CK1 is used for the programming pulse.

There are four important points to note about low-end PICmicro® MCU programming. The first is when _MCLR is active at 13 volts. The program counter is set to the configuration register, which is different from the mid-range devices. The configuration register also requires a considerably longer pulse to program than the standard addresses.

Second, just pulsing the OSC1 pin can be used to implement a "fast verify" (Fig. 13-4).

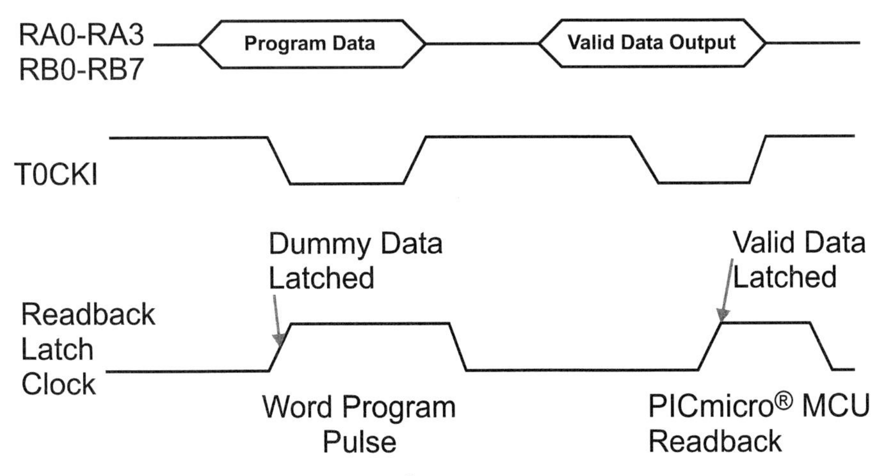


**Figure 13-3** Low-end PICmicro® MCU programming waveform

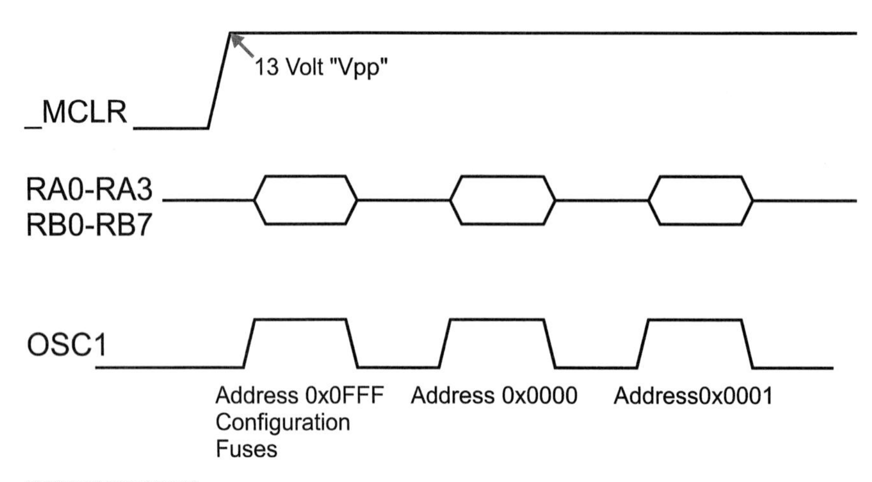


**Figure 13-4** **Low-end PICmicro "Fast Verifying" waveform**

Each time that OSC1 is pulsed, data at the current address will be output and then increment the PICmicro® MCU's program counter. Figure 13-4 shows the fast verify right from the start with the configuration fuse value output first for verify before the contents of the program memory.

Past the end of the low-end PICmicro® MCU's program memory is four bytes of EPROM words that can be used for serial number or application code version information. These four words, known as the *ID Location (IDLOCS)*, cannot be accessed by the PICmicro® MCU during application execution.

The last point is that the configuration fuses should *always* be programmed last. Thus, the configuration information is skipped over when burning the program memory, *_MCLR* cycled, the programming is started over, and the configuration fuses programmed with the actual value.

The reason for programming the configuration fuses last is to ensure that the code-protect bit of the configuration register is not reset (enabled) during program memory programming. If code protection is enabled, then data read back will be scrambled during programming, which makes code verification impossible.

# Mid-Range and ICSP Serial Programming

*In-Circuit Serial Programming (ICSP)* is one of the big advantages of the PICmicro® MCU microcontrollers. This feature allows you to program PICmicro® MCUs after they have been assembled into the application circuit, which eliminates one manufacturing step or eliminates the need to buy specialized sockets and handling equipment for different devices. The ICSP features of the PICmicro® MCU allow very simple programmers to be built. The El Cheapo, presented later in this chapter, is an example of a very simple PICmicro® MCU programmer that you can build very inexpensively that will allow new users to program the PICmicro® MCU.

The first feature of in-circuit serial programming, allowing the PICmicro® MCU to be programmed in circuit, is something only a "card stuffer" would appreciate. Being able to program parts after they have been soldered to a circuit board eliminates the need for costly (and unreliable) programmer sockets and extra device handling, which can result in bent pins or reversed components being put into the placement equipment and onto the boards. This gives the card manufacturer an opportunity to provide a cheaper service to their customers and provide the flexibility of shipping product with the latest-possible software burned into the product.

In-circuit serial programming is one of the reasons why the PICmicro® MCU is as popular as it is (other reasons include the MPLAB IDE and the wide availability of different PICmicro® MCUs part numbers and features from a number of different sources). This section introduces In-Circuit Serial Programming, explains how it is implemented for the mid-range PICmicro® MCUs, and how programming works. The following sections cover different ICSP programmers and the design and assembly of the El Cheapo and YAP-II programmers. ICSP for the low-end (which have it) and mid-range parts are shown in Table 13-4.

This section covers EPROM-based mid-range devices. The following sections cover ICSP programming for the low-end devices with ICSP capability, the Flash based mid-range devices, and the PIC17Cxx and PIC18Cxx ICSP methods. The information in this section is prerequisite material for in-circuit serial programming of these devices.

To program and read data, the PICmicro® MCU must be put into *Programming mode* by raising *_MCLR* to 13 to 14 volts, and pulling the Data and Clock lines low for several ms. Once the PICmicro® MCU is in Programming mode, data can then be shifted in and out using the Clock line.

A note on the *Vpp* line; when the programming voltage is applied to the _MCLR pin, it is important that up to 50 mA must be supplied on the *Vpp* circuit. This is relatively high and the best way I have found to provide this voltage is using a 78L12 regulator with two silicon diodes used to shift up the regulator's ground reference (Fig. 13-5). The two diodes will shift up the GND reference by 0.7 volts because of the pin junction voltage. This shift will pull up the 78L12's output to allow the PICmicro® MCU to go into Programming mode.

*Vpp* is at 5 volts and it requires 20 to 50 mA. Thus, either a 78L05 regulator, or a Zener diode regulator (as used in the El Cheapo) can be used to supply power to the PICmicro® MCU being programmed.

Transistor physical switches can be used to turn on and off the *Vpp* and *Vpp* voltages. If *Vpp* is not being driven, internal pull downs in the PICmicro® MCU will pull its *_MCLR*

**TABLE 13-4 Mid-Range PICmicro® MCU ICSP Programming Pins**

| PIN | 12C5XX | 16C50X | 18 PIN MID | 28 PIN MID | 40 PIN MID |
|---|---|---|---|---|---|
| 1 Vpp | 4 _MCLR | 4 _MCLR | 4 _MCLR | 1 _MCLR | 1 _MCLR |
| 2 Vdd | 1 Vdd | 1 Vdd | 14 Vdd | 26 Vdd | 11,32 Vdd |
| 3 GND | 8 Vss | 14 Vss | 5 Vss | 8,21 Vss | 12,31 Vss |
| 4 DATA | 7 GPO | 13 RBO | 13 RB7 | 28 RB7 | 40 RB7 |
| 5 CLOCK | 6 GP1 | 12 RB1 | 12 RB6 | 27 RB6 | 39 RB6 |

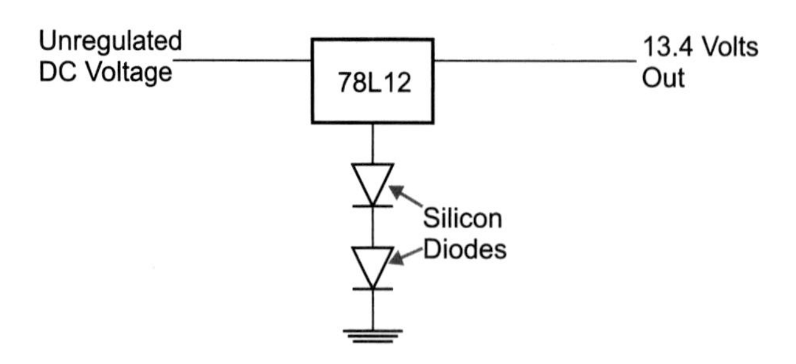


**Figure 13-5** Programmer voltage supply

pin to ground, which eliminates the need for a ground driver on the reset line. Putting the PICmicro® MCU into Programming mode requires the data waveform shown in Fig. 13-6.

When *_MCLR* is driven to *Vpp*, the internal program counter of the PICmicro® MCU is reset. The PICmicro® MCU's program counter is used to keep track of the current program memory address in the EPROM that is being programmed. When it is set to 0x02000, the ID locations of the PICmicro® MCU will be programmed and address 0x02007 points to the PICmicro® MCU's configuration fuses.

Data is passed to and from the PICmicro® MCU using a synchronous data protocol. A six-bit "command" is always sent before data is transferred. The commands (and their bit values and data) is listed in Table 13-5.

Data is shifted in and out of the PICmicro® MCU using a synchronous protocol. Data is shifted out, least-significant bit first, on the falling edge of the clock line. The minimum period for the clock is 200 ns with the data bit centered (Fig. 13-7), which is sending an *increment address* command.

When data is to be transferred, the same protocol is used, but a 16-bit transfer (LSB first) follows after 1 $\mu$s has passed since the transmission of the command. The 16 bits consist of the instruction word shifted to the left by one. This means the first and last bits of the data transfer are always zero.

Before programming can start, the program memory should be checked to ensure it is blank. This is accomplished by simply reading the program memory (*read data* command) and comparing the data returned to 0x07FFE. After every compare, the PICmicro® MCU's program counter is incremented (using the *increment address* command to the size

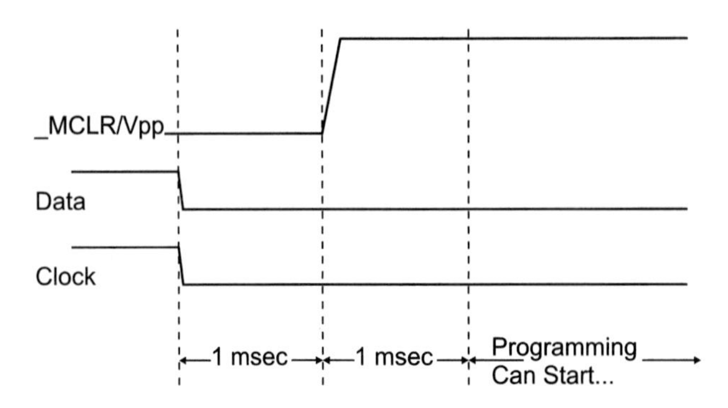


**Figure 13-6** Programmer initialization

**TABLE 13-5 Mid-range PICmicro® MCU EPROM ISCP Programming Commands**

| COMMAND | BIT PATTERN | DATA | COMMENTS |
|---|---|---|---|
| Load data | 0b0000010 | 0,14 Bits Data, 0 | Load word for programming |
| Begin Programming | 0b0001000 | none | Start Programming Cycle |
| End Programming | 0b0001110 | none | End Programming Cycle after 100 msec |
| Increment Address | 0b0000110 | none | Increment the PICmicro's Program Counter |
| Read Data | 0b0000100 | 0,14 Bits Data, 0 | Read Program Memory at Program Counter |
| Load Config | 0b0000000 | 0x07FFE | Set the PICmicro's Program Counter to 0x02000 |

of the device's program memory. Once the program memory is checked, the PICmicro® MCU's program counter is "jumped" to 0x02000 (using the *load configuration* command) and then the next eight words are checked for 0x07FFE.

For different mid-range devices, I use Table 13-6 to keep track of different program memory sizes.

To program an EPROM program memory instruction word, a *load data* command (followed by the instruction value) is sent to the PICmicro® MCU, followed by a begin *programming* command. After at least 100 ms has passed, an *end programming* command is sent. This sequence is known as a *programming cycle*. After each programming cycle, the contents are read back and compared to the expected value.

This process is repeated up to 25 or until the program memory is correct. If the program memory is correct, then three times the number of programming cycles needed to get the correct value are repeated to ensure that the instruction word is not marginally programmed.

This is a bit confusing. An example would be a PICmicro® MCU program memory word that requires four programming cycles before the correct data is returned. After the correct data has been returned, an additional 12 programming cycles (three times the four cycles required to get the correct value) are repeated to *overprogram* the instruction address to ensure the instruction is properly programmed in.

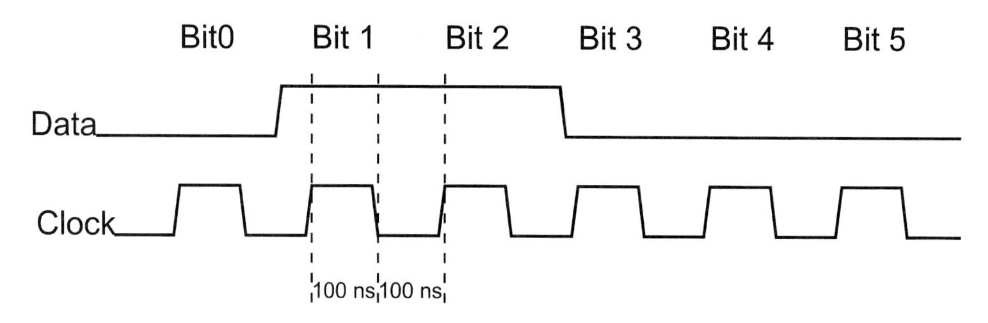


**Figure 13-7** Programmer command-6 bits

**TABLE 13-6 Mid-Range PICmicro® MCU RPROM Sizes By Part Number Mask**

| DEVICE | SIZES |
|---|---|
| PIC16Cx1 | 1k |
| PIC16Cxx0 | 0.5k |
| PIC16Cxx1 | 1k |
| PIC16Cxx2 | 2k |
| PIC16Cx3 | 4k |
| PIC16Cx4 | 2k |
| PIC16Cx5 | 4k |
| PIC16Cx6 | 8k |
| PIC16Cx7 | 8k |

Once a memory location has been correctly programmed, the PICmicro® MCU's program counter can be incremented. If there is nothing to program at a memory location or if the value is 0x03FFF, you can simply send an increment to skip to the next address and ignore programming the instruction completely.

The configuration memory and ID locations are programmed the same way after sending a *load configuration* command.

The process for programming program memory could be blocked out with the pseudocode:

```
ICSPProgram()            //  Program to be burned in is in an array of
{                        //   addresses and data

int PC = 0;              //  PICmicro® MCU's program counter
int i, i j k;
int retvalue = 0;


  for (i  = 0; (i - PGMsize) && (retvalue == 0); I++) {

    if (PC ! = address[i]) {
      if ((address[I] >= 0x02000) && (PC < 0x02000)) {
        LoadConfiguration(0x07FFE);
        PC = 0x02000;
      }

    for (; PC < address[i]; PC++)
      IncrementAddress();

    for (i = 0; (i < 25) && (retvalue != data[I]); I++) {
      LoadData(ins[i] << 1); //  Programming Cycle
      BeginProgramming();
      Dlay(100usec);
      EndProgramming();
      Retvalue = ReadData();
    }

    if (i == 25)
      retvalue = -1;         //  Programming Error
    else {
```

```
        retvalue = 0;          //  Okay, Repeat Programming Cycle 3x
        for (k = 0; k < (j * 3); k++){
          LoadData(ins[i] << 1);
          BeginProgramming();
          Dlay(100usec);
          EndProgramming();
        }  //  endif
      }  //  endif
    }  //  endfor
}  //  end ICSPProgram
```

After the program memory has been loaded with the application code, *Vpp* should be cycled off and on, then the PICmicro® MCU's program memory is read out and compared against the expected contents. When this verify is executed, *Vpp* should be cycled again with *Vpp* a minimum voltage (4.5 volts) and then repeated again with *Vpp* at a maximum voltage (5.0 volts) value.

When this verify is executed at voltage margins, the PICmicro® MCU is said to be *production programmed*. If the margins are not checked, then programming operation is said to be *prototype programmed*. Most hobbyist programmers (including the two presented in the book and the Microchip PICSTART Plus) are prototype programmers because they cannot margin *Vpp* when the value is checked.

## PIC12C50x AND PIC16C505 SERIAL PROGRAMMING

If you've read through the "low-end programming" section of this chapter, you have learned that at least 17 pins are required to program the PIC16C5x devices. This might have caused you to question how programming is carried out in the PIC12C50x and PIC16C505 devices, which don't even have 17 pins. These devices use the serial EPROM ICSP programming protocol used by the mid-range PICmicro® MCUs with two small modifications to take into account the low-end device's characteristics.

Like the traditional parallel-programmed low-end devices, when the PIC12C50x and PIC16C505 first enters Programming mode (with the data and clock pins pulled low, followed by _MCLR driven with 13+ volts), the PICmicro® MCU's program counter is set to 0x0FFF, which is the configuration register address.

I recommend that the configuration fuses for the PIC12C50x and PIC16C505 devices should not be changed during the first programming pass because enabling code protection will scramble the programmed data being read back when other instructions are programmed.

The PIC12C50x and PIC16C505 use a 12-bit instruction word. When data is passed to the PICmicro® MCU, the upper three bits (instead of the upper one) is zero and ignored by the device as it is programmed. The first bit sent is still *0*, with the least significant bit (LSB) of the instruction word following.

A simple way to calculate the 16 data bits to be programmed into the PIC12C50x and PIC16C505 microcontrollers from the instruction is to save the instruction in a 16-bit variable and shift it up (to the left) by one bit.

The commands available for programming the PIC12C50x and PIC16C505 have a six-bit header and optional 16-bit instruction or configuration fuse data word. They are listed in Table 13-7.

The programming algorithm is identical to that described in the previous section with, of course, the configuration fuses ignored at the start of programming. Then the operation

**TABLE 13-7 PIC12C5xx And PIC16C505 Programming Commands**

| COMMAND | BITS |
|---|---|
| Load data | 000010 +0, data(12), 000 |
| Read data | 000100 +0, data(12), 000 |
| Increment PC | 001000 |
| End programming | 001110 |

is reset to program the configuration fuse once the instructions are burned into the PICmicro® MCU and have been verified as being correct.

## FLASH PROGRAMMING

I realize that the programming operations for the PIC12C5xx, PIC16C505, and EPROM mid-range devices seem complex, but they are designed so that relatively simple programmers can be used to burn applications into PICmicro® MCU devices. Microchip uses a modified version of this programming algorithm for the PIC16F8x Flash-based parts. Along with the programming algorithm being much simpler, the actual programming circuit is much easier to implement and has resulted in some ridiculously simple programmers being developed for them. For these reasons (as well as the electrically erasable program memory), the PIC16F8x family has become a very popular choice for hobbyists to start learning about the PICmicro® MCU.

The programming circuit should be familiar to you (Fig. 13-8).

Electrically, the programming voltages are basically the same as what is required for the mid-range devices. One difference, however, is the voltage and current required for *Vpp*. For the mid-range parts, up to 50 mA is required for EPROM programming. Because the PIC16F8x parts have built-in EROM data and Flash *Vpp* generator, this circuit will provide adequate voltage and current to program and engage the data and program memory, meaning that very little current is required from the programmer's *Vpp* voltage.

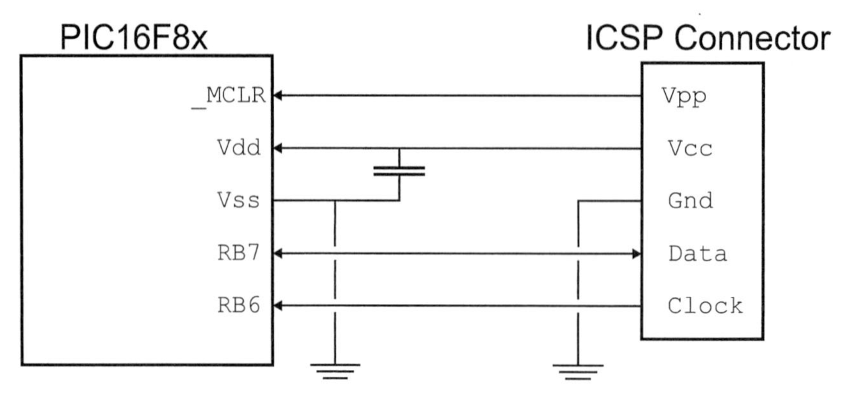


**Figure 13-8** PIC16F8x ICSP connection

**TABLE 13-8 Mid-Range Flash PICmicro® MCU Programming Commands**

| COMMAND | BITS | DATA |
|---|---|---|
| Load Configuration | 000000 | 07FFE In |
| Load Data for Program Memory | 000010 | Word × 2 Going In |
| Load Data for Data Memory | 000011 | Byte × 2 Going In |
| Read Data from Program Memory | 000100 | Word × 2 Going Out |
| Read Data from Data Memory | 000101 | Byte × 2 Going Out |
| Increment PICmicro® MCU's PC | 000110 | none |
| Begin Programming | 001000 | none |
| Bulk Erase Program Memory | 001001 | none |
| Bulk Erase Data Memory | 001011 | none |

The *Vpp* voltage that is applied just turns on the built-in programmer circuit and puts the PICmicro® MCU into Programming mode. For this reason, the actual current drain is very minimal (tens of $\mu$As) and a *Vpp* voltage as low as 10 volts will enable the *Vpp* generator and put the PICmicro® MCU into Programming mode.

The same data packet format is used for the PIC16F8x as was used for the mid-range EPROM parts, but the commands and how they work is slightly different. Table 13-8 lists the different commands.

The data, as in the mid-range EPROM part, is always 16 bits with the first and last bit always equal to zero. Data is always transferred LSB first using the same timings as specified earlier in the chapter for the mid-range parts. When I have designed PICmicro® MCU programmers, I multiply the data word by two (or shift it to the left by one) to provide a 16-bit word with the first and last bit equal to zero and the data word in between.

The programming cycle for the PIC16F8x is:

1. *Load data* command (000010 + data word * 2)
2. *Begin programming* command (001000).
3. Wait 10 ms.

No *end programming* command is required, but the 10-ms delay makes programming Flash parts somewhat slower.

For most simple ("development") programmers, *Vpp* cannot be set to minimum (4.5 volts) or maximum (5.5 volts), so the margined voltage verify steps are skipped or the contents are just checked once at nominal voltage.

Configuration memory is programmed the same way with the *load configuration* command (with 0x07FFE as data to avoid incorrect data writes taking place) executed before the memory writes. The ID locations are set by convention, at addresses 0x02000 to 0x02003. The 16-bit *__IDLOCS* value is written to each of the four ID locations with the least-significant four bits written to each location for a 16-bit ID value. The configuration register is at address 0x02007, like the EPROM mid-range parts, with the register set using the same programming cycle as given previously.

The only issue left is how to erase the program memory before programming. Like the EPROM devices, the PIC16F8x program in operation converts specific *1*s in memory to *0*s. Like with the EPROM device's ultraviolet erase, the PIC16F8x erase step loads *1*s in all of the memory locations.

This could be done using the *bulk erase* commands, but I prefer to use the Microchip-specified erase for code-protected devices. This operation will erase all Flash and EEPROM memory in the PICmicro® MCU device—even if code protection is enabled.

1. Apply *Vpp*.
2. Execute load configuration (0b0000000 + 0x07FFE)
3. Increment the PC to the configuration register word (send 0b0000110 seven times).
4. Send command 0b0000001 to the PICmicro® MCU.
5. Send command 0b0000111 to the PICmicro® MCU.
6. Send begin programming (0b0001000) to the PICmicro® MCU.
7. Wait 10 ms.
8. Send command 0b0000001.
9. Send command 0b0000111.

Notice that two undocumented commands (0b0000001 and 0b0000111) are used in this sequence.

When this sequence is followed, all data within the PIC16F8x PICmicro® MCU is erased and the device is ready for a new application to be burned into it.

## PIC17Cxx ICSP PROGRAMMING

Depending on your familiarity with the PIC17Cxx devices, seeing that they have ICSP capability is probably surprising to you. It was to me because I thought they were parallel programmed, as covered in a later section of this chapter. PIC17Cxx devices have the capability to write to their own EPROM program memory, which gives them some in-circuit programming capability that can be exploited in your applications.

The capability of a PIC17Cxx application to write to program memory is enabled when *_MCLR* is driven by more than 13 volts and a *tablwt* instruction is executed. When *tablwt* is executed, the data loaded into the TABLATH and TABLATL registers is programmed into the memory locations. This instruction keeps executing (it doesn't complete after two cycles, as it would if the TBLPRH and TBLPTRL registers were pointing outside the internal EPROM) until it is terminated by an interrupt request or *_MCLR* reset.

To perform a word write, the following mainline process would be used:

1. Disable TMRO interrupts.
2. Load TABPTRH and TABPTRL with the address.
3. Load TABLATH or TABLATL with the data to be stored.
4. Enable a 1,000-$\mu$s TMRO delay interrupt (initialize TMRO and enable TMRO interrupt).
5. Execute *tablwt* instruction with the missing half of data.

**6** Disable TMRO interrupts.
**7** Read back data, check for match.
**8** If no match, return error.

The interrupt handler for this process can just consist of resetting the TOIF flag and executing a *retfie* instruction.

To enable internal programming, _MCLR has to be "switched" from five volts (*Vdd*) to 13 volts. The recommended Microchip circuit is shown in Fig. 13-9. This circuit will drive the PIC17Cxx's *_MCLR* pin at 5 volts until *RA2* is pulled low. When *RA2* is pulled low, the voltage driven in to *_MCLR* will become 13 volts (or *Vpp*). The programming current at 13 volts is a minimum of 30 mA.

Notice that Fig. 13-9 shows an RS-232 level converter and implies that the PICmicro® MCU can be programmed in circuit. This implication is wrong because code (known as *boot code*) must be burned into the PICmicro® MCU to take advantage of the EPROM self-program capability of the part. The boot code is a host-interface application that reads data and then programs it at the specified address. This code must be burned into the PICmicro® MCU before ICSP can execute. The typical boot code for a PIC17Cxx PICmicro® MCU would:

**1** Establish communication with programming host.
**2** If no communication link established jump to application code.
**3** Enable *Vpp* (*RA2* = 0)
**4** Wait for host to send instruction word address.
**5** Program in the word.
**6** Confirm word programmed correctly.
**7** Loop back to 4.

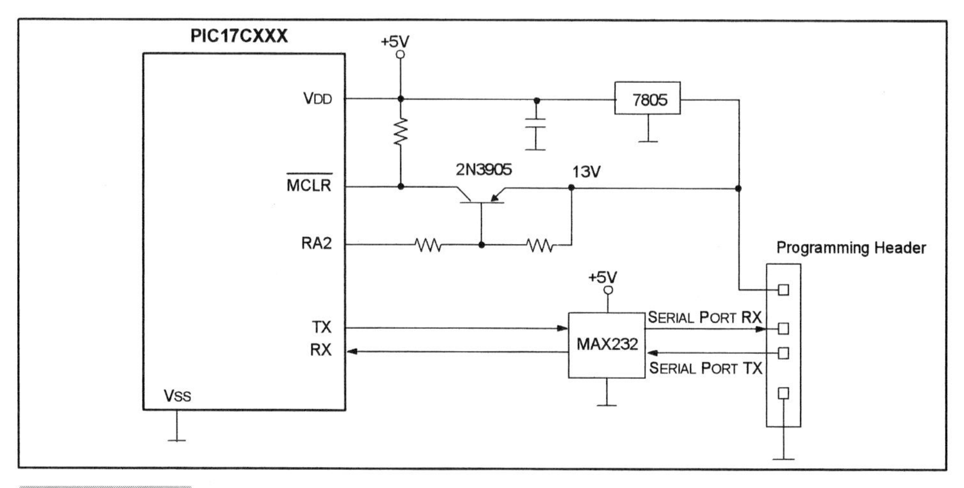


**Figure 13-9** PIC17Cxxx in circuit serial programming schematic

In this process, you will probably want to program as few instruction locations as possible. This is caused by the need for programming in the boot code initially. If this code has to be programmed in, then you might as well program in the application at the same time, leaving the boot code for programming serial numbers or calibration values. This is really the optimal use of the self-program capabilities of the PIC17Cxx devices.

If you do use the self-programming capability for applications, then I recommend that the jump to the boot code be overwritten with zeros (which changes them into *nop* instructions). The interrupt handler used to end the *tablwt* instruction should be *nop*ped as well. This will prevent any possibility that the boot code will be inadvertently enabled or incorrect interrupt handlers accessed.

After the application has been stored in the PIC17Cxx EPROM and the boot code jump and any interrupt vector code overwritten with zeros, the PIC17Cxx will boot with the application code and avoid the boot programming code all together.

Notice that in this case, the boot code space will be unusable. Thus, care must be taken to ensure that sufficient space is available for the application code to be loaded. As well, the application must be written in such a way that the program memory resources used by the boot code must not be specified by the application. PIC17Cxx ICSP, sample boot code, and operations are shown in Chapter 16.

# PIC17Cxx Programming

When a PIC17Cxx is programmed using a parallel programmer, it is wired as shown in Figure 13-10. Notice that PORTB and PORTC are used to transfer data 16 bits at a time and PORTA is used for the control bits that control the operation of the programmer. The *_MCLR* pin is pulled high to 13 volts, as would be expected, to put the PICmicro® MCU into Programming mode.

Although the programming of the PIC17Cxx is described as being in *parallel*, a special boot ROM routine executes within the PICmicro® MCU, which accepts data from the I/O ports and programs the code into the PICmicro® MCU. To help facilitate this, the test line, which is normally tied low, is pulled high during application execution to ensure that the programming functions can be accessed. The clock, which can be any value from 4 MHz to 10 MHz, is used to execute the boot ROM code for the programming operations to execute.

To put the PICmicro® MCU into Programming mode, the test line is made active before *_MCLR* is pulled to *Vpp* and then 0x0E1 is driven on PORTB to command the boot code to enter the programmer routine (Fig. 13-11).

To end Programming mode, *_MCLR* must be pulled to ground 10 ms or more before power is removed from the PICmicro® MCU. *Test* should be de-asserted after *_MCLR* is pulled low.

When programming, the *RAO* pin is pulsed high for at least 10 instruction cycles (10 $\mu$s for the PICmicro® MCU running at 4 MHz) to load in the instruction address, followed by the PICmicro® MCU latching out the data (so that it can be verified). After the data has been verified, *RAO* is pulsed high for 100 $\mu$s to program the data. If *RA1* is low during the *RA0* pulse, then the PICmicro® MCU program counter will be incremented. If it goes high during the pulse, the internal program counter will not be incremented and the instruction word contents can be read back in the next RA1 cycles without having to load in a new address.

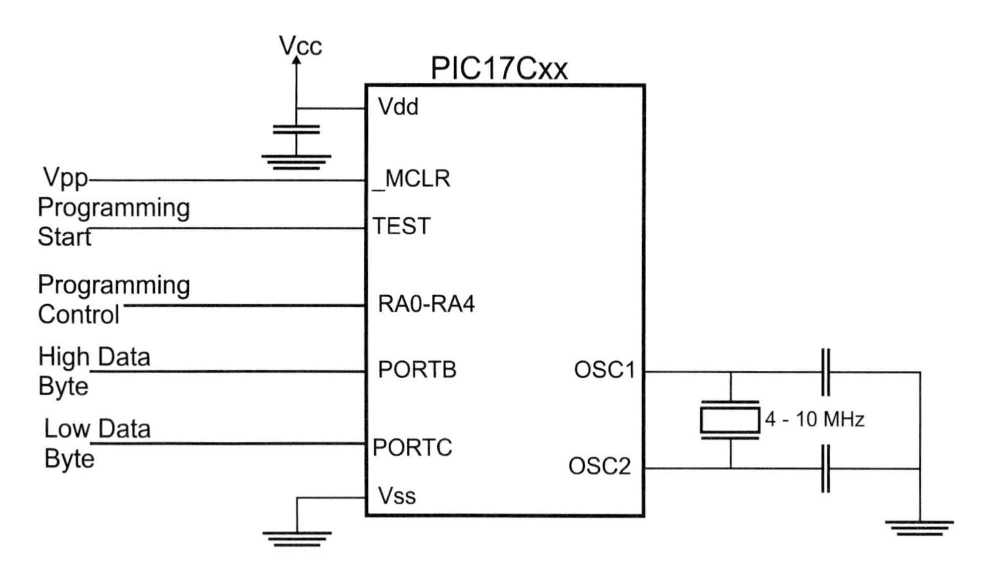


**Figure 13-10** PIC17Cxx parallel programming connections

The latter operation is preferred and looks like the waveforms shown in Figure 13-12.

This waveform should be repeated until the data is loaded or up to 25 times. Once it is programmed in, then three times the number of programming cycles must be used to "lock" and "overprogram" the data. This process is similar to that of the other EPROM parts.

Writing to the specified addresses between 0x0FE00 and 0x0FE0F programs and verifies the configuration word. To program (make 0) one of the configuration bits, its regis-

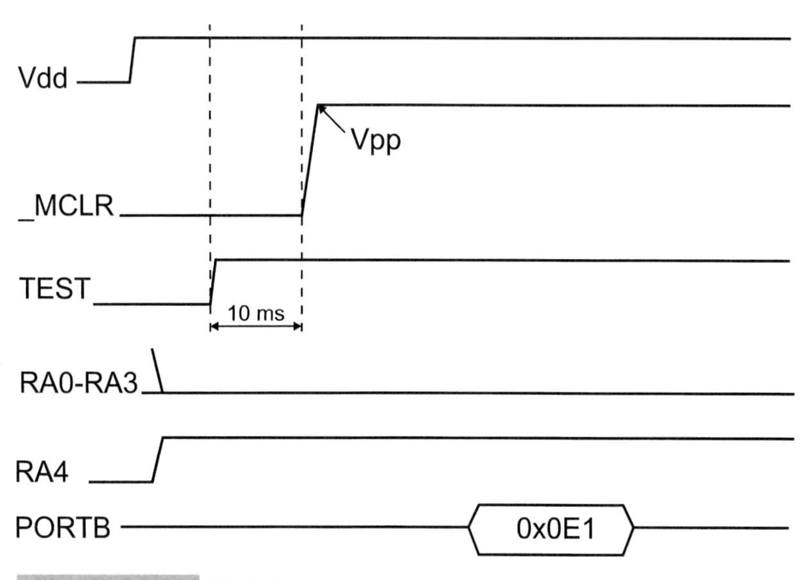


**Figure 13-11** PIC17Cxx parallel programming connections start up

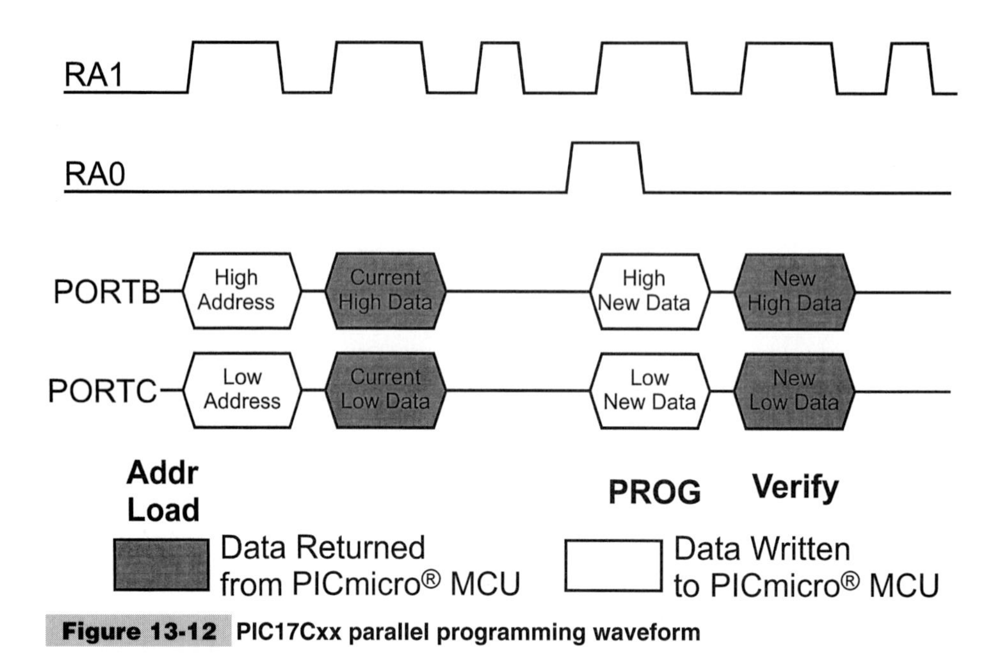


**Figure 13-12** PIC17Cxx parallel programming waveform

ter is written to. Reading back the configuration word uses the first three RA1 cycles of Fig. 13-12 at either 0x0FE00 or 0x0FE08. Reading 0x0FE00 will return the low byte of the configuration word in PORTC (0x0FF will be in PORTB) and reading 0x0FE08 will return the high byte in PORTC. The configuration bits for the PIC17Cxx are defined in Tables 13-9 and 13-10.

Notice that configuration bit addresses must be written in ascending order. Programming the bit in nonregister ascending order can result in unpredictable programming of the configuration word as the processor mode changes to a Code-Protected mode before the

**TABLE 13-9 PIC17Cxx Configuration Bits**

| ADDRESS | BIT |
|---|---|
| FOSCO | Bit |
| FOXC1 | 0x0FE01 |
| WDTPSO | 0x0FE03 |
| WDTPS1 | 0x0FE03 |
| PMO | 0x0FE05 |
| PM1 | 0x0FE06 |
| PM2 | 0x0FE0F |

**TABLE 13-10 PIC17Cxx Configuration Bit Definition**

| PM2:PM0 | PROCESSOR MODE |
|---|---|
| 111 | Microprocessor Mode |
| 110 | Microcontroller Mode |
| 101 | Extended Microcontroller Mode |
| 000 | Code Protected Microcontroller Mode |
| | |
| WDTPS1:WDTPS0 | Watchdog Timer and Postscaler Mode |
| 11 | WDT Enabled, Postscaler = 1:1 |
| 10 | WDT Enabled, Postscaler = 256:1 |
| 01 | WDT Enabled, Postscaler = 64:1 |
| 00 | WDT Disabled, 16 bit Overflow Timer |
| | |
| F0SC1:F0SC0 | Oscillator Mode |
| 11 | External Oscillator |
| 10 | XT Oscillator |
| 01 | RC Oscillator |
| 00 | LF Oscillator |

data is loaded in completely. This issue is important to watch out for in all PICmicro® MCU programming. The configuration fuses must be programmed last, with any code protection programmed into the PICmicro® MCU as the last possible programming operation.

# PIC18Cxx Programming

Like the PIC17Cxx, the PIC18Cxx has the capability to "self program" using the table read and write instructions. In the PIC18Cxx, this capability is not only available within applications, but is used to program the device right from the start, without the need for specialized boot ROM or ICSP interface code, unlike the PIC17Cxx. This ability to run instructions from an external device makes me wonder about how easy it would be to create an emulator for the PIC18Cxx using a standard part and no specialized hardware.

To program the PIC18Cxx, instructions are downloaded into the PICmicro® MCU after setting the *_MCLR* pin to $V_{PP}$ (13 to 14 volts, as in the other EPROM PICmicro®

MCUs) with both *RB6* and *RB7* low. Passing instructions (which contain the program data) to the PICmicro® MCU is accomplished by first sending a four-bit special instruction, followed by an optional 16-bit instruction.

The four-bit special instruction is sent most-significant bit first and can either specify that an instruction follows or that it is a mnemonic for a *tblrd* or *tblwt* instruction (Table 13-11).

The data transmission looks like that shown in Fig. 13-13. The four-bit *nop* code is transmitted first, followed by the 16-bit instruction.

If the instruction is a table operation, then the special instruction code can be used instead of the *nop* to simplify the data transfer. At the end of Fig. 13-13, bit pattern 0b01101 (*TBLWT* *-) is sent to the PICmicro® MCU.

Although the table reads and writes only require four bits, to carry out the program operation, 16 bits always follow the mnemonic (just as if it were an *nop*) for data transfer. This avoids the need to explicitly load and unload the table latch registers using instructions. In Fig. 13-14, the *tblwt* * instruction (write to table and don't change TBLPTR) is shown.

After the first 20-bit sequence, a second 20-bit sequence is executed to allow the programming operation to complete (this is what is meant by the 2 in the "Cycles" in Table 13-11). The PICmicro® MCU ignores the second sequence of 20 bits and the initial sequence is processed. Reading data from the PICmicro® MCU's program memory is accomplished in exactly the same way.

To set up a table read or write, first the TBLPTR has to be initialized. This is done using standard *movlw* and *movwf* instructions. For example, to program address 0x012345 with 0x06789, the data sequence is written to the PIC18Cxx:

```
Mnemonic        Instruction/Data
nop             movlw  UPPER 0x012345
nop             movwf  TBLPTRU
```

**TABLE 13-11 PIC18Cxx Programming "Mnemonics"**

| SPECIAL INSTRUCTION | MNEMONIC | INSTRUCTION OPERATION | CYCLES |
|---|---|---|---|
| 0000 | nop | Shift in Next Instruction | 1 |
| 1000 | TBLRD * | Read Table | 2 |
| 1001 | TBLRD *+ | Read Table, Increment TBLPTR | 2 |
| 1010 | TBLRD *− | Read Table, Decrement TBLPTR | 2 |
| 1011 | TBLRD +* | Increment TBLPTR, Read Table | 2 |
| 1100 | TBLWT * | Write Table | 2 |
| 1101 | TBLWT *+ | Write Table, Increment TBLPTR | 2 |
| 1110 | TBLWT *− | Write Table, Decrement TBLPTR | 2 |
| 1111 | TBLWT +* | Increment TBLPTR, Write Table | 2 |

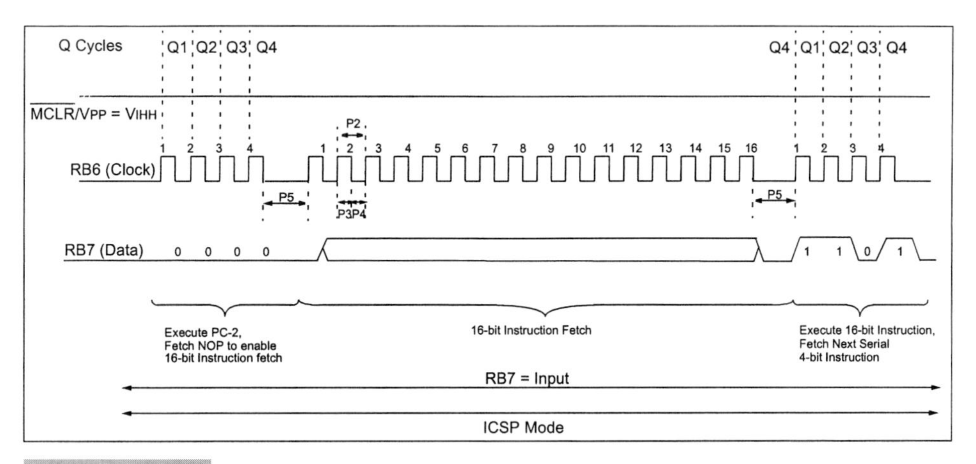


**Figure 13-13** Serial instruction timing for 1 cycle 16-bit instructions

```
nop               movlw  HIGH 0x012345
nop               movwf  TBLPTRH
nop               movlw  LOW 0x012345
nop               movwf  TBLPTRL
tblwt *           0x06789
```

This method of programming and reading back the contents of the PIC18Cxx program memory looks like it would be quite easy to implement. It avoids one of the biggest drawbacks of being able to quickly program mid-range PICmicro® MCUs. This drawback is the need to increment the program counter for each instruction address. In the PIC18Cxx, the program destination can be programmed explicitly (as in the previous sequence), instead of having to be stepped through.

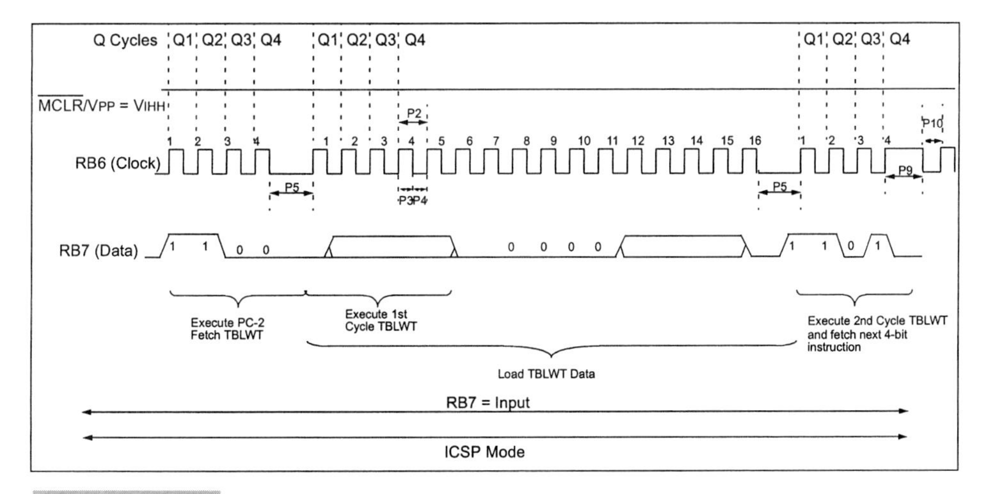


**Figure 13-14** TBLWT instruction sequence

# PICSTART Plus and PRO MATE II

Microchip, as part of its developer support, offers a number of programmers and programming options for its PICmicro® MCU and serial EPROM products. These programmers and options are reasonably priced and integrate seamlessly to MPLAB for direct application programming, eliminating possible problems with moving hex files between applications.

The basic Microchip programmer is the PICSTART Plus (Fig. 13-15). I have owned and used one since they first came out. The PICSTART Plus, which can be referred to as *PSP* or *PS+,* is a development programmer that connects to a PC via an RS-232 cable. The programmer itself consists of a small box with a ZIF socket for programming all of the different DIP PICmicro® MCUs.

The PICstart Plus can only be controlled from the MPLAB IDE. The latest versions of the software integrate very well (Fig. 13-16). When *PICstart Plus → Enable* is selected from the MPLAB IDE's top pull-down line, the memory contents are displayed, along with a PICSTART Plus' control box and a dialogue box showing how the configuration fuses will be set. The configuration values are set automatically from the values specified by the *__CONFIG* statement in the assembler source code. To "burn" an application into a part, normally all that has to be done at this stage is to click on *program*.

In earlier versions of the MPLAB IDE, the configuration register values would not be transferred automatically into these dialog boxes. But, the latest versions of the MPLAB IDE (including the version on the CD-ROM), do update the programmed configuration settings properly when the programmer is enabled.

With PICSTART Plus, you should be aware of a few things. First, as designed, it will only program DIP parts. This isn't a problem for hobbyists and pin-through-hole prototypes, but it can be a problem for SMT parts. An inexpensive ICSP cable can be made with a connector to connect the PICSTART Plus to the circuit in which the chip is soldered to sidestep this problem. Also, sockets are available that converts the pin-through-hole device spacing into a surface-mount socket.

You will have to deal with firmware revisions. Inside the PICSTART Plus is a 17C44 that controls the operation of the PICSTART Plus. Periodically, you will have to update this code. To do this, a windowed (OTP) 17C44 will have to be bought and programmed by the PICSTART Plus and then used to replace the one already inside it. This operation seems to be required about once a year.

Notice that the PICSTART Plus is a development programmer; as such, it does not check the contents of a programmed part at low voltage for prototyping operations. Microchip will not consider a PICmicro® MCU to be "production" programmed by the PICSTART Plus and will not respond to field problems if the PICSTART Plus is used.

The PICSTART Plus package consists of the PICSTART Plus, a power supply, an RS-232 9-pin straight-through male-to-female cable. A sample 16F84 and CD-ROM containing data sheets, the MPLAB IDE, and applications notes are also included to help you get started. As I write this, the list price of the PICSTART Plus is $200 (USD).

If you are interested in buying one, remember that Microchip often sells the PICSTART Plus at their seminars for a "sale price" of $50 less than the retail price (and it includes

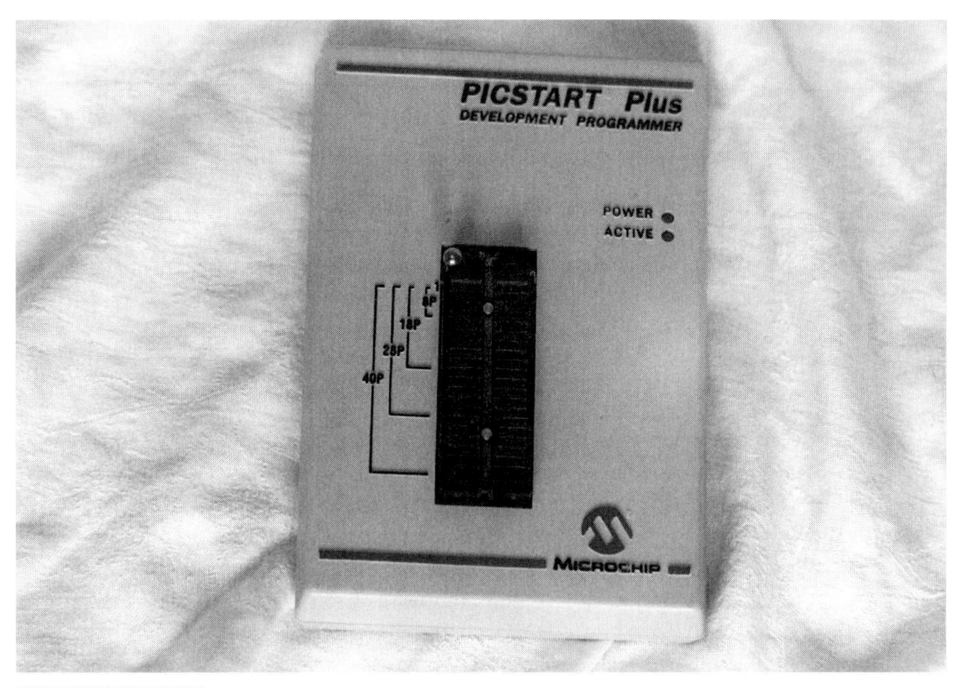


**Figure 13-15** Microchip "PICSTART Plus" programmed

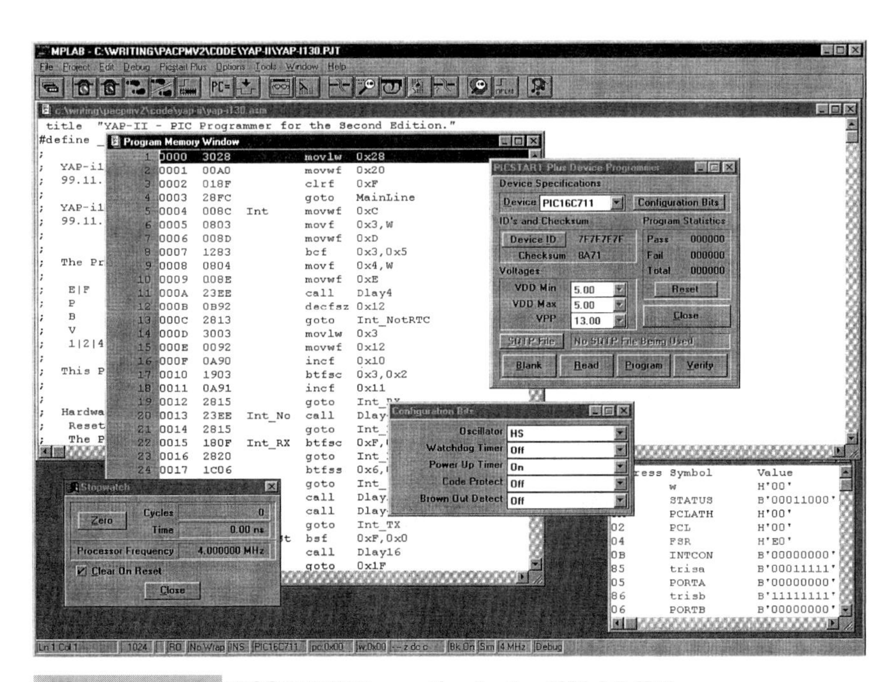


**Figure 13-16** PICSTART Plus active in the MPLAB IDE

admission to the seminar). Going to the seminar and buying a PICSTART Plus is an excellent way to start getting seriously into the PICmicro® MCU.

If you require a production programmer or need the ability to program a surface-mount PICmicro® MCU, then you should look at the PRO MATE II programmer by Microchip. This programmer is much more flexible than the PICstart Plus and offers the following additional features:

- Executes from the MPLAB IDE or the MS-DOS command line.
- Can work in a stand-alone mode.
- Provides production low-voltage verify.
- Interchangeable sockets for PTM, SMT, and ICSP cabling.
- Also programs Microchip serial EPROMS.
- Can serialize parts.

These additional features do come at a price, however. The PRO MATE II is about $1000 (USD), which might make it less attractive for hobbyists or companies that want to see what the PICmicro® MCU is all about before making substantial investments in it.

The PRO MATE II supports a 5-pin ICSP cable adapter that can be bought separately. This cable provides power, *Vpp,* programming clock, and data signals to a connector built onto a product. The connector is an IDE 5x1 connector with pins spaced 0.100 apart. The ICSP cable connects to PICmicro® MCU's application circuits using the conventions shown in Table 13-12.

To connect a PICmicro® MCU that has been put into an application circuit, the interface shown in Fig. 13-17 should be used.

The PICmicro® MCU must be isolatable from the application circuit. The diode on the *_MCLR/ Vpp* pin and the "breakable connections" on *Vpp,* RB7 and RB6 isolate the PICmicro® MCU. These breaks are best provided by unsoldered 0-$\Omega$ resistors or unconnected jumpers in the circuit. This has to be done because the ICSP specification will only provide 50 mA for *Vpp* and has 1-k resistors in the data and clock lines to protect the driver circuits.

The 17Cxx parts can also be ICSP programmed using RS-232 and the PRO MATE II's ICSP cable will support this.

The PICSTART Plus and PRO MATE II are not the fastest PICmicro® MCU programming solutions on the market today. As pointed out later in this chapter, other vendor's products

**TABLE 13-12 Microchip "ICSP" Pin Definition**

| PIN | 12C5xx | 16Cxx | 16Fxx |
|---|---|---|---|
| 1. Vpp | -MCLR/Vpp | -MCLR/Vpp | -MCLR/Vpp |
| 2. Vdd | Vdd | Vdd | Vdd |
| 3. Vss | Vss | Vss | Vss |
| 4. DATA | GPO | RB7 | RB7 |
| 5. CLOCK | GP1 | RB6 | RB6 |

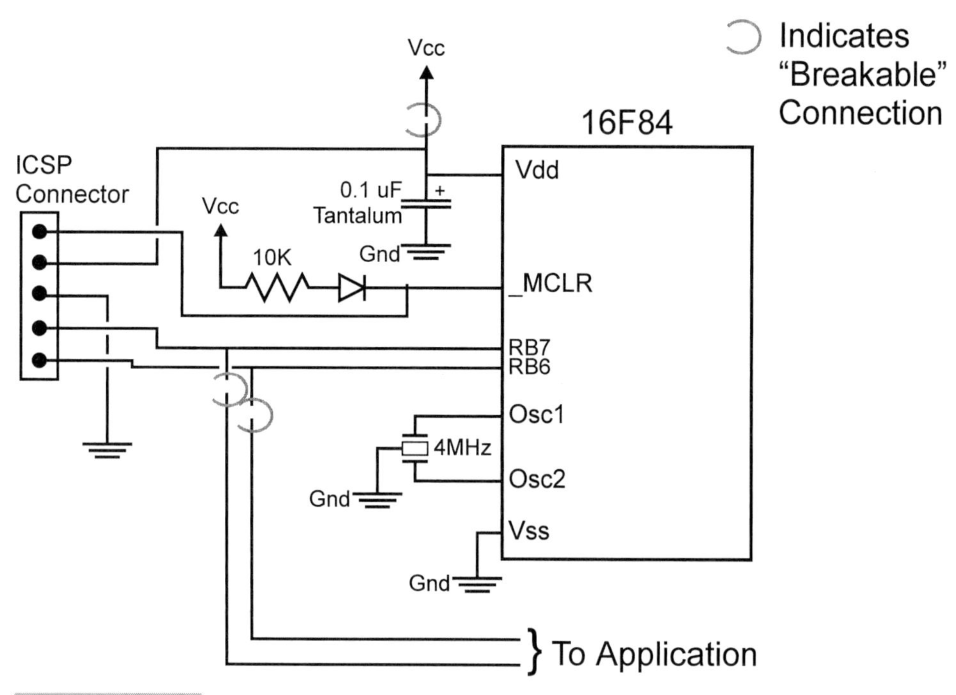


**Figure 13-17** 16F84 ICSP circuit

perform the programming operations lightning fast (e.g., in less than eight seconds for a PIC16F84). The advantages of the Microchip programmers is the seamless interface to MPLAB, along with Microchip's excellent support in ensuring that development software is available for new parts before the parts are available.

# The El Cheapo

Over the past year, I have been working on a simple PC parallel-port programmer that is compatible with a variety of PCs and will work for both Flash and EPROM program memory-based PICmicro® MCU microcontrollers. The result of four redesigns is the El Cheapo (Fig. 13-18), which can be used in virtually any PC with MS-DOS or Microsoft Windows. The programmer itself will allow you to program all of the different PICmicro® MCU part numbers used in the experiments and projects presented in this book, with the exception of the PIC17Cxx and PIC18Cxx.

I am pleased to be able to include a PC board for the El Cheapo in this book. This section covers building it and using it to program different PICmicro® MCU part numbers. This PC board is designed to handle 8-, 14-, 18-, 28-, and 40-pin low-end and mid-range PICmicro® MCUs that can be programmed using the two-wire ICSP synchronous data protocol (presented previously in this chapter). The basic circuit design is shown in Fig. 13-19 which shows how an 18-pin PICmicro® MCU is wired into the circuit.

**Figure 13-18** The "El Cheapo" PICmicro® programmer

The bill of materials for the programmer is shown in Table 13-13. Notice that for the 2.5-mm power connector (J1), I have included a Digi-Key part number to make it easier for you to find.

These parts are quite easy to find in most electronic stores. In fact, you will probably be able to find most of the components in your local Radio Shack. If you have problems finding any of these components, please let me know and I'll see if I can point you in the right direction. Notice that the 2N7000 N-channel MOSFET transistor can be any small-signal N-channel MOSFET. The maximum "on" current through the transistor is 41 mA.

The circuit can be broken up into four major subsystems. First is the power supply. The input DC power is passed through the 78L12 to provide the programming voltage ($V_{PP}$) for the PICmicro® MCUs. This voltage must be at least 13 volts. To shift the voltage output of the 78L12 from 12 volts, I have added two silicon diodes to the ground reference of the 78L12. Each silicon diode will "bump up" the ground reference by 0.7 volts. The two of them raise the ground reference to 1.4 volts. With it, the 78L12's voltage output is raised to 13.4 volts above the El Cheapo's ground reference.

To provide the 5-volt $V_{PP}$ required by the PICmicro® MCU, I used a Zener diode and resistor to regulate down the 13.4 volts to 5.1 volts. When I checked the programming specifications of the various PICmicro® MCU devices that I wanted this programmer to handle, I found that the maximum $V_{PP}$ during programming was 40 mA. This value determined what resistor would be used in the circuit. For a voltage drop of 8.3 volts with 40 mA, a 207-Ω resistor would be best. I used a 220-Ω resistor simply because I have a lot of them

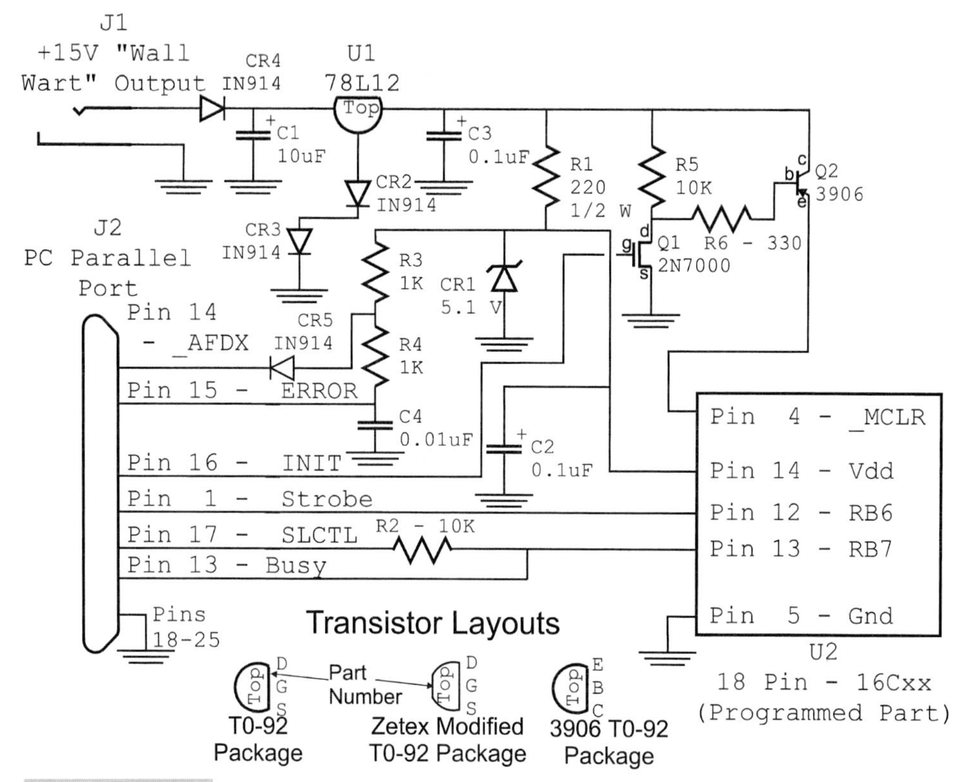


**Figure 13-19** "El Cheapo" circuit schematic

around. The 8.3-volt drop at 40 mA dissipates 0.33 watts of power, which is why I have specified a 1/2-watt resistor. If no load is on the 5.1-volt power supply, this 40 mA is passed through the Zener diode. Doing the calculation again, the total power dissipated by the Zener diode is 0.24 watts. To be on the safe side, this diode and the resistor should be rated for at least half a watt.

The five-volt power supply is designed so that if it is shorted out, only 40 mA will be supplied by it. This allows the PICmicro® MCU to be put and pulled out without switching or disconnecting the power supply. The power supply seems to be a circuit that many people want to change, thinking they have something more efficient.

Please do not change or modify this circuit simply because it does work very well and will protect the circuit against short circuits or other upsets. I know of people that have used bench supplies and one poor soul that used his PC's +5 and +12 Volt power supplies and ended up burning them out because he didn't know what he was doing. The circuit shown here can take a lot of abuse. If too much current is drawn, the 78L12 will shut down, protecting you and the circuitry to which it is connected.

The second subsystem in this programmer is the *Vpp*-control circuit, consisting of the dual transistor switch of Q1 and Q2. Q1 converts the five-volt logic signal from the PC's parallel port to a 0- to 13.4-volt signal that controls the PNP transistor at Q2. This circuit

**TABLE 13-13 "El Cheapo" Bill of Material**

| PART | DESCRIPTION | COMMENTS |
|---|---|---|
| U1 | 78L12 | |
| Q1 | 2N7000 N-channel MOSFET | Note Zetex modified TO-92 package pinout in Fig. 13-19 |
| Q2 | 2N3906 | Labeled *3906* in Fig. 13-19 |
| CR1 | 5.1 V, 1/2 watt Zener | |
| CR2–CR5 | 1N914 silicon diodes | Any silicon diode can be used in this circuit |
| C1 | 10 $\mu$F electrolytic | 35+ volts |
| C2, C3 | 0.1 $\mu$F | Any type of capacitor can be used |
| C4 | 0.01 $\mu$F | Any type of capacitor can be used |
| R1 | 220 $\Omega$, 1/2 watt | |
| R2, R5 | 10 K, 1/4 watt | |
| R3–R4 | 1 K, 1/4 watt | |
| R6 | 330 $\Omega$ power socket | |
| J1 | 2.5-mm power socket | Digi-Key part number: SX1152-ND |
| J2 | DB25-F PC board-mounted socket | |
| U2 | 18-pin DIP socket | Can be a ZIF socket. Use 3M Texttool P/N 218-3341-00-0602R |
| U3 | 14-pin DIP socket | Can be a ZIF socket. Use 3M Texttool P/N 214-3341-00-0602R |
| P28 | 14-pin SIP socket | Can be cut down from a 28-pin DIP socket |
| P40 | 40-pin DIP socket | See text regarding socketing options |
| PC board | P/N 136437-4 | Provided with the book |
| Power | +14-volt AC/DC | The power supply must source at least 250 mA. Output must match J1 |
| PC I/F cable | DB-25M to DB-25M | Straight-through parallel port switch cable. See text |

probably seems a bit unwieldy and unnecessarily complex, but $V_{PP}$ will require up to a 50-mA source to program EPROM parts. This circuit will switch the 13.4 volts of regulated voltage and allow the maximum current supplied by the 78L12 to be passed to the PICmicro® MCU being programmed.

The RC delay circuit (consisting of R3, R4, C3, and CR5) is used to provide an external delay to the PC to program the PICmicro® MCU. The delay itself is on the order of

100 $\mu$s. Five volts is provided by the El Cheapo's power supply and is controlled by pin 14 of the PC's parallel port. The RC network will delay the action and allow the PC to poll the error (pin 15) line of the parallel port to get a (relatively) constant delay that is independent of the PC's operating speed and internal architecture.

The diode (CR5) is used to prevent the RC delay from being driven by the PC's parallel port and only allow it to be driven by the El Cheapo's power-supply circuitry. As indicated elsewhere, there are essentially two types of parallel-port drivers. The most common type used today is an open-drain driver with a weak (1 mA max.) pull up. But, older PCs and most add-on cards use TTL totem-pole drivers that can source and sink current. These drivers can charge C3, leading to the El Cheapo software "finding" that the RC delay is working properly. By placing the diode on the pin 14 line, current cannot be passed from the PC into the RC delay circuitry.

Other simple programmers have been built without this circuit, but the RC delay subsystem ensures that the El Cheapo can be used with virtually all the PCs available on the market today.

The last subsystem in the El Cheapo is the PICmicro® MCU socket and programming data pins. Figure 13-19 shows just an 18-pin PICmicro® MCU socket. As can be seen in Fig. 13-18, there are actually three sockets, which allow the programming of five different *Pin-Through Hole (PTH)* PICmicro® MCU configurations.

The PC board design is shown in Figs. 13-20 and 13-21. Figure 13-20 is the *overlay,* indicating where parts are to be placed on the board and their orientation. Figure 13-21 is the bottom side copper used on the board and the lettering on it is the mirror image of what is going to be displayed because the stencils made for the PC board are made from the "top

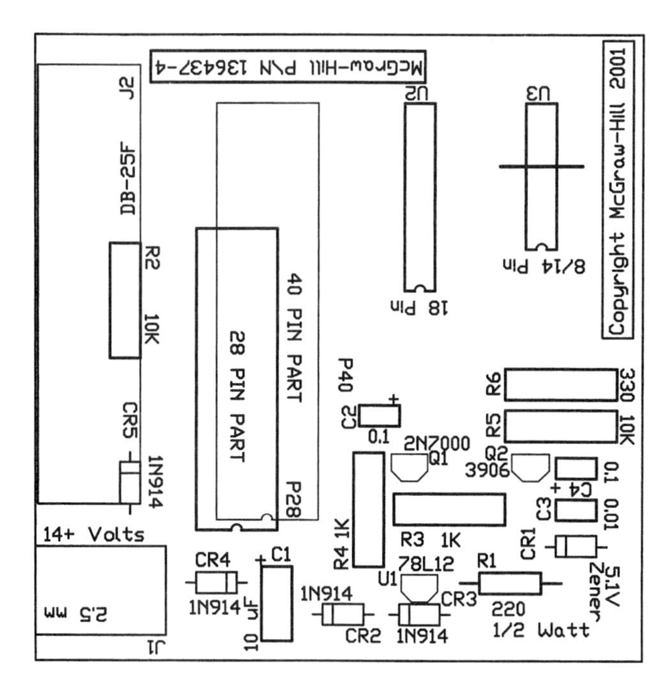


**Figure 13-20** "Overlay" for El Cheapo PCB

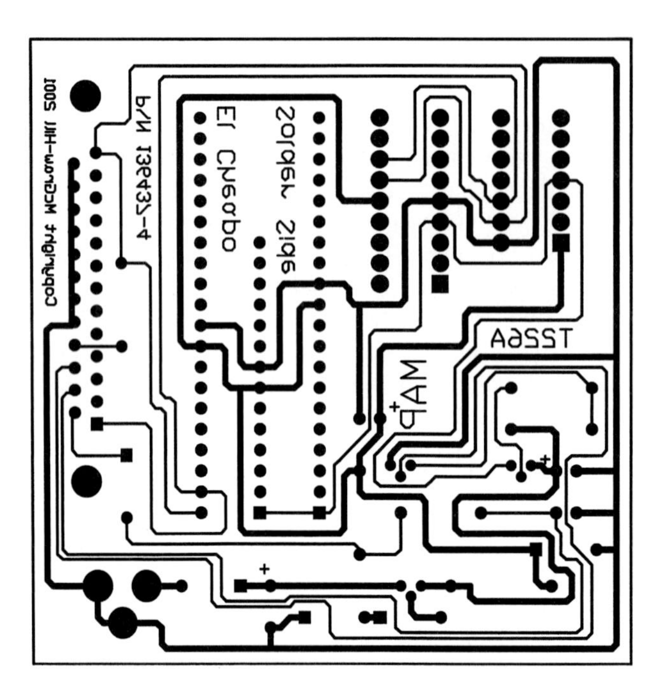

**Figure 13-21** El Cheapo bottom side copper pattern

down." When you look at the actual PC board that came with the book, all of its lettering will be readable (without using a mirror).

The circuit is quite easy to assemble on the PC board, but I want to make a few comments. First, this is a single-sided board. To lay out the board without any jumpers on the topside, some traces run in between 0.100″ parts and between the transistors' outside pins. Be careful when you are soldering in the components that the traces are not shorted to the pins.

Next are the three or five polarized components. Be sure that the transistors and the 78L12 regulator are inserted properly. If you buy Zetex-"modified TO-92" package transistors, remember that the flat side with the labeling should match the flat side on the PC board's overlay silkscreen.

The power connector (J1) is a 2.5-mm center-tap power connector and the AC/DC (wall wart) power converter that you use with it should have a 2.5-mm center-tap mating connector on it. These connectors are very common. As indicated in "Introduction to Electronics" on the CD-ROM, it prevents arcing when the power connector is applied to the circuit.

Notice that the socket is a female socket and the cable connecting the El Cheapo to your PC is a DB-25 male to DB-25 male straight-through cable. Do not use a male socket on the El Cheapo and a DB-25 male to DB-25 female cable; this will reverse the pins going into the programmer and prevent it from working.

The sockets used for the eight, 14, and 18 PICmicro® MCU positions can either be standard DIP sockets or *ZIF (Zero Insertion Force)* sockets. The holes in the PC board are large enough to support ZIF sockets. I recommend that you use them. I realize that a single ZIF socket will cost three or four times what the other sockets on the board cost, but they will make using the programmer a lot easier.

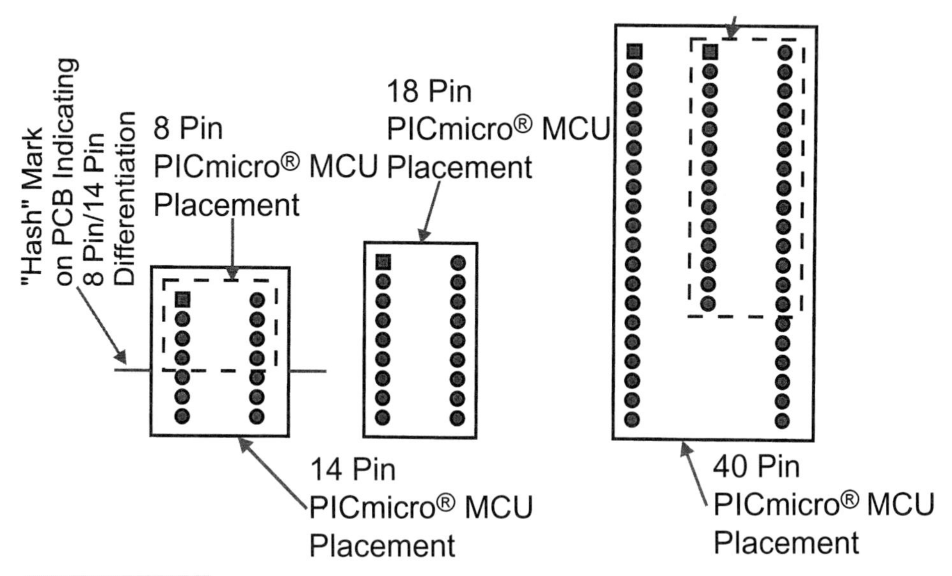


**Figure 13-22** El Cheapo part socketing options

The 8-/14-pin socket is used to program both eight- and 14-pin devices, as does the 40-pin part and 28-pin part. They are combined as shown in Fig. 13-22.

Although a simple 14-pin DIP socket can be used for the 8-/14-pin PICmicro® MCU devices, the 28 and 40-pin PICmicro® MCUs use a slightly different arrangement. The 28-pin PICmicro® MCUs are built as 0.300″ skinny DIPs. The combined socket uses a cut-up 40-pin socket along with a sip connector that was cut from another socket. This socket is not wired for a ZIF socket.

Soldering the circuit together is not that difficult. Even novices should be able to do it in less than an hour. Along with the parts to be installed on the board, you will also require the tools listed in Table 13-14.

**TABLE 13-14 "El Cheapo" Assembly Tools**

| TOOL | DESCRIPTION |
|---|---|
| Soldering Iron | Thin Tip Soldering Iron |
| Solder | Thin Diameter Resin Core Flux Solder |
| Side/Diagonal Cutters | Small |
| Digital Multi-Meter | For Testing the El Cheapo during Assembly |
| Bared 20 Gauge Wire | For Testing the El Cheapo during Assembly |
| PC | Loaded with El Cheapo Software |

Notice that these tools are of small size and should be designed for electronics work only.

When building the El Cheapo, I suggest that you do it in the order of the following list and stop and check after each point.

1 First test the power connector (J1) with the AC/DC power supply that you have selected. This is done by plugging in the power connector into the power supply, plugging the power supply into a wall socket, and measuring the voltage at the connector. The two end pins of the connector should be checked with the terminal away from the connector hole. The output voltage must be at least 14.5 volts for the El Cheapo to work properly. The actual output voltage of the power supply should be between 14.5 and 16.0 volts for the power supply to work properly. Too high and you will find that the 78L12 (U1) will get very hot during programming and could shut down.
2 Next, solder the connector and the CR4 diode onto the board. This diode is used to ensure that the power is positive. It cannot be depended upon to provide rectified AC output. After soldering J1 and CR4 to the board, check the voltage at C1 to ensure that the input voltage is greater than 13.75 volts. If the voltage is less than 13.75 volts, then you will have to find another AC/DC power supply.
3 Wire in the 13.4-volt power supply. In doing this, solder in C1, U1, CR2, CR3, and C4. When finished, check the voltage output with the AC/DC power supply connected to the El Cheapo board. The output should be between 13.0 and 14.0 volts. The two diodes, CR2 and CR3, are used to boost the voltage output from 12 volts to more than 13 volts to ensure that the EPROMs will be properly programmed.
4 Now, wire in the 5.1-volt power supply by soldering in R1, C2, and CR1. Notice that R1 and CR1 should be capable of dissipating 0.5 watt. When no PICmicro® MCU is in any of the sockets, 40 mA will flow through CR1 and R1. This translates to 320 mW or more of power, which requires the larger parts to dissipate the heat. The output should be checked with the DMM. Look for 5.1 volts at pin 1 and pin 14 of the 14-pin socket.
5 Solder in J2. This will be required for the following building steps. As noted, be sure that the connector is female (with holes for accepting pins).
6 With J2 in, the Reset control circuit will be soldered in. Solder in Q1, Q2, R5, and R6 ensuring that you get the transistor polarities correct before soldering in the parts. To test the reset-control circuit, connect the El Cheapo to the PC, power, and follow option 2 (reset) of the El Debug program.
7 Next, solder in R3, R4, C3, and CR5. This provides a hardware delay that will be used by the programmer software. This circuit is tested using option 3 (RC delay) of the MS-DOS El Debug application.
8 The last electronic component to install is R2. Once this is done, check the operation of the Clock and Data pins using El Cheapo.
9 Finally, install the PICmicro® MCU sockets. If ZIF sockets are to be used for 14- and 18-pin parts, then be sure that the parts are "open" when they are soldered in. This avoids problems with the pins being soldered into a position where they are stressed and the socket pin cannot open properly. The 28- and 40-pin socket is created by cutting the top strut from a 40-pin socket to allow the 14-pin SIP socket to be installed inside it with the same level for pin 1. The 14-pin SIP socket is cut down from a 28- or 40-pin socket and soldered in between. When installing the 14-pin socket, be sure that the pins are orientated in such a way that the PICmicro® MCU will be fully seated when it is installed.

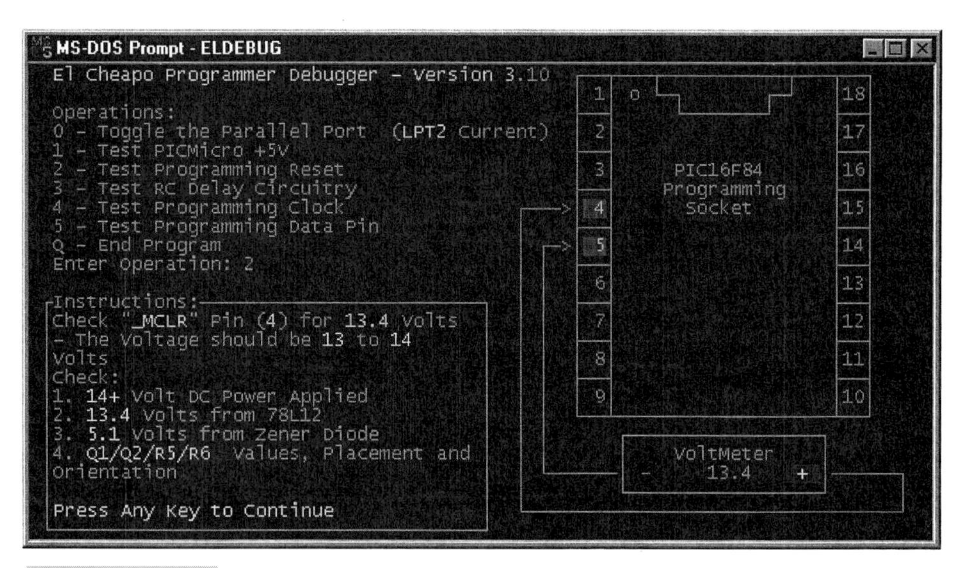


**Figure 13-23** "El Debug" for the "El Cheapo"

To help you debug the application, the Windows Interface provides you with a series of steps to follow, or you can use the El Debug MS-DOS application. A screen shot of this program is shown in Fig. 13-23. This tool is referenced in the assembly instructions given above. The application program provides a graphic of the 18-pin socket (the El Cheapo was originally developed as a simple PIC16F84 programmer) for you to check the actual pin values. As indicated, the three sockets on the PC board are wired together in such a way to provide control for five different PICmicro® MCU configurations. The 18-pin socket is the only one that doesn't provide a two-device interface, which is why I continued its use here.

If you are running from MS-DOS (or the MS-DOS prompt in Microsoft Windows), you can use the ELCHEAPO.EXE application. This command-line program is invoked with the parameters specified and explained (Table 13-15):

```
elcheapo ProgName[.hex] [/1|/2|/3] [/b] [/v] [/e]
```

When you have built your El Cheapo and tested it with El Debug, try programming a PICmicro® MCU (PIC16F84) with a test file and verifying it with both the PICmicro® MCU in and out of the sockets. You should find that with the PICmicro® MCU in the socket, the El Cheapo will pass verify; with it out of the socket, it should fail. Once you can program a part and verify it and fail an empty socket, you have finished building and debugging the programmer. It's ready for use.

One comment for Windows NT and Windows 2000 users: to operate the El Cheapo from a MS-DOS prompt, you will have to have administrator access rights to the system. This application writes directly to the PC's I/O ports. To do this, you must have your ID set up with the correct access.

**TABLE 13-15 "El Cheapo" Command Line Parameters**

| PARAMETER | DESCRIPTION |
|---|---|
| FileName.hex | The intel Format Hex File to Program into PICmicro that was produced by MPASM/MPLAB–this is required for programming or verifying a PICmicro |
| [/1\|/2\|/3] | LPT Port that Programmer Hardware is connected to (LPT1 is the default) |
| /b | Do Blank Check on the Device. If Requested, FileName is Optional. Cannot be Present with "/c", "/r" or "/v" |
| /v | Verify the Program Memory Contents. Cannot be Present with "/b", "/c" or "/r" |
| /e | Program an EPROM PICmicro® MCU |

Along with the MS-DOS prompt, I have created a Microsoft Windows GUI interface in Visual Basic. Figure 13-24 is an early version of the application. To use this application, the file, parallel port and the device is specified using typical windows GUI controls followed by clicking on the operation you want to use. Instructions for installing this application can be found on the CD-ROM.

You will have to have administrator access rights to load this program (and the DLL used to access the parallel ports) onto your PC if you are running Windows NT or Windows 2000.

The El Cheapo programmer operation itself is not very fast. It takes about thirty seconds to erase and program a PIC16F84. This is in contrast to other simple programmers that can burn a PIC16F84 in less than 10 seconds. What it loses in speed it makes up for in robustness.

As indicated, this programmer is designed to work on virtually all PCs. Two specific aspects of the design accomplished this robustness. The parallel-port pins that are specified do not use any advanced or programmable features. The RC delay circuitry will ensure that the El Cheapo will time the application's access to the PC and not require that changes will have to be made of the software.

If you do have problems with the El Cheapo, then perform the following steps before contacting me (because I'll just ask you to do them before I respond to you):

1. Be sure that the parts are soldered down and orientated.
2. Check the parallel cable and be sure that it is straight through and male to male and J2 is a female PC board-mount DB-25 connector.
3. Use El Debug and ensure that each aspect of the programmer is working.
4. Check for any error messages (such as "Unable to Access Port Registers") and check that you have administrator access rights to the PC.
5. Test the programmer with a known-good Flash PICmicro® MCU (ideally, a PIC16F84) with it placed in and out of the socket.

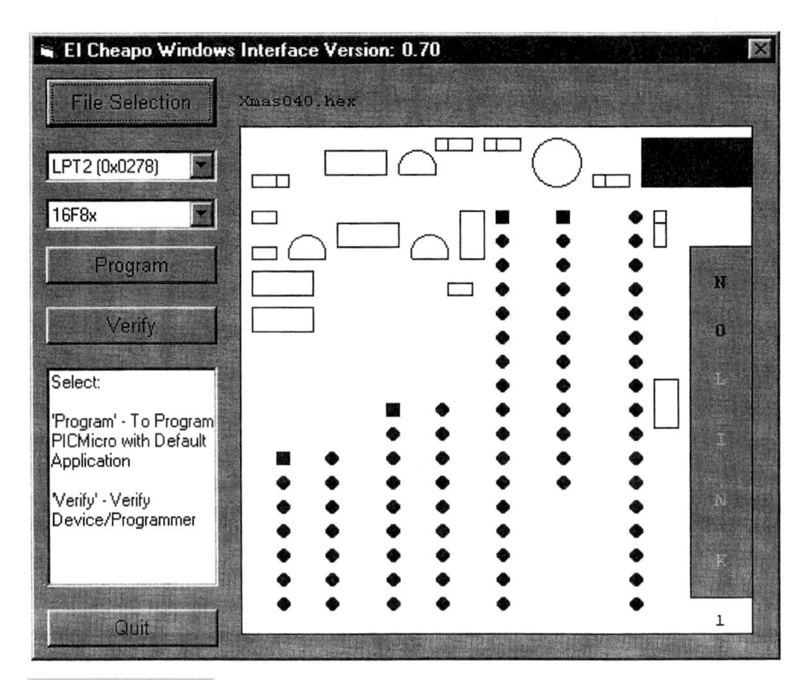


**Figure 13-24** "El Cheapo" windows interface

# The YAP-II

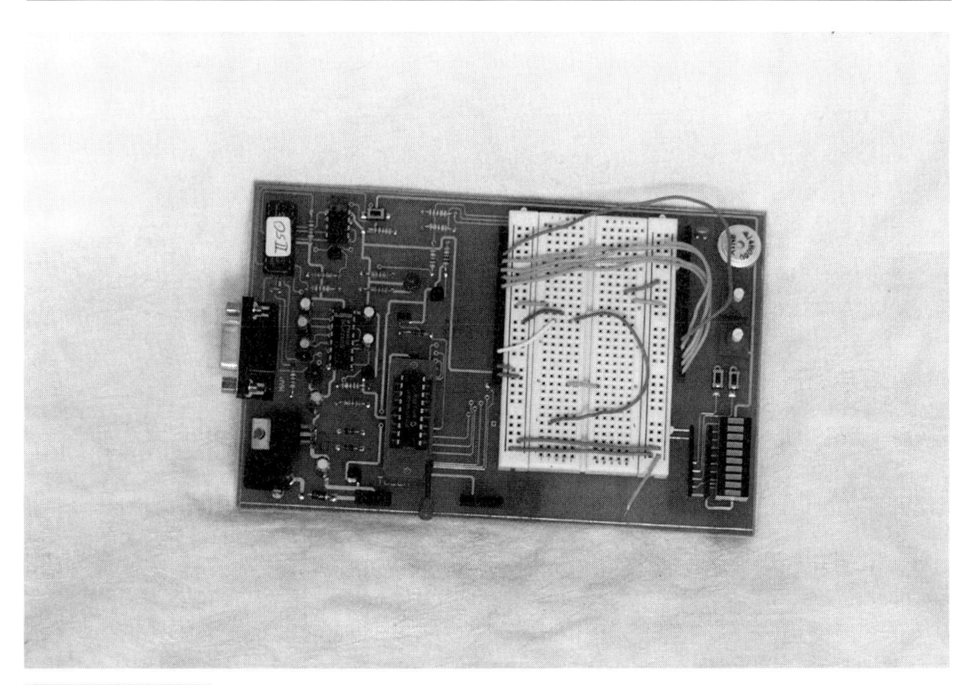

**Figure 13-25** YAP-II programmer with PCB

When I wrote the first edition of this book, I spent an unreasonable amount of time trying to create a programmer. As I was working through the book, I came up with three different programmers, using a PC's parallel port and serial port. Each one had some strengths and weaknesses. Usually, the programmers were very inexpensive, but none of them was able to run on a reasonably wide variety of PCs.

My final solution to the problem was the *YAP (Yet Another Programmer)*. This programmer used a PIC16C61 with an RS-232 interface that took a downloaded hex file and programmed it into target PICmicro® MCU as the file was downloaded. The programmer worked quite well although it only runs at 1200 bps, so it is somewhat slower than other devices out there. The reason for 1200 bps was to ensure that there would be enough time for the worst-case programming of EPROM parts.

As included in the El Cheapo description previously, the problem was caused by the different PCs accessing and timing accesses to their I/O ports differently. Rather than try to figure out a way in the software to time the application, which appeared very difficult because of the operation of the PC, I looked at having the card itself time the programming operation. The El Cheapo provides this timing function with a built-in RC network that is used to delay parallel-port operations and provide a basic ping function to ensure that the programmer is connected to the PC and running properly.

The YAP approached the problem from a different perspective and used the PC's RS-232 ports as the basic interface, not only to the programmer, but for programming applications as well. This had three advantages over the other methods tried. The first was the use of an I/O port with standard timings (while the communication voltages would have to be translated from RS-232 protocol to CMOS/TTL), the incoming data rate could be used to time data going into the PICmicro® MCU. The second advantage was that the ability of PICmicro® MCU application to interface to standard ASCII terminal emulators and not require custom PC software that would have to be debugged in parallel with the PICmicro® MCU application. Lastly, many people who wanted to learn about the PICmicro® MCU, but were not running a MS-DOS or Microsoft Windows PC could run the YAP on their hardware to develop their own applications.

Once I had this concept, I created the following specifications for the YAP programmer:

1. Able to program all mid-range parts (EPROM and Flash-based program memory).
2. Able to take the programming signals and use them in ICSP applications.
3. Allow any RS-232-equipped host PC or workstation to program PICmicro® MCUs.
4. Allow serial communications between the host PC or workstation and the executing application for debugging applications "on the fly."

The result was the YAP, which really wasn't a bad design, but fell short in a number of areas. These included problems with the reset circuit that made programming EPROM parts unreliable, selecting parts that were difficult for people to find were expensive, and creating a form factor that made building sample applications more difficult than it should have been.

After writing the first edition of the book, HyperTerminal (the terminal emulator shipped with Windows 95 and later operating systems) had a problem in which data files would not be transmitted with a trailing ASCII carriage return and line-feed character string. The code had to be modified to accept either these two characters or just a carriage return character.

The original YAP did have some strong points: its ease of use and reliability for different PCs and workstations. I also created a Visual Basic interface that make using the YAP much easier than running it from a basic terminal emulator. Once the problem with the line-ending characters was resolved, the programmer itself has been downloaded and built by a number of people. Wirz Electronics has sold a large number of built and tested units with very few complaints from people.

For this edition, I wanted to enhance the YAP in the following areas:

1 Make programming more reliable for EPROM parts.
2 Program PIC12C5xx and PIC16C505 PICmicro® MCUs, as well as 28- and 40-pin parts.
3 Eliminate some of the difficult-to-find and expensive parts.
4 Provide a better form factor for hobbyists and people learning the PICmicro® MCU.

The result, the YAP-II (Fig. 13-25), is presented here and has been available for sale from Wirz Electronics for a number of months before this book's printing. The circuit itself has been simplified somewhat and the actual PC board has been laid out to include a built-in "breadboard" and a set of sample devices that you can interface a PICmicro® MCU to test applications simply.

This is not to say that the YAP-II (or original YAP) circuit is easy to build. As I write this, I do know of eight people who have successfully built their own YAPs without PC boards or made their own PC boards for the programmer. If you are new to electronics, you will probably want to build the El Cheapo first. When you become more comfortable with the PICmicro® MCU and electronics, you might want to tackle this project. If you are interested in buying a YAP-II, check the information in the appendix about buying an assembled and tested unit (along with a PIC16F84 and wires for building your own applications).

The YAP-II really is a PICmicro® MCU application, in which the PICmicro® MCU communicates with a host system via RS-232 and provides some interesting interfaces other devices (which includes a second PICmicro® MCU with a synchronous serial interface and a high-voltage, up to 50 mA current, control). For this reason, I present it as a project, with the same kind of description that you will see in Chapter 16.

The schematic for the YAP-II is shown in Fig. 13-26 and is the basic application circuit. Attached to it on the PC board that I have designed for it is a set of I/O accessories, which are shown in Fig. 13-27.

The parts for building the YAP-II are quite straight forward and are listed in Table 13-16.

The only part that you might have difficulty getting is the 2106A P-channel MOSFET. This device can be substituted for other P-Channel MOSFETs, its crucial parameters are its ID (on maximum current) of 280 mA and low internal resistance ($R_{ds}$) of 5 Ω.

The source code for the YAP-II (YAP-II.ASM) can be found in the *code\YAP-II* subdirectory of the *PICmicro* directory.

The PC board designed for this circuit is a two-layer board. The top and bottom layers are shown in Figs. 13-28 and 13-29, respectively. The silkscreen overlay information is shown in Fig. 13-30. The CD-ROM section provides the YAP-II Gerber files and other files so that you can build the circuit or have the PC board made from you at a quick-turn PC board manufacturer.

Notice that in the overlay layer, the part number references have a *2* added to them (e.g., *R8* in Fig. 13-26 is *R28* in Fig. 13-30). This change is because I placed multiple PC board

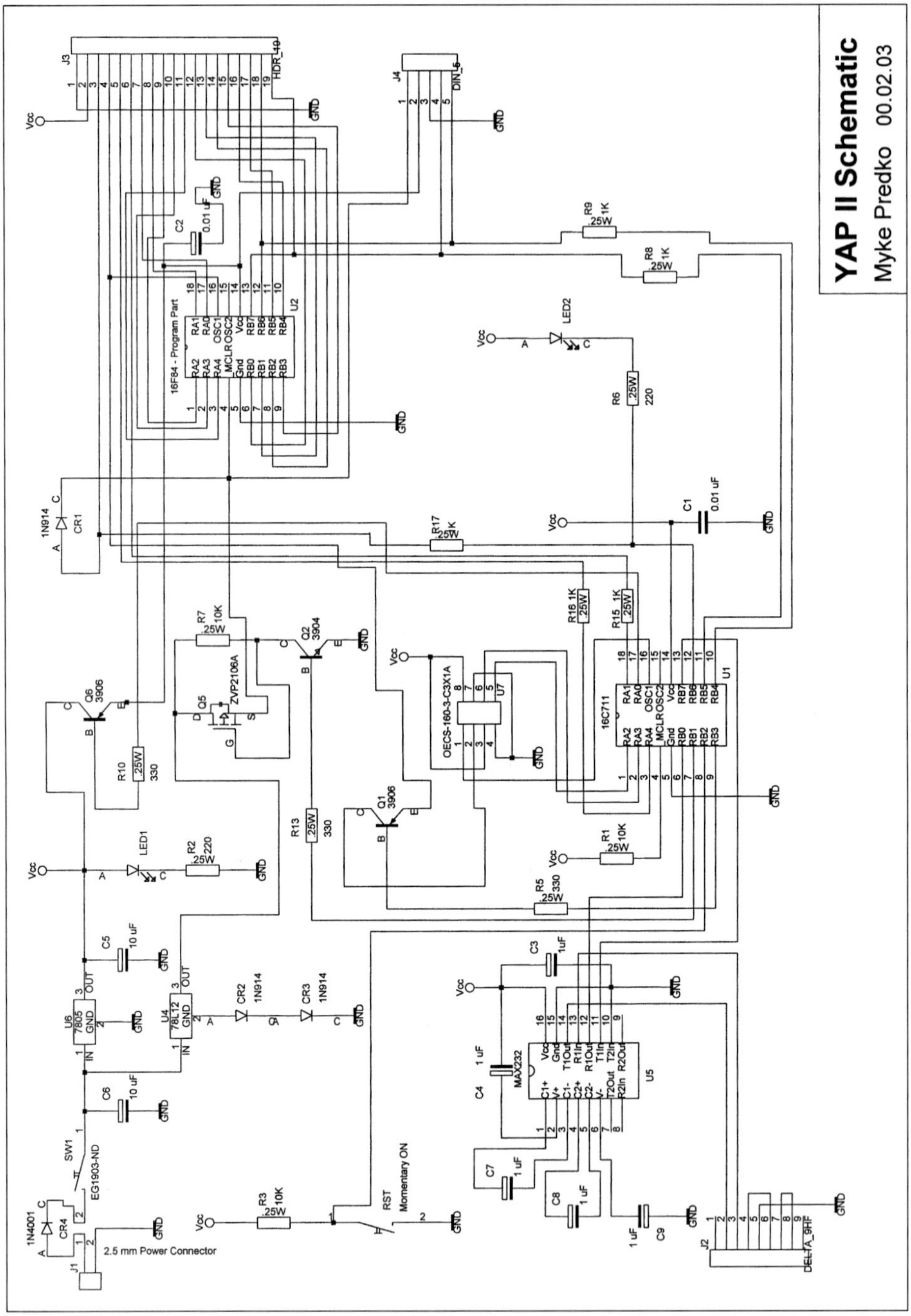

YAP II Schematic

Myke Predko 00.02.03

**Figure 13-26** YAP II schematic

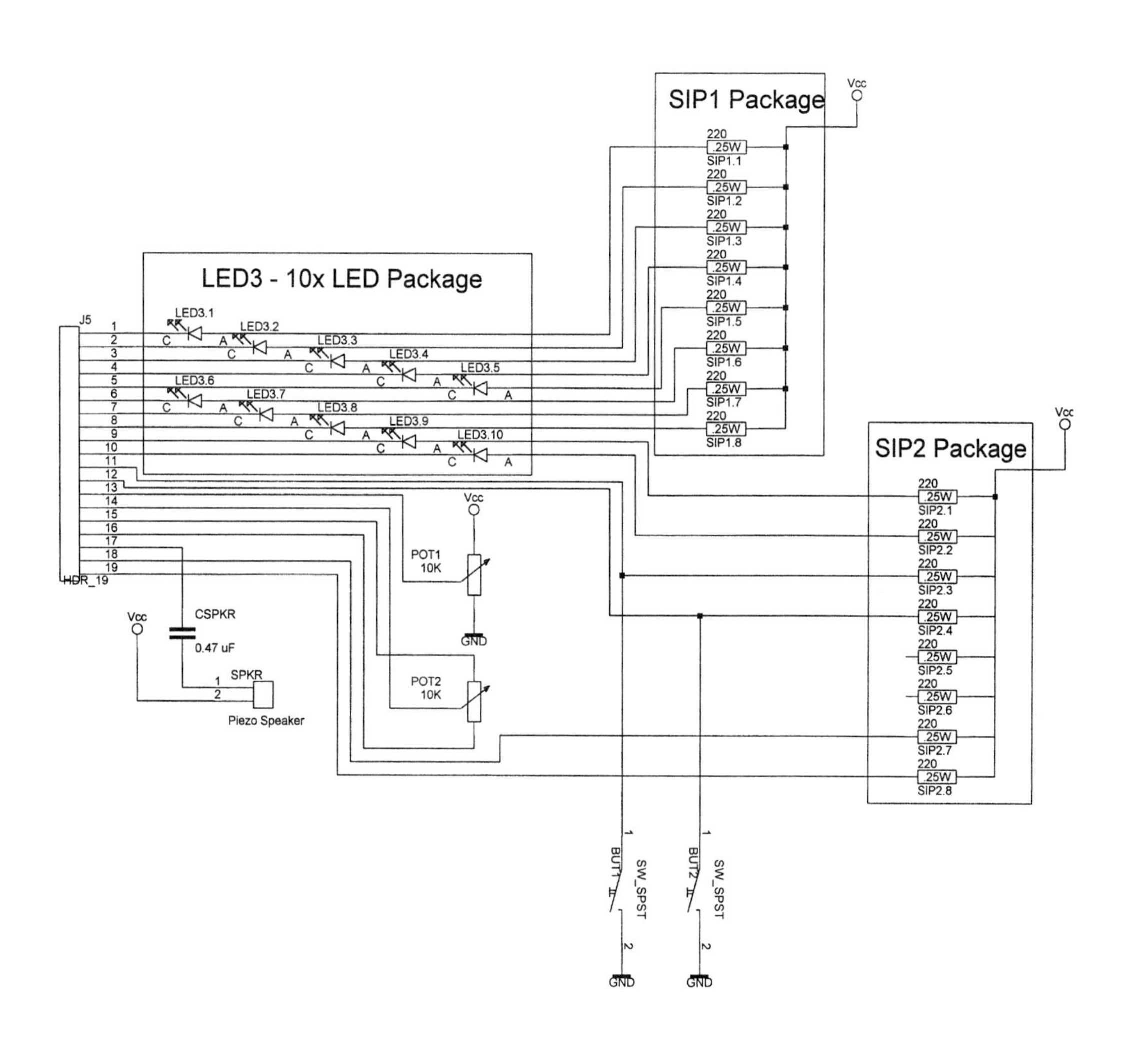


**Figure 13-27** YAP-II accessories

**TABLE 13-16 "YAP-II" Bill of Materials**

| PART | DESCRIPTION/COMMENTS | DIGI-KEY P/N |
|---|---|---|
| U1 | PIC16C711-20/P<br>Programmed with YAP-II Software | |
| U2 | 18 Pin Socket/ZIF Socket | |
| U4 | 78L12 | |
| U5 | MAX232 | |
| U6 | 7805 | |
| U7 | ECS Programmable Oscillator<br>OECS-160-3-C3X1A | XC307-ND |
| CR1–CR3 | 1N914 Silicon Diode | |
| CF4 | 1N4001 Silicon Diode | |
| LED1–LED2 | Red LED with 0.100″ Lead Spacing | |
| LED3 | 10x Red LED Bargraph | |
| Q1, Q6 | 2N3906 PNP Bipolar Transistor | |
| Q2 | 2N3904 NPN Bipolar Transistor | |
| Q6 | 2106A P-Channel MOSFET | ZVP2106A-ND |
| R1, R3, R7 | 10K, 1/4 Watt | |
| R2, R6 | 220 Ohm, 1/4 Watt | |
| R5, R10, R13 | 330 Ohm, 1/4 Watt | |
| R8, R9, R15–R17 | 1K, 1/4 Watt | |
| POT1–POT2 | 10K, Single Turn PCB Mount POT | 3310Y-1-103-ND |
| SIP1–SIP2 | 220 Ohm x9 Common Pin SIP | |
| C1–C2 | 0.01 $\mu$F, Any Type | |
| C3–C4, C7–C9 | 1 $\mu$F, Any Type | |
| C5–C6 | 10 $\mu$F Electrolytic | |
| CSPKR | 0.47 Tantalum | |
| SPKR | Piezo Speaker | P9922-ND |
| J1 | SPDT PCB Mount Switch | EG1903-ND |
| J2 | 9 Pin Female PCB Mount D-Shell | |
| J3, J5 | 19x1 PCB Mount Socket Strip | |
| J4 | 5x1 PCB Mount Socket Strip | |
| RST, BUT1–BUT2 | Momentary On PCB Mount Switch | CKN9013-ND |
| Misc. | PCB Board, Serial Cable, Power Supply | |

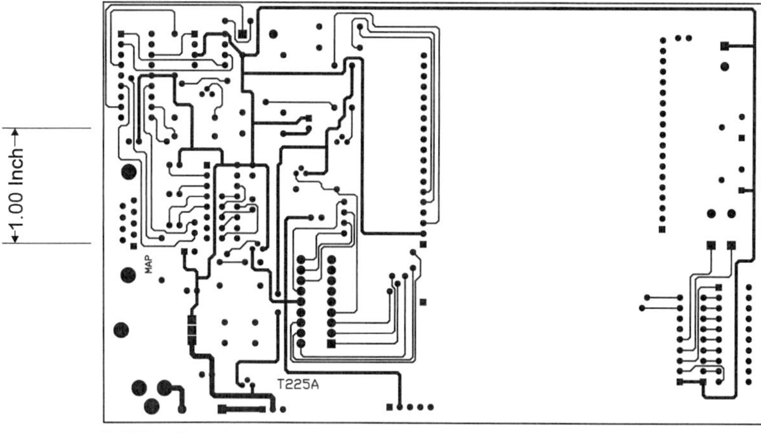


**Figure 13-28** YAP-II top layer design

images on one card and the PC board design system does not allow multiple parts with the same part number on the PC board.

The basic circuit of the YAP-II is a PIC16C711 running at 16 MHz (from the dual-output programmable oscillator) communicating to a host system via an RS-232 interface. As mentioned elsewhere, I hate making up my own cables, so the circuit is designed to be used with a standard straight-through 9-pin RS-232 cable. Although my circuit uses a 9-pin D-shell, you can use whatever method of connections you are most comfortable with. The RS-232 interface is essentially a three-wire RS-232 connection.

The RS-232 interface application code executing in the PICmicro® MCU uses TMR0 to provide an interrupt at three times the incoming data rate. The RS-232 interface code is designed to buffer the incoming serial data and indicate when the current byte being sent has completed. This interface is used to allow programming operations to occur in the

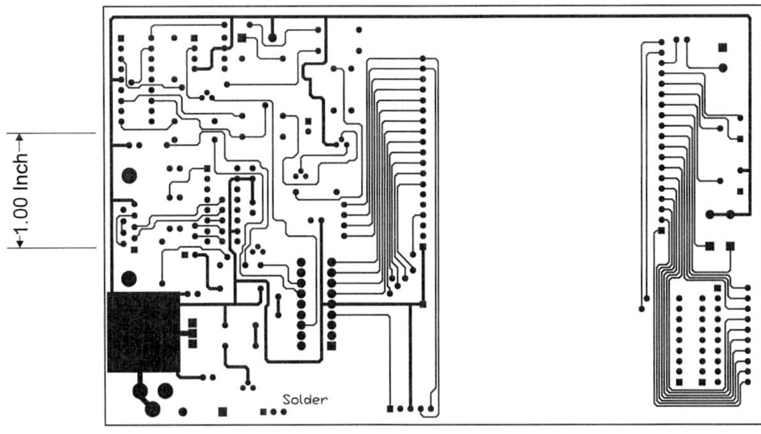


**Figure 13-29** YAP-II bottom layer design

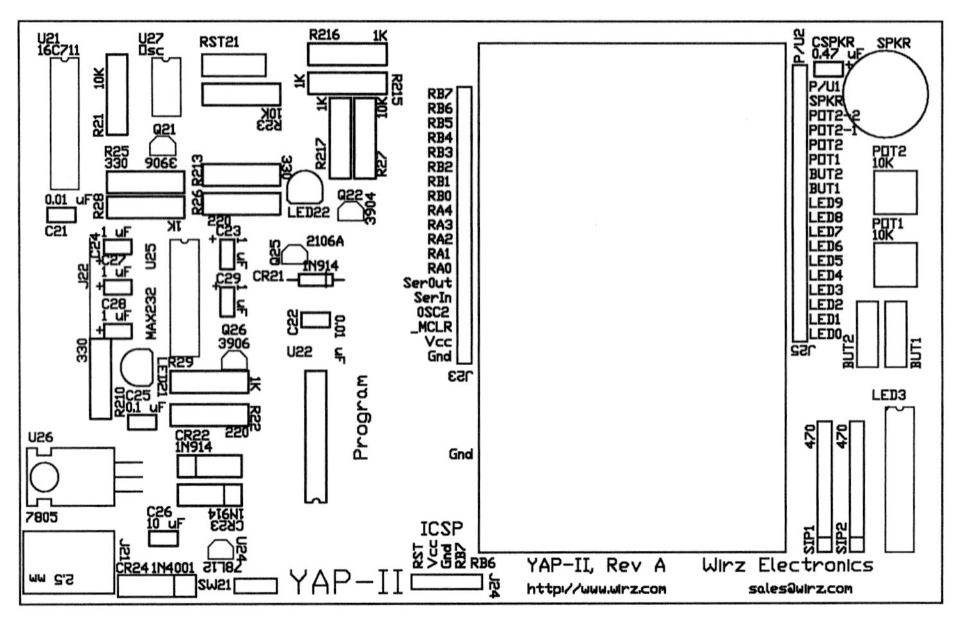


**Figure 13-30**

foreground while serial I/O is occurring in the background. When data is being received, a new programming operation is initiated every four instructions.

The power supply is quite straight forward. A 7805 provides up to 1 A of current at 5 volts, and a 78L12 and two 1N914 diodes provide 13.4 volts at up to 100 mA. In the power-supply circuit, notice that I have included a 1N4001 diode to ensure that negative voltages cannot damage the circuit. For the power source, use a wall-mounted AC/DC converter with an output of at least 14 volts and 500 mA. Wall power adapters with these specifications can usually be bought from discount stores for five dollars or less.

The programming interface circuit consists of three transistors, one diode and seven resistors. Transistor Q6 (along with R10) provides a switched power supply to U2 or the part to be programmed. Transistors Q2 and Q5 (along with R7 and R13) provide a control to the 13.4-volt power supply to the *Vpp* pin. The reset circuit is also driven by U1's RB6 pin, which provides reset voltages when the programmed part is run in a circuit.

Resistors R8 and R9 provide the Data and Clock interfaces to the part being programmed. The resistors are used to provide protection to U1's pins. To initiate a programming operation, the U1 pins on R8 and R9 (RB5 and RB4, respectively) are pulled low. Then +13.4 volts is applied to the PICmicro® MCU in the socket at U2 or connected to the ICSP port at J4. Once the programming voltage has stabilized, instructions are sent to the PICmicro® MCU being programmed.

The programmed parts reset can also be controlled by U1's RB6 for allowing the PICmicro® MCU in the socket to execute. When the PICmicro® MCU in the U2 socket is to execute, U1's RB3 is pulled low, turning on the gate to the programmable oscillator.

The U2 socket is designed to provide a method of programming the PICmicro® MCU and allowing it to execute freely (clocked by the programmable clock, U7). The U2 socket

**TABLE 13-17 "YAP-II" 19 Pin Interface**

| PIN | FUNCTION | U1 CONNECTION | U2 CONNECTION |
|---|---|---|---|
| 1 | Gnd | | |
| 2 | Vcc | | |
| 3 | _Reset | RB6/1K Resistor | No Direct Connect |
| 4 | YAP-II Oscillator | | |
| 5 | U1 Serial In | RA4/1K Resistor | |
| 6 | U1 Serial Out | RA1/1K Resistor | |
| 7 | U2 RA0 | | RA0 |
| 8 | U2 RA1 | | RA1 |
| 9 | U2 RA2 | | RA2 |
| 10 | U2 RA3 | | RA3 |
| 11 | U2 RA4 | | RA4 |
| 12 | U2 RB0 | | RB0 |
| 13 | U2 RB1 | | RB1 |
| 14 | U2 RB2 | | RB2 |
| 15 | U2 RB3 | | RB3 |
| 16 | U2 RB4 | | RB4 |
| 17 | U2 RB5 | | RB5 |
| 18 | U2 RB6 — Programming Clock | RB4/R9 | RB6 |
| 19 | U2 RB7 — Programming Data | RB5/R8 | RB7 |

itself is connected to a 19-pin interface (J3) that can be connected to circuits on the breadboard attached to the YAP-II PC board or to the second 19-pin socket, which provides the built-in accessory interface. The PICmicro® MCU 19-pin interface is defined in Table 13-17.

Notice that pins 18 and 19 are directly connected to U2 and are used to program the PICmicro® MCU in U2 from U1. There are 1-K resistors between the U2 and U1 pins, but an active driver should never be on J3's pin 18 and 19 when you are trying to program the part in the U2 socket. If there is an active driver (and this can be an LED on a pull up), U1 will be unable to "overpower" it because of the 1-K current limiting resistors on the ICSP Clock and Data lines.

Along with the circuit necessary to program the PICmicro® MCU in the U2 socket, I have also included an ICSP-compatible connector at J4. This connector is defined in Table 13-18.

This J4 connector can be used with J3 to program PICmicro® MCUs that are different than 18 pins. A 28-pin device could be programmed by wiring it into the YAP-II's breadboard and providing $V_{PP}$ from the ICSP connector.

Along with the PICmicro® MCU interface, I have also included a set of accessories to allow new users to try out new applications very quickly, without having to find parts and

**TABLE 13-18 "YAP-II" ICSP Connector Pin Definition**

| PIN | FUNCTION |
|---|---|
| 1 | Vpp—Connected to PICmicro® MCU programming pin |
| 2 | Vcc |
| 3 | Gnd |
| 4 | ICSP Data |
| 5 | ICSP Clock |

figure out how to wire them in. As you will see, these features greatly simplify the wiring of the experiments.

J5 is a 19-pin connector, like J3, that provides an interface to LEDs, buttons, potentiometers, a speaker, and some pull ups. Ten LEDs are built into a bargraph display that are soldered into the board. These LEDs are pulled up by a 220-Ω resistor. To turn them on, they have to be pulled to ground. Two pulled-up buttons are also available, along with a potentiometer that acts like a voltage divider. The second potentiometer has all three connections passed to the J5 connector, so different circuits can be built with it. The piezo speaker is connected to J5 through a 0.47-$\mu$F capacitor so that the driver is isolated from the speaker (and any transients coming from it). Finally, there are two pull-ups for convenience's sake. The pin out of J5 is shown in Table 13-19.

As I have designed the YAP-II's PC board, a breadboard is placed between J3 and J5. As you work through the experiments, you will see that for each experiment, I have included a wiring diagram for both a breadboard version of the experiment, as well as a YAP-II. These graphics are similar to those in Fig. 13-31.

In Fig. 13-31, notice that I have connected the *Vcc* and Gnd connections from the 19-pin connectors directly to the two top sideways rails on the breadboard and jumper the connections to the bottom two rails. This gives me convenient power that will make the application wiring quite a bit easier to do when you are creating applications.

In the original YAP, I created what I thought was a very minimal command set. The basic commands consisted of programming and verifying commands for EPROM, Flash-based PICmicro® MCUs, and execution control. The problem with these commands is that they didn't support different-sized PICmicro® MCUs very well. I designed the original YAP for PICmicro® MCUs with 1024 instructions of program memory.

The original YAP was well designed for the PIC16F84 and PIC16Cx(x)1 part numbers, but not very many others. The YAP-II is designed for a wider range of PICmicro® MCUs with varying program memory sizes.

In the YAP-II, I've further simplified the command set so that a single character is sent as a command, followed by a carriage return. The 11 commands are listed in Table 13-20.

The interface itself is designed to run at 1200 bps. This speed was chosen as the fastest standard speed in which the time it takes to receive four bytes (from a hex file), giving an instruction for programming and perform a programming operation in parallel. Running the interface at 1200 bps makes the YAP somewhat slower than other PICmicro® MCU programmers, but the tradeoff would be to add an external buffer memory, which would

**TABLE 13-19 "YAP-II" Accessory Connector Pin Definition**

| PIN | FUNCTION |
|---|---|
| 1 | LED1 |
| 2 | LED2 |
| 3 | LED3 |
| 4 | LED4 |
| 5 | LED5 |
| 6 | LED6 |
| 7 | LED7 |
| 8 | LED8 |
| 9 | LED9 |
| 10 | LED10 |
| 11 | BUT1 |
| 12 | BUT2 |
| 13 | POT1 |
| 14 | POT2 Wiper |
| 15 | POT2 — Connection 1 |
| 16 | POT2 — Connection 2 |
| 17 | Speaker |
| 18 | Pull Up 1 |
| 19 | Pull Up 2 |

add to the cost of the device. The programming operation is explained in more detail in this section.

When demonstrating how the commands work, I have provided screen shots of HyperTerminal operating with the data shown on the display. HyperTerminal operation is explained elsewhere, but to connect to the YAP-II, HyperTerminal should be set up with a direct connect to the YAP-II at 1200 bps with an 8-N-1 data format.

The interface will convert lowercase ASCII to upper case and ignores all characters, except for the ones listed in Table 13-20 and ASCII *Backspace* (0x008) and *Enter* (0x00D).

The *ping* command is designed for advanced interfaces (such the Visual Basic interface presented later) to check and see if the YAP is connected and working properly. After sending ASCII *A* (0x041), followed by an ASCII *Enter* (0x00D), the YAP-II returns a carriage-return/line-feed string. This instruction is simply used to check the interface without having to parse the:

```
"<== Invalid"
```

message that is returned for invalid commands.

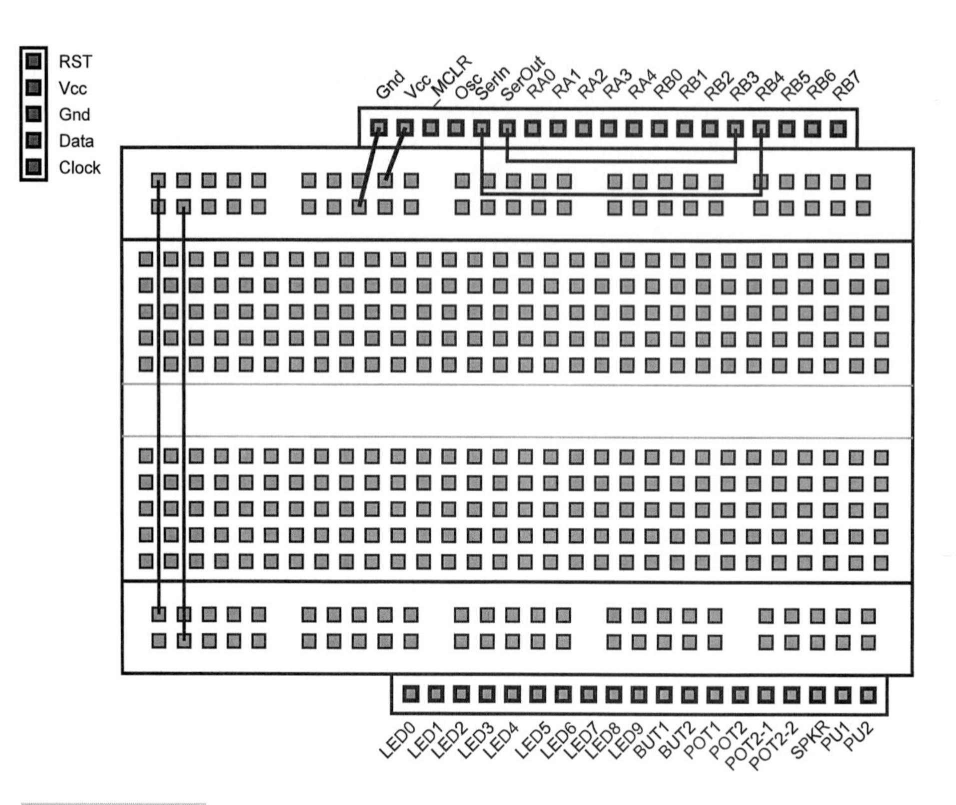


**Figure 13-31** YAP breadboard for "basicRS"

**TABLE 13-20 "YAP-II" Commands**

| COMMAND | OPERATION |
|---|---|
| A | "Ping" — Return Nothing but Carriage Return/Line Feed |
| B | Reset the Program Counter to 0 |
| C | Clear the Contents of Flash Memory |
| D | Dump 256 Instructions |
| | Increment "Read" Program Counter by 256 |
| E | EPROM Part Programming — "Text Send" Hex File |
| F | Flash Part Programming — "Text Send" Hex File |
| G | Get Eight Instructions Starting at Address 0x02000 |
| 1, 2, 4, 8 | Run PICmicro® MCU at the Specified Speed |

During programming, the PICmicro® MCU uses its built-in program counter to track where operations are occurring. This program counter is shadowed within the YAP-II to track where it is executing. I use the shadowed program counter to track the offset of the 256 instructions last returned by the YAP-II during the *dump* instruction. To reset it, the *B* command is used.

*C* clears the contents of Flash program memory using the Microchip-specified *all clear* instruction. This is the same process as given previously in this chapter.

1 Apply *Vpp*.
2 Execute load configuration (0b0000000+ 0x07FFE)
3 Increment the PC to the configuration register word (send 0b0000110 seven times).
4 Send command 0b0000001 to the PICmicro® MCU.
5 Send command 0b0000111 to the PICmicro® MCU.
6 Send *begin programming* (0b0001000) to the PICmicro® MCU.
7 Wait 10 ms.
8 Send command 0b0000001.
9 Send command 0b0000111.

I separated this from the program command to allow the clear to be confirmed with the *dump* (*D*) command.

The first version of the YAP had a *verify* command that compared downloaded data from the host computer to the contents of the PICmicro® MCU in the programming socket. In the YAP-II, I have dispensed with this command, instead pulling down the contents of the PICmicro® MCU and sending it to the PC host. The *dump* command sends 256 instructions to the host computer in a format of 16 instructions per line (each one taking up five bytes). The results of the *dump* command are shown in Fig. 13-32.

The dump operation takes about 11 seconds for each 256 instructions. This translates to about five and a half minutes for a device with 8,192 instructions of program memory (and about half a minute for a 1024-instruction PICmicro® MCU). This is approximately the same speed as will be required for programming, so a complete operation of blank check, program, and verify can take as long as 20 minutes for an 8K PICmicro® MCU. This is why I generally do a cursory blank check and no verify in the large devices.

This change was added to allow host computers to do their own blank check and verify on PICmicro® MCU microcontrollers of varying program memory sizes. The original YAP was designed to only work with PICmicro® MCUs that had 1024 instructions of program memory space. By downloading data at 256 instructions at a time, different program memory sizes can be supported. It can also support the PIC12C5xx and PIC16C505, which are low-end devices, but use the same programming protocol as the mid-range devices and program the configuration fuses as the first byte.

Programming the PICmicro® MCU is accomplished by downloading the MPASM-produced hex file into the YAP-II after specifying either the *E* (EPROM) or *F* (Flash) commands. The YAP-II will decode the hex file format and check that the results of the programming operation are correct for each instruction. When the file has finished being transmitted, if an error occurred in the programming, the first instance of problems will be returned with the expected (*E*) and actual (*A*) values displayed and the address where the error occurred.

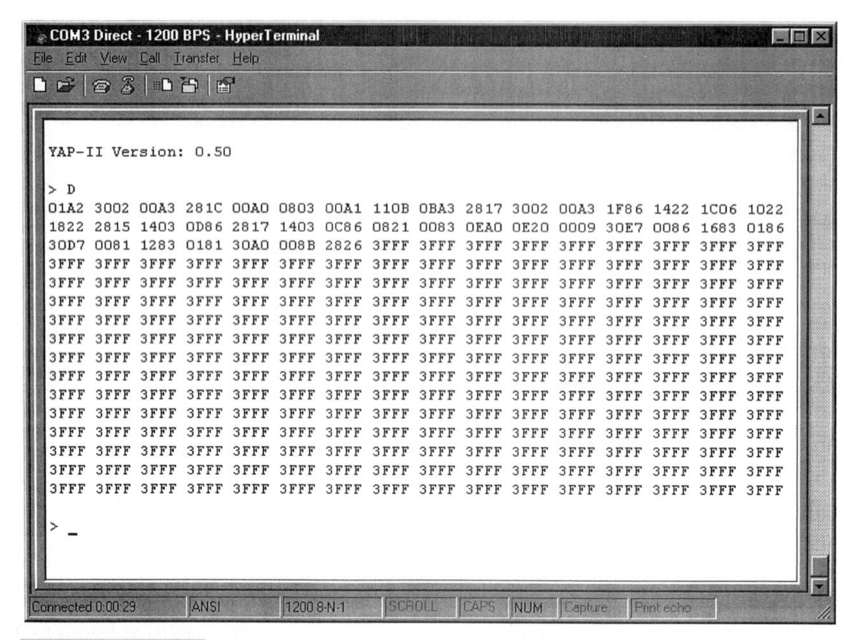


**Figure 13-32** YAP-II "dumping" initial 256 instructions

The programming operations wait for the data file for one minute before timing out. If you don't want to go through with the programming operation, then send a 0x003 (*Ctrl-C*) character to the YAP-II to stop it.

The programming operations are identical to what is described, except that there is no final verify operation (except using the *dump* command). When each line of the hex file comes in, the address to program is compared against the current value of the shadowed program counter. If there is a difference between the two values, the PICmicro® MCU's program counter is incremented while the next six bytes are coming in. These six bytes take 50 ms to come in, which is just enough to increment the PICmicro® MCU's program counter 50 times. If the difference between the current PICmicro® MCU program counter and the specified address is greater than 50, then an error (jump) will be flagged by the YAP-II and programming will stop. Generally, this is not a major concern because most applications are written with instruction addresses written consecutively and no space is left between the modules in the application.

When the configuration fuses are programmed (at address 0x02007), the *load configuration* command will be used, which changes the internal program counter to 0x02000 and avoids the need to repetitively increment the PICmicro® MCU's program counter.

This is not true for data EEPROM initialization. This data is located at address 0x02100 of the hex file, requires a different programming algorithm, and cannot be accessed. If the *de* directive is used in your source file, the application will return a jump error.

The configuration memory (at 0x02007) is read for verify by using the *G* command. The eight bytes starting at 0x02000 are dumped onto the screen (in the same format as the *D* command), showing you not only the configuration fuses, but the ID locations as well.

The last four commands specify the operating speed for the PICmicro® MCU. On the card is an ECS 16-MHz dual-output programmable oscillator. The primary, 16-MHz output is passed to the PIC16C711, which controls the operation of the YAP-II. The secondary output, which has a programmable divisor that is passed to the part in the programming socket when one of these commands is executed.

When *1, 2, 4,* or *8* is passed to the YAP-II, the part in the programming socket has power applied to it, along with the programmable oscillator output and finally the *_MCLR* pin goes high. During execution, serial data can be passed to and from the PICmicro® MCU with the application stopped when a *Ctrl-C* (0x003) character is received by the YAP-II.

When the *Ctrl-C* character is received, you will see a funny string of characters overwriting the running message (Fig. 13-33). This is caused by invalid serial line data being inadvertently sent by the serial pass through option described in the previous paragraph when the programmed PICmicro® MCU's *_MCLR* line becomes active.

This is an *artifact* as to how the PIC16C711 YAP-II controller is operating and there is no way to prevent it, except to eliminate the pass-through option. When *Ctrl-C* is received, the serial interface changes mode with the value of the *SerIn* line being passed to the host without a full byte being sent. I have chosen to leave this in because I normally use the YAP-II with the Visual Basic front end and it masks the invalid characters returned.

The PIC16C711 software used in the YAP-II is quite complex. Although it is based on the original YAP application, substantial changes have been made to the operation of the code. Put another way, if you compare the code for the YAP-II to the original YAP, you will discover that the interface subroutines are the same, but the mainline logic is substantially different.

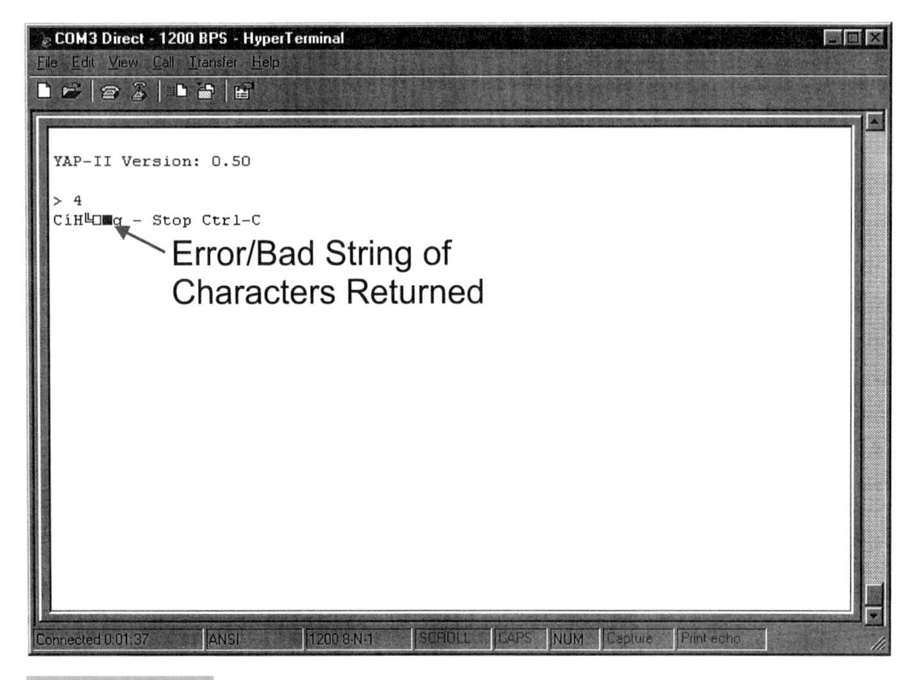


**Figure 13-33** YAP-II stopped showing bad string

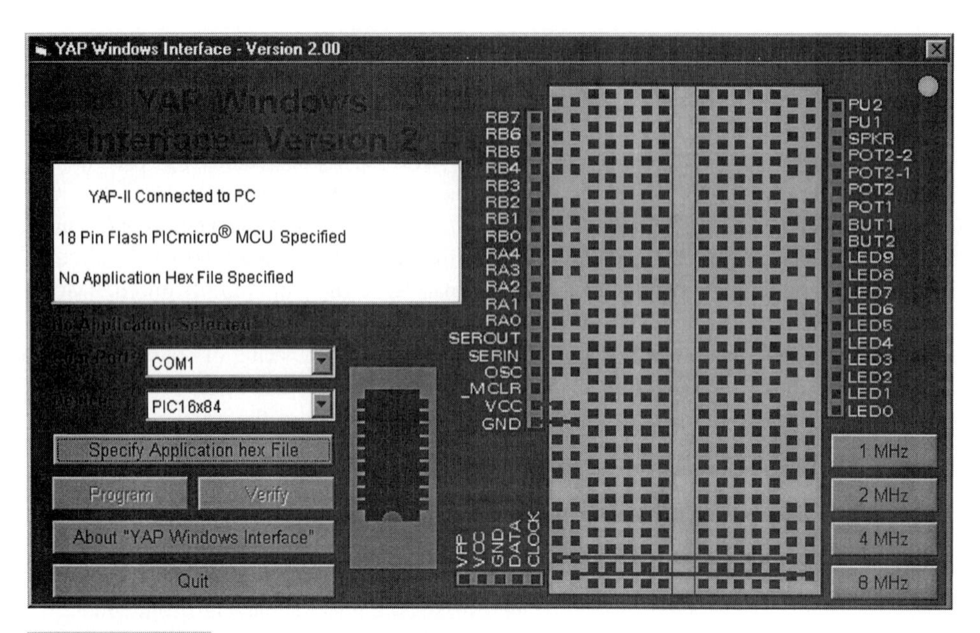


**Figure 13-34** **YAP-II windows interface**

Also in the YAP-II application code, because I was using a PIC16C711 (which has a lot more file registers instead of the YAP's PIC16C61), I was not as restricted in the number of variables that I used in the application. This makes the code somewhat easier to read and follow.

The Visual Basic front end (Fig. 13-34) provides a graphical interface for you to operate the YAP-II. There is an installation package in the *code\YAP-II\vb* subdirectory of the *PICmicro* directory, as well as instructions on how this application is to be installed. Check my web page for updated installation packages.

This Visual Basic front end provides similar functions to the El Cheapo front end. The interface port (COM1, COM2, or COM3) can be selected from the front window, as well as the PICmicro® MCU to work with and the hex file to program into it. The Visual Basic front end continually pings for a YAP-II on the specified interface port and indicates when one is available. To simplify the Visual Basic front end, the primary parameters, COM port, hex file, and PICmicro® MCU target is saved on your PC's hard file so that when you start up again, you do not have to re-enter these parameters.

When the application is executing, a separate terminal emulator display allows you to send and receive executing data from your application.

When you are working with the YAP-II, I recommend that you initially wire the two top and bottom rails with *Vcc* and ground, as I've shown in the Windows Interface (Fig. 13-34). This will give you convenient power connections as you build a test application circuit.

The YAP-II will program all mid-range and the PIC12C5xx and PIC16C505 PICmicro® MCUs. If you are going to experiment with using the YAP-II with the PIC12C5xx and the PIC16C505, then be sure that the first instruction word is 0x0FFF in the hex file in which you program into the devices. The YAP Windows interface will modify the hex file data sent to the YAP-II automatically to provide this feature.

When I first created the YAP-II (and I was quite excited about the prospect of being able to program PIC12C5xx parts with it), I tried it out with a variety of parts. I had an unexpected error when I programmed it: the memory location at address 0x0000 was incorrectly programmed. My initial reaction was to be mad at myself because I remembered that the first word accessed in the PIC12C5xx and PIC16C505 contains the configuration fuses. When I looked at what I was programming, I discovered that the expected word had bit three of the first instruction reset, which enabled the PICmicro® MCU's program memory code protection. This obviously scrambled the data returned from the first program and caused the application to fail.

# Third-Party Programmers

The first edition of this book stated that a real "cottage industry" has grown up around creating PICmicro® MCU programmers. There are literally hundreds of different designs and different software interfaces, which make choosing a programmer difficult. Choosing the "right" programmer for you can be a frustrating experience—especially if you try different designs, only to discover problems.

Two types of programmers on the market today can be used for the PICmicro® MCU. The first is the commercial, "professional" programmer. These programmers can usually program a lot more than just a PICmicro® MCU and will program the devices exactly according to the Microchip specifications. The downside to these programmers is their price; they are often 500 dollars or more for the "base unit" and additional money must be spent for adapters designed for specific parts. These devices are best suited for manufacturing sites or commercial development sites, where designs using more than one microcontroller device type are produced.

For the hobbyist or the small business that does not want to invest a lot of money into equipment, there are a plethora of different low-cost programmers. The El Cheapo and YAP-II presented here fit into this category, as well as the Microchip PICSTART Plus. The microEngineering Labs' EPIC programmer (Fig. 13-35) also belongs here.

Another product that you should be aware of when you are starting to learn about the PICmicro® MCU is Labtronics International's PICFun and PICFun2 development boards (Fig. 13-36). This board contains a serial interface PICmicro® MCU programmer, along with a number of LEDs, buttons, potentiometers, and external interfaces.

When looking at a low-cost programmer, consider the following characteristics:

1 What are the supported PICmicro® MCU devices?
2 What is the interface and how is the application timed?
3 How are the configuration fuses programmed?
4 What operating system does it run under?

Notice that I did not list *cost* or *programming time* as a characteristic that you should consider when buying PICmicro® MCU programmer. *Cost* is a very subjective term that can be very misleading. A very low cost programmer can become very costly if all the

**Figure 13-35** microEngineering Labs' EPIC programmer

"bells and whistles" are added to it. As I was writing this book and planning for the El Cheapo to be included with it, I looked at listing the costs of all the parts needed to build the El Cheapo and run through the experiments. I found that retail prices for the El Cheapo with all of the parts on its board (including ZIF sockets), the cables, a wall-mounted

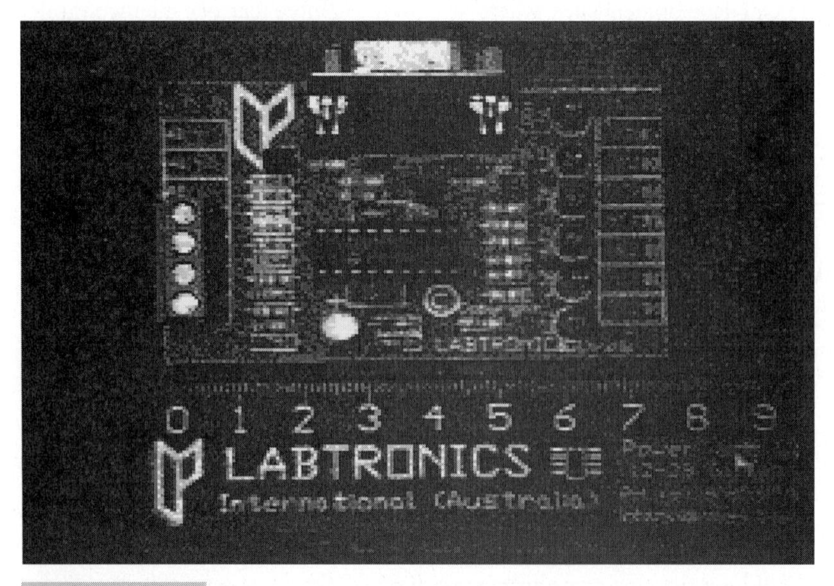


**Figure 13-36** "PICFun" PICmicro® MCU programmer and development kit

AC/DC adapter, 5-volt power supply, a sample PIC16F84, a breadboard, and the parts needed to work through the experiments, the cost would be approximately $100. It makes you feel like the El Cheapo doesn't live up to its name.

One hundred dollars for a complete PICmicro® MCU experimenter's kit is not unreasonable, however.

The programmer's speed should not be considered a major consideration because, depending on how the programmer is built, it might work very quickly with one host PC and not at all in another. This is why I went through several revisions of the El Cheapo and YAP-II to be sure that the programmers were as device independent as possible. As I write this, the first 800-MHz PCs are becoming available and 1 GHz PCs are not far away. Coupled with many sub-500-dollar PCs with Celeron processors running under 400 MHz, the performance range on modern systems is staggering. If a parallel-port programmer does not have a circuit on board to time itself, you will have to contact the manufacturer (or the developer) for patches and new applications to run on it when you get a new PC or update the operating system.

You should be aware of the different PICmicro® MCU devices that are supported by the programmer you are looking at. I tend to design my projects with only ICSP-enabled devices. This simplifies the types of programmers needed to a very small set. Notice that the 17Cxx ICSP is not compatible with the other device's ICSP; programming of the boot ROM is required before the application can be loaded into the PICmicro® MCU.

One issue that isn't often addressed is whether EPROM and Flash devices can be programmed. Many people assume that because the ICSP pins and packet protocol for the two program memory types is the same, a programmer that can do Flash can also do EPROM. This is not the case because of the EPROM's 50 mA $V_{PP}$ source requirements. For the El Cheapo and the YAP-II, the $V_{PP}$ circuits are seemingly more complex than they should be to be able to supply the 50 mA $V_{PP}$ for EPROM programming. Very few low-cost devices have the capability of programming EPROM devices. The El Cheapo, YAP-II, and meLab's EPIC are among the few that do have the appropriate $V_{PP}$ current-drive capabilities.

If you are new to working with the PICmicro® MCU, I highly recommend not buying a programmer in which you have to set the configuration fuses. Chances are that when you first start working with the PICmicro® MCU, you will have a lot to learn and remember. Don't try to remember fuse settings. You will invariably forget and end up with a PICmicro® MCU that doesn't run after a simple application "tweak" and you have no way to find out what it is.

This problem invariably happens when you are under stress already with an assignment or project deadline due. Not having to worry about what the configuration fuses are set to (other than in the source code) eliminates one variable and potential problem for you.

Along with PC speeds getting faster and faster, you also have to consider the operating system that is used with PC. Many programmers are designed to run under MS-DOS. Although there is an MS-DOS prompt under Windows 3.1x/95/98/NT/2000, you cannot count on it being able to access all of the hardware in the system. This is of considerable concern for programmers that use the parallel port; if you do not have administrator access in Windows NT/2000, you probably cannot use the programmer without your network administrator granting a session-special access. Linux offers MS-DOS support, but this might limit the I/O ports that are accessible from it.

The safest bet is to only use a programmer that accesses the PC via the RS-232 serial port. RS-232 support and drivers are available under all operating systems without requiring special access rights.

If you are going to use a programmer that accesses to your PC via the parallel port, I suggest that you buy a second parallel-port adapter if you are using the primary parallel port for a printer. Doubling up the functions on a PC's parallel port is time consuming and probably won't work that well with different software applications "claiming" the hardware as their own. An ISA, or PCI parallel-port adapter card can be purchased new for less than 15 dollars. You will feel like the card is worth it after just a few minutes of working with it, instead of sharing a port with other devices.

The information provided in this section should be available from the programmer manufacturer. If it isn't, then look at another programmer. Over the years, I have had enough problems with balky PC adapters. Avoid any potential problems and find a programmer that is well supported by an established company or has hardware interfaces that do not require software upgrades as time goes on.

# 14

# EMULATORS

## CONTENTS AT A GLANCE



Microcontroller emulators can be extremely useful tools in finding problems within completed application code. Unfortunately, they are often not used for this purpose. Instead, they are used as an integral part of the development process. The danger with using an emulator as a development tool (putting in changes during debug) can result in an application that is not fully thought out and can have design problems that are not readily evident.

Four types of emulators are available for the PICmicro® MCU. Although a number of low-cost emulators are available (including the EMU-II presented later in this chapter), the full device emulator (like the MPLAB ICE-2000) is the best device on the market to use. This introduction to emulators goes through the different types and explains the differences between them. I also present my own design for an emulator.

The first type of emulator is a simple hardware interface emulator that is controlled by a PC or workstation (Fig. 14-1). In this type of emulator, the simulator code in the PC runs the PICmicro® MCU application and accesses the emulator pod's I/O pins.

This type of emulator is very inexpensive, but probably is the least accurate of the three types in terms of timing and electrical behavior of the I/O pins. Using a PC with a PIII

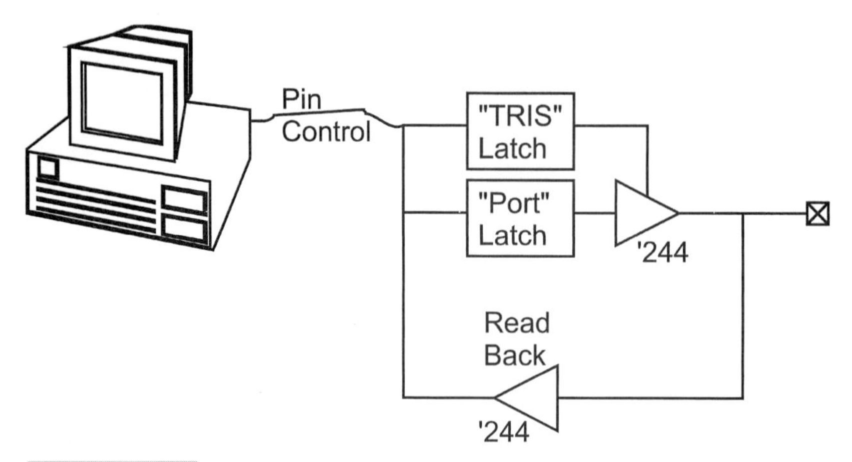


**Figure 14-1** Simple I/O port emulator

processor and a high-speed interface, the actual speed of the application will be approximately the same as what the PICmicro® MCU would produce, although there would probably be some significant differences in edge-to-edge timing.

The pin operation can be difficult to properly simulate using discrete chips, but a CMOS PLD could be designed to accurately model the PICmicro® MCU's pin behavior. This problem could be eliminated by writing a small PICmicro® MCU application that allows a PC to interface with it and remotely control the operation of the I/O ports (Fig. 14-2).

The second type of emulator is presented later in this chapter (the EMU-II). Recently, the PIC16F87x family of PICmicro® MCUs has become available where the internal program memory can be read from and written to. The EMU-II uses this feature to load and execute code in the PICmicro® MCU (Fig. 14-3). This type of emulator is often called a *downloader.*

This type of emulator can be created quite inexpensively, but the limited program memory of the PICmicro® MCU means that relatively simple operations can only be implemented.

Each PICmicro® MCU is designed with emulator features built in. The PIC16F87x series of PICmicro® MCUs have built-in serial-port features that allow custom hardware to an asynchronous serial port to access the different registers and functions of the PICmicro® MCU (Fig. 14-4).

The only issue I have with this method is the selection of I/O pins (most notably RB3, RB6, and RB7) for the emulator functions. These pins are part of the only full eight-bit port available in the 18-pin PICmicro® MCUs, which means that often the application cannot run the full, unchanged application with the emulator. This is especially true if PORTB is used for full eight-bit I/O or with PORTB built-in hardware features.

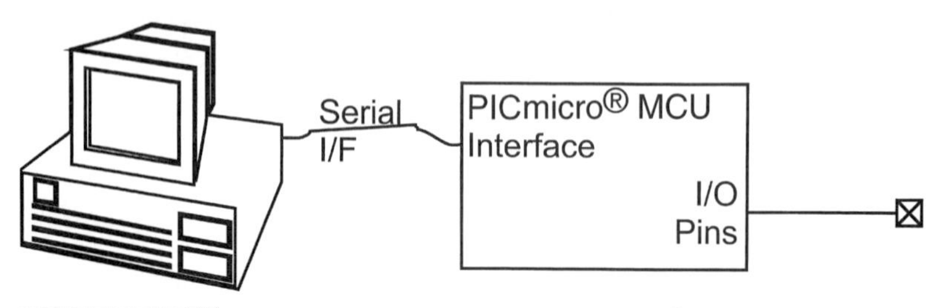


**Figure 14-2** Simple emulator using a PICmicro® MCU for I/O

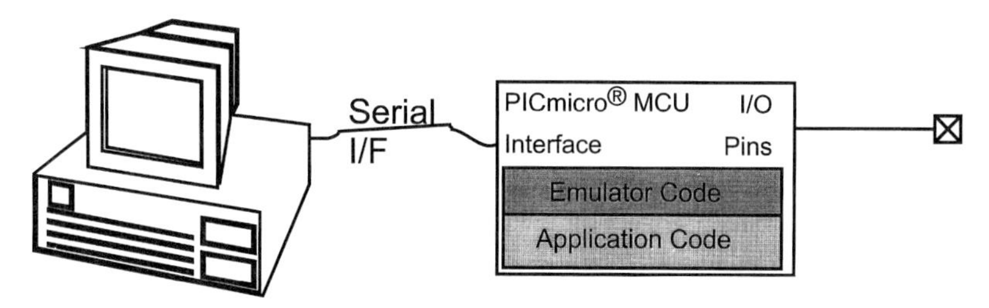


**Figure 14-3** **Emulator using internal program memory**

The last type of emulator is the most expensive, but it does address all of the issues raised by the other methods that have been presented. Built into every PICmicro® MCU chip are the I/O pins, which can be used by external hardware to control the operation of the PICmicro® MCU, as well as provide a separate memory to execute out of. To provide access to these I/O pins, a special type of package (known as a *bondout chip*) is used. The bondout chip package is used to interface to the application and the emulator so that if any damaging voltages or currents are driven into the emulator by the application, this is the only part that has to be replaced. The cost for a bondout chip is on the order of $500, which, although expensive, is a lot cheaper than replacing a $2,000 emulator.

The block for this type of emulator is shown in Fig. 14-5.

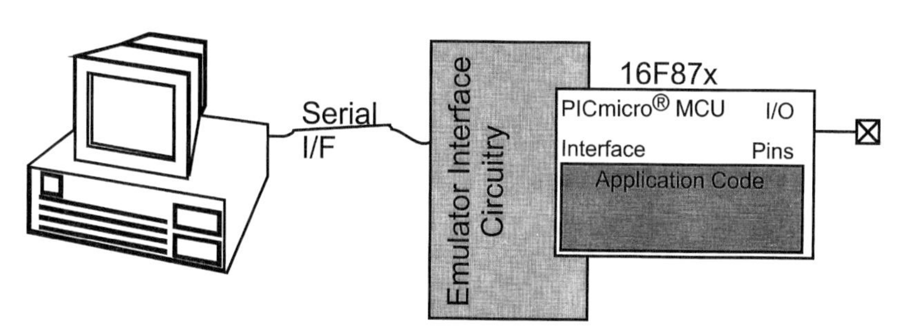


**Figure 14-4** **16F87x built-in emulator**

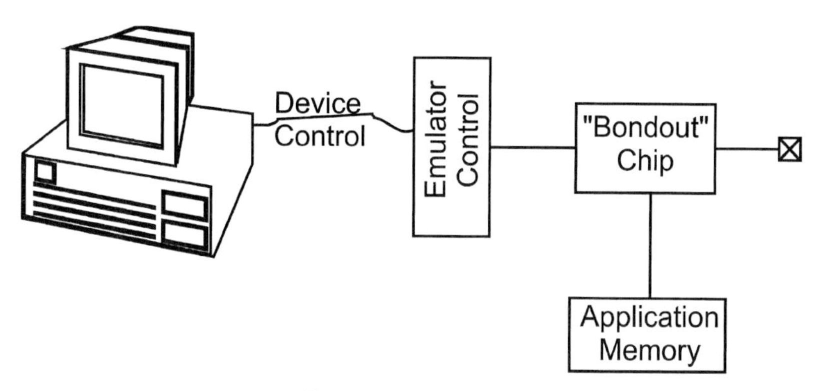


**Figure 14-5** **PICmicro® MCU emulator using a "bondout" chip**

Personally, I do not believe that emulators are absolutely required for debugging PICmicro® MCU (or any other microcontroller) applications. They are, however, excellent tools to have when you are first working with the PICmicro® MCU because they will show what is happening from both the hardware and the software's perspective.

# MPLAB ICE-2000

I can only think of a very few electronic-development products that I have been able to work with and would consider truly useful or something that I would aspire because it is so useful. Development tools usually come in two flavors: very expensive with a few deficiencies, but, because you paid a lot of money for the tool, these problems are addressed. The other type is the low-cost development tool that has a number of limitations and you tend to just have to live with the problems. Microchip is a company that has avoided these two generalizations: the MPLAB IDE is probably one of the best software application-development tools that is available on the market today—and it is free to download from the Internet. The PICSTART Plus is an excellent reasonable-cost programmer that integrates seamlessly to the MPLAB IDE and is continually updated for new products, so your investment is never lost as new products become available. Microchip's PICmicro® MCU-development tools bridge the gap between the high- and low-cost tools, taking advantage of the best features of both.

Microchip has recently complemented their list of affordable, high-quality tools with the addition of the MPLAB ICE-2000 emulator (Fig. 14-6). Although the cost will seem high (the hardware necessary to start working with the MPLAB ICE-2000 costs about $2,000), this is actually thousands of dollars less than the previous PICMaster system and is much better designed.

**Figure 14-6** **MPLAB-ICE with a processor module**

The MPLAB ICE-2000 consists of a roughly 7″ × 6″ box (pod) that is about 3/4″ deep (it is surprisingly small) and connects to an unregulated DC power source, a PC's parallel port, and a processor module. The processor module is a PCMCIA-like (PC card) circuit with a 15″ ribbon cable leading from it to a product connector that is connected to a device adapter or transition socket. This hardware organization is shown in Fig. 14-7 and should give you an idea of how simple the connections are. What you cannot see in the diagrams is the small tripod, which allows you to position the emulator in such a way that there is no mechanical stress on the PICmicro® MCU socket, resulting in a much better electrical connection.

The MPLAB-ICE 2000 has the following list of impressive features and can emulate the entire PICmicro® MCU line with the current exception of the PIC18Cxx.

- Programmable internal clockable to clocking devices from 32 kHz to 40 MHz
- Full 2.0- to 5.5-volt operating voltage operation
- Breakpoints on execution address, internal device address, register contents or external events
- Complex breakpoints can be built out of four different breakpoint sources
- 32 KB of trace memory
- Oscilloscope/logic analyzer input/output
- Processor modules/device adapters/transition sockets for all classes of PICmicro® MCUs

The programmable clock is a very nice feature to have because it gives you "what if" capability to look at how the application could be implemented for different clocking schemes. This can be especially useful for looking at different ways to implement applications.

The different breakpoint options bear a few comments and their usefulness is probably not readily apparent when you first look at the capability. As you work through applications, you will find events that you want to understand happening "in the middle" of the application and can be very hard to trigger on.

An example of this would be a PICmicro® MCU application that is communicating with another device and seems to have a problem in the fourth communications packet. In

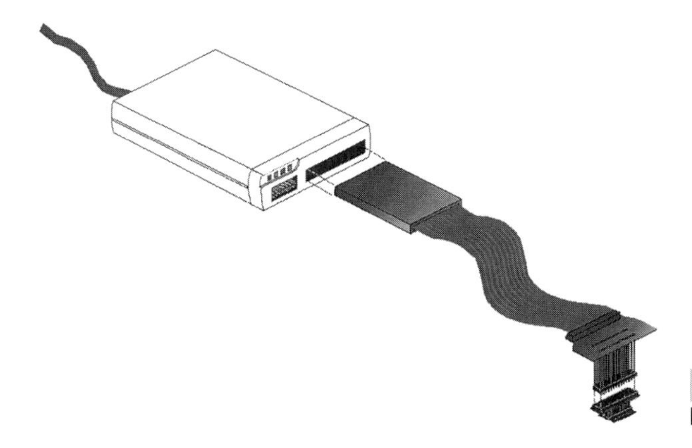

**Figure 14-7** MPLAB-ICE connections

a traditional system, to trigger on this, you would either need a very long delay or you would have to develop some hardware to look for the specific event (I've done both over the years). With the MPLAB-ICE 2000, you can set up where you want to trigger algorithmically and then wait for the event to happen to look at where the problem is.

Many interfaces cannot stand to have the processor stopped for a human to single step through to see what the problem is. The reasons for this not being possible can be the specific timing of the interface or the timeouts built into the communication protocol. To avoid these issues, you could single step both communicating devices or you could use the trace buffer in the PICmicro® MCU to see what happened at a specific point of time.

I really like the trace buffer feature because it allows me to see exactly what has happened and I can identify very specifically what the problem is and how it is manifested. When I make my proposed changes to the code to fix the problem, I can run the changes through the trace buffer and see how the changes affect the operation of the application.

Further enhancing your ability to observe problems are the logical analyzer and oscilloscope interfaces that are available on the front of the MPLAB-ICE 2000. These connectors (which are not discernible in the diagrams used in this section) allow you to either trigger the emulator's trace buffer remotely or use the complex triggering capability of the MPLAB-ICE 2000 to trigger other pieces of test equipment. The connectors can be used for external devices to trigger the MPLAB-ICE 2000 to see what is happening in the code at a specific external event.

The last point about processor modules, device adapters, and transition sockets being available for every "class" of PICmicro® MCU is important to understand. As covered elsewhere in the book, there are literally hundreds of different PICmicro® MCU devices. To provide a processor module for each and every PICmicro® MCU part number would be economically unfeasible for Microchip to produce, for distributors to keep in stock, and for you to buy. Instead, you should be looking for the supported device that best fits your requirements.

As I write this, the supported devices include:

- PIC12C5xx
- PIC12C67x
- PIC14000
- PIC16C505
- PIC16C52, PIC16C54, PIC16C55, PIC16C56, PIC16C57, and PIC16C58
- PIC16C55x
- PIC16C62x
- PIC16C6x
- PIC16C6x2
- PIC16C71x, PIC16C72, PIC16C73A, and PIC16C74A
- PIC16C773, PIC16C774, and PIC16C777
- PIC16F877
- PIC17C4x
- PIC17C5x

Looking at this list and comparing it to the devices that I use in this book, you will see some missing parts, most notably the PIC16F84. To emulate a PIC16F84 application, I would probably use a PIC16C556 or a PIC16C711 processor module and create the code in such a way that can be easily ported between the emulated device and the actual hardware.

This is what I mean when I say that the MPLAB-ICE 2000 supports "classes" of PICmicro® MCUs. You should be able to find a supported part that is very close to the part that you want to use in your application.

As indicated, the cost of an MPLAB-ICE 2000 capable of working with a single PICmicro® MCU is about $2,000, with support for another PICmicro® MCU costing about $500. The system is developed so that damaged components (device adapters, transition sockets, and even processor modules) can be easily replaced and not damage the MPLAB-ICE 2000 emulator pod, the most significant part of your investment.

This might seem expensive, but remember that time is money and your clients will not be happy paying for your time as you flail around looking for ways to figure out why a byte in the middle of a 10-byte packet is incorrect. No debug or failure-analysis tool will save you and your customers money like the MPLAB-ICE 2000.

# The PICMaster

When I was writing the first edition of this book, I was very lucky to be lent a PICMaster for my use. Although I didn't use it to debug any of the applications, I did get to spend enough time with the PICMaster to understand what it is capable of doing. To summarize my feelings at the time, I said "the PICMaster is the Cadillac of PICmicro® MCU *InCircuit Emulators (ICEs).*"

This statement was primarily made because of two reasons:

1. The integration of the PICMaster with the MPLAB development environment. The PICMaster works exactly like the MPLAB simulator (but with more features), which minimizes the learning required to use it.
2. The PICMaster was designed for (and by engineers). As I worked with the unit, I didn't find anything that I would change (and no wishes for more functions).

Along with being able to stop at events, like the MPLAB-2000, once the event has completed, you have an instruction flow record and with the ability to trigger an oscilloscope to see what is happening "outside" of the PICmicro® MCU.

Like any emulator, the PICMaster can be used as a "what if" tool. If you were working on a project and wanted to try something new or take a different approach to a problem, you would use this tool.

A few points to note about the PICMaster. The device itself consists of a PC ISA card connected to the PICMaster box, which is connected to a probe (which is plugged into the PIC socket of the application). The ISA card has an address switch that must be set to a valid, unused address range (as explained in the manual) before it can run properly.

I found the pod to be quite difficult to work with in some cases. The problems were poor connections to the socket on the PC board and the weight of the pod, which I found tended to pull the connections out if either device was at a different height. Last, the relative stiffness of the cable between the emulator box and the pod meant there was quite a bit of "fiddling" to place the pod in such a way that there was no uneven weight or strain on the pod and the connection to the emulated PICmicro® MCU socket. Despite this problem, the PICMaster would be a part of my "dream" PIC development system of three years ago.

The PICMaster has been superseded by the MPLAB-2000 emulator. In the last two years since this book has been released, PICMasters and their pods have been on sale, and trade-ins have been offered by Microchip. Although the PICMaster is an excellent tool, I remember all too well the problems I had with configuring the ISA card so that it wouldn't conflict with any other devices in my PC.

# MPLAB-ICD

If you are beginning with the PICmicro® MCU and want the most cost-effective in-circuit emulator solution available, you might want to look at the Microchip MPLAB-ICD emulator. This product takes advantage of the built-in features of MPLAB and the PIC16F87x devices to provide a low-cost emulator that can be used to develop your own applications. The product itself consists of a small module board that can be connected directly into an application or use a small header board that can be used to replace the PICmicro® MCU in an application.

The MPLAB-ICD kit (Figs. 14-8 and 14-9) consists of the module board, the header board, and a demo board that can be used to test applications and learn about the PICmicro® MCU. In fact, using the demo board, most of the experiments in this book could be executed with little modification, except for the changes in the source code that are outlined in this section.

The version of MPLAB that comes with this book is enabled for MPLAB-ICD. To enable MPLAB-ICD, select it from the *Options Development Mode* dialog box of MPLAB (Fig. 14-10). After MPLAB-ICD is selected, you will get the MPLAB-ICD execution dia-

**Figure 14-8** MPLAB-ICD interface card

**Figure 14-9** MPLAB-ICD prototype card

log box (Fig. 14-11). This dialog box indicates the current mode of MPLAB-ICD, as well as the execution parameters.

From the MPLAB-ICD execution box, you can select the configuration fuse options of the PICmicro® MCU that are executing with the MPLAB-ICD interface. For MPLAB operation, the following configuration fuse selections must be made:

1. No code protection
2. No *Low-Voltage Programming (LVP)*
3. Debug interface enabled

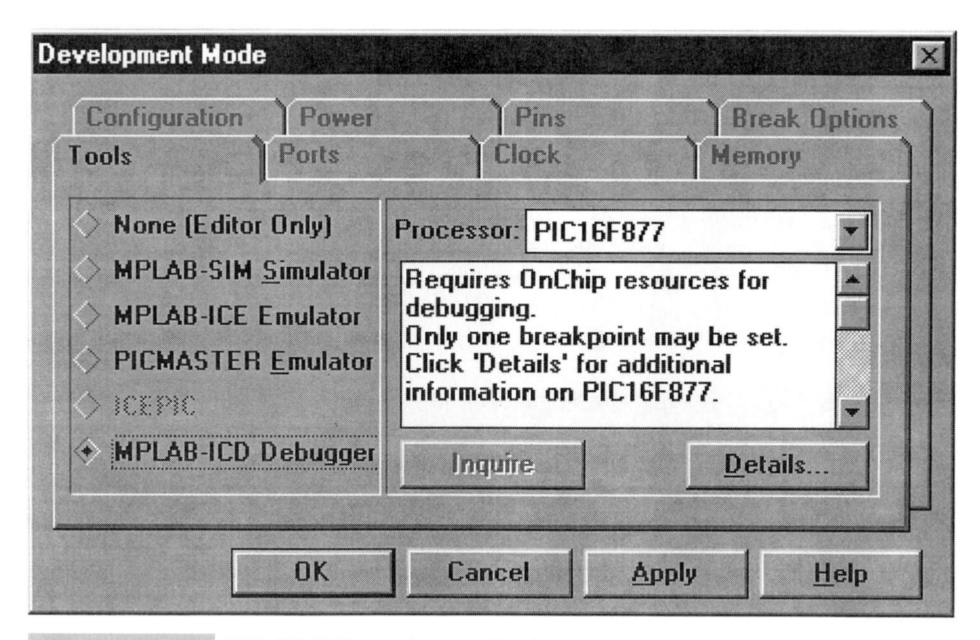


**Figure 14-10** MPLAB-ICD emulator selection

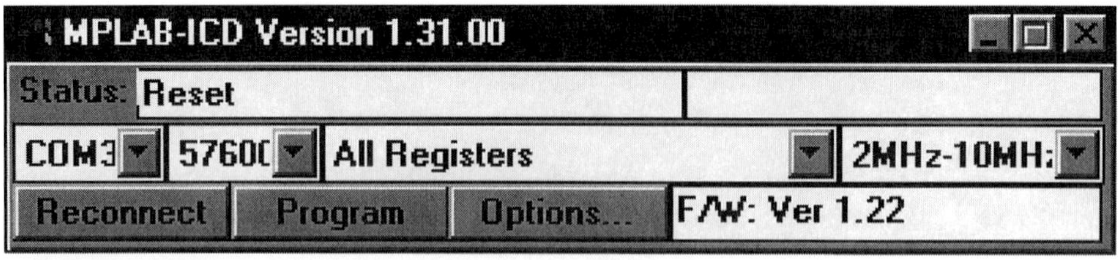

**Figure 14-11** **MPLAB-ICD emulator interface**

**4** Internal program memory writes enabled
**5** Brown out reset disabled
**6** Watchdog timer disabled

These selections should be made within the application using the *__CONFIG* directive of MPASM. The 72-ms power-up delay timer and oscillator selection is up to you and the application.

Along with the configuration fuse selections, you must follow some rules in your application. First, the code must start with an *nop* instruction to allow the ICD software to insert its own instruction. Along with this, the program memory from 0x01F00 to 0x01FFF cannot be used because this space is reserved for the MPLAB-ICD's monitor program.

It is important to remember that MPLAB-ICD is simply an interface to the PICmicro® MCU. The PICmicro® MCU itself is programmed by the module board and interfaces to it using the ICSP connection during execution to allow MPLAB to monitor execution and interface with the PICmicro® MCU's registers and program memory. Along with the considerations made for the application code, you should also be aware of some considerations when designing the hardware aspects of the application.

When you are using MPLAB-ICD, you will find that a number of functions that execute quite quickly in the MPLAB simulator (or MPLAB-ICE 2000) are breathtakingly slow in MPLAB-ICD. These functions include "building" the project and "single stepping." *Building* the project includes programming the PICmicro® MCU with the application (the programming operations can be observed in the MPLAB-ICD emulator dialog box). *Single stepping* includes returning the contents of all the hardware I/O registers, as well as the contents of the file registers. This operation cannot be limited to just the registers that are changed because there is no space to store the previous contents of the registers for a compare to see what has been updated.

After using MPLAB-ICD for a while, you'll discover that you rely on strategically placed breakpoints more than single stepping to monitor the execution of an application.

To perform the MPLAB-ICD functions, a Flash-program memory-equipped PICmicro® MCU is used with the MPLAB-ICD module board. The PIC16F87x parts are currently the only parts that are used with MPLAB-ICD, which makes the emulator excellent for learning about the PICmicro® MCU. If you are porting an application to another PICmicro® MCU, you will find that the limited number of PICmicro® MCUs will make it necessary to track the pins and file registers that are available in the device that you want to use.

My first suggestion is that you do not use PORTB, except for some simple digital I/O. I/O pins RB3, RB6, and RB7 interface directly to the MPLAB-ICD module board and prevent its use as an eight-pin I/O register. When I have tested applications on MPLAB-ICD, I have used PORTC for eight-bit ports. When I build the application for the final circuit, I change this over to PORTB.

Notice that the MPLAB-ICD module board also provides the programming voltage on the PICmicro® MCU's *_MCLR* pin. When designing your application, be sure that *_MCLR* can be driven directly by the MPLAB-ICD module board and the reset circuitry is protected by a diode. The demo board that comes with MPLAB-ICD has some very large current-limiting resistors for this function, but I would recommend using a diode for your own applications that are going to use an MPLAB-ICD device.

One of the interesting aspects of MPLAB-ICD is the ability to simply add the RJ-45 connector that is used with the MPLAB-ICD module to your application and run it with a PIC16F87x built into it. The connector is a standard AMP part. Notice that the MPLAB-ICD module card is powered from the application; an extra 70 mA at 5 volts should be available for the module card's power requirements.

The excellent support Microchip provides its customers became very evident when I was involved in the final proofreading of this book. I had been using MPLAB-ICD on and off for the previous six months. Although a few aspects of it made it clumsy to use, I didn't think about checking the latest firmware version of the PICmicro® MCU on the module card. I found out that I was running Version 1.22 and the latest version was 2.04. Fortunately, I was able to get an updated copy of the code and burned it into the PICmicro® MCU on the module board. I was up and running in about one business day.

The MPLAB-ICD is an excellent tool to learn about the PICmicro® MCU. In fact, if you were looking at what to buy as you set up your own PICmicro® MCU lab, I would recommend buying the MPLAB-ICD before a PICSTART Plus or other tools. MPLAB-ICD is inexpensively priced and gives you many of the basic functions of a "professional" in-circuit emulator (including source-code debugging). As well, the demo board is an excellent introduction to interfacing to the PICmicro® MCU and has almost all of the capabilities of the YAP-II or EMU-II boards presented in this book.

# The EMU-II

When I first heard about the MPLAB-ICD (and PIC16F877) capabilities, I was quite excited as I thought this would be an excellent way of helping me to debug my applications. After using it for a while, I ended up putting the tool on a shelf because of its slow response to user inputs and the single breakpoint built into it. This is not to say that it isn't a good emulator, it actually is quite good, especially considering the cost but I found that its limitations (especially in the area of response speed) were more than what I am comfortable with.

I also wanted to improve on the "EMU" project I presented in the first edition of the book to make it more useable for different applications. The problems with the "Emu" was the relatively slow emulation speed (about 10 KHz or one hundred or more times less than what I use for the projects presented in this book) and its propensity to "lock up" during execution that I was never able to find.

The root cause problem with this emulator was the need for an external 12C serial EEPROM to store the application code. To execute the application, I had to download

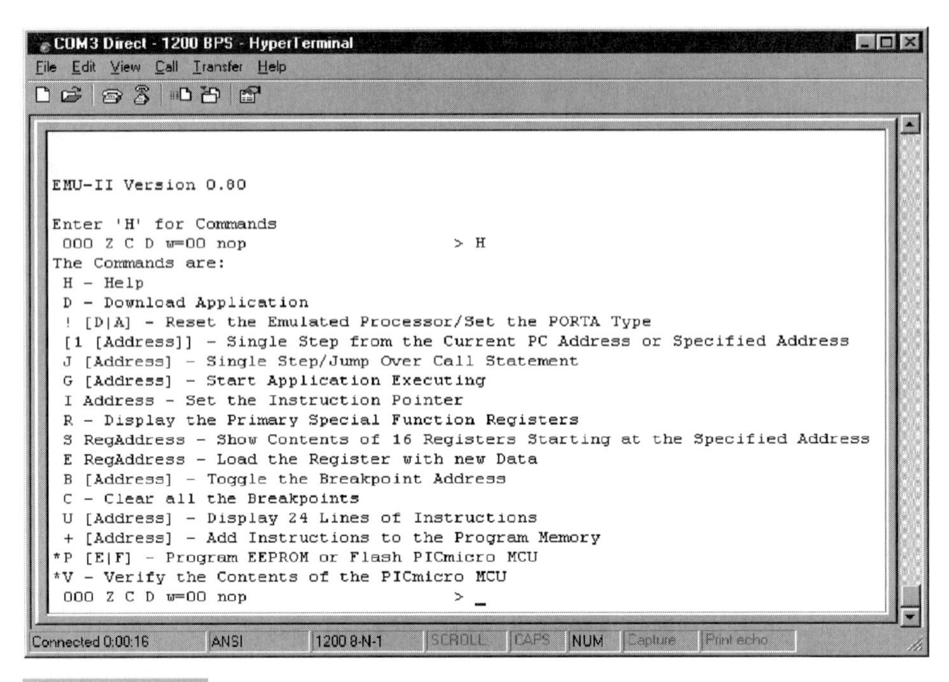


**Figure 14-12** EMU-II "help" display

each instruction from the EEPROM into the PICmicro® MCU and interpret it (much the same way the Basic Stamp interprets PBASIC tokens). This resulted in code that was amazingly complex (and not to mention very difficult to completely debug) and its execution was very slow.

To come up with a better emulator, I wanted to use the internal debug features of the PIC16F877 for my own emulator to see if I could improve upon the MPLAB-ICD implementation and come up with a product that I would be comfortable with. The final result is definitely not what I initially envisioned when I started the project, but it did turn into one of the most interesting applications in this book and something that I probably learned the most on.

Even if you are not going to build this application, I urge you to read through this write up as there will be some information in here that I think you will find interesting and useful. As well, I have provided some macros that could be useful in your own applications.

The issues I had with MPLAB-ICD that I wanted to improve upon were:

1. Slow Response to Reset and Single Step
2. The Single Breakpoint
3. No way of changing the Program Counter from the MPLAB interface

The resulting application worked similarly to the YAP-II in that I provided an intelligent RS-232 interface to a Terminal Emulator (as well as a Visual Basic front end) that is completely self contained. The different commands available to the EMU-II are listed in

(Fig. 14-12). Like the YAP-II, the EMU-II runs at 1200 bps and takes applications as hex files. The maximum application size is up to 1,792 (0x0700) instructions.

The EMU-II provides the same functions as higher cost emulators with the added features:

- The EMU-II is not device or operating system specific. Any PC, Mac or Workstation with a Terminal Emulator program and a serial port capable of running at 1200 bps can be used with the EMU-II
- Pins 6 and 7 of PORTB are usable, allowing 18 pin PICmicro® MCU applications to be emulated and programmed directly into parts without having to debug on PORTC or splitting up the PORTB functions
- Provisions for a built-in PICmicro® MCU programmer is included in the application
- Up to eight breakpoints can be specified during application execution

Unsupported features and potential limitations of the EMU-II to be aware of include:

- Applications must start with a "nop" to allow an Emulator execution breakpoint (this is the same as what MPLAB-ICD requires)
- The PIC16F877 reset pin is passed to the application but consists of just an unbounced 10K pullup and pull down switch
- The 4 MHz clock available to the EMU-II is not distributed to the application circuit
- It is not recommended that power provided on the board is distributed to the application circuit
- Built in Assembly of Instructions is not supported. There is a Patch ("+") command, however
- There is no protected Program Memory for the application. For this reason, writes to Flash/Data EEPROM are not recommended except using the built-in functions described below
- Applications to be emulated are limited to 1,792 instructions in size
- The EMU-II cannot single step from a "return," "retlw" or "retfie" instruction
- Variable memory in Banks 2 and 3 should not be accessed

The first order of business was to come up with a circuit to start from. The initial circuit was very similar to the YAP-II's and I designed a PCB that was based on the YAP-II's. I jokingly said that the circuit was almost identical to the YAP-II—the two ends are the same only the middle was changed. This is actually quite true for the PCB that I designed for the board (the Gerber files of which can be found on the CD-ROM).

What I didn't expect was the amount of difficulty I would have getting the emulator working in a way that I was comfortable with.

The final circuit I came up with is shown in Figure 14-13. What is not shown in Figure 14-13 is that either a PIC16F877 or PIC16F876 can be used for driving the EMU-II. To allow either part to be used in the emulator, I avoided using PORTD and PORTE (which are not available in the 28 pin PIC16F876) for control pins. The PCB, which is presented below, is designed to accept either a PIC16F877 or PIC16F876. The reason for doing this was some initial problems with getting initial PIC16F876 parts when I was prototyping the application.

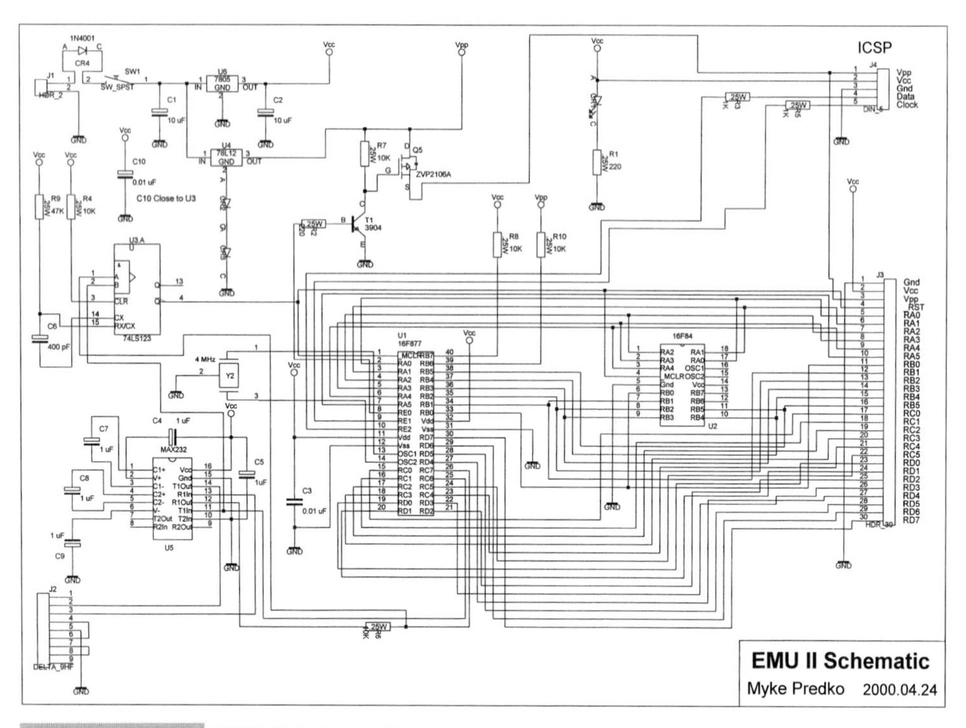


**Figure 14-13** EMU-II Schematic

There is a nineteen pin connector that includes ten LEDs, two buttons, two potentiometers, a Speaker and some pull ups that are wired exactly the same way as the YAP-II's built in I/O devices.

This circuit can be built quite easily on a prototyping card or breadboard. Following the YAP-II PCB design, I have created a PCB design for the EMU-II which has its topside copper shown in Figure 14-14, bottomside copper in Figure 14-15 and overlay in Figure 14-16. The Gerber files for this board is available on the CD-ROM.

When you look at this circuit, you should notice four important differences between it and the YAP-II. They are:

1 The use of a PIC16F877/PIC16F876 instead of a PIC16C711
2 A 4 MHz Ceramic Resonator is used instead of a programmable oscillator
3 The processor signals passed to the breadboard area are different
4 The Addition of "U4," which is labeled "16F84" and is a socket in which eighteen pin "ScotchFlex" DIP connectors can be used to pass the emulated signals from the EMU-II to a development board
5 A number of PIC16F877/PIC16F876 I/O pins are unavailable in the EMU-II

The EMU-II is designed to emulate an 18 pin PICmicro® MCU as closely as possible. The I/O signals from the PIC16F877/PIC16F876 are available to the breadboard in exactly the same fashion as the YAP-II. This has resulted in some restrictions in how the PIC16F877/PIC16F876 is used:

**TABLE 14-1 Bill of Material for the EMU-II**

| PART | DESCRIPTION |
| --- | --- |
| U1 | PIC16F877-04/P or PIC16F876-04/SP |
| U2 | Eighteen Pin (ZIF) DIP Socket |
| U3 | 74LS123 Single Shot |
| U4 | 78L12 +12. Volt Regulator in TO-92 Package |
| U5 | MAX232 RS-232 Interface |
| U6 | 7805 +5 Volt Regulator in TO-220 Package |
| T1 | 2N3904 NPN Transistor |
| Q5 | ZVP2106A P-Channel MOSFET Transistor |
| CR1 | Red LED |
| CR2, CR3 | 1N1414 Silicon Transistors |
| CR4 | 1N4001 Silicon Transistor |
| Y2 | 4 MHz Ceramic Resonator with Internal Capacitors |
| R1 | 10K, 1/4 Watt Resistor |
| R2, R7 | 220 Ohm, 1/4 Watt Resistor |
| R3 | 330 Ohm, 1/4 Watt Resistor |
| C1–C2 | 10 $\mu$F, 16 Volt Electrolytic Capacitors |
| C3 | 0.01 $\mu$F Tantalum Capacitors |
| C4 –C8 | 1.0 $\mu$F Capacitor, Any Type |
| SW1 | SPST Switch |
| SW2 | "Momentary On" SPST Switch |
| J1 | 2.5 mm Power Socket |
| J2 | 9 Pin Female "D-Shell" Connector |
| J3 | 19x1 Socket |
| J4, J6 | 5x1 ICSP Connector |
| Misc. | PCB Board, Wiring |

1. The USART port is dedicated to the EMU-II function. The RCSTA, TXSTA, RCREG, TXREG and SPBRG registers should not be accessed in any way from an application.
2. Bringing RC3 – RC5 to the 19 pin product connector means that the MSSP hardware can be used with the application. But SPI "Slave Mode" cannot be implemented because RA5 (the "_SS" pin) is not passed to the "emulated" part.
3. Other than MSSP functions, no other PORTC functions are to be accessed from the application. PORTC controls the operation of the programming reset circuitry and inadvertent writes to the hardware could potentially damage the EMU-II or the application circuit it is connected to.

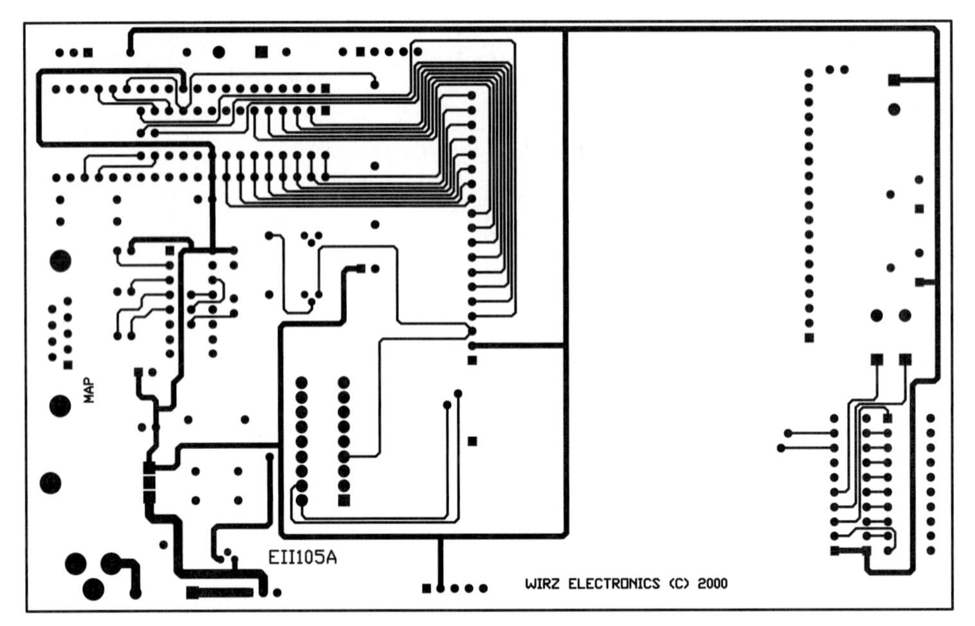


**Figure 14-14** EII105A Top layer

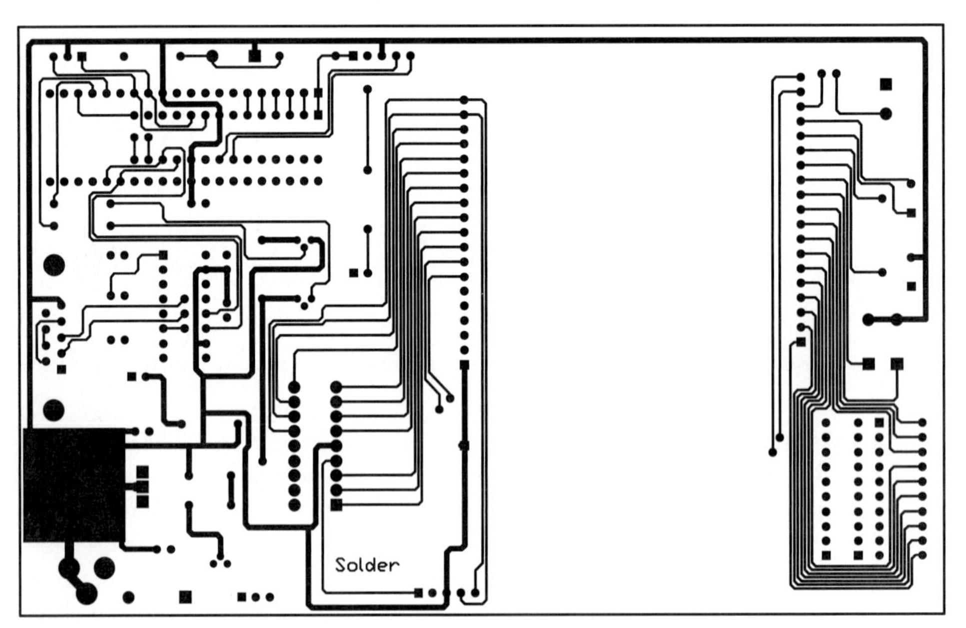


**Figure 14-15** EII105A Bottom Layer

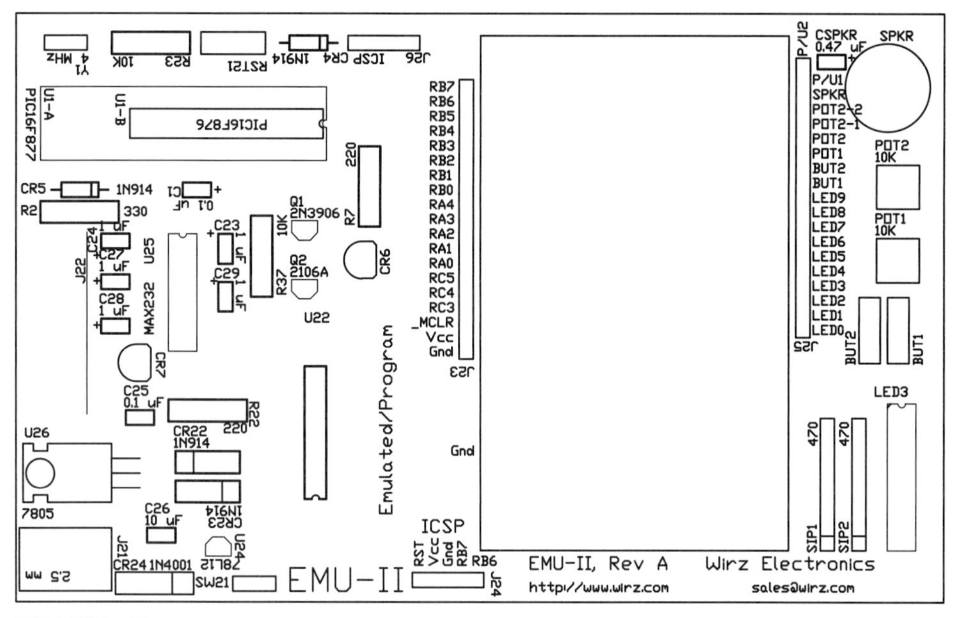


**Figure 14-16** EII105A Overlay

**4** Note that none of the connections from the PIC16F877/PIC16F876 to the application circuit are protected. The reason for this is to allow the EMU-II to interface directly without any unexpected impedances which will affect the operation of the application.

The reason for the single frequency option for this board was to avoid the need for changing the delay constants (or timers) when the user selected a different operating frequency. 4 MHz was chosen as the constant frequency because it is the one chosen for most applications.

"U4" is left unpopulated and is in fact a socket for wiring an eighteen pin emulated PICmicro® MCU into the application's circuit. This is done by using an eighteen pin "ScotchFlex" ribbon cable between two eighteen pin DIP connectors. Note that the $V_{CC}$, OSC1, and OSC2 pins are left floating (unconnected) on the U4 connector to avoid contention problems with the EMU-II. "_MCLR" will be driven high by the EMU-II and will be toggled low when the "reset" command is initiated.

Once the EMU-II is built, it is connected to a PC (of any type) via an RS-232 cable and powered by 14+ Volts from a wall mounted AC/DC power converter. To interface with the EMU-II, a terminal emulator set to 1200 bps, 8-N-1 data protocol is used. No hardware handshaking or XON/XOFF control protocol is required for the interface.

The commands to be sent to the EMU-II are similar to that of the YAP-II but allow for hexadecimal address and data parameters as well as alphanumeric register parameters. Table 14-2 lists the different commands available from the EMU-II.

**TABLE 14-2 Command Set for the EMU-II**

| COMMAND | PARAMETERS | COMMENTS |
|---|---|---|
| H | | Display List of Commands |
| D | | Download Application Hex File |
| ! | [D I A] | Reset Emulated Part with "A"nalog or "D" PORTA I/O Pins |
| 1 | [Address] | Single Step starting at current Program Counter or Specified Address – If "Enter" – If "Enter"pressed/"Carriage Return" Sent without a command then Single Step from current Program Counter is invoked |
| J | [Address] | Single Step starting at current Program Counter or Specified Address – If Instruction to single step is a subroutine "call", then Execution resumes at the instruction after the call instruction – If breakpoint is encountered within the subroutine, then Execution will stop at that breakpoint |
| G | [Address] | Start Executing at current Program Counter or specified Address – If breakpoint is encountered, then Execution will stop at that breakpoint |
| I | Address | Set new Program Counter value |

Note that "Parameters" in square brackets ("["/"]") are optional. All numeric data in hexadecimal and register addresses can either be a hex address or a register name.

The operation of the application is quite straightforward with the only non-intuitive operation being the application download operation. To load a new application into the EMU-II's Program Memory, carry out the following steps:

1 Enter "D"/CR ("Enter" if a PC is being used). The EMU-II will return with a message saying that Program Memory is being cleared.
2 When the EMU-II requests the application to be downloaded, use the Terminal Emulator to send the application's hex file as a "textfile." This operation will be picked up by the EMU-II and the application code will be stored into the Program Memory devoted to the operation.

The Download operation will not typically have any feedback as to the status of the operation. This can make the operation worrisome but note that after the application hex file has been sent, the EMU-II will poll incoming serial data for five seconds to ensure the host

**TABLE 14-2 Command Set for the EMU-II *(Continued)***

| COMMAND | PARAMETERS | COMMENTS |
|---|---|---|
| R | | Display the Primary Special Function Registers in the PICmicro® MCU |
| S | Register | Display the contents of 16 Registers starting at "Register" Address |
| E | Register | Display and optionally change the contents of the specified Register |
| B | [Address] | Toggle a breakpoint at either the specified address or the current Program Counter |
| C | | Clear all the Breakpoints in the Emulator |
| U | [Address] | Disassemble the 22 instructions starting at either the current Program Counter or the specified address |
| 1 | [Address] | Load hex values that are to be entered into Program Memory either Starting at the current Program Counter or the specified address |
| P | [E I F] | Program an EPROM or Flash PICmicro® MCU from the contents of the EMU-II's Program Memory |
| V | | Verify the contents of a PICmicro® MCU with the contents of the EMU-II's Program Memory |

system (PC) download was not pre-empted by another task. No characters should be entered in the PC keyboard until the EMU-II prompt (which will be at address 0) has been displayed.

Register namers have been built into the EMU-II to allow for some symbolic application debuggings listed in Table 14-3.

The following notes apply to the registers:

**(1)** Registers and I/O Ports are Not Available in the PIC16F876
**(2)** Register is known as "OPTION_REG" in Microchip Documentation/Tools
**(3)** No special functions devoted to these registers, 0x000 always returned upon register read
**(4)** Registers used by the EMU-II. These registers, along with the registers EEPROM registers (EEDATA, EEDATH, EEADR, EEADRH, EECON1 and EECON2), should never be accessed except using the functions listed below.

As indicated in the notes above, the USART and EEPROM registers should never be accessed by an application. Instead, the functions listed in Table 14-4 should be used for se-

**TABLE 14-3 Registers Supported by the EMU-II**

| BANK0 | | BANK1 | |
|---|---|---|---|
| ADDRESS | REGISTER NAME | ADDRESS | REGISTER NAME |
| 0x000 | INDF | 0x080 | INDF |
| 0x001 | TMR0 | 0x081 | OPTION (2) |
| 0x002 | PCL | 0x082 | PCL |
| 0x003 | STATUS | 0x083 | STATUS |
| 0x004 | FSR | 0x084 | FSR |
| 0x005 | PORTA | 0x085 | TRISA |
| 0x006 | PORTB | 0x086 | TRISB |
| 0x007 | PORTC | 0x087 | TRISC |
| 0x008 | PORTD (1) | 0x088 | TRISD (1) |
| 0x009 | PORTE (1) | 0x089 | TRISE (1) |
| 0x00A | PCLATH | 0x08A | PCLATH |
| 0x00B | INTCON | 0x08B | INTCON |
| 0x00C | PIR1 | 0x08C | PIE1 |
| 0x00D | PIR2 | 0x08D | PIE2 |
| 0x00E | TMR1L | 0x08E | PCON |
| 0x00F | TMR1H | 0x08F | Zero (3) |
| 0x010 | TCON1 | 0x090 | Zero (3) |
| 0x011 | TMR2 | 0x091 | SSPCON2 |
| 0x012 | TCON2 | 0x092 | PR2 |
| 0x013 | SSPBUF | 0x093 | SSPADD |
| 0x014 | SSPCON | 0x094 | SSPSTAT |
| 0x015 | CCPR1L | 0x095 | Zero (3) |
| 0x016 | CCPR1H | 0x096 | Zero (3) |
| 0x017 | CCP1CON | 0x097 | Zero (3) |
| 0x018 | RCSTA (4) | 0x098 | TXSTA (4) |
| 0x019 | TXREG (4) | 0x099 | SPBRG (4) |
| 0x01A | RCREG (4) | 0x09A | Zero (3) |
| 0x01B | CCPR2L | 0x09B | Zero (3) |
| 0x01C | CCPR2H | 0x09C | Zero (3) |
| 0x01D | CCP2CON | 0x09D | Zero (3) |
| 0x01E | ADRESH | 0x09E | ADRESL |
| 0x01F | ADCON0 | 0x09F | ADCON1 |

rial communications with the PC Host and for accessing the Data EEPROM. The Program Memory EEPROM must never be accessed. The serial port registers are enabled for non-interrupt communications at 1200-8-N-1 and should not be modified in any way.

Applications written for generic PICmicro® MCU applications can be debugged using the EMU-II with very little modification. The EMU-II was designed to minimize the need for developing applications that had to be modified for both EMU-II operation and actual application operation.

**TABLE 14-4 Built in Interface Functions**

| ADDRESS | FUNCTION NAME | DESCRIPTION |
|---|---|---|
| 0x07B0 | Serial Poll | If Character Received and not yet read, return with the carry flag set |
| 0x07C0 | SerialRead | Wait until a character has been received and return it in "w" |
| 0x07D0 | SerialWrite | Send the Character in "w" out serially to the host |
| 0x07E0 | EERead | Read the Data EEPROM at the Address specified in FSR |
| 0x07F0 | EEWrite | Write the Data EEPROM with the value in "w" at the Address specified in FSR |

To create applications that can be debugged on the EMU-II, the following rules must be followed:

1 "nop" as the first instruction at address 0x0000
2 The maximum size of the application is 1,791 (0x06FF) instructions
3 Variables should start at 0x020 rather than 0x00C as is possible in some devices
4 Use the USART and EEPROM functions listed above and do not access the Special Function Registers that control these functions directly
5 For variables that are accessed from either Bank 0 or Bank 1, use address range 0x070 to 0x07E

Ideally, applications should not access any registers in Bank 2 or Bank 3 as the EMU-II state variables are stored in these banks along with the EEPROM access control registers.

The "nop" instruction at the start of the application is used to allow a "goto EMU-II Reset" instruction which jumps to the EMU-II emulator application when the EMU-II first boots. This goto instruction is tested for in the EMU-II software and is treated like a "nop" during application execution.

As noted above, applications cannot be larger than 1,791 (0x06FF) instructions. The reason for this size is that the 256 instructions at address 0x0700 to 0x07FF is used by the EMU-II to provide an interface for the breakpoint handlers as well as the USART/data EEPROM functions listed above. As will be discussed below, when a breakpoint is "set" in the application, it jumps to the handler address. By placing the handler vectors in the first page, a single "goto" instruction is required and the PCLATH registers do not have to be modified by the breakpoint operation.

It is recommended that all applications specify the configuration flags to be used by the final PICmicro® MCU application using the:

```
_CONFIG
```

statement in the application source code. The config and_IDLOCS data will be stored within the EMU-II and will be programmed into a PICmicro® MCU using the Program (and Verify) commands.

I originally wanted this application to take advantage of the built in debug features demonstrated by MPLAB-ICD. The MPLAB-ICD's operation and the PIC16F87x built-in debug features are not documented by Microchip so I did a bit of "hacking" to find out what was going on inside them and how they worked.

Actually, the "Hacking" just consisted of taking a PIC16F877 that had been programmed by the MPLAB-ICD and looking at what was put in it. I found that there was a code, which is very similar to an interrupt handler, placed at addresses 0x01F00 to 0x01FFF. This matched what I expected based on the requirements for the MPLAB-ICD—in the specifications Microchip indicates that no application code can be placed in the PICmicro® MCU at addresses 0x01F00 to 0x01FFF. (Along with leaving 0x01F00 to 0x01FFF alone, address 0 must have a "nop" instruction.)

To look at this code, I used my PICStart Plus to read the contents of the PIC16F877 and then dumped the "Unassembled" code into a text file. Once this was done, I then spent some time and labeled the code and made sure it could be reassembled properly (i.e., loaded back into a PIC16F877 and tested with MPLAB-ICD). The resulting application code is "ICD.ASM" and can be found in the "code\ICD" subdirectory of the *PICmicro* directory.

This code is copyright Microchip and cannot be distributed in any products. If you are planning on using this code in your own product you must contact Microchip first and get their permission. I have reproduced it here so you can see what is going on inside the PICmicro® MCU controlled by the MPLAB-ICD card.

When you look at this code, you will see that it has the entry point of 0x01F00—I will call this the "debug interface" as that is what it is; it provides an interface between the executing application in the PICmicro® MCU and the MPLAB-ICD. Execution jumps to this address any time the PICmicro® MCU is reset or the previous command has completed executing. When execution jumps to this address, the context registers (the w, STATUS, FSR and PCLATH registers) are saved and the contents of 0x018E and 0x018F are saved.

Registers 0x018E and 0x018F are reserved and used for passing information between the executing program and the debug interface. Upon debug interface entry at 0x01F00, these registers normally contain the address execution stopped at. Upon return to the executing application, these registers are loaded with a command, the context registers are restored and a "return" instruction is executed.

The debug interface communicates using a synchronous serial protocol with the MPLAB-ICD. RB7 is a bi-directional I/O pin and RB6 is a clock provided by the MPLAB-ICD. The first clock cycle is a synching bit and is followed by a thirty-two bit data transfer. The data transfer is a bit unusual in that when the clock line rises, the debug interface drives a bit from 0x018E and 0x018F onto the data line and when the clock line falls, a new bit for 0x018E and 0x018F are read from the data line. Using this method, sixteen bits of data are input and output using one clock.

The 0x018E and 0x018F registers are used either to write to the PICmicro® MCU or used as command to the internal debug hardware. When the command is passed to the executing hardware, registers 0x018E and 0x018F are written to before the context registers are restored and the "return" instruction executed.

From what I could tell, there are four instructions passed from the MPLAB-ICD to the PICmicro® MCU via this interface. They are listed in Table 14-5.

**TABLE 14-5 MPLAB-ICD Reserved Address Instruction**

| 0x018E | 0x018F | Command |
|---|---|---|
| 0x000 | 0x000 | Reset the PICmicro® MCU |
| 0x040 | 0x000 | Run to breakpoint address added to the 0x04000 instruction value |
| 0x05F | 0x000 | Run Freely |
| 0x07F | 0x000 | Single Step |

The "Run to breakpoint address" executes from the current program counter value and stops after it executes the specified instruction.

If an instruction to the executing hardware instead of the debug hardware is to be passed, then a command that does not have bit 6 of 0x018E is passed and this instruction is jumped to a handler address where the command (such as read the contents of "w") is executed.

In observing this code and experimenting with my own, I was able to make the following observations about the built in hardware debugger of the PIC16F877:

1 Execution jumps to 0x01F00 after reset or after a command has finished executing
2 The first time "return" is executed, the debug interface ends and the application code executes again using the command in 0x018E and 0x018F
3 Hardware resources may or may not be available in the PICmicro® MCU. I found that most resources are not available except for:
   – EEPROM read/write interface
   – RB6 and RB7 (via Bank2 and Bank3)
4 In the Application Code, RB6 and RB7 are not available for I/O
5 Changes to Hardware Registers take effect after the "return" is executed
6 Reset is required to enable the internal debug hardware

This set of rules was developed using both the MPLAB-ICD and monitoring the data being passed back and forth with an oscilloscope as well as experimenting with the Emu-II hardware.

My original plan was to implement the EMU-II using the built in debug functions of the PIC16F87x and providing "protected" EMU-II application program memory from the address range 0x01000 to 0x01FFF. While I was somewhat successful, I discovered that there were three problems with this approach. They were:

1 No hardware resources of the PICmicro® MCU (i.e., the USART) except for the EEPROM/Flash read/write registers could be accessed while the PICmicro® MCU was in "Debug" mode. Serial communications relied upon "Bit Banging" functions.
2 Only one breakpoint is available in the application. If I were to implement multiple breakpoints, I would have to put in code like:

```
movwf   _w              ; Save the "w" Register
swapf   PCLATH, w       ; Save the current PCLATH Register
movwf   _pclath
```

```
movlw    HIGH BreakpointVector
movwf    PCLATH             ; Setup the Jump to the Next page
goto     (BreakpointVector & 0x07FF) | ($ & 0x01800)
```

This would make setting single step stop points at skip instructions (which have two possible stop points) impossible as well as making it impossible to set breakpoints for "goto" loops which were less than six instructions from their destinations.

**3** At the end of an operation, the PICmicro® MCU has to be reset (to put it back into internal debug mode). Using a 74123 "single shot" to toggle reset, I was able to accomplish this, but I felt the circuit was "kludgy."

After experimenting with the built in debugging mode of the PIC16F87x, I settled on the design that I am presenting in this section which does not utilize the built-in debugging features of the PICmicro® MCU. The resulting application is surprisingly fast (especially considering it works at 1200 bps) and I think very useful for new application developers.

This application is probably the most complex of anything presented in the book (as well as being far and away the longest). Despite having 6,400 instructions to work with, I found that I had to come up with some new ways of doing things in order to fit in all the functions that I wanted. Ideally, I would like to implement a line assembler as part of the EMU-II package—but for that, I would require 1,000 or more free instructions in program memory. There are four aspects of this application that I would like to bring to your attention as I believe they are noteworthy and may help you out in your own applications.

The first is how the text table reads are implemented in the EMU-II application. For most other applications, I pass a text message number to a "SendMsg" subroutine which reads through a large table of ASCIIZ messages to find the specific ASCIIZ message. When the selected message is found, it is output via the serial port until the ending NUL (0x000) character is encountered.

For the EMU-II, I had planned on implementing a large number of text messages to help guide the user (I wasn't expecting to have an application like the "YAP-II Windows Interface" to help the user). The message table that was developed was 2,888 ASCII characters long. If this table was used directly in the application, it would take up 45 percent of the total I had available.

The solution to this problem was to "compress" the tables so that two seven bit ASCII characters could be placed in a single fourteen bit instruction location. This would eliminate the two "retlw" instructions and take advantage of the PIC16F87x to read its own program memory.

To do this, I started with a typical table declaration (the "compress.asm" file in the code\EMU-II subdirectory of the *PICmicro* directory) like:

```
MsgTable                                ; List of Messages
Msg0                                    ; Start with \r\n (New Line)
dt          0x00D, 0x00A, 0
Msg1                                    ; Put in the "Prompt" Message:
dt          " > ", 0
Msg2                                    ; Backspace
dt          8, " ", 8, 0
Msg3                                    ; Introductory Message
dt          "Enter 'H' for Commands",   ; 0x00D, 0x00A, 0
Msg4                                    ; "Help" Message
dt     "The Commands are:", 0x00D, 0x00A
```

```
dt      "H - Help", 0x00D, 0x00A
dt      " D - Download Application", 0x00D, 0x00A
dt      " ! [D|A - Reset the Emulated Processor/Set the PORTA Type"
dt      0x00D, 0x00A
dt      " [1 [Address]] - Single Step from the Current PC Address or"
dt      " Specified Address", 0x00D, 0x00A
dt      "J [Address] - Single Step/Jump Over Call Statement", 0x00D
dt      0x00A
dt      "G [Address] - Start Application Executing", 0x00D, 0x00A
dt      "I Address - Set the Instruction Pointer", 0x00D, 0x00A
dt      "R - Display the Primary Special Function Registers", 0x00D
dt      0x00A
dt      "S RegAddress - Show Contents of 16 Registers Starting at"
dt      "the Specified Address", 0x00D, 0x00A
dt      "E RegAddress - Load the Register with new Data", 0x00D
dt      0x00A
dt      "B [Address] - Toggle the Breakpoint Address", 0x00D, 0x00A
dt      "C - Clear all the Breakpoints", 0x00D, 0x00A
dt      "U [Address - Display 24 Lines of Instructions", 0x00D
dt      "0x00A
dt      " + [Address] - Add Instructions to the Program Memory"
dt      0x00D, 0x00A
dt      "P [E|F] - Program EEPROM or Flash PICmicro® MCU", 0x00D
dt      0x00A
dt      "V - Verify the Contents of the PICmicro® MCU", 0x00D, 0x00A
dt      0
```

and added the two statements:

```
;#CompStart
```

and

```
;#CompEnd
```

before and after the table. These keywords were used to indicate that the table between them is to be compressed with the first (least significant address) character being used as the least significant seven bits of the instruction. The second (Fig. 14-7—most significant address) character being used as the most significant seven bits of the instruction.

I then wrote the Visual Basic application "DT Compress" (which can be found in the "code\DT Compress" subdirectory of the *PICmicro* directory) which converts the table between the ";#CompStart" and ";#CompEnd" statements to the compressed code similar to what's shown below:

```
;#CompStart

MsgTable        ; List of Messages
Msg0            ; Start with \r\n (New Line)
  dw 0x050D
Msg1            ; Put in the "Prompt" Message:
  dw 0x01000, 0x0103E
Msg2            ; Backspace
  dw 0x0400, 0x0420
Msg3            ; Introductory Message
  dw 0x02280, 0x03A6E, 0x03965, 0x013A0, 0x013C8, 0x03320, 0x0396F, 0x021A0
  dw 0x036EF, 0x030ED, 0x0326E, 0x06F3, 0x0A
Msg4     ; "Help" Message
```

```
dw 0x03454, 0x01065, 0x037C3, 0x036ED, 0x03761, 0x039E4, 0x030A0, 0x032F2
dw 0x06BA
dw 0x0100A, 0x01048, 0x0102D, 0x032C8, 0x0386C, 0x050D
dw 0x02220, 0x016A0, 0x02220, 0x03BEF, 0x0366E, 0x030EF, 0x01064, 0x03841
dw 0x03670, 0x031E9, 0x03A61, 0x037E9, 0x06EE
dw 0x0100A, 0x01021, 0x0225B, 0x020FC, 0x0105D, 0x0102D, 0x032D2, 0x032F3
dw 0x01074, 0x03474, 0x01065, 0x036C5, 0x03675, 0x03A61, 0x03265, 0x02820
dw 0x037F2, 0x032E3, 0x039F3, 0x0396F, 0x029AF, 0x03A65, 0x03A20, 0x032E8
dw 0x02820, 0x0294F, 0x020D4, 0x02A20, 0x03879, 0x06E5
dw 0x0100A, 0x018DB, 0x02DA0, 0x03241, 0x03964, 0x039E5, 0x02EF3, 0x0105D
dw 0x0102D, 0x034D3, 0x033EE, 0x032EC, 0x029A0, 0x032F4, 0x01070, 0x03966
dw 0x036EF, 0x03A20, 0x032E8, 0x021A0, 0x03975, 0x032F2, 0x03A6E, 0x02820
dw 0x01043, 0x03241, 0x03964, 0x039E5, 0x01073, 0x0396F, 0x029A0, 0x032F0
dw 0x034E3, 0x034E6, 0x03265, 0x020A0, 0x03264, 0x032F2, 0x039F3, 0x050D
dw 0x02520, 0x02DA0, 0x03241, 0x03964, 0x039E5, 0x02EF3, 0x016A0, 0x029A0
dw 0x03769, 0x03667, 0x01065, 0x03A53, 0x03865, 0x0252F, 0x036F5, 0x01070
dw 0x03B4F, 0x03965, 0x021A0, 0x03661, 0x0106C, 0x03A53, 0x03A61, 0x036E5
dw 0x03765, 0x06F4
dw 0x0100A, 0x01047, 0x020DB, 0x03264, 0x032F2, 0x039F3, 0x0105D, 0x0102D
dw 0x03A53, 0x03961, 0x01074, 0x03841, 0x03670, 0x031E9, 0x03A61, 0x037E9
dw 0x0106E, 0x03C45, 0x031E5, 0x03A75, 0x03769, 0x06E7
dw 0x0100A, 0x01049, 0x03241, 0x03964, 0x039E5, 0x01073, 0x0102D, 0x032D3
dw 0x01074, 0x03474, 0x01065, 0x03749, 0x03A73, 0x03AF2, 0x03A63, 0x037E9
dw 0x0106E, 0x037D0, 0x03769, 0x032F4, 0x01072, 0x050D
dw 0x02920, 0x016A0, 0x02220, 0x039E9, 0x03670, 0x03CE1, 0x03A20, 0x032E8
dw 0x02820, 0x034F2, 0x030ED, 0x03CF2, 0x029A0, 0x032F0, 0x034E3, 0x03661
dw 0x02320, 0x03775, 0x03A63, 0x037E9, 0x0106E, 0x032D2, 0x034E7, 0x03A73
dw 0x03965, 0x06F3
dw 0x0100A, 0x01053, 0x032D2, 0x020E7, 0x03264, 0x032F2, 0x039F3, 0x016A0
dw 0x029A0, 0x037E8, 0x01077, 0x037C3, 0x03A6E, 0x03765, 0x039F4, 0x037A0
dw 0x01066, 0x01B31, 0x02920, 0x033E5, 0x039E9, 0x032F4, 0x039F2, 0x029A0
dw 0x030F4, 0x03A72, 0x03769, 0x01067, 0x03A61, 0x03A20, 0x032E8, 0x029A0
dw 0x032F0, 0x034E3, 0x034E6, 0x03265, 0x020A0, 0x03264, 0x032F2, 0x039F3
dw 0x050D
dw 0x022A0, 0x02920, 0x033E5, 0x03241, 0x03964, 0x039E5, 0x01073, 0x0102D
dw 0x037CC, 0x03261, 0x03A20, 0x032E8, 0x02920, 0x033E5, 0x039E9, 0x032F4
dw 0x01072, 0x034F7, 0x03474, 0x03720, 0x03BE5, 0x02220, 0x03A61, 0x06E1
dw 0x0100A, 0x01042, 0x020DB, 0x03264, 0x032F2, 0x039F3, 0x0105D, 0x0102D
dw 0x037D4, 0x033E7, 0x032EC, 0x03A20, 0x032E8, 0x02120, 0x032F2, 0x035E1
dw 0x037F0, 0x03769, 0x01074, 0x03241, 0x03964, 0x039E5, 0x06F3
dw 0x0100A, 0x01043, 0x0102D, 0x03643, 0x030E5, 0x01072, 0x03661, 0x0106C
dw 0x03474, 0x01065, 0x03942, 0x030E5, 0x0386B, 0x034EF, 0x03A6E, 0x06F3
dw 0x0100A, 0x01055, 0x020DB, 0x03264, 0x032F2, 0x039F3, 0x0105D, 0x0102D
dw 0x034C4, 0x03873, 0x030EC, 0x01079, 0x01A32, 0x02620, 0x03769, 0x039E5
dw 0x037A0, 0x01066, 0x03749, 0x03A73, 0x03AF2, 0x03A63, 0x037E9, 0x039EE
dw 0x050D
dw 0x015AA, 0x02DA0, 0x03241, 0x03964, 0x039E5, 0x02EF3, 0x016A0, 0x020A0
dw 0x03264, 0x024A0, 0x039EE, 0x03974, 0x031F5, 0x034F4, 0x0376F, 0x01073
dw 0x037F4, 0x03A20, 0x032E8, 0x02820, 0x037F2, 0x03967, 0x036E1, 0x026A0
dw 0x036E5, 0x0396F, 0x06F9
dw 0x0150A, 0x01050, 0x022DB, 0x0237C, 0x0105D, 0x0102D, 0x03950, 0x033EF
dw 0x030F2, 0x0106D, 0x022C5, 0x02950, 0x026CF, 0x037A0, 0x01072, 0x03646
dw 0x039E1, 0x01068, 0x024D0, 0x036C3, 0x031E9, 0x037F2, 0x026A0, 0x02AC3
dw 0x050D
dw 0x02B2A, 0x016A0, 0x02B20, 0x03965, 0x03369, 0x01079, 0x03474, 0x01065
dw 0x037C3, 0x03A6E, 0x03765, 0x039F4, 0x037A0, 0x01066, 0x03474, 0x01065
dw 0x024D0, 0x036C3, 0x031E9, 0x037F2, 0x026A0, 0x02AC3, 0x050D
dw 0x00
;#CompStart
```

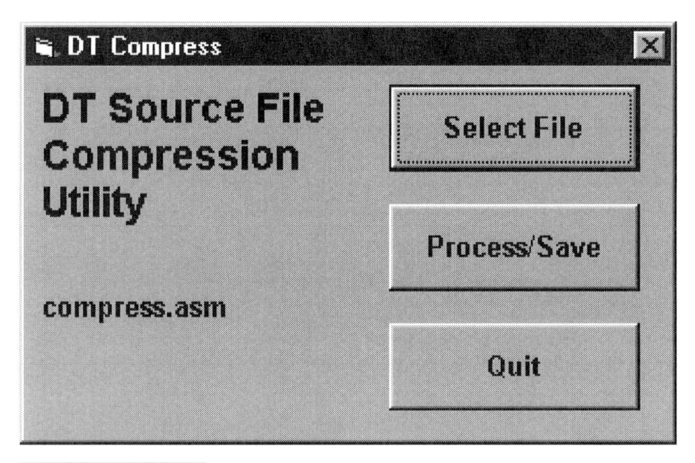


**Figure 14-17** DT compress

Admittedly, this is a lot harder to understand than the ASCII text above (especially with the high order byte having to be shifted down by seven to read it), but it cuts the instruction requirements in half for the application text files.

"DT Compress" is a Windows GUI application (shown in Figure 14-17 that simply allows a user to select an assembler (".asm") file and then converts it into a compressed include (".inc") file. Depending on the amount of text to compress, the application may take a few seconds to run (it takes five seconds to do the EMU-II "compress.asm" file on my 300 MHz Pentium-II PC).

To read and output the compressed file, I created the subroutine:

```
SendMsg                               ; Call Here for sending a specific
                                      ; message string to the Serial port.
                                      ; The Message Number is in "w"
 movwf     MsgTemp ^ 0x0100           ; Save the Message Number
 clrf      MsgOffset ^ 0x0100         ; Reset the
 clrf      (MsgOffset 1 1) ^ 0x0100
 clrf      SMCount ^ 0x0100           ; Clear Count of Output Bytes

MT_Loop1                              ; Loop Here Until "MsgTemp" is == 0
 movf      MsgTemp ^ 0x0100, f        ; Is "MsgTemp" Equal to Zero?
 btfsc     STATUS, z
  goto     MT_Loop2                   ; Yes, Start Displaying Data
 bcf       STATUS, C                  ; Calculate Address of Next Word to
 rrf       (MsgOffset + 1) ^ 0x0100, w      ; Display
 addlw     HIGH MsgTable
 movwf     EEADRH ^ 0x0100
 rrf       (MsgOffset 1 1) ^ 0x0100, w      ; Setup Carry Correctly for
 rrf       MsgOffset ^ 0x0100, w      ; Increment
 addlw     LOW MsgTable
 movwf     EEADR ^ 0x0100
 btfsc     STATUS, C
  incf     EEADRH ^ 0x0100, f         ; If Carry Set, Increment to Next Page
EEPReadMacro                          ; Now, Do the Program Memory Read
 rlf       EEDATA ^ 0x0100, w         ; Start with the Odd Byte
 rlf       EEDATH ^ 0x0100, w
 btfss     MsgOffset ^ 0x0100, 0      ; Odd or Even Byte?
```

```
 movf       EEDATA ^ 0x0100, w       ; Even Byte
 andlw      0x07F                    ; Convert to ASCII 7 Bits
 btfsc      STATUS, z
  decf      MsgTemp ^ 0x0100, f      ; Decrement the Value Count
 incf       MsgOffset ^ 0x0100, f    ; Point to the Next Byte in the String
 btfsc      STATUS, Z
  incf      (MsgOffset 1 1) ^ 0x0100, f
 goto       MT_Loop1

MT_Loop2                            ; Have Correct Offset, Now, Display the
                                    ; Message
 bcf        STATUS, C               ; Calculate Address of Next Word to
 rrf        (MsgOffset + 1) ^ 0x0100, w      ; Display
 addlw      HIGH MsgTable
 movwf      EEADRH ^ 0x0100
 rrf        (MsgOffset + 1) ^ 0x0100, w      ; Setup Carry Correctly for
 rrf        MsgOffset ^ 0x0100, w   ; Increment
 addlw      LOW MsgTable
 movwf      EEADR ^ 0x0100
 btfsc      STATUS, C
  incf      EEADRH ^ 0x0100, f      ; If Carry Set, Increment to Next Page
EEPReadMacro                        ; Now, Do the Program Memory Read
 rlf        EEDATA ^ 0x0100, w      ; Start with the Odd Byte
 rlf        EEDATH ^ 0x0100, w
 btfss      MsgOffset ^ 0x0100, 0   ; Odd or Even Byte?
  movf      EEDATA ^ 0x0100, w      ; Even Byte
 andlw      0x07F                   ; Convert to ASCII 7. Bits
 btfsc      STATUS, z
  goto      MT_End                  ; Zero, Yes
SendCharMacro
 incf       MsgOffset ^ 0x0100, f   ; Point to the Next Byte in the String
 btfsc      STATUS, z
  incf      (MsgOffset + 1) ^ 0x0100, f
 incf       SMCount ^ 0x0100, f
 goto       MT_Loop2

MT_End                              ; Finished sending out the Table Data
  movf      SMCount ^ 0x0100, w     ; Return the Number of Bytes Sent
EmuReturn
```

In this code there are three macros. The "EEPReadMacro" (along with the "EEP-WriteMacro") is used to access the Flash program memory of the PIC16F87x.

"SendCharMacro" is used to poll the transmit holding register empty interrupt request flag (TXIF of PIR1) and send the byte when the holding register is open. The macro code is:

```
SendCharMacro Macro
  bcf     STATUS, RP1
 ifndef Debug
  btfss   PIR1, TXIF
   goto   $ - 1
 else
  goto    $ + 3                ; Put in a Skip over the "nop" to save
  nop                          ; a Mouse Click
 endif
  movwf   TXREG                ; Send the Byte
  bcf     PIR1, TXIF           ; Reset the Interrupt Request Flag
  bsf     STATUS, RP1
endm
```

and it should be noted that if the label "Debug" is defined, the polling loop is ignored because in MPLAB, the USART hardware is not simulated and execution will never fall out of this loop.

There are two other things to notice about this macro. The first is in the code that executes when the "Debug" label is defined. I kept two instructions to match the btfss/goto instructions of the polling loop but I jump over the second one to save a mouse click when I'm single stepping through the application. This might seem like a petty place to save a mouse click or two, but "SendCharMacro" is used a lot in this application and when single stepping through the application the skipping over the instructions seems to reduce the number of mouse clicks significantly.

The second point to notice about this macro is that it changes the operating bank from two to zero and then back to two. The EMU-II application has all its variables in banks two and three of the PICmicro® MCU. This allows the user to access almost all the registers (except for the USART specific ones) in banks zero and one without affecting the operation of the EMU-II in any way. Executing out of bank two primarily was a new experience for me and if you look at the application code, you will see a lot of statements like:

```
xorwf (InstrPtr + 1) ^ 0x0100, w
```

in which I have to remember to keep the banks straight. This wasn't terribly hard to do, but did require a paradigm shift for me (or suffer through innumerable "Warning, register is not in bank 0" messages).

The third macro relates to the second point I wanted to discuss about the EMU-II. Instead of using the "call" and "return" instructions in the EMU-II, I used two macros, "EmuCall" and "EmuReturn," which I wrote to implement a subroutine call that does not access the built in program counter stack. The reason for writing these subroutines was to avoid the possibility that the subroutine calls in the EMU-II application code would affect the emulated application.

The EmuCall and EmuReturn macros are:

```
EmuCall Macro Address           ; Stackless Emulator Call
 local ReturnAddr
  movwf  tempw ^ 0x0100         ; Save the Call Value in "w"
  incf   FSR, f
  movlw  LOW ReturnAddr         ; Setup the Return Address
  movwf  INDF
  incf   FSR, f
  movlw  HIGH ReturnAddr
  movwf  INDF
  movlw  HIGH Address           ; Jump to the Specified Address
  movwf  PCLATH
  movf   tempw ^ 0x0100, w      ; Restore "w" before doing it
  goto   (Address & 0x07FF)     ($ & 0x01800)
ReturnAddr
  movf    tempw ^ 0x0100, w     ; restore the "w" from the Subroutine
 endm

EmuReturn Macro                 ; Return from the Macro Call
  movwf   tempw ^ 0x0100        ; Save the Temporary "w" Register
  movf   INDF, w                ; Get the Pointer to the Return Address
  movwf  PCLATH
  decf   FSR, f                 ; Point to the Low Byte of the Return
```

```
Address
  movf   INDF, w
  decf   FSR, f
  movwf  PCL                    ; Jump to the Return Address
 endm
```

and will save the return address in a data stack implemented with the FSR register. This address is then used to return to the calling section of code.

These macros are reasonably efficient, but it should be noted that they do affect the state of the zero flag in the STATUS register and they do take up a number of instructions. The number of instructions taken up by the subroutine calls is why I created other macros, like "EEPReadMacro" and "SendCharMacro" which actually require fewer instructions to implement the required function than the EmuCall macro.

The last aspect of the application that I would like to bring to your attention is how I implemented the breakpoints for application single stepping and breakpoints. As I pointed out above, if I were to use multiple instructions for breakpoints then code like:

```
btfss  PIR1; TXIF      ; Poll until USART Free to Send a
 goto  $ - 1           ; Character
movwf  TXREG           ; Output the character in "w"
```

will not be able to be stepped through.

The approach I took was to create a single step (and "breakpoint") mechanism that would not have problems with these situations. By limiting application size to one page, I can use a single "goto" instruction for implementing the return to the EMU-II application code.

For example, if I was single stepping at the "btfss" instruction in the example above, the EMU-II code would put in the breakpoints shown below:

```
 btfss   PIR1, TXIF    ; Poll until USART Free to Send a
  goto   NextStep      ; Was "goto $ - 1"
goto     SecondStep    ; Was "movwf TXREG"
```

Now, depending on the value of "TXIF," execution will return to the EMU-II code vial "goto NextStep" or "goto SecondStep," which in either case is located in the instruction code area 0x0700 to 0x07FF. "NextStep" and "SecondStep" are separate from each other in order for the correct new Program Counter value to be noted and recorded by the EMU-II application.

The "NextStep," "SecondStep" and breakpoint code is similarly designed and uses the following instructions:

```
Step#                           ; "#" is from 0 to 8 for Breakpoints
  movwf    _w                   ; Save the "w" Register
  movf     STATUS, w            ; Save STATUS and Change to Bank 2
  bsf      STATUS, RP1          ; Execute from Page 2 in EMU-II
  bcf      STATUS, RP0
  movwf     _status ^ 0x0100
  movf     PCLATH, w            ; Save the PCLATH Registers
  movwf    _pclath ^ 0x0100
  movlw    # * 2                ; Save the Breakpoint Number
  gotom    StopPoint
AddressIns#
  dw       0x03FFF              ; Address of the Breakpoint/Single Step
  dw       0x03FFF              ; Instruction at Breakpoint/Single Step
```

The two words at the end of the breakpoint are used to save the address where the breakpoint was stored and what was the original instruction. The breakpoint address is used to update the EMU-II's Program Counter along with replacing the "goto Step#" instruction with the application's original instruction.

Before any type of execution, all the enabled breakpoints are put into the application program memory. Upon completion of execution, the application program memory is "scrubbed" for all cases of breakpoint "gotos" and they are restored with the original instructions.

Note that when setting up single step breakpoints, the next instruction or destination of a "goto" or "call" is given the "goto NextStep" and "goto SecondStep" breakpoints. This is possible for all instructions instead of "return," "retlw" and "retfie." The reason for these three instructions to get an error message during single stepping is that the destination cannot be taken from the processor stack. Instead of putting a breakpoint at potentially every address in the application, I decided to simply prevent single stepping at these instructions.

As applications may be halted by pressing the "Reset" button in the application, when the EMU-II first boots up, the scrub operation takes place to ensure that there are not any invalid "gotos" left in the application.

There are a few things to watch out for with breakpoints and interrupts. For most application debug, I do not recommend setting breakpoints within the interrupt handler. The reason for this is to avoid any missed interrupt requests or having multiple requests queued up in such a way that the application's mainline never returns. I originally thought that this was a limitation of the EMU-II, but I tried some experiments with MPLAB-ICD and found

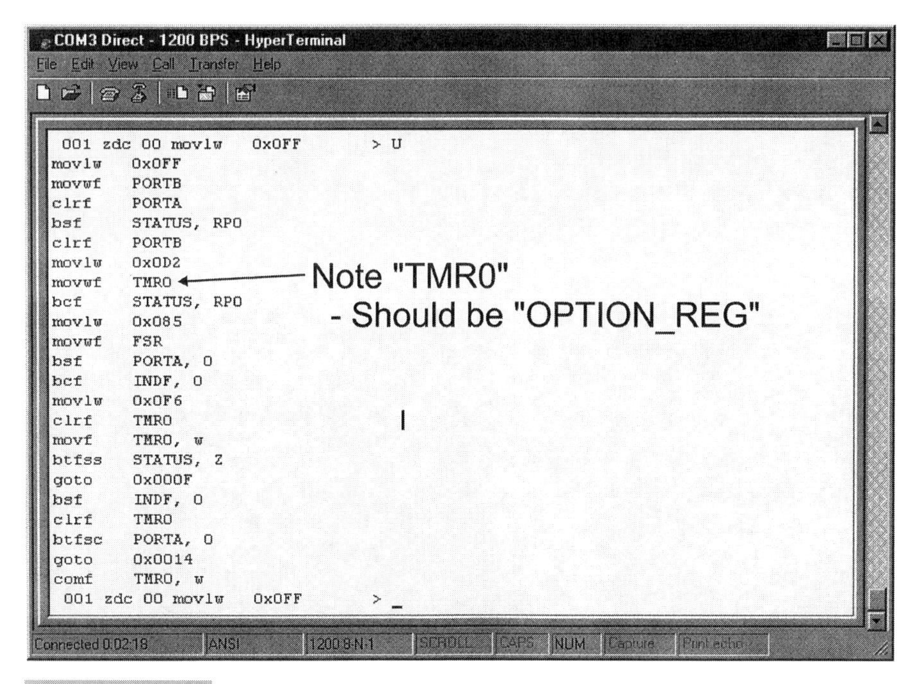


**Figure 14-18** EMU-II unassemble display

that it also has similar issues. Interrupt handlers should always be debugged as thoroughly as possible using a simulator so as to not miss or overflow on any interrupt events and requests.

This is not to say that simple applications (such as just waiting for a TMR0 overflow) cannot be used with the EMU-II. In testing out the application, I did work through a number of experiments with interrupt handlers without problems—there is one point I should make to you and that is the breakpoints should **NEVER** be enabled in both an interrupt handler and mainline code. The reason for saying this is that if an interrupt breakpoint is executed while a mainline breakpoint is being handled by the EMU-II, then the mainline breakpoint context registers will be changed to the values of the interrupt handler. As well, the execution of the application may become erratic.

If you have debugging an application that requires breakpoints in both the interrupt handler and mainline code, then I recommend setting only one at a time and using the "C" (breakpoint all clear) command before setting the next breakpoint.

The EMU-II includes a simple disassembler for reviewing the source code. A typical "Unassembled" application is shown in Figure 14-18 and there are two things I want to point out about the disassembled function.

The first point to make about the disassembled code is the lack of useful labels. The disassembled code in Figure 14-18 probably doesn't look too bad, but compare it against the source (from the "Experiments" chapter):

```
  movlw    0x0FF
  movwf    PORTB                      ; Turn off all the LED's
  clrf     PORTA                      ; Use PORTA as the Input

  bsf S    TATUS, RP0                 ; Have to go to Page 0 to set Port
                                      ; Direction
  clrf     TRISB & 0x07F              ; Set all the PORTB bits to Output
  movlw    0x0D2                      ; Setup the Timer to fast count
                                      ; Put in Divide by 8 Prescaler for 4x
  movwf    OPTION_REG & 0x07F         ; Clock
  bcf       STATUS, RP0               ; Go back to Page 0

  movlw    TRISA                      ; Have to Set/Read PORTA.0
  movwf    FSR

Loop

  bsf      PORTA, 0                   ; Charge Cap on PORTA.0
  bcf      INDF, 0                    ; Make PORTA.0 an Output
  movlw    0x0100 2 10                ; Charge the Cap
  clrf     TMR0                       ; Now, Wait for the Cap to Charge
Sub_Loop1                             ; Wait for the Timer to Reach 10
  movf     TMR0, w                    ; Get the Timer Value
  btfss    STATUS, Z                  ; Has the Timer Overflowed?
   goto    Sub_Loop1                  ; No, Loop Around again

  bsf      INDF, 0                    ; Now, Wait for the Cap to Discharge
  clrf     TMR0                       ; and Time it.
Sub_Loop2                             ; Just wait for PORTA. 1 to go Low
  btfsc    PORTA, 0
   goto    Sub_Loop2

  comf     TMR0, w                    ;Get the Timer Value
```

This is an excellent example of why I prefer only using source code enabled development tools. Trying to see the function of the application from Figure 14-18 is just about impossible, but when you look at the source, the function that it implements, Potentiometer measuring code with an RC delay circuit, is quite obvious.

The second problem with what is pointed out in Figure 14-18 is that the disassembler doesn't "know" what bank is currently executing. In Figure 14-18, you should see that the "TRISB" register, when it is enabled for all output is referenced as "PORTB" (the fifth line of the code). This is not a big problem and one that I can usually work my way through without any problems.

What I find to be very confusing in Figure 14-18 is the identification of "TMR0" when I want "OPTION_REG" (or "OPTION" as it is displayed by the EMU-II). As you step through the application, you will discover that the instruction on the prompt line will display the instruction based on the state of the "RP0" bit of the emulated device's STATUS register.

While application code can be downloaded to the EMU-II, it cannot be uploaded into a PC. This was not implemented specifically to discourage the practice of modifying an application in the emulator and then uploading the .hex file into the host PC and replicating the application from this "source." This is a very dangerous practice and should be avoided at all costs to prevent the proliferation of executable code without supporting application code.

When you work through the "Experiments" in the next chapter, you will find that some of the code is written for the EMU-II. You will find that you will be able to execute the PIC16F84 "Experiment" code directly on the EMU-II without making any changes to the source.

The EMU-II is probably the most involved application that you will find in this book. I am pleased with the way in which it came out and I think it is a very interesting project and tool to have while learning about the PICmicro® MCU. I don't think that it is adequate as a professional development tool due to the lack of a source code interface, but for very simple applications this emulator can be an invaluable tool for you to learn about the PICmicro® MCU.

# Other Emulators

While there are a number of very good PICmicro® MCU emulators designed and marketed by third parties, there hasn't been the same explosion of designs as with PICmicro® MCU programmers. The reason for this really comes down to the complexity of work required to develop an emulator circuit along with the difficulty of developing a user interface for them. The EMU-II is a very simple example of how an emulator could work with a very basic user interface that does not contain many of the features I would consider critical for using an emulator in a professional environment.

A professional product would require a "bond out" chip and a source/listing file interface for the user to use it effectively. Other features would include providing a high-speed interface to avoid the time required to download application hex files into the EMU-II.

If you are interested in designing your own full emulator, then you will have to contact your local Microchip representative to find out about entering into a "Non-Disclosure

Agreement" ("NDA") with them to learn more about the options Microchip has for developing emulators for their products. Microchip does make "bondout" chips available for all of the PICmicro® MCU products, but technical information about them is considered proprietary and not for general release.

Partial emulators (like the EMU-II) are still a lot of work to get running, but designing them does give you a much better appreciation of how the PICmicro® MCU works. If you are interested in designing your own, please take a look at the code in the EMU-II to see how the various issues of executing an application from within a PICmicro® MCU "monitor" are handled.

Having said there are few commercial emulators available, I should point out that there are a number of products that you can buy. These commercial emulators provide wide ranges of services and can be bought for a few hundred dollars up to a thousand dollars or more for a "full" bondout chip based complete system.

# 15

# EXPERIMENTS

## CONTENTS AT A GLANCE

**Analog Input/Output**



**I/O with Interrupts**



**Serial I/O**



**Debugging**



I'm an experimenter by nature. As I work through computer and electronic projects, I often create small applications to understand a new microcontroller or help clarify features in devices that I already am familiar with. This chapter goes through many of the different aspects of the PICmicro® MCU on an experimental basis to help you to understand how applications execute in the processor and how hardware interacts with the PICmicro® MCU.

The experiments are ordered in such a way that you will develop your understanding of the PICmicro® MCU, starting with software operations, followed by actual hardware applications. The purpose is to demonstrate the operation of the PICmicro® MCU processor before hardware interfacing (which requires processor operation) with prerequisite information provided for each application.

In the experiments, I have tried to be as flexible as possible. The circuits will execute on a PIC16F8x, a YAP-II, or an EMU-II and can be easily modified for the MPLAB-ICD and other third party protyping and test tools. The experiments should also work on a large range of PICmicro® MCU devices, including those of different families. Along with schematic diagrams, I have included wiring diagrams for small breadboards and the YAP-II/EMU-II. I am hoping that this variety will make it easier for you to work through the experiments and try them out on different devices and PICmicro® MCU architectures.

# Tools and Parts

In the 45 experiments that follow, a number of modes tools and parts are required to build, assemble, and test the different applications. Table 15-1 consists of a complete kit of parts, but you will need a number of other things to successfully build the experiments. This section covers the background materials and then the actual parts that are required to implement the experiments.

First and foremost, you will need a PC running Microsoft Windows 95 or later. Hardfile space will be required to load the MPLAB IDE, the experiments and applications, as well as provide space for your own applications. The features of the PC include:

- Intel Pentium II or equivalent processor running at 200 MHz (or faster).
- 64 MB (or more) of free memory.
- SVGA display capable of displaying 1024 by 768 (or more) pixels and 2 MB (or more) of VRAM.
- CD-ROM reader.
- Mouse.
- Serial port.
- Parallel port.
- Microsoft Windows 95/98/NT/2000/Me installed on the harddrive.
- Internet connection.
- Microsoft Internet Explorer or Netscape Navigator Web browsers.
- Adobe Acrobat file reader.
- 500 MB (or more) of free harddrive space for the MPLAB IDE and applications.

A PC with these features is not very expensive (actually, as I write this, a new one with these features can be bought for less than $500). Once you have the PC, you should install the MPLAB IDE software and application code from the CD-ROM that comes with the book. The instructions are found in the HTML files on the CD-ROM.

Along with the PC, you should have a PICmicro® MCU programmer. The El Cheapo programmer board included with this book will program all the parts presented in this chapter. For the experiments, I used the Microchip PICSTART Plus, which connects to the PC via a serial port.

The experiments are designed to be built and tested using nothing more than a digital multimeter. You might want to get a logic probe and an oscilloscope to help understand what is happening at the pin level as the applications execute as well as debug your applications.

For the experiments themselves, you should have a 5-volt power supply, a breadboard, and a collection of wires that can be used to wire parts together. I have designed the experiments to use a small breadboard and the 5-volt power supply presented in the "Introduction to Electronics" on the CD-ROM. No experiment requires more than 100 mA, so a 9-volt battery connected to a 78L05 can be used as a quick and dirty power supply. Only *Pin-Through-Hole (PTH)* parts are used, so you should not need a soldering iron or any special PICmicro® MCU sockets.

For wire, you can either buy a "kit" of prestripped wires or a roll of 20-gauge solid-core wire. To wire the circuits, I have included suggested wiring diagrams that can be used to wire the applications together. Follow these diagrams as much as possible and keep the wiring as neat as possible (i.e., use the shortest wires for a particular location as is feasible) to minimize the opportunity for incorrect wiring and to make it easier to find your problems.

For handtools, you will probably need a pair of wire clippers, wire strippers, needlenose pliers, and a small flat-bladed screwdriver. The flat-bladed screwdriver is to be used to pry chips up from the breadboard. A ZIF socket could be used. It will eliminate the need for the screwdriver, but it can be quite expensive and difficult to properly plug into a breadboard. Other than adding wires to the momentary on switches, a soldering iron is not required.

Next, you should have a work area. This can be a desktop with enough space for your PC (including display, keyboard, and mouse), as well as the programmer you want to use, a static-safe assembly area, an oscilloscope, and a DMM. If at all possible, avoid putting the PC and programmer on a different bench than the prototyping area because it will make trying new things or changing code more difficult.

To summarize, the following tools are required to create the experiments in this chapter:

- PC
- PICmicro® MCU programmer
- Digital multimeter
- +5-volt application power supply
- Breadboard
- 20-gauge solid-core wire
- Wire clippers
- Wire strippers
- Needlenose pliers
- Small flat-bladed screwdriver
- Listed PICmicro® MCUs
- Listed electronic parts

Optional tools:

- Logic probe
- Oscilloscope

- Logic analyzer
- Ultraviolet EPROM eraser
- PICmicro® MCU emulator

In the experiments, I use the following PICmicro® MCU devices:

- PIC16F84-04/P
- PIC16C505-20/JW
- PIC16C711-20/JW
- PIC16C73B-20/JW
- PIC16F877-04/P

The PIC16F84 is an 18-pin PICmicro® MCU Flash base that can be electrically erased and reprogrammed by the programmer. The PIC16C505 and PIC16C73B are EPROM based and require an ultraviolet light source to erase them. Notice that I have specified the JW versions of the EPROM PICmicro® MCUs, which allows them to be erased. Expect to pay up to $20 apiece for the JW parts. The PIC16F877 (or the PIC16F876) can be used for most of the experiments and can be combined with an emulator such as the EMU-II or MPLAB-ICD.

The JW parts are somewhat more expensive than the *One-Time Programmable (OTP) versions* of the same part. In fact, depending on the source, they can be upwards of three times more expensive. When programming them, be sure that you do not enable the code protect function of the chips. As I have indicated elsewhere in the book, the code-protect configuration fuses are protected from ultraviolet erasure by metal layers over the EPROM cells and can be almost impossible to clear.

The ultraviolet eraser that I use is the Walling Co. Datarase ][, which can erase four EPROM pin-through hole devices simultaneously. The eraser is reasonably inexpensive, running from $30 for a basic unit to $50 with a built-in timer. I recommend buying the version with the timer because I've forgotten parts in the eraser for a few days. Although doing so won't damage the expensive JW parts, it will reduce the useful life of the (expensive) ultraviolet bulb in the eraser.

This list is drastically reduced if the YAP-II is used instead of the breadboard solution. With some judicious shopping, you should be able to get all these parts except for the microcontrollers for less than $10. In most of the experiments, I have included wiring diagrams for both the breadboard and YAP-II solutions.

This list contains a few components that should be mentioned. For some applications, I have specified a 10-LED bargraph display instead of 10 individual LEDs. The reasons for this should be obvious because it is easier to wire a 20-pin DIP package to a breadboard than 10 individual LEDs. The LED bargraph is used in any application that requires more than two LEDs.

The second one is the momentary-on switches. I have taken normal PC board mount momentary-on switches and soldered wires onto them and then coated the wire with "Weldbond" (Fig. 15-1). This will make adding the switches to your breadboard experiments quite easy.

**TABLE 15-1 Parts Needed To Implement The Experiments**

| PART | DESCRIPTION |
|---|---|
| 3x PIC16F84-04/P | 18-pin flash-based PICmicro® MCU. Three parts required for PowerUp experiment |
| 3x PIC16F877-04/P | Optional. Used instead of PIC16F84. Part of EMU-II or MPLAB |
| 1x PIC16C505-JW | Used to test the different oscillator configurations |
| 1x PIC16C711-JW | Used for built-in ADC |
| 1x PIC16C73B-JW | Used for built-in ADC/USART |
| 1x 4-MHz ceramic resonator | 4 MHz with built-in capacitors |
| 1x 4-MHz crystal | 4-MHz parallel-cut crystal |
| 2x Crystal capacitors | 27- to 33-pF ceramic disk capacitors |
| 1x 10-K potentiometer | 10-K single-turn/PC board-mount potentiometer |
| 2x 0.1-$\mu$F capacitors | 0.1-$\mu$F tantalum or ceramic disk |
| 5x 1.0-$\mu$F capacitors | 1.0-$\mu$F tantalum/electrolytic capacitors |
| 2x Switch | Momentary on |
| 2x Red LED | 5-mm red LEDs with 0.100" lead spacing |
| 1x 10 LED bargraph display | DIP package bargraph display |
| 10x LED current-limiting resistor | 220 Ohm, 1/4 watt |
| 2x 330 Ohm | 330 Ohm, 1/4 watt |
| 1x 1-K resistor | 1 K, 1/4 watt |
| 2x 2.2-K resistors | 2.2 K, 1/4 watt |
| 2x 3.3-K resistors | 3.3 K, 1/4 watt |
| 2x 4.7-K resistors | 4.7 K, 1/4 watt |
| 2x Pull ups | 10-K, 1/4-watt resistor |
| 2x NPN transistors | 2N3904 in TO-92 package |
| 1x PNP transistor | 2N3906 in TO-92 package |
| 1x DS275 | DS275/DS1275 |
| 1x MAX232 | Maxim MAX232 or equivalent |
| 1x Female 9-pin D-shell | DB-9F |

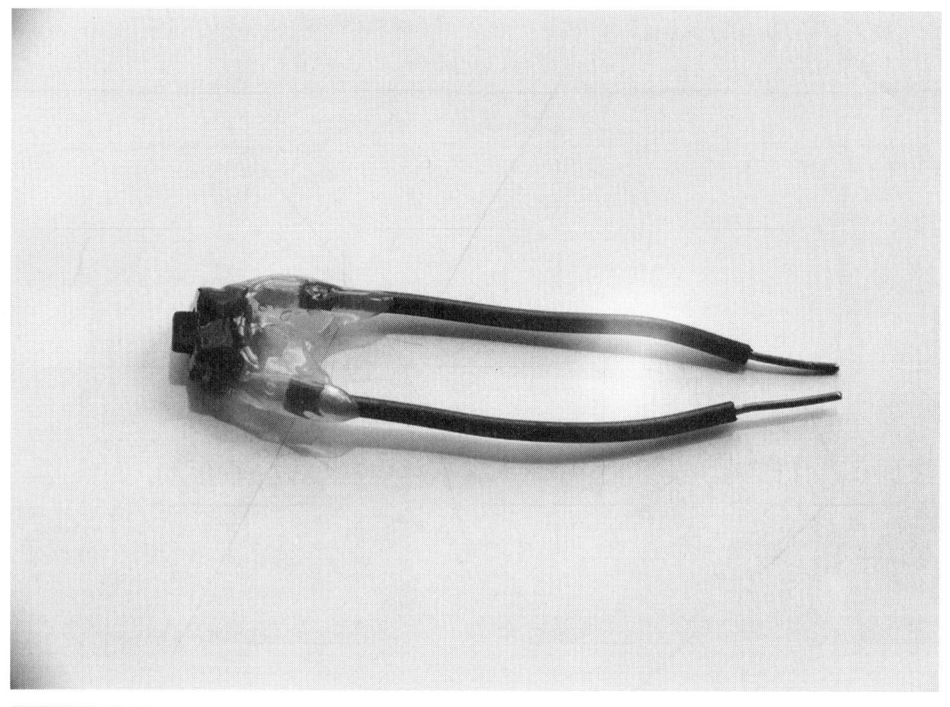

**Figure 15-1** Breadboard switch

# PICmicro® MCU Processor Execution

The previous chapters cover the different issues regarding PICmicro® MCU application programming at length. Before I start into programming actual PICmicro® MCU hardware, I would like to first introduce you to many of the issues concerning developing your application software and watching it execute in the MPLAB simulator.

Installing the MPLAB IDE into a Microsoft Windows PC is very easy. It simply consists of executing an .EXE file that performs all of the decoding information for you. The CD-ROM contains instructions to install the MPLAB IDE on your PC.

When I work with the MPLAB IDE, I'm not that adventurous with how I place the windows on the MPLAB IDE desktop. For most applications, I put the application's source code window at the top of the desktop, with the *Stopwatch* in the lower left hand corner and the PICmicro® MCU register's *Watch* windows to the *Stopwatch*'s right. This arrangement is shown in Fig. 15-2.

The *Toolbar* is the collection of buttons below the MPLAB IDE pulldowns. A number are built into the MPLAB IDE which can be selected from:

1 Project
2 Editor
3 Assembler
4 Simulator
5 Simulator/Editor

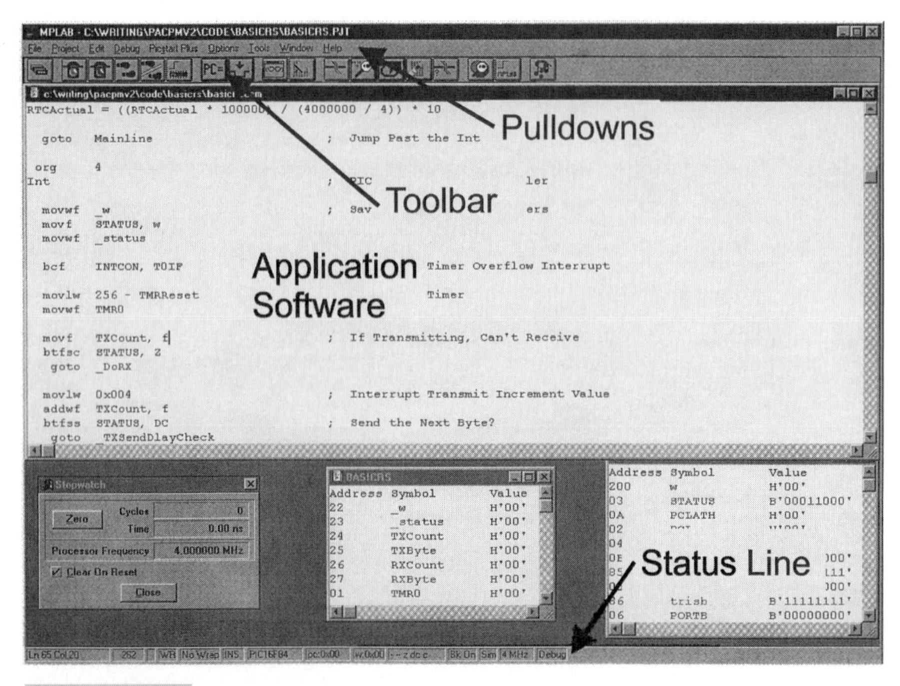


**Figure 15-2** MPLAB IDE labeled desktop

The *Toolbar* is set to *Simulator* because this provides a basic interface to execute applications in either the simulator or using one of the emulators that can access the MPLAB IDE directly. The *Simulator* toolbar is shown in Fig. 15-3 with the primary buttons labeled. Along with this description, if you move your mouse pointer over a button, its definition will be displayed on the left field of MPLAB's STATUS line.

The *Stopwatch* (Fig. 15-4) is used to monitor the number of cycles executed within an application. The *Processor Frequency* is set from the *Options* pull down and then clicking on *Processor* and *Clock Frequency.* This process might seem a bit cumbersome (and difficult to remember at times, resulting in you scrambling around trying to find the function), but remember that it should only have to be done once in an application. Normally, when I invoke the *Stopwatch,* I have it automatically reset when the simulated PICmicro® MCU is reset.

As covered later in this chapter, the *Stopwatch* is used to control the operation of stimulus files by determining the step or cycles (in the *Stopwatch* window) value. Resetting or making the *Stopwatch* zero will cause the step used for the stimulus file to also be reset to zero, which can cause problems if you are timing a loop and expecting a certain input from a stimulus file.

The watch window allows you to monitor the current values of different registers active in the application. The register data can be presented in hexadecimal, decimal, octal, binary, or ASCII formats, whatever is most appropriate for the data. Along with being displayed in a specific format, the number of bytes (up to four) can be displayed. The *Properties* window makes this specification when the register is being added (Fig. 15-5). Registers can either be selected by their address or label (the label is much easier to work with).

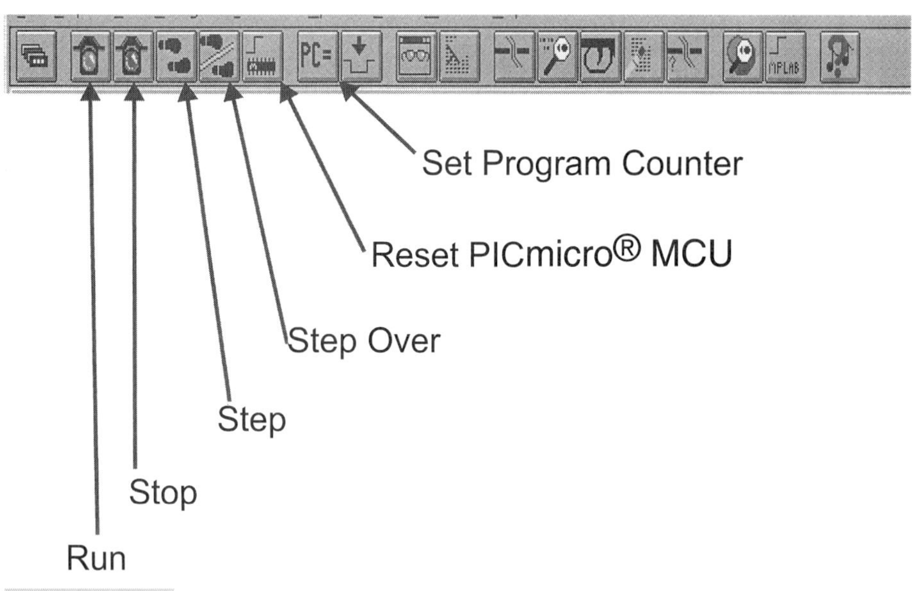


**Figure 15-3** The MPLAB IDE debug toolbar

I tend to create some basic watch windows for specific PICmicro® MCUs and reuse them every time the PICmicro® MCU is used in an application. In this chapter, you will see a lot of the 16F84.WAT file, in which I have defined the basic context registers, INTCON, and the I/O registers (Fig. 15-6). When an application uses variables, I create a unique watch window for it and make it specific to the application (like BASICRS, shown in Fig. 15-5).

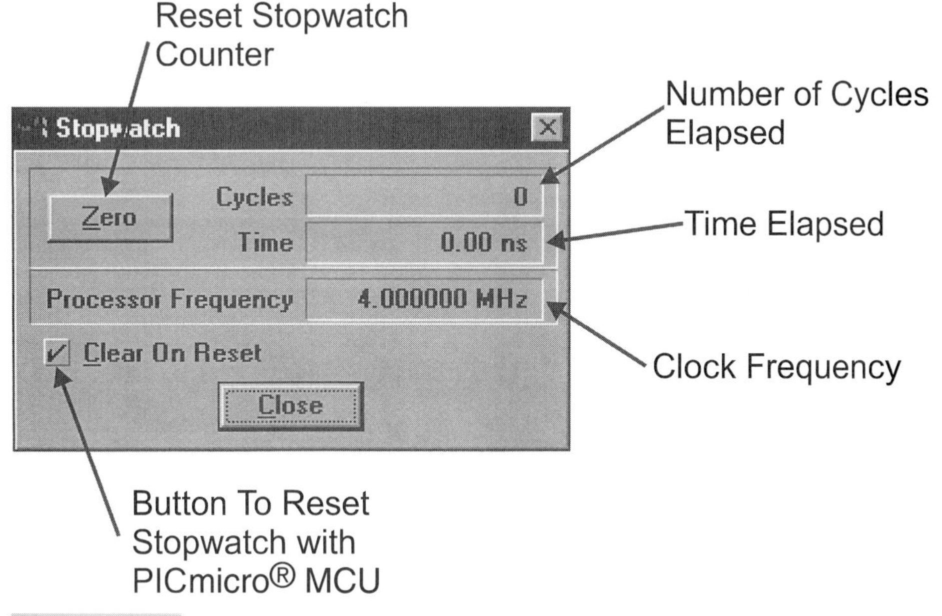


**Figure 15-4** The MPLAB IDE stopwatch

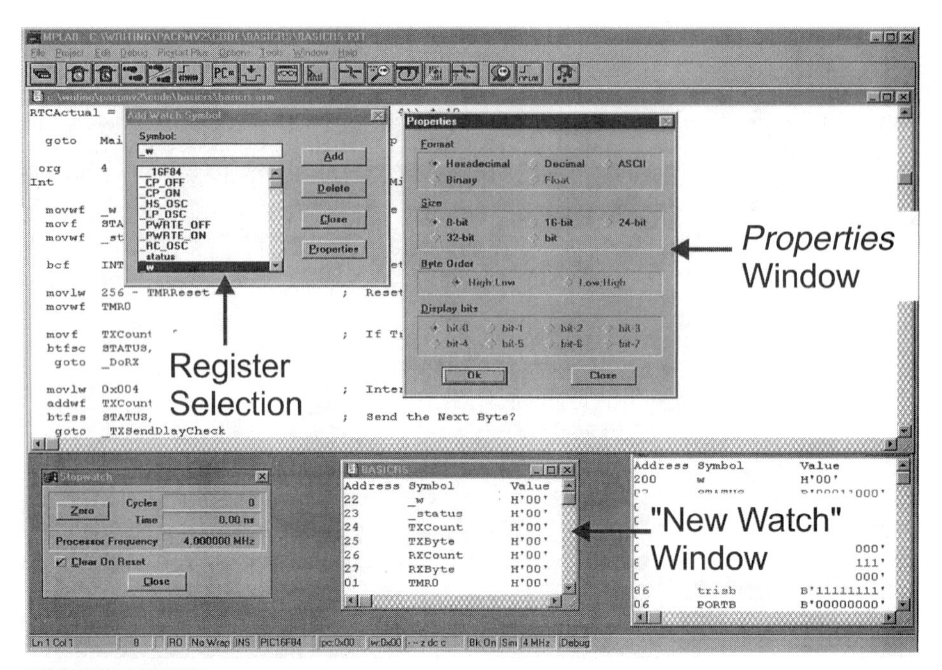


**Figure 15-5** The MPLAB IDE watch window create

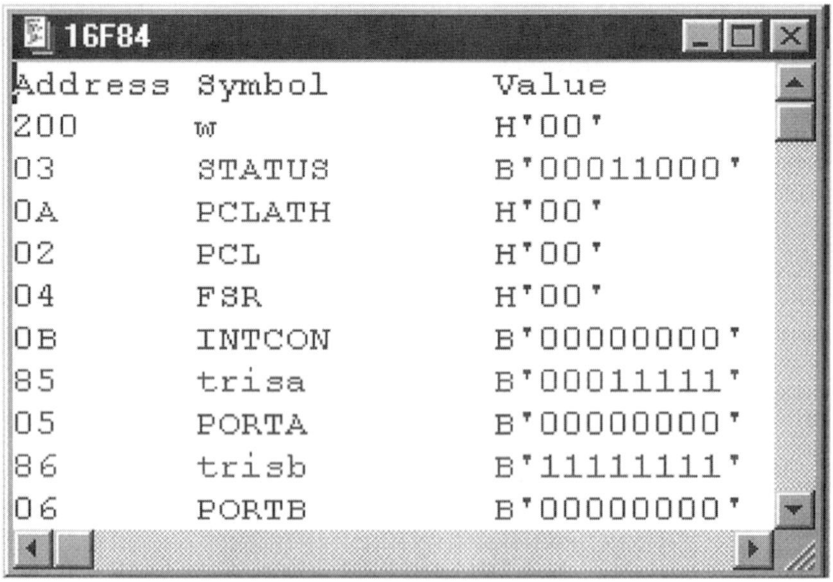

16F84

| Address | Symbol | Value |
|---|---|---|
| 200 | W | H'00' |
| 03 | STATUS | B'00011000' |
| 0A | PCLATH | H'00' |
| 02 | PCL | H'00' |
| 04 | FSR | H'00' |
| 0B | INTCON | B'00000000' |
| 85 | trisa | B'00011111' |
| 05 | PORTA | B'00000000' |
| 86 | trisb | B'11111111' |
| 06 | PORTB | B'00000000' |

**Figure 15-6** PIC16F84 register watch window

I have included the subdirectory *code\ProcWat* in the *PICmicro* directory, which contains the processor specific watch windows that I use.

This background information should give you a better idea of how the MPLAB IDE works and some of its features. As I go through the experiments, I point out how the different MPLAB IDE features that are required for simulating the different applications are used.

## FIRSTAP: DIRECT REGISTER ADDRESSING AND CREATING YOUR FIRST MPLAB PROJECT

After reading through the book up to this point, I hope that you don't feel that you now have to work up the nerve to start working with the PICmicro® MCU. I realize that I have presented you with a lot of information on interfacing to the PICmicro® MCU, programming it, and working with the MPLAB IDE, but I believe that once you start working with the three elements, you will be able to see how everything comes together. This first experiment walks you through a simple application and explains how to use the MPLAB IDE to create the application, assemble it, and simulate it.

Notice in the "simulator only" experiments that I specify the PIC16F84 only and do not include emulator "hooks." This hardware specification should not be limited to this part for you in your experiments and applications. The code can be used in different devices with the changes shown later in this chapter.

FIRSTAP.ASM is the source code for the first experiment that I am going to work with. The application itself is very simple and is located on the CD-ROM in the *code\FirstAp* subdirectory. The code itself repeatedly adds two values together, using a counter, and then finishes by going into an endless loop. The source code itself is quite simple and follows the conventions set out in the template presented earlier in the book.

```
 title "FirstAp - First PICmicro MCU Application"
 ;
 ;  This Application is a First PICmicro MCU Application to demonstrate how
 ;   MPLAB works. The Application code itself simply Adds two Variables
 ;   together inside a loop.
 ;
 ;
 ;  Hardware Notes:
 ;   Simulated PIC16F84 Running at 4 MHz
 ;
 ;  Myke Predko
 ;  99.12.23
 ;
   LIST R=DEC
   INCLUDE "p16f84.inc"

 ;  Register Usage
  CBLOCK 0x020;  Start Registers at End of the Values
 i, j, k
  ENDC

  PAGE
  __CONFIG _CP_OFF & _XT_OSC & _PWRTE_ON  & _WDT_OFF

 ;  Mainline of FirstAp

   org    0
```

```
  movlw  1                 ;  Initialize the Variables
  movwf  i
  movlw  2
  movwf  j
  movlw  3
  movwf  k

;  For k = 0 to 5         //  Loop 5x
  movlw  5
  movwf  k
Loop
  movf   i, w              ;  Add "i" and "j" and Put Result in "j"
  addwf  j, f
  decfsz k, f              ;  Decrement "k" until it's equal to Zero
   goto  Loop

  goto   $                 ;  Loop Forever when Done

  end
```

Before starting up the MPLAB IDE, I recommend that you read through this section so you have an understanding of what I am going to do. The first time you start MPLAB, there will be a lot to absorb while working through the first experiment. If you haven't installed the MPLAB IDE and the other utilities on the CD-ROM, please take this opportunity to go to the "CD-ROM" Appendix and follow the instructions for installing the code. It should only take a few moments. Once this is done, you can follow along with me and load your first application.

One of the first things I recommend for a new application developer is to copy the MPLAB IDE (and WordPad and other often-used utilities) onto your desktop. These procedures only take a few seconds and will make launching the MPLAB IDE very simple and quick. The sequence of operations can be done with other applications as well.

The procedures presented here were carried out on my Microsoft Windows 98 PC, but they work similarly to that of other Windows-based PCs. After the MPLAB IDE has been installed on your PC, you will have to first *copy* it in from the *Start* taskbar. As shown in Fig. 15-7, find the MPLAB IDE from *Start, Programs* and click on the right mouse button with the mouse pointer over *MPLAB.* This will bring up the pull-down menu shown in Fig. 15-7.

In the full screen-shot diagrams, I am using a 1024-by-768 pixel Windows desktop. This is a useful resolution for working with the MPLAB IDE because it allows you to get the source file on with a reasonable size and also have *Watch, Stopwatch,* and *Asynchronous Debug* windows active on the desktop at the same time. In the experiments, when I start using a new or different feature of the MPLAB IDE, I show what it looks like as a screen shot.

After left clicking on *Create Shortcut,* a new copy of the MPLAB IDE, called *MPLAB (2)* will show up on the bottom of the list (Fig. 15-8). In this figure, I moved my mouse pointer over *MPLAB (2)* to highlight it in the program list.

When the copy has been made, you can now *drag* (hold down the mouse's left button and move it with the symbol) *MPLAB (2)* onto your desktop. You should see an outline moving with the mouse pointer, which you can place on the desktop. Your PC's desktop should look something (although probably less cluttered) like mine (Fig. 15-9).

With the MPLAB IDE copied to your desktop, you can now start it up by double left clicking on the MPLAB icon or the MPLAB label. Do this and you will get the MPLAB IDE desktop (Fig. 15-10), although an MPLAB logo bitmap will come up first and leave after the MPLAB IDE desktop has been displayed and set up.

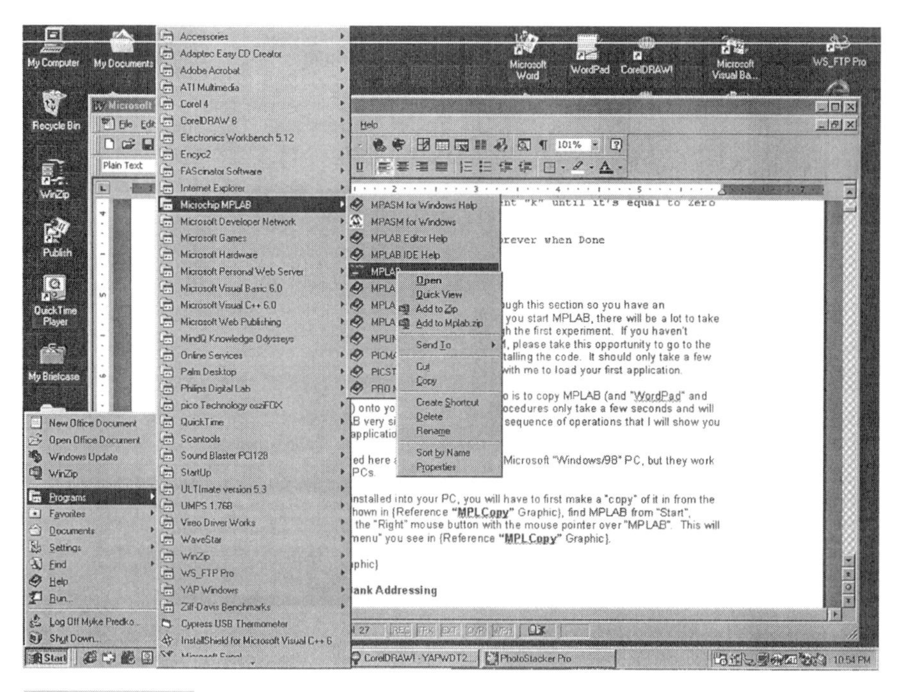


**Figure 15-7** The MPLAB IDE copy operation

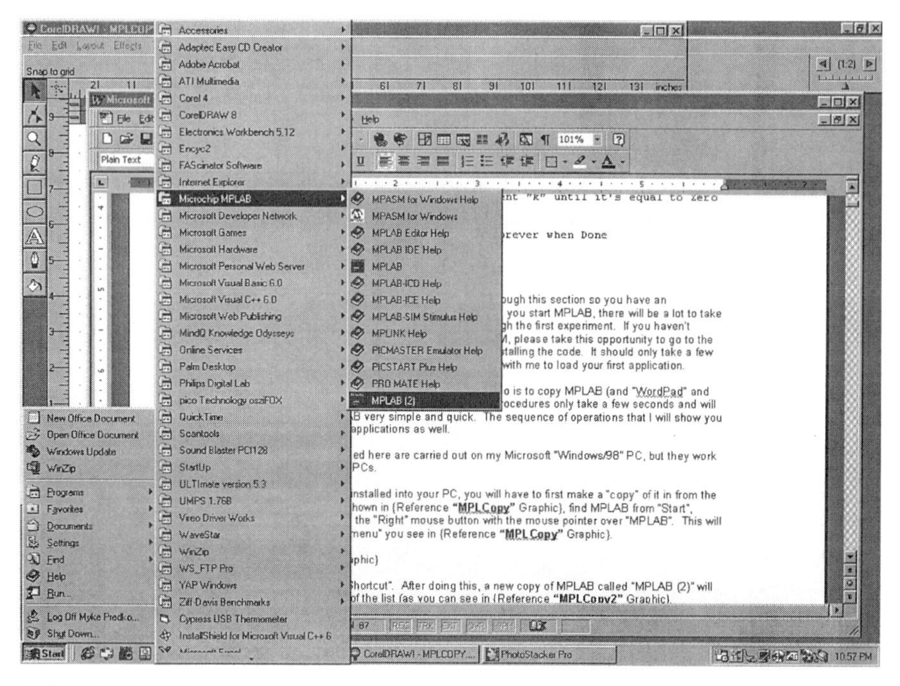


**Figure 15-8** The MPLAB IDE copy saved in selection

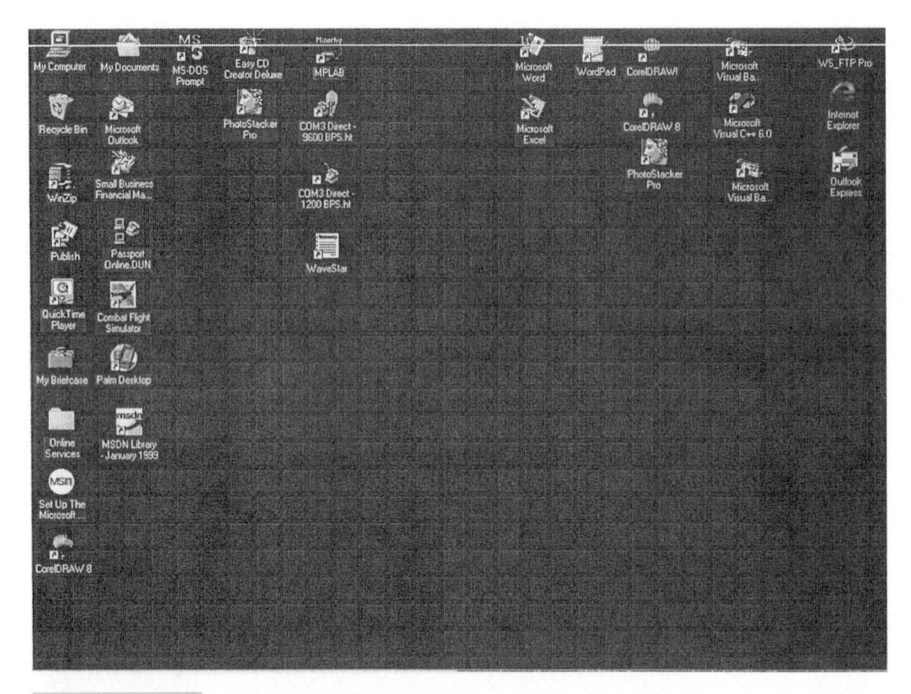


**Figure 15-9** The MPLAB IDE copy moved onto desktop

**Figure 15-10** The MPLAB IDE desktop after startup

With the desktop active, you will now have to select the PICmicro® MCU to use. To do this, click on *Options* (on the top line of the MPLAB desktop) and then click on *Development Mode...* (Fig. 15-11) to bring up the Development mode selection window (Fig. 15-12). For this chapter, unless you are using a PIC16F87x for the experiments or an EMU-II, the processor selected should be a PIC16F84. As well, the MPLAB-SIM simulator should also be selected (Fig. 15-12).

For virtually all of the experiments that have hardware connected to them, I used a 4-MHz ceramic resonator. Because of the four-to-one ratio between the clock and the PICmicro® MCU processor, a 4-MHz clock will result in a 1-$\mu$s instruction cycle period, which is easy to work with for timing an application. To set this clock speed, click on the *Development Mode* dialog box's *Clock Tab* and enter *4 MHz* and an *XT* oscillator. Once this is done, the dialog box shown in (Fig. 15-13) will be displayed.

Once this is done, the MPLAB IDE is ready for you to start working through the experiments. In the applications, I use a number of different tools. The procedures outlined can be repeated to change the processor or the execution frequency.

Now that the MPLAB IDE is set up, you can now create a project. Left click on *Project* and then *New* (Fig. 15-14) to bring up the *New Project* dialog box (Fig. 15-15). When the *New Project* dialog box comes up, it will show the location of the MPLAB IDE files. This should be changed to *PICmicro\code\FirstAp* (Fig. 15-16) so that the correct source file is picked up.

For every application in this book, I included a separate subdirectory that is named according to the application. I highly recommend that you continue this practice because it will allow you to keep track of the different applications easily. As shown in later experiments

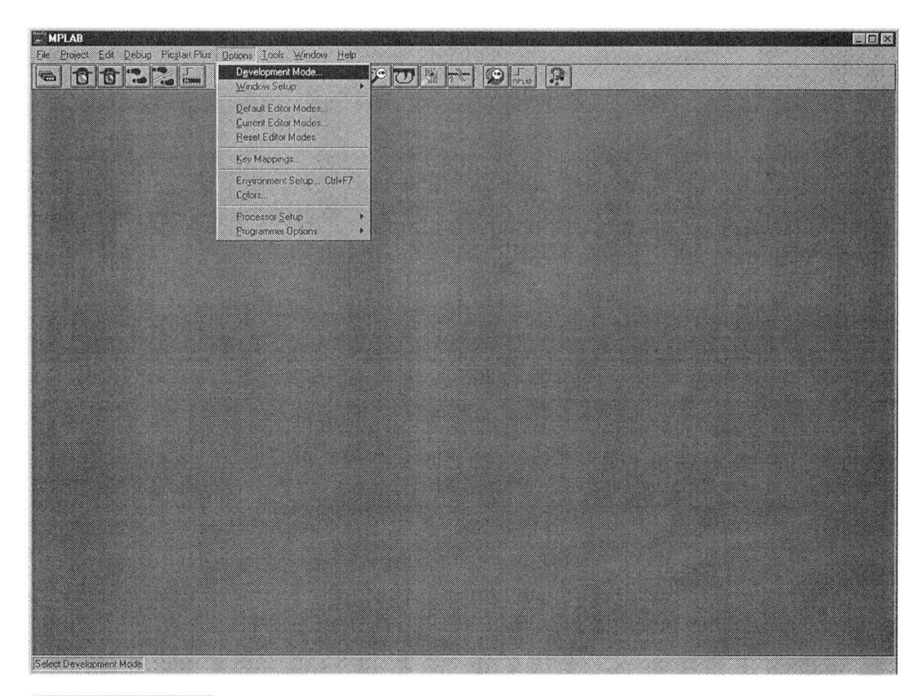


**Figure 15-11** The MPLAB IDE PICmicro® MCU selection

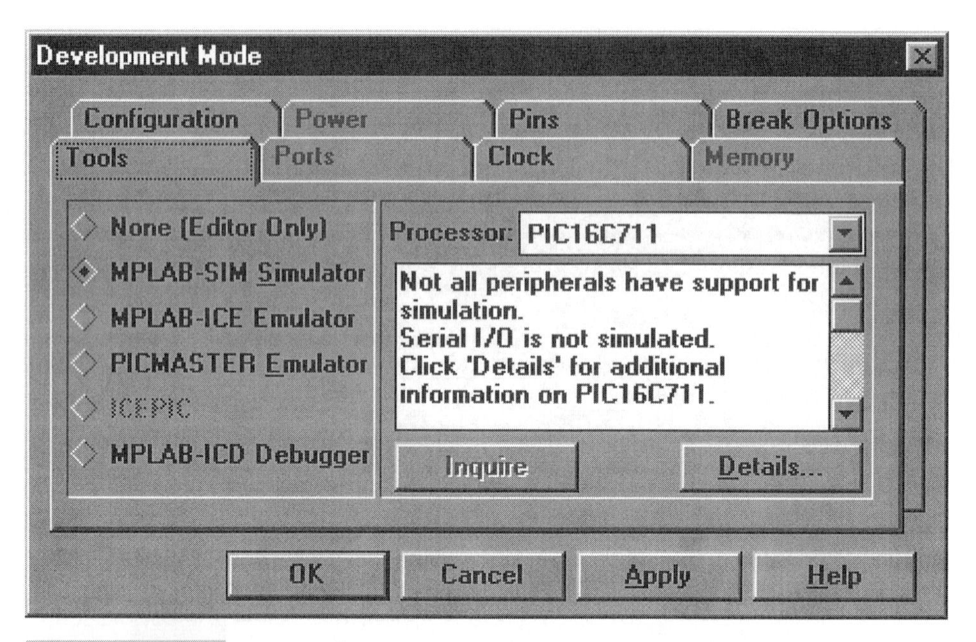


**Figure 15-12** The MPLAB IDE development mode selection

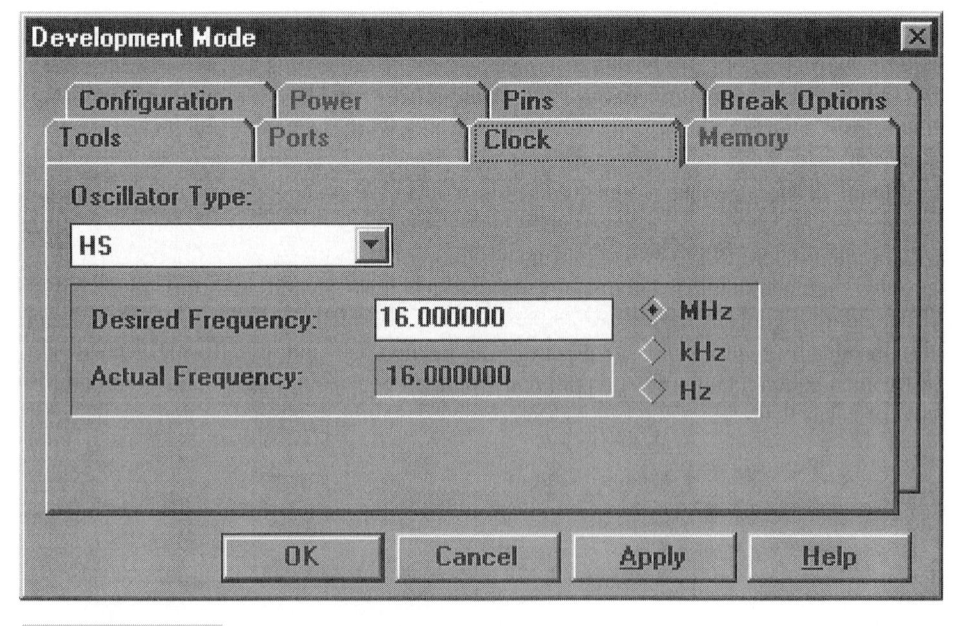


**Figure 15-13** Setting the MPLAB-simulated clock

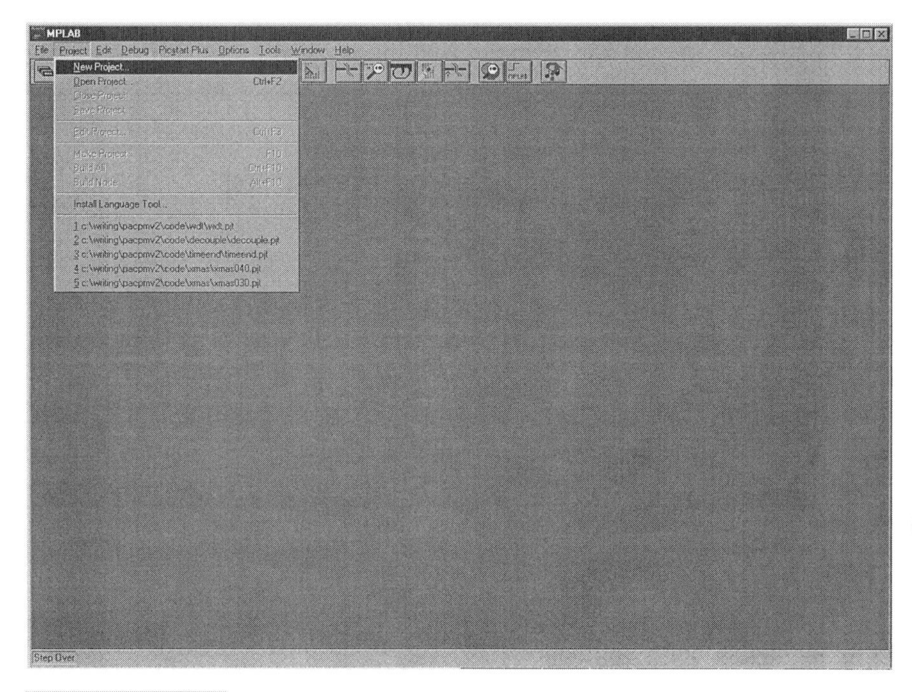


**Figure 15-14** Creating a new MPLAB IDE project

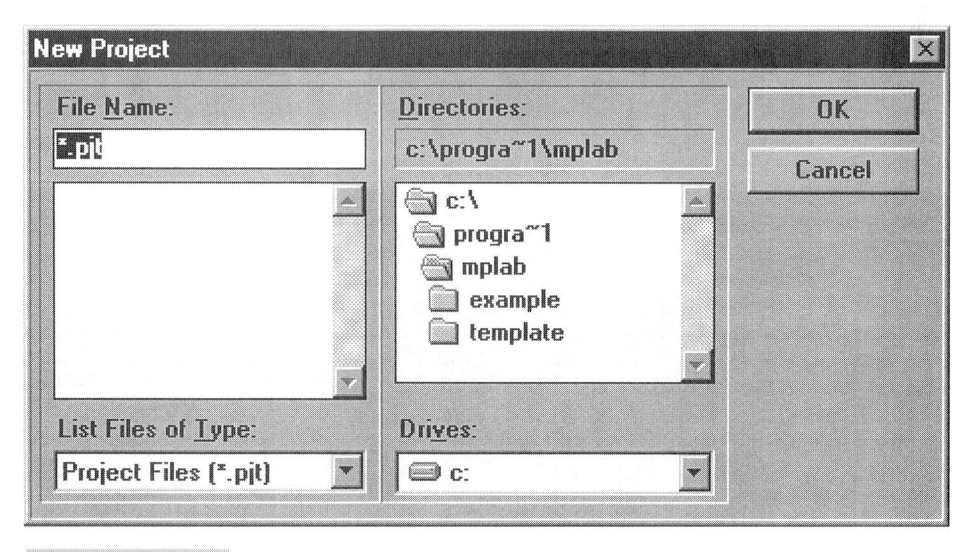


**Figure 15-15** New MPLAB IDE project subdirectory select

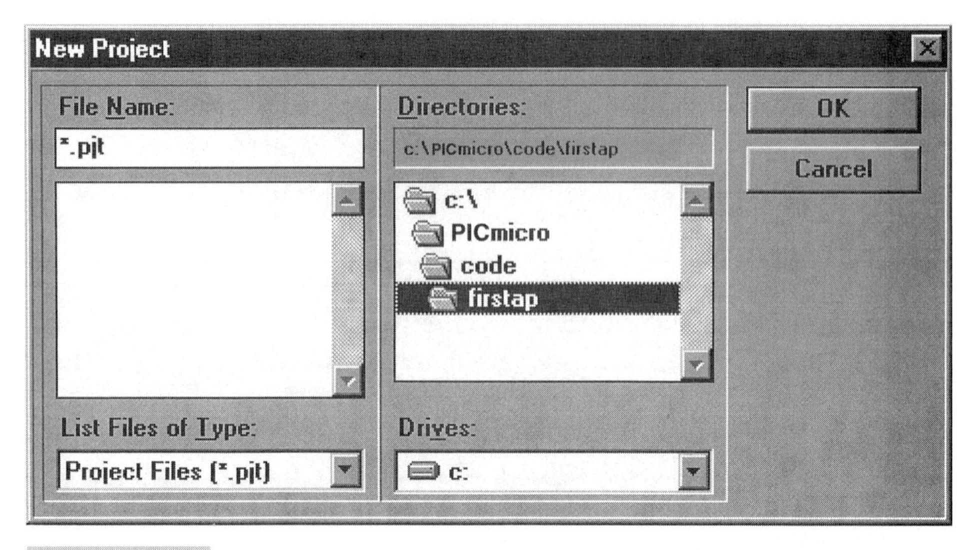


**Figure 15-16** Selecting the MPLAB IDE project directory

and projects, I keep track of different versions of an application, not by changing the subdirectory, but by changing the version number of the source code in the application subdirectory.

A name for the project must be selected for the experiments. This is the same as the source-code file name, as well as the subdirectory. In this case, it is *FirstAp* (Fig. 15-17). Clicking on *OK* after specifying the project name will bring up the *Edit Project* window shown in Fig. 15-18.

Left click on the *Add Node* and select the source file (FIRSTAP.ASM), as shown in Fig. 15-19.

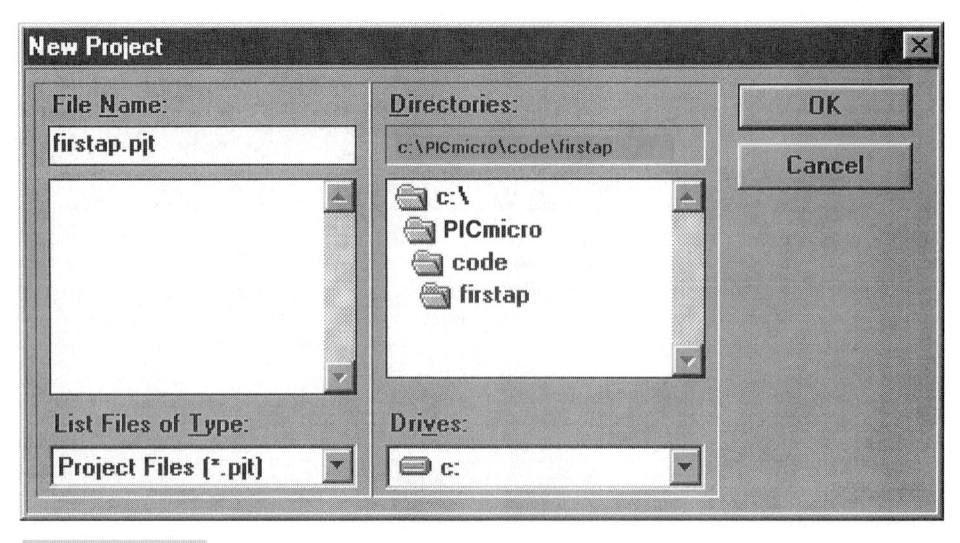


**Figure 15-17** Selecting the MPLAB IDE project name

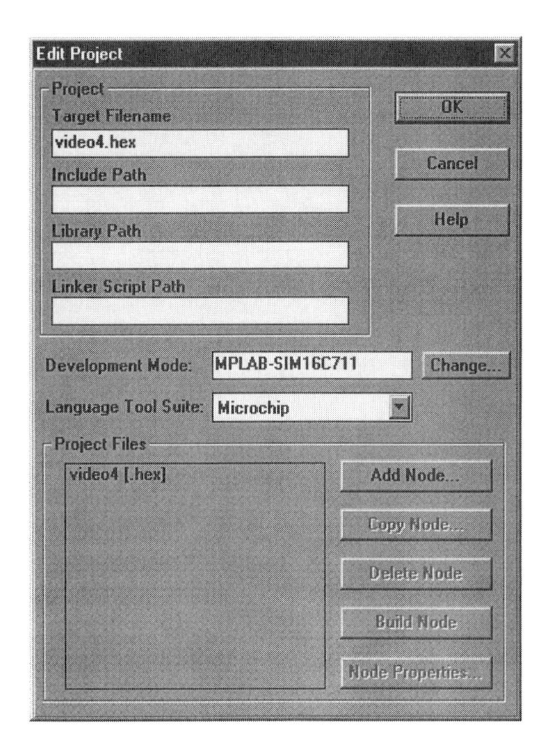


**Figure 15-18** Selecting the MPLAB source files

With the source file selected, click on *OK* and then *OK* in the *Edit Project* window. You have just created the project!

The process probably seems very complex, but after you do it a few times, it will become second nature for you. It really only takes a few seconds to create new projects using the process outlined here.

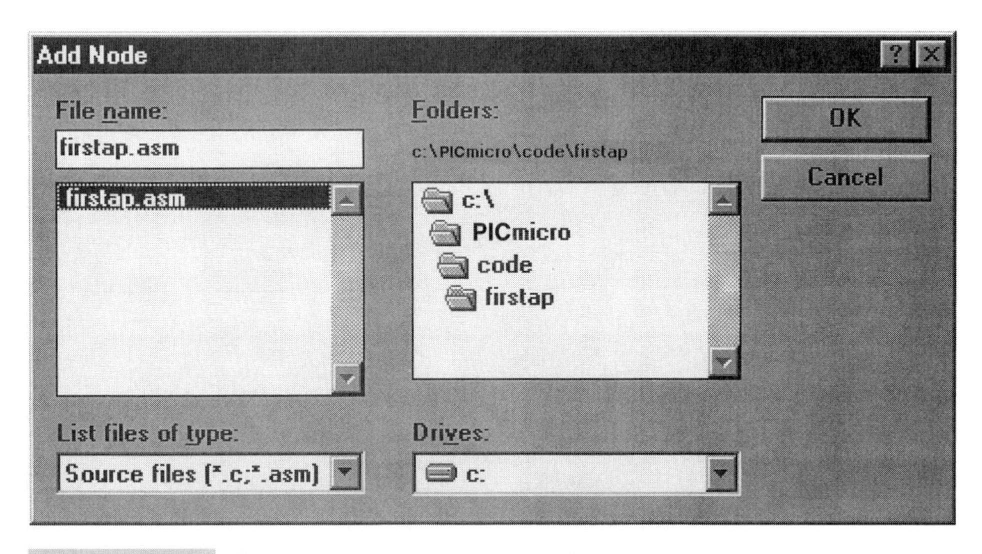


**Figure 15-19** Specifying the project source file

Now, you can load the source file. From the *File* pull-down at the top line of the MPLAB IDE desktop, click on *Open* and select FIRSTAP.ASM. *Open* will start at the subdirectory in which the project is located. For new projects, I will create a new subdirectory for them and copy the TEMPLATE.ASM file into it as the basis for an application. You will get a screen that looks like (Fig. 15-20). I normally will "pull" the right edge of the source file over to the right edge of the desktop to allow me to write source code that is wider than 80 columns.

With the desktop set up with the source file, I now edit the source into the application. For the experiments, I have created the source code so that you only have to create a project (using the guide listed previously).

After the code has been edited, it can be assembled ("built") by pressing *Ctrl-F10*. You could also go through clicking on *Project* and then *Build All,* but the *Ctrl-F10* key sequence is much faster. Once the code assembles cleanly (no errors, warnings, or messages), you are ready to start simulating the application.

FirstAp (listed previously) really does nothing useful, except function as an initial application that you can use to build a project for and then see how the MPLAB simulator works. After you have set up the application, you should display the PIC16F84 (or whatever PICmicro® MCU you are going to be using for the experiments) base registers. As indicated, in the *ProcWat* subdirectory, I have created a number of watch windows that are specific to the devices presented in this book. Click on *Window* and then *Load Watch Window . . .* to load an existing watch window. Find 16F84.WAT in *PICmicro\code\procwat* and resize it and put it in the bottom right corner of the MPLAB IDE desktop (Fig. 15-21).

I've set up the experiments to use the processor watch windows so that you can create simple watch windows to display the registers (including variables) that are not included

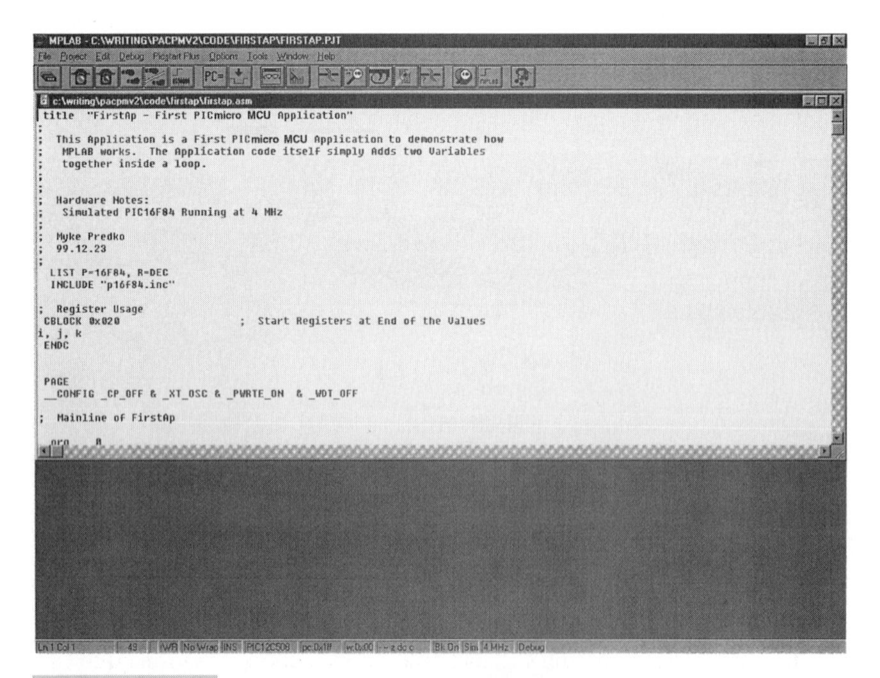


**Figure 15-20** Editing a file in the MPLAB IDE

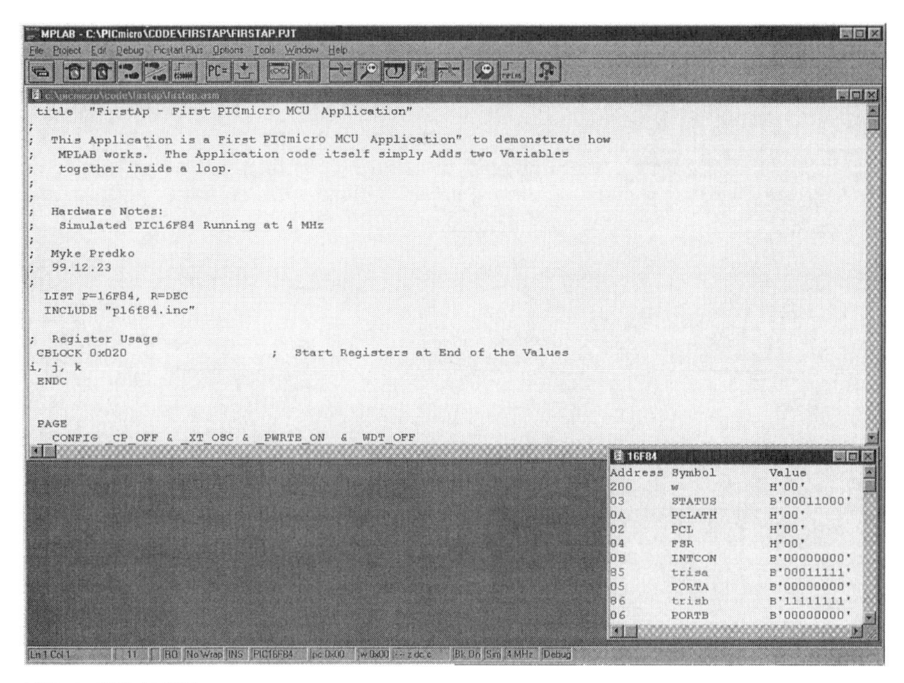


**Figure 15-21** PICmicro® MCU processor watch file selected

in the base processor registers. For FirstAp, these registers are i, j, and k. To create a new watch window for these labeled file registers, click on *Window* on the MPLAB IDE's top line, followed by *Watch* and then *New Watch Window,* and you will get the two new windows on the MPLAB IDE desktop (Fig. 15-22). When they first come up, the *Add Watch Symbol* window will be placed over the *Watch_1* window. To allow me to more easily observe what is being added (and the format it is in), I usually move the *Add Watch Symbol* window to another location (Fig. 15-22).

To add a variable, add it in the *Symbol* of the *Add Watch Symbol* window. If the symbol has been correctly identified, it should also be listed in the list below the symbol entry point. Symbols can be added directly from this list as well. Figure 15-23 shows how j is added to the list. Note that it shows up in the *Watch_1* window when the operation has been successfully created.

There could be two reasons for the label not being known to the MPLAB IDE watch window. The first (and most obvious) is that you have spelled it wrong or specified the capital characters in the label incorrectly. The second reason might be that you are setting up the unique watch window before the application has been assembled. If this is the case, then only the built-in registers will be available to select from. In this case, you will have to assemble the source code successfully before selecting any watch window file registers.

When the variables in the watch window are selected, you can click on *Properties* in the *Add Watch Symbol* window and you can select what kind of data is displayed in the *Watch_1* window. The *Properties* window that comes up is shown in Fig. 15-24. Along with selecting how the data is to be displayed, the size and order of the data can be displayed. This chapter will use registers in watch windows with different parameters specified.

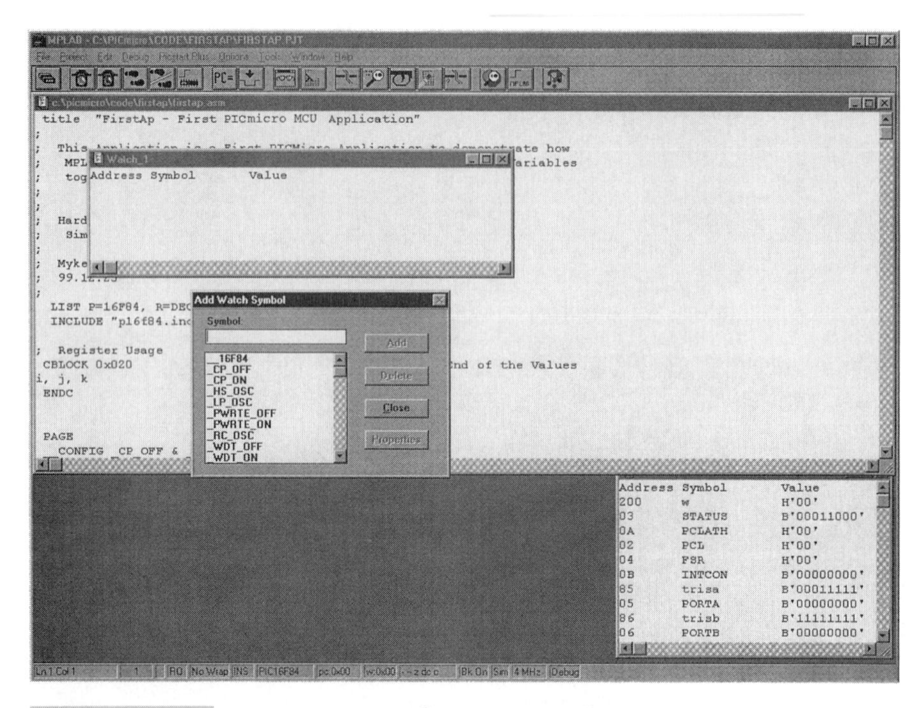

**Figure 15-22** Adding PICmicro® MCU file register watch

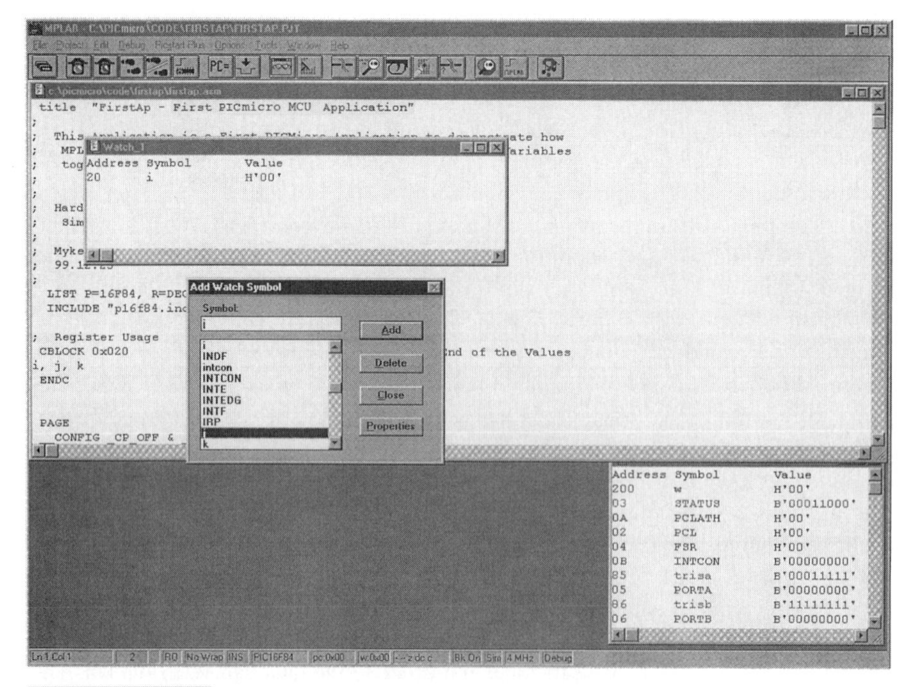

**Figure 15-23** Adding PICmicro® MCU file register

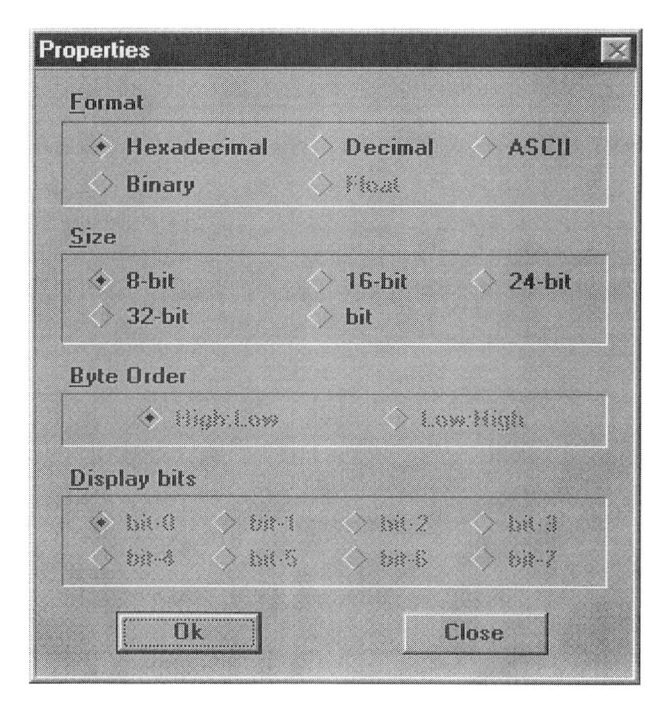


**Figure 15-24** **Setting file register watch parameters**

For FirstAp, click on *OK* without selecting any different parameters from the default (hex-value display and single byte) and add *k* to the *Watch_1* window. You should have now added all three variables (*i, j,* and *k*) to the *Watch_1* window. Press *Close* on the *Add Watch Symbol* window.

You should now just have the *Watch_1* window on the desktop. Resize it and place it on an unused part of the desktop (I normally put it in the bottom middle). By clicking on the icon in the upper left corner of the *Watch_1* window, you will get a selection of different options (including *Insert* and *Delete*), click on *Save* and put in the source file name that is being simulated (*FirstAp,* in this case). When this is done, the MPLAB IDE desktop will look like that shown in Fig. 15-25.

The last aspect of the MPLAB IDE to set up before you are ready to start simulating is to select the correct MPLAB toolbar for working with an application. The toolbar can be selected by clicking on the leftmost icon (a series of overlapped boxes) on the toolbar line (which is the line of icons below the pull-downs) until the toolbar matches that shown in Fig. 15-26. This toolbar is optimized for using the MPLAB simulator and is the one that most people seem to use. The purposes of the different *Toolbar* icons are shown in Fig. 15-26. I have not labeled all of them, just the ones that will be of most use to you as you start working with the MPLAB IDE.

Figure 15-27 shows the desktop that I typically use to simulate applications. You might have some preferences that work better for you. By all means, set up the desktop as you prefer for your own applications. As I have noted, this is *my* preference and I use it to demonstrate the experiments in this chapter.

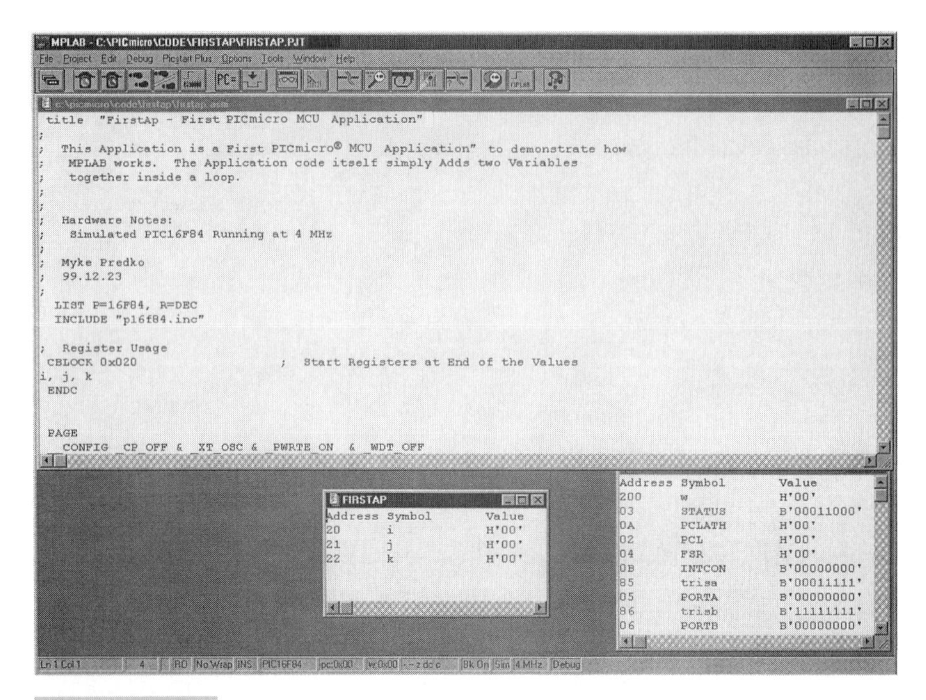


**Figure 15-25** MPLAB IDE desktop ready for simulation

With the MPLAB project set up, the source file loaded and assembled, and the watch windows specified and placed on the MPLAB IDE desktop, you are finally ready to start simulating your application. Click on the *Reset the Application* icon, which will reset the simulated PICmicro® MCU processor and highlight the first instruction in the application (at the processor reset, address zero), as shown in Fig. 15-27.

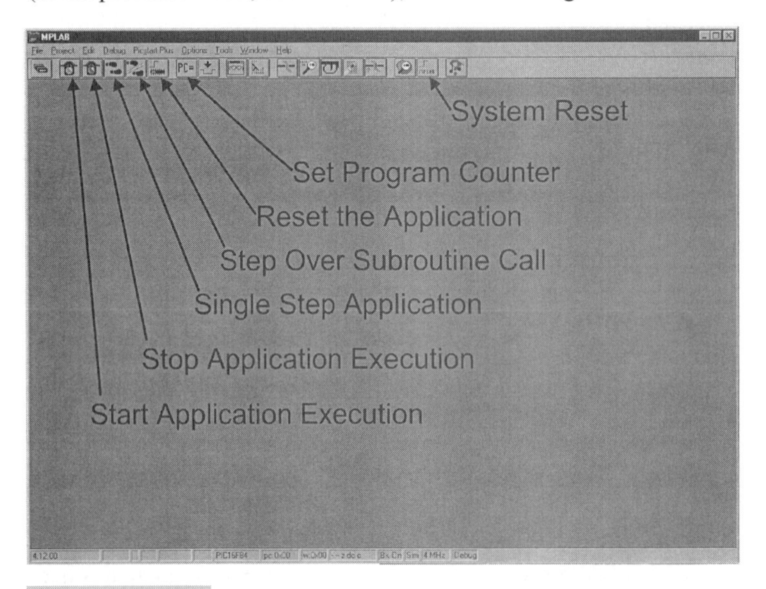


**Figure 15-26** MPLAB simulation buttons labeled

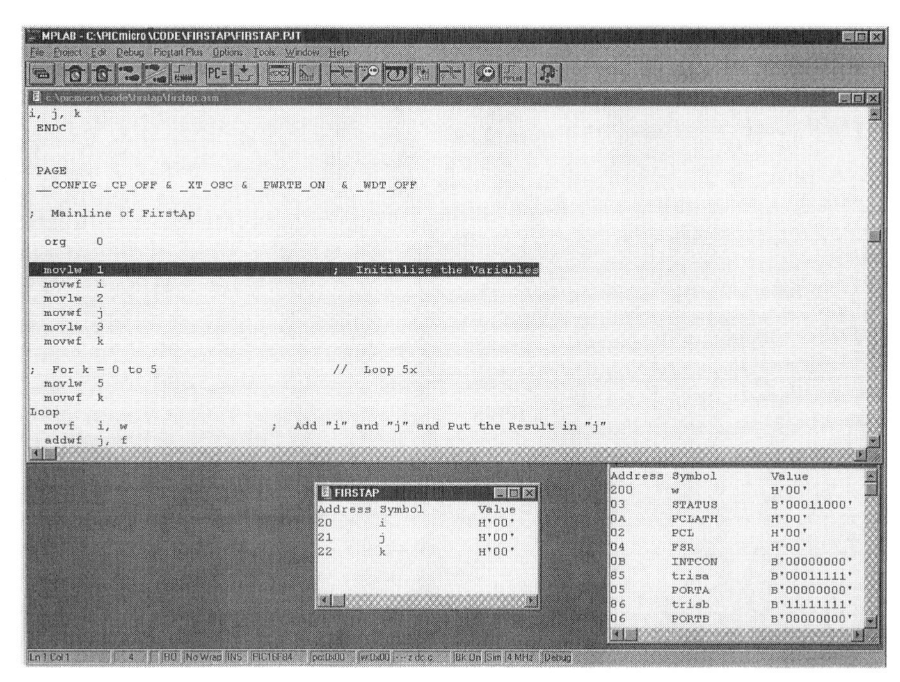


**Figure 15-27** The MPLAB IDE desktop after PICmicro® MCU reset

From here, you can start single stepping (by clicking on either one of the single-step icons on the toolbar). As you work through the application, you will see the values in the watch windows change with the execution of the code. For the first few single steps, the values of *i, j,* and *k* in the *FIRSTAP* watch window will be updated with their initial values. Next, *Loop* is entered and *j* will be incremented each time through (because *i* is equal to one when it is added to it). As you work through the loop, you will see *j* increase by one and *k* decrease by one until it is equal to zero. Then, execution "falls out" of the loop code and executes the endless loop (the *goto $* instruction).

Before single stepping through the loop code, I want to show one last feature of the MPLAB simulator, the breakpoint. Move the MPLAB editor cursor on the line that has the *goto $* and click on the right button. When you do this, you will get a pull-down menu (Fig. 15-28). Click on *Break Point(s)* and the pull down will disappear, but the *goto $* line will be highlighted. Now, click on the *Start Application Execution* toolbar icon and the simulated PICmicro® MCU will execute to the breakpoint that you just set.

When the breakpoint has been reached (Fig. 15-29), the loop code will have executed five times, *j* will be 7, and *k* will be 0—all as expected.

Despite the number of screen shots and the depth of information provided here, this explanation of the features of the MPLAB IDE is pretty cursory. As you work through the experiments in this chapter, you should get a better understanding of how the MPLAB simulator works. I also provide you with a better idea of what you can do to find problems with your applications in this easy-to-use interface before you program a PICmicro® MCU and can't figure out why your application doesn't work as expected.

You can speed up the MPLAB IDE by moving the mouse while the simulation is executing to a breakpoint. This is a known problem in the MPLAB IDE and might be fixed in

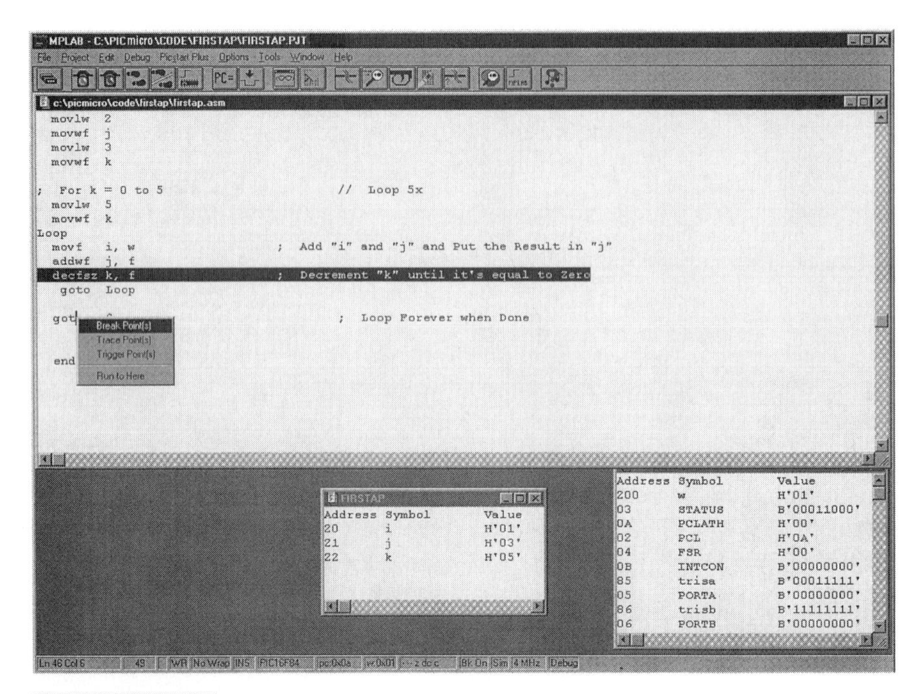


**Figure 15-28** Setting the MPLAB simulator breakpoint

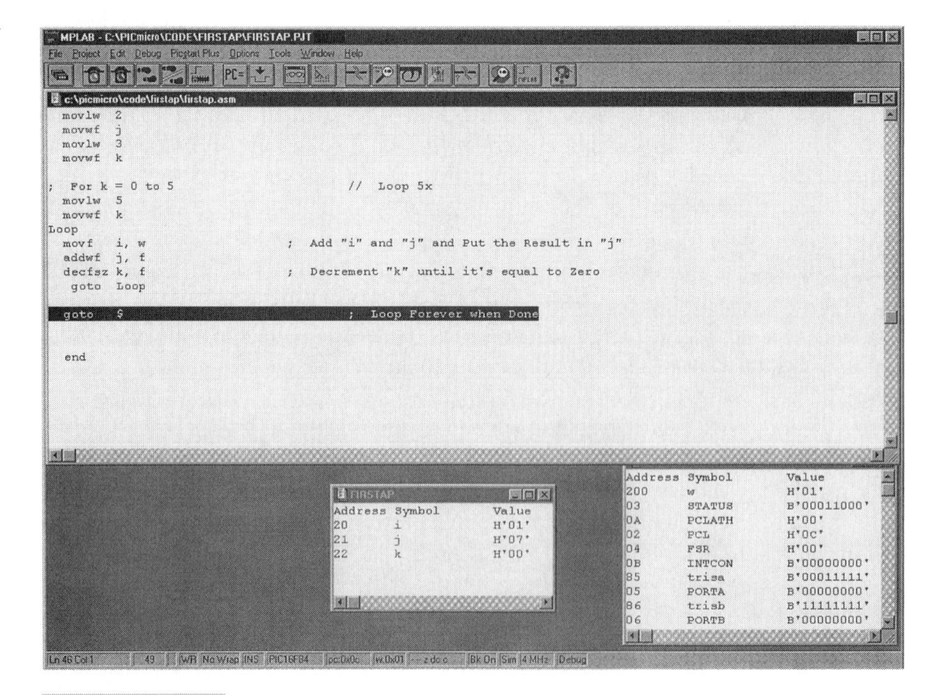


**Figure 15-29** FirstAp at endless-loop breakpoint

the version that comes with the book (MPLAB IDE version 5.11 is originally being shipped on the book's CD-ROM). If you have a copy of the MPLAB IDE version 5.11 (or earlier), you should move the mouse back and forth to speed up the execution of the application by 10 times (or more). Also some utilities that can be downloaded from the Internet speed up the MPLAB IDE and increase the keyboard/mouse return speed.

## REGADDR: REGISTER BANK ADDRESSING

Before the PICmicro® MCU can do anything really useful, you will have to understand how to access the TRIS and OPTION registers by using their direct write instructions or, for the mid-range PICmicro® MCUs, be able to change the register set to Bank1 and write to the registers from there. The RegAddr application shows how this can be done with an application that is used later in the chapter to turn on a simple LED. The source file can be found in the *\code\RegAddr* subdirectory of the *PICmicro* directory.

```
    title  "RegAddr - Turn on a LED"

  ;
  ;  This is the First Program to be Burned in and Run in a PICmicro® MCU.
  ;
  ;  The Program simply sets up Bit 0 of Port "A" to Output and then
  ;   Sets it Low.
  ;
  ;  Hardware Notes:
  ;   Reset is tied through a 4.7K Resistor to Vcc and PWRT is Enabled
  ;   A 220 Ohm Resistor and LED is attached to PORTB.0 and Vcc
  ;
  ;  Myke Predko
  ;
  ;  99.11.13 - Updated for MPLAB Operation
  ;
  ;  96.05.17 - Created
  ;
   LIST R=DEC
   INCLUDE "p16f84.inc"

  ;  Registers

  __CONFIG _CP_OFF & _WDT_OFF & _XT_OSC & _PWRTE_ON

   PAGE
  ;  Mainline of RegAddr

   org 0

   clrf   PORTB             ;  Clear all the Bits in Port "B"

   bsf    STATUS, RP0       ;  Have to go to Page 0 to set

                            ;  Port Direction
   bcf    TRISB ^ 0x080, 0  ;  Set RB0 to Output
   bcf    STATUS, RP0       ;  Go back to Page 0

 Finished
   goto  $

   end
```

When I created this application in the MPLAB IDE, I just used the *16F84. WAT* window so that I could monitor how the TRISB and PORTB registers are modified. Figure 15-30 shows the MPLAB IDE set up that I used.

Clearing the PORTB register is accomplished with the *clrf* instruction. The PORTB register is located in Bank 0 (which is what the PICmicro® MCU is using upon power up) and the TRISB register is located in Bank 1.

To access TRISB, the RP0 bit of the STATUS register must be set using the bit set instruction (*bsf*). Once this is done, only the registers in Bank 1 and the shadowed registers from Bank 0 can be accessed. Changing the access back to Bank 0 is accomplished by simply resetting the RP0 bit.

Rather than changing the bank select bit (RP0), the *tris* instruction can be used to set the TRISB register contents.

So, instead of:

```
bsf     STATUS, RP0        ;  Goto Bank 1
bcf     TRISB ^ 0x080, 0   ;  Make RB0 an Output Bit
bcf     STATUS, RP0
```

the following two instructions could be used:

```
movlw   0x0FE               ;  RB0 is Output/Everything else is input
tris    PORTB
```

Changing these three lines of code and assembling them, you will get the line:

```
"Warning 224: Use of this Instruction is Not Recommended"
```

in the ASSEMBLY STATUS register and it would point to the line the *tris PORTB* instruction statement is on. This warning indicates that using the *tris* (and *option*) instruction

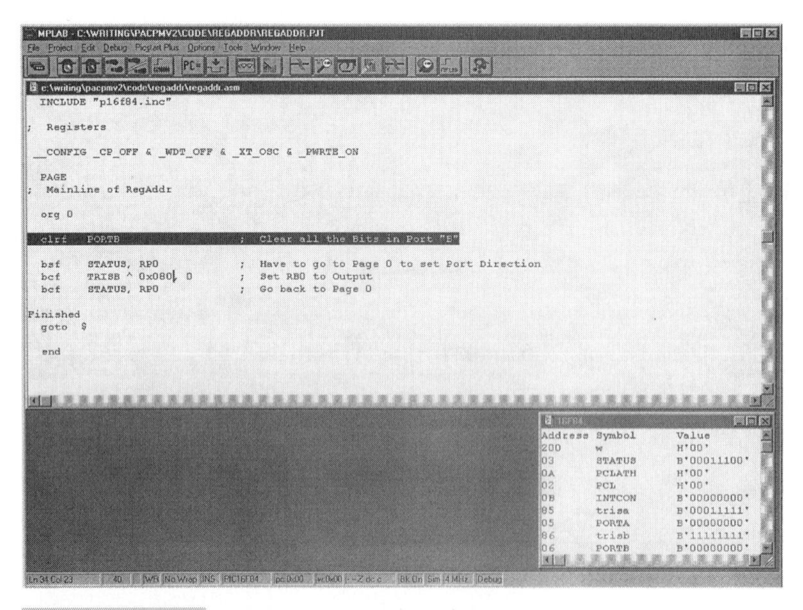


**Figure 15-30** **Screen shot of RegAddr project in operation with watch window**

is not recommended on mid-range PICmicro® MCUs. As stated elsewhere in the book, I also don't recommend the use of this instruction because it does not allow access to all the I/O registers (PORTD and PORTE) in some PICmicro® MCUs.

Even though PORTB will show all of its bits as *0* before TRISB.0 is set to *0,* they are actually in Input mode and are responding to the inputs that they are given. To change the input state, you would have to either create a stimulus file with all the bits set to *1* or use the *Asynchronous Stimulus* dialog to set the state or specific I/O pins. Changing the input values of simulated I/O pins in the MPLAB IDE is examined later in this chapter.

## STATUS: THE STATUS REGISTER

A good understanding of the Processor Status register flags (zero, carry, and digit carry) is crucial to being able to develop and debug applications. These registers are primarily used to provide a simple hardware interface to a previous operation. These bits are *positive active,* which means that the condition is true when they are set (equal to *1*).

The Zero flag is set when the result of an operation is zero. Actually, it might be more accurate to say that the Zero flag is set when the data leaving the ALU is equal to zero. I have put in this clarification because of the ubiquitous PICmicro® MCU zero register check *movf Reg, f*. In this instruction, the contents of the registers are run through the ALU and stored back into the register (Fig. 15-31).

The Carry flag is normally set/reset after an *addition/subtraction* or *rotate* instruction. For addition (as well as subtraction, which consists of "negative addition"), the Carry flag is set when the result is greater than 0x0FF.

Subtraction is another matter because of its operation. Remembering that subtraction is actually negated addition:

```
subwf Reg, w
```

is actually:

```
w = Reg - w
  = Reg + ( w ^ 0x0FF ) + 1
```

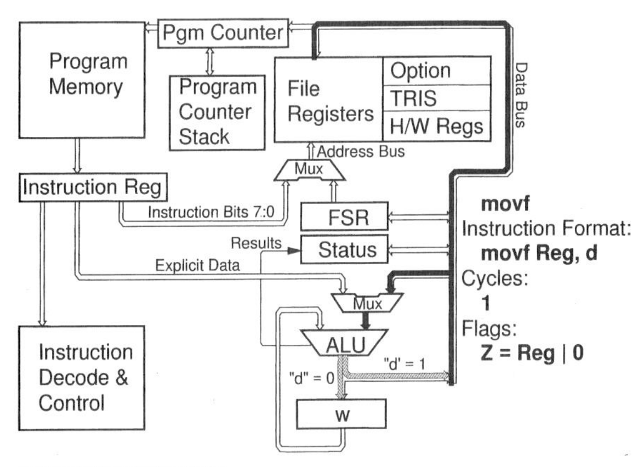


**Figure 15-31** "movf Reg, f" PICmicro® MCU instruction execution

You can see that the Carry flag is reset when *Reg* is less than w, but it is set when *Reg* is greater than or equal to *w*. This can be reasoned out by noting if *w* is greater than *Reg*, then its negative will be less than 0x0100 − *Reg*. If this is true, the sum of *Reg* and 0 − *w* will be less than 0x0100, (which is the first sum that will set the Carry flag).

It has been pointed out to me that the Carry flag in subtraction is really a Positive flag; it is set when the result is not less than zero. I find that changing the way I think about Carry when a subtraction instruction (*subwf, sublw,* or *addlw 0−*) is executed, makes predicting how the Carry flag will behave easier.

The Carry flag is also used integrally by the Rotate instructions (*rlf* and *rrf*). The instructions rotate the data from the Carry flag, through the register and back out to the Carry flag.

The Digit Carry flag works in exactly the same manner as the Carry flag, except that it only operates on the lowest four bits (also known as the least-significant nybble). The Digit Carry flag is only changed during addition/subtraction instructions (and not in rotates).

The program presented here (located in *\code\Status* subdirectory of the *PICmicro* directory) is designed to demonstrate how the flags work during various arithmetic operations.

```
 title  "Status - Showing How the Status Flags Work"
 ;
 ;  This Program plays around with the Execution Status Flags to
 ;   show how they are changed by different operations.
 ;
 ;  The Flags are:
 ;   STATUS Bit 0 = "C"  - The "Carry" Flag
 ;   STATUS Bit 1 = "DC" - The "Digit Carry" Flag
 ;   STATUS Bit 2 = "Z"  - The "Zero" Flag
 ;
 ;  Myke Predko
 ;  96.05.13
 ;
 ;  Updated:
 ;  99.11.01 - Corrected the "Clear Flags" Macro
 ;           - Updated with Additional Comments to Explain how the
 ;              Instructions Execute
 ;           - Updated with "Rotate" Instructions to Show how the
 ;              Carry Flag is used with them
 ;           - "nops" added after Status Flags Updates to allow
 ;              user to observe STATUS Flag States outside of
 ;              "ClearFlags" Macro
 ;
 ;  Hardware Notes:
 ;  Simulated 16F84
 ;  No I/O
 ;  Use "status.pjt" along with "status.wat"
 ;
 LIST R=DEC
 INCLUDE "p16f84.inc"

 ;  Macros
ClearFlags Macro        ;  Clear the Processor Status Flags
  movlw  0x0F8
  andwf  STATUS, w
  movwf  STATUS
 endm
 ;  Registers
```

```
 CBLOCK 0x020
i, j                    ;  Registers to Operate on
 ENDC

__CONFIG _CP_OFF & _WDT_OFF & _RC_OSC & PWRTE_ON

  PAGE
;  Mainline of Status

  org 0

  movlw  H'80'          ;  Set "i" to 0x080
  movwf  i
  movlw  8
  movwf  j

;  Clear the Status Flags before doing any operations

 ClearFlags

  movf   i, w           ;  Zero Flag not Set (w != 0)
  nop

  xorwf  i, w           ;  Zero Flag Set After Instruction
  nop                   ;   ((i ^ i) == 0)

  movf   i, w           ;  Add to Show Carry and Zero being Set
  addlw  1              ;   Add one - No Flags Set (w = 0x081)
  nop

  addwf  i, w           ;   ((0x081 + 0x080) == 0x0101)
  nop                   ;   0x0101 => CarrySet
                        ;   0x0101 => w = 1/Zero Reset

 ClearFlags

  rlf    i, w           ;  Load "Carry" with the Bit 7 of "i" and
                        ;   Shift i to the left by 1 (w == 0x000)
  rlf    j, w           ;  Shift j to the left by 1 (w == 0x010)
  nop                   ;   with carry inserted as the LSB (w == 0x011)

 ClearFlags

  movf   j, w           ;  Now, Show the Digit Carry Flag Being
  addwf  j, w           ;   Set ((0x008 + 0x008) == 0x010)
  nop

 ClearFlags

  movf   i, w           ;  Now, Set all three Status Bits
  subwf  i, w           ;   0x080 - 0x080  = 0x080 + (0x080 ^ 0x0FF) + 1
  movwf  i              ;                  = 0x080 + 0x07F + 1
  nop                   ;                  = 0x0100
                        ;  0x0100 => Carry Set
                        ;  0x0F + 1 => Digit Carry Set
                        ;  0x0100 =.> Zero Result in "w"/Zero Set
 ClearFlags

  movf   i, f           ;  Now, just show the zero flag set
  nop                   ;   ( i == 0)

Finished                     ;  Finished, Just Loop Around Forever
  goto   $

  end
```

When working with this application, I used the 16f84.WAT file for monitoring the w and STATUS registers. For *i* and *j*, I created a small, separate watch window (STATUS.WAT). The window that I used is shown in Fig. 15-32.

Throughout the program, you will see that I have put the three instruction statements:

```
movlw  0x0F8
andwf  STATUS, w
movwf  STATUS
```

into a macro. These three instructions will clear the PICmicro® MCU Processor Status flags before the next arithmetic operation to better illustrate how the flags are affected before each arithmetic operation.

When the first edition of this book came out, I cleared the STATUS registers using the two-instruction macro:

```
movlw  0x0F8
andwf  STATUS, f
```

This was in error because if the STATUS register in the PICmicro® MCU is used as the destination of an arithmetic or bit-wise operation, the bits will be set according to the result of the operation, not the actual result. To set the result, they have to be explicitly written to using the *movwf STATUS* or *bcf* and *bsf* instructions.

In the program, the first set of instructions is used to initialize the two variables (*i* and *j*) that are used in the program. After stepping through these instructions, you will see that the Zero flag is not changed.

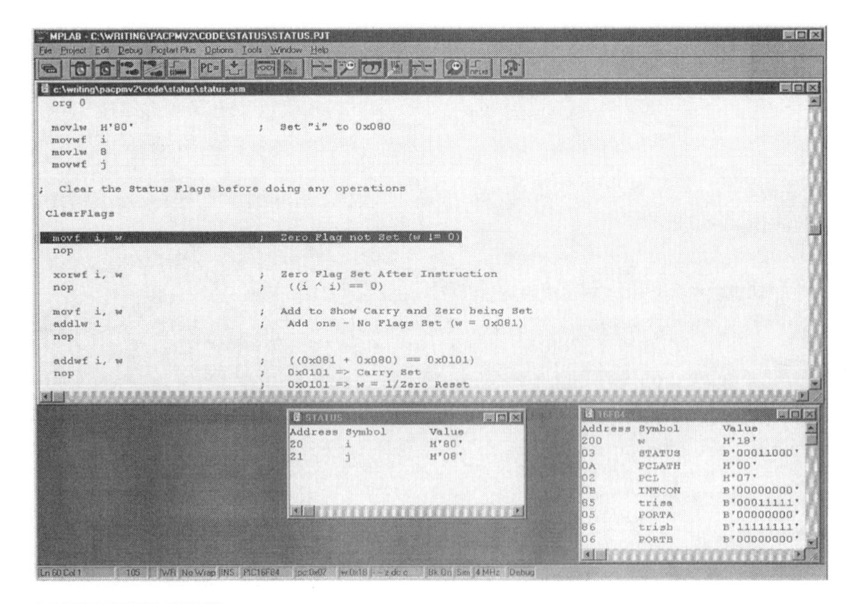


**Figure 15-32** Screen shot of status project in operation with watch windows

Either the STATUS.WAT watch window can be used along with the status project or you can monitor the Z DC C flags at the bottom line of the MPLAB IDE. On the MPLAB IDE's bottom line, these three bits are continually displayed with active flags set in upper case.

After the variables are initialized, the first operation is to load the w register with the contents in the register addressed at *i*. As noted, *i* is not an actual variable in the traditional programming sense; it is an address in the file register space. Once *w* is loaded, it is XORed with the same value to produce zero. After this operation, the Zero flag will be set and no other flags will be changed (set).

Next, the contents of the file register space at *i* are added to themselves after *i* has been incremented. Because *i* currently has 0x080, the result will be 0x0101. 0x0101 is greater than an eight-bit number and the setting of the Carry flag will signify the overflow. The least-significant eight bits are put into the result (the w register) of the *addwf* instruction. This means that one is put into the w register with the Zero flag reset at the same time. The Digit Carry flag is reset by this operation.

The operation of the Carry flag during shifting is then shown with the two *rlf* instructions. In the first *rlf* (*rlf i, w*), 0x080 is shifted to the left as it is loaded into w. The result stored in w is 0x000 because a ClearFlags macro was invoked before the *rlf* instruction and the reset carry is stored in the least-significant bit of w. Bit 7 of *i* (which was set), is moved into the Carry flag.

When the second *rlf* instruction is executed (*rlf j, w*), the original value 0x08 is shifted up by one and stored into the w register and the contents of carry are moved into the least-significant bit of the shifted value. After *rlf j, w* executes, w will be loaded with 0x011 because carry was set by the previous instruction. Because the most-significant bit of *j* was reset, the carry bit will be reset after this instruction.

Next, the value of *j* (8) is doubled in w by adding its value to itself. The result, 16 (or 0x010 in hex), has an overflow into the next highest nybble. When this happens, the Digit Carry flag is set. The Zero and Carry flags are not set because the result is not equal to zero and the sum is less than 0x0100.

The operation of loading *i* and then subtracting it from itself is going to take some explanation. When the *subwf* operation is executed, we have the following situation:

```
w = 0x080 - w
  = 0x080 - 0x080
```

But, as I've shown before, subtraction is actually addition of a negative. This means that the actual operation is addition of a negative, so the operation becomes:

```
w = 0x080 - (( 0x080 ^ 0x0FF ) + 1 )
  = 0x080 + ( 0x07F + 1 )
  = 0x080 + 0x07F + 1
  = 0x0100  (0x000 is actually stored in w)
```

You must be aware of three things happening here. First, the lower eight bits of the result are equal to zero (which means that the Zero flag is set). The result will be greater than 0x0FF, so the Carry flag will be set. And, if you look at the lowest nybble, 0x0F plus 1 is 0x010, so the Digit Carry flag will be set. Therefore, all of the PICmicro® MCU processor flags will be set after subtracting something from itself.

You might want to execute this operation over several times with different values to prove that this is true. Once this operation is completed, the result is stored in *i*.

Finally, the contents of the register at address *i* is run through the ALU to test for zero value. Because it is zero, the Zero flag will be set.

With this program, you should get a good idea of how the PICmicro® MCU processor flags work. It would be wise to play around with the different instructions and values and observe how the flags are affected. You might also want to modify it when you develop new applications on your own and are not sure how the flags will behave.

## ARITH: ARITHMETIC OPERATIONS

The purpose of the previous experiment was to introduce you to the STATUS register and how it is updated by the various instructions. This experiment goes a bit further and shows how the two arithmetic instructions (add and subtract) execute and affect the Status flags in different situations.

The source code for this experiment, ARITH.ASM can be found in the *code\Arith* subdirectory of the *PICmicro* directory. It uses the ClearFlags macro from the previous experiment to allow you to observe changes to the flags more clearly.

```
 title  "Arith - Showing How Arithmetic Operations Work"
;
;  This simply supplies an "addlw" and "sublw" instruction for
;   the user to observe how these operations execute.
;
;  Myke Predko
;  99.12.29
;
;  Hardware Notes:
;  Simulated 16F84
;  No I/O
;
 LIST R=DEC
 INCLUDE "p16f84.inc"

;  Macros
ClearFlags Macro             ;  Clear the Processor Status Flags
        movlw    0x0F8
        andwf    STATUS, w
        movwf    STATUS
        endm
 __CONFIG _CP_OFF & _WDT_OFF & _RC_OSC & _ PWRTE_ON

  PAGE
;  Mainline of Arith

  org 0
Loop

 ClearFlags

  movlw  0x012               ;  Load a Value in "w"

  nop                        ;  <--- Breakpoint Here
                             ;  <--- Change the contents of "w"
  addlw  0x088               ;  Perform the Addition Operation
  nop                        ;  <--- Breakpoint Here for STATUS Check
```

```
ClearFlags

 movlw  0x012              ;  Load a Value in "w"

 nop                       ;  <--- Breakpoint Here
                           ;  <--- Change the contents of "w"
 sublw  0x088              ;  Perform the Subtraction Operation

 nop                       ;  <--- Breakpoint Here for STATUS Check

 goto   Loop

 end
```

Using the procedures outlined earlier in this chapter, set up the Arith project and load the 16F84.WAT file to show the hardware registers inside the PICmicro® MCU. Of these registers, the only ones that we are really concerned with for this experiment are the w and STATUS registers.

To help show you how this works and allow you to experiment with the values passed to the add and subtract operations, when you set up your Arith project, I would like you to add the register Modify window as shown in Fig. 15-33. This window allows you to modify register contents directly without having to changing the source code. I use it fairly extensively in this experiment and you can use it in other experiments as well.

To add this window, left click on the *Window* pulldown and then click on the *Modify . . .* selection to put the window in the middle of the MPLAB IDE desktop. When this is done, move the *Modify* window to the bottom left corner of the desktop as in Fig. 15-33.

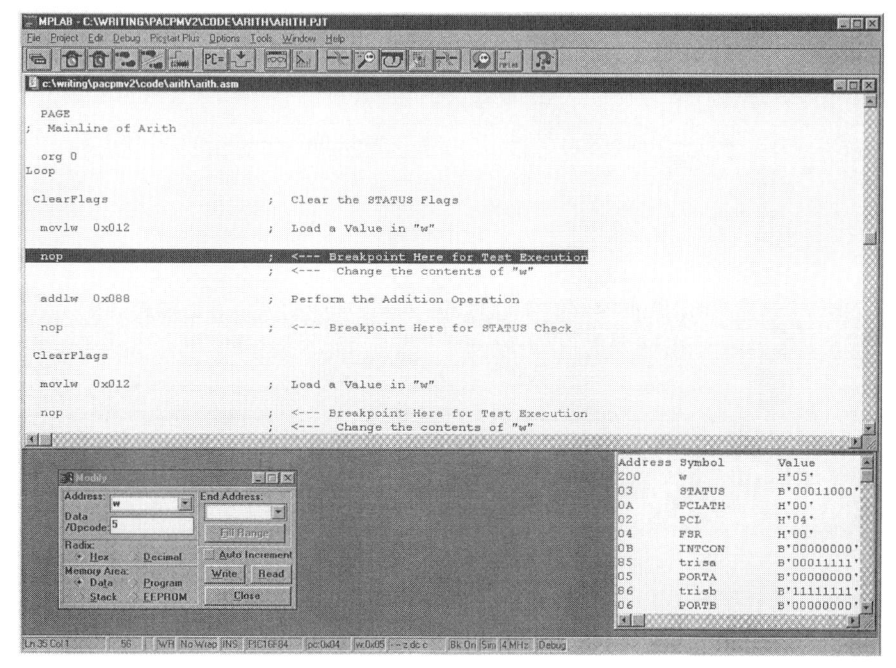


**Figure 15-33** The MPLAB IDE desktop for Arith application

Next, set a breakpoint at each of the four *nop* instructions in the ARITH.ASM file. This will force execution to stop and allow you to load new values into w or look at the STATUS bits as a result of the Add or Subtract operation.

When you get to the *nop* before the arithmetic statement, you can load it with a hex value using the Modify window to test out what will happen. For this experiment, enter *w* as the *Address* in the *Modify* window with the data being a hex value that you want to experiment with.

For example, if the first time through, you might want to see what happens when five is added to 0x088. To test this, execute (click on the *green light* or single step) to the first *nop* and then enter *w* into the *Address* box of the Modify window, put *5* into the *Data* box and click on *Write*. When this is done, the MPLAB IDE desktop will look like Fig. 15-33.

Executing to the next *nop* will add 5 to 0x088 and the result will be saved in w as 0x08D and none of the three execution Status flags (Zero, Carry, and Digit Carry) will be set. This can be repeated for different values.

The operation of addition is very straightforward. What I really wanted to do was focus on the subtraction instruction. *sublw* is quite useful, but I don't recommend that you use it until you are very familiar with the PICmicro® MCU architecture and the subtraction instructions.

In most processors, a subtract operation provides the parameter that is taken away from the contents of the accumulator. For example, the typical subtract instruction could be shown as:

```
sub    0x012                 ;  Accumulator = Accumulator - 0x012
```

In the PICmicro® MCU, the *sublw* (and the *subwf*) instruction provides the value that has the contents of the accumulator taken away from it. Using the format used for the typical subtract instruction, *sublw,* executes as:

```
sublw  0x012                 ;  Accumulator = 0x012 - Accumulator
```

This reverse operation can be a bit hard to understand and think of when you are writing your applications. I have seen people that program their PICmicro® MCU applications with adding negatives to avoid the difficulty of remembering to reverse the parameters for the subtraction operation.

The complexity of the *sublw* instruction is compounded by its adding of the negative. To save circuits within the PICmicro® MCU's ALU, the negative is calculated for the contents of the accumulator using the adder rather than implementing a separate subtractor:

```
Negative = 0 - Positive
         = 0 + Positive ^ 0x0FF + 1
```

So, the actual subtraction instruction is:

```
sublw   0x012        ;  Accumulator = 0x012 - Accumulator
                     ;              = 0x012 + Accumulator ^ 0x0FF + 1
```

This operation is simply not something you will remember easily nor be comfortable working with when you are first learning the PICmicro® MCU.

To see how the *sublw* instruction works, jump to the *nop* instruction before the *sublw* 0x088 and start changing the values in w subtracted from 0x088. The value 0x088 was chosen because the effects of *less than* or *greater than* values can be easily seen.

To help show what is happening, I created Table 15-2 to try out different values and show what is actually happening to get the calculated result for "sublw 0x088."

In this table, the basic operation of the Carry flag can be seen as being set if the result of the subtraction instruction is equal to or greater than zero. If you look down the "Operation" column, you can see that the result is actually greater than 0x0FF (255) and cannot be fully stored in the eight-bit destination (w for the *sublw* instruction). This should reinforce the notion that the Carry flag is a Positive flag after subtraction or the inverted *Borrow flag* that Microchip has recently describing the Carry flag after subtraction.

The Digit Carry flag's operation is somewhat more difficult to predict when you are working with subtraction operations. As a general rule, you can say that it will be set if the lower nybble result is also positive or zero. To demonstrate this, I created Table 15-3 to show how this works (and can be confirmed in the *Arith* application).

Subtraction, with the result saved in w, is often used for data comparison. By remembering the rules that if carry is set, then the result is not negative and that the value being subtracted is loaded into w first, you should not have any problems adding comparison code to your applications. An example of the comparison statement:

```
if (A >  7)
  A = A + 1;
```

could be implemented in PICmicro® MCU assembly as:

```
movlw  7              ;  Load "w" with the Value to be Subtracted
subwf  A, w           ;  Subtract the Test Value by "w"
btfss  STATUS, C      ;  If Carry is Set, Execute the Increment
 goto  $ + 2          ;  Carry Not Set, Skip Over the
incf   A, f           ;  Execute if Condition is True
```

**TABLE 15-2 Different PICmicro® MCU Subtraction Operations Showing Carry Flag Behavior for "sublw 0×088"**

| "W" | OPERATION | ZERO | CARRY | DC | ACTUAL OPERATION |
|---|---|---|---|---|---|
| 0x010 | 0x088 – 0x010<br>= 0x088 + 0x010<br>^ 0x0FF + 1<br>= 0x089 + 0x0EF<br>= 0x0178 | 0 | 1 | 1 | 0x088 – 0x010<br>= 0x078 |
| 0x088 | 0x088 – 0x088<br>= 0x088 + 0x088<br>^ 0x0FF + 1<br>= 0x089 + 0x077<br>= 0x0100 | 1 | 1 | 1 | 0x088 – 0x088<br>= 0x000 |
| 0x099 | 0x088 – 0x099<br>= 0x088 + 0x099<br>^ 0x0FF + 1<br>= 0x089 + 0x066<br>= 0x0EF | 0 | 0 | 0 | 0x088 – 0x099<br>= 0x0EF |

**TABLE 15-3 Different PICmicro® MCU Subtraction Operations Showing Digit Carry Flag Behavior for "sublw" 0×088**

| "W" | OPERATION | ZERO | CARRY | DC | ACTUAL OPERATION |
|---|---|---|---|---|---|
| 0x003 | 0x088 − 0x003<br>= 0x088 + 0x003<br>^ 0x0FF + 1<br>= 0x089 + 0x0FC<br>= 0x0185 | 0 | 1 | 1 | 0x088 − 0x003<br>= 0x085 |
| 0x008 | 0x088 − 0x008<br>= 0x088 + 0x008<br>^ 0x0FF + 1<br>= 0x089 + 0x0F7<br>= 0x0180 | 0 | 1 | 1 | 0x088 − 0x008<br>= 0x080 |
| 0x00B | 0x088 − 0x00B<br>= 0x088 + 0x00B<br>^ 0x0FF + 1<br>= 0x089 + 0x0F4<br>= 0x017D | 0 | 1 | 0 | 0x088 − 0x00B<br>= 0x07D |

or "optimized" to:

```
movlw  7                ;  Load "w" with the Value to be Subtracted
subwf  A, w             ;  Subtract the Test Value by "w"
btfsc  STATUS, C        ;  If Carry is Set, Execute the Increment
 incf  A, f             ;  Execute if Condition is True
```

The Arith experiment can be repeated for the other instructions in the PICmicro® MCU, including the bitwise logical operations and the *rotate* instructions. For the high-end PICmicro® MCUs, you might also want to repeat this experiment for the *multiplication* instruction to understand how it works.

# Jumping Around

Execution change is probably an area that you didn't expect that you would have to learn with the PICmicro® MCU. In most processors, the *jump* instructions and their operation is quite straightforward and usually do not require the additional support of the PICmicro® MCU. Most processors have conditional *jump* instructions that can execute anywhere in the processor's instruction space.

With the PICmicro® MCU, we are not quite that lucky. From the perspective of conditional jumping, there are no specific instructions for executing a jump based on status register contents, but there is the mechanism to conditionally execute based on the state of any bit in the PICmicro® MCU. The PICmicro® MCU's capabilities are actually quite a bit more powerful and flexible than what you would get with a processor with conditional jumps built in. The problem with them is that they are somewhat more difficult to learn to use effectively. The "inter-page" jumping is really not that unusual for processors. For example, the 8088 used in the IBM PC has short, intrasegment, and intersegment jumps—each one giving the application programmer different options for changing execution to a different location anywhere in the system's memory. In the PICmicro® MCU, writing jumps that go across page boundaries and require changes to PCLATH and the STATUS registers (depending on the processor architecture) are not that difficult, but some rules should be followed when doing them.

## MIDGOTO: MID-RANGE JUMPING BETWEEN PROGRAM MEMORY PAGES

You will have to consider jumping between program memory "pages" when your assembly-language code becomes larger than 2,048 (0x0800) instructions. In the mid-range PICmicro® MCUs, jumping between the pages is accomplished by setting the PCLATH register before executing a *goto* or *call* instruction or changing the PICmicro® MCU's program counter via the PCL register. This feature might seem intimidating when you first start working with it, but as you gain experience with the PICmicro® MCU, it will actually be quite easy to work through.

To show how jumping between program memory pages is accomplished, I created the MidGoto application, which is located in the *code\MidGoto* subdirectory of the *PICmicro* directory on your PC.

```
 title  "MidGoto - Low-End Jumping Around."
;
;  In this Application, Jumps Between Device Pages is
;   Demonstrated.
;
;
;  99.12.25 - Created for the Second Edition
;
;  Myke Predko
;
  LIST R=DEC
  INCLUDE "p16c73b.inc"     ;  <-- Note the Changed Processor
;  Registers

;  Macros
MyLGoto MACRO Label
  movlw   HIGH Label
  movwf   PCLATH
  goto    Label & 0x07FF     ;  Jump to Label Without Page Selections
 endm

 __CONFIG _CP_OFF & _WDT_OFF & _XT_OSC & _BODEN_OFF & _PWRTE_ON

  PAGE
;  Mainline of LowGoto

 org     0

   goto   Page0Label         ;  Goto an Address Within the Same Page

Page0Label                   ;  Label in Page 0
```

```
  MyLGoto Page1Label

 org     0x0800
Page1Label                    ;  Label in Page 1
  MyLGoto Page0Label

  end
```

This application first jumps to a label within the current page (0) and then does an interpage jump using the MyLGoto macro. This macro sets up the PCLATH register before executing the *goto* instruction.

To allow the interpage jumps to be meaningful, I wanted to use a PICmicro® MCU with more than one code page. One of the most basic devices for doing this is the 16C73B, which is a 28-pin mid-range PICmicro® MCU with a number of ADC inputs. For some of the experiments, you'll see that I change the PICmicro® MCU that is worked with to a different device.

In the mid-range (and higher-end) PICmicro® MCUs, the lower eight bits of the program counter are available in PCL and the upper bits use a PCLATH (and, optionally, a PCLATU) register. When a *goto* or *call* instruction is executed or PCL is updated, the contents of PCLATH are loaded into the high bits of the PICmicro® MCU's program counter, along with the new address.

To demonstrate how this works, after you have simulated the application and watched the PCLATH register change in the 16C73B watch window, comment out the line that updates PCLATH (*movwf PCLATH*) in the MyLGoto macro, rebuild, and then step through the code.

As you work through the code, you'll see that the label in the second page (Page1Label) is never reached and the application apparently is resetting itself (because execution jumps back to address 0 each time the MyLGoto Page1Label macro is encountered). This problem is quite common with people working with going between pages for the first time.

To avoid these problems, the MyLGoto macro will allow the operation to be completed properly, but with a few caveats. The first is: this macro only works for the *goto* instruction. If you want to have interpage *call* instructions, use the code I've presented elsewhere in this chapter as well as the rest of the book. Secondly, this macro should never be used with conditional *skip* instructions. As noted in the next experiment, only nonconditional jumps should be done between pages as you are learning the PICmicro® MCU. The third caveat is that you will get a Message if you are jumping from a page in which a PCLATH page-selection bit is set, to an address where the page-selection bit is reset. This isn't a big problem. I didn't want to add the code to eliminate this error because it would cause an incorrect address to be used with the *goto,* which could be confusing when an application is being debugged. Instead, the macro *gotos* the address within the destination page.

Last, this method is probably the most efficient and straightforward code in which to jump between pages. If you want to pass the contents of w to an address in another page (either via *goto* or *call*), instead of setting or resetting the PCLATH bits individually, you could use the macro:

```
MyLGotoSave Macro Label      ;  Jump to "Label" with contents of "w"
                             ;   unchanged
  movwf   _wGoto             ;  Temporarily Save the Contents of "w"
  movlw   HIGH Label         ;  Update PCLATH
```

```
  movwf  PCLATH
  movf   _wGoto              ;  Restore "w"
  goto   (Label & 0x07FF) | ($ & 0x01800)
 endm
```

This macro will change the value of the Zero flag in the STATUS register. Other than that, it can be used without modification and w will be passed correctly to the label. No "page jump" messages will be created by this macro, regardless of the source and destination address.

For the low-end PICmicro® MCUs, the page-selection bits are located in the STATUS register and, as shown in the next experiment, conditional assembly will have to be used to set and reset these bits.

## LOWGOTO: LOW-END JUMPING BETWEEN PROGRAM MEMORY PAGES

To round out the issues with regards to jumping around in the PICmicro® MCU, I thought I would present how it is done in the low-end PICmicro® MCUs. The low-end devices do not have a PCLATH register, and the page-selection bits are in the STATUS register, which has to be set or reset (according to the page that the address being jumped to). Changing the STATUS page-selection bits in the low-end devices is really only slightly different than doing it in the mid-range's PCLATH register writes, with the majority of the issues being in how the subroutine and table calls work.

Even though this experiment is the first low-end code-specific application that I have presented in this book, I don't think you'll see too many differences in how low-end PICmicro® MCU applications are written, compared to applications written for mid-range devices. This experiment (and the previous one) actually demonstrates the major differences between the two devices. Along with showing how the interpage jumps work in the low-end PICmicro® MCU processors, I also show you something else regarding how low-end PICmicro® MCU applications work.

The LowGoto application is located in the *code\LowGoto* subdirectory of the *PICmicro* directory and is very simple, with only seven instructions executed to jump within the initial page (page 0) and then jumping to and from page 1.

```
 title  "LowGoto - Low-End Jumping Around."
;
;  In this Application, Jumps Between Device Pages is
;   Demonstrated.
;
;
;  99.12.25 - Created for the Second Edition
;
;  Myke Predko
;
  LIST R=DEC
  INCLUDE "p16c5x.inc"         ;  <-- Note PICmicro MCU is PIC16C56

;  Registers

;  Macros
```

```
MyLGoto MACRO Label
 if ((Label & 0x0200) != 0)
  bsf    STATUS, PA0
 else
  bcf    STATUS, PA0
 endif
 if ((Label & 0x0400) != 0)
  bsf    STATUS, PA1
 else
  bcf    STATUS, PA1
 endif
  goto   Label & 0x01FF     ;  Jump to Label Without Page Selections
 endm

 __CONFIG _CP_OFF & _WDT_OFF & _XT_OSC

  PAGE
;  Mainline of LowGoto

 org     0
  goto   Page0Label         ;  Goto an Address Within the Same Page

Page0Label                  ;  Label in Page 0

  MyLGoto Page1Label

 org     0x0200
Page1Label                  ;  Label in Page 1
  lgoto   Page0Label

  end
```

In this application, I have changed the default processor from the PIC16F84 and I'm now using the PIC16F56. The PIC16C56 is a low-end PICmicro® MCU with 1,024 instructions of EPROM program memory. This is actually two instruction pages and works well for this experiment and for those that show how interpage calls and program counter updates work in the low-end devices.

In this experiment, first notice what happens when you click on the *PICmicro® MCU Reset* icon in the MPLAB IDE. If you have set up your LowGoto project as I have shown so far in the chapter, you'll see that the black "highlight" line is displayed on the first *goto Page0Label*. This is actually what you would expect, based on your experience with the PIC16F84 and other mid-range devices.

Even though this is what you are used to, it is not 100% correct. To show what I mean, add the Stopwatch to the application by clicking on *Window* and then *Stopwatch* . . . The timer should be put into the lower left corner of the MPLAB IDE desktop (Fig. 15-34).

The Stopwatch is a timer that can be added to your applications for seeing how many instructions operations require. I use it a lot in the later experiments, but I wanted to use it here to demonstrate what is happening with the low-end PICmicro® MCU's reset.

Notice in Fig. 15-34 that after clicking on the *Reset Processor* icon, in the MPLAB IDE toolbar, the number of cycles shown executed is *1*. This shouldn't seem right because before the first instruction is executed, the timer should be reset to 0.

The reason for this discrepancy is how reset works in the low-end PICmicro® MCU processors and how I have presented that they should be used. In the low-end devices, the reset vector is always the last address of program memory, not the first, as in the other

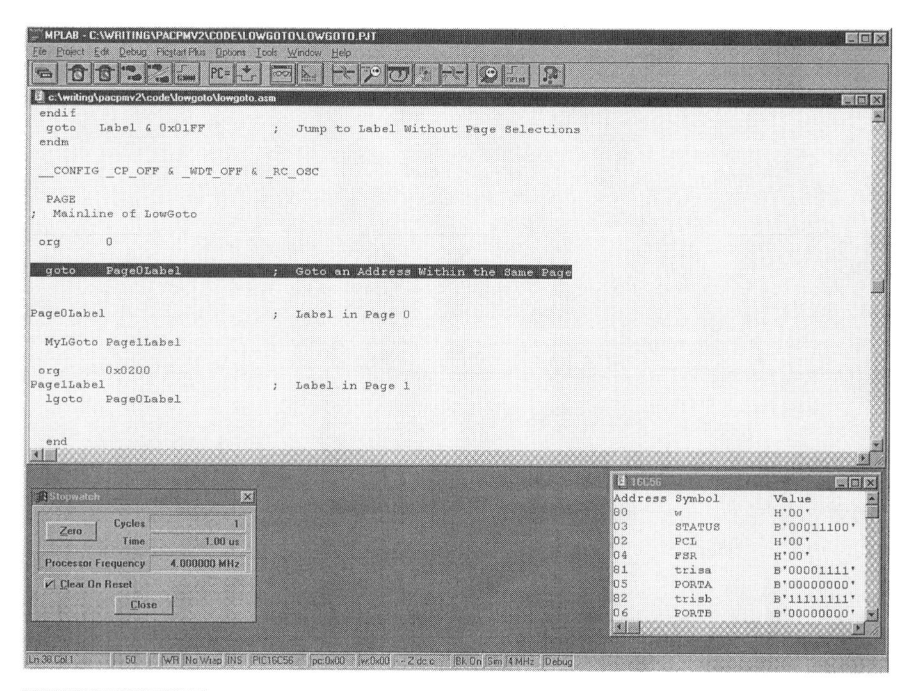


**Figure 15-34** LowGoto reset desktop

PICmicro® MCU processor families. To make the devices "appear" more common to the other processor families (with the mid-range family in particular), ignore the last instruction. Thus, when all the bits are set, the resulting instruction *xorlw 0x0FF,* inverts the bits in the w register. This is not an issue because the value in w is "undefined" at power up anyway.

If you are to bring up the *Program Memory* window by clicking on *Window, Program Memory,* and then the *Reset Processor* icon, you should see the desktop shown in Fig. 15-35. In this case, notice that the *Stopwatch* Window indicates that at reset, the instruction count is at *0,* and the program counter is pointing to the last instruction in program memory, which is just what is expected. Clicking on a single-step icon will cause execution to jump to address 0 (which has the first *goto* instruction) and the *StopWatch* window to increment to *1,* which is what was observed originally.

Having gone through what is happening at reset, the actual code itself should actually be pretty anti-climactic. Like MidGoto, the first instruction jumps to a label within the first page. From this label, it jumps to a label in the second page. Unlike MidGoto, the label in the second page uses the *lgoto* "pseudo-instruction" built into the MPLAB assembler instead of the MyLGoto macro. If you look at the *Absolute Listing* or the *Program Memory* Windows, you will see that the code created by MyLGoto and *lgoto* are identical.

For both cases, the *PA0* and *PA1* bits of the STATUS register are set according to where the destination label is located. The *PA0* is set according to the state of bit 9 of the destination address and *PA1* is set according to the state of bit 10 of the destination address. In the MyLGoto macro, instead of setting or resetting these bits according to the label address, I could have used the code:

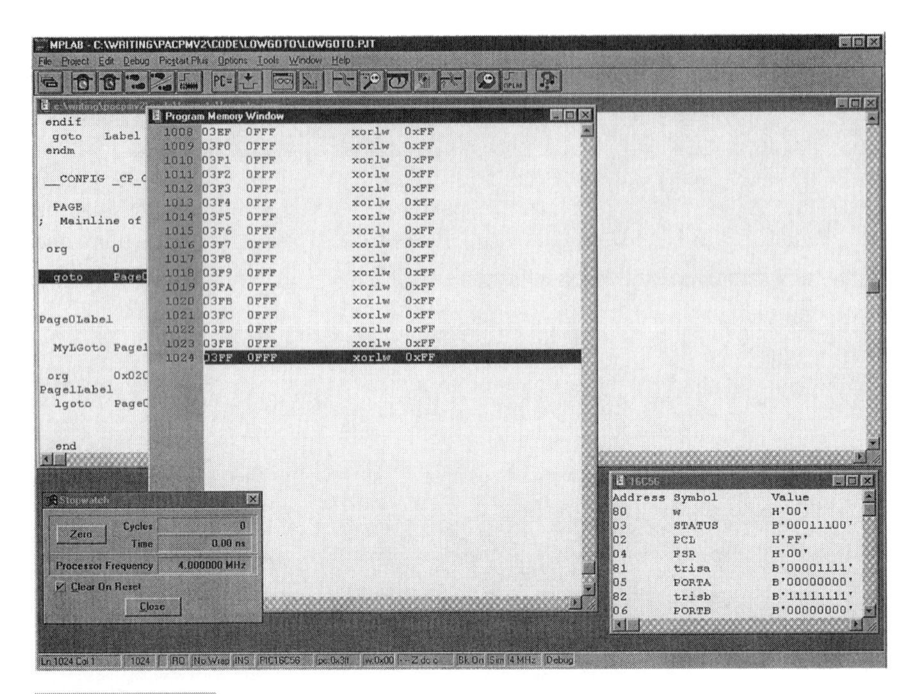


**Figure 15-35** LowGoto reset showing program memory

```
movlw   (1 << PA0) | (1 << PA1)
iorwf   STATUS, w                  ;  Set PA0 and PA1
andlw   ((HIGH Label) & 0x006) << 4) 10x09F
movwf   STATUS
goto    (Label & 0x01FF) | ($ & 0x0600)
                                   ;  Jump to the Address within the Bank
```

which loads in the contents of the STATUS register, sets *PA0* and *PA1,* and then loads them with the value directly from *Label.* After *PA0* and *PA1* are correct, I then jump to the address within the instruction bank. This method takes twice as many instructions as the method that I (and apparently Microchip) used to no advantage. I would also say that the method shown is suboptimal to the macro's version because it is also much harder to understand exactly what is happening.

Using *lgoto* is something that you might want to consider in your applications when you are definitely jumping between banks. The *lcall* pseudo-instruction is not recommended for use because it does not restore the state of the *PA0* and *PA1* bits (or PCLATH in the mid-range) after return. This is covered in more detail in a later experiment.

Also, the *lgoto* pseudo-instruction should not be used after a conditional *skip* instruction as adds extra instructions (which are not "seen" in the source code) that will affect the operation of *skip*. Ideally, conditional branching should not occur across pages.

## CONDJUMP: CONDITIONAL JUMPING

Conditional jumping in the PICmicro® MCU is carried out differently than in any other eight-bit processor that I know of. Instead of having instructions that are based on the state of one of the execution STATUS bits (zero, carry, and digit carry), the conditional instructions are based on the state of any bit accessible to the processor. This is quite a profound method of operation and one that you really have to sit down and think about because it can allow you to create applications that are startling in their efficiency and their ability to work through complex comparisons.

Elsewhere in the book, I have described the "skip on bit condition" instructions. This experiment demonstrates their use in a few cases. It also covers how operations can be written to avoid taking up extra cycles and instructions for operations that just require one instruction. The three conditional cases in CONDJUMP.ASM might seem to be quite limited in their scope, but used with the STATUS register experiments presented earlier in the chapter and with variable flags, it is really all you have to be aware of for most of your applications.

When I created the CondTest application, it was created as an assembly-language version of the C application:

```
Main()
{

int  i, j, k;
bit  TestFlag;

  i = 5;                        //  Initialize the Variables
  j = 7;
  TestFlag = 0;

  if (I < j)                    //  Set TestFlag Based on "i" and "j"
    TestFlag = 1;

  if (TestFlag == 1)            //  If "TestFlag" set, Jump
    goto LongJump;

  while (1 == 1);               //  Loop Forever

   :

LongJump:                       //  Label in another Page

  for (k = 0; k < j; k++ )      //  Repeatedly increment "I"
    i = i + 1;

  while (1 == 1);               //  Loop Forever

}
```

The CondJump code demonstrates an eight-bit comparison, a bit test, and a counting loop that you can use in your own applications. The source code can be found in *code\CondJump,* which is in the *PICmicro* directory of your PC. Notice that the code is written for a 16C73B, which has more than one page of instructions. This was done to show you how to implement interpage conditional jumps.

```
 title  "CondJump - Conditional Jumping"
;
;  This application shows how conditional jumping can be
;   implemented in a variety of different situations,
;   including interpage jumps.
;
;  Myke Predko
;  99.12.29
```

```
;
;  Hardware Notes:
;  Simulated 16C73B
;  No I/O
;
 LIST R=DEC
 INCLUDE "p16c73b.inc"      ;  <-- Note PIC16C74B

;  Registers
 CBLOCK 0x020
Flags
i, j
 ENDC
#define TestFlag Flags, 0  ;  Define a File Register Flag

 __CONFIG _CP_OFF & _WDT_OFF & _RC_OSC & _PWRTE_ON & RODEN_ON

  PAGE
;  Mainline of CondJump

  org 0

  clrf   Flags              ;  No Flag is Set

  movlw  5                  ;  Setup Test Variables
  movwf  i
  movlw  7
  movwf  j

;  if (i < j)
;   TestFlag = 1;

  movf   j, w               ;  Subtract "j" from "i"
  subwf  i, w               ;   And Look at Carry Result
  btfsc  STATUS, C
   goto  $ + 2
  bsf    TestFlag

;  if (TestFlag == 1)
;   goto LongJump

  movlw  HIGH LongJump      ;  Set up PCLATH for the Long Jump
  movwf  PCLATH
  btfsc  TestFlag
   goto  LongJump & 0x07FF
  movlw  HIGH $             ;  Restore PCLATH to the Current
  movwf  PCLATH             ;   Page

  goto   $                  ;  Endless Loop if TestFlag == 0

 org     0x0800
LongJump

;  for (j = 0; j < 7; j++ )
;    i = i + 1;

  incf   i, f               ;  Increment "i", "j" Times
  decfsz j, f
   goto  $ - 2

  goto   $

  end
```

To explain the code, I work through each test individually, looking back to the original C source and comparing it to the source code presented previously.

The variable initialization should not be of any surprise to you. In this code, I load *i* with 5 and *j* with 7 and clear the FLAGS register, clearing TestFlag at the same time.

The first comparison is executing the next instruction if *i* is less than *j*. This comparison was covered in previous experiments and in the text. It has the basic form:

```
  movf   Parameter1, w
  subwf  Parameter2, w
  btfss|c  STATUS, Z|C      ;  "Test" in Table Below
   goto  PastTrue
;  Code Executed if Condition is "True"
PastTrue
```

With *Parameter1, Parameter2,* and whether or not the Carry or Zero flags is specified according to the table below:

| CONDITION | PARAMETER1 | PARAMETER2 | TEST |
|---|---|---|---|
| A == B | A | B | btfss STATUS, Z |
| A == B | A | B | btfsc STATUS, Z |
| A > B | A | B | btfsc STATUS, C |
| A >= B | B | A | btfss STATUS, C |
| A < B | B | A | btfsc STATUS, C |
| A <= B | A | B | btfss STATUS, C |

In CondJump, I followed this format instruction explicitly, but I could have simplified it to:

```
  movf   j, w              ;  Subtract "j" from "i"
  subwf  i, w              ;   And Look at Carry Result
  btfss  STATUS, C
   bsf   TestFlag
```

because only one instruction is executed conditionally.

The next conditional jump is interesting for two reasons: jumping to another page and jumping based on the condition of the flag. In a traditional processor, a *jump on flag* condition would probably involve the operations:

```
  if (((Flag & (1 << Bit)) != 0)
    goto Label;
```

which, in PICmicro® MCU assembler, would look like:

```
  movf   Flags, w
  andlw  1 << Bit
  btfsc  STATUS, Z
   goto  Label
```

and is not that efficient compared to the code that can be generated. The PICmicro® MCU itself can access the bits in the processor, so the test on the bit can be done in one instruction and not by ANDing the test register with a mask value:

```
btfsc  TestFlag
 goto   Label
```

which is much simpler and faster.

This simple bit test and jump (or execute a single conditional instruction) can also be applied to the various hardware register bits in the PICmicro® MCU. In fact, when one of the processor STATUS condition bits is tested and a jump is made from its state, this is exactly what is happening. By eliminating the need to isolate a bit, as shown in the inefficient code, the PICmicro® MCU can carry out some very fast and efficient operations.

Interpage conditional jumps are somewhat difficult to conceptualize. Although I recommend avoiding them as much as possible, sometimes they have to be done. The format used here, where I initialize PCLATH before the test and jump and then reset it after the test and jump, seems to be the simplest way to do it. Because *movlw* and *movwf* do not change the processor STATUS registers, they can be used before the conditional *skip* instruction.

For the comparison operation that is the first conditional jump, the code could be changed to:

```
  movf   Parameter1, w
  subwf  Parameter2, w
  movlw  HIGH PastTrue
  movwf  PCLATH
  btfss|c STATUS, Z|C       ;  "Test" in Table Below
   goto  (PastTrue & 0x07FF) | ($ & 0x01800)
  movlw  HIGH $
  movwf  PCLATH
;  Code Executed if Condition is "True"
PastTrue
```

if the *Code Executed if Condition is "True"* is across a page boundary.

Another way of doing this is to jump "around" the *goto "PastTrue",* as in:

```
  movf   Parameter1, w
  subwf  Parameter2, w
  btfsc|s STATUS, Z|C        ;  "Test" in Table Above
   goto  $ + 4
  movlw  HIGH PastTrue
  movwf  PCLATH
  goto   PastTrue
;  Code Executed if Condition is "True"
PastTrue
```

But this code only saves one instruction and requires you to think through the negative condition to jump over. Personally, I prefer the first method that doesn't require any "negative" thinking to work through.

The last conditional operation is the *for* loop at the end of the application. In the PICmicro® MCU, this can most efficiently be done with the *decfsz* instruction, as is shown in CondJump, rather than something like:

```
  clrf   k                   ;  k = 0
```

```
ForLoop
  movf   j, w                 ;  is "k" == "j"?
  subwf  k, w
  btfsc  STATUS, Z
   goto  ForEnd

;  Execute Code Here

  incf   k, f                 ;  Increment Counter
  goto   ForLoop              ;  Repeat the Test
ForEnd
```

which actually performs the same function, but requires quite a few more instructions. To implement a loop for a set number of loops, the number is loaded into a variable. Then the *decfsz* instruction is used until it is equal to zero.

The only advantage I can see with using this format is if the intermediate values of the counter are required. The intermediate value in the *decfsz for* loop can be found using the instruction snippet:

```
  movlw  LoopNumber           ;  Final Value of "k"
  subwf  k, w
```

In these two instructions, the current value is subtracted from the number of loops (*LoopNumber*) to be executed. The result would be the same as *k* in the *forloop* analog code for a specific loop.

# Data

Data manipulation in the PICmicro® MCU is quite similar to that of other processors. This might be quite a surprise to you because of the use of the Harvard architecture in the PICmicro® MCU. Actually, the PICmicro® MCU has the same basic addressing modes (immediate, direct, and indexed) that you would find in other processors.

The following programs will give you an idea of how the PICmicro® MCU can access data. The applications also show how the PICmicro® MCU architecture works to actually require fewer instructions than other processors to carry out various tasks.

## VARMANI: VARIABLE MANIPULATION

The most basic form of addressing is by not specifying an address at all. Instead, just the value to use is provided as part of the instruction. This is known as *immediate addressing*. *Direct addressing* is when the address of the register to be used is actually part of the instruction. Both of these forms of addressing are specified explicitly in the program instructions. Even if the program or the data is changed, these values remain constant.

In direct-addressed instructions; the address in the instruction depends on the bank information in the STATUS register (*RP0* bit). Banks and registers are covered elsewhere and demonstrate to show how the bank select bits in the STATUS register are used to access the PORTB and TRISB registers. Both of these registers have the same address within the different banks.

VarMani carries out a number of immediate and direct instructions and can be found in the *\code\Varmani* subdirectory of the *PICmicro* directory.

```
 title  "VarMani - Variable Manipulation."
;
;  This Program is meant to be run in the Simulator to show how
;   Variables can be manipulated and accessed.
;
;  Myke Predko
;  96.05.08
;
;  99.11.01 - Updated for MPLAB
;
;  Hardware Notes
;  Simulated 16F84
;  No I/O
;  Use "varmani.pjt" along with "varmani.wat"
;
  LIST R=DEC
  INCLUDE "p16f84.inc"

;  Registers
 CBLOCK 0x020
i, j
 ENDC

 __CONFIG _CP_OFF & _WDT_OFF & _RC_OSC & _PWRTE_ON

  PAGE
;  Mainline of VarMani

  org 0

  movlw  3                    ;  Initialized the Variables
  movwf  i
  movlw  5
  movwf  j

;  i = i + 1  -  Show how Register is Incremented, Return Result

  incf   i, f              ;  Increment and Store the Result

;  j = i + 2  -  Now, Add two to a Register and Store it in another one

  movlw  2                    ;  Get the Value to Add
  addwf  i, w                 ;  Add the Register to it
  movwf  j                    ;  Store the Added Value
  movf   i, w                 ;  Now, do the Addition a different way
  addlw  2                    ;  Load then Add, opposite to above
  movwf  j

;  if ( i == j ) then i = i + 1  -  Do a simple Comparison with an Add

  movf   i, w                 ;  Subtract i from j and if the result
  subwf  j, w                 ;   is equal to zero, increment i
  btfsc  STATUS, Z
   incf  i, f
```

```
Finished                    ;  Finished, Just Loop Around Forever
  goto  $

  end
```

After assembling the code, you can begin to single step through the program. In Fig. 15-36, you can see that I used the processor-specific watch window and a variable watch window to observe what happens as the application executes.

The first block of code is the file register initialization. To initialize a register, w is first loaded with the initial value to be put into the file register (using the *movlw* instruction). The contents of w are then put into the file register (using the *movwf* instruction). If a register is to be initialized to zero, a *clrf* instruction can be used.

In some processors that have a common program and variable memory (such as the Intel 8086), variables can be saved as initialized variables and set to specific values when the application is loaded into memory. The PICmicro® MCU cannot take advantage of this for two reasons. First, the program and variable memory is separate; the program memory contents can in no way influence the variable data except by excuting instructions. Second, the application is "burned" into the device. If the contents of shared program and variable memory were changed by the application, the value will be different the next time it runs.

Next, incrementing a register is demonstrated. As can be seen, the statement:

```
i = i + 1
```

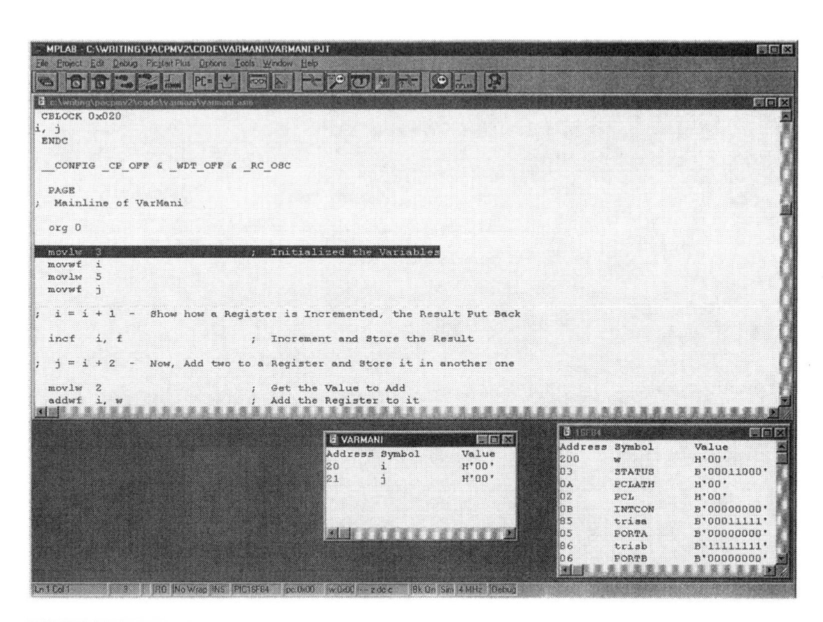


**Figure 15-36** Screen shot of Varmani project in operation with watch windows

can be neatly reduced to:

```
incf i, f
```

Many processors have instructions that can carry out a simple increment. What makes the PICmicro® MCU special is when the original high-level statement puts the result back into one of the parameter variables.

For example:

```
j = j + 3
```

In a traditional processor (e.g., the Motorola 68HC11) the following code would be required:

```
ldaa   j          ;  Load accumulator with the contents of "i"
adda   3          ;  Add three to the value
staa   j          ;  Store the result of the operation
```

In the PICmicro® MCU, the equivalent series of instructions is:

```
movlw   3         ;  Get the value to Add
addwf   j, f      ;  Add the contents of "w" to the value in
                     "j"
```

Along with the PICmicro® MCU code taking fewer instruction addresses than the 'HC11 (two versus eight for the three instructions), the PICmicro® MCU also only requires eight clock cycles versus the 'HC11's 32 (based on four *E-Clocks* for every address access cycle).

The next operation ($j = i + 2$) is carried out in two different ways in the PICmicro® MCU code. After single stepping through each of them, you will see that both get the same result. The first method adds $i$ to an immediate value and the second adds an immediate value to the w register (which was loaded with the value at address $i$). Either method is valid and neither has advantages over the other.

The last block of code is used to simulate a short block of high-level conditional code. The snippet of code compares two values and if they are equal (subtracting one from the other results in zero), one of the two values is incremented.

The comparison (subtraction) is carried out and the result is put into w, so neither source variable is changed by the operation. The results of the comparison are stored in the STATUS register (Zero, Carry, and Digit Carry flags). The zero flag is checked for zero (both values being equal). If they aren't, the next instruction is skipped (the variable increment).

Now, if the high-level statement executed on the condition was more than one instruction long, the condition check and skip code would change to:

```
movf    i, w           ;  t one value for the comparison
subwf   j, w           ;  Compare it to the other, Result in "w"
btfss   STATUS, Z
 goto   IF_Skip        ;  Else, Skip over the conditional code
  .
  :                    ;  Code to be executed if "i" == "j"
IF_Skip
```

## VARARRAY: SIMULATING A VARIABLE ARRAY

One of the most useful data constructs is the array. In PICmicro® MCU assembly-language terms, an array can be thought of as "indexed addressing." The FSR register is used to access a software-specified address (i.e., an address loaded and manipulated in the FSR by software during execution time). Indexed addressing differs from immediate or direct addressing in that the address is dynamically created during program execution.

The VarArray application (in the *code/VarArray* subdirectory of the CD-ROM or the *PICmicro* directory of your PC), shows examples of how array reads and writes can be accomplished.

```
 title  "VarArray - Simulating an Array."
;
;  This Program shows how a single-dimensional array can be implemented
;   on a PICmicro® MCU using the "FSR" and "INDF" Registers.
;
;  99.11.13 - Updated for Second Edition
;
;  Myke Predko
;  96.05.10 - Original Version
;
  LIST R=DEC
  INCLUDE "p16f84.inc"

;  Registers
i     equ 12
Array equ 13                    ;  Four Bytes of an Array

 __CONFIG _CP_OFF & _WDT_OFF & _RC_OSC & _PWRTE_ON

  PAGE
;  Mainline of VarArray

  org 0

  movlw  3                 ;  Initialized the Variables
  movwf  i
  movlw  Array             ;  Set up the Array Pointer to Start
  movwf  FSR

  movlw  'm'               ;  Initialize the Array
  movwf  Array             ;  Put in "myke".
  movlw  'y'
  movwf  Array + 1
  movlw  'k'
  movwf  Array + 2
  movlw  'e'
  movwf  Array + 3

;  i = Array[ 2 ] - Get the third element in the Array

  movlw  2                 ;  Move the Array to the Character
  addwf  FSR, f

  movf   INDF, w           ;  Get the Character at third Element
  movwf  i                 ;   and Store it

  movlw  2                 ;  Restore Pointer to the start of the
  subwf  FSR, f            ;   Array

;  Array[ 0 ] = 'M' - Change the high byte of the Array
```

```
    movlw  0                      ;  Move the Array to the Character
    addwf  FSR, f

    movlw  'M'                    ;  Store the New Value
    movwf  INDF

    movlw  0                      ;  Restore Pointer to the start of the
    subwf  FSR, f                 ;   Array

  Finished                        ;  Finished, Just Loop Around Forever
    goto  $
    end
```

The first part of the program is (as always) the variable initialization. Figure 15-37 shows how I set up the VarArray project with two watch windows, the 16F84.WAT, for the basic processor registers and the VARARRAY.WAT window, which shows the array and the *i* variable.

When I set up the *Array* line in the VARARRAY.WAT file (after selecting the *Window -> New Watch Window*), I set the *Properties* to *ASCII* for 32 bits (Fig. 15-38). The data can only be specified as displayed with the most-significant byte first, which means that the bytes in the array is displayed "backwards."

Once the array is defined in RAM, the general form to access an element of the array is:

```
    movlw   element#              ;  Get the Address of the Element to be
    addwf   FSR, f                ;   Accessed

    read/write Array element      ;  Access the Array Element

    movlw   element#              ;  Restore Array Pointer (FSR) back to
    subwf   FSR, f                ;   the Start of the Array
```

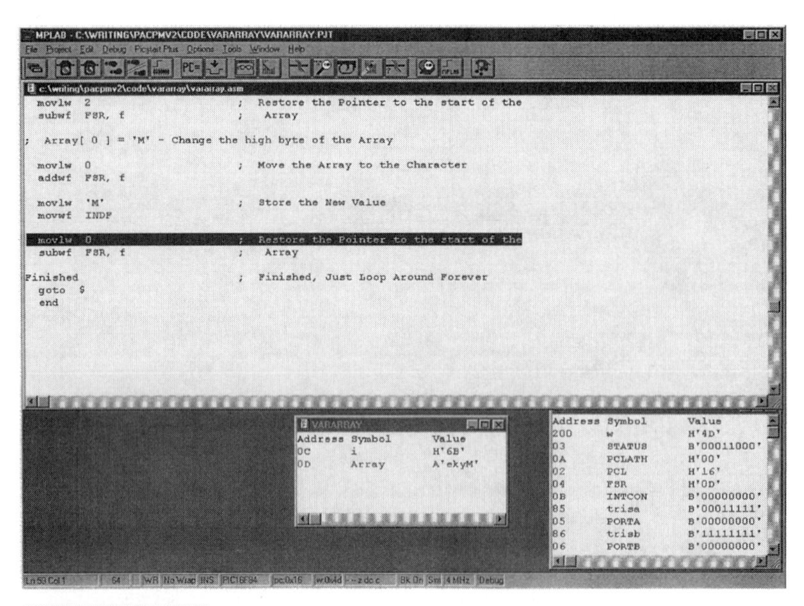


**Figure 15-37** **Screen shot of VarArray project in operation with watch windows**

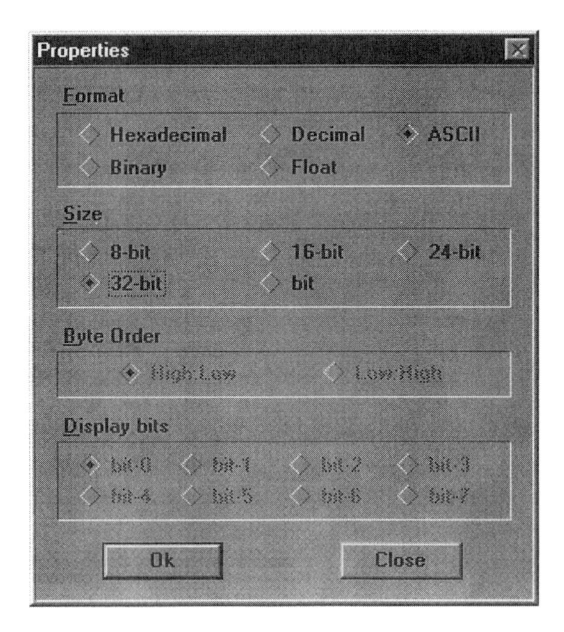


**Figure 15-38** **Screen shot of setting the Array properties for the VarArray project**

Before and after the array element access, the FSR register is kept at the start of the array so that the program has a constant starting point. Notice that in the second access (writing *M* to the first element of the array) has an add and subtract of zero to the FSR. Obviously, this could have been optimized (removal of the *movlw/addwf* and *movlw/subwf* statements), but all the instructions were left in to show that the first element of an array is given the index of zero.

I always use a first array index of zero instead of one. Even though in cases like the second access, where I add and subtract zero from the index register (FSR) to access the first element, I feel that doing this is more efficient and consistent with other basic programming operations.

The index into the array can also be another variable (not a constant, as shown in VarArray).

For example, if you wanted to create the assembler code for:

```
k = Array[ i ];
```

the following assembler would be used (it is assumed that the FSR is already loaded with the address of the start of *Array*):

```
movf    i, w            ;  Update the FSR with the index "i"
addwf   FSR, f

movf    INDF, w         ;  Read and save the Element j[ i ]
movwf   k

movf    i, w            ;  Restore FSR to Point to Start of "j"
subwf   FSR, f
```

Multidimensional arrays are best created with the various dimensions as a multiple of two. This isn't an absolute requirement, but it will make coding the array easier.

For example, the PICmicro® MCU assembly code for:

```
j = Array[ i ][ 2 ]
```

where *Array* is a 4 × 4 two-dimensional array, would be:

```
bcf     STATUS, C           ;  Get i x 4 as the first Dimension Value
rlf     i, w
movwf   ArrayTemp
bcf     STATUS, C
rlf     ArrayTemp, f

movlw   2                   ;  Get the 2nd Dimensional Offset
addwf   ArrayTemp, f        ;    And add it to the First Dimension

movf    ArrayTemp, w        ;  Add Total Offset to the Current Array
addwf   FSR, f              ;   Pointer

movf    INDF, w             ;  Get Byte Pointed at Array[ i ][ 2 ]
movwf   j

movf    ArrayTemp, w        ;  Restore FSR to its previous value
subwf   FSR, f
```

When you are using arrays indexed by variables, be sure that all indices are always within the array size limits. Your software must ensure that the FSR never points outside of the array boundaries.

## STACKOPS: SIMULATING A STACK FOR ARITHMETIC OPERATIONS

When a high-level language converts a complex arithmetic expression, i.e.:

```
i = ( j + 2 ) + ( k * 3 )
```

in PICmicro® MCU assembly language, it can stack and execute the instructions in a "push/pop" format, such as:

```
push    j                   ;  j + 2
push    2
add

push    k                   ;  k * 3
push    3
mul

add                         ;  ( j + 2 ) + ( k * 3 )

pop     i                   ;  i = ( j + 2 ) + ( k * 3 )
```

The actual stack operations are usually inserted in the PICmicro® MCU assembly code as macros.

The PICmicro® MCU doesn't have a traditional stack pointer along with push/pop operations, but they can be simulated easily using the StackOps application, which is located in the *code\StackOps* subdirectory or in the *PICmicro* subdirectory.

```
 title  "StackOps - Using a Stack during Arithmetic Calculations"
;
;  This Program shows how to use a stack for saving intermediate values
;   during complex mathematical operations.
;
;  99.11.13 - Updated for Second Edition
;           - Deleted Table in Original
;
;  Myke Predko
;  96.05.10
;
  LIST R=DEC
  INCLUDE "p16f84.inc"

;  Registers
a        equ 12              ;  Registers to Operate on
bi       equ 13              ;  "b" is an MPASM pseudo-Op
c        equ 14
d        equ 15
Stack    equ 16              ;  Stack for Operations (two bytes)

 __CONFIG _CP_OFF & _WDT_OFF & _RC_OSC & _PWRTE_ON

  PAGE
;  Mainline of StackOps

  org 0

  movlw  3                   ;  Initialized the Variables
  movwf  a
  movlw  5
  movwf  bi
  movlw  7
  movwf  c
  movlw  9
  movwf  d

  movlw  Stack               ;  Set up the Stack Pointer
  movwf  FSR

;  d = ((( a << 1 ) + bi ) << 1 ) + (( a >> 1 ) - c )

  rlf    a, w                ;  Get values in the 1st set of
  addwf  bi, w               ; Brackets
  movwf  INDF
  rlf    INDF, f
  incf   FSR, f

  rrf    a, w                ;  Get Values in 2nd set of Brackets
  movwf  INDF                ;  Do the Temporary Storage
  movf   c, w                ;  Subtract "c" from the Shifted Value
  subwf  INDF, w

  decf   FSR, f              ;  Finally, Add the Two Values together
  addwf  INDF, w             ;   and Store
  movwf  d

;  c = (( a << 1 ) - ( bi - d )) + ( 37 + d )

  movf   a, w                ;  Get the First Value
  movwf  INDF
  rlf    INDF, f             ;  #### - See Text for Comments about
                             ;   this Line
```

```
    incf    FSR, f

    movf    d, w            ;  Now, do the next Value
    subwf   bi, w           ;  Do the Subtraction
    movwf   INDF            ;  Store on the Stack
    incf    FSR, f

    movlw   37              ;  Get the Last Value First
    addwf   d, w
    movwf   INDF
    incf    FSR, f

    decf    FSR, f          ;  With the values in Parenthesis,
                            ;   Put them together
    decf    FSR, f          ;  Get Subtract Value First
    movf    INDF, w
    incf    FSR, f          ;  Subtract it From the Last Value
    subwf   INDF, w
    decf    FSR, f          ;  Add the Result to the First Value
    decf    FSR, f
    addwf   INDF, w
    movwf   c               ;  Store the Result

  Finished                  ;  Finished, Just Loop Around Forever
    goto    $

    end
```

When I created the project, I created the StackOps watch window, which shows the values for the four variables and the stack (Fig. 15-39). You can see that I was able to display the stack contents in hex format using a 32-bit variable with the *Low:High* parameter that I wasn't able to use with the ASCII array variable of the VarArray project.

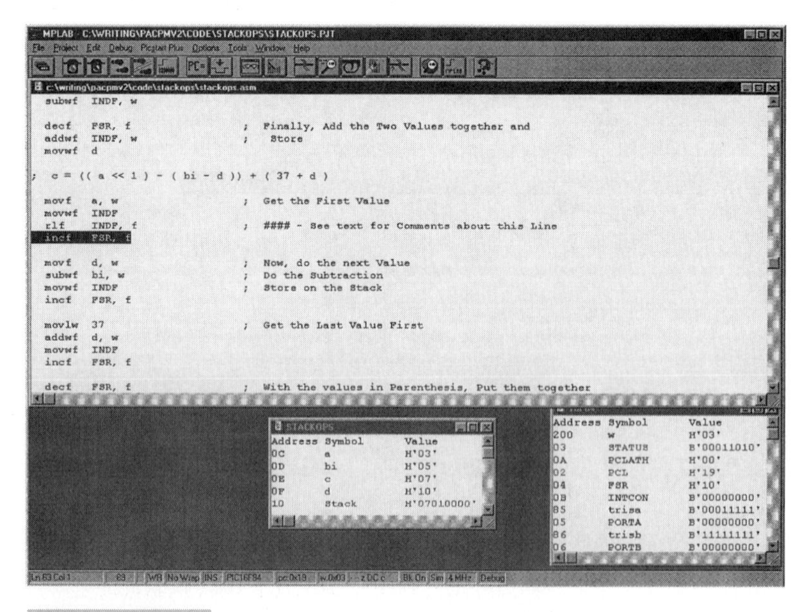


**Figure 15-39** Screen shot of setting the StackOps project with watch windows

Looking over StackOps, you might feel the code is needlessly complex (and I'd have to agree with you in regard to this application). Stack-based operations are important in applications written in high-level languages with many complex operations or different data formats (e.g., 8-, 16-, 32-bit, or floating-point numbers), which need multiple temporary storage registers.

The very complex mathematical case using a stack for arithmetic operations has advantages because much fewer temporary file registers will be required—only what is required will be used. For simpler arithmetic calculations, a stack is not the best way to go.

Although most compilers must be able to handle very complex operations, they should also be able to optimize statements like:

```
j = j + 1
```

into:

```
 incf     j, f
```

rather than process through:

```
  push   j
  push   1
  add
  pop    j
```

This code has a problem that you should have noticed when you simulated the application. On line 62 (the one with the comment *#### - See Text for Comments on this Line*), the rotate left instruction (*rlf*) is executing with carry set, which means that the least-significant bit of the result is set. If this were a rotate right instruction (*rrf*), then the most-significant bit would be set.

In most high-level language applications, this is not correct because when shifting values, you are typically multiplying or dividing by two. In this case, the result will not be exactly what you are expecting. Always be sure that you clear the Carry flag (using the *bcf STATUS, C* instruction) before executing a rotate instruction unless the Carry flag is loaded with a value that you want to be shifted into the result.

# Subroutines

A basic feature of virtually all processor and programming languages is the ability to jump to a routine after saving the address execution can return to after the routine has completed. This rather cumbersome statement is designed to encompass all of the different ways that subroutines are implemented in different devices. Although the PICmicro® MCU is quite conventional in its operation, some other devices do not have *call* and *return* instructions. Instead, the *goto* instruction saves the return address in a register and the application can choose whether or not to return to it.

I have not included any experiments with subroutine calls across pages because this is demonstrated earlier in the chapter with regard to jumping between pages. However, there are a couple of comments that I do want to make with respect to interpage calls. First, don't use the *LCALL* special instruction provided by Microchip in the MPASM assembler. Instead, you can use something like my LMCALL and LLCALL macros, which modify the processor page-selection bits before and after the call to a subroutine.

The mid-range subroutine LMCALL macro updates the PCLATH register before and after a call to ensure that PCLATH is correct for the new address and afterward, regardless of where execution is currently located, the high-order address bits are set correctly as well. The mid-range LMCALL macro subroutine call is:

```
LMCALL macro Label
 variable mask=0

 if (((Label & 0x0800) != 0) & (($ & 0x0800) == 0))
  bsf     PCLATH, 3
mask = mask | 0x0800
 else
 if (((Label & 0x0800) == 0) & (($ & 0x0800) != 0))
  bcf     PCLATH, 3
mask = mask | 0x0800
 else
  nop
 endif
 endif

 if (((Label & 0x01000) != 0) & (($ & 0x01000) == 0))
  bsf     PCLATH, 4
mask = mask | 0x01000
 else
 if (((Label & 0x01000) == 0) & (($ & 0x01000) != 0))
  bcf     PCLATH, 4
mask = mask | 0x01000
 else
  nop
 endif
 endif

  call   Label ^ mask

 if (((Label & 0x0800) != 0) & (($ & 0x0800) == 0))
  bcf     PCLATH, 3
 else
 if (((Label & 0x0800) == 0) & (($ & 0x0800) != 0))
  bsf     PCLATH, 3
 else
  nop
 endif
 endif

 if (((Label & 0x01000) != 0) & (($ & 0x01000) == 0))
  bcf     PCLATH, 4
 else
 if (((Label & 0x01000) == 0) & (($ & 0x01000) != 0))
  bsf     PCLATH, 4
 else
  nop
 endif
 endif

 ENDM
```

and the low-end LLCALL macro subroutine call is located in the *code\macros* subdirectory of the *PICmicro* subdirectory.

If you are going to use subroutines in different pages of a PICmicro® MCU, you should be sure that none of the code within the subroutine is in different pages (i.e., the subroutine "straddles" a page boundary). It's easy to have this happen if you have a large application that is larger than one page. The assembler will give you a warning about this problem, but it is quite easy to avoid by checking the listing file and setting *org* statements at page-boundary starts to minimize the opportunity for subroutines to go over the page boundary.

The following experiments concentrate on passing data to and from a subroutine and the problems with the reduced program stack of the PICmicro® MCU.

## FIRSTCAL: PASSING DATA THROUGH REGISTERS

The simplest method of passing data to a subroutine is through a register. In the PICmicro® MCU, this can be done by either passing a byte parameter between the calling code and the subroutine using the w register or using a temporary file register common to both the caller and the routine. As shown in this experiment, passing data to a subroutine using w is always safe, but be careful to return parameters if w is to be used.

FirstCal, located in the *code\firstcal* subdirectory of the CD-ROM or in the *PICmicro* directory of your PC, shows how both of these methods would be implemented.

```
 title  "FirstCal - Passing Subroutine Parms via Registers"
;
;  This Program shows how to Pass Parameters back and forth between the
;   Mainline and Routines using Registers.
;
;  99.11.13 - Updated for Second Edition
;
;  Myke Predko
;  96.05.10
;
  LIST R=DEC
  INCLUDE "p16f84.inc"

;  Registers
i        equ 12              ;  Registers to Operate on
j        equ 13
Temp     equ 14

 __CONFIG _CP_OFF & _WDT_OFF & _RC_OSC & _PWRTE_ON

  PAGE
;  Mainline of FirstCal

  org 0

  movlw  3                   ;  Initialized the Variables
  movwf  i
  movlw  5
  movwf  j

;  j = increment ( i )

;  Pass Parameters using the "w" Register

  movf   i, w                ;  Get value to Increment
  call   increment_w         ;  Increment using the 'w' Register
  movwf  j
```

```
;  Pass Parameters using Temporary Registers

  movf   i, w               ;  Get the Value to Increment
  movwf  Temp
  call   increment_Temp     ;  Increment the Temp Value
  movf   Temp, w            ;  Store the Incremented Value
  movwf  j
Finished                    ;  Finished, Just Loop Around Forever
  goto   $

;  Increment Subroutines

increment_w                 ;  Increment the value in "w"
  addlw  1                  ;  Add one to Value in "w"
  return

increment_Temp              ;  Increment the Temporary Value
  movlw  1
  addwf  Temp, f
  return

  end
```

The *FirstCal* code executes the operation:

```
i = increment( j )
```

two different ways.

The first method uses w to store the input parameter and return the result. This works well and is quite efficient, but it does have one major and one minor drawback (the minor one is explained in the following paragraph). The major limitation is that passing data in w means that only one byte can be passed in and out of the routine at any given time. Some functions have no problems with this limitation; for others, one byte simply is not sufficient.

The second method stores the input and output parameters in the Temp subdirectory. This method works quite well, but does require extra code, cycles, and file registers that the first method doesn't require. This is the preferred method for more than one byte of parameter data. Using file registers to pass data back and forth eliminates the limitation of using w to pass a single-byte parameter back and forth. But, this method introduces its own problem; when calling multiple nested routines there is the issue of having correct common parameter registers for the different routines.

Elsewhere, the book notes that the *return* instruction is valid for low-end PICmicro® MCUs—even though the instruction does not exist for the parts and I warned you that this could cause problems when porting code between architectures. To show what I mean, I modified FirstCal to run in a 16C54 (a low-end processor) and saved it as FirstC54. The most-significant change made to the code is in the increment_w subroutine in which I changed the *addlw 1* instruction by saving the value in w into *Temp* and adding one to *Temp*.

The code is:

```
 title  "FirstC54 - Passing Subroutine Parms via Registers"
;
;  This Program is the same as "FirstCal" except the device
;   has been changed from a mid-range to a low-end.
;
;  99.11.13 - Created
;
;  Myke Predko
```

```
;
  LIST R=DEC                 ;  <-- Note Processor is PIC16C54
  INCLUDE "p16c5x.inc"

;  Registers
i        equ 12              ;  Registers to Operate on
j        equ 13
Temp     equ 14

 __CONFIG _CP_OFF & _WDT_OFF & _RC_OSC

  PAGE
;  Mainline of FirstCal

  org 0

  movlw  3                   ;  Initialized the Variables
  movwf  i
  movlw  5
  movwf  j

;  j = increment( i )
;  Pass Parameters using the "w" Register

  movf   i, w                ;  Get value to Increment
  call   increment_w         ;  Increment using the 'w' Register
  movwf  j

;  Pass Parameters using Temporary Registers

  movf   i, w                ;  Get the Value to Increment
  movwf  Temp
  call   increment_Temp      ;  Increment the Temp Value
  movf   Temp, w             ;  Store the Incremented Value
  movwf  j

Finished                     ;  Finished, Just Loop Around Forever
  goto   $

;  Increment Subroutines

increment_w                  ;  Increment the value in "w"
  movwf  Temp                ;  - Save in "Temp" for Update
  movlw  1                   ;  Add one to Value in "w"
  addwf  Temp, w
  return

increment_Temp               ;  Increment the Temporary Value
  movlw  1
  addwf  Temp, f
  return

  end
```

When I simulated the code, I found that upon return from *increment_w,* the value in w was zero. The problem is that MPASM converts return into a *retlw 0* instruction. This is not a trivial problem to find. As shown in Fig. 15-40, the code assembles without any errors, warnings, or messages. You will only find the problem by carefully single-stepping through the application and then noticing that the return value in w is always zero. Once you see this, you will have to check the low-end instruction set and discover that 0x0800 is actually the:

```
retlw 0
```

instruction.

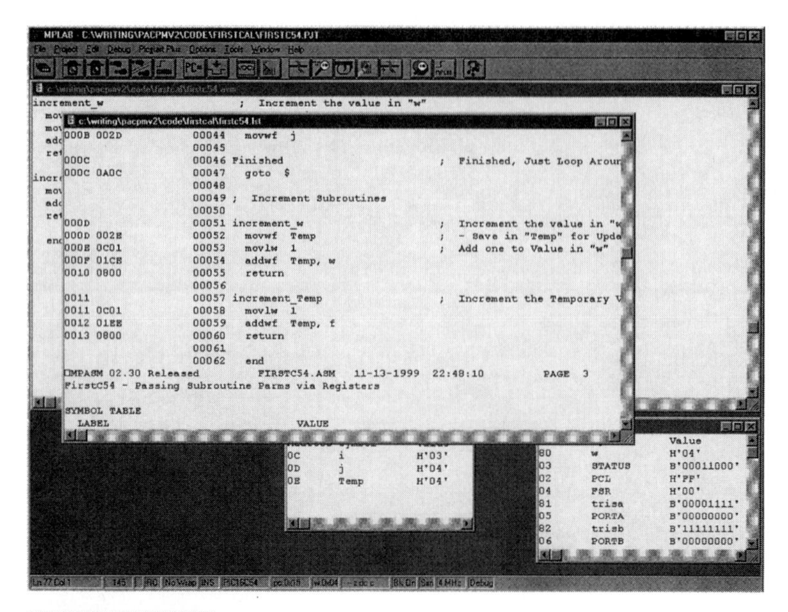

**Figure 15-40** Screen shot of setting the FirstC54 project with listing file

If you are going to be "porting" applications or code between the mid-range and low-end devices, you might want to keep this in mind and never return data in w, except by using a definite *retlw* instruction.

## STAKCALL: PASSING DATA ON A STACK

Passing parameters to and from a subroutine on a stack eliminates the problems in passing parameters in common registers that might be re-used during "nested" calls.

```
 title  "StakCall - Passing Subroutine Parms via Stack"
;
;  This Program shows how to set up a pseudo-Stack and use it to pass
;   Parameters between the Mainline and Routines.
;
;  99.11.13 - Updated for Second Edition
;
;  Myke Predko
;  96.05.10
;
 LIST R=DEC
 INCLUDE "p16f84.inc"

;  Registers
i        equ 12              ;  Registers to Operate on
j        equ 13
Stack    equ 14              ;  Start of the virtual Stack

 __CONFIG _CP_OFF & _WDT_OFF & _RC_OSC & _PWRTE_ON
```

```
  PAGE
;  Mainline of StakCall

  org 0
  movlw  4                  ;  Initialized the Variables
  movwf  i

  movlw  Stack              ;  Set up the Stack
  movwf  FSR

;  j = Increment( i )

;  Get the value of i + 1 and put it in "j"

  incf   FSR, f             ;  Make Space for the Sum

  movf   i, w               ;  Get First Value to Add and store on
  movwf  INDF               ;   the Stack
  incf   FSR, f

  call   Increment          ;  Increment the Value

  decf   FSR, f             ;  Get and Store the Sum
  movf   INDF, w
  movwf  j

Finished                    ;  Finished, Just Loop Around Forever
  goto   $

;  Increment Subroutine

Increment                   ;  Add the Two Values on the Stack

  decf   FSR, f             ;  Get the First Value and Increment it
  incf   INDF, w

  decf   FSR, f             ;  Store the result
  movwf  INDF

  incf   FSR, f             ;  Reset the Stack Pointer

  return

  end
```

This application can be found in the *code\StakCall* subdirectory of the *PICmicro* directory. When I simulated the file, I used the standard format shown in Fig. 15-41 with the *Stack* variable displayed as a 32-bit hex decimal.

In StakCall, notice that the FSR always points to the next available stack element and that, for returned parameters, I make space for them before loading the input parameters on the stack. Thus, the stack is always ready to take a new value without any of the previous values being affected.

When determining the number of file registers to allocate for parameter passing, simply count the number of parameters (both input and output) for each routine and sum the "deepest" path to get the total stack required. Chances are that you'll find that you'll need fewer stack file registers than the various parameter registers required if each subroutine is to have parameter variables.

Although I feel that implementing a PICmicro® MCU stack is more trouble than it's worth in arithmetic operations, they do offer advantages for the user in the area of multiple or nested subroutines.

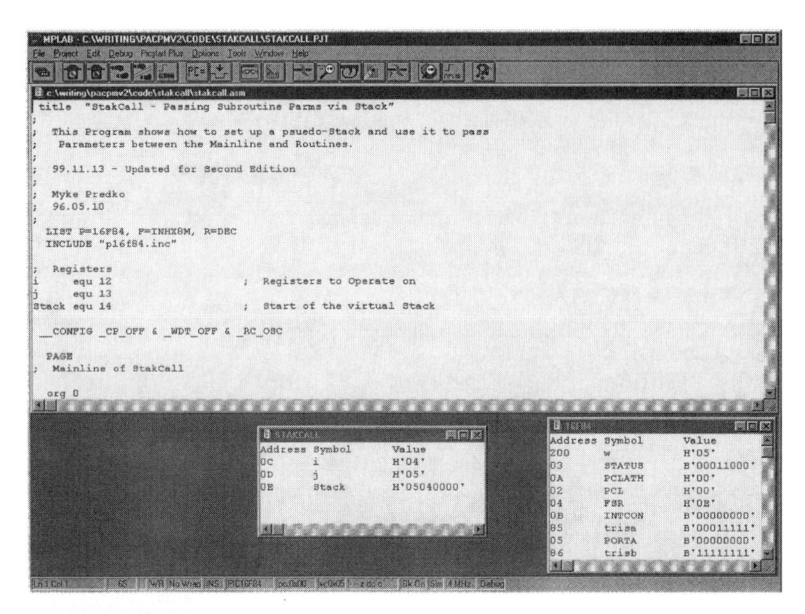


**Figure 15-41** Screen shot of setting the StakCall project with the Stack watch file

## CALLBUP: BLOWING UP THE PROGRAM COUNTER STACK

The PICmicro® MCU's Harvard architecture means that the PICmicro® MCU uses a separate *LIFO (Last In-First Out)* register stack for storing the program counter when execution changes to subroutines or interrupts. Because the stack is separate from the control store and file registers, it is quite limited.

This limitation means, in the mid-range devices, that no more than eight levels of subroutines and interrupt-handler subroutines can be executed. For most programming styles, this is not a problem. It is an issue if you are going to implement a recursive subroutine (which is a subroutine that can call itself).

What happens when more than eight subroutines are called? To answer this, I have created the following program which can be found in the *\code\CallBUp* subdirectory of the *PIC micro* directory:

```
 title  "CallBUp - Blowing up the PC Stack."
;
;  This Program calls a bunch of routines to show how the Stack can
;   be exceeded and an incorrect address returned
;
;  99.11.13 - Updated for Second Edition
;
;  Myke Predko
;  96.05.13
;
  LIST R=DEC
  INCLUDE "p16f84.inc"

;  Registers

 __CONFIG _CP_OFF & _WDT_OFF & _RC_OSC & _PWRTE_ON
```

```
  PAGE
;  Mainline of CallBUp

  org 0

  call   Prog1              ;  Now, Just Call Subroutines

Finished                    ;  Finished, Just Loop Around Forever
  goto   $

;  Subroutines
;  The Subroutines consist of:
;  Prog_at_Label
;    calling Prog_at_label_plus_one
;    return

Prog1

  call   Prog2

  return

Prog2

  call   Prog3

  return

Prog3

   :

Prog10

  return                    ;  At End of "Call" Chain, Return

  end
```

After assembling this program, you can single step through clicking on the Step button (and not the Step Over button), to see how execution goes. In MPLAB, as you step through the application, you will receive a "ding" in *Prog8* when the *call Prog9* instruction is executed. This is your only indication that the stack is overflowing. As you keep stepping through the program, you will get another "ding" at the *call* instruction to *Prog10*. In the MS-DOS command-line MPSIM simulator, a true error message will be brought up to indicate that the stack has overflowed.

Even though you have received a "ding" to indicate that there is a problem, keep stepping through until you reach the *return* statement in Prog10. At this point, the program counter stack elements that had the return address back into *Prog1* has been overwritten and execution from *Prog2* will return to *Prog9*. This will go on forever and you will never return to *Prog1* or the mainline of the application.

What happened is that the program counter stack has been overwritten and the original initial contents have been lost.

To avoid this situation, I always check all the subroutines used and ensure that the eight levels of program counter stack is never exceeded in the mid-range (two for the low end and 16 for the PIC17Cxx). To do this, I find the deepest level in the mainline, add it to the deepest subroutine level plus one for the interrupt handler, and add the depth of any subroutine calls from the interrupt handler itself.

For the most part, the limited program counter stack of the PICmicro® MCU is not a great liability. As I've noted, the only programming construct that this limitation makes not recommended for the PICmicro® MCU is recursive subroutines.

# Table Data

Tables are excellent tools to use in the PICmicro® MCU because they can provide the ability to save and retrieve strings easily, as well as give you some conditional execution options that you might not have expected. Successfully writing to the *PCL (Program Counter Low)* register requires a good understanding of how the PICmicro® MCU's program counter is architected. It also requires you to understand how PCLATH and the *PA0* to *PA1* bits (in the low end) are used to select an arbitrary address in a 256-instruction address block within the PICmicro® MCU.

Looking over the first edition, I feel like the section on accessing tables by modifying PCLATH was deficient because I focused primarily on presenting tables in the first 256 instructions of a PICmicro® MCU to avoid having to alter PCLATH. To be honest, I cringe at what I wrote for the first edition because it does not emphasize that tables can be loaded anywhere in a PICmicro® MCU's program memory.

The following experiments show how tables can be created from a much more general perspective and if issues arise from how the table is placed (and potentially going over a 256-instruction "boundary").

## TABLE0: MID-RANGE TABLE CALLING

The Table0 application (which can be found in the *code\table0* subdirectory of the *PICmicro* directory) consists of a simple table loop that reads a five ASCII byte table at three different locations. They are in the first 256 instructions, in the second 256 instructions, and straddling the second and third 256-instruction boundary of the PICmicro® MCU's program memory. If the *table* calls execute correctly, each byte returned is stored in the *Dest* array until a *0* is returned, at which time execution exits from the current loop.

Bring your attention to two different aspects of Table0. First, I load *Dest* in what seems to be "backwards." Because of the limitations of the watch windows in MPLAB, I store the data from high byte to low byte so that the information is more "man readable."

The second issue to notice is how I try to ensure that PCLATH is correct. Because both Table2 and Table3 start in the second block of 256 instructions, I save the table offset first, then set up PCLATH to point to the second block of 256 instructions and then change PCL to the correct address.

```
 title  "Table0 - Mid-Range Page0 Table Calling."
;
;  An invalid Table Read is implemented.
;
;
;  99.11.14 - Updated for Second Edition
;
;  Myke Predko
;  96.05.17
;
  LIST R=DEC
  INCLUDE "p16f84.inc"
```

```
;  Registers
 CBLOCK 0x020
i,Dest:4, Temp
 ENDC

 __CONFIG _CP_OFF & _WDT_OFF & _RC_OSC & _PWRTE_ON

  PAGE
;  Mainline of Table0

 org     0

  movlw  ' '                 ;  Initialize "Dest" so it can be seen
  movwf  Dest + 0
  movwf  Dest + 1
  movwf  Dest + 2
  movwf  Dest + 3

;  Table Read, Get values in Table until they equal Zero.

  clrf   i                   ;  Use "i" as the Index
  movlw  Dest + 3            ;  Point to the Destination
  movwf  FSR
Table1_Loop                  ;  Loop Around Here until finished
  movf   i, w                ;  Get the Index into the Table
  incf   i, f                ;  Increment the Table Index
  call   Table1              ;  Call the Table
  iorlw  0                   ;  At Table End?
  btfsc  STATUS, Z
   goto  Table1_Skip
  movwf  INDF                ;  No - Store the Returned Value
  decf   FSR, f              ;  Point to Next Position in "Dest"
  goto   Table1_Loop         ;   Array
Table1_Skip

  clrf   i                   ;  Same as Previous, but Different Page
  movlw  Dest + 3            ;   Table
  movwf  FSR
Table2_Loop
  movf   i, w
  incf   i, f
  call   Table2
  iorlw  0
  btfsc  STATUS, Z
   goto  Table2_Skip
  movwf  INDF
  decf   FSR, f
  goto   Table2_Loop
Table2_Skip

  clrf   i                   ;  Same as Previous, but Table
  movlw  Dest + 3            ;   "Straddling" 256 Instruction
  movwf  FSR                 ;   Boundary
Table3_Loop
  movf   i, w
  incf   i, f
  call   Table3
  iorlw  0
  btfsc  STATUS, Z
   goto  Table3_Skip
  movwf  INDF
  decf   FSR, f
  goto   Table3_Loop
Table3_Skip

  goto   $                   ;  All Done, Go to Infinite Loop
```

```
Table1
  addwf  PCL, f            ;  Get the Table Offset
  retlw  't'
  retlw  'a'
  retlw  'b'
  retlw  '1'
  retlw  0                 ;  End the String

 org     0x0100            ;  Second 256 Instruction Jump
Table2
  movwf  Temp              ;  Save the Table Offset
  movlw  HIGH $
  movwf  PCLATH
  movf   Temp, w           ;  Jump to the Table Address
  addwf  PCL, f
  dt     "TAB2", 0

 org     0x01FA            ;  This Table goes over a 8 Bit (256
Table3                     ;   Address) Boundary
  movwf  Temp              ;  Save the Table Offset
  movlw  HIGH $
  movwf  PCLATH
  movf   Temp, w           ;  Jump to the Table Address
  addwf  PCL, f
  dt     "tab3", 0

  end
```

Now, when simulating the application, you will find that Table1 and Table2 seem to execute correctly and *Dest* is loaded up with *tab1* and *TAB2* (as is expected). The problem occurs when you start executing in *Table3_Loop.*

If you are single stepping through the application, you will discover that execution will jump to Table2, after returning the first character of Table3 (the *t*), return *A,* and then *B.* When the fourth character of the table is read (what should be the *3,* or since the Table2 seems to be called) execution jumps to 0x0308 and the MPLAB display looks like that shown in Fig. 15-42.

The errors in execution are caused by the table straddling the second and third 256-instruction block boundary. When PCL is updated after PCLATH is already set to the second 256-instruction block, the jump is invalid because it should be to the third.

This is a very difficult problem to understand if you have programmed a PICmicro® MCU and during a table read, a problem like this happens. When the 0x0308 address jump occurs, execution can be almost literally anything. The PICmicro® MCU will most likely become totally nonresponsive and seem to have "locked up."

This is the biggest potential problem when working with tables. The only way you will really find it is by simulating the application and working through each table jump to try and find a case where execution does a "branch to the boonies" like what happens with *Table3*.

By changing the code in the Table3 subroutine to:

```
Table3
  movwf  Temp              ;  Save the Table Offset
  movlw  HIGH Table3_Start
  movwf  PCLATH
  movf   Temp, w           ;  Jump to the Table Address
  addlw  LOW Table3_Start
  btfsc  STATUS, C
   incf  PCLATH, f
```

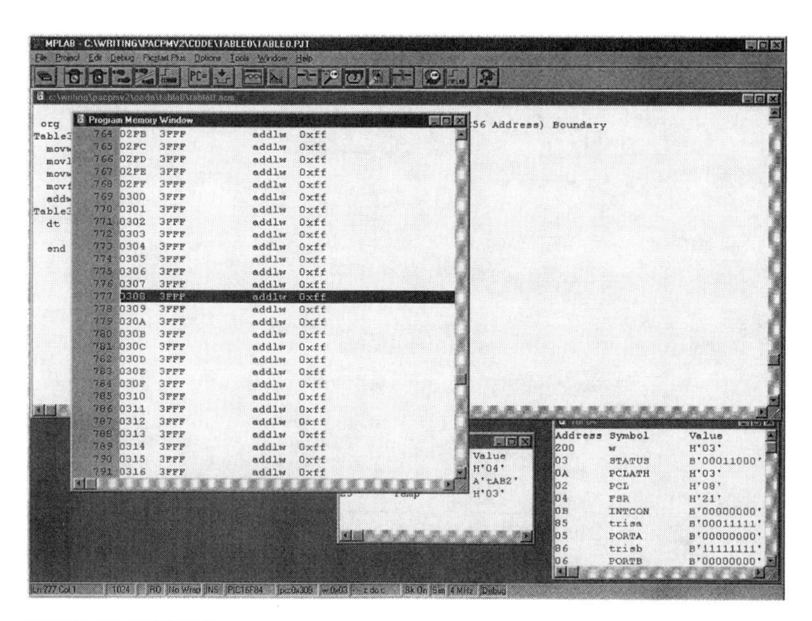


**Figure 15-42** **Screen shot of invalid tabe jump from Table0 application**

```
   movwf  PCL
 Table3_Start
   dt     "tab3", 0
```

the problem is eliminated.

The "fixed" code snippet is my general-case solution to supporting tables in and across different pages.

In Table0, I first use the individual *retlw* instructions followed by the *dt* directive to define the tables. In the first case, I used the individual *retlw* instructions to allow you to watch the jumps and returns happening in the simulator. The *dt* directive makes it easier to write the application (and for ASCII data, it makes the code very easy to read), although the MPLAB simulator might not jump to the *dt* line in the source code and you cannot see what individual byte is being returned.

## ARBTABLE: MID-RANGE TABLES LONGER THAN 256 ENTRIES

Sometimes you will have to create a table that is longer than 256 entries. The table will have to be accessed using a 16-bit variable that requires two file registers for the index into the table, rather than the eight bits required for the small table in the previous experiment. Because two bytes are used for the offset into the table, PCLATH may have to be changed after it is initialized. The source code can be found in the *\code\ArbTable* subdirectory of the *PICmicro* directory

```
  title  "ArbTable - Tables Longer than 256 Entries."
 ;
 ;  This Program demonstrates how a table that is longer than
 ;   than 256 entries (and therefore, jumps over the 256 "addwf PCL"
 ;   Page boarder).
 ;
```

```
;  99.11.14 - Updated for the Second Edition
;
;  Myke Predko
;  97.02.19
;
  LIST R=DEC
  INCLUDE "P16F84.inc"

 __CONFIG _CP_OFF & _WDT_OFF & _RC_OSC & _PWRTE_ON

;  Registers
 CBLOCK 0x020
TableOff:2                  ;  Offsets into the Table
Temp
 ENDC

  PAGE

;  Mainline of ArbTable

  org 0

  movlw  9                  ;  Get the 10th Message
  call   GetMSG

  goto   $

GetMSG                      ;  Figure Out where the Specified
                            ;   Message is
  movwf  Temp               ;  Save the Message Count

  clrf   TableOff           ;  Reset the Table Values
  clrf   TableOff + 1

GM_Loop

  movf   Temp, f            ;  If Temp == 0, then Stop Search
  btfsc  STATUS, Z
   goto  GM_End

  call   MSGTable           ;  get the Value in the Table

  iorlw  0                  ;  Are we at the End of a Message?
  btfsc  STATUS, Z
   decf  Temp, f            ;   Yes - Decrement Temp

  incf   TableOff, f        ;  Point to the Next Table Element
  btfsc  STATUS, Z
   incf  TableOff + 1, f

  goto   GM_Loop

GM_End                      ;  TableOff Now Points to the
                            ;   First Character in the Specified
  return                    ;   Message

MSGTable                    ;  Place all the Messages Here
  movlw   HIGH TableStart   ;   Setup PCLATH for the Table
  addwf  TableOff + 1, w
  movwf  PCLATH
  movlw  TableStart & 0x0FF;  Figure out the Offset
  addwf  TableOff, w
  btfsc  STATUS, C          ;  If Necessary, Increment PCLATH
   incf  PCLATH, f          ;   to get the Correct 256 Address Page
  movwf  PCL                ;  Update the PC
```

```
TableStart                      ;  Table Data
Msg0
  dt      "This is the First Message", 0
Msg1
  dt      "This is the Second Message", 0
Msg2
  dt      "This is the Third Message", 0
Msg3
  dt      "This is the Forth Message", 0
Msg4
  dt      "This is the Fifth Message", 0
Msg5
  dt      "This is the Sixth Message", 0
Msg6
  dt      "This is the Seventh Message", 0
Msg7
  dt      "This is the Eighth Message", 0
Msg8
  dt      "This is the Ninth Message", 0
Msg9
  dt      "This is the Tenth Message", 0

  end
```

When *GetMSG* returns, the 16-bit variable *TableOff* will be pointing to the first entry in the appropriate message.

The big thing to notice about this program over the previous one is that a new PCL and PCLATH are calculated before the *MSGTable* entry can be retrieved. In the previous experiment, the offset of the desired table element is simply added to the offset of the current PCL. As can be seen, this is not very hard to implement, but you must follow some rules to ensure that both PCL and PCLATH have the correct values in the table.

The eight instructions used in *MSGTable*:

```
  movlw  HIGH TableStart      ;   Setup PCLATH for the Table
  addwf  TableOff + 1, w
  movwf  PCLATH
  movlw  TableStart & 0x0FF   ;  Figure out the Offset
  addwf  TableOff, w
  btfsc  STATUS, C            ;  If Necessary, Increment PCLATH
   incf  PCLATH, f            ;   to get the Correct 256 Address Page
  movwf  PCL                  ;  Update the PC
TableStart                    ;  Table Data Start
```

can be used as a "general" case for jumping to tables of any size anywhere in a mid-range PICmicro® MCU's program memory. These eight instructions, along with a 16-bit *TableOff* variable could be considered a general-case table-call algorithm.

## SMALLTBL: LOW-END TABLE CALLING AND PLACEMENT

Tables in the low-end PICmicro® MCUs work very similarly to that of the tables in the mid-range (and high-end) devices, but with one important difference. SmallTbl (which is located in the *code\smalltbl* subdirectory of the *PICmicro* directory) is a simple application that you can use to examine tables in the low-end devices.

The SmallTbl code is very simple:

```
 title  "SmallTbl - Low-End Table Calling."
;
;  In this Application, A Page 1 Table Read
;   is Implemented on a Low-End Device
;
;
;  99.12.13 - Created for the Second Edition
;
;  Myke Predko
;
  LIST R=DEC
  INCLUDE "p16c5x.inc"       ;  <-- Note PIC16C56 used

;  Registers
 CBLOCK 0x010
i, Dest:4                    ;  Variables for Use with Monitoring
the Table
 ENDC

 __CONFIG _CP_OFF & _WDT_OFF & _RC_OSC

  PAGE
;  Mainline of SmallTbl
 org     0

  clrf   Dest                ;  Clear Destination for Each Execution
  clrf   Dest + 1
  clrf   Dest + 2
  clrf   Dest + 3

;  Table Read, Get values in the Table in Page 1 until they equal Zero.

  clrf   i                   ;  Use "i" as the Index
  movlw  Dest + 3            ;  Point to the Destination
  movwf  FSR                 ;  Look at FSR after this Instruction
Table_Loop                   ;  Loop Around Here until finished
  movf   i, w                ;  Get the Index into the Table
  incf   i, f                ;  Increment the Table Index
  call   Table               ;  Call the Table
  iorlw  0                   ;  At Table End?
  btfsc  STATUS, Z
   goto  Table_Skip
  movwf  INDF                ;  No - Store the Returned Value
  decf   FSR, f              ;  Point to Next "Dest" Array Position
  goto   Table_Loop
Table_Skip

  goto   $                   ;  All Done, Go to Infinite Loop

 org     0x00FA              ;  Table in Page 0
Table                        ;  Change "org" to 0x00FB
  addwf  PCL, f
  dt     "TAB1", 0

  end
```

In the SmallTbl application, notice that I placed the table at the absolute end of the first 256 instructions of the program memory's page 0. This was done to allow the maximum amount of space for the mainline code without having to place the table in page 1 of the simulated PIC16C56's program memory.

When setting up this application in the MPLAB IDE, use the 16C56.WAT file for monitoring the PICmicro® MCU's internal registers and create a watch file of your own to observe *i* and *Dest*. For *Dest,* set the watch *Properties* to *32 Bit ASCII* to observe the table as was done in the previous experiments.

With the MPLAB project set up, after assembling the SmallTbl application, single step through it and watch the *Dest* get filled with the string *TAB1* before a 0x000 character is returned and execution jumps to *Table_Skip,* where execution goes into a loop.

To calculate the last instruction that can be used for the table (and then using it in the *org* statement to locate it), I took the total number of instructions (five table entries and the *addwf PCL, f* instruction) for the table subroutine (Table) and subtracted them from 256. The 250 decimal value (0x0FA) was used as the address for the table.

To show that it is the last address, try changing the *org 0x00FA* statement to *org 0x00FB*. In this case, execution will never reach *Table_Skip.* When the fifth element of the table is to be retrieved by the call to Table, you'll discover that execution jumps to address zero. This is not unexpected because the last address, PCL, which is *0x00FB* (251), has five added to it, resulting in a value of *0x0100.*

The low-end PICmicro® MCUs cannot call or change the PICmicro® MCU's program counter (via updating PCL) to the second half of a page because the ninth bit of the address cannot be written to by either of these methods. It is possible to *goto* the second half of a low-end PICmicro® MCU page because the ninth bit of the address is specified in the instruction. You must be sure that your tables do not go over the first 256 instructions of a page boundary or you will find the execution jumping to unexpected addresses.

For this experiment, I used the simulated PIC16C56 instead of the PIC16C54 because the PIC16C56 has a 1-KB program memory space that is two low-end PICmicro® MCU pages. I did this because I wanted to show how an interpage call is accomplished in the low-end PICmicro® MCUs. The Table subroutine's *org* statement can be changed to 0x0200 to show how a call to another page can be accomplished.

After changing the *org* statement and rebuilding the application, you will get the message:

```
Crossing page boundary -- ensure page bits are set.
```

pointing to the *call Table* instruction. If you try executing the application, you'll find that the *call Table* instruction actually calls address zero of the PICmicro® MCU's program memory. The problem is that *Table* is in Page 0 and the *PA* bits of the STATUS register (which get added to the address of the *call, goto,* or *PCL* update) are still pointing to page 0. The problem is that Table has been relocated to page 1 by changing the *org* statement.

To ensure that the *call* instruction changes execution to the correct page, the *PA* bits of the STATUS register have to be set correctly. With the *call* instruction in page 0 and the destination being in page 1, just the *PA0* bit has to be set before the *call* can be executed. Once the subroutine returns to the location where it is called, the *PA0* bit has to be reset to ensure that jumps and calls within the current page are correct.

The *call Table* instruction becomes:

```
bsf    STATUS, PA0        ;  "Call" is to Page 1
call   Table              ;  Call Table Subroutine
bcf    STATUS, PA0        ;  Restore STATUS Page Bits to Page 0
```

which will still give you the same message. To eliminate it, you will have to reset the address bits that specify the new address. This is done by XORing the new address with the start of its page (0x0200 for page 1 of the low-end PICmicro® MCUs):

```
bsf     STATUS, PA0          ;  "Call" is to Page 1
call    (Table & 0x01FF) ($ & 0600)
bcf     STATUS, PA0          ;  Restore STATUS Page Bits to Page 0
```

With this done, the application will assemble cleanly and execute properly with Table relocated to the PICmicro® MCU's Page 1.

There is one last thing I want to point out to you and that is the operation of the FSR register in the low-end PICmicro® MCUs. To illustrate this, reset the simulated PICmicro® MCU and before single stepping through the application, change the contents of the FSR register to 0x000.

To change the contents of the FSR register, double click on *FSR* in the 16C56 watch window. A dialog box like that in Fig. 15-43 will appear, allowing you to modify the FSR (and any other) register. In the Data/Opcode entry, put in *00* (as shown in Fig. 15-43) and click on *Write*. After clicking on *Write,* you'll see that the value of the *FSR* in the watch window becomes *H'E0',* along with the value in the *Modify* dialog box changing to *E0.*

This will seem strange, but remember that for the low-end PICmicro® MCUs, the FSR sets any bits that are not used to access registers in specific banks. In the 16C56, only one bank is used for registers, so only a total of 32 register addresses are required. The most significant three bits are not required to access memory.

As you single step through the code, you'll see that FSR is loaded with the last address of *Dest,* which should be 0x014, but actually becomes 0x0F4 when the FSR is updated. This is another example of the unneeded FSR bits staying set after being loaded with a specific value. If the *FSR* was to be read back into the w register or used as part of an arithmetic or bitwise instruction, you will find that it will have these bits set as well.

In other low-end PICmicro® MCUs with differing file register sizes, the set bits of the FSR will change according to the number of register banks in the processor. Because there are a maximum of four banks in the low-end PICmicro® MCUs, Bit 7 of the FSR will be set. Only a maximum of 128 register addresses are spread over four 32-bit banks in the microcontrollers.

This is a quirk of the low-end PICmicro® MCUs that I mentioned earlier in the book. I felt it was important to point it out explicitly because, if you use low-end devices, I'm sure it will bite you. Actually, I'm pretty sure that the only way for you to remember that the FSR works this way is for you to forget this aspect of the low-end PICmicro® MCU family and have to debug the problem just as I did.

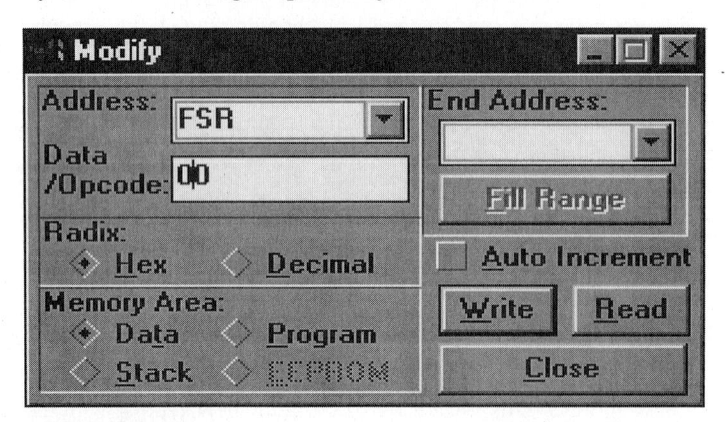


**Figure 15-43** **The MPLAB IDE register *Contents Change* dialog box**

## STATEMC: EXECUTION CONTROL STATE MACHINES

Execution state machines became popular when the first read-only programmable memories became available as a method to algorithmically control digital devices without using a microprocessor. The external input data values are combined with the current state of the memory to determine what the outputs will be. A block diagram of this is shown in Fig. 15-44.

A typical example application for the state machine is controlling traffic lights. The state machine's read-only memory is loaded from a table. The inputs are listed, along with the current state to get expected outputs for the next states.

For a traffic light, I have assumed that there are two timers (one second long and 15 seconds long), which are started by the state machine as one of its inputs. The timers are reset by the application. When they time out, they provide new inputs to the state machine. The timers are started by a rising edge on the RST lines. When they overflow (reach their delay values), a *1* is output onto the O/F lines. The block diagram of the state machine can be modified to what is shown in Fig. 15-45.

The inputs and outputs for the traffic-light state machine are shown in Table 15-4.

The table data is then converted to address and corresponding data. In this table, you should be able to see how the address output (which is actually the address input) is dependent on the state of the timers. When the timers overflow, they advance to the next state.

As you would expect, the actual hardware and I/O states for a state machine is somewhat more complex. The code presented here for the traffic lights only handle them in one direction and do not take into account such variables as cars wanting to turn left or pedestrians pressing a crossing button.

State machines can be implemented in the PICmicro® MCU quite simply using Tables to decide which routine to jump to next. For example, if you were keying off a file register value (e.g., 0, 1, 2, . . .), you could use the following code:

```
movf    Reg, w              ;  Get the Index
addwf   PCL, f              ;  Jump to the State Machine Table
                            ;  Entry
goto    State0              ;  Reg = 0
goto    State1              ;  Reg = 1
goto    State2              ;  Reg = 2
:                           ;  ... And so on ...
```

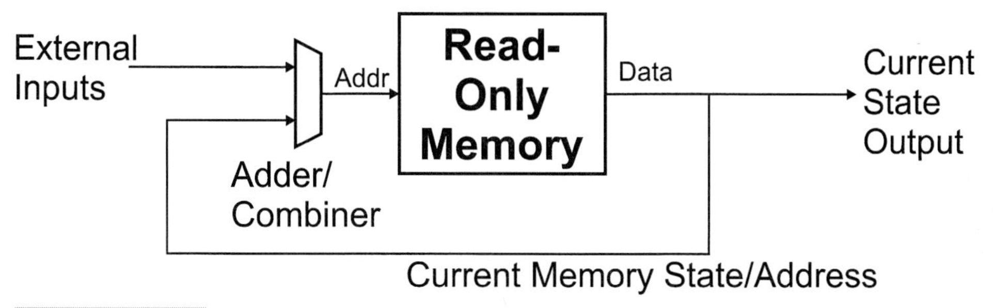


**Figure 15-44** Hardware state machine implementation

**TABLE 15-4 Traffic Light State Machine Inputs And Outputs**

| ADDRESS | 1 SEC O/F | 15 SEC O/F | ADDROUT | 1 SEC RST | 15 SEC RST | LIGHTS |
|---|---|---|---|---|---|---|
| Start | x | x | Start + 1 | 1 | 0 | None |
| Start + 1 | 0 | 0 | Start + 1 | 0 | 0 | None |
| Start + 1 | 1 | 0 | Green | 0 | 1 | Green |
| Green | 0 | 0 | Green | 0 | 0 | Green |
| Green | 0 | 1 | Yellow | 1 | 0 | Yellow |
| Yellow | 0 | 0 | Yellow | 0 | 0 | Yellow |
| Yellow | 1 | 0 | Red | 0 | 1 | Red |
| Red | 0 | 0 | Red | 0 | 0 | Red |
| Red | 0 | 1 | Green | 0 | 1 | Green |

There are three advantages to using a table in this situation:

1. Fewer instructions are required for the decision-making process.
2. Each different jump condition takes the same number of cycles. This could be important for some applications.
3. The table jumps are a lot easier to understand by looking at just the code.

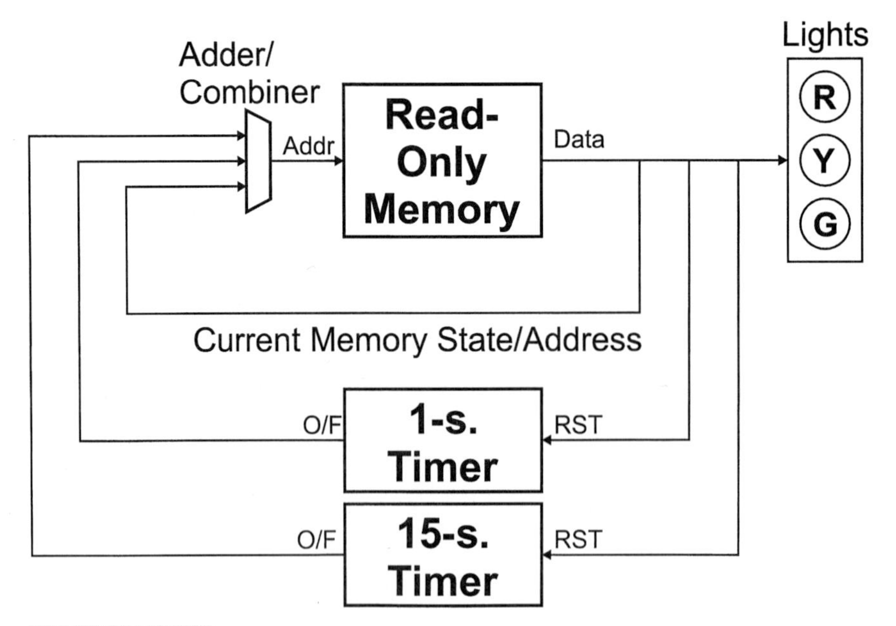


**Figure 15-45** Traffic-light state machine implementation

I define a *state machine* as a program that uses a single value to execute different functions the program. This value is set according to the previous state and current environmental conditions (e.g., "if RB0 is Set, increment the State and execute the response to RB0 being Set").

State machines are a form of *nonlinear programming*. Nonlinear programming includes different methods of programming when the traditional *if/else/endif* structure of coding is not used. To demonstrate this in the PICmicro® MCU, I created the StateMC application, which is located in the *code\StateMC* subdirectory in the *PICmicro* directory.

```
title  "StateMC - Demonstrating a State Machine"
;
;  This Program is Demonstrates how a State Machine could work with
;   the PICmicro MCU Architecture.
;
;  99.11.14 - Updated for Second Edition
;
;  Myke Predko
;  96.05.14
;
  LIST R=DEC
  INCLUDE "P16F84.inc"

;  Registers
 CBLOCK 0x020
i                           ;  General Counter
state                       ;  Returned Value
Temp                        ;  Temporary Storage Variable
 ENDC

 __CONFIG _CP_OFF & _WDT_OFF & _RC_OSC & _PWRTE_ON

  PAGE
;  Mainline of StateMC

  org 0

  clrf   i                  ;  Initialize Variables
  clrf   state

  clrf   PORTB              ;  Setup PortB
  bsf    STATUS, RP0
  clrf   TRISB ^ 0x080
  bcf    STATUS, RP0

;  Now, Execute the Program

Loop                        ;  Return Here every Execution
  movlw  1                  ;  Check the Least Significant Bit
  andwf  PORTB, w           ;   of PORTB
  movwf  Temp
  bcf    STATUS, C          ;  Now, Shift over the state
  rlf    state, w           ;   Variable
  addwf  Temp, w            ;  Add the Least Significant Bit of
                            ;  PORTB
  addwf  PCL, f             ;  Jump to the Correct State
  goto   State0             ;   Execution Vector
  goto   State0
  goto   State10
  goto   State11
```

```
  goto   State2
  goto   State2

;  State Routines...

State0                       ;  Increment i to 4
  incf   i, f

  movlw  4
  subwf i, w                 ;  is "i" greater than 3?
  btfsc STATUS, C
   incf state, f             ;  Yes, Increment the State
                             ;  Variable
  goto  Loop                 ;  Execute the State value again

State10                      ;  Increment the LSB of PORTB if
                             ;    it's == 0
  movlw  1
  addwf  PORTB, f

  goto   Loop

State11                      ;  Shift Over PORTB by one until
                             ;   Carry Set

  bcf    STATUS, C
  rlf    PORTB, f

  btfsc  STATUS, C           ;  Is the Carry Set?
   incf  state, f            ;  Yes, Go to the Next State

  goto   Loop

State2                       ;  Reset Everything and Restart the
  clrf   I                   ;   Program
  clrf   state

  goto  Loop

  end
```

I realize that StateMC is a pretty simple example of a state machine, but it does show how the state is changed with different conditions and the program progresses forward. This application might not seem so simple–especially considering that some of the operations and execution will be quite unexpected. In fact, this can be a problem with state machines—the operation becomes even more confusing as you modify the application over time and see simple changes that can be made. In StateMC, an example of this would be how I increment and shift the output value in PortB when there are other ways to do this (that are not quite so complex).

State machines are particularly useful in low-end PICmicro® MCUs, where the two-level stack may be a hindrance to traditional programming methods.

# Playing with the Hardware

With a clearer understanding of how to program the PICmicro® MCU, simulate applications, and debug them, I want to go on to actually developing circuits with the PICmicro® MCU. All of the rest of the experiments in this chapter involve "burning" an application into a PICmicro® MCU or using an emulator to test the code and circuit. To keep the wiring complexity down, I only run the experiments at 4 MHz, use simple buttons and pots for inputs, and LEDs for outputs. I have also tried to avoid complex software interfaces.

All of the circuits are designed for use with a simple "core circuit" and have been tested on 16F84s and 16F877s, except for the "different oscillator" experiment, which requires a PIC16C505. The software used for these applications can be used with MPLAB-ICD or the EMU-II, as well as the YAP-II. The 16F84 and 16F877 are the current favorites of most people creating their first applications because they can be reprogrammed very easily—and in the circuit. These circuits can all be built on prototyping breadboards to make the building of the various experimental circuits easier.

To help facilitate building the circuits, I use the same "core circuits" for both the 16F84 (Fig. 15-46) and the 16F877 (Fig. 15-47). Notice that I specify a 4-MHz ceramic resonator (such as an ECS ZTT-4.00MG) with internal 30-pF capacitors built in to eliminate the need for external capacitors in the oscillator circuit. I also use a 10-K resistor for the _MCLR pull up and a 0.1-$\mu$F tantalum capacitor to debounce the PICmicro® MCU.

When putting them on a breadboard, you can either use a *ZIF (Zero Insertion Force)* socket or "reference" the location where the PICmicro® MCU goes into the circuit using a wire across the breadboard. Figure 15-48 shows how the core circuit is assembled onto a breadboard, along with the photograph of the baseboard (Fig. 15-49).

As shown in Fig. 15-48, a wire is put across the breadboard to indicate where to put the PICmicro® MCU in. Notice that this wire is on the "bottom" (away from pin 1) of the PICmicro® MCU. It allows the PICmicro® MCU to be repositioned after it has been pulled out and programmed. When I set up this circuit, I put the ceramic resonator away from the PICmicro® MCU so that the PORTA I/O pins can be used in the application. I placed the PICmicro® MCU at the edge of the board to allow it to be easily pried up for reprogramming. Once you've set up your core circuit, you will not have to change it for any of the experiments (except for "different oscillators").

For the applications, I have assumed the simple 15 volt power supply that is presented in the "Introduction to Electronics" pdf file on the CD-ROM is used. For your own applications, you can use a bench supply or a 5 volt regulator powered by a 9 volt alkaline

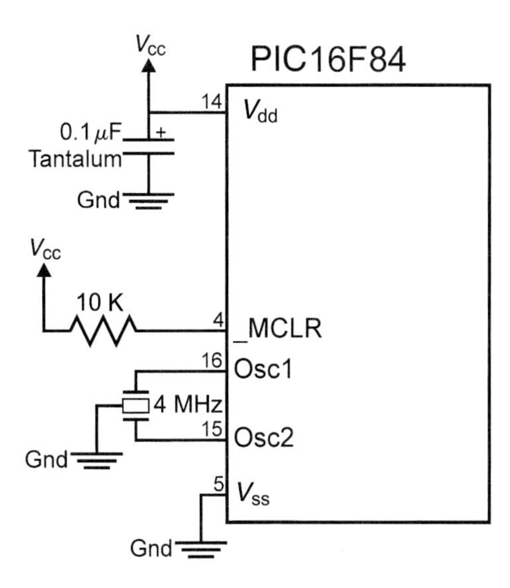


**Figure 15-46** PIC16F84 Core Circuit

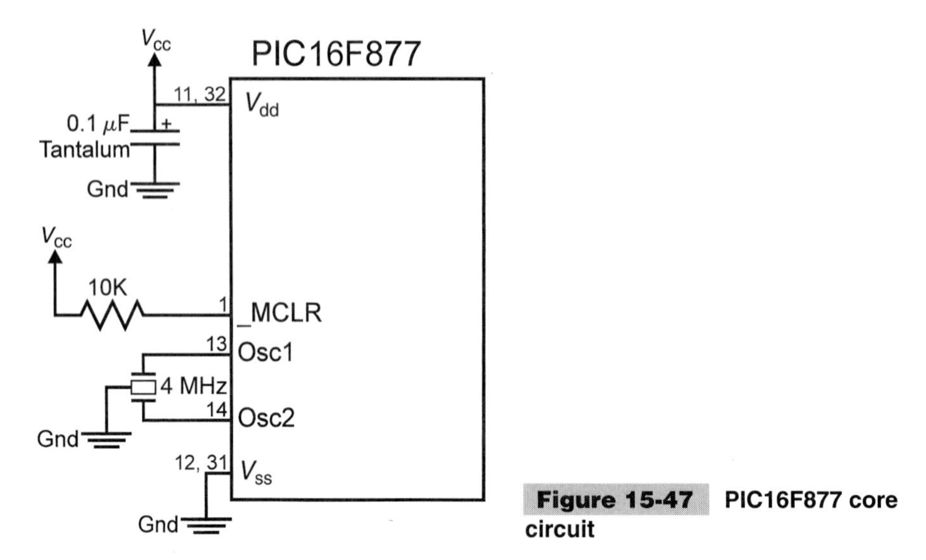


**Figure 15-47** PIC16F877 core circuit

transistor radio battery. The only requirements I would put on the power supply is the need for a fuse or other mechanism (such as an internal "crowbar") to protect you and the circuit in the case of short circuits.

Figure 15-50 shows how a Microchip-compatible ICSP connector can be added to the circuits to avoid having to pull the PICmicro® MCU from the breadboard to reprogram it. The *breakable connections* are connections that can be disconnected during programming to avoid any contention issues with the application circuit's loads. If the YAP or YAP-II is used for the experiments, only the RB6 and RB7 breakable connections have to be installed because the YAP-IIs control the reset/$V_{pp}$ circuit directly. These "'breakable' con-

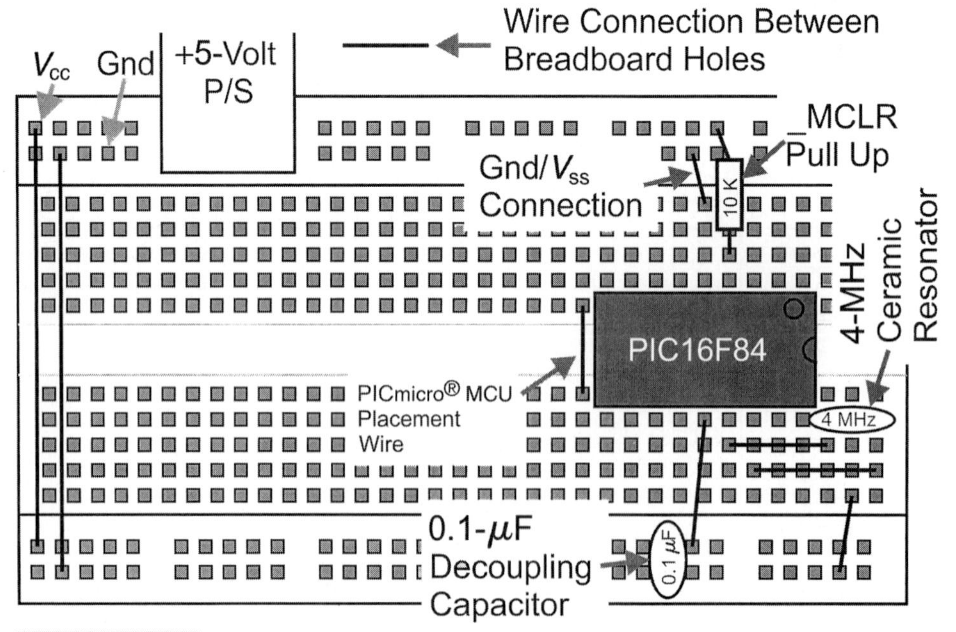


**Figure 15-48** PIC16F84 breadboard core circuit

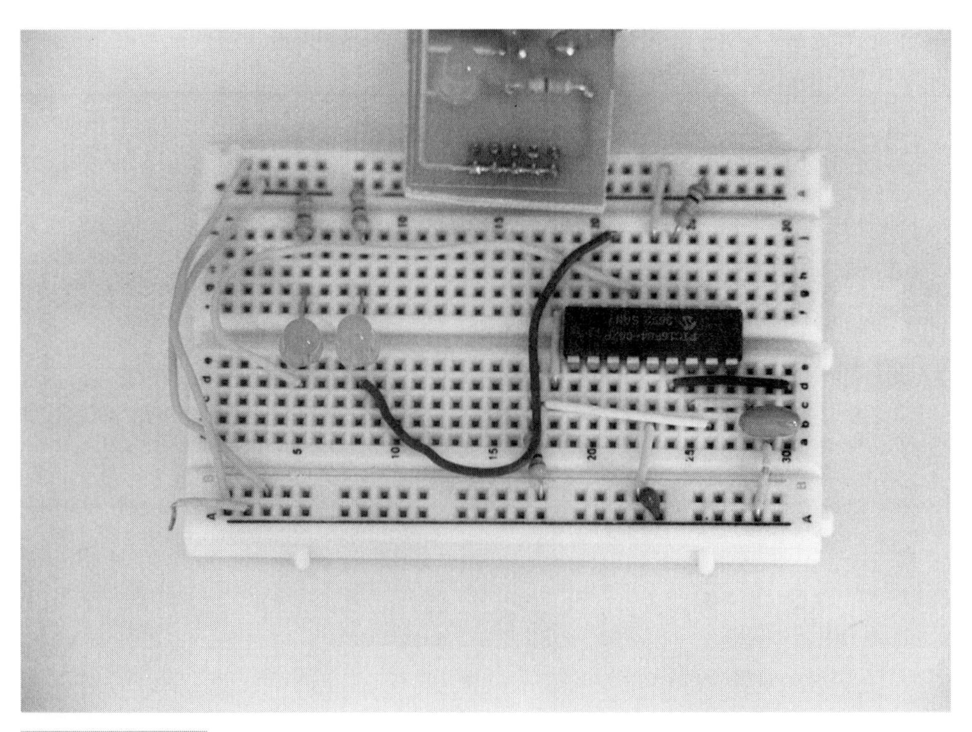

**Figure 15-49** Two LED application wiring example

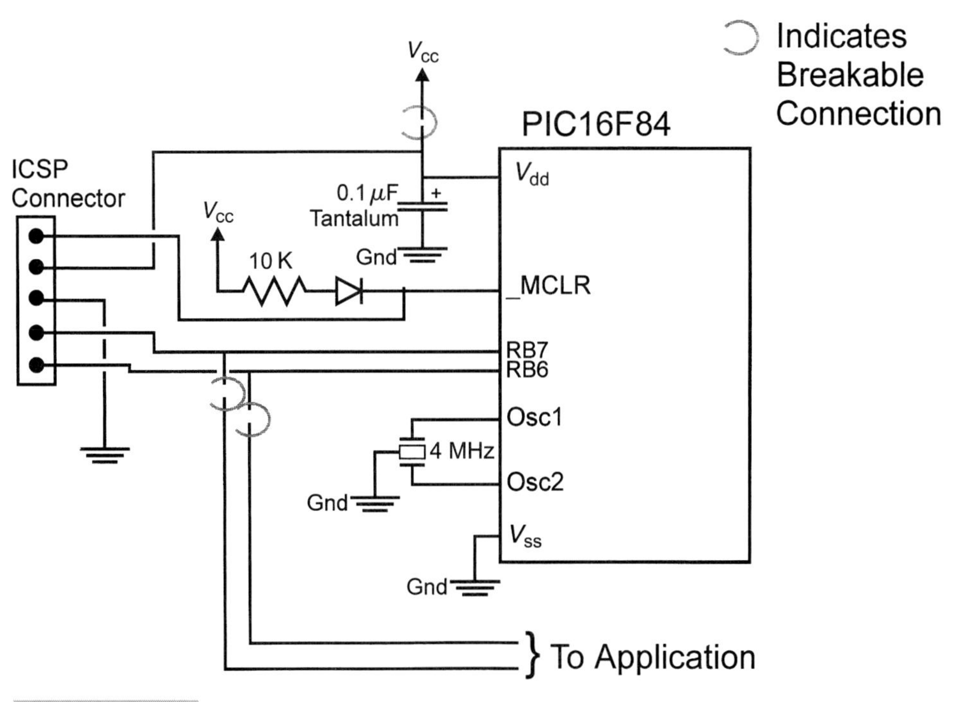


**Figure 15-50** PIC16F84 ICSP circuit

nections" can be as simple as disconnecting the wires attached to RB6 and RB7 during programming.

## LEDON: TURNING ON A LIGHT-EMITTING DIODE (LED)

In the first edition of this book, the first hardware application was simply turning on an LED. This edition is expanded to controlling the LED by a button connected to a PICmicro® MCU and turning on the LED when the button is pressed (Fig. 15-51). I have deliberately kept this application very simple to show how looking at what the hardware has to do to make decisions that will simplify the software. Also, it shows how a hardware application can be very easily simulated with "live" input in MPLAB.

The bill of materials for Ledon is given in Table 15-5.

The third column of Table 15-5 (and the ones that follow) indicates whether or not the part is required if the YAP-II (or EMU-II) is used for experiment development and testing. This column is in all of the applicable experiments in this chapter except the few in which I do not recommend the YAP-II or EMU-II is used to simplify the circuit development.

For this application, I could show how it executes by using the pseudo-code:

```
ledon()
{

  PORTB.0 = High;             //  Make RB0 High so LED is Off
  TRISB.0 = Output;           //  Make RB0 Output for LED Driving

  while (1 == 1)              //  Loop Forever
    PORTB.0 = PORTA.0;

}  //  end ledon
```

This simple C application pseudo-code translates into the very simple assembly-

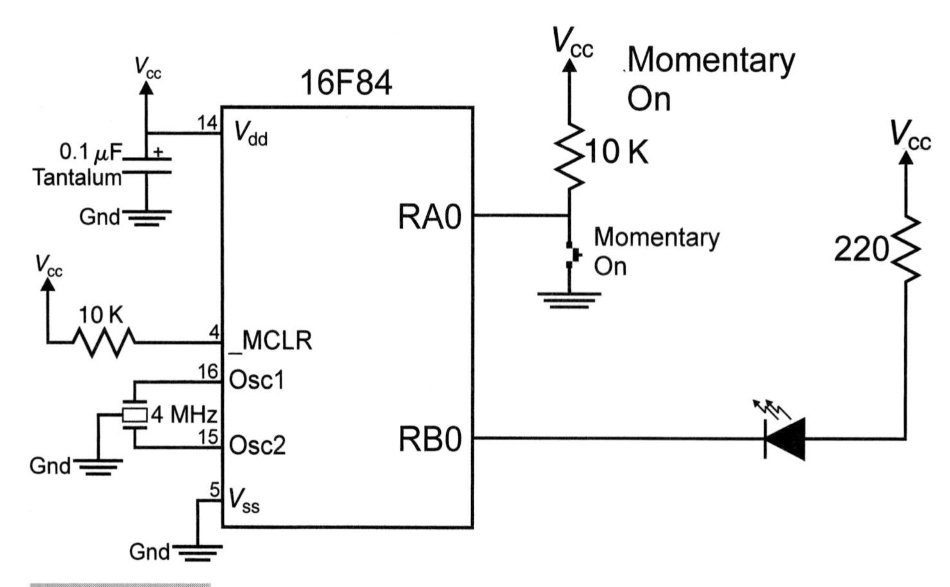


**Figure 15-51** Ledon: turning on an LED using a button as a control

**TABLE 15-5 Bill Of Materials For The "Ledon" Experiment**

| PART | DESCRIPTION | REQUIRED ON YAP-II |
|---|---|---|
| PICmicro® MCU | PIC16F84-04/P | In Socket |
| Vdd/Vss decoupling capacitor | 0.1 $\mu$F tantalum | No |
| _MCLR pull up | 10 K, 1/4 watt | No |
| 4-MHz ceramic resonator | 4 MHz with built-in capacitors | No |
| RA0 Pull up | 10 K, 1/4 watt | No. BUT1 used |
| RA0 Switch | Momentary on | No. BUT1 used |
| RB0 LED | Red LED | No. LED0 |
| RB0 LED resistor | 220 Ohm, 1/4 watt | No. LED0 |

language application code that is located in the *code\ledon* subdirectory of the *PICmicro* subdirectory:

```
  title  "ledon - Turn on a LED when a Button is Pressed"
 ;
 ;  This is the First Program to be Burned in and Run in a PICmicro MCU.
 ;
 ;  The Program simply sets up Bit 0 of Port "A" to Output and then
 ;   Sets it Low when RA0 is pulled low.
 ;
 ;  Hardware Notes:
 ;   _MCLR is tied through a 4.7K Resistor to Vcc and PWRT is Enabled
 ;   A 220 Ohm Resistor and LED is attached to PORTB.0 and Vcc
 ;   A 10K pull up is connected to RA0 and it's state is passed to
 ;    RB0
 ;
 ;  Myke Predko
 ;  99.12.03
 ;
   LIST R=DEC
  ifdef __16F84
   INCLUDE "p16f84.inc"
  endif
  ifdef __16F877
   INCLUDE "p16f877.inc"
  endif

 ;  Registers

  ifdef __16F84
  __CONFIG _CP_OFF & _WDT_OFF & _XT_OSC & _PWRTE_ON
  else
  __CONFIG _CP_OFF & _WDT_OFF & _XT_OSC & _PWRTE_ON & _DEBUG_OFF &
_LVP_OFF & _BODEN_OFF & _WRT_ENABLE_ON & _CPD_OFF
  endif

   PAGE
 ;  Mainline of ledon

  org     0

   nop                          ;  "nop" is Required for Emulators
```

```
    bsf     PORTB, 0            ;  Make the LED on PORTB.0 "off"
    bsf     STATUS, RP0         ;  Goto Bank 1 to set Port Direction
    bcf     TRISB ^ 0x080, 0    ;  Set RB0 to Output
    bcf     STATUS, RP0         ;  Go back to Bank 0

  Loop
    movf    PORTA, w            ;  Simply Transfer PORTA.0 to PORTB.0
    movwf   PORTB

    goto    Loop

   end
```

In the application code, I first load PORTB bit zero with a *1* so that when it is changed to an output pin, the LED will be off. In the circuit, I connect the LED to a 220-Ohm pull up so that if the PICmicro® MCU pin is connected to is high, there is no current path and the LED will be off.

In the application, I choose RA0 and RB0 for the button and LED pins that the PICmicro® MCU is connected to for more than just the reason that they were the first pins in the two I/O ports. I choose them because I could pass the data from the button to the LED very efficiently. The button is wired to a pull up. When it is pressed, it pulls the line low (or to a *0*). When this value is passed to the output pin, the LED is turned on.

I could have placed the button pin and the LED pin with other pins in the PICmicro® MCU, but if I did that I would have to change the simple *movf PORTA, w* and *movwf PORTB* instructions to:

```
   btfsc  inputPORT, inputPIN     ;  Is the Pin High or Low?
    goto  outputHIGH              ;   High

   bcf    outputPORT, outputPIN   ;  It's Low, turn on the LED

   goto   outputDONE

 outputHIGH                       ;  Input High - turn off the LED

   bsf    outputPORT, outputPIN

 outputDONE
```

This is an example of what I consider to be effective PICmicro® MCU optimization. By looking at the inputs and outputs, you can come up with some amazing simplifications in terms of hardware and software. Although this example might seem simplistic, it is actually very representative of the kind of application improvements you can get by planning which pins you are going to use for I/O.

For all of the experiments and projects presented in this book, as well as any of your own that are "burned" into PICmicro® MCUs and run in circuits, I stress the importance of simulating the application before trying it out on any hardware. Any problems you find in the MPLAB IDE are much easier to find and fix, compared to the work that will have to be done in the actual hardware.

For this application, I recommend the use of an asynchronous input, which allows you to change the value of a PICmicro® MCU input pin. This feature is turned on by clicking on a button on a dialog box that is added to your project desktop in the MPLAB IDE when you click on *Debug,* followed by *Simulator Stimulus,* and then *Asynchronous Input.* For the Ledon project, Fig. 15-52 shows the *Asynchronous Input* dialog box in the lower center of the desktop.

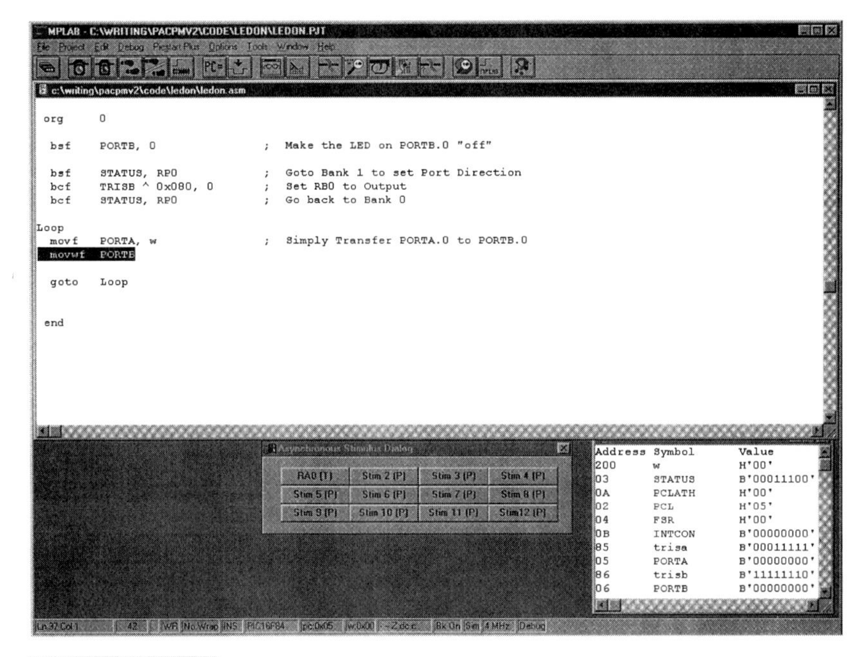


**Figure 15-52** Ledon MPLAB IDE project

When the *Asynchronous Input* dialog box first comes up, none of the buttons are set to any pins or operations. First, the pin must be selected. This is done by moving your mouse over the pin you want to use (*Stim 1* for Ledon), then right clicking the mouse. You will get the pull down menu shown in Fig. 15-53. Click on *Assign Pin* . . . and then select the pin that you want to use. For ledon, this is RA0 and the selection is shown in Fig. 15-54.

Once the pin is selected, then the operation has to be selected. This is done by right clicking on the pin again in the *Asynchronous Input* dialog box (Fig. 15-53) and selecting the operation type for the button. The default for asynchronous input pins is *P* for *Pulse* (which pulses the input and returns it to the original state). The other options are to set the pin *Low* or *High*. For this application (and most of the ones that I work with) I toggle the state of the pin when the button is pressed.

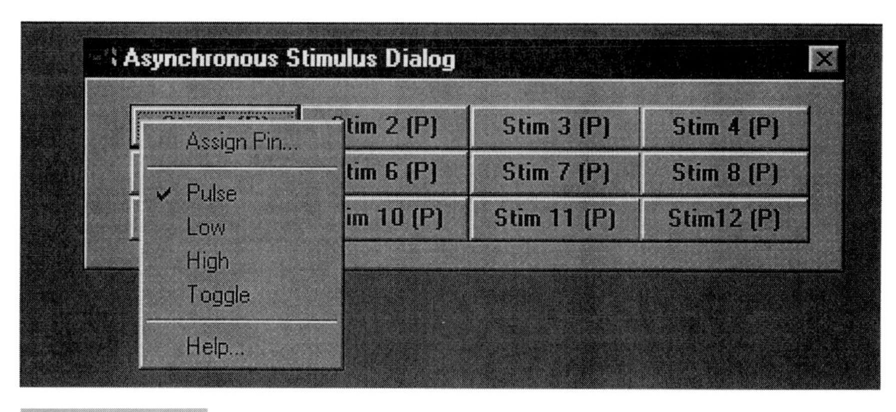


**Figure 15-53** Asynchronous input: selecting operation

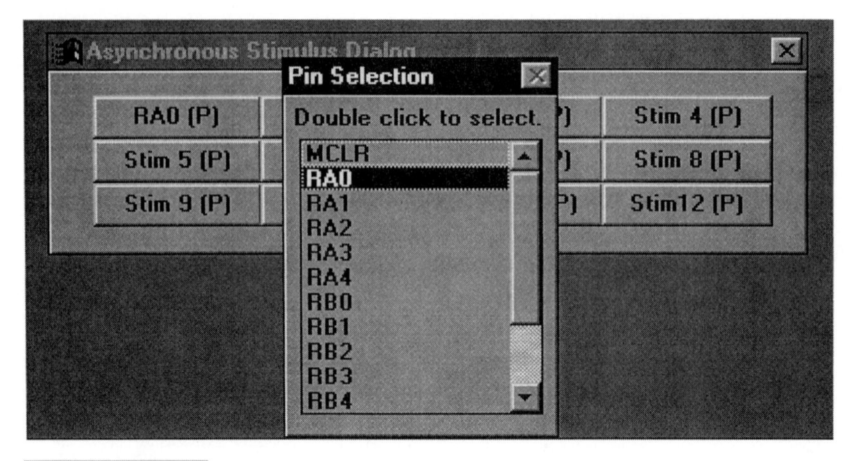


**Figure 15-54** Asynchronous input: selecting the pin

If you experiment with the different pins, you'll find that *P* doesn't seem to do anything. The pulse operation actually changes the state of the input pin and then returns to the original state before the application starts executing again (or while it is executing, if *Asynchronous Input* is clicked on while an instruction is executing). The pulse input is meant for TMR0 or an interrupt source pin, rather than a straight I/O pin, which will not display the action of the "pulse," as implemented in the MPLAB simulator.

With the *Asynchronous Input* dialog box set up to toggle the state of RA0, you can now single step through the application to watch how it executes. After resetting the output, you will see that the RA0 input pin will be low and that this value will be passed to RB0. After executing the Loop code, click on the *RA0* button in the *Asynchronous Input* to see it change to a high (or 1) and then step through the application some more to see how the high RA0 value is passed to RB0.

*Asynchronous Input* is a very handy tool to use while you are simulating an application. It eliminates the need to build a stimulus file or to change the contents of a register to see how input changes results in changing execution in the application.

With the simulation done, you are now ready to build the application. This experiment application, like all of the ones that follow in this chapter, is designed to have its circuit built on a breadboard, or it can use the YAP-II to simply connect it. For a Breadboard, the core circuit is enhanced (Fig. 15-55). If you are using a YAP-II to test the application, simply add the two wires (Fig. 15-56). In either case, or building the circuit on another prototyping system, it will only take a few moments and will be the basis for the next few experiments.

With the circuit built, you are finally ready to program the application into a PICmicro® MCU. It is indicated for all applications that you follow these steps to ensure that the application is programmed with the correct code:

1 Load the Ledon application hex file from the *code\ledon* subdirectory of your PC's PICmicro® MCU directory.
2 Perform a *Verify* operation with the Ledon application with the programmer's PICmicro® MCU socket empty. The results of the verify should be a failure with the first or second address value returning 0x00000.
3 Program the PICmicro® MCU.
4 Verify the PICmicro® MCU. This operation should pass.

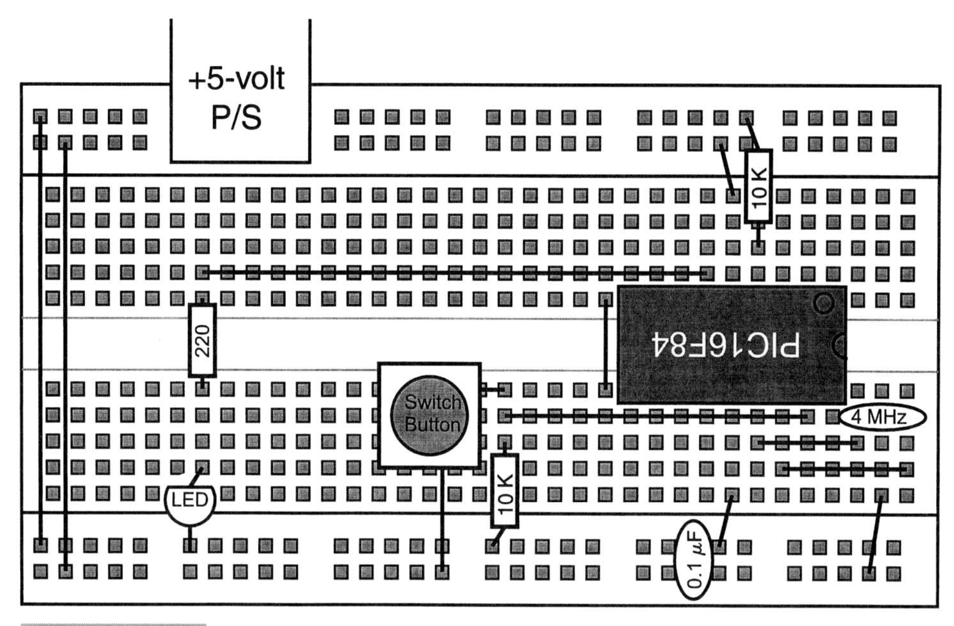


**Figure 15-55** Ledon PIC16F84 breadboard circuit

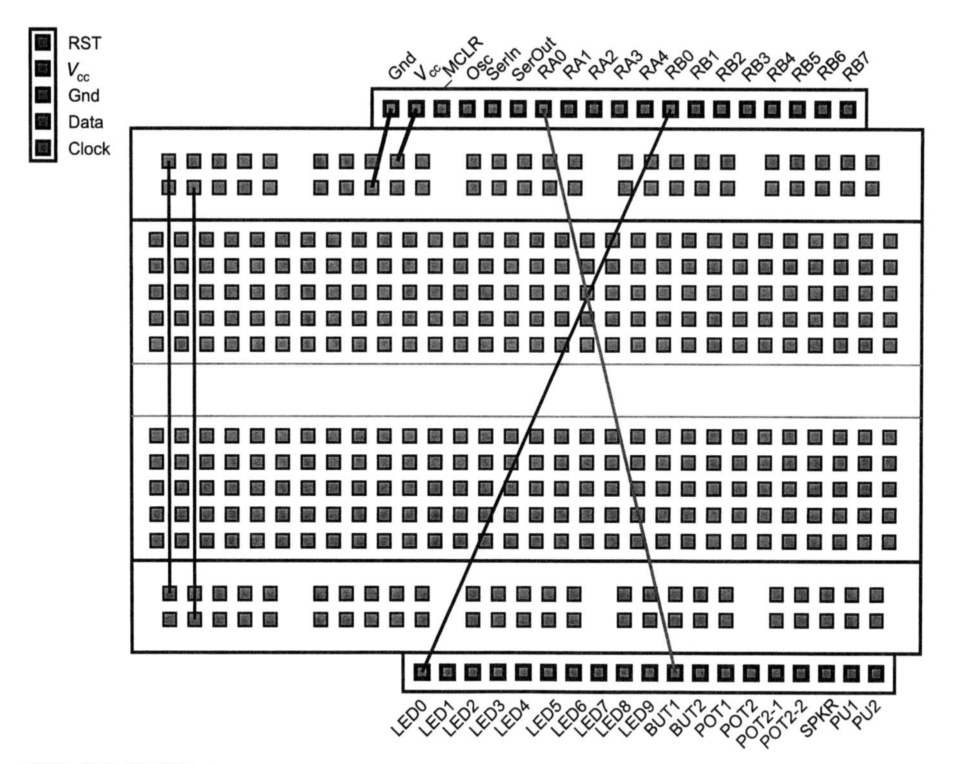


**Figure 15-56** EMU-II/YAP-II breadboard for the Ledon application

This series of steps will verify that the programmer is able to detect the absence of the PICmicro® MCU by not successfully verifying the application. Once this is done, you can program a PICmicro® MCU and be reasonably sure that the contents of the device are programmed properly.

With the circuit built and the PICmicro® MCU programmed and installed in the circuit, you can now apply power ($V_{cc}$ is 5.0 volts) and press on the button. Each time you press the button, you should see the LED turn on. Although this experiment doesn't really do all that much, I think it is an excellent way to get started in PICmicro® MCU application development. It will give you some confidence that later experiments and projects (including your own) will be successful.

## CALCULATING CURRENT REQUIREMENTS/ CHECKING EXPERIMENTALLY

With the Ledon circuit working, I wanted to see how good an estimate I could make on checking the current requirements for the circuit. With the current known, I can go back and determine the power being used by the circuit using the formula:

$$Power = V * I$$

Where $V$ is the voltage applied to the circuit (+5 volts for the experiments presented in this chapter).

Earlier in the book, I stated that I didn't think that my PICmicro® MCU current estimations would be very accurate. When I was developing the power supply specification for the application, I "derated" the calculated current value by 25 to 100%. In this experiment, I wanted to check how useful this derating value is and whether or not I can accurately predict how much current the application really requires.

When I look at the PIC16F84 datasheet, I can see that at 4 MHz and the XT oscillator specified, the PICmicro® MCU requires a typical intrinsic current of 1.8 mA and a maximum intrinsic current of 4.5 mA. This means that when the LED is off (no current flowing through it), anywhere from 1.8 to 4.5 mA would flow through the circuit.

When the LED is turned on, the current passing through the PICmicro® MCU will be increased by the current that is being "sunk" through the LED. For my typical LED circuits, I assume that the LED has a voltage drop of 0.7 volts (the same as any silicon diode) with a maximum current of 20 mA. To provide this current, I placed a 220-Ohms current-limiting resistor in series with the LED.

The 220-Ohms resistor was chosen by using Kirchoff's law, which states that the voltage applied to a circuit is equal to the voltage drops within it. If 5.0 volts was applied to the circuit and the LED has a voltage drop of 0.7 volts, then the resistor has 4.3 volts across it. Knowing that the LED must have a maximum of 20 mA flowing through it, I used Ohm's law to calculate the resistance:

$$\begin{aligned} R &= V/I \\ &= 4.3 \text{ volts} / 20 \text{ mA} \\ &= 215 \text{ Ohms} \end{aligned}$$

I used a 220-Ohms resistor because it is a standard value. The actual current flowing through the LED/resistor and "sunk" by the PICmicro® MCU is then:

$$I = V/R$$
$$= 4.3 \text{ volts}/220 \text{ Ohms}$$
$$= 19.5 \text{ mA}$$

Therefore, when the LED is on, the total current passing through the PICmicro® MCU is 1.8 mA typical intrinsic current plus 19.5 mA of LED current. For the power supply of this circuit, I would probably derate this by 50% in real life, so I would have to provide a 30-mA 5-volt power supply for this circuit.

I used this value for the total current used by the application when the LED is on. I realize that I am not including the current through the momentary on switch, but this will be 10 $\mu$A (according to Ohm's Law) and it really doesn't change the total current required by the application in an appreciable manner.

With the application designed and the total current estimated, it is now time to do an empirical check on what the PICmicro® MCU actually requires. To do this experiment, I used the same circuit as the previous experiment, except that I "broke" the ground connection between the PICmicro® MCU and ground and wired in my DMM set on the *mA* reading (Fig. 15-57). With this circuit, I can now check the actual current drawn by the Ledon circuit.

If you are using the YAP-II for this circuit, I suggest that you build this circuit on the prototype board with the PICmicro® MCU separate (although you can use the built-in LEDs and switches) so that you can monitor the current.

When I first turned on the power to this circuit (LED off), I found that the current passing from the PICmicro® MCU to ground was 1.4 mA. This is a bit lower than the "typical" value quoted by Microchip, but only off by 400 $\mu$A. This might have been a part variation, but it was still within the minimum derating tolerance that I set for PICmicro® MCU circuits (25%) and at a level that is difficult to measure with most low-cost hand held DMMs.

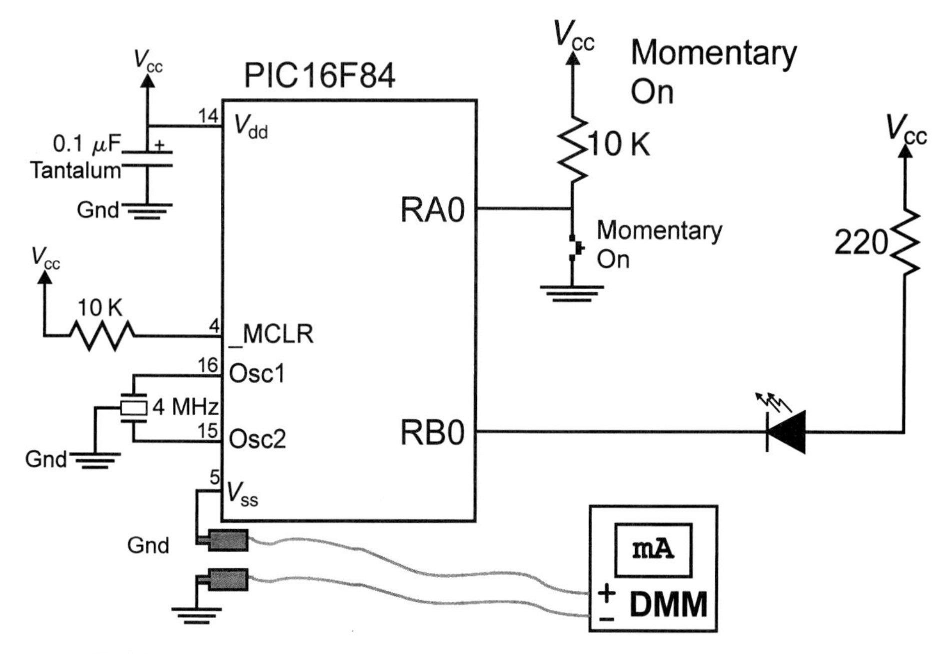


**Figure 15-57** Checking the current draw by the Ledon circuit

When I turned on the LED, I found that the current jumped to 14.1 mA—28% lower than what I calculated. This was the first time that I had really checked the current drawn by an LED connected to a PICmicro® MCU and I was surprised at the difference, which was larger than I had expected.

Looking at the operation of the circuit, I found that one of my basic assumptions was incorrect. When I assumed that the LED had a 0.7-volt drop, I was wrong. The actual voltage drop across the LED was 2.12 volts and the actual drop across the resistor was 2.6 volts, which changes the current calculations dramatically. Going back to Ohm's law for the resistor, the current flowing through the resistor/LED combination is actually 11.8 mA.

Adding 11.8 mA to the observed intrinsic current of 1.4 mA, the total current expected is really 13.2 mA, a difference of only 6% from the observed current of 14.1 mA. When I replaced the LED with a 1N914 silicon diode, I found that the actual current from the PICmicro® MCU through ground jumped to 18.7 mA, which is very close to my original calculation.

This was the first time I had ever measured the voltage across a LED and I guess I should have done it sooner because I knew LEDs are not built from silicon like "normal" diodes. Instead, they are built with other materials that aren't as electrically efficient as silicon, but do emit light in the visible spectrum (silicon emits light in the infrared spectrum). These materials have a much higher P/N junction voltage drop, as observed here.

The question remains after working through this experiment: if I was able to observe a 6% difference between my calculations, why do I still want to derate this value by up to 100%? The answer lies in the efficiency and stability of most power supply circuits. As the load approaches the rated maximum current supplied by a power supply, the power level could sag or have extra ripple that will affect the performance of the application (i.e., it will become intermittent).

As stated elsewhere, power supply problems are probably the most difficult to find and fix. So, by derating the circuit's current requirements, I have added a safety margin for the power supply that will help it work properly—even when the application is drawing its full load.

## DEBOUNCE: BUTTON PRESS WITH DEBOUNCE

Depending on the quality of the switch that you used for the first PICmicro® MCU hardware experiment (turning on an LED), you might have noticed that the LED flickered while you were pushing down on the switch. This is caused by intermittent contacts being made in the switch. The original switches that I used for electronics had a lot of problems with this situation because they're weren't very good quality. This is one reason why I suggest that you use switches that "click" when they are pressed and released; the clicking action ensures a good contact and it "wipes" the switch's contacts clean.

Even with this capability built within the switch, if you were to observe the action of the switch, you would probably see some "bouncing" (Fig. 15-58). The jagged edges on the oscilloscope plot are the bounces that occur within the switch when it is pressed. For this experiment, I used the same circuit as the Ledon experiment (Fig. 15-59) and the oscilloscope picture shown in (Fig. 15-58) was taken at RA0.

The bill of material for Debounce is shown in Table 15-6. This circuit can be built on a breadboard (Fig. 15-60) or using the YAP-II with the wiring shown in Fig. 15-61.

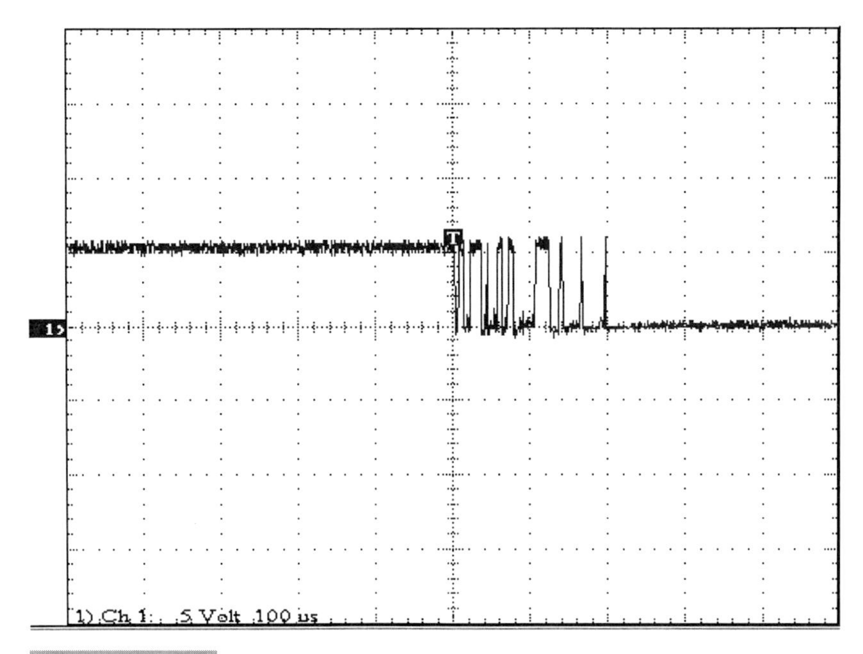


**Figure 15-58** Oscilloscope picture of a switch bounce

To eliminate the switch bounce, code must be written to wait for the bouncing to stop for a set amount of time. As indicated elsewhere in the book, the normal interval of time to wait for the bouncing to stop is 20 ms. This is done by continually polling the switch until it has remained in the same state for 20 ms. This operation is shown in the pseudo-code:

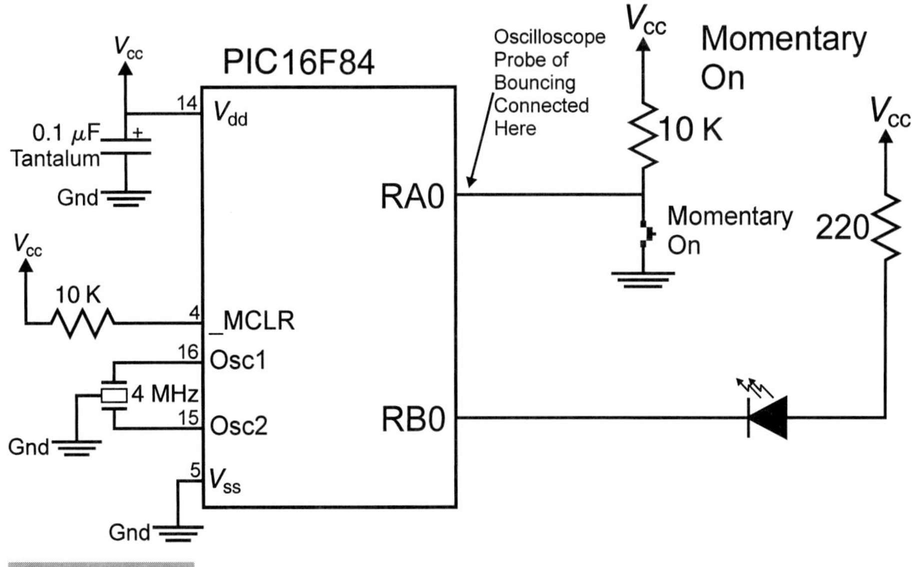


**Figure 15-59** Debounce circuit

**TABLE 15-6 Bill Of Materials For The "Debounce" Experiment**

| PART | DESCRIPTION | REQUIRED ON YAP-II |
|---|---|---|
| PICmicro® MCU | PIC16F84-04/P | In socket |
| Vdd/Vss decoupling capacitor | 0.1 μF tantalum | No |
| _MCLR pull up | 10 K, 1/4 watt | No |
| 4-MHz ceramic resonator | 4 MHz with built-in capacitors | No |
| RA0 Pull up | 10 K, 1/4 watt | No. BUT1 used |
| RA0 Switch | Momentary on | No. BUT1 used |
| RB0 LED | Red LED | No. LED0 |
| RB0 LED resistor | 220 Ohm, 1/4 watt | No. LED0 |

```
Debounce_Down:                    //  Wait for the Switch to be held down
                                  //    for 20 msec
  while (Switch == Up);           //  Wait for the Switch to go down

  for (Dlay = 0; (Switch == Down) && (Dlay < 20msec); Dlay++);
                                  //  Poll the Switch while Incrementing
                                  //    the Delay Timer

  if (Dlay < 20msec)
    goto Debounce_Down;           //  If 20 msecs has NOT passed, wait for
                                  //    Switch Down and Repeat Process

//  When Execution reaches here, the switch is "Down" and Debounced
```

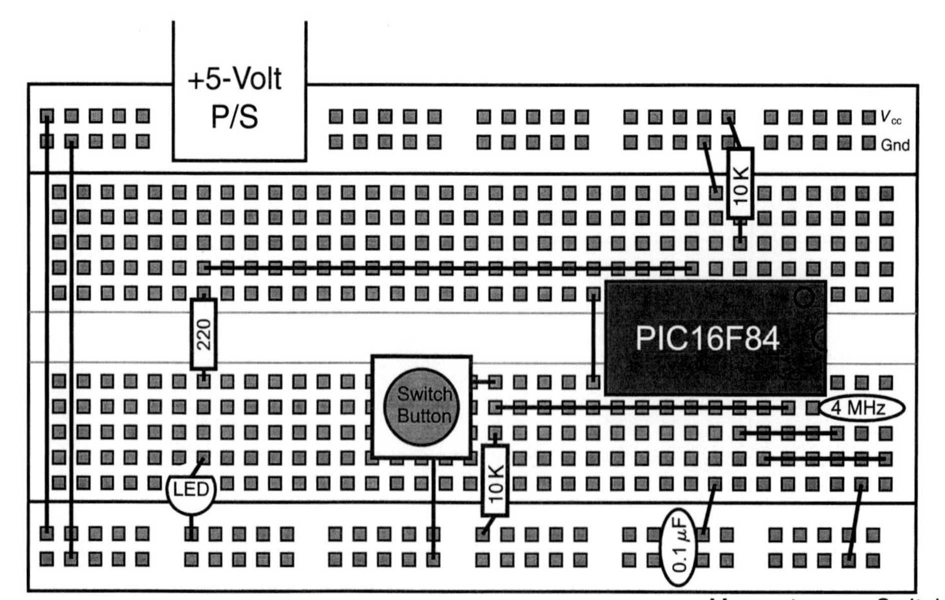


**Figure 15-60** Debounce 16F84 breadboard circuit

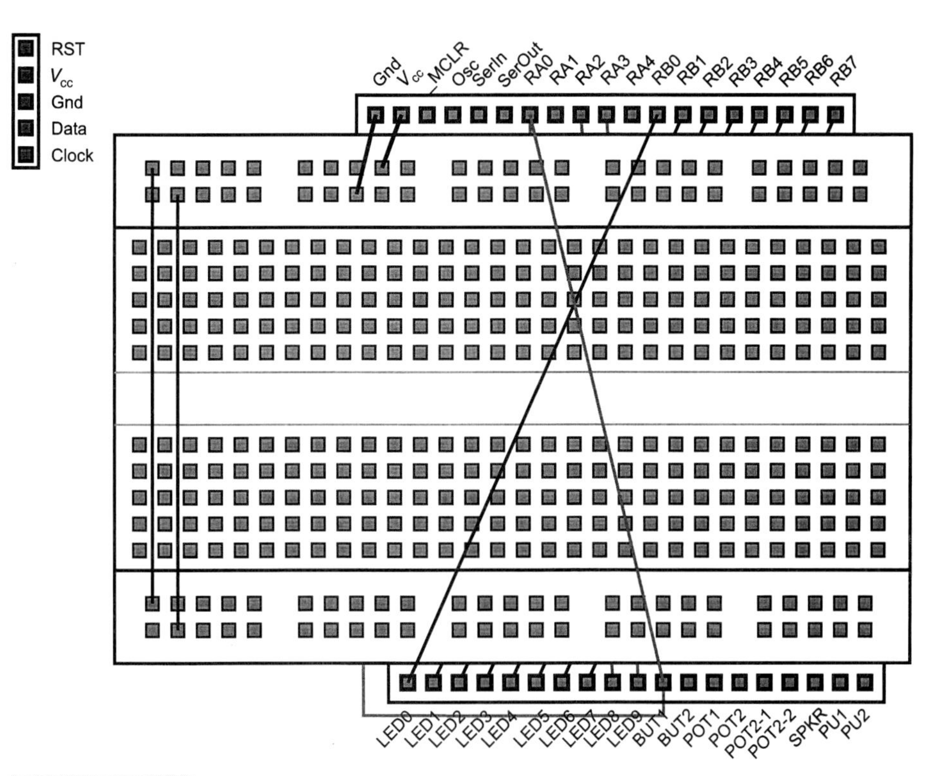


**Figure 15-61** EMU-II/YAP-II breadboard for Debounce

The actual PICmicro® MCU assembly code that debounces a switch "down" for a PIC16F84 running at 4 MHz with the button at RA0 is:

```
Debounce_Down

  btfsc  PORTA, 0
   goto  Debounce_Down      ;  Wait for Button "Down"

  movlw  0x0100 - 0x0C4     ;  Initialize Dlay for a 20 msec
  movwf  Dlay               ;   Delay
  movlw  0x0100 - 0x00A
  movwf  Dlay + 1

  bcf    STATUS, Z          ;  Make Sure that Zero is Reset

Debounce_Down_Loop
  incfsz Dlay, f
   goto  $ + 2
  incf   Dlay + 1, f
  btfsc  PORTA, 0
   goto  Debounce_Down       ;  No - Loop Around Again
  btfss  STATUS, Z           ;  Zero Flag Set (20 mSecs Past?)
   goto  Debounce_Down_Loop

;   When Execution reaches here, the switch is "Down" and Debounced
```

One aspect of this code that might be confusing is how I came up with the values *0x0100 - 0x0C4* and *0x0100 - 0x00A* for the initial values in *Dlay* and how the delay count works in *Debounce_Down_Loop*. The code in *Debounce_Down_Loop* takes eight instruction cycles to execute for a 20-ms delay and a 4-MHz clock speed, which means it has to execute 2,500 times.

2,500 in decimal is 0x009C4. Because the *incf* instruction is used for the upper byte, I added one cycle to it to ensure that it increments the lower byte 9 * 256 times through and once for 0x0C4 (196 decimal) times.

The three instructions for the delay count:

```
incfsz Dlay, f
 goto  $ + 2
incf   Dlay + 1, f
```

always execute in the same number of cycles and only set the STATUS register's Zero flag when the 2,500 loops through *Debounce_Down_Loop* have executed. This code could have been executed differently, specifically using decrements, but I wanted to try something different. The advantage of this code is that it allows the loop to take a total of eight cycles, which is evenly divisible by 20,000.

This code is useful for checking for button down, but the button up also has to be checked. In the following Debounce source code snippet, in the switch up and down, the Debounce application first checks for the button to be debounced as up, before waiting for it to go down. When the switch has been debounced up, then the *Debounce Down* routine is invoked, which waits for the switch to be debounced down, at which point PORTB is complemented, to change the state of the LED output on RB0.

```
Loop                          ;  Loop Here
 Debounce Up                  ;  Wait for Key to Go Up
 Debounce Down                ;  Wait for Key to Go Down
  comf   PORTB, f             ;  Toggle the PORTB Value
  goto   Loop
```

In this source code, you'll see that I invoke *Debounce* with the parameters *Up* and *Down,* which is the macro version of the button debounce code I presented previously. I used a macro for this function simply because it is repeated. By providing a macro for essentially the same code, I avoid having to write it twice and, more importantly, I avoid the opportunity for errors between the same functions to be repeated.

The source code for Debounce, which is located in the *code\debounce* subdirectory of the *PICmicro* directory is:

```
 title  "debounce - Debounce Button input and Toggle LED"
;
;  This Application waits for the button on RA0 to be released
;   (for more than 20 msecs) and then when it has been pressed and
;   stays constant for 20 msecs, the LED on RB0 is toggled.  The
;   Process then repeats.
;
;  Hardware Notes:
;   _MCLR is tied through a 4.7K Resistor to Vcc and PWRT is Enabled
;   A 220 Ohm Resistor and LED is attached to PORTB.0 and Vcc
;   A 10K pull up is connected to RA0 and it's state is passed to
;    RB0
```

```
;
;  Myke Predko
;  99.12.04
;
  LIST R=DEC
 ifdef __16F84
  INCLUDE "p16f84.inc"
 endif
 ifdef __16F877
  INCLUDE "p16f877.inc"
 endif

;  Registers
 CBLOCK 0x020
Dlay:2                      ;  Delay Value
 ENDC

Up   EQU 1                  ;  Flag Value for Debounce "Up"
Down EQU -1                 ;  Flag Value for Debounce "Down"

;  Macros
Debounce MACRO Direction

 if (Direction < 0)         ;  Going Down
  btfsc  PORTA, 0
 else
  btfss  PORTA, 0           ;  Wait for Button Released
 endif
   goto  $ - 1

  movlw  0x0100 - 0x0C4     ;  Initialize Dlay for a 20 msec
  movwf  Dlay               ;   Delay
  movlw  0x0100 - 0x00A
  movwf  Dlay + 1

  bcf    STATUS, Z          ;  Make Sure that Zero is Reset

  incfsz Dlay, f
   goto  $ + 2
  incf   Dlay + 1, f
 if (Direction < 0)
  btfsc  PORTA, 0
 else
  btfss  PORTA, 0           ;  Button Still Released?
 endif
   goto  $ - 11             ;  No - Loop Around Again
  btfss  STATUS, Z          ;  Zero Flag Set (20 mSecs Past?)
   goto  $ - 6
 ENDM                       ;  End the Macro

 ifdef __16F84
 __CONFIG _CP_OFF & _WDT_OFF & _XT_OSC & _PWRTE_ON
 else
 __CONFIG _CP_OFF & _WDT_OFF & _XT_OSC & _PWRTE_ON & _DEBUG_OFF &
_LVP_OFF & _BODEN_OFF & _WRT_ENABLE_ON & _CPD_OFF
 endif

  PAGE
;  Mainline of ledon

 org     0

  nop

  bsf    PORTB, 0           ;  Make LED on RB0 "off" Initially
```

```
    bsf     STATUS, RP0         ;  Goto Bank 1 to set Port Direction
    bcf     TRISB & 0x07F, 0    ;  Set RB0 to Output
    bcf     STATUS, RP0         ;  Go back to Bank 0

  Loop                          ;  Loop Here

   Debounce Up                  ;  Wait for Key to Go Up

    nop                         ;  Location for Stopping after
                                ;    "Up" Debounce
   Debounce Down                ;  Wait for Key to Go Down

    comf    PORTB, f            ;  Toggle the PORTB Value

    goto    Loop

   end
```

I want to point out a few things in the Debounce source code. First is the addition of the *nop* instruction between the two Debounce macro invocations. This instruction is used as an MPLAB simulator breakpoint to test the code. You cannot put a breakpoint at a macro invocation and there would be no way to easily check whether the Debounce Up macro invocation worked correctly unless you were to input a high value on the switch at RA0, wait 20 ms, and then input a low value. The *nop* instruction allows you to "hang" a breakpoint between the two macro invocations during simulation in source code.

To test this application, I used the asynchronous input, just as was used in the Ledon experiment. To test the debounce capability of the Debounce macro, after I had toggled RA0 high, I waited a few cycles (anywhere from 10 to several millisecond's worth) to test the operation of Debounce. In my PC (300 MHz Pentium II), waiting a full 20 ms takes about two minutes. If you have a slower PC, you might want to reduce the loop values while you are simulating the application.

In the Debounce application, notice that I debounce the button high (not pressed) before debouncing it low. This is done to ensure that a definite break occurs between button presses. This check is normally required for most applications that process button inputs.

The Debounce macro isn't bad for applications that require a single button's input to debounce before proceeding with an application. For applications that cannot spend large periods of time waiting for button inputs, the debounce code will have to be interrupt based, as shown in a later experiment.

## PINCHG: CHANGING AN OUTPUT BIT VALUE INADVERTENTLY

When I looked at the issue of how the PICmicro® MCU's I/O pins worked, with regard to inadvertently setting or resetting them in the first edition, I have to admit that I really came up with a contrived situation. I always wanted to show how this could happen in real life, but I couldn't come up with a good example.

I hadn't come up with a good, real-life example until I was working on the updated YAP-II programmer. When I was doing an initial version of the programmer, I used just MOSFET transistors for the inverters because they are very simple to work with and don't require the driver-to-base current-limiting resistor of bipolar transistors. When I built the circuit, I accidentally substituted a 3904 bipolar NPN transistor for a 2N7000 N-channel MOSFET and found that the $V_{pp}$ driver shut off when I started to program the part.

After a bit of probing around the board, I found that the 3904 had been placed in the circuit incorrectly (without the normal 330-Ohm current-limiting resistor) and the voltage on the output pin was only 0.8 volts when the 3904 was on. When another pin was set using the *bsf* instruction, the read of the pin the 3904 was connected to returned a *0* because the 3904 was pulling it down. Knowing that, the *bsf* instruction executes as:

```
bsf PORT#, PIN#        ; PORT# = PORT# | (1 << PIN#)
```

When the pin with the incorrect transistor was connected to it was read, even though it was driving a *1*, returned a *0* because the output is below the voltage threshold for a *1* (which is approximately one half $V_{dd}$ for most PICmicro® MCU pins).

The fix for this problem was to replace the incorrect part with the correct one—or I could have added a 330-Ohm resistor between the PICmicro® MCU's pin and the transistor. I tried the second solution and confirmed that it worked.

This experience led to the Pinchg application, which is located in the *code\pinchg* subdirectory the *PICmicro* directory. The source is very simple:

```
 title  "PinChg - Change the State of an LED Inadvertently"
#define _version "1.00"
;
;  Update History:
;
;  99.11.23 - Created
;
;  This Application turns on a LED at RB0.  An input switch
;   at RA0 is then polled and when it is pressed, a LED at
;   RB4 is then turned on using the "bsf" instruction.  The
;   LEDs will be driven from PICmicro® MCU's I/O Ports directly
;   (no current limiting resistor and to Ground), so when
;   the button is pressed, the LED at RB0 should go out.
;
;  Myke Predko
;
;  Hardware Notes:
;  16F84 Running at 4 MHz
;  RA0 - Button Pulled up to Vcc and Active when pressed
;  RB0, RB1 - LED controlled by a NPN/N-Channel FET Transistor
;   with a 220 Ohm Pull Up
;
 LIST R=DEC
 ifdef __16F84
  INCLUDE "p16f84.inc"
 else
 ifdef __16F877
  INCLUDE "p16f877.inc"
 endif

;  Variable Register Declarations

;  Macros

 ifdef __16F84
 __CONFIG _CP_OFF & _WDT_OFF & _XT_OSC & _PWRTE_ON
 else
 __CONFIG _CP_OFF & _WDT_OFF & _XT_OSC & _PWRTE_ON & _DEBUG_OFF &
_LVP_OFF & _BODEN_OFF & _WRT_ENABLE_ON_ & _CPD_OFF
 endif
```

```
 org     0
Mainline

  nop

  clrf  PORTB              ;  Initialize PortB to Nothing On
  bsf   STATUS, RP0
  movlw 0x0FC              ;  Set RB0 & RB4 as Outputs
  movwf TRISB ^ 0x080
  bcf   STATUS, RP0

  bsf    PORTB, 0          ;  Turn on RB0 LED

Loop
  btfsc PORTA, 0           ;  Wait for Button to be Pressed
   goto Loop

  bsf   PORTB, 1           ;  Turn on RB1 LED
                           ;  #### - RB0 LED Turned Off

  goto   $                 ;  Loop Forever

 end
```

After you have assembled this circuit, try it out in the simulator using a programmable button for stimulus. By doing this, you should find that pin RB0, once it is set, it is never reset—even after the button (RA0 going low) has been pressed. This is because the simulator cannot pull this pin low.

To actually see the problem, you will have to build the circuit shown in Fig. 15-62. This circuit uses two 3094s that do not have current-limiting resistors between the PICmicro® MCU's pin and their bases to replicate the problem.

The bill of materials for Pinchg is shown in Table 15-7. The breadboard circuit is shown in Fig. 15-63 and the YAP-II circuit is shown in Fig. 15-64. This circuit is somewhat more

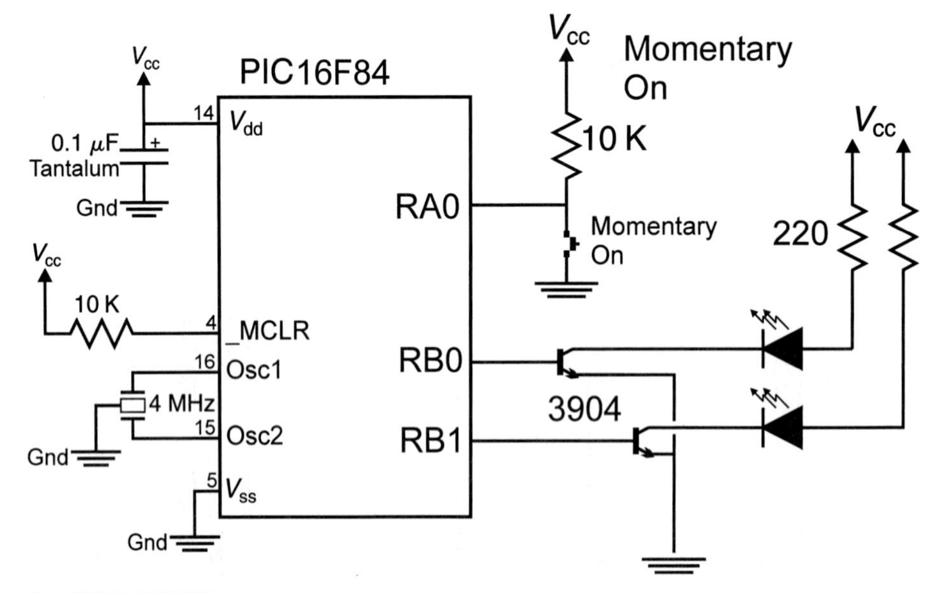


**Figure 15-62** **Pinchg: inadvertently turn off one LED while turning on another**

**TABLE 15-7 Bill Of Materials For The "DEBOUNCE" Experiment**

| PART | DESCRIPTION | REQUIRED ON YAP-II |
|---|---|---|
| PICmicro® MCU | PIC16F84-04/P | In socket |
| Vdd/Vss decoupling capacitor | 0.1 $\mu$F tantalum | No |
| _MCLR Pull up | 10 K, 1/4 watt | No |
| 4-MHz ceramic resonator | 4 MHz with built-in capacitors | No |
| RA0 Pull up | 10 K, 1/4 watt | No. BUT1 used |
| RA0 Switch | Momentary On | No. BUT1 used |
| Transistors | 2N3904 | Yes |
| RB0, RB1 LEDs used | Red LEDs | No. LED0/LED1 |
| RB9, RB1 LED resistors | 220 Ohm, 1/4 watt | No. LED0/LED1 used |

complex than what was built previously, but it will only require 10 minutes or so to assemble—even if you are starting from scratch. Notice that if you are working with the EMU-II or YAP-II, you only have to supply the transistors.

When you apply power to the circuit (start the programmed PIC16F84 in the YAP-II executing), you should see that the LED connected to RB0 through the 3904 is lit. Look-

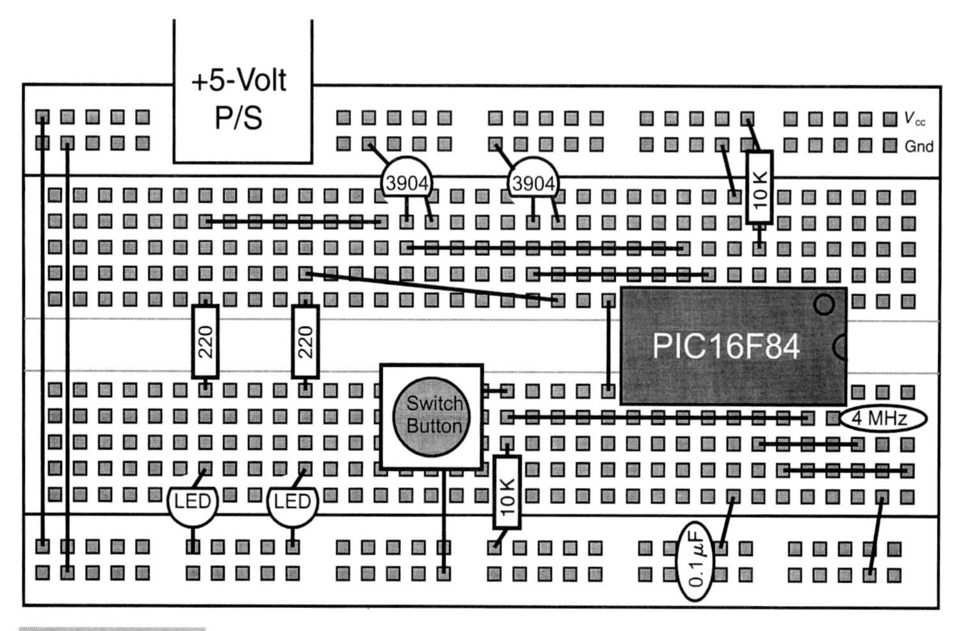


**Figure 15-63** Pinchg 16F84 breadboard circuit

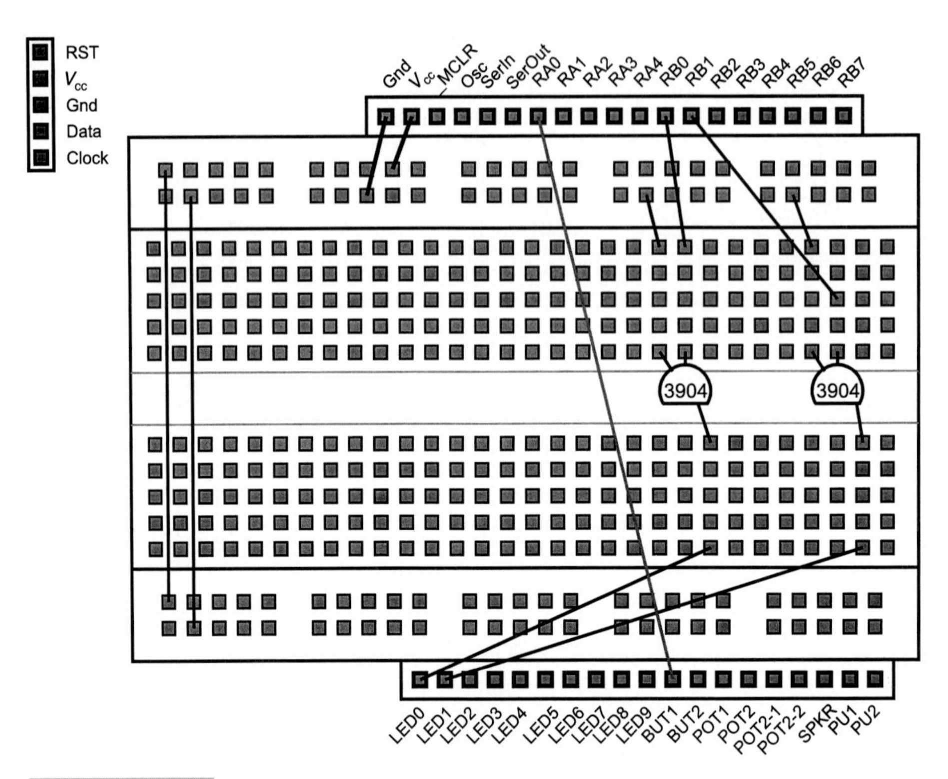


**Figure 15-64** EMU-II/YAP-II breadboard for Pinchg

ing at the previous source code, you'll see that the LED is enabled using a *bsf* instruction. When the button is pressed, you'll see the LED at RB1 light, but the LED at RB0 will shut off.

To understand what is happening, reset the PICmicro® MCU and do a voltage measurement on RB0. You will see that the voltage output with the 3904 in place is somewhere between 0.8 and 1.0 volts. This value is less than the voltage threshold for a *1* (which is 0.48 $V_{dd}$, according to the PIC16F84's data sheet). When it is read during the *bsf PORTB, 1* instruction, a *0* is returned for this bit and this value is then written back.

In the oscilloscope picture (Fig. 15-65), the transition point for *RB0* going low and *RB1* going high is when the *bsf PORTB, 1* instruction is executed.

Finding this cause of the inadvertent turning off of $V_{pp}$ in the YAP-II prototype took me about two minutes. This is not an attempt to brag, but rather to point out how I looked for the problem. When the problem occurred, instead of going over the code, I took a look at the voltage level of the pin when it was active. This led me to go back over the parts that I had soldered into the board. From there, I was able to find the incorrect part.

When you experience a situation where an output changes unexpectedly, you should first check its voltage level when it is active and be sure that it is above the PICmicro® MCU's switching voltage threshold. At first, it might seem to make more sense to look at the code, but in some situations (such as this), a hardware problem is the root cause of a pin change. These problems are fast and usually easy to fix.

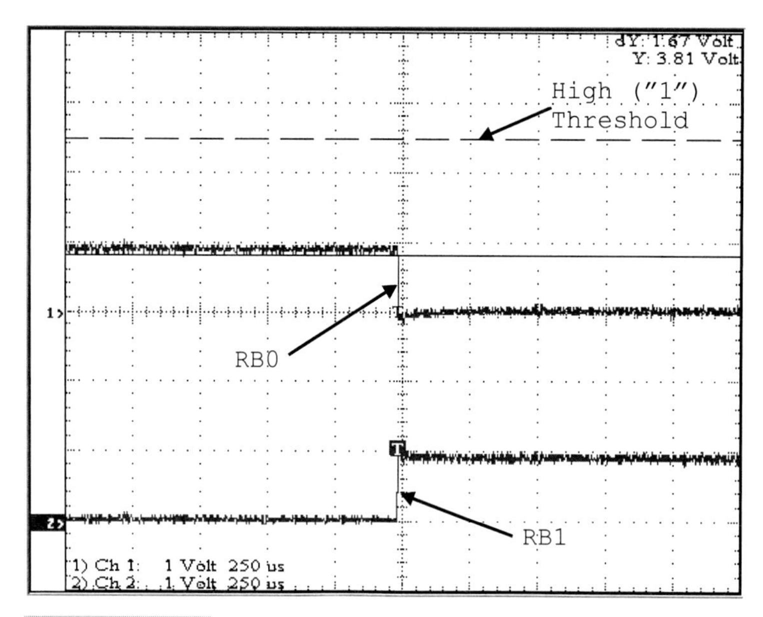


**Figure 15-65** Waveform on two LEDs when button pressed

## TIMEEND: TMR0 DELAY THAT NEVER ENDS

There is an awful lot to remember about the PICmicro® MCU. Later in this chapter is a small application for you to debug. Also, an appendix provides you with a table of responses to find and fix different problems to help get yourself out of a mess. Up to this point in the book, I have given you a lot of little hints and things to beware of. These really are not that easy to remember unless you have some examples to work through. This experiment illustrates one aspect of TMR0 and the prescaler that you should be aware of.

The TimeEnd application is designed to use TMR0 to delay 1 ms after reset and turn on an LED on RB0. This application is easy to set up and run. The parts list is shown in Table 15-8.

The application circuit can be built on either a breadboard (Fig. 15-66) or using the YAP-II (Fig. 15-67).

Once you've built the circuit, you can program a 16F84 with the TimeEnd application (located in the *code\TimeEnd* subdirectory of the *PICmicro* directory).

```
 title  "TimeEnd - Loop While Waiting for TMR0 to End"
;
;  This Uses TMR0 with a 4x Prescaler to Provide a 1 msec
;   Delay.  The Access instruction will cause the Timer to
;   Never Reach Zero
;
;
;  Hardware Notes:
;   PIC16F84 Running at 4 MHz
;   _MCLR is Pulled Up
;   PortB0 Pulled up and Connected to LED
;
```

**TABLE 15-8 Bill Of Materials For The "Timeend" Experiment**

| PART | DESCRIPTION | REQUIRED ON YAP-II |
|---|---|---|
| PICmicro® MCU | PIC16F84-04/P | In socket |
| Vdd/Vss decoupling capacitor | 0.1 $\mu$F tantalum | No |
| _MCLR pull up | 10 K, 1/4 watt | No |
| 4-MHz ceramic resonator | 4 MHz with built-in capacitors | No |
| RB0 LED | Red LED | No. LED0 used |
| RB0 LED resistor | 220 Ohm, 1/4 watt | No. LED0 used |

```
;  Myke Predko
;  99.06.22
;
  LIST R=DEC
 ifdef __16F84
  INCLUDE "p16f84.inc"
 endif
 ifdef __16F877
  INCLUDE "p16f877.inc"
 endif

;  Register Usage
 CBLOCK 0x020                   ;  Start Registers at End of the SFRs
 ENDC
```

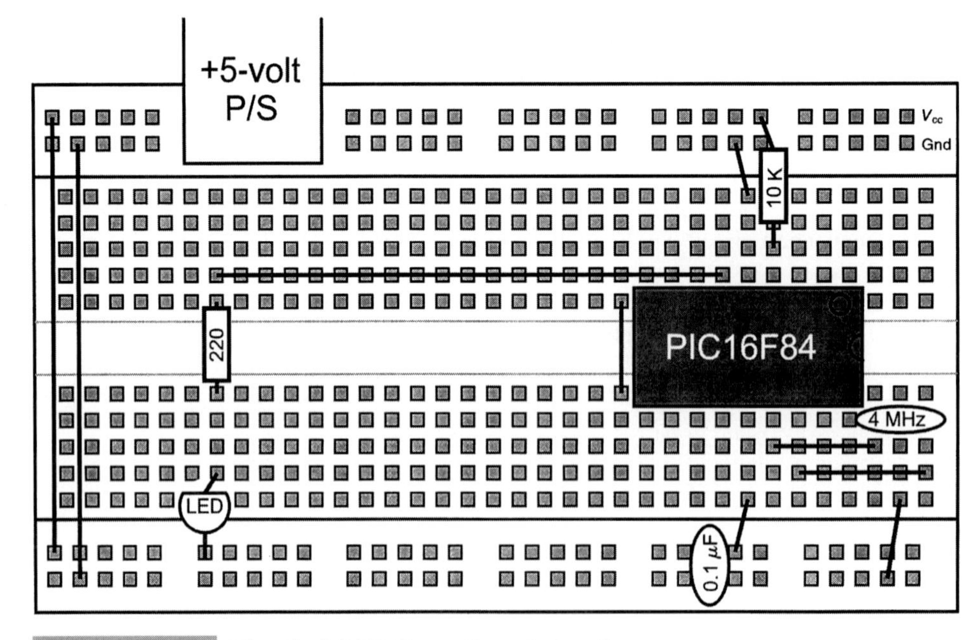


**Figure 15-66** TimeEnd 16F84 breadboard circuit

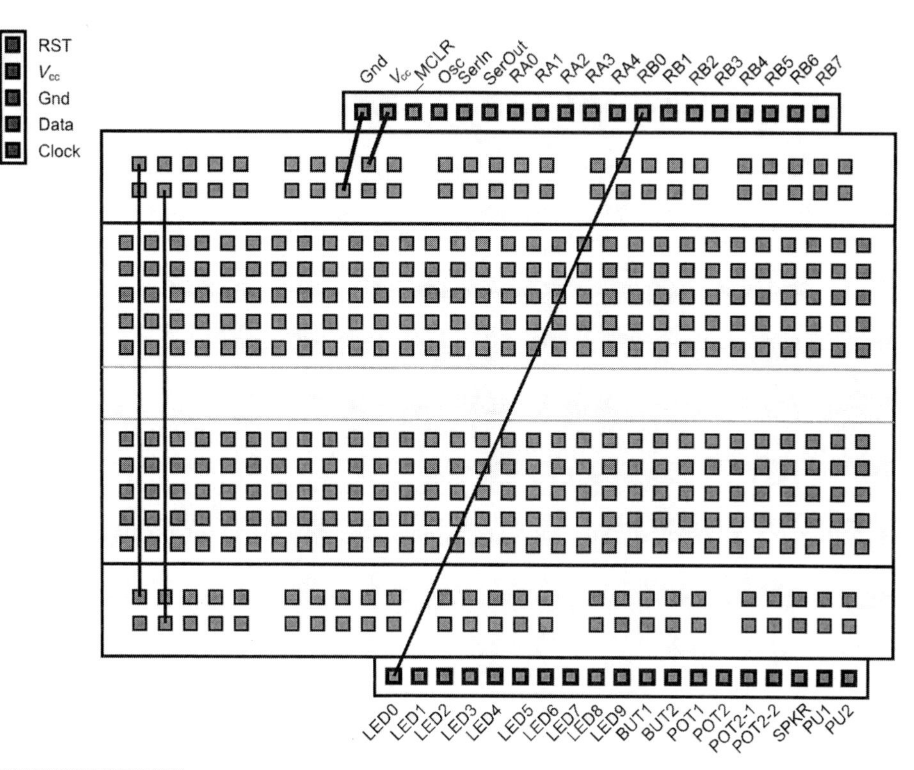


**Figure 15-67** EMU-II/YAP-II breadboard for TimeEnd

```
  PAGE
  ifdef __16F84
  __CONFIG _CP_OFF & _WDT_OFF & _XT_OSC & _PWRTE_ON
  else
  __CONFIG _CP_OFF & _WDT_OFF & _XT_OSC & _PWRTE_ON & _DEBUG_OFF &
_LVP_OFF & _BODEN_OFF & _WRT_ENABLE_ON_ & _CPD_OFF
  endif
 ;  Mainline of TimeEnd

   org    0

   nop

   bsf    PORTB, 0          ;  Turn RB0 off
   bsf    STATUS, RP0
   clrf   TRISB ^ 0x080
   movlw  0x0D1             ;  Set TMR0 to Prescaler and 4x
   movwf  OPTION_REG ^ 0x080
   bcf    STATUS, RP0

   movlw  (1024 - 1000) / 4 ;  Reset TMR0 to Wait 1 msec
   movwf  TMR0

 Loop
   movf   TMR0, f           ;  Does TMR0 Equal 00?  (1 msec Passed)
   btfss  STATUS, Z
    goto  Loop
```

```
    bcf     PORTB, 0            ;  Turn RB0 ON

    goto    $                   ;  Loop Forever When Done

    end
```

The code is very simple, sets up the prescaler to be used with TMR0 to divide by four, and has TMR0 driven by the instruction clock. Once this is done, TMR0 is loaded with *6* and waits for the timer to overflow to zero. The only problem is that when you build the circuit and program a PICmicro® MCU with TimeEnd, the application "sits there" and doesn't turn on the LED.

When you look at the source, you probably won't see anything amiss. The problem only becomes apparent when you simulate the application and look at the value of TMR0 as the PICmicro® MCU executes through *Loop*. Each time *Loop* executes, TMR0's value stays the same at *6*.

Figure 15-68 shows what you should be seeing when you simulate the application. Along with the standard processor registers, I added a TIMEEND watch window for the OPTION register, as well as TMR0. As the code executes, TMR0 stays at value *6* and never changes, no matter how long it is executed.

The problem with the application is the *movf TMR0, f* instruction just after the *Loop* label. This instruction moves the contents of TMR0 through the ALU to test the value against zero, conditionally set the Zero flag, and then stores this value back into TMR0. When the value is stored back in TMR0, the prescaler is reset.

If you check the number of cycles used by *Loop*:

```
Loop
  movf   TMR0, f               ;  Does TMR0 Equal 00?  (1 msec Passed)
  btfss  STATUS, Z
   goto  Loop
```

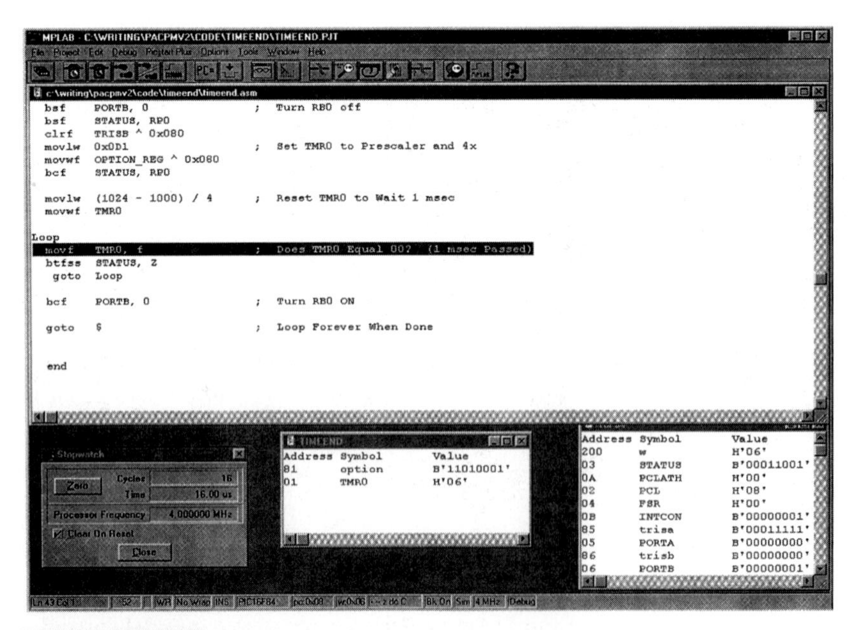


**Figure 15-68** MPLAB screenshot of TimeEnd experiment

you'll come up four instruction cycles each time *Loop* executes. This is a problem because the TMR0 prescaler is set to be a divide-by-four counter. Each time *Loop* executes, the prescaler is reset, never passing a clock tick to TMR0 to increment it. As a result, TMR0 never changes value and *Loop* becomes an endless loop.

The easiest way to fix the problem is to change the *movf TMR0, f* instruction to *movf TMR0, w* with the value in TMR0 being stored in the w register instead of back into itself. When you program the PICmicro® MCU with the updated code, you'll find that it works without any problems.

There might be a question as to how likely this problem is. If you're going to use TMR0, chances are that you will use it with an interrupt, so there would be no need to poll it at all. This is true for mid-range devices, but what about the low-end?

Using TMR0 to set up delays in the low-end PICmicro® MCUs is a fairly common operation. The problem shown here causes problems for many new application developers. The easiest way to avoid this is to *never* write to the TMR0 register, except when setting it originally. If you have a value in w that you don't want changed, then it should be saved before polling TMR0.

## DECOUPLE: POWER/DECOUPLING PROBLEMS

The Decouple application (which uses the circuit shown in Fig. 15-69) is an attractive LED display that demonstrates the need for decoupling the PICmicro® MCU's power ($V_{dd}$ and $V_{ss}$) pins. The circuit itself is not terribly hard to build on a breadboard, but it does require quite a few wires (Fig. 15-70).

If the YAP-II is used, then to demonstrate the operation of the application, the PICmicro® MCU will have to be taken out of the socket after programming and put into the built-in breadboard (Fig. 15-71). When the PICmicro® MCU is put into the breadboard, you will be responsible for providing power, clocking, and reset.

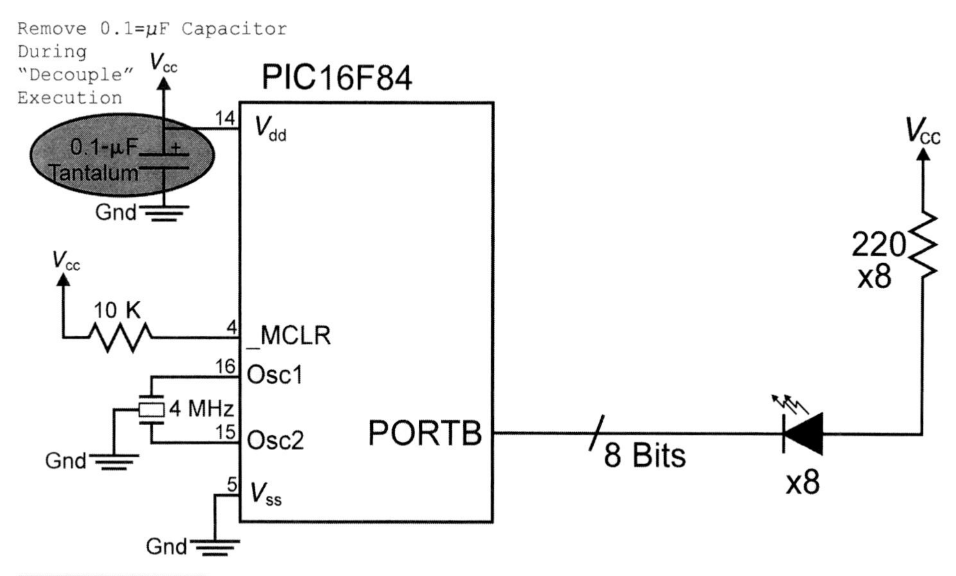


**Figure 15-69** **Decouple: Hammer LEDs on PortB to maximize internal current transients**

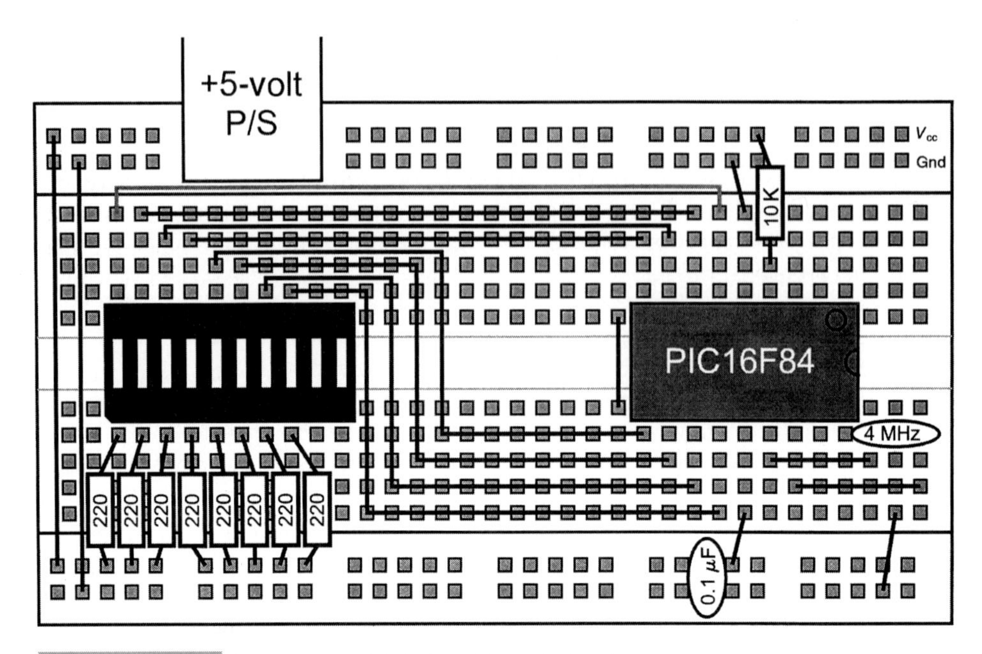


**Figure 15-70** Decouple 16F84 Breadboard circuit

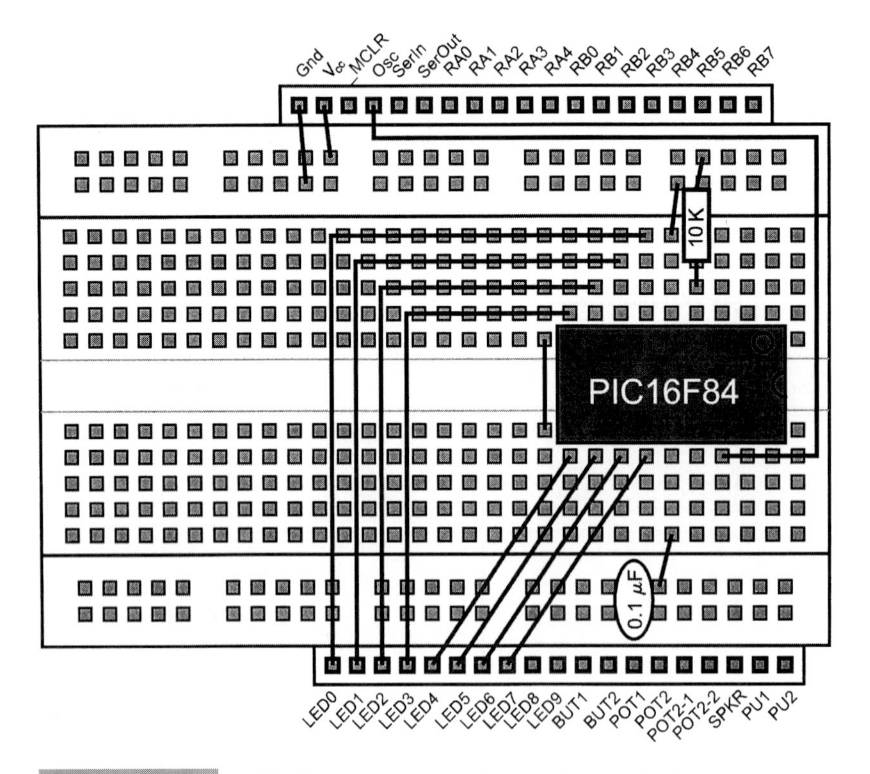


**Figure 15-71** YAP-II breadboard for Decouple application

Figure 15-71 shows a separate 10-K resistor connected to the PICmicro® MCU's _MCLR pin instead of one of the"pull ups built into the YAP-II. One of the built-in pull-ups could have been used, but to try and keep the diagram as uncluttered as possible, I show a 10-K pull up connected between _MCLR and $V_{cc}$.

When I built the circuits for myself (Fig. 15-70), notice that I used a 10-LED bargraph display. This display is a lot easier to put into an application than multiple LEDs, is much neater, and does not require any lead forming. The part is somewhat more expensive than eight (or 10) individual LEDs, but it pays for itself in simplifying the application's wiring.

The parts shown in Table 15-9 are required for Decouple.

This application is not terribly hard to build, although it is the most complex thus far in this chapter. Notice that in the wiring diagrams (Figs. 15-70), the 10-LED bargraph display has a notch on it. The side with the notch is the anode. It must be connected to the positive voltage supply for the application to work.

When building a circuit like this on a breadboard with a large number of wires leading to a similar location (or have to go between circuits, as the wires between the *PICmicro* and LED display do), be sure that you leave enough space for the wires and to keep track of what is underneath. Thinking first will save you some work later when adding additional circuits to an application. I've been caught enough on this (and having to rebuild parts of a prototype circuit) that I always spend more than a few moments ensuring that space is left for later circuits to be added to the application.

The application code (located in the *code\Decouple* subdirectory of the *PICmicro* directory) flashes a varying number of LEDs on and off. The application starts with all eight LEDs connected to PORTB turned off and then starts turning on a single LED and incrementing the number of LEDs while turning off the LEDs between changes to the number of LEDs turned on. The purpose of this application is to induce as large current transients into the PICmicro® MCU as possible to see the effects of removing the decoupling capacitor connected to the $V_{dd}$ pin of the PICmicro® MCU.

```
  title  "Decouple - Decoupling Effects on a PICmicro MCU's Operation"
 #define nDebug
 ;
 ;  This Code "Hammers" a PICmicro MCU's PortB with Changing Values
 ;   to investigate the effects of a Decoupling Capacitor
```

**TABLE 15-9 Bill Of Materials For The "DECOUPLE" Experiment**

| PART | DESCRIPTION | REQUIRED ON YAP-II |
|---|---|---|
| PICmicro® MCU | PIC16F84-04/P | In socket |
| Vdd/Vss decoupling capacitor | 0.1 μF tantalum | Yes |
| _MCLR Pull Up | 10 K, 1/4 watt | (see text) |
| 4-MHz ceramic resonator | 4 MHz with built-in capacitors | No |
| 10-LED display | Bargraph red | No. LED0/LED7 used |
| 8 PORTB LED resistors | 220 Ohm, 1/4 watt | No. LED0/LED7 used |

```
;
;
;  Hardware Notes:
;   PIC16F84 Running at 4 MHz
;   _MCLR is Pulled Up
;   All 8 bits of PortB are Pulled up and Connected to LEDs
;
;  Myke Predko
;  99.12.22
;
  LIST R=DEC
  INCLUDE "p16f84.inc"

;  Register Usage
 CBLOCK 0x020                ;  Start Registers at End of the SFRs
Dlay:2
BValue
 ENDC

 PAGE
 __CONFIG _CP_OFF & _XT_OSC & _PWRTE_ON  & _WDT_OFF
                             ;  Note that the WatchDog Timer is OFF

;  Mainline of Decouple
  org    0

  movlw  0x0FF
  movwf  PORTB
  movwf  BValue

  bsf    STATUS, RP0         ;  Make All 8 PortB Bits Output
  clrf   TRISB ^ 0x080
  bcf    STATUS, RP0

Loop                         ;  Loop Here

  call    Delay

  bcf    STATUS, C           ;  Change PORTB

  btfss  BValue, 7
   bsf   STATUS, C
  rlf    BValue, w
  movwf  PORTB
  movwf  BValue

  call   Delay

  movlw  0x0FF               ;  Turn OFF LEDs
  movwf  PORTB

  goto   Loop

Delay                        ;  Delay 1/5 Seconds
  clrf   Dlay
  clrf   Dlay + 1
 ifndef  Debug
  decfsz Dlay, f
   goto  $ - 1
  decfsz Dlay + 1, f
   goto  $ - 3
 else
  nop
  nop
```

```
  nop
  nop
 endif

  return

  end
```

This experiment is not appropriate for use in an emulator. This is because of the intermittent nature of the application's execution. If an emulator is run without the correct decoupling capacitors, it could reset or "lock up," resulting in your being unable to see what is actually happening.

One aspect of this code that I want to bring to your attention is the Delay subroutine. When the two eight-bit variables that are used to count down are cleared, as in Decouple, the delay is approximately 200,000 instruction cycles. For a PICmicro® MCU running at 4 MHz, this translates to a fifth of a second delay that is useful for many applications.

After the circuit is built and the PICmicro® MCU programmed, you can run it and watch the LEDs "move" across the display. When the LEDs are turned on and off, the amount of current drawn through the PICmicro® MCU is changed. In current "draw" experiment earlier in the chapter, I have found that a visible-light LED and a 220-Ohm resistor will require approximately 12 mA. Thus, the LEDs are turning on and off, the current being passed to them changes from zero to 100 mA, in 12-mA increments. Changing the number of LEDs that are turned on was done to allow me to visually observe where the application is executing, as well as to change the current being passed through the PICmicro® MCU.

After a few minutes, remove the 0.1-$\mu$F decoupling cap and watch what happens. From my experience, one of three things can happen:

1. The PICmicro® MCU seems to reset itself at halfway (or a little beyond halfway) and start the application over.
2. The PICmicro® MCU locks up.
3. The application runs without any problems.

In the devices I tested this application with (several different PIC16F84s, PIC16F877s, PIC16C84s, and PIC16C711s), I got a range of results. When I created this experiment for the first edition, only the PIC16C84 was tested and I continuously got the first result (the PICmicro® MCU seemed to reset after five LEDs were lit and start the application up again with only one LED lit).

When I repeated this experiment with the other PICmicro® MCU part numbers, I got the varying results listed. These results were not consistent for any one part number (when different parts were tried). Actually, I was most surprised that the first PIC16F84 that I tried with this application didn't reset itself (as the PIC16C84, used in the first edition did). This caused me to retest the application on the different part numbers.

The obvious conclusion to the results of this experiment is that a decoupling capacitor on the PICmicro® MCU's $V_{dd}$ pin is crucial to proper operation of the device for widely varying current loads. When laying out your application, always be sure that you keep the decoupling capacitor as close to $V_{dd}$ as possible.

The proximity of the decoupling capacitor to the $V_{dd}$ pin of the PICmicro® MCU seems to be the crucial parameter when working with any digital device. As a matter of course, you should always place a decoupling capacitor by all active devices to ensure that varying current loads do not affect their operation or the operation of devices connected in the application circuit.

If you are working through the experiments one by one, you might want to leave this circuit intact because it will be used as a base for the next experiment.

## WDT: THE WATCHDOG TIMER

If you are new to microcontrollers in general and the PICmicro® MCU specifically, you might be unsure as to what the *Watchdog Timer (WDT)* offers for your application. A very common informal experiment that is performed is to enable the watchdog timer and see what happens to the application. The result is usually a problem that is difficult to characterize and find by looking externally to the PICmicro® MCU.

To demonstrate how the watchdog timer works, I created the WDT.ASM application, which can be found in the *code\WDT* subdirectory of the *PICmicro* directory of your PC.

```
 title  "WDT - Demonstrate PICmicro MCU Reset using WatchDog Timer"
 ;
 ;  This Code Puts a changing value into a PICmicro MCU's PortB after
 ;   loading PortA bits 2 and 3 with the _TO and _PD bits.
 ;
 ;  The PICmicro MCU should be reset during execution by the operation
 ;   the Watchdog timer, which will cause the _TO and _PD Display
 ;   To Change.
 ;
 ;
 ;  Hardware Notes:
 ;   PIC16F84 Running at 4 MHz
 ;   _MCLR is Pulled Up
 ;   All 8 bits of PortB are Pulled up and Connected to LEDs
 ;   PORTA.2 is Pulled up and Connected to a LED for _PD
 ;   PORTA.3 is Pulled up and Connected to a LED for _TO
 ;
 ;  Myke Predko
 ;  99.12.23
 ;
  LIST R=DEC
  INCLUDE "p16f84.inc"


 ;  Register Usage
 CBLOCK 0x020                  ;  Start Registers at End of the Values
BValue
Dlay:2
 ENDC

 PAGE
 __CONFIG _CP_OFF & _WDT_ON & _XT_OSC & _PWRTE_ON
```

```
; Mainline of WDT
  org    0

  nop

  movlw   0x0FF
  movwf   PORTB
  movwf   BValue

  movlw   0x00C
  movwf   PORTA

  bsf     STATUS, RP0
  clrf    TRISB ^ 0x080     ; Make All 8 PortB Bits Output
  movlw   0x013             ; Make RA2 and RA3 Outputs
  movwf   TRISA ^ 0x080
  bcf     STATUS, RP0

  rrf     STATUS, w         ; Get the Status Bits
  xorlw   0x00C             ; Invert _PD and _TO
  andlw   0x00C             ; Clear the Other Bits
  movwf   PORTA             ; Set the LEDs Appropriately

Loop        ; Loop Here

  call    Delay

  bcf     STATUS, C         ; Change PORTB
  btfss   BValue, 7
   bsf    STATUS, C
  rlf     BValue, w
  movwf   PORTB
  movwf   BValue

  call    Delay

  movlw   0x0FF             ; Turn OFF LEDs
  movwf   PORTB

  goto    Loop

Delay                       ; Delay 1/5 Seconds
  clrf    Dlay
  clrf    Dlay + 1
  decfsz  Dlay, f
   goto   $ - 1
  decfsz  Dlay + 1, f
   goto   $ - 3

  return

  end
```

The application itself is based on the previous (Decouple) application and requires the addition of two LEDs (and their current-limiting resistors) on RA2 and RA3 (Fig. 15-72). If you still have the Decouple application wired, then you can simply add two 220-Ohm resistors and connections from RA2 and RA3 (Fig. 15-73). The YAP-II circuit is quite easy to assemble (Fig. 15-74).

The bill of material for the WDT experiment is shown in Table 15-10.

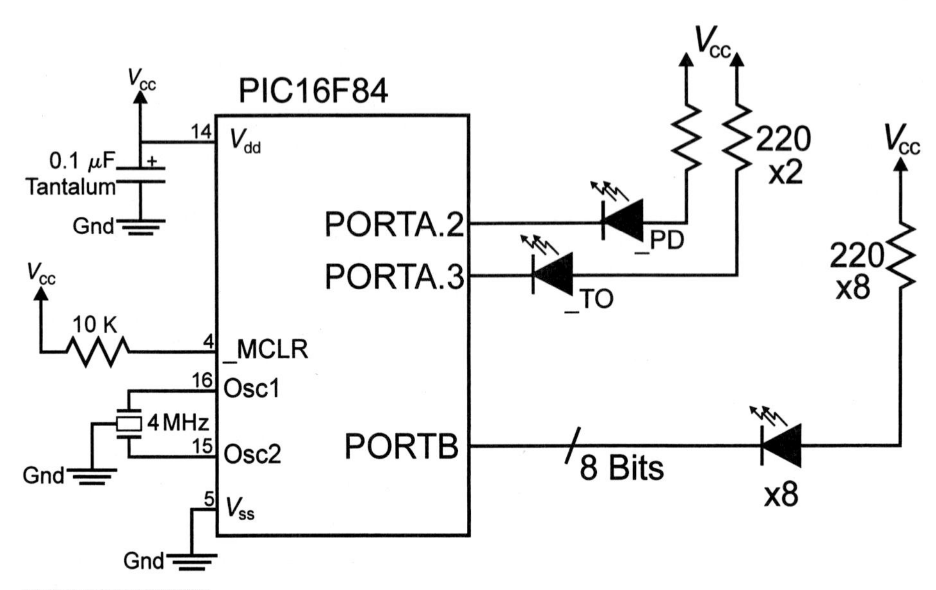


Figure 15-72 WDT: Monitor PD and TO during watchdog timer reset

In the previous experiment, if you were using the YAP-II, I recommended that you put the 16F84 onto the breadboard and wire to it directly to allow you to easily remove the 0.1-$\mu$F decoupling capacitor. If this is the circuit you are using, then two wires, from RA2 and RA3 can be connected to the built-in bargraph LED (Fig. 15-75).

When you first run the application, notice that the two LEDs that were added to RA2 and RA3 are on initially. The application code, being based on Decouple will start flash-

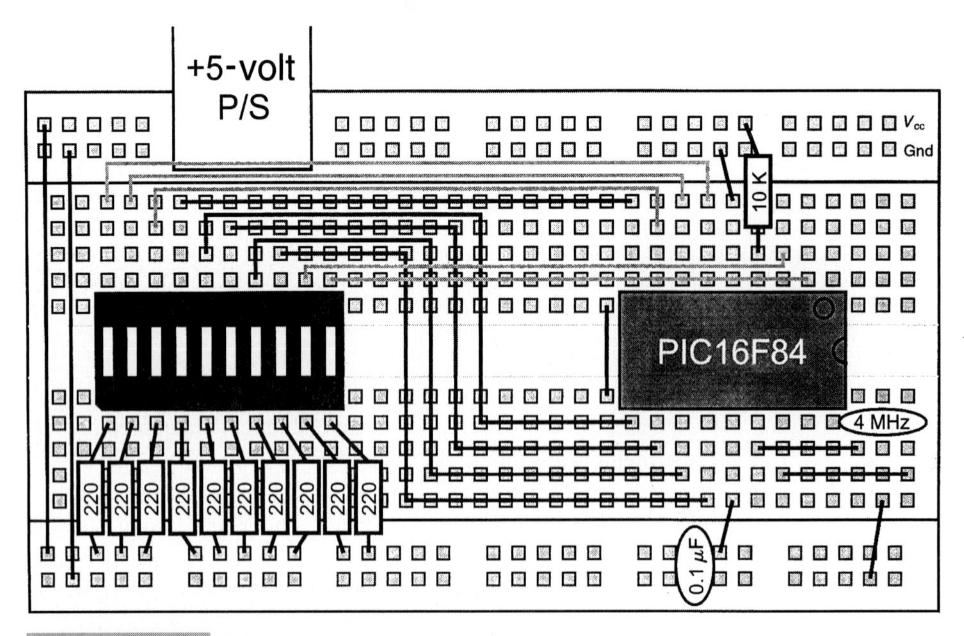


Figure 15-73 WDT 16F84 breadboard circuit

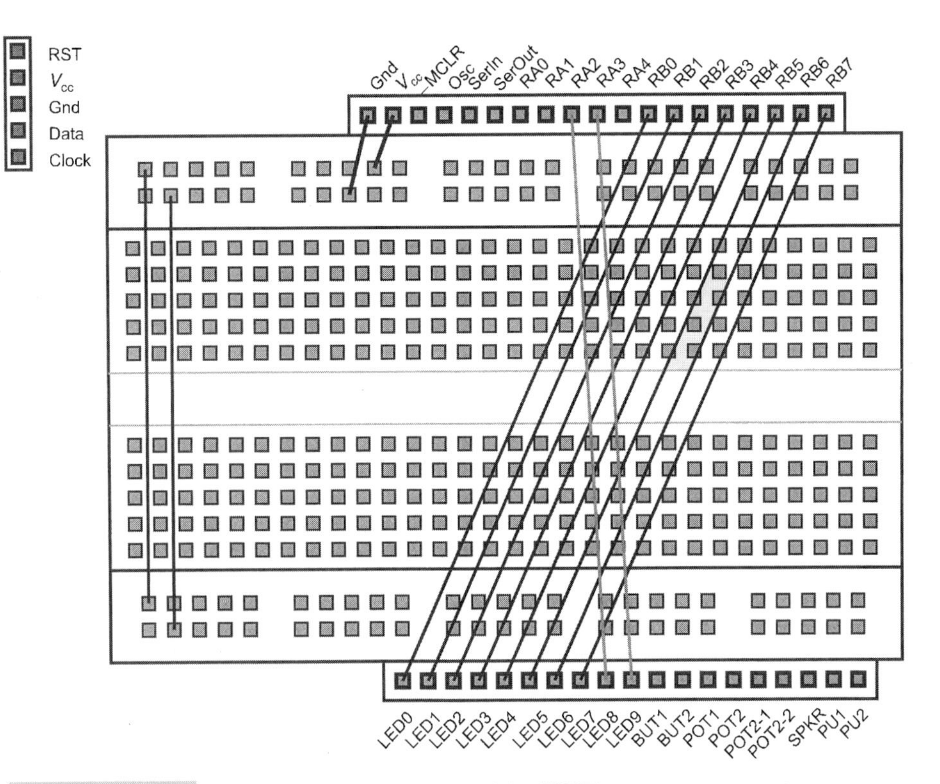


**Figure 15-74** EMU-II/YAP-II breadboard for WDT

ing the LEDs in the same, increasing pattern. After a few seconds (about the same number as the Decouple application with the decoupling capacitor pulled out), the PICmicro® MCU will reset and start executing again. Notice that when the PICmicro® MCU resets, the LED connected to RA2 (which outputs the value of *_PD*) is turned off. This indicates that the PICmicro® MCU has been reset by a watchdog timer time out.

Also notice, that the only thing done to enable the watchdog timer in the application code is to just change the *_WDT_OFF* configuration fuse parameter to *_WDT_ON*. When

**TABLE 15-10 Bill Of Materials For The "WDT" Experiment**

| PART | DESCRIPTION | REQUIRED ON YAP-II |
|---|---|---|
| PICmicro® MCU | PIC16F84-04/P | In Socket |
| Vdd/Vss decoupling capacitor | 0.1 μF tantalum | Yes |
| _MCLR Pull Up | 10 K, 1/4 watt | See Text |
| 4-MHz ceramic resonator | 4 MHz with built-in capacitors | No |
| 10-LED display | Bargraph red LED | No. LED0 used |
| 10 PORTB LED resistors | 220 Ohm, 1/4 watt | No. LED0 used |

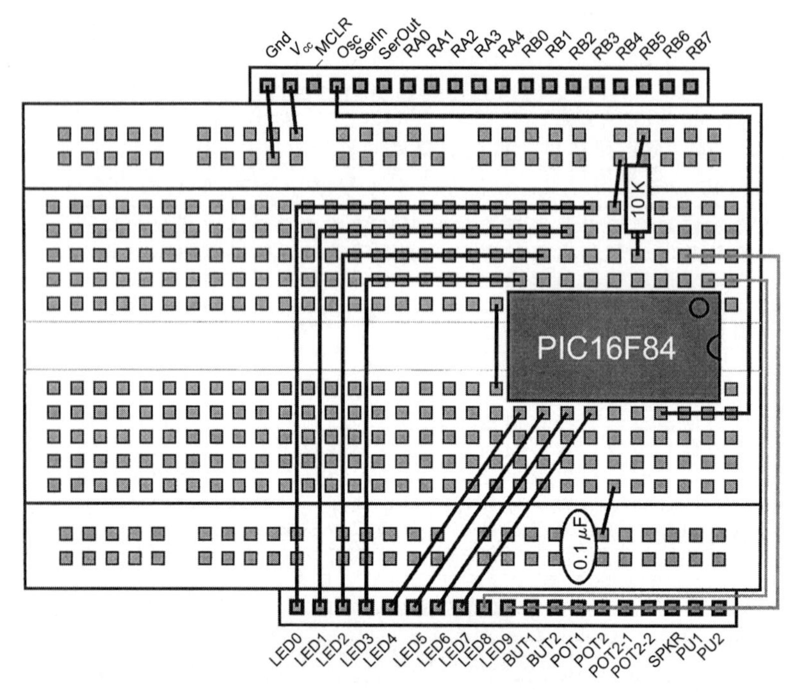


**Figure 15-75** **YAP-II breadboard for circuit WDT using Decouple's wiring**

the watchdog timer is enabled and the PICmicro® MCU powers up, the OPTION register is loaded with all 1s, which means that the prescaler is dedicated to the watchdog timer and it is running at the maximum value (which is 127). Because the nominal watchdog timer time-out value is 18 ms, the actual time-out interval becomes:

$$\begin{aligned} \textit{Watchdog timer time out} &= \textit{Nominal time-out interval} * \textit{Prescaler} \\ &= 18 \text{ ms} * 127 \\ &= 2.286 \text{ seconds} \end{aligned}$$

or the 2.3 seconds that is normally quoted by Microchip as the maximum time for the watchdog timer time-out interval.

To prevent the watchdog timer from resetting the PICmicro® MCU, you have to add the *clrwdt* instruction somewhere within the Loop code to reset the watchdog timer (and the prescaler). There should only be one *clrwdt* instruction in the code (to avoid an application that is running amok to reset it accidentally) and it should be reset after about 50% of the nominal time-out interval has executed. This 50% value is a general rule to ensure that additional code added to the application does not cause a problem with the timer and any variances in the watchdog timer circuit do not cause a short interval to result in an unwanted reset.

Proper watchdog timer application specification is a bit of a "black art." I have not found many applications that really require it. The PICmicro® MCU is quite a tough little device from the perspective of electrical interference. The only situations where I would look toward enabling it is in a high-noise environment, such as a TV set (the flyback transformer can generate large EMFs), manufacturing floors, automotive applications and avionics. Personally, I have never had an application upset by what I would consider to be a noise problem when the PICmicro® MCU has been properly decoupled.

Some software issues occur with regard to using the watchdog timer. First, it is an excellent tool to ensure that your application software doesn't run amok because of problems with the source code. Actually, this is what I would consider to be the most likely use for the watchdog timer. If you have created a very complex application that you are not comfortable with (or have used a compiler that isn't as robust as you think it should be), then "protecting" the user with the watchdog timer is an excellent idea. When using the watchdog timer, the *clrwdt* instruction should be surrounded by *return* instructions to prevent an application that is running incorrectly to step into it accidentally.

Having said all this, you are probably not thinking of even using the watchdog timer for your applications, but there is always the chance that you will accidentally. For many PICmicro® MCUs (like the PIC16F84), enabling the watchdog timer is accomplished by leaving a bit set in the configuration fuses. In these cases, if you forget to put the *_WDT_OFF* parameter into your *__CONFIG* statement, then you will inadvertently enable the watchdog timer. When this happens, and you are new to the PICmicro® MCU, finding the problem is just about impossible. As mentioned previously, the problem can appear to be the same as if the decoupling capacitor was forgotten or incorrectly wired.

Thus, if you have an application that seems to start okay, but resets itself, you should probably take the PICmicro® MCU out of the application and check it in your programmer to see if the watchdog timer is enabled. For some programmers (such as those presented in this book), you will have to go back to the source and be sure that the *_WDT_OFF* parameter is specified.

## POWERUP: REGISTER POWER-UP VALUES

One of the biggest traps that new PICmicro® MCU developers fall into when creating their first applications is to not initialize variables correctly. An uninitialized variable can cause you a lot of problems to find and debug. When I was creating the example projects, an improperly initialized variable stymied me for several days because I wasn't looking at the problem with the simulator properly.

This mistake is compounded by the operation of the MPLAB simulator that initializes all the file registers to zero (0x000) upon power up. You can have an application simulate correctly. Because you are depending on a file register to be set to zero, the programmed PICmicro® MCU will not work correctly. This is a very frustrating and difficult problem to find.

Another problem that some people will encounter with their applications is that if they are running out of space, they look at power-up values for opportunities to eliminate the code used to initialize the variables. If a file register has the same power-up value (either *0x000* or *0x0FF*) multiple times, the assumption is made that it will *always* have this power-up value. When this assumption is made, the developer takes advantage of this and does not initialize the variable. This assumption is not correct and will cause problems after the application is "finished" and "released."

To show the PICmicro® MCU's power-up values, I created the PowerUp application (located in the *code\PowerUp* subdirectory in the *PICmicro* directory of your PC). This application initially displays the contents of the w register upon power up. Each time that a button is pressed, the contents of a file register (from 0x030 to 0x040) are displayed. The code was largely taken from WDT and Debounce. As initially implemented, the code caused me some problems with how this application worked right from the start.

```
 title  "PowerUp - List Power Up Values of "w" and 16 Registers"
#define nDebug
;
;  This Application displays the contents of "w" upon power up
;   (Reset) and as the button on RA0 is pressed, the contents of
;   the file registers from 0x030 to 0x040 are displayed on LEDs
;   connected to PORTB.  LEDs on PortA bits 2 & 3 display the
;   least significant two bits of the file register address (and
;   are both on for the "w" display.
;
;
;  Hardware Notes:
;   PIC16F84 Running at 4 MHz
;   _MCLR is Pulled Up
;   PORTA.0 is Pulled up and Connected to a Momentary "On" Switch
;   All 8 bits of PortB are Pulled up and Connected to LEDs
;   PORTA.2 is Pulled up and Connected to a LED for _PD
;   PORTA.3 is Pulled up and Connected to a LED for _TO
;
;  Myke Predko
;  99.12.26
;
  LIST R=DEC
 ifdef __16F84
  INCLUDE "p16f84.inc"
 endif
 ifdef __16F877
  INCLUDE "p16f877.inc"
 endif

;  Register Usage
 CBLOCK 0x020                ;  Start Registers at End of the SFRs
Dlay:2
Temp
 ENDC

Up        EQU 1                 ;  Flag Value for Debounce "Up"
Down      EQU -1                ;  Flag Value for Debounce "Down"

;  Macros
Debounce MACRO Direction

 if (Direction < 0)             ;  Going Down
  btfsc  PORTA, 0
 else
  btfss  PORTA, 0               ;  Wait for Button Released
 endif
   goto  $ - 1

 ifndef Debug
  movlw  0x0100 - 0x0C4         ;  Initialize Dlay for a 20 msec
  movwf  Dlay                   ;   Delay
  movlw  0x0100 - 0x00A
  movwf  Dlay + 1

  bcf    STATUS, Z              ;  Make Sure that Zero is Reset
```

```
 incfsz Dlay, f
  goto  $ + 2
 incf   Dlay + 1, f
if (Direction < 0)
 btfsc  PORTA, 0
else
 btfss  PORTA, 0          ;  Button Still Released?
endif
  goto  $ - 11            ;  No - Loop Around Again
 btfss  STATUS, Z         ;  Zero Flag Set (20 mSecs Past?)
  goto  $ - 6
else
 nop                      ;  movlw  0x0100 - 0x0C4
 nop                      ;  movwf  Dlay
 nop                      ;  movlw  0x0100 - 0x00A
 nop                      ;  movwf  Dlay + 1

 nop                      ;  bcf    STATUS, Z

 nop                      ;  incfsz Dlay, f
 nop                      ;   goto  $ + 2
 nop                      ;  incf   Dlay + 1, f
 nop                      ;  btfsc  PORTA, 0
 nop                      ;   goto  $ - 11
 nop                      ;  btfss  STATUS, Z
 nop                      ;   goto  $ - 6
endif

endm                      ;  End the Macro


PAGE
ifdef __16F84
__CONFIG _CP_OFF & _WDT_OFF & _XT_OSC & _PWRTE_ON
else
__CONFIG _CP_OFF & _WDT_OFF & _XT_OSC & _PWRTE_ON & _DEBUG_OFF &
_LVP_OFF & _BODEN_OFF & _WRT_ENABLE_ON & _CPD_OFF
endif

;  Mainline of PowerUp
 org    0

 movwf  PORTB             ;  Save the Contents of "w" into PortB
 clrf   PORTA             ;  Turn on Both PortA LEDs to Indicate "w"

 bsf    STATUS, RP0
 clrf   TRISB ^ 0x080     ;  Make All 8 PortB Bits Output
 movlw  0x013             ;  Make RA2 and RA3 Outputs/RA0 Input
 movwf  TRISA ^ 0x080
 bcf    STATUS, RP0

 movlw  0x030             ;  Start Displaying the Data at Address 0x020
 movwf  FSR

Loop                      ;  Loop Here

 Debounce Up              ;  Wait for Key to Go Up

 nop                      ;  Location for Stopping after
                          ;   "Up" Debounce
 Debounce Down            ;  Wait for Key to Go Down

 rlf    FSR, w            ;  Get the Current Address
 movwf  Temp
 rlf    Temp, w           ;  Shift it up by 2 to Display LSBs in RA2/RA3
```

```
  xorlw  0x0FF          ;  Invert the Value
  movwf  PORTA

  comf   INDF, w        ;  Get the Value at FSR
  movwf  PORTB          ;  Output it
  incf   FSR, f         ;  Point to the Next Value
  movf   FSR, w         ;  Displayed 16 File Registers?
  xorlw  0x040
  btfss  STATUS, Z
   goto  Loop

  goto   $              ;  When Finished, Infinite Loop

 end
```

The circuit used for this application (Fig. 15-76) is an enhancement to WDTs. If you still have the circuit wired, just add a 10-K resistor and a momentary-on switch. If you are building the circuit from scratch, the breadboard circuit is shown in Fig. 15-77. The YAP-II circuit is shown in Fig. 15-78.

The bill of materials for PowerUp are shown in Table 15-11. As indicated, I originally used code WDT and Debounce to create the application. When I built the circuit originally, I found that after pressing the button to look at the first address (0x030), the display would change value to turn on the two LEDs at RA2 and RA3 and then display 0x011 on the LED display. This was quite frustrating for me. I modified the Debounce macro to support simulation better (i.e., not wait the 20 ms for the debounce circuit) and try to find the problem. Out of frustration, I also tried a different PICmicro® MCU.

The problem was that I based the code on WDT and I forgot that I had enabled the watchdog timer. Although I did not spend a lot of time trying to find the problem, I was embarrassed that I spent any at all. I should have noticed that the value was changing a few moments after the button was pressed, which should have told me that the watchdog timer was timing out.

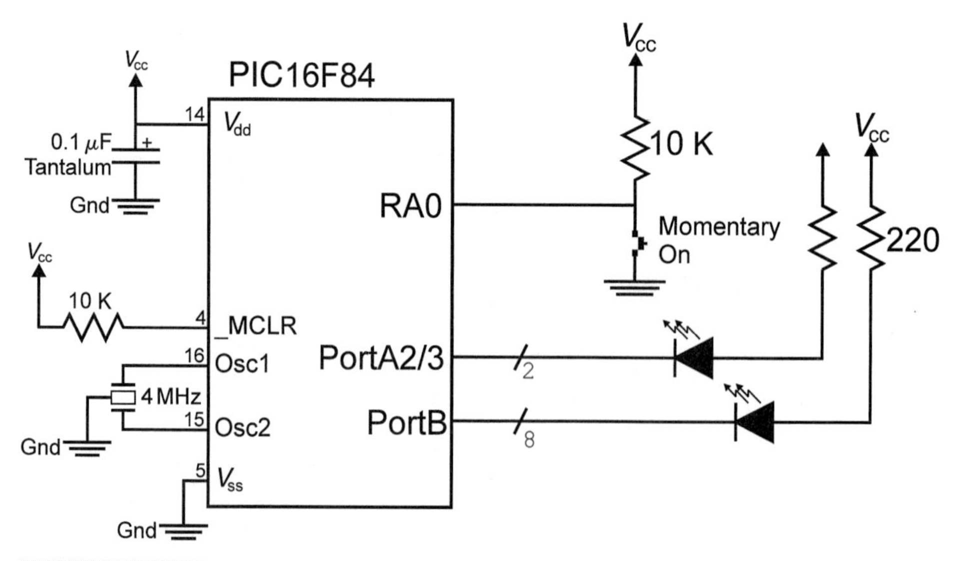


**Figure 15-76** PowerUp: circuit showing the power-up values of w and file registers

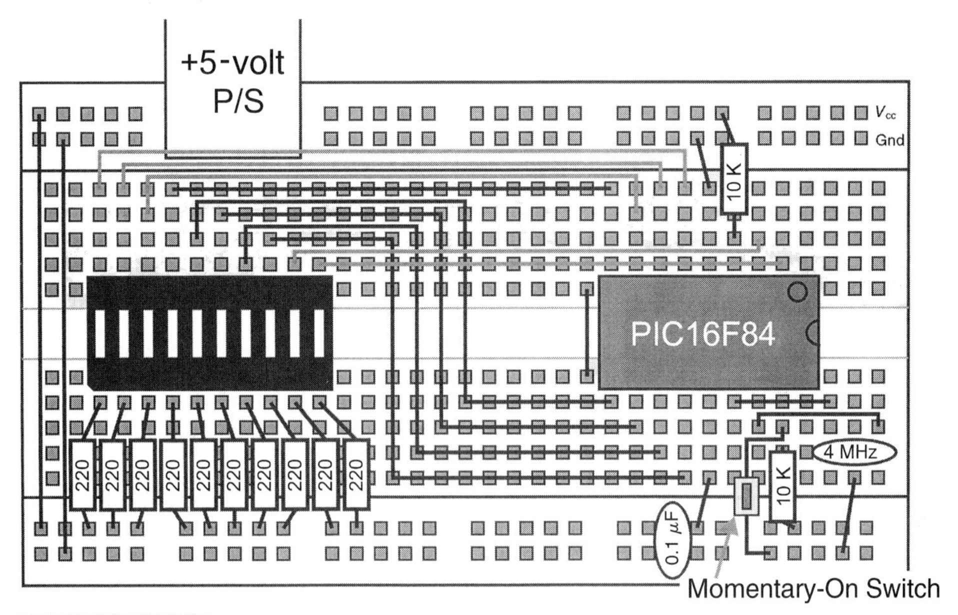


**Figure 15-77** PowerUp PIC16F84 breadboard circuit

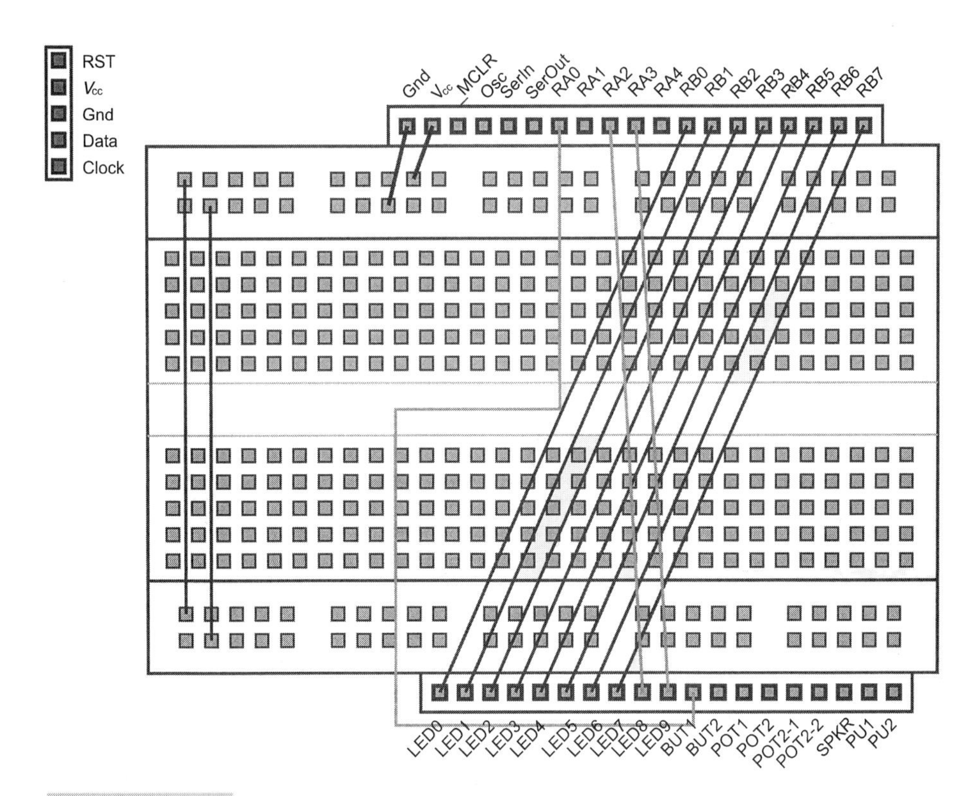


**Figure 15-78** YAP-II breadboard for PowerUp

**TABLE 15-11 Bill Of Materials For The "POWERUP" Experiment**

| PART | DESCRIPTION | REQUIRED ON YAP-II |
|---|---|---|
| PICmicro® MCU | PIC16F84-04/P | In Socket |
| Vdd/Vss decoupling capacitor | 0.1 $\mu$F tantalum | No |
| _MCLR pull up | 10 K, 1/4 watt | No |
| 4-MHz ceramic resonator | 4 MHz with built-in capacitors | No |
| 10-LED display | Bargraph red LED | No. LED0 used |
| 10 PORTB LED resistors | 220 Ohm, 1/4 watt | No. LED0 used |
| 10-K pull-up resistor | 10 K, 1/4 watt | No. BUT1 used |
| Momentary-on push button | Momentary on | No. BUT1 used |

With the application running, I tried it on three different PIC16F84s and I documented the power-up values in Table 15-12. For the first two PICmicro® MCUs, I ran the application twice (after waiting about two minutes between power ups to let the file register lose all charge).

Looking at the results, you can see that the w register seems to power up with a value of *0*x*000*. If the FILE register address bit 2 is reset, *0*x*0FF* tends to be the power-up value of a file register and when bit 2 is set, the file register power-up value tends to be *0*x*000*. You should also notice that between power-up sequences, the contents of file registers might be different because of the state that the flip flops within the register "settle" during power up. In the first edition of the book, when I ran this experiment with a 16C84, the values returned were dramatically different.

If you have a PICmicro® MCU emulator, you should be able to see these random variable memory values by simply reading them out at power up.

What conclusion should you draw from this? The contents of a PICmicro® MCU file register are unknown at power up and should be treated accordingly. Also notice that this experiment was run with only three PICmicro® MCUs, each of the same part number and each from the same lot code. Different PICmicro® MCU part numbers will yield different results in this experiment.

By assuming that the contents of the file registers are unknown at power up, you should always initialize all variables before they are used.

I realize that the contents of w were consistently zero (0x000), but this is also not something that you can assume to be true in all cases. It is especially not true in the case of PICmicro® MCUs that have a built-in oscillator and the calibration value is set upon execution of address 0x000 of the program memory.

## RESET: RESET

The previous experiment covered how the PICmicro® MCU's file registers power up to different values. This experiment uses this aspect of the PICmicro® MCU to understand

**TABLE 15-12 Bill Of Materials For The "POWERUP" Experiment**

| REGISTER | PIC16F84#1 | | PIC16F84#2 | | PIC16F84#3 |
|---|---|---|---|---|---|
| | Run #1 | Run #2 | Run #1 | Run #2 | Run #1 |
| "w" | 0x000 | 0x000 | 0x000 | 0x000 | 0x000 |
| 0x030 | 0x0FF | 0x0FF | 0x0FF | 0x0FF | 0x0FF |
| 0x031 | 0x0FF | 0x0FF | 0x0FF | 0x0FF | 0x0FF |
| 0x032 | 0x0FF | 0x0FF | 0x0EF | 0x0F7 | 0x0FF |
| 0x033 | 0x0FF | 0x0FF | 0x0FF | 0x0FF | 0x0FF |
| 0x034 | 0x002 | 0x002 | 0x080 | 0x080 | 0x000 |
| 0x035 | 0x000 | 0x000 | 0x040 | 0x0C0 | 0x000 |
| 0x036 | 0x000 | 0x000 | 0x001 | 0x001 | 0x000 |
| 0x037 | 0x000 | 0x000 | 0x000 | 0x000 | 0x000 |
| 0x038 | 0x0BF | 0x0BF | 0x0FF | 0x0FF | 0x0FF |
| 0x039 | 0x0FF | 0x0FF | 0x0F7 | 0x0F7 | 0x0FF |
| 0x03A | 0x0FF | 0x0FF | 0x0FF | 0x0F7 | 0x0FD |
| 0x03B | 0x0FF | 0x0FF | 0x0FF | 0x0FF | 0x0FF |
| 0x03C | 0x000 | 0x000 | 0x004 | 0x004 | 0x000 |
| 0x03D | 0x000 | 0x000 | 0x000 | 0x002 | 0x000 |
| 0x03E | 0x000 | 0x000 | 0x000 | 0x000 | 0x000 |
| 0x03F | 0x020 | 0x040 | 0x000 | 0x080 | 0x000 |

whether or not the PICmicro® MCU is powering up for the first time. In the Reset experiment, the values of four bytes are checked and execution occurs, depending on whether or not the four bytes are set to the specific values.

The circuit used is similar to the last few experiments and uses PORTB as an 8-LED driver and RA2 and RA3 for driving an auxiliary LEDs. When *Reset* first executes from power up, the values of *_TO* and *_PD* are output onto the LEDs on RA2 and RA3 and the eight LEDs on PORTB are turned off. After *_MCLR* is pulled low and then released, the RA2 and RA3 LEDs are given the initial values of *_TO* and *_PD* and the LEDs on PORTB are lit. The circuit that I used is shown in Fig. 15-79 and has a momentary-on switch on _MCLR to activate the PICmicro® MCU's reset. The circuit can be built on either a breadboard (as shown in Fig. 15-80 or using the YAP-II, as shown in Fig. 15-81). If the YAP-II is used, then the execution reset button on the board is used instead of a separate reset switch connected to the PICmicro® MCU's _MCLR pin.

The bill of materials for Reset is shown in Table 15-13. The RESET.ASM application can be found in the *code\Reset* subdirectory of the *PICmicro* directory on your PC's hardfile:

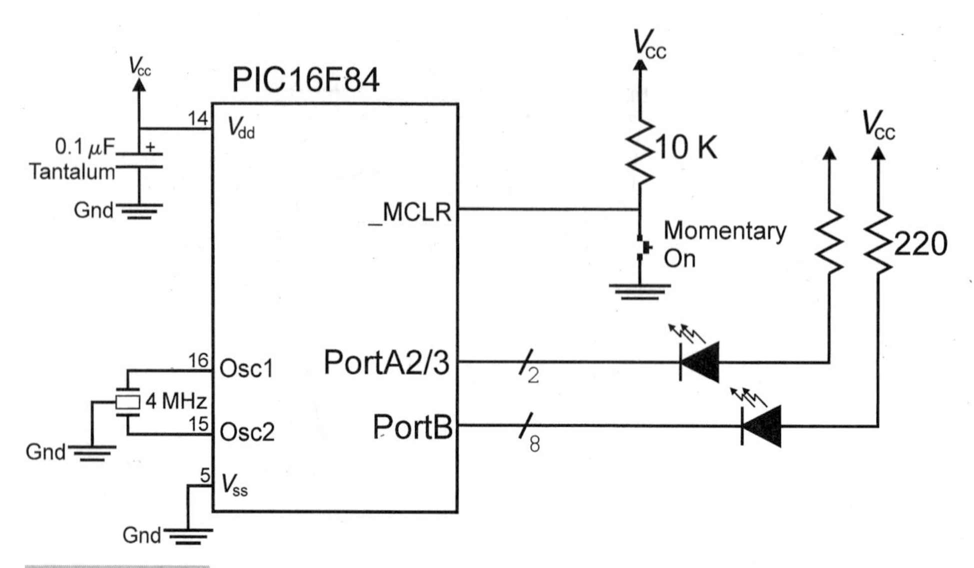


**Figure 15-79** Reset: circuit showing when _MCLR is toggled instead of a power up

```
  title  "Reset - Turn on PORTB LEDs if _MCLR Reset"
;
;  This Application checks four bytes to see if they are at an
;   expected value and turns on all the LEDs on PORTB if this is
;   true, else the LEDs are turned off.
;
;
```

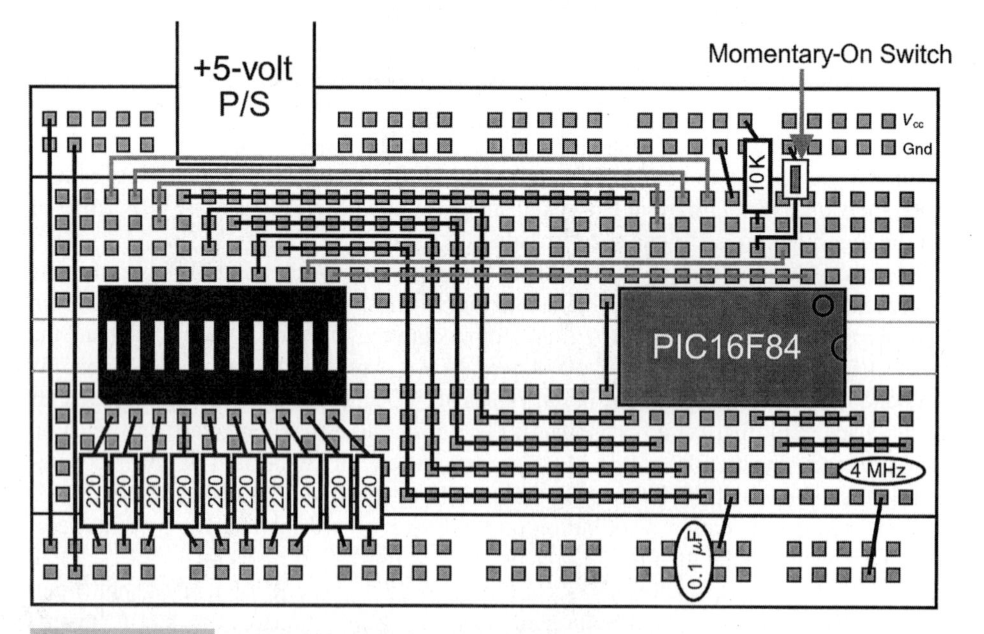


**Figure 15-80** Reset PIC16F84 breadboard circuit

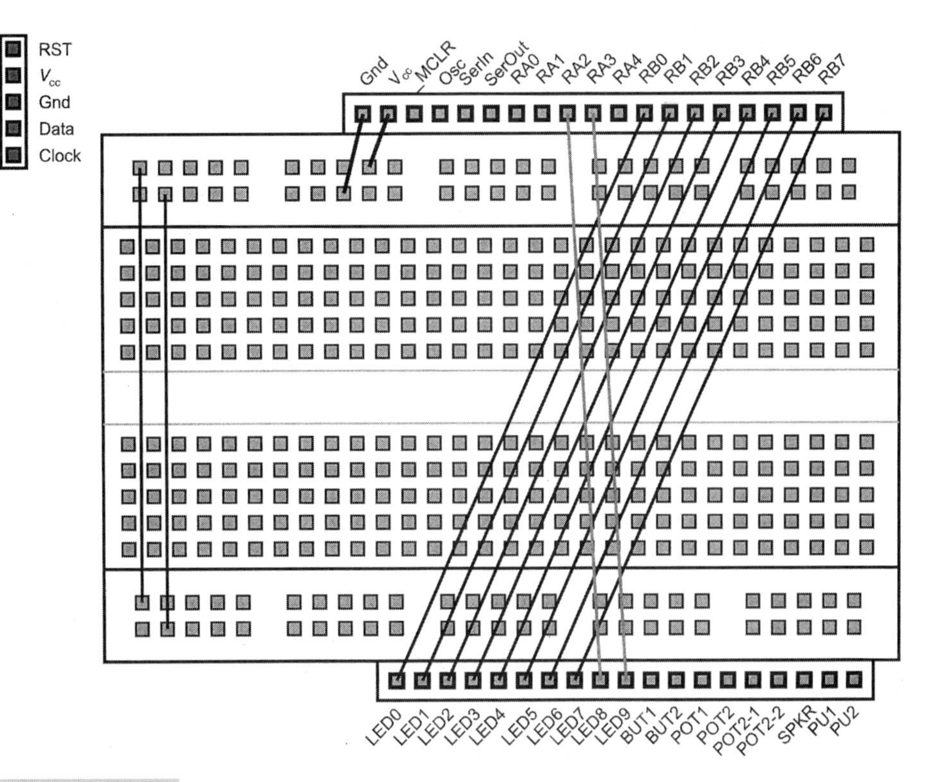


**Figure 15-81** EMU-II/YAP-II breadboard for reset

```
;  Hardware Notes:
;   PIC16F84 Running at 4 MHz
;   _MCLR is Pulled Up
;   All 8 bits of PortB are Pulled up and Connected to LEDs
;   PORTA.2 is Pulled up and Connected to a LED for _PD
;   PORTA.3 is Pulled up and Connected to a LED for _TO
```

**TABLE 15-13 Bill Of Materials For The "RESET" Experiment**

| PART | DESCRIPTION | REQUIRED ON YAP-II |
|---|---|---|
| PICmicro® MCU | PIC16F84-04/P | In Socket |
| Vdd/Vss decoupling capacitor | 0.1 $\mu$F tantalum | No |
| _MCLR pull up | 10 K, 1/4 watt | No |
| Momentary-on push button | Momentary on | No |
| 4-MHz ceramic resonator | 4 MHz with built-in capacitors | No |
| 10-LED display | Bargraph red LED | No. LED0/LED10 used |
| 10 PORTB LED resistors | 220 Ohm, 1/4 watt | No. LED0/LED10 used |

```
;
;  Myke Predko
;  99.12.26
;
  LIST R=DEC
 ifdef __16F84
  INCLUDE "p16f84.inc"
 endif
 ifdef __16F877
  INCLUDE "p16f877.inc"
 endif

;  Register Usage
 CBLOCK 0x020                   ;  Start Registers at End of the Values
Check:4                         ;  Four Byte Check Value
 ENDC


 PAGE
 ifdef __16F84
 __CONFIG _CP_OFF & _WDT_OFF & _XT_OSC & _PWRTE_ON
 else
 __CONFIG _CP_OFF & _WDT_OFF & _XT_OSC & _PWRTE_ON & _DEBUG_OFF &
 _LVP_OFF & _BODEN_OFF & _WRT_ENABLE_ON & _CPD_OFF
 endif

;  Mainline of Reset
  org    0

  nop
  rrf    STATUS, w              ;  Save the Current STATUS Value
  xorlw  0x0FF                  ;  Invert to Show Value
  movwf  PORTA

  bsf    STATUS, RP0
  clrf   TRISB ^ 0x080          ;  Make All 8 PortB Bits Output
  movlw  0x013                  ;  Make RA2 and RA3 Outputs/RA0 Input
  movwf  TRISA ^ 0x080
  bcf    STATUS, RP0

  movf   Check, w               ;  Check for First/Subsequent PICmicro MCU
  xorlw  0x0FF                  ;   Reset
  btfss  STATUS, Z
   goto  FirstTime              ;  No Match - Subsequent Reset
  movf   Check + 1, w           ;  Check Second Byte
  xorlw  0x000
  btfss  STATUS, Z
   goto  FirstTime
  movf   Check + 2, w
  xorlw  0x0AA
  btfss  STATUS, Z
   goto  FirstTime
  movf   Check + 3, w
  xorlw  0x055
  btfss  STATUS, Z
   goto  FirstTime

  clrf   PORTB                  ;  Match - Turn on LEDs

  goto   $                      ;  Finished - Infinite Loop

FirstTime                       ;  No Match, Set up Values/Turn off LEDs

  movlw  0x0FF
  movwf  PORTB                  ;  Turn Off LEDs
```

```
  movwf  Check              ;  Save the Check Values
  clrf   Check + 1
  movlw  0x0AA
  movwf  Check + 2
  movlw  0x055
  movwf  Check + 3

  goto   $                  ;  When Finished, Infinite Loop

 end
```

This application is quite straightforward to program into a PICmicro® MCU and build the circuit for. Without looking too closely at the code, I suggest that you put in a programmed PICmicro® MCU and test its operation. You should see that the two LEDs that show off *_TO* and *_PD* are lit (which is expected because these bits are set on power up) and then the remaining LEDs connected to PortB are off. When you press the reset momentary-on switch, you'll see that all the LEDs go out, then all 10 light. This is exactly what is expected.

The code should be very straightforward to understand, but when you look at it, you might wonder why the check code is not:

```
  movf   Check, w
  xorwf  Check + 1, w
  xorwf  Check + 2, w
  xorwf  Check + 3, w
  btfss  STATUS, Z
   goto  FirstTime
```

This code is a bit more than one third of the size of the code that I used in the Reset application code and executes in the same number of cycles each time. This would seem like a reasonable optimization of the code. This code tests the four Check bytes together and jumps to zero if the result is not equal to zero, which is essentially what the code in RESET.ASM does.

I didn't use the suggested optimization because there is a good chance that it would not work properly in this application. Looking at the previous experiment, notice that there seems to be register values that have a predisposition of being set to specific values. To make matters worse, they are in groups of four and the groups are 0x0FF or 0x0000, which equals 0x000 when they are XORed together.

The code that I use in RESET.ASM checks each bit in the four *Check* variables to ensure that the full-value *0x055AA00FF* is present. The suggested optimization really only compares eight bit sets to zero, which is not as rigorous as the method that I used. With the code that I used, only one pattern out of 4.3 billion is valid. Using the results of the PowerUp experiment, the actual chance of this pattern coming up is significantly higher than that.

You might be concerned about the differing number of cycles that can cause a reset. In the comparison code that is used in RESET.ASM, FirstTime can execute 13, 17, 21, or 25 cycles. I would argue that this is not important because it is at the start of the application and the maximum difference of 12 cycles is insignificant, compared to the delay needed to power up and start the PICmicro® MCU's processor. This potential delay is based on the PICmicro® MCU's reset and clock circuitry when the conditions are correct for the PICmicro® MCU to start its oscillator and then start the processor.

Many of the later PICmicro® MCUs have a *PCON* register, which has a _POR bit that is reset upon power-up reset and is set after a _MCLR reset. This hardware function is very simple to create code to poll, does not require any file registers, and is actually more reliable than the method used in RESET.ASM.

If you power down the Reset PICmicro® MCU circuit and power it back up after a few seconds (rather than a few minutes), you will find that the PICmicro® MCU will turn on the PortB LEDs, as if it were a _MCLR reset. This is because of the flip flops within the PICmicro® MCU staying at their current state with very small charges left in their circuits. For the operation to be reliable, I found that the PICmicro® MCU should be powered off for at least two minutes.

## TMR0: TMR0 SET UP WITH PRESCALER

As I work through the later experiments and the projects, I will be using the TMR0 and prescaler hardware a lot. So far in the experiments, I have only simply presented the prescaler as part of the watchdog timer and TMR0 as a device that can be held with a specific value without changing. In this experiment, I want to spend some time showing how the timer works and how the prescaler can be used to change its operation. This experiment will also give you an idea of the speeds the PICmicro® MCU operates at. For this experiment, you might want to look at the data output with an oscilloscope, but it is not necessary if you have a logic probe.

The TMR0 application (which can be found in the *code\TMR0* subdirectory of the *PICmicro* directory on your PC) is quite simple. A TMR0 does enable TMR0 to run in from the processor's clock and then poll the TMR0 value and write it out to PORTB, so the value of TMR0 can be checked externally.

```
 title  "TMR0 - Demonstrate the Operation of TMR0 with Prescaler"
;
;  This Code sets up TMR0 to run from the Instruction Clock and
;   uses the Prescaler to divide the incoming clock.
;
;
;  Hardware Notes:
;   PIC16F84 Running at 4 MHz
;   _MCLR is Pulled Up
;   All 8 bits of PortB are Pulled up and Connected to LEDs
;
;  Myke Predko
;  99.12.26
;
  LIST R=DEC
 ifdef __16F84
  INCLUDE "p16f84.inc"
 endif
 ifdef __16F877
  INCLUDE "p16f877.inc"
 endif

;  Register Usage
 CBLOCK 0x020                 ;  Start Registers at End of the Values
 ENDC

 PAGE
 ifdef __16F84
```

```
 __CONFIG _CP_OFF & _WDT_OFF & _XT_OSC & _PWRTE_ON
 else
 __CONFIG _CP_OFF & _WDT_OFF & _XT_OSC & _PWRTE_ON & _DEBUG_OFF &
 _LVP_OFF & _BODEN_OFF & _WRT_ENABLE_ON & _CPD_OFF
 endif

;  Mainline of TMR0
  org     0

  nop
  bsf     STATUS, RP0
  clrf    TRISB ^ 0x080      ;  Make All 8 PortB Bits Output
  movlw   0x0F8 ^ ((1 << T0CS) | (1 << PSA))
  addlw   0                  ;  add prescaler values to the option register
  movwf   OPTION_REG ^ 0x080;  Load the Option Register Value
  bcf     STATUS, RP0

Loop                         ;  Loop Here

  comf    TMR0, w            ;  Output the TMR0 Value
  movwf   PORTB

  goto    Loop

  end
```

The instructions that set up the OPTION register to use the PICmicro® MCU's TMR0 clock input and the prescaler for TMR0 will seem a bit cumbersome and hard to understand. Normally, all of the bits in the OPTION register are set, to change its operation for specifying TMR0 as using the instruction clock input, as well as directing the prescaler to TMR0, bits have to be reset. As well, I clear the prescaler's value. To be honest, instead of creating the cumbersome instruction pair:

```
 movlw 0x0F8 ^ ((1 << T0CS) | (1 << PSA)
 addlw 0
```

I simply use the constant value 0x0D0 as:

```
  movlw   0x0D0
  movwf   OPTION_REG ^ 0x080
```

which is a lot easier to read. If it is found before a *movwf OPTION_REG ^ 0x080* instruction, then I know that I am setting up the OPTION register.

I used the more cumbersome format to show how the OPTION register bits are set for this application. The next instruction, the *addlw 0* is used to set the prescaler to something other than one to one, the reason for which will become obvious.

This application is yet another one that uses PORTB for LED output. When you build it and run the PICmicro® MCU, all eight LEDs connected to PORTB will appear to be on. Figure 15-82 shows the schematic for the circuit and Figs. 15-83 and 15-84 are used for the breadboard and YAP-II circuits, respectively.

The bill of materials for the circuit is shown in Table 15-14. When you first run the circuit, you will probably think that TMR0 is actually 0 and is not changing state. Actually, if you check the I/O pins with a logic probe or an oscilloscope, you will discover that the pins are actually switching, with the most-significant bit (RB7) switching at a rate of just less than 2 kHz (Fig. 15-85). If you check the other PORTB pins, you will find that they are switching at higher frequencies with RB1 switching at a rate of 125 kHz.

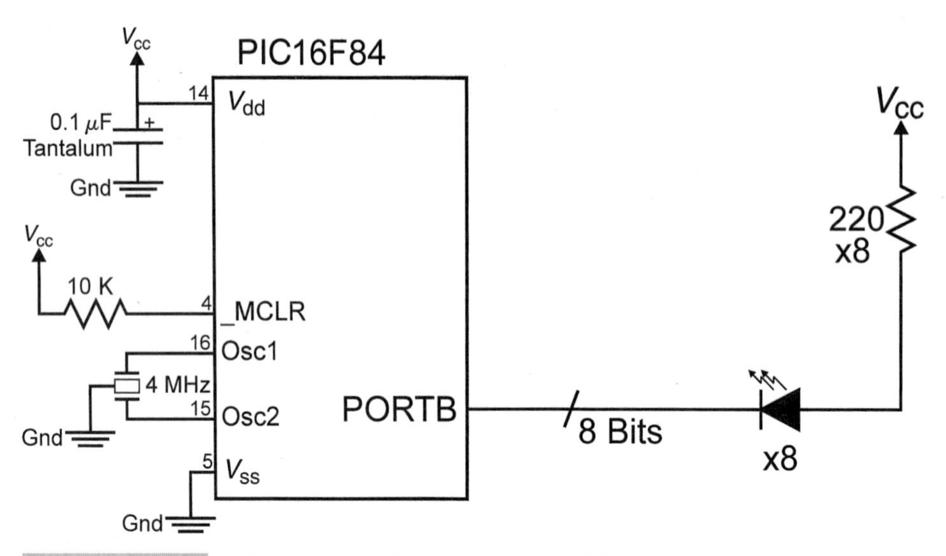


**Figure 15-82** TMR0: display TMR0 values on LEDs

When you look at the I/O pins, you should notice that RB0 does not switch at all. This is because the Loop code in TMR0.ASM takes four instruction cycles. Each time through, the least-significant bit of TMR0 is always at the same value for each instruction in the loop and will never change relative to the output. This is something that you can see if you simulate the application.

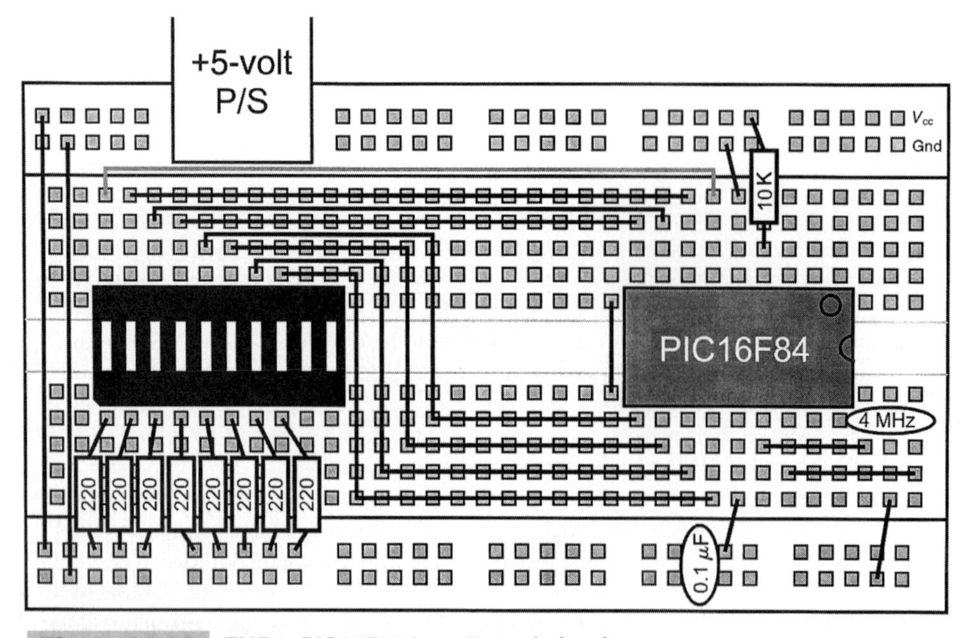


**Figure 15-83** TMR0: PIC16F84 breadboard circuit

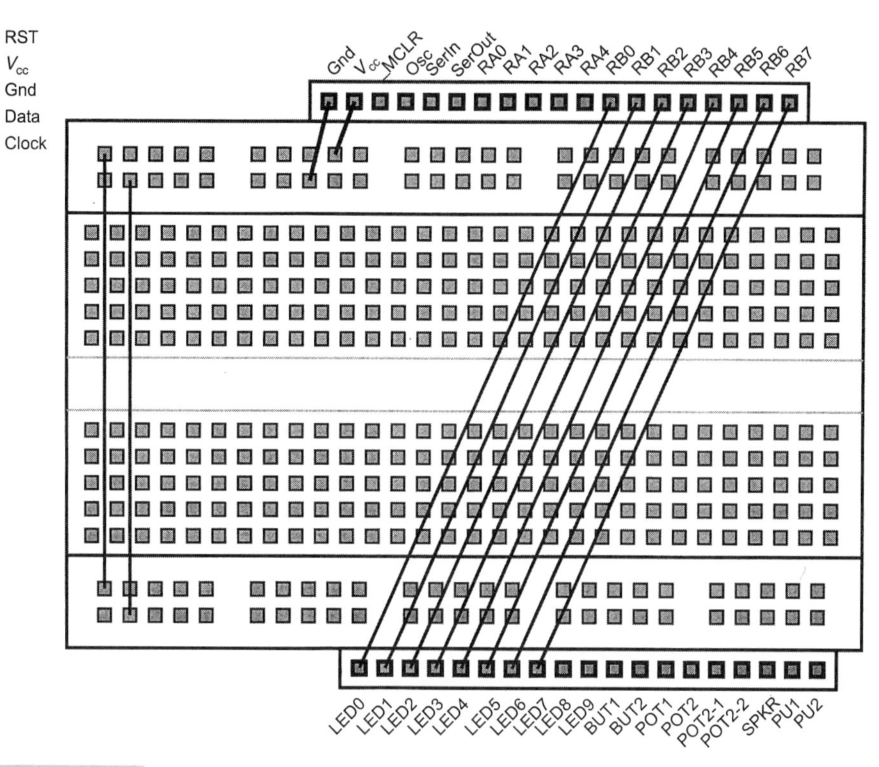


**Figure 15-84** EMU-II/YAP-II breadboard for TMR0

To actually see TMR0 changing state, change the:

```
addlw 0
```

instruction in the code to:

```
addlw 7
```

**TABLE 15-14 Bill Of Materials For The "TMR0" Experiment**

| PART | DESCRIPTION | REQUIRED ON YAP-II |
|---|---|---|
| PICmicro® MCU | PIC16F84-04/P | In Socket |
| Vdd/Vss decoupling capacitor | 0.1 $\mu$F tantalum | No |
| _MCLR pull up | 10 K, 1/4 watt | No |
| 4-MHz ceramic resonator | 4 MHz with built-in capacitors | No |
| 10-LED display | Bargraph Red LED | No. LED1/LED7 used |
| 8 PORTB LED resistors | 220 Ohm, 1/4 watt | No. LED0/LED7 used |

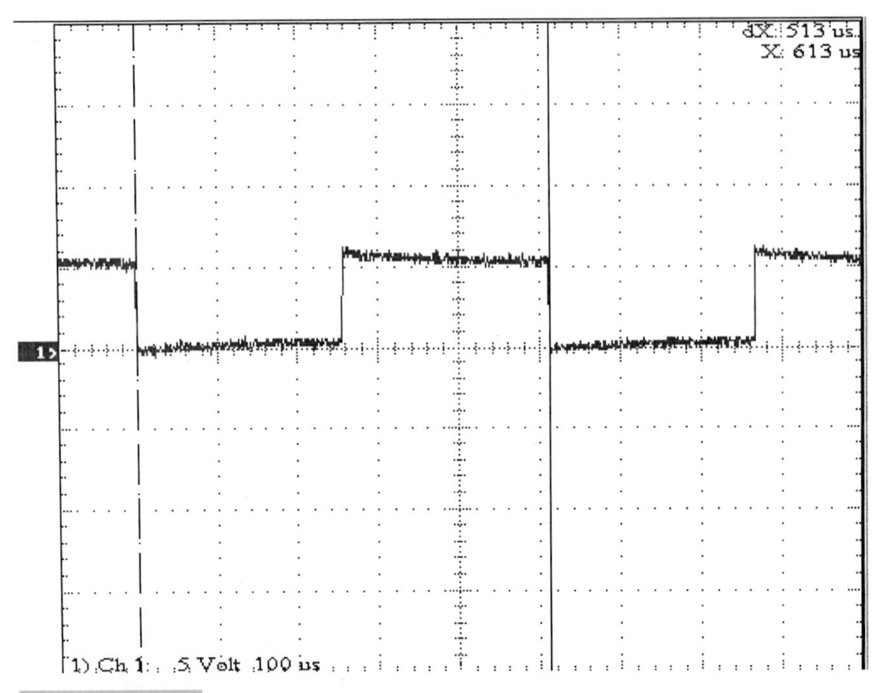


**Figure 15-85** TMR0 prescaler equals 0 RB7 waveform

which will change the prescaler from a one to one to one to 128. When the code has been reassembled and put into a PICmicro® MCU that is in a circuit, the LED connected to RB7 will flash noticeably. If you put an oscilloscope on RB7, you will find that it is flashing at a rate of 16 times per second (16 Hz).

The RB7 LED flashes should not be individually observable by you. Instead, you should see a flashing blur because 16 Hz is at about the limit of your ability to discriminate between different events. If you were to slow down the flashing any more, you would be able to distinguish individual on and off events. If the flashing was any faster (as is the case with RB6), the LED will just appear to be on all the time. However, it will not be at the maximum brightness possible because the signal passed to the LED is actually a PWM signal with a 50% duty cycle.

This application should give you an appreciation of how fast the PICmicro® MCU actually executes. Note that for changes to the TMR0's most-significant bit to be noticeable, a prescaler of 256 had to be assigned, which divides the RB7 by 65,356 times. This application shows just how fast instructions execute inside of the PICmicro® MCU. It should also help you to understand when I say that getting the maximum speed out of the PICmicro® MCU is not always crucial—especially when it increases the power required by the microcontroller and increases its electrical emissions.

## RANDOM: RANDOM NUMBER GENERATOR

The code in RANDOM.ASM application (located in the *code\Random* subdirectory of the *PICmicro* directory on your PC) can be used in other applications. Every time that the button in the Random circuit (Fig. 15-86) is pressed, a new random number is displayed on the eight LEDs connected to PORTB.

The circuit is quite easy to build (especially considering the previous experiments). The parts list is shown in Table 15-15.

The circuit can be built on a breadboard (Fig. 15-87) or using the YAP-II (Fig. 15-88).

The code itself takes advantage of the Debounce macro and the TMR0 set up provided in the previous experiment.

```
 title  "Random - Produce a Random Number Generator with TMR0"
;
;  This Code Displays TMR0 on RB0 every time the Button on RA0
;   is Pressed.
;
;
;  Hardware Notes:
;   PIC16F84 Running at 4 MHz
;   _MCLR is Pulled Up
;   All 8 bits of PORTB are Pulled up and Connected to LEDs
;   PORTA.0 is Pulled up with a Momentary On Pulling to Ground
;
;  Myke Predko
;  99.12.26
;
  LIST R=DEC
 ifdef __16F84
  INCLUDE "p16f84.inc"
 endif
 ifdef __16F877
  INCLUDE "p16f877.inc"
 endif
```

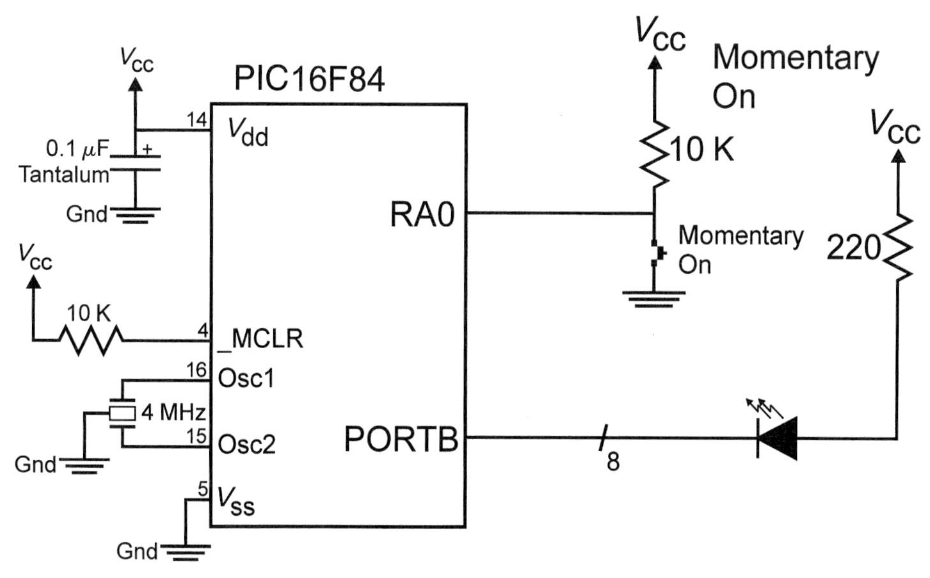


**Figure 15-86** Random: generating a random number using a debounced button

**TABLE 15-15 Bill Of Materials For The "RANDOM" Experiment**

| PART | DESCRIPTION | REQUIRED ON YAP-II |
|---|---|---|
| PICmicro® MCU | PIC16F84-04/P | In Socket |
| Vdd/Vss decoupling capacitor | 0.1 $\mu$F tantalum | No |
| _MCLR pull up | 10 K, 1/4 watt | No |
| 4-MHz ceramic resonator | 4 MHz with built-in capacitors | No |
| 10-LED display | Bargraph Red LED | No. LED0/LED7 used |
| 8 PORTB LED Resistors | 220 Ohm, 1/4 watt | No. LED0/LED7 used |

```
;  Register Usage
 CBLOCK 0x020                ;  Start Registers at End of the Values
Dlay:2                       ;  Delay Value
 ENDC

Up       EQU 1               ;  Flag Value for Debounce "Up"
Down     EQU -1              ;  Flag Value for Debounce "Down"

;  Macros
Debounce MACRO Direction

 if (Direction < 0)          ;  Going Down
  btfsc  PORTA, 0
```

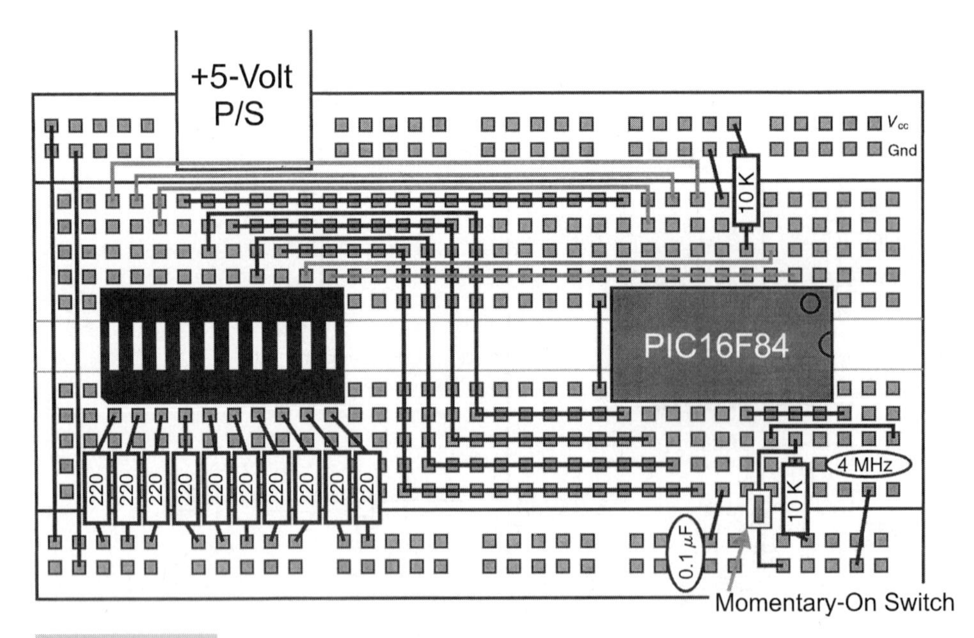


**Figure 15-87** Random PIC16F84 breadboard circuit

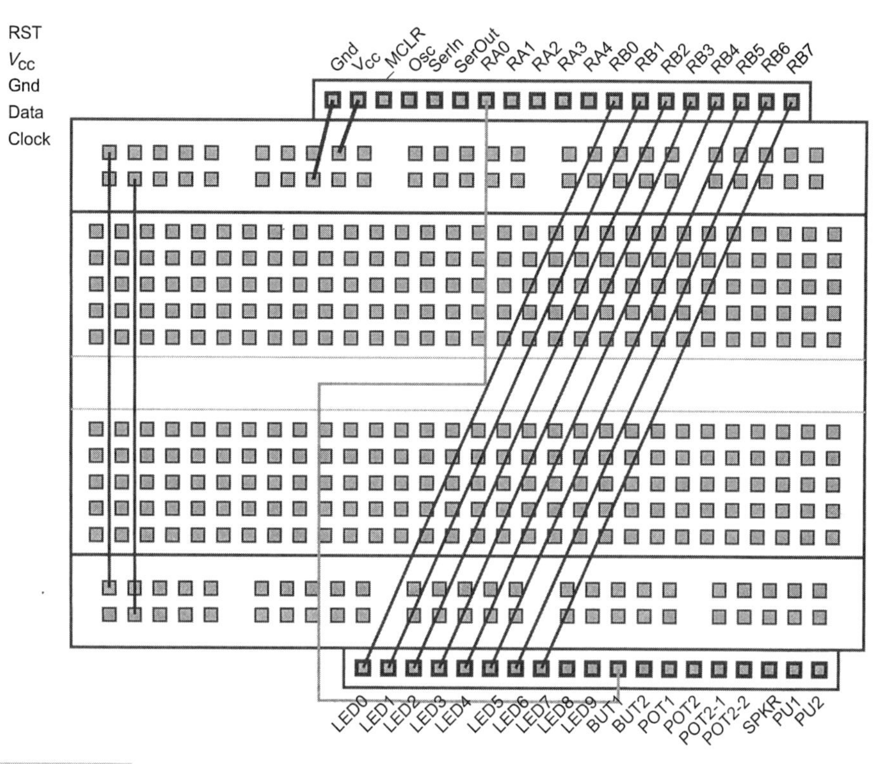


**Figure 15-88** EMU-II/YAP-II breadboard for Random

```
else
 btfss  PORTA, 0              ;  Wait for Button Released
endif
  goto  $ - 1
 movlw  0x0100 - 0x0C4        ;  Initialize Dlay for a 20 msec
 movwf  Dlay                  ;   Delay
 movlw  0x0100 - 0x00A
 movwf  Dlay + 1

 bcf    STATUS, Z             ;  Make Sure that Zero is Reset

 incfsz Dlay, f
  goto  $ + 2
 incf   Dlay + 1, f
if (Direction < 0)
 btfsc  PORTA, 0
else
 btfss  PORTA, 0              ;  Button Still Released?
endif
  goto  $ - 11                ;  No - Loop Around Again
 btfss  STATUS, Z             ;  Zero Flag Set (20 mSecs Past?)
  goto  $ - 6

endm                          ;  End the Macro

PAGE
ifdef __16F84
```

```
 __CONFIG _CP_OFF & _WDT_OFF & _XT_OSC & _PWRTE_ON
 else
 __CONFIG _CP_OFF & _WDT_OFF & _XT_OSC & _PWRTE_ON & _DEBUG_OFF &
 _LVP_OFF & _BODEN_OFF & _WRT_ENABLE_ON & _CPD_OFF
 endif

                              ;  Note that the WatchDog Timer is OFF
;  Mainline of Random
  org     0

  nop

  movlw   0x0FF               ;  Turn off all the LEDs initially
  movwf   PORTB

  bsf     STATUS, RP0
  clrf    TRISB ^ 0x080       ;  Make All 8 PortB Bits Output
  movlw   0x0D0               ;  Assign Prescaler of 1:1 to TMR0
  movwf   OPTION_REG ^ 0x080;    Load the Option Register Value
  bcf     STATUS, RP0

Loop                          ;  Loop Here

 Debounce Up                  ;  Wait for Key to Go Up

  nop                         ;  Location for Stopping after
                              ;   "Up" Debounce
 Debounce Down                ;  Wait for Key to Go Down

  comf    TMR0, w             ;  Output the TMR0 Value
  movwf   PORTB
  goto    Loop

 end
```

As indicated at the start of this experiment, the code simply starts off TMR0 with the prescaler assigned, but using a 1 : 1 value. Thus, TMR0 is updated once every two instruction cycles. The *Loop* code waits for the button on RA0 to go high. When it is pressed (and debounced), the current value in TMR0 is output onto the eight LEDs.

To demonstrate the random nature of the application, I pressed the button on RA0 16 times and recorded the results (Table 15-16).

As you can see in this list, the numbers are reasonably random. Actually, I was a bit surprised that the same value didn't come up twice. The statistical "birthday test" can be applied to this situation to see how many button presses have to be made before there is a 50% chance of two numbers being the same.

To compute this, if you assume that the TMR0 value returned can be any number from 0 to 255, there are 256 different opportunities for different values. For the second random number generated to be not equal to the first, the chances are 255/256. For the third number not to be equal to the first or second, the chances are 255/256 times 254/256. The ultimate product is multiplied to each time the button is pressed. This is computed until there is a 50% chance for the number pressed not to be equal to anything before it.

I've always found statistics to be the science of the negative, rather than the positive. If you've taken statistics, you'll know what I'm talking about.

In any case, 18 presses need to occur until there is a 50% chance of a repeated random number. It took me eight additional presses (for a total of 24) to get 0x0A9, which matches the fourteenth press.

**TABLE 15-16 Returned Values From The "Random" Experiment**

| BUTTON PRESS | TMR0 VALUE |
|---|---|
| 1 | 0x043 |
| 2 | 0x0BA |
| 3 | 0x075 |
| 4 | 0x012 |
| 5 | 0x0C3 |
| 6 | 0x0D2 |
| 7 | 0x049 |
| 8 | 0x000 |
| 9 | 0x05A |
| 10 | 0x0E8 |
| 11 | 0x0C6 |
| 12 | 0x044 |
| 13 | 0x06C |
| 14 | 0x0A9 |
| 15 | 0x0C4 |
| 16 | 0x0F9 |

## SLEEP: SLEEP

When you look at the Microchip PICmicro® MCU datasheets, you will probably be under the impression that Sleep mode is not very complex. Personally, I find that the different operating modes of Sleep are quite complex. Understanding all of the different things that can happen can be confusing and difficult to work through. This experiment will give you an idea of how Sleep works in the PICmicro® MCU. I hope this experiment gives you some things that you can think about when you are going to use Sleep in your own applications.

The circuit that I have used for this experiment is one of the more complex in this chapter (in terms of wiring). Figure 15-89 is the schematic diagram of the circuit. Figures 15-90 and 15-91 are the wiring diagrams for the breadboard and YAP-II versions, respectively.

The bill of materials for Sleep is shown in Table 15-17. The SLEEP.ASM application code follows. This code is located in the *code\Sleep* subdirectory of the *PICmicro* directory.

```
 title  "Sleep - Demonstrate the PICmicro® MCU Modes of 'Sleep'"
;
;  This Application Displays a Counter on PORTB (if code is
;   set up) and shows the current "_TO" and "_PD" bits of the
;   STATUS Register.
;
;
;
```

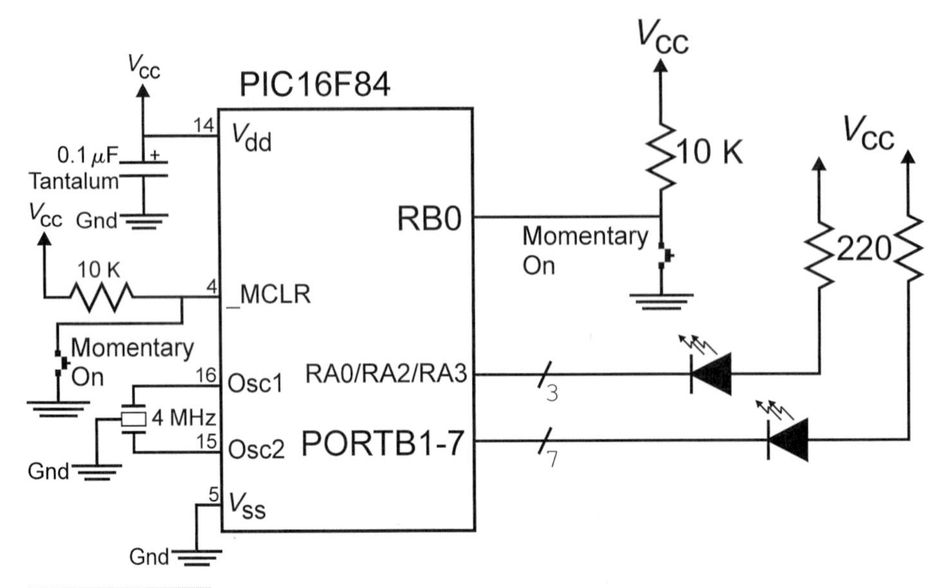


**Figure 15-89** Sleep: looking at a Sleeping PICMicro® MCU

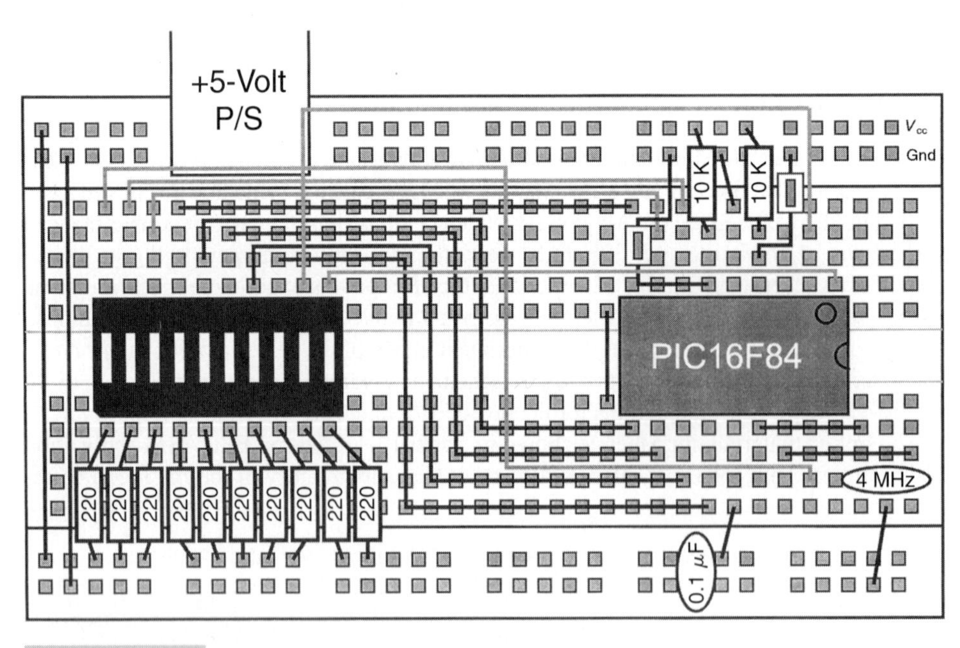


**Figure 15-90** Sleep PIC16F84 breadboard circuit

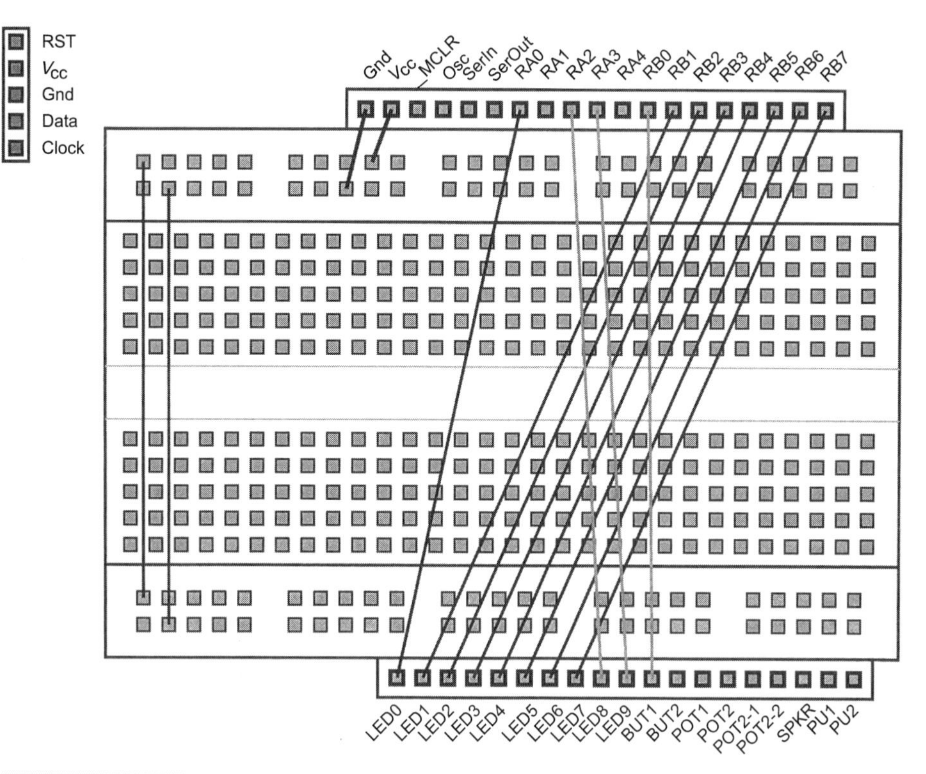


**Figure 15-91** EMU-II/YAP-II breadboard for Sleep

**TABLE 15-17 Bill Of Materials For The "SLEEP" Experiment**

| PART | DESCRIPTION | REQUIRED ON YAP-II |
|---|---|---|
| PICmicro® MCU | PIC16F84-04/P | In Socket |
| Vdd/Vss decoupling capacitor | 0.1 $\mu$F Tantalum | No |
| _MCLR pull up | 10 K, 1/4 watt | No |
| Momentary-on push button | Momentary on | No |
| 4-MHz ceramic resonator | 4 MHz with built-in capacitors | No |
| 10-LED display | Bargraph red LED | No. LED0/LED10 used |
| 10 PORTB LED resistors | 220 Ohm, 1/4 watt | No. LED0/LED10 used |
| Momentary-on push button | Momentary on | No. BUT1 used |
| 10-K pull up | 10 K, 1/4 watt | No. BUT1 used |

```
;  Hardware Notes:
;   PIC16F84 Running at 4 MHz
;   _MCLR is Pulled Up with a Button on it
;   RB0 is Pulled up with a Momentary On Switch Pulling to Ground
;   All 7 remaining bits of PortB are Pulled up and Connected to LEDs
;   PORTA.0 is Pulled up and Connected to a LED for LSB of the Counter
;   PORTA.2 is Pulled up and Connected to a LED for _PD
;   PORTA.3 is Pulled up and Connected to a LED for _TO
;
;  Myke Predko
;  99.12.28
;
  LIST R=DEC
 ifdef __16F84
  INCLUDE "p16f84.inc"
 endif
 ifdef __16F877
  INCLUDE "p16f877.inc"
 endif

;  Register Usage
 CBLOCK 0x020                ;  Start Registers at End of the Values
Check:4                      ;  The Four Bytes of the Check
Counter                      ;  Count Value
 ENDC

 PAGE
 ifdef __16F84
 __CONFIG _CP_OFF & _WDT_OFF & _XT_OSC & _PWRTE_ON
 else
 __CONFIG _CP_OFF & _WDT_OFF & _XT_OSC & _PWRTE_ON & _DEBUG_OFF &
 _LVP_OFF & _BODEN_OFF & _WRT_ENABLE_ON & _CPD_OFF
 endif
                             ;  Note that the WatchDog Timer is ON

;  Mainline of Sleep
  org  0
Loop                         ;  Loop Here Each Time Executes

  nop

  bsf   STATUS, RP0
  movlw 0x012                ;  RA0, RA2 & RA3 are Outputs
  movwf TRISA ^ 0x080
  movlw 0x01                 ;  RB1 to RB7 are Outputs
  movwf TRISB ^ 0x080
  bcf   STATUS, RP0

  rrf   STATUS, w            ;  Get the _TO/_PD Bits
  xorlw 0x0FF
  andlw 0x0FE                ;  Make Sure LED at RA0 is Clear
  movwf PORTA                ;   Display on RA2/RA3

  movf  Check, w             ;  Is this a Reset or First Time Through?
  xorlw 0x0FF
  btfss STATUS, Z
   goto FirstTime
  movf  Check + 1, w
  xorlw 0x000
  btfss STATUS, Z
   goto FirstTime
  movf  Check + 2, w
  xorlw 0x0AA
```

```
    btfss STATUS, Z
     goto FirstTime
    movf  Check + 3, w
    xorlw 0x055
    btfss STATUS, Z
     goto FirstTime

    incf  Counter, f

    comf  Counter, w           ;  Output the Counter Value
    movwf PORTB                ;  Output the High Seven Bits
    andlw 1
    iorwf PORTA, f             ;  Output the Low Bit

    bsf   INTCON, INTE         ;  Enable a Pin Interrupt

    sleep                      ;  Execute the Sleep Instruction
    nop

    clrwdt                     ;  Clear WDT To Reset 2.3 seconds later

    goto Loop

  FirstTime                    ;  Set up the Count to See Differences

    movlw 0x0FF
    movwf Check
    movlw 0x000
    movwf Check + 1
    movlw 0x0AA
    movwf Check + 2
    movlw 0x055
    movwf Check + 3

    movlw 0x00                 ;  Turn on all the LEDs
    movwf PORTB
    bcf   PORTA, 0

    clrf  Counter              ;  Reset the Counter

    bsf   INTCON, INTE         ;  Allow Buttons to Reset

    sleep                      ;  Go to Sleep
    nop

    goto  Loop                 ;  Display the New Data

    end
```

I have refrained from describing how the application works. Instead I would like you to spend a few moments working through it and try to figure out what you expect when the application runs and the buttons are pressed. When you have done that, build the circuit and try it out. Chances are that you'll be surprised at its operation.

Certain aspects of this application can be simulated, but not all. You should be able to guess some of the things that occur during execution (especially with regard to the watchdog timer), but I'm sure some behaviors exhibited by the application will surprise you.

When this application starts, you should see all 10 LEDs light up and then start incrementing with a 2.3-second interval. If you press the button connected to RB0, you will see all the LEDs turn on (although if you check the LEDs with an oscilloscope or a logic probe, you will see that they are actually toggling very quickly). This will only stop when you press the button connected to _MCLR. When the button at _MCLR is pressed, all the

LEDs will go off and the incrementing will continue as before. When the application first powers up, the LEDs that are used to display _TO and _PD are probably on and stay on, except for after the button on _MCLR being pressed.

Chances are that you probably missed a few of the things that happen. When you think about it, what is actually happening probably doesn't make sense.

The beginning with all the LEDs turned on should not be surprising because this is very similar to the code used in the Reset experiment. The first time through, the power-up values of the Check file registers are unknown. Execution jumps to the code that turns on all the LEDs and waits for a watchdog timer reset. After the watchdog timer overflows and resets the PICmicro® MCU, then when the Check file registers are compared to expected values, they are the same and the eight counter output LEDs are incremented.

What should be surprising to you is how the _TO and _PD indicator LEDs behave. When the application powers up, they are both lit (which means that _TO and _PD are set), which is expected because in the PICmicro® MCU documentation, both bits are specified as being set when the PICmicro® MCU first powers up. But, the _TO bit should be turned off after a watchdog timer overflow and reset because it is specified to work this way. In this application, the operation of the *sleep* instruction overrides the watchdog timer setting and sets the _TO bit.

Pressing the button connected to RB0 probably gives you a surprising result (especially if you didn't think to look at the I/O pins with a logic probe or an oscilloscope until I suggested it). When you press reset, notice that the _PD bit is now reset (upon execution return from the _MCLR reset, the _PD LED is off until the next timeout). This indicates that a *sleep* instruction has executed.

The operation of the PICmicro® MCU after the RB0 is set should not be surprising because the *sleep* instruction ends after an interrupt request is received. In the PICmicro® MCU, an interrupt request is made when an enabled IF flag is set. By setting the INTE flag of the INTCON register, when the button on the RB0/INT pin is pressed, an interrupt request is passed to the processor. This takes the PICmicro® MCU out of the Sleep mode. When the PICmicro® MCU exits sleep because of the interrupt, it continues to execute (e.g., the *nop* instruction and the ones that follow) and then jump to the loop address (address zero). At no time in this process is the INTF flag reset. When a *sleep* instruction is encountered again, execution will continue because an interrupt request is pending. Thus, the PICmicro® MCU code will execute continuously.

Putting a logic probe on the ceramic resonator while *sleep* is executing can show this. When the LEDs are simply incrementing every 2.3 seconds because of the watchdog timer overflow, you will see that the oscillator is only active for a brief interval. When the button at RB0/INT is pressed and the LEDs are updating continuously, the oscillator will be running continuously as well.

This check with a logic probe also shows something else about this application. Most of the time, it is shut down. This is a bit of a profound statement, but it is true. It also means that by using the input interrupts, the *sleep* instruction, and the watchdog timer, you can set up delays that poll the outside world in your application and use much less power than they would normally require. This method of creating PICmicro® MCU applications will respond to external inputs quickly enough as to be instantaneous for human users.

This feature can be used to create battery-powered PICmicro® MCU applications that do not have on/off switches, providing a very professional "feel" to the application. For these applications, the power supply or voltage regulator must have a very low current drain during low-current-load operations. You will probably find that the 78L05 (my first choice for low-current PICmicro® MCU applications) uses too much current and some other regulator will have to be selected.

## DIFFOSC: DIFFERENT OSCILLATORS

The PICmicro® MCU has a number of different oscillator options that you can take advantage of for your applications. In this experiment, I wanted to look at three of the internally generated options to give you an idea of how the different ones work and their accuracy. When I implemented this experiment, I wanted to introduce you to a device with an internal oscillator, so I decided on the PIC16C505, which is a 14-pin device with the option of an internal oscillator.

Before describing the circuit, I wanted to show you the source code for the experiment, DIFFOSC.ASM. It in the *code\DiffOsc* subdirectory of your PC's *PICmicro* directory.

```
 title  "DiffOsc - Increment PORTC Once a Minute"
;
;  This Application Tests the time keeping Ability of the 16C505
;   With Different Clocks by updating the PORTC Value once every
;   1,000,000 Cycles (or One Second).
;
;
;  Hardware Notes:
;   PIC16C505 Running at 4 MHz with Different Clocks
;   _MCLR is Pulled Up
;   All 6 bits of PORTC are Pulled up and Connected to LEDs
;   PORTC.5 ("TOCK1") I/O set to I/O
;
;  Myke Predko
;  99.12.29
;
  LIST P=16C505, R=DEC
  INCLUDE "p16c505.inc"

;  Register Usage
 CBLOCK 0x010                  ;  Start Registers at End of the Values
Dlay:3                         ;  Counter for 1,000,000 Cycles
 ENDC

 PAGE
; __CONFIG _MCLRE_ON & _CP_OFF & _WDT_OFF & _XT_OSC
 __CONFIG _MCLRE_ON & _CP_OFF & _WDT_OFF & _IntRC_OSC_RB4EN

;  Mainline of DiffOsc
  org    0

  movwf  OSCCAL                   ;  Save the Oscillator Calibration Value

  movlw  0x0FF ^ (1 << TOCS)      ;  Turn Off TOCK1 (RC5) Pin Input
  option
  movlw  0x0FF                    ;  Turn Off the LEDs on PORTC
  movwf  PORTC

  movlw  0x0C0                    ;  Turn on the PORTC Pins
  tris   PORTC
```

```
Loop                             ; Loop Here Every 60,000,000 Cycles
  call   Delay
  decf   PORTC, f                ; "Increment" the LED output
  goto   Loop
Delay                            ; Delay 1,000,000 - 5 Cycles
  movlw  4                       ; Loop 4x
  movwf  Dlay + 2
DelayLoop
  movlw  High 0x0C44E            ; 249,993 Cycles Delay
  movwf  Dlay + 1
  movlw  Low 0x0C44E
  movwf  Dlay
  decf   Dlay, f                 ; Delay 5 Cycles for Each Loop
  btfsc  STATUS, Z
  decfsz Dlay + 1, f
    goto $ - 3
  decfsz Dlay + 2, f             ; Loop 4x
   goto  DelayLoop
  goto   $ + 1                   ; Have to Add 8 Cycles
  goto   $ + 1
  goto   $ + 1
  goto   $ + 1
  return
 end
```

Notice the similarity of this application, which is written for the low-end processor in the PIC16C505, to the mid-range processor of the other experiments' code. I only had to make four changes in this code to convert it from a PIC16F84 application to a 16C505. Notice that I did not put in the code for the different processors or emulators. This is because I wanted to demonstrate the PIC16C505's built-in oscillator.

The first was to use PORTC instead of PORTB for the application. In the other counting applications, I used PORTB in the mid-range devices. In the 16C505, PORTB is used for the Oscillator and Reset (_MCLR) pins. By using PORTC, these issues were avoided. The second change was also PORTC related; the T0CS bit of the OPTION register was reset.

The third change was to add the *movwf OSCCAL* instruction as the first instruction in the application. I put this instruction in even when I was working with an external oscillator so that I wouldn't forget it.

The last change was the only PIC16C505 specific and that was the use of a file register for the low byte of the five-cycle delay instead of saving the value in w and then decrementing it. The latter method could not be used because the low-end processor does not have the *addlw* instruction, which is needed to implement the w decrement (i.e., *addlw 0x0FF*).

These changes are remarkably minor and do not require a large amount of work to carry them out successfully. This is true for many ports of mid-range applications to low-end devices; other than some minor hardware and instruction tweaking, the applications can be passed between devices in literally a matter of minutes.

The most crucial part of the application is the loop code with the delay subroutine. I used the five-cycle delay loop for the 250-ms primary delay and then worked around it to execute four times to provide an even 1,000,000 cycles (for one second).

When calculating delays like this, it is important to remember two things. First, add 256 to the high byte, 49,998, which, is 0x0C34E. Notice in the code that I used 0x0C44E. This ensures that the least-significant byte is run through the loop the correct number of times. Second, remember that the last cycle through requires four cycles instead of five. This aspect of the delay code is only crucial for timed applications like this, but it is something to remember.

The circuit itself takes on three forms (Figs. 15-92 through 15-94). In the three cases, I have kept the circuits the same, except for differences in oscillators. For the crystal, notice that two 30 pF capacitors (27 or 33 pF can be used) are required while for the ceramic resonator these capacitors are built in.

The bill of materials is shown in Table 15-18. The wiring diagram for the ceramic resonator version is shown in Fig. 15-95. I have included a YAP-II version of the wiring diagram.

I did not include a test for the RC oscillator simply because it cannot be counted on for clocking accuracy within 20% (200,000 ppm). Unless you are willing to "tune" the clock, it is not very useful in a precision circuit. The RC clock circuit is cheap and reliable, but it is not something you want to work with in most applications (especially with the internal oscillator available).

When I tested the different clocks, I ran them each for 16 minutes and compared them against a digital clock. I chose 16 minutes because the seconds are binary equivalent of: 0b01111000000 and I could easily check this pattern against the clock. Table 15-19 records the differences in seconds and calculated *Parts Per Million (ppm)* between the clock and the PICmicro® MCU circuit. For each clock type, I used two different crystals, ceramic resonators, and two different PIC16C505 to see what kind of variance in the values I could expect.

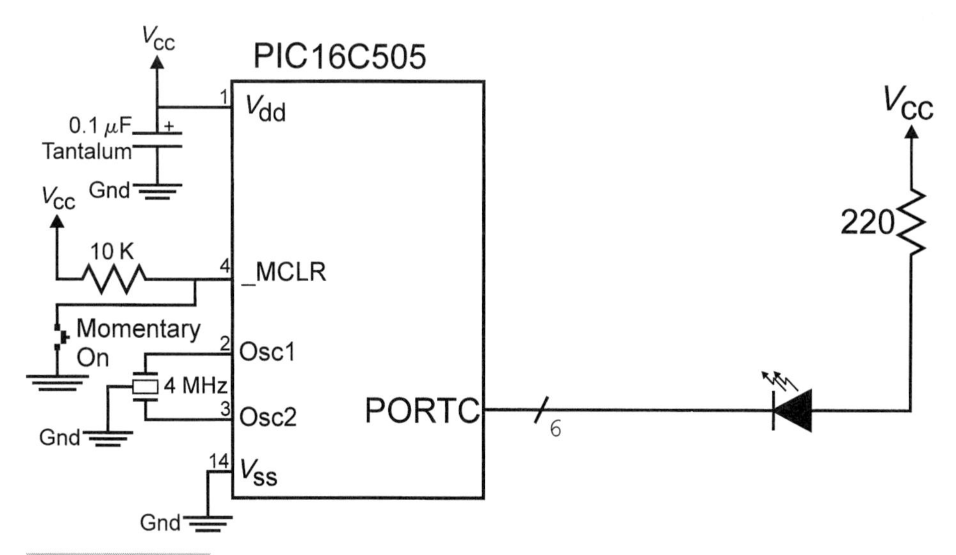


**Figure 15-92** **DiffOsc: looking at the accuracy of different PICMicro® MCU clocks (resonator)**

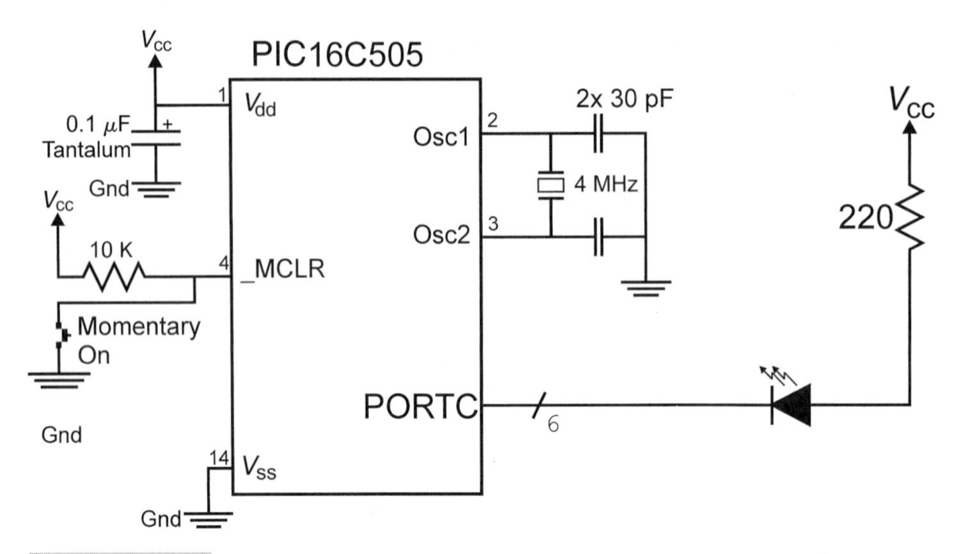


**Figure 15-93** DiffOsc: looking at the accuracy of different PICMicro® MCU Clocks (crystal)

From this table, the obvious conclusion is that a crystal is the way to go if accuracy is required. Over 16 minutes, they had the same accuracy as the digital clock for a 0% or 0 ppm error rate. These results are more a result of luck and a relatively low sample space. I suspect that timing errors would begin to creep in if the PICmicro® MCUs were run for more than 16 minutes, but the errors would still be very low. The higher cost of crystals, their fragility, and the need for external capacitors are why I prefer using ceramic resonators in my applications.

You might be surprised at the relatively high error rates with the internal RC clocks in the 16C505s. These rates are 1.4% and 2.1%, which is at the high end of the specification,

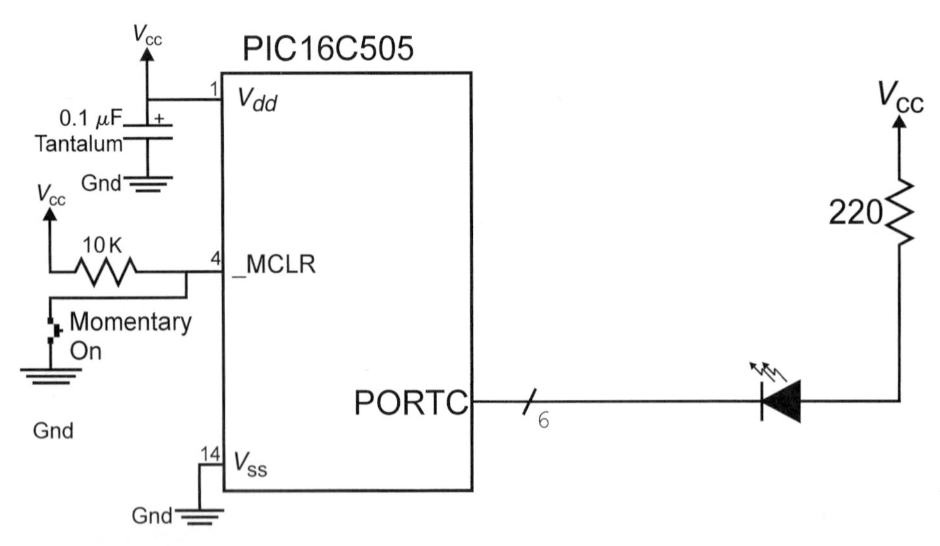


**Figure 15-94** DiffOsc: looking at the accuracy of different PICMicro® MCU Clocks (internal)

**TABLE 15-18 Bill Of Materials For The "DIFFOSC" Experiment**

| PART | DESCRIPTION |
|---|---|
| PICmicro® MCU | PIC16C505-04/JW |
| Vdd/Vss decoupling capacitor | 0.1-$\mu$F tantalum |
| _MCLR pull up | 10 K, 1/4 watt |
| 4-MHz ceramic resonator | 4 MHz with built-in capacitors |
| 4-MHz crystal | 4-MHz parallel cut |
| 2 27- to 33-pF capacitors | Any type |
| 10-LED display | Bargraph red LED |
| 6 PORTB LED resistors | 220 Ohm, 1/4 watt |

but still good enough to allow the PICmicro® MCUs to be used for asynchronous serial communication (as shown in the next chapter). When I use the 16C505 or the 12C5xx parts, I always use the built-in oscillators because they allow simpler circuit wiring and the use of all possible pins in an application.

## EEPROM: ACCESSING EEPROM DATA MEMORY

Accessing the built-in EEPROM data memory in the 16F8x PICmicro® MCUs is not something that I naturally think of when creating my own applications. This memory can be useful in a variety of different situations, from saving user-interface preferences to con-

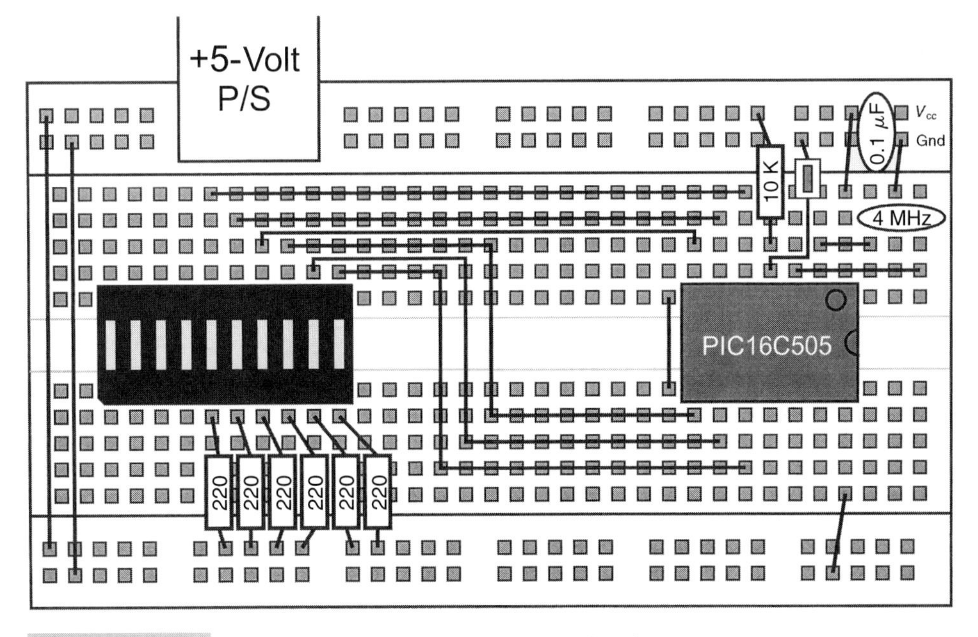


**Figure 15-95** DiffOsc PIC16C505 breadboard circuit

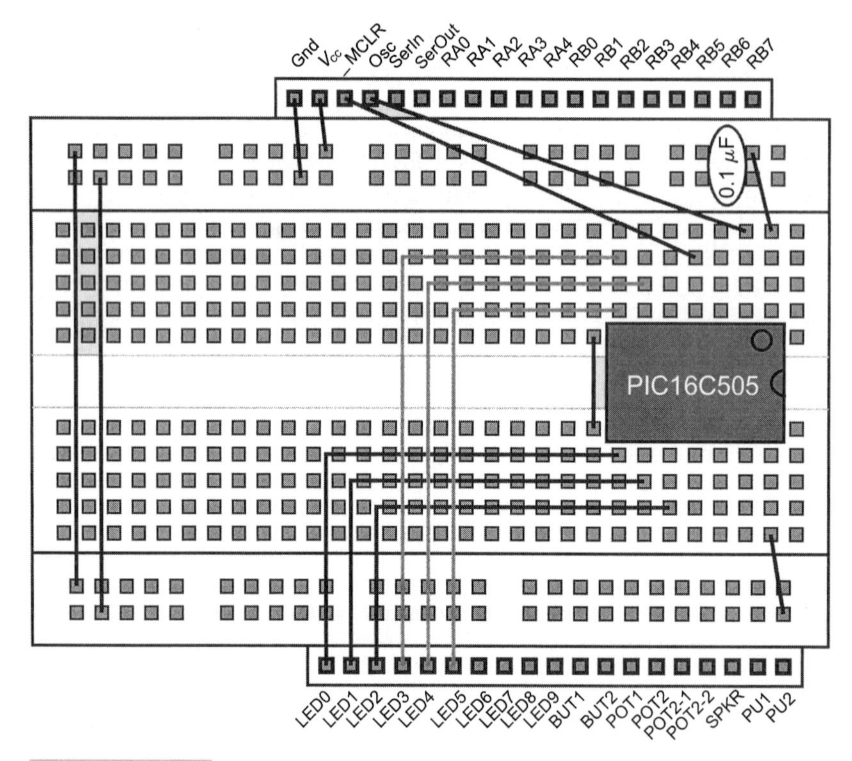


**Figure 15-96** YAP-II breadboard for DiffOsc application

stants and even simple program data. Chapter 16 presents a couple of applications that use the EEPROM memory. Before getting to them, I wanted to present you with a simple example of how EEPROM is accessed and how it can be manipulated in the source code and the MPLAB simulator.

The EEPROM.ASM application, which is located in the *code\EEPROM* subdirectory in the *PICmicro* directory checks the first four addresses of the 16F84's data EEPROM. If it is set to a predetermined pattern (0x055AA00FF), it reads the fifth byte, displays it, and increments it. If the predetermined pattern is not discovered, then it is loaded into data

**TABLE 15-19 Results Of The "DIFFOSC" Experiment**

| PICmicro® MCU Clock | Error in 16 Minutes (960 Seconds) | ppm Error |
|---|---|---|
| Ceramic resonator #1 | 3 Seconds | 3,125 |
| Ceramic resonator #2 | 4 Seconds | 4,167 |
| Crystal #1 | 0 | 0 |
| Crystal #2 | 0 | 0 |
| PIC16C505 #1 Internal | 14 Seconds | 14,583 |
| PIC16C505#2 Internal | 21 Seconds | 21,875 |

EEPROM and the counter byte is loaded with 0x000. After the predetermined pattern is loaded into EEPROM, the PORTB LEDs are flashed to indicate that the correct data was not encountered.

The circuit is not that unusual to the Reset application's circuit (Fig. 15-97). As will be covered later in the text, I had some problems with the momentary-on switch on reset bouncing and causing problems with the application. To avoid this, you might want to add a 0.1-$\mu$F filter capacitor on the _MCLR line to prevent these problems.

The bill of materials for the EEPROM application is shown in Table 15-20. The breadboard circuit (minus the optional filter capacitor on _MCLR) is shown in Fig. 15-98. The YAP-II circuit (Fig. 15-99) does not require any special treatment of reset because the YAP drives this itself with the Reset button.

This application could be implemented with the EMU-II, but the built in "eeread" and "eewrite" routines would have to be used instead. This is also true for MPLAB-ICD.

The code is somewhat different from the Reset applications because of the data being stored in EEPROM instead of file registers.

```
 title  "EEPROM - Show Contents of an EEPROM Counter on Reset"
;
;  This Application checks four bytes in the Data EEPROM to see
;   if they are at an expected value and then increments an
;   EEPROM counter and displays it.  If the Data EEPROM Check
;   bytes are not true, then they are set and the LEDs on PORTB
;   are flashed.
;
;  Flash Memory is Organized as:
;  Byte 0 - 0x0FF
;  Byte 1 - 0x000
;  Byte 2 - 0x0AA
;  Byte 3 - 0x055
;  Byte 4 - EEPROM Counter
```

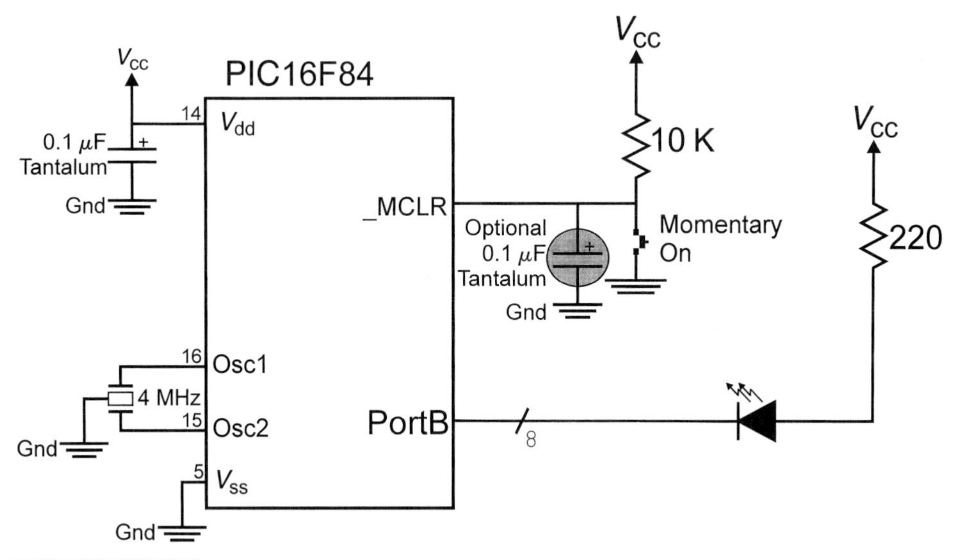


**Figure 15-97** EEPROM: circuit displaying EEPROM counter on _MCLR or power up

**TABLE 15-20 Bill Of Materials For The "EEPROM" Experiment**

| PART | DESCRIPTION | REQUIRED ON YAP-II |
|---|---|---|
| PICmicro® MCU | PIC16F84-04/P | In Socket |
| Vdd/Vss decoupling capacitor | 0.1-$\mu$F tantalum | No |
| _MCLR pull up | 10 K, 1/4 watt | No |
| 4-MHz ceramic resonator | 4 MHz with built-in capacitors | No |
| 10-LED display | Bargraph red LED | No. LED0/LED7 used |
| 8 PORTB LED resistors | 220 Ohm, 1/4 watt | No. LED1/LED7 used |
| Momentary-on push button | Momentary on | No |
| Optional 0.1-$\mu$F capacitor | 0.1-$\mu$F tantalum | No |

```
;
;  Hardware Notes:
;   PIC16F84 Running at 4 MHz
;   _MCLR is Pulled Up with a Momentary On Pull Down Switch
;   All 8 bits of PortB are Pulled up and Connected to LEDs
;
;  Myke Predko
;  99.12.28
;
  LIST R=DEC
  INCLUDE "p16f84.inc"
```

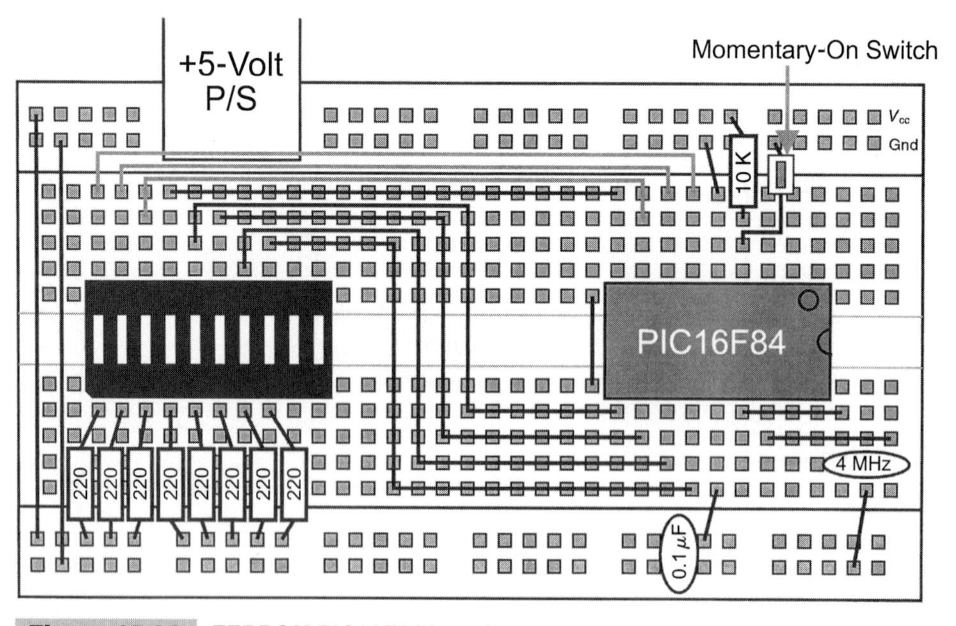


**Figure 15-98** EEPROM PIC16F84 breadboard circuit

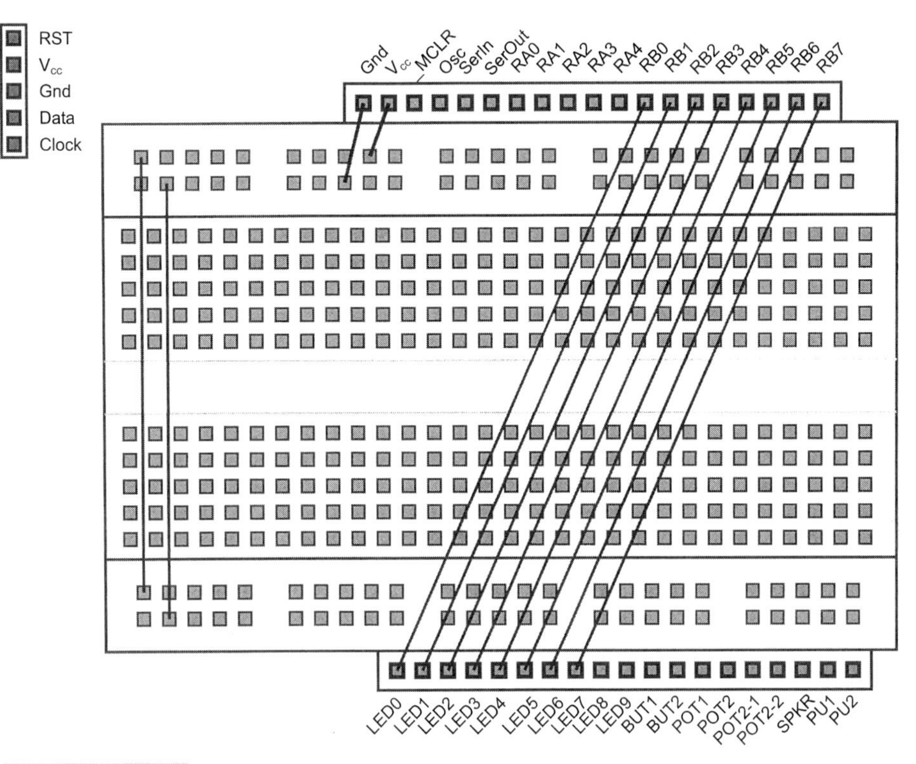


**Figure 15-99** YAP-II breadboard for EEPROM

```
;  Register Usage
 CBLOCK 0x020                 ;  Start Registers at End of the Values
Dlay:2                        ;  Two Bytes for Flashing Delay
 ENDC

 PAGE
 __CONFIG _CP_OFF & _XT_OSC & _PWRTE_ON  & _WDT_OFF

;  Mainline of EEPROM
  org    0

  bsf    STATUS, RP0
  clrf   TRISB ^ 0x080        ;  Make All 8 PortB Bits Output
  bcf    STATUS, RP0

  movlw  EECON1               ;  Use FSR To point to EECON1
  movwf  FSR                  ;  To Avoid Going Back and Forth

  movlw  0                    ;  Look for "Check" Bytes in Data EEPROM
  movwf  EEADR
  bsf    INDF, RD             ;  Read and Compare the First Byte
  movf   EEDATA, w
  xorlw  0x0FF
  btfss  STATUS, Z
   goto  WrongEEPROM          ;  Not 0x0FF - Reset the EEPROM and Continue
  movlw  1
  movwf  EEADR
  bsf    INDF, RD             ;  Read and Compare the Second Byte
```

```
  movf   EEDATA, w
  xorlw  0x000
  btfss  STATUS, Z
   goto  WrongEEPROM
  movlw  2
  movwf  EEADR
  bsf    INDF, RD           ;  Read and Compare the Third Byte
  movf   EEDATA, w
  xorlw  0x0AA
  btfss  STATUS, Z
   goto  WrongEEPROM
  movlw  3
  movwf  EEADR
  bsf    INDF, RD           ;  Read and Compare the Fourth Byte
  movf   EEDATA, w
  xorlw  0x055
  btfss  STATUS, Z
   goto  WrongEEPROM

;  Check Data is Correct - Display Current Contents and Increment Them

  movlw  4
  movwf  EEADR
  bsf    INDF, RD
  comf   EEDATA, w          ;  Get the Complemented Data for LEDs
  movwf  PORTB

  xorlw  0x0FF              ;  Increment and Store the Byte Back
  addlw  1
  movwf  EEDATA
  call   EEWrite            ;  Write to the EEPROM Data

  goto   $                  ;  Finished, Endless Loop

WrongEEPROM                 ;  Rewrite the Contents of Data EEPROM
  movlw  0                  ;  Write the First Byte
  movwf  EEADR
  movlw  0x0FF
  movwf  EEDATA
  call   EEWrite

  movlw  1                  ;  Write the Second Byte
  movwf  EEADR
  movlw  0x000
  movwf  EEDATA
  call   EEWrite

  movlw  2                  ;  Write the Third Byte
  movwf  EEADR
  movlw  0x0AA
  movwf  EEDATA
  call   EEWrite

  movlw  3                  ;  Write the Fourth Byte
  movwf  EEADR
  movlw  0x055
  movwf  EEDATA
  call   EEWrite

  movlw  4                  ;  Reset the Counter
  movwf  EEADR
  movlw  0x000
  movwf  EEDATA
  call   EEWrite
```

```
Loop                          ;  Flash the LEDs in this Case

  call   Delay                ;  Delay 200 msecs

  comf   PORTB, f             ;  Complement the Contents of PORTB

  goto   Loop

;  Subroutines

Delay                         ;  Delay 200 msecs

  clrf   Dlay
  clrf   Dlay + 1
  decfsz Dlay, f
   goto  $ - 1
  decfsz Dlay + 1, f
   goto  $ - 3

  return

EEWrite                       ;  Write EEData into EEPROM at Address
EEADR

  bsf    STATUS, RP0
  bcf    INDF, WRERR          ;  Make Sure Write Error Bit is Reset
  bsf    INDF, WREN           ;  Enable the Write
  movlw  0x055                ;  #### - Required Write Sequence
  movwf  EECON2 ^ 0x080       ;  ####
  movlw  0x0AA                ;  ####
  movwf  EECON2 ^ 0x080       ;  ####
  bsf    EECON1 ^ 0x080, WR;  ;  #### - End of Required Write Sequence

  btfss  INDF, EEIF           ;  Wait for Finished Interrupt Request
   goto  $ - 1

  bcf    STATUS, RP0          ;  Return to Bank 0
  bcf    INDF, WREN
  bcf    INDF, EEIF
  return

 org     0x02100              ;  EEPROM Set Here
  de     0x0FF, 0x0FF, 0x0FF, 0x0FF, 0x0FF

 end
```

At the end of the application, I included two lines (*org 0x02100* and *de 0x0FF, . . .*) that I haven't mentioned before. For programming, the data EEPROM is located at address 0x02100. To program it, the programmer increments the PICmicro® MCU's program counter to 0x02100 and then writes the byte into the EEPROM as if it were program memory. If you look at the hex file produced for the EEPROM.ASM, you will see that these five bytes are specified at address 0x04200, which is twice 0x02100, and is what you should expect. These two lines are included into the application to ensure that the EEPROM is put into the initial state when the application starts so that I could test its behavior.

The El Cheapo and YAP-II cannot program a PIC16F84's data EEPROM memory from the hex file. If you are using a El Cheapo or a YAP-II with your application, you will have to comment out this line before downloading the hex file. Once this is done, if you want to test the application multiple times, then you will have to change the comparison values written into the data EEPROM.

As indicated, I had some problems with noise on the PICmicro® MCU's _MCLR line when I was testing how the increment function worked. As I originally set up the application,

I just used a pull up with a momentary-on switch to reset the PICmicro® MCU. I have used this circuit in a number of other applications without any problems. But when there was a bounce in the _MCLR reset, the EEPROM read/write was interrupted, which caused a problem with the PICmicro® MCU locking up. I added the 0.1-$\mu$F capacitor to the reset circuit and the problems went away.

The conclusion that I get from this is that when working with EEPROM, the Reset should be driven by a power supply or another device to avoid bounces that could cause problems with accesses.

The data EEPROM is simulated in the MPLAB IDE and you can work through the application before you test it on a part. To do this, I created the rather busy MPLAB desktop shown in Fig. 15-100. On the desktop, along with the standard PIC16F84 registers, I also included a watch window for the data EEPROM-specific registers (the EEPROM watch window), the window for the EEPROM and the *Change Program Counter* control.

The EEPROM window displays the contents of the simulated EEPROM memory. If Data EEPROM is changed by the application, then the contents of this window will be changed appropriately. When the application is reassembled, the contents of this window will change, resetting the simulated processor will not change the contents of the simulated data EEPROM memory. If a PICStart Plus is connected to the MPLAB IDE, the contents of the hex file (which are set by the *org 0x02100/de 0x0FF, . . .* instructions) will be used and not the currently simulated values in the EEPROM window.

For this application, when I simulated it, I wanted to test the operation of the application, but without waiting for the Delay subroutine to return after 200 simulated ms. Instead

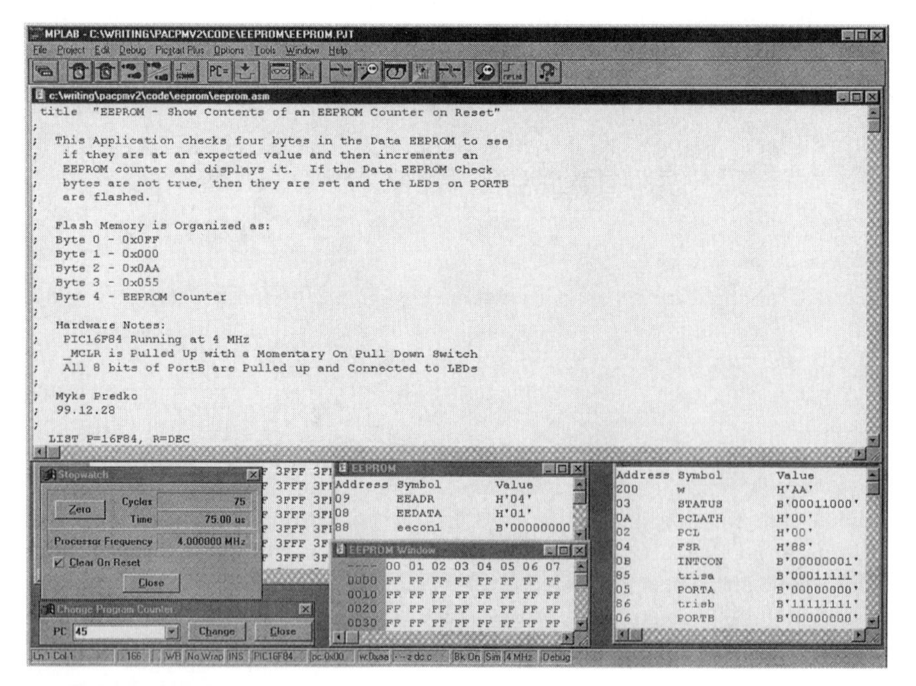


**Figure 15-100** EEPROM MPLAB desktop

of using the Debug and conditional code, I used another feature of MPLAB and that is the ability to change the program counter. The *Change Program Counter* dialog box allows you to change the simulated PICmicro® MCU's program counter by just clicking on *Change*.

This feature was useful in this application because I only needed to skip over one address. As can be seen in Fig. 15-100, I loaded it with 0x045, which is the address after *call Delay*. As I was simulating the application, after updating PORTB to avoid the long delay, all I had to do was click on *Change* to jump over the *call Delay* instruction. Using the *Change Program Counter* dialog box in this method is only this simple when there is only one location to jump over. If multiple addresses are required, then I recommend using the Debug *#define* and conditional code.

## SHORT: THE SIMPLEST PRACTICAL PICmicro® MCU APPLICATION POSSIBLE

The title of this experiment probably seems somewhat arrogant, but I think that I have created a small application that does not make a liar out of me. This application is only two instructions long and I think it shows off some obscure aspects of the PICmicro® MCU and how PICmicro® MCU programming works. It also provides you with a simple application that can be used in a variety of situations to check on the "health" of a PICmicro® MCU before an application is programmed in.

The application itself simply flashes an LED on RB7. The circuit that I used (Fig. 15-101) requires the parts listed in Table 15-21.

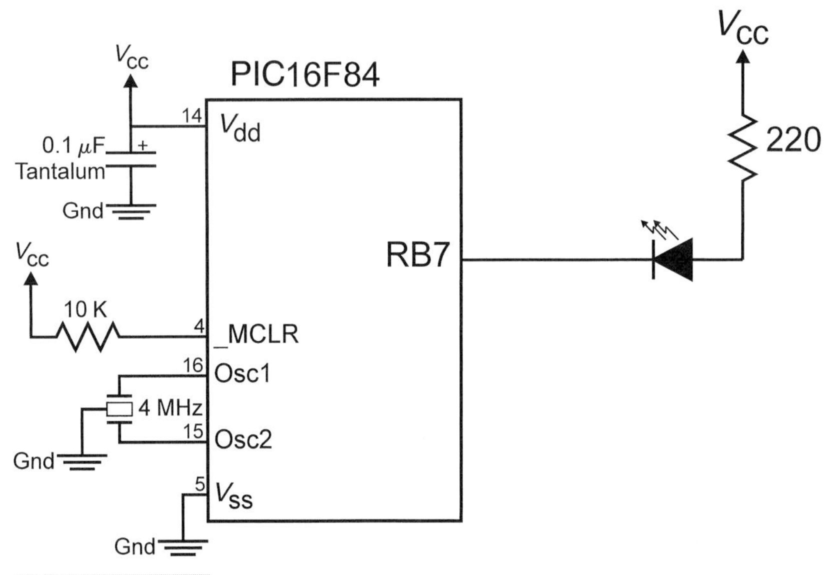


**Figure 15-101** **Short: circuit for shortest possible application**

**TABLE 15-21 Bill Of Materials For The "SHORT" Experiment**

| PART | DESCRIPTION | REQUIRED ON YAP-II |
|---|---|---|
| PICmicro® MCU | PIC16F84-04/P | In Socket |
| Vdd/Vss decoupling capacitor | 0.1-$\mu$F tantalum | No |
| _MCLR pull up | 10 K, 1/4 watt | No |
| 4-MHz ceramic resonator | 4 MHz with built-in capacitors | No |
| RB7 LED | Red LED | No. LED0 used |
| RB7 LED Resistor | 220 Ohm, 1/4 watt | No. LED0 used |

The circuit itself is pretty easy to assemble, either on a breadboard (Fig. 15-102) or on the YAP-II breadboard (Fig. 15-103). In either case, the circuit will take just a few moments to assemble. The only thing to watch out for is when you program the PICmicro® MCU with the application code. If the programming is going to occur in circuit (which is what happens with the YAP-II), then the LED should be disconnected from RB7 until after it has been programmed and it is ready to execute.

The application code itself is honestly only two instructions long and will cause the LED to flash at about seven and a half times per second when the PICmicro® MCU is running at 4 MHz. The application itself is located in the *code\short* subdirectory of the *PICmicro* subdirectory.

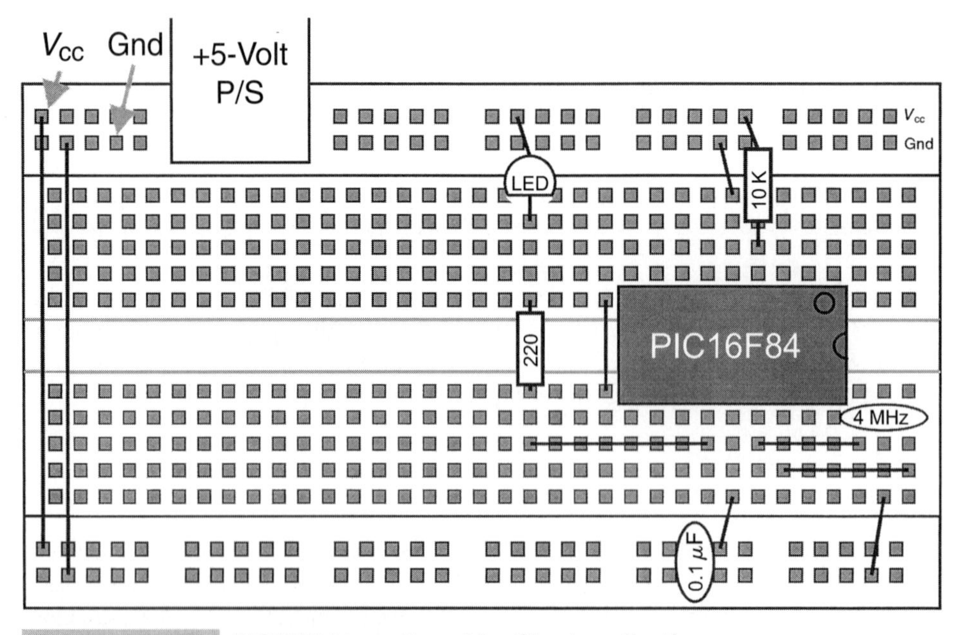


**Figure 15-102** PIC16F84 breadboard for Short application

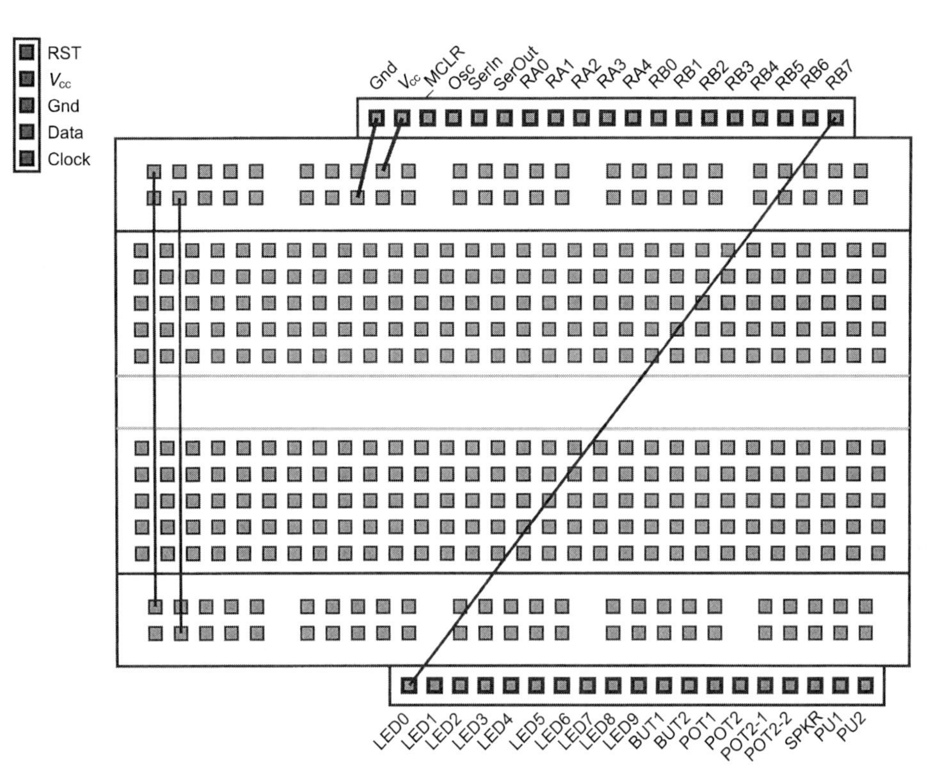


**Figure 15-103** YAP-II breadboard for Short

The two instructions of the application code are:

```
tris    PORTB             ;  Save "w" in TRISB
xorwf   PORTB, f          ;  XOR PORTB with the contents of "w"
```

I'm not going to initially reveal the entire source code just yet because I want to cover some of its aspects later in this section. In any case, the unique application code is only these two instructions long, but they take advantage of two features of the PICmicro® MCU and its processor.

Elsewhere in the book, I mention that the *tris* instruction should never be used in the mid-range PICmicro® MCUs. I use the *TRIS* instruction in this application simply because it is the method of changing PORTB's *TRIS* register in the fewest number of instructions. If I were to update TRISB "properly," I would use the code:

```
bsf     STATUS, RP0       ;  Access Bank 1
movwf   TRISB ^ 0x080     ;  Store the Contents of "w" in TRISB
bcf     STATUS, RP0       ;  Return to Bank 0
```

This application code takes advantage of the fact that unprogrammed program memory addresses are set to *1*s. This is a function of EPROM and flash memory; when it is erased, the bits in the memory are all set to *1*s. For the mid-range PICmicro® MCUs, the instruction *addlw 0x0FF* has the bit pattern 0x03FFF, which is all of the bits set in an instruction word.

As the code executes through the unprogrammed program memory, it is adding 0x0FF (or -1) to the contents of the w register. Upon power up, the contents of w are undefined, but as the code executes through the unprogrammed instructions, the value within it will be continually incremented.

The other aspect of the PICmicro® MCU that this application takes advantage of is that when the program counter reaches the end of program memory, it resets and continues executing from address zero.

These two features of the PICmicro® MCU means that the application code, although only two instructions long actually executes as if it were:

```
Loop                            ;  Loop Back Here
  call   Dlay1019               ;  Delay Same Cycles executing
                                ;   "addlw 0x0FF" 1022x
  addlw  2                      ;  Equivalent result as "addlw 0x0FF" 1022x
  tris   PORTB                  ;  Save "w" in TRISB
  xorwf  PORTB, f               ;  XOR PORTB with the contents of "w"
  goto   Loop
```

The effective adding of two each time through the loop is why I have put the LED on RB7, the most-significant bit in the PORTB register. As the w register is incremented by two, the most-significant register will be toggled at the slowest rate. The LED could be put on any output from RB1 to RB7, but you will find that as you use the least-significant bits, the rate at which the LED flashes increases to the point where it cannot be observed.

The actual source code is:

```
 title  "SHORT - Is this the Shortest Possible Application?"
#define nYAPProg
;
;  Look at a Two Instruction Application.  Continually Update
;   w (with the "unprogrammed" ADD 0x0FF) and then use the
;   value for "PORTB" and "TRISB".
;
;  Hardware Notes:
;   16F84 Running at 4 MHz
;   Reset is tied directly to Vcc and PWRT is Enabled.
;   PortB is used for Output
;
;
;  Myke Predko
;
;  99.10.26 - "Short" Created
;
  list R=DEC
  include "p16f84.inc"

 __CONFIG _CP_OFF & _XT_OSC & _PWRTE_ON  & _WDT_OFF

 ifndef YAPProg
 org     0x03FE               ;  Start Application at Program End
 else
 org     0x0                  ;  Load Instructions to Program Memory
                              ;   End so there isn't a large "jump"
 variable i = 0               ;   in instructions the YAP can't handle
 while (i < 0x0100)
  dw     0x03FFF
```

```
i = i + 1
 endw
 while   (i < 0x0200)
  dw     0x03FFF
i = i + 1
 endw
 while (i < 0x0300)
  dw     0x03FFF
i = i + 1
 endw
 while (i < 0x03FE)
  dw     0x03FFF
i = i + 1
 endw
 endif

  tris   PORTB              ;  Save "w" in TRISB
  xorwf  PORTB, f           ;  XOR PORTB with the contents of "w"

  end
```

This code runs through the PICmicro® MCU's program memory repeatedly, adding −1022 to the contents of w (which is the same as adding positive two when the eight-bit register is taken into account). Next, the new value for w is used for the TRISB register and then XORed with the contents of PORTB to change PORTB's output value.

The operation of the application can be observed by enabling the *Program Memory* window (from the *Window* pull down on MPLAB's top pull downs) as is shown in Fig. 15-104. Once you have brought this up, you can single step through the application (the two instructions explicitly programmed, along with the *addlw 0x0FF* instructions) to see that the five-instruction simulation written before the source code is actually correct.

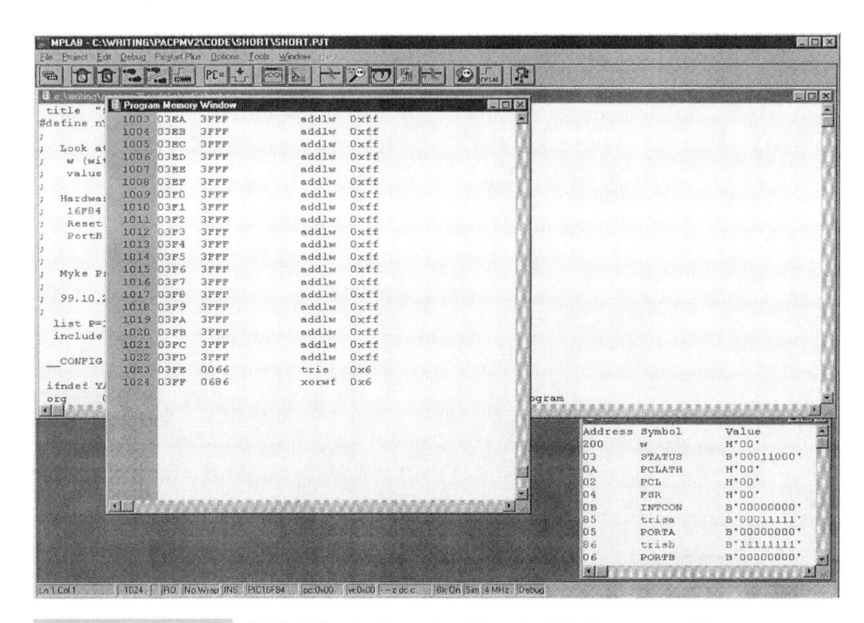


**Figure 15-104** MPLAB window for Short with Program Memory window active

If you are going to use a programmer other than the YAP-II, then you can leave the:

```
#DEFINE nYAPProg
```

statement just the way it is. If you are going to use the YAP-II, then the *#DEFINE* statement should be changed to:

```
#DEFINE YAPProg
```

In either case, the two instructions (*tris PORTB* and *xorwf PORTB, f*) will be at the same locations in the program memory (addresses 0x03FE and 0x03FF, respectively). The difference is that in the YAP-II case, the hex file will be built with the 0x03FFF (*addlw 0x0FF*) instructions declared explicitly. The reason for doing this is that the YAP-II updates the PICmicro® MCU's program counter as it is programming the part while data is coming in. The YAP-II can only increment the program counter by 50 instructions between the time it gets the address and when it has to start saving data into the PICmicro® MCU. The YAP-II will return an error if a jump between the current program counter and the destination program counter is greater than 50. This problem is avoided by explicitly loading the memory with the unprogrammed instructions.

A more obvious solution to this problem would be to put the two instructions of the application at the start of the PICmicro® MCU's memory space instead of at the end. The reason I don't do this is because of the nature of this experiment.

This experiment is actually an application. The purpose of this application might not seem clear when you first see it, but it is designed for the cases where you are unsure whether or not an OTP PICmicro® MCU (which can only be programmed once) will run in a circuit or if it can be programmed. By placing these two instructions at the end of the program memory, another application can be programmed into the PICmicro® MCU with very little worry that the two instructions of this application will affect the operation of the final application.

When this application is used with a PICmicro® MCU soldered into a circuit, then the PORTB interfaces should not be assembled in the board in order to prevent errant operation of the final application circuit. The operation of the application can be checked with a logic probe or an oscilloscope. The 7.5-Hz LED flashing on RB7 will probably be too fast to observe on a DMM.

Once you are ready to "burn" in the final application, you could overwrite the two instructions of this application with two *nop* instructions (all zeros) to ensure that this application does not affect the operation of any others. These two actions make absolutely sure that the two instructions of this application will not affect any other operations on the board.

Having the *addlw 0x0FF* instruction as the equivalent for an unprogrammed program memory word is only available on the mid-range PICmicro® MCUs. Thus, this application and the trick of the two instructions will only work on the mid-range devices and not the low-end, 17Cxx, or 18Cxx PICmicro® MCUs. For these other devices, a more explicit test of the PICmicro® MCU's operation will have to be created. Remember that in all the PICmicro® MCUs, *nop* is when all the bits in an instruction word are programmed (reset or *0*s). This allows you to put a PICmicro® MCU execution test (like this application) in "patch" memory and then erase it when the final application is to be burned into the part.

No EMU-II version is included for this application because the code put into the PICmicro® MCU's program memory will cause execution to stop at addresses 0X0700.

# Analog Input/Output

When I took Electromagnetics and Transmission Theory in university, I wished for nothing more than the world to consist of just ones and zeros. Unfortunately, the world just doesn't work that way; varying voltage levels are available inside the PICmicro® MCU's $V_{dd}$ and $V_{SS}$ voltage range to the outside in terms of positive and negative voltages. As well, the different speeds and waveforms that these voltages are presented with can make your life more difficult.

The next sections provide some practical examples of how analog I/O interfacing with the PICmicro® MCU is accomplished using both built-in hardware devoted to the task and using the digital I/O functions of the PICmicro® MCU to perform the same tasks.

## ADCLESS: MEASURING RESISTANCE VALUES WITHOUT AN ADC

To measure resistance values without an ADC, a simple RC network can be used with the PICmicro® MCU (Fig. 15-105). This method has been taken from the Parallax Basic Stamp 2. It is quite easy to do and works quite well, although a few concerns have to be worked through for this method to be used in an application.

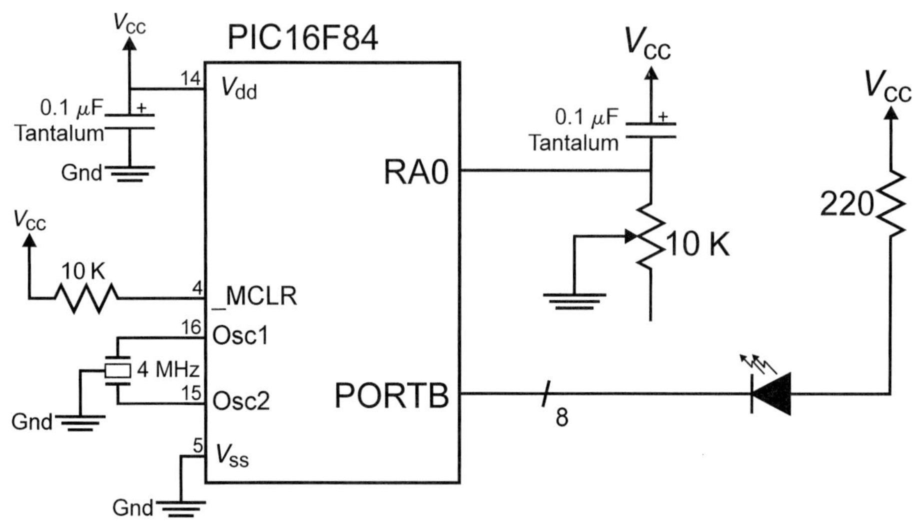


**Figure 15-105** ACDLESS: reading a potentiometer position without an ADC

To measure the resistance (assuming the capacitor is of a known value), the PICmicro® MCU first charges the capacitor to 5 volts (or its nominal output) using the I/O pin in Output mode. Once this is done, the pin changes to Input mode and waits for the capacitor to discharge through the potentiometer. Looking at this operation on an oscilloscope, the waveform produced by the circuit looks like that in Fig. 15-106. The charge cycle and discharge cycle can clearly be seen in this figure.

From basic electronic theory, the time required for the capacitor to charge is:

$$Time = R * C * ln(V_{end}/V_{start})$$

where the $V_{start}$ and $V_{end}$ are the starting and ending voltages that we are interested in. For the PICmicro® MCU, we would be interested in the capacitor voltage starting at $V_{dd}$ (after being charged by the PICmicro® MCU to 5 volts) and then waiting for the capacitor to discharge to the input transition point (2.5 volts in the PICmicro® MCU).

Because the capacitor value, voltages, and the capacitor discharge time are known, the formula can be rearranged to find $R$:

$$R = Time/[C * ln(V_{end}/V_{start})]$$

Therefore, by controlling the voltage applied to the network and knowing the value of the capacitor, you can determine the value of the resistor.

The code used to test the analog I/O uses the following logic:

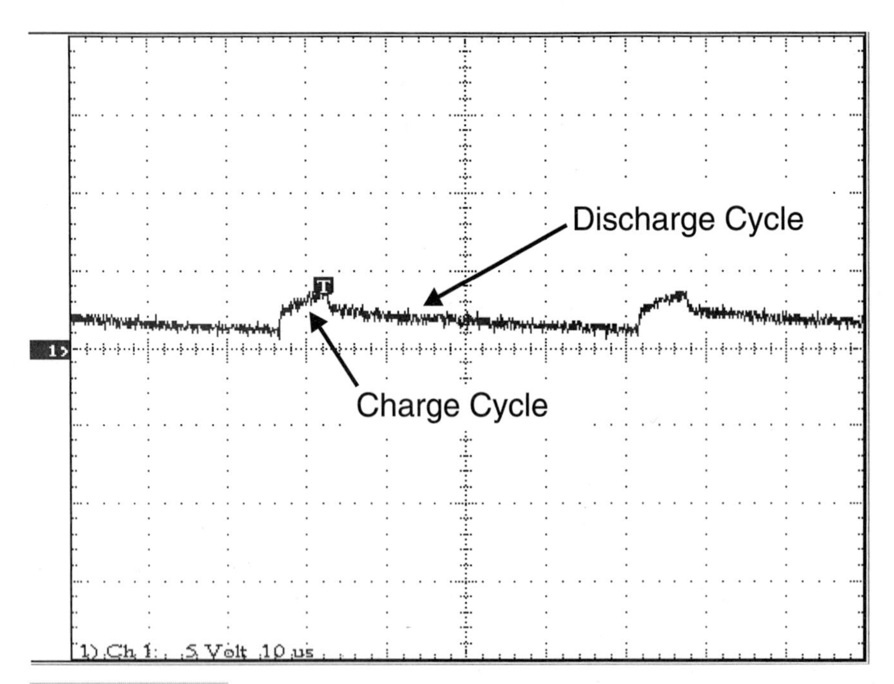


**Figure 15-106** Oscilloscope picture for ADCLess operation

```
int PotRead()                  //  Read the Resistance at the I/O Pin
{

int i;

  TRIS.Pin = Output;           //  Set the Output Mode
  Pin = 1;                     //  Output a "1" to Charge the Capacitor
  for (i = 0; i < 5usec, i++ );
  TRIS.Pin = Input;            //  Now, Time How Long it Takes for the
  TMR0 = 0;                    //   the Capacitor to Discharge through
  while (Pin == 1);            //   the Potentiometer

  return TMR0;                 //  Return the TMR0 Value for the
                               //   Discharge Time

} //  end PotRead
```

This code is unique because no file registers are used for the timing, which is totally done within the PICmicro® MCU hardware. TMR0 does not have to be used. Instead, the w register could be incremented within the *while (Pin == 1)* loop.

The source for this experiment is ADCLESS.ASM, which is located in the *code\ADC-Less* subdirectory of the *PICmicro* directory.

```
 title  "ADCLess - Reading a Resistor Value without an ADC"
;
;  This Program copies the "RCTIME" instruction of the Parallax Stamp.
;  A resistor value is read repeatedly and displayed.
;
;  This program is a modification of PROG17.ASM
;
;  Hardware Notes:
;   PIC16F84 running at 4 MHz
;   Reset is tied directly to Vcc and PWRT is Enabled.
;   A 10K Pot along with a 0.1uF Cap and 100 Ohm Series Resistor on
;    PORTA.0
;   A 220 Ohm Resistor and LED is attached to all the PORTB.7:0
;
;  Application Updated: 99.12.26 for 4 MHz PIC16F84.
;
;  Myke Predko
;  96.06.02
;
  LIST R=DEC
 ifdef __16F84
  INCLUDE "p16f84.inc"
 endif
 ifdef __16F877
  INCLUDE "p16f877.inc"
 endif
;  Registers

 ifdef __16F84
 __CONFIG _CP_OFF & _WDT_OFF & _XT_OSC & _PWRTE_ON
 else
 __CONFIG _CP_OFF & _WDT_OFF & _XT_OSC & _PWRTE_ON & _DEBUG_OFF &
 _LVP_OFF & _BODEN_OFF & _WRT_ENABLE_ON & _CPD_OFF
 endif

  PAGE
;  Mainline of ADCLess
```

```
 org      0

  nop

  movlw  0x0FF
  movwf  PORTB              ;  Turn off all the LED's
  clrf   PORTA              ;  Use PORTA as the Input

  bsf    STATUS, RP0        ;  Have to go to Page 0 to set Port
Direction
  clrf   TRISB & 0x07F      ;  Set all the PORTB bits to Output
  movlw  0x0D2              ;  Setup the Timer to fast count
  movwf  OPTION_REG & 0x07F;    Put in Divide by 8 Prescaler for 4x
                            ;   Clock
  bcf    STATUS, RP0        ;  Go back to Page 0

  movlw  TRISA              ;  Have to Set/Read PORTA.0
  movwf  FSR                ;    - Use FSR instead of Changing RP0

Loop

  bsf    PORTA, 0           ;  Charge Cap on PORTA.0
  bcf    INDF, 0            ;  Make PORTA.0 an Output
  movlw  0x0100 - 10        ;  Charge the Cap
  clrf   TMR0               ;  Now, Wait for the Cap to Charge
Sub_Loop1                   ;  Wait for the Timer to Reach 10
  movf   TMR0, w            ;  Get the Timer Value
  btfss  STATUS, Z          ;  Has the Timer Overflowed?
   goto  Sub_Loop1          ;   No, Loop Around again

  bsf    INDF, 0            ;  Now, Wait for the Cap to Discharge
  clrf   TMR0               ;   and Time it.
Sub_Loop2                   ;  Just wait for PORTA.1 to go Low
  btfsc  PORTA, 0
   goto  Sub_Loop2

  comf   TMR0, w            ;  Get the Timer Value
  movwf  PORTB

  goto   Loop               ;  Get another Time Sample

  end
```

The circuit itself is relatively easy to build. The breadboard version is shown in Fig. 15-107 and the YAP-II version is shown in Fig. 15-108. Notice that the potentiometer is wired somewhat differently than what you are probably used to. It is not used as a voltage divider, but as a variable resistance path to ground for the charge in the capacitor. The bill of materials for ADCLess is shown in Table 15-22.

I have not spent a lot of time simulating this experiment. The reason is because the operation of the RC network (with varying resistances) cannot be easily simulated by MPLAB. A stimulus file could produce simulated delays, but I decided to go ahead and try out the application directly.

If you work through different capacitors for the RC network in this experiment, you'll discover how dependent the circuit is on the capacitor value. I found that after trying four different capacitors, I got four different "upper limits," with some going beyond 0x0FF (the limit that can be returned by the eight-bit TMR0) and one having a maximum value of 0x046. This leads to the biggest problem with this circuit, which is its dependency on the parts used.

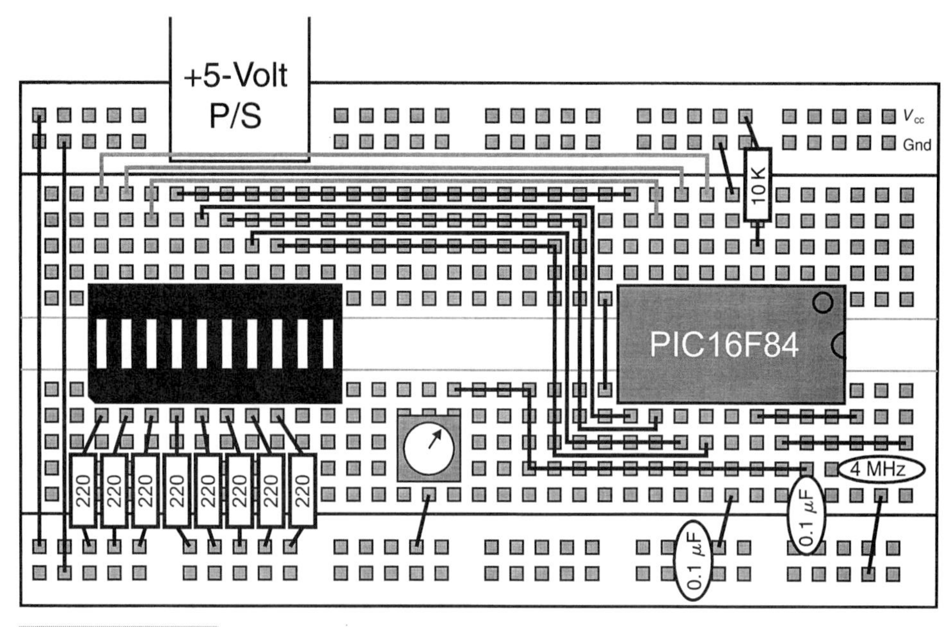


**Figure 15-107** ADCLess PIC16F84 breadboard circuit

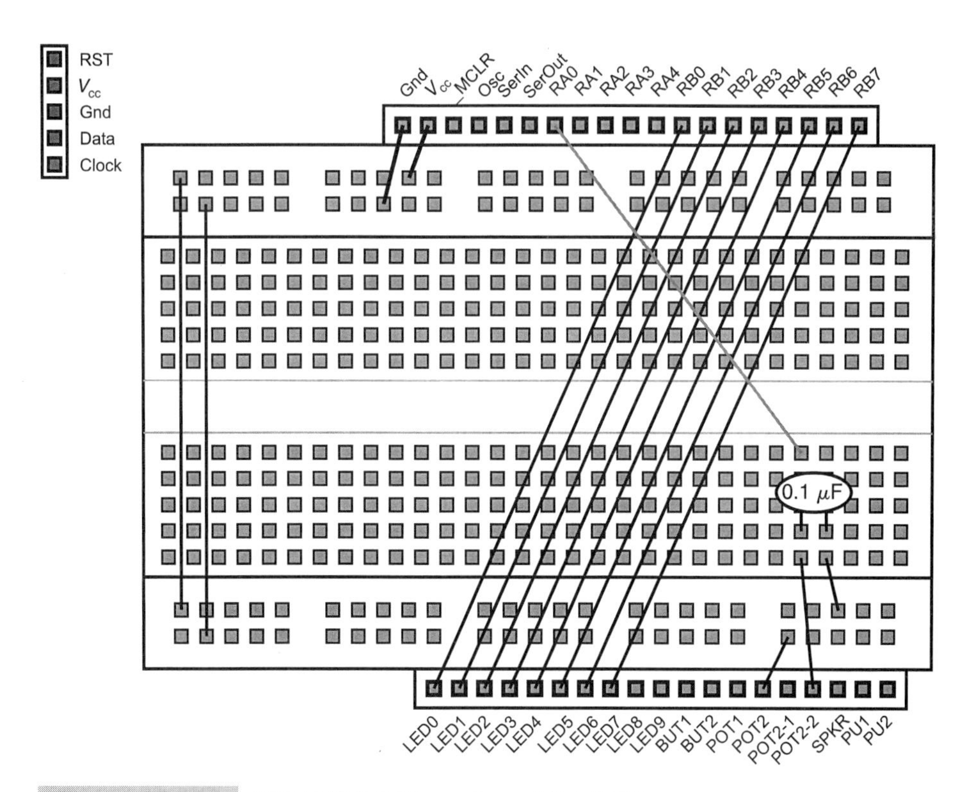


**Figure 15-108** EMU-II/YAP-II breadboard for ADCLess

| TABLE 15-22 Bill Of Materials For The "ADCLESS" Experiment | | |
|---|---|---|
| PART | DESCRIPTION | REQUIRED ON YAP-II |
| PICmicro® MCU | PIC16F84-04/P | In Socket |
| Vdd/Vss decoupling capacitor | 0.1-$\mu$F tantalum | No |
| _MCLR pull up | 10 K, 1/4 watt | No |
| 4-MHz ceramic resonator | 4 MHz with built-in capacitors | No |
| 10-LED display | Bargraph red LED | No. LED0/LED7 used |
| 8 PORTB LED resistors | 220 Ohm, 1/4 watt | No. LED1"/LED7 used |
| 10-K Potentiometer | Single-turn/PC board mount | No. Pot2 used |
| 0.1-$\mu$F tantalum capacitor | 0.1-$\mu$F tantalum | Yes |

Because of the variance to the capacitor value, I would not recommend this circuit for crucial resistance measurements. Yes, a precision cap and power supply and characterizing the timer values from the PICmicro® MCU would provide more accurate results, but like using precision parts for an RC oscillator, this is not reasonable for volume production and would require application "tuning".

In the Basic Stamp, a "scale" value is specified and the result is returned as a fraction of this value. Although this is better than trying to match parts, it still requires some extra work to "tune" the scale value to the individual circuit. Chances are, if this operation is costed out, using an external ADC connected to the PICmicro® MCU or using a PIC16C7x (which has a built-in ADC) it would cost less than finding the "scale" value in a manufacturing setting.

You probably won't observe another problem with this circuit unless you put an oscilloscope or DMM on the RA0 pin. When the resistor is set to a very low value, the RA0 pin is essentially connected to ground. Many people put in a 100-Ohm resistor to prevent a dead short to ground, but this increases the time for the capacitor to fully charge. This experiment does not include the 100-Ohm current-limiting resistor between RA0 and the RC network so that you can observe this problem.

In the PICmicro® MCU, the amount of current that a pin is able to source is limited to about 20 mA. So, not using this resistor will not cause serious problems with the PICmicro® MCU sourcing too much current to work properly when a low resistance is specified. Personally, I would want to have a 100-Ohm current-limiting resistor on this pin to avoid any potential problems, but if you can guarantee that the potentiometer will never have a resistance less than several hundred ohms, it is not required.

The advantages of this circuit and software are its simplicity and few PICmicro® MCU advanced resources used. Although not providing accuracy, the circuit does provide excellent repeatability that can be very useful in many applications.

It is not an optimal application because interrupts must be disabled during a resistance-value read and the limits to the circuit must first be established according to the value of the capacitor used in the RC network.

One last thing to notice about this circuit is the use of FSR to point to TRISA (for the switching between output and input in the A/D function). FSR is loaded with the address of TRISA (0x085). Using INDF, TRISA is accessed directly without having to change the default bank (the usual *bsf STATUS, RP0* instruction). As discussed elsewhere in the text, this is possible because the FSR register provides up to eight bits of addressing, which eliminates the need for the RP0 bit when it is addressing registers. This method of using FSR might not be advisable in all applications because dedicating it to an individual function limits its usefulness elsewhere in the application.

## ADC: USING THE BUILT-IN ADC

The resistance-measurement method from the previous experiment is interesting, but I would not consider it to be the best way of measuring resistance values. Personally, I would consider the built-in *Analog-to-Digital Converter (ADC)* of the PIC16C7x family to be a much better solution. The PIC16C7x ADC will return an exact representation of the analog voltage without having to time an RC rise/fall. Also, the value returned will not have to be scaled, as it would with the previous RC network solution.

One argument against using ADC-equipped PICmicro® MCUs is the lack of flash program memory-based PICmicro® MCUs available. This argument has become less significant over time with the availability of low-cost ADC-equipped PICmicro® MCUs. These PICmicro® MCUs are actually much lower in cost than the flash-equipped PICmicro® MCUs. In any case, as time goes on, more and more PICmicro® MCUs will be available with flash program memory and built-in ADCs. As I write this, the only flash program memory PICmicro® MCUs with ADCs available are the PIC16F876 and PIC16F877, but a number are in the development queue.

This experiment introduces the PIC16C711, which, as I write this, is the low-cost ADC-equipped 18-pin PICmicro® MCU. The function provided by the experiment is identical to what was produced by the previous experiment: a potentiometer's position is read and displayed by a PICmicro® MCU. The circuit (Fig. 15-109) lacks the RC capacitor and probably costs a dollar or two less for parts than the 16F84-equipped previous experiment. The bill of materials for the ADC application is shown in Table 15-23.

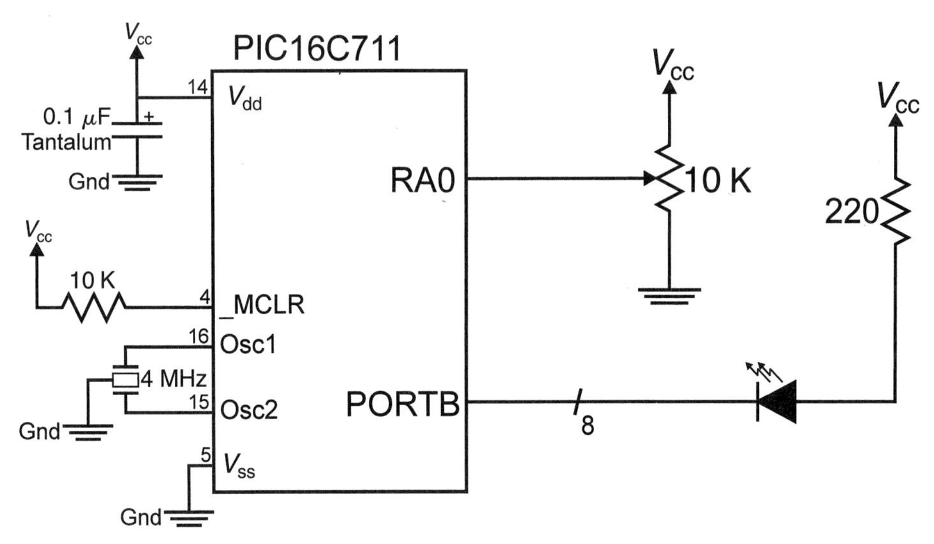


**Figure 15-109** ADC: reading a potentiometer position with an ADC

**TABLE 15-22 Bill Of Materials For The "ADC" Experiment**

| PART | DESCRIPTION | REQUIRED ON YAP-II |
|---|---|---|
| PICmicro® MCU | PIC16C711-04/JW | In Socket |
| Vdd/Vss decoupling capacitor | 0.1-$\mu$F tantalum | No |
| _MCLR pull up | 10 K, 1/4 watt | No |
| 4-MHz ceramic resonator | 4 MHz with built-in capacitors | No |
| 10-LED display | Bargraph red LED | No. LED0/LED7 used |
| 8 PORTB LED resistors | 220 Ohm, 1/4 watt | No. LED0/LED7 used |
| 10-K Potentiometer | Single-turn/PC board mount | No. Pot1 used |

The circuit is fairly easy to build and should be familiar to you if you've worked through the previous experiments. The breadboard version's wiring is shown in Fig. 15-110 and the YAP-II version is shown in Fig. 15-111. The YAP-II wiring diagram shows an external potentiometer attached to the YAP-II's breadboard, the POT1 potentiometer could be wired to RA0, and the external potentiometer eliminated from the circuit.

The application code, ADC.ASM, located in the *code\ADC* subdirectory of the *PICmicro* directory, should appear straightforward, but a few comments follow the code.

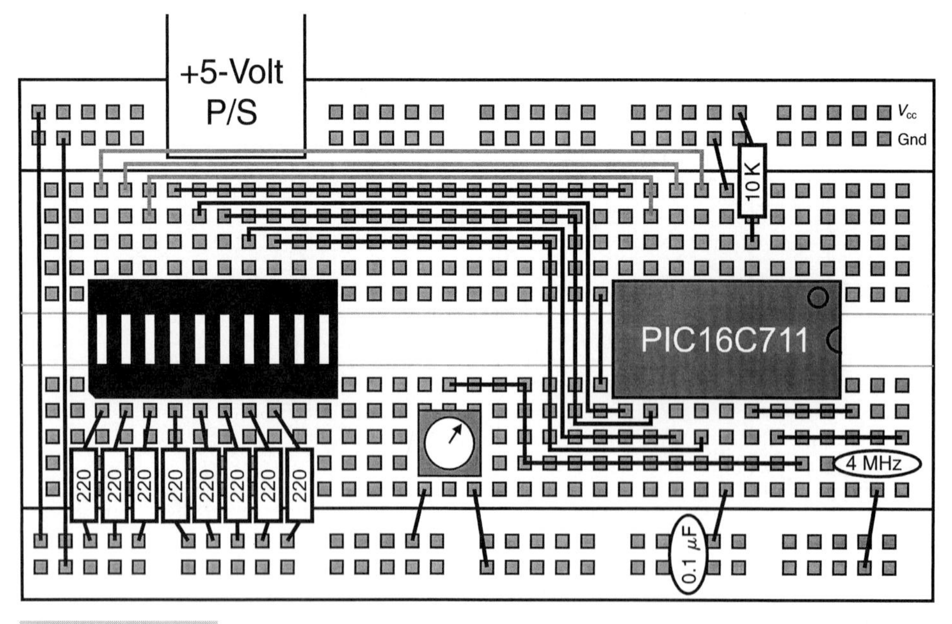


**Figure 15-110** ADC PIC16C711 breadboard circuit

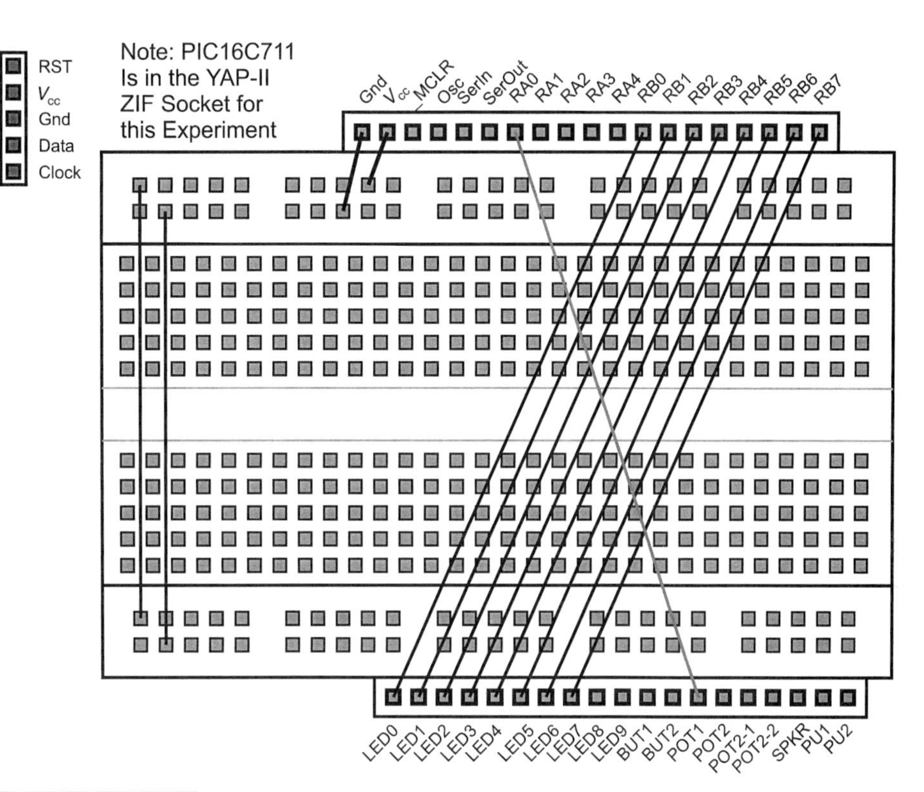


**Figure 15-111** YAP-II/EMU-II breadboard for ADC

```
 title  "ADC - Reading a Resistor Value with an ADC"
;
;  This Program Uses the ADC built into a PIC16C711 and
;  Reads an ADC Value and displays it on eight LEDs.
;
;  Hardware Notes:
;   PIC16C711 running at 4 MHz
;   Reset is tied directly to Vcc and PWRT is Enabled.
;   A 10K Pot Wired as a Voltage Divider on PORTA.0
;   A 220 Ohm Resistor and LED is attached to all the PORTB.7:0
;
;  Myke Predko
;  99.12.27
;
  LIST R=DEC
 ifdef __16C711
  INCLUDE "p16c711.inc"        ;  <-- Note PIC16C711 Used
 endif
 ifdef __16C877
  INCLUDE "p16f877.inc"
 endif
 endif

;  Registers

 ifdef __16C711
 __CONFIG _CP_OFF & _WDT_OFF & _XT_OSC & _PWRTE_ON
```

```
  else
  __CONFIG _CP_OFF & _WDT_OFF & _XT_OSC & _PWRTE_ON & _DEBUG_OFF &
_LVP_OFF & _BODEN_OFF & _WRT_ENABLE_ON & _CPD_OFF
  endif

   PAGE
 ;  Mainline of ADC

  org     0

   nop

   movlw  0x0FF
   movwf  PORTB              ;  Turn off all the LED's
   clrf   PORTA              ;  Use PORTA as the Input

   bsf    STATUS, RP0
   clrf   TRISB & 0x07F      ;  Set all the PORTB bits to Output
  ifdef __16C711             ;  Make Sure all ADC Bits are Output
   clrf   ADCON1 ^ 0x080     ;  Make RA0 to RA3 ADC input
  else
   movlw  B'00000000'        ;  For 16F877, Make 10 Bit ADC "Left
   movwf  ADCON1 ^ 0x080     ;   Justified"
  endif
   bcf    STATUS, RP0        ;  Go back to Page 0

   movlw  0x081              ;  Setup ADCON0 for ADC Conversion
   movwf  ADCON0             ;   ADCS1:ADCS0 - 10 for /32 Clock
                             ;   Unimplemented - 0
                             ;   CHS1:CHS0 - 00 for RA0/AN0
                             ;   Go/_Done - 0
                             ;   ADIF - 0
                             ;   ADON - 1

 Loop

   movlw  3                  ;  Wait 12 usec for ADC to Charge
   addlw  0x0FF              ;  Take One Away to Setup the Charge
   btfss  STATUS, Z
    goto  $ - 2

   bsf    ADCON0, GO         ;  Turn on the ADC
   btfsc  ADCON0, GO         ;  Wait for it to Complete
    goto  $ - 1

   bsf    STATUS, RP0
  ifdef __16C711
   comf   ADRES, w           ;  Get the Timer Value
  else
   comf   ADRESH, w          ;  Read Most Significant 8 Bits in
  endif                      ;   PIC16F877
   bcf    STATUS, RP0
   movwf  PORTB

   goto   Loop               ;  Get another Time Sample

   end
```

The first comment is regarding the ADC set up. I explicitly set the ADC inputs for RA0 through RA3—even though these are the default settings of the PICmicro® MCU on power up. This is done by writing 0x000 to ADCON1. In the other PIC16C711 applications presented in this book, you'll notice that I really don't use the ADC function of the PIC16C711. Instead, I just use the PIC16C711 as a cheap EPROM-based PICmicro® MCU. In these cases, before writing to the TRIS registers of the I/O port registers, I write *0x003* to ADCON1, which turns off the ADC inputs and allows the PORTA pins to be used only as digital I/O pins.

If you are going to use an emulator and/or another ADC-equipped PICmicro® MCU for this experiment, you would have to be sure that you set the ADCON1 register correctly for the part number that you are going to use. This application can be used with a PIC16F877, and conditional code has been included to select between the PIC16C711 and the flash program memory part that is used for emulators.

Notice that in the application code, I support the PIC16C711 and the PIC16F877. If another processor was used (which would potentially have different ADCON0 or ADCON1 register bit definitions), the correct Microchip include file would not be loaded. Like the other experiments that use the PIC16F877, the application would fail because of references to registers that were not defined by an include file. This application actually implements a method of ensuring that the supported parts are the only ones the application is built for in the MPLAB IDE.

Next, the setting of the ADCON0 register, like trying to properly set up the OPTION register, can be somewhat difficult to get the correct I/O bits. The ADC.ASM application code provides the different bit values, which are probably quite straightforward, except for the */32* setting. The */32* is the A/D conversion clock select bit, which allows for a full 4.2-$\mu$s analog-to-digital conversion (the "typical" time required is 4.0 $\mu$s).

With the ADC hardware set up and enabled, the code then enters a loop in which it waits 13 $\mu$s for the ADC's input capacitor to stabilize at the input voltage and then performs the conversion by setting and polling the *Go/_Done* bits.

In this application, the ADC is very straightforward and quite easy to work with. The actual code is quite compact (it only requires 21 instructions) and is very easy to set up. As for the ADC input, the PIC16C711 is very tolerant of all signals, as long as they are not greater than $V_{dd}$ and do not change at a rate of 30 kHz (or faster). Coupled with the elimination of the scaling or tuning step required by the method shown in the previous experiment, it should be easy to see why I consider using a built-in ADC to be superior.

## VLADDER: RESISTOR LADDER OUTPUT

Although a single reference voltage produced by a voltage divider might be useful for some applications, a variable-voltage output is much more useful for many other applications. This experiment shows how multiple analog voltages (approximately 0.55 volts apart) can be produced by the PICmicro® MCU using the digital I/O pins with a voltage divider.

The output voltage is determined from a voltage divider that has the formula:

$$V_{out} = V_{cc} * [R_n/(R_s + R_n)]$$

Where $R_n$ is the resistance between the "tap" and $V_{cc}$ and $R_s$ are the resistance between the tap and ground.

A variable-resistance voltage divider can be implemented on the PIC as a resistor ladder, like the $V_{Ref}$ circuit of the 16C62x devices (Fig. 15-112). Depending on which PORTB output is active, the $R_s$ resistance can be varied to change the output voltage of the circuit. The $R_s$ resistance is varied by changing when the PORTB I/O pin is active. All the port pins have been loaded with a *0*. When one of them is put into Output mode, the circuit is grounded at this point. The resistance includes all of the resistors between it and the analog voltage output.

The bill of materials for the VLadder application is shown in Table 15-24. Wiring this circuit is actually quite a challenge. I found that when I redid it for the second edition, I had

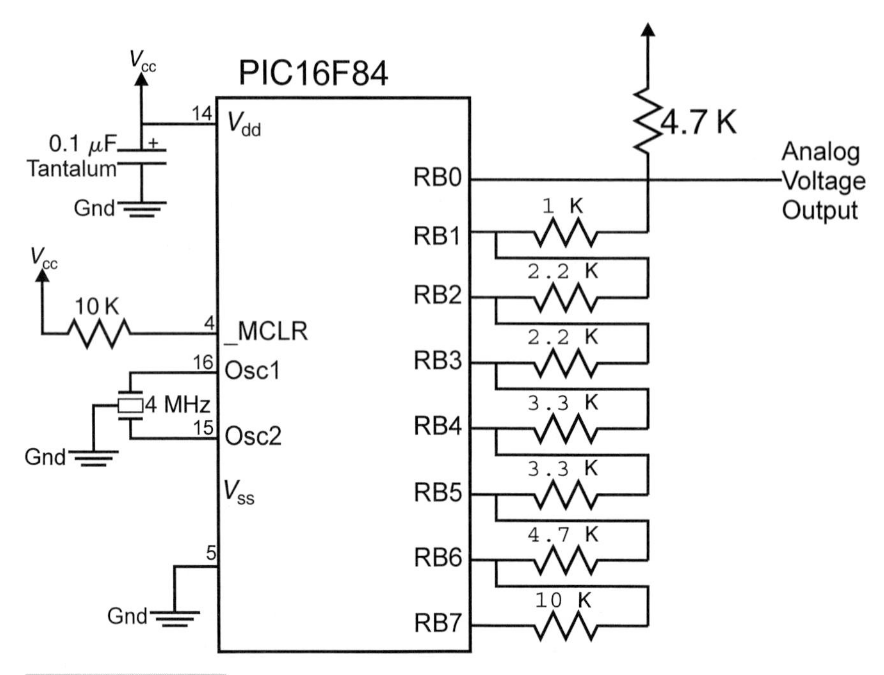


**Figure 15-112** **VLadder: outputting an analog signal using a voltage ladder**

a number of difficulties getting it right (described further in this section). For the breadboard version of the application, the wiring is shown in Fig. 15-113 and the YAP-II wiring is shown in Fig. 15-114. Getting the wiring right is not trivial. In many ways, this was the most difficult application for me to specify the wiring to keep it simple enough for you to follow and build your own circuit.

**TABLE 15-24 Bill Of Materials For The "VLADDER" Experiment**

| PART | DESCRIPTION | REQUIRED ON YAP-II |
|---|---|---|
| PICmicro® MCU | PIC16F84-04/P | In Socket |
| Vdd/Vss decoupling capacitor | 0.1-µF tantalum | No |
| _MCLR pull up | 10 K, 1/4 watt | No |
| 4-MHz ceramic resonator | 4 MHz with built-in capacitors | No |
| 2 4.7-K resistors | 4.7 K, 1/4 watt | Yes |
| 1-K resistor | 1 K, 1/4 watt | Yes |
| 2 2.2-K resistors | 2.2 K, 1/4 watt | Yes |
| 2 3.3-K resistors | 3.3 K, 1/4 watt | Yes |
| 10-K resistor | 10 K, 1/4 watt | Yes |

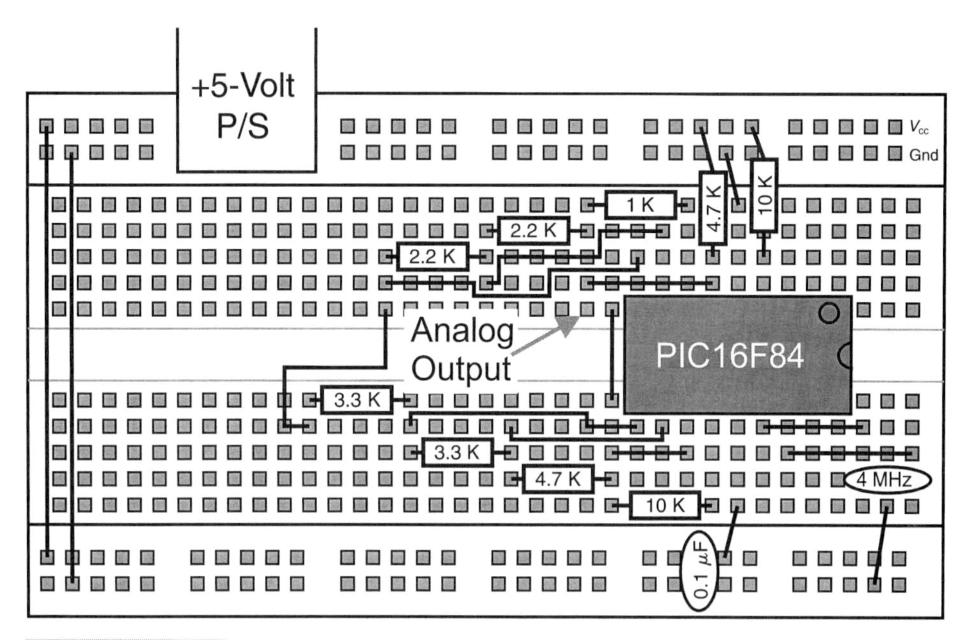


**Figure 15-113** VLadder PIC16F84 breadboard circuit

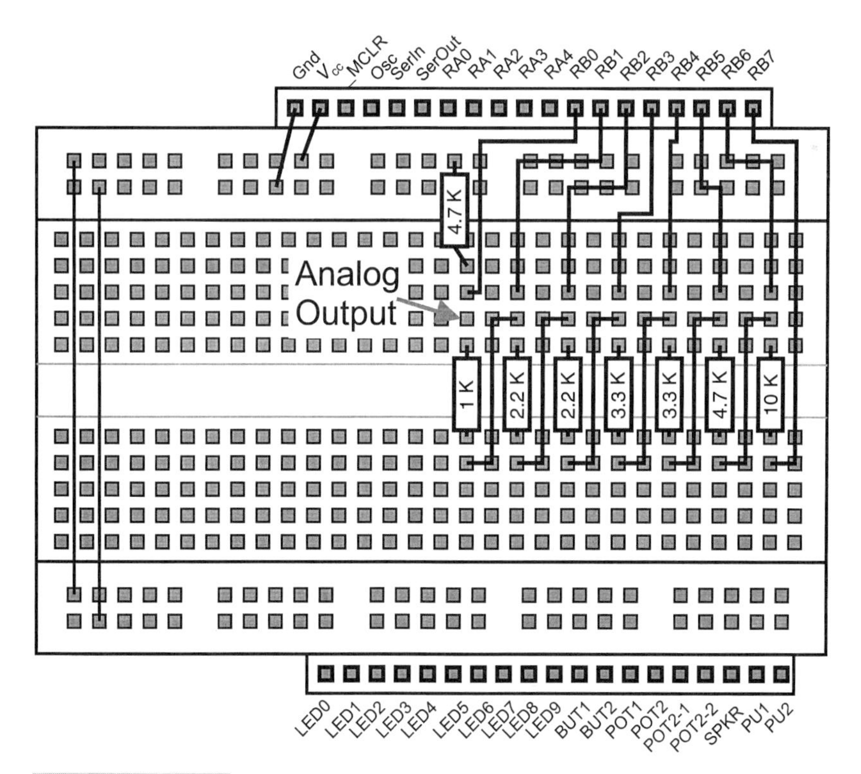


**Figure 15-114** EMU-II/YAP-II breadboard circuit for VLadder

When I originally created this experiment (for the first edition), I rearranged this formula to:

$$R_n = (V_{out}/V_{cc}) * R_s/[1 - (V_{out}/V_{cc})]$$

By choosing a value for $R_s$, (for example, 10 K), you can easily calculate the value of $R_n$ for a given $V_{out}$.

I went to all this trouble with calculating the formula to ensure that I can get reasonable linearity in $V_{out}$ for different bit outputs. With the resistor ladder connected to Port B, we can have nine different voltages. In selecting these voltages, I have tried to space them evenly 0.55 volts apart.

With the circuit shown in Fig. 15-112, you can get nine different voltages. First, when all the output bits are turned off, only a pull up is left connected to the analog voltage output for a maximum voltage output. If RB0 is in Output mode and pulling the output line to ground, this is the low value for the application. $R_s$ values can be added to the circuit by outputting a zero (low voltage) on a pin (terminating the resistor ladder at that point). The remaining seven voltages are selected by grounding intermediate resistors in the ladder using this method.

It is very important to remember the minimal current (load) output drive characteristics of this circuit. A resistor ladder such as this is very poor at maintaining the output voltage if it has to source or drive current. If there is any type of tangible current flow (more than a few $\mu$A) outside of the voltage divider, the output voltage will be changed from what you expect. The best way to avoid this is to "buffer" the resistor ladder output using an op amp (Fig. 15-115).

When I originally created this circuit, I calculated specific values so that the output would be linear. The same value could be used for each resistor in the ladder, but if these were plugged into the PICmicro® MCU, the result would be the voltage output shown in Fig. 15-116. This figure plots what I would consider to be the ideal voltage ramp.

This output obviously deviates significantly from the desired (linear) output. Actually, finding the correct resistor values is not that difficult. Using the rearranged formula for $R_n$, you can plug in an $R_s$ value of 10 K, a $V_{cc}$ of 5 volts, and then figure how to go from 0 to $V_{max}$ in nine steps.

When I created this application for the first edition of the book, I used a PIC16C84 (which is different from the PIC16F84 that is used for this edition) and I was able to calculate values very straightforwardly to get a linear output. This was not possible for the PIC16F84. I found that I had to experiment a bit to get the values specified in Fig. 15-112. These values will produce the reasonable linear steps shown in Fig. 15-117.

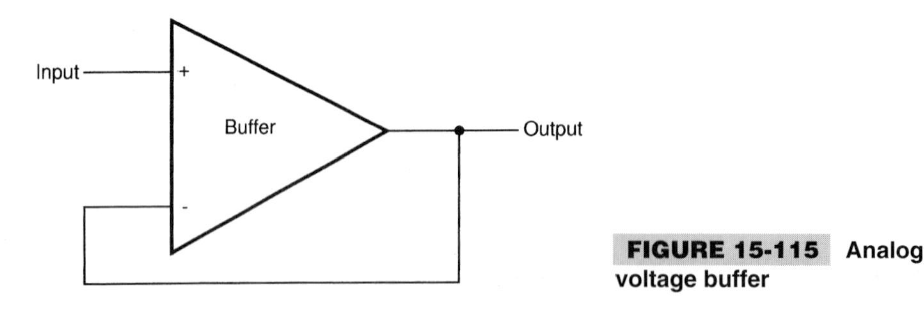


**FIGURE 15-115** Analog voltage buffer

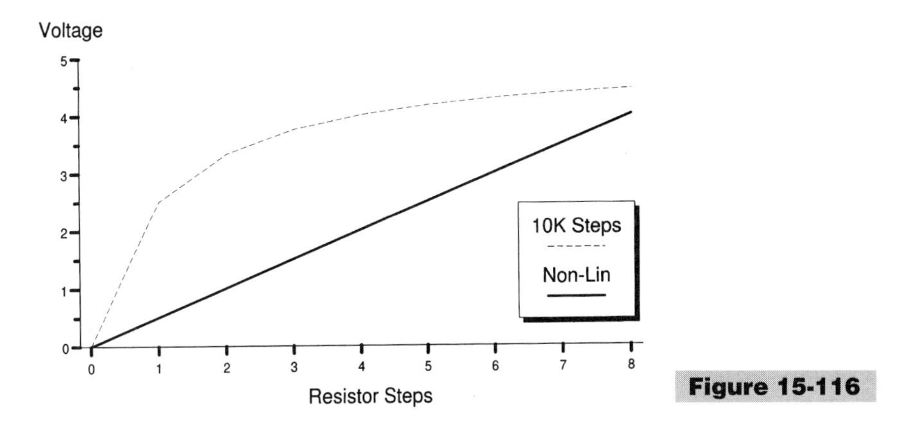


**Figure 15-116**

Despite the complexity of wiring this application, one of the really nice things about this circuit is that it can be checked for wiring errors without the PIC installed. This is done by hooking up your DMM and checking the resistances between the I/O pins.

If you're curious, you could rewire the circuit using a constant value for each step in the ladder. If you repeat the experiment with the voltmeter, you will find the asymptotic curve predicted in the previous graph.

The PICmicro® MCU application code (VLADDER.ASM) is very simple and is located in the *code\VLadder* subdirectory of the *PICmicro* directory.

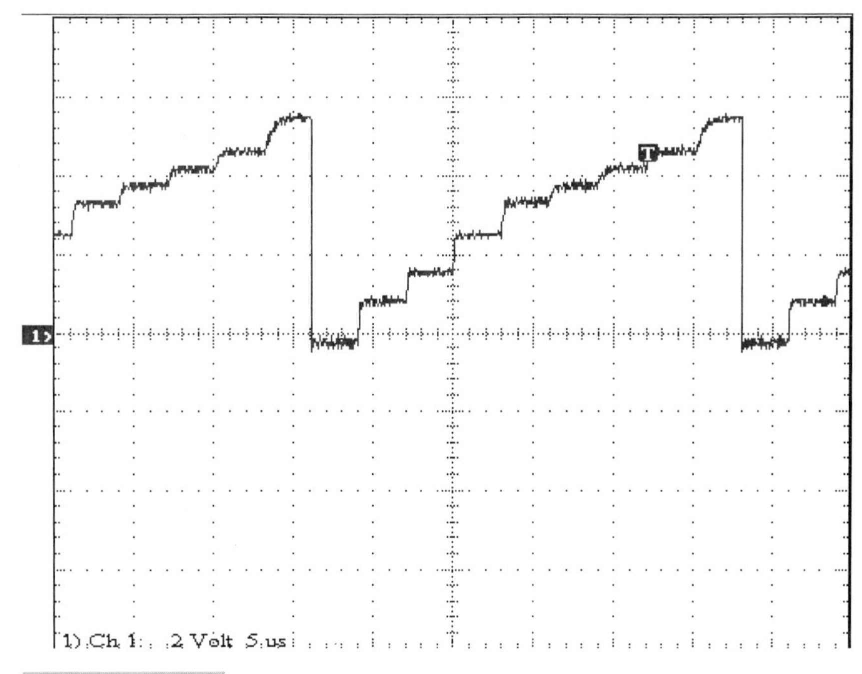


**Figure 15-117** Voltage output from VLadder

```
 title  "VLadder - Resistor Ladder Analog Output."
#define DMM
;
;  This Program Runs through a Saw Tooth Analog Output from the
;  16C84.  The Output is generated by a Resistor Ladder attached
;  to PORTB.  To set a particular voltage, a bit is output to 0 volts.
;
;  Hardware Notes:
;   PIC16F84 Running at 4 MHz
;   Reset is tied directly to Vcc and PWRT is Enabled.
;   The Resistor Ladder is attached to PORTB.7:0
;
;   A 4.7K Resistor between PORTB.0 and Vcc (Output is taken from here
;    as well)
;   A 1K Resistor between PORTB.0 and PORTB.1
;   A 2.2K Resistor between PORTB.1 and PORTB.2
;   A 2.2K Resistor between PORTB.2 and PORTB.3
;   A 3.3K Resistor between PORTB.3 and PORTB.4
;   A 3.3K Resistor between PORTB.4 and PORTB.5
;   A 4.7K Resistor between PORTB.5 and PORTB.6
;   A 10K Resistor between PORTB.6 and PORTB.7
;
;  Updated: 99.12.27 - For Second Edition
;
;  Myke Predko
;  96.06.27
;
  LIST R=DEC
 ifdef __16F84
  INCLUDE "p16f84.inc"
 endif
 ifdef __16F877
  INCLUDE "p16f877.inc"
 endif

;  Registers
 CBLOCK 0x020
Count, Counthi, Countu
 ENDC

 ifdef __16F84
 __CONFIG _CP_OFF & _WDT_OFF & _XT_OSC & _PWRTE_ON
 else
 __CONFIG _CP_OFF & _WDT_OFF & _XT_OSC & _PWRTE_ON & _DEBUG_OFF &
_LVP_OFF & _BODEN_OFF & _WRT_ENABLE_ON & _CPD_OFF
 endif

  PAGE
;  Code for VLadder

  org    0

  nop

  clrf   PORTB

  movlw  TRISB              ;  Setup the TRIS Values
  movwf  FSR

  bcf    STATUS, C          ;  Use the Carry as the Skip Value

Loop                        ;  Loop Around Here to Output Sawtooth

 ifdef   DMM                ;  Just Dlay if only a DMM Available
  call   Delay              ;   for Seeing the Output
 endif
```

```
  rlf     INDF, f

  goto    Loop

 ifdef   DMM
Delay

  clrf    Count              ;  Display for one Second
  clrf    Counthi
  movlw   6
  movwf   Countu
Dlay
  decfsz Count, f
   goto  Dlay
  decfsz Counthi, f
   goto  Dlay
  decfsz Countu, f
   goto  Dlay

  return
 endif

 end
```

The code for this application is designed with the DMM conditional code. If the *DMM #define* is used, the one-second Delay subroutine is called so that you can watch the voltage change on a DMM. If *DMM* is not defined, then the application will run with the voltage changing once every 3 $\mu$s and the waveform can be observed on an oscilloscope, like the oscilloscope picture shown in Fig. 15-117.

As you step through the code (be sure that you don't define *DMM* when doing this or you'll be clicking your mouse a lot), watch the zero shift through TRISB, which causes PORTB to act like a programmable open-collector output and change the effective voltage divider resistance. It is important that this code takes advantage of the value in the Carry flag, not changing between Loop iterations. This was important when specifying the Delay code to ensure that no add or subtract instructions (which could modify the Carry flag) were used.

When I created this circuit for the second edition, I must confess that I had some significant problems. The first and most obvious one was that the schematic presented in the first edition was not correct and did not match the application code presented with it. This experiment is correct and has been tested with a number of different PICmicro® MCUs.

I also had a lot of problems replicating what I saw in the first edition with this edition. It took me several hours to reconcile the actual results, but I can say confidently that the circuit diagrams presented here will work with the application code to produce a high-frequency sawtooth. The information contained in this experiment was used as a base to produce the NTSC composite video output described in Chapter 16.

## PWMOUT: PWM VOLTAGE OUTPUT

One of the most interesting ways to generate an analog voltage is through the use of a PWM signal and a low-pass filter, like the circuit shown in Fig. 15-118. The PWM output is used to partially charge a capacitor. The on-time average or duty cycle of the PWM is the determining value of the analog output.

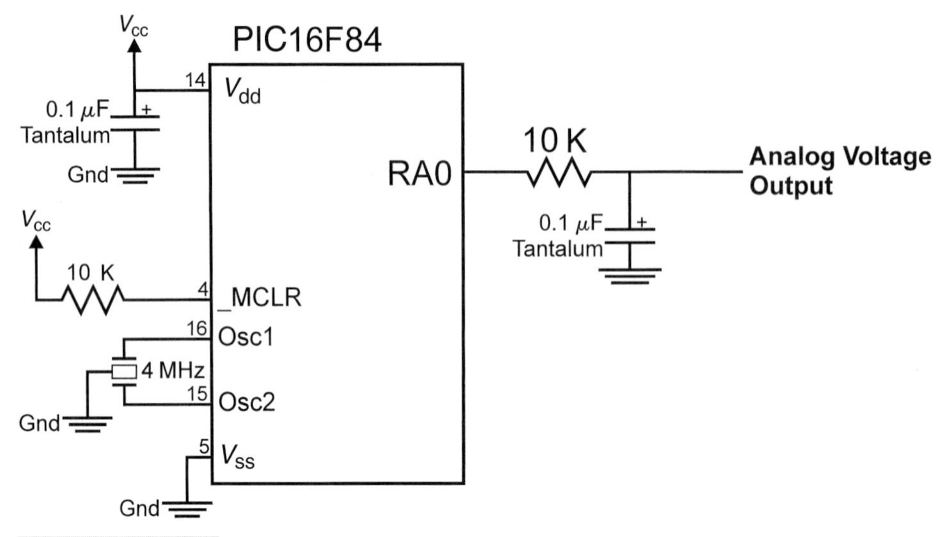


**Figure 15-118** PWM: outputting an analog signal using a PWM

Providing a PWM analog output is quite easy to wire together. It just requires a resistor and capacitor added to a PICmicro® MCU I/O pin for analog voltage output). The actual code within it is much more difficult to work with to provide a full range of analog outputs. As I'll show in this application, the voltage output is somewhere between the extremes and does not have a small value "granularity" for the output.

The bill of materials for this circuit is shown in Table 15-25. It can also be assembled quite quickly, using the breadboard wiring diagram shown in Fig. 15-119. If you are working with the YAP-II, you will have to add a resistor and capacitor to the breadboard using the wiring diagram shown in Fig. 15-120.

The code itself, although quite simple, is designed to provide a PWM at a 20-kHz frequency (50-μs period). To observe the operation of this experiment, you will need an oscilloscope capable of at least a 500-kHz bandwidth. The code for this experiment is PWM.ASM, which is located in the *code\PWM* subdirectory of the *PICmicro* directory.

**TABLE 15-25 Bill Of Materials For The "PWMOUT" Experiment**

| PART | DESCRIPTION | REQUIRED ON YAP-II |
|---|---|---|
| PICmicro® MCU | PIC16F84-04/P | In Socket |
| Vdd/Vss decoupling capacitor | 0.1-μF tantalum | No |
| _MCLR pull up | 10 K, 1/4 watt | No |
| 4-MHz ceramic resonator | 4 MHz with built-in capacitors | No |
| 10 K resistor | 10 K, 1/4 watt | Yes |
| 0.1-μF tantalum capacitor | 0.1-μF tantalum | Yes |

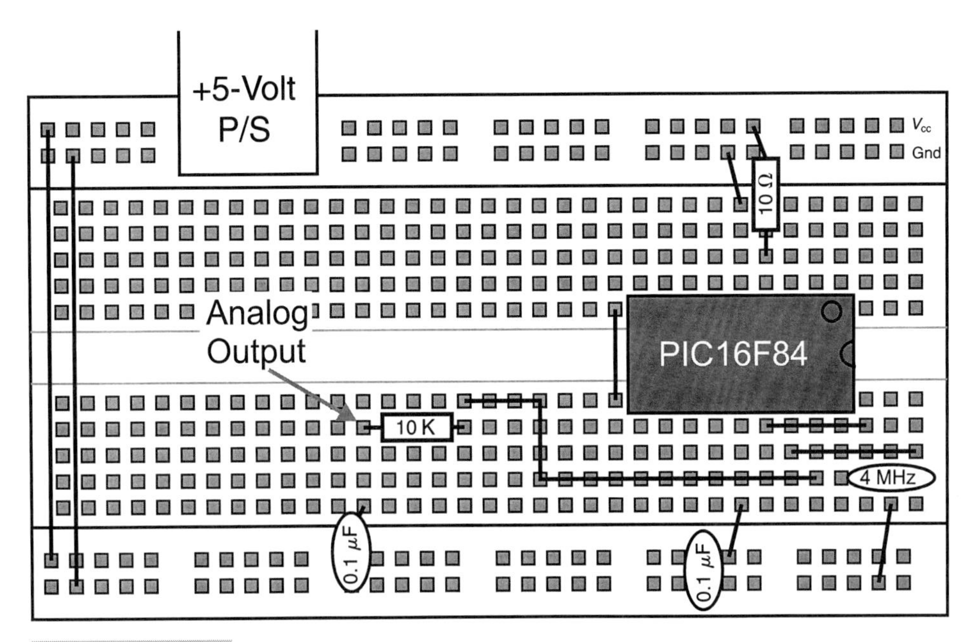


**Figure 15-119** PWM PIC16F84 breadboard circuit

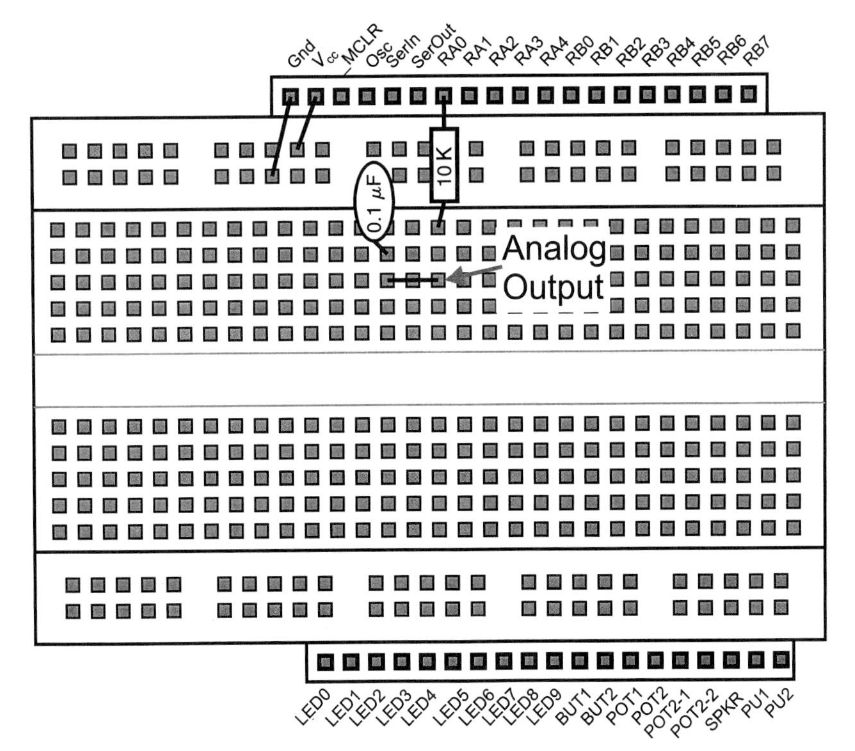


**Figure 15-120** EMU-II/YAP-II breadboard circuit for PAM

```
 title  "PWMOut - Output a PWM Analog Voltage"
;
;  This Application simply Outputs an Analog Voltage by Driving a
;   Resistor/Capacitor ("RC") Network with a PWM signal.
;
;  The PWM runs at 20 kHz, with a Duty Cycle range from 0 to 100%
;   Demonstrated.
;  This is an Actual PWM Duty Cycle Range of 0 to 50 Instructions/usecs
;
;  Hardware Notes:
;   This application runs on a 16F84 executing at 4 MHz
;   _MCLR is tied through a 4.7K Resistor to Vcc and PWRT is Enabled
;   A 10K Resistor/0.1 uF Capacitor Network is connected to RA0 to Show
;    the PWM in Operation
;
;  Myke Predko
;  99.12.27
;
  LIST R=DEC
 ifdef __16F84
  INCLUDE "p16f84.inc"
 endif
 ifdef __16F877
  INCLUDE "p16f877.inc"
 endif

;  Registers
 CBLOCK 0x020
LoopNumber
PWMOn                          ;  PWM "On Value"
PWMOff
 ENDC

#define PWM PORTA, 0           ;  LED on PORTB.0

 ifdef __16F84
 __CONFIG _CP_OFF & _WDT_OFF & _XT_OSC & _PWRTE_ON
 else
 __CONFIG _CP_OFF & _WDT_OFF & _XT_OSC & _PWRTE_ON & _DEBUG_OFF &
_LVP_OFF & _BODEN_OFF & _WRT_ENABLE_ON & _CPD_OFF
 endif

  PAGE
;  Mainline of pwmout

 org     0

  nop

  movlw  2                     ;  Start the PWM with Nothing
  movwf  PWMOn

  clrf   LoopNumber            ;  Output Each Voltage Level for 10
                               ;   Iterations
  bcf    PWM                   ;  Make the PWM "off" Initially
  bsf    STATUS, RP0           ;  Goto Bank 1 to set Port Direction
  bcf    PWM                   ;  Set PWM Pin to Output
  bcf    STATUS, RP0           ;  Go back to Bank 0

Loop                           ;  Loop Here and Output the PWM

  bsf    PWM                   ;  Turn on the PWM

  movf   PWMOn, w              ;  Calculate Delay Time
  sublw  12
  movwf  PWMOff
```

```
    decf    PWMOn, w          ;  Display the Data Here
    addlw   0x0FF             ;  Take One Away
    btfss   STATUS, Z
     goto   $ - 2

    bcf     PWM               ;  Turn OFF PWM

    incf    LoopNumber, f

    btfsc   STATUS, Z         ;  Looped 256x?
;    btfsc  LoopNumber, 0     ;  Looped 1x?
     goto   NewLoop

  OffLoop

    decf    PWMOff, w         ;  No, Just Output It
    addlw   0x0FF
    btfss   STATUS, Z
     goto   $ - 2

    goto    Loop

  NewLoop                     ;  Increment the Output Value

    clrf    LoopNumber

    incf    PWMOn, w          ;  Increment the Value
    xorlw   8
    btfsc   STATUS, Z         ;  At 8?
     movlw  2 ^ 8             ;   Yes - Reset
    xorlw   8
    movwf   PWMOn

    movlw   3                 ;  "NewLoop" takes 3 Cycles of Time
    subwf   PWMOff, f

    goto    OffLoop


  end
```

In this application, the high voltage for the on period is output. Then the low or off interval is calculated by subtracting the length of the on period from 50 $\mu$s (for a 20-kHz PWM).

In the code, I use a four-instruction-cycle delay loop for the delay. A total of 12 (and a half) of these loops is required for one iteration of the PWM. This provides up to 12 PWM output voltage levels to select from. The actual number available to the application is only six because of the overhead needed in calculating the off time and the time required to calculate a new value to show a change in the PWM's output.

In this application, I stay at the same voltage level for 256 PWM cycles (50 ms long each), which means that the voltage is active for a total of 12.8 ms. This is why I said at the start of the experiment that an oscilloscope is needed to observe the execution of the application. Figure 15-121 shows how this application outputs less than a 1-volt value, based on the minimum time (of eight instruction cycles) on. Figure 15-122 shows the slightly less than three volts output, based on the 28 maximum cycle on.

With a bit of code tweaking, these values could be set to full-volt values with the delays evenly divided by 10. The 20-kHz PWM frequency was chosen because it is a typical PWM frequency for motors and other devices as it is above the range of human hearing.

The problem with this PWM frequency is that there are only 50 $\mu$s in which the signal can be output. For a PICmicro® MCU running at 4 MHz, this requires a lot of careful plan-

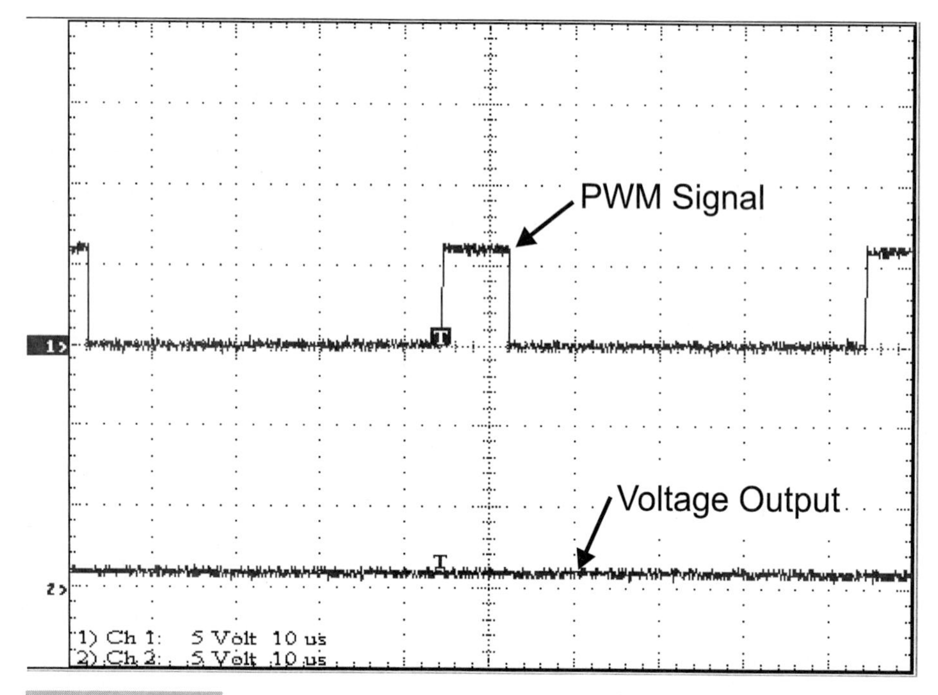


**Figure 15-121** Low PWM voltage output

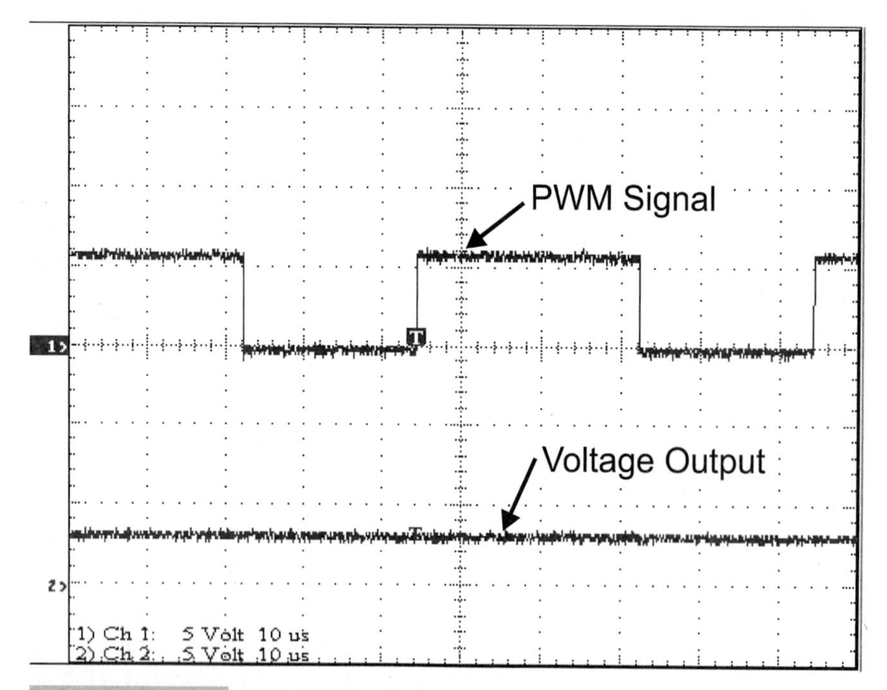


**Figure 15-122** High PWM voltage output

ning to get the frequency set up correctly for the application. As you can see in this experiment, creating a general case for the application is difficult to get a wide range of outputs and a small granularity or difference between voltage steps. The code also has the disadvantage that it requires the entire resources of the PICmicro® MCU.

An obvious solution to some of these problems is to reduce the actual speed of operation by an order of magnitude or so. The problem is that motor drivers and other mechanical controls require a frequency that is above human hearing. To be on the safe side, this should be at least 20 kHz, which is difficult to produce for the PICmicro® MCU in software. Many of these problems can be alleviated by running the PICmicro® MCU at a much higher frequency than the 4 MHz of this application. Running at 20 MHz, would provide a minimum of 30 steps to the PWM and a much better range of outputs.

A much better solution is to use the CCP module with PICmicro® MCUs set up with TMR2 and TMR3. This hardware provides a much wider range of possible duty cycles (and output voltages) and a much smaller granularity than what can be achieved within software. The hardware also has the advantage of not requiring any processor cycles.

Last, you should remember that the RC low-pass filter, which provides the output to the analog voltage, should drive a very high input impedance device. Ideally, this should be an op amp set up as a zero-gain buffer that is able to drive other circuits without affecting the operation of the PWM-driven RC.

# I/O with Interrupts

Thus far, the experiments have focused on the basic hardware that is built into both the low-end and mid-range PICmicro® MCUs. These experiments are very important because they will help you to understand how the PICmicro® MCU works in a variety of situations and how the PICmicro® MCU responds to different hardware inputs.

The next series of experiments look at using the mid-range PICmicro® MCU's interrupt capability to enhance applications and, in many cases, simplify the code and operation of the application. This probably seems surprising to you, but interrupts can make applications much easier to work with. Chapter 16 shows how a serial LCD interface board can be built without interrupts or serial interface hardware, but the code is quite complex and was difficult to accurately time.

Chapter 16 also covers a lot more of the different interrupt sources and operation than this chapter. This chapter introduces how interrupts work in the mid-range PICmicro® MCUs and how handlers are written for them.

## CYLON: TIMER/INTERRUPT HANDLER WITH CONTEXT SAVING

Throughout the book, I have pointed out the advantages of using the TMR0 interrupt source to carry out background tasks. Later, this chapter shows a TMR0 interrupt button debounce application that debounces button input without affecting the operation of the mainline code. To help introduce interrupts and the TMR0 interrupt, I have created a little interrupt handler that moves two lit LEDs back and forth in kind of a Cylon Eye, after an eighth of a second delay.

The name "Cylon Eye" comes from the old television show, *Battlestar Gallactica*. This was a knock off of *Star Wars* and featured the last human survivors in the universe running away from a race of menacing robots that clanked around a lot and scanned everything with a single eye that moved back and forth. The show really wasn't very good, but just about every teenager in the 1970s (myself included) was glued to it because the only other TV science fiction options were reruns of *Star Trek* and *Lost in Space*.

*Battlestar Gallactica* is long gone, but the eye movement has somehow remained and is often used in many applications as an indicator that a circuit is working properly. I wanted to create the code so that it could be ported into a PICmicro® MCU application that supervises the operation of a larger circuit. The Cylon Eye application uses PORTB, connected to a series of LEDs, with TMR0 providing an interrupt request every 1/16 second. The 1/16 second is the maximum delay for a PICmicro® MCU running at 4 MHz. This leaves the mainline or foreground execution open to application monitoring code, which can stop the eye if any problems are discovered.

The circuit itself simply consists of eight LEDs connected to PORTB of the core circuit (Fig. 15-123). For the breadboard application, I used a bargraph LED to allow simple continuity of the display (Fig. 15-124) and the YAP-II connection just uses eight LEDs in its bargraph (Fig. 15-125). The bill of materials for the application is shown in Table 15-26.

The application code is CYLON.ASM, which is located in the *code\Cylon* subdirectory of the *PICmicro* directory. Along with the context-saving registers, two variables are required for the programmable delay and tracking the direction where the eye is moving.

```
  title  "Cylon - Output a Cylon Eye in the Background"
 #define nDebug
 ;
 ;  This Application uses TMR0 to Move a "Cylon Eye" back and forth
 ;   Across eight LEDs connected to PORTB.
 ;
```

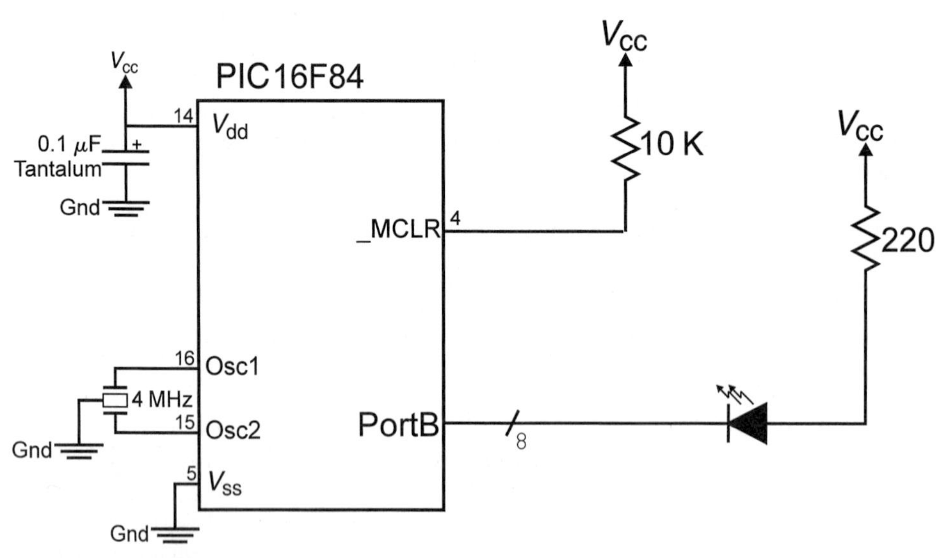


**Figure 15-123** **Cyclon: circuit with moving LEDs**

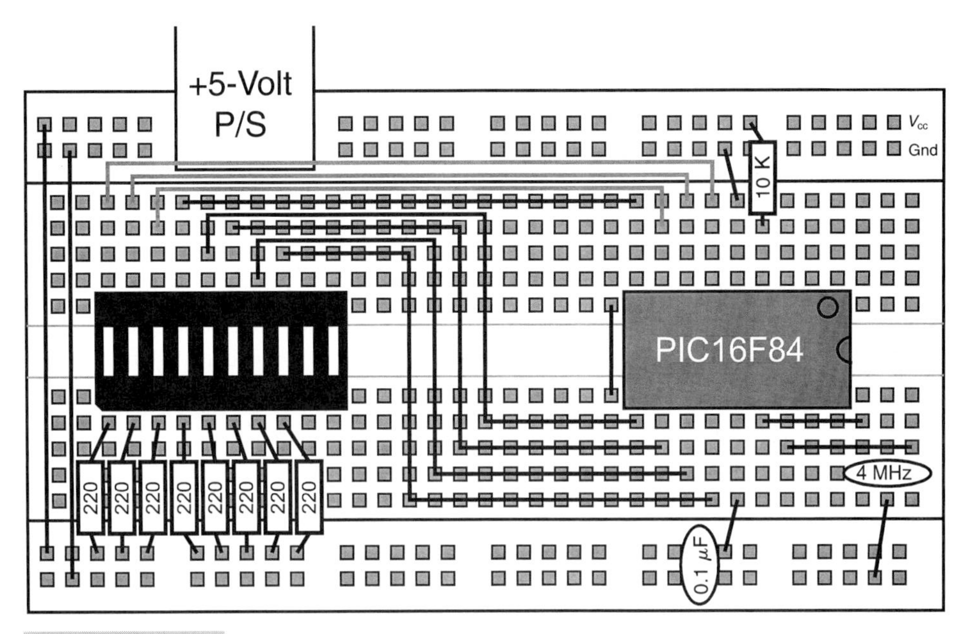


**Figure 15-124** Cyclon PIC16F84 breadboard circuit

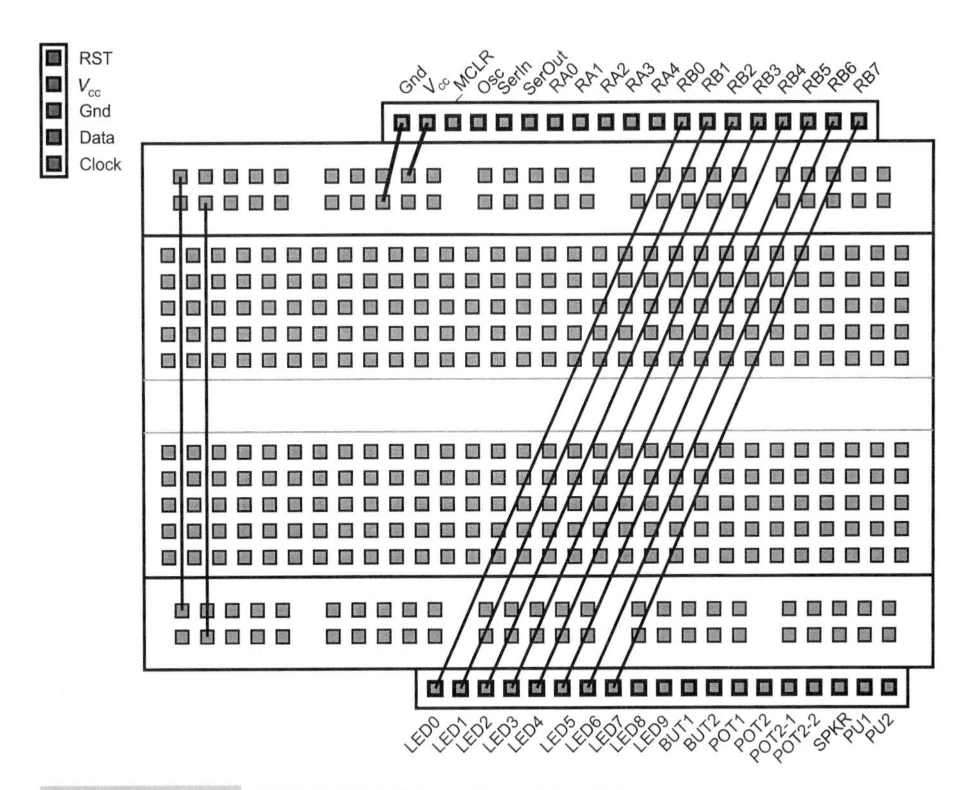


**Figure 15-125** EMU-II/YAP-II breadboard for Cylon

**TABLE 15-26 Bill Of Materials For The "CYLON" Experiment**

| PART | DESCRIPTION | REQUIRED ON YAP-II |
|---|---|---|
| PICmicro® MCU | PIC16F84-04/P | In Socket |
| Vdd/Vss decoupling capacitor | 0.1-$\mu$F tantalum | No |
| _MCLR pull up | 10 K, 1/4 watt | No |
| 4-MHz ceramic resonator | 4 MHz with built-in capacitors | No |
| RB0 LED | 10-LED bargraph | No. LED0/LED7 used |
| RB0 LED resistor | 220 Ohm, 1/4 watt | No. LED0/LED7 used |

```
;  Hardware Notes:
;   This application runs on a 16F84 executing at 4 MHz
;   _MCLR is tied through a 10K Resistor to Vcc and PWRT is Enabled
;   A 220 Ohm Resistor and LED is connected between PORTB.0 and Vcc
;
;  Myke Predko
;  99.12.28
;
  LIST R=DEC
 ifdef __16F84
  INCLUDE "p16f84.inc"
 endif
 ifdef __16F877
  INCLUDE "p16f877.inc"
 endif

;  Registers
 CBLOCK 0x020
_w, _status                    ;  Context Register Save Values
Direction                      ;  0 for Up, !0 for Down
Count                          ;  Count the Number of Times Through
 ENDC

 ifdef __16F84
 __CONFIG _CP_OFF & _WDT_OFF & _XT_OSC & _PWRTE_ON
 else
 __CONFIG _CP_OFF & _WDT_OFF & _XT_OSC & _PWRTE_ON & _DEBUG_OFF &
_LVP_OFF & _BODEN_OFF & _WRT_ENABLE_ON & _CPD_OFF
 endif

  PAGE
;  Mainline of cylon
 org     0

  nop

  movlw  2                     ;  Setup the Count
  movwf  Count

  goto   Mainline

 org     4
Int                            ;  Interrupt Handler

  movwf  _w                    ;  Save Context Registers
  movf   STATUS, w             ;  - Assume TMR0 is the only enabled
  movwf  _status               ;    Interrupt
```

```
  bcf     INTCON, T0IF       ;  Reset the Interrupt Flag

  decfsz Count, f            ;  Execute Once Every Two times
   goto  IntEnd

  movlw  2                   ;  Reset the Counter
  movwf  Count

  btfss  PORTB, 7            ;  At the Top?
   bsf   Direction, 0

  btfss  PORTB, 0            ;  At the Bottom?
   bcf   Direction, 0

  btfsc  Direction, 0        ;  Going Up?
   goto  Down                ;  No, Down

Up                           ;  Moving the LEDs Up

  bsf    STATUS, C           ;  Set the Status Flag
  rlf    PORTB, f            ;   Shift the Data Up

  goto   IntEnd

Down

  bsf    STATUS, C           ;  Shift the Data Down
  rrf    PORTB, f

IntEnd

  movf   _status, w          ;  Restore the Context Registers
  movwf  STATUS
  swapf  _w, f
  swapf  _w, w

  retfie

Mainline                     ;  Setup PWM And Monitor it, Updating "PWMOn"

  clrf   Direction
  movlw  0x0E7               ;  Start in the Middle
  movwf  PORTB

  bsf    STATUS, RP0         ;  Goto Bank 1 to set Port Direction
  clrf   TRISB ^ 0x080       ;  PORTB is Output
 ifdef Debug
  movlw  0x0D0               ;  Debug - Minimum Timer Delay
 else
  movlw  0x0D7               ;  Normal Operation - Maximum Timer
 endif
  movwf  OPTION_REG ^ 0x080
  bcf    STATUS, RP0         ;  Go back to Bank 0

  clrf   TMR0                ;  Start the Timer from Scratch

  movlw  (1 << GIE) + (1 << T0IE)
  movwf  INTCON              ;  Enable Interrupts

Loop                         ;  Loop Here

  goto   Loop                ;  Let Interrupt Handler Work in the
                                Background

 end
```

Notice the use of the prescaler for the application and the *Count* variable for delaying the output. I use the prescaler set to the maximum delay for the TMR0 input (including using the instruction clock as the TMR0 input source) to delay the interrupts as long as possible. In this configuration, the delay is 1/64,000 of the instruction clock. For a PICmicro® MCU running at 4 MHz, this works out to an interrupt interval of 16 times per second. I used the prescaler on this application because it simplified the *Count* variable's operation.

By using the prescaler for TMR0 in this application, I am not indicating that I feel that the watchdog timer should not be used with this application. If the eye would be used to indicate the health of a larger circuit, I would presume that the use of the watchdog timer would be used to indicate if the PICmicro® MCU has had a problem in its execution. In this case, the prescaler could be assigned to the watchdog timer and the *Count* variable increased to 16 bits to perform the same function.

The use of the *Count* variable is a result of the maximum delay being put on TMR0 as not being acceptable for the application. With the 1/16-second maximum delay, I found the movement of the eye to be somewhat manic. By placing the *Count* variable and counting down from two, I have delayed the operation to something reasonable. For PICmicro® MCUs that are clocked with higher frequencies, this variable can be given a higher value to maintain the speed of the eye going back and forth. As well, if the prescaler is to be used with the watchdog timer, then *Count* could be used to increase the number of times TMR0 overflows before the eye's position is updated.

I originally used the prescaler set to the largest value to minimize the impact of the interrupt handler on the mainline code. Eliminating the prescaler input will result in fewer instructions available to the mainline code for system monitoring.

## TMR0INT: SIMULATING INPUT PIN INTERRUPT WITH TIMER PIN INPUT

Some time ago, the question came up on the PICList asking whether or not TMR0 could be used as an interrupt input pin, if it were set up in such a way that the next transition would cause an Overflow condition and request that the interrupt be handled. I discovered that it is indeed possible to use TMR0 as an interrupt input pin. Using TMR0 in this fashion is something that could be very desirable in the cases when no more interrupt source pins are left available or if a low-end PICmicro® MCU is being used. Rather than polling an input, a latch (like TMR0) could be used to detect a transition. This experiment shows how TMR0 can be set up to request an interrupt after the first incoming data transition.

The circuit itself is very simple as you can see in the schematic (Fig. 15-126) and the breadboard assembly drawing (Fig. 15-127). The YAP-II circuit is also quite simple (Fig. 15-128). The only unusual aspect of the circuit is that I have specified that a switch is connected to the pull down connected to the PICmicro® MCU's _MCLR pin to allow you to easily reset the application. The reason for this will become obvious when you work with it.

The bill of materials (Table 15-27) for the experiment is also quite modest. The code itself is also very simple and is known as *TMR0Int*. It is located in the *code\TMR0Int* subdirectory of the *PICmicro* directory.

```
 title  "TMR0Int - Treat TMR0 Input like an Interrupt Input"
;
;  This Application uses the TMR0 Input Pin (RA4 in the mid-range
;   PICmicro MCUs) as an interrupt source.  When the input data changes,
;   TMR0 overflows, which causes an interrupt request.
;
;  Hardware Notes:
;   This application runs on a PIC16F84 executing at 4 MHz
;   _MCLR is tied through a 10K Resistor to Vcc and PWRT is Enabled
```

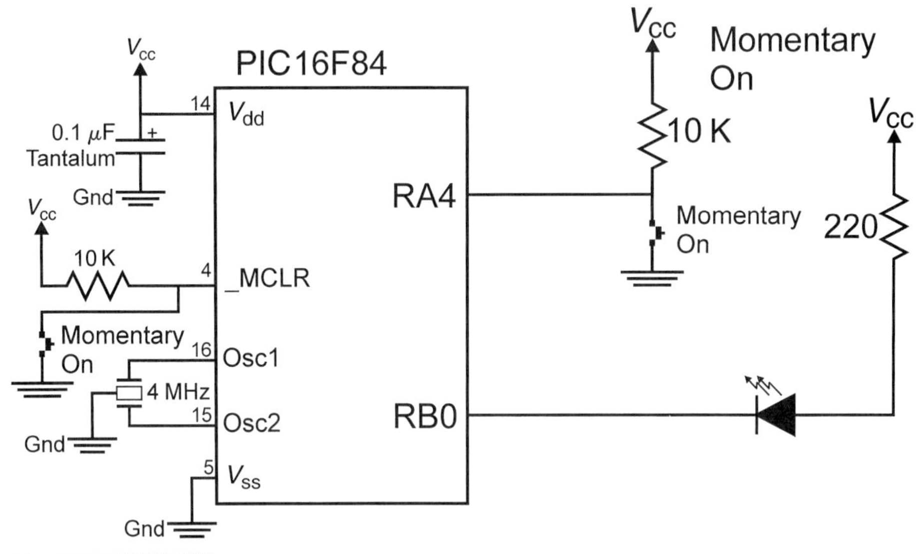


Figure 15-126 TMR0Int: using TMR0 as an interrupt source

```
;   A 10K Pull-Up and a Momentary "On" Switch is Connected to RA4
;   A 220 Ohm Resistor and LED is connected between PORTB.0 and Vcc
;
;  Myke Predko
;  99.12.28
;
  LIST R=DEC
```

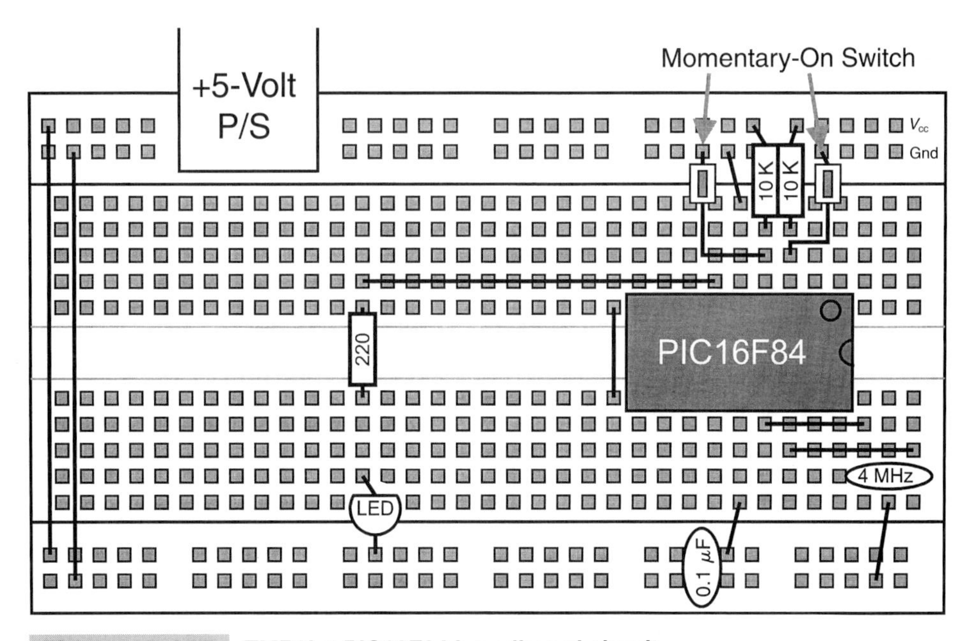


Figure 15-127 TMR0Int PIC16F84 breadboard circuit

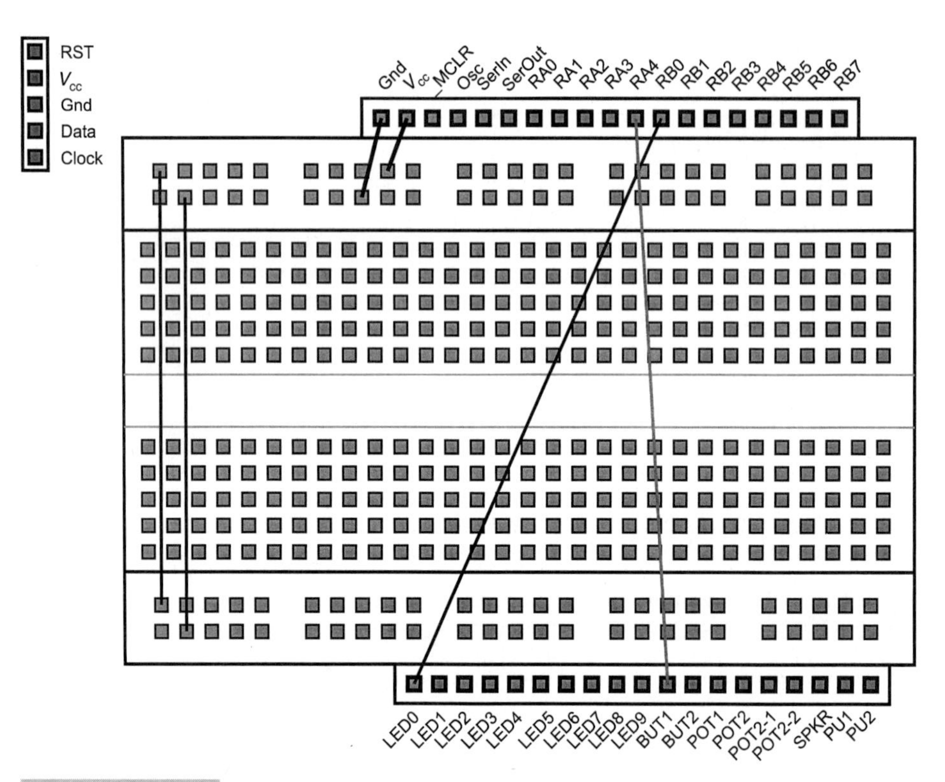


**Figure 15-128** EMU-II/YAP-II breadboard for TMP0Int

**TABLE 15-27 Bill Of Materials For The "TMR0INT" Experiment**

| PART | DESCRIPTION | REQUIRED ON YAP-II |
|---|---|---|
| PICmicro® MCU | PIC16F84-04/P | In Socket |
| Vdd/Vss decoupling capacitor | 0.1-$\mu$F tantalum | No |
| _MCLR pull up | 10 K, 1/4 watt | No |
| Momentary-on switch for _MCLR | Momentary on | No |
| 4-MHz ceramic resonator | 4 MHz with built-in capacitors | No |
| RB0 LED | Red LED | No. LED0 used |
| RB0 LED resistor | 220 Ohm, 1/4 watt | No. LED0 used |
| 10-K resistor RA4 pull up | 10 K, 1/4 watt | No. BUT1 used |
| Momentary-On switch for RA4 | Momentary On | No. BUT1 used |

```
  ifdef __16F84
   INCLUDE "p16f84.inc"
  endif
  ifdef __16F877
   INCLUDE "p16f877.inc"
  endif

 ;  Registers
  CBLOCK 0x020
 _w, _status                  ;  Context Register Save Values
  ENDC

  ifdef __16F84
  __CONFIG _CP_OFF & _WDT_OFF & _XT_OSC & _PWRTE_ON
  else
  __CONFIG _CP_OFF & _WDT_OFF & _XT_OSC & _PWRTE_ON & _DEBUG_OFF &
_LVP_OFF & _BODEN_OFF & _WRT_ENABLE_ON & _CPD_OFF
  endif

   PAGE
 ;  Mainline of cylon

  org     0

   nop

   movlw  0x0FF
   movwf  PORTB

   goto   Mainline

  org     4

 Int                          ;  TMR0 has Overflowed - New Input

   bcf    INTCON, T0IF        ;  Reset the Interrupt Flag

   bcf    PORTB, 0            ;  Turn on the LED

   retfie

 Mainline                     ;  Setup TMR0 Interrupt

   bsf    STATUS, RP0         ;  Goto Bank 1 to set Port Direction
   bcf    PORTB, 0            ;  Enable RB0 for Output
   bcf    STATUS, RP0         ;  Go back to Bank 0

   movlw  OPTION_REG          ;  Point to the Option Register
   movwf  FSR
   clrf   TMR0                ;  Reset the Timer

   movlw  (1 << GIE) | (1 << T0IE)
   movwf  INTCON              ;  Enable Interrupts

   movlw  0x0C0               ;  Make TMR0 Driven by the Instruction Clock
   movwf  INDF

   movlw  0x0FF
   movwf  TMR0

   bsf    INDF, T0CS          ;  Now, Make TMR0 Driven Externally

 Loop                         ;  Loop Here

   goto   Loop                ;  Let Interrupt Handler Work in the
                                 Background

  end
```

This code will seem somewhat conventional, except for my passing the address of the OPTION register to FSR and updating it from there, rather than the usual way of updating the register while Bank 1 is being accessed. I used this code because the time required to set TMR0 and then jump to Bank 1 would be much more difficult to time than the code I used.

In the example code, TMR0 is initially given the value of *0x0FF*. It waits to be updated by the instruction clock. As indicated earlier in the book, TMR0 has a two-cycle clock counter (synchronizer) that must have two inputs before it passes one on to TMR0. This synchronizer, like the prescaler, is reset any time TMR0 is written to. Thus, it will overflow and request an interrupt two instructions after it is set. For this application, I want TMR0 to overflow after the first external input. To set up this condition, I want one instruction between the instruction when TMR0 is loaded with 0x0FF and when I change its input source to RA4.

My solution is to use the INDEX register (FSR) to point to OPTION and change the source (by resetting the T0CS bit) using the *bcf* instruction. In the past, I have used the *option* instruction, which is what I would do if I were implementing this code in a low-end PICmicro® MCU. The code here avoids using the not recommended *option* for the mid-range PICmicro® MCUs.

When using this code in low-end PICmicro® MCUs, remember to not use the *movf TMR0, f* instruction. Instead, use the *movf TMR0, w* instruction. Both will set the STATUS register's Zero flag, but the first instruction will reload TMR0 with the current value. If it is 0x0FF, it will reset the two-cycle timer, which means that two valid interrupt inputs are required to change the state. Worse yet, if TMR0 is polled before the two interrupt inputs are received, the TMR0 will never reset and increment for the interrupt source.

This application is very focused on the single task of latching the LED connected to RB0 "on" when the input on RA4 goes low. For this reason, I didn't bother with a complicated interrupt handler (in fact, I dispensed with context register saving and restoring). In a "real" application, the interrupt-handler code would probably carry out the code for setting TMR0 as an interrupt source after acknowledging that the interrupt has been received.

## LEDPWM: TIMER0 INTERRUPT USED FOR LED PWM DIMMING

To demonstrate how TMR0 and interrupts can be used to control the brightness of a LED by producing a PWM signal, this application's circuit is the simplest in the book. The circuit (Fig. 15-129) only interfaces to one LED, which will get brighter, turn off and repeat using the LEDPWM code that is located in the *code\ledpwm* subdirectory of the *PICmicro* directory. The bill of materials for LedPWM is shown in Table 15-28.

The code used for LedPWM is reasonable simple and explained in quite a bit of detail in Chapter 6. Upon reset, the PICmicro® MCU's RB0 pin is enabled for output and the TMR0 interrupt is enabled. Once this is done, an incrementing LED power PWM signal is set and is passed to an interrupt handler, which outputs a pulse of varying timing widths every 1 ms.

```
 title  "LedPWM - Show an LED Changing Brightness"
;
;  This Application simply waits for the TMR0 interrupt handler to
;   occur and after the LED PWM "ON" is complete, decrement the "ON"
;   value.  This Application repeats endlessly with the LED getting
;   lighter before turning off and starting over.
;
```

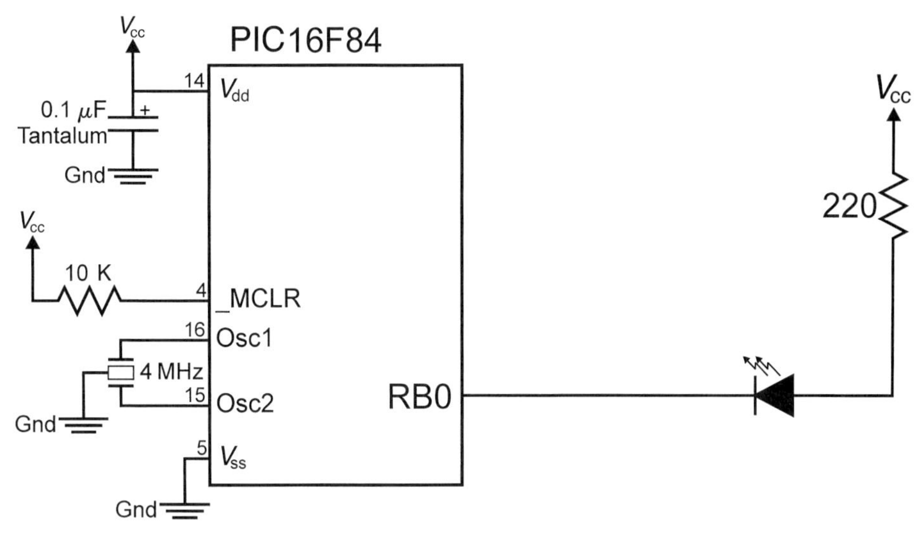


**Figure 15-129** **LedPWM: Varying the brightness of a LED using a PWM signal**

```
;  Hardware Notes:
;   This application runs on a 16F84 executing at 4 MHz
;   _MCLR is tied through a 4.7K Resistor to Vcc and PWRT is Enabled
;   A 220 Ohm Resistor and LED is attached to PORTB.0 and Vcc
;   A 10K pull up is connected to RA0 and it's state is passed to
;    RB0
;
;  Myke Predko
;  99.12.14
;
  LIST R=DEC
 ifdef __16F84
  INCLUDE "p16f84.inc"
 endif
 ifdef __16F877
  INCLUDE "p16f877.inc"
 endif
```

**TABLE 15-28 Bill Of Materials For The "LEDPWM" Experiment**

| PART | DESCRIPTION | REQUIRED ON YAP-II |
|---|---|---|
| PICmicro® MCU | PIC16F84-04/P | In Socket |
| Vdd/Vss decoupling capacitor | 0.1-μF tantalum | No |
| _MCLR pull up | 10 K, 1/4 watt | No |
| 4-MHz ceramic resonator | 4 MHz with built-in capacitors | No |
| RB0 LED | Red LED | No. LED0 used |
| RB0 LED resistor | 220 Ohm, 1/4 watt | No. LED0 used |

```
;  Registers
 CBLOCK 0x020
_w, _status                  ;  Context Register Save Values
PWMOn:2                      ;  PWM "On Value"
PWMDouble: 2                 ;  Divide PWM down for Slowing Down
 ENDC

#define PWM PORTB, 0         ;  LED on PORTB.0

 ifdef __16F84
 __CONFIG _CP_OFF & _WDT_OFF & _XT_OSC & _PWRTE_ON
 else
 __CONFIG _CP_OFF & _WDT_OFF & _XT_OSC & _PWRTE_ON & _DEBUG_OFF &
_LVP_OFF & _BODEN_OFF & _WRT_ENABLE_ON & _CPD_OFF
 endif

  PAGE
;  Mainline of ledpwm

 org    0

  nop

  clrf   PWMDouble
  clrf   PWMDouble + 1

  goto   Mainline

 org    4
Int                          ;  Interrupt Handler

  movwf  _w                  ;  Save Context Registers
  movf   STATUS, w           ;   - Assume TMR0 is the only enabled
  movwf  _status             ;     Interrupt

  btfsc  PWM                 ;  Is PWM O/P Currently High or Low?
   goto  PWM_ON
  nop                        ;  Low - Nop to Match Cycles with High

  bsf    PWM                 ;  Output the Start of the Pulse

  movlw  6 + 6               ;  Get the PWM On Period
  subwf  PWMOn, w            ;  Add to PWM to Get Correct Period for
                             ;   Interrupt Handler Delay and Missed
                             ;   cycles in maximum 1024 usec Cycles

  goto   PWM_Done

PWM_ON                       ;  PWM is On - Turn it Off

  bcf    PWM                 ;  Output the "Low" of the PWM Cycle

  movf   PWMOn, w            ;  Calculate the "Off" Period
  sublw  6 + 6               ;  Subtract from the Period for the
                             ;   Interrupt Handler Delay and Missed
                             ;   cycles in maximum 1024 usec Cycles

  goto   PWM_Done

PWM_Done                     ;  Have Finished Changing the PWM Value

  sublw  0                   ;  Get the Value to Load into the Timer
  movwf  TMR0

  bcf    INTCON, T0IF        ;  Reset the Interrupt Handler

  movf   _status, w          ;  Restore the Context Registers
  movwf  STATUS
```

```
  swapf   _w, f
  swapf   _w, w

  retfie

Mainline                    ;  Setup the PWM And then Monitor it,
                            ;   Updating "PWMOn"
  bsf     PORTB, 0          ;  Make the LED on PORTB.0 "off"
                            ;   Initially
  bsf     STATUS, RP0       ;  Goto Bank 1 to set Port Direction
  bcf     TRISB ^ 0x080, 0  ;  Set RB0 to Output
  movlw   0x0D1             ;  Setup TMR0 with a 4x prescaler
  movwf   OPTION_REG ^ 0x080
  bcf     STATUS, RP0       ;  Go back to Bank 0

  clrf    TMR0              ;  Start the Timer from Scratch

  movlw   (1 << GIE) + (1 << T0IE)
  movwf   INTCON            ;  Enable Interrupts

Loop                        ;  Loop Here

  btfsc   PWM               ;  Wait for PWM to go Low
   goto   $ - 1

  incfsz TMR0, w            ;  Wait for TMR0 to Equal 0x0FF
   goto   $ - 1

  movf    PWMDouble, f      ;  Decrement PWM Double
  btfsc   STATUS, Z
   decf   PWMDouble + 1, f
  decf    PWMDouble, f

  rrf     PWMDouble + 1, w  ;  Divide by 4
  movwf   PWMOn + 1
  rrf     PWMDouble, w
  movwf   PWMOn

  rrf     PWMOn + 1, w
  rrf     PWMOn, w

  addlw   255 - 0x0FE       ;  Get the High limit
  addlw   0x0FE - 0x00E + 1 ;  Add Lower Limit to Set Carry
  btfss   STATUS, C         ;  If Carry Set, then Lower Case
   movlw 0x0FC              ;   Carry NOT Set, Reset the Character
  addlw   0x00E             ;  Add Lower Limit to restore the Character
  movwf   PWMOn             ;  Save the New Value

  btfss   PWM               ;  Wait for the PWM to Go High
   goto   $ - 1
  goto    Loop

 end
```

The PWM signal-generating code in the interrupt handler can be described using the pseudocode:

```
Interrupt PWMOutput()       //  When Timer Overflows, Toggle "On" and
                                "Off"
{                           //   and Reset Timer to the correct delay for
                                   the Value
  if (PWM == ON) {          //  If PWM is ON, Turn it off and Set Timer
    PWM = off;              //   Value
    TMR0 = PWMPeriod - PWMOn;
```

```
  } else {                    //  If PWM is off, Turn it ON and Set Timer
    PWM = ON;                 //   Value
    TMR0 = PWMOn;
  }  //  end if

  INTCON.T0IF = 0;            //  Reset Interrupts

}  //  End PWMOutput TMR0 Interrupt Handler
```

This operation first outputs a high on the PWM pin for the specified number of TMR0 ticks, up to 250. When TMR0 overflows and requests an interrupt, the PWM is set to low and TMR0 is loaded with the number of ticks remaining for the 250-instruction cycle PWM signal period.

The generation of the TMR0 initial value is somewhat unusual and might be somewhat difficult to follow. I suggest that you go back to the "Pulse-Width Modulation (PWM) I/O" section of Chapter 6 to see how the *add* and *two* subtracts work for loading TMR0 with the correct value.

I recommend that you build this application first and try it before reading on. The circuit is quite simple on a breadboard (Fig. 15-130) and is dead simple using the YAP-II (Fig. 15-131). For either method, you will probably spend more time waiting for the device to be programmed than building the circuit.

When you run the application, you should see the LED connected to RB0 increase in brightness to a certain point and then turn off and start again. The entire process takes about one second. If you were to look at the PWM signal sent to the LED, you would see something like that shown in Fig. 15-132, transitioning to a high or low value with short pulses. A mostly low signal will translate to the LED being very bright.

It will be probably surprising to you to discover that the most difficult code that I had to write for the experiment was the code used to update PWMOn (which is used to select the PWM signal's pulse width). When I originally created this application, I wanted to simply

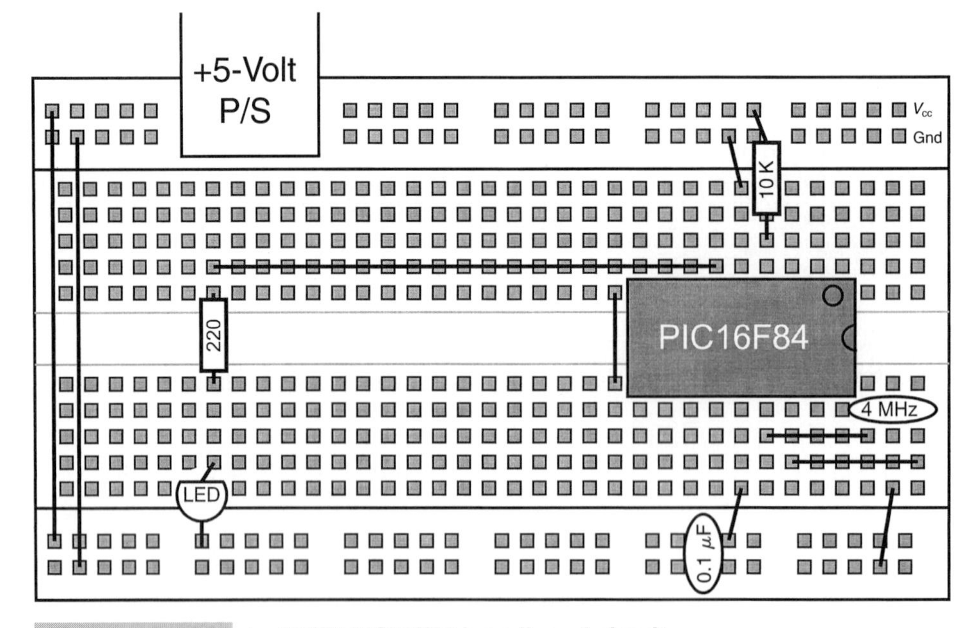


**Figure 15-130** LedPWM PIC16F84 breadboard circuit

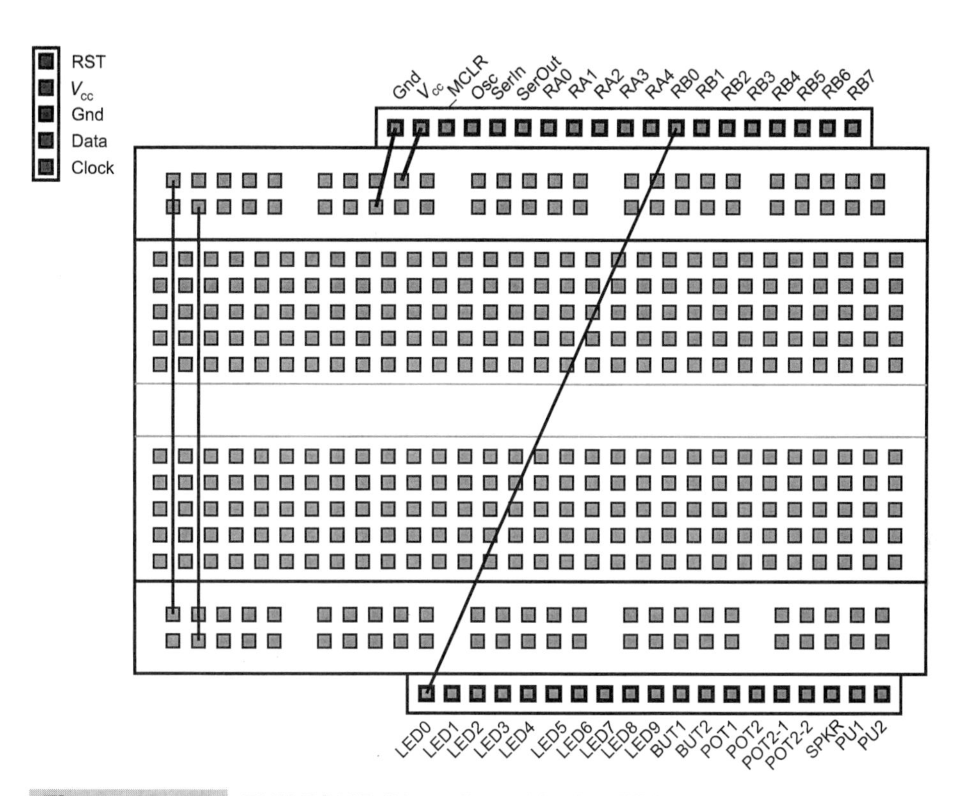


**Figure 15-131** EMU-II/YAP-II breadboard for LedPWM

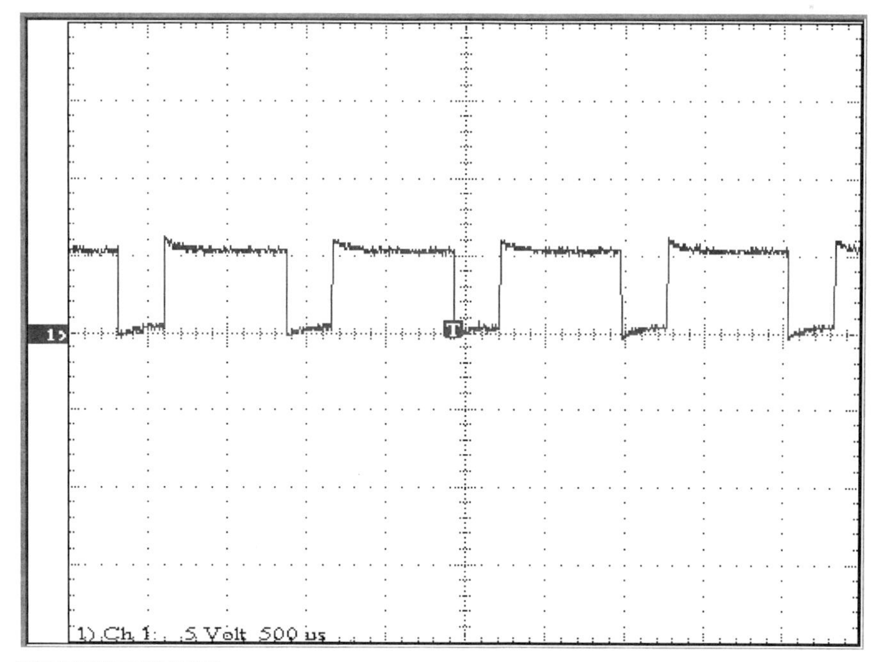


**Figure 15-132** LedPWM application LED drive PWM signal

decrement the PWMOn variable to make the off time of the output PWM signal longer as time goes on, which, in turn, makes the LED brighter as time goes on. To do this, I originally wanted to use code that can be shown as the pseudocode:

```
while (1 == 1) {             //  Loop Forever
  while (PWM == ON);         //  Wait for Pulse to Finish
  PWMOn = PWMOn - 1;         //  Decrement the Pulse Width
  while (PWM == off);        //  Wait for the "Off" to Finish
} //  end while
```

which simply waits for an interval between pulses in which the *PWMOn* value could be changed. The following assembly-language code was created for this function:

```
Loop                        ;  Return here forever
  btfsc  PWM                ;  Loop while PWM == ON
   goto  $ - 1
  decf   PWMOn, f           ;  Decrement the Pulse Width
  btfss  PWM                ;  Loop while PWM == off
   goto  $ - 1
  goto   Loop               ;  Loop Again
```

When I simulated the application, the output seemed to be correct for 10 ms (the length of time in which I simulated the operation for 10 PWM pulses). When I programmed a PICmicro® MCU and tried out the application, I found that the LED flashed briefly on power up and then stayed dark. When I put an oscilloscope probe on the LED, I found that the PWM signal was high with a few short pulses.

I then simulated the application beyond the first 10 ms and discovered that when the value put into TMR0 was very high (i.e., 0x0FF), a new interrupt request from TMR0 would be received before the interrupt handler had returned to the mainline (Loop) code. Execution jumped immediately back into the interrupt handler and changed the polarity of the PWM output pin without giving the mainline code a chance to recognize the change. To fix this, I then monitored the values of *PWMOn* and ensured that it never became less than 0x00C, at which point *PWMOn* was reset to 0x0F4 to restart itself.

This change to the code did fix the problem, but the LED flashed on and off very quickly (about four times per second). To make the display a bit more user friendly, I decided to use a 16-bit counter for the *PWMOn* value (this is *PWMDouble*) and shift it down to slow down the actual PWM display. I had to do this twice (divide the value by four) to get an appropriate value. After *PWMDouble* is divided by four, I check the range and reset the output to a high value to make the full on time a bit longer. You can see the results of this code in the "Loop" section at the end of the *LedPWM* application.

This was not the best way to change *PWMOn*. Along with being quite complex, requiring three additional file registers for temporary values, it also took me a long time to figure out the exact values for the range check and reset for the code. A much better way would have been to simply wait four ms (the time for four pulses to be output) and then change *PWMOn*. This method would not require the range checking of the method that I used because the updates are not dependent on the changes in the PWM bit.

The code to implement the changing PWM by simply delaying by four ms is:

```
Loop                          ;  Loop Here

  decf   PWMOn, f             ;  Decrement the PWM Value

  movlw  4                    ;  Delay 4 msecs
  movwf  Dlay
  movlw  0x020
  addlw  0 - 1                ;  Decrement the Value
  btfsc  STATUS, Z            ;   Using the 5 Cycle Delay Loop
   decfsz Dlay, f
    goto $ - 3

  goto   Loop
```

The application source code for *LedPWM* using this method is called *lpbetter* of the *code\ledpwm* subdirectory.

This method uses less than 40 percent of the code of the original method and half of the file registers. It also took me about two minutes to make the changes to the original application to come up with lpbetter.

I've left the original application in place as a monument to stubbornness. To get the original LedPWM code working, I spent almost two hours trying different tricks before settling on the 16-bit counter, the shift down by four, and the range check and update. Instead of doing all this work, I really should have been thinking of a better way to implement the code.

The lesson I want to impart to you is: don't try plowing through a troublesome application to make it work. Take a few moments and reflect on your goal and see if there is a better way to achieve it.

My problem was, that I wanted to be clever and use the interrupt handler's 1-ms operation for the high and low of the PWM interrupt to provide a counter for changing the PWM signal-control variable (*PWMOn*). The much more efficient method of implementing the changing *PWMOn* value was to simply put in my own delay counter in the mainline and ignore what was happening in the interrupt handler.

Experiences like this are lessons, not mistakes. Don't get mad at yourself if you miss an obvious opportunity like what I have shown here. Instead, remember it for the next time you get in the situation where you are struggling to get an application to work and try to think up a better way to do it. Chances are that you'll come up with something pretty spectacular.

## INTDEB: DEBOUNCING INPUTS WITH INTERRUPTS

When using interrupts for debouncing applications, the decision has to be made on the interface to the mainline. In some applications, it might be appropriate to initiate the action of the button read. In others, the interrupts built into the PICmicro® MCU can be used to provide a debounced button function that runs in the background and can be checked as required. The IntDeb application uses the RB0/INT pin and TMR0 to provide a background-debounced button input that can be polled by the mainline application as required.

The circuit for IntDeb is a bit unusual because I have split the eight bits output by the PICmicro® MCU between PORTA and PORTB. The reason for doing this is expediency because PORTB.0 is required for the button interrupt input. As shown in the application code, this does not cause a major problem for the application. The schematic for the IntDeb experiment is shown in Fig. 15-133 and the bill of materials is shown in Table 15-29.

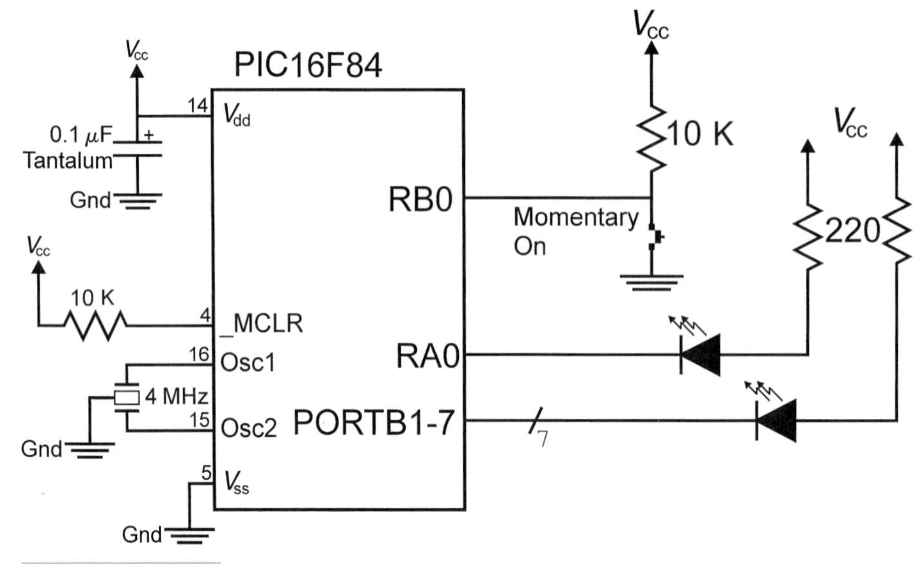


**Figure 15-133** IntDeb: counting using a debounced button

The need for the RB0/INT pin to be used as an input makes wiring the application somewhat different from the other applications. The least-significant bit displayed has been moved from RB0 to RA0. Moving the bit means that one of the wires will have to be moved from one side of the PICmicro® MCU to the other (Fig. 15-134), which is the wiring diagram for the breadboard version. The YAP-II wiring is shown in Fig. 15-135.

The INTDEB.ASM code, located in the *code\IntDeb* subdirectory of the *PICmicro* directory is:

```
 title  "IntDeb - Register Contents Int Debounce."
#define nDebug
;
```

**TABLE 15-29 Bill Of Materials For The "INTDEB" Experiment**

| PART | DESCRIPTION | REQUIRED ON YAP-II |
|---|---|---|
| PICmicro® MCU | PIC16F84-04/P | In Socket |
| Vdd/Vss decoupling capacitor | 0.1-µF tantalum | No |
| _MCLR pull up | 10 K, 1/4 watt | No |
| 4-MHz ceramic resonator | 4 MHz with built-in capacitors | No |
| 8 220-Ohm resistors | 220 Ohm, 1/4 watt | No. LED0/LED7 used |
| 10-LED bargraph | 10-LED bargraph | No. LED0/LED7 used |
| 10 K | 10 K, 1/4 watt | No. BUT1 used |
| Momentary-on switch | Momentary On | No. BUT1 used |

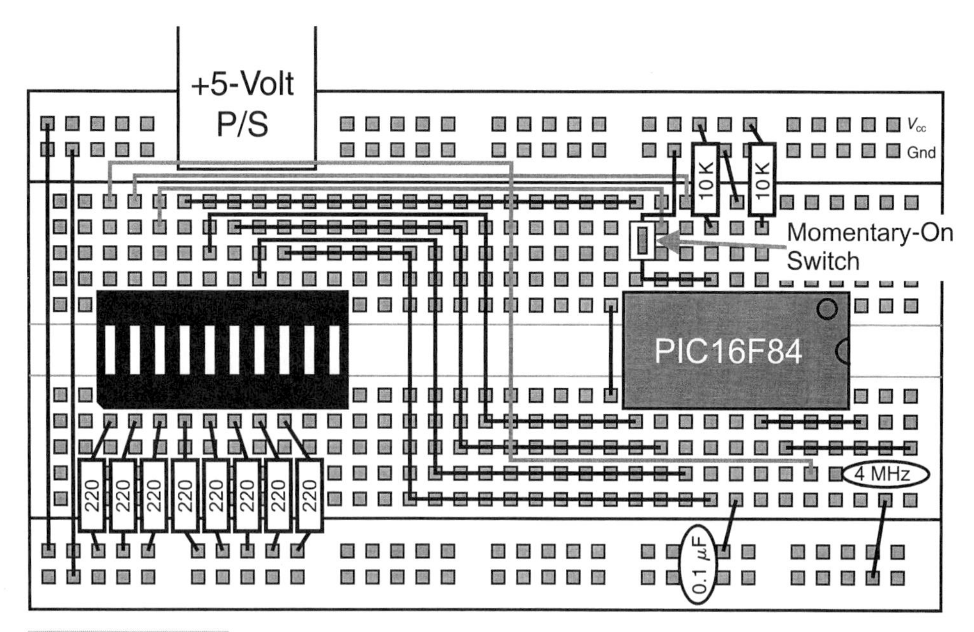


**Figure 15-134** IntDeb PIC16F84 breadboard circuit

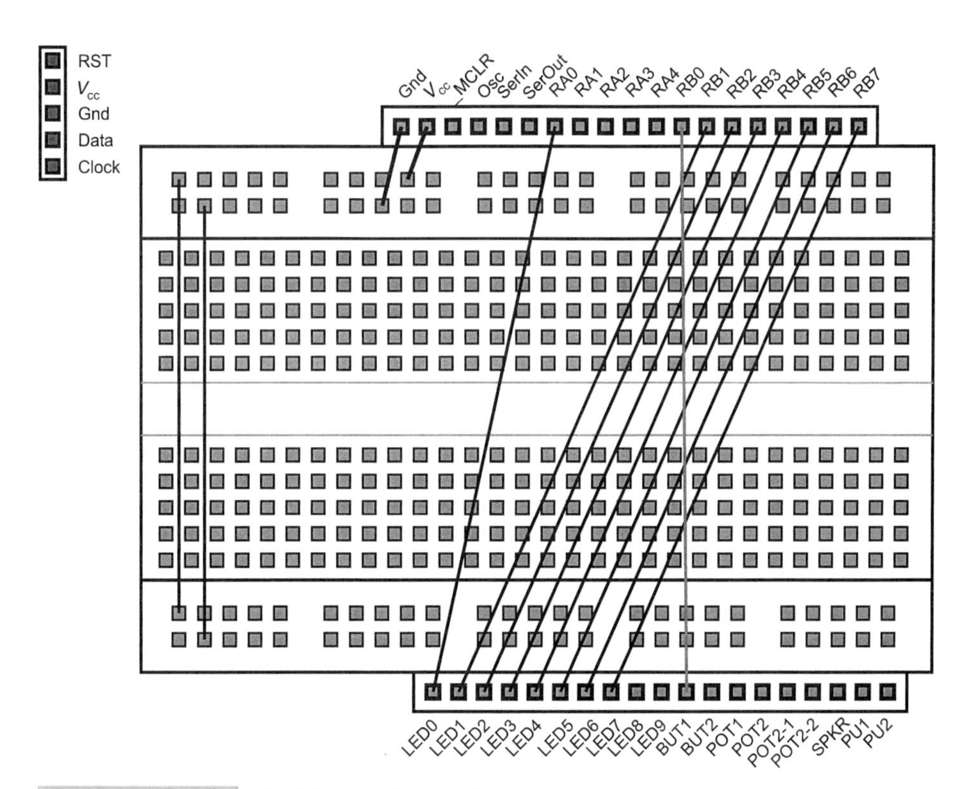


**Figure 15-135** EMU-II/YAP-II breadboard for IntDeb

```
 ;  This is Program reads the value in a RAM Register and outputs it
 ;  inverted onto PORTB (which has LEDs to Display the Value).  All the
 ;  RAM Registers are Read and Displayed.  A button is used as the
 ;  instigator of the next value read.  The FSR is Copied into the LEDs
 ;  to Display the current Register Being Displayed.
 ;
 ;  This program is a modification of PROG18.ASM to use the Interrupt
 ;   Handler to Debounce the Button Input.
 ;
 ;  Hardware Notes:
 ;   Reset is tied directly to Vcc and PWRT is Enabled.
 ;   A 4.7K Pullup and Switch Pull-Down is attached to PORTB.0
 ;   A 220 Ohm Resistor and LED is attached to PORTB.7:1
 ;   A 220 Ohm Resistor and LED is attached to PORTA.0
 ;
 ;  Updated for the Second Edition: 99.12.28
 ;
 ;  Myke Predko
 ;  97.02.22
 ;
   LIST R=DEC
  ifdef __16F84
   INCLUDE "p16f84.inc"
  endif
  ifdef __16F877
   INCLUDE "p16f877.inc"
  endif

 ;  Registers
  CBLOCK 0x020
 Flags
 Reg                            ;  Register to Display
  ENDC

 #define ButUp   Flags, 0       ;  Flags Indicating Button State
 #define ButDown Flags, 1
  ifdef __16F84
  __CONFIG _CP_OFF & _WDT_OFF & _XT_OSC & _PWRTE_ON
  else
  __CONFIG _CP_OFF & _WDT_OFF & _XT_OSC & _PWRTE_ON & _DEBUG_OFF &
_LVP_OFF & _BODEN_OFF & _WRT_ENABLE_ON & _CPD_OFF
  endif

   PAGE
 ;  Mainline of IntDeb

   org 0

   nop

   clrf   Flags                 ;  No Button Pressed Yet
   clrf   Reg

   goto   MainLine

   org 4                        ;  Interrupt Handler Address

 Int

   btfss  INTCON, TOIF          ;  Do we have a Timer Overflow?
    goto  Int_Switch            ;   No - Handle the Switch

   bcf    INTCON, TOIF          ;  Yes, Reset Timer Interrupt Request

   movlw  0x001                 ;  Assume the Button is Up
   btfss  PORTB, 0              ;  Is the Button Up or Down?
    movlw 0x002                 ;   - If PORTB is Low, Button Down
```

```
  movwf   Flags

  goto    Int_End

Int_Switch                    ;  Interrupt on the Switch - Reset Timer

  bcf     INTCON, INTF        ;  Reset the Interrupt Request

  clrf    Flags               ;  Indicate that Nothing is Valid

  movf    PORTB, 0            ;  What is the Button State?
  andlw   1
  bsf     STATUS, RP0
  bcf     OPTION_REG ^ 0x080, INTEDG
  btfsc   STATUS, Z           ;  Determine the Actual Edge
   bsf    OPTION_REG ^ 0x080, INTEDG
  bcf     STATUS, RP0

  clrf    TMR0                ;  Going to Wait for another Key Press

Int_End

  retfie
 PAGE
MainLine                      ;  Mainline of reading the File Registers

  movlw   0x0FF
  movwf   PORTB               ;  Turn off all the Indicator LED's
  movwf   PORTA

  movf    PORTB, w
  andlw   1                   ;  Set 0 if Pin Down

  bsf     STATUS, RP0
  bcf     TRISA ^ 0x080, 0    ;  RA0 is Output
  movlw   1                   ;  RB0 is Input/Interrupt
  movwf   TRISB ^ 0x080       ;  Set PORTB.7:1 bits to Output
 ifdef Debug
  movlw   0x090               ;  Zero Prescaler for TMR0 if Debug
  btfsc   STATUS, Z
   movlw 0x0D0                ;  Pin down, Go for the Rising Edge
 else
  movlw   0x096               ;  Setup Prescaler for TMR0 to 32.7 msec
  btfsc   STATUS, Z
   movlw 0x0D6                ;  Pin down, Go for the Rising Edge
 endif
  movwf   OPTION_REG ^ 0x080
  bcf     STATUS, RP0

  clrf    TMR0                ;  Reset TMR0

  movlw   (1 << GIE) | (1 << T0IE) | (1 << INTE)
  movwf   INTCON              ;  Setup the Interrupt Delays

Loop                          ;  Loop to Here for Each Register

  btfss   ButUp               ;  Wait for the Button to be Debounced UP
   goto   $ - 1

  btfss   ButDown             ;  Wait for Button be to Debounced Down
   goto   $ - 1

  incf    Reg, f              ;  Increment the Counter

  comf    Reg, w              ;  Display the Counter Value
  movwf   PORTB
  movwf   PORTA

  goto    Loop

 end
```

There shouldn't be too many surprises in this application for you (although a big one is described in the following paragraphs). The mainline code sets up the interrupts, along with the prescaler for TMR0, to cause an interrupt after 32.7 ms. I used this interval for the debounce interval (if no button changes have occurred within 32.7 ms, then the *Flags'* variable bits are set appropriately). The reason for this interval is that it is greater than the general rule of 20 ms and did not require any different values to be set in TMR0. I could simply reset it to restart the delay.

The interrupt handler code should not be surprising either. It could be written out in pseudo-code as:

```
interrupt RB0Debounce()      //  Debounce the Button on RB0
{
  if (T0IF != 0) {           //  Timer Interrupt (32.7 msec Delay)
    if (Button == 0) {       //  Button Pressed Down
      ButUp = 0; ButDown = 1;
    } else {                 //  Button Released Up
      ButUp = 1; ButDown = 0;
    }
    TMR0 = 0;                //  Reset the Timer
    T0IF = 0;
  } else {                   //  Button State Changed
    if (Button = 0)          //  Wait for Next Button Based on Current State
      OPTION.INTEDG = 1;     //  Button Down, Wait for going Up
    else
      OPTION.INTEDG = 0;
    INTF = 0;                //  Reset Button State
    TMR0 = 0;                //  Reset the Timer
  }
}                            //  end RB0Debounce
```

The advantages of this method are that it can be very crisply defined and written in a high-level language (this pseudo-code could be ported directly to C) and it does not affect the operation of the mainline code or even other interrupts. If you look back at the interrupt handler in IntDeb, you should be able to match this code very clearly to the pseudo-code.

In this application, track the current state of the button and change the INTEDG flag of the OPTION register accordingly. For button debounce routines that are interrupt based, I would prefer to use the Interrupt on Port Change function of the PICmicro® MCU's PORTB pins and ignore this. In this case, I could have created two variables that are set with the PORTB input states when TMR0 overflowed and requested an interrupt. The interrupt handler itself would be a bit simpler and more than one button could be debounced within an application.

When writing the counter value out to the LEDs, notice that I arranged it so that the least-significant bit (which is lost in PORTB because of the RB0/INT pin) is replicated in PORTA. Thus, I don't have to change the bit number of the count register, which simplifies the code and the application quite a bit. Obviously, this action is further simplified by not using any of the other PORTA pins. Even if they were, the code to update RA0 based on the least-significant bit of the counter (*Reg*) would be:

```
comf    Reg, w
movwf   PORTB             ;  Write Most Significant 7 Bits of "Reg"
movf    PORTA, w          ;  Clear the LSB of PORTA
andlw   0x01E             ;   Before setting if LSB of Reg requires it
```

```
btfss  Reg, 0             ;  If LSB Set, then Output Reset for LED On
 iorlw 1                  ;  LSB Reset, Set output
movwf  PORTA
```

which is not very difficult to implement and does not require any thinking on how to pass register bits back and forth. This is what I mean when I say that optimization occurs during application design, not after the circuit is created and the code is being written.

The problem with the interrupt handler is that I didn't bother to put in context register saves or restores. In the IntDeb application, where the button *Flags* values are polled without executing any other code, this is not a problem. But if this button-debounce code was to be used in another application that executed code, buttons could be pressed resulting in interrupt request flags being set. This will result in some very difficult to find and debug problems.

I want to impress you with the idea that when you create code that could be reused, try to make it as generic as possible. In this case, the context saves are not required, so I skipped the seven instructions or so and two file registers that are required for the function. The problem occurs if I want to use the code in another application. When it fails intermittently, the last place I will look is in the "debugged" code that I took from another source.

# Serial I/O

If you've read through Chapter 7, you will very definitely get the idea that I believe that the best method to interface to the PICmicro® MCU from a PC (or workstation) is serially, using RS-232. This standard, available in a wide variety of devices, is quite simple to wire—even though the electrical standard seems difficult to connect with.

The following experiments show how the PICmicro® MCU can be simply wired to a PC serially and interfaced using either built-in hardware or the I/O pins using bit-banging serial algorithms. Last, it shows how a three-wire RS-232 connection can be created that allows a PC to poll a serial port to determine whether or not a PICmicro® MCU is connected and operating.

## TRUERS: ASYNCHRONOUS SERIAL I/O USING THE BUILT-IN USART

Connecting a PICmicro® MCU to a PC or other RS-232 device can be daunting, but you can use some PICmicro® MCU and external hardware to make developing the software and hardware interface easier. The next three experiments introduce three of these methods, which were described earlier in the book. This experiment introduces what could be considered to be the "proper" way to connect a PICmicro® MCU to another device using RS-232. The next two experiments show some ways in which RS-232 connections can be implemented with less than the "proper" way of doing things.

In all of my RS-232 applications, I try to work with a three-wire RS-232 interface only. This interface consists of data transmission, data reception, and a signal ground. I avoid the handshaking lines as much as possible to avoid having to work through some of the more-complex aspects of the RS-232 conventions.

The circuit used for this application is shown in Fig. 15-136. It consists of a PIC16C73B

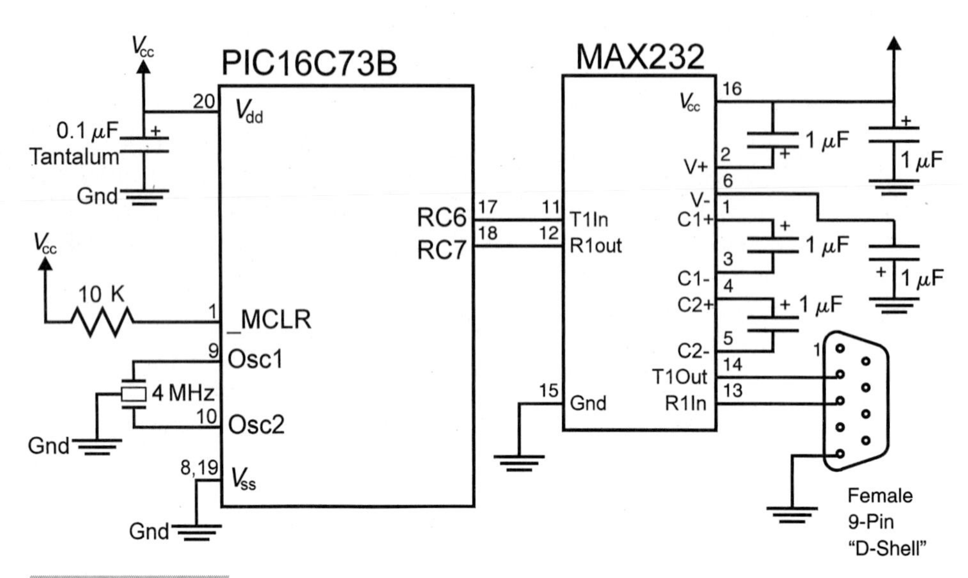


**Figure 15-136** **TrueRS: Proper RS-232 PICmicro® MCU/serial communications**

connected to a Maxim MAX232 level-translator chip. The PIC16C73B has a built-in *USART (Universal Synchronous/Asynchronous Receiver Transmitter)* that can be used for serial communications.

The MAX232 converts TTL/CMOS data levels into RS-232 data levels with negative voltage for the mark (*1*) and a positive voltage for the space (*0*). The only drawback to the MAX232 is the requirement for five 1.0-µF capacitors, which complicates the wiring of the circuit. Some RS-232 level converters have the capacitors built in (simplifying the wiring), but these chips generally cost three or four times that of the cost of the MAX232. This gap is widened even further when "knock off" MAX232s, which cost a fraction of the "true" MAXIM MAX232s, are used.

This circuit can be wired to a breadboard by following the wiring diagram in Fig. 15-137.

The circuit is actually quite easy to build, although you should spend a few moments thinking about the capacitor placement–especially if you have large capacitors. The capacitors can be either electrolytic or tantalum. The deciding factor should be the lowest price. The bill of material for the TrueRS project is shown in Table 15-30.

Notice the female 9-pin D-shell with three wires coming out of it. As you can see in Fig. 15-138, I have made up a connector with 20-gauge solid wire for use in multiple applications. This little gimmick can be very useful when you are working with RS-232 to help debug your applications. After soldering on the wire, I cover the connections with WeldBond for "strain relief" to eliminate the possibility of the wires working loose from the solder joints.

The TRUERS.ASM application code (located in the *code\TrueRS* subdirectory of the *PICmicro* directory) is actually one of the shortest applications shown in this book. The reason for its brevity is the use of the built-in hardware USART. Instead of timing the I/O accesses (as is done in the next three experiments) in code, the USART hardware takes care of the data transfer with Interrupt Request flags being set when the operations are complete.

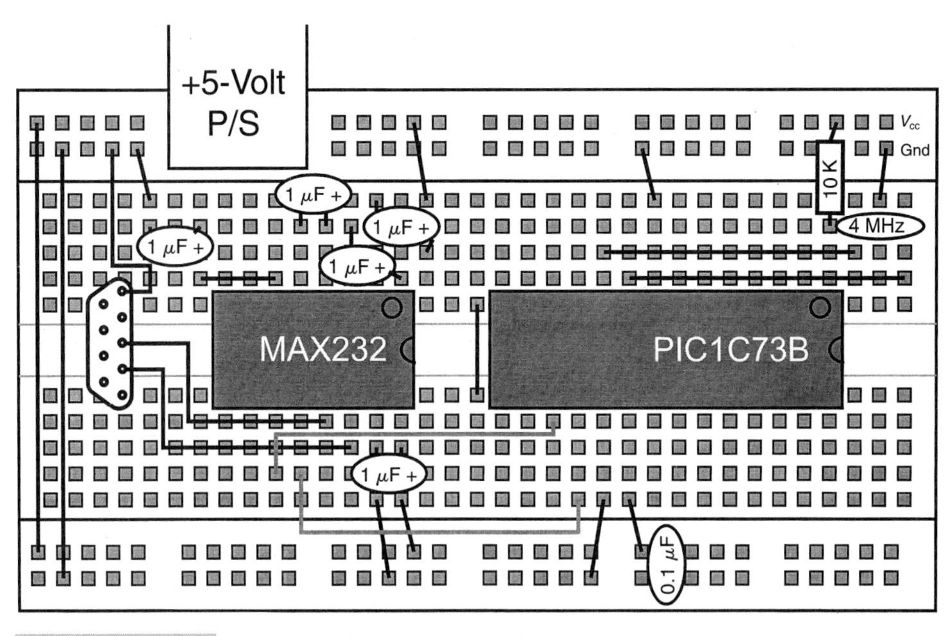


**Figure 15-137** TrueRS PIC16C73B breadboard circuit

```
 title  "TrueRS - Simple 3-Wire RS-232 Application"
;
;  This Application Implements a simple "3 Wire" RS-232 Link
;   to another Computer.  The built in USART is used with a
;   MAX232 Protocol translation chip.
;
;
;  Hardware Notes:
;   PIC16C73B Running at 4 MHz
;   _MCLR is Pulled Up
;   RC7 - Serial Receive Pin
;   RC6 - Serial Transmit Pin
;
```

**TABLE 15-30 Bill Of Materials For The "TRUERS" Experiment**

| PART | DESCRIPTION |
|---|---|
| PICmicro® MCU | PIC16C73B-04/JW |
| Vdd/Vss decoupling capacitor | 0.1-μF tantalum |
| _MCLR pull up | 10 K, 1/4 watt |
| 4-MHz ceramic resonator | 4 MHz with built-in capacitors |
| MAX232 | Maxim MAX232 or equivalent |
| 5 1.0-μF capacitors | 1.0-μF tantalum/electrolytic capacitors |
| Female 9-pin D-shell | DB-9F |

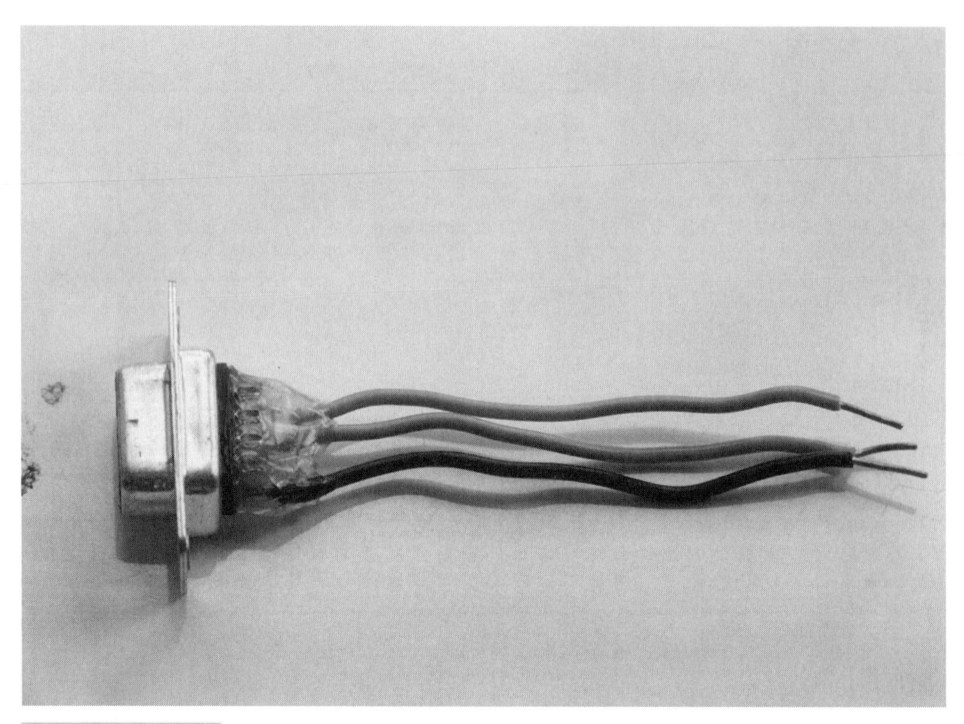

**Figure 15-138** Breadboard 9-pin DSHELL connector for RS-232

```
;  Myke Predko
;  99.12.31
;
  LIST P=16C73B, R=DEC
  INCLUDE "p16c73b.inc"

 PAGE
 __CONFIG _CP_OFF & _XT_OSC & _PWRTE_ON  & _WDT_OFF ^ _BODEN_ON

;  Mainline of TrueRS
  org    0

  bsf    STATUS, RP0        ;  Enable the Serial Port
  movlw  (1 << TXEN)        ;  Enable Serial Transmission
  movwf  TXSTA ^ 0x080
  movlw  51                 ;  Make the Serial Port Run at 1200 bps
  movwf  SPBRG ^ 0x080
  bcf    STATUS, RP0
  movlw  (1 << SPEN) | (1 << CREN)
  movwf  RCSTA              ;  Enable Serial I/O

Loop
  btfss  PIR1, RCIF         ;  Wait for Received Character Set
   goto  Loop               ;   - Nothing There
  movf   RCREG, w           ;  Get the Received Character
  bcf    PIR1, RCIF         ;  Clear the Character Present Flag

  addlw  255 - 'z'          ;  Get the High limit
  addlw  'z' - 'a' + 1      ;  Add Lower Limit to Set Carry
```

```
    btfss  STATUS, C          ;  If Carry Set, then Lower Case
     addlw h'20'              ;   Carry NOT Set, Restore Character
    addlw  'A'                ;  Add 'A' to restore the Character

    movwf  TXREG              ;  Send the Character

    goto   Loop

  end
```

In this application, after a byte is received (at 1200 bps), it is checked to see if the ASCII character is in the range of "a" to "z." If it is, the byte is converted into an uppercase ASCII representation of the byte. To test out the application, I used HyperTerminal, set at 1200 bps with no hardware handshaking to send characters.

This application cannot be easily simulated by the MPLAB IDE. The MPLAB IDE has not been programmed with a lot of the hardware peripheral functions, such as the USART, the synchronous serial port, or the ADC. To implement applications that use these functions, you will have to read through the Microchip datasheets and experiment with them to see how they work. The Register Stimulus files are used to show how an application responds to hardware I/O. The basic test of whether or not the hardware works in the first place has to be tested using applications like this one.

## BASICRS: SIMULATED ASYNCHRONOUS SERIAL I/O HARDWARE WITH PIN STIMULUS FILE

My favorite part for interfacing digital devices via RS-232 is the Dallas Semiconductor DS275. This "voltage stealing" device acts like the resistor/transistor circuit presented in the next experiment. Instead of producing its own negative voltage (like the MAX232, presented in the previous experiment), it takes the negative voltage from the idle transmit line and uses it for a valid negative voltage for its own output.

The DS275 is very easy to use and wire into your application. Its only real deficit is that it is more expensive than the MAX232 or the resistor/transistor circuit presented in the other RS-232 experiments.

The BASICRS.ASM application uses a three-times clock poll of the incoming data and uses this polling to observe when the data is coming in. The receive "synchs" from this "low" data. This is done by assuming that when the *Start* bit is discovered, by waiting one cycle (which is three times the data rate), polling will now take place in the middle of the bit (33% to 67% from the start of the bit).

This method works very well, allows applications to run in the "foreground," and treats the data coming in and out like a USART built into the PICmicro® MCU.

The BASICRS.ASM source code is located in the *code\BasicRS* subdirectory of the *PICmicro* directory.

```
 title  "BasicRS - Simple 3-Wire RS-232 Application"
 ;
 ;  This Application Implements a simple "3 Wire" RS-232 Link
 ;   to another Computer.  A Interrupt Based Bit Banging
 ;   Routine is used instead of a USART.
 ;
 ;
 ;
```

```
;  Hardware Notes:
;   PIC16F84 Running at 4 MHz
;   _MCLR is Pulled Up
;   RC7 - Serial Receive Pin
;   RC6 - Serial Transmit Pin
;
;  Myke Predko
;  99.12.31
;
  LIST R=DEC
 ifdef __16F84
  INCLUDE "p16f84.inc"
 endif
 ifdef __16F877
  INCLUDE "p16f877.inc"
 endif

;  Registers
 CBLOCK 0x020
Dlay:2
_w, _status
TXCount, TXByte
RXCount, RXByte
 ENDC

#define TX PORTB, 4
#define RX PORTB, 3


 PAGE
 ifdef __16F84
 __CONFIG _CP_OFF & _WDT_OFF & _XT_OSC & _PWRTE_ON
 else
 __CONFIG _CP_OFF & _WDT_OFF & _XT_OSC & _PWRTE_ON & _DEBUG_OFF &
_LVP_OFF & _BODEN_OFF & _WRT_ENABLE_ON & _CPD_OFF
 endif

;  Mainline of TrueRS
  org    0

  nop

 variable TMRDlay, TMRReset, PreScaler, PreScalerDlay, TMRActual, RTCAc-
tual, TMRError

TMRDlay = 4000000 / (1200 * 3 * 4)
TMRDlay = TMRDlay - 8
PreScalerDlay = 0
PreScaler = 2

TMRReset = TMRDlay / PreScaler
TMRActual = TMRReset * PreScaler
TMRError = ((TMRDlay - TMRActual) * 100) / TMRDlay
RTCActual = TMRActual + 9
RTCActual = ((RTCActual * 100000) / (4000000 / 4)) * 10

  goto   Mainline          ;  Jump Past the Interrupt Handler

 org     4
Int                        ;  PICmicro MCU Interrupt Handler

  movwf  _w                ;  Save the Context Registers
```

```
  movf    STATUS, w
  movwf   _status

  bcf     INTCON, TOIF       ;  Reset the Timer Overflow Interrupt

  movlw   256 - TMRReset     ;  Reset the Timer
  movwf   TMR0

  movf    TXCount, f         ;  If Transmitting, Can't Receive
  btfsc   STATUS, Z
   goto   _DoRX

  movlw   0x004              ;  Interrupt Transmit Increment Value
  addwf   TXCount, f
  btfss   STATUS, DC         ;  Send the Next Byte?
   goto   _TXSendDlayCheck

  bsf     TXCount, 2         ;  Want to Increment 3x not Four for each
                             ;  byte

  bsf     STATUS, C
  rrf     TXByte, f

  movf    PORTB,  w          ;  Send Next Bit
  andlw   0x0FF ^ (1 << 4)
  btfsc   STATUS, C
   iorlw  1 << 4
  movwf   PORTB              ;  Cycle 12 is the Bit Send

  goto    _IntEnd

_TXSendDlayCheck             ;  Don't Send Bit, Check for Start Bit

  btfss   TXCount, 0         ;  Bit Zero Set (Byte to Send)?
   goto   _TXNothingtoCheck

  goto    $ + 1              ;  Send Bit at Start of Cycle 12

  movlw   0x004              ;  Setup the Timer to Increment 3x
  movwf   TXCount

  bcf     TX                 ;  Output the Start Bit

  goto    _IntEnd

_TXNothingtoCheck            ;  Nothing Being Sent?

  movf    TXCount, w
  xorlw   0x004              ;  Zero (Originally) TXCount?
  btfsc   STATUS, Z
   clrf   TXCount

  movf    TXCount, w         ;  Sent 10 Bits?
  xorlw   0x09C
  btfsc   STATUS, Z
   clrf   TXCount            ;  Yes, Clear "TXCount" for Next Byte

  goto    _IntEnd

_DoRX


;  RXCount Bit 1  Bit 0
;              0      0  - Waiting for Character to Come In
;              0      1  - Receiving Character
;              1      0  - Have Received Valid Character
;              1      1  - Error Receiving, Clear Buffer/RXCount
;
```

```
  movlw  0x004              ;  Check for Bit?
  addwf  RXCount, f
  btfss  STATUS, DC
   goto  _RXNo              ;  Nothing to Check for (Yet)

  movf   RXCount, w         ;  At the End?
  xorlw  0x091
  btfsc  STATUS, Z
   goto  _RXNo

  bcf    STATUS, C          ;  Read the Current Bit
  btfsc  RX
   bsf   STATUS, C
  rrf    RXByte, f
  bsf    RXCount, 2         ;  Start Counting from 4

  goto   _IntEnd            ;  Finished Receiving Byte

_RXAtEnd                    ;  Check Last Bit
  btfss  RX
   goto  _RXOverrun         ;  Not Valid - Error

  movlw  2                  ;  Valid - Save Value
  movwf  RXCount

  goto    _IntEnd

_RXNo                       ;  No Bit to Receive

  movf   RXCount, w
  xorlw  0x09D              ;  Read with Stop Bit?
  btfsc  STATUS, Z
   goto  _RXAtEnd

  btfsc  RXCount, 0         ;  Something Running?
   goto  _IntEnd            ;   - Yes, Skip Over

  btfss  RXCount, 3         ;  Checking Start Bits?
   goto  _RXStart

  btfsc  PORTB,  3
   bcf   RXCount, 3         ;  Nothing - Keep Waiting

  bsf    RXCount, 0         ;  Mark it has Started

  btfss  RXCount, 1         ;  Something Already Saved?
   goto  _IntEnd

_RXOverrun                  ; Error - Mark the Overrun

  movlw  0x003
  movwf  RXCount

  goto   _IntEnd

_RXStart                    ;  Check for Low

  btfsc  PORTB,  3
   bcf   RXCount, 2         ;  Don't Have a "Start" Bit

:Finished with the Receive Interrupt Code

_IntEnd

  movf   _status, w         ;  Restore the Context Registers
  movwf  STATUS
  swapf  _w, f
  swapf  _w, w
```

```
  retfie
Mainline

  bsf    TX                   ;  Initialize the Output Port
  bsf    STATUS, RP0
  bcf    TX
  movlw  0x0D0 + PreScalerDlay  ;  Set up the Timer
  movwf  OPTION_REG ^ 0x080
  bcf    STATUS, RP0

  call   Delay                ;  Delay 200 msecs before Sending

  clrf   TXCount              ;  Make Sure Nothing is Happening
  clrf   RXCount              ;   On Boot

  bsf    INTCON, T0IE         ;  Initialize the Interrupt Handler
  bsf    INTCON, GIE

  movlw  256 - TMRReset       ;  Start the Timer Going
  movwf  TMR0

  movlw  "*"                  ;  Send out a Start Character
  call   NRZSend

Loop

  call   NRZReceive           ;  Wait for a Serial Byte to be Received

  addlw  255 - 'z'            ;  Get the High limit
  addlw  'z' - 'a' + 1        ;  Add Lower Limit to Set Carry
  btfss  STATUS, C            ;  If Carry Set, then Lower Case
   addlw h'20'                ;   Carry NOT Set, Restore Character
  addlw  'A'                  ;  Add 'A' to restore the Character

  call   NRZSend

  goto   Loop

NRZSend                       ;  Send the Value in "w" Serially
  movf   TXCount, f           ;  Wait for Previous Data Sent
  btfss  STATUS, Z
   goto  $ - 2                ;  Counter = 0 when Sent
  movwf  TXByte
  bsf    TXCount, 0           ;  Indicate there is Data to Send
  return

NRZReceive                    ;  Wait for a New Character to be Received
  btfss  RXCount, 1           ;  Bit 0 Set when Data Received
   goto  $ - 1
  movf   RXByte, w            ;  Get the Received Byte
  clrf   RXCount
  bcf    STATUS, C            ;  Carry Reset - No Error
  return

Delay
  clrf   Dlay
  clrf   Dlay + 1
  decfsz Dlay, f
   goto  $ - 1
  decfsz Dlay + 1, f
   goto  $ - 3

  return

 end
```

Notice that in the code, I do not receive anything while transmitting. This is because when data is being sent, the DS275 is also passing the sent signal back through the RX lines. This results in the code actually "seeing" its own data coming back as input data and, in the application, converting it and sending it back out. Sending the data back out causes it to be received again. The application begins to operate as if it were in some kind of feedback loop, with no new data getting sent and the current byte never ending. Normally, when I write this algorithm (it is used in the YAP-II), I allow simultaneous (full duplex) sending and receiving. This is not possible in this application because of the operation of the DS275.

To test out the code, I created the stimulus file BASICRS.STI, which follows. This file is also located in the *code\BasicRS* subdirectory of your PC's PICmicro® MCU directory.

```
!  SimpRS.STI - Stimulus File for SimpRS.ASM
!
!  Send an "a" and See How it Gets Processed
!
Step    RB3
1       1                !  Initial Conditions
100     0                !  Start Bit
933     1                !  Bit 0
1767    0
2600    0
3433    0
4267    0                !  Bit 4
5100    1
5933    1
6767    0
7600    1                !  Stop Bit
```

This stimulus file was useful for testing the code, but not for finding the problem with the DS275 passing the sent data back to the receiver. This stimulus file consists of breaking an ASCII *a* into bits and then passing each bit to the application.

When sending ASCII data, I usually pass data this way because it is easier to increment time by one bit and keep the value the same, rather than calculate multiple bits and compress everything together.

If I were to add a second ASCII byte, I would probably start it at Step 10000, copy in the current byte step values with a *1* in front of them, and change the bit values according to the new byte. This eliminates a lot of the grudge work of creating a complex stimulus file in calculating new offsets.

Stimulus files are enabled in the MPLAB IDE by clicking on *Debug, Simulator Stimulus, Pin Stimulus,* and then *Enable . . .* (Fig. 15-139). Once *Enable . . .* is clicked on, you will be given a dialog box from which to select the stimulus file that you want to use. Once the file is selected, it is loaded into memory and used every time the application code is simulated.

Watch out for two things with stimulus files. First, if you reset the stopwatch during simulation, your stimulus file will be picked up as if you had reset the PICmicro® MCU. This always trips me up when I am debugging an application like BasicRS, which has crucially timed loops that are processing input.

Second, if you change the stimulus file and resave it, you have to reload it (using the procedure outlined previously). This trips up a lot of people as they work through their applications and make changes to the stimulus file to better test the application. By forgetting to reload the stimulus file, they will probably start looking at their application code for problems that don't exist.

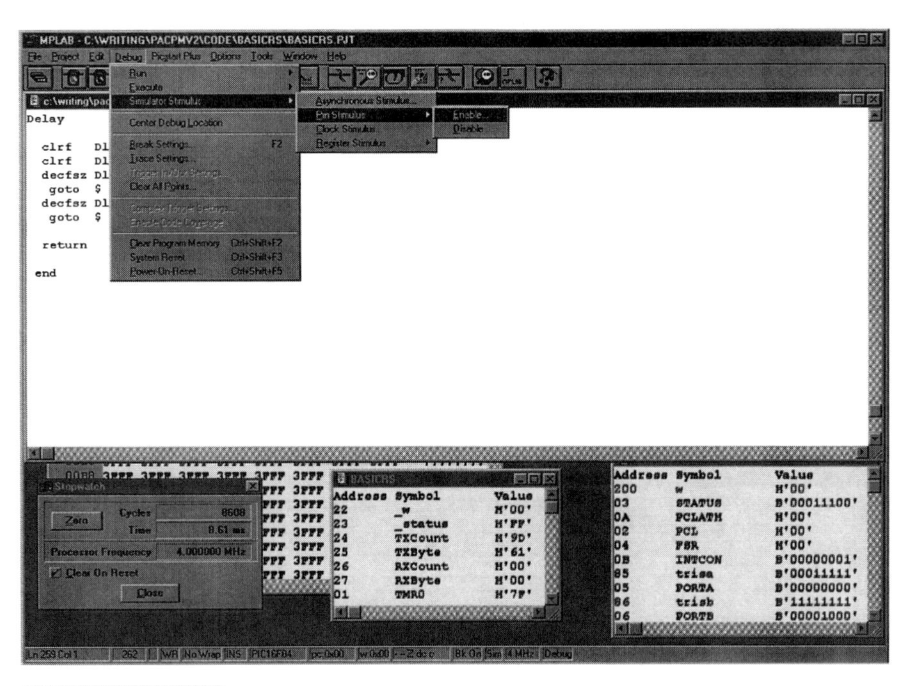

**Figure 15-139** BasicRS stimulus-enable desktop view

The circuit that I used for this application is shown in Fig. 15-140. The breadboard wiring diagram is shown in Fig. 15-141, along with the YAP-II wiring diagram in Fig. 15-142. Notice that the YAP-II circuit takes advantage of the built-in RS-232 level converter. The bill of materials for the experiment is quite modest (Table 15-31).

This was the last experiment that I completed, so, of course, it was the one that had the most problems. I was trying to get this chapter finished on New Year's Eve 1999 so that I

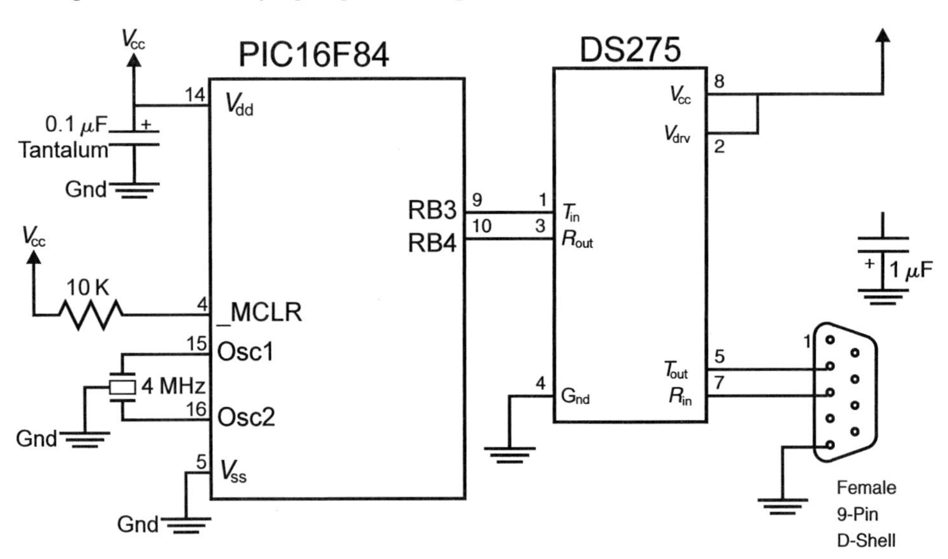


**Figure 15-140** BasicRS: simple RS-232 PICmicro® MCU/serial communications

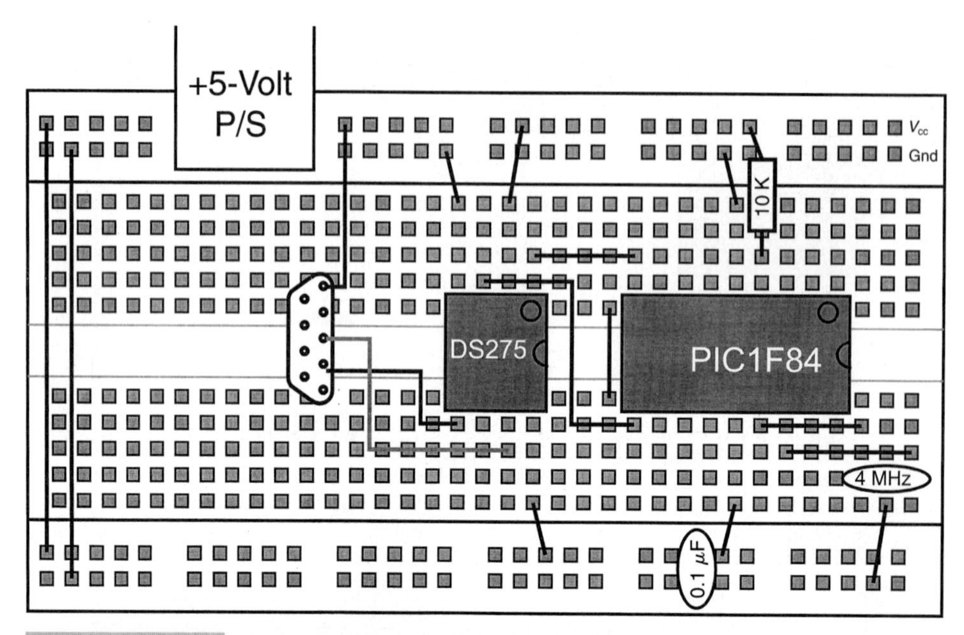


Figure 15-141 BasicRS PIC16F84 breadboard circuit

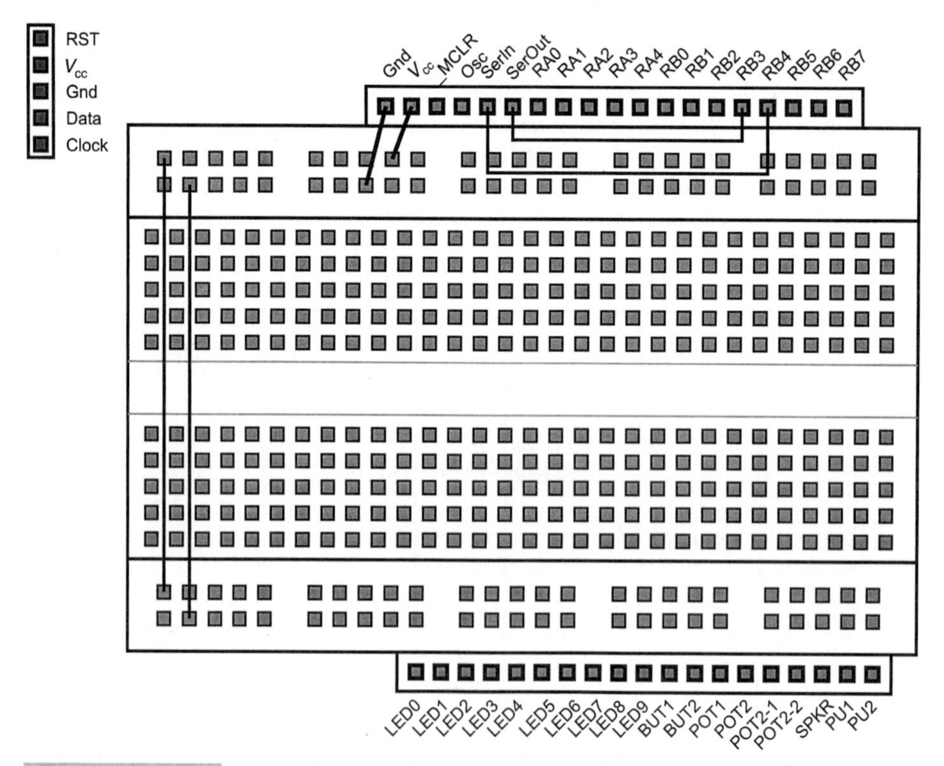


Figure 15-142 EMU-II/YAP-II breadboard for BasicRS

**TABLE 15-31 Bill Of Materials For The "BASICRS" Experiment**

| PART | DESCRIPTION | REQUIRED ON YAP-II |
|---|---|---|
| PICmicro® MCU | PIC16F84-04/P | In Socket |
| Vdd/Vss decoupling capacitor | 0.1-μF tantalum | No |
| _MCLR pull up | 10 K, 1/4 watt | No |
| 4-MHz ceramic resonator | 4 MHz with built-in capacitors | No |
| DS275 | DS275 | No. SERIN/SEROUT |
| Female 9-pin D-shell | DB-9F | No. SERIN/SEROUT |

could save the current manuscript on CD-ROM and avoid any Y2K problems without losing anything and we were starting to get ready to move. This resulted in a lot of pandemonium and three problems with getting the application running. I learned a few things about being in a stressful situation and feeling like you just *have* to get things done.

First, keep your perspective. The crucial aspect of getting the manuscript onto CD-ROM was just that I should not feel like my "back is up against a wall" because something wasn't finished. I should have made a CD-ROM of the current manuscript and then looked at the problem again when things calmed down. CD-ROMs cost only $1.50. If I was able to get the application running before Y2K and written up, I only would have wasted an extra dollar or so and produced a "Frisbee" for my kids to play with.

When I was able to realize that the important thing was to ensure that I would be ready for the deadline and not have my back against the wall, my blood pressure went down a lot and I was able to find and fix the problems.

As stated, I have implemented this bit-banging algorithm on a number of different systems, but it turned out that I had not done any with the DS275. This caused me some problems trying to figure out exactly what the problem was. It was compounded by my not waiting a full *stop* bit's width before sending data back to the PC. This was a problem because the PC's UART could not interpret the data coming back and simply ignored the characters. Because I was using HyperTerminal with it, the errors were not reported back to me.

I was able to find the issues with the *stop* bit with an oscilloscope. When I was able to see that the *stop* bit of the data going back to the PC was only 520 us long (it should be at least 833 us for 1200 bps), then I was able to go back to the code and fix the problem.

Next, I discovered that after I sent the first character, data would be sent repeatedly. This is because of the negative voltage stealing of the DS275. The problem was that as data was being transmitted, it was also being received and resent over and over again in a digital feedback loop. Normally, as stated, when have I implemented this algorithm in a PICmicro® MCU in the past, it was always with a MAX232, so the possible feedback angle wasn't possible. To fix this problem, I stopped polling the incoming data while a byte transmit was occurring.

Last, I discovered that I miswired the DS275. Instead of connecting the $V_{cc}$ Pin (8) to my breadboard $V_{cc}$, I accidentally connected it to ground. I found this when I was trying to figure out the other problems. I discovered that the voltage level was very low (not between zero and five volts, as I expected from my previous experiences with the DS275). This was the keystone to all of the symptoms. When this was fixed, everything fell into place.

The problem was calming down enough to go back and check every connection against the wiring diagram that I produced for it (Fig. 15-141). Once I did this, I was able to find the wiring problem and then worked back to the other two problems. I had them fixed in time to save the manuscript on CD-ROM for Y2K.

## SIMPRS: BIT-BANGING ASYNCHRONOUS SERIAL I/O TEST

As indicated elsewhere in this book, there are other ways to implement an RS-232 without using any chips. This experiment introduces you to a three-resistor/single-transistor method that I do use occasionally (mostly when I don't have any MAX232 or DS275 chips lying around), along with a very simple bit-banging asynchronous serial I/O algorithm. The hardware and software presented here can be used with a low-end PICmicro® MCU and provide a very cheap and easy RS-232 serial communications capability for it.

The circuit shown in Fig. 15-143 is really not much more complicated than the "core circuit" presented at the start of the chapter. Along with the basic PIC16F84 core, just a few, very inexpensive components have been added to give the PICmicro® MCU a negative voltage stealing RS-232 interface. When the host (connected to the female 9-pin D-shell) is not transmitting, it is sending a *mark,* which is a negative voltage. This voltage is used as a negative bias for the transmit back. When a *space* (positive voltage) is to be sent, then the PNP transistor is turned on, driving a positive voltage to the host.

This circuit works very well, except for two things. First, anything that the host sends to the PICmicro® MCU will be "echoed" back because the transmit line is connected to the

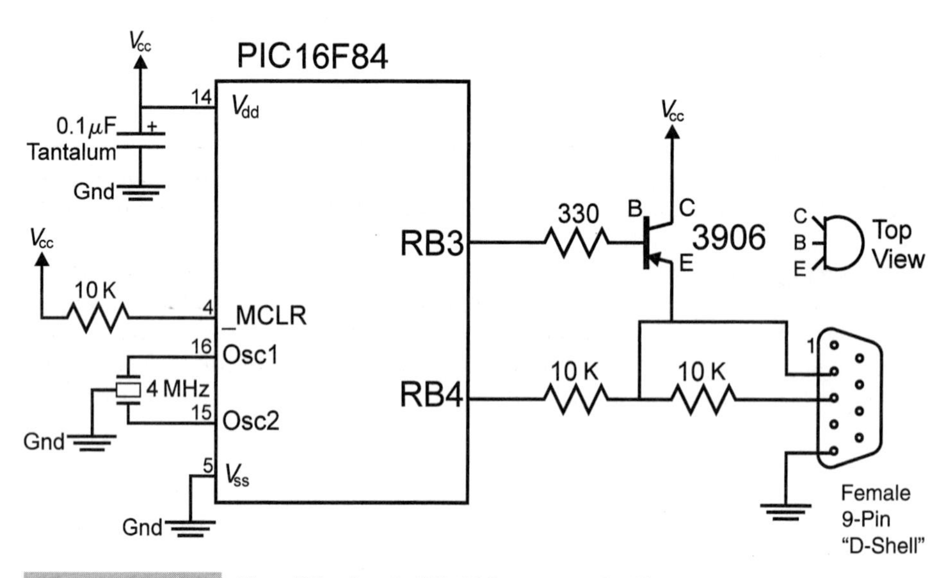


**Figure 15-143** SimpRS: simple RS-232 communications

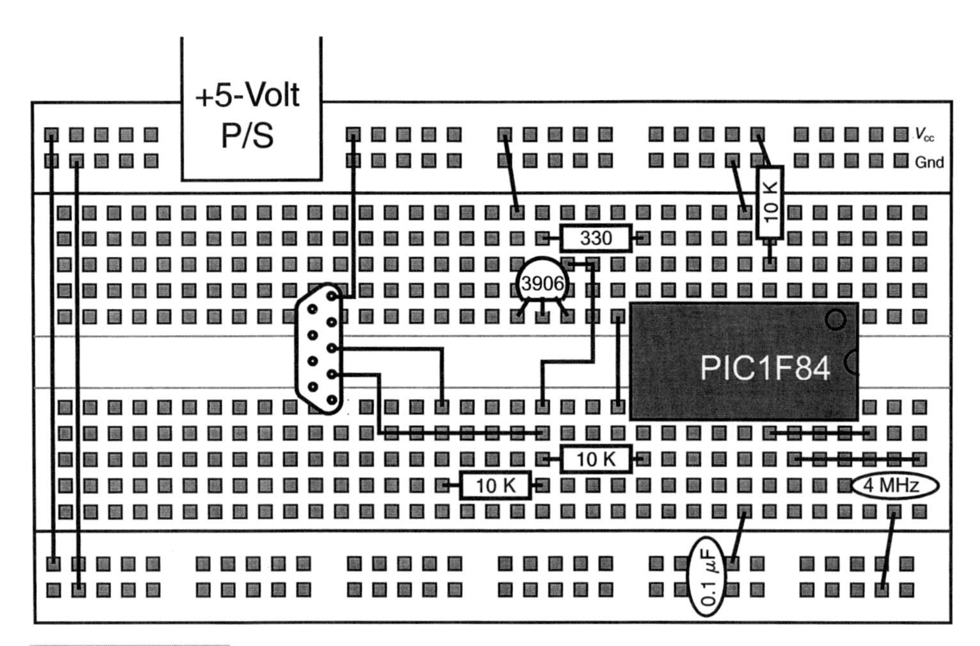


**Figure 15-144** SimpRS PIC16F84 breadboard circuit

receive line. Thus, the host software will have to be tolerant of the characters coming back. This aspect of the circuit can be used to advantage, as shown in the next experiment. The second point with this circuit is that it will pass negative-received data back, but the transmitted data will be positive. This can make things a bit confusing when the PICmicro® MCU software and stimulus files are created and tested.

Earlier in the book, when describing RS-232, I showed the RS-232 level-conversion circuit using a MOSFET transistor, but this circuit shows a 3906 PNP transistor. The bipolar transistor has been specified because it is very cheap and easy to find. Just about any small-signal PNP, NPN, N-channel MOSFET, or P-channel MOSFET can be used, with the appropriate changes made to the software to account for any polarity differences required for the device's operation.

The breadboard circuit wiring (shown in Fig. 15-144) is quite easy to wire and does not require any special components. The bill of materials for SimpRS is shown in Table 15-32. The SIMPRS.ASM application is similar to 3RS.ASM in that once a character is received, if it is ASCII *a* to *z,* it is converted to upper case. SIMPRS.ASM is located in the *code\SimpRS* subdirectory of the *PICmicro* directory.

```
 title  "SimpRS - Simple PICmicro MCU RS-232 Interface"
;
;  This Application executes a "Bit-Banging" RS-232 Interface
;   using a Simple Resistor/Transistor Interface.  This
;   Application Returns a "Capitalized" ASCII Input
;
;
```

**TABLE 15-32 Bill Of Materials For The "BASICRS" Experiment**

| PART | DESCRIPTION |
|---|---|
| PICmicro® MCU | PIC16F84-04/P |
| Vdd/Vss decoupling capacitor | 0.1-μF tantalum |
| _MCLR pull up | 10 K, 1/4 watt |
| 4-MHz ceramic resonator | 4 MHz with built-in capacitors |
| 2 10-K | 10 K, 1/4 watt |
| 330-Ohm resistor | 330 Ohm, 1/4 watt |
| 3906 transistor | 2N3906 transistor |
| Female 9-pin D-shell | DB-9F |

```
;  Hardware Notes:
;   PIC16F84 Running at 4 MHz
;   _MCLR is Pulled Up
;   PORTB.3 is the Transmit Output
;   PORTB.4 is the RS-232 Input
;
;  Myke Predko
;  99.12.30
;
  LIST R=DEC
  INCLUDE "p16f84.inc"

;  Register Usage
 CBLOCK 0x020                 ;  Start Registers at End of the Values
Byte, Count                   ;  Variables for RS-232
Dlay                          ;  Dlay Count
 ENDC

#define  TX PORTB, 3
#define  RX PORTB, 4


 PAGE
 __CONFIG _CP_OFF & _WDT_OFF & _XT_OSC & _PWRTE_ON
 endif

;  Mainline of SimpRS
  org    0

  nop

  bsf    TX                   ;  Start Sending a "1"
  bsf    STATUS, RP0
  bcf    TX                   ;  Enable TX for Output
  bcf    STATUS, RP0

Loop

  btfss  RX                   ;  Wait for a Start Bit
   goto  $ - 1

  call   HalfBitDlay          ;  Wait 1/2 a Bit
```

```
   btfss  RX                ;  Make Sure Bit is Still Low
    goto  Loop

   movlw  8
   movwf  Count
 RXLoop                     ;  Loop Here to Read in the Byte

   call   BitDlay           ;  Wait a Full Byte

   bcf    STATUS, C         ;  Set Carry Accordingly
   btfss  RX                ;  Bit High or Low?
    bsf   STATUS, C

   rrf    Byte, f           ;  Shift in the Byte

   goto   $ + 1             ;  Make 11 Cycles in Loop
   goto   $ + 1

   decfsz Count, f
    goto  RXLoop

   call   BitDlay           ;  Make Sure there is a "Stop" Bit

   btfsc  RX
    goto  Loop              ;  Not High, Then No Byte

   movf   Byte, w           ;  Change Byte to Upper Case
   addlw  255 - 'z'         ;  Get the High limit
   addlw  'z' - 'a' + 1     ;  Add Lower Limit to Set Carry
   btfss  STATUS, C         ;  If Carry Set, then Lower Case
    addlw h'20              ;   Carry NOT Set, Restore Character
   addlw  'A'               ;  Add 'A' to restore the Character
   movwf  Byte

   movlw  10                ;  Send Upper Case Back with Start and Stop
   movwf  Count
   bcf    STATUS, C

 TXLoop

   btfsc  STATUS, C         ;  Send the Bit in "Carry"
    goto  $ + 4
   nop                      ;  Send a "Low"
   bcf    TX
   goto   $ + 3
   bsf    TX                ;  Send a "High"
   goto   $ + 1             ;  6 Cycles in Loop

   call   BitDlay           ;  Wait a Bit

   bsf    STATUS, C         ;  Shift Out the Next Bit into Carry
   rrf    Byte, f

   decfsz Count, f          ;  11 Intrinsic Delays in TXLoop
    goto  TXLoop

   goto   Loop

 BitDlay                    ;  Delay 833 - 15 Cycles (including
                            ;   Call/Return)
   movlw  204
   addlw  0x0FF             ;  Take 1 Away from the Loop
   btfss  STATUS, Z
    goto  $ - 2

   goto   $ + 1

   return
```

```
HalfBitDlay                     ;  Delay (833 - 15) / 2 Cycles

  movlw  100
  addlw  0x0FF                  ;  Take 1 Away from the Loop
  btfss  STATUS, Z
   goto  $ - 2

  return

 end
```

To receive a character, the application polls the input line for a high voltage that indicates a space or *start* bit on the line. After the high has been received, the code waits for half a bit and checks the value again to ensure that a glitch wasn't received. If the bit is good, the data bits are shifted in with one bit period in between them. The Receive algorithm could be written as:

```
char RSReceive()                        //  Receive a character
{

int  i;
int  StartFlag = 0;                     //  Set when Data Received
char Byte;

  while (StartFlag == 0) {              //  Wait for Valid Read
    while (StartFlag == 0) {
      while (RX == Mark);               //  Wait for a "Start Bit"
      HalfBitDlay();                    //  Wait half a bit
      if (RX == Space)
        StartFlag = 1;                  //  Still Space, not a glitch
    }

    for (i = 0; i < 9; i++) {           //Read in the Bits
      Byte = (Byte >> 1) + RX;          //  Add in the Bit
      BitDlay();                        //  Delay to the Next Bit
    }

    if (RX != Mark)                     //  Is there a Stop Bit?
      StartFlag = 0;                    //   No - Ignore Byte
  }

  return Byte;

}  //  end RSReceive
```

When you look back at the serial receive code, you probably will be hard pressed to see the complexity I've put in this pseudo-code, but the PICmicro® MCU assembly language actually carries out the functions described here. If a glitch is received or if the *stop* bit is not active, then the byte is "thrown out" and RSReceive continues to wait for a valid serial character.

Notice that this routine only returns when a character has been received and will not return under any other circumstances. This might be unacceptable in some PICmicro® MCU applications. A timer (counter) could be implemented to time out the wait for data to be sent to the PICmicro® MCU and other functions executed before returning to the serial receive code.

The serial transmission function is much simpler, really just consisting of a timed shift out, based on the least-significant bit. For the transmit function, I do fool around with the data a bit to ensure that *start* and *stop* bits are shifted out as part of the data. This function could be modeled as:

```
RSTransmit(char Data)               //  Transmit a Byte of Data
{
int  i;
int  Byte;

  Byte = 0x0FE00 + (Data << 1);    //  Set up Data to Shift Out

  for (I = 0; I < 10; I++) {       //  Shift out the Data
    TX = Byte & 1;                 //  Shift out the LSB
    Byte = Byte >> 1;              //  Shift down the Data for the
    BitDlay();                     //   Next Bit
  }

}  //  end RSTransmit
```

Notice that I only use one BitDlay subroutine. In the case of the receive code, I have added four number cycles to the loop to make it equal to the transmit code. By doing this, the software can be used for a variety of different data speeds. Both routines take 11 cycles to execute. To change them to a 9600-bps routine, I just have to change BitDlay to execute in 93 cycles, instead of the 822 cycles that it executes in for the application.

The last point to touch on for this application is how it interfaces to the host. Figure 15-145 shows a screen shot of a 1200-bps HyperTerminal session that is communicating with the SimpRS application. In the screen shot, you can see the lower-case letter being sent to the PICmicro® MCU and the PICmicro® MCU returning with the upper-case character. To test the application, I have also sent numbers, special characters, and upper-case characters to ensure that there are no problems with the application.

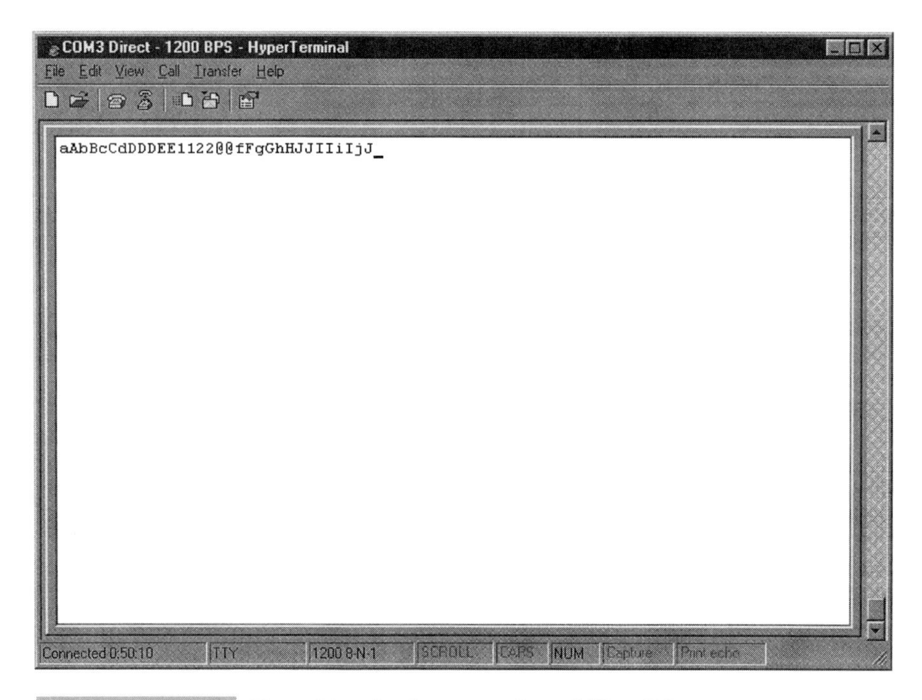


**Figure 15-145** HyperTerminal screen shot of SimpRS

In the screen shot shown in Fig. 15-145, you have to remember that as far as the host (which is a PC, in this case is concerned), the lower-case character has come back from the PICmicro® MCU and is not echoed within the HyperTerminal application. For some applications this can be a problem and can be confusing for the user.

For other applications, such as the one that follows, it can be used to advantage.

## 3RS: DETECTING A PICmicro® MCU USING A 3-WIRE RS-232 INTERFACE

I wanted to end the experiments with one that demonstrates how easy it is to interface a PICmicro® MCU with a Microsoft Windows application. If you take a look at the code for both the PC and the PICmicro® MCU, you'll probably feel like the application is quite complex. But as I go through the operation, you will see how easy it is to interface a PICmicro® MCU serially to PC applications.

The experiment itself uses the PICmicro® MCU application hardware covered in the previous experiment (Fig. 15-143) is used to implement a serial connection detect protocol, as covered earlier in the book. The circuit wiring diagrams for the application are presented in the previous experiment. The bill of materials for 3RS is shown in Table 15-33.

The serial data connect uses the "Ping" described earlier in the book, in which an echoed (by hardware) 0x0FF character is changed on its receive to something different. Figure 15-146 shows this operation with the output byte of 0x0F0 changed to 0x090. In this application, a data byte of 0x0FF is changed to another character. 0x0FF was chosen because it is not a valid ASCII symbol.

In the PICmicro® MCU application, data byte 0x0FF is waited for. If it is received, the next byte (which should be 0x0F0) is "modified" by changing the character that is echoed back to the host computer (a PC, in this case). An LED connected to PORTA.2 (RA2) lights if the connection is active (i.e., 0x0FF characters are being received). The 3RS.ASM application can be shown as the pseudo-code:

**TABLE 15-33 Bill Of Materials For The "3RS" Experiment**

| PART | DESCRIPTION |
|---|---|
| PICmicro® MCU | PIC16F84-04/P |
| Vdd/Vss decoupling capacitor | 0.1-$\mu$F tantalum |
| _MCLR pull up | 10 K, 1/4 watt |
| 4-MHz ceramic resonator | 4 MHz with built-in capacitors |
| 220-Ohm resistor | 220 Ohm, 1/4 watt |
| LED | LED |
| 2 10-K resistors | 10 K, 1/4 watt |
| 330-Ohm resistor | 330 Ohm, 1/4 watt |
| 3906 transistor | 2N3906 transistor |
| Female 9-pin D-shell | DB-9F |

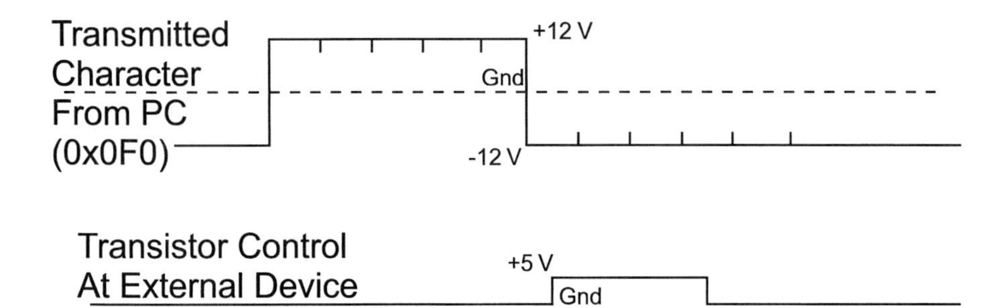


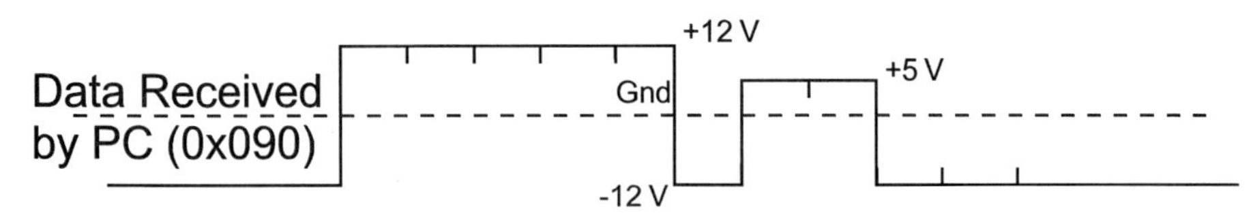


**Figure 15-146** RS-232 Ping using transistor/resistor TTL/CMOS voltage-conversion circuit

```
main()                      //  3RS Pseudo-Code, Used to
{                           //   Test and Establish Link to PC Host

int  i;
char Data;

  LED = off;                //  No Connection, LED Off

  while (1 == 1) {          //  Loop Forever

    for (i = 0; (i < 600msec) && (SERIN == MARK);  i++);
                            //  Poll Serial In Line for "Start" Bit
    HalfBitDlay();
    if (SERIN == MARK)      //  Timeout - Nothing Received
      LED = off;            //  Turn off the LED/No Connection
    else {                  //  Something Received

      for (I = 0; I < 8; I++) {
        Data = (Data >> 1) + (SERIN << 7);
        BitDlay();
      }                     //  Read Incoming Byte

      if (Data = 0x0FF){    //  0x0FF Received
        LED = on;           //  Turn on LED to Indicate Data Incoming

        for (i = 0; (i < 600msec) && (SERIN = MARK); i++);
                            //  Poll Serial Line for Second Byte

        HalfBitDlay();
        if (SERIN == MARK)//  Timeout - Nothing Received
          LED = off;        //  Turn off the LED
        else {              //  Send Back "Synch"

          for (i = 0; i < 5; i++ )
            BitDlay();      //  Wait Past First Four Bits (+ Stop)

          SEROUT = SPACE;  //  Change the Data Going Out
```

```
          For (I = 0; I < 3; I++)
            BitDlay();      //  Wait for 3 Bits

          SEROUT = MARK;    //  Don't Change Anything Else

          BitDlay();        //  Don't Change the Stop Bit

        }                   //  End, Second 0x0FF Received
      }                     //  End, First 0x0FF Received
    }                       //  End, First Byte Received
  }
} //  end 3RS
```

The actual PICmicro® MCU assembly-language code is 3RS.ASM, which is located in the *code\3RS* subdirectory of the *PICmicro* directory.

```
 title  "3RS - Simple 3-Wire Communication Protocol"
;
;  This Application Implements a simple Communication
;   protocol.  The Link is assumed to be down unless the
;   the character 0x0FF is received.  When this character
;   is received, then the next character (which is also
;   0x0FF), is received, the PICmicro MCU, Changes bits 4
;   through 6 from "1" to "0" and sends it back to the
;   "host".  The "RSReceive" and "RSTransmit" were taken
;   from "SimpRS".
;
;
;  Hardware Notes:
;   PIC16F84 Running at 4 MHz
;   _MCLR is Pulled Up
;   PORTB.3 is the Transmit Output
;   PORTB.4 is the RS-232 Input
;   PORTA.2 is an LED indicating the Link is Up
;
;  Myke Predko
;  99.12.30
;
  LIST R=DEC
  INCLUDE "p16f84.inc"

;  Register Usage
 CBLOCK 0x020                  ;  Start Registers at End of the Values
Byte, Count                    ;  Variables for RS-232
Dlay:2                         ;  Dlay Count
OnCount, OffCount              ;  Reset On/Off Counts
 ENDC

#define  TX  PORTB, 3
#define  RX  PORTB, 4
#define  LED PORTA, 2

 PAGE
 __CONFIG _CP_OFF & _WDT_OFF & _XT_OSC & _PWRTE_ON

;  Mainline of SimpRS
  org    0

  nop

  bsf    LED                   ;  LED is "Off"
  bsf    TX                    ;  Start Sending a "1"
  bsf    STATUS, RP0
  bcf    LED                   ;  Enable the LED Output
```

```
  bcf    TX                  ; Enable TX for Output
  bcf    STATUS, RP0

Loop

  clrf   Dlay                ; Put in a 1/2 Second Delay
  clrf   Dlay + 1

  btfsc  RX                  ; Wait for a Start Bit
   goto  HaveBit             ; If Line High, Have the Start Bit
  goto   $ + 1
  goto   $ + 1
  decfsz Dlay, f             ; Loop 64K Times
   goto  $ - 5
  decfsz Dlay + 1, f
   goto  $ - 7

  bsf    LED                 ; No Start Bit, Start All Over

  goto   Loop

HaveBit

  call   HalfBitDlay         ; Wait 1/2 a Bit

  btfss  RX                  ; Make Sure Bit is Still Low
   goto  Loop

  movlw  8
  movwf  Count

SynchRXLoop                  ; Loop Here to Read in the Byte

  call   BitDlay             ; Wait a Full Byte

  bcf    STATUS, C           ; Set Carry Accordingly
  btfss  RX                  ; Bit High or Low?
   bsf   STATUS, C

  rrf    Byte, f             ; Shift in the Byte

  goto   $ + 1               ; Make 11 Cycles in Loop
  goto   $ + 1

  decfsz Count, f
   goto  SynchRXLoop

  call   BitDlay             ; Make Sure there is a "Stop" Bit

  bsf    LED                 ; Turn Off LED Just in Case

  btfsc  RX
   goto  Loop                ; Not High, Then No Byte - Start Over

  movf   Byte, w
  xorlw  0x0FF               ; Is the Value Received 0x0FF?
  btfss  STATUS, Z           ; No, LED off, Keep Waiting
   goto  Loop

  bcf    LED                 ; Indicate the Link is Up

  clrf   Dlay                ; Put in a 1/2 Second Delay
  clrf   Dlay + 1

  btfsc  RX                  ; Wait for a Start Bit
   goto  Have2Bit            ; If Line High, Have the Start Bit
  goto   $ + 1
  goto   $ + 1
  decfsz Dlay, f             ; Loop 64K Times
   goto  $ - 5
```

```
  decfsz Dlay + 1, f
   goto  $ - 7

  bsf    LED               ;  No Start Bit, Start All Over

  goto   Loop

Have2Bit

  call   HalfBitDlay       ;  Wait 1/2 a Bit

  bsf    LED

  btfss  RX                ;  Make Sure Bit is Still Low
   goto  $ - 1

  bcf    LED

  movlw  9
  movwf  Count
  movlw  5                 ;  Start at Bit 4 Coming Back
  movwf  OnCount
  movlw  3
  movwf  OffCount

RXLoop                     ;  Loop Here to Read in the Byte

  call   BitDlay           ;  Wait a Full Byte
  decfsz OnCount, f
   goto  $ + 6             ;  No - Decrement and Continue
  decfsz OffCount, f       ;  Keep Bit Off?
   goto  $ + 7
  bsf    TX                ;  Put it On
  bsf    OnCount, 7        ;  Yes, Make Sure Never Goes Off Again
  goto   $ + 7

  nop                      ;  Not at Turn Off Yet
  goto   $ + 1
  goto   $ + 4

  bcf    TX                ;  Output a Low Bit
  bsf    OnCount, 0        ;  Make Sure it Happens Again
  nop

  decfsz Count, f
   goto  RXLoop

  goto   Loop

BitDlay                    ;  Delay 833 - 15 Cycles (including
                           ;   Call/Return)
  movlw  204
  addlw  0x0FF             ;  Take 1 Away from the Loop
  btfss  STATUS, Z
   goto  $ - 2

  goto   $ + 1

  return

HalfBitDlay                ;  Delay (833 - 15) / 2 Cycles

  movlw  100
  addlw  0x0FF             ;  Take 1 Away from the Loop
  btfss  STATUS, Z
   goto  $ - 2

  return

 end
```

To understand this code, I recommend going back over the pseudo-code and try to follow the data coming in. I tried to avoid using any definite values in the pseudo-code, simply because they can be confusing (especially with the inverted serial input and positive serial output).

You might also want to go over the previous experiment (SimpRS) to better understand the operation of the serial data reception. The code I used here is a simple bit-banging protocol in which after the middle of the *start* bit has been established, I delay one bit cycle before reading in the next bit.

After writing the application, I created the stimulus file (3RS.STI), which sends two 0x0FF characters so that I could test the PICmicro® MCU application before connecting it to the PC host.

```
!  3RS.STI - Stimulus File for 3RS.ASM
!
!  Send two 0x0FF and see how the Application Responds
!
Step      RB4
1         0                  !  Initial Conditions
100       1                  !  Start Bit
933       0                  !  Bit 0
1767      0
2600      0
3433      0
4267      0                  !  Bit 4
5100      0
5933      0
6767      0
7600      0                  !  Stop Bit
          !
10100     1                  !  Start Bit
10933     0                  !  Bit 0
11767     0
12600     0
13433     0
14267     0                  !  Bit 4
15100     0
15933     0
16767     0
17600     0                  !  Stop Bit
```

I have used Microsoft's Visual Basic as the application development environment for the host code in this experiment. I have included the Visual Basic source (3RS VISUAL BASIC.FRM) and project (3RS VISUAL BASIC.VBP). The required files are located in the *code\3RS\vb* subdirectory of the *PICmicro* directory. The Basic source code is:

```
' 3RS Visual Basic
'
'  This Application Polls a Serial Port for an Active
'  PICmicro MCU Connected to it.
'
'  When "Start" is Clicked, 0x0FF, followed by 0x0FF is
'  Sent out of the Serial Port.  The two Values are
'  checked for a Valid Return (which means the Second
'  0x0FF is Scrambled by the PICmicro MCU).
'
'  If nothing is returned, then the "No Connect"
'   (Label(0)) is Active, if 0x0FF and no change, then
```

```
'   (Label(1)) is Active, if 0x0FF is returned and
'   Mangled, then (Label(2)) is Active.
'
'  99.12.30
'  Myke Predko
Dim oldCombo1 As Integer    '  Flag of Old Serial Port
```

```
Private Sub Combo1_Click()
'  Combo1 Box is Clicked on.  Can Only Change if "Start" is in
'   Command2
  If (Command1.Caption = "Start") Then
    oldCombo1 = Combo1.ListIndex + 1
  End If

End Sub
```

```
Private Sub Command1_Click()
'  "Quit" Button

  If (Command2.Caption = "Stop") Then
    MSComm1.PortOpen = False
  End If

  End

End Sub
```

```
Private Sub Command2_Click()
'  "Start"/"Stop" Button is Active

  If (Command2.Caption = "Start") Then
    MSComm1.CommPort = Combo1.ListIndex + 1
    On Error GoTo invalidCommPort
    MSComm1.PortOpen = True
    Timer1.Enabled = True
    Command2.Caption = "Stop"
  Else                        '  Stop the Action
    MSComm1.PortOpen = False
    Timer1.Enabled = False
    Command2.Caption = "Start"
  End If

  Exit Sub

invalidCommPort:              '  Notify of Invalid Selection
  MsgBox "Selected CommPort Could NOT be Opened.  Try Another", vbExcla-
mation, "Invalid CommPort"

  Exit Sub

End Sub
```

```
Private Sub Timer1_Timer()
'  1/2 Second "Ping" Timer

  MSComm1.Output = Chr$(&HFF)

  If (MSComm1.InBufferCount <> 0) Then
    temp$ = MSComm1.Input
    If (temp$ = Chr$(&HFF)) Then
      If (Label1(2).BackColor <> &HFFFF&) Then
        Label1(1).BackColor = &HFFFF&
        Label1(0).BackColor = &HFFFFFF
        Label1(2).BackColor = &HFFFFFF
      End If
    Else                    '  Data Scrambled
```

```
      Label1(2).BackColor = &HFFFF&
      Label1(0).BackColor = &HFFFFFFF
      Label1(1).BackColor = &HFFFFFFF
    End If
  Else                     '  Nothing Received
    Label1(0).BackColor = &HFFFF&
    Label1(1).BackColor = &HFFFFFFF
    Label1(2).BackColor = &HFFFFFFF
  End If
End Sub
```

```
Private Sub Form_Load()
  Combo1.AddItem "COM1", 0
  Combo1.AddItem "COM2"
  Combo1.AddItem "COM3"
  Combo1.ListIndex = 0

  Label1(0).BackColor = &HFFFF&
  Label1(1).BackColor = &HFFFFFFF
  Label1(2).BackColor = &HFFFFFFF

End Sub
```

The Visual Basic application is available in both source-code format and as a package that can be installed on your PC without requiring compilation. Execute SETUP.EXE in the *code\3RS\vb\Package* subdirectory of the *PICmicro* directory to load the 3RS Visual Basic executable application code, along with all the necessary device drivers for the RS-232 interface.

I always feel like the source code in Visual Basic (and any Visual development environment) is only half the story. In Fig. 15-147, the dialog box or form that I created for this application is shown and has a "combo" box to select the serial (COM). These controls are used with Start/Stop, Quit, and Button controls and three text boxes that indicate the current state of the link (which are all visible). In the active form (Fig. 15-148), you can't see the Timer and MSComm serial-port interface that I have added to control the application.

The timer even handler (known as the *Timer1* control in the 3RS Visual Basic application) is probably the most important control in the Visual Basic application because it is responsible for periodically sending a *0x0FF* out on the specified serial port. *Timer1_Timer* executes every time that *Timer1* overflows, which happens every 500 ms, and sends a *0x0FF* character out on the specified COM port. If a character is received, it is checked as either *0x0FF* or something else (in which case the Active box is made yellow). If nothing is received, the *No Connect* box is yellow. If *0x0FF* is received, then *Connect* is returned.

This application violates one of the basic tenants that I try to establish for connecting the PICmicro® MCU to host computers via RS-232. A terminal emulator (like HyperTerminal) cannot be used for debugging the PICmicro® MCU application itself. Despite this, I did not have any trouble testing and debugging the interface.

If you have trouble with the interface, the *ping* character should be changed from *0x0FF* to something easy to add (e.g., *0x061,* ASCII *a*) to see how the application works. If this is done to help debug the application, be sure that characters are sent within a half second. Otherwise, the PICmicro® MCU application will time out and the second character will not be changed (essentially the process is happening all over).

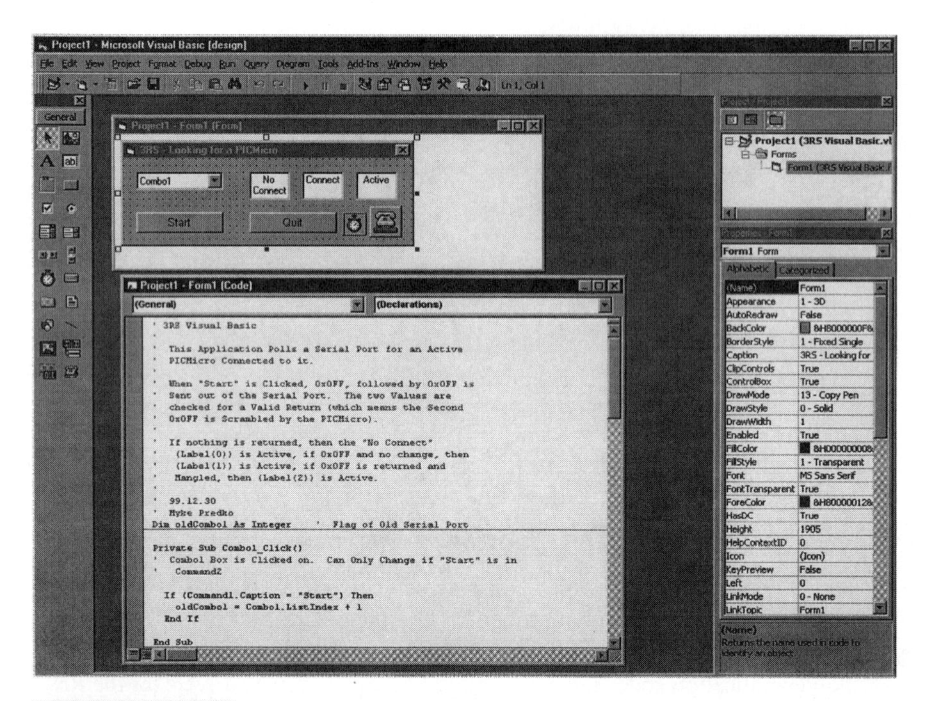


**Figure 15-147** Screen shot of 3RS VB desktop

# Debugging

I suspect that as you've worked through the experiments, you have made more than a few wiring or PICmicro® MCU programming mistakes along the way. If so, you have had to try and figure out what you did wrong. Chances are, you weren't very efficient at finding the problems and you might have spent a few hours tearing your hair out trying to find them.

**Figure 15-148** Screen shot of 3RS VB active form

Earlier in the book I presented the idea of performing failure analysis on any problems that you encounter. The steps I presented are:

1 Characterize the problem.
2 Hypothesize what the problem is.
3 Test your hypothesis.
4 Apply the fix according to your hypothesis and evaluating whether or not the application's operation changes as you expected.

The final experiment contains a number of problems that you will very likely encounter in real life. The only difference between the experiment and real life is that in the experiment, I explain how you can find the problems.

## DEBUG: AN APPLICATION WITH SOME SUBTLE PROBLEMS

After working through more than 40 experiments, I'm sure that you have experienced a few problems along the way with programming the PICmicro® MCU or building a circuit incorrectly. I'm also pretty sure that you spent a lot of time backtracking, trying to figure out the actual problem. For the last experiment, I want to give you a bit of a test to see how efficient you are at finding problems and fixing them.

The bill of materials for Debug is shown in Table 15-34. For this experiment, I would like you to build the circuit shown in Fig. 15-149 using the breadboard layout shown in Fig. 15-150.

Once the circuit is built, load your PIC16F84 with DEBUG.ASM (which is located in the *code/debug* subdirectory of the *PICmicro* directory):

```
 title  "debug - An application with a few problems"
;
;  This is an application to demonstrate how insidious some problems
;   can be.  This application should be burned into a PICmicro® MCU after
;   assembly to see if the problems built into it can be found.
;
;  The application is *supposed* to turn on a LED at RB1 and wait for
;   a button to be pressed.  When it is, an LED at RB0 should be turned
;   on as well.
;
;  Hardware Notes:
;   PIC16F84 running at 4 MHz
;   _MCLR is tied through a 4.7K Resistor to Vcc and PWRT is Enabled
;   A 220 Ohm Resistor and LED is attached to PORTB.0/PORTB.1 and Vcc
;   A 10K pull up is connected to RA0 with a Momentary on Switch
;
;  Myke Predko
;  99.12.07
;
  LIST R=DEC
  INCLUDE "p16f84.inc"

;  Registers
 __CONFIG _CP_ON & _WDT_ON & _XT_OSC & _PWRTE_ON

  PAGE
;  Mainline of debug
```

**TABLE 15-34 Bill Of Materials For The "3RS" Experiment**

| PART | DESCRIPTION |
|---|---|
| PICmicro® MCU | PIC16F84-04/P |
| Vdd/Vss decoupling capacitor | 0.1-μF tantalum |
| _MCLR pull up | 10 K, 1/4 watt |
| 4-MHz ceramic resonator | 4 MHz with built-in capacitors |
| RA0 pull up | 10 K, 1/4 watt |
| RA0 switch | Momentary on |
| RB0, RB1 LED | Red LED |
| RB0, RB1 LED resistor | 220 Ohm, 1/4 watt |

```
 org     0

  nop

  movlw  0x001              ;  LED at RB1 is On/RB0 is Off
  movwf  PORTB

  bsf    STATUS, RP0        ;  Goto Bank 1 to set Port Direction
  movlw  0x0FC              ;  Set RB0/RB1 to Output
  movwf  TRISB ^ 0x090
  bcf    STATUS, RP0        ;  Go back to Bank 0
Loop

  btfsc  PORTA, 0           ;  Wait for RA0 Button to be Pressed
   goto  Loop
```

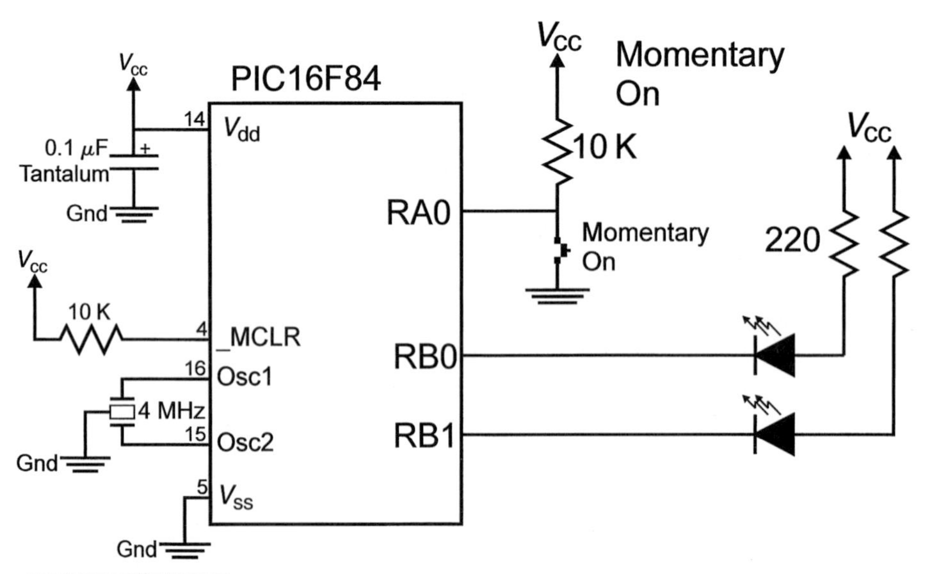


**Figure 15-149** **Debug: turn on LED on RB1 at power up, LED on RB0 when button pressed**

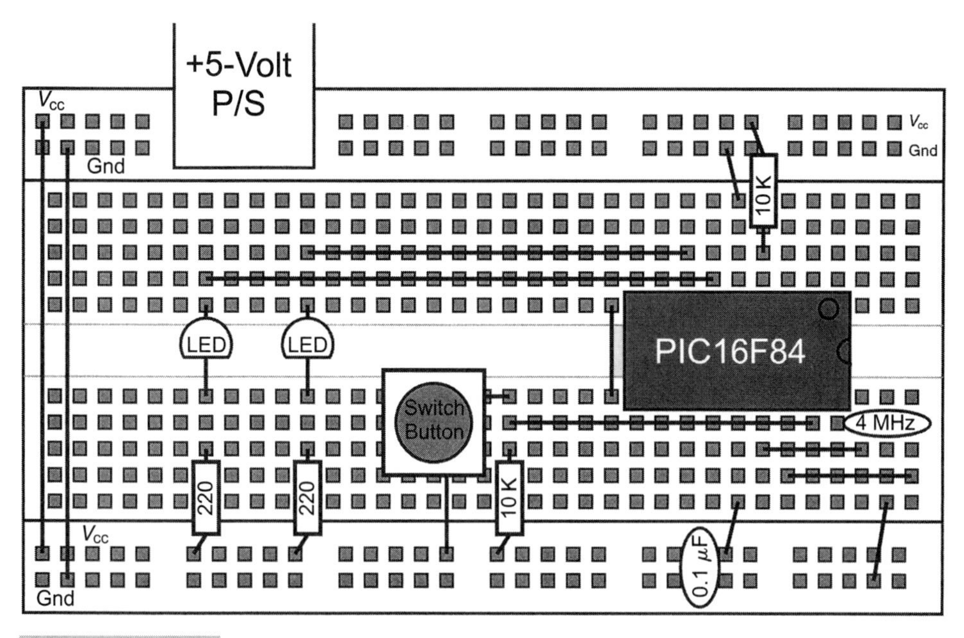


**Figure 15-150** Debug PIC16F84 breadboard circuit

```
    clrf    PORTA              ;  Set RB0 = RB1 = 0 for Both LEDs on
    goto    Loop               ;  Loop Forever

  end
```

This application is designed to turn on an LED at RB1 and wait for the button at RA0 to be pressed (pulled down low). When the button is pressed, the LED at RB0 will be turned along with the button at RB1.

Without simulating this application, I would like you to build it and put in a programmed PIC16F84. Next, I would like you to try and get the application to execute properly. If you have read through the preceding chapters and worked through all the experiments, you should be able to get this application running without too many problems.

So that you don't cheat, I've asked that this book be laid out in such a way that you can't see my comments on finding the problems on the following pages. When you think you have found all of the problems, turn the page to see the problems that I put in the application and how I think you should have approached finding them.

How did you do? You should have found two wiring errors, three definite errors with the source code, and one questionable instruction. The problems are very representative of what you will see both from your own applications and designs that you pick up from other people.

Really, the biggest question should be: how efficient were you at finding the problems? I would expect that someone very experienced with the PICmicro® MCU could find them all within about five minutes. If you spent an hour or more trying to find all the problems, don't feel bad about it. Finding and eliminating problems is one of the biggest skills that you will have to learn when working with the PICmicro® MCU and the software.

After building the circuit shown in Fig. 15-150, when you apply power to the circuit, neither LED will turn on. First, check:

**1** Power
**2** Reset
**3** Clocking

If you built the circuit correctly, you will find that these three potential problems are nonissues. Power is going to the correct pins, Reset is a 10-K pull up on the _MCLR pin and the ceramic resonator is wired correctly. If you check the ceramic resonator with an oscilloscope or a logic probe, you will find that it is oscillating correctly at 4 MHz.

With these three problems out of the way, where do you go next? Personally, I would start by looking at the pins to which the LEDs are connected. With a logic probe, check to see if either one is in Output mode (indicated by the high or low LED and audible tone of the logic probe) after disconnecting the LEDs.

In both cases, you will find that neither pin is in Output mode. You have just characterized the first problem with the application: I/O pins are not in Output mode. What is your hypothesis?

For this problem, I would say that the PORTB TRIS register is not getting written to correctly. To confirm this problem, I would add the P16F84.WAT file and start simulating the application (Fig. 15-151).

In Fig. 15-151, notice that I have single stepped through the application to the instruction after the write to the TRISB register. At this point, the TRISB register is equal to 0b011111111, which indicates that all the bits are in Input mode and none is in Output mode. This is unexpected because the value *0x0FC* (0b011111100) has been written into w and should have been stored in TRISB.

To figure out what is happening, you should re-run the simulation and look at how the *movwf TRISB ^ 0x090* instruction executes. If the TRISB register is being written to, then the whole *trisb* line should turn red, but it doesn't. This indicates that the write to the register is not successful. In this case, you will probably check to see that the RP0 is in the right state (it is). Next, you should check the *File Register* window to see if the write has gone awry. As you can see in Fig. 15-152, address 0x016 should have the value *0x0FC,* which is not expected.

For some reason, the code is writing *0x0FC* to address 0x016 instead of address 0x006 of Bank 1. Remembering that all the file registers of Bank 0 are shadowed in Bank 1, and knowing that RP0 was set correctly, you can deduce that the write was actually to address 0x096 and not 0x086 (which is the address of TRISB).

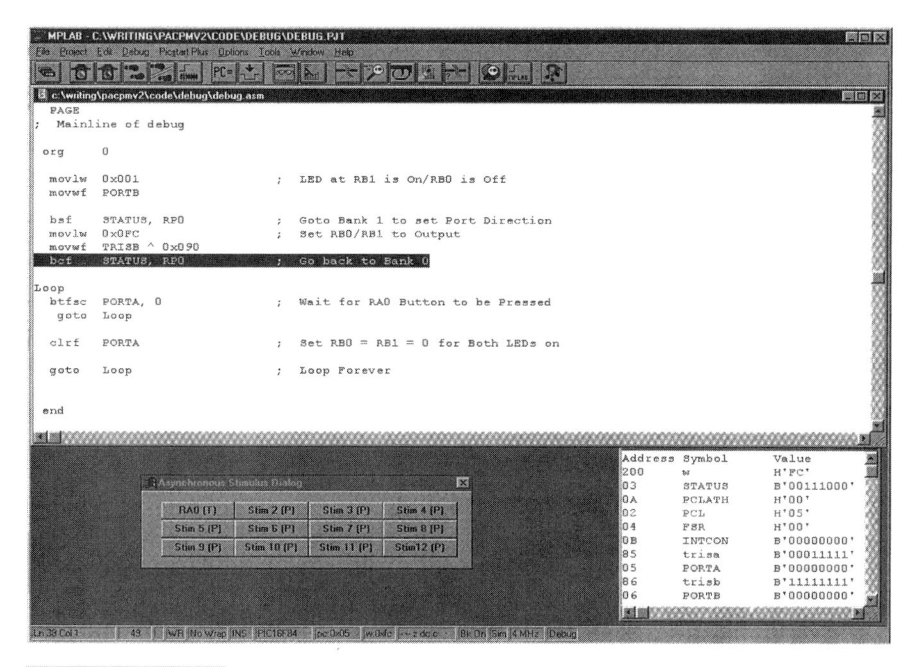


**Figure 15-151** "Debug" MPLAB simulation

Going back over the code, you'll see that the write is:

```
movwf TRISB ^ 0x090
```

and not

```
movwf TRISB ^ 0x080
```

as is required. This typo is a very common mistake that you will probably make more than a few times as you work with the PICmicro® MCU. Now, you can hypothesize that the RB0 and RB1 bits are not going into Output mode because the address being written to is incorrect.

Now, you can make the change to the source, simulate it, and confirm that the *0x0FC* value is going into TRISB and then program the PIC16F84 with the updated code to see if

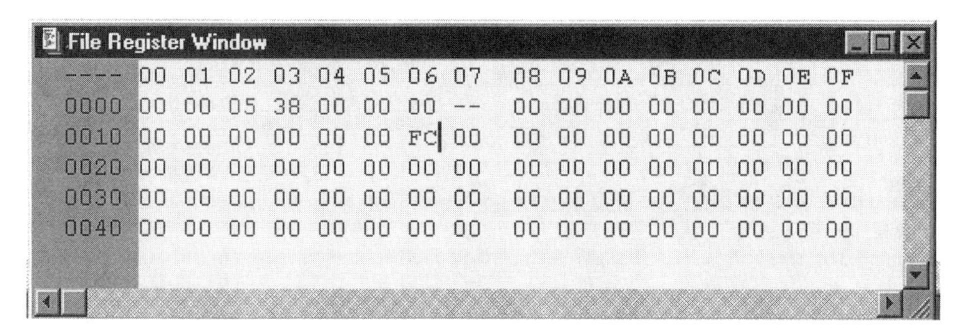


**Figure 15-152** MPLAB PIC16F84 file register display

this fixes the problem. If you put a logic probe on the PICmicro® MCU in the circuit, you'll see that these pins are now going into Output mode. But, if you connect the LEDs as shown, you'll find that they still don't light.

To find problems with LEDs, I always first check to see if they can light. I often find that I install them backwards and don't check the application closely enough. The LEDs can be checked by disconnecting the PICmicro® MCU connection and connecting the LED circuit to ground. When the connection to ground is made, the LEDs should light.

In this case, they won't. The obvious check (reversing the LEDs) will fix the problem. This fix is shown in Fig. 15-153. This type of error is very common with published graphics (and I hope and pray that I haven't made any in the circuit diagrams presented in this book).

Now, when you apply power, the LED on RB1 lights, but the LED on RB0 still does not light. If you disconnect the LED and check the output value on RB0, you'll see that it remains high—even after the button is pressed. After checking the switch and pull-up connections (which are correct in Fig. 15-153), you will have to go back to the source code and simulate some more to see if you can find the problem.

As you can see in Fig. 15-151, I have set up RA0 as an asynchronous input that toggles each time the button is pressed. As you simulate the code, you will see that bit zero of PORTB never changes even though you have changed RA0 from a high to a low and back again.

The problem here is another typo. Change the line:

```
clrf PORTA
```

to:

```
clrf PORTB
```

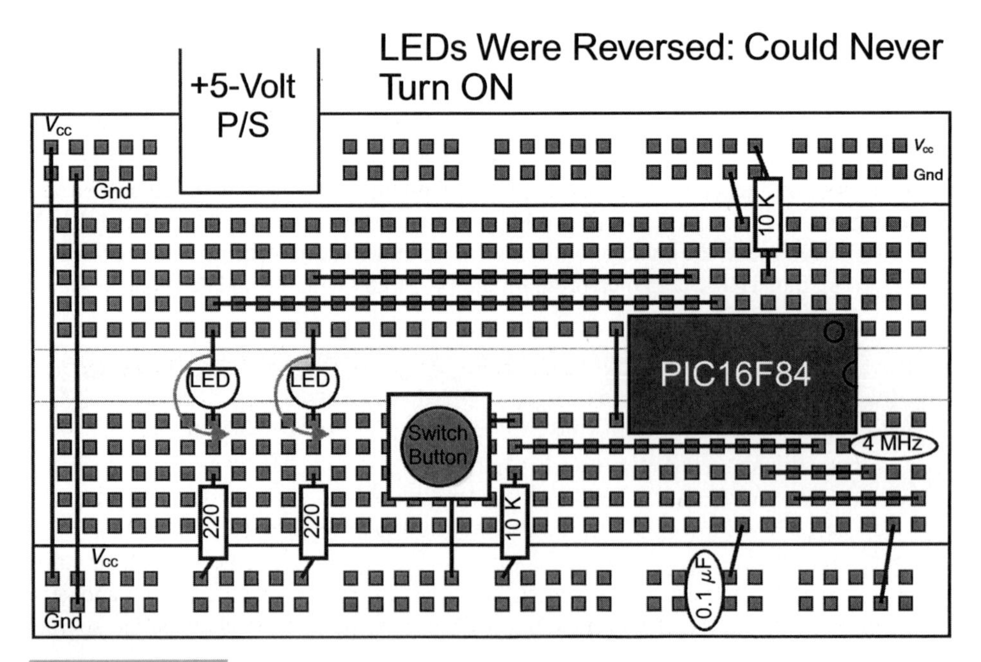


**Figure 15-153** Fixed debug PIC16F84 breadboard circuit

and see what happens. In the simulator, you will see RB0 now going low. When you program the PICmicro® MCU with the new code, you will see that the application is now working as you expected . . .

Well, almost. One strange problem is that a few moments after the button is pressed, the LED turns off. It will turn on again after the button is pressed, but periodically, it will turn back off.

At this point, most people would say that it is a reset problem. If you try to simulate the application, chances are you won't see any problem with the code—even if you simulate for a long period of time, up to a second or so (which takes an unreasonable amount of time in MPLAB to get to).

What's going on? The problem is that the watchdog timer is enabled. In most PICmicro® MCUs, leaving the *WDT* bit set in the configuration fuses enables the watchdog timer. In the DEBUG.ASM code, I deliberately enabled it because I thought that leaving it out would be more noticeable than enabling it.

The most common problem with the watchdog timer and other configuration fuse-enabled features (such as the PIC16F87x's LVP and Debug mode) is that they are enabled when the bit is set. Forgetting to put in parameter settings for all of the configuration fuses can lead you to some problems where the device won't work or won't work 100% properly (as in this case). As covered elsewhere in the book, you should be sure that you know all of the configuration fuse options for the PICmicro® MCU part that you are using and be sure that they are set properly.

Once the *_WDT_OFF* parameter is added to the *__CONFIG* statement of the DEBUG.ASM source file, the application will run properly.

Well, almost. I am still not happy with how the application ends. The last instruction in the application is:

```
goto Loop
```

and if I had created this application from somebody else's code, I would be very suspicious of it. In a typical application, after a button is pressed and acknowledged, I would think that the code would continue on, or get stuck in a hard loop in which it does not jump to any other instructions. Jumping back to *Loop,* code that has already executed will be repeated over and over again. In this application, this instruction is not a problem, but I would feel more comfortable with:

```
goto $
```

which does not execute the polling for the button and does not clear PORTB repeatedly.

With the fixes outlined, the DEBUG.ASM code becomes DEBUGFIX.ASM (which is located in the same subdirectory as DEBUG.ASM). In DEBUGFIX.ASM, the lines with problems are commented out with the corrected lines inserted afterward:

The problems outlined here are pretty common for applications in magazines, books, and over the Internet. You will also make these problems for yourself, as well as when you develop your own applications. Along with finding other people's problems, this process for finding and debugging problems can be applied to modifying circuits and getting them working—even if you have used different parts or have changed the source code.

The purpose of this experiment has really been to show you that if you do encounter problems, you should be able to puzzle them out for yourself and fix them without getting angry at the source or giving up on the project completely.

I could have thrown in a few more problems for you that have been covered elsewhere in this chapter and this book. These problems include noninitialized variables and variables that are overlayed on hardware I/O registers. You will have to watch for these problems when you are debugging an application that isn't working properly.

# 16

# PROJECTS

## CONTENTS AT A GLANCE

When I wrote the first edition of this book, I wanted to provide a good range of sample applications for the reader to work through and use for a basis for their own applications. With this goal, I presented the applications as introductions to the PICmicro® MCU and different interfacing methods.

This was actually quite a big miscalculation on my part. It turned out that most readers wanted to build the applications just as they were. The limited circuit information caused problems for a number of readers who did not have the experience in designing their own PICmicro® MCU applications and felt that I did not provide enough information to help them wire the applications. To remedy this, I have provided full schematics for each project, along with pointers to the CD-ROM, where the code can be found.

I am also pleased to offer twice as many sample applications as were present in the original book. To help you sort through them, I have grouped them according to the PICmicro® MCU family to which they belong. Each project has its own HTML page with links to the source code and other pertinent information. In previous books, people have had problems deciphering schematics, so I have included adobe Acrobat "pdf" versions of the schematics on the CD-ROM.

The projects themselves are generally quite simple and are designed to be built and debugged in the space of a weekend. I have designed the different projects to use a variety of different control, display, and power-supply methods. As I describe them, I point out their pros and cons, as well as describe their suitability for other applications.

# Low-End Devices

As I've gained insight and expertise into working with the different PICmicro® MCU devices, I've found that many people use the different PICmicro® MCU architecture families in ways that I consider to be inappropriate. No where is this more obvious than in regard to how low-end devices are used.

Most of the published applications that I see in magazines and journals use low-end devices in applications I consider them poorly suited for. As indicated earlier in this book, I consider the low-end best suited for applications that have the following characteristics:

1. No interrupts
2. No complex dialog-based user I/O
3. No complex interfaces
4. Minimal subroutines

This does not mean that sophisticated applications cannot be implemented with the low-end PICmicro® MCUs, just that the code cannot be very long or require interfaces that are best suited for interrupts or the peripherals of the mid-range devices.

I have targeted the example low-end applications in this chapter to use the 12C5xx and 16C505 PICmicro® MCUs, which are very low cost (less than one dollar each in reasonably low quantities) and do not require external clocks or resets.

## TRAINCTL: MODEL-TRAIN TRAFFIC-LIGHT CONTROL USING A HALL-EFFECT SENSOR

As my kids have gotten older, I have been reintroduced to many of the hobbies and toys that I had when I was young. Many of these activities have changed in ways that would be unimaginable for me as a kid 30 years ago (for example, *Hot Wheels* cars can be bought with computer chips inside them). Despite this, I was disappointed to find that model trains are virtually unchanged from when I was a kid; many of the products sold are identical to what I had 30 years ago.

The DCC protocol which has recently become available, allows digital commands to be "broadcast" on the rails to engines, rolling stock, switches, and accessories. DCC is very expensive to get as a starter set and can be very difficult ("fiddly" is the word I hear most often) to set up and use.

I would like to create a computer-controlled system that is somewhere between the two. The flexibility of a computer system should be able to be added to the train system simply, for relatively low cost.

This project (Fig. 16-1) is one of the first of my experiments toward this type of system. It uses the 18-VAC auxiliary power supply built into most electric train systems to detect the presence of a train and change a three-color light display until the train has passed.

The circuit shown here interfaces and controls AC power with triacs. I have to very strongly caution you that although the techniques to determine component values could be

**Figure 16-1** Model train traffic light controller

used to develop applications that control household mains wiring; the component values cannot be used with 110- to 120- or 220-VAC power. You should not attempt to use the information contained in this project if you are the least bit unsure what you are doing or are not fully aware of the dangers involved.

The circuit shown in this application is designed only for the 18 VAC that is available from model train transformers for driving accessories. Using 18 volts results in very little danger of electrical shock, but short circuits can result in high currents that can burn you or cause a fire if you are not careful with the circuit.

Despite these seemingly dire warnings, the circuit I came up with to control lights is actually very simple (Fig. 16-2).

The bill of materials for the application is shown in Table 16-1. The circuit has a 5.1-volt 50-mA power supply that also provides half-wave-rectified power from the 18-VAC source. This sounds like a pretty fancy specification, but it is actually produced by a very simple circuit (Fig. 16-3).

In this circuit, the Zener diode limits the voltage across it to 5.1 volts. This voltage is used to power the PICmicro® MCU that controls the circuit. The silicon diode allows current to pass in only one direction. These two diodes have a combined voltage drop of 5.8 volts. Because I wanted to provide 50 mA to the PICmicro® MCU, I used this information to calculate the resistor value needed.

Using Kirchoff's and Ohm's laws:

$$
\begin{aligned}
V_{ac} &= V_{zener} + V_{diode} + V_r \\
18 \text{ volts} &= 5.1 \text{ volts} + 0.7 \text{ volts} + V_r \\
V_r &= 18 - 5.8 \text{ volts} \\
&= 12.2 \text{ volts at } 50 \text{ mA} \\
R &= V/I \\
&= 12.2/50 \text{ mA} \\
&= 244\ \Omega
\end{aligned}
$$

I used a 220-Ω resistor in my circuit because it is a standard value.

With 50 mA of current going through the 220-Ω resistor, the maximum power dissipated is:

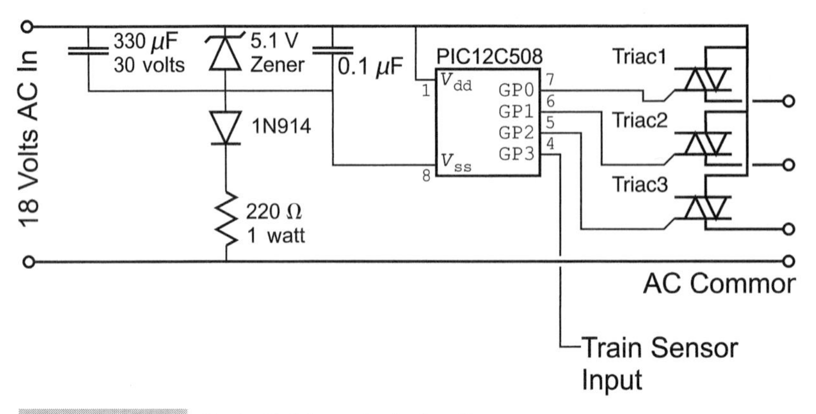


**Figure 16-2** **Train light-control circuit**

**TABLE 16-1 Bill Of Materials**

| PART | DESCRIPTION |
|---|---|
| PIC12C508 | PIC12C508-JW |
| 5.1 Zener | 5.2 Volt, 1/2 Watt Zener Diode |
| 1N914 | 1N914 Silicon Diode |
| TRIAC1-TRIAC3 | Forward Current 6 Amp, 25mA Gate Current, Radio Shack Part Number 276-1000 |
| 220 Ohm | 220 Ohm, 1 Watt Resistor |
| 0.1 $\mu$F | 0.1 $\mu$F Capacitor (any type) |
| 330 $\mu$F | 330 $\mu$F, 35 Volt Electrolytic Capacitor |
| Train Sensor | Hall Effect Sensor (See Text) |
| Misc. | Prototype PCB, Wiring, Screw Terminals for Electrical Connections |

$$\begin{aligned} P &= V * I \\ &= 12.2 \text{ volts} * 50 \text{ mA} \\ &= 0.61 \text{ watts} \end{aligned}$$

This is actually a significant amount of power and this is why I specified the 1-watt resistor in the bill of materials.

The maximum power going through the diode is:

$$\begin{aligned} P &= V * I \\ &= 0.7 \text{ V} * 50 \text{ mA} \\ &= 0.035 \text{ Watts} \end{aligned}$$

which is actually quite low. A small-signal diode like the 1N914 can be used safely.

The large (330 $\mu$F) capacitor is used to maintain an even voltage to the PICmicro® MCU–even when the diode isn't conducting. This cap is very effective, as I measured a 10-mV ripple on the PICmicro® MCU's $V_{dd}$, relative to $V_{ss}$ (or what I call *logic Gnd*). The 0.1-$\mu$F cap is used to decoupling the PICmicro® MCU and should be as close to the $V_{dd}$ and $V_{ss}$ pins as possible.

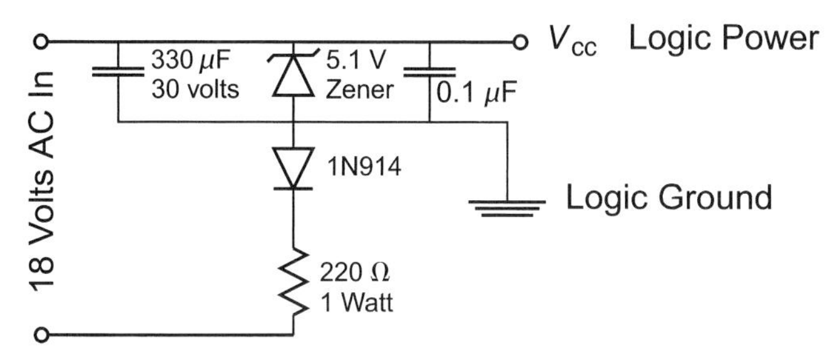


**Figure 16-3** Train 5-volt power supply

Triacs are interesting devices that come under the heading of *Thyristors,* which can be used to switch AC signals on and off. TRIACS do not rectify the AC voltage because they consist of two *Silicon-Controlled Rectifiers (SCRs),* which allow the AC current to pass without any clipping. A typical circuit for triacs is shown in Fig. 16-4.

Triacs do not allow AC current to pass unless their gates are biased relative to the two AC contacts. To do this, I pull the gates to logic Gnd by the PICmicro® MCU. As noted in Table 16-1, the current required to close the triacs is 25 mA, which can be sunk by most PICmicro® MCUs without any problems.

The load in this project are colored "grain of wheat" bulbs that are designed for model-train applications and can handle the 18-VAC accessory drive provided by the model-train transformer. For this application's circuit, I have put three of these circuits in parallel to control three separate lights in a traffic-light configuration.

To detect the train, I used a hall-effect switch with magnets glued to the bottom of the train and its cars. A *hall-effect switch* is a clever device in which if a current passing through a piece of silicon is deflected by a magnetic field. The output changes state as shown in Fig. 16-5.

The hall-effect switch that I used is an open-collector device and requires a pull up on its output, which can either be provided by an external resistor or use the internal port pull ups of the PICmicro® MCU. You'll find that you will have to play around with which magnet pole works best. For the parts I used, I found the "North Pole" and top edge of the hall-effect sensor worked best.

Other sensors could be tried in this circuit, which is why I left it somewhat vague. Along with hall-effect sensors, light-dependent resistors, or even simple switches could be used for the application.

The application code is very simple; it's barely more than 100 instructions and it turns the green light on until the sensor detects the passage of a train. When this happens, the green is turned off and the amber light is turned on for a second. After the second, the amber light is turned off, the red light is turned on, and the hall-effect sensor is polled. The red light will stay on until the hall-effect sensor has not returned "on" for 10 seconds. This code, although very simple, very nicely simulates the behavior of traffic lights with a traffic sensor.

TrainCtl The application, which is found in the *code\trainctl* subdirectory of the *PICmicro* directory, can be described using the pseudo-code:

```
main() {                                    // Train Control Application

 red = on;                                  // Initially Stop Traffic
 dlay(5 seconds);
```

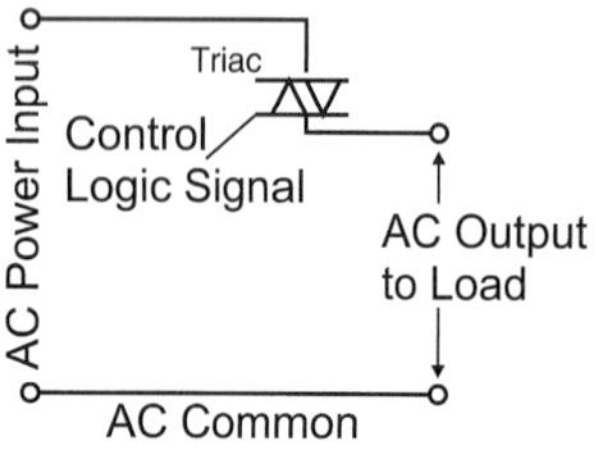

Triac
Control Logic Signal
AC Power Input
AC Output to Load
AC Common

**Figure 16-4** **Typical triac AC-control circuit**

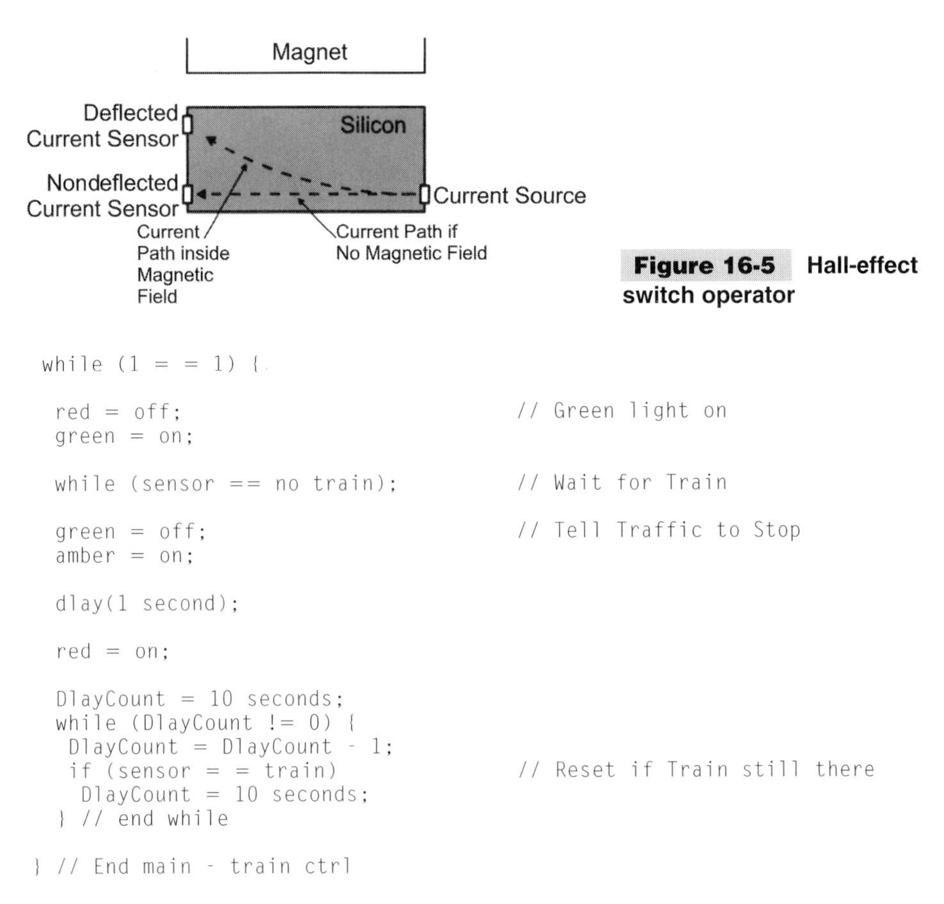


**Figure 16-5** Hall-effect switch operator

```
while (1 = = 1) {

  red = off;                       // Green light on
  green = on;

  while (sensor == no train);      // Wait for Train

  green = off;                     // Tell Traffic to Stop
  amber = on;

  dlay(1 second);

  red = on;

  DlayCount = 10 seconds;
  while (DlayCount != 0) {
   DlayCount = DlayCount - 1;
   if (sensor = = train)           // Reset if Train still there
    DlayCount = 10 seconds;
  } // end while

} // End main - train ctrl
```

My prototype circuit was built on a simple proto-board (shown at the start of this section). For the 18-VAC lines, I used 20-gauge solid-core signal wires. To facilitate easy insertion/removal into a circuit, I used screw terminals for the power and light signals.

As a final word of warning and caution; this type of construction is adequate for the model-train 18-VAC accessory drive; it is not acceptable nor safe if you are going to work with 110-VAC or 220-VAC home main wiring.

## SLI: SERIAL LCD INTERFACE

One of the most useful tools you can build for yourself is a simple LCD interface for your projects. The first edition presented a serial LCD interface application that demonstrated how an application written for a mid-range PICmicro® MCU could be ported to a low-end device. It had the interesting capability of sensing the timing of the incoming signals (and its polarity), which made it useful for some applications where a carriage-return character (which was used for timing) was first sent. It was less useful in applications where the interface was connected to a device that was already existing and the data coming out was arbitrary and probably not a carriage return. In the latter cases, the serial LCD interface could not be used.

**Figure 16-6**

The Serial LCD Interface presented here is a completely new design and is used a lot from that presented in the first edition. This application was written exclusively for a low-end PICmicro® MCU (the PIC16C505, specifically) and was designed to show off the following characteristics of the part and application design:

1. The built-in 4-MHz oscillator of the PIC16C505.
2. The built-in reset of the PIC16C505.
3. Using the _MCLR pin of the PIC16C505 for serial input.
4. Combining PORTC and PORTB of the PIC16C505 as an eight-bit I/O bus.
5. Combining switch inputs with bi-directional pins connecting the PICmicro® MCU to the LCD.
6. Providing a timed, interrupt-like interface to read incoming serial data.

I chose the PIC16C505 because when the internal 4-MHz oscillator and internal reset are used, the PICmicro® MCU provides 11 I/O pins and one input pin. The 11 I/O pins are perfect for bi-directional I/O on an LCD and the single input can be used for the serial input. To set the operating mode (speed and signal polarity), I provided two switches consisting of two pins that can be shorted with a jumper. The circuit that I came up with is shown in Fig. 16-7.

The bill of materials for the Serial LCD Interface is shown in Table 16-2.

When I built my prototype, I used point-to-point wiring on a simple phenolic prototyping card with the nine-pin D-shell connector pressed onto the card and the mounting holes tied to the card for stability. I mounted the LCD's 14-by-one connector in such a way that it covers most of the card, including the two pin headers, used to select the speed and polarity of the incoming data.

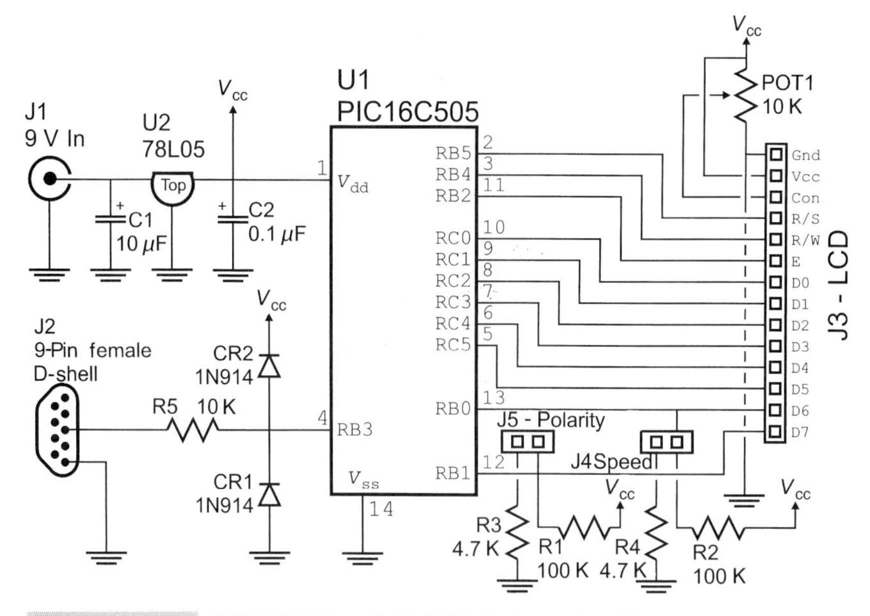


**Figure 16-7** PIC16C505 serial LCD interface circuit

| TABLE 16-2 "SLI" Bill Of Materials | |
|---|---|
| **PART** | **DESCRIPTION** |
| U1 | PIC16C505/JW |
| U2 | 78L05 |
| C1 | 10 μF, 16 Volt Electrolytic |
| C2 | 0.1 μF 16 Volt Tantalum |
| R1, R2 | 100 K, 1/4 Watt |
| R3, R4 | 4.7 K, 1/4 Watt |
| R5 | 10 K, 1/4 Watt |
| Pot1 | 10 K, Single Turn Potentiometer |
| CR1, CR2 | 1N914 Silicon Diode |
| J1 | 9 Volt Battery Connector |
| J2 | 9 Pin "Female" D-Shell with "Solder Tail" Connections |
| J3 | 14x1 Connector |
| J4 | 2x1 Pin Header, "Speed" Switch |
| J5 | 2x1 Pin Header, "Polarity" Switch |
| Misc. | Prototype Board, Wire |

Notice that I have included a simple 5-volt power supply, consisting of a 10-$\mu$F capacitor and a 78L05. This power supply provides 100 mA for the circuit and allows me to use a 9-volt radio alkaline battery to power the Serial LCD Interface.

To run the PIC16C505 with the internal oscillator and internal reset, I used the *__CONFIG* statement in the source code:

```
__CONFIG _CP_OFF & _WDT_OFF & _IntRC_OSC_RB4EN & _MCLRE_OFF
```

This statement allows RB4 and RB5 to be used as I/O pins instead of OSC2 and OSC1, respectively, and allows RB3 to be used as a digital input (and not _MCLR/$V_{pp}$). To allow RB5 (which is normally T0CKI, the TMR0 input pin) to be used as an I/O, I had to reset the T0CS bit of the OPTION register, which was accomplished by the two instructions:

```
movlw  0x0FF ^ (1 << TOCS)
option
```

The serial input includes two clamping diodes, which ensure that the input voltage on the PICmicro® MCU does not exceed any of the specified voltages. When I first implemented this circuit, I only had one clamping diode (CR1 in Fig. 16-7) on RB3. When RS-232 signals were sent to the Serial LCD Interface with only one clamping diode, I found that the input to the PICmicro® MCU ranged from 0.2 to 10 volts and placed the PICmicro® MCU periodically in Program mode.

When the second clamping diode (CR2) was added, the voltage swing on RB3 was reduced from 12 to 4.7 volts and PICmicro® MCU stopped going into Program mode.

The source code for this application is located in the *code\SLI* subdirectory in the *PICmicro* directory. The final version that is used with the application is SLI04.ASM.

The two switches caused some problems with the LCD. When I first started working with them, I was surprised to find that the LCD, which did not have any of its I/O pins driven by the PICmicro® MCU, was holding down the lines. To allow the PICmicro® MCU to read the switch values (which are either pulled up by 100-k resistors or pulled down by 4.7-k resistors), the E bit (pin 6) of the LCD has to be pulled down low. To do this, I used the three statements:

```
clrf  PORTB                  ;  See If PORTB Initialized Helps
movlw  0x0FF ^ ((1 << 5) + (1 << 4) + (1 << 2))
tris  PORTB                  ;  Drive the LCD Control Bits
```

before the pin read to force the E, R/S, and R/W lines of the LCD low and then perform the switch read. The need to hold E down to read the switches was a surprise and not something that I had expected.

When writing to the LCD, the eight I/O bits are shared between PORTB and PORTC with the two most-significant bits of the LCD's eight I/O bits as PORTB bits zero and one. To shift up the bits, I used the code:

```
movf  char, w  ; Get the Character to Send to the LCD
movwf  Dlay
```

```
movwf   PORTC    ; Store the Least Significant bits in PORTC
rlf     Temp, w ; Convert the Most Significant bits to PORTB's
rlf     Temp, f ; Least Significant Bits
rlf     Temp, w
rlf     Temp, w
andlw   0x003
movwf   PORTB    ; Output the Data to PORTB
```

This code should be explained a bit because it will not be expected—especially with all of the *rlf* statements that seem to continually load w with the same value.

If you look in the appendices, you will see that the two instructions:

```
rlf  Register, w
rlf  Register, f
```

will do a *rotate* on the contents of Register. The first instruction will load the Carry flag with the contents of the most-significant bit. The second instruction will shift up the contents of Register while passing the most-significant bit (which was loaded into the Carry flag in the previous instruction) in the least-significant bit of Register.

The next two instructions (both of which are *rlf Register, w*) will first load the contents of the most-significant bit into Carry and then load the rotated register into w. These four instructions (along with the following *andlw 0x003* instruction) moves the most-significant two bits of the byte to write to the LCD into the two least-significant bits of PORTB.

Another way to do this is:

```
movf    char, w     ; Get the Character to Send to the LCD
movwf   Dlay
movwf   PORTC       ; Store the Least Significant bits in PORTC
swapf   Dlay, f     ; Move the Most Significant Nybble down
rrf     Dlay, f     ; Shift Bits down by 2
rrf     Dlay, w
andlw   0x003
movwf   PORTB       ; Output the Data to PORTB
```

With the circuit and interface code working, I was then able to focus on the serial input read and then output to the LCD. As indicated, I wanted to create an interface that would simulate the operation of a TMR0 overflow interrupt that polls the serial line three times for each bit.

As I envisioned the Serial LCD Interface, I wanted it to work for 9600 bps and 1200 bps. A 9600-bps data rate has a bit period of 104.167 $\mu$s (which I usually round off to 104 $\mu$s) and the 1200-bps data rate has a bit period of 833.333 $\mu$s (which I round off to 833 $\mu$s).

Using these timings, I had to come up with code that would poll either once every 35 $\mu$s (35 instruction cycles) for 9600 bps or 278 $\mu$s for 1200 bps. Polling every 35 $\mu$s would give me an effective data rate of 105 $\mu$s for 9600 bps for an error of 0.8 percent. The 278-$\mu$s polling rate gave me an effective data rate of 1,199.041 bps with an error of 0.08 percent to an ideal of 1200 bps. In either case, this error rate is not low enough to result in incorrect data being read in.

To actually poll the data, I created a macro that would check for data already coming in. If it was, or if a *start* bit was detected, it would call a serial processing subroutine. This macro only required three cycles (allowing up to 32 cycles to execute "outside" of it) to check the incoming data:

```
SerCheck Macro              ; Check Serial Data
  btfss    SerActive        ; Reading to be Done?
    btfss  SerIn
  call     RSSerialPos      ; Yes - Check the RS-232 Line Input
 endm
```

In the code, the SerActive flag is set to indicate that data is coming in and to jump to the serial read routine. If this flag is not set, then if the incoming data line is a 0 (*start* bit), then the serial read routine is called to start reading the incoming data.

The serial read routine can be written in high-level pseudo-code as:

```
SerialRead()                        // Read Incoming Serial Data
{

 Temp = inp(PORTB);                 // Get the Contents of the Serial Data
                                    //  for Checking
 if (SerActive == 0)                // Is this a Start Bit?
   if ((Temp & (1 << SerInBit)) == 0) {

     Dlay( 1/3 Bit );               // Delay 1/3 Bit

     if ((inp(PORTB) & (1 << SerInBit)) == 0) {

     SerActive = 1;                 // Not a Glitch, Read in Byte

     BitCount = 9;                  // Going to Read through 9 Bits

     Dlay( 2/3 Bit );               // Delay Full Bit

    }
    } else ;                        // Not Low - Glitch?
 else {                             // Else, Reading in Data


  BitCount = BitCount - 1;          // Decrement the Bit Count
  if (BitCount != 0)                // Read Bit?
    RXData = (RXData >> 1) + ((Temp & (1 <<SerInBit)) != 0);
  else {                            // Last Bit - Is there a Stop Bit?

   SerActive = 0;                   // In any Case, not Reading Bits

   if ((inp(PORTB) & (1 << SerInBit)) != 0)
     SaveData(RXData);              // Valid Stop Bit - End

  }

  Dlay( 1/3 Bit );                  // Delay Remaining Bit

 }

} // End SerialRead
```

In this code, notice that under all circumstances, a delay occurs for at least one third of a bit. In the application code, this meant that any time SerialRead was called, at least a 2/3-bit period is used in the subroutine. This leaves only a 1/3-bit period to actually execute mainline code. This is not a terrible hardship. In the nine bits required for each bit, up to five characters could be sent to the LCD with a polling routine on the Busy flag to wait for the LCD to complete writing to its display.

In the *start* bit, notice that I end up waiting a full bit period before returning to the mainline code. This code performs two glitch checks and ensures that the data poll occurs in the 1/3-bit period that is closest to the middle of the bit. As indicated, the possible bit-time errors are anywhere from 0.08 to 0.8 percent, which means that any kind of movement from this position should keep the poll well within the bit. I capture the serial data at the start of the serial input routine to ensure that it is polled at the same "location" during the application's execution.

The assembly code itself is quite a bit more complex. The most important aspect of it is the delay code. In the previous pseudo-code above, I label delays simply like *Dlay (1/3 bit),* but in the actual application code, it is quite a bit more complex. In the application code, I first debugged the application to run properly at 9600 bps. To do this, I had to ensure that the line is polled at exactly a 1/3-bit period. I counted through the number of cycles for each action and then put in a delay to ensure that the number of cycles between the SerCheck polling macros was exactly 35.

To ensure that this wouldn't be a problem, I checked that everything else executed in less than 35 cycles and used the following macro to delay the required number of cycles:

```
Delay Macro Value                   ; Figure Out Delay Values
  variable Count, Remainder
Count = (Value - 4) / 3
Remainder = Value - ((Count * 3) + 4)
  movlw  Count + 1
  movwf  Dlay
  decfsz Dlay, f
    goto  $ - 1
if (Remainder == 2)
  goto  $ + 1
endif
if (Remainder == 1)
  nop
endif
endm
```

This macro will provide valid delay code from four to 771 instruction cycles. When you look through the code, you will see statements like:

```
Delay 35 - 18
```

which was my shorthand way to record the number of cycles total to be delayed, and subtracted from it, the number of cycles already executed. I have never used this method to provide precise delays before and I found it to be quite easy to track. The only issue I had with it was ensuring that the Delay macro was accurate to the cycle.

Once I had RS-232 input (which is essentially the complement of TTL/CMOS NRZ serial input) running at 9600 bps, I then went ahead and made the changes to the code that would allow the switch selection between 9600 and 1200 bps. I was looking ahead at doing this. To accomplish it, I simply selected between two different time delays. The code I normally used looked something like:

```
btfsc  SpeedSet        ; Delay 1/3 Cycle
  goto  Speed3P
Delay 35 - 17 - 4
  goto  Speed4P
```

```
Speed3P
  Delay 278 - 17 - 4
    nop
Speed4P
```

which executes a delay specific to the speed that the application is running at. If the SpeedSet flag is set, then the application is running at 1200 bps, otherwise 9600 bps. In the previous code, the *- 4* operator is used to include the four cycles that were added to select the actual data speed.

With the application running for RS-232, I put the SerialRead code into a macro with the ability to select positive (TTL/CMOS) or negative (RS-232) inputs. This further adds to the complexity of trying to see what is happening inside of the source code. If you look over the different copies of the application code, you will see how this macro was developed.

You will also see that I put the serial input and LCD output code in two different pages in the source code. I did this because there was not enough space for both to reside in the same page. Even though the actual application is just a few instructions longer than 512 (the low-end PICmicro® MCU's page size), it cannot be placed in one page. To simplify the application, I placed the code for each type of polarity in its own page.

Normally, when I am creating an application, I would consider placing the code all together and putting in the comparisons between positive and negative incoming serial data. I didn't in this case because there is no 512-instruction PIC16C505 derivative and I had more than enough program memory space to implement the application this way.

Many serial LCD interfaces on the market today have the capability to accept LCD instructions and LCD data. The project given here has this capability as well. To send an LCD instruction, the data character with the ASCII code 0x0FE has to be sent first. The next byte received will be written to the LCD as an instruction, rather than as a data byte.

The most significant problem I had with getting this application running (once I had figured out the switch problem) was at 1200 bps with the Busy flag polling. I incorrectly left the state of the R/S flag set after a data byte write to the LCD and when polling the value back, a data read was accidentally initiated.

This was a very confusing problem to find because there was no problem at 9600 bps. I eventually found the problem by looking at the state of the LCD pins when the E strobe was active. In each data transfer, the first action was correct and the data byte was being written to the LCD. However, when I went through all of the LCD's bits, I discovered that RS was in the wrong state to poll the Busy flag.

## ULTRA: ULTRASONIC LCD RANGING

With the Serial LCD Interface completed, I wanted to interface with it using a PICmicro® MCU to see how useful it would be. This project is a distance-measuring tool that uses the Polaroid 6500 module and transducer (Fig. 16-8). This resulting project can measure distances using the same ultrasonic transducer that many cameras use to set their lenses. The Polaroid 6500 is very easy to work with, although there are a couple of things to watch out for.

The PICmicro® MCU used with this project is a PIC12C509. This application was also an experiment to find out how well two PICmicro® MCUs with internal RC oscillators could be used to communicate with each other. The "Positive" NRZ signal input to the Serial LCD Interface (the previous project) was used without any type of signal conditioning. The actual communications was not perfect, but does give me some insight into how the internal oscillators of the PICmicro® MCUs work and their relative accuracy.

**Figure 16-8** Polaroid 6500 ultrasonic range finder

The ultrasonic transducer chosen for this application is the *Polaroid 6500,* which was first developed in the mid-1970s for use with Polaroid autofocusing cameras. The operation of the device is quite simple: a controller outputs a pulse to the module (which can be up to 100 ms) long and then measures the time for the echo to come back. This can be simply accomplished by two wires connected to a microcontroller.

The black disk (Fig. 16-8) is the ultrasonic transducer, which both outputs the ultrasonic signal (which can be heard as a "click") and receives the echo when the sound waves bounce off the first object that they come to. The ultrasonic pulse is triggered by the INIT line of the *Polaroid 6500.* Its ECHO line is driven high to indicate that the echoed signal has been received. The time required for the signal to return to the transducer is proportional to the distance from the transducer to an object in its line of sight. The 6500 module is specified to work for ranges from six inches to 35 feet with an error of 1%. The INIT enable and ECHO return are shown in Fig. 16-9 for an object placed 16 feet from the transducer. On the oscilloscope picture, I have marked the pulse edges with "cursor" lines.

Before going too far into this project, I should mention a few things. The first is a warning. When the ultrasonic signal is sent, an incredible amount of energy is generated. At the transducer, 400 volts is produced and about 1.2 amps is drawn from the power source. This is important to realize because nowhere in the Polaroid (or any other) documentation did I find a warning about electrical shocks.

When I was first experimenting with the device and aiming it at different walls of my workshop, I inadvertently touched the two contacts on the transducer. After doing that, I discovered that the transducer was dangling by its wires and my right hand was about four feet behind me. A few seconds later the pain followed. I received a pretty good shock between my right hand's index finger and thumb.

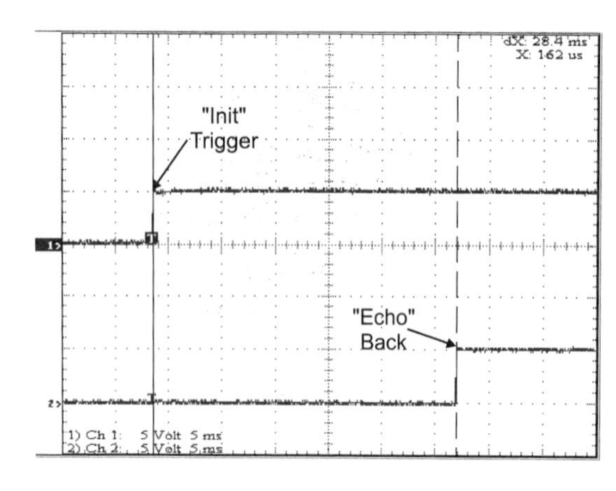


**Figure 16-9** Init/Echo signal to/from the 6500

For this reason, I insist that you do not use both hands when handling the operating transducer. When you do handle the transducer, put your left hand in your pocket to ensure that there is no chance for a current path through your heart.

The voltage and current the device produces should not be capable of severely hurting you, but you should take precautions and treat the driver module and transducer with respect when power is being applied to them.

The 6500 has a nine-conductor connector that is designed to have a flat "flex" cable connected to it. One of the first things that I did with the connector was to cut it off, unsolder the connections, and put wires into them (Fig. 16-8). This was one of the smartest things that I did because it really made prototyping and experimenting with the 6500 very easy.

To experiment with the Polaroid 6500, I used a PIC12C509 to take advantage of its internal 4-MHz oscillator and reset. With these functions taken care of, all I required of it was three wires: two to interface to the Polaroid 6500 and one to send serial data to the Serial LCD Interface. The circuit that I used is shown in Fig. 16-10.

The bill of materials for the project is shown in Table 16-3.

Use a 1-amp rated 5-volt bench power supply. The Polaroid 6500 is capable of drawing a large amount of current when it is activated. The reasonably large supply and the very large (1,000-$\mu$F electrolytic) filter capacitor is to ensure that the operation of the Polaroid 6500 does not cause voltage surges that would affect the operation of the circuit or power supply.

The ULTRA02.ASM code (found in the *code\Ultra* subdirectory of the *PICmicro* directory) is only 156 instructions long. I choose the PIC12C509 and not the PIC12C508 simply because I had more JW PIC12C509s on hand when I built this project than PIC12C508s. Either PICmicro® MCU will work fine without any problems in this application.

This application was really an experiment to learn more about the Polaroid 6500 and the ability of two internally clocked PICmicro® MCUs to communicate with each other. To keep the application very simple, I built it on a breadboard, which only takes five to 10 minutes for all five components to be wired together and to the power supply.

Figure 16-10 shows how I implemented the interface wiring for the Polaroid 6500. After removing the flex socket that comes with the Polaroid 6500, I discovered that the connector vias were in the pattern shown in the schematic, with pin 1 on the left side and pin 9 on the right side. The lower pins, halfway between the upper row are the even pins. The upper row contains the odd pins. The pin out of the Polaroid 6500 is shown in Table 16-4.

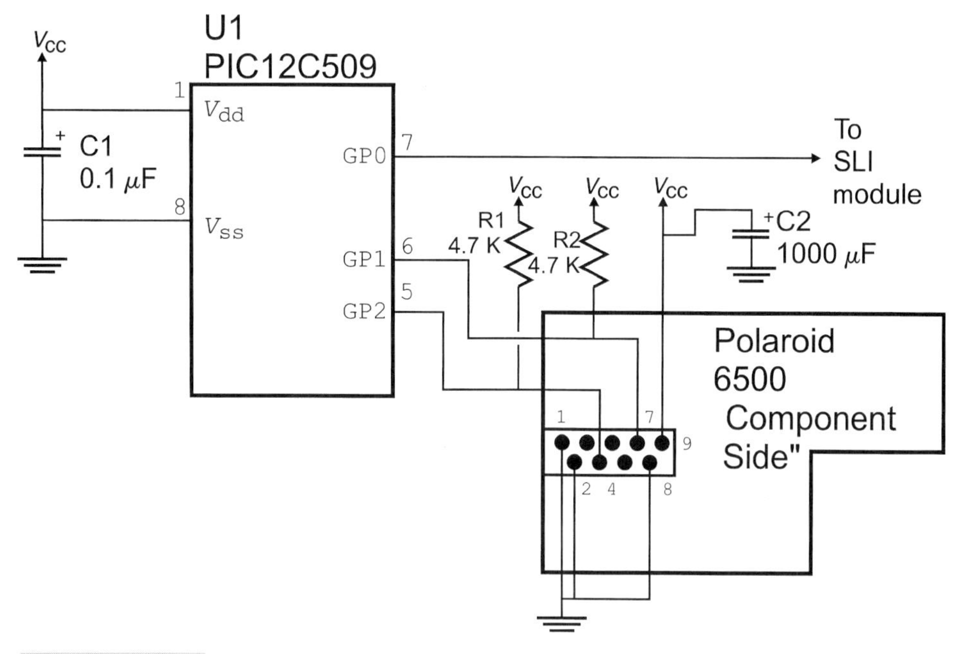


**Figure 16-10** Polaroid 6500 ultrasonic distance measuring

The operation of the 6500 is very straightforward; the INIT pin is typically asserted for up to 100 ms and the controlling processor waits for the ECHO to come back. If the surface the ultrasonic burst reflects off of is too close or too far away, then ECHO will never become asserted after INIT.

The 6500 is specified to work at a minimum distance of 16 inches. This can be reduced to about six inches by asserting the BINH line. By asserting BINH, any ringing of the transducer will be masked.

BLNK will ignore any reflected inputs at specific times after INIT is asserted. This is useful if the 6500 could detect multiple objects. By blanking out the receiver when echoes are being received, a more accurate reading of the furthest object from the Polaroid 6500 can be made, which helps avoid fluctuating readings.

**TABLE 16-3 "Ultra" Bill Of Materials**

| PART | DESCRIPTION |
|---|---|
| U1 | PIC12C509/JW |
| R1, R2 | 4.7 K, 1/4 Watt |
| C1 | 0.1 µF Tantalum |
| C2 | 1000 µF Electrolytic |
| Polaroid 6500 | Polaroid 6500 Ultrasonic Distance Finder with Transducer |
| Misc. | Breadboard Prototyping System, Wiring, 1 Amp 5 Volt Power Supply |

**TABLE 16-4 Polaroid 6500 Pinout**

| PIN | FUNCTION | COMMENTS |
|---|---|---|
| 1 | Ground | Connected to the Cathode of 1000 $\mu$F Filter cap |
| 2 | BLNK | When Asserted, the 6500 will "Blank" out any reflected signal |
| 3 | N/C | |
| 4 | INIT | When Asserted, the 6500 will output the Ultrasonic "Burst" |
| 5 | N/C | |
| 6 | OSC | 49.4-Khz Oscillator |
| 7 | ECHO | Asserted after the Received Signal Returned |
| 8 | BINH | When Asserted, the 6500 will ignore any Transducer "Ringing" |
| 9 | Vcc | Connected to the Anode of 1000 $\mu$F Filter Cap |

For any inputs that are going to be driven, you must pull up the lines with 4.7-k resistors. This is especially important when wiring the 6500 to a microcontroller, which has an open drain I/O pin (like RA4 in the mid-range PICmicro® MCUs), or weak MOSFET pull ups, such as on the mid-range PICmicro® MCU's PORTB.

This explanation of the 6500 probably seems quite simplistic. On the CD-ROM included with this book, I have included a .pdf file of the Polaroid datasheet for your use. The 6500 can be purchased from Wirz Electronics (the address is given in the Appendices).

When I work with devices that have attached wires, I typically put on a good coating of Weldbond or hot-melt glue to provide strain relief for the solder joints. I did put Weldbond on the transducer wires and the interface wiring to be sure that as I handled the card, I wouldn't break any joints. The transducer wires do have a hole in the card to provide some measure of strain relief.

I found that until the Weldbond hardened (which took three days), the Polaroid 6500 stopped working because the uncured Weldbond broke down and provided a relatively low resistance path (compared to the transducer) for the 400-volt output signal. If you are going to use the 6500 and are planning to glue the interface wires to the Polaroid 6500 to provide strain relief, be sure that you understand the electrical characteristics of the glue, both when it is a liquid and after it has cured.

After all these warnings and related problems, you might not be feeling very good about the Polaroid 6500, but it's a tough little critter. When I first added the wires, I got $V_{cc}$ and Gnd mixed up. Although the chips on the Polaroid 6500's card got quite hot, the circuit did work afterward without any problems. As well, there was the problem with the uncured Weldbond and my modest attempts at jump starting my heart. After all of these mishaps, the 6500 module worked perfectly. I really haven't had a moment's problem with it, aside from the self-inflicted ones.

To simplify the operation of the Polaroid 6500, I drove it with BINH and BLNK pulled to ground. This makes measurements of less than 4 feet a bit unreliable (i.e., jumping up and down, depending on the first object to echo back to the Polaroid 6500). Other than this, the application works very well.

Once I had created a simple application to test the operation of the Polaroid 6500 and allow me to get oscilloscope readings (Fig. 16-9), I then wanted to convert the distance measurement to feet and inches and display it on an LCD.

To determine the distance between the Polaroid 6500's transducer and another object, the time between the initiation of the INIT pulse and the ECHO return is measured. Using the PIC12C509, running the internal 4-MHz clock, this time was measured in 148-$\mu$s increments.

The 148-$\mu$s increment probably seems like an unusual measurement, but it is based on the speed of sound for ultrasonic frequencies. Assuming that the speed of sound at sea level for a 40-kHz signal is 1,127 feet per second (13,523 inches per second), the flight time from the transducer to an object one inch away and back (twice this distance) is 147.9 $\mu$s. Using a 148-$\mu$s check results in a 0.068% error, which should be insignificant in the actual measurement. It allows the PIC12C509 to be used with the internal oscillator without modification.

The Polaroid 6500 is specified to a maximum distance of 35 feet (420 or 0x01A4 inches). In the application code, I wait for the ECHO line to become active, incrementing a distance counter once every 148 $\mu$s and indicating that there is No Echo if the counter reaches 420.

The code that does this distance and No Echo check is:

```
 clrf  Distance             ; clear the Distance Calculation
 clrf  Distance + 1

 bsf   Init                 ; Toggle the "Init" Pin High

EchoLoop                    ; Loop Here until

 movf  Distance, w          ; At 35' (Max Distance)?
 xorlw  0x0A4               ;  = 420?
 btfsc  STATUS, Z
  decfsz Distance + 1, w    ; Low byte == 0x0A4, High Byte == 1?
    goto $ + 2
      goto NoEcho           ;  - No, Give up

 Delay 148 - 12             ; Make sure each Loop is 148 usecs (1 ")

 incf  Distance, f          ; Increment the Distance Count
 btfsc  STATUS, Z
  incf  Distance + 1, f

 btfss  Echo                ; Wait for "Echo"
  goto  EchoLoop

 bcf   Init                 ; Turn off "Init" Pin
```

This code bears a few comments. In the code after the *EchoLoop* label, I first compare the two-byte *Distance* variable to be equal to 0x01A4. This is done by first setting the Zero flag if the lower byte is 0x0A4 and then optionally decrementing the upper byte using the *decfsz* instruction to check if it is equal to one as well. This code is quite efficient, although it is probably difficult to walk through if you are new to the PICmicro® MCU. It only works if the value being checked for has a high byte of *1*.

The Delay macro is the same one that I developed for the Serial LCD Interface project.

Once a valid distance has been determined, I then convert the value into feet and inches. To convert the distance into feet, I divide it by 12 using the code:

```
 clrf  Feet                   ; Clear the Number of Feet to Display

ConvertLoop

 movlw  12                    ; Is the Value > 12?
 subwf  Distance, w
 btfss  Distance + 1, 0       ; Rolled Over to Second Byte?
  btfsc STATUS, C             ; Negative Value?
   goto $ + 2
 goto  HaveFeet               ; Remaining Distance is Less than 12

 btfss  STATUS, C             ; Shift Down?
  decf  Distance + 1, f

 movwf  Distance              ; Save the New Value

 incf  Feet, f                ; Add to the Foot Counter

 goto  ConvertLoop

HaveFeet                      ; "Feet" is Correct, Distance has Inches
```

This code divides a two-byte variable by a single-byte constant (12 in this case). Elsewhere in the book, I have presented you with other division by constant algorithms, but I used the repeated subtraction one because I knew that the maximum distance would only be 35 feet. This results in a maximum number of cycles through the loop of 420 instruction cycles (420 $\mu$s in this application).

In this code, notice that I save the result in w before jumping to *HaveFeet.* If the result is greater than zero, then the new value is stored back into *Distance.* When *HaveFeet* is executed, the *Feet* variable has the number of feet found in the *Distance,* and the lower byte of *Distance* has the remainder in inches. These two values are then sent to the Serial LCD Interface.

The NRZ serial output routine is a very standard bit-banging output routine with the Carry flag used to indicate the bit value after the data has been shifted.

When using the Polaroid 6500 PIC12C509 distance-measuring circuit initially with the Serial LCD Interface, I found that there were a number of errors in the transmitted data. When I looked at the data being sent on an oscilloscope, I found that the bit periods were in error by about 1.7 percent. By programming another PIC12C509, I found that I could reduce the bit period error to about zero, at which point the data presented on the LCD display was perfect.

After doing this, I realized that I probably should have tried to calibrate the built-in oscillator in the PCI16C505 built into the Serial LCD Interface. I suspect that the part that I used in the Serial LCD Interface also has a reasonably high error, which added to the error of the PIC12C509. By choosing a different device, I was able to decrease the combined error to the point where the data was accurate under all circumstances.

## KEY: SWITCH MATRIX KEY INTERFACE

The typical example for interfacing a microcontroller to a switch matrix keyboard is a $4 \times 4$ keypad (Fig. 16-11). Although this can be useful in some applications, often only a

**Figure 16-11** Switch Matrix Keyboard interface

full QWERTY keyboard will do. This project shows how a switch matrix keyboard could be attached to a PICmicro® MCU.

A *switch matrix* keyboard is a series of switches wired in rows and columns. They can be read by pulling up each row individually and then tying down a column to ground and seeing if a switch is pulling a row to ground through the column (Fig. 16-12).

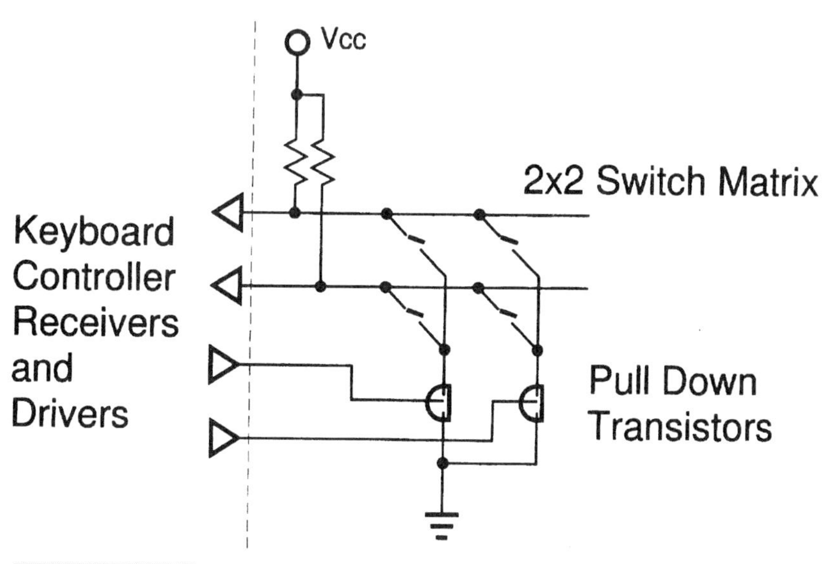


**Figure 16-12** Switch Matrix Keyboard connections

The pull-down transistors shown can be either discrete transistors or PIC I/O pins individually enabled to a *0* state to simulate the pull down to ground.

Figure 16-12 shows that when one of the switches is closed and the column it is connected to is pulled down to ground, the receiver will sense a low. If the switch was open and the column was pulled down to ground, the pull-up would cause the PICmicro® MCU to sense a high value. This method of switch sensing is analogous to having multiple open collector outputs on a single pulled-up line.

To "scan" the keyboard, each pull down is individually enabled in series. If any time an input value is low, the key's switch at that row/column address is closed and the key is pressed.

With this scheme, consider a few issues:

1. How should the keyboard be wired to the controller?
2. What about switch bouncing?
3. What about multiple keys pressed at the same time?
4. What about *Shift/Ctrl/Alt/Function* key modifiers?

In developing the first software application for the project Key1, I had to understand these issues and have a plan to deal with them. The source for the Key application is located in the *code\Key* subdirectory of the *PICmicro* directory.

Notice that these applications were designed to work with a keyboard that I found for a dollar in a surplus store in Toronto five years or so ago. The keyboard was manufactured by General Instruments for the Texas Instruments' TI-1000 personal computer that was last built more than 20 years ago. No visible part number is on the device itself, except for the Texas Instruments logo. I doubt you will be able to find this keyboard, but this still makes this application and the process I went through to decode the keyboard useful as a reference for whatever keyboards or keypads you want to work with.

The schematic of the circuit I used is shown in Fig. 16-13. To assemble my prototype, I used a *vector board,* which is a PC board that has a series of copper strips on the backside, to make a circuit, wires (or components) are brought to a specific strip. The strip can be cut to allow multiple circuits to use the same strip. This method of prototyping is somewhat obscure (which is why it is not mentioned in "Introduction to Electronics" on the CD-ROM), but it can be very useful when single-bussed devices, like the keyboard (or an LCD) is being prototyped in the circuit. I have used the vector board in a few of the projects simply because it is reasonable for the applications.

The bill of materials for the circuit is shown in Table 16-5. For my prototype, I used a 9-volt alkaline radio battery to the circuit to power it. My original intention was to use the keyboard interface with the Serial LCD Interface (presented elsewhere in this chapter) to make a simple 1200-bps RS-232 TTY terminal. Actually, this can still be done using these two components.

Trying to figure out how a keyboard should be wired to the PICmicro® MCU when you don't have any information is not as daunting a task as it would appear to be. When I bought the keyboard used for this project, its only identifying feature with regard to electrical connections was a strip indicating pin 1 on a 15-pin IDC connector.

The first thing that I did was set up a matrix. Using a DMM, I "beeped" out every key with the two different connector pins. With this information, I created the matrix shown in Table 16-6.

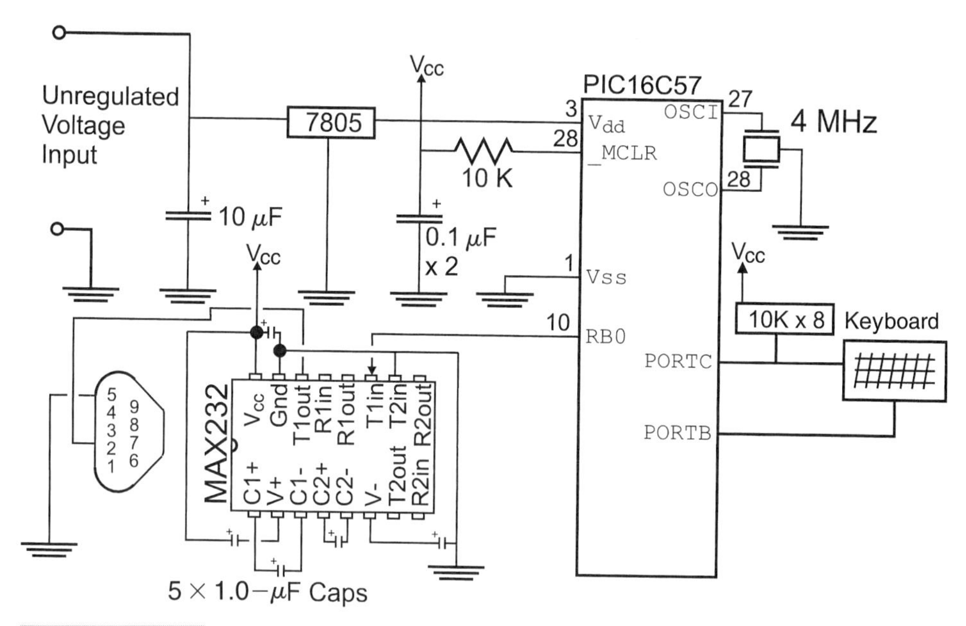


**Figure 16-13** Simple keyboard interface

Once this was done, I manipulated the table until I could get a good understanding how the keyboard was wired and the best way to attach it to the PICmicro® MCU.

The design point I decided upon was setting up eight rows (or register bits) for each column. I defined the *row* as where I put the pull up and the *column* as the pin I pull to ground.

**TABLE 16-5 "Key" Bill of Materials**

| PIN | DESCRIPTIONS |
|---|---|
| Part | Connected to Negative Pin of 1000 µF Filter cap |
| PIC16C57 | PIC16C57-JW |
| 7805 | 78L05 in a TO-92 Package |
| MAX232 | Maxim MAX232 RS-232 Interface Chip |
| 4 Mhz | 4 Mhz Ceramic Resonator |
| 10 K | 10 K, 1/4 Watt Resistor |
| 10 K | 10 K common tap, 9 pin "SIP" |
| 10 µF | 10 µF, 35 Volt Electrolytic Capacitor |
| 2x 0.1 µF | 0.1 µF, 16 Volt Tantalum Capacitor |
| 5x 1.0 µF Caps | 1.0 µF, 16 Volt Capacitors (any type) |
| RS-232 Connector | 9-Pin "Male" D-Shell Connector |
| Misc. | Prototype Card, Hook Up Wire, 15x1 IDC Connector for Keyboard |

**TABLE 16-6 Example Keyboard Matrix**

| PIN | 1 | 2 | 3 | 4 | 5 | 6 | 7 | 8 | 9 | 10 | 11 | 12 | 13 | 14 | 15 |
|---|---|---|---|---|---|---|---|---|---|---|---|---|---|---|---|
| 1 | | | | | “V” | “R” | | | “4” | | “M” | “J” | “F” | “7” | “U” |
| 2 | | | | | “C” | “E” | | | “3” | | “,” | “k” | “D” | “8” | “1” |
| 3 | | | | | “X” | “W” | | | “2” | | “.” | “L” | “S” | “9” | “O” |
| 4 | | | | | | Ctrl | | | Fctn | | “=” | “ ” | Shft | | Ent |
| 5 | | | | | | | “B” | “Z” | | | | | | | |
| 6 | | | | | | | “T” | “Q” | | | | | | | |
| 7 | | | | | | | | | “5” | | “N” | “H” | “G” | “6” | “Y” |
| 8 | | | | | | | | | “1” | | “/” | “;” | “A” | “0” | “P” |
| 9 | | | | | | | | | | Caps | | | | | |
| 10 | | | | | | | | | | | | | | | |
| 11 | | | | | | | | | | | | | | | |
| 12 | | | | | | | | | | | | | | | |
| 13 | | | | | | | | | | | | | | | |
| 14 | | | | | | | | | | | | | | | |
| 15 | | | | | | | | | | | | | | | |

The data was transformed into Table 16-7, where the rows and columns are the pin numbers on the connector.

With this information, I was ready to specify the wiring. At this point, it was simply wiring the rows and columns to the PICmicro® MCU. The rows have the pull up as noted previously and connected to PORTC while the columns are pulled down within PORTB.

**TABLE 16-7 Example Keyboard Matrix**

| COLUMN | ROW-> | 5 | 6 | 9 | 11 | 12 | 13 | 14 | 15 |
|---|---|---|---|---|---|---|---|---|---|
| 1 | | “V” | “R” | “4” | “M” | “J” | “F” | “7” | “U” |
| 2 | | “C” | “E” | “3” | “,” | “K” | “D” | “8” | “I” |
| 3 | | “X” | “W” | “2” | “.” | “L” | “S” | “9” | “O” |
| 4 | | | Ctrl | Func | “=” | “ ” | Shft | | Enter |
| 7 | | “B” | “T” | “5” | “N” | “H” | “G” | “6” | “Y” |
| 8 | | | “Q” | “1” | “/” | “;” | “A” | “0” | “P” |
| 10 | | | Caps | | | | | | |

I used a PIC16C57 for the keyboard and RS-232 interface because it had more than enough I/O pins to handle the 15 pins required by the keyboard. I outputted the NRZ serial data (through a MAX232) to my PC's RS-232 HyperTerminal terminal emulator so that I could monitor what was coming out (and, if needed, send debug information as well).

To read the keys, I used the algorithm:

```
KeyPreviously = 0;                    // Nothing Currently read

While (1 == 1)}

  Dlay4ms();                          // Delay for key Debouncing

  if (KeyPreviously = 0)}

  KeyCount = 0                        // Reset # of Times through Loop

  For i = 0; (i < # of columns) && (KeyPreviously == 0);i++)

   Scan Column                        // Check each Column and Set
                                      // KeyPreviously

 }else{                               // Else Key Previously pressed

  if (KeyPreviously Still Pressed)}

   KeyCount = KeyCount + 1;           // Increment the Actual Count

   if (KeyCount == 5)                 // Do we have the First Press?
    Send (KeyPreviously);

   else
    If KeyCount = 128                 // Have we Waited 1 Sec?
     Send KeyPreviously

    else
     if (KeyCount = 192)              // Have we Waited another 1/2 Sec?
      Send (KeyPreviously);
      KeyCount = 128                  // Reset for Next AutoSend

     }
  else                                // Key was lifted
   KeyPreviously = 0                  //  Start all over

 }
 goto   Loop
  }
```

This algorithm handles the key bouncing by requiring that five consecutive polls 4 ms apart "sense" the key being pressed. When the bit goes high, the keyread is reset until the next keypress pulls down the row and the counting resumes.

Multiple key presses (which can obviously occur when more than two keys in the same column are pressed at the same time) are resolved within the Scan Column routine. It is resolved by taking the lowest active bit in the column.

The Send routine looks up the ASCII code to send by reading the value in a row/column table.

Notice that the key modifiers (*Ctrl, Func,* etc.) are always masked off in Key1. I didn't bother to put in the modifiers because I didn't have an application that required the keyboard (and being lazy . . .). They can be implemented easily by adding new tables for each modifier and then before calling the table to look up the value, adding an offset to the correct table values.

Notice that didn't bother debouncing the key modifiers because they are not the action that initiates the action of sending the keys. They are just used to ensure that the proper keys are sent.

The code presented here (KEY1.ASM) could very easily be ported to a mid-range PICmicro® MCU. The advantage is that the TMR0 interrupt could be used to initiate the scan, allowing the keyboard read code to be run totally in the background. Another advantage of using a mid-range device is the ease in which the TRIS bits could be rotated directly within software, rather than keeping track of the current *TRIS* value, as I have to do in this application.

Along with porting the application to another PICmicro® MCU, you can also very easily change the tables I created to interpret the different "scan codes" to another switch matrix keyboard. This application is a bit "high end." If you would use a four-by-four switch pad, this code could be used quite easily within one port, rather than the two presented here. One last enhancement that could be made to this application is to use the built-in pull ups on PORTB of most mid-range PICmicro® MCU devices. This would avoid the need for the nine-pin common tap *Single In-Line Package (SIP)* resistor that I used in this application.

# Mid-Range Devices

By far the most popular PICmicro® MCU family is the mid-range. The family is well designed for sophisticated single-chip applications using the interrupt-enabled processor. Advanced I/O features can simplify applications and avoid the need for bit-banging applications. The mid-range devices, with these features, often allow simpler applications than what might be ordinarily possible.

The following example projects display the feature of the mid-range family that I take the most advantage of: the interrupt capability. As I've said elsewhere, this allows more sophisticated applications for the PICmicro® MCU and, in many cases, simplifies the final applications. When I use interrupts in an application, notice that I rarely use more than one interrupt source at a time. When I do, the second one is almost always the TMRO interrupt to provide timing functions in the projects.

The projects presented in the following sections not only are used to demonstrate different I/F and code methods, but also give you some concrete examples of how the different peripheral I/O functions work in an application.

## CLOCK: ANALOG CLOCK

This was one of my first PICmicro® MCU applications. It turned out to be very successful. In fact, it has spent about five years on my desk at work keeping just about perfect time. When I was making up the list of projects for this book, I had to try and remember

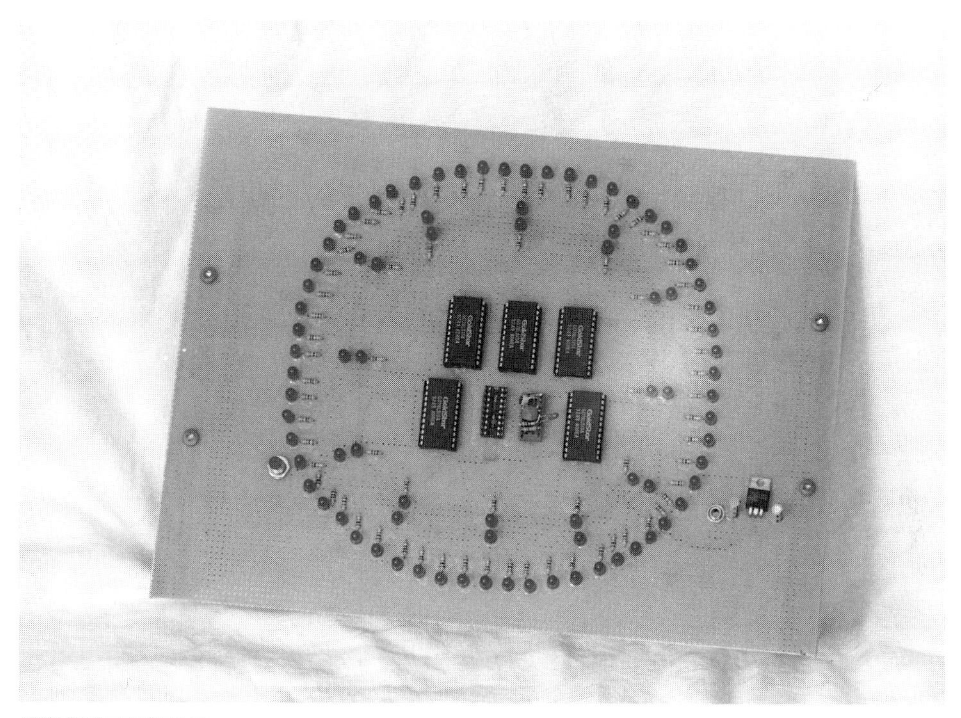

**Figure 16-14**

why I didn't include it in the first edition of this book. For the life of me, I have no idea why it didn't make it. Maybe because it was at work and was such a part of the office that I never thought anything of it.

This unique timepiece can be wire-wrapped together in about an afternoon. Seventy-two LEDs (or 84, if you built it the same way I did and used two LEDs for the hours) provide a digital replica of an analog clock. For the application, I used a 32.768-kHz quartz watch crystal and a 2.8-volt lithium PC back-up battery to ensure that I don't have to reset the clock if power is lost.

The circuit utilizes five 74LS154 four to 16 demultiplexers (Fig. 16-15) to drive the LEDs. One "minute" LED and one "hour" LED is displayed at any time from PORTB and PORTA or the PIC16LF84, respectively.

The bill of materials for the Analog Clock is shown in Table 16-8. For my prototype, I point-to-point wired the circuit. For your version, I recommend that you either design and embed the PC board for the circuit or wirewrap it. Using point-to-point wiring, it took me about a day's worth of effort to build this circuit. If you are going to wirewrap the circuit, I recommend that you wrap directly to the LED posts, which will save you a lot of time and make your life a lot easier.

I specified a "low-voltage" PICmicro® MCU for this circuit because I've included a simple battery backup circuit using a 2.8-volt lithium PC backup battery. If power is ever lost to the circuit (I powered my prototype from a wall-mounted AC/DC power adapter), the lithium battery will keep the PICmicro® MCU running until power is restored.

The lithium battery will run the PICmicro® MCU for a very long time because the

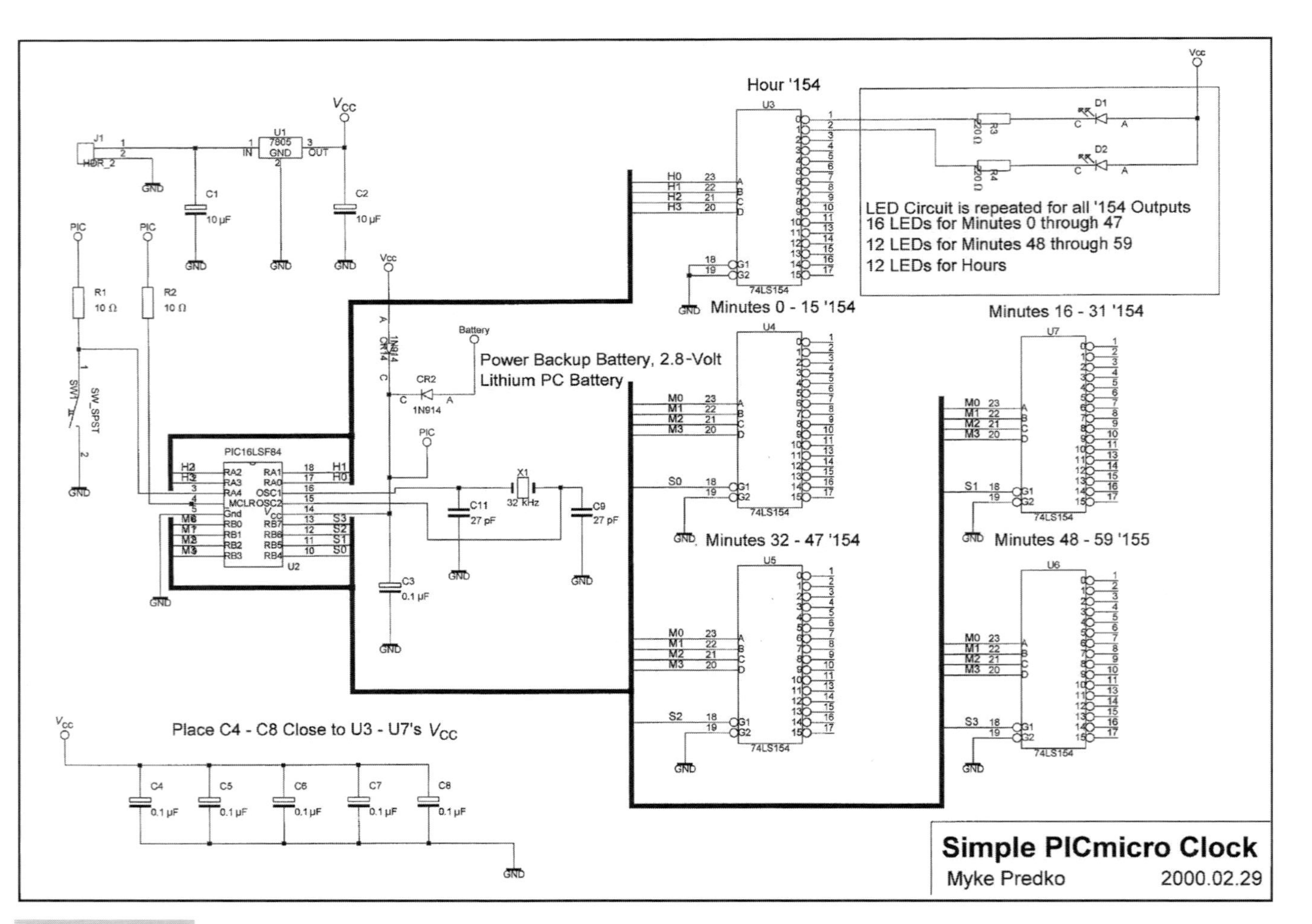


**Figure 16-15** "Analog" clock schematic

**TABLE 16-8 "Clock" Bill of Materials**

| PART | DESCRIPTION |
|---|---|
| U1 | 7805 +5 Volt Regulator in a TO-220 Package |
| U2 | PIC16LSF84-04/P |
| U3 - U7 | 74LS154 4 to 16 Demultiplexor |
| X1 | 32.768 KHz Watch Crystal |
| CR1 - CR2 | 1N914 Silicon Diodes |
| D1 - D72 (D84) | Red LEDs |
| R1 - R2 | 10 K, 1/4 Watt Resistors |
| R3 - R74 | 220 Ohm, 1/4 Watt Resistors |
| C1 - C2 | 10 $\mu$F, 35 Volt Electrolytic Capacitors |
| C3 - C8 | 0.1 $\mu$F, 16 Volt Tantalum Capacitors |
| J1 | 2.5 mm Power Plug |
| SW1 | Momentary On Push Button Switch |
| Battery | 2.8 Volt Lithium PC Backup Battery |
| Misc. | Prototype Board, Wire Wrap Sockets, Wire, DC Power Supply |

PICmicro® MCU's current consumption is on the order of microamps and the 2.1 volts of the PIC power output from the PICmicro® MCU is insufficient to cause a parasitic drain in the 74LS154s. This is a nice feature to have if you have occasional power outages.

To keep the time, I allow TMR0 to run continuously with a divide-by-64 prescaler that overflows once every second. In the software, the T0IF bit is continually polled. If it is set, then it is reset and the button is checked for being pressed. If it isn't, the seconds counter is updated with, if necessary, the minutes and hours.

The entire application is quite simple and could be written out in pseudo-code as:

```
main()                              // Simple PICmicro MCU Clock
{

int Seconds = 0;
int Minutes = 0;
int TimeInc = 1;

 OPTION_REG = TMR0 - Internal Clock/TMR0 - Prescaler/Prescaler = 64;

 PORTB = 0x0E0;                     // PORTB outputs "Minutes"
 TRISB = 0;                         // PORTB All Output
 PORTA = 0;                         // PORTA is the "Hours" Counter
 TRISA = 0x010;                     // RA4 is an Input Pin

 TMR0 = 0;                          // Wait 1 Second before Starting

 while (1 == 1) {                   // Loop Forever

  while ((INTCON & (1 << T0IF)) == 0);
```

```
  T0IF = 0;                          // One Second has Gone By

  if (RA4 == 0) {                    // The Button has been Pressed
   TimeUpdate(TimeInc);              // Increment the Time
   TimeInc = ((TimeInc << 1) + 1) ^ 0x03F;
   Seconds = 0;
  } else {                           // Update the Second Counter
   TimeInc = 1;                      // Button isn't pressed
   Seconds = Seconds + 1;
   if (Seconds > 59) {               // Minute has gone by
    Timeupdate(TimeInc);
    Seconds = 0;
   }
  }
 }
} // End Simple PICmicro® MCU Clock
```

This code has actually been updated for this edition of the book and uses the clock button time-set algorithm that I first created for *Handbook of Microcontrollers.* When the button is first pressed (it is polled once per second), then the time is incremented by one minute. After incrementing the time, the increment value is shifted up by one and has one added to it to 0x03F (63 decimal). When 63 decimal is added to the *Minutes,* then it will roll over and increment the hours.

Using this algorithm, the clock can roll over 12 hours in just 20 seconds–a lot faster than if you were to hold an Incrementing Minutes button down. As well, it offers nice capabilities to accurately set the clock because the incrementing period changes according to how long the button is pressed. This button set code is a nice algorithm that is easy to implement.

The most complex part of the application is the Timeupdate subroutine. This routine is complex because the upper four bits of PORTB are used to select which 74LS154 is enabled for output (with the lower four bits selecting the 16-bit output of the '154). The code for updating the time uses the *Minutes* variable. It adds the passed value to it appropriately and updates both the minutes and hours, depending on the result.

```
Timeupdate                    ; Add the contents of "w" to PORTB
                              ; ("Minutes")

 addwf  Minutes, f            ; Save the Update

 movlw  60                    ; Are the Minutes Rolled Over?
 subwf  Minutes, w
 btfss  STATUS, C
  goto  TUMinutes             ; No, Just Save the New Value

 clrf   Minutes               ; Reset the Minutes

 incf   PORTA, w              ; Do the Hours get Incremented?
 Andwf  0x00F
 xorlw  0x00C
 btfsc  STATUS, Z             ; Are we Up to 12?
  movlw  0x00C                ; Yes - Reset to Zero
 xorlw  0x00C                 ; Convert the Time Back
 movwf  PORTA

TUMinutes                     ; Convert Minutes to LED Positions
 movlw  0x00F                 ; Get the 16 Minute selection
```

```
 andwf  Minutes, w
 movwf  Temp
 call   Get154               ; Get the enable for the specific '154
 iorwf  Temp, w              ; Combine with the Minutes for the '154
 movwf PORTB

 return
Get154                       ; Use the MS 2 Bits of "Minutes" to
 movlw HIGH Get154Table      ;  Get the '154 Enable Setting
 movwf PCLATH
 swapf Minutes, w
 andlw 2
 addlw LOW Get154Table       ; Compute the Table Address
 btfsc STATUS, C
  incf PCLATH
 movwf PCL
Get154Table
 retlw 0x0E0                 ; Minutes 0 - 15
 retlw 0x0D0                 ; Minutes 16 - 31
 retlw 0x0B0                 ; Minutes 32 - 47
 retlw 0x070                 ; Minutes 48 - 59
```

This code takes a maximum of 34 instruction cycles to execute. This code, with the code presented previously means that the vast majority of the time in this application, the PICmicro® MCU spends processing is spent polling on the T0IF flag instead of calculating the new LED display value. This is important to avoid any possibility that the T0IF flag being set is missed (and the time is lost).

## XMAS: CHRISTMAS DECORATION WITH BLINKING LIGHT AND MUSIC

If you look through the hobbyist electronics magazines in December, you are sure to find a number of flashing-light PICmicro® MCU applications. These applications usually use a number of LEDs to flash on and off in a seemingly random pattern. These lights are often put in a Christmas tree decoration or a stand-alone decoration as is shown in Xmas Project (Fig. 16-16), which has 15 flashing LEDs and can play a tune.

This is the third time I've created a lighted Christmas tree with a PICmicro® MCU. In the first case, I handwired all the connections (which made it truly horrific to wire). The second version used a predesigned PC board, but the design was lost in a hardfile crash (before I started religiously backing up files onto CD-ROM) and I wrote the software for an early PICmicro® MCU compiler, which never worked extremely well. I ended up updating it just for the light application (and the "tune generator" that went with it). Also with the second version, I created an unnecessarily complex tune generator, which requires two separate PC applications to run before the code can be built with the "tune" installed.

This third version avoids the problems of the other two and, by using LEDs with leads soldered to them, the build time for the entire project is just a few hours (not including decoration paint drying time).

The application itself supports up to 16 LED outputs, a speaker with software that can drive a 128-note tune, and a built-in power supply so that the application can be run from a wall-mounted AC/DC adapter. The schematic is shown in Fig. 16-17.

The bill of materials for the project is shown in Table 16-9. Notice a couple of things in this application. First, there is no on/off switch for power. I left this off because I assumed

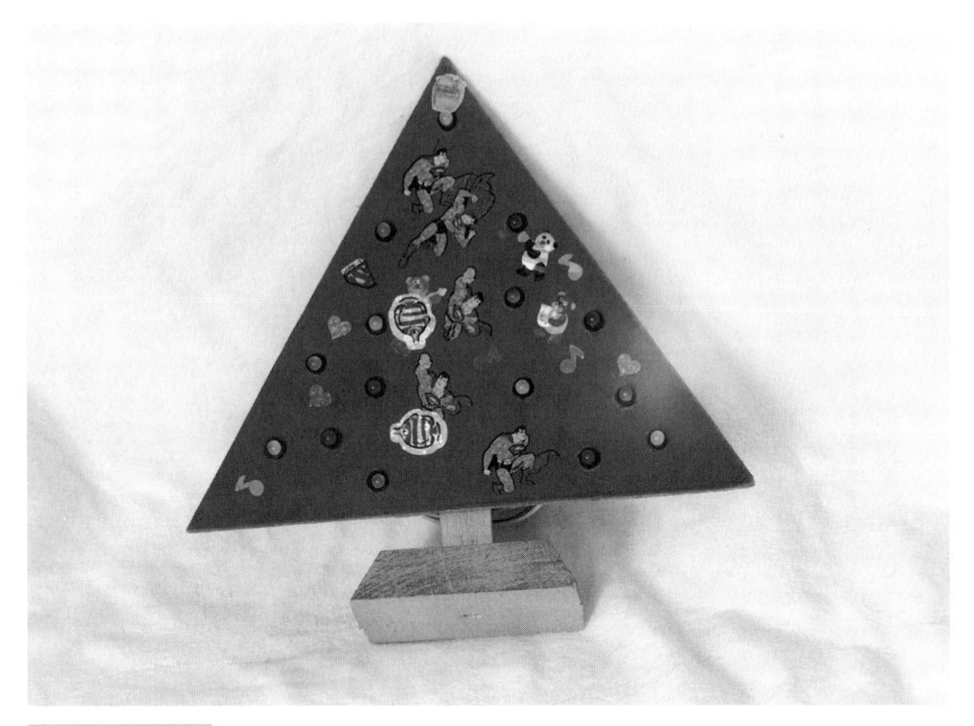

**Figure 16-16**

that once power was applied, the decoration would run continuously. Next, I do not interconnect the 74LS374's, which are used as shift registers, because ample pins are available for the application and randomizing can be done in software. Last, I used a 3.58-MHz ceramic resonator instead of my typical 4.0 MHz one because I ran out of 4-MHz ceramic resonators. This change did not result in a major problem with timing the application, as is described in the following paragraphs. The macro calculator was used to calculate delays and a 4-MHz ceramic resonator or crystal could be substituted in its place.

The software used to run the application could be considered to be a simple multitasker with two "processes." The crucial process is the tune process, which plays a tune programmed into the program memory. The other process is a pseudo-random number generator using a linear feedback shift register that runs in the foreground. The application can be modeled as:

```
main ( )                        // Xmas Tree high level code
{

int TuneDelay = 1;
int TuneIndex = 0;
int Data = 0x0FFFF;

 PORTA = output;                // Set Up Outputs
 OPTION = TMR0 Instruction Clock | TMR0 Prescaler;
                                // Set Up Timer
 INTCON = GIE | T0IE;           // Enable Timer Interrupt

 While (1 = = 1) {              // Loop Forever Updating LEDs
```

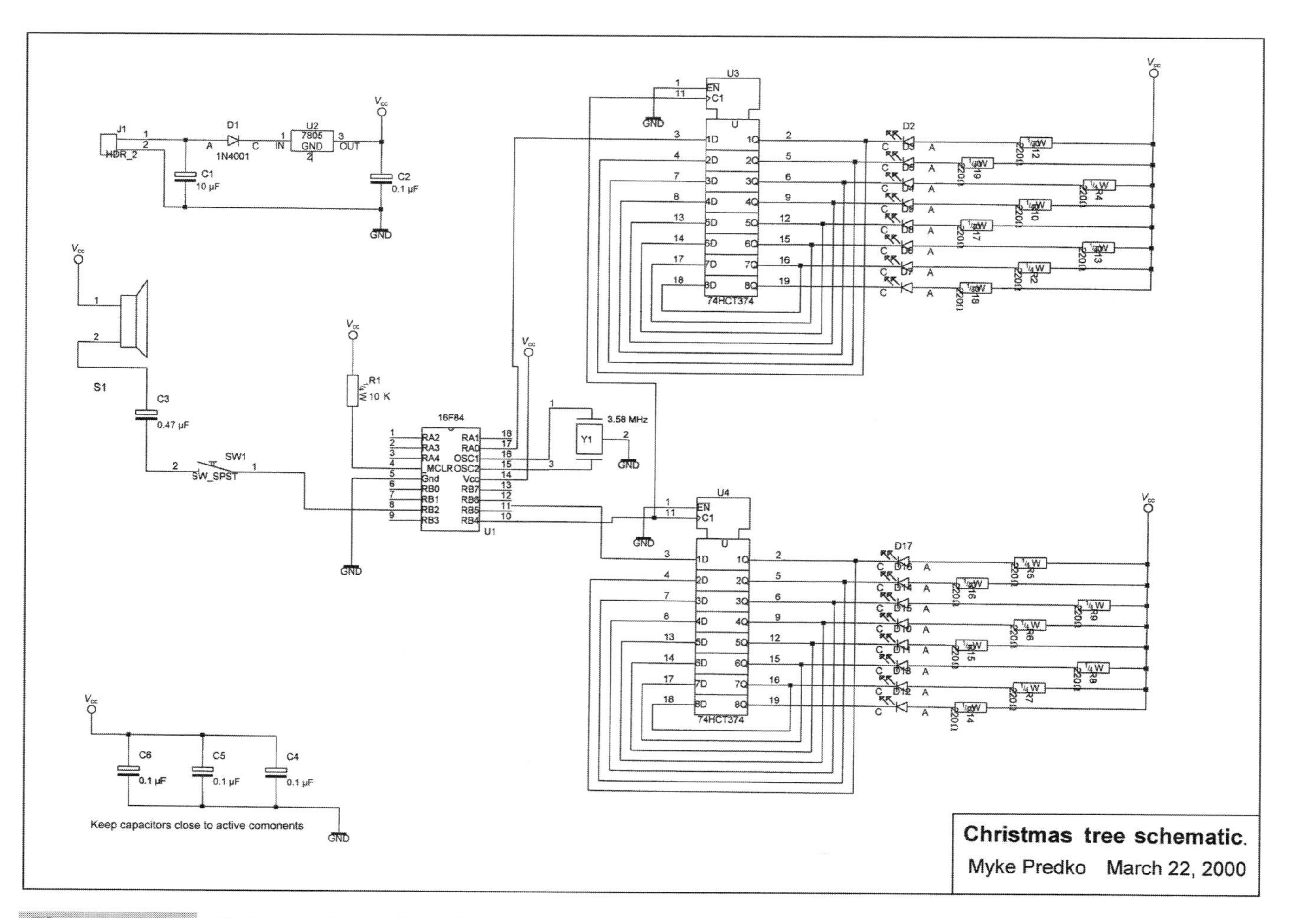


Figure 16-17 Christmas tree schematic

**TABLE 16-9 "Xmas" Bill of Materials**

| PART | DESCRIPTION |
|---|---|
| U1 | PIC16F84-P/04 |
| U2 | 7805 |
| U3 - U4 | 74LS374 |
| Y1 | 3.58 MHz Ceramic Resonator |
| D1 | 1N4001 Silicon Diode |
| D2 - D17 | Leaded, Multi-colored LEDs |
| C1 | 10 $\mu$F, 16 Volt Electrolytic Capacitor |
| C2, C4-C6 | 0.1 $\mu$F, 16 Volt Tantalum Capacitors |
| C3 | 0.47 Capacitor (Any Type) |
| R1 | 10 K, 1/4 Watt Resistor |
| R2-R17 | 220 $\Omega$,1/4 Watt Resistors |
| SW1 | Single Pole, Single Throw Switch |
| SPEAKER | Piezo Electric Speaker (Digi-Key Part Number P9922-ND) |
| J1 | 2.5 mm Power Connector |
| Misc. | PCB Board, Wall Mounted 8 Volt, 500 mA AC/DC Power Supply |

```
   dlay();                        // Delay 1 second

   Data = (data << 1) + (Data.15 ^ Data.12);
                                  // Create Pseudo Random Number
   ShiftOut(Data);                //  and Shift it out to the LEDs

 }
} // End Xmas Mainline

Interrupt TMR0Int()               // Timer0 Interrupt Handler
{

 TMR0 = Note(TuneIndex);          // Reset TMR0 to Delay Again
 T0IF = 0;

 TuneDelay = TuneDelay - 1;
 if (TuneDelay = = 0) {           // If Tune Delay Finished, Output a

  TuneIndex = TuneIndex - 1;      //  New Note
  if Note(TuneIndex) == 0)        // At End of the Tune?
   Tune index = 0;                // Yes, Reset and Start Again

  TuneDelay = Delay(TuneIndex);

 } else                           // Toggle Note
  if (Note(TuneIndex) != Pause)
     SpkrOutput = SpkrOutput ^ 1;

} // End TMR0 int.
```

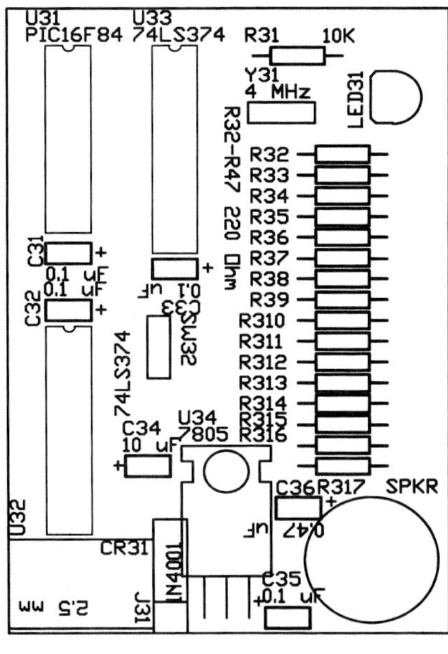


**Figure 16-18** Christmas tree PC board overlay

The table (listed as the *Note* and *Delay* functions in this code) contains the "tune" with notes and duration programmed in using the Note Add macro. This macro will allow the development of the code needed for a tune very quickly. Just transcribe the tune into the note range described in the following paragraphs and program a PICmicro® MCU.

After getting a PC board built, you will be able to solder the application together in less than an hour. The PC board design is shown in Figs. 16-18 through 16-20.

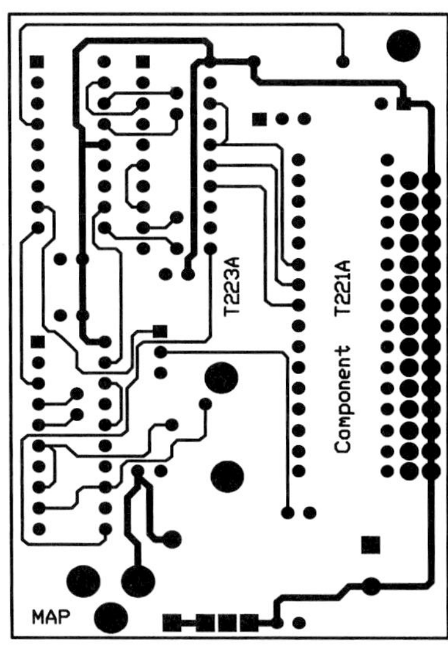


**Figure 16-19** Christmas tree top-side copper PC board

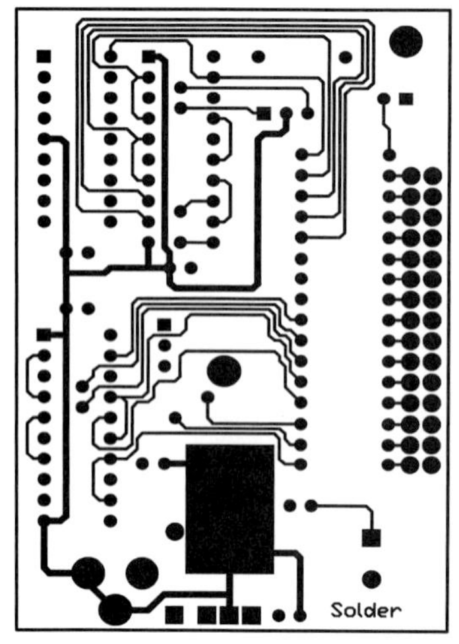


**Figure 16-20** Christmas tree top-side copper PC board

For my application, I created the wooden Christmas tree shown in Fig. 16-22. The PC board has two 0.125″ holes drilled in it so that you can mount it to a display. When I mounted my board to the Christmas tree, I simply screwed it to the back of the tree using small wood screws (Fig. 16-21).

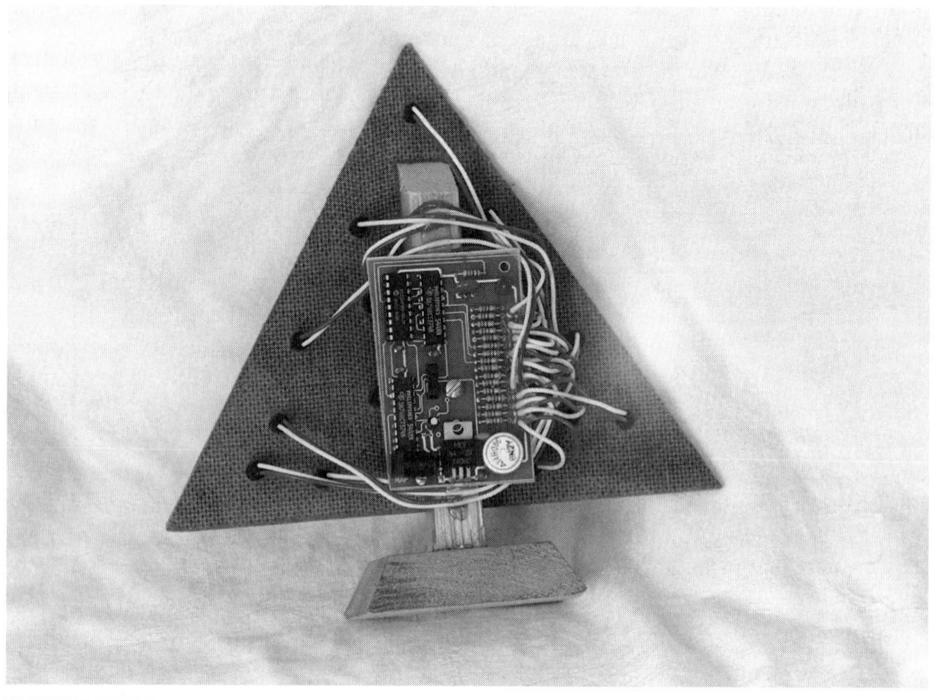

**Figure 16-21** Backside of Christmas tree display

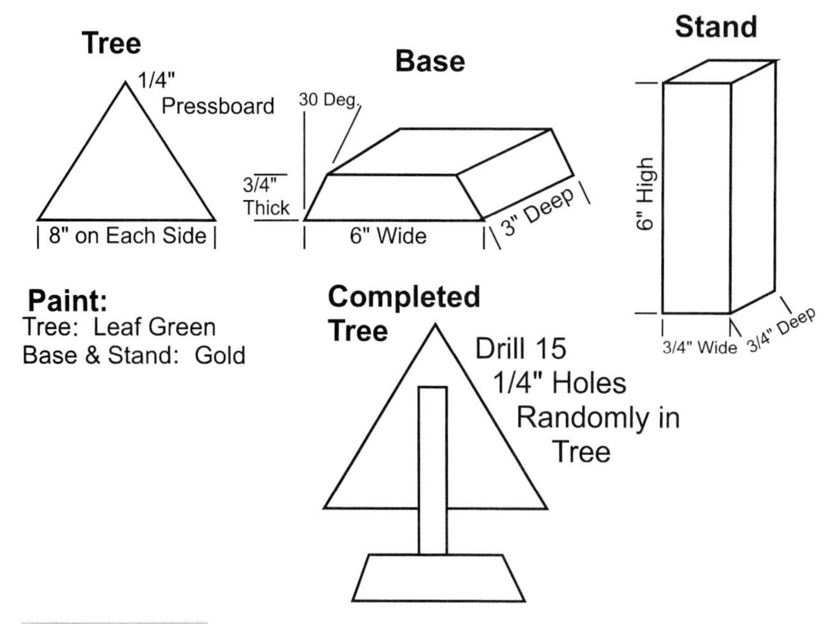


**Figure 16-22** Wood pieces for Christmas tree display

When I assembled this project, I tried to keep the LED wires away from the PC board to allow removal of the PICmicro® MCU for reprogramming, as well as allow me to fix any potential problems (there weren't any).

For previous Christmas tree projects, I laboriously soldered leads onto LEDs and then glued them into holes on the tree. For this project, I was able to find LEDs that were presoldered onto leads and glued into panel-mount plastic carriers. These parts are a bit more expensive than buying LEDs and wires, but the time saved made the project quite enjoyable, rather than a particularly odious chore.

For my Christmas tree, I used a piece of scrap pressboard cut into an equilateral triangle 8 inches per side and 3/4-inch thick piece of wood for the stand (Fig. 16-22). I painted the triangle forest green, and the base and the stand gold. To assemble the tree, I simply glued the triangle to the stand and screwed the stand to the base.

After the glue and paint had dried, I inserted the LEDs into the holes and soldered their leads to the PC board. The total time for me working on the tree was about two hours, with a total elapsed time of three days (mostly for the paint and glue to dry).

When the tree was completed, it seemed to be a bit "bare." To help dress it up, I bought a few sheets of small, sparkly stickers showing various toys and Christmas decorations and gave them and the tree to my 4 1/2-year-old daughter to decorate. This was a lot of fun for her. As a bonus, if enamel paint is used with a smooth surface for the tree triangle, the stickers can be peeled off easily and the tree "redecorated" the next year.

Creating tunes is a relatively simple chore. Once you have chosen the song you would like the application to play, you will have to "transpose" it into the notes provided in Table 16-10.

When specifying the time for the note, the delay is given in 1/16th beats. To specify a tune, the notes are specified, along with the number of 1/16th beats. For example, the first line of "O, Christmas Tree," which is:

**TABLE 16-10 "Xmas" Bill of Materials**

| NOTE | LABEL | FREQUENCY | COMMENTS |
|---|---|---|---|
| B | lnB | 494 | B Below Middle "C" |
| C | nC | 523 | C - Middle "C" |
| C Sharp | nCS | 554 | |
| D | nD | 587 | |
| D Sharp | nDS | 622 | |
| E | nE | 659 | |
| F | nF | 699 | |
| F Sharp | nFS | 740 | |
| G | nG | 784 | |
| G Sharp | nGS | 831 | |
| A | Na | 880 | |
| A Sharp | nAs | 923 | |
| B | nB | 988 | |
| C | hnC | 1047 | One Octave Above Middle C |
| C Sharp | hnCS | 1109 | |
| D | hnD | 1175 | |
| D Sharp | hnDS | 1245 | |
| E | hnE | 1319 | |
| Pause | nP | N/A | Pause a Sixteenth Beat |

```
"Oh  Christ-mas  tree, Christ-mas  tree, how lovely are..."
 1/8D 3/16G 1/16G 1/8A  3/16B 1/16B 3/8B 1/8B 1/8A  1/4C...
```

is written out as:

```
nD,2,nG,3,nG,1,nG,6,nA,2,nB,3,nB,1,nB,6,nB,2,nA,2,nB,2,hnC,4...
```

The application is timed for 72 beats per minute, which allows me to do a pretty good rendition of "O, Christmas Tree."

Other songs can be very easily programmed in. To do this, I recommend buying a cheap beginner's piano book with Christmas carols (or other songs) and transcribing the notes. This may be surprising, but the most difficult aspect of this application for me was to find and transcribe a version of "O Christmas Tree" that had notes that fell in the proper range for the application.

The pseudo-random formula:

```
Data = (data << 1) + (data.15 ^ data.12)
```

shifts the current value up by one bit and then fills in the least-significant bit with the most-significant bit and another XORed together. This seems unbelievably simple (and it is), but it does a very good job of randomizing the data as a very simple linear-feedback shift register.

The two 74LS374s are wired as synchronous serial shift registers. I use this format for a variety of different applications to shift data serially out to a parallel I/O. To shift the data out, notice that I perform a shift on the data with the carry up from the lower byte being used as the LSB of the upper (the lower byte's LSB is the pseudo-random value described previously.

The XMAX040.ASM application code is located in the *code\Xmas* subdirectory of the PC's hard file *PICmicro* directory. If you decide to put in your own tune, change the name of the file so that you don't overwrite the original application.

## FANCTRL: SIMPLE FAN SPEED CONTROLLER

As I was writing this book, I got the question about how to use the PIC12CE6xx *Analog-to-Digital Controller (ADC)*. My immediate reply was that it was the same as for the other ADC-equipped PICmicro® MCUs. Then I got the question "How are the 'eight pin' features, like the Internal RC oscillator, the OSCCAL register, and TMR0 input (T0CKI) pins handled?" This project is a result of these questions, as well as looking at implementing a PWM application in PICmicro® MCUs that do not have the TMR2 and CCP hardware built in. This project is very fast to build and a good way to experiment with controlling DC motors (not just fans) using the PICmicro® MCU.

The application circuit is very simple and only eight parts have to be wired together. The schematic is shown in Fig. 16-24.

The bill of materials needed for the fan controller is shown in Table 16-11. I built my prototype on a fully drilled PC board prototyping card and it took me less than half an hour to solder together the various components. I did this because the circuit is so simple that if I had made any mistakes, I could rewire and add components very easily. Once the circuit was tested, I used Weldbond to hold the wires to the backside.

Notice in the photograph (Fig. 16-23) of the project that I used a 2.5-mm power plug for the input where I simply soldered some 20-gauge solid core wire to the power plug and wired it to the screw terminals used for the power connection.

The fans that I used were 12-VDC muffin fans that I bought at a local surplus store for a few dollars each. As I indicated previously, the fans could be replaced with 12-VDC electric motors. Fans are nice because their operation will give you immediate feedback as to the output speed in which they are running. The application circuit with the software provided in this project will provide an output PWM duty cycle signal anywhere from 1.5 to 90.3%. The reasons why I could not go from 0 to 100% duty cycle are described in the software description that follows.

The potentiometer is used with the PICmicro® MCU's ADC to provide the operating level for the fan. I used the GP0 pin for the analog voltage input with the remaining five pins as digital I/O. The potentiometer will provide voltages anywhere from 0 to $V_{cc}$ (5 volts).

The motor control simply consists of an IRF510 N-channel MOSFET driver transistor. This transistor is quite inexpensive and can handle up to 5.6 amps. It has a 10-volt input and an on resistance (between the drain and source) of 0.54 Ω. This transistor is used as a switch for the PWM output. Figure 16-25 shows the PWM signal for a slightly less-than-

**Figure 16-23** PIC12CE673 fan motor controller

half $V_{cc}$ analog input. Figure 16-26 shows the PWM running at the maximum duty cycle of 90.3%. In both Fig. 16-25 and Fig. 16-26, the top line is the PWM signal out of the PICmicro® MCU and the lower line is the analog voltage input.

The actual application code is very simple and only requires 42 instructions. It was de-

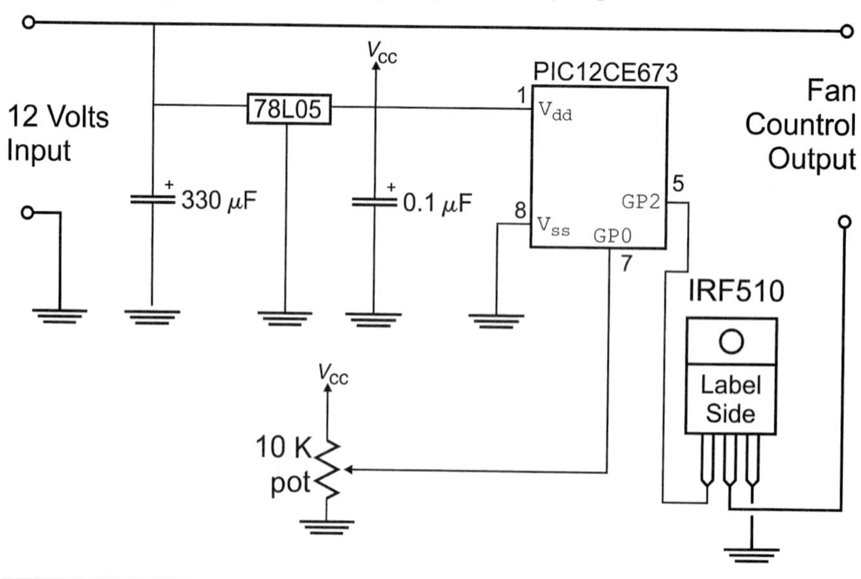


**Figure 16-24** Very simple tan-control project

**TABLE 16-11 "Fanctrl" Bill of Materials**

| PART | DESCRIPTION |
|---|---|
| PIC12CE673 | PIC12CE673-JW |
| 78L05 | 78L05 in a TO-92 Package |
| IRF510 | International Rectifier IRF510 in a TO-220 Package |
| 0.1 $\mu$F | 0.1 $\mu$F, 16 Volt Tantalum Capacitor |
| 330 $\mu$F | 330 $\mu$F, 50 Volt Electrolytic Capacitor |
| 10 K | 10 K PCB Mount Single turn Potentiometer |
| Misc. | 2x Two Screw Terminal Connectors, PCB Prototyping Board, Wiring, "Muffin Fan" |

signed to output a 10-kHz PWM signal, with the ADC being polled in between each ADC "pulse." This polling and the software associated with it are the reason why there is a longer inactive period at full speed that can be seen in Fig. 16-26.

The pseudo-code for the application is:

```
main()                          // Fan Control Application
{

int ADCON = 32;
int ADCoff = 1;
int i;
```

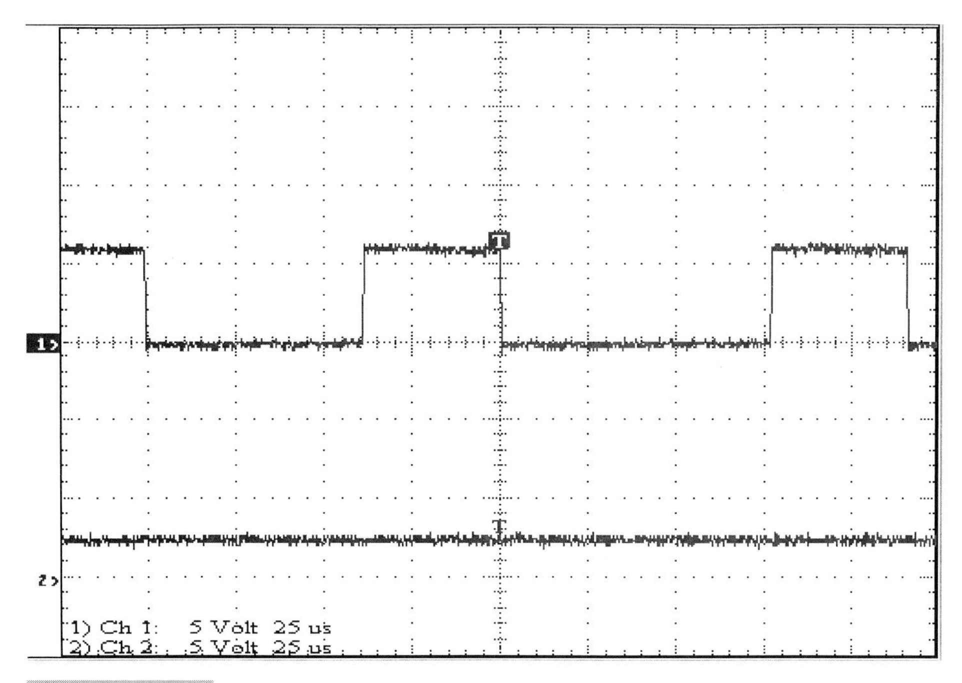


**Figure 16-25** Fan-control project running at half speed

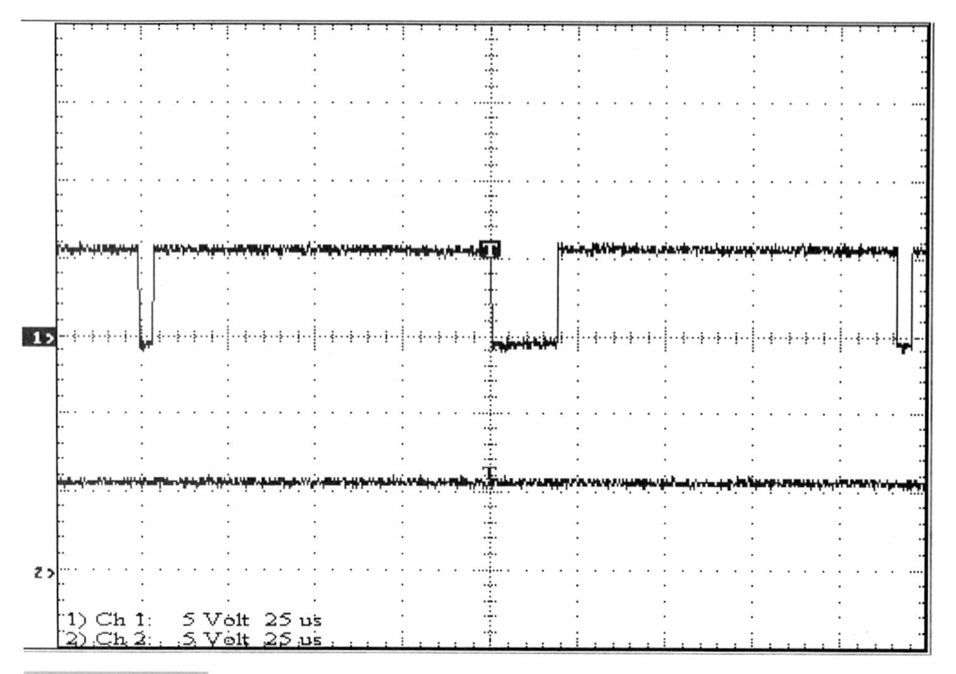


**Figure 16-26** Fan-control project running at full speed

```
OSCCAL = CalibrationValue;

ADCON1 = 6;                     // All I/O Pins Digital Except for GP0
ADCON0 = 0x041;                 // Enable the ADC

TRISIO = 0x03F ^ (1 << 2);  // GP2 is the PWM Output

while (1 == 1) {                // Loop Forever

 GP2 = 1;                       // Output the PWM Signal while ADC
                                //  Capacitor is Stabilizing to the Pot
                                //  Input
 for (i = 0; i < ADCON; i++ );
 GP2 = 0;
 for (i = 0; i < ADCoff; i++ );

 ADCON0 = ADCON0 | Go;          // Start the ADC Operation

 GP2 = 1;                       // Output the PWM Signal while ADC
                                //  is processing the Pot Input
 for (i = 0; i < ADCON; i++ );
 GP2 = 0;
 for (i = 0; i < ADCoff; i++ );

 ADCON = (ADRES>>3) + 1;    // Get the ADC Value and Scale it for
 ADCoff = 33 - ADCON;       //  the Application

}
} // End Fan Control Application
```

In this application, notice that I output the PWM signal twice for each ADC sample loop. During the first PWM signal output, the ADC's input capacitor is allowed to stabilize. During the second PWM output, the ADC operation is allowed to occur. I set the ADC with the divide-by-eight TAD clock.

The actual assembly-language code is not much more complex and is located in the *code\fanctrl* subdirectory of the *PICmicro* directory.

In the application, I want to bring a few things to your attention. The first is that the PWM output operation, which was coded in assembly language as:

```
bsf     PWMOut                  ; Output the First Value
decfsz  ADCON,  f
 goto   $ - 1
bcf     PWMOut
decfsz  ADCoff, f
 goto   $ - 1
```

This code takes 99 instruction cycles under all cases and I have written the ADC "scaling" so that the value will always be between one and 32. By making the value "one based" instead of "zero based," I avoided the problem where if one of the ADC time values was equal to zero, it would loop 256 times.

The second issue is the scaling routine, which takes more cycles that I would have liked:

```
rrf    ADRES, w                 ; Read the ADC result
movwf  ADCON
rrf    ADCON, f                 ; Convert to a 1-32 Value
rrf    ADCON, f
movlw  0x01F
andwf  ADCON, f
incf   ADCON, f
```

The problem is that the ADCON scaling has to be:

```
ADCON = (ADRES >> 3) + 1;
```

for the PWM code to work properly.

In the mid-range architected eight-pin PICmicro® MCUs, the calibration value is located at the last address in memory, just the same as in the low-end architected eight-pin parts. The difference is, the mid-range parts calibration value is part of a subroutine and consists of a *retlw calibration* instruction instead of the *movlw calibration* of the low-end devices.

To access the calibration value and program the OSCCAL register, I used the code:

```
bsf     STATUS, RP0
ifndef  Debug
  call    0x03FF                ; Read the Calibration Value
else
  nop
endif
errorlevel 1, -219
  movwf  OSCCAL ^ 0x080
errorlevel 1, +219
  bcf   STATUS, RP0
```

which I am sure is definitely more complex than you would have expected. When testing the application in MPLAB, the calibration subroutine (the *retlw calibration value* instruction) is not simulated. Thus, a call to the last address in program memory (address 0x03FF) will actually be to a 0x03FFF (*addlw 0x0FF*) instruction, which will then roll over and start executing from address 0 (the reset vector) again.

To avoid getting into an endless loop in the simulator, I put in the Debug Define flag. When simulating, I ignore the calibration value read all together.

I was surprised to find that writing to OSCCAL returned the invalid register address warning in MPLAB. To suppress the message, I used the *errorlevel* directive and then enabled it after the instruction (to catch any other problems in my code).

With the hardware built and application code written, I tried the code out on two types of fans. For a standard DC fan, the application ran perfectly. The second fan had a "tachometer" output and had a very noticeable high-frequency "whine." This whine was caused by the PWM signal turning on and off the power to the fan. To fix this problem, the application would have to be re-written for a higher-frequency PWM output (on the order of 20 kHz is the general rule).

To get the higher PWM frequency, the application's clock would have to be increased. To keep the application simple, I elected to use the PICmicro® MCU's built in 4-MHz RC oscillator. If I were to increase the operating frequency, I would have had to use an external oscillator.

## IRTANK: TV I/R REMOTE-CONTROL ROBOT

Most engineers enjoy playing around with robots. I'm sure if you were to question their appeal, you'd hear a lot of comments about their interest in digital-control theory, artificial intelligence, and electro-mechanical interfaces. Personally, I think most robots are built so that their creators have something to boss around.

This chapter presents two different robot applications that can be controlled by a PIC. This section shows how to use the PICmicro® MCU with a TV remote control to send commands to a tracked vehicle. Towards the end of the chapter, I go through a programmable servo controller.

The genesis of this project was an article in *Electronics Now* (earlier in this chapter, I made the comment about how hobbyist magazines always have Christmas decorations in their November/December issues. If you watch them, you'll also notice a number of different robot designs throughout the year. The project was for a simple "differential drive" robot that could be controlled by a TV remote control. After looking through the article, I decided to build my own robot, but I could improve upon a few things (Fig. 16-27).

The robot in the article used a single *InfraRed (I/R)* receiver (typically used with TV remote controls) to control the robot. Two motors (for the robot's left and right sides) were switched on and off by an H-bridge motor control made out of discrete transistors.

The magazine robot used a PIC 16C5x for control. I felt that by using a mid-range part, I would have the advantage of interrupts to handle the incoming commands (which I refer to as *data packets*).

Most (if not all) I/R TV remotes use a Manchester encoding scheme in which the data bits are embedded in the packet by varying the lengths of certain data levels. Figure 16-28 shows a theoretical perspective of the data and Fig. 16-29 shows the actual output from an infrared receiver.

The normal signal from an I/R receiver circuit is high when nothing is coming (line idle)

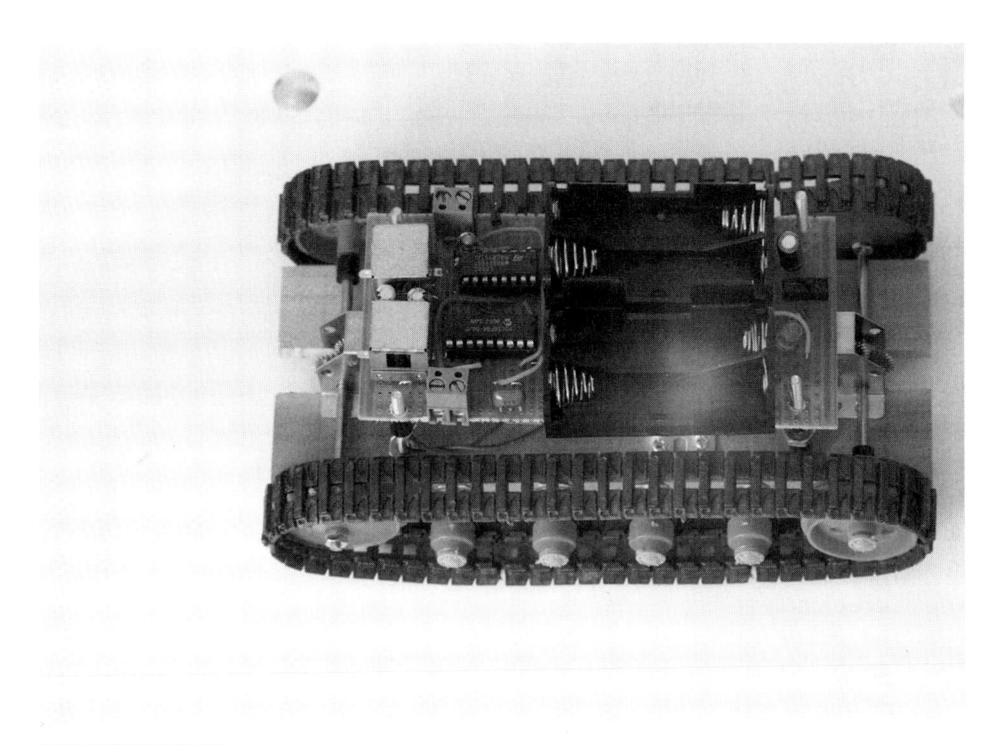

**Figure 16-27** I/R controller "tank"

and then goes low with a "leader" signal to indicate that data is coming in. The data consists of a bit synch. When it completes, the bit value is transmitted as the length of time before the next bit synch. The I/R robot is designed for Sony-brand TV remotes, which have 12 data bits and a 40-kHz carrier. The timings are as given in Table 16-12 (and use a base timing $T$ of 550 $\mu$s).

To read the incoming signal, I used the following state machine code in an interrupt handler that differentiated between TMR0 overflows and PORTB change on interrupt:

```
interrupt IRTankInt()         // Handle Incoming IR Serial Data
{

 if (INTCON.TOIF != 0) {
                              // TMRO Overflow - Timeout Indicating Invalid
                              //  Data Being Received
  BitCount = 0;               // Prepare for the Next Packet
  State = 0;
```

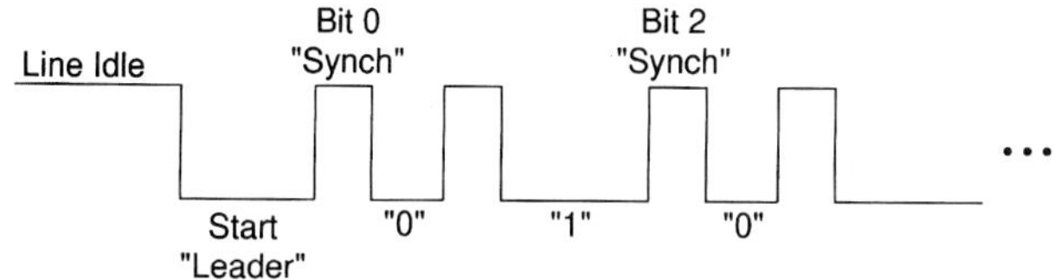


**Figure 16-28**

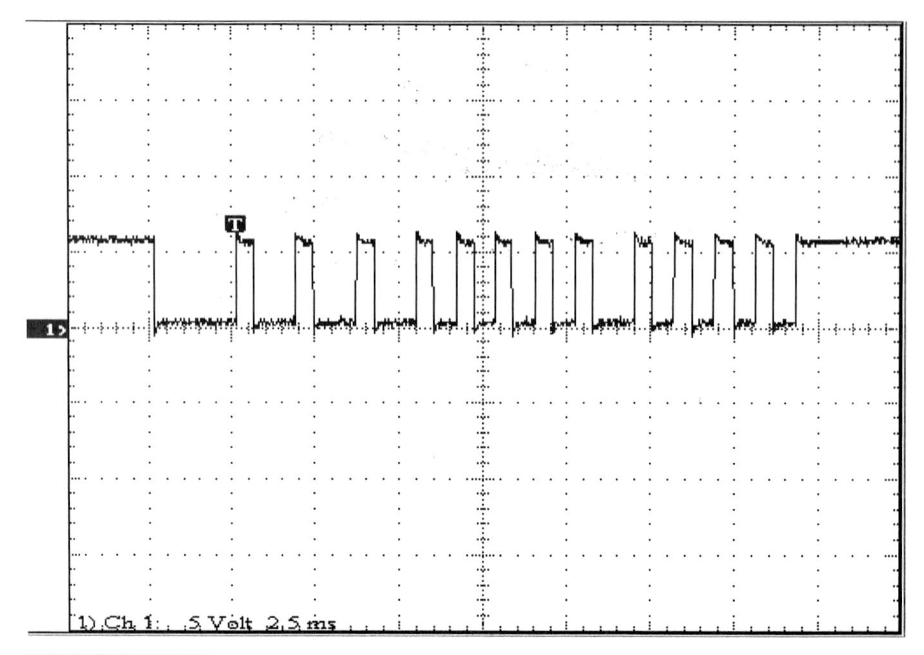


**Figure 16-29** Scope view of TV I/R remote-control input

```
 INTCON = 1 << RBIE;       // Wait for Change on PORTB Interrupt Request
                           //  Also Reset "TOIF"
} else {                   // Else, Handle Incoming Bit Transition

 Temp = PORTB;             // Save the Changed State in PORTB to Reset
                           //  Interrupt Request
 INTCON = (1 << RBIE) + (1 <<TOIE);
                           // Reset and Enable TMR0 and RB Interrupts
 switch(State) {           // Handle PORTB Change Based on Current State
  case 0:                  // Expect Leader/Synch Coming In

   TMR0 = 5msec;           // Leader Valid for 2.2 msec - Put in Timeout

   State = State + 1;

   Value = 0;              // Haven't received anything yet

   break;
```

**TABLE 16-12 Sony I/R Remote Control Timing**

| FEATURE | "T" TIMING | ACTUAL LENGTH |
|---|---|---|
| Leader | 4T | 2.20 msec |
| Synch | T | 0.55 msec |
| "0" | T | 0.55 msec |
| "1" | 2T | 1.10 msec |

```
    case 1:                        // End of Leader/Get "Synch"

     TMR0 = 1 msec;                // "High" Synch Pulse is Valid for 0.55 msec

     BitCount = 12;                // Expect twelve Incoming Bits

     State = State + 1;

     break;

    case 2:                        // Start of Synch Bit - Come back here for
                                   // Each One
     TMR0 = 2 msec;                // "Low" Value can be valid for 1.10 msec

     State = State + 1;

     break;

    case 3:                        // Bit Ended - Use TMR0 to Get Value

     Value = Value >> 1;

     If (TMR0 < 0.8 msec)
        Value.msb = 1;
                                   // If Less than 0.80 msec has passed, then
                                   //  "1" was received
     BitCount = BitCount - 1;
     if (BitCount == 0) {
      DataFlag = 1                 // Flag Mainline that Data is Available

      INTCON = 1 << RBIE;

     }
   } // end Switch
  } // end if
} // End IRTankInt
```

This code has the advantage of being able to discriminate between different manufacturer's I/R codes (the differences lay in the length of time during the transitions, the number of bits, and the I/R carrier frequency).

As stated earlier, one of the features I didn't like about the magazine article robot was its use of a single I/R receiver. I was concerned that this would not provide adequate reception in a crowded room, where the robot could be turned in different directions relative to the remote control.

For my robot, I decided to use two I/R receivers pointing 180° apart (this can be seen in the photograph; the I/R receivers are the two square metal cans at the front of the robot). This gives almost 360° coverage.

By doing this, a new problem came up: arbitration. I found that one receiver might not receive the transmitted signal, or that it would be a 40-kHz cycle (or two) ahead or behind the other receiver. The arbitration scheme used is quite simple; the first receiver to transmit the leader to the PICmicro® MCU will be the one that is "listened to" for the remainder of the data packet.

The application code is located in the *code\IRTank* subdirectory of the *PICmicro* directory. The application has been updated from the first edition's code to use a PIC16F84 instead of a PIC16C84.

IRTANK.ASM was based on Test9A from the first edition. This was the last application in the series of programs written to understand how TV remote controls work and then build up the understanding into the final application. One thing that I am proud of this application code is, when I did it, nothing (like an oscilloscope) was used to debug it. I was able to figure out the actual times by writing simple applications that wrote out signal timing and I was able to differentiate the different codes that were being sent. Once I had enough information to understand how the remote control worked, I was able to write the application.

The I/R receiver the code was developed using a LiteOn 40-kHz I/R remote-control receiver module (Digi-Key part number LT1060-ND). This receiver was connected to a PICmicro® MCU that had a number of LEDs on PORTB/PORTA to try and decode what was happening.

Once I had the infrared interface worked out, I thought I was away to the races. Unfortunately, this did not turn out to be the case.

As a robot platform, I bought two Tamiya tracked vehicle parts kits (Item 70029). Each one has a single electric motor and gearbox. Buying two gave me a motor and gearbox for each side's tracks. Running the motors, I found that they used about 250 mA, which was within the stated current ranges of the H drive used in the magazine robot's circuit. I then went off happily building the magazine circuit's H drive.

After hooking up the recommended circuit, nothing happened, except that the transistors got very hot (one actually exploded). This lead to (literally) several weeks of asking questions and experimenting with different circuits (I even pulled out some of my old transistor textbooks; when I did that, I knew I was really desperate) until I determined that I had a circuit that wouldn't work for this application.

To make a long story short, I decided to see what others had done before me. Looking up robotics web sites (such as the Seattle Robotics Society), I discovered the L293D chip. This is a single IC that provides the H drive function for two motors (making this application much easier to wire than the original circuit, which used eight discrete transistors and resistors).

The L293D consists of four drivers (with clamping diodes) and is designed to be used as a two-motor H-drive control. Figure 16-30 shows how one half of the 293D can be used to control a motor.

Now, with everything working properly, I created the circuit shown in Fig. 16-31. The controller to be used is a universal remote-control set to Sony TV. For the electronics, I used the bill of materials shown in Table 16-13.

Four AA alkaline batteries provided power for my prototype. Using nickel-cadmium batteries, I found the robot to be very sluggish. Looking for the cause of the sluggish performance, I found that the tracks and axles were binding because they weren't exactly square.

The Tamiya-tracked vehicle kit provides a piece of hardwood to put on the various pieces. This is an extremely hard piece of wood and it didn't take well to non-precision parts placement. I ended up using the piece from the second kit and had better results.

A much better solution for this project is to buy a Tamiya wall-hugging mouse kit (item 70068-1300), which contains two motors, two gear boxes, and a plastic chassis to keep everything aligned. In this case, the robot worked a lot better (giving my kids something to terrorize our cats with).

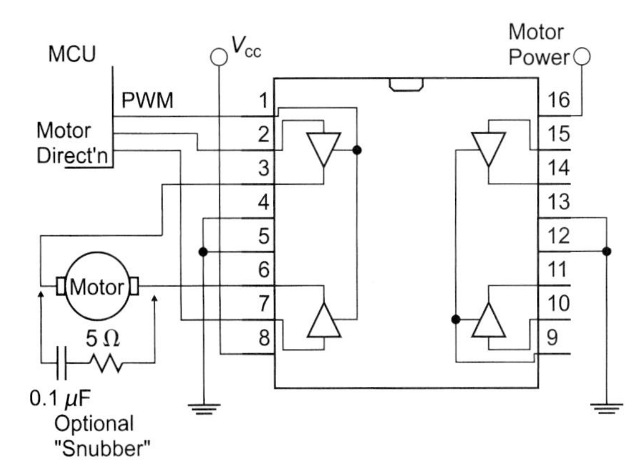


**Figure 16-30** Wiring a motor to the 293D

## IRBETTER: ADDENDUM I/R ROBOT

Sometime after I completed the I/R-controlled robot (and the kids were well on their way to demolishing it), the question came up on the PICList on alternative methods of reading infrared remote-control transmissions. The question was about sampling and "learning" the received data packet, rather than comparing each incoming bit to a predetermined value. The question sparked the idea for representing the codes inside the PICmicro® MCU in other ways.

My first attempt at this was to count the numbers of ones during a sample period knowing that a characteristic pattern of ones and zeros would be sampled.

This was the beginning of IRLCD_2 in subdirectory *IRBetter* in the *PICmicro* directory. The program used the following schematic on a breadboard.

Notice that this circuit uses the serial-LCD interface described earlier in this chapter. Using the LCD really made debugging this project/experiment a lot easier and it really showed me the use of simple interfaces. The YAP's built-in serial port could be used to make this project very simple to wire and interface to.

The main body of the code, where the I/R stream is read/sampled is:

```
 movlw    0x0A0             ; Setup the Timer Interrupt
 movwf    INTCON

Loop                        ; Loop Here for Each Update of Screen

 movlw    200               ; Wait for the Time Out
 subwf    IntCount, w
 btfss    STATUS, Z
  goto    Loop              ; Has NOT timed out

 movlw    200               ; Can we Display?
 subwf    ReadCount, w
 btfsc    STATUS, Z
  goto    Loop_Reset        ; Reset the Count Values

 movf     ReadCount, w      ; Now, Display what was read in

 clrf     IntCount          ; Clear the Display Values
 clrf     ReadCount
```

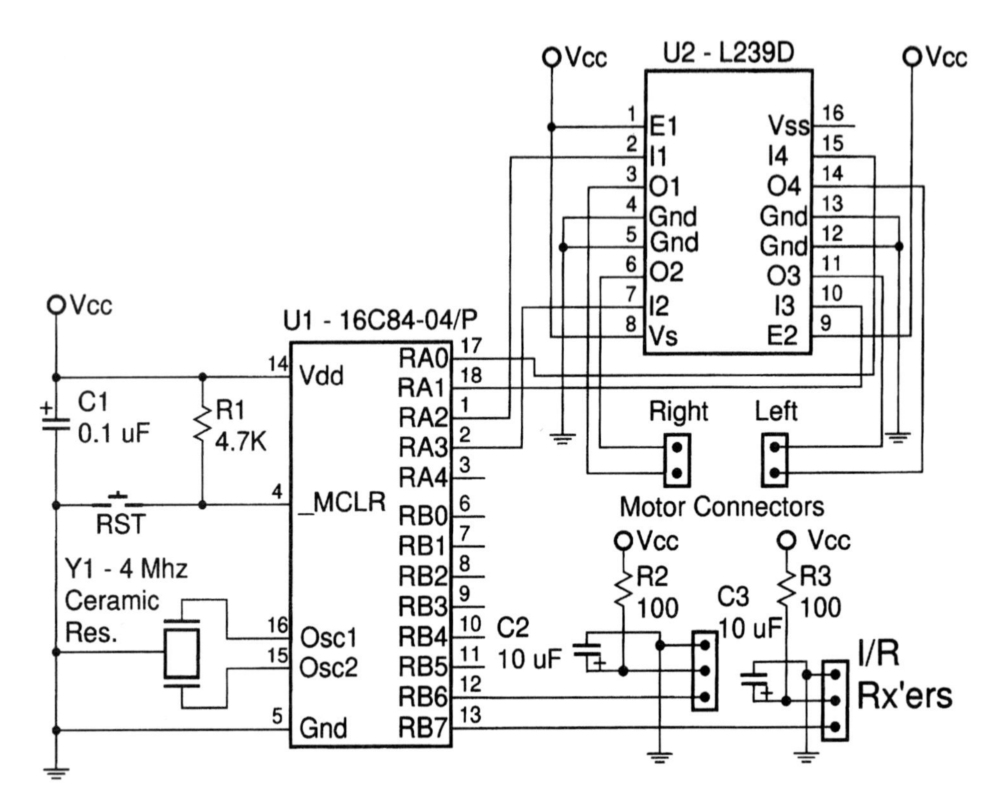


**Figure 16-31** I/R controlled robot schematic

```
 call       DispHex               ; Display the Hex Value

 movlw      0x08E                 ; Reset the Cursor for Writing
 call       WriteINS

 goto       Loop                  ; Wait for the Next Loop Around

Loop_Reset

 clrf       IntCount              ; Reset the Values
 clrf       ReadCount

 goto       Loop

Int                               ; Interrupt, Check I/R Input

 movwf      _w                    ; Save the Context Registers
 swapf      STATUS, w
 movwf      _status

 bcf         INTCON, T0IF         ; Clear the Timer Interrupt

 incf       IntCount              ; Increment the Count Register

 btfsc      PORTB, 6              ; Increment the Read Value?
  incf      ReadCount
```

**TABLE 16-13 "I/R Tank" Bill of Materials**

| PART | VALUE | COMMENTS |
|---|---|---|
| U1 | PIC16F84 | |
| U2 | 293D | |
| R1 | 10 K 1/4 Watt | |
| R2, R3 | 100 Ohm, 1/4 Watt | |
| C1 | 0.1 $\mu$F Tantalum | Tantalum should be used to provide best filtering possible |
| C2, C3 | 10 $\mu$F, 35V Electrolytic | |
| Y1 | 4 MHz Ceramic Resonator | ECS ATT-4.00MG |
| RST | Momentary On Push Button | |
| I/R Rx'ers | LiteOn 40 KHz I/R Remote Control Receiver Module | Digi-Key Part Number LT1060-ND |
| Board | "Vero Board" | Assembly Board is not critical |
| Power | 6 Volts Alkaline Batteries | |

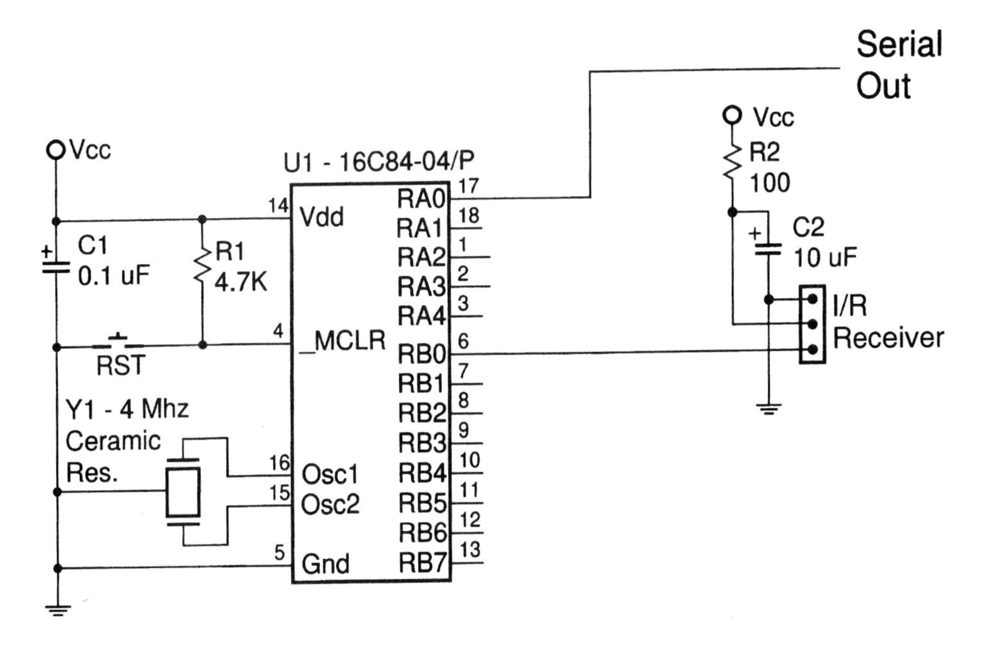


**Figure 16-32** Test circuit for better operation of the PICmicro® MCU I/R receiver

```
movlw     256 - 25           ; Reset the Timer
movwf     TMR0

swapf     _status, w         ; Restore the Context Registers
movwf     STATUS
swapf     _w
swapf     _w, w

retfie
```

This code simply counts the number of ones and stores it in *ReadCount* for a given amount of time. The theory behind this method of sampling was that the dead space between packets would be read along with the data and the result would combine them.

The actual value returned from the program wasn't very repeatable (as was expected). For example, five tries with the *1* key from a universal remote programmed with Sony codes produced these results:

```
0x09F
0x09D
0x08C
0x09D
0x09D
```

Generally, the results from this program were repeatable about 60% of the time. This might have been acceptable, except for the poor discrimination of this method. For example, the codes for *2* and *3* are 0x081 and 0x082, respectively. The problem lies in the fact that the two codes have the same number of ones and zeros. The code may pick up the differences, but I didn't find this to be the case.

So, the code for reading, the I/R packet was changed to:

```
 clrf     IntCount           ; Reset the Counters
 clrf     ReadCount

GetPack                      ; Get the Next Packet Coming In

 movlw    0x088              ; Wait for Port Change Interrupt
 movwf    INTCON

Loop                         ; Loop Here for Each Update of the Screen

 movlw    150                ; Wait for 25 msec of Data from I/R
 subwf    IntCount, w
 btfss    STATUS, Z
  goto    Loop               ; Has NOT timed out

 clrf     INTCON             ; No more interrupts for a while
 movf     ReadCount, w       ; Get the Read in CRC

 clrf     IntCount           ; Reset for the Next Packet
 clrf     ReadCount

 call     DispHex            ; Now, Display the Character

 movlw    0x08E              ; Reset the Cursor
 call     WriteINS

 goto     GetPack            ; Wait for the Next I/R Packet

Int                          ; Interrupt, Check I/R Input
```

```
 movwf    _w                    ; Save the Context Registers
 swapf    STATUS, w
 movwf    _status

 movlw    0x020                 ; Just wait for a Timer Interrupt
 movwf    INTCON

 movlw    256 - 20              ; Reset the Timer
 movwf    TMR0

 incf     IntCount              ; Increment the Count Register

 bcf       STATUS, C            ; Now, Figure out what to Add to LSB
 btfsc    PORTB, 6              ; Is the Incoming Value Set?
  goto    Int_Set

 btfsc    ReadCount, 5          ; Do we Update the Value coming in?
  bsf     STATUS, C
 goto     Int_End

Int_Set                         ; Incoming Set
 btfss    ReadCount, 5          ; Is the Current Bit Set?
  bsf     STATUS, C             ; No, Turn on the Incoming Bit

Int_End

 rlf ReadCount                  ; Shift Over with New Input Data

 swapf    _status, w            ; Restore the Context Registers
 movwf    STATUS
 swapf    _w
 swapf    _w, w
 retfie
```

which can be found in IRLCD_3.

The fundamental changes were that the sampling started after the Leader was received and the "1"s and "0"s were treated as the inputs to a linear feedback shift register. Elsewhere in the book, I have discussed how linear feedback shift registers work. For this Code, an 8-bit LFSR was used to produce *Cyclical Redundancy Check (CRC)* codes. In this case, the input wasn't the high bit of the shift register, it is the input from the I/R receiver.

Using this code, the CRC codes shown in Table 16-14 were generated from the Sony I/R transmitter.

The interrupt-handler code waits for a port change interrupt (the I/R line going low from its nominal state of *1*). Once that happens, the line is sampled every 200 $\mu$s and a CRC is generated from each sample. After 150 samples (30 ms), the CRC is output serially in Hex format (i.e., sending the high nybble followed by the low one).

The CRC generated is rock solid (none of the 60% repeatability that I had with just sampling bits). I don't know if I'm going to go back and update my I/R robot code (lack of initiative more than anything else), but this is clearly a much more elegant and robust method to handle I/R codes.

I did a limited amount of checking for invalid code rejection by reprogramming my universal remote with Panasonic and RCA codes. The CRCs generated were different from the Sony ones.

I was never pleased with the XORing used to create the CRC in this code. I felt that it was too confusing to understand. After some thought, I came up with the idea that if I used

**TABLE 16-14 Experiment CRC Results From Sony I/R Transmitter**

| KEY | CODE |
|---|---|
| Power | 0x052 |
| Vol+ | 0x05E |
| Vol– | 0x0BB |
| Ch+ | 0x0DC |
| Ch– | 0x062 |
| "1" | 0x07A |
| "2" | 0x08D |
| "3" | 0x033 |
| "4" | 0x01F |
| "5" | 0x04E |
| "6" | 0x072 |
| "7" | 0x0CC |
| "8" | 0x0B9 |
| "9" | 0x023 |

the same bit number for the PORT input bit as the CRC "tap," I could simplify the CRC generator (from *bcf STATUS, C* to *rlf ReadCount*) to:

```
bcf       STATUS, C
movf      PORTB, w            ; Get the Value Read in
xorwf     ReadCount, w        ; XOR it with the current
andlw     0x040               ; Clear all the bits but the two
btfss     STATUS, Z           ;  we're interested and if not = 0
 bsf      STATUS, C           ;  then make LSB of the CRC = 1
rlf       ReadCount
```

This code only improves the original by two addresses, but it sure is a lot easier to understand. This is IRLCD_4.ASM in the *IRBetter* subdirectory.

When I did this, I did cheat a bit and changed the CRC tap (and not the line coming in, which meant that I would have to change the LCD code). But, it still ran very well, with unique CRCs generated for each of the different keys of the Sony mimicking universal remote.

I guess the moral of this whole escapade is that tremendous improvements in your code (in terms of size and effort requirements) can be made if you look at a problem from a different direction. The code literally took less than six hours to develop and debug (compared to more than two weeks for the I/R receiver code for the robot). The code takes up about a third of the space of the infrared read algorithm used in the robot, and only uses two 8-bit variables, compared to the seven of the robot's code. This is a tremendous improvement!

Philosophically, this experiment is actually restructuring the application (reading an I/R transmitter) to best fit the PICmicro® MCU. The data read is now totally eight bit, as opposed to the 12/16 bits that had to be handled in the original application.

## THERMO: ELECTRONIC THERMOMETER WITH SEVEN-SEGMENT LED DISPLAYS

One of the most popular PICmicro® MCU projects I have ever created is this electronic thermometer (Fig. 16-33). Unfortunately, I could have done a bit better job in documenting it—I have received a lot of e-mails over the years asking for more information on the thermometer and how to build it. This section has been updated significantly from the first edition to help make building this circuit easier and result in fewer questions for me.

The application itself uses the resistor/capacitor network described elsewhere in the book to determine the resistance value of a thermistor. The temperature output is displayed on three seven-segment LEDs. The circuit is driven by a PIC16F84 using a 1-MHz crystal. This circuit has a number of possible inaccuracies, so I have included the ability to set a calibration value (stored in the PIC16F84's data EEPROM), which is used to allow the temperature range to be set accurately.

The thermistor that I used was bought from Radio Shack. Although the part is widely available in North America, it cannot be ordered under the Radio Shack brand name elsewhere in the world. I have described the operation of the thermistor so that you can find the equivalent part numbers in your own location.

One comment/caveat about the application, *despite* feedback that I have received from people who have built this circuit on their own: this is not a precision instrument. Despite the inclusion of the capability of calibrating the output temperature, you should not assume

**Figure 16-33** Digital Thermometer

that this thermometer can be used for precision operations. I am putting in this warning because I know of at least two people who have used this circuit to control the temperature within their barns. Although both people have said the project worked extremely well for them, if you are looking for accurate temperature readings, use the DS1820 digital thermometer, which has an interface application, presented later in this chapter.

In terms of actual applications, this circuit has apparently been used as the basis for a thermostat in a chicken-hatching incubator without any ill effects to the chicks. I suspect that this circuit is reasonably good around a small set range, although I would be very wary of the accuracy of results returned over a wide temperature range.

The code displays the current temperature in degrees Celsius. The Thermistor was bought from Radio Shack (part number 271-110). This part is a 10-K thermistor with *Negative Temperature Coefficient (NTC)* of −3.85%. Thus, for each degree Celsius that the thermistor is raised, the resistance within the thermistor drops by 3.85%. The base temperature is 25° Celsius and the response to different temperatures can be shown in Table 16-15.

For the thermometer, I created the circuit shown in Fig. 16-34. My prototype was built on a phenolic prototyping board that was bought as part of a prototype box.

The bill of materials for this project is shown in Table 16-16. The seven-segment LED displays used are of a common cathode type (which is to say that the cathodes of all the LEDs within the display are connected to common pins) that take up a 14-pin 0.300″ DIP pattern. The pinout of the conventional seven-segment LED displays is shown in Fig. 16-35. When I first created this application, I couldn't find the "standard" reference to the display, so I came up with my own standard (which is almost right). Figure 16-35 includes a table to allow you to convert between my number convention and the universal letter convention. *DP* represents the display's *Decimal Point.*

To drive the LEDs, I control the connection through the common cathode to ground us-

**TABLE 16-15 Resistance Versus Temperature For Radio Shack Thermistor**

| TEMPERATURE | ACTUAL RESISTANCE |
|---|---|
| 20 C | 12.079 K |
| 21 C | 11.631 K |
| 22 C | 11.200 K |
| 23 C | 10.785 K |
| 24 C | 10.385 K |
| 25 C | 10.000 K |
| 26 C | 9.615 K |
| 27 C | 9.245 K |
| 28 C | 8.889 K |
| 29 C | 8.547 K |
| 30 C | 8.218 K |

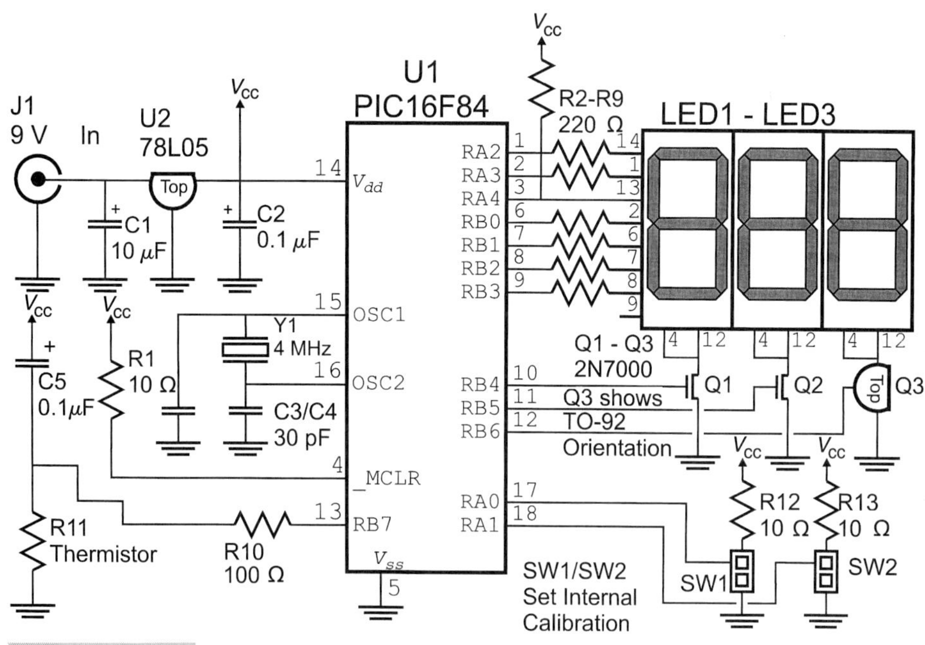


**Figure 16-34** Digital thermometer schematic

**TABLE 16-16 Digital Thermometer Bill of Materials**

| PART | COMMENTS |
|---|---|
| U1 | PIC16F84-04/P |
| U2 | 78L05 |
| Y1 | 1.00 MHz Crystal |
| C1 | 10 µF Electrolytic |
| C2, C5 | 0.1 µF Tantalum |
| C3, C4 | 30 pF |
| R1, R12, R13 | 10 K, 1/4 Watt |
| R2 - R9 | 200 Ohm, 1/4 Watt |
| R10 | 100 Ohm Resistor |
| R11 | 10 K, −3.85% NTC Thermistor (Radio Shack Part Number 271-110 |
| Q1 - Q3 | 2N7000 N-Channel MOSFET in TO-92 Package |
| LED1 - LED3 | 7 Segment Common Cathode LEDs |
| J1 | 9 Volt Radio Battery Connector |
| SW1 - SW2 | 2x1 0.100 Pin Headers |

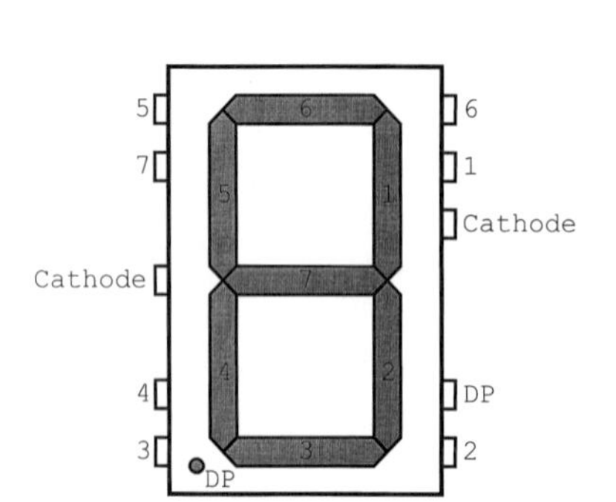


Note: the following table converts segment numbers to conventional letters:

| Number | Letter |
|---|---|
| 1 | B |
| 2 | C |
| 3 | D |
| 4 | E |
| 5 | F |
| 6 | A |
| 7 | G |

**Figure 16-35** 7-segment Common-cathode LED pinout

ing a 2N7000 N-channel MOSFET. Each segment is toggled through 1/500 second to display the three-digit temperature on the display.

The schematic shows (and for my prototype I used) a 9-volt alkaline radio battery. Instead of a 9-volt battery, an AC/DC wall adapter can be used for the circuit. Notice that I included a SPST switch to turn on and off power to the circuit.

Each LED segment is connected to a PIC pin via a 220-Ω resistor, except for the segment connected to RA4, in which the 220-Ω resistor is attached to $V_{dd}$ and the PIC pin pulls it low to turn off the LED. This is to avoid the issue of RA4 not being able to drive positive signals and it ensures that there is no possibility for high currents to be sunk by RA4. This was done to avoid both the PICmicro® MCU being burned out and to minimize current consumption when RA4 is pulling the line low.

In my application, I wired the segments as:

RA2 is Connected to Segment 6
RA3 is Connected to Segment 5
RA4 is Connected to Segment 1—Note the Comments Above
RB0 is Connected to Segment 7
RB1 is Connected to Segment 4
RB2 is Connected to Segment 3
RB3 is Connected to Segment 2

The DP pin of the LED display was left unconnected.

The reason for using these values was strictly to make the wiring easier. If you look up at the photograph at the top of this section, you'll see that the seven current-limiting resistors are placed between the PIC and the LED displays. All of LED display connections use the seven I/O pins on the left side of the PICmicro® MCU.

When creating an application like this, you really have to plan ahead. In this application, the best example of this is wiring the seven left-side PICmicro® MCU I/O pins directly to the current-limiting resistors and then to the seven-segment LED displays to avoid having to resort to very complex wiring. To help simplify the wiring even further, I also put the switching transistors along the "bottom" of the seven-segment LED displays to allow a common ground to be run to each one. Even with this planning, I still had to move around a couple of components and wire them on the backside of the board to get everything to work together. The final mess of wires is shown in Fig. 16-36.

An useful strategy in the application-development process is to ensure that each subsystem in the application is tested before the final application is created. The source code for the Thermo application (THERMO.ASM) is located in the *code\Thermo* subdirectory of the *PICmicro* directory, along with a number of small, focused test programs to verify each facet of the design and assembly.

For the digital thermometer, I wrote Test1 (in subdirectory *Prog32*). This program simply cycles through the individual segments of the display to ensure that they are wired correctly. This application was used to develop the tables used to reference different values for each display.

Next, TEST6 can be used to test out the application. This code will allow you to start checking the function of the thermistor/capacitor measuring circuit. The program uses the formula from elsewhere in the book,

$$Time = R * C * -ln(V_{end}/V_{start})$$

to measure the resistance of the thermistor. To initially calculate the actual temperature, I used idealized components and had the temperature looked up from a table (because the thermistor resistance versus temperature is a nonlinear function and I didn't want to use an extensive mathematical formula). The table relates the resistance value read (where the time taken for the capacitor to discharge is proportional to the resistance) to

**Figure 16-36** Backside of the digital thermometer

an actual temperature. The table values relate back to idealized components (i.e., exact values).

In the actual application, I provide a constant value that will shift the actual value returned into an accurate temperature value. This constant changes the previous formula to:

$$Time = Constant * R * C * -ln(V_{end}/V_{start})$$

TEST6 is also significantly different from TEST1 in how it displays values on the LEDs. TEST1 was unable to light any more than one display at a time. TEST6 uses an interrupt handler to toggle between the displays (at 2 kHz) to give the appearance that they are all working at the same time. The interrupt handler makes the multiplexing between displays to be transparent to the mainline code. All it has to do is load up the values to be displayed for each digit and the interrupt handler takes care of the rest.

Once the program starts to run, you will see that it displays a value for the current temperature or an error message (*uuu* or ^^^) for too cold and too hot, respectively.

While the circuit is up and running, you might want to fool around with it a bit. For example, you could put the thermistor between your fingers and watch the temperature go up or put it into a refrigerator/freezer and watch it go down. In doing this, you will discover that the thermistor-based thermometer will seem a lot faster than most mercury-based ones. That is because the thermistor has a lot smaller thermal mass than a mercury thermometer and it can reach its environment's temperature faster.

As you look at the displayed temperature, you will probably notice two things. First, the temperature is probably wrong. Because you are using components that are not perfect (ideal), their values are somewhat off from the exact specified values. Second, you will probably see the temperature creep up if the thermistor is close to the PICmicro® MCU as the PICmicro® MCU warms up from use. Both of these problems can be overcome by setting the calibration constant in the PICmicro® MCU.

Instead of working with a constant, I could have added a trimmable part in series with the resistor like a potentiometer and then changed the total resistance. The downside to this method is that the trim parts can drift themselves—either by temperature, material breakdown over time or by the circuit being knocked around.

TEST7 uses SW1 and SW2 to allow the user to change the calibration constant and shift the temperature displayed in different values and have them stored into the PICmicro® MCU's EEPROM. Although I call these controls *switches,* they are actually twin posts with one side connected to ground.

When the connectors are shorted to ground, the PICmicro® MCU pins to which they are connected will be pulled to ground and used to change the constant value. I put in a 1/3-second delay to allow for debouncing and give a delay between updating the value and disconnecting the connection to the switch.

A significant feature of the THERMO.ASM application code is how the data is stored in the EEPROM. On power up, two data-check bytes are checked for the values *0*x*0AA* and *0*x*055*. These data-check bytes are used to indicate that the value in the data EEPROM byte is actually correct. On the first power up of the PICmicro® MCU with this program, these memory locations are invalid and a separate set of code is executed to initialize the checksum bytes and the data EEPROM bytes to an idealized constant. Using the check bytes along with value byte means that if you shut off the power to the PICmicro® MCU. When you come back later, the value will still be there and usable.

When I was adding the EEPROM code, I encountered the most-significant problem in debugging the code that I would experience in this application. Every time I would run the application in hardware, only one digit would be displayed and only the first digit of the EEPROM data check was written by the application. This was very confusing because, when I simulated the application, I found that it seemed to be running correctly, but it wouldn't when put into real hardware.

After much and protracted debugging, I found the problem was with the initialization of the segment variables. I had copied one line from the previous, but hadn't changed its value to the correct variable name. This cost me about three weeks of part-time debugging to find the problem.

The problem with the simulating that I was doing was that I had put in a Debug define and had jumped over a three-second set-up delay. On the system that I was working with at the time (a 50-MHz 486), this simulation literally took more than 45 minutes to execute. When I finally gave in and allowed the simulator to run through the full 45 minutes, the problem was obvious and easily fixed.

The lesson was that I didn't ensure that the variable initialization was correct before I looked for other problems. Also, I was totally naive when looking at the problem and not expecting everything to be correct.

Once I had gotten the testing applications working, I combined them into the PROG32.ASM application that was shipped with the first edition (and put on my web page). This application contains the temperature-reading and displaying code from TEST6, along with the EEPROM read and write code from TEST7 to provide a calibration value that can be used to correct the delay value to get an accurate temperature.

The calibration value itself is 16 bits long and is multiplied by the actual delay value. The high byte of the resulting 16-bit number is used as the corrected delay value. Using the high byte is the same as dividing the result by 256 (which is what the calibration value is based on). Doing the calculation this way eliminates the requirement to provide a division routine or floating-point routines as part of the calibration.

This method of implementing fractions is covered elsewhere in the book, but I want to go through it again because it relates to the problem of providing a constant fraction value to this application.

For example, you could find 30% percent of an eight-bit number two different ways. First would be to multiply by 3 and divide by 10. To get 30% of 123, the operations would be:

$$\begin{aligned} 30\% \text{ of } 123 &= (123 * 3)/10 \\ &= 369/10 \\ &= 36 \text{ or } 37 \text{ (depending on rounding)} \end{aligned}$$

The problem with this method in the PICmicro® MCU is that division cannot be implemented easily. Instead of multiplying by a fraction, a constant value for a particular denominator is used. For the PICmicro® MCU, a denominator of 256 or 65,536 (0x0100 or 0x010000) is probably the best way to do this.

Using this method, instead of multiplying 123 by three and then dividing by 10, 123 can be multiplied by the fractional value of 256 and then divided by 256 to get the actual value. This is:

```
30% of 123 = [123 * (30% of 256)]/0x0100
           = (123 * 77)/0x0100
           = 9,471/0x0100
           = 0x024FF/0x0100
           = 0x024
           = 36
```

The seven-segment LED driver code and hardware specified within this application can be used for a variety of purposes. Different character sets (i.e., hex codes), and additional and different displays can be implemented easily by modifying cut and pasting this code into another application.

## MARYATOY: MARYA'S TOY, AN ADDENDUM TO THE ELECTRONIC THERMOMETER

After I built the LED thermometer, my daughter, who was 18 months old at the time, was absolutely fascinated with its LED displays. In fact, I had a lot of problems trying to keep her from playing with it. The obvious solution for me was to create a toy of her own that had lights and buttons that would respond to her inputs.

This was also a good chance for me to experiment with other types of LED displays. I used a 15-segment alpha-numeric display. This display is very similar to the seven-segment displays of the Electronic Thermometer, except that it has a lot more segments (Fig. 16-37). In this diagram, I have shown how I numbered each of the segments.

```
 ---1---
|\  |  /|
6 91011 2
|  \|/  |
 -7- -8-
|  /|\  |
5121314 3
|/  |  \|
 ---4---      Dot (15)
```

The most-significant part of this project was figuring out how to drive these displays. Ideally, I wanted to use a simple PICmicro® MCU (a PIC16C84 at the time).

Actually, this was quite easy to do because the experience I got from doing a "snowman" display that was the precursor to the Christmas Tree Display and Digital Thermometer (presented in this chapter).

The circuit I came up with (originally) is shown in Fig. 16-38. Six 15-digit displays were wired in parallel, like the three seven-segment displays of the digital thermometer, with each digit driven from an output of the 74LS374. To select which display is active, I used a 74S138, rather than a single transistor. This way, up to eight displays could be handled without additional components.

Conceptually, the wiring of the display is very simple. In actuality, when you wire it, you'll feel like you are going blind. I suggest that you buy a display with multiple digits that just have multiple common-cathode connections for each of the digits within the dis-

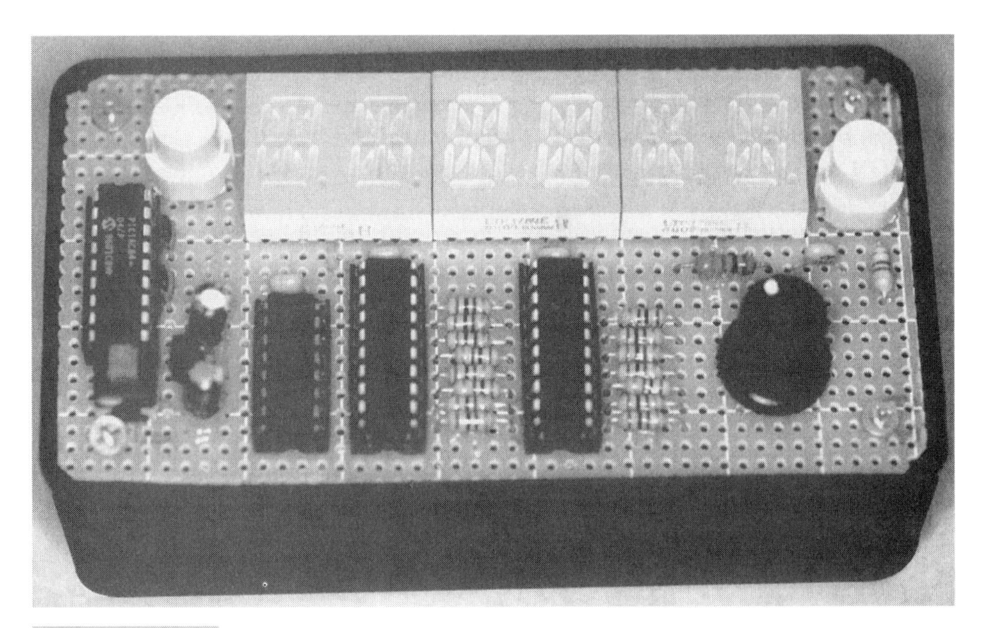

**Figure 16-37** Maryatoy prototype

play. For my prototype, I used displays with two digits built in (and two common cathodes). The bill of materials for this project is given in Table 16-17.

Despite essentially halving the amount of wiring, I still found that it took me a whole afternoon to point-to-point wire the displays to the current-limiting resistors connected to the 74LS374s. The displays that I used have their pins on the top and bottom of the packages, which makes daisy chaining the wiring quite difficult to do (I was able to do it in the digital thermometer relatively easily).

The mainline code simply updates a six-byte array called *Disp,* which consists of the ASCII codes (from 0x020 to 0x05F) that are currently displayed on the 15-segment LEDs. TMR0 was enabled along with its interrupt request and each time it overflows (every 512 instruction cycles, 512 $\mu$s because the PICmicro® MCU is running with a 4-MHz clock), the character in each Disp element is output on its respective 15-segment LED display. The 512-$\mu$s interval between digit displays gives an overall display frequency of 325 Hz, which is flicker free and provides an acceptably bright output. I say that the display is "acceptably bright" because the output is essentially a PWM with a duty cycle of 1/6 (16%) of a total possible output.

I found that the interrupt handler required 162 instruction cycles of the 512 available. I am mentioning this because you should be aware of the 30% overhead that the display interrupt handler operation places on the PICmicro® MCU's execution. This was something to be aware of when I ported the code to PicBasic in the next section.

For the second edition, I updated PROG1.ASN (located in the *code\MaryaToy* subdirectory of the *PICmicro* directory) to work with a PIC16F84 and also took out the functions that I didn't use in the original.

When I built the original, it was my plan to poll the buttons and read the potentiometer (as part of an RC network) to display the alphabet, as well as numbers. I didn't do this be-

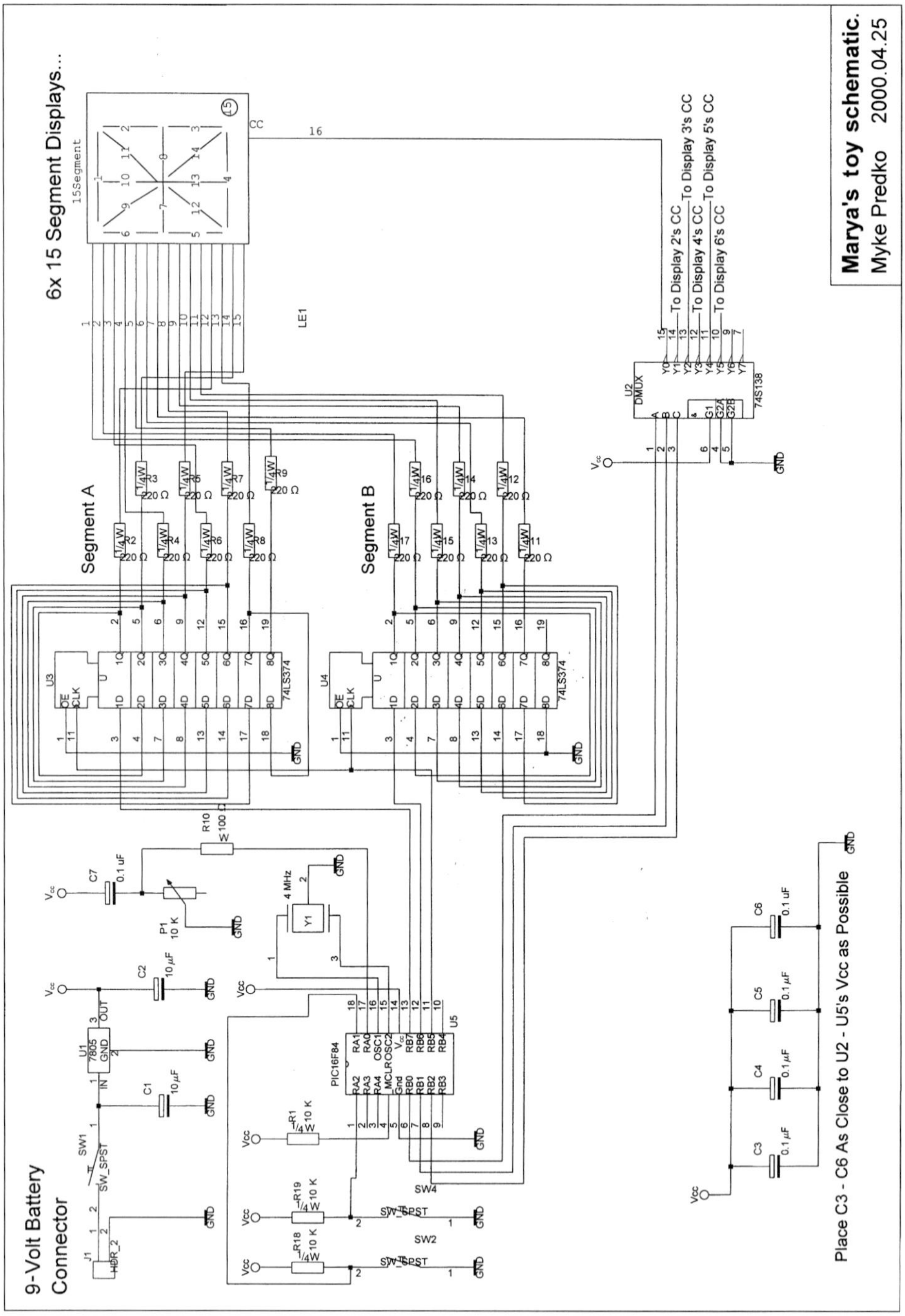

6x 15 Segment Displays...
15Segment
LE1
CC
16
Segment A
Segment B
U3
74LS374
U4
74LS374
R2
R3
R4
R5
R6
R7
R8
R9
R10
100 Ω
R11
R12
R13
R14
R15
R16
R17
220 Ω
1/4W
U2
DMUX
74S138
To Display 2's CC
To Display 3's CC
To Display 4's CC
To Display 5's CC
To Display 6's CC
Marya's toy schematic.
Myke Predko 2000.04.25
9-Volt Battery Connector
J1
HDR_2
SW1
SW_SPST
U1
7805
C1
10 μF
C2
10 μF
C7
0.1 uF
P1
10 K
Y1
4 MHz
U5
PIC16F84
R1
10 K
R18
10 K
R19
10 K
SW2
SW4
C3
C4
C5
C6
0.1 μF
Place C3 - C6 As Close to U2 - U5's Vcc as Possible
GND


**Figure 16-38** Maryatoy schematic

**TABLE 16-17 Marya's Toy Bill Of Materials**

| PART | DESCRIPTION |
|---|---|
| U1 | 7805 +5 Volt Regulator |
| U2 | 74S138 3 to 8 Demultiplexor |
| U3 - U4 | 74LS374 |
| U5 | PIC16F84-04/P |
| Y1 | 4-MHz Ceramic Resonator with Internal Capacitors |
| 15 Segment | 6x15 Segment Alpha-Numeric LED Displays |
| R1, R18, R19 | 10 K, 1/4 Watt Resistors |
| R2 - R9 | 220 Ohm, 1/4 Watt Resistors |
| R10 | 100 Ohm, 1/4 Watt Resistors |
| R11 - R17 | 220 Ohm, 1/4 Watt Resistors |
| POT1 | 10-K Single Turn Potentiometer |
| C1 - C2 | 10 $\mu$F Electrolytic Capacitors |
| C3 - C7 | 0.1 $\mu$F Capacitors, Any Type |
| SW1 | SPST Power Switch (Push Button Switched used for Prototype) |
| SW2, SW4 | Momentary On Push Button Switches |
| J1 | 9-Volt Alkaline Battery Connector |
| Misc. | Prototype Board, Wiring, Project Case |

cause my daughter wasn't very interested in it—even though it had the letters "MARYA'S TOY" run across the LEDs. For some reason, she was always a lot more interested with the digital thermometer. The effort to add the additional functions didn't seem to be worthwhile until I started working on the second edition of the book and revisited this project.

## MARYABAS: PICBASIC "PORT" OF "MARYA'S TOY"

After updating the Marya's Toy project for the second edition, I wanted to use it as a test bed to test out the capabilities of PicBasic, from the perspective of:

1. Understanding how easy it is to develop complex applications in PicBasic using built-in PICmicro® MCU hardware features.
2. Measure the performance of PicBasic compared to straight assembly-language programming.
3. Mix a reasonably complex interrupt handler written in assembly language with straight PicBasic code.
4. Evaluate the ease in which new features can be added to an application.

My overall impressions of PicBasic from this project are very favorable, but you should be aware of a few things. Also, this project turned out to be more than I bargained for in terms of code development, debugging, and in terms of difficulty. I would rate this application as being quite difficult to implement and, when it comes right down to it, quite advanced in terms of the skills I had to apply to get it working.

If you look at the source code on the CD-ROM, you will probably be surprised at these statements. This was a difficult application to create, but I cover the different problems that I encountered and how I solved them.

For this application, I wanted to use a PICmicro® MCU with a built-in ADC, rather than rely on the RC network (as I originally proposed for this project). I used a PIC16C711, which has a built-in ADC. I wanted to see how easily the ADC registers could be accessed from PicBasic. I was happy to find that to use the PIC16C711, I only had to specify the processor in the *compile* statement and the correct libraries would be automatically specified and loaded.

The original circuit was modified slightly. The R/C network was changed to a simple potentiometer voltage divider and the button wired to RA1 was relocated to RA3 because the PIC16C711 cannot just have RA0 as an analog input without RA1 also being an analog input as well. The updated circuit is shown in Fig. 16-39.

The bill of materials for the circuit is shown in Table 16-18. The PicBasic application source code and PicBasic compiled assembly-language code (which can be run from MPLAB) is located in the *code\maryabas* subdirectory of the *PICmicro* directory.

To create the application code, first replicate the functions that I had in the original assembly-language application. This application simply displays a scrolling "MARYA'S TOY" string on the six 15-segment LED displays. The first part of the implementation plan that I had for this application was to copy in the interrupt handler from the previous project exactly and just use PicBasic to create the display information.

Along with leaving the interrupt handler in assembler, I would also include the tables used to store the different ASCII characters. I wanted to keep the assembly-language interrupt handler as separate from the PicBasic application as possible; the only interface would be the six-byte Disp array, which contained the ASCII codes for each display. The PicBasic mainline would provide the interface functions.

As it turned out, this was an excellent approach to implementing the application, which eventually included providing the same initial scrolling display, polling the two buttons, and displaying either the alphabet or the numbers zero through nine (the potentiometer selects the initial digit).

Implementing the initial capabilities is where I first ran into significant problems. It was my intention to simply use the interrupt-handler code from the original project, put the PicBasic handler prefix and suffix that I show in Appendix D and add some mainline code to have TMR0 driven from the instruction clock and enable the TMR0 interrupt request. I expected this to be a half-hour exercise and it turned into eight hours of frustrating debugging.

When I tried to compile this simple application, I found that I received *Error Line ###: Syntax Error* and *Bad Token ";"* error statements from the assembler and a lot of gyrations through different formats for data until I tried deleting the tables to look at the absolute smallest problem (and, if I could isolate it, send the code to meLabs for them to take a look at). When I deleted the tables, I found that the errors completely went away.

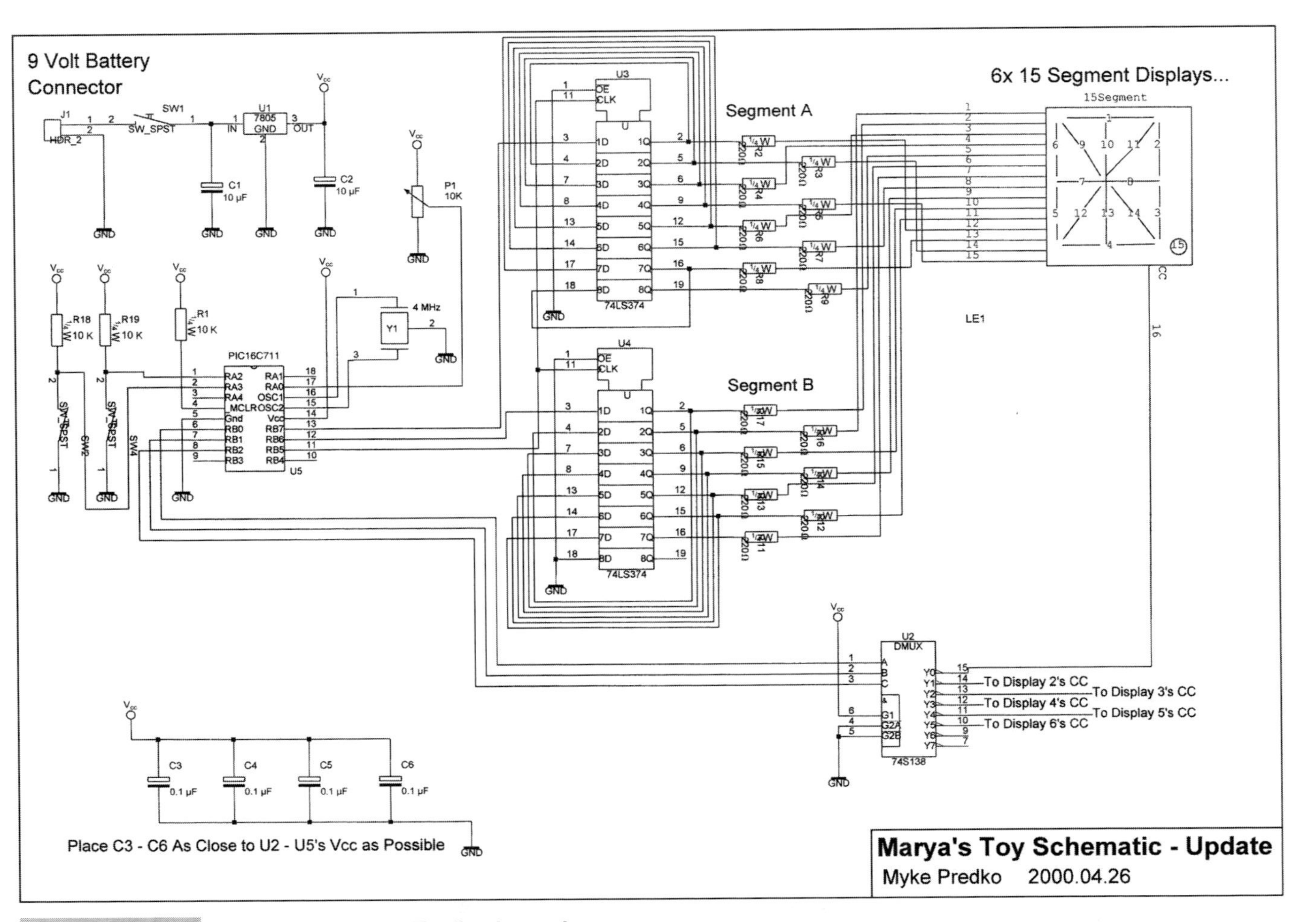


**Figure 16-39** Updated "Marya's Toy" schematic

**TABLE 16-18 Updated Marya's Toy Bill Of Materials**

| PART | DESCRIPTION |
|---|---|
| U1 | 7805 +5 Volt Regulator |
| U2 | 74S138 3 to 8 Demultiplexor |
| U3 - U4 | 74LS374 |
| U5 | PIC16C711-TW |
| Y1 | 4 MHz Ceramic Resonator with Internal Capacitors |
| 15 Segment | 6x 15 Segment Alpha-Numeric LED Displays |
| R1, R18, R19 | 10 K, 1/4 Watt Resistors |
| R2 - R9 | 220 Ohm, 1/4 Watt Resistors |
| R11 - R17 | 220 Ohm, 1/4 Watt Resistors |
| POT1 | 10 K Single Turn Potentiometer |
| C1 - C2 | 10 $\mu$F Electrolytic Capacitors |
| C3 - C6 | 0.1 $\mu$F Capacitors, Any Type |
| SW1 | SPST Power Switch (Push Button Switched used for Prototype) |
| SW2, SW4 | Momentary On Push Button Switches |
| J1 | 9 Volt Alkaline Battery Connector |
| Misc. | Prototype Board, Wiring, Project Case |

Thinking that I had put the data in the wrong format for the tables, I then went back and started adding the table code back in. I started by adding the table information a line at a time to find the problem. After about 10 lines, I added blocks of 16 table lines (instructions and comments) at a time and I didn't get the errors until I had restored the tables to their original size.

Playing around with deleting individual lines, I found that if I deleted four lines from the total, the errors would go away. Needless to say, this was pretty strange until I remembered a compiler that I worked with years before that had a limited amount of buffer/heap space. This compiler could only process so much source code until it started producing strange errors.

As an experiment, I restored the tables to their original size and deleted all of the comments to the second table (for the second 74LS374). Amazingly enough, the error messages disappeared. After a bit of analysis, I found that the PicBasic (version 2.21) could only support about 0x0A00 (or 2560) bytes of assembler source code. Anything more than this amount caused the problems. This is something to remember if you are adding assembler code to your own PicBasic applications.

This was not a trivial problem to find, characterize, and overcome. If you are new to programming, I doubt you would have been able to find this problem and work out a solution for it. One of the recommendations that I would have from this is to avoid embedding assembly-language code in your PicBasic applications unless you are comfortable with debugging assembly-language code and have some experience with programming in different environments. This is not a terrible restriction because, as discussed at the end of this section, I found the compiled instructions produced by PicBasic to be very efficient. There are few reasons why you would want to use assembly language with PicBasic.

Once I got the assembly-language interrupt handler compiled by PicBasic and the "mainline" code written, I wanted to test it out. The first time I programmed a PIC16C711, I used an meLab's EPIC programmer. The application seemed to be programmed in properly (I could verify it and get an error from verifying an empty socket), but as you probably guessed, the application did nothing.

This lead to the next major problem: trying to figure out how to simulate the application to see where the problem was. I decided to produce an MPASM-compatible assembler file from PicBasic and simulate it from MPLAB.

When you look at PicBasic compiler output assembly-language file, you will find that it loads and invokes a number of different macros. This makes single stepping through the application just about impossible. It is literally impossible to set breakpoints, except at assembly-language instructions written explicitly in the source code.

My bypass for this problem was to display the MPLAB-produced assembly-language listing file and debug from there. Figure 16-40 shows what the listing file looks like. Simulating from the listing file can be quite difficult and disorientating unless you know exactly what you are looking at. I found that if I specified the C command-line parameter for PicBasic, which forces the source lines as comment lines in the assembler file, it was a lot easier to follow and find specific locations in the code.

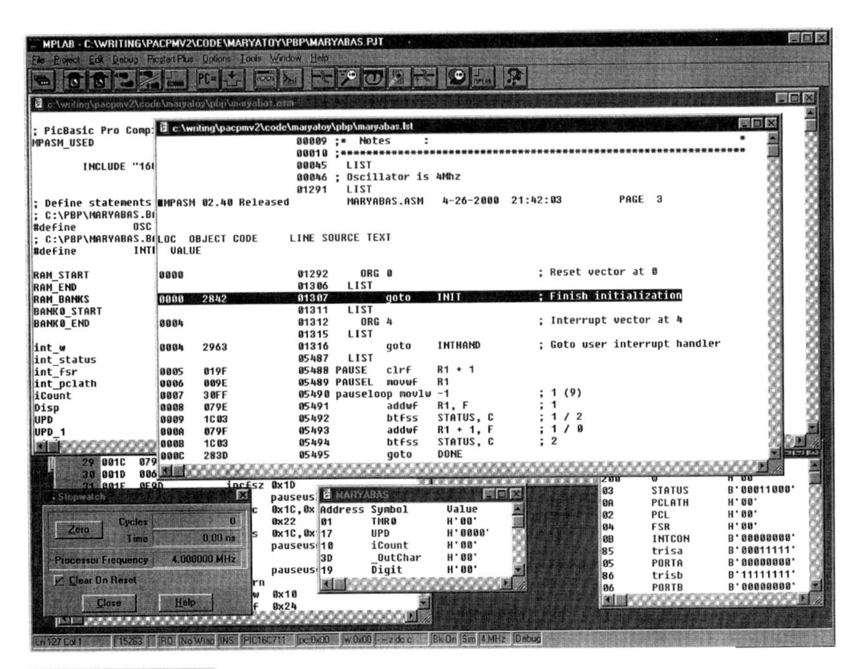


**Figure 16-40** MPLAB view of a simulated PicBasic application

To compile the code, I used the command-line statement:

```
pbp -p16c711 -c -ampasm maryabas
```

which probably seems like a handful to create and remember, but you can easily put the statements in a batch file and avoid having to repeatedly type them in.

The display did not work because I forgot to put the shift data, clock, and display select pins in Output mode. I was able to quickly find this problem once I had figured out how to simulate the application.

I also found that my experiences with writing compilers for the PICmicro® MCU were useful for seeing what the compiler produced. In this application, I spent quite a bit of time looking at the instructions that PicBasic produces from the source code. It is probably just as efficient as anything an experienced PICmicro® MCU assembly-language programmer would create. There are some caveats to this statement, as presented when describing the PicBasic application code.

With the PicBasic version of Marya's Toy at the same level of operation as the assembly-language version (both were displaying the scrolling text string MARYA'S TOY), I wanted to take a look at the differences in the two applications. The most obvious thing to check was the number of instructions required for both applications. The assembly-language version uses 336 instructions and the PicBasic version uses 345—a difference of 2.7%. The obvious conclusion is that PicBasic is very efficient in developing its assembler instructions. As indicated, when I simulated the assembly-language output, I found the code to be quite logical and efficient, which bears out the conclusion reached by the comparison of the number of instructions.

The only execution difference that I noticed between the two versions is that in the PicBasic version, the scrolling was somewhat slower. When I looked at the reason for this, I explicitly added a 125—ms delay between character scrolls in the PicBasic source code. This delay is stretched out by the 30% overhead of the 15-segment LED display interrupt-handler overhead. When I wrote the assembly-language code and was debugging it, I decreased the mainline delay so that the characters would scroll at a rate of one character every 125 ms—even with the interrupt handler operating.

With the scrolling speed increased, I wanted to go on and add polling of the two buttons. When one of the buttons is pressed, another display (either the alphabet or the first 10 numbers) is displayed and its position could be specified by turning the potentiometer.

When I originally implemented the button polling, I used the PicBasic *button* function and specified the information as if it were Parallax Stamp PBASIC. In doing this, I found that the operation of the *button* function is actually quite a bit different between the two languages. In PicBasic, *button* polls the specified button. If it is in the target state, it performs a debounce poll of the button for 10 ms before jumping to the destination. The Stamp *button* function is designed to poll the button in a loop-around increment a count each time the button is down (and clear it if it is up).

For the application, I really wanted to implement a 1-ms loop in which each of the two buttons would be polled. To do this, I went away from the PicBasic *button* function and wrote my own Stamp *button* equivalent code:

```
if PORTA.2 = 0 then         ' Debounce Button2
 Button1 = Button1 + 1      ' Count Number of Times Down
```

```
else
 Button1 = 0                  ' If Up, Reset
endif
if Button1 >= 15 then Destination
```

In my code, each time the button is polled (which happens once every 1.5 ms nominally), if it is down (the bit equal to zero), I increment a counter. If it is high, then the counter is reset. When the counter reaches a preset value (*15* for the example code), the button is determined to be reset and execution jumps to the *Destination* label.

In this application, after one of the buttons is pressed, the application jumps to either an alphanumeric display that uses the potentiometer to specify what portion of the string is to be displayed. The code that I created for the Alphabet display is quite elegant:

```
Alphabet:                           ' Put in the Alphabet

 LoopCount = 0
 Button2 = 0                        ' Alphabet is From Button1

Loop_2:

 ADCON0.2 = 1                       ' Start ADC Operation

 pause 1                            ' delay 1 msec

 if PORTA.3 = 0 then                ' Debounce Button2
  Button2 = Button2 + 1
 else
  Button2 = 0                       ' If Up, Reset
 endif
 if Button2 >= 15 then Numbers

 OutPos = ADRES                     ' Read the Results of the ADC

 OutPos = OutPos */ 21              ' Get the Scaled Value

 for i = 0 to 5                     ' Update the Alphabet
  lookup OutPos, ["ABCDEFGHIJKLMNOPQRSTUVWXYZ"], j
'                  01234567890123456789012345
  Disp[i] = j                       ' Get the Next Character
  OutPos = OutPos + 1               ' Increment to the Next Character
 next i

 goto Loop_2
```

Before starting the button-polling loop (which executes once per second), I set the GO/_DONE bit of the ADCON0 register of the PIC16C711. This starts the analog-to-digital conversion on the specified pin (RA0 in this case). After the 1-ms delay (the *pause 1* statement), the other button is polled in my debounce routine and execution jumps to the different display routine if more than 20 ms has passed with the switch not changing. Notice that I do not check both switches, just the one for the other function. This was done to avoid the need to wait for the pressed button to be debounced high. Instead, it is assumed that only one button is pressed at any time.

If the button has not been pressed, requiring the jump to the other display routine, I read the potentiometer's position from the ADC and scale it for the display. The scaling statement:

```
OutPos = OutPos */ 21
```

probably requires some explanation.

When displaying the alphanumeric characters in this application, I wanted to ensure that all six displays were in use at all times. Thus, the last five positions of each string cannot be used as the starting position of the display. For the alphabet, instead of being able to start with 26 different letters, this application can only start with 21 (the last five are in displays other than the leftmost one).

If you were to scale a value between 0 and 21 in real life, you would use a calculator and divide by the fraction of 21/256. The result's fractional value would be taken off the result and an integer between 0 and 20 (just what you want) would remain. Put mathematically, the formula would be:

$$OutPos = \text{int}(OutPos * 21/256)$$

For example, a value of *142* from the ADC would become:

$$\begin{aligned} OutPos &= \text{int}(OutPos * 21/256) \\ &= \text{int}(142 * 21/256) \\ &= \text{int}(2{,}982/256) \\ &= \text{int}(11.648) \\ &= 11 \end{aligned}$$

The result is correct, but the process taken to get there requires real numbers, which the PICmicro® MCU simply does not work with very well. This operation could be done in PicBasic, but it would be terribly expensive (in terms of the number of instructions) to complete the task.

A better way to calculate fractions is required. The best way that I have found is by multiplying by the maximum value of the new range and discarding the lower eight bits of the product. Put mathematically, this is:

$$OutPos = (OutPos * 21) >> 8$$

and it can be demonstrated using the previous example as:

$$\begin{aligned} OutPos &= (OutPos * 21) >> 8 \\ &= (142 * 21) >> 8 \\ &= 2{,}982 >> 8 \\ &= \text{0x0BA6} >> 8 \\ &= \text{0x0B} \\ &= 11 \end{aligned}$$

Multiplication by a constant is very easy in the PICmicro® MCU. Shifting a value to the right by eight bits is accomplished by simply discarding the lower eight bits. Elsewhere, the book shows how this formula can be implemented very easily in PICmicro® MCU assembler.

PicBasic has an operator that makes this operation even simpler than the formula above. The */ operator multiplies two numbers together and lops off the least-significant byte (eight bits) exactly in the manner required for this scaling operation.

I am showing you this to get you used to the idea of looking for tricks to make your applications more efficient and avoid trying to implement applications using the same techniques as you would have used for your high-school math homework.

Going through the "calculator process" will remove any chances for PicBasic to produce code that is just as efficient as assembler and will become a lot more work for you. If you are going to use a high-level language, like PicBasic, be sure that you understand it reasonably thoroughly and you can take advantage of special instructions and operations, like the ones shown here.

The other point in using PicBasic (and the Alphabet code) is to limit the number of *lookup* and *lookdown* statements in your application. This is not to say that *lookup* and *lookdown* are not implemented efficiently. They are actually quite efficiently implemented with table operations in PicBasic. However, they are so useful when programming that you will want to use a lot of them. This will require a lot of space in the PICmicro® MCU, which has little to begin with.

In the Alphabet code, you might have thought it would be better to do something like:

```
Alphabet:                           ' Put in the Alphabet

 for i = 0 to 5                     ' Update the Alphabet
  lookup i, ["ABCDEFGHIJKLMNOPQRSTUVWXYZ"], j
'             01234567890123456789012345
  Disp[i] = j                       ' Get the Next Character
 next i

 LoopCount = 0
 Button2 = 0                        ' Alphabet is From Button1

Loop_2:

 ADCON0.2 = 1                       ' Start ADC Operation

 pause 1                            ' delay 1 msec

 if PORTA.3 = 0 then                ' Debounce Button2
  Button2 = Button2 + 1
 else
  Button2 = 0                       ' If Up, Reset
 endif
 if Button2 >= 15 then Numbers

 OutPos = ADRES                     ' Read the Results of the ADC

 OutPos = OutPos */ 21              ' Get the Scaled Value

 for i = 0 to 5                     ' Update the Alphabet
  lookup OutPos, ["ABCDEFGHIJKLMNOPQRSTUVWXYZ"], j
'                  01234567890123456789012345
  Disp[i] = j                       ' Get the Next Character
  OutPos = OutPos + 1               ' Increment to the Next Character
 next i

 goto Loop_2
```

where the display is given an initial alphabet value before the ADC is read. The problem with this code is that it doubles the amount of space required for the lookup table. By itself, one lookup is not a lot (about 35 instructions) and it makes implementing a table very easy. However, two lookups require more space: 70 instructions (7% of the total available in the PICmicro® MCU).

Along with using more space than necessary in the PICmicro® MCU's program memory, the first lookup is probably not desired because the correct location for the alphabet display is not known and the display will flash with the arbitrary string that is sent to the 15-segment LED displays. Even though the flash is only for 1 ms, it will probably be picked up by a user and will not be that attractive. The code used will set the correct starting point for the display immediately following the press of the button from the previous display.

In case you are wondering; the comment line that follows the lookup table *012345678901234567890123456* is just my way to track of all the offsets for the different characters within the lookup table and ensure that I do not forget any table elements. It is a good idea to add little tools like this to ensure that everything is correct, in place, and providing an easy visual way to check actual to expected values.

I have gone on *ad nauseams* about application efficiency because in a high-level language like PicBasic, good planning pays off the most. In the PICmicro® MCU assembler, tables can be accessed multiple times, but in the PicBasic compiler each lookup—even if it is identical with another one, will produce code for a unique table. Many other examples of this can be found and avoided, making the compiled PicBasic code just as efficient as well-written PICmicro® MCU assembler if you look at the situation and see if there are other ways to implement the function.

To close, here are a few comments about PicBasic. First, it is a very efficient compiler in terms of code size and execution speed. I would not hesitate to recommend it for new PICmicro® MCU users to learn how to program. PicBasic has extremely rich functions and capabilities that will allow you to develop very sophisticated applications quickly and efficiently. Internal PICmicro® MCU features and registers can be accessed in line without regard to banks or bit values. In many ways, PicBasic is an optimal blending of a high-level language while retaining assembly language's ease of accessing specific registers and bits.

The biggest problem I found with PicBasic is the difficulty in which code can be simulated (and, by extension, emulated). For this reason, I do not recommend it for applications that are very complex or have mixed-in assembler. For large projects, I recommend that you look at a C compiler that produces .COD files that can be used "natively" with MPLAB.

## EMAIL: CONNECTING THE PICmicro® MCU TO THE INTERNET

This was the last project for me to create and I wasn't sure if I would be able to complete it. The first issue was that I was not sure that I would receive working Seiko S7600A TCP-protocol stack chips in time. Working chips became available in February of 2000, which did not provide a lot of "runway" for the project's completion. Third, the S7600A is only available as a relatively small surface-mount chip. I was nervous about creating a debugged PC board in time for the manuscript's release. Third, I was scared to do the project; originally, I wanted to create a small webserver and copy the work that has been done by a number of other people on the Internet. To summarize, I was expecting the development effort to be considerable.

Surprisingly enough, I was wrong on all three points and I created the project given here in two days.

The two-day figure might not be considered accurate because it does not reflect the extensive research I did before designing, building, and coding the project. I spent about a week researching Internet protocols and the best interfaces for the job. I had originally planned to implement the webserver totally within the PICmicro® MCU and I wanted to be sure that I understood the requirements for implementing the code before I started.

This research pointed me toward the Seiko S7600A chip, which provides a modem interface, TCP stack, and a PPP engine for logging onto an ISP's server. Don't worry if you don't really understand what these acronyms are (much less understand what they mean). The Seiko S7600A is a complete Internet interface, except for a modem interface and processing logic. To execute, it requires an external clock and can interface to the controlling processor using a serial or parallel data protocol. As shown later in this section, the software to interface to it is quite easy to create. The chip itself uses just about no power during normal operation.

The S7600A is about a 1/3″ square in a 44-pin TQFP package. Although it is not terribly difficult to interface to, I wanted to avoid having to create a PC board design for the book. In hopes that I could hack into a prebuilt board with the S7600A built on it, I bought the Seiko iChip Designer's kit from Mouser Electronics. This kit consists of a PC ISA card with a S7600A built on it and glue hardware to allow it to interface to your PC as a prototype card address device.

When I got the kit, I felt like all my problems were gone. The board has a single S7600A mounted on it, along with the ISA interface and a 25 × 2 connector that can be used for wire wrapping. I was wrong about all my problems being gone, but a significant one had been eliminated and I had an easy way to interface to the chip in a prototyping environment.

The major problem resulting from using this chip and development board for my prototype was that the chip requires no more than 3.3 volts to operate. Although I did not have any problems finding a PICmicro® MCU that could run at this voltage level, I did have considerable problems finding a widely available 3.3-volt power regulator chip and a single-voltage RS-232 interface that could be powered by 3.3 volts.

I finally decided on using the PIC16LF877, the Linear Technology LT1121-3.3CN voltage regulator and Linear Technology LT1331 single-voltage RS-232 interface chip. The circuit that I created is surprisingly simple (Fig. 16-41).

The bill of materials for this project is shown in Table 16-19. I wirewrapped the three-pin through-hole chips (and analog components) to the board and straddled the solder connectors of the 9-pin D-shell connectors to the edges of the card. The 2.5-mm power connector was glued to the surface of the card to avoid having to stress relieving the wire connections. The solder mask used on the iChip Designer's kit ISA card is quite dark and, unfortunately, I could not get a photograph of the completed circuit with sufficient contrast to the black chips for you to see how the circuit was arranged.

You might want to take a look at the current consumption of the three primary chips in this circuit. The PICmicro® MCU will require about 3.8 mA, the LT1331 will require somewhere about 10 mA, and the S7600A requires an astounding 300 $\mu$A. In contrast, the modem I bought for this project requires 800 mA to operate.

The total of approximately 14 mA means that this circuit could be battery powered or, at the very least, battery backed up. One of the example webservers that can be down-

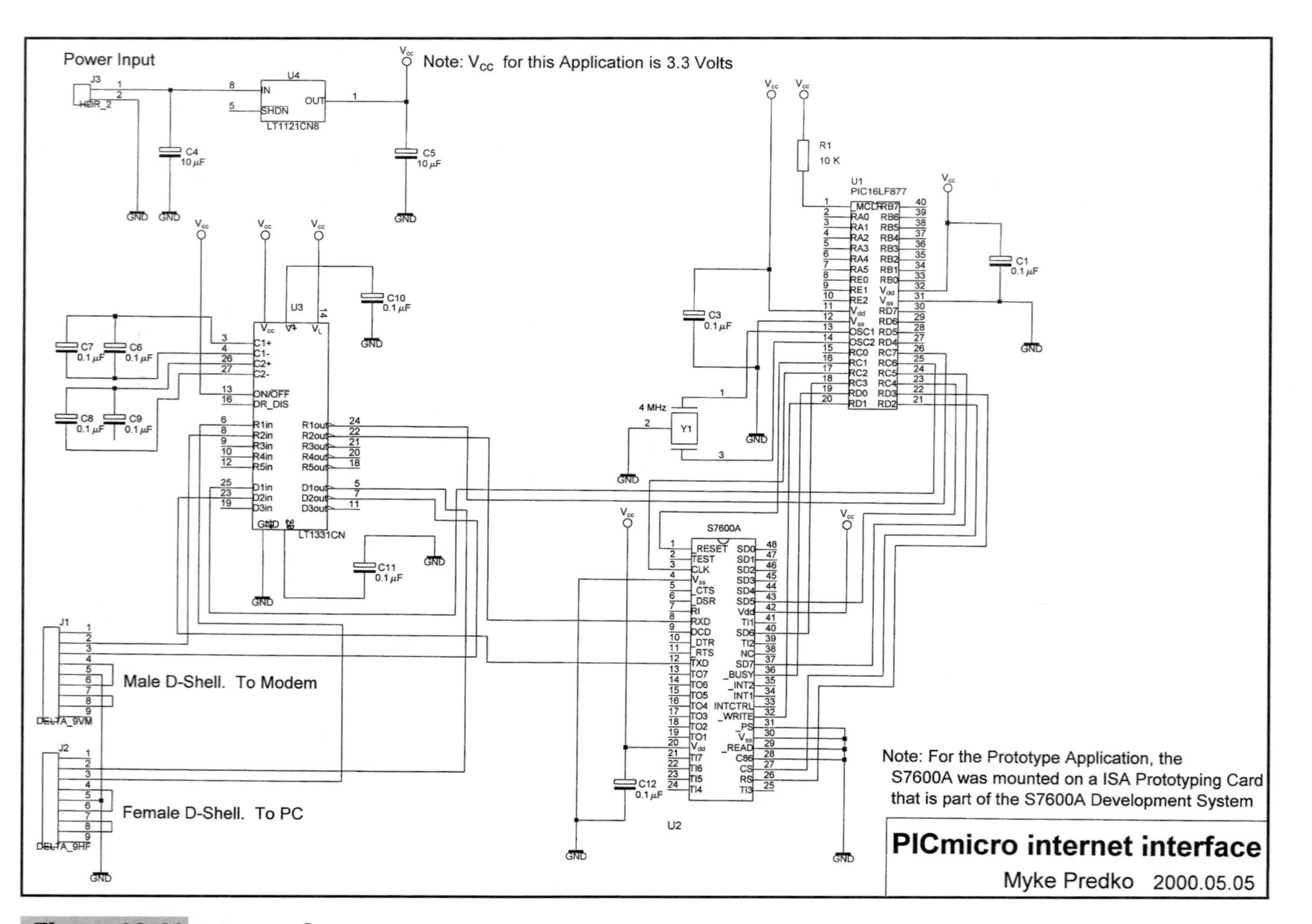


**Figure 16-41** PICmicro® MCU based mail server

**TABLE 16-19 PICmicro® MCU Internet Interface Bill Of Materials**

| PART | DESCRIPTION |
|---|---|
| U1 | PIC16LF877-04/P |
| U2 | Seiko S7600A TCP Protocol Stack |
| U3 | LT1331 Single Voltage RS-232 Interface |
| U4 | LT1121-3.3CN 3.3 Volt Regulator |
| Y1 | 4 MHz Ceramic Resonator with Internal Capacitors |
| R1 | 10 K, 1/4 Watt Resistor |
| C1, C3, C6-C12 | 0.1 $\mu$F, 16 Volt Tantalum Capacitors |
| C4 - C5 | 10 $\mu$F, 35 Volt Electrolytic Capacitors |
| J1 | Male 9 Pin D-Shell Connector |
| J2 | Female 9 Pin D-Shell Connector |
| J3 | 2.5 mm Power Socket |
| Misc. | Seiko "iChip Designer's Kit" ISA prototyping card, wire wrap wire, wire wrap sockets, External PC Modem |

loaded from the Internet uses the positive voltage current provided by the modem to power it, which is entirely reasonable. This can give you options where power is unreliable, or thieves might try to disable an Internet-based alarm by removing power from the circuit.

When I designed the circuit, I wanted to monitor the PICmicro® MCU interface and connection to the Internet using a PC. To facilitate this, I used the PICmicro® MCU's USART and communicated to the PC using RS-232. This is the purpose of the female nine-pin connector. The male nine-pin connector is wired directly to an external modem for interfacing to an *Internet Service Provider (ISP)*. To allow the PICmicro® MCU keep up with the Internet connection, I specified a data rate of 9600 bps for the S7600 interface to the Internet and a 19,200 bps interface to the PC. Choosing these data rates meant that the PC would always be updated with the current Internet information without missing any characters that were exchanged.

After looking at Fig. 16-41, you are probably wondering why I stated that this circuit is "surprisingly simple." To explain this comment, I should explain my typical experiences with wire wrapping. Normally, I work with boards that are so complex that if they are wire wrapped, not only can you not see the underside of the board, but the board actually rests on wire-wrap wires when it is put down on a table. This board requires less than 20 interchip wire-wrap connections and I was able to wrap the circuit in less than a half hour (with the chip sockets already in position).

The S7600A requires a clock to sequence data operations within the chip and time-asynchronous serial operations with the modem. For most of the applications that I saw from the Internet, this clock was produced by a signal that was the processor's clock divided down. For my application, I used TMR2 and CCP1 to produce a 250-kHz clock from the 4-MHz operating clock.

To interface to the S7600A, either an eight-bit parallel interface or a three-line serial interface is used to read and write its registers. These registers are used to control the basic chip operation, modem access, PPP connection, PAP connection, and TCP/IP. These registers are very easy to use and can be largely set up automatically, with no intervention on your part.

When a data byte is transferred, first an eight-bit address is transmitted, followed by the data byte. Figure 16-42 shows the timing used to write to the S7600A. First, the register address is transmitted (with RS low), followed by the data byte (RS high). CS is used to enable the S7600A and to monitor the other I/O lines. WRITEX (what I have called *_WRITE* in this application's circuit and code) is a negative active line, which indicates that the S7600A should accept the byte being transferred to it.

BUSYX (what I've called *_BUSY* in my application) is low when the S7600A is processing information and should be polled before passing any new commands to the S7600A from the PICmicro® MCU.

The serial interface data rate is restricted to be no less than 0.526 times the S7600A clock speed; to simplify the application, I decided to transfer data at 250 kHz using the SSP hardware. To implement this, I did a few things that you should know about. First, from looking at the schematic, you will see that the nonclocked serial interfaces use PORTD. The reason for this was to allow me to directly change the pin states with a write to a single port, instead of individually changing bits.

This ease of passing data to the S7600A is easily seen in the macros that I created to send data to the S7600A. To send a constant byte to a S7600A register, I use the macro:

```
S7600WriteI MACRO SAddress, SData    ; Write Constant
  btfss _BUSY                        ; Wait for S7600A to be able
   goto $ - 1                        ;  to Accept New Command
  movlw WriteAddr
  movwf PORTD
  movlw SAddress
```

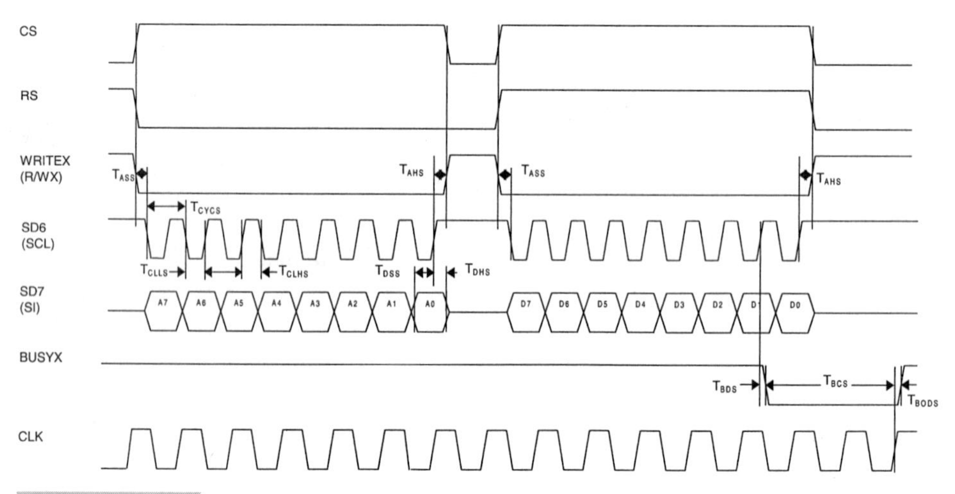


**Figure 16-42** Serial interface write timing

```
  call   S7600ASend
  movlw  WriteData
  movwf  PORTD
  movlw  SData
  call   S7600ASend
  movlw  NoTrans
  movwf  PORTD
endm
```

The *WriteAddr, SAddress,* and *NoTrans* constants are the PORTD pins controlling CS, RS and _WRITE.

The WriteData subroutine simply pulls the SCL (clock) line low and then transfers the byte in w and returns the read in data in w. This allows it to "pull double duty" in sending and receiving data without requiring a separate read subroutine. WriteData is:

```
S7600ASend                      ; Send 8 Bits from the SSP
  bcf    SDO                    ; Pull Down the Clock Line
  bsf    SSPCON, SSPEN          ; Enable the SSP Port
  bcf    PIR1, SSPIF
  movwf  SSPBUF                 ; Send the Character in "w"
  btfss  PIR1, SSPIF            ; Wait for SSP Transfer to
    goto  $ - 1                 ;  Complete
  movf   SSPBUF, w              ; Read Data Sent from S7600A
  bcf    SSPCON, SSPEN
  bsf    SDO                    ; Restore the Clock Line High
  return
```

The Serial Read (Fig. 16-43) is similar, but probably doesn't operate as you would expect. After sending the register address to read, a dummy write occurs, followed by the actual write. The expected waveform would have been to send the register address followed by the register content read.

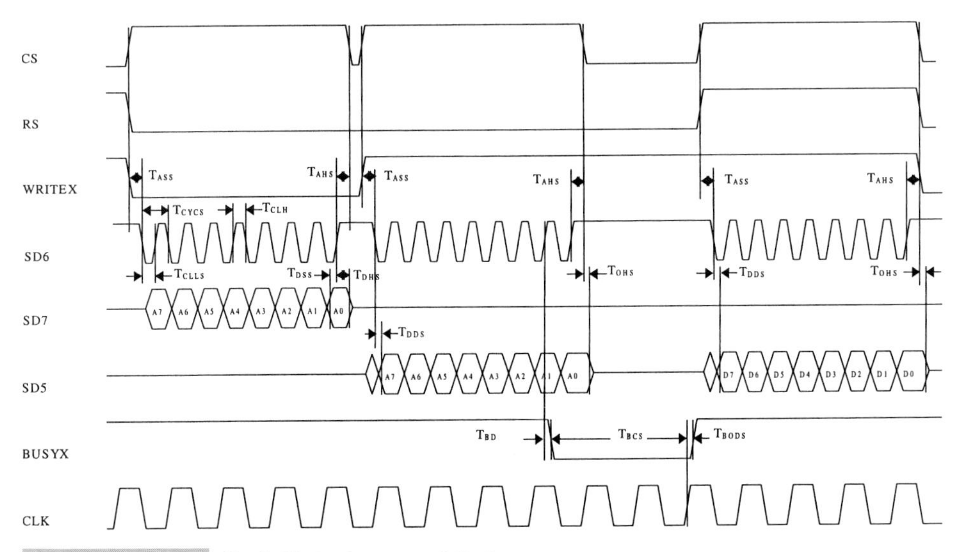


**Figure 16-43** Serial interface read timing

To read a byte, the following macro is used:

```
S7600Readw MACRO SAddress       ; Return Contents of Register
 btfss _BUSY                    ; Wait for S7600A to be able
  goto $ - 1                    ;  to Accept New Command
 movlw WriteAddr
 movwf PORTD
 movlw SAddress                 ; Output the Address
 call  S7600ASend
 movlw ReadSetup
 movwf PORTD
 movlw 0x0FF                    ; Do the Dummy Read
 call  S7600ASend
 btfss _BUSY                    ; Wait for S7600A to have byte
  goto $ - 1                    ;  Ready for Read
 movlw ReadDo
 movwf PORTD
 movlw 0x0FF
 call  S7600ASend               ; Send Also Returns Shifted in
 movwf ReceiveTemp              ; Save Received Byte
 movlw NoTrans
 movwf PORTD
 movf  ReceiveTemp, w           ; Return the Received Byte
endm
```

To log onto an ISP server using the S7600A, I carried out the following actions in the PICmicro® MCU application:

1. Reset the S7600A (send 0x001 to register 0x001)
2. Load the S7600A's baud-divisor register for 9600 bps
3. Load the S7600A's 1-kHz clock-divisor register
4. Enable the direct link to the serial port (write 0x000 to register 0x008)
5. Send *AT* and wait for the modem to reply with *OK*. This action is repeated five times or until the modem replies with *OK*. Multiple data transmissions were done to allow the modem to sense the 9600 bps (it was a 57.6-kbps modem) coming in.
6. Send *ATDT #######*, where ####### is the phone number of my ISP
7. When *Connected 9600* was output from the modem, I disabled the direct control of the modem.
8. Next, the S7600A was instructed to initiate PPP communications and control the modem. This was accomplished by writing *0x021* to register 0x008.
9. Next, PPP was enabled by sending *0x00A* to 0x062 (try to log on 10 times) and then *0x060* to register 0x060. Finally, the logon information (the userid and password) are sent with first their length, the strings and a Null character when done.
10. *0x062* is written to register 0x060 to enable PPP communications. Register 0x060 is polled until bit 0 is set to ensure the application is executing

Upon completion of step 10, the PICmicro® MCU is logged onto the Internet. The PICmicro® MCU's IP address and the host's IP address can be read out of the S7600A to verify that the link is active.

I originally wanted to replicate the other applications out on the Internet and make a web server out of this application. When I did the research preparing for this project, I decided that this would be one of the less-useful applications I could think for it. Instead, I thought that a periodic e-mail sent from the PICmicro® MCU could be a lot more useful.

To send e-mail, once you have established the connection, link to your ISP's *Mail Transport Agent (MTA)*. When this is done, you can start sending e-mail. How this is done (and what is the TCP port you have to use) is different for different systems and will be something that you will have to find out from your ISP. I found that by just calling my ISP and explaining what I wanted to do got me all the help I ever could have used. I guess to work for an ISP, you have to be something of a "propeller head."

To start sending the e-mail, send the string to a TCP socket:

```
HELO PICEMAIL
```

which notifies the server that you are going to access the server's e-mail system. The server will respond with something like:

```
250 mailrelay.passport.ca Hello PICEMAIL.passport.ca Please to meet you
```

After every data string, be sure to end the string with a ASCII *Carriage Return* (0x00D).

Next, indicate where the e-mail is coming from:

```
mail from joelb@passport.ca
```

For these experiments, I used my son's e-mail (he just uses it to download "Our Lady Peace" MP3, anyway) and had them send the e-mails to my ID.

The server next has to be told where the e-mail is going:

```
rcpt to: myke@passport.ca
```

and will respond with:

```
250 myke@pasport.ca... Recipient ok
```

Now, you can indicate that the text of the e-mail follows by sending:

```
data
```

The server will respond with:

```
354 Enter mail, end with "." on a line by itself
```

From here, you can send your e-mail, with each field or paragraph delineated by an ASCII *Carriage Return.* At the end of the message, place the . (period) character on a line by itself. An example of an e-mail message is:

```
From: PIC16F877
Subject: I'm Working!
Myke, you will be pleased to know that I am working
.
```

The *From:* and *Subject:* lines are pulled out by the e-mail receiver (Outlook Express, in my case) and displayed to the recipient. The blank line between the *Subject* and the body of text seems to be needed by Outlook Express for it to be able to figure out how to display the entire message.

The webserver will respond with:

```
250 $$$###### Message accepted.
```

where *$$$######* is a mixed character and text string that is used by the server to keep track of the e-mail.

When I did this, I was thinking of the possibilities for the application. I can see a lot more for the e-mail program, rather than the webserver. The reason for this is the cost of the webserver. To implement the webserver, you must have a dedicated phone line (the S7600A only works with modems and telephone systems), which will cost you somewhere in the neighborhood of $30 per month. Next, you will have to have access to an ISP all of the time. Considering that there are over 700 hours in 11 months, the costs for the always up access will probably be another $50 to $100 per month.

This $80 to $130 per month is probably not that expensive for a company, but, for an individual, it is probably prohibitive (at least I found this to be the case). Instead, this circuit is ideal for periodically logging onto an ISP, sending an e-mail (or receiving one), and then logging off.

## PCTHERM: RS-485 MASTER (PC WITH PSP)/SLAVE REMOTE (PIC16HV540) THERMOMETER INTERFACE

In this book, I've talked about the pre-eminence of the PC as the device you want to connect the PICmicro® MCU to for different applications. Throughout the book, I have presented a number of ways to interface the PICmicro® MCU to the PC via RS-232. In the El Cheapo programmer circuit, I have a parallel port interface (and one in which you can see the difficulties in interfacing to the PC's parallel port). I have not yet shown how the PICmicro® MCU can be connected to the ISA bus. As well, I have described RS-485, but I have not yet presented an application that uses it. This project wraps up these two "missing" interfaces, along with a Visual Basic PC interface to provide a remote temperature sensor for your PC (Figs. 16-44 and 16-45).

This application consists of three quite simple pieces that work together to provide the dialog box shown in Fig. 16-46 on your PC to allow you to monitor outside temperature from the comfort of your PC. The dialog box was written in Visual Basic and interfaces to an ISA card that has a PIC16F877 with its *Parallel Slave Port (PSP)* enabled. This PIC16F877 uses its USART to communicate to a remote PIC16HV540 that interfaces with a Dallas Semiconductor DS1820 temperature sensor via RS-485. This entire application uses less than 620 lines of code (including comments). Although this seems like the application was quite simple, I had a number of problems that I had to work through.

The first card that I created was the ISA card that the PIC16F877 resided upon. Figure 16-44 shows that the circuit (Fig. 16-47) is quite simple to implement with only five chips. The circuit itself uses two 74LS85 value comparators and a 74LS138 for address selection. The RS-485 interface is implemented with a 75176 and the PC's +12-volt power supply is passed to the remote sensor for its power. In the photograph (Fig. 16-44), you can see that other (un-

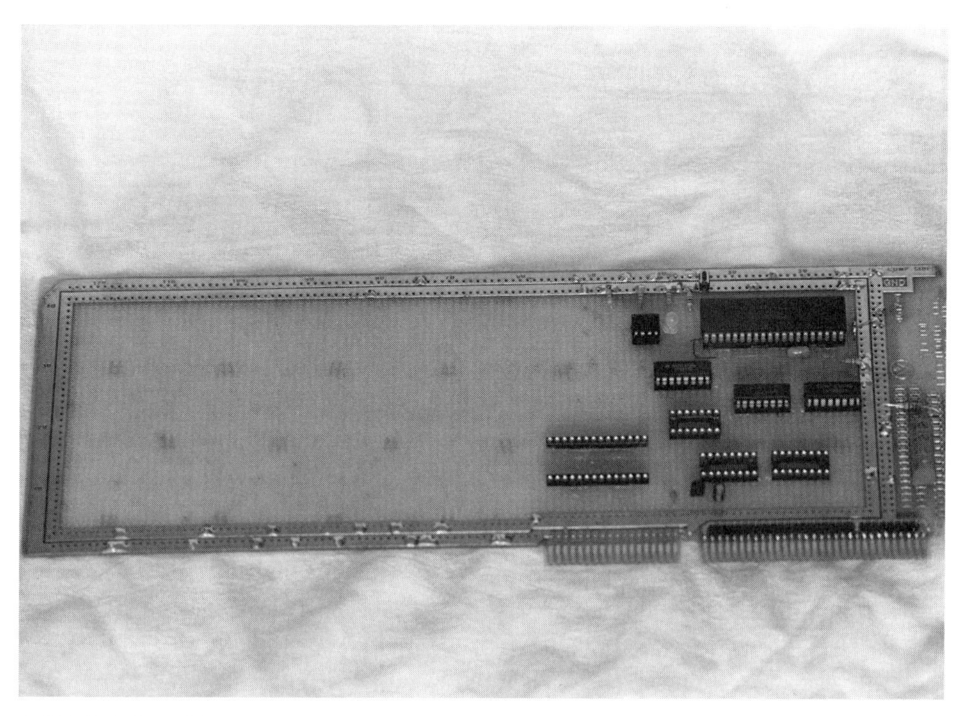

**Figure 16-44** PIC16F877 ISA RS-485 Interface

**Figure 16-45** PIC16HV540 Remote Temperature Sensor (DS1820 on far left)

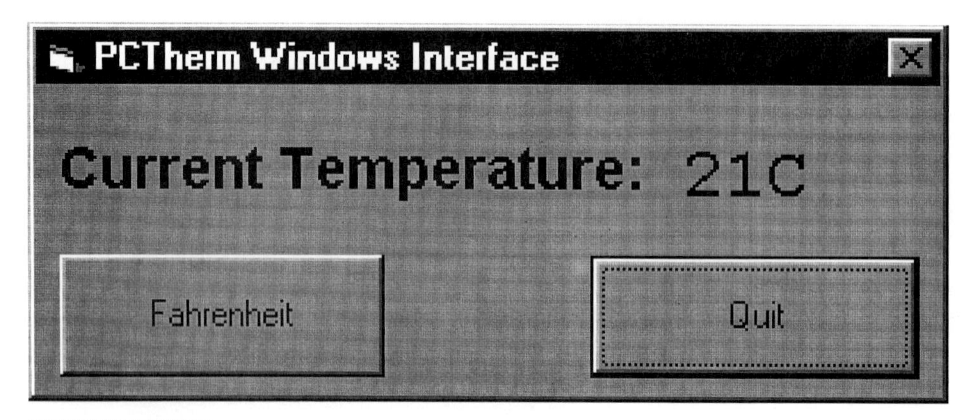


**Figure 16-46** PCThermVisual Basic interface

populated) sockets are on the board. This ISA prototype board was taken from previous projects simply because it is quite expensive and I wanted to reuse it, rather than spend $100 on a new one.

The bill of materials for this circuit is shown in Table 16-20. A number of aspects of this circuit will seem unusual. I use the PICmicro® MCU itself to drive the address that is compared to when writes to the PSP port occur. The circuit was originally an ISA prototype card that I created for *PC Ph.D.* Rather than rewire the circuit, I used the prototype card circuit. This circuit uses the PICmicro to output the most-significant seven bits of the PC's I/O port address. The lower three bits are decoded and selected by the 74LS138. This method of decoding the PC's I/O port addresses allows me to select different addresses according to my own requirements. For this application, the ISA card presented here was the only optional card, so I used address 0x0300 (the first prototype adapter block address) for this application.

Notice that, in this circuit, the received data driver control (_RE) of the 75176 is always pulled low so the driver is active. I did this so that the PICmicro® MCU's RX pin (RC7) is always being driven. The LED was added to provide a visual indicator when the ISA card received valid data.

When I originally built this card (for the PC Ph.D.), I used an old 80386-processored PC that I had bought for $30. This PC had I/O slots that were a standard 3/4" apart. For this project, I used a different PC that only had 0.5" between slots. Thus, I had to reposition some cards in the system, as well as cut down the ends of the wirewrap pins. If you have to do this, please be careful. Clean all of the cut pins off of the prototyping card before puting it into your PC. You do not want any of them falling into the PC and shorting out anything on the PC's motherboard. As another safety precaution, only put in and take out this card when power is off.

The ISA card's PIC16F877 source code is PCTHERM2.ASM and can be found in the *code\pctherm* subdirectory of the *PICmicro* directory. This project does not have to use the PIC16F877, any 40-pin mid-range PICmicro® MCU will work (because they all have built-in USARTs and PSPs).

To communicate with this card, I wanted to use a PIC16HV540 (high voltage) PICmicro® MCU to take the 12-volt power from the ISA card and convert it to five volts for the remote application. In doing this, I wanted the PIC16HV540 to power the 75176 RS-485

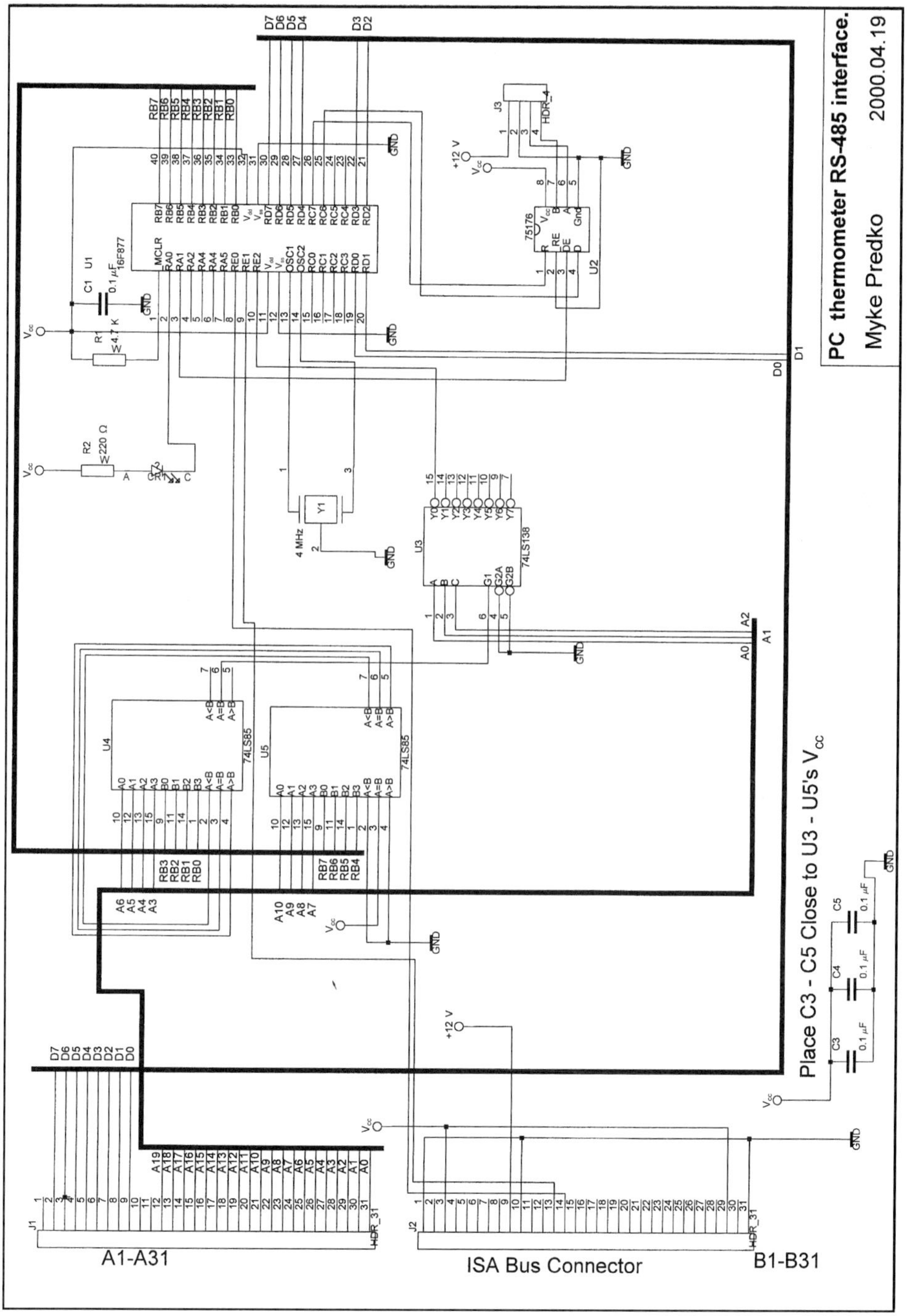

A1-A31
ISA Bus Connector
B1-B31
J1
J2
HDR_31
U4
U5
74LS85
U3
74LS138
U1
16F877
U2
75176
J3
HDR_4
R1 4.7 K
R2 220 Ω
C1 0.1 μF
Y1
4 MHz
+12 V
GND
Place C3 - C5 Close to U3 - U5's $V_{CC}$
C3 0.1 μF
C4 0.1 μF
C5 0.1 μF
PC thermometer RS-485 interface.
Myke Predko
2000.04.19


Figure 16-47

**TABLE 16-20 PC Thermometer ISA Card Bill of Materials**

| PART | DESCRIPTION |
|---|---|
| U1 | PIC16F877-04/P |
| U2 | 75176 RS-422/RS-485 Interface |
| U3 | 74LS138 3 to 8 Demultiplexor |
| U4 - U5 | 74LS85 Value Comparators |
| CR1 | Red LED |
| Y1 | 4 MHz Ceramic Resonator |
| R1 | 4.7 k, 1/4 Watt Resistor |
| R2 | 220 Ohm, 1/4 Watt Resistor |
| C1 | 0.1 $\mu$F, Tantalum Capacitor |
| C3 - C5 | 0.1 $\mu$F, Capacitor, Any Type |
| J3 | 4x1 Screw Terminal |
| Misc. | ISA Prototyping Boards (with J1 and J2 Built in), Wire Wrap Sockets, Wire Wrap Wire |

interface and the DS1820 digital thermometer. When I started working with this circuit, I discovered that the PIC16HV540 did not have the current drive capabilities required for the 75176 (especially when it was driving the line low when the ISA card's 75176 was driving it high). The solution to this problem was to add a 7805 to provide power for it and the DS1820. The final circuit (Fig. 16-48) was built using point-to-point wiring.

The bill of materials for the Remote Card is shown in Table 16-21.

The PIC16HV540 source code (REMPIC.ASM) is located in *code\PCTherm* subdirectory of the *PICmicro* directory. The code can be converted to any other 18-pin PICmicro® MCU without modification. You might also want to use a PIC16C505 and have it powered from the 7805.

The reason for using the +12-volt power from the PC to the remote card was to avoid very long wire voltage drops, which would cause problems with the voltage from the PC being high enough to reliably run the PICmicro® MCU and the other parts on the card. Using 100-feet of four-conductor telephone cable, I found a 0.35-volt drop in the voltage.

To debug the application, before it was installed in the PC, I connected the ISA card to a bench +5-volt power supply and the remote card to a +12-volt power supply. I then connected them with two conductors of the four-conductor telephone cable. Notice that in both Figs. 16-47 and Fig. 16-48 that I have put the two 4-by-1 screw terminals in the same orientation. This was to avoid issues with keeping track of the wiring. The convention I used for the wiring is given in Table 16-22.

This is no different from how a real telephone is connected but the voltage signals are radically different, so if you are going to install this cable along side telephone cable, be sure you keep track of which cable is which. Notice that plugging phones into this cable

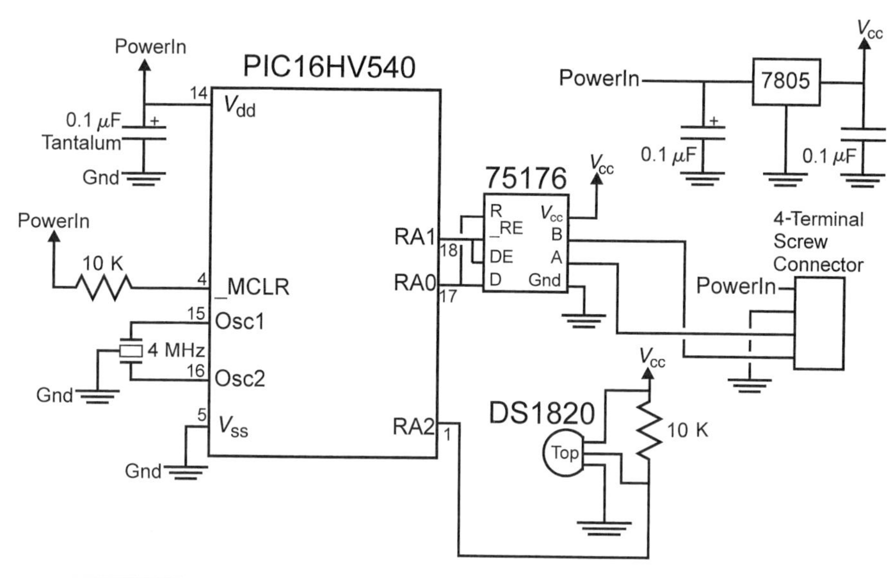


**Figure 16-48** PIC16HV540 PICmicro® MCU thermometer

or a PC into a phone line could damage either device (which could be very expensive if phone company equipment is damaged).

For the data communication, I used 1200-bps NRZ serial communications with some special 75176 control timings to ensure that the two PICmicro® MCUs do not get into contention with one another. The communication's flow for the two PICmicro® MCUs is shown in Fig.16-49. When the communication is working correctly, neither device will be pulling the RS-485 line low. When I introduced RS-422 and RS-485, I did notice that multiple devices could drive at the same time. I found that in this application, because of the

**TABLE 16-21 PC Thermometer Remote PICmicro® MCU Card Bill Of Materials**

| PART | DESCRIPTION |
|---|---|
| PIC16HV540 | PIC16HV540-JW |
| 7805 | 7805 +5 Volt Regulator |
| 75176 | 75176 RS-422/RS-485 Interface |
| DS1820 | DS1820 TO-92 Package |
| 4 MHz | 4 MHz Ceramic Resonator |
| 3x 0.1 μF | 0.1 μF, Tantalum Capacitors |
| 2x 10 K | 10 K, 1/4 Watt Resistors |
| Connector | 4 Terminal Screw Connector |
| Misc. | Prototype Card, Sockets, Hook up Wire" |

**TABLE 16-22 PC Thermometer Wire Connectors**

| WIRE COLOR | FUNCTION |
|---|---|
| Red | +12 Volts |
| Black | Ground |
| Yellow | "A" Pin (Positive RS-485 Voltage) of 75176 |
| Green | "B" Pin (Negative RS-485 Voltage) of 75176 |

method of providing power to the remote PICmicro® MCU, I could not get enough current to reliably pull the data line low.

This method of communication works quite well, as you can see in the oscilloscope pictures shown in Figs. 16-50 and 16-51. The top line in both diagrams shows the actual RS-485 positive voltage signals. Figure 16-50 shows the signals that the PIC16F877 master works with. You can see the 75176 driver being turned off (the lower line) when the temperature data from the remote PICmicro® MCU is expected. During this time, the line remains driven and valid data is received by the PIC16F877.

Figure 16-51 shows the incoming *ping* character (*P,* ASCII 0x050) on the 75176 data line, after which it drives the line, waits 15 ms, and drives the current temperature back on the line. When the transmission has been completed, the remote PIC16HV540 stops driving the RS-485 line (at which point, the PIC16F877 resumes driving the line high).

After the remote PICmicro® MCU has responded to the PIC16F877, it polls the DS1820 for the current temperature. This temperature is transmitted back to the PIC16F877 the next time the *ping* character is received. The PIC16F877 polls the remote device once every second, so there is plenty of time to read the current temperature from the DS1820.

If the remote PICmicro® MCU is disconnected or cannot respond in any other way, the PIC16F877 will experience the data shown in Fig. 16-52. When the RS-485 voltage goes to the "half" voltage level, the 75176 will convert that to a low with periodic voltage

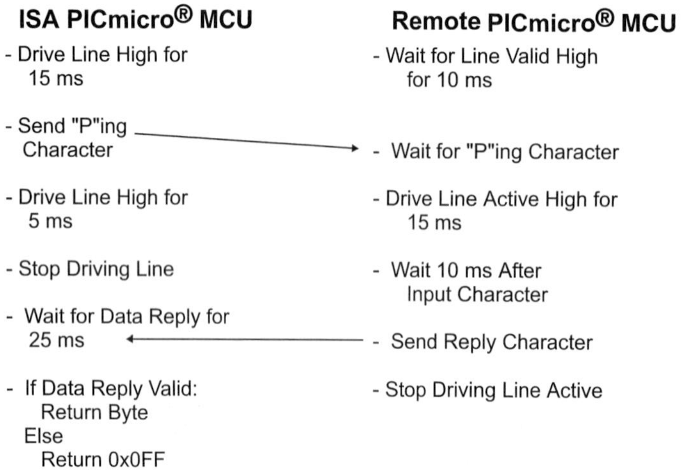

**ISA PICmicro® MCU**

- Drive Line High for 15 ms
- Send "P"ing Character →
- Drive Line High for 5 ms
- Stop Driving Line
- Wait for Data Reply for 25 ms ←
- If Data Reply Valid: Return Byte Else Return 0x0FF

**Remote PICmicro® MCU**

- Wait for Line Valid High for 10 ms
- Wait for "P"ing Character
- Drive Line Active High for 15 ms
- Wait 10 ms After Input Character
- Send Reply Character
- Stop Driving Line Active

**Figure 16-49** PCTherm data flow

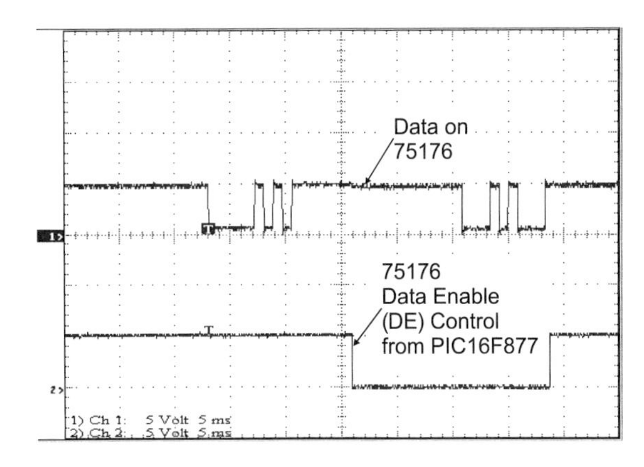


**Figure 16-50** PICTherm ISA PICmicro® MCU data transfer

spikes. The voltage spikes are caused by noise on the RS-485 lines differing enough for the differential input drivers to interpret them as high-voltage values. These spikes are the reason for why I poll the line for 10 ms before accepting a low-voltage level as a valid NRZ serial character in the remote PICmicro® MCU.

When the low data is received by the PIC16F877's USART, the data is interpreted as an invalid character (it is not in 8-N-1) format and the *FERR (Framing ERRor)* bit of the RCSTA" register is set. This bit indicates that when a high value was expected for the *stop* bit, a low was received. To clear this error, I reset the USART before transmitting the *ping* character.

As indicated, this is not a terribly hard circuit to build. I recommend that you do test the two cards and their connections on a bench using the power supplies outlined previously to avoid the need to power up and power down the PC to debug the application. Another reason to do this is that probing an ISA card in the PC is actually quite difficult—especially if other cards are present in the system.

On the bench, with the two cards hooked up and working correctly, you should see the LED flash on and off at a rate of once per second. I was not able to get an LED connected to the PIC16HV540 to work because of the low-current source and sink capabilities of the chip. Once the LED is flashing (and you can stop it by disconnecting the remote PICmicro® MCU), you can test it out in your PC.

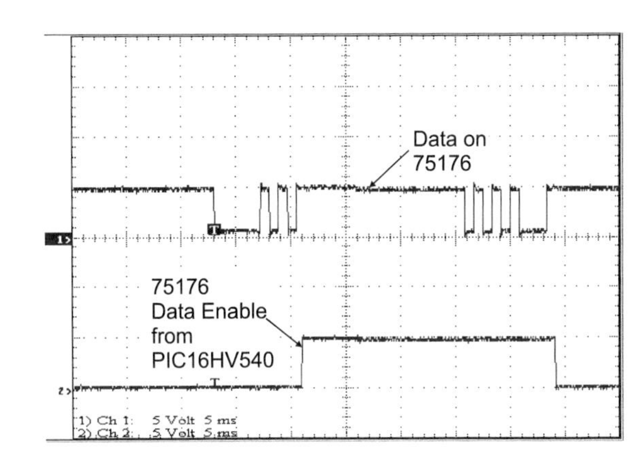


**Figure 16-51** Remote temperature-sense data transfer

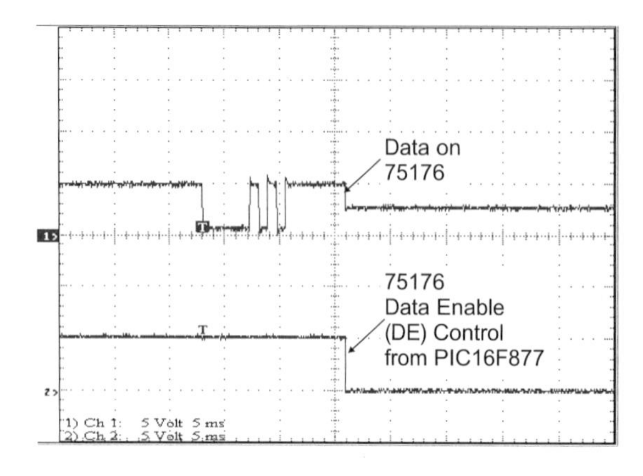


**Figure 16-52** **No-reply ISA PICmicro® MCU data transfer**

When I first tested the ISA card and connections in the PC, I used MD-DOS *Debug,* rather than going directly to the Visual Basic application. To use Debug, start up your PC either in MS-DOS mode or with an MS-DOS prompt active. From the MS-DOS prompt, key in Debug, followed by pressing the *Enter* key. You should get the - prompt.

From the -, enter in *i 300* so that the input line looks like:

```
-i 300
```

after pressing *Enter,* debug will respond with the byte data at the I/O port address. If nothing is addressed at this location, *FF* (0x0FF) is returned.

Once you see values other than 0x0FF when you check the address, you are ready to work with the Visual Basic application code. The code for this application is very simple. Just a timer polls I/O port 0x0300 (where the PSP is located) once every 777 ms. Watch because the DLL has to be loaded to allow you to access the internal I/O ports of the PC. For this application, I used the same DLL as with the El Cheapo. If you are working with the El Cheapo, you should have DLPORTIO already loaded into your PC, so you can run the Visual Basic interface without modification.

The Visual Basic source code project is PCTHERM MICROSOFT WINDOWS INTERFACE.VBP, which is located in the *code\PCTherm\vb* subdirectory of the *PICmicro* directory. A packaged and compiled version of this code is located in the *code\PCTherm\vb\Package* subdirectory. To load the application, execute SETUP.EXE.

If you want to create your own Visual Basic applications that will access your PC's I/O ports, you can use the DLPORTIO DLL as well. As indicated elsewhere, this package (PORT95NT.EXE is the installation file) will allow you to access I/O ports in Windows 95/98/NT, automatically detect which operating system is in use, and load the correct DLL for you. Notice that if you are using Windows/NT, you will have to have administrator access to the PC.

When you are creating a Visual Basic Application, to access the DLL routines for the I/O Port reads and writes, simply add a "module" to the project. Figure 16-53 shows the Visual Basic development screen for the PCTherm Visual Basic source code. When you are ready to start working with the I/O ports, single left click on your *Project* and then right click on it. You will be given a pull down. Select *Add Modules*.

From here, add the DLPORTIO.BAS module (which will probably be installed in your

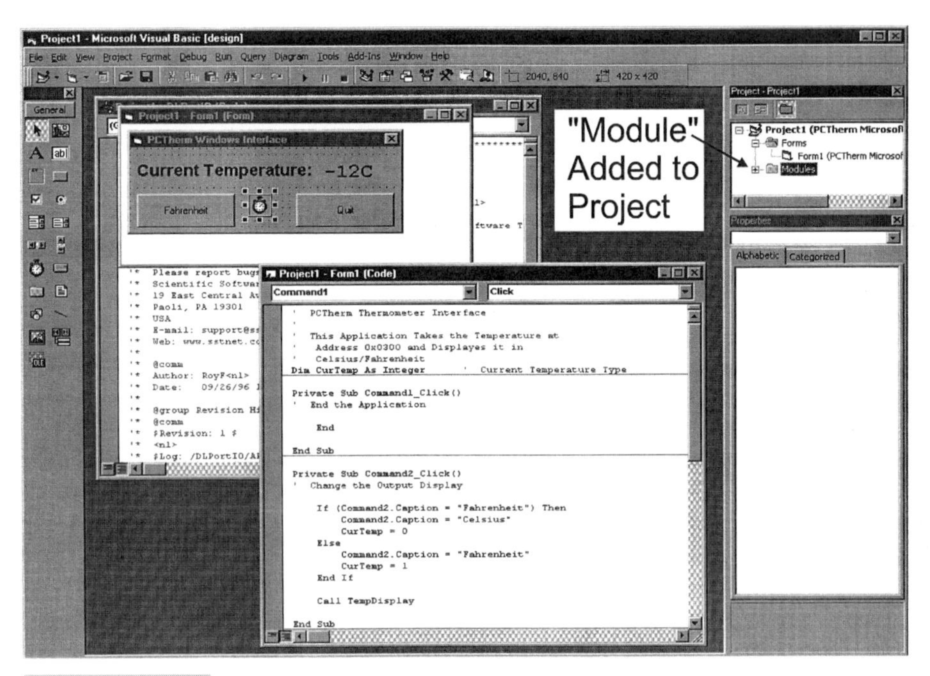


**Figure 16-53** PCTherm Visual Basic module addition

*c:\Program Files\DLPORTIO\API* subdirectory), which has the following I/O subroutines and functions that can be accessed from within your Visual Basic source code to access the I/O ports within the PC. Each function takes a "long" data type as the address and passes the type of data specified in the subroutine or function name. The list of subroutines and functions are:

```
Function Name
DlPortReadPortUchar
DlPortReadPortUshort
DlPortReadPortUlong

Subroutine Name
DlPortReadPortBufferUchar
DlPortReadPortBufferUshort
DlPortReadPortBufferUlong
DlPortWritePortUchar
Dl PortWritePortUshort
DlPortWritePortUlong
DlPortWritePortBufferUchar
DlPortWritePortBufferUshort
DlPortWritePortBufferUlong
```

## SERVO R/C SERVO PROGRAMMER/CONTROLLER

This project (Fig. 16-54) really demonstrates what kind of complete applications can be implemented with the PICmicro® MCU. The servo controller project uses a PIC16C71 to provide a text user interface with an LCD and allows the user to control up to four servos,

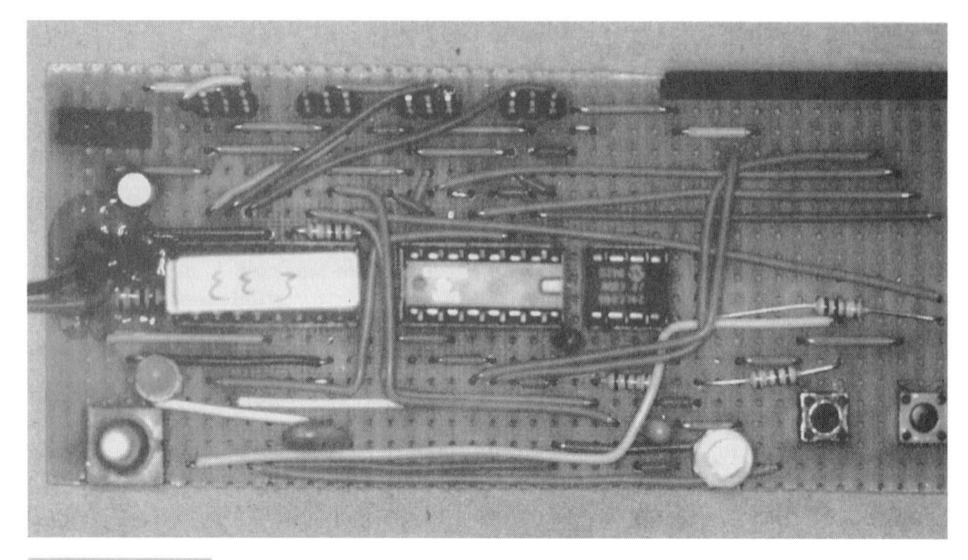

**Figure 16-54** **PICmicro® MCU servo controller board**

develop a sequence of events for the servos to run, and allow the user to save a sequence for later execution. Pretty good for an 18-pin device.

This project is designed to control the servos used in "armature" robots or mechanical displays that require moving parts. The servos can either be controlled individually or sequenced (Fig. 16-55).

I apologize for the cluttered schematic, but I wanted to ensure that all of the devices to which the PICmicro® MCU is interfacing can be shown in the graphic. Notice that a lot of the lines are actually $V_{cc}$ or Gnd. If all of the power lines were taken off the schematic, it would be a lot simpler. The bill of materials is shown in Table 16-23.

In the schematic, I did not include the power for the 74LS174. Pin 16 is connected to $V_{cc}$ and pin 8 to ground. Actually, when I first developed the circuit (and the schematic), I just assembled it on a *vector* board (as shown in the photograph). A Vector board is a good prototyping tool for this project because of the repeated power and grounds that are easily bussed.

The user interface consists of a potentiometer, two buttons and a 16-character by two-line LCD. Making it all work together is a menu routine that takes a source message, puts it on the screen, and then handles user input via a pot position and buttons that select the action.

The initial screen looks like:

```
Servo Controller
>Pos< Pgm Run
```

The button at RB4 is the *Select* button, which will move the cursor (the > and <) between the actions on the lower line of the LCD. When the *Enter* button (at RA2) is pressed, the display program is ended and the cursor position is returned to the caller.

The pot is used to select arbitrary values (i.e., the servo position). This was the first time I had worked with the PIC16C7x analog inputs (the pot was used as a voltage divider).

The code for reading the pot value is really quite simple:

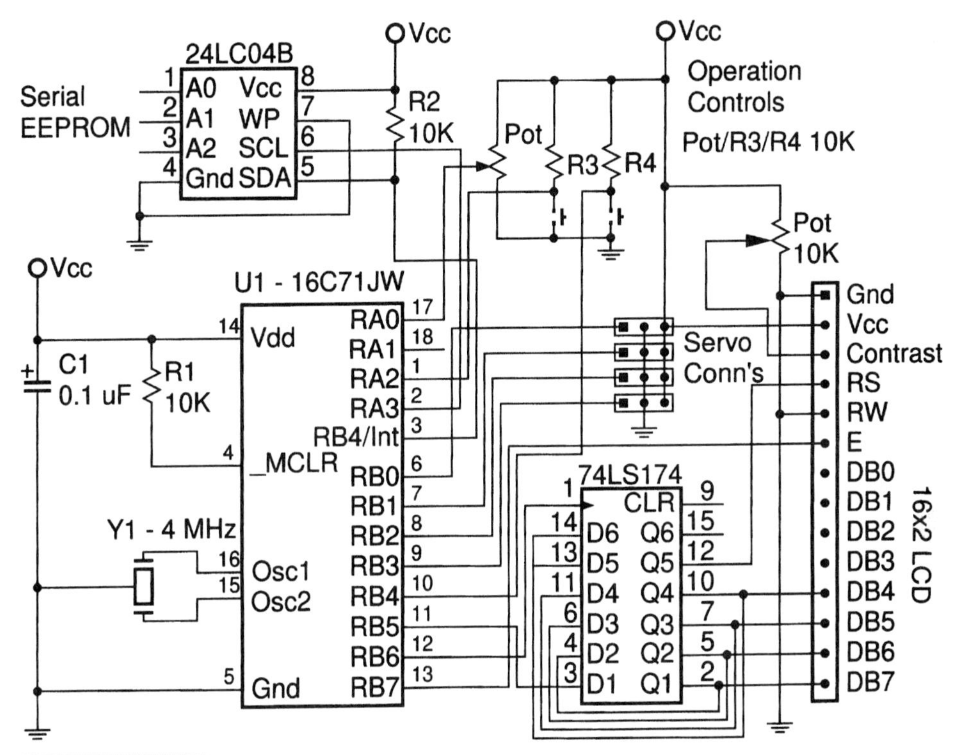


**Figure 16-55** PICmicro® MCU servo controller circuit

```
ADCON1 = 2;                      ; Just RA0/RA1 for Analog I/P

ADCON = 0x041;                   ; Enable A/D Conversions for RA0
 :
ADCON.GO = 1;                    ; Start the A/D Conversion
while (ADCON.GO == 1);           ; Wait for the A/D to complete

ADTemp = ADRES;                  ; Read the A/D result
```

Using a pot to specify exact values is kind of tricky and will require some practice and patience. The optimum solution would be using a multi-turn pot instead of the single-turn one that I used (you should also put on a knob to make turning the pot easier on the fingers). I went with the pot in the first place because the servo position and program delay values are not really precision operations. Adding the *goto* and *print* functions to the program specification were an afterthought.

The user interface is used to provide a nonlanguage-based programming environment. With the *Pot, Select,* and *Enter* buttons, you can specify an immediate servo position, enter a program, or run the program (either single step or running with 20-ms nominal steps). I believe that this is a very efficient interface for such applications.

The servo interface is actually embarrassingly simple. I used radio-control model servos for this project, which rely on a 1- to 2-ms pulse (the duration specifies the position) every 20 ms. The 20-ms cycle is a natural for a TMR0-based interrupt handler. When invoked, the

**TABLE 16-23 Servo Controller Bill Of Materials**

| P/N | DESCRIPTION/COMMENTS |
|---|---|
| U1 | PIC16C711-JW |
| U2 | 24LC04B/I2C Serial EEPROM |
| U3 | 74LS174/Hex "D" Flip Flop |
| Y1 | 4 MHz Ceramic Resonator |
| C1 | 0.1 $\mu$F, Capacitor (Any Type)/As Close as Possible to U1 |
| C2 | 0.1 $\mu$F, Capacitor (Any Type)/As Close as Possible to U2/Not shown on Schematic |
| C3 | 0.1 $\mu$F, Capacitor (Any Type)/As Close as Possible to U3/Not shown on Schematic |
| R1 - R4 | 10 K 1/4 Watt |
| R5 - R6 | 10-K Potentiometers/Part Number not given on Schematic |
| SW1 - SW2 | "Momentary On" Pushbutton Switches/Part Number not given on Schematic |
| – | "Vector" Board or other Prototyping Card for Schematic |
| – | 8, 16 and 18 Pin Sockets for U2, U3 and U1 Respectively |
| – | 1 of 14x1 0.100" Socket for LCD |
| – | 4 of 3x1 0.100" Pin Connectors for Servos |
| – | 5 Volt, 500-mA Power Supply |

interrupt handler outputs a pulse of 1 ms to each servo and then loops with a counter. When the counter value is greater than the value for a particular servo, the pulse for that servo is turned off. The counter continues to loop until a full 2 ms (to allow full travel of a servo) is complete. TMR0 is then reset to an 18-ms delay (so that the whole cycle is 20 ms long).

The mainline program is responsible for updating the servo positions from user input or the application "program" in the serial EEPROM.

Notice that for this project, the servo position granularity is such that there are 50 steps from stop to stop. Because of the parallel control of the servos, the count loop takes quite a long period of time to check each servo value to see if it has to be updated. The number of steps can be increased by either using a faster PICmicro® MCU clock or by using fewer servos and taking out the code used to support the unneeded devices.

The last major sub-block of this project is the I2C serial EEPROM. The 24LC04B provides 4 kbits or 512 bytes of EEPROM. For each program instruction, two bytes are used, in the format shown in Table 16-24.

When the program is running, it can be stopped manually by pressing the *Enter* button.

The PICmicro® MCU itself communicates with the serial EEPROM by behaving as an I2C master. The 24LC04B is an I2C *slave device,* which means it responds to instructions

**TABLE 16-24 Servo Controller I2C Memory Data**

| BITS | FUNCTION |
|---|---|
| 15–10 | Instruction Check Sum |
| 9–8 | Instruction Type |
| | – Set Servo Position |
| | – Goto Location |
| | – Delay n/10 Seconds |
| | – Print Character on LCD |
| 7–0 | Servo Position/Goto Location/Delay/Display Character |

directed to it by a master (these instructions are prefaced by the I2C control byte, as described elsewhere in the book).

There are two things I would like to point out to you in this application. First is with the 24LC04B. The bit 9 of the byte address is in the "control" byte (the least-significant eight bits are passed in the next byte). The other thing is that the A0 to A2 pins on the EEPROM package are not connected to the chip inside. Thus, the 24LC04B cannot be used with other EEPROM devices (which typically use the A0 to A2 bits to differentiate each other when the control byte is sent) on the I2C bus because of the danger of contention (two devices each trying to transmit data).

The actual code for communicating with the EEPROM is quite simple. As is laid out in SERVOEE3.ASM, it makes reading and writing to the EEPROM simply just subroutine calls. The code presented here has been modified, used as an .INC file, and presented in the appendices.

By adding together the LCD/Pot and Button interface to the servo controller and serial EEPROM, you really end up with a project that is more than the sum of its parts. I think you'll find the interface intuitive and easy to work with. This was my two boys' first experience in programming (they were eight and 10 when this project was first done) and they were able to master the four-instruction program environment very easily.

One of the things that made the project a lot more accessible was mounting the servos on Lego blocks (if your household has kids, chances are you have a few hundred pounds of the stuff) as shown in Fig. 16-56.

## ADDENDUM: SIMMSTICK SERVO CONTROLLER

A lot of PICmicro® MCU development tools (the programmers and emulators presented in this book are some of them) are available for the PICmicro® MCU. The SimmStick tool can improve the time needed to build your own applications by providing a basic framework for wiring up your PICmicro® MCU. To show how easy it is to setup one of these for a project, I decided to redo one of my more complex projects (the servo controller) on a SimmStick (Fig. 16-57).

The whole operation went quite easily. I only had to cut two traces going from RA0 and RA1 and wired RA0 to a 10-K potentiometer on the SimmStick board. Once this was done, I ran wires from RA3 and RA4 to the serial EEPROM on board the SimmStick. If I

**Figure 16-56**

wasn't using the built-in A/D on the 16C71, I probably would not have even done this; I just would have changed the code to use whatever pins the SimmStick had already wired for this purpose to be compatible with the Basic Stamp 1.

Once the traces were cut and wires were put in to change the EEPROM use, I built the SimmStick for the servo controller application. I stopped to ensure that the PICmicro® MCU would run (basically, put the board into a SimmStick backplane, apply power to it, and check Osc2 to be sure that it was running).

Using wirewrap techniques, I then put on the 74LS174 that's used as a shift register for the LCD and the two control buttons.

Once this was completed, I then created a simple Vector board wiring of the application.

When this project was originally done, the documentation for the SimmStick was not complete. I had to manually trace out the wiring on the card to figure out how to wire the circuit for this application. Total time to do this and build the application was about five hours. Since the first edition of the book was written, the documentation on the SimmStick has been improved significantly and this work can be done without having to trace all of the wiring on the card to figure out where everything goes.

Today, I would estimate the work to take less than an hour to decide where to make the cuts and to manually add the circuits.

If you compare the SimmStick version of the servo controller to the original version,

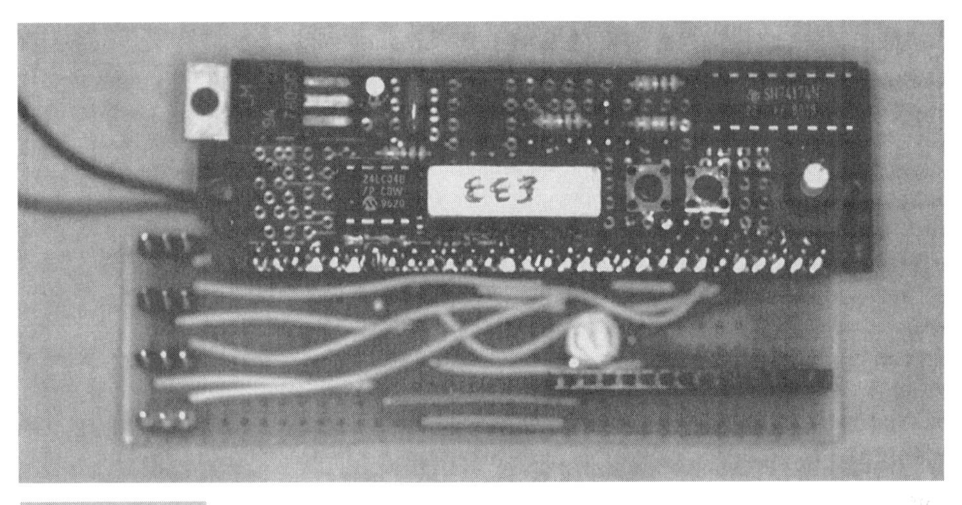

**Figure 16-57**

you'll see that the wiring is quite a bit simpler and the whole package is reduced significantly in size. I highly recommend the SimmStick for developing applications because of the ability to simplify wiring (basically eliminating the PICmicro® MCU set up, like the oscillator and power supply wiring from being an issue).

## MIC-II: SINGLE CHIP CONTROLLER

The one project I did not get any feedback on in the first edition was the MIC. This software application provides a complete debugging and development system for a simulated microcontroller with a built-in UART in a 1024-instruction PICmicro® MCU. I would have thought that the sheer audacity of such a device would have made a number of people stand up and take notice. Along with the MIC-II application itself, I have also included two sample applications to give you an idea of what can be done with it.

The original was not without some problems. For the second edition, I have taken a look at the original design and worked through some different ways to do things. In the original design, I had trouble figuring out how to implement labels and still fit everything within the 1024 instructions. The version presented here not only has resolved these shortcomings, but also has 94 instructions of program memory left over!

The reasons for the improvements is largely because of taking another look (almost four years later) at the original design with ideas on how to do better serial interfacing and processing data within the PICmicro® MCU. The second version (the MIC-II presented here), runs two times faster than the original and implements all the features that I wanted to add when I created the MIC originally.

The Parallax Basic STAMP was actually how I first got interested in the PICmicro® MCU microcontrollers. I was fascinated in how an 18-pin microcontroller could be used to load and execute PBASIC applications. The actual hardware was very simple and quite efficient, although I felt there were three shortcomings with the Basic Stamp.

The first shortcoming was the need for an external chip to store the program. When I first started to work with the PICmicro® MCU, I really got interested in working with the

PIC16C84. This was an earlier version of the PIC16F84 that is featured in Chapter 15. It has 64 bytes of data EEPROM that can be used to store applications.

The second shortcoming with the Parallax Basic Stamp was the need to compile applications on a host computer and the limited debugging capabilities of the interface. This is not to say that the MS-DOS IDE interface is not efficient, however, it requires an IBM PC running Microsoft MS-DOS or Windows connected to the Basic Stamp, which seemed limiting. At the time I was working with a number of different manufacturer's high-end UNIX workstations and I wanted to see more PICmicro® MCU applications available for them.

The last concern I had about the Basic Stamp was the relatively slow execution speed (1,000 instructions per second for the BS1 and 2,000 instructions per second for the BS2). I felt that using a PICmicro® MCU with an internal data EEPROM could run much faster and require less space.

I decided to see what I could do with a simulated/emulated microprocessor executing its own assembly-language instructions. I am very happy with what I have been able to implement with the MIC-II.

The original MIC project served as the basis for the YAP-II programmer and EMU-II emulator presented elsewhere in this book. This project that gave me the realization that the PICmicro® MCU is capable of much more than just being a programmable piece of logic that fits into a circuit; it can also be used to interface intelligently with the person using it. As shown with this project (as well as some of the others in this book), you can see that surprisingly complex interfaces can be created using quite a small PICmicro® MCU.

Like all projects, the first order of business was to decide what I would call the finished product. *MIC* jumped right out at me. Whereas the original *PIC* was an acronym for *Peripheral Interface Controller,* I thought the *MIC* would be a good acronym for *Myke's Interfacing Controller.* The slight difference in the second word is important because this microcontroller application can interface to the user without any special host software for controlling or programming.

With the name out of the way, I then began to try to decide what the emulated device's architecture would be. I don't know if the way I did it was putting the cart before the horse (I've never designed a computer architecture before), but I based the design on what I wanted the instructions to look like when they were displayed for the programmer.

To take advantage of the hardware features, I designed the microcontroller to use a similar Harvard architecture to the PICmicro® MCU. This means that program memory (which is the PIC16F84's 64 bytes of EEPROM data memory) is separate from the 11 file registers (including the PIC16F84's PORTB).

With this basis, a simple monitor program, with its own assembler and disassembler, to be built into the PICmicro® MCU. Along with handling source code, the monitor program would control application single stepping, program execution, and modifying and handling instructions. Obviously, this is a lot to ask for in only 1,024 instructions.

The design I came up with is a true eight-bit processor. All data paths are eight bits wide and there is only a maximum of eight address bits. The architecture that I came up with is shown in Fig. 16-58. This architecture is able to handle four addressing modes and can provide four symbolic labels within the application.

A sample application circuit could be similar to the one shown in Fig. 16-59. In this circuit, the MIC-II uses the bill of materials shown in Table 16-25.

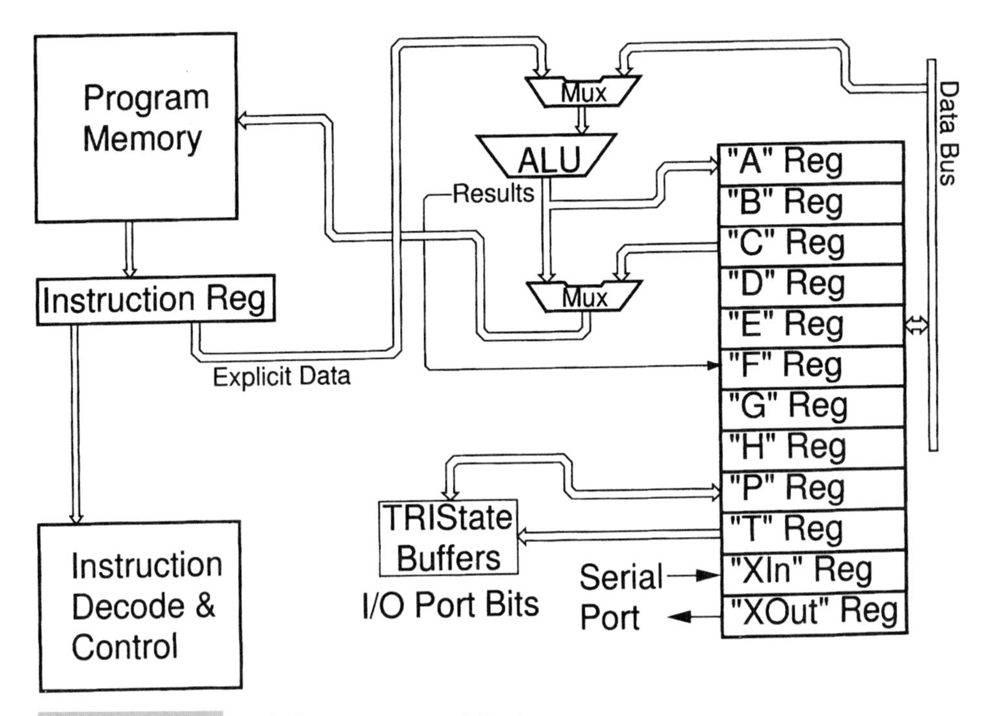


**Figure 16-58** MIC-II processor architecture

To develop this application, I used a YAP-II that just required pins 5 and 8, as well as pins 6 and 7, on the top connector to be wired together to provide the serial connection. This turned out to be a very fast and efficient way to implement and test the application, which is largely a software based, instead of a combination of software and hardware.

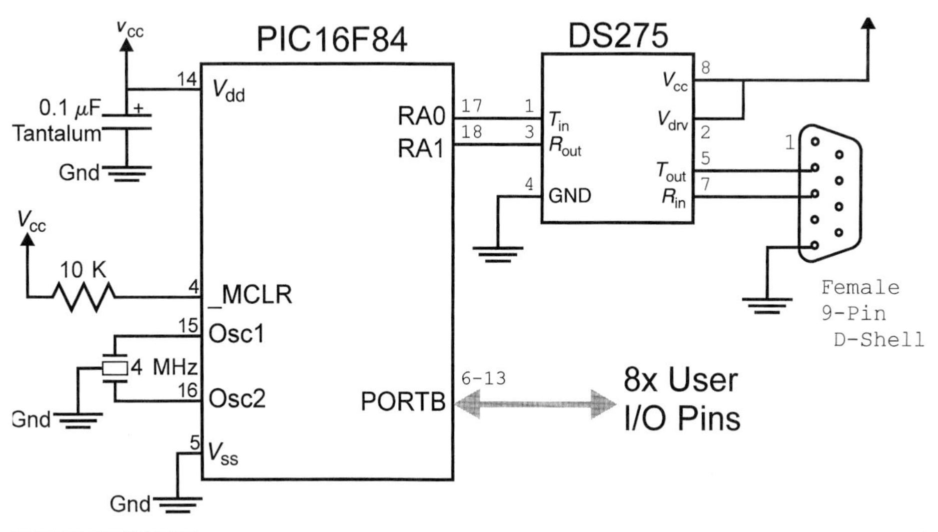


**Figure 16-59** MIC-II circuit using Ds275 RS-232 level conversion

**TABLE 16-25 MIC-II Bill Of Material**

| PART | DESCRIPTION |
|---|---|
| PIC16F84 | PIC16F84-04/P |
| DS275 | Dallas Semiconductor DS275 |
| C1 | 0.1-$\mu$F Tantalum Capacitor |
| R1 | 10 K, 1/4 Watt Resistor |
| Y1 | 4-MHz Ceramic Resonator with Built in Capacitors |
| J1 | Female 9-Pin D-Shell Connector |
| Misc. | Board, +5-Volt Power Supply, Wiring |

To make the instructions more intuitive, I wanted the instructions to be reference to the accumulator at all times. I felt that this would also greatly simplify the work required to implement the assembly language. I wanted to create instructions in the format:

```
+77
```

which is the addition operation:

```
Accumulator = Accumulator + 0x077
```

With this format, I decided upon 10 instructions for the MIC (Table 16-26).

**TABLE 16-26 MIC-II Instructions**

<table>
<tr><th>INSTRUCTION</th><th>DESCRIPTION</th></tr>
<tr><td><</td><td>Load the Accumulator</td></tr>
<tr><td>></td><td>Store the Accumulator in Register</td></tr>
<tr><td>+</td><td>Add to Contents of Accumulator (Store Result in Accumulator)</td></tr>
<tr><td>−</td><td>Subtract from Contents of Accumulator (Store Result in Accumulator)</td></tr>
<tr><td>|</td><td>OR with Contents of Accumulator (Store Result in Accumulator)</td></tr>
<tr><td>&</td><td>AND with Contents of Accumulator (Store Result in Accumulator)</td></tr>
<tr><td>^</td><td>XOR with Contents of Accumulator (Store Result in Accumulator)</td></tr>
<tr><td>/</td><td>Shift Register to the Right by 1 (Store Result in Accumulator)</td></tr>
<tr><td>#</td><td>Skip the Next Instruction if the Specified Accumulator Bit is Set</td></tr>
<tr><td>@</td><td>Jump to the specified Address, Store Address + 1 in “B”</td></tr>
</table>

Each one of these instructions has a parameter that is addressed in one of four modes. The four modes are: immediate, direct, indirect, and label. The 10 instructions gave me all the necessary functions I could think of (with some caveats).

A *shift left* of the contents of a register would be accomplished by adding a value to itself:

```
< Reg                    ; Load Accumulator with Reg
+ Reg                    ; Reg + Reg = Reg << 1
```

*Branch on condition* is implemented by moving the MIC's STATUS register contents into the accumulator (A Register) and then testing the condition of a bit within the register. A *skip on carry set* would be:

```
< F                      ; Load the Accumulator with the Flags Register
# 0                      ; Skip if Carry Bit Set
```

You'll also notice there isn't a *call* instruction. Instead, any time a *goto* (@) instruction is executed, the incremented program counter is stored in B. Doing subroutines this way saved code and file registers needed for a program counter stack. A traditional subroutine *return* instruction would be implemented using the two instructions:

```
< B                      ; Get the return value
> C                      ; Change the Program Counter
```

This subroutine return will be expanded upon in the following examples. With the instruction set specified, I then decided on how to do the register addressing.

I decided upon four modes: immediate, direct, indirect, and label. The first three can be easily shown with the example instructions:

```
+ 37                  ; Add 0x037 to contents of Accumulator
+ D                   ; Add contents of the "D" register to Accumulator
+ [D]                 ; Add contents of register addressed by "D" to the
                      ;  Accumulator
```

If you look back at the instructions, you'll see that each instruction takes data in these formats and acts on the value they represent. Notice that when specifying hex values where the most-significant nybble is in the range of A to F, that you should put in a leading zero to the value or the PICmicro® MCU will assume that the A to F registers are being accessed and code the instruction accordingly.

The application has four labels (A through D), which are assigned during application development. When the application is being passed to the MIC, a label is assigned at a specific address using the *!* directive to identify the address as a label (instead of a register). Accessing this address from within the source code uses the *!* letter format. For example:

```
@ !A          ; Jump to Address for Label "A"
```

The label is stored as an address in the Data EEPROM. When the application references the label, its value is referenced to a register. Loading the accumulator with a label value is accomplished using the instruction:

```
< !A          ; Load Accumulator with the Address of Label "A"
```

**TABLE 16-27 MIC-II Register Definitions**

| REGISTER | FUNCTION |
|---|---|
| A | Accumulator (the destination for all arithmetic operations) |
| B | "Goto" Return Address |
| C | Program Counter |
| F | Flags Register |
| D, E, G, H | General Purpose Registers |

In specifying the registers, I decided upon eight Base registers and three special-purpose registers. The base registers (A–H) can all be written to and read from, but some special purposes are assigned to some of the registers (Table 16-27).

The Program Counter (C) can only contain values between 0x000 and 0x01D. The last two possible addresses (up to 30 instructions can be programmed into the MIC as 16-bit words) are used to store the label addresses, as well as provide a check for a valid application in the EEPROM.

If a value outside of these limits is written to the C register, it is loaded with zero as a default.

The F (FLAGS) register contains the result of arithmetic STATUS, along with the serial receive/transmit status. This register is continually updated after each instruction or as the serial interface receives or transmits data.

The Flags bits are arranged as shown in Table 16-28.

The high four bits are the complement of their low nybble counterpart to allow the *skip on* bit set to work for all conditions.

Notice that the FLAGS register cannot be written to—either from the user interface or the application. If the register is written to in the application a *no* instruction halts execution. Thus, a *breakpoint* instruction in the MIC-II is:

**TABLE 16-28 MIC-II Flag Register Definition**

| BIT | FUNCTION |
|---|---|
| 0 | Carry Flag, Set on Addition > 0x0100 or Subtraction < 0 |
| 1 | Zero Flag, Set on Arithmetic Result = 0 |
| 2 | Receiver Byte Waiting |
| 3 | Transmitter Free |
| 4 | Not Carry Flag |
| 5 | Not Zero Flag |
| 6 | Not Receiver Byte Waiting |
| 7 | Not Transmitter Free |

```
> F   ; Invalid Write to Flags Register
```

An *nop* in the MIC-II is a *goto* (@) instruction pointing to the next instruction in the program memory.

The *UART (Universal Asynchronous Receiver/Transmitter)* built into the MIC-II is a three times sampling NRZ software handler using the PIC16F84's TMR0. The timer algorithm (and code) used is essentially identical to the YAP-IIs. When the application is running (using the *R*) command, a *Ctrl-C* (ASCII 0x003) byte will stop the application running or you can insert the *breakpoint* instruction (> *F*) as described.

P, T, and X are the three special-purpose registers. P and T are the PORTB and TRISB registers of the PICmicro® MCU, which operate identically as to how they work in the PICmicro® MCU. When a *0* is written to a T bit, the corresponding PORTB bit will be put into Output mode. When a *1* is written to a T bit, the corresponding PORTB bit will be in Input mode.

The X register, when written to, transmits the byte. When read from, it reads the last received character (0x000, if there isn't an unread character). Data should not be written to until the Transmitter Free bit is set; while data is being sent, the Transmitter Free bit will be reset. When a character is available for reading, the Receiver Byte Waiting bit will be set. Reading from the X register will reset this bit. As indicated, the monitor program will poll the incoming data. When a *Ctrl-C* is received, execution will end.

With the instructions, architecture, and peripheral devices defined, I then turned my attention toward the user interface. As I indicated, this interface consists of an NRZ serial interface running at 1200 bps with an 8-N-1 data packet format. Any commercially available RS-232 level translator or the YAP-II can be used to interface to the PC.

When the MIC-II is executing, it behaves as if it is an emulator controlling the processor. The prompt that I came up with for the user is:

```
[Label:...] PC ACC Flags Ins > _
```

Where:

```
!Label   Is the Label at the current Address
PC       Program Counter
ACC      Accumulator
Flags    Zero and Carry Flags
Ins      The Disassembled Instruction at the Program Counter
```

You will probably notice the similarity of this prompt to the one I used for the EMU-II. The different commands that can be entered into the MIC-II are shown in Table 16-29.

To set the label positions, first specify the address where you want the label. To do this, assign the desired address to the C register. Next, put in the *!Label* command on the next line. Figure 16-60 shows the process for running the MIC-II application code in a PIC16F84 that is in a YAP-II socket. After starting the MIC-II application running at 4 MHz, I change the program Counter C to address 4.

With the instruction at address 4 displayed, I then enter *!A* to indicate that the A label should move to this address. The final line of Fig. 16-60 shows Label A moved to address 4.

Labels are used in the application as in any traditional programming language. Jumping to B from the current execution is accomplished by referencing the label in the instruction as shown in the code example:

**TABLE 16-29 MIC-II Commands**

| COMMAND | DESCRIPTION |
|---|---|
| Nothing/1 | Single Step the MIC-II Processor. Execute one instruction and update the Program Counter ("C" Register) |
| R | Run the MIC-II at full speed from the current Program Counter ("C" Register) location |
| Register | Return the Contents of the Specified Register |
| Register = Constant | Set the Register to a Specific Value |
| !Label | Specify the Label Position |
| Instruction Parameter | Load the instruction with parameter into the PIC16F84's data EEPROM |

```
@ !B                    ; jump on to Label "B"
  :
B:
```

Here is a simple test program that you might want to try. It is designed to have four LEDs connected to the PIC16F84's RB0 to RB3 pins and will increment them and then restart. I have written it with the labels on the start of the line. As shown, to put in a la-

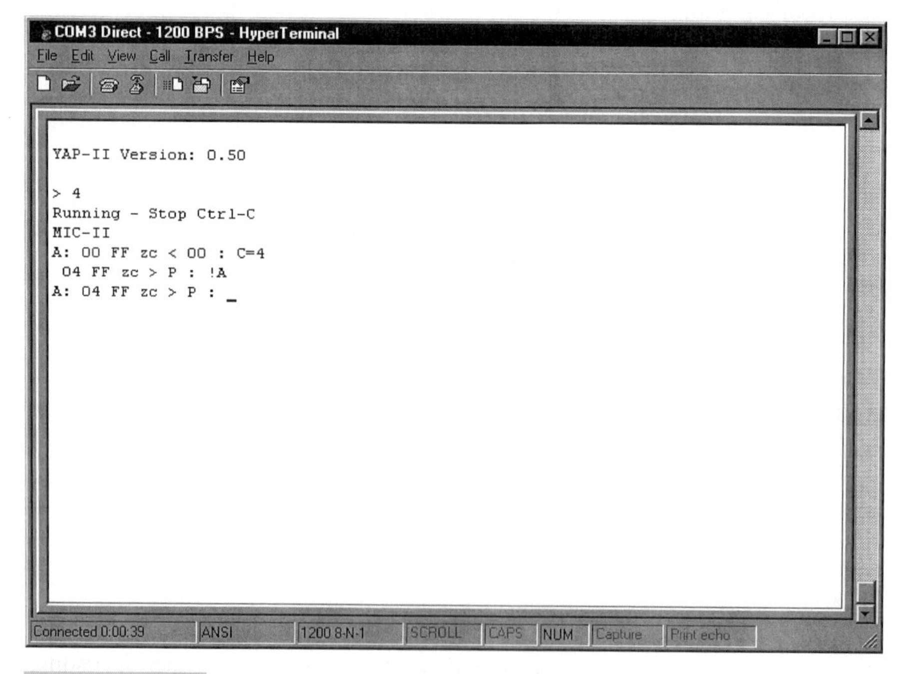


**Figure 16-60** MIC-II label-assignment operation

bel, you will have to enter in the *!Label* command before the instruction. This command does not increment the C register, so you can enter the instruction right after it.

```
A: < 0                    ; Enable all the Outputs
   > T
   > D                    ; Use "D" as the Output Counter
   ^ OFF                  ; Invert the Data to Turn OFF the LEDs
   > P
   < 10                   ; Loop Sixteen Times before Restarting
   > E
B: < D                    ; Increment the LED Display
   + 1
   > D
   ^ OFF                  ; Invert the Count for the LEDs
   > P
   < E                    ; Decrement the Loop Counter
   - 1
   > E
   < F                    ; Check for being Done
   # 1                    ; Skip Next if "Zero" is Set
   @ !B                   ;  Not Zero, Repeat at "B"
   @ !A                   ; Start All Over Again
```

When this application runs, 184 instructions are executed before looping back to the start and beginning again. When I watched this on my oscilloscope (probing RB3 because this cycles once during the execution), it took 19.4 ms for an average execution speed of approximately 9,500 instructions per second.

If you want to watch the LEDs change, you could try the following application. This code calls a subroutine at *C:*, which loops 256 times to provide a delay in the application so that it can be observed.

```
A: < 0                    ; Enable all the Outputs
   > T
   > D                    ; Use "D" as the Output Counter
   ^ OFF                  ; Invert the Data to Turn OFF the LEDs
   > P
B: < D                    ; Increment the LED Display
   + 1
   > D
   ^ OFF                  ; Invert the Count for the LEDs
   > P
   @ !C                   ; Jump to Label "C"
   @ !B                   ; Loop Back to "B" Forever
C: < B                    ; Save the Return Address
   > H
   < 0                    ; Loop 256 Times
   > G
D: < G                    ; Increment the Counter
   + 1
   > G
   < F                    ; Check to See if Zero Set (256x)
   # 1
   @ !D
   < H                    ; Return to Caller
   > B
```

This application is shown entered into the MIC-II in Fig. 16-61. To produce the listing, I changed the program counter (C register) value to the next instruction.

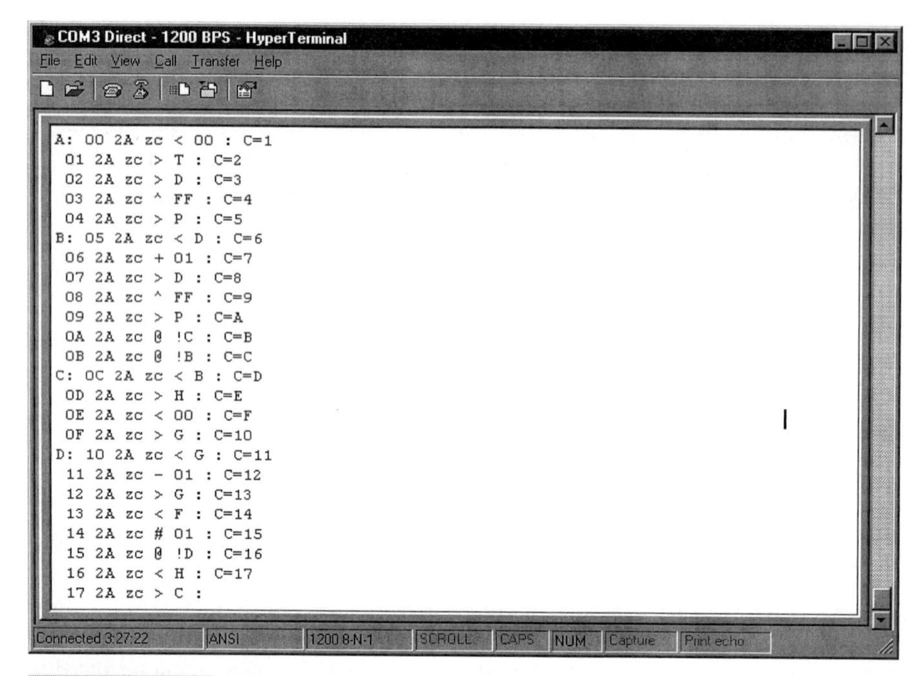


**Figure 16-61** MIC-II subroutine example

In the application's subroutine, notice that I save the value in the B register into the H register. I did this because within the subroutine, I perform a *goto* instruction, which changes the value of the B register to the address following the *@ !D* instruction. This is not the return address for the subroutine. When I first coded the application, I forgot to do this, but I was able to find the problem by stepping through the application (after changing the value loaded into G into 0x0FE or 254 to enable me to quickly execute the code in the subroutine's loop).

## VIDEO: NTSC VIDEO OUTPUT

Throughout the book, I have used LEDs extensively to provide user feedback on the status of an application or its input. I have also used LCDs to display text data for users to allow complex operations to be explained to the user, rather than relying on panels with text or instruction books. For home electronics, one of the output devices that you are probably most familiar with is the cathode-ray tube of your TV, which is used to provide I/O for the TV itself, your VCR, DVD player, and maybe your stereo.

If you've looked at the different PICmicro® MCU projects that are available on the Internet, you should not be surprised to find that a number of different projects are available for the PICmicro® MCU, in which it can be used to drive NTSC video output. In this project, I would like to introduce you to *NTSC (National Television Standards Committee)* composite video and a PICmicro® MCU application that will show you how to process PICmicro® MCU ADC data, along with moving data. Although the hardware is very simple and the software is also quite simple, this was one of the most challenging projects in the book.

I do not address other standards, such as PAL or SECAM, but the generation of composite video for these standards is simple and the application should be reasonably easy to port to these systems.

This project demonstrates how ADC data can be captured while driving an NTSC composite video output, along with computing the position of a bouncing ball. If you've looked further in the book, you'll see that the circuit that I created for this application is extremely simple, but the code probably took me the most time to develop. This was probably the most challenging project for me to develop in this book because of the stringent timing required by NTSC and the challenges I had in developing the simple displays used in this application.

Before explaining the application, I should first introduce you to composite NTSC video, as well as two warnings first.

First, for this project, I used a $10 television that I bought at a garage sale and used a $1.50 video modulator that I bought at a surplus store to convert the composite video that the PICmicro® MCU microcontroller circuit produces. This signal is passed to the TV using a standard 75-Ω coaxial cable.

I did not modify the TV in any way (such as providing a bypass to the video preamp from the tuner) and I don't recommend that you do this, either. Potentially lethal voltages are available inside a TV, stored in capacitors. These voltages can always be present—even if the TV is turned off and not plugged in.

Although the circuit here should not produce any voltages or currents that could damage a TV, I don't recommend that you hook this circuit up to the family's large-screen TV (or any TV that you care about). Instead, you should look for an old 12″ black-and-white TV that you can pick up for a few bucks. If flames come out of your home-entertainment system, I'm not to blame.

The second warning is that the circuit and software presented here is essentially a video transmitter. The modulator will convert the composite video to a frequency that could be picked up by your neighbor's TVs on channel 3 (which they are probably using for cable converters, VCRs, DVD players, etc.). Please be sensitive to whether or not you are interfering with their reception (or you might get a visit from your local FCC representative).

If you find that there are any problems with your home TVs or radios while this circuit is in operation, shut it down and use a different modulator or cable set up. These problems will manifest themselves as *snow* (white spots randomly on the screen) or audio static. You might find that you are unable to use this application without causing problems. In this case, go on to the next project.

With the warnings out of the way, let's look at the NTSC composite video signal produced by this application. After the tuner in your TV set has demodulated an incoming signal, the actual video information, called *composite video* is passed to the video drive electronics. If you were to look at the composite video signal on an oscilloscope, you would see something like that shown in Fig. 16-62.

Figure 16-62 identifies two features that you will have to become familiar with. The first is the *vertical synch,* a series of specialized pulses that "tells" the video drivers to reset the *raster* (the electron beam that travels across the CRT) to move to the top left corner of the screen. After the vertical synch is sent, video line data is sent. Each line of the data is a corresponding line on the TV display.

The start of each line of video data is identified with a horizontal synch pulse. During each line, data is output as an analog voltage. The description I'm going to give here is for black-and-white composite video with no colorburst information.

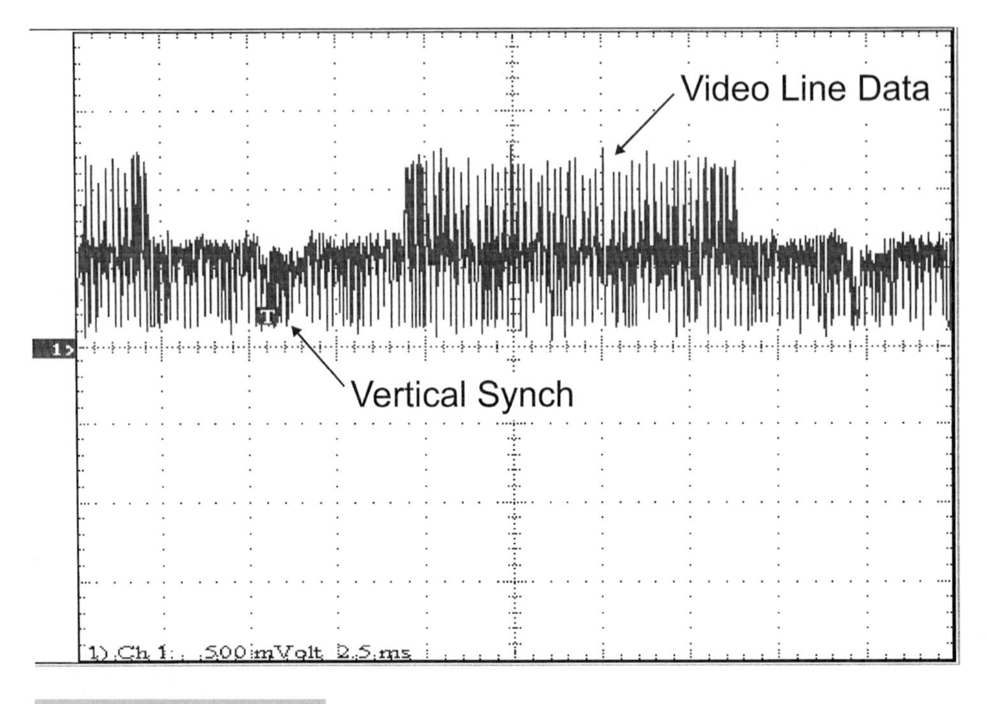


**Figure 16-62** Single NTSC field

The *colorburst* is a 3.579-MHz sine wave that is output after the horizontal synch to allow the video circuitry to latch onto the phase of the color signal sent to the TV. Along with the brightness (luminance) information, which consists of DC voltage levels, an analog signal is attached to the signal, as well for the color (chrominance) value for what is displayed on the TV's screen. The color displayed is dependent on the phase of the 3.579-MHz chrominance signal. By changing the phase of the signal, the color is changed on the display. As you can imagine, producing and detecting the changing phase of the 3.579-MHz chrominance signal is nontrivial. I doubt it could be done by a PICmicro® MCU because of the need for both a high-speed digital-to-analog converter, as well as a signal that constantly changes the phase according to the current color being displayed.

Composite video gets its name from the fact that it combines three different signals (the vertical synch, horizontal synch, and video data) all in one line. Special circuitry in the TV (*CRT,* as it is called for most of this section) splits this information out to control how the output is displayed.

The period of time between vertical synchs is known as the *video field*. For NTSC composite video (which this circuit creates), 59.94 fields are displayed each second. One complete frame of video consists of two fields, with the raster scan line of one field overlapping the other (known as *interlaced video*). The output produced by the project shown here produces the appropriate timing for the output data to repeat over two interlaced lines.

As part of the vertical synch, a number of "dummy" horizontal lines are passed at the same time and are known as *vertical blanking*. When the vertical synch is recognized within the TV and the raster moves to the top of the screen to begin scanning again, the CRT guns are turned off to prevent any spurious signals from being driven on the screen. The vertical synching operation is shown in Fig. 16-63.

The analog voltage output used for synch pulses and CRT control are always at a level below the video data black level. These are often known as "blacker than black." The normal synch level is at 0.4 volts and the active synch pulse is at 0 volts.

The vertical blanking before the vertical synch consists of six or seven half lines (at 31.8 $\mu$s long), followed by six negative half lines (these are the vertical synch pulses), and then another six or seven half lines, followed by 10 or 11 full lines (63.5 $\mu$s long).

With the raster gun now pointing to the top of the CRT, data can be output a line at a time. One of the lines, taken from this circuit, is shown in Fig. 16-64.

Each line is 63.5 $\mu$s with a horizontal synch pulse to indicate where the line starts, followed by the data to be output on the line. The data output ranges from 0.48 volts (black) to 1.20 volts (white). The voltages in between are gray. Figure 16-64 shows the different voltage levels for the horizontal synch, the front porch, the back porch, black, and white.

Notice that approximately 100 mV of noise is on each of the signals. This is largely caused by the prototyping construction that I used for this application and the poor ground that I have for it. In a "professional" application, I would expect that the noise on the line would be on the order of 10 mV (or less). For this application (and the very cheap, used TV set that I used), the 100 mV of noise did not cause a problem with image stability (a big consideration for video applications) or the brightness output of different parts of the signal.

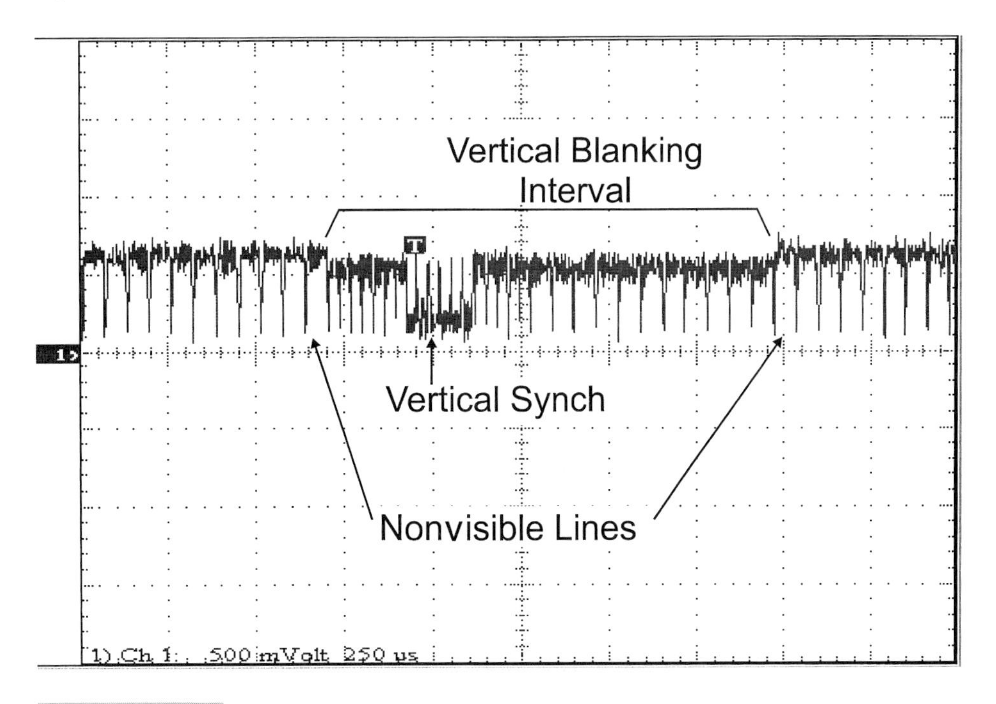


**Figure 16-63** NTSC vertical synch signals

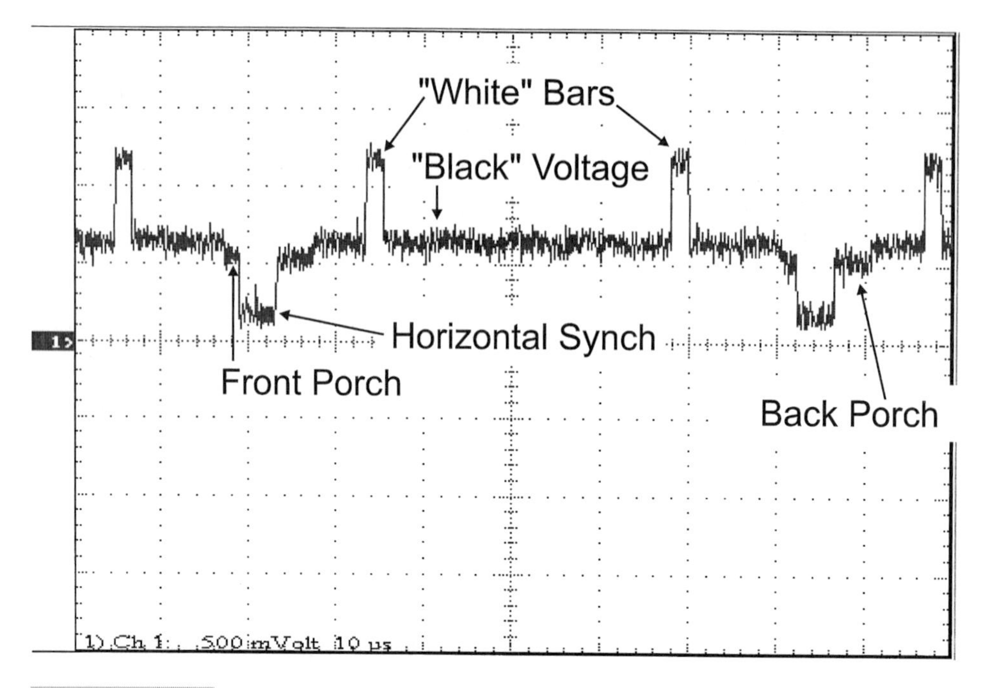


**Figure 16-64** Single NTSC line

The front porch and back porch are 1.4 and 4.4 $\mu$s in length, respectively, and are at 0.40 volts. The synch pulse itself is a zero voltage active for 4.4 $\mu$s. These signals (with their voltage levels) must be present in any video signal. For the 53.3 $\mu$s after the synch pulse, the voltage level must be at 0.48 volts to 1.20 volts. If the output dips below 0.48 volts, the TV (or CRT) might interpret the signal as a new horizontal synch with terrible results (in terms of the output display).

For the TV to accept the composite video, the quoted signal lengths must be adhered to as closely as possible. Failure to have the same number of cycles on different lines will result in a "broken" screen, which is described at length later in this project.

The actual circuit used to create the composite video is almost unbelievably simple and is shown in Fig. 16-65. The circuit itself is probably simpler than many of the circuits that were designed for the experiments presented in Chapter 15.

The bill of materials for the project is listed in Table 16-30. The composite video voltage output was produced by placing different I/O pins in Output mode with an output of *0*. The circuit is designed so that only one pin can be enabled as output at any one time. When a pin is in Output mode, the pin connected is pulled down and allows current to pass through the resistor connected to the pin. This added resistance changes the resistance of the voltage-divider output and changes the voltage output from it.

The circuit works very well and provides a very fast specific digital to analog conversion. In the oscilloscope, pictures of the composite video signals shown in this section, I used the circuit presented in Fig. 16-65.

When I built this project, I used the SimmStick prototyping system to build the circuit. I used a SimmStick prototyping card because I had used up all the DT101 cards that I had on hand. Notice that I put the 10-k potentiometer on the backside of the prototyping card

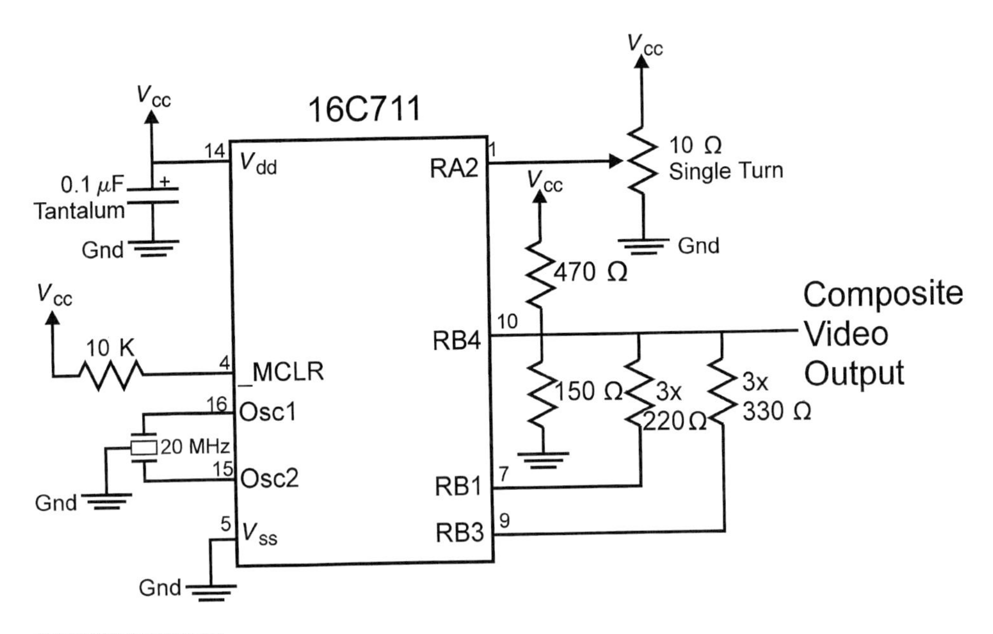


**Figure 16-65** Video generator circuit

so that I could access it while both the PICmicro® MCU and modulator card were connected in the DT003 backplane card.

One of the advantages of the SimmStick in applications like this is that the unregulated voltage is available with five-volt regulated voltage for the PICmicro® MCU. I

**TABLE 16-30 Video Generator Bill of Materials**

| PART | DESCRIPTION |
|---|---|
| PIC16C711 | PIC16C711-JW |
| 20 MHz | 20 MHz Ceramic Resonator with Built in Capacitors |
| 0.1 μF | 0.1 μF Tantalum Capacitor |
| 10 K | 10 K, 1/4 Watt Resistor |
| 10 K Pot | 10 K, 1 Turn Potentiometer |
| 470 | 470 Ohm, 1/4 Watt Resistor |
| 150 | 150 Ohm, 1/4 Watt Resistor |
| 3x220 | Three 220 Ohm, 1/4 Watt Resistors wired in Parallel |
| 3x330 | Three 330 Ohm, 1/4 Watt Resistors wired in Parallel |
| Misc. | Prototype board, wiring, +5 and 15 Volt Power Supply, Video Modulator, Video Modulator Power Supply, 75 Ohm Co-Ax cabling, TV set |

used this feature instead of having to come up with a dual power-supply circuit for the application. (The video modulator I used required at least 12 volts to run).

The video modulator was hot-melt glue attached to the prototyping card and the power, ground, and composite video output passed from the SimmStick edge connector. When I first was setting up the project, I used a small Tyco toy video camera as a sample composite video source. An RCA socket was glued to the video modulator, using hot-melt glue and the shield soldered to the modulator's case for ground and the signal line passed to the modulator's input. I originally created this SimmStick card for an Atmel AVR composite video output for the *Handbook of Microcontrollers* and I've used it on a number of projects since.

You might think that all the science went into the code development, but I should explain how the ADC built into the circuit works.

Looking at the circuit (Fig. 16-65) for this project, you will probably have a few questions. First concerns my use of a ceramic resonator instead of a crystal for the PICmicro® MCU clock. This is probably surprising—especially in light of the harping I've done about the importance of an accurate clock that I've done so far in this section. When working with a stand-alone video generator, like this circuit, the crucial parameter is to ensure that the timing is perfectly accurate, relative to the various signals and reasonably accurate to the specifications.

This is described in more detail, but most modern (which is to say built within the last 30 years) TV sets are able to work with a relatively wide range of input timing parameters. The reason for putting in this tolerance is not to make life easy for people like us, but to make the TV set insensitive to changes within itself as the components age. I am continually amazed at the reliability of TV sets (and consumer electronics in general). One of the reasons for this reliability is the ability of the circuit's designs to continue operating even though their components have degraded. This built-in tolerance makes the lives of experimenters like us much easier.

The second thing you will probably notice in Fig. 16-65 and Table 16-30 is the use of resistors in parallel. The three 220-Ohm resistors in parallel result in a resistance of 73 Ohms, which is close to one half of the 150 Ohms built into the voltage-divider circuit. The three 330-Ohm resistors in parallel have an equivalent resistance of 110 Ohms. Both of these resistances were calculated as part of the composite video output voltage for the 0-, 0.40-, 0.48-, and 1.20-volt outputs needed for the composite video output. The actual values needed was 107 Ohms, which are not available as "standard" values. By applying parallel resistance theory, I was able to approximate this value quite closely, instead of having to rely on standard values.

The composite video output is driven directly into the video modulator. Most video modulators have a capacitor in series with the input to negate any DC voltage offsets. To avoid any problems with RC delays in the composite video signal, I tried to keep the output impedance as close to 75 Ohms as I could (without having to resort to buffering the analog output from the PICmicro® MCU). This is the reason for the relatively low resistances in the digital-to-analog circuit in this project. Normally, when producing analog voltages using a voltage-divider circuit (as I use in this project), I would tend to use values in the tens of kilohms. TV circuitry is normally designed to work with low characteristic-impedance signals, so resistors in the 10-K range would have problems providing crisp changes in voltage levels to the video modulator with its input capacitor.

To find the resistor levels, I first calculated the proper values in a circuit that connected the resistors to ground directly (and not through an open-drain I/O pin). I then wired the circuit (Fig. 16-65) and found that the voltages had changed somewhat (actually, I just had to change the resistor always pulling to ground to 150 Ohms).

After calculating the expected values for the resistors, I then put the proposed circuit on a breadboard and wrote a small application to test out the voltage output. This application simply pulled RB1, RB3, and RB4 to ground and allows me to monitor the voltage levels output by the voltage divider. The application that I used was *VLadder* application taken from Chapter 15. Using this application with an oscilloscope allowed me to verify the voltage levels output from the composite video generator circuit.

The I/O pins RB1, RB3, and RB4 might seem strangely arbitrary, but they were used because they simplified the wiring of my prototype on the SimmStick. Depending on the method used to build your own circuit, you might want to change the specified I/O pins to keep your wiring as simple as possible.

With the circuit built, I then created VIDEO1.ASM, which can be found in the *code\video* subdirectory of the *PICmicro* directory. Video1 creates the various signal timings to produce the composite video output for the two fields that make up a frame. As indicated previously, 59.94 NTSC composite video fields are sent each second. Each frame consists of two of these fields, each of the second field's raster placed in between the rasters of the first field. This creates a "richer" display with fewer visible raster scan lines than if a single field was used for a frame.

To start the vertical blanking interval, a set of half lines is sent. This line is 31.8 $\mu$s in length with a 2.2-$\mu$s synch at its start. For the first field of the frame, seven of these lines are sent and six are sent in the second field. Next, a reverse image is sent as the vertical synch, with 2.2-$\mu$s porch voltage and the remaining 29.6 $\mu$s at zero voltages. Six vertical synchs are output for both fields of the frame. Finally, 10 (for the first field) or 11 (for the second field) half lines are sent. For this application, which runs the PICmicro® MCU at 16 MHz (with a 250-ns instruction cycle time), 127 instruction cycles were used for each line.

After the vertical blanking interval, 243 video data lines were sent using the specifications I outlined previously. To produce a 63.5-usec line period, 254 instruction cycles were used for each line.

Video1 was relatively easy to create and I was able to get it working in an evening with just a few minutes of debug because I had done a lot of simulation of the application before trying it out in hardware and driving out to a TV. The biggest issue I had was determining the location of the box so that it would be easily seen on the TV.

Next, I started work on VIDEO2.ASM, which set up an inverted U-shaped "playing field" on the TV's display, along with a PICmicro® MCU generated "paddle." The position of this paddle is determined by the voltage input from the 10-k potentiometer into the PICmicro® MCU's ADC. Debugging this program was considerably more challenging than Video1's.

The initial problem I had was with displaying the playing field. This involved creating two routines. The first was used to generate the top line of the display. The second was to create the two vertical bars of the field. Although not terribly hard to do, a fair amount of work was involved in determining values that would result in the playing field properly displayed on my TV's screen.

Next, I did the paddle's position. To read and process the paddle's position, I initiated the PICmicro® MCU's ADC in the first half line of the vertical blanking interval. In the second half line, I read out the ADC and then converted the value read into a value that could be displayed.

To do this, I had to know how wide the field was. This is why I did the playing field first. The playing field size that I settled on was 131 instruction cycles wide (32.75 $\mu$s) with eight-instruction-cycle (two $\mu$s) wide bars on each side. I wanted to have a bar that was about one sixth of the field wide, so I decided to make it 20 instruction cycles wide. This left 111 instruction cycles to move the bar. Because I wanted to use the code:

```
decfsz PaddleCount, f
  goto  $ - 1
nop
```

to provide a consistent three-instruction-cycle delay to move the paddle, this gave me a granularity of thirty seven positions for the paddle on the field.

To scale the eight-bit output of the ADC to the 0 to 37 range of the paddle, I multiplied the eight-bit output by 37 and used the top byte for the paddle position. This value was then multiplied by three and then used to "locate" the paddle on the screen.

Creating the software to move the paddle accurately was a significant challenge for me (and I still don't feel that I have it 100% right). I found that there was five cases that I had to be able to handle for the paddle to appear to move linearly to the potentiometer input. These cases were:

1. Against the left wall.
2. One position from the left wall.
3. Two or more positions from the left wall and two or more positions from the right wall.
4. One position from the right wall.
5. Against the right wall.

I found that each case had to be accurately timed, which required most of an evening for me to do.

The final code, which is shown in the photograph at the start of this section, is VIDEO3.ASM. This code incorporates a bouncing ball on the display. This ball moves diagonally until it hits a boundary of the playing field. The obvious original intention of this application was to create a *Breakout*-type game. The ball itself is three lines by three instruction cycles in size. Each time a new frame (two fields) is displayed, the position of the ball is updated by the size of the ball. A 16-bit variable, called *BallPos,* is used to track the X and Y position of the ball on the screen, as well as the X and Y direction the ball is moving in. Each time the ball "hits" a playing field wall, the direction changes. The initial position and direction of the ball is determined by the TMR0 value.

To be able to handle the ball's position, I separated each set of three lines of the playing field. The ball itself, if it is within these lines when the PICmicro® MCU generates them, is displayed on the line.

Getting the timing correct for the ball was a daunting task and took me two evenings to do. After getting each case (the same as the paddle plus the case of no ball) to execute in 63.5 $\mu$s (or 254 instruction cycles), the movement of the ball seemed to be appropriate.

Unfortunately, in creating the multiple lines for the ball to be displayed, I used an awful lot of space. This application uses 899 of the 1024 instructions available in the PICmicro®

MCU and does not leave enough space for collision detection between the ball and the paddle, or the ball and other features on the screen. To follow through and create a game from this application, I would want to use a 2048-instruction PICmicro® MCU, like the PIC16C712, instead of the 1024-instruction PIC16C711, which I used.

In the work that was done for this project, I was able to reach a number of conclusions. First, the PICmicro® MCU is not well suited to generate video signals. Electrically, this was not a huge issue and "gray" voltage levels could be added to the voltage-divider output quite easily.

The significant problems were in the software used by the project. Although the PICmicro® MCU is running at a reasonably fast speed (16 MHz), this translated to being able to place a "feature" on the display arbitrarily with a best-case accuracy of three instruction cycles (750 ns). Because 53.3 $\mu$s are available in each line for placing features, the resolution was only 71 positions on each line. This positioning accuracy is even further decreased when the playing-field width of 131 instruction cycles is available with the ball being the smallest feature displayable on the screen with up to 43 different positions. Using a PIC18Cxx running at 40 MHz would improve the granularity of the ball position to about 107 different positions for each line of the playing field. Ideally, I would like to see something approaching 256 different positions for the display to be as accurate as possible.

With different cases, timing the application is a major headache and a difficult piece of work. This was especially true when I was calculating the ball's position with two playing field borders to also display at the same time. Ideally, a PICmicro® MCU video generator application would only have one feature to put on any line at a given time.

This circuit could be used as the basis of a PICmicro® MCU-based video data overlay circuit. Instead of generating the video timing signals within the PICmicro® MCU, synch separators could be used to indicate the start of a field and a line for the PICmicro® MCU to then count from and drive a signal on top of the input signal. An obvious example of an overlay circuit is the text generator used at the end of TV shows to display the credits over some other signal.

To get accurate positioning of the overlay, you will probably have to use some kind of *Phase-Locked Loop (PLL)* circuit to ensure the PICmicro® MCU's clock is "locked" on the source generator's and that the overlay appears on the display at the same position each time. This isn't a terribly hard circuit to design and could be useful for placing cross hairs or arrows on a video signal to indicate where problems are or features to look at.

# PIC17Cxx Devices

In the 17Cxx application presented in the following sections, I wanted to present what I felt are the 17Cxx family's strong points. These strong points include advanced instructions for data processing, and the ability to write to its own program memory. Personally, I consider the ability to access external memory and memory-mapped I/O devices as the family and architecture's strongest point. The following application demonstrates how memory is accessed, as well as comment on how difficult it is to program, relative to the other PICmicro® MCU devices.

## PIC17DEV: PIC17Cxx MEMORY INTERFACE DEMONSTRATION CIRCUIT AND MONITOR PROGRAM

I didn't start this project until quite late in my development schedule, although I was quite comfortable with the design and what I was going to do with it—especially because I designed the board as part of the description on project design in "Introduction to Electronics" on the CD-ROM. Although I was comfortable with what I was doing with the hardware, I should have been more nervous about the software. This was my first "true" PIC17Cxx application . . . and I only budgeted a day to do it. I can't tell you how surprised I was at 10 P.M. on a Sunday night to have the application finished, along with this write up.

The board presented here provides an external interface to two 8-KB SRAMs, as well as the modified reset circuit used to work the built-in programming capability of the PIC17Cxx. The software provided for this application is a simple monitor program that allows you to access file registers inside the PIC17C44 that is used for the application, the SRAM and the EPROM memory. Along with being able to read and write this memory, I also provided a method of calling subroutines at arbitrary points within the PIC17Cxx execution space. With two simple changes to the software, this project could be used as a PIC17Cxx emulator and application-development test tool.

The application circuit is shown in Fig. 16-66 and uses an RS-232 interface as the primary I/O function of the board.

The bill of materials for the PIC17Cxx Memory Interface Demonstration Circuit is given in Table 16-31.

A two-layer PC board was designed for this project. Figures 16-67 and 16-68 show the top and bottom layers of the PC board, respectively, and Figure 16-69 shows the topside silkscreen overlay.

Notice that I designed the board with a 14-by-1 connector (J3 in the schematic) to provide I/O to all the unused PIC17Cxx pins (PORTA and PORTB), as well as a small prototyping area in the far corner of the board. When you design a first-pass PC board, it is always a good idea to include an area to add new circuits or, as in this case, to have a prototyping area "play around" with different aspects of the design.

The PC board was almost perfect from the work that I did in Appendix C. The only error on it was two shorted traces on the backside. Once the offending lines were cut and a wire was added to put in the correct signal path, the board worked without any problems. The PC board design given in this section has this problem corrected.

I found this problem when I was doing a visual inspection of the boards. It is always a good idea to do this because your board layout software and your own visual inspection of the board data files will miss such problems. Shorted traces usually manifest themselves as "crosses" and are very easy to repair and add a wire to the board.

For the software, I wanted to follow the conventions I established with the other projects, such as the Emu-II, which have an intelligent interface. The interface and options that I came up with are shown in Fig. 16-70.

The six program options should be pretty simple to understand. The only surprise was my splitting up the EPROM program and SRAM Write. I did this to split out the two different memory writes and not try to develop a single routine that did everything (and probably didn't work).

When I designed the board, I decided that I would run the PIC17C44 in Extended Microcontroller mode. This means that the PICmicro® MCU boots out of its internal

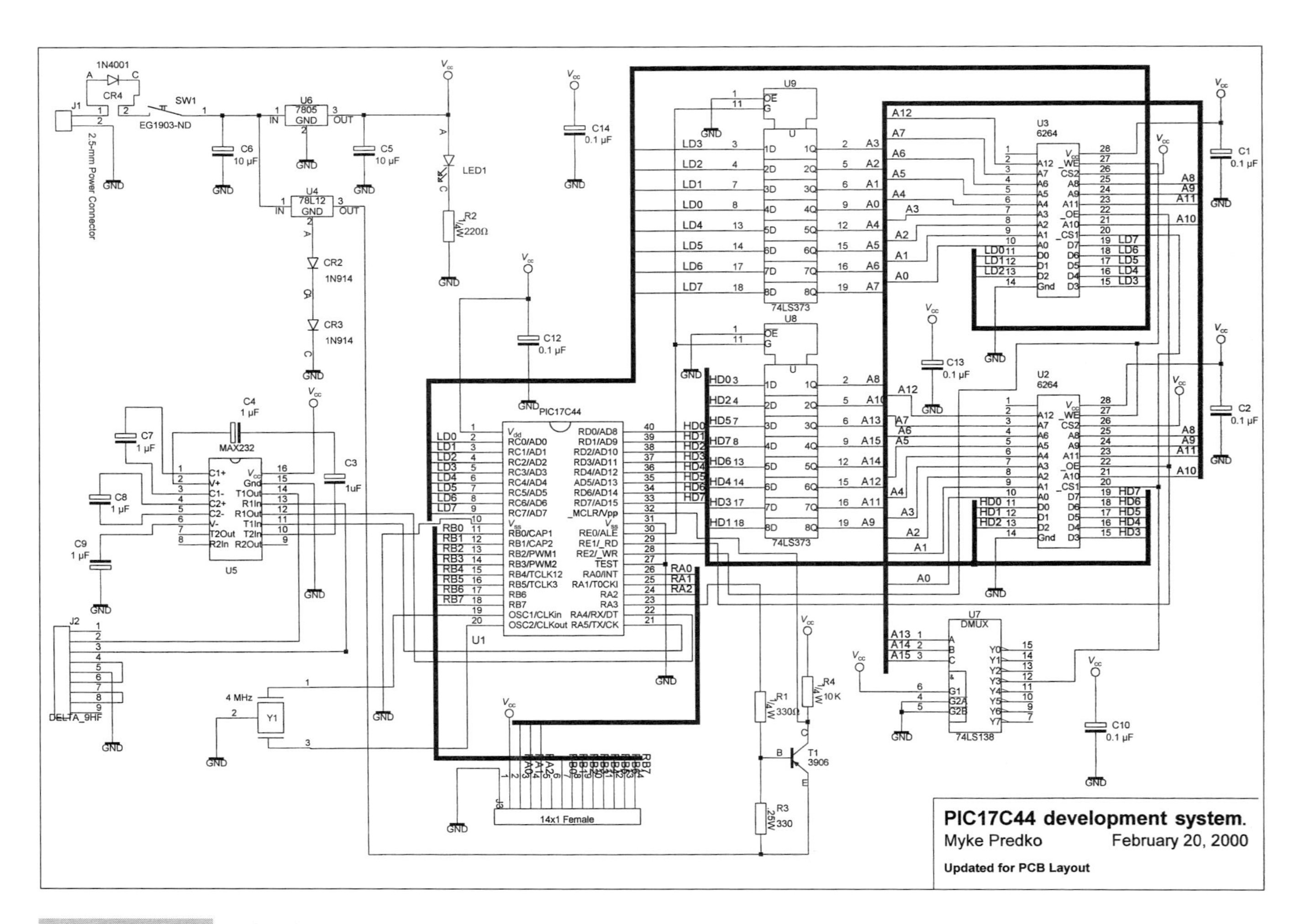


**Figure 16-66** PIC17C44 development system

**TABLE 16-31 PIC17C44 Development System Bill of Materials**

| PART | DESCRIPTION |
|---|---|
| U1 | PIC17C44/JW |
| U2 - U3 | 6264 150 nsec Access Time 8Kx8 SRAMs |
| U4 | 78L12 +12 Volt Regulator in a TO-92 Package |
| U5 | MAXIM MAX232 RS-232 Interface |
| U6 | 7805 +5 Volt Regulator in a TO-220 Package |
| U7 | 74LS138 Demultiplexor |
| U8 - U9 | 74LS373 Eight Bit Latches |
| T1 | 2N3906 PNP Transistor |
| Y1 | 4 MHz Ceramic Resonator with Built in Capacitors |
| CR1 | 5 mm Red LED |
| CR2 - CR3 | 1N914 Silicon Diodes |
| CR4 | 1N4001 Silicon Diode |
| R1, R3 | 330 Ohm, 1/4 Watt Resistors |
| R2 | 220 Ohm, 1/4 Watt Resistor |
| R4 | 10 K, 1/4 Watt Resistor |
| C1 - C2, C10, C12 - C14 | 0.1 $\mu$F 16 Volt Tantalum Capacitors |
| C3 - C4, C7 - C9 | 1.0 $\mu$F 16 Volt Capacitors, Any Type |
| C5 - C6 | 10 $\mu$F 35 Volt Capacitors, Any Type |
| J1 | 2.5 mm Power Connector |
| J2 | 9 Pin Female "D-Shell" |
| J3 | 24x1 Female Connector or wire-wrap posts |
| Misc. | PCB or Prototype Card and Sockets. 15+ Volt DC Power Source |

EPROM, but it can also access external memory. The PIC17Cxx has 8,192 words of EPROM, so I used a 74LS138 demultiplexer to select the SRAM. The program memory in this application is broken up as shown in (Table 16-32).

The difference between *application code* and *free EPROM* is that the application code is the code written for the application and I didn't want there to be the opportunity for it being overwritten. For this reason, I prevent the user from writing to it (Fig. 16-71).

In Fig. 16-71, the first word write attempt fails because it is in the software-protected

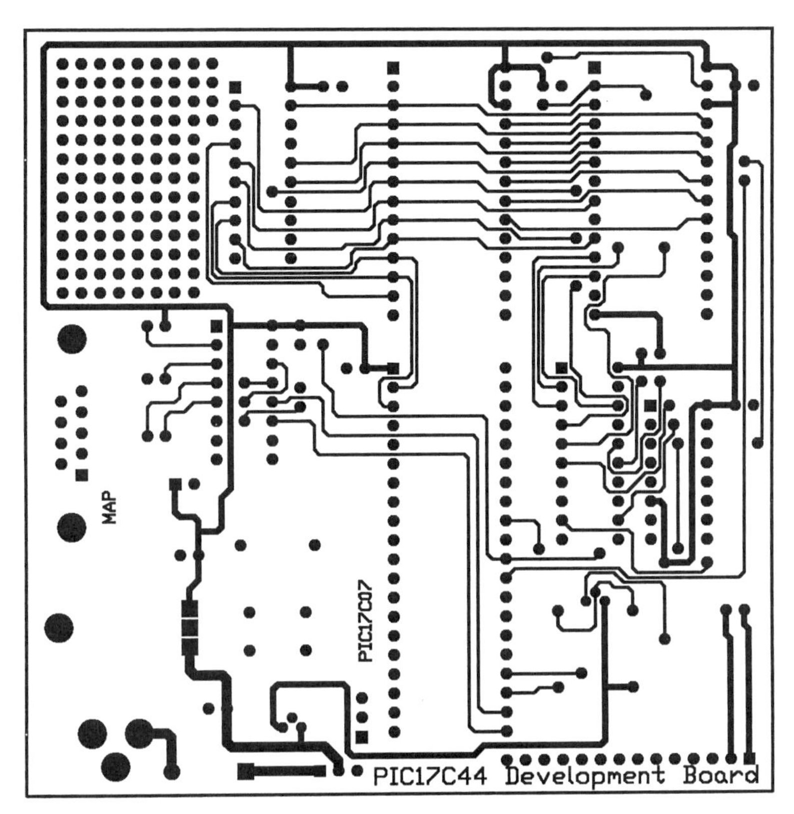


PIC17C07 Top Layer

**Figure 16-67**

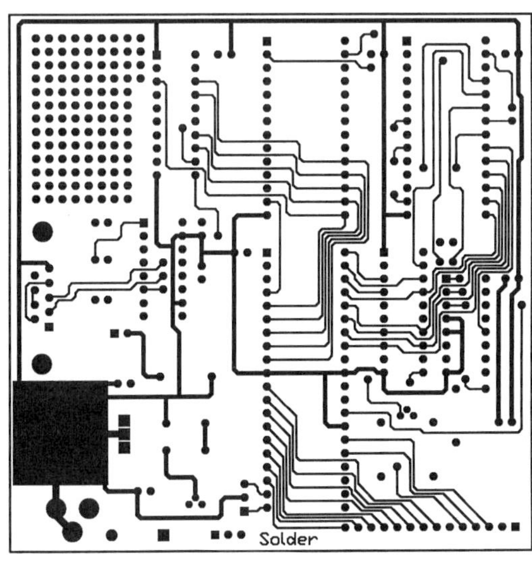


PIC17C07 Bottom Layer

**Figure 16-68**

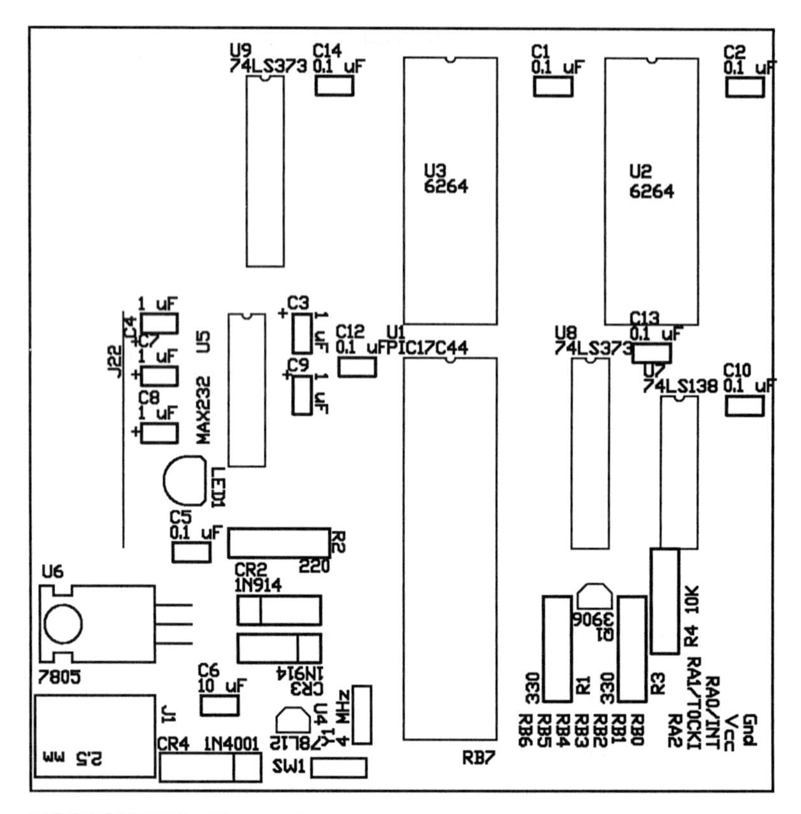


**Figure 16-69**

Application Code space. Next, I write a *return* instruction to the EPROM, followed by a test word (0x0AA55). To see that the information "took," I dumped the contents of program memory, starting at 0x0800. The two valid commands were accepted and overwrote the EPROM.

Notice that the unprogrammed EPROM returns with all bits set. This is to be expected. As I have explained elsewhere in the book, programming an EPROM bit is the process of changing it from a *1* to a *0*. Figure 16-71 shows this to good advantage.

Writing to SRAM has a similar interface. The only difference being that the valid address range for SRAM write is 0x06000 to 0x07FFF.

In the PIC17Cxx, PORTA does not have a DDR register, the PIC17Cxx equivalent of the other PICmicro® MCU families' TRIS register. The bits in this register are normally used for special functions (RA4 and RA5 are used by the USAR, for example) or for input. The exception to this is RA3 and RA2, which are open-drain drivers and can either be floating or pulled down. To enable $V_{pp}$ in this application, I pull down RA3, which turns on T1 and drives the $V_{pp}$ (13.4 volts) programming voltage to the PICmicro® MCU for it to perform its internal program operation.

When I created the software for this application, I discovered three things. First, it

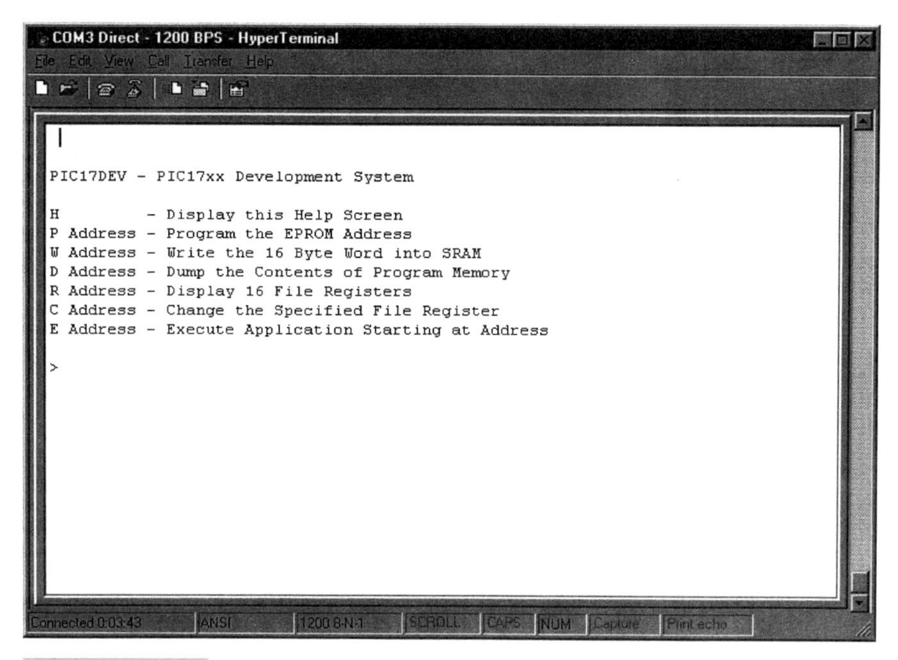


**Figure 16-70** PIC17C44 development board options

wasn't that hard to convert my mental thought processes from the other PICmicro® MCU families. I just had to remember that *movf Register, w* was implemented as:

```
movfp Register, WREG
```

and that instead of STATUS, I had to use the ALUSTA register to check the Carry, Digit Carry, and Zero flags.

Second, I discovered that if I treated the bank registers the same way as I did with the mid-range PICmicro® MCUs, with just changing the active register when it is required, working with the differences in the architecture was quite simple. In Table 16-33, I listed the different banks, along with the XOR values and instructions to make the changes for both the PIC17Cxx and the mid-range PICmicro® MCUs.

Last, I discovered that I had to recreate all my "typical" pieces of code to implement dif-

**TABLE 16-32 PIC17C44 Development System Memory Access Mapping**

| MEMORY TYPE | STARTING ADDRESS | ENDING ADDRESS |
|---|---|---|
| Application Code | 0x00000 | 0x007FF |
| "Free" EPROM | 0x00800 | 0x01FFF |
| SRAM | 0x06000 | 0x07FFF |

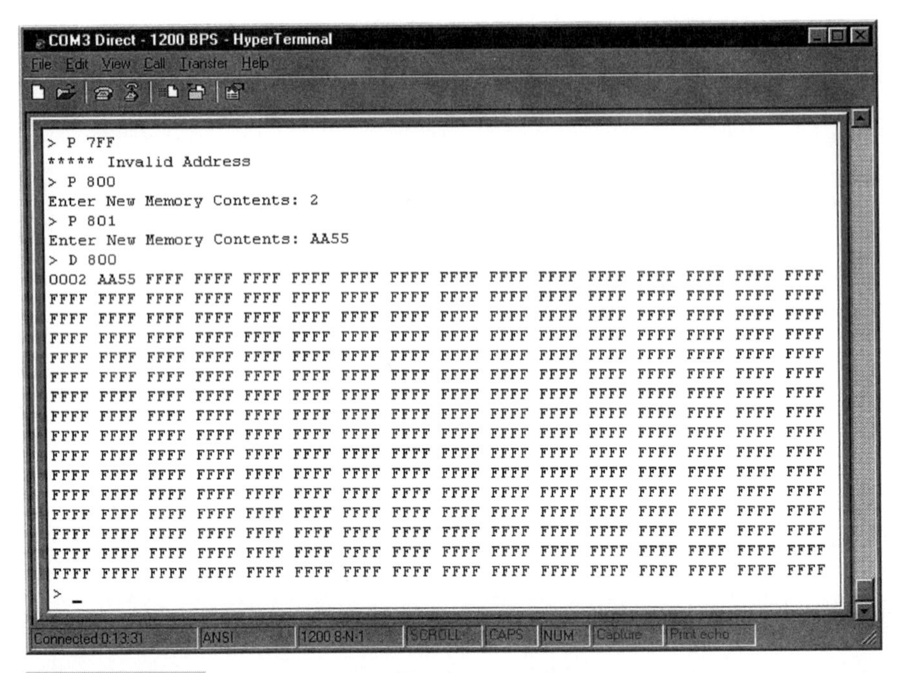


**Figure 16-71** **PIC17C44 EPROM writes and display**

ferent functions. As I have said over and over again, when I create applications, I try to "steal" from other applications as much as possible. In this application, I found that I had to recreate many of my favorites because they could not be taken directly from my existing mid-range application code.

For example, to poll the RCIF interrupt-request bit to wait for an incoming USART character, I had to create the code:

```
movlb    1                         ; Poll RXIF
btfss    PIR ^ 0x0100, RCIF
 goto    $ - 1
movlb    0
movfp    RCREG, WREG
```

which wasn't very hard, but did require some thinking and looking through data sheets to correctly implement the code.

The enhanced instruction set did allow me to more efficiently implement my string search routine. For most of my applications that have to transmit text data to a user, I set up a large dt table, with each entry terminated by an ASCII Null (0x000) character. To implement this in the PIC17Cxx, I used the code:

```
SendMSG
 movwf   SMIndex                   ; Save the Index Value
 clrf    SMOffset, f               ; Clear the Offset Counter
 clrf    SMOffset + 1, f
SMLoop1
 movfp   SMIndex, WREG             ; Is the Index equal to zero?
 iorlw   0
```

**TABLE 16-33 PIC17C44 Development System Bank Operations**

| BANK | PIC17CXX | "XOR" | MID-RANGE CODE | "XOR" |
|---|---|---|---|---|
| 0 | for Address 0x010-0x017: movlb 0 for Address 0x020-0x0FF movlr 0 | 0 | bcf STATUS, RP0 bcf STATUS, RP1 | 0 |
| 1 | for Address 0x010-0x017: movlb 1 for Address 0x020-0x0FF movlr 1 | 0x100 | bsf STATUS, RP0 bcf STATUS, RP1 | 0x080 |
| 2 | for Address 0x010-0x017: movlb 2 for Address 0x020-0x0FF movlr 2 | 0x200 | bcf STATUS, RP0 bsf STATUS, RP1 | 0x0100 |
| 3 | for Address 0x010-0x017: movlb 3 for Address 0x020-0x0FF movlr 3 | 0x300 | bsf STATUS, RP0 bsf STATUS, RP1 | 0x0180 |

```
 btfsc   ALUSTA, Z
  goto   SMLoop2                 ; Yes, Display the message
 movlw   LOW OutputTable
 addwf   SMOffset, w
 movwf   SMPDest
 movlw   HIGH OutputTable
 addwfc  SMOffset + 1, w
 movwf   SMPDest + 1
 call    GetTable
 iorlw   0                       ; Zero Returned for the Offset?
 btfsc   ALUSTA, Z
  decf   SMIndex, f              ; Yes, Decrement the Index Counter
 infsnz  SMOffset, f             ; Increment the Offset
  incf   SMOffset + 1, f
 goto    SMLoop1

  :                              ; #### - String Transmit Code
  :                              ; #### - Deleted
SMEnd                            ; Finished Displaying Message

 return
```

```
GetTable
 movfp    SMPDest + 1, PCLATH
 movfp    SMPDest, PCL
OutputTable                   ; Help Screen
  dt      0x00D, 0x00A, 0x00D, 0x00A
  dt      "PIC17DEV   - PIC17xx Development System", 0x0D, 0x0A
  dt      "H          - Display this Help Screen", 0x0D, 0x0A
  dt      "P Address - Program the EPROM Address", 0x0D, 0x0A
  dt      "W Address - Write the 16 Byte Word into SRAM", 0x0D, 0x0A
  dt      "D Address - Dump Contents of Program Memory", 0x0D, 0x0A
  dt      "R Address - Display 16 File Registers", 0x0D, 0x0A
  dt      "C Address - Change Specified File Register", 0x0D, 0x0A
  dt      "E Address - Execute Application Starting at Address",
  dt      0x00D, 0x00A, 0x00D, 0x00A, 0
Msg1                          ; Prompt Message
  dt      "> ", 0
Msg2                          ; New Line
  dt      0x00D, 0x0A, 0
```

In this code, I was able to use the *addwfc* and *infsnz* instructions to simplify the operations needed to search for the specified string somewhat.

To provide the *execute* function, I set up a table of three instruction *lcall* table elements and multiplied the least-significant eight bits of the new address by three before jumping to the table element where a PCLATH and *lcall* setup was located. The code itself is quite simple:

```
LongJump                               ; Jump Somewhere in Memory
 movfp      SaveAddress, WREG
 mullw      3                          ; Multiply the Offset by 3
 movfp      PRODH, WREG
 addlw      HIGH LongJumpTable
 movwf      PCLATH
 movfp      PRODL, WREG
 addlw      LOW LongJumpTable
 btfsc      ALUSTA, C
  incf      PCLATH, f
 movwf      PCL
LongJumpTable
 variable k = 0
 while (k < 0x0100)                    ; Put in the Jump Tables
 movfp      SaveAddress + 1, PCLATH
 lcall      k                          ; Call 16 Bit Address
 return
k = k + 1
 endw
```

but when it is expanded (in the *while* statement), you will discover that it actually takes up 778 instructions because of the need for 256 table elements (one for each possible combination of the least-significant byte of the destination address).

For this function, I could have implemented a simple PCLATH/PCL update, but I wanted to implement a *call* so that if this application was used as a downloader or simple emulator, the user interface would be the central focus. In the EMU-II for a similar function, I used the ability of the PIC16F877 to write to flash memory and update the instruction, instead of relying on a large table (as I did in this application).

For my first application, I would probably give myself a *A*−. I was able to implement quite a complex function in a relatively small amount of time. Probably the biggest inefficiency in the code is the replication of the *data value* input function for the three write functions (write to a register, write to SRAM and EPROM program) when I should have implemented one subroutine. Overall, I would say that I wasn't very frugal with my use of program memory.

I guess you could argue that with 8K instructions, I never had to worry about being frugal as I learned to program the PIC17Cxx. As well, I was able to implement these functions with over 6K instructions left. But I like to produce the most efficient code that I can and there are definitely opportunities here that I should (sometime) go back and capitalize on.

If I was going to make this application into something I would consider for sale, like the EMU-II, I would have relocated all the variables from the start of Bank0 to the end of Bank1. This would allow users to write their own applications and not worry about interfering with the interface variables.

Last, I would go back and look at how I implemented the EPROM program code. The code itself:

```
 org     0x00010
TMR0Int                                ; Timer Interrupt Request
                                       ; Acknowledge
 bcf    INTSTA, T0IF                   ; Reset Interrupt

 retfie

  :

 movfp   SaveAddress, TBLPTRL          ; Point to the Memory being
 movfp   SaveAddress + 1, TBLPTRH      ; written to

 bcf     PORTA, 3                      ; Turn on Programming Voltage

 movlw   HIGH ((100000 / 5) + 256)     ; Delay 100 msecs for
 movwf   Dlay                          ; Programming Voltage to Stabilize
 movlw   LOW ((100000 / 5) + 256)
 addlw   0x0FF
 btfsc   ALUSTA, Z
  decfsz  Dlay, f
  goto   $ - 3

 movlw   HIGH (65536 - 10000)          ; Delay 10 msecs for EPROM Write
 movwf   TMR0H
 movlw   LOW (65536 - 10000)
 movwf   TMR0L

 bsf     T0STA, T0CS                   ; Start up the Timer

 movlw   1 << T0IE                     ; Enable Interrupts
 movwf   INTSTA
 bcf     CPUSTA, GLINTD

 tlwt    0, SaveData                   ; Load Table Pointer with Data
 tlwt    1, SaveData + 1
 tablwt  1, 0, SaveData + 1            ; Write the Data In
 nop
 nop

 clrf    INTSTA, f                     ; Turn Off Interrupts
 bsf     CPUSTA, GLINTD
```

```
movlw   2
call    SendMSG

bsf     PORTA, 3
```

is not terribly inefficient, especially in this application, but I would have liked to spend some more time understanding if the 100-ms $V_{pp}$ stabilize delay is really needed. Also, I used a 10-ms delay for the programming operation when Microchip specifies that 1 ms should be sufficient. Both these values are about eight times more than what I would consider to be optimal. I have added a number of instructions (the *bcf INTSTA, T0IF* and *nops*) that I am not sure are absolutely necessary.

Despite the concerns I have about this application, it is actually quite close to being in a state where it could be offered as a product. The two big additions that I would like to see are the inclusion of an assembler/disassembler and the ability to download a hex file. With this capability (along with some enhanced application-execution options), this project would be essentially a PIC17C44 emulator.

Not bad for a day's work.

# PIC18Cxx Devices

To finish off the example applications, I wanted to demonstrate how the PIC18Cxx is used in a circuit and how code is written for it. When I created the application, along with the *Real-Time Operating System (RTOS)* and RTOS application in the next chapter, only engineering samples were available (although MPLAB had been updated to handle code written for it). With this in mind, you should check both the Microchip web site, as well as mine for updated PIC18Cxx information and tools.

As stated elsewhere in this book, I expect the PIC18Cxx to become very popular in the coming years—eventually becoming the most popular PICmicro® MCU device, taking away the mid-range crown. The reasons for making this prediction include the much larger program memory and register spaces and the ease of application development resulting from the PIC18Cxx's less "quirky" architecture. The PIC18Cxx should be a much easier device to develop high-level compilers for than the low-end, mid-range, and PIC17Cxx devices.

I believe that the following application demonstrates this and gives you an idea of how the PIC18Cxx can be used in your application, instead of the other PICmicro® MCU devices. As I write this, the PICStart Plus and PROMate II are the only programmers on the market that can program them. This probably will have changed when the book has gone into print (fall of 2000) and should not be an issue in the coming years as third-party programmers for the PIC18Cxx become more common.

## FUZZY: FUZZY LOGIC FAN GOVERNOR

To finish off the projects, I wanted to take a look at using the PICmicro® MCU as a digital controller (Figs. 16-72 and 16-73). Elsewhere in this book, I have noted that the eight-bit data word and limited mathematical capabilities instructions decrease the PICmicro® MCU's attractiveness as a DSP. The PICmicro® MCU has the ability to implement fuzzy-logic control, which requires significantly less computational power than DSP algorithms. Along with using fuzzy logic for processing, I wanted to use built-in hardware features of

**Figure 16-72** PIC18Cxx fuzzy logic fan controller

**Figure 16-73** Fan blade sensor

the PICmicro® MCU wherever possible, instead of relying on bit-banging interface methods. This design philosophy allowed the hardware interfaces to be created quickly and easily modified when I was debugging the application. For this application, I wanted to use a PIC18Cxx to test how easily code could be taken from mid-range PICmicro® MCU applications and ported to the PIC18Cxx. In many ways, this application was an experiment, with some interesting results, especially in the developing of fuzzy-logic applications and what I learned about them.

The circuit used for this project was taken from the PIC12C673-based motor controller that I presented earlier in the chapter. The project was built on a small prototyping card (Figs. 16-74 and 16-75), as you can see at the start of this chapter. Using point to point wiring, it took me about five hours to wire the application, with a few problems, as outlined below.

The bill of materials for this project is shown in Table 16-34. When I originally built this board, I used a Dallas Semiconductor DS-275 "voltage-stealing" RS-232 interface. This chip had problems because the reflected data on the TX line, returning the data sent on the receiving line (a function of this type of RS-232 interface) caused problems with the FUZZYtech tool during the fuzzy-logic development. To get the RS-232 interface to work correctly, the MAX 232 was substituted. Unfortunately, installing this chip was a major exercise in point-to-point wiring. The resulting circuit is half hanging into the DS275 socket and half hanging in air. Wiring this chip in took me at least as long as building the entire circuit.

To measure the fan's speed, I used an OPB804 slotted opto-isolator switch. This device consists of an infrared LED providing a signal to a photo transistor, which turns on when the infrared signal is received. When the infrared signal is blocked, the photo transistor is off. I recommend that you prototype the circuit in the top right corner of Fig. 16-76 to better understand its operation.

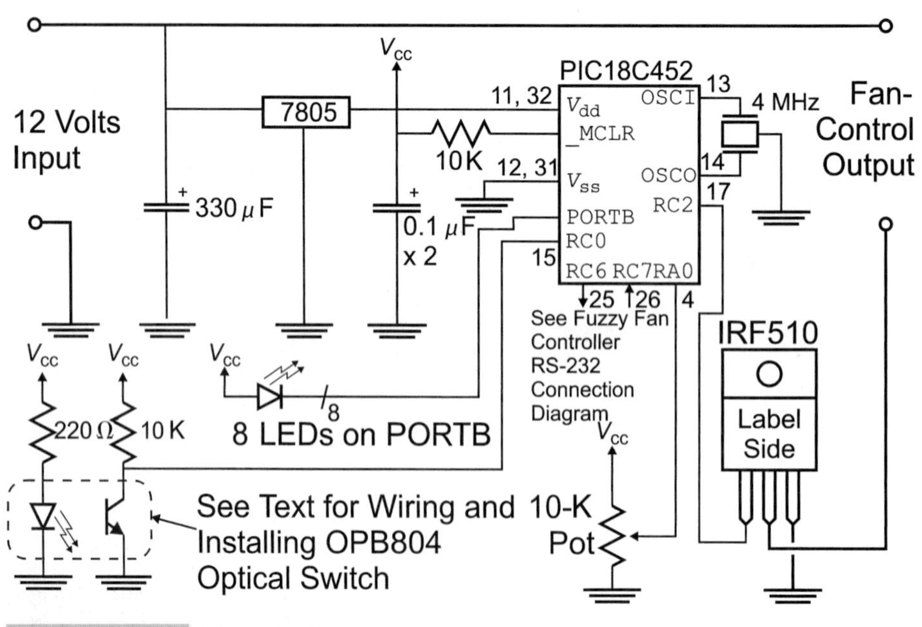


**Figure 16-74** **Fuzzy logic fan controller**

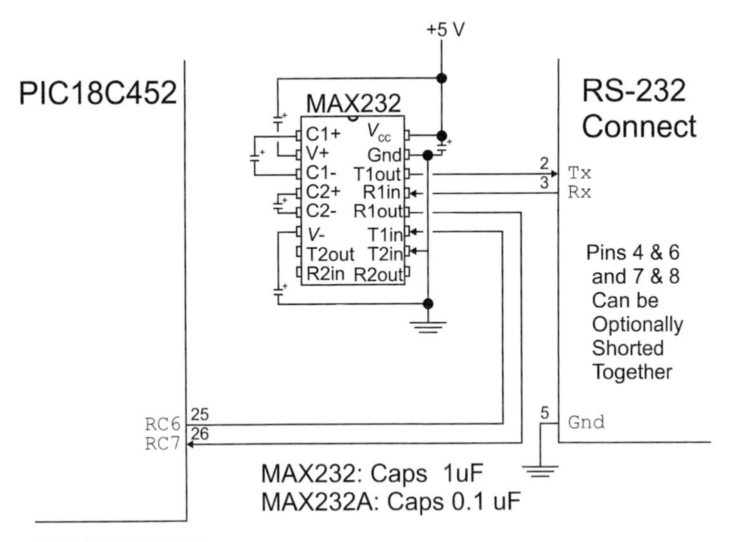


**Figure 16-75** Fuzzy fan controller RS-232 connections

**TABLE 16-34 Fuzzy Logic Motor Controller Bill of Materials**

| PART | DESCRIPTION |
|---|---|
| PIC18C452 | PIC18C452-JW |
| 7805 | 7805 in TO-220 package |
| MAX232 | MAX232 |
| IRF510 | IRF510 N-Channel MOSFET in TO-220 Package |
| OPB804 | OPB804 “Slotted OPTO Isolator Switch” or Equivalent |
| 4 MHz | 4-MHz Ceramic Resonator with Built in Capacitors |
| 2x10 K | 10 K, 1/4 Watt |
| 220 | 220 Ohm, 1/4 Watt |
| 10 K pot | 10 K, Single Turn Potentiometer |
| 330 $\mu$F | 330 $\mu$F, 35 Volt Electrolytic Capacitor |
| 2x 0.1 $\mu$F | 0.1 $\mu$F, 16 Volt Tantalum Capacitors |
| 5x 1.0 $\mu$F | 1.0 $\mu$F, 35 Volt Electrolytic Capacitors |
| 8x LED | 8 individual LED’s or 10x “bar graph” LED display |
| Misc. | Prototyping board, 8x screw terminal, 9 pin female D-shell connector wiring, 12-volt wall mounted AC/DC power supply. |

To use the OPB804 in this application, carefully cut a notch in the side of the 12-volt fan that you are going to use for this application. The notch cannot be cut so far as to intersect the path of the fan blades and prevent them from turning when the opto-isolator switch is installed. For my prototype, I used a nibbling tool and checked the clearance after every cut. For the fan that I used, I found that I could safely cut out enough material to allow the LED and phototransistor to be within 0.800″ (in the stock OPB805, the distance between the LED and phototransistor is 0.150″).

A couple of comments about the OPB804; this part is not crucial to the operation of the application. If you are going to substitute it for another part, I do have some comments about it, however. First, only use an infrared slotted switch. I realize that there are visible-light switches available, but these should be avoided because of the chance for ambient light reflected off of the fan blades to become "confused." Second, be sure that the slotted switch that you use will work with the LED and phototransistor directly apart and that a direct path lies between the LED and the phototransistor. Last, test the LED and potentiometer on the prototype PC board before it is mounted on the fan; you don't want to have glued it onto the fan, only to discover it doesn't work.

With the notch cut in the fan, carefully cut away the plastic bridge that connects the LED and phototransistor. No signals or circuitry are contained within the bridge. When soldering the LED and phototransistor to the prototype board, be sure that they are as square as possible.

You will find that as the distance between the LED and phototransistor increases, the gain of the transistor decreases. When I originally built my prototype, I had an LED connected to the phototransistor, like the top right corner of Fig. 16-76. When the phototransistor PC board was tested, I found that its gain had decreased markedly and the output voltage was always above the PICmicro® MCU's logic threshold, making counting im-

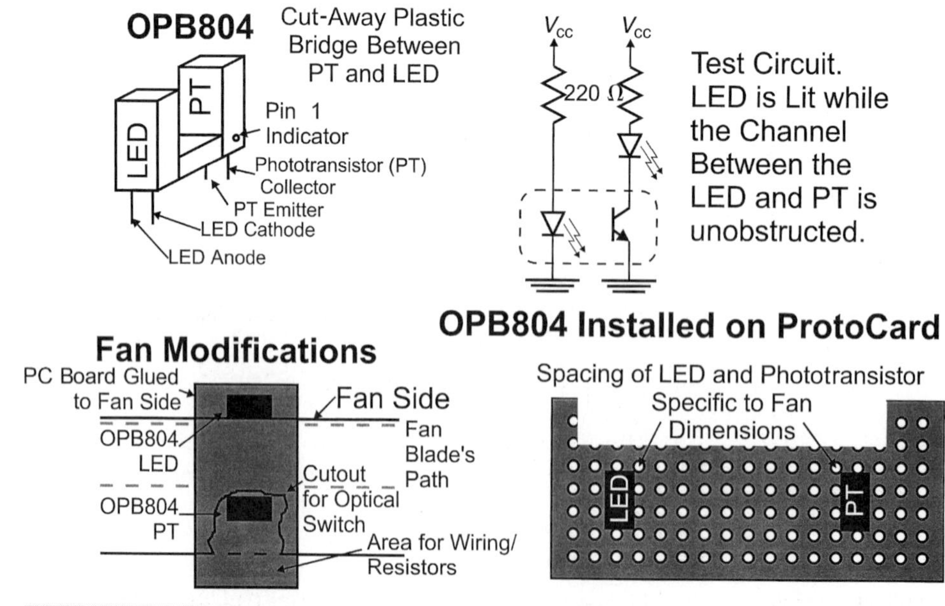


**Figure 16-76** Fuzzy fan-controller fan-speed sensor

possible. Using the 10-k pull-up on the OPB804 provided good-quality signals that I could use to count the speed to the fan's rotation.

As indicated at the start of this section, I wanted to take advantage of the PICmicro® MCU's built-in interfacing hardware for this application. Like the PIC12C673-based fan control, I used the PICmicro® MCU's built-in ADC and a potentiometer to read the set speed.

To set the power output to the fan, I used the PWM hardware built into TMR2 and the CCP. One of the problems I had with the PIC12C673-based fan control was not being able to implement a 20-kHz PWM output signal. This would have been a problem with this application because 20 kHz is a 50-ms signal, which would only allow 50 PWM output levels. I used the four-times clock multiplier built into the PIC18C452 to provide a 250 nsec instruction cycle and give TMR2 the capability of driving out a 20-kHz PWM with 200 duty cycle levels available.

As an aside, after building this circuit and writing preliminary software for it, I tried it out with the fan that "squealed" with the 10-kHz PWM control. In this case, there was no squeal and the fan was almost silent throughout the full speed range.

To measure the fan speed, I counted the output from the OPB804 infrared opto-isolator slotted switch with the 10-k pull up. I used TMR1 because it can be configured as a 16-bit counter and I had no idea what the speed range of the fan would be (and, consequently, how many blades would pass by in a second).

To better understand what the fan's speed range would be, I created a simple application to show the analog voltage output by the potentiometer and read by the ADC. This value was scaled to a value between 0 to 199 for easy comparison with the TMR2 CCP PWM control value.

To scale the 10-bit ADC's output into an 8-bit maximum 199 decimal value, I outputted the ADC values in Left-Justified mode, which allowed me to read the most-significant eight bits and ignore the least-significant two. To convert the eight-bit value that has the range of 0 to 255 (decimal), I multiplied it by 200 (using the *mullw* instruction) and only used the most-significant byte. This is the same as multiplying the value by the fraction 200/256 and is very fast to calculate in the PICmicro® MCU.

By using the Left-Justified mode and the multiplication trick, the code required for scaling the 10-bit ADC result to 0 to 199 was just:

```
movf    ADRESH, w
mullw   200
movf    PRODH, w
```

This type of trick will not work for all applications. Knowing that these capabilities exist in the PICmicro® MCU and planning on the numeric ranges that will be required for the application, you can take advantage of them in your design. It basically comes down to making your own luck.

When I created the application to find the fan speed, I started working with the ADC and PWM duty cycle and output (via RS-232) both the ADC value as a percentage (which was proportional to the PWM duty cycle) and the fan speed. Table 16-35 lists the results, along with what I would consider to be the idealized fan speed. The fan speeds are in blades past per minute (RPM).

For this application, I decided upon a maximum fan speed of 800 blades past per second because the maximum speed was somewhere about 850. The top speed of 800 blades past

**TABLE 16-35 Fuzzy Logic Motor Controller Fan Speed Table**

| DUTY CYCLE | FAN SPEED | IDEAL |
|---|---|---|
| 0 | 0 | 0 |
| 10 | 360 | 80 |
| 20 | 530 | 100 |
| 30 | 600 | 240 |
| 40 | 660 | 320 |
| 50 | 700 | 400 |
| 60 | 720 | 480 |
| 70 | 750 | 560 |
| 80 | 770 | 640 |
| 90 | 785 | 720 |
| 100 | 800 | 800 |

per second would give me a margin of control for speeding up the fan. Because I used seven blades on the fan, this translates to 114 revolutions per second (6,840 RPM).

When I produced this chart, I realized something I hadn't before about fuzzy logic. Along with controlling a process or device, it can also be used to linearize its response. For this application, the goal was to provide an output using a single input. This (rather belatedly) became the purpose of this project: provide a fuzzy-logic fan-motor linearized control, as well as govern speed based on the ADC input value.

To create the fuzzy-logic rules for this application, I used FUZZYtech in a similar manner as I used with the prototype board that comes with it (and described elsewhere in this book). The reason for doing this was based on my assumption that the processes to be controlled were similar. This was an invalid assumption because the temperature-control 10-Hz PWM was much more easily controlled using a PWM output than a motor running at a few thousand RPMs.

To develop a fuzzy-logic application, you must create an application that has the structure shown in the pseudo-code:

```
main()                          // Fuzzy Logic Application
{

 initializeIO( );               // Initialize the I/O

 while (1 == 1) {
  dlay();                       // Delay for the Sample time
  output = fuzzyMod(input);
 }
} // End Fuzzy Logic Application
```

I used this structure for the application. Changing the ADC-to-speed output application into this format was quite simple, but I did run into a few glitches along the way.

First was with the serial communications. When I first connected the fan-controller hardware to FUZZYtech, I continually got the error "multiple data input," which I didn't understand. This caused FUZZYtech to continually shut down and have to be rebooted. When you read the FUZZYtech manual, you will find that it does not handle invalid serial data very well and this is the recommended procedure if invalid data is received.

This was perplexing to me because I used both the manual and the fuzzy-logic temperature controller as guidelines for the application. I was able to match the fan controller's RS-232 output to the temperature controller's board very closely. I looked at the signal at the PICmicro® MCU. As far as I could tell, it was perfect. After trying to discover the cause of this, I did what I always do in these cases: I slept on it. Sometime the next day, I remembered the operation of the DS275 and that it was a "voltage stealer" that uses the RX voltage for its TX. If the RX line is positive, the TX line will also be positive. I was able to confirm this with an oscilloscope and I found that every byte from the PC was "echoed" back to it. To prevent this problem, I replaced the DS275 with the MAX232. After this (laborious) switch, the FUZZYtech software performed better.

But it didn't perform perfectly. Occasionally, I got the message "invalid input." In the FUZZYtech manual, it documents the serial data stream as being the decimal representative of the value with an optional + or − at the start, followed by the ASCII decimal representation of the number. Valid numbers are:

```
 123
+123
-123
+1.23
-1.23
+1.23E-4
```

In the last example, the *E* represents "ten to the power of."

I found that if I always sent data with a leading + sign and three data digits, the "invalid input" message would go away. Despite having a specific format for input, the data from FUZZYtech could be in any of the previous formats. Thus, although the output handler is quite simple, the input data handler is quite complex.

I simplified the serial input data handler by not having exponents and by truncating off fractional data to handle the integer values of the FUZZYtech input.

When I first designed this application, I copied the temperature controller almost exactly. This was a mistake based on how the two applications work. The input for the fuzzy controller was the difference between the input control and the motor-control PWM. This value, from −20 to +20, was converted into the data string and passed to FUZZYtech. The output was a PWM range based on the error rate.

When I started the application, I found that I could not adequately regulate the speed. Although not a fully out-of-control situation, I did have wild oscillations that could not be dampened by changing the rules. The difference between the two systems was the amount of inertia that was being controlled in the thermometric example. Temperature changes occurred over many seconds. In the fan controller, changes in speed could be implemented in less than a second. Although the fuzzy-logic controller could prevent the oscillations from becoming unbounded, a 20% to 30% change in the fan's speed occurred each second.

The solution to this problem was to change the operation of the fuzzy logic system. Although I still retained the input as the difference between the set speed and the actual speed, the output was changed to output differences in the PWM duty cycle from the actual PWM duty cycle. The application code was modeled as:

```
main()                          // Fuzzy Logic Fan Motor Controller
{

 initialize();                  // Initialize the Hardware
 PWM = 50 percent;              // Initialize the PWM to a 50% duty
                                //  cycle
 While (1 == 1) {
   Dlay ( );
   Speed-error = ADCinput - countACT;
   PWM = PWM + FuzzyMod(speed-error);
   If (PWM <0)                  // Keep PWM within Valid Range
     PWM = 0;
   else if (PWM > 200)
     PWM = 200;
 }
} // End Fuzzy Logic Fan Motor Controller
```

The *speed-error* FUZZYtech input was given the four rules shown in Fig. 16-77. In this set of rules, I identified difference ranges (or "patches") for too fast, too slow and just right. These were matched with the output rules shown in Fig. 16-78. For the input, I used a range of −20 to +20, where this value is the difference in counts, to a maximum of 200 between the ADC input (desired speed) and the counter input (actual speed).

The change in PWM output is set to the range of −20 to +20. Each increase of one digit resulted in about a 1% change of fan speed.

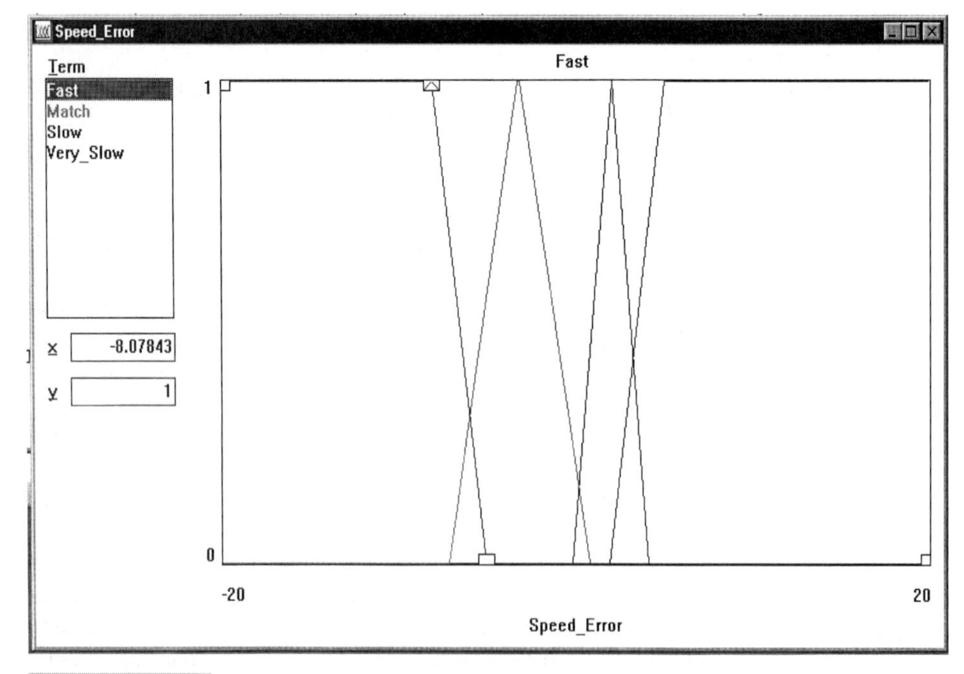


**Figure 16-77** **Fuzzy fan-control rules for speed errors**

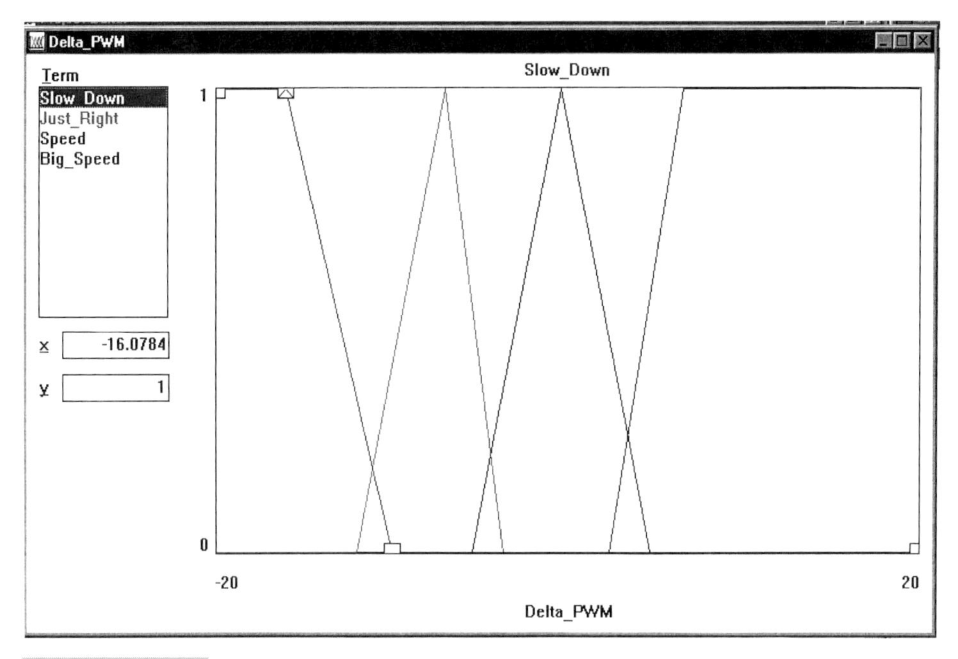


**Figure 16-78** Fuzzy fan-control PWM rules

When I started this fuzzy-logic system with FUZZYtech, I found an immediate improvement in the performance of the system. Rather than oscillating between 20 and 30% from the target, I found that the motor was regulated to within 8%.

Although 8% was a lot better, I thought I could do better. My primary strategies were to cut the delay interval by four (down to a four-times-per-second sample) and "tuning" the input and output rules. I found that these simple changes resulted in the system being able to keep the fan motor running to within 2% (although it occasionally extends to 3% or 4%. In Fig. 16-79, the light-colored line is the actual speed error. The darker line is the Delta_PWM or change in speed-control PWM duty cycle. The changes in the speed-control PWM duty cycle are in 0.5-percent duty-cycle increments because the total PWM range is 0 to 199.

For this application, this error rate is quite acceptable. You might find that for your own applications, this error rate is either too stringent or too loose. One of the biggest advantages to using FUZZYtech that I can see is the graphical performance output that can be observed easily.

The next step in the process was to compile the fuzzy logic system and place the code into the application. I did this to see how well the fuzzy rules would perform in a stand-alone system and the accuracy of Microchip's statement that the PIC18Cxx was source-code compatible with the PIC16Cxx (the mid-range).

I compiled the application and attempted to "include into the fuzzy2" source. Fuzzy2ASM is located in the *code/fuzzy* subdirectory of the *PICmicro* directory. After attempting to get the combined source to assemble without any errors, I gave up with more than 200 errors and 100 warnings left to resolve (and probably many more as I worked through the application).

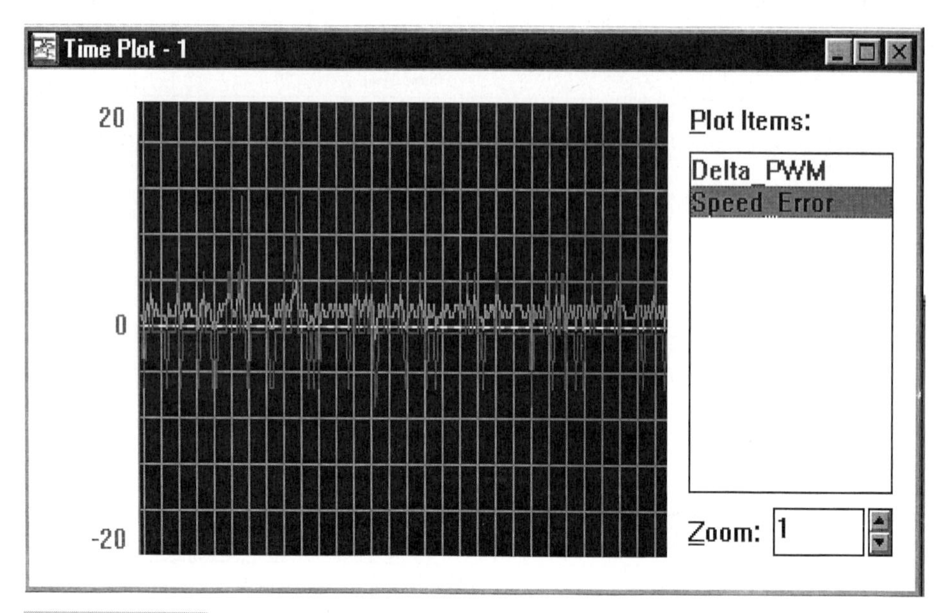


**Figure 16-79** **Actual fuzzy fan-control performance**

Looking at the errors and warnings, I came to the conclusion that the PIC18Cxx is very source-code compatible with the mid-range PICmicro® MCUs. The problem was the source code was not very well written for my style of coding as well as the PIC18Cxx, requiring bank changes and accessed registers using labels that are no longer used (such as *RTCC* for *TMRO*).

If you look at the fuzzy2ASM source code, you will see that the code works like mid-range source, except for three differences. The first is that I didn't have to worry about register banks for the hardware interface registers. I also kept all of the file register variables in the first 128 addresses, which meant the BSR register never had to be accessed in the application.

Second, notice the lack of *goto* instructions. Instead, I used the *BRanch Always (BRA)* because it requires just one instruction word. It can access all the instructions in the applications easily.

The third change was to replace branches on Status flag conditions to branch instructions. An example, you'll see in the source code is replacing:

```
btfsc  STATUS, C
 bra   Label
```

with:

```
bc      Label
```

I found myself often going to the two-instruction format (out of habit), but it was easy to go back and put in the conditional branch statements before I was ready to build the code. This is something that you will have to watch (as well as avoiding the *goto* instruction in the PIC18Cxx).

# 17

# REAL TIME OPERATING SYSTEMS

## CONTENTS AT A GLANCE



One of the most-useful application programming environments that can be used in a microcontroller is the *Real-Time Operating System (RTOS)*. When properly used, an RTOS can simplify application development and, by compartmentalizing tasks, the opportunity for errors is significantly reduced. This option has not been available to the PICmicro® MCU until the initial availability of the PIC18Cxx. This PICmicro® MCU family allows access to the processor stack, which can then be modified with different task data. This feature is not available in other PICmicro® MCUs and will be used in the PIC18Cxx Real-Time Operating System presented in this chapter.

The best definition I can come up with for a RTOS is: *A background program that controls the execution of a number of application subtasks and facilitates communication between the subtasks.*

An important point about *subtasks* (which are usually referred to as just *tasks*): if more than one task can be running at one time, the operating system is known as a *multitasking operating system* (which I usually refer to as a *multitasker*). Each task is given a short "time slice" of the processor's execution, which allows it to execute all or part of its function before control is passed to the next task in line.

I usually call a *subtask* a *task* or *process*. I also blur the difference between an RTOS and an *OS (Operating System).*

I tend to lump both OS and RTOS together is because the central *kernel* (the part of the operating system that is central to controlling the operation of the executing tasks) is really the same for both cases. The difference comes into play with the processes that are loaded initially along with the kernel. In a PC's operating system, the console I/O, command interpreter, and file system processes are usually loaded with the kernel. Everything has been optimized to run in a PC environment (which means respond to operator requests). In an RTOS, the actual application tasks are loaded in with the kernel, with priority given to tasks that are crucial to the operation of the application.

You might be a bit suspicious of a RTOS after what I've written. After all, you probably have a PC running Windows 95 or Windows NT and you are probably familiar with problems working with different pieces of hardware or software applications. I would be surprised if somebody reading this book had never had a problem with Windows not coming up properly, crashing when you least expect it, or hanging up and not responding to input (or displaying a "blue screen of death"). Along with these problems, these operating systems require literally hundreds of megabytes on a hard file to operate. With this background, you're probably wondering why I would suggest a multitasking operating system for an eight-bit microcontroller with only a few thousand instructions of program memory.

Another question you might ask is "What features does the PICmicro® MCU PIC18Cxx have that would make me feel like I would want to invest the time and effort into developing a multitasking operating system for it?"

To answer this question, I would look at it from the negative and ask what the PIC18Cxx doesn't have compared to the PC. The smaller "system environment" of the PIC18Cxx is what is important. It makes an RTOS very appropriate for this type of device. The PIC18Cxx does not have a file system, arbitrary amounts of memory (including virtual memory models) or the sophisticated user interface (unless you want to provide one yourself in your application). As well, in the microcontroller's case, you can specify the interfaces and hardware, rather than create standards that might have to be bent when new technology becomes available or interpreted incorrectly by other developers.

To make matters worse for the PC, literally millions of lines of code have been written for the Windows operating system, drivers, and applications. When more code is written, the greater the chance that incompatibilities between the system, drivers, and applications will occur. This makes it more likely that problems will occur when everything is working together.

In a small microcontroller, the team of people developing the application and the interface software is very small. The chances for these incompatibilities not being detected and passed on to the end user is much more remote.

In previous books, when I've described RTOS, I've used the example of a space-station control system and a nuclear reactor control system. Both are fairly complex applications with important safety aspects associated with them. For this book, I want to present a web surfer's PC and how the different features and software interconnect to provide a method to connect to the Internet. Instead of having multiple computers, each one providing one function, the PC makes functions (known as *resources*) available to the different applications.

In a real system, the different tasks (which are represented by different boxes in Fig. 17-1), would be given priorities. This reflects the importance of messaging and operation of the specific tasks relative to the others. Thus, if high-priority and medium-priority tasks are waiting to execute, the high-priority task will run first. In most PCs, the modem would be given a specific priority because it is the resource that is the most constrained by speed that it can operate at and the demands by the different applications for the available bandwidth.

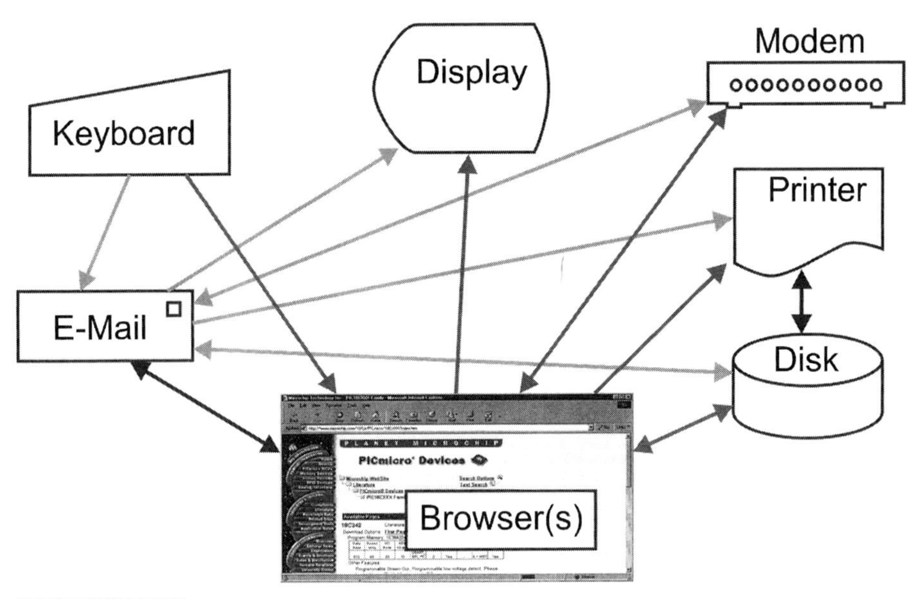


**Figure 17-1** PC hardware and software used while Web surfing

In periods of inactivity (between the surfer's keystrokes with no web pages being currently downloaded), low or medium tasks are executed because the high-priority tasks are not required. Low-priority tasks are normally interactions with the operator because if a task takes a long time to execute (in computer terms), a human working with it probably won't notice any delays in the computer's operation.

All of the tasks can be arranged in a diagram to show how they communicate via messages.

In Fig. 17-1, the arrows represent the directions in which the messages move. An important concept about messaging in an RTOS is that each message is initiated by a task. Normally, each task is waiting (blocking) on a message, waiting to respond to it, and execute the request that is part of the message. After the task has executed its function, it then can respond to the conditions by sending messages to other tasks.

For example, to send e-mail, the "Modem" task could execute the code:

```
ModemSend()                    // "Modem" Data Packet Send Routine
{

char * Packet;                 // "Packet" to be Sent

  while ( 1 == 1 ) }           // Loop forever
    while (GetMsg() == 0);     // Wait for a Message to Send Out
    while (ModemBusy != 0);    // Wait for the Modem to be Available
    Packet = ReadMsg();        // Read the Sent Packet
    SendMsg(Packet);           // Send the Packet over the Internet
    AckMsg();                  // Acknowledge the Original Message
  } // endwhile
} // End ModemSend
```

This task will wait for a packet to be passed to it that will have to be sent to the Internet. If the modem is already active (i.e., a previous packet is being sent or a packet is being received), it will wait for the modem to become free before sending the packet. Once the packet is in the queue to be sent by the modem, the *ModemSend* task acknowledges the original request to send the packet and waits for the next request to come in.

This is a grossly oversimplified example of what actually happens, but should give you the idea that tasks send request messages to other tasks to request that a function be performed. The receiving tasks will only process the request when they are able to and acknowledge that they have done this when the resources to do so are free.

One aspect of tasks and processes in RTOS might not be readily apparent. Tasks should only access one hardware device. So the *SendModem* task in this example only accesses the modem-output functions. To access other hardware in the system, the RTOS application will send messages to the appropriate tasks that control the different hardware.

This is an important philosophical point about access and RTOS because, obviously, applications could be written to access multiple hardware interfaces. This allows code re-use or modification and allows multiple software developers to work on an application without having to understand how the other interfaces work. At first, it might be unnatural for you to think in terms of simple, single devices, but once you get the hang of it, you'll be amazed at how easy it is to work with an RTOS.

The last aspect of RTOS presented here is the *semaphore*. The semaphore is a flag that is controlled by the operating system, which can be used to restrict access to a resource (a hardware device, data, or even another task) until the "owning" task has finished with it.

Semaphores are used in such situations as controlling access to an operator console. In the *SendModem* task, a semaphore could be used to indicate that the modem hardware is available for a packet transmission or that the modem is involved in a data packet transfer.

I have thrown a lot of information and concepts to you in this introduction to RTOS. I'm sure that many of these concepts have never been presented to you before. If it doesn't make a lot of sense or if you can't see how something is implemented, don't worry. As I go through the rest of the chapter, I show how these concepts are implemented in the PIC18Cxx.

# RTOS01: Simple PIC18Cxx RTOS

When I originally set out to develop the RTOS for the PIC18Cxx, I had a number of ideas. This was based on the RTOS that I have written for other devices. The philosophy was to keep it quite simple. My goals were:

1. Required program memory for the application to be less than 256 instructions.
2. Fast task-switch capabilities.
3. Sixteen bytes or less of overhead file registers would be required.
4. Eight tasks with an initial *AllTask*.
5. The tasks would execute as if they were the only application in the PICmicro® MCU.
6. Each task could have at least 32 bytes of variable space.
7. The RTOS would support at least one subroutine call.
8. Interrupts would be supported.
9. Application tasks would be assembled with the RTOS code.
10. A single OS entry point.

The result is located in the *code\rtos* subdirectory of the *PICmicro* directory. This is a reasonably full-function RTOS with a simple application that increments a series of LEDs on a PIC18C452's PORTB. The application code was developed to test the different functions of the PICmicro® MCU. I describe the application in more detail in this section and a more complex application is presented at the end of this chapter.

To achieve a very small program memory requirement for the RTOS, I wanted to ensure that I had optimized the application as much as possible for the PIC18Cxx. I was able to write and debug the code over the space of about 12 hours, but the design of the RTOS was done over the space of a couple of weeks. This time spent in design and doodling resulted in fairly small code, but, what I feel is a reasonably elegant RTOS that is well suited to the PIC18Cxx and offers good "protection" for the application developer.

Because the RTOS was the first "full" application I implemented on the PIC18Cxx, this was to be a learning experience for me. It probably sounds risky to do an entire operating system as a first application, but I actually used this occasion to ensure that I understood how the PIC18Cxx worked and how the instructions execute. Making this more of an adventure was my use of an interim upgrade to MPLAB and engineering sample PIC18Cxx. Production-level PIC18Cxx's become available when I was proofreading this book.

In terms of measuring against requirements, the final RTOS weighs in at 748 instructions, requires only eight bytes of overhead to run eight tasks and executes a task switch in about 750 $\mu$s with the PICmicro® MCU running at 4 MHz. It might sound like I blew the RTOS size specification, but because of the relatively large size of program memory available to the PIC18Cxx (I used a PIC18C452, which has 16,384 instructions), I do not consider this issue to be significant.

When you write RTOS applications, you will discover that the actual application code is quite small. The test application that is included with the RTOS only requires 26 instructions to create an application that starts up TMR0 to interrupt once every 16.384 ms and uses a counter to count down 64 TMR0 interrupts and increment the LEDs.

When I first learned about RTOS, I was told that the maximum task switching time should be 500 $\mu$s. In this RTOS, I used 750 $\mu$s at 4 MHz. I don't consider this to be a major concern because a 4-MHz clock is actually quite modest. To meet the 500-$\mu$s requirement, I could simply turn on the four times clock PLL built into the PIC18Cxx's oscillator circuit or add a faster clock. This might sound somewhat facetious, but I have found over the years with reasonably simple microcontroller RTOS and properly designed and written applications, switching times of two or three milliseconds can be tolerated without any problems by most applications.

When tasks are created for this RTOS, they execute as if they are the only task running in the PICmicro® MCU. To help facilitate this, I made the following assumptions about how the application tasks would work. This results in a somewhat simplified view of the PIC18Cxx, but one that is more than offset by the capabilities offered by the RTOS.

The basic assumption I made was that the application would only access the first 63 file registers of the access bank. When I created the RTOS, I wanted to avoid global variables as much as possible. Instead I wanted to take advantage of the RTOS' ability to send messages back and forth.

Sixty-three file registers is actually quite generous and will allow you to develop tasks for most applications with very few problems. The number of file registers is specified when the task is started. For many applications, you will discover that quite a few tasks require only one or two file registers that will allow you to run the RTOS quite easily in the PIC18Cxx devices with 512 file registers.

At the top of the memory space in the PIC18Cxx's with 1,536 file registers, you could put in some global variables, but I suggest that you place them starting at address 1,024 to ensure that none of the tasks can access them.

When the RTOS executes, you will notice that the file registers dedicated to the task start at the beginning of the access bank and even in the PICmicro® MCU's file register memory itself. This was done to further simulate the perception that the task was the only code running in the PICmicro® MCU.

The 17 special-function registers that are specific to each task are located in the second 128 bytes of the access bank (Table 17-1).

The limitations are the lack of the third index register (FSR2) and restricting the stack depth so that only three pushes could not be used by the application (because it is changed during the RTOS execution).

When a task is not executing, these 24 registers are read out of the PICmicro® MCU's special-function register area and copied into what I call the *Task Information Block (TIB)*. This data structure, which is provided for each task, consists of the 24 special-function registers and the variable file registers.

To make moving data back and forth between the task information block simpler in the RTOS, I copy the 24 bytes of the task's special-function registers into address 0x060 to 0x077 of the access bank. This allows me to set or restore the special-function registers without hav-

**TABLE 17-1 Real Time Operating System "Context Registers"**

| REGISTER | COMMENTS |
|---|---|
| WREG | |
| STATUS | |
| BSR | |
| PCLATH | |
| PCLATU | |
| TBLPTRL | |
| TBLPTRH | |
| TABLAT | |
| PRODL | |
| PRODH | |
| FSR0L/FSR0H | |
| FSR1L/FSR1H | |
| TOSL/TOSH/TOSU | Top Three Program Counter "Pushes" on the Stack |

ing to cross-bank boundaries (which simplified the development of the RTOS considerably).

This is probably the largest area of inefficiency in the RTOS. If the task information block's special-function register area would pass data to and from the saved task information block, rather than the access bank copy, quite a few instructions and cycles could be saved in the RTOS execution. Although to be honest, even if this was done, both the code size and task-switch parameters would probably still exceed the targets that I gave for them.

The ability to access the program counter's stack in the PIC18Cxx is what sets this architecture apart from the others and makes the RTOS possible. For this application, I only save the top three elements of the stack, and not the total 31. The primary reason is to save file registers. If all eight possible tasks saved all 31 program counter stack elements, then 744 bytes would be required. By cutting the saved value down to only the top three, I only require 72 file registers to save the tasks' stack.

Saving the top-three program counter stack elements means that the task can be nested down two levels of subroutines when the RTOS is active. Thus, the task's mainline can call a subroutine and that subroutine can call another subroutine. Recursive subroutines cannot be implemented at all in the RTOS.

Watch out for when a new task is started. When the TaskStart macro is invoked, it pushes the starting address onto the stack. I did this because it allowed me to simply copy data from one task information block into a new one that I have set up for it. It also meant that I would not have to make space in any other registers when the RTOS is invoked.

The first byte of the task information block is the *task byte.* This byte is used to indicate to the RTOS what is the status of the specific task. The task byte is defined in Table 17-2. The interrupts supported for the task to wait for are listed in Table 17-3.

**TABLE 17-2 Real Time Operating System "Task Byte"**

| BITS | FUNCTION |
|---|---|
| 7–6 | Task Operating Mode |
| | 11 Able to execute |
| | 10 Task is waiting for a message to be sent |
| | 01 Task has sent a message and is waiting for it to be acknowledged |
| | 00 Task is waiting for an interrupt request |
| 5–4 | Task priority |
| | 11 Highest |
| | : |
| | 00 Lowest |
| 3–0 | Task operating indicator, specific to the "task operating mode bits" |

**TABLE 17-3 Real Time Operating System Interrupt List And Numbers**

| INTERRUPT NUMBER | INTERRUPT SOURCE | REGISTER |
|---|---|---|
| 0 | TMR1 | PIR1 |
| 1 | TMR2 | PIR1 |
| 2 | CCP1 | PIR1 |
| 3 | SSP | PIR1 |
| 4 | TX | PIR1 |
| 5 | RC | PIR1 |
| 6 | AD | PIR1 |
| 7 | PSP | PIR1 |
| 8 | CCP2 | PIR2 |
| 9 | TMR3 | PIR2 |
| 10 | LVD | PIR2 |
| 11 | BCL | PIR2 |
| 12 | RB | INTCON |
| 13 | INT0 | INTCON |
| 14 | TMR0 (16.384 ms delay) | INTCON |
| 15 | | Used by the RTOS |

All of the possible PIC18Cxx interrupt-request sources are accounted for, except for the INT1 request. This request was left off because it is put in a register that doesn't follow the conventions of the other bits and I wanted to have a special-purpose value that is used to flag the RTOS that an operating-system request is required.

The interrupts are run in Normal, not Priority mode. Priority mode must never be enabled with the RTOS because the code is designed for a single-entry point for interrupts and code. Enabling interrupts will cause a second-entry point that will result in task data being saved incorrectly when the interrupt request is acknowledged. When using interrupts with this RTOS, ensure you only access the E bits to set them to enable the interrupt request.

When an interrupt request is acknowledged by the RTOS, the E flag is also reset by the RTOS, except in the case of TMR0. For all interrupts, except for TMR0, the tasks will have to enable the interrupts and reset the hardware after the request has been acknowledged. This means setting up the hardware and setting the E flag to enable the interrupt.

TMR0 is enabled within the RTOS and will interrupt the application once every 16.384 ms. This interrupt can be waited on, but TMR0IE should never be reset by a task. The purpose of the TMR0 interrupt is to stop an application from taking up too many processor cycles and "starving" the other tasks. If you have a task with a large amount of data processing or I/O polling, I suggest that you place this at as low a priority as possible to allow other tasks to execute.

During a task's execution, the INTCON GIE bit can be reset for crucially timed code or a high-priority operation. I do not recommend disabling interrupts for more than 20 instruction cycles as the longer interrupts are disabled, the better chance an interrupt will be "missed" or overwritten. The RTOS code only responds to one interrupt at a time and responds to them in the order given. Any delays in acknowledging an interrupt can result in problems later in the application.

The RTOS requests are listed in Table 17-4, along with the registers (and bits) that are affected. Each request is actually a macro, which was done to eliminate the need for keeping a reference for how to call the RTOS correctly.

When the RTOS is booted, it will start *AllTask*. This task has two purposes. The first is to start up the application code. To do this, it invokes the TaskStart macro to start up Task1 at a priority of *1*. AllTask has a priority of *0*. Once Task1 is executing, it will just loop as the lowest-priority task in the RTOS, providing a lowest level of functionality for the RTOS. The code that I used is:

```
_AllTask                        ; Always Return Here
 TaskStart Task1, 1, 63         ; Start Application Code
_AllTask_Loop
 TaskNext                       ; Jump to Next Active Task
  bra _AllTask_Loop
```

The AllTask looping might not seem like an important function, but it actually is because, when the RTOS code is active, I disable interrupts. If all the other tasks in the PICmicro® MCU are waiting on interrupt requests or messages, then AllTask will be the only one capable of executing when it executes with interrupts enabled. The application is basically just the *bra* instruction jumping back to a previous TaskNext RTOS request. Any pending interrupts can be acknowledged and processed.

Task1 is the first task of the application code. This standard label is used to indicate the application code. Note that I have given it a priority of *1* and the full 63 file registers for variables. *1* is the lowest priority that your application should have to ensure that AllTask performs its function properly.

When I created the RTOS, I wanted to use it with a simple application. As I mentioned, this application simply waits on the timer and updates LED values once per second. The circuit that I used is shown in Fig. 17-2 and is part of the RTOS01.ASM application that is found in the *code\rtos* subdirectory of the *PICmicro* directory.

The bill of materials for the project is given in Table 17-5. The actual application code is:

```
Task1

  CBLOCK     0x000                  ; Put the Variables Starting at 0x000
LEDTask                             ;

Counter                             ; Number of Times Executing
LEDValue
  ENDC

  clrf       Counter, 0             ; Clear the Variables
  movlw      0x0FF
  movwf      LEDValue

TaskStart    LEDTaskStart, 1, 2     ; Put in the LED Executing Task
  movwf      LEDTask                ; Save the Task Number of LEDTask
```

**TABLE 17-4 Real Time Operating System Task Request Ddefinition**

| RTOS | INPUT | OUTPUT | MACRO |
|---|---|---|---|
| Start Task | Starting Address<br>Priority (03)<br>file register<br>requirements | Carry Reset, OK and new Task #<br>in “WREG” | TaskStart Address, Priority, Size |
| | | Carry Set, NO Available Tasks<br>or more than 63 Task Variables | *Address, Priority* and *Size* are all constants |
| Next Task | None | None | TaskNext |
| Wait for Interrupt | Interrupt Number<br>in WREG | None | IntWait interrupt<br>*Interrupt* is a constant |
| Send Message | Task Number in WREG<br>2 Byte Message in<br>FSR0L/FSR1L | None | MsgSend TaskNumber<br>*TaskNumber* is a variable |
| Wait for Message | None | 2 Byte Message in FSR0L/FSR1L<br>TaskNumber in WREG | MsgWait |
| Read Message | Task Number in WREG | Carry Reset, 2-byte message<br>in FSR0L/FSR1L<br>Carry set, no message to get | MsgRead TaskNumber<br>*TaskNumber* is a variable |
| Ack Message | Task Number in WREG | Carry reset, operation complete<br>Carry set, no message to acknowledge | MsgAck TaskNumber<br>*TaskNumber* is a variable |

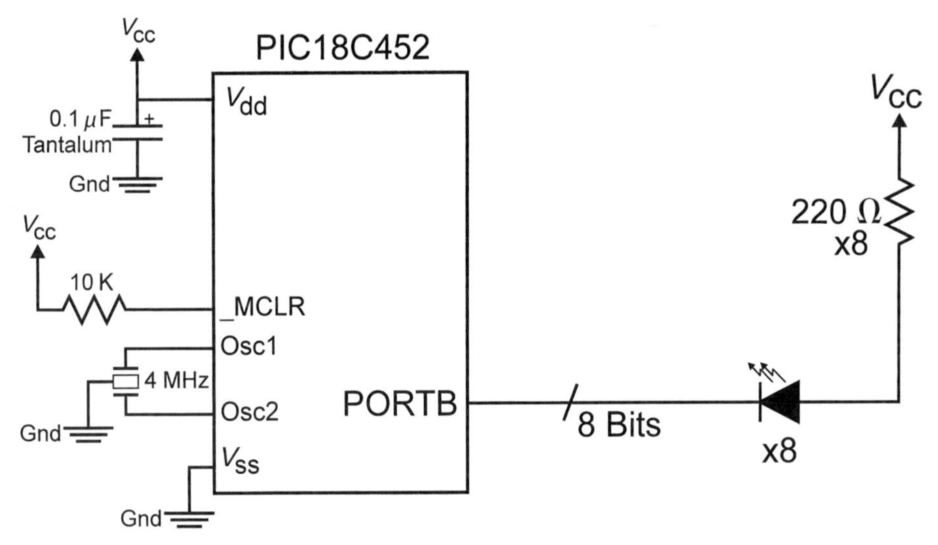


**Figure 17-2** RTOS01 test circuit. Increment LEDs once per second

```
  movff      LEDValue, FSR0L      ; Save the LED Task Value
MsgSend      LEDTask
  nop

Task1Loop                         ; Loop Here Until 64x Past

  IntWait    14                   ; Wait for the Timer0 Interrupt

  incf       Counter, f, 0

  movlw      64                   ; Done 64x?
  xorwf      Counter, w, 0
  btfss      STATUS, Z, 0
   bra       Task1Loop            ; If Not, Loop Around Again
  movwf      Counter, 0

  decf       LEDValue, f, 0       ; Increment the Value displayed
```

**TABLE 17-5 Real Time Operating System Example Application Bill Of Materials**

| PART NUMBER | DESCRIPTION |
|---|---|
| PIC18C452 | PIC18C452-10/JW |
| Capacitor | 0.1-μF tantalum capacitor |
| R1 | 10-K, 1/4-watt resistor |
| R2 to R9 | 220-Ohm, 1/4-watt resistor |
| Y1 | 4-MHz ceramic resonator with built-in capacitors |
| Misc | Prototyping board, +5-volt power supply, wiring |

```
  movff      LEDValue, FSR0L
 MsgSend     LEDTask

  bra        Task1Loop

LEDTaskStart                        ; Make PORTB Output and Wait for
                                    ;   Messages

 CBLOCK 0x000                       ; Variables
LEDTaskNumber
 ENDC

  clrf       TRISB, 0

LEDTaskLoop                         ; Loop Here for Each Character

 MsgWait
   movwf     LEDTaskNumber, 0       ; Save the Task Number

 MsgRead     LEDTaskNumber
   movff     FSR0L, PORTB           ; Update the LED Value with the Message

 MsgAck LEDTaskNumber               ; Acknowledge the Message

   bra LEDTaskLoop
```

In this code, Task1 first enables LEDTaskStart which puts all the bits of "PORTB" into "output mode" and then waits for a message to update the LEDs. Task1 saves the task number returned for LEDTaskStart and then sends a message to the LEDTaskStart to turn off all the LEDs. Next, Task1 waits for the 16.384-ms timer to overflow 64 times (which takes approximately one second) and then updates the LED value and sends a message to LEDTaskStart.

This is a very simple application, but there are a few things to notice. In this application, I have saved the LEDTaskStart locally in Task1. For many other applications where a task provides a central resource to the complete application, you will probably want to make the task number available globally within the PICmicro® MCU.

To do this, the variable should be placed at the end of the file registers, to ensure there isn't a conflict with any task variables. Each task has its own unique bank-select register, so access can be made by one task without affecting any of the others.

In the code, notice that I explicitly access the variables using the access bank (the *, 0* at the end of variable-/register-access instructions). This is optional and applications do not need to use this convention unless the file registers outside the address bank are accessed.

In the previous code, notice that I placed an *nop* instruction after the first MsgSend macro invocation in Task1. The reason for doing this is to provide a place to hang a breakpoint. You cannot place a breakpoint at a macro invocation in source code, so I put in the *nop* to allow my to break after the MsgSend to be able to go back and take a look at what happened in the application. This seems to be a trick that not many people know about. It can avoid frustration later when you have an application with a lot of macros with straight instructions in between them and you want to understand what is happening.

Last, the MsgRead macro invocation in the LEDTaskStart is unnecessary. This was done to test the function and the MsgWait will return the same information.

This is the fourth RTOS that I have presented in my books and it has one important difference from the others. In the previous RTOSes (for the Motorola 68HC05 and Intel 8051), I put a lot of emphasis on placing the RTOS and task variables where I could easily find them and translate them to debug the application.

For this PIC18Cxx RTOS, I placed the emphasis on just making the Task's variable space easy to look at. This made creating and debugging the RTOS somewhat more difficult because I had to leave the thread of the debugging to move the *Program Memory* window back and forth to see what was going on.

The *File Register* window in MPLAB is shown in Fig. 17-3 and you can see how only the first 192 file registers can be shown with the source file. These 192 file registers are enough for the variables used within the tasks and the 24-byte special-function register block, as well as the first two task's (AllTask and Task1) task information blocks. Looking at Fig. 17-3, finding specific values, other than the first three variables of Task1, at addresses 0x000 to 0x002, is not easy to do.

I made a compromise when designing the application. For people who are developing applications, this display is not a problem (in fact, it works very well to see what is happening in a specific task).

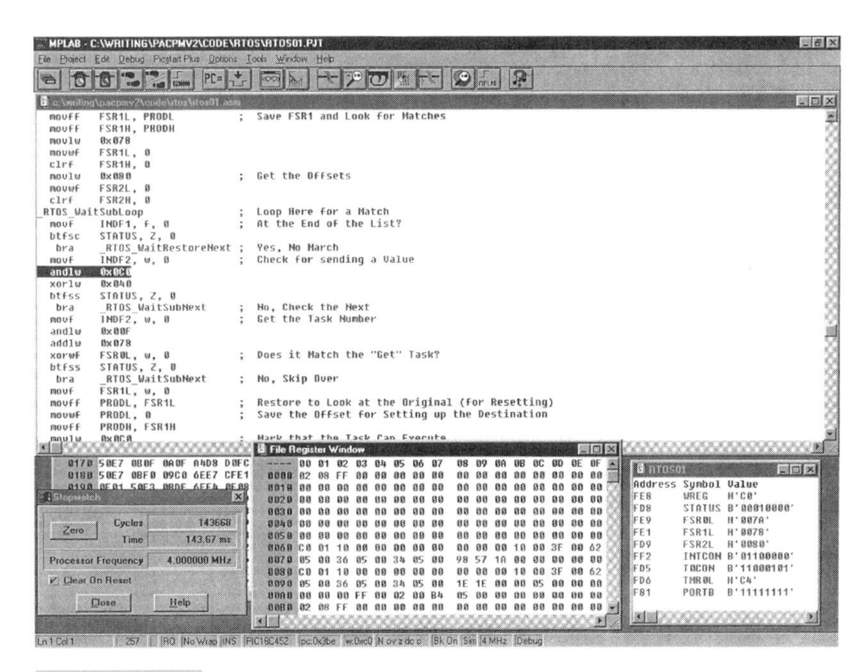


**Figure 17-3** Screen shot of MPLAB debugging the real-time operating system

# RTOS02: Multitasking Application Example

The first example application in this book is a digital clock and thermometer that runs on a PIC18C452, with a Dallas Semiconductor DS1820 temperature sensor and a Hitachi 44780-based two-line LCD. The circuit used for this application is shown in Fig. 17-4.

If you have read the *Handbook of Microcontrollers,* this application will probably look pretty familiar. I used this application to demonstrate how different microcontrollers would implement the same application. For this book, I am using this application to show how a typical microcontroller application can be implemented using an RTOS to control the application's execution. The bill of materials is listed in Table 17-6.

The application code interfaces with three hardware devices: the LCD, DS1820, and a button is used to set the time. The circuit itself is pretty easy to wire because I arranged the PORTB pins to be wired directly across to the LCD. The LCD's pin 14, if it lines up with pin 40 of the PIC18C452, will allow mostly straight-through wiring to the LCD.

The DS1820 has a 10-K pull up on its line. It uses one of the I/O pins as a simulated open-drain driver. The DS1820 will pull down the line when data is read back. By simulating the open drain and putting the I/O pin in Input mode while the DS1820 is driving the line low, there is no opportunity for bus contention between the PICmicro® MCU and the DS1820.

When I created this application, I used a 4-MHz ceramic resonator simply because it was already on the breadboard. I had left it wired for a PIC16F877 and the core circuit was

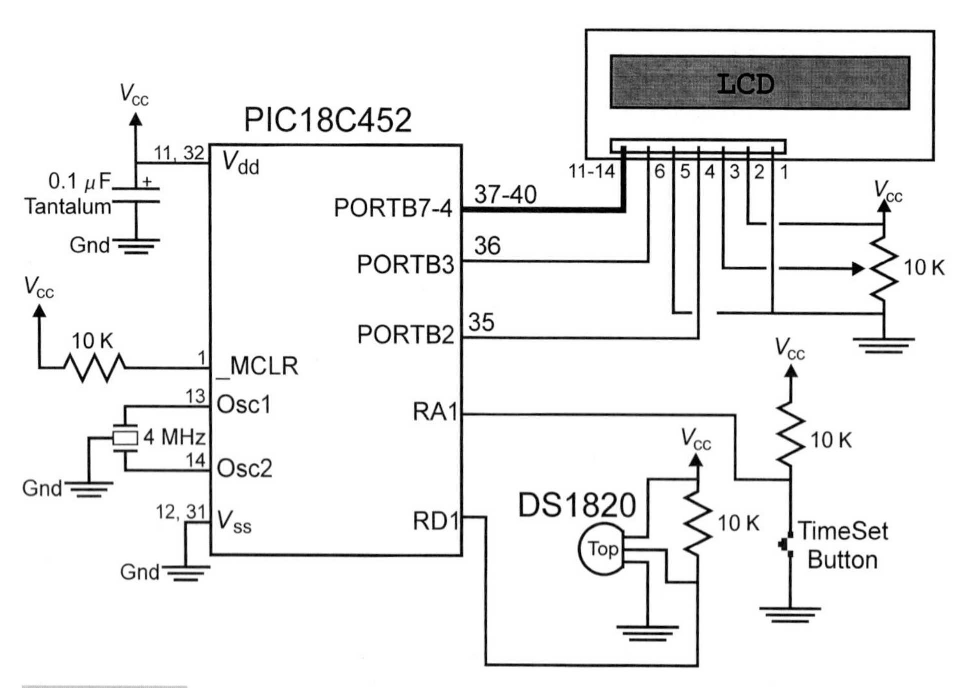


**Figure 17-4** PIC18Cxx digital clock/thermometer

**TABLE 17-6 Real Time Operating System Digital Clock/ Thermometer Bill Of Materials**

| PART | DESCRIPTION |
|---|---|
| PIC18C452 | PIC18C452-10/JW |
| DS1820 | DS1820 in TO-92 package |
| 4 MHz resonator | 4 MHz ceramic or 4-MHz crystal and capacitors |
| 3 10-K resistors | 10 K, 1/4 watt |
| 10-K potentiometer | 10 K single turn |
| 0.1-$\mu$F capacitor | 0.1 $\mu$F tantalum |
| Button | Momentary-on button |
| LCD | 2x16, Hitachi 44780 based LCD |
| Misc | Prototyping card, wiring, +5-volt power supply |

identical to what I needed for the PIC18C452. The same pinout is a useful feature of the PIC18C452 because it allows you to easily replace a mid-range PICmicro® MCU with the PIC18Cxxx without changing the connections. If you are going to use this application as a clock, you might want to put in a 4-MHz crystal (with external capacitors) to get the most accurate timing possible.

With the circuit designed, I then looked at how I would architect the application software. Because this code would be running under the RTOS, I wanted to ensure that I could develop it without using any more than the available resources.

I always find creating a simple block diagram of how I expect the tasks to execute to be invaluable when I am designing an RTOS application. For this application, I was able to create the task block diagram shown in Fig. 17-5 and allow only five tasks in total to be created.

The AllTask is built into the RTOS. Its function is to start the Task1 task executing and provide a task that never blocks and is able to execute instructions that aren't masked for interrupts. Task1 starts the other tasks and then initializes the LCD. Once the temperature task (TempTask) or the TimeTask has a message for it, it updates the LCD with the appropriate data.

Notice that in Task1, when the LCD is set up, I use the IntWait request to wait on the 16.384-ms TMR0 interrupt interval to provide long delays for the LCD. For short delays, I use the TaskNext RTOS request, which has the operating-system check for other tasks and return. As indicated in the previous sections, this executes in about 700 $\mu$s, which is more than long enough delay to allow LCD short-delay commands to execute.

The 16.384-ms TMR0 interrupt built into the RTOS is used by TimeTask, ButtonTask, and TempTask to initiate the requests. Each task spends most of its time doing nothing and just waiting for a specific delay. When you look through the application, you will probably be surprised to find that this 16.384-ms interval is used for the real-time clock. The advantages of using this interrupt interval is that, regardless of what kind of delays occur occur from other tasks, each task will be interrupted every 16.384-ms (on average). When you first single step through the application, you'll find that the first time TimeTask is in-

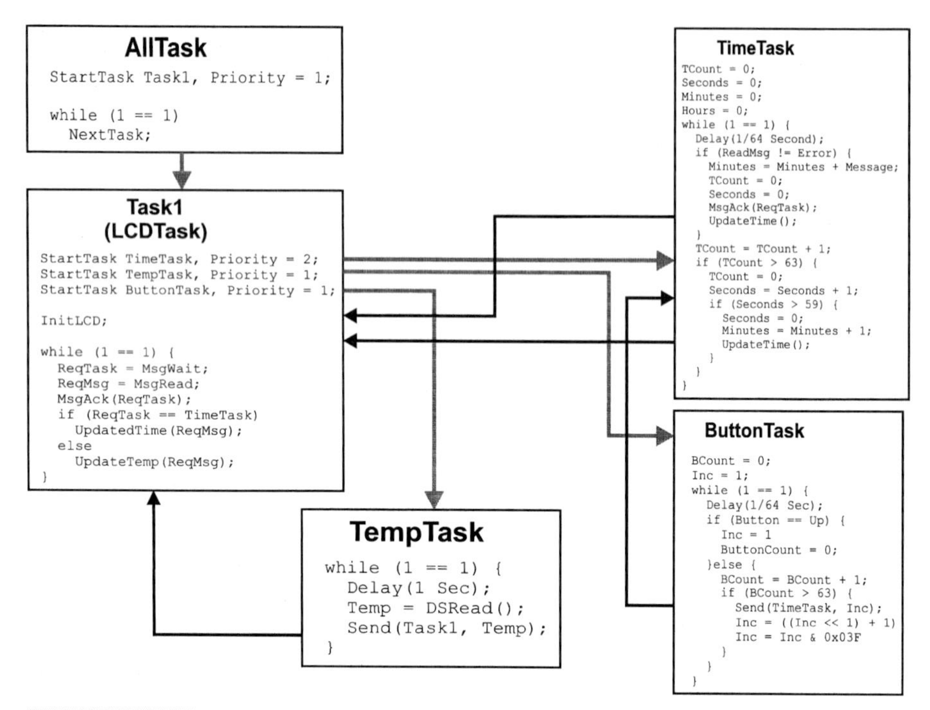


**Figure 17-5** PIC18Cxx digital clock/thermometer tasks

terrupted, almost 19 ms has passed, but the average will work out to 16.384, allowing this interval to be used as a regular clock.

When you look at the TimeTask code, you will discover that I count the number of 16.384-ms intervals and add the delay until the total is greater than one second. Each time the TMR0 interrupt allows TimeTask to execute, the code executes as:

```
TotalTime = TotalTime + 0x04000;  // Increment Second Fraction by
                                  //   16,384
if (TotalTime > 1000000) {        // One Second has Past?
  TotalTime = TotalTime - 1000000;// Yes, take 1,000,000 from the
                                  //   total
  Seconds = Seconds + 1;
  if (Seconds > 59) {             // Increment the Minute if
                                  //   Appropriate
 :
 ,
```

The 0x04000 that is added to *TotalTime* each time TMR0 overflows is 16,384 decimal, the number of microseconds in hex. When the value is more than one million, one million is subtracted from the total and the process repeats. Each time the process repeats, the *TotalTime* overflow after subtracting one million from it is added to the total. In Table 17-7, the TotalTime and number of TMR0 interrupts are shown for waiting for the initial second delays.

To ensure that the TMR0 interrupt is never missed by TimeTask, I have set it to the highest priority of any task in the PICmicro® MCU. If an interrupt is missed by one of the other tasks (the temperature or button task), it is not crucial. If there is an opportunity for

**TABLE 17-7 Digital Clock/Thermometer Application Time Update With TMR0 Intervals**

| TOTALTIME START | TOTALTIME END | TMR0 INTERVALS |
|---|---|---|
| 0 | 1,015,808 | 62 |
| 15,808 | 1,015,232 | 61 |
| 15,232 | 1,014,656 | 61 |
| 14,656 | 1,014,080 | 61 |

the TMR0 interrupt to be missed by TimeTask, the clock will not be accurate and "lose" time the longer it executes.

When you read through the code, you'll see that the LCDTask does the conversion of the time and temperature before they are displayed on the LCD. For this application, the sending task number is used to determine which message should be processed.

One aspect of the RTOS messaging to the Task1 that I am pleased with is the ability of the RTOS MsgWait to return the task number of the transmitting tasks. I found this to be very useful in this application because it allowed me to respond immediately to incoming messages. If I hadn't included this feature, I would have to poll through each available task to find what was sending messages to it. This polling could be a problem with TimerTask because I would like to respond to that task first so that I will always respond before the next timer "tick." To also ensure that TimerTask never misses an interrupt, I acknowledge the message before processing it and updating the time.

Before sending a message to the LCD task to notify it that it's time to update the display, the ButtonTask is polled to see if it is sending a message to TimeTask. If an increment command is sent to the TimeTask, the seconds information is deleted and the minutes is updated with the latest increment value.

The increment value is kept at one while the time set button is released and increments by powers of two (to a maximum of 63) for each second the button is pressed. Each second the button is pressed, a message is sent to TimeTask, indicating that it should increment the minutes count by the increment value. The time increment value is increased by the next power of two (to a maximum value of 63) or reset back to one.

By pressing the button, the increment value is increased by a power of two. So, after one second, it will increase the minute counter variable by three minutes (not one); after two seconds, it will increase the minute counter variable by seven minutes; and after three seconds, by 15. When an increment variable value of 63 is reached, *increment* won't be incremented any more and every second, the hour will be updated. Using this algorithm, it's possible to run through a total of 24 hours in less than 30 seconds, with quite a bit of control over the time to which you are trying to arrive.

If you are a digital clock manufacturer, I hope you remember where you first saw this! I would be happy to license this algorithm to you for a modest fee.

The temperature sensor used in this application, the Dallas Semiconductor DS1820, is a rather interesting little beast. It is available in a three-pin transistor (TO-92) style package and only requires a 10-K pull-up resistor for the driving signal. I have included a PostScript format of the datasheet on the CD-ROM.

The data sent and received used a Manchester encoded signal, in which the controlling device pulls down the line for a specific amount of time to write to the DS1820. A *0* is a low pulse of 15 to 60 $\mu$s (30 $\mu$s is typical) and a *1* is a low pulse greater than 1 $\mu$s and less than 15. The write operation is shown in Fig. 17-6.

To read data from the DS1820, the microcontroller pulls down the line for approximately 1 $\mu$s and releases it, at which time the DS1820 holds the line low for a specific length of time to write a bit on the line. If the DS1820 holds the line low for less than 15 $\mu$s, then a *1* is being sent. If the line is low for greater than 15 $\mu$s, then a "0" is being transmitted. I found it best to poll the line 3 or 4 $\mu$s after the PIC18Cxx stops driving the line.

Figure 17-7 shows a read of two *1*s followed by two *0*s, to give you an idea of what data looks like on the single line.

Data is passed between the controlling device and the DS1820 eight bits at a time, with the least-significant bit first. Before any command, a reset pulse of approximately 500 $\mu$s is output to the line. When the line goes high, the DS1820 responds by pulling the line low for about 60 æs after the line has gone high. After the reset pulse, the DS1820 should not be accessed again for 1 ms.

Figure 17-8 shows the reset pulse and the DS1820 response for a temperature read operation. Notice that the DS1820 response is at a slightly higher voltage level than the PICmicro® MCU initiated reset. This is an example of bus contention, which is caused by the PICmicro® MCU still driving the line high and the DS1820 trying to pull it down low. In the following code and on the CD-ROM, the PICmicro® MCU's I/O pin is not driving the line for the 1 ms of the reset operation to allow the DS1820 to drive the line without interference from the PIC18C452.

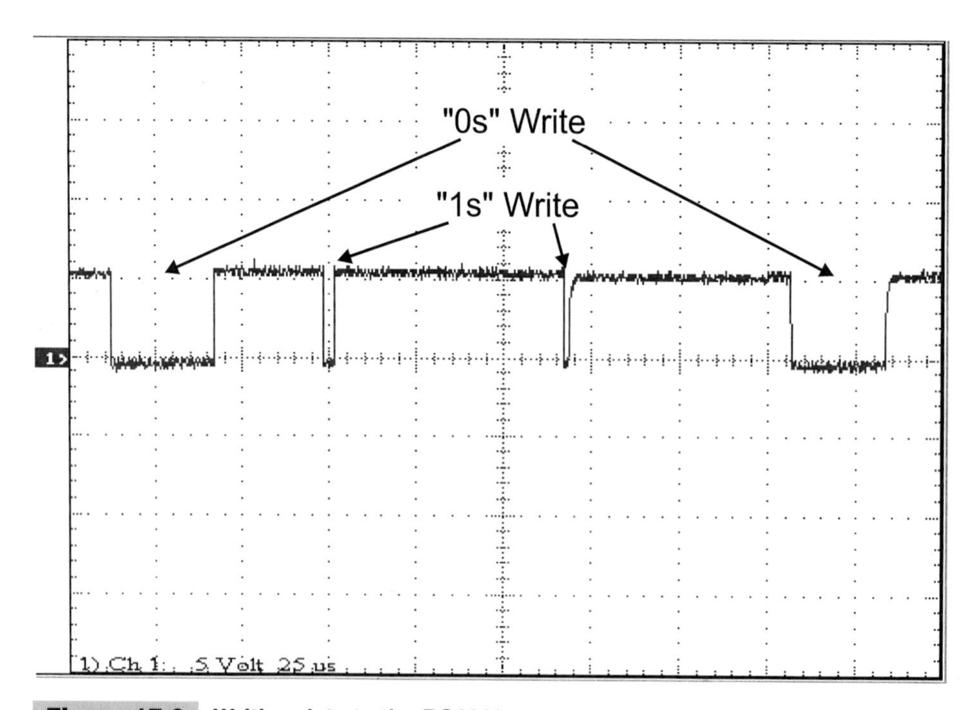


**Figure 17-6** Writing data to the DS1820

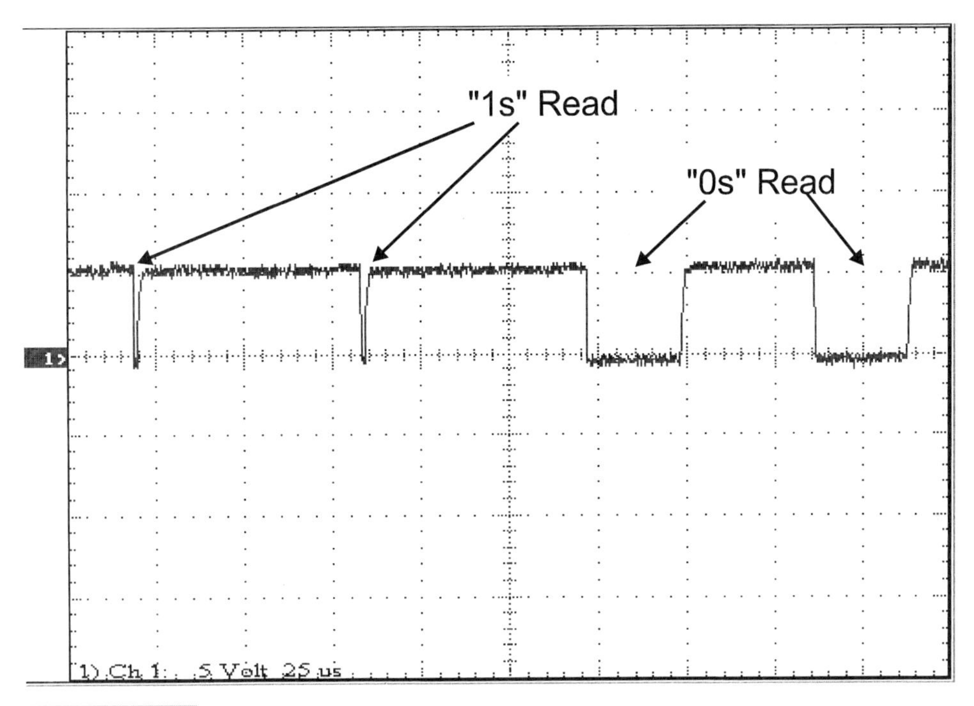


**Figure 17-7** Reading data from the DS1820

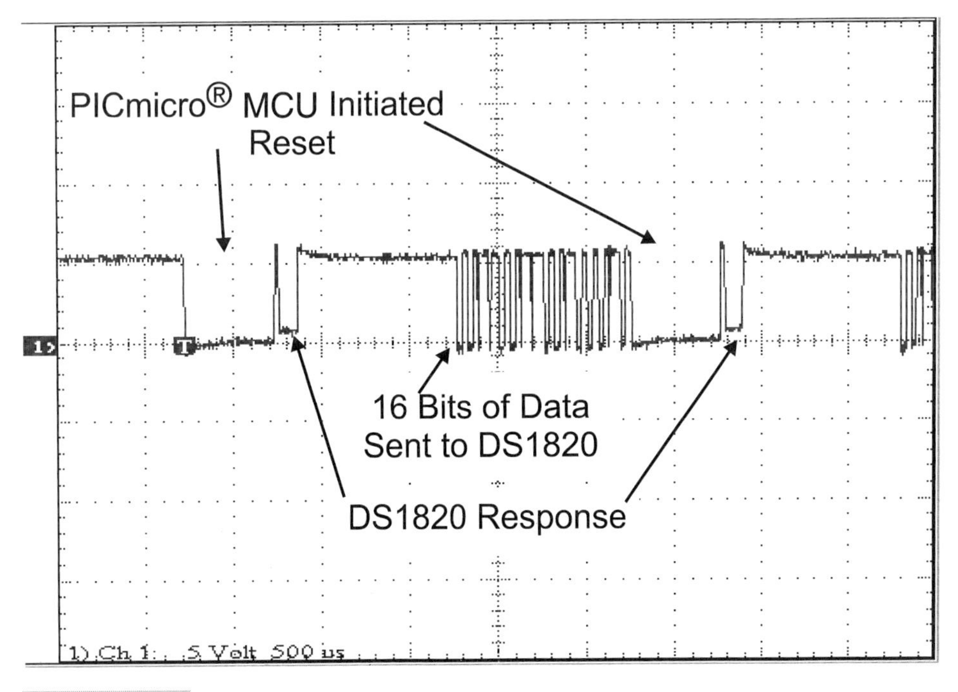


**Figure 17-8** DS1820 Reset with data send following

When I first coded this application, I created three subroutines to operate the DS1820 from the DS87C520 with this application. The first routine is DSReset, which pulls the line down for 480 μs and then waits 1 ms before returning to its caller.

```
DSReset                          ; Reset the DS1820
  bcf       DS1820               ; Hold the DS1820 Low for 500 μs to reset
  movlw     125
  addlw     0x0FF                ; Add -1 until Reset is Equal to Zero
  btfss     STATUS, Z
   bra      $ – (2 * 2)          ; 4 Cycles for Each Loop
  bsf       DS1820
  bcf       DSTRIS
  movlw     0                    ; Wait 1 msec before sending a command
  addlw     0x0FF
  btfss     STATUS, Z
   bra      $ – (2 * 2)
  bsf       DSTRIS
  bsf       DS1820
  return
```

The transmit routine, DSSend, was coded as a simple subroutine with no hooks used for the RTOS.

```
DSSend                           ; Send the Byte in "WREG" to the DS1820
  movwf     FSR0L, 0
  clrf      FSR0H, 0
  movlw     8
DSSendLoop
  bcf       INTCON, GIE          ; Make Sure Operation isn't interrupted
  bcf       DS1820               ; Drop the DS1820's Control Line
  rrcf      FSR0L, f, 0          ; Load Carry with Contents of the
  btfsc     STATUS, C, 0
   bsf      DS1820               ; If "1" Sent, Restore After 4 Cycles
  bsf       FSR0H, 3, 0          ; Loop for 24 Cycles
  decfsz    FSR0H, f, 0
   bra      $ – (2 * 1)
  bra       $ + (2 * 1)          ; Put in a Full 30 Cycle Delay
  bsf       DS1820               ; The Line is High
  bsf       INTCON, GIE          ; Restore the Interrupts
  bsf       FSR0H, 3, 0          ; Loop Another 24 Cycles for Execution Delay
  decfsz    FSR0H, f, 0
   bra      $ – (2 * 1)
  addlw     0x0FF                ; Subtract 1 from the Count
  btfss     STATUS, Z, 0
   bra      DSSendLoop
  return                         ; Finished, Return to Caller
```

Notice in both DSSend and DSRead (the data receive routine from the DS1820) that I first mask interrupts. This was done to ensure that the TMR0 interrupt does not interrupt the timed data transfers when the DS1820 line is low. If the TMR0 interrupt occurs after the data bit is transferred, then there won't be any problems and execution can pick up just where it left off.

Reading data from the DS1820 was accomplished by pulsing the line low and then seeing how long it would stay low.

```
DSRead                           ; Receive the Byte from the DS1820 and put in
  movlw 8                        ; "WREG"
DSReadLoop
  bcf       INTCON, GIE          ; Make Sure Operation isn't interrupted
  bcf       DS1820               ; Drop the DS1820's Control Line
```

```
bsf       DSTRIS            ; Turn Port into a simulated Open Drain Output
nop
bsf       STATUS, C, 0      ; What is Being Returned?
btfss     DS1820
  bcf     STATUS, C, 0      ; If Line is high, a "1"
rrcf      FSR0L, f, 0       ; Shift in the Data bit
bsf       INTCON, GIE       ; Can Interrupt from here
clrf      FSR0H, 0
decfsz    FSR0H, f, 0
  bra     $ - (2 * 1)
bsf       DS1820
bcf       DSTRIS
bsf       DS1820
addlw     0x0FF             ; Loop Around for another Bit
btfss     STATUS, Z, 0
  bra     DSReadLoop
movf      FSR0L, w, 0       ; Return the Byte Read in
return                      ; Finished, Return to Caller
```

To read data from the DS1820, I carried out the following instruction process:

1 Reset the DS1820.
2 Send 0CCh followed by 044h to begin the temperature sense and conversion.
3 Wait 480 $\mu$s for the temperature conversion to complete.
4 Send another reset to the DS1820.
5 Send 0CCh and 0BEh to read the temperature.
6 Wait 100 $\mu$s before reading the first byte in the DS1820.
7 Read the first (SP0) byte of the DS1820.
8 Wait another 100 $\mu$s before reading the second (SP1) byte of the DS1820.

The DS1820 has a unique serial number burned into it, which allows multiple temperature sensors (and other devices using the Dallas Semiconductor one-wire protocol) to be placed on the same pulled-up bus. To avoid first having to read the serial number out of the device and then referencing it each time that a command is being sent to it, a 0CCh byte is sent first. This indicates to the DS1820 that it is to send the temperature without checking for a valid serial number being sent.

I found that for the DS1820 to work properly, I had to use the typical values for the data writes and not extremes in the specification. When I first started working with the DS1820, I used the minimums (15 $\mu$s for a *0*) because I wanted to keep the time the microcontroller was running with interrupts masked to a minimum. When I did this, I found that there were problems with the DS1820 not recognizing correct data. After changing the timings to the typical values in the datasheets, the DS1820 seemed to respond properly. But instead of a single long pulse for each bit returned, I received a valid mix of short pulses and longer pulses.

Using the RTOS and ensuring that most operations are keyed off of the timer, rather than instruction delays, means that this application should be easily transferred to another microcontroller and RTOS combination. The only issue that you would find in porting this RTOS application to another microcontroller is with the DS1820 code and how it is timed in the application.

In *Handbook of Microcontrollers,* I implemented this application on a variety of microcontrollers, including on a Motorola 68HC05 in an RTOS that I had written for it and in *Programming and Customizing the 8051 Microcontroller* I did the same thing with an

8051-derivative microcontroller. This application is quite a bit easier to implement with a real-time operating system because it avoids the need to control access to the LCD, which must be considered if the timer interrupt occurs while the LCD was being updated with the temperature.

Possible solutions to this problem include making a queue for LCD requests or providing timing to avoid the possibility of updating the time on the LCD while the updating the temperature. This work is completely avoided by using an RTOS, which provides a built-in method of serializing operational requests for the application developer.

# 18

# IN CLOSING

When you write a book about microcontrollers, it is easy to think in terms of the advances made in miniaturization, cost effectiveness, and the remarkable reliability of today's devices. There is also a strong allure to write about how efficient a certain type of interface or code algorithm is and how it was derived. The ease of use of development tools is also an attractive topic, as is how easy it is to build your own programmer followed by the simple rules needed to create applications. When I look over this book, I see there is one aspect of the PICmicro® MCU chips I haven't written anything about. That is how much fun they are to work with.

Working with the PICmicro® MCU and creating your own applications can be an enjoyable experience because you are developing everything from the ground up. It always reminds me of my early experiences with computers, where there was always the chance to try different things or redesign a feature of an application to learn something new or see if you could make it better (faster, simpler, or more reliable). Along with this are the challenges of just trying to get the application running, such as to having to squeeze in five extra instructions when no more program memory available. These problems can be at times frustrating, but, for some reason, I only remember the thrill of solving the problem.

I hope to hear from you and find out what kind of applications you have developed on your own. I am interested in hearing from you on what you have learned about the PICmicro® MCU and finding out what I can do better—most of the changes in this edition are a result of the feedback I have gotten from the first edition.

Good luck and I hope you have a lot of fun working with the PICmicro® microcontroller.

Myke
Toronto, Canada
September 2000

APPENDIX A

# GLOSSARY

If this book is your first experience with microcontrollers, I'm sure there are a lot of terms I've thrown at you that are unfamiliar to you. I've tried to give a complete list of all the acronyms, terms, and expressions that might be unfamiliar to you. Acronyms are expanded before they are described.

**Accumulator** Register used as a temporary storage register for an operation's data source and destination. In the PICmicro® MCU, this is known as *w*.

**Active Components** Generally, integrated circuits and transistors. Active components require external power to operate.

**ADC (Analog-to-Digital Converter)** Hardware devoted to converting the value of a DC voltage into a digital representation. See *DAC*.

**Address** The location a register, RAM byte, or instruction word is located at within its specific memory space.

**ALU (Arithmetic Logic Unit)** Circuit within a processor that carries out mathematical operations.

**Amps** A measure of current. One amp is the movement of one Coulomb of electrons in one second.

**Analog** A quantity at a fractional value, rather than a binary, one or zero.

**Anode** Positive terminal of a polarized device. See *Cathode*.

**AND** Logic gate that outputs a *1* when all inputs are a *1*.

**Array** Collection of variables that can be addressed by an arbitrary index.

**ASCII (American Standard Character Interchange Interface)** The bit-to-character representation standard most used in computer systems.

**ASCIIZ** A string of ASCII characters ended by a *null* (0x000) byte.

**Assembler** A computer program that converts assembly-language source to object code. See *Cross Assembler.*

**Assembly Language** A set of word symbols use to represent the instructions of a processor. Along with a primary instruction, parameters are used to specify values, registers, or addresses.

**Asynchronous Serial** Data sent serially to a receiver without clocking information. Instead, data-synching information for the receiver is available inside the data packet or as part of each bit.

**Bare Board** See *Raw Card.*

**BCD (Binary Code Decimal)** Using four bits to represent a decimal number (zero to nine).

**BGA (Ball Grid Array)** A chip solder technology that provides connection from a chip to a bare board via a two-dimensional grid of solder balls (typically 0.050″ from center to center).

**Binary Numbers** Numbers represented as powers of two. Each digit is two, raised to a specific power. For example, 37 Decimal is $32 + 4 + 1 = 2^4 + 2^2 + 2^0 =$ 00010101 binary. Binary can be represented in the forms: *0b0nnnn, B'nnnn',* or *%nnnn,* where *nnnn* is a multidigit binary number composed of *1*s and ***0***s.

**Bipolar Logic** Logic circuits made from bipolar transistors (either discrete devices or integrated onto a chip).

**Bit Banging** Simulating hardware interface functions with software.

**Bit Mask** A bit pattern that is ANDed with a value to turn off specific bits.

**Bounce** Spurious signals in a changing line. Most often found in mechanical switch closings.

**Burning** See *Programming.*

**Bus** An electrical connection between multiple devices, each using the connection for passing data.

**Capacitor** Device used to storing an electrical charge. Often used in microcontroller circuits to filter signals and input power by reducing transient voltages. The different types include ceramic disk, polyester, tantalum, and electrolytic. Tantalum and electrolytic capacitors are polarized.

**Cathode** Negative terminal of a polarized device. See *Anode.*

**Ceramic Resonator** A device used to provide timing signals to a microcontroller. More robust than a crystal, but with poorer frequency accuracy.

**Character** A series of bits used to represent an alphabetic, numeric, control or other symbol or representation. See *ASCII.*

**Chip Package** The method in which a chip is protected from the environment (usually either encased in either ceramic or plastic) with wire interconnects to external circuitry (see *PTH* and *SMT*).

**CISC (Complex Instruction Set Computer)** A type of computer architecture that uses a large number of very complete instructions, rather than a few short instructions. See *RISC.*

**Clock** A repeating signal used to run a processor's instruction sequence.

**Clock Cycle** The operation of a microcontroller's primary oscillator going from a low voltage to a high voltage and back again. This is normally referenced as the speed in which the device runs. Multiple clock cycles are normally used to make up one instruction cycle.

**CMOS Logic** Logic circuits made from N-channel and P-channel *MOSFET (Metal-Oxide Silicon Field-Effect Transistors)* devices (either discrete devices or integrated onto a chip).

**Comparator** A device that compares two voltages and returns a logic *1* or *0,* based on the relative values.

**Compiler** A program that takes a high-level language source file and converts it to either assembly-language code or object code for a microcontroller.

**Concatenate** Joining two pieces of data together to form a single contiguous piece of data.

**Contiguous** When a set amount of data is placed together in memory and can be accessed sequentially using an index pointer, it is said to be *contiguous.* Data is *noncontiguous* if it is placed in different locations that cannot be accessed sequentially.

**Control Store** See *Program Memory.*

**Constant** Numeric value used as a parameter for an operation or instruction. This differs from a variable value that is stored in a RAM or register memory location.

**Coulomb** Charge produced by $1.6 \times 10^{19}$ electrons.

**CPU (Central Processing Unit)** What I refer to as the microcontroller's "processor."

**Cross Assembler** A program written to take assembly-language code for one processor and convert it to object code while working on an unrelated processor and operating system. See *Assembler.*

**Crystal** A device used for precisely timing the operation of a microcontroller.

**Current** The measurement of the number of electrons that pass by a point in a second each second. The units are amperes, which are coulombs per second.

**DAC (Digital-to-Analog Converter)** Hardware designed to convert a digital representation of an analog DC voltage into that analog voltage. See *ADC.*

**DCE (Data Communications Equipment)** The RS-232 standard by which modems are usually wired. See *DTE.*

**Debounce** Removing spurious signals in a "noisy" input.

**Debugger** A program used by an application programmer to find the problems in the application. This program is normally being run on the target system that the application runs on.

**Decimal Numbers** Base-10 numbers used for constants. These values are normally converted into hex or binary numbers for the microcontroller.

**Decoupling Capacitor** Capacitor placed across $V_{cc}$ and ground of a chip to reduce the effects of increased/decreased current draws from the chip.

**Demultiplexer** Circuit for passing data from one source to an output specified from a number of selections. See *Demux, Multiplexer,* and *Mux.*

**Demux** Abbreviation for *Demultiplexer.*

**Digital** A term used to describe a variety of logic families, where values are either high (*1*) or low (*0*). For most logic families, the voltage levels are either approximately 0 volts or approximately 5 volts, with an input switching level somewhere between 1.4 and 2.5 volts.

**Driver** Any device that can force a signal onto a net. See *Receiver.*

**DTE (Data Terminal Equipment)** The RS-232 standard to which the IBM PC's serial port is wired. See *DCE.*

**D-Shell Connectors** A style of connector often used for RS-232 serial connections, as well as other protocols. The connector is *D* shaped to provide a method to polarize the pins and ensure that they are connected the correct way.

**Duty Cycle** In a pulse-wave modulated digital signal, the duty cycle is the fraction of time that the signal is high over the total time of the repeating signal.

**Edge Triggered** Logic that changes, based on the change of a digital logic level. See *Level Sensitive.*

**Editor** Program located on your development system that is used to modify application source code.

**EEPROM (Electrically Erasable Programmable Read-Only Memory, sometimes referred to as *Flash*)** Nonvolatile memory that can be erased and reprogrammed electrically (i.e., it doesn't require the UV light of EPROM).

**Emulator** Electrical circuit connected to the development system that allows the application to be executed under the developer's control. Thus, it is possible to observe how it works and make changes to the application ("what if?").

**EPROM (Erasable Programmable Read-Only Memory)** Nonvolatile memory that can be electrically programmed and erased using ultraviolet light.

**Event Driven Programming** Application code that waits for external events to process before executing.

**External Memory** RAM or ROM memory attached to the PIC17Cxx I/O ports, providing additional Variable memory I/O functions or application space.

**FIFO (First In, First Out)** Memory that will retrieve data in the order in which it was stored.

**Flash** A type of EEPROM. Memory that can be electrically erased in blocks, instead of as individual memory locations. True Flash is very unusual in microcontrollers, many manufacturers describe their devices as having *Flash,* when, in actuality, they use EEPROM.

**Flip-flop** A basic memory cell that can be loaded with a specific logic state and read back. The logic state will be stored as long as power is applied to the cell.

**Floating** The term used to describe a pin that has been left unconnected and is floating relative to ground.

**$F_{osc}$** PICmicro® MCU clock frequency.

**Frequency** The number of repetitions of a signal that can occur in a given period of time (typically one second). See *Period* and *Hertz.*

**FTP (File Transfer Protocol)** A method of transferring files to/from the Internet.

**Functions** A subroutine that returns a parameter to the caller.

**Fuzzy Logic** A branch of computer science in which decisions are made on partially on data, rather than on or off data, such as digital logic uses. These decisions are often made to control physical and electronic systems. See *PID.*

**Ground** Negative voltage to microcontroller/circuit. Also referred to as $V_{ss}$.

**GUI (Graphical User Interface)** Often pronounced "Gooey." A GUI is used, along with a Graphical Operating System (such as Microsoft Windows), to provide a simple, consistent computer interface for users that consists of a screen, keyboard, and mouse.

**Harvard Architecture** Computer processor architecture that interfaces with two memory subsystems, one for instructions (control store) memory and one for variable memory and I/O registers. See *Princeton Architecture.*

**Hertz (Hz)** A unit of measurement of frequency. *One Hertz* means that an incoming signal is oscillating once per second.

**Hex Numbers** A value from 0 to 15 that is represented using four bits or the characters *0* through *9* and *A* through *F.*

**High-Level Language** A set of English (or other human language) statements that have been formatted for use as instructions for a computer. Some popular high-level languages used for microcontrollers include C, BASIC, Pascal, and Forth.

**Horizontal Synch** A pulse used to indicate the start of a scan line in a video monitor or TV set.

**Hysteresis** Characteristic response to input that causes the output response to change based on the input. Typically, it is used in microcontroller projects to debounce input signals.

**Index Register** An eight- or 16-bit register that can have its contents used to point to a location in variable storage, control store, or the microcontroller's register space. See *Stack Pointer.*

**Inductor** Wire wrapped around some kind of form (ceramic, metal or plastic) to provide a magnetic method of storing energy. Inductors are often used in oscillator and filtering circuits.

**Infrared** A wavelength of light (760 nanometer or longer) that is invisible to the human eye. Often used for short-distance communications.

**Interpreter** A program that reads application source code and executes it directly, rather than compiling it.

**Interrupt** An event that causes the microcontroller's processor to stop what it is doing and respond.

**Instruction** A series of bits, which are executed by the microcontroller's processor to perform a basic function.

**Instruction Cycle** The minimum amount of time needed to execute a basic function in a microcontroller. One instruction cycle typically takes several clock cycles. See *Clock Cycles.*

**In-System Programming** The ability to program a microcontroller's program memory while the device is in the final application's circuit without having to remove it.

**I/O Space** An address space totally devoted to providing access to I/O device-control registers.

**I2C (Inter-InterComputer communication)** A synchronous serial network protocol that allows microcontrollers to communicate with peripheral devices and each other. Only common lines are required for the network.

**kHz** An abbreviation for measuring frequency in thousands of cycles per second.

**Label** An identifier used within a program to denote the address location of a control store or register address. See *Variable.*

**Latency** The time or cycles required for hardware to respond to a change in input.

**LCD (Liquid Crystal Display)** A device used to output information from a microcontroller. Typically it is controlled by a Hitachi 44780 controller, although some microcontrollers contain circuitry for interfacing to an LCD directly without an intermediate controller circuit.

**LED (Light-Emitting Diode)** Diode (rectifier) device that will emit light of a specific frequency when current is passed through it. When used with microcontrollers, LEDs are usually wired with the anode (positive pin) connected to $V_{cc}$ and the microcontroller I/O pin sinking current (using a series 200- to 470-Ω resistor) to allow the LED to turn on. In typical LEDs in hemispherical plastic packages, the flat side (which has the shorter lead) is the cathode.

**Level Conversion** The process of converting logic signals from one family to another.

**Level Sensitive** Logic that changes based on the state of a digital logic signal. See *Edge Triggered.*

**LIFO (Last In, First Out)** Type of memory in which the most recently stored data will be the first retrieved.

**Linker** A software product that combines object files into a final program file that can be loaded into a microcontroller.

**List Server** An Internet server used to distribute common-interest mail to a number of individuals. Also known as a *ListServ.*

**Logic Analyzer** A tool that graphically shows the relationship of the waveforms of a number of different pins.

**Logic Gate** A circuit that outputs a logic signal based on input logic conditions.

**Logic Probe** A simple device used to test a line for either being high, low, transitioning, or in a high-impedance state.

**Macro** A programming construct that places a string of characters (and parameters) into a previously specified block of code or information.

**Manchester Encoding** A method to serially send data that does not require a common (or particularly accurate) clock in which the data is encoded by pulse lengths.

**Mask Programmable ROM** A method of programming a memory that occurs at the final assembly of a microcontroller. When the aluminum traces of a chip are laid down, a special photographic mask is made to create wiring, which will result in a specific program being built into a microcontroller's program memory.

**Master** In microcontroller and external device networking, a *master* is a device that initiates and optionally controls the transfer of data. See *Multimaster* and *Slave.*

**Matrix Keyboard** A set of pushbutton switches wired in an X/Y pattern to allow button states to be read easily.

**MCU** Abbreviation for *Microcontroller.*

**Memory** A circuit designed to store instructions or data.

**Memory Array** A collection of flip-flops arranged in a matrix format that allows consistent addressing.

**Memory-Mapped I/O** A method of placing peripheral registers in the same memory space as RAM or variable registers.

**MHz** An abbreviation for measuring frequency in millions of cycles per second.

**Microwire** A synchronous serial communications protocol.

**MIPS (Millions of Instructions Per Second)** This acronym should really be: "Misleading Indicator of Performance" and should not be a consideration when deciding which microcontroller to use for an application.

**Monitor** A program used to control the execution of an application inside of a processor.

**MPU** Abbreviation for *microprocessor.*

**ms** One thousandth of a second (0.001 seconds). See *ns* and *μs.*

**Multimaster** A microcontroller networking philosophy that allows multiple masters on the network bus to initiate data transfers.

**Multiplexer** Device for selecting and outputting a single stream of data from a number of incoming data sources. See *Demultiplexer, Demux,* and *Mux.*

**Mux** Abbreviation for *Multiplexer.*

**Negative Active Logic** A type of logic where the digital signal is said to be *asserted* if it is at a low (*0*) value. See *Positive Active Logic.*

**Nesting** Placing subroutine or interrupt execution within the execution of other subroutines or interrupts.

**Net** A technical term for the connection of device pins in a circuit. Each net consists of all the connections to one device pin in a circuit.

**Net, The** A colloquial term for the *Internet.*

**NiCad (Nickel-Cadmium) Batteries** These batteries are rechargeable, although typically provide 1.2 volts per cell output, compared to 1.5 to 2.0 volts for standard "dry" or alkaline radio batteries.

**NMOS Logic** Digital logic where only N-channel MOSFET transistors are used.

**Noise** High-frequency variances in a signal line that are cause by switch bounce or electrical signals picked up from other sources.

**NOT** Logic gate that inverts the state of the input signal (*1* NOT is *0*).

**ns** One billionth of a second (0.000000001 seconds). See *μs* and *ms.*

**NTSC (National Television Standards Committee)** The standards organization responsible for defining the TV signal format used in North America.

**Object File** After assembly or high-level language compilation, a file is produced with the hex values (op codes), which make up a processor's instructions. An object file can either be loaded directly into a microcontroller or multiple object files can be linked together to form an executable file that is loaded into a microcontroller's control store. See *Linker.*

**Octal Numbers** A method of representing numbers as the digits from *0* to *7.* This method of representing numbers is not widely used, although some high-level languages, such as C, have made it available to programmers.

**One's Complement** The result of XORing a byte with 0x0FF, which will invert each bit of a number. See *Two's Complement.*

**Op Codes** The hex values that make up the processor instructions in an application.

**Open Collector/Drain Output** An output circuit consisting of a single transistor that can pull the net it is connected to ground.

**OR** Basic logic gate, when any input is set to a *1,* a *1* is output.

**ORT (Ongoing Reliability Testing)** A continuing set of tests that are run on a manufactured product during its life to ensure that it will be as reliable as originally specified.

**Oscillator** A circuit used to provide a constant-frequency repeating signal for a microcontroller. This circuit can consist of a crystal, ceramic resonator, or resistor-capacitor network to provide the delay between edge transitions. The term is also used for a device that can be wired to a microcontroller to provide clocking signals without having to provide a crystal, capacitors, and other components to the device.

**Oscilloscope** An instrument that is used to observe the waveform of an electrical signal. The two primary types of oscilloscopes in use today are the *analog oscilloscope,* which "writes" the current signal onto the phosphers of a CRT. The other common type of oscilloscope is the *digital storage oscilloscope,* which saves the analog values of an incoming signal in RAM for replaying on either a built-in CRT or a computer connected to the device.

**OTP (One-Time Programmable)** This term generally refers to a device with EPROM memory encased in a plastic package that does not allow the chip to be exposed to UV light. Notice that EEPROM devices in a plastic package might also be described as *OTP* when they can be electrically erased and reprogrammed.

**Parallel** Passing data between devices with all the data bits being sent at the same time on multiple lines. This is typically much faster than sending data serially.

**Parameter** A user-specified value for a subroutine or macro. A *parameter* can be a numeric value, string, or pointer, depending on the application.

**Passive Components** Generally resistors, capacitors, inductors, and diodes. Components that do not require a separate power source to operate.

**PCA (Printed Circuit Assembly)** A bare board with components (both active and passive) soldered onto it.

**PC Board (Printed Circuit Board, PCB)** See *Raw Card.*

**PDF Files** Files suitable for viewing with Adobe PostScript.

**Period** The length of time that a repeating signal takes to go through one full cycle. The reciprocal of frequency.

**PID (Parallel Integrating Differential)** A classic method to control physical and electronic systems. See *Fuzzy Logic.*

**Ping** The operation of sending a message to a device to see if it is operating properly.

**Pod** The term given for a hardware circuit that interfaces some piece of equipment to another circuit.

**Poll** A programming technique in which a bit (or byte) is repeatedly checked until a specific value is found.

**Pop** The operation of taking data off of a stack memory. See *Push.*

**Positive Active Logic** Logic that becomes active when a signal becomes high (*1*). See *Negative Active Logic.*

**PPM (Parts Per Million)** A measurement value. An easy way to calculate the PPM of a value is to divide the value by the total number of samples or opportunities and multiplying by 1,000,000. 1% is equal to 10,000 PPM and 10% is equal to 100,000 PPM.

**Princeton Architecture** A computer processor architecture that uses one memory subsystem for instructions (control store) memory, variable memory, and I/O registers. See *Harvard Architecture* and *Von Neumann.*

**Program Counter** A counter within a computer processor that keeps track of the current program execution location. This counter can be updated by the counter and have its contents saved/restored on a stack.

**Program Memory** Or *program store.* Memory (usually nonvolatile) devoted to saving the application program for when the microcontroller is powered down. Also known as *control store.*

**Programming** Loading a program into a microcontroller program memory, also referred to as *burning.*

**PROM (Programmable Read-Only Memory)** Originally an array of "fuses" that were "blown" to load in a program. Now *PROM* can refer to EPROM memory in an OTP package.

**PTH (Pin-Through Hole)** Technology in which the pins of a chip are inserted into holes drilled into a printed circuit card before soldering.

**Pull Down** A resistor (typically 100 to 500 Ω) that is wired between a microcontroller pin and ground. See *Pull Up.*

**Pull Up** A resistor (typically 1 to 10 K) that is wired between a microcontroller pin and $V_{cc}$. A switch pulling the signal at the microprocessor pin might be used to provide user input. See *Pull Down.*

**Push** The operation of putting data onto a stack memory. See *Pop.*

**PWB (Printed Wiring Board)** See *Raw Card.*

**PWM (Pulse-Width Modulation)** A digital output technique where a single line is used to output analog information by varying the length of time that a pulse is active on the line.

**RAM (Random-Access Memory)** Memory that you can write to and read from. In microcontrollers, virtually all RAM is *Static RAM (SRAM),* which means that data is stored within it as long as power is supplied to the circuit. *Dynamic RAM (DRAM)* is very rarely used in microcontroller applications. *EEPROM* might be used for nonvolatile RAM storage.

**Raw Card** Fiberglass board with copper traces attached to it, which allows components to be interconnected. Also known as *PCB, PC Board, PWB,* and *Bare Board.*

**RC** A resistor/capacitor network used to provide a specific delay for a built-in oscillator or reset circuit.

**Receiver** A device that senses the logic level in a circuit. A receiver cannot drive a signal.

**Recursion** A programming technique where a subroutine calls itself with modified parameters to carry out a task. This technique is not recommended for microcontrollers that have a limited stack like the PICmicro® MCU.

**Register** A memory address devoted to saving a value (such as RAM) or providing a hardware interface for the processor.

**Relocatable** Code written or compiled in such a way that it can be placed anywhere in the control store memory map after assembling and running without any problems.

**Resistor** A device used to limit current in a circuit.

**Resistor Ladder** A circuit that consists of a number of resistors, which can be selected to provide varying voltage-divider circuits and output differing analog voltages.

**Reset** Placing a microcontroller in a known state before allowing it to execute.

**RISC (Reduced Instruction Set Computer)** A philosophy in which the operation of a computer is sped up by reducing the operations performed by a processor to the absolute minimum for application execution and making all resources accessible by a consistent interface. The advantages of RISC include faster execution time and a smaller instruction set. See *CISC.*

**ROM (Read-Only Memory)** This type of memory is typically used for control store because it cannot be changed by a processor during the execution of an application. *Mask-programmable ROM* is specified by the chip manufacturer to build devices with specific software as part of the device and cannot be programmed "in the field."

**Rotate** A method of moving bits within a single or multiple registers. No matter how many times a *rotate* operation or instruction is carried out, the data in the registers will not be lost. See *Shift.*

**RS-232** An asynchronous serial communications standard. Normal logic levels for a *1* is −12 volts and for a *0,* +12 volts.

**RS-485** A differential pair, communications standard.

**RTOS (Real-Time Operating System)** A program that controls the operation of an application.

**Scan** The act of reading through a row of matrix information for data, rather than interpreting the data as a complete unit.

**Serial** Passing multiple bits using a serial line one at a time. See *Parallel.*

**Servo** A device that converts an electrical signal into mechanical movement. Radio-control modeler's servos are often interfaced to microcontrollers. In these devices, the position is specified by a 1- to 2-ms pulse every 20 ms.

**Shift** A method of moving bits within a single or multiple registers. After a *shift* operation, bits are lost. See *Rotate.*

**Simulator** A program used to debug applications by simulating the operation of the microcontroller.

**Slave** In microcontroller networking, a device that does not initiate communications, but does respond to the instructions of a master.

**SMT (Surface-Mount Technology, SMD)** Technology in which the pins of a chip are soldered to the surface of a printed-circuit card.

**Software** The term used for the application code that is stored in the microcontroller's program memory. Some references might use the term *firmware* for this code.

**Source Code** Human-readable instructions used to develop an application. Source code is converted by a compiler or assembler into instructions that the processor can execute. It is stored in a .HEX file.

**SPI** A synchronous serial communications protocol.

**Splat** Asterisk (*). Easier to say, spell, and funnier than "asterisk."

**SRAM (Static Random-Access Memory)** A memory array that will not lose its contents while power is applied.

**Stack** LIFO memory used to store program counter and other context register information.

**Stack Pointer** An index register available within a processor that is used for storing data and updating itself to allow the next operation to be carried out with the index pointing to a new location.

**State Analyzer** A tool used to store and display state data on several lines. Rather than requiring a separate instrument, this is often an option available in many *Logic Analyzers.*

**State Machine** A programming technique that uses external conditions and state variables to determine how a program is to execute.

**String** Series of ASCII characters saved sequentially in memory. When ended with 0x000 to note the end of the string, known as an *ASCIIZ string.*

**Subroutines** A small application program devoted to carrying out one task or operation. Usually called repeatedly by other subroutines or the application mainline.

**Synchronous Serial** Data transmitted serially, along with a clocking signal, which is used by the receiver to indicate when the incoming data is valid.

**Task** A small, autonomous application, similar in operation to a subroutine, but it can execute autonomously to other application tasks or mainline.

**Timer** A counter incremented by either an internal or external source. It is often used to time events, rather than counting instruction cycles.

**Traces** Electrical signal paths etched in copper in a printed circuit card.

**Transistor** An electronic device by which current flow can be controlled.

**Two's Complement** A method to represent positive and negative numbers in a digital system. To convert a number to a two's complement negative, it is complemented (converted to one's complement) and incremented.

**UART (Universal Asynchronous Receiver/Transmitter)** Peripheral hardware inside a microcontroller used to asynchronously communicate with external devices. See *USART* and *Asynchronous Serial.*

**USART (Universal Synchronous/Asynchronous Receiver/Transmitter)** Peripheral hardware inside of a microcontroller used to synchronously (using a clock signal either produced by the microcontroller or provided externally) or asynchronously communicate with external devices. See *UART* and *Synchronous Serial*

**$\mu$s** One millionth of a second (0.000001 seconds). See *ns* and *ms.*

**UV Light (UltraViolet Light)** Light at shorter wavelengths than the human eye can see. UV light sources are often used with windowed microcontrollers with EPROM control store to erase the contents of the control store.

**Variable** A label used in an application program that represents an address, which contains the actual value to be used by the operation or instruction. Variables are normally located in RAM and can be read from or written to by application software.

**$V_{cc}$** Positive power voltage applied to a microcontroller/circuit. Generally 2.0 to 6.0 volts, depending on the application. Also known as $V_{dd}$.

**$V_{dd}$** See $V_{cc}$.

**Vertical Synch** A signal used by a monitor or TV set to determine when to start displaying a new screen (field) of data.

**Vias** Holes in a printed circuit card.

**Volatile** RAM is considered to be *volatile* because when power is removed, the contents are lost. EPROM, EEPROM, and PROM are considered to be *nonvolatile* because the values stored in the memory are saved—even if power is removed.

**Voltage** The amount of electrical force placed on a charge.

**Voltage Regulators** A circuit used to convert a supply voltage into a level useful for a circuit or microcontroller.

**Volts** A unit of voltage.

**Von Neumann, John** The chief scientist responsible for the Princeton architecture.

**$V_{ss}$** See *Ground.*

**Wait States** Extra time added to an external memory read or write.

**Watchdog Timer** A timer used to reset a microcontroller upon overflow. The purpose of the watchdog timer is to return the microcontroller to a known state if the program begins to run errantly (or amok).

**Wattage** The measure of power consumed. If a device requires one amp of current with a one-volt drop, one watt of power is being consumed.

**Word** The basic data size used by a processor. In all the PICmicro® MCU families, the word size is eight bits.

**XOR** A logic gate that outputs a *1* when the inputs are at different logic levels.

**ZIF (Zero Insertion Force)** ZIF sockets allow the plugging/unplugging of devices without placing stress upon the device's pins.

**.ZIP Files** Files combined together and compressed into a single file using the PKZIP program by PKWARE, Inc.

APPENDIX B

# USEFUL TABLES AND DATA

## CONTENTS AT A GLANCE

# Physical Constants

**TABLE B-1 Physical Constants**

| SYMBOL | VALUE | DESCRIPTION |
|---|---|---|
| AU | 149.59787 x (10^6) km<br>92,955,628 miles | Astronomical Unit (Distance from the Sun to the Earth) |
| c | 2.99792458 x (10^8) m/s<br>186,282 miles/s | Speed of Light in a Vacuum |
| e | 2.7182818285 | |
| Epsilon-o | 8.854187817 x (10^−12) F/m | Permittivity of Free Space |
| Ev | 1.60217733 x (10^−19) J | Electron Volt Value |
| g | 32.174 ft/sec^2<br>9.807 m/sec^2 | Acceleration due to gravity |
| h | 6.626 x (10^−34) Js | Planck Constant |
| k | 1.380658 x (10^−23) J/K | Boltzmann Entropy Constant |
| me | 9.1093897 x (10^−31) kg | Electron Rest Mass |
| mn | 1.67493 x* 10^−27) kg | Neutron Rest Mass |
| mp | 1.67263 x (10^−27) kg | Proton Rest Mass |
| pc | 2.06246 x (10^5) AU | Parsec |
| pi | 3.1415926535898 | Ratio of circumference to Diameter of a circle |
| R | 8.314510 J/(K * mole) | Gas Constant |
| sigma | 5.67051 x (10^−8) W/(m^2 * K^4) | Stefan-Boltzmann Constant |
| $\mu$ | 1.66054 x (10^−27) grams | Atomic Mass Unit |
| mu-o | 1.25664 x (10^−7) N/A^2 | Permeability of Vacuum |
| Mach 1 | 331.45 m/s<br>1087.4 ft/s | Speed of Sound at Sea Level, in Dry Air at 20C |
| None | 1480 m/s<br>4856 ft/s | Speed of Sound in Water at 20C |

# Audio Notes

Notes Around Middle "C". Note that Octave above is twice the note frequency and Octave below is one half note frequency.

**TABLE B-2 Audio Notes**

| NOTE | FREQUENCY |
|---|---|
| G | 392 Hz |
| G# | 415.3 Hz |
| A | 440 Hz |
| A# | 466.2 Hz |
| B | 493.9 Hz |
| C | 523.3 Hz |
| C# | 554.4 Hz |
| D | 587.3 Hz |
| D# | 622.3 Hz |
| E | 659.3 Hz |
| F | 698.5 Hz |
| F# | 740.0 Hz |
| G | 784.0 Hz |
| G# | 830.6 Hz |
| A | 880.0 Hz |
| A# | 932.3 Hz |
| B | 987.8 Hz |

# "Touch-Tone" Telephone Frequencies

**TABLE B-3 "Touch-Tone" Telephone Frequencies**

| FREQUENCY | 1209 Hz | 1336 Hz | 1477 Hz |
|---|---|---|---|
| 697 Hz | 1 | 2 | 3 |
| 770 Hz | 4 | 5 | 6 |
| 852 Hz | 7 | 8 | 9 |
| 941 Hz | * | 0 | # |

# Electrical Engineering Formulas

Where:

V = Voltage
I = Current
R = Resistance
C = Capacitance
L = Inductance

# DC Electronics Formulas

Ohm's Law:

$$V = IR$$

Power:

$$P = VI$$

Series Resistance:

$$Rt = R1 + R2 \ldots$$

Parallel Resistance:

$$Rt = 1/((1/R1) = (1/R2) \ldots)$$

Two Resistors in Parallel:

$$Rt = (R1 * R2) / (R1 + R2)$$

Series Capacitance:

$$Ct = 1 / ((1/C1) + (1/C2) \ldots)$$

Parallel Capacitance:

$$Ct = C1 + C2 \ldots$$

Wheatstone Bridge:

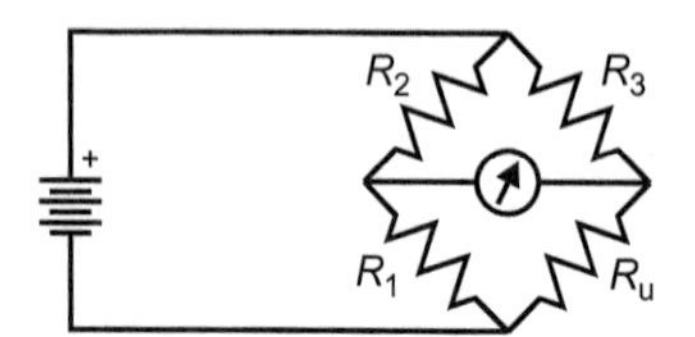


$$R_u = R_1 \times R_3/R_2$$

When No Current Flow is in the Meter

**Figure B-1** **Wheatstone bridge operation**

# AC Electronics Formulas

Resonance:

frequency = $1 / (2 * pi \sqrt{(L * C)})$

RC Time Constant:

Tau = R * C

RL Time Constant:

Tau = L / R

RC Charging:

V(t) = Vf * (1 − e^ (−t/Tau))

i(t) = if * (1 − e^ (−t/Tau))

RC Discharging:

V(t) = Vi * e^ (−t/Tau)

i(t) = ii * e^ (−t/Tau)

Transformer Current/Voltage:
Turns Ratio = Number of Turns on Primary ("p") Side / Number of Turns on Secondary ("s") Side

Turns Ratio = Vs / Vp = Ip / Is

Transmission Line Characteristic Impedance:

Zo = SQRT(L / C)

# Mathematical Formulas

Frequency = Speed / Wavelength

For Electromagnetic Waves:

Frequency = c / Wavelength

Perfect Gas Law:

PV = nRT

## BOOLEAN ARITHMETIC

Identify Functions:

A AND 1 = A

A OR 0 = A

Output Set/Reset:

A AND 0 = 0

A OR 1 = 1

Identity Law:

A = A

Double Negation Law:

NOT(NOT(A)) = A

Complementary Law:

A AND NOT(A) = 0

A OR NOT(A) = 1

Idempotent Law:

A AND A = A

A OR A = A

Commutative Law:

A AND B = B AND A

A OR B = B OR A

Associative Law:

(A AND B) AND C = A AND (B AND C) = A AND B AND C

(A OR B) OR C = A OR (B OR C) = A OR B OR C

Distributive Law:

A AND (B OR C) = (A AND B) OR (A AND C)

A OR (B AND C) = (A OR B) AND (A OR C)

De Morgan's Theorem:

NOT(A OR B) = NOT(A) AND NOT(B)

NOT(A AND B) = NOT(A) OR NOT (B)

Note:

AND is often Represented as Multiplication, nothing between the Terms, or the "." or "*" characters between them.

OR is often Represented as Addition with "+" between Terms.

NOT is indicated with a "-" or "!" character before the Term. "~" is usually used to indicate a multi-bit, bitwise inversion.

# Mathematical Conversions:

**TABLE B-4 Mathematical Conversions**

1 Inch = 2.54 Centimeters

1 Mile = 1.609 Kilometers

1 Ounce = 29.57 Grams

1 US Gallon = 3.78 Liters

1 Atmosphere = 29.9213 Inches of Mercury
= 14.6960 Pounds per Square Inch
= 101.325 kiloPascals

10,000,000,000 Angstroms = 1 Meter

1,000,000 Microns = 1 Meter

Tera = 1,000 Giga

Giga = 1,000 Mega

Mega = 1,000 Kilo

Kilo = 1,000 Units

Unit = 100 Centi

Unit = 1,000 Milli

1 Hour = 3,600 Seconds

1 Year = 8,760 Hours

# ASCII

The ASCII Definition uses the least significant eight bits of a byte to specify a character as shown in Table B-5.

**TABLE B-5 ASCII Character Table**

| 3–0 | I 6–4 –> | 000 | 001 | I | 010 | 011 | 100 | 101 | 110 | 111 |
|---|---|---|---|---|---|---|---|---|---|---|
| V | I | Control | | I | Characters | | | | | |
| 0000 | I | NUL | DLE | I | Space | 0 | @ | P | ` | p |
| 0001 | I | SOH | DC1 | I | ! | 1 | A | Q | a | q |
| 0010 | I | STX | DC2 | I | " | 2 | B | R | b | r` |
| 0011 | I | ETX | DC3 | I | # | 3 | C | S | c | s |
| 0100 | I | EOT | DC4 | I | $ | 4 | D | T | d | t |
| 0101 | I | ENQ | NAK | I | % | 5 | E | U | e | u |
| 0110 | I | ACK | SYN | I | & | 6 | F | V | f | v |
| 0111 | I | BEL | ETB | I | ` | 7 | G | W | g | w |
| 1000 | I | BS | CAN | I | ( | 8 | H | X | h | x |
| 1001 | I | HT | EM | I | ) | 9 | I | Y | I | y |
| 1010 | I | LF | SUB | I | * | : | J | Z | j | z |
| 1011 | I | VT | ESC | I | + | ; | K | [ | k | { |
| 1100 | I | FF | FS | I | ' | < | L | \ | I | I |
| 1101 | I | CR | GS | I | – | = | M | ] | m | } |
| 1110 | I | SO | RS | I | . | > | N | ^ | n | ~ |
| 1111 | I | SI | US | I | / | ? | O | _ | o | DEL |

# ASCII Control Characters

The ASCII Control Characters were specified as a means of allowing one computer to communicate and control another. These characters are actually commands and if the BIOS or MS-DOS display or communications APIs are used with them they will revert back to their original purpose. As I note below when I present the IBM Extended ASCII characters, writing these values (all less than 0x020) to the display will display graphics characters.

Normally, only “Carriage Return”/“Line Feed” are used to indicate the start of a line. “Null” is used to indicate the end of an ASCIIZ string. “Backspace” will move the cursor back one column to the start of the line. The “Bell” character, when sent to MS–DOS will cause the PC’s speaker to “beep”. “Horizontal Tab” is used to move the cursor to the start of the next column that is evenly distributed by eight. “Form Feed” is used to clear the screen.

**TABLE B-6 ASCII Control Character Definitions**

| HEX | MNEMONIC | DEFINITION |
|---|---|---|
| 00 | NUL | "Null"—Used to indicate the end of a string |
| 01 | SOH | Message "Start of Header" |
| 02 | STX | Message "Start of Text" |
| 03 | ETX | Message "End of Text" |
| 04 | EOT | "End of Transmission" |
| 05 | ENQ | "Enquire" for Identification or Information |
| 06 | ACK | "Acknowledge" the previous transmission |
| 07 | BEL | Ring the "BELL" |
| 08 | BS | "Backspace"—Move the Cursor on column to the left |
| 09 | HT | "Horizontal Tab"—Move the Cursor to the Right to the next "Tab Stop" (Normally a column evenly divisible by eight) |
| 0A | LF | "Line Feed"—Move the Cursor down one line |
| 0B | VT | "Vertical Tab"—Move the Cursor down to the next "Tab Line" |
| 0C | FF | "Form Feed" up to the start of the new page. For CRT displays, this is often used to clear the screen |
| 0D | CR | "Carriage Return"—Move the Cursor to the leftmost column |
| 0E | SO | Next Group of Characters do not follow ASCII Control conventions so they are "Shifted Out" |
| 0F | SI | The following Characters do follow the ASCII Control conventions and are "Shifted In" |
| 10 | DLE | "Data Link Escape"—ASCII Control Character start of an Escape sequence. In most modern applications "Escape" (0x01B) is used for this function |
| 11 | DC1 | Not defined—Normally application specific |
| 12 | DC2 | Not defined—Normally application specific |
| 13 | DC3 | Not defined—Normally application specific |
| 14 | DC4 | Not defined—Normally application specific |
| 15 | NAK | "Negative Acknowledge"—the previous transmission was not properly received |
| 16 | SYN | "Synchronous Idle"—If the serial transmission uses a synchronous protocol, this character is sent to ensure the transmitter and receiver remain synched |
| 17 | ETB | "End of Transmission Block" |
| 18 | CAN | "Cancel" and disregard the previous transmission |

*(continued)*

**TABLE B-6 ASCII Control Character Definitions (*Continued*)**

| HEX | MNEMONIC | DEFINITION |
|---|---|---|
| 19 | EM | "End of Medium"—Indicates end of a file. For MS—DOS files, 0x01A is often used instead |
| 1A | SUB | "Substitute" the following character with an incorrect one |
| 1B | ESC | "Escape"—Used to temporarily halt execution or put an application into a mode to receive information |
| 1C | FS | Marker for "File Separation" of data being sent |
| 1D | GS | Marker for "Group Separation" of data being sent |
| 1E | RS | Marker for "Record Separation" of data being sent |
| 1F | US | Marker for "Unit Separation" of data being sent |

# IBM Extended ASCII Characters

The additional 128 characters shown in Figure B-3 can do a lot to enhance a character mode application without having to resort to using graphics in MS-DOS. These enhancements include special characters for languages other than English, engineering symbols and simple graphics characters. These simple graphics characters allow lines and boxes in applications can be created.

| Hex | 0x | 1x | 2x | 3x | 4x | 5x | 6x | 7x |
|---|---|---|---|---|---|---|---|---|
| x0 | 0 | ► 16 | SP 32 | 0 48 | @ 64 | P 80 | ` 96 | p 112 |
| x1 | ☺ 1 | ◄ 17 | ! 33 | 1 49 | A 65 | Q 81 | a 97 | q 113 |
| x2 | ☻ 2 | ↕ 18 | " 34 | 2 50 | B 66 | R 82 | b 98 | r 114 |
| x3 | ♥ 3 | ‼ 19 | # 35 | 3 51 | C 67 | S 83 | c 99 | s 115 |
| x4 | ♦ 4 | ¶ 20 | $ 36 | 4 52 | D 68 | T 84 | d 100 | t 116 |
| x5 | ♣ 5 | § 21 | % 37 | 5 53 | E 69 | U 85 | e 101 | u 117 |
| x6 | ♠ 6 | ▬ 22 | & 38 | 6 54 | F 70 | V 86 | f 102 | v 118 |
| x7 | • 7 | ↨ 23 | ' 39 | 7 55 | G 71 | W 87 | g 103 | w 119 |
| x8 | ◘ 8 | ↑ 24 | ( 40 | 8 56 | H 72 | X 88 | h 104 | x 120 |
| x9 | ○ 9 | ↓ 25 | ) 41 | 9 57 | I 73 | Y 89 | i 105 | y 121 |
| xA | ◙ 10 | → 26 | * 42 | : 58 | J 74 | Z 90 | j 106 | z 122 |
| xB | ♂ 11 | ← 27 | + 43 | ; 59 | K 75 | [ 91 | k 107 | { 123 |
| xC | ♀ 12 | ∟ 28 | , 44 | < 60 | L 76 | \ 92 | l 108 | ¦ 124 |
| xD | ♪ 13 | ↔ 29 | - 45 | = 61 | M 77 | ] 93 | m 109 | } 125 |
| xE | ♫ 14 | ▲ 30 | . 46 | > 62 | N 78 | ^ 94 | n 110 | ∼ 126 |
| xF | ☼ 15 | ▼ 31 | / 47 | ? 63 | O 79 | _ 95 | o 111 | △ 127 |

**Figure B-2** IBM PC extended ASCII Set 0-0x07F

| Hex | 8x | 9x | Ax | Bx | Cx | Dx | Ex | Fx |
|---|---|---|---|---|---|---|---|---|
| x0 | Ç 128 | É 144 | á 160 | ░ 176 | └ 192 | ╨ 208 | α 224 | ≡ 240 |
| x1 | ü 129 | æ 145 | í 161 | ▒ 177 | ┴ 193 | ╤ 209 | β 225 | ± 241 |
| x2 | é 130 | Æ 146 | ó 162 | ▓ 178 | ┬ 194 | ╥ 210 | Γ 226 | ≥ 242 |
| x3 | â 131 | ô 147 | ú 163 | │ 179 | ├ 195 | ╙ 211 | π 227 | ≤ 243 |
| x4 | ä 132 | ö 148 | ñ 164 | ┤ 180 | ─ 196 | ╘ 212 | Σ 228 | ⌠ 244 |
| x5 | à 133 | ò 149 | Ñ 165 | ╡ 181 | ┼ 197 | ╒ 213 | σ 229 | ⌡ 245 |
| x6 | å 134 | û 150 | ª 166 | ╢ 182 | ╞ 198 | ╓ 214 | µ 230 | ÷ 246 |
| x7 | ç 135 | ù 151 | º 167 | ╖ 183 | ╟ 199 | ╫ 215 | τ 231 | ≈ 247 |
| x8 | ê 136 | ÿ 152 | ¿ 168 | ╕ 184 | ╚ 200 | ╪ 216 | Φ 232 | ° 248 |
| x9 | ë 137 | Ö 153 | ⌐ 169 | ╣ 185 | ╔ 201 | ┘ 217 | Θ 233 | • 249 |
| xA | è 138 | Ü 154 | ¬ 170 | ║ 186 | ╩ 202 | ┌ 218 | Ω 234 | · 250 |
| xB | ï 139 | ¢ 155 | ½ 171 | ╗ 187 | ╦ 203 | █ 219 | δ 235 | √ 251 |
| xC | î 140 | £ 156 | ¼ 172 | ╝ 188 | ╠ 204 | ▄ 220 | ∞ 236 | ⁿ 252 |
| xD | ì 141 | ¥ 157 | ¡ 173 | ╜ 189 | ═ 205 | ▌ 221 | φ 237 | ² 253 |
| xE | Ä 142 | ₧ 158 | « 174 | ╛ 190 | ╬ 206 | ▐ 222 | ∈ 238 | ■ 254 |
| xF | Å 143 | ƒ 159 | » 175 | ┐ 191 | ╧ 207 | ▀ 223 | ∩ 239 | 255 |

**Figure B-3** IBM PC extended ASCII set 0x080-0x0FF

APPENDIX C

# PICmicro® MCU APPLICATION DEBUGGING CHECKLIST

In the first edition of this book, I had an appendix entitled "Hints" and in it I gave ten suggestions for designing and building PICmicro® MCU applications. Because of the changes to the format of the second edition (with more specific application debug information), I felt it would be useful to change this appendix into a checklist for what to do if something isn't working.

The table below is a list of possible causes for problems with your PICmicro® MCU application along with a number of things to check and some suggestions of what the problems are. The "Problems" are the top dozen problems that I have with my own PICmicro® MCU applications as well as questions I get from other people.

These suggestions should be taken as a list of things to try out and characterize your problem before you try and implement a fix. As I have said throughout the book, you should be working toward doing "Failure Analysis" instead of "Debug". The list of suggestions here should point you in that direction with each point under "Check" being used to characterize the problem so you can make your hypothesis of what is the problem and what should be done to fix it.

When you look at the potential reasons for problems, you should notice that I never suggest that a PICmicro® MCU or another silicon part as the ultimate cause of the problem. Instead, the problems are almost all identified as a specification, interfacing or software problem with the fix being relatively easy. PICmicro® MCUs and other chips are very reliable and you can conceivably go years without seeing a bad one or have one that fails through normal use and abuse. The last thing you should do is simply replace a part because it's something easy to try that could fix the problem.

**TABLE C-1 Debugging Checklist**

| PROBLEM | POTENTIAL CAUSES | CHECK |
|---|---|---|
| PICmicro® MCU application does not start | 1 No/Bad power | 1a Be sure that $V_{dd}$ is between 4.5 volts and 5.5 volts relative to $V_{ss}$. |
| | | 1b Be sure $V_{dd}$ ripple is less than 100 mV. |
| | 2 No/Bad reset | 2a Be sure _MCLR is pulled up from 4.5 volts to 5.5 volts. |
| | | 2b Be sure disabled _MCLR pin is not pulled below ground. |
| | 3 Missing/bad decoupling capacitor | 3a Check for 0.01-$\mu$F to 0.1-$\mu$F capacitor close to PICmicro® MCU's $V_{dd}$ pin. |
| | 4 Part orientation | 4a Check that PICmicro® MCU part is installed correctly. |
| | | 4b Be sure the PICmicro® MCU is not getting very hot. |
| | 5 Oscillator not running | 5a Check both the OSC1 and OSC2 pins with an oscilloscope or logic probe. |
| | | 5b If internal oscillator, check configuration fuses for correct setting. |
| | | 5c Check for present and correct capacitors. |
| | 6 Device programming incorrect | 6a Check/verify device programming. |
| | | 6b Look for I/O pins being set high or low. |
| | 7 Watchdog timer enabled | 7a Check I/O pins for changing between input output states. |
| | | 7b Check actual configuration fuse value. |
| | 8 Uninitialized variable/value incorrect | 8a Check variable initialization at application start. |
| | | 8b After resetting the simulated PICmicro® MCU, load file registers with a random value (such as 0x05A). |

**TABLE C-1 Debugging Checklist *(Continued)***

| PROBLEM | POTENTIAL CAUSES | CHECK |
|---|---|---|
| | 9 Interrupt handler NOT allowing execution exit from handler | 9a Simulate interrupt hander and be sure that execution can return to mainline before next interrupt request is acknowledged.<br>9b Be sure that correct Interrupt flag (IF) is reset in handler. |
| | 10 Variable address overlayed onto a hardware I/O register | 10a Be Sure that the variable *CBLOCK* statement is in the file register area of the PICmicro® MCU and not in the hardware I/O Area. |
| | 11 Outputs switching too fast to see | 11a Probe the outputs using a logic probe or oscilloscope. |
| PICmicro® MCU device seems to rest itself unexpectedly | 1 Watchdog timer enabled | 1a Check configuration fuse values.<br>1b Check for I/O pins changing state with reset. |
| | 2 High internal current and inadequate decoupling | 2a Check for correlation to to reset with changes in load drawn by PICmicro® MCU.<br>2b Check for power-supply "sags" when the load is drawn. |
| | 3 Check for a "noisy" power supply | 3a Check for greater than 100-mV ripple from power supply. |
| | 4 Execution jumps past application end | 4a Check code for subroutine without *return* instruction or table that is accessed past its end. |
| | 5 Uninitialized variable/value incorrect | 5a Check variable initialization for missed variable.<br>5b Set variables to random values (such as 0x05A) before starting simulation to find problem. |
| Peripheral hardware not active | 1 Pin programming not correct | 1a Check register access prerequisites.<br>1b Check TRIS registers For values that prevent peripheral operation. |

*(continued)*

**TABLE C-1 Debugging Checklist *(Continued)***

| PROBLEM | POTENTIAL CAUSES | CHECK |
|---|---|---|
| | 2 Incorrect part number PICmicro® MCU | 2a Check to see if the part being actually used has hardware.<br>2b Check to see that part being used has hardware registers that match source. |
| | 3 Hardware switching too fast to observe | 3a Check the hardware using a logic probe or an oscilloscope. |
| No Output mode for I/O pin | 1 Incorrect TRIS specification | 1a Check values saved in TRIS registers.<br>1b Check for inadvertent execution.<br>1c Float pin (disconnect from circuit) to see if pin is actually in Output mode with a logic probe. |
| | 2 If peripheral hardware built into pin, check for activation | 2a This problem might not be apparent in MPLAB simulation because peripherals are often not modeled.<br>2b If Pin Shared with T0CKI in 12C5xx or 16C505, check for correct state of OPTION register. |
| | 3 If dual–use pin, check for output capability | 3a If pin shared with _MCLR in 12C5xx or 16C505, then no output capabilities are built in.<br>3b If pin shared with T0CKI in 12C5xx or 16C505, check for correct state of OPTION register. |
| Output pin not changing state | 1 Pin not in Output mode | 1a Float pin (disconnect from circuit) to see if pin is actually in Output mode with a logic probe.<br>1b Check causes for no Output mode for I/O pin. |
| | 2 Pin being held by high current source/sink | 2a Float pin and see if state changes are possible with pin disconnected.<br>2b Look for shorts to $V_{cc}$/Gnd.<br>2c Look for missing/incorrect resistors or components. |

*(continued)*

**TABLE C-1 Debugging Checklist *(Continued)***

| PROBLEM | POTENTIAL CAUSES | CHECK |
|---|---|---|
| | 3 Output changing state too quickly to be observed. | 3a Check output with a logic probe or an oscilloscope. |
| Pin changes state unexpectedly | 1 Look for *bcf*, *bsf* or *movf*/*movwf* instruction combinations that could reset the pin | 1a Check value written to I/O port.<br>1b Check computed values that are used to modify pin values.<br>1c Look for saved port values that are incorrect or inappropriate. |
| | 2 Look for hardware that "backdrives" the pin | 2a Float pin (disconnect from circuit) to see if state is incorrect.<br>2b Check output-enable pins of tri-state drivers on the pin's net. |
| | 3 Variable address overlayed onto a hardware I/O register | 3a Be sure that the variable *CBLOCK* statement is in the file register area of the PICmicro® MCU and not in the hardware I/O area. |
| Output timing not as expected | 1 Delay calculations incorrect | 1a Check to see if the calculations match the actual output.<br>1b Use the assembler calculator to calculate delays and match to developer values. |
| | 2 Interrupt handler active during timed output | 2a Check for enabled interrupts.<br>2b Put *bcf INTCON, GIE* before timed code and *bsf INTCON, GIE* after |
| | 3 Check instruction timings | 3a Note that *goto, call,* return, and PCL modifications require two instruction cycles. |
| Register values incorrect/change unexpectedly | 1 Check for interrupt handler active | 1a Look for instances in the interrupt handler when register is changed.<br>1b Mask interrupt handler during crucial periods of register operation. |

*(continued)*

**TABLE C-1 Debugging Checklist *(Continued)***

| PROBLEM | POTENTIAL CAUSES | CHECK |
|---|---|---|
| | | 1c Use another register in the interrupt handler and update mainline's version, as appropriate. |
| | 2 Be sure that variables are not located in hardware register space | 2a Check actual register address from listing file to hardware register addresses.<br>2b Be sure that variables are in memory space "above" the hardware registers for all PICmicro® MCU family devices on which the Application runs. |
| | 3 Variable registers in "shadowed" memory space | 3a Check file register addresses with "Shadowed" registers.<br>3b Mark unused shadow Registers as BADRAM. |
| LED not lighting | 1 LED polarity incorrect | 1a Short PICmicro® MCU pin to ground to ensure LED can light. |
| | 2 Check PICmicro® MCU pin for not changing to output | 2a Check PICmicro® MCU pin as specified previously. |
| | 3 PWM active with setting that turns off LED | 3a Check PWM output with an oscilloscope or logic probe.<br>3b Check for PWM code active. |
| | 4 PICmicro® MCU not working | 4a Check the PICmicro® MCU, as specified. |
| | 5 Output changing too fast to observe | 5a Check the output using a logic probe or an oscilloscope. |
| Button: no response | 1 Pin pullup/pull down incorrect | 1a Check the wiring of the button to the PICmicro® MCU, $V_{cc}$ and Gnd.<br>1b Check the operation of the PICmicro® MCU's internal pull ups. |

*(continued)*

**TABLE C-1 Debugging Checklist *(Continued)***

| PROBLEM | POTENTIAL CAUSES | CHECK |
|---|---|---|
| | 2 Pin in Output mode | 2a Check to be sure that the PICmicro® MCU I/O pin is in Input mode.<br>2b Look for inadvertent changes to the TRIS register. |
| | 3 Output changing to quickly to be observed | 3a Check pin output using a logic probe or an oscilloscope. |
| Button: strange response | 1 Poor debounce | 1a Check for multiple button presses recognized by software.<br>1b Check voltage levels on hardware to ensure button press is within 0.2 volts from $V_{cc}$ or $G_{nd}$. |
| | 2 Interrupt handler response incorrect | 2a Check the interrupt handler's operation with the input conditions. |
| LCD: no output | 1 Check wiring | 1a Check ground is on pin 1.<br>1b Check data pins.<br>1c Be Sure that R/W line is held low during writes. |
| | 2 Check contrast | 2a Contrast different for different LCDs. |
| | 3 Check timing | 3a Be sure that LCD E strobes are a minimum of 450 ns in width.<br>3b Be sure signals do not change during "E" strobes. |

# APPENDIX D

# RESOURCES

## CONTENTS AT A GLANCE

**Web Sites of Interest**



# Microchip

Microchip's corporate headquarters is:

**Microchip Technology Inc.**
2355 W. Chandler Blvd.
Chandler, AZ 85224
Phone: (480) 786-7200
Fax: (480) 917-4150

Their website ("Planet Microchip") is:

**http://www.microchip.com**
and contains a complete set of data sheets in .pdf format for download as well as the latest versions of MPLAB.

Look for the "World Wide Sales and Distribution" page on the website for technical information for your region.

**Reference Information and Tools**
As I indicated above, the microchip website, known as "Planet Microchip" has a wealth of information on PICmicro® MCU products, application notes and articles that can be downloaded free of charge.

Also on the website is

**http://buy.microchip.com**
that is Microchip's on line ordering system. From this website, you will be able to order any of the tools presented in this book. For buying actual PICmicro® MCU parts, you will

have to contact one of the distributors listed below.

Microchip puts on a series of seminars throughout the world every year. For a nominal cost, you will be treated to a tutorial about the PICmicro® MCU, what's new, the latest datasheets in .pdf format on CD-ROM as well as the opportunity to contact other individuals in your area that are working with the PICmicro® MCU. At these meetings, Microchip will make various tools available at reduced ("sale") costs. I highly recommend going to these seminars, as they are excellent ways to meet your "FAEs" ("Field Application Engineers") and find out what is new in the PICmicro® MCU world.

# Contacting the Author

I can be contacted by sending an email to **"myke@passport.ca"** or visit my web site at:

**http://www.myke.com**

For technical questions or suggestions, please contact me through the PICList so that the information can be made available to a much larger audience.

# Buying Project Kits Presented in this Book

In the first edition of this book, I didn't do a very good job of explaining what would be available and when the projects would be available as kits. Other than the programmers and emulator presented in this book, there are no plans for kits to be made available of the different projects that I go through in this book.

I have included raw card information on all the projects that I have presented PCBs for. To minimize my costs (as well as yours) in having the boards built at quick-turn assembly houses, I have incorporated the PCBs on "panels" with multiple images. The "Gerber" and "Drill" files are included on the CD-ROM and are presented in the "PCB Boards" appendix.

As I write this, plans are in place for the YAP-II and EMU-II to be available from Wirz Electronics (see below) as assembled and tested products. Unassembled kits of these tools will not be made available. The reason is mostly for self preservation; these circuits are quite complex to build (and may require some parts that are difficult for you to find), and while a number of people have successfully built them—I have spent a lot of time supporting them.

The "El Cheapo" raw card is different because of its relative simplicity. While there are no plans at the current time to offer it as a kit, this will probably not always be true. Check my site periodically for El Cheapo kit information as well as component supplier suggestions. If you have problems finding the parts necessary to build the El Cheapo board shipped with this book, please send me an email and I'll see what I can do to point you in the correct direction.

# PICmicro® MCU Books

When the first edition of this book came out, it was one of the PICmicro® MCU books available on the market. Along with these books, Microchip has an excellent set of datasheets that are available on CD-ROM (although there are books available for each part number). In the three years since the first edition, the following books have become available:

**Design with PIC Microcontrollers**
Author: JB Peatman
ISBN: 0-13-759259-0

**PICTUTOR**
Author: J Becker
URL: http://www.matrixmultimedia.co.uk/picprods.htm

**Programming and Customizing the PIC Microcontroller**
Author: M Predko
ISBN: 0-07-913645-1

**PIC'n Techniques**
Author: D Benson
ISBN: 0-9654162-3-2

**PIC'n Up The Pace**
Author: D Benson
ISBN: 0-9654162-1-6

**Serial PIC'n**
Author: D Benson
ISBN: 0-9654162-2-4

**Easy PIC'n**
Author: D Benson
ISBN: 0-9654162-0-8

**An Introduction To PIC Microcontrollers**
Author: RA Penfold
ISBN: 0-85934-394-4

**Practical PIC Microcontroller Projects**
Author: RA Penfold
ISBN: 0-85934-444-4

**A Beginners Guide to the Microchip PIC—2nd Edition**
Author: N Gardner
ISBN: 1-899013-01-6

**PIC Cookbook**
Author: N Gardner
ISBN: 1-899013-02-4

# Useful Books

Here are a collection of books that I have found useful over the years for developing electronics and software for applications. Some of these are hard to find, but definitely worth it when you do find them in a used bookstore.

## THE ART OF ELECTRONICS—1989

The definitive book on electronics. It's a complete engineering course wrapped up in 1125 pages. Some people may find it to be a bit too complex, but just about any analog electronics question you could have will be answered in this book. I find that the digital information in this book to be less complete.
ISBN: 0-521-37095-7

## BEBOP TO THE BOOLEAN BOOGIE—1995

Somewhat deeper in digital electronics (and less serious) than "The Art of Electronics", Clive Maxwell's introduction to electronics stands out with clear and insightful explanations of how things work and why things are done the way they are. I bought my copy when it first became available in 1995 and still use it as a reference when I'm trying to explain how something works. It distinguishes itself from other books by explaining Printed Wiring Assembly technology (PCB Boards, Components and Soldering). This book complements "The Art of Electronics" very nicely.
ISBN: 1-878707-22-1

## THE ENCYCLOPEDIA OF ELECTRONIC CIRCUITS—VOLUME 1 TO 6

Rudolf Graf's Encyclopedia series of Electronic Circuits is an excellent resource of circuits and ideas that have been cataloged according to circuit type. Each book contains thousand of circuits and can really make your life easier when you are trying to figure out how to do something. Each volume contains an index listing circuits for the current volume and the previous ones.
Volume 1 ISBN: 0-8306-1938-0
Volume 2 ISBN: 0-8306-3138-0
Volume 3 ISBN: 0-8306-3348-0
Volume 4 ISBN: 0-8306-3895-4
Volume 5 ISBN: 0-07-011077-8
Volume 6 ISBN: 0-07-011276-2

## CMOS COOKBOOK—REVISED 1988

In "CMOS Cookbook," Don Lancaster introduces the reader to basic digital electronic theory. Also explaining the operation of CMOS gates, providing hints on soldering and prototyping, lists common CMOS parts (along with TTL pin out equivalents) and provides a number of example circuits (including a good basic definition of how NTSC video works). The update by Howard Berlin has made sure the chips presented in the book are still available. In the 1970s, Don Lancaster also wrote the "TTL Cookbook" (which was also updated in 1998), but I find the "CMOS Cookbook" to be the most complete and useful for modern applications.
ISBN: 0-7506-9943-4

## THE TTL DATA BOOK FOR DESIGN ENGINEERS—TEXAS INSTRUMENTS

I have a couple of 1981 printed copies of the second edition of this book and they are all falling apart from overuse. The Texas Instruments TTL data books have been used for years by hundreds of thousands of engineers to develop their circuits. Each datasheet is complete with pinouts, operating characteristics and internal circuit diagrams. While the data books are not complete for the latest "HC" parts, they will give you just about everything you want to know about the operation of small scale digital logic. The latest edition I have references for was put out in 1988 and are no longer in print, but you can pick them up in used book stores for relatively modest prices.
ISBN: N/A

## PC PHD—1999

This Book/CD-ROM package was written to give a clear introduction to the PC, from a "bottoms up" hardware perspective as well as an explanation of how code works in the PC. Along with explaining the architecture, I also provide over twenty applications that will help you to understand exactly how MS-DOS and Windows code executes in the PC and how hardware is accessed using the various interfaces available within the PC.
ISBN: 0-07-134186-2

## PC INTERFACING POCKET REFERENCE—1999

This Book is designed as an easy to use pocket reference for programmers and engineers working on the PC. Along with detailing the PC's architecture, the Intel 8086 and later microprocessors are described. The instruction sets used in the processor are listed along with addressing and value information. The information is useful for all PCs from the first 8088s to the most modern multi-Pentium III systems.
ISBN: 0-07-135525-1

## THE PROGRAMMER'S PC SOURCE BOOK—2ND EDITION 1991

Thom Hogan's 850 page book is just about the best and most complete reference that you can find anywhere on the PC. This book basically ends at the '386 (no 486, Pentiums of

any flavor, PCI, Northbridge, Southbridge or SuperIO or any ASICs of any type), but if you need a basic PC reference that explains, BIOS, all the "Standard" I/O, DOS and Windows 3.x Interfaces, this is your book. Look for it at your local used bookstore and if they have a second one, let me know - my copy is falling apart. The only problem with this book is there are no later editions.
ISBN: 1-55615-118-7

## THE EMBEDDED PC'S ISA BUS: FIRMWARE, GADGETS AND PRACTICAL TRICKS—1997

Ed Nisley's book is an almost complete opposite to the previous two books. Where the other books' focus is on documenting the innards of the PC, Nisley's shows you how to practically interface to the PC's "Industry Standard Architecture" ("ISA") bus and if you follow through the book you will end up with an LCD graphic display. Theory, register addresses and programming information is available in this book, but it is presented as the author works through the projects. This book is a resource that you can go back to and look at actual 'scope photographs of bus accesses or discussions on how to decode bus signals. There are a lot of great tricks in this book that can be used for many different PC interfacing applications.
ISBN: 1-5739-8017-X

## HANDBOOK OF MICROCONTROLLERS—1998

I wrote this very thick book as an introduction and complete reference package for modern eight bit embedded microcontrollers. As well as providing technical and processor programming information on the: Intel 8051, Motorola 68HC05, Microchip PICmicro® MCU, Atmel AVR and Parallax Basic Stamp, I have also provided datasheets, development tools and sample applications on the included CD-ROM. To help with your future applications, I explore interfacing to RS-232, I2C, LCD and other devices and I devote a fair amount of space to such advanced topics as Fuzzy Logic, Compilers, Real Time Operating Systems (I have included a sample one for the 68HC05) and Network Communications.
ISBN: 0-07-913716-4

## IBM PC AND ASSEMBLY LANGUAGE AND PROGRAMMING—FOURTH EDITION 1997

This is an excellent introduction to assembly language programming with a fairly low level approach concentrating on Microsoft's "MASM" and Borland's "TASM." "Debug.com" is used exclusively as a debug tool, which makes this book reasonably inexpensive to get involved with. I bought the first edition in 1983 when I first started working with the PC and I have kept up with the new editions over the years (largely because the older books fell apart from overuse).
ISBN: 1-1375-6610-7

## THE C PROGRAMMING LANGUAGE—SECOND EDITION 1988

Brian W. Kernighan, Dennis M. Ritchie's classic text explaining the "C" programming language has not lost any of it's usefulness since it's first introduction. This book has prob-

ably been used by more students in more colleges and universities than any other has. Despite the fact that the book was written originally for a programming course, the writing is crisp, sharp and easily understandable. This is a book I've probably owned five copies of with a couple being worn out and a few growing legs and walking out of my office.
ISBN: 0-13110-362-8

## PICList INTERNET LIST SERVER

I have found Internet "List Servers" to be an invaluable technical resource with MIT's "PICList" being the most valuable resource available for the PICmicro® MCU. The purpose of a List Server is to provide an address that you can send a message to and have it relayed to other subscribers. This means that a large amount of mail can be sent and received in a short period of time (the "PICList" typically processes fifty to one hundred e-mails per day) to a large number of people—as I write this the "PICList" has over two thousand subscribers.

List Servers are really wonderful things with it possible to get answers to questions within literally minutes after posing them (although hours afterward is probably more typical). There are a great deal of very knowledgeable people who can answer questions on a very wide variety of subjects (and have opinions on more than just the current subject does).

Having a great deal of people available makes the list essentially a community. This is a worldwide community and you have to try to be sensitive to different people's feelings and cultures. In thinking of this, I have created the following set of guidelines for sending email to List Servers. The genesis of these guidelines came about after the PICList went through a period of time in which the same type of message was being sent and people complaining about it. There is also a PICList "FAQ" that is available for when you subscribe to the list to help you understand the particular "netiquette" that is required when dealing with the PICList.

I think these are pretty good guidelines for any list server or news group and I suggest that you try to follow them as much as possible to avoid getting into embarrassing situations (or having your ancestry questioned because you made a gaff). After the guidelines, there are instructions for subscribing to the PICList.

1. Don't subscribe to a List and then immediately start sending questions. Instead, wait a day or so to get the hang of how messages are sent and replied to on the List and get a "feel" for the best way of asking questions.
2. Some lists send an email sent to them back to the author (while others do not). If you receive a copy of your first email, don't automatically assume that it is a "bounce" (wrong address) and resend it. In this case, you might want to wait a day or so to see if any replies to show up before trying to resend it. Once you've been on the list for a while, you should get an idea of how long it takes to show up on the list and how long it takes to get a reply.
3. If you don't get a reply to a request, don't get angry or frustrated and send off a reply demanding help. There's a good chance that nobody on the list knows exactly how to solve your problem. In this case, try to break down the problem and ask the question a different way.

**4** I mentioned above that you maybe able to get replies within minutes; please don't feel that this is something that you can count on. Nobody on the PICList is paid to reply to your questions. The majority of people who reply are doing so to help others. Please respect that and don't badger and help out in anyway that you can.

**5** If you are changing the "Subject" line of a post, please reference the previous topic (ie put in "was: '. . .' "). This will help others keep track of the conversation.

**6** When replying to a previous post, try to minimize how much of the previous note is copied in your note and maximize the relevance to your reply. This is not to say that none of the message should be copied or referenced. There is a very fine balance between having too much and too little. The sender of the note you are replying to should be referenced (with their name or ID).

My rule of thumb is, if the original question is less than ten lines, I copy it all. If it is longer, then I cut it down (identifying what was cut out with a "SNIP" Message), leaving just the question and any relevant information as quoted. Most mail programs will mark the quoted text with a ">" character, please use this convention to make it easier for others to follow your reply.

**7** If you have an application that doesn't work, please don't copy the entire source code into an email and post it to a List. As soon as I see an email like this I just delete it and go on to the next one (and I suspect that I'm not the only one). Also, some lists may have a message size limit (anything above this limit is thrown out) and you will not receive any kind of confirmation.

If you are going to post source code: keep it short. People on the list are more than happy and willing to answer specific questions, but simply copying the complete source code in the note and asking a question like "Why won't the LCD Display anything" really isn't useful for anybody. Instead, try to isolate the failing code and describe what is actually happening along with what you want to happen. If you do this, chances are you will get a helpful answer quickly.

A good thing to remember when asking why something won't work, make sure you discuss the hardware that you are using. If you are asking about support hardware (ie a programmer or emulator), make sure you describe your PC (or workstation) setup. If your application isn't working as expected, describe the hardware that you are using and what you have observed (ie if the clock lines are wiggling, or the application works normally when you put a scope probe on a pin).

**8** You may find a fantastic and appropriate web page and want to share it with the List. Please make it easier on the people in the list to cut and paste the URL by putting it on a line all by itself in the format:

http://www.awesome-pic-page.com

**9** If you have a new application, graphic, or whatever that takes up more than 1 K you would like to share with everyone on the List, please don't send it as an attachment in a note to the List. Instead, either indicate that you have this amazing piece of work and tell people that you have it and where to request it (either to you directly or to a web server address). Many list servers if a large file is received may automatically delete (thrown into the "bit bucket") it and you may or may not get a message telling you what happened.

If you don't have a web page of your own or one you can access, requesting somebody to put it on their web page or ftp server is acceptable.

**10** Many of these List Servers are made available, maintained and/or moderated by a device or product manufacturer. Keep this in mind if you are going to advertise your own product and understand what the company's policy on this is before sending out an advertisement.

The PICList is quite tolerant of advertisements of *relevant* products. If you are boarding puppies or have something equally non-PICmicro® MCU related, find somewhere else to advertise it.

**11** Putting job postings or employment requests *may* be appropriate for the PICList (like the previous point, check with the list's maintainer). However, I don't recommend that the rate of pay or conditions of employment should be included in the note (unless you want to be characterized as cheap, greedy, unreasonable or exploitive).

**12** "Spams" are sent to every list server occasionally. Please do not "reply" to the note even if the message says that to get off the spammer's mailing list just "reply." This will send a message to everyone in the list. If you must send a note detailing your disgust, send it to the spam originator (although to their ISP will probably get better results).

**NOTE:** There are a number of companies sending out bogus spams to collect the originating addresses of replying messages and sell them to other companies or distributors of addresses on CD-ROM. When receiving a spam, see if it has been sent to you personally or the list before replying—but beware if you are replying to the spam, you may be just sending your e-mail address for some company to resell to real spammers. I know it's frustrating and, like everyone else, I'm sure you would like to have all spammers eviscerated, but if you want to minimize how much you are bothered in by spams in the future, you just have to ignore any spams that are sent to you.

**13** Following up with the previous message, if you are going to put in pointers to a list server, just put a hyperlink to the list server request email address, **NOT TO THE LIST SERVER ITSELF.** If you provide the address to the list server, Spammers can pull the link from your page and use it as an address to send Spams to. By not doing this, you will be minimizing the opportunity for spammers to send notes to the list.

**14** Off topic messages, while tolerated will probably bring lots of abuse upon yourself, especially if you are belligerent about it. An occasional notice about something interesting or a joke is fine as long as it is not offensive.

If you feel it is appropriate to send an off topic message, make sure that you put "[OT]:" in the subject line, some members of the list use mail filters and this will allow them to ignore the off topic posts automatically.

Eventually a discussion (this usually happens with off topic discussions) will get so strung out that there is only two people left arguing with each other. At this point stop the discussion entirely or go 'private.' You can obtain the other person's e-mail address from the header of the message—send your message to him or her and not to the entire list. Everyone else on the list would have lost interest a long time ago and probably would like the discussion to just go away (so oblige them).

**15** Posts referencing Pirate sites and sources for "cracked" or "hacked" software is not appropriate in any case and may be illegal. If you are not sure if it is okay to post the latest software you've found on the 'Web, then DON'T until you have checked with the owners of the software and gotten their permission. It would also be a good idea to indicate in your post that you have the owner's permission to distribute cracked software.

A variety of different microcontrollers are used in "Smart Cards" (such as used with Cable and Satellite scrambling) or video game machines and asking how they work will probably result in abusive replies at worst or having your questions ignored at best. If you have a legitimate reason for asking about smart cards, make sure you state it in your email to the list.

**16** When you first subscribe to a list, you will get a reply telling you how to unsubscribe and indicate the note type. **DON'T LOSE THIS NOTE.** In the past, people having trouble unsubscribing have sent questions to the list asking how and sometimes getting angry when their requests go unheeded. If you are trying to unsubscribe from a list and need help from others on the list, explain what you are trying to do and how you've tried to accomplish it.

**17** When working with a list server, do *not* have automated replies set. If they are enabled then, all messages sent by the server to you will be replied to back to the list server. This is annoying for other list members and should be avoided.

**18** If you're like me and just log one once or twice a day, read all the notes regarding a specific thread before replying. When replying to a question that has already be answered look for what you can add to the discussion, not reiterate what's already been said.

**19** Lastly, please try to be courteous to all on the PICList. Others may not have *your* knowledge and experience or they may be sensitive about different issues. There is a very high level of professionalism on the lists, please help maintain it. Being insulting or rude will only get the same back and probably have your posts and legitimate questions ignored in the future by others on the list who don't want to have anything to do with you.

To put this succinctly: **"Don't be offensive or easily offended."**

To subscribe to the PICList, send an email to

listserv@mitvma.mit.edu

with the message:

subscribe piclist *your name*

In the body of the message.

*Save* the confirmation message; this will give you the instructions for signing off the list as well as instructions on how to access more advanced PICList List Server functions.

To sign off the list, send a note to the same address (listserv@mitvma.mit.edu) with the message:

signoff piclist

When signing off the PICList make sure that you are doing it from the ID that you used to sign on to the list.

Once you have subscribed to the PICLIST, you will begin receiving mail from

piclist@mitvma.mit.edu

Emails can be sent to this address directly or can be replied to directly from your mailer.

The list archive is available at:

http://www.iversoft.com/piclist/

and it has a searchable summary of the emails that have been sent to the PICList.

# My Favorite PICmicro® MCU Web Sites

When I have a problem, here is the list of sites that I turn to try and find the answers in what other people have already done. Most of these sites are dedicated to the PIC16F84 (and the PIC16C84), but there is a lot of useful information, code and circuits that you can get from these sites. This list is current as of February 2000 and is changing all the time. The CD-ROM contains the latest set of links as of the publishing date and my web page may be updated with additional information.

As I write this, there is somewhere in the neighborhood of one thousand web pages devoted to the PICmicro® MCU with different applications, code snippets, code development tools, programmers and other miscellaneous information on the PICmicro® MCU and other microcontrollers. To try and simplify your own searching, I have tried to include web sites that either have unique information and applications or ones that have substantial links to other sites.

Alexy Vladimirov's outstanding list of PICmicro® MCU resource pages. Over 700 listed as of February 2000.

**http://www.geocities.com/SiliconValley/Way/5807/**

Bob Blick's web site. Some interesting PICmicro® MCU projects that are quite a bit different than the run of the mill.

**http://www.bobblick.com/**

Scott Dattalo's highly optimized PICmicro® MCU math algorithms. The best place to go if you are looking to calculate Trigonometric Sines in a PICmicro® MCU.

**http://www.dattalo.com/technical/software/software.html**

Along with the very fast PICmicro® MCU routines, Scott has also been working on some GNU General Purpose License Tools designed to run under Linux. More information on these tools can be found in Appendix H. The tools can be downloaded from:

**http://www.dattalo.com/gnupic/gpsim.html**
**http://www.dattalo.com/gnupic/gpasm.html**

Marco Di Leo's "PIC Corner." Some interesting applications including information on networking PICmicro® MCUs and using them for cryptography.

**http://members.tripod.com/~mdileo/**

Dontronics Home Page. Don McKenzie has a wealth of information on the PICmicro® MCU as well as other electronic products. There on lots of useful links to other sites and it is the home of the SimmStick.

**http://www.dontronics.com/**

Fast Forward Engineering. Andrew Warren's page of PICmicro® MCU information and highly useful question answer page.

**http://home.netcom.com/~fastfwd/**

Steve Lawther's list of PICmicro® MCU Projects. Interesting PICmicro® MCU (and other microcontroller) projects.

**http://ourworld.compuserve.com/homepages/steve_lawther/ucindex.htm**

Eric Smith's PIC Page. Some interesting projects and code examples to work through.

**http://www.brouhaha.com/~eric/pic/**

Rickard's PIC-Wall. Good site with a design for PICmicro® MCU based composite video game generator.

**http://www.efd.lth.se/~e96rg/pic.html**

PicPoint—Lots of good projects to choose from including 5 MB free to anyone that wants to start their own PICmicro® MCU web page.

**http://www.picpoint.com/**

MicroTronics—Programmers and Application reviews.

**http://www.eedevl.com/index.html**

Home page to the "Pic 'n Poke" development system. This system includes an animated simulator that is an excellent tool for learning how data flows and instructions execute in the PICmicro® MCU microcontroller.

**http://www.picnpoke.com/**

# Periodicals

Here are a number of magazines that do give a lot of information and projects on PICmicro® MCUs. Every month, each magazine has a better than 50% chance of presenting a PICmicro® MCU application.

## CIRCUIT CELLAR INK

Subscriptions:
P.O. Box 698
Holmes, PA
19043-9613

1(800)269-6301

Web Site: **http://www.circellar.com/**
BBS: (860)871-1988

## POPTRONICS

Subscriptions:
Subscription Department
P.O. Box 55115
Boulder, CO

1(800)999-7139

Web Site: **http://www.gernsback.com**

## MICROCONTROLLER JOURNAL

Web Site: **http://www.mcjournal.com/**

This is published on the Web.

## NUTS & VOLTS

Subscriptions:
430 Princeland Court
Corona, CA
91719

1(800)-783-4624

Web Site: **http://www.nutsvolts.com**

## EVERYDAY PRACTICAL ELECTRONICS

Subscriptions:
EPE Subscriptions Dept.
Allen House, East Borough,
Wimborne, Dorset,
BH21 1PF
United Kingdom

+44 (0)1202 881749

Web Site: **http://www.epemag.wimborne.co.uk**

# Web Sites of Interest

While none of these are PICmicro® MCU specific, they are a good source of ideas, information, and products that will make your life a bit more interesting and maybe give you some ideas for projects for the PICmicro® MCU.

## SEATTLE ROBOTICS SOCIETY

**http://www.hhhh.org/srs/**

The Seattle Robotics Society has lots of information on interfacing digital devices to such "real world" devices as motors, sensors and servos. They also do a lot of exciting things in the automation arena. Most of the applications use the Motorola 68HC11.

## LIST OF STAMP APPLICATIONS (L.O.S.A)

**http://www.hth.com/losa.htm**

The List of Parallax Basic Stamp Applications will give you an idea of what can be done with the Basic Stamp (and other microcontrollers, such as the PICmicro® MCU). The list contains projects ranging from using a Basic Stamp to give a cat medication to providing a simple telemetry system for model rockets.

## ADOBE PDF VIEWERS

**http://www.adobe.com**

Adobe .pdf file format is used for virtually all vendor datasheets, including the devices presented in this book (and their datasheets on the CD-ROM).

## "PKZIP" AND "PKUNZIP"

Web Site: **http://www.pkware.com**

PKWare's "zip" file compression format is a "Standard" for combining and compressing files for transfer.

## HARDWARE FAQS

Web Site: **http:paranoia.com/~filipg/HTML/LINK/LINK_IN.html**

A set of FAQs (Frequently Asked Questions) about the PC and other hardware platforms that will come in useful when interfacing a microcontroller to a Host PC.

**http://www.innovatus.com**

Innovatus has made available "PICBots," an interesting PICmicro® MCU simulator which allows programs to be written for virtual robots which will fight amongst themselves.

## PART SUPPLIERS

The following companies supplied components that are used in this book. I am listing them because they all provide excellent customer service and are able to ship parts anywhere you need them.

## DIGI-KEY

Digi-Key is an Excellent Source for a wide range of Electronic Parts. They are reasonably priced and most orders will be delivered the next day. They are real lifesavers when you're on a deadline.

Digi-Key Corporation
701 Brooks Avenue South
P.O. Box 677
Thief River Falls, MN
56701-0677

Phone: 1(800)344-4539 (1(800)DIGI-KEY)
Fax: (218)681-3380

**http://www.digi-key.com/**

## AP CIRCUITS

AP Circuits will build prototype bare boards from your "Gerber" files. Boards are available within three days. I have been a customer of theirs for several years and they have always produced excellent quality and been helpful in providing direction to learning how to develop my own bare boards. Their website contains the "EasyTrax" and "GCPrevue" MS-DOS tools necessary to develop your own Gerber files.

Alberta Printed Circuits Ltd.
#3, 1112-40th Avenue N.E.
Calgary, Alberta
T2E 5T8

Phone: (403)250-3406
BBS: (403)291-9342
Email: staff@apcircuits.com

**http://www.apcircuits.com/**

## WIRZ ELECTRONICS

Wirz Electronics is a full service Microcontroller component and development system supplier. Wirz Electronics is the main distributor for projects contained in this book and will sell assembled and tested kits of the projects. Wirz Electronics also carries the "SimmStick" prototyping systems as well as their own line of motor and robot controllers.

Wirz Electronics
P.O. Box 457
Littleton, MA
01460-0457

Toll Free in the USA & Canada: 1(888)289-9479 (1(888)BUY-WIRZ)
Email: sales@wirz.com

**http://www.wirz.com/**

## TOWER HOBBIES

Excellent source for Servos and R/C parts useful in homebuilt robots.

Tower Hobbies
P.O. Box 9078
Champaign, IL
61826-9078

Toll Free Ordering in the USA & Canada: 1(800)637-4989
Toll Free Fax in the USA & Canada: 1(800)637-7303
Toll Free Support in the USA & Canada: 1(800)637-6050
Phone: (217)398-3636
Fax: (217)356-6608
Email: orders@towerhobbies.com

**http://www.towerhobbies.com/**

## JAMECO

Components, PC Parts/Accessories, and hard to find connectors.

Jameco
1355 Shoreway Road
Belmont, CA
94002-4100

Toll Free in the USA & Canada: 1(800)831-4242

**http://www.jameco.com/**

## JDR

Components, PC Parts/Accessories, and hard to find connectors.

JDR Microdevices
1850 South 10th St.
San Jose, CA
95112-4108

Toll Free in the USA & Canada: 1(800)538-500
Toll Free Fax in the USA & Canada: 1(800)538-5005
Phone: (408)494-1400
Email: techsupport@jdr.com
BBS: (408)494-1430
Compuserve: 70007,1561

**http://www.jdr.com/JDR**

## NEWARK

Components—Including the Dallas Line of Semiconductors (the DS87C520 and DS275 is used for RS-232 Level Conversion in this book).

Toll Free in the USA & Canada: 1(800)463-9275 (1(800)4-NEWARK)

**http://www.newark.com/**

## MARSHALL INDUSTRIES

Marshall is a full service distributor of Philips microcontrollers as well as other parts.

Marshall Industries
9320 Telstar Avenue
El Monte, CA
91731
1(800)833-9910

**http://www.marshall.com**

## MOUSER ELECTRONICS

Mouser is the distributor for the Seiko S7600A TCP/IP Stack Chips.

Mouser Electronics, Inc.
958 North Main Street
Mansfield, Texas 76063
Sales: (800) 346-6873
Sales: (817) 483-6888
Fax: (817) 483-6899
Email: sales@mouser.com

**http://www.mouser.com**

## MONDO-TRONICS ROBOTICS STORE

Self-proclaimed as "The World's Biggest Collection of Miniature Robots and Supplies" and I have to agree with them. This is a great source for Servos, Tracked Vehicles, and Robot Arms.

Order Desk
Mondo-tronics Inc.
524 San Anselmo Ave #107-13
San Anselmo, CA
94960

Toll Free in the USA & Canada: 1(800)374-5764
Fax: (415)455-9333

**http://www.robotstore.com/**

APPENDIX E

# PICmicro® MCU PRODUCT AND SERVICE SUPPLIERS

CONTENTS AT A GLANCE



Even though Microchip has done an excellent job at providing a wide range of products for the PICmicro® MCU, a lot of other companies have excellent products and services for the PICmicro® MCU that you can take advantage of. These companies provide products and services with a very wide range of capabilities and price ranges.

In this appendix, I have included a number of companies that provide products and services in a wide number of areas that you may want to contact for your own needs and services.

I do not have any interest in any of the companies listed here, and in many cases, I have not seen their products. As well, I cannot guarantee that all the phone numbers, addresses and Internet email address and URLs will change between the time that the information was provided to me and the publication of this book. This means that you should be careful before ordering any products or services that are listed here.

## PROGRAMMERS

YAP-II Programmer and Development Kit—Wirz Electronics
PICFun2—Labtronics International (Australia)
"Introduction to 16C84," programmer and components—DIY
General 16Cxx Programmer—DIY
DonTronics DT.001 Programmer—DonTronics
HTH Programmer—High Tech Horizon
PIC-1a Programmer—ITU Technologies
WARP-3 Programmer—ITU Technologies
WARP-3—Newfound Electronics, DonTronics
WARP-17—Newfound Electronics
5xer Programmer—ITU Technologies
PIC16CXX Programmer—Parallax Inc.
PROBYTE Programmer—PROBYTE Oy
PICProg V3 Production Programmer—Telesystems
ProPic—Tato Computadores
Linux Command Line Interface for PICStart Pro—Cosmodog, Ltd.

## DEVELOPMENT TOOLS

Pic'N Poke Software Simulator/Development Tool—Bubble Software
Fuzzy Logic Development System—Byte Craft Limited
CCS Software Prototyping Board—CCS
CESSIM Development Environment—ITU Technologies
Linux Software for Parallel Port Programmers—Nexus Computing
Programmer Adapters for various PIC Package types—Parallax, Inc.
SimmStick—SiStudio
PICkle—SuperComputing Surfaces, Inc.

## COMPILERS

MPC "C" Compiler—Byte Craft Limited
Jal Freeware PICmicro® MCU Compiler
PCB DOS "C" Compiler (PIC16C5xx and 12000)—CCS
PCM DOS "C" Compiler (PIC16C6xx, 7xx, 8x, 92x, and 14000)—CCS
PCW Professional Package for Windows (all PIC16Cxx chips)—CCS
PicBasic—micro-electronics Labs. Inc.
ANSI C Compiler for the PIC—HI-TECH Software
PIC Assembler—Parallax Inc.

## EMULATORS

Emu-II Emulator—Wirz Electronics, DonTronics, Interface Products (Pty) Ltd.

## APPLICATIONS

PIC-Based Closed Caption Decoder—Brouhaha Computer Mercenary Services
Radio Based Data Collection System—Cedardell Inc.

DIX: Digital Intelligent Cross Switch—CherryTronics B.V.
CAPS: CherryTronics Automatic Power Switch—CherryTronics B.V.
COPS: CherryTronics OS/2 Power Switch—CherryTronics B.V.
TELEPHONEGUARD—CherryTronics B.V.
Disk Drive Test Power Supply—Diamond Mountain Engineering, Inc
Single and Dual Dice—DIY
Unipolar Stepper Motor Controller—DIY
Servo Motor Driver—DIY
PICPlus (tm) Microcontroller Board—E-Lab Digital Engineering, Inc
Alarm and Monitoring System—Engenharia Mestra de Sistemas, sociedade limitada
Unipolar 3, 6, or 8 Wire Stepper Motors—Fisher Automation, Inc.
DC Load Switching—Fisher Automation, Inc.
Motor Control—JS Controls
DigiTemp temperature sensor system for DOS and Linux—Nexus Computing
BASIC Stamp—Parallax, Inc.
PROBYTE Tools—PROBYTE Oy
GPS Based Timing Receiver—Rack and Stack Systems
Mini Mods—Solutions Cubed
SLI-OEM Interface—Wirz Electronics, DonTronics, Interface Products (Pty) Ltd.
PIC16C57accessory board, serial LCD Interface IC's, a Serial Keypad Encoder and a Serial-to-Parallel Printer Interface IC—E-Lab

## PUBLICATIONS

Byte Craft Quarterly Newsletter—Byte Craft Limited
PIC Datasheets, magazine articles, and application notes in Russian—ORMIX Ltd.
Easy PIN'n Beginner's Guide—DonTronics
Scott Edward's PIC Source Book/Disk—DonTronics
The Personal Computer Technical Reference

## CONSULTANTS

Aengineering Co.
AmberDrew Ltd.
Brouhaha Computer Mercenary Services
Cinematronics
CherryTronics B.V.
DASCOR
Datalink, The Engineering Company
Design Concepts, Inc.
E-Lab Digital Engineering, Inc.
Ed Edmondson Consulting
Embedded Research
Engenharia Mestra de Sistemas, sociedade limitada
Fast Forward Engineering
HyperDyne Systems
Interface Products (Pty} Ltd.

Iversoft, Inc.
Jones Computer Communications
Nelson Research
Nexus Computing
Oak Valley Development
ORMIX Ltd.
Pragmatix
Precision Design Services
PROBYTE Oy
Rack and Stack Systems
Rochester MicroSystems, Inc.
Rocolec
Solutions Cubed
SuperComputing Surfaces, Inc.
Tato Computadores
Telelink Communications
Trinity Electronics Systems Ltd
Ultrawave Designs

## MICROCHIP REPRESENTATIVES

Nelson Research
ORMIX Ltd.
Pipe-Thompson Technologies Inc.
PROBYTE Oy

## THE COMPANIES

### Aengineering Co.

We design non-contact sensors for OEMs and users, especially sensors used for sensing people. Successful commercial designs include: lavatory faucet and toilet no-touch controls, intrusion alarms, aiming and focusing devices, and many others. Ultra-low cost consumer product designs make ingenious use of PICmicro® Microcontrollers.

3300 S Fox Spi Rd.
Langley, WA
98260-8010

Phone: (360)730-2058
Fax: (360)730-2058
Email: optoeng@whidbey.com
**http://www.whidbey.com/optoinfo**

### AmberDrew Ltd.

AmberDrew specialises in the development of Software and Hardware solutions in Process Monitoring and Control using embedded controllers based on PICmicro® MCUs and PCs.

Thistle Lodge,
Alltewn, Pontardawe
Swamsea, South Wales, UK
SA8 3A!

Phone: +44 (0)1792 862912
Fax: +44 (0)1792 862912
Email: efoc@cyberstop.net

**Brouhaha Computer Mercenary Services**
Embedded System Design Consulting. A few PICmicro® MCU project Designs Available free from Web Page.

142 North Milpitas Boulevard, Suite 379
Milpitas, CA
95035

Phone: (408)263-3894

**http://www.brouhaha.com/~eric/pic/**

**Bubble Software**

The "PICNPoke" software runs under Windows 95 and has been created for people who have no assembly language programming experience with PICs, or any processor for that matter. If you want to learn about this fascinating subject in a fun and interesting way, then this is the place to start. You begin with one of the most popular processors around today—the PIC16F84.

8 Westminster Crt.
Somerville, Victoria, Australia
3912

Phone: +61 3 59775792
Email: sales@picnpoke.com

**http://www.picnpoke.com**

**Byte Craft Limited**

Byte Craft Limited is a software development company specializing in PC-based code development tools for 8-bit microcontrollers. Byte Craft products include a line of C cross-compilers and a C pre-processor that adds fuzzy logic capability to any C compiler.

The MPC Code Development System is a C cross-compiler targeted to all Microchip PICmicro® MCU families. It comes with a development environment, built-in macro assembler and the Byte Craft BClink Optimizing Linker which runs a final optimization pass on the entire program.

MPC features include:

- Interrupt context save and restore macros for 12-, 14- and 16-bit cores
- Compiler-generated bank-switching
- Sbility to call a function located anywhere in memory in the 12-bit core
- C support for hardware registers, including setting configuration fuses
- Extensive set of header files that describe the resources of individual PICmicro® MCUs
- Named address space to help you organize the allocation of memory
- Library source (16-bit math, delay functions, asynchronous serial support, 4 and 8 channel A/D, EEPROM access)
- Source-level debugging with Microchip's PICMASTER emulator system and MPLAB-SIM simulator system, Advanced Transdata, Tech-Tools Mathias Clearview and iSystem.

421 King Street North
Waterloo, Ontario, Canada
N2J 4E4

Phone: (519)888-6811
Fax: (519)746-6751
Email: **info@bytecraft.com**

**http://www.bytecraft.com**

**Cedardell Inc.**

A radio based data collection/delivery system, using a multi-loop architecture. System can be connected to a PC or a microprocessor based control system via a RS-232 link. 16 bit Windows support DLL provided, sample Windows management application provided. Currently used in a security system.

2919 17 St, #210
Longmont, CO
80503

Phone: (303)651-2442
Fax: (303)651-2426
Email: edtodd@sni.net

**http://www.sni.net/cedardell**

### CherryTronics B.V.

DIX—An intelligent power saving device for computer systems which diminishes power-consumption of PC's, equipped with Advanced Power Management (APM) with more than 99% in Standby Mode.

CAPS—A device used for remotely powering on and off computer systems (used for the unattended exchange of data/software).

COPS—A device for automatic powering on and off computer systems by telephone-line without answering the call; information is transferred toll-free.

TELEPHONEGUARD—A device which prevents unauthorized use of one's telephone line by illegal callers (a very hot item in our region of the world).

Haarstraat 25-a
Ammerzoden, Holland
5324 AM

P.O. Box 39
Ammerzoden, Holland
5324 ZG

Phone: +31 73 5991098
Fax: +31 73 5994712
Email: mdekkers@inter.nl.net
Email: b.janssen@inter.nl.net
Email: infor@cherrytronics.nl

**http://www.cherrytronics.nl**

### Cinematronics

Consultant specializing in the design and fabrication of electronic devices for the film/video and entertainment industry. Custom design of lighting controllers, motion control devices and remote control electronics. Cinematronics also designs PICmicro® MCU based electronic devices for consumer electronics.

344 Dupont St., Suite 304
Toronto, Ontario, Canada
M5R 1V9

Phone: (416)927-7679
Fax: (416)927-7679
Email: cinetron@passport.ca

**Cosmodog, Ltd.**

PICPpicp, is a Linux-based open source command line interface to the PICSTART Plus programmer, and picdasm, a PICmicro® MCU disassembler. picp supports all devices supported by MPLAB, and does it swiftly and cleanly from the command line. picdasm takes Intel of Motorola hex files and re-generates PICmicro® MCU assembler source files, making all possible label substitutions for the specified processor, like STATUS or RB0.

415 West Superior Street
Chicago, IL
60610-3428

Phone: (312)440-0344
Fax: (312_440-0355
Email: **info@cosmodog.com**

**http://www.cosmodog.com**

**Custom Computer Services, Inc.**

CCS has created a line of powerful C compilers that are designed specifically for the Microchip PIC16Cxx devices to make development fast, easy and efficient. Standard C operators and the special built in functions are optimized to produce very efficient code for the bit and I/O functions normally required for these microcontrollers.

P.O. Box 53008
Brookfield, WI
53008

Phone: (262)797-0455 extension 35.
Fax: (414)781-3241
Email: ccs@execpc.com

**http://www.ccsinfo.com/PICC.html**

**Dascor**

Dascor specializes in developing turnkey signal conditioning and PICmicro® MCU based data acquisition systems for the OEM market. Auxiliary products include High and Low Speed Data Loggers and Earthquake sensor/alarm systems. Auxiliary services include product design in mixed analog and digital modes, PCB layout and packaging, Programming in a variety of languages, and contract manufacturing.

P. O. Box 46-2885
Escondido, California
92046-2885

Phone: (760)796-7785 voice mail and fax
Fax: (760)796-7785
Email: support@dascor.com

**www.dascor.com**

**Datalink, The Engineering Company**
Datalink provides Custom designed, Hardware and Software solutions for a variety of environments including Industrial.

Zurichstrasse 25
Maur, Zurich, Switzerland
CH 8124

Phone: ++41 1 887 6400
Fax: ++41 1 887 6414
Email: info@datalink.ch

**Design Concepts, Inc.**
Design Concepts designs and engineers products for a variety of industries. The Industrial Designers, Mechanical and Electrical Engineers, and Prototype Specialists can turn ideas into one-of-a-kind or mass produced products. PICmicro® MCU projects include a consumer use hand tool with a yearly production of two million units and an industrial use labeling machine with a yearly production of five hundred units.

5301 Buttonwood Drive
Madison, Wisconsin
53718

Phone: (608)221-2623
Fax: (608)221-2133
Email: larryg@design-concepts.com

**http:www.design-concepts.com**

**Diamond Mountain Engineering, Inc.**
We sell a programmable power supply based on a PIC16C64. It is designed for disk drive testing with a 5 and 12 Volt output rated at 3 Amps. It features programmable linear ramping from 10 microseconds to 10 seconds in three ranges. The voltage range is from 0 to 6 and 0 to 16 Volts. The fast PIC16C64 allows positive, negative and bi-polar spikes from 10 microseconds to 10 seconds.

3123 Whipple Road
Union City, CA
94587-1218

Phone: (510)487-9530
Fax: (510)487-9531
Email: mschwabe@ricochet.net

**Diy**

Manufacturers of Electronic Kits using PICmicro® MCU IC: All source code supplied on floppy Disk.
Products:

- Single and dual dice
- Unipolar Stepper Motor Driver
- Servo Motor Driver
- "Introduction to 16C84," programmer and all components
- General 16Cxx Programmer

DIY Electronics
P.O. Box 88458
Sham Shui PO
Hong Kong

Phone: 852-2720 0255
Fax: 852-2725 0610
Email: diykit@hk.super.net

**http://kitsrus.com**

**DonTronics**

DIY PCB kits including Microchip Technology PICmicro® MCU based micro products, FED and MEL PICmicro® MCU Basic Compilers, FED PICmicro® MCU Basic Interpreters, CCS C Compilers, Square 1 Electronics EasyPIC'n Beginners Guide, Newfound PIC Programmers DonTronics DT.001 Programmers, Nuts and Volts Magazine, Scott Edward's PICmicro® MCU Source Book/Disk, SiStudio SimmStick, and Wirz Electronics SLI-LCD Interface and Emu PIC 18 Pin Emulator.

P.O. Box 595
Tullamarine 3043 Australia

Phone: 613+9338-6286
Fax: 613+9338-2935
Email: don@dontronics.com

**http://www.dontronics.com**

**E-Lab Digital Engineering, Inc.**

E-Lab manufacturers an accessory board for the PIC16C57, including an LCD interface and A/D addition, as well as expanded I/O. We also manufacture the EDExx Family of IC's, including serial LCD Interface IC's (EDE702, EDE701, EDE700), a Serial Keypad Encoder (EDE1144), and a Serial-to-Parallel Printer Interface IC (EDE1400) that are of use to a great many PICmicro® MCU enthusiasts.

In addition, we offer the 'EasyASIC™' IC creation service, in which we create 'custom' IC's to client's specifications using PICmicro® MCU microcontrollers.

1932 Hwy. 20
Lawton, IA
51030

Phone: (816) 257-9954
Fax: (816) 257-9945

**http://www.elabinc.com**

### Ed Edmondson Consulting

Custom embedded systems and software with Microchip PICmicro® MCU series and Atmel AVR series controllers. No job too small or too large.

200 East Jack Street, Suite 15
PO Box 1407
Mabank, Texas
75147-1407

Phone: (903)887-4783
Fax: (903)887-5762
email: eaejrphd@aol.com

### Embedded Research

Embedded Research offers engineering services, both hardware and software, for the embedded controller market. We specialize in PICmicro® MCU based designs.

P.O. Box 92492
Rochester, NY
14692

Phone: (716)359-3941
Email: gmdsr@vivanet.com

**http://www.vivanet.com/~gmdsr**

### Engenharia Mestra de Sistemas, sociedade limitada

Engenharia Mestra makes alarm systems for all types of clients, from correctional facilities to the regular home. We have four types of panels in production, and we also make custom systems, as well as act in consulting.

Rua Guaiauna, 439
Sao Paulo, SP, Brazil
03631-000

Phone: 55-11-218.5008
Fax: 55-11-217.6610
Email: mestra@u-netsys.com.br

**Fast Forward Engineering**

Contract design of PIC-based prodcuts, specializing in RF systems, high-volume consumer electronics, automotive electronics, and remote telemetry and control. Software-only services are available, as well as start-to-finish handling of the entire design process. Consulting/training services are also available.

Certified by Microchip as one of their Microchip Technology Consultant Program Members.

1320 Kilby Lane
Vista, CA
92083-5043

Phone: (760)724-9600
Fax: (760)724-0042
Email: fastfwd@ix.netcom.com

**http://www.geocities.com/SiliconValley/2499**

**Fisher Automation, Inc.**

Output Interface Boards Available:

#1—Control of unipolar 5, 6, or 8 wire step motors (up to 2 amps/coil up to 30 VDC) from only two pins.

#2—General on-off switching of DC loads. Four loads can be controlled up to 2 Amps each to 60 VDC each.

150-45 12 Rd.
Whitestone, NY
11357

Phone: (718)767-8250
Fax: (718)767-8251
Email: fishfam@pipeline.com

**http://www.plasticsnet.com/mbr/fishfam**

**High Tech Horizon**

HTH distributes Parallax PICmicro® MCU and Basic Stamp products in Scandinavia.

Asbogatan 29 C
Angelholm, Sweden
S-262 51

Phone: +46 431 41 00 88
Fax: +46 431 41 00 88
Email: info@hth.com

**http://www.hth.com**

### HI-TECH Software

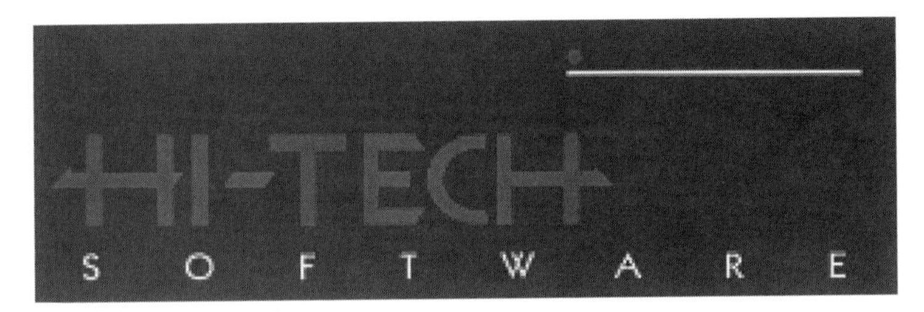


HI-TECH produces an ANSI-C compiler for PICmicro ® MCU series microcontrollers—includes floating point, long arithmetic, compact code, complete with assembler and linker.

HI-TECH Software
PO Box 103
Alderley QLD 4051
AUSTRALIA

HI-TECH Software LLC
Suite 105
7830 Ellis Road
Melbourne, FL 32904
USA

sales : sales@htsoft.com
support : support@htsoft.com

**http://www.htsoft.com/**

### HyperDyne Systems

HyperDyne Systems is a small design firm located 20 miles south of Detroit Michigan. We specialize in application specific controls which essentially take the place of large PLCs. Application specific controls are designed specifically to do one job which, depending on the application, can reduce the cost of both setup time and training. Our current products include early warning systems, high security RF systems and online calibration systems.

10222 Loma Circle
Grosse Ile, Michigan
48138

Phone: (734)675-2386
Fax: (734)675-2242
Email: jmb@webbernet.net

**Interface Products (Pty) Ltd.**
Supplier of microcontroller development tools, custom electronic design and manufacturing. Remote automation, monitoring and control across commodity networks. Positioning support for GPS assisted airborne remote sensing; stocklist of Garmin equipment. Business level Internet services for engineering professionals. Distributor for DonTronics, Silicon Studio and Wirz Electronics microcontroller products.

2nd Floor, Quinor Court, 81 Beit Street
New Doorfontein, Gauteng, South Africa
P.O. Box 15775, DOORNFONTEIN, 2028 South Africa

Phone: +27 (11) 402-7750
Fax: +27 (11) 402-7751
Email: info@ip.co.za

**http://www.ip.co.za**

**ITU Technologies**
ITU Technologies is a leading supplier of PICmicro® MCU development tools. We provide the highest quality tools at the most affordable prices! Our PIC-n-GO package is a perfect introductory package for only $45 kit or $65 assembled. This includes our PIC-1a programmer, PIC16F84 microcontroller and an assembler.

3477 Westport Ct.
Cincinnati, OH
45248

Phone: (513)574-7523
Fax: (513)574-4245
Email: sales@itutech.com

**http://www.itutech.com**

### Iversoft, Inc.

Programming services and product consulting.

40 Samoset Ave. 1st Fl.
Mansfield, MA
02048

Phone: (508)337-9926
Fax: (508)337-9926
Email: anick@iversoft.com

**http:www.iversoft.com**

### Jal Freeware Compiler

Jal is a freeware compiler for the 16c84, 16f84, SX18 and SX28. Jal is a simple, Pascal-like language. Libraries are provided for delays, serial communication, LCD interface etc.

Primay Contact: Wouter van Ooijen
Email: **wf@xs4all.nl**

**http://www.xs4all.nl/~wf/wouter/pic/jal**

### Jones Computer Communications

Embedded systems design and programming in assembler and/or C. Strong background in networks, serial communications and radio.

509 Black Hills Dr.
Claremont, CA
91711

Phone: (909)621-9008
Fax: On Request
Email: lee@frumble.claremont.edu

### JS Controls

Civils laboratory force testing equipment, manufacture, service and calibration. Currently offering PIC16C84 based DC motor speed controller:

- 200V shunt wound motors
- 3 digit LED rpm readout
- Push-Button stop/start and speed control
- non-volatile memory for speed set-point

8 Nolloth St., Groeneweide
Boksburg 1460, Republic of South Africa

Phone: Johannesburg 893-4154
Fax: Johannesburg 893-4154
Email: jsand@pixie.co.za

**Nelson Research**

Consulting and design services specializing in Microchip PICmicro® MCU microprocessors. Listed with Microchip as a Master Consultant. Nelson Research is a consulting firm specializing in electronics design including analog, digital, and embedded microprocessors. We support the product line with the full line of development tools and have experience in creating a variety of products for widely varying customers. With the associates we have available we can handle projects ranging from one day reviews to major development efforts.

130 School Street, P.O. Box 416
Webster, MA
01570-0416

Phone: (508)943-1075
Fax: (508)949-2914
Email: L.Nelson@ieee.org

**http://www.ultranet.com/~NR**

**Newfound Electronics**

WARP-3 is a high featured, low cost development programmer for all PIC16Cxx and derivative devices. WARP-3 is a DOS based programmer and suitable for use with anything from an XT to a Pentium PC. Please see the web site for further WARP-3 details and of also the low cost, fully featured "production" programmer to follow.

WARP-17 is a low cost development programmer for the PIC17C4x devices. All configuration options are supported and the software allows easy code and configuration changes. The WARP-17 is the ideal programmer to evaluate the increased power of the PIC17C4x family.

14 Maitland St.
Geelong West
Victoria 3218
Australia

Phone: ++61 3 5224 1833
Fax: ++61 3 5224 1833
Email: **newfound@pipeline.com.au**

**http://www.new-elect.com**

**Nexus Computing**

Nexus Computing specializes in the development of small embedded microprocessor products using Microchip PICmicro® MCU and Motorola 8 and 16 bit processors.

Email: nexus@eskimo.com

**http://www.eskimo.com/~nexus**

### Oak Valley Development

Oak Valley Development is a group of consultants that specializes in microprocessor controlled electro-mechanic and analog electronics. We have complete engineering resources to do all phases of a project from prototype to high volume production.

P.O. Box 41473
San Jose, CA
95160

Phone: (408)489-9623
Fax: (408)268-6184
Email: paul@oakvalley.com

**http://www.oakvalley.com**

### ORMIX Ltd.

Low-cost development programmers COMPIC-1 and COMPIC-5X for Microchip MCUs and Serial EEPROMs, connectored to the Serial Port. COMPIC-1 works without an external power supply, COMPIC-5X has two ZIF Sockets. Excellent software flexibility—new devices can be added to the text configuration file by end user. Very simple and friendly user interface, built-in HEX editor, serialization of parts possible. Low price.

Kr. Barona, 136
Riga, Latvia
LV-1012

Phone: (371)-7310660
Fax: (371)-2292823
Email: avlad@mail.ormix.riga.lv

**http://www.ormix.riga.lv/eng/index.htm**

### Parallax, Inc.

The Parallax PIC16Cxx Programmer can program many PICmicro® MCU devices and is constantly updated for newer PICmicro® MCUs throughout the year. We also offer many adaptors which plug directly into the programmer for use with various package types available in the PIC16Cxx series.

3805 Atherton Rd., Suite #102
Rocklin, CA
95675

Toll Free in the USA & Canada: 1(888)512-1024
Phone: (916)624-8333
Fax: (916)624-8003

Email: info@parallaxinc.com

**http://www.parallaxinc.com**

## THE PERSONAL COMPUTER TECHNICAL REFERENCE

The Technical Reference website contents range from low level machine code tricks for several processors to command references, tips and trick for large applications. You will find detailed information (not just links) on Scenix and PICmicro® microcontrollers, the IBM PC parallel port, electronic construction, and much more. We focus on the actual engineering data and details, not on canned, commercial solutions to common problems.

The TechRef is designed as a "mass mind" for technical information. Every page on the TechRef has a small form attached to the bottom that allows members to add their own comments, links, and entire web pages to the site. All members are allowed and encouraged to add their own links and comments to each existing page or to add new pages of their own. Non-members are also excluded from some "insider" content.

Information that you have in your mind right now, will be gone in a very short time. TechRef membership allows you to organize and record (privately if you wish) that information and to see information on the same subject from others. Advertisements are welcomed, personal information is fine.

You can even add your own pages and links! We call this TANSTAAFL web hosting. We don't charge, but we do:

- reserve the right to edit in order to improve readability and organization. (We will credit you and let you know what we have done)
- reserve all publication rights (non-exclusive). We intend to sell the (public NOT private) data on the techref site as a book and CD. Authors who make significant contributions will be financially rewarded.

1452 View Pointe Ave
Escondido, CA
92027

Phone: (619)652-0593
Email: jamesnewton@geocities.com

**http://techref.homepage.com**

### Pipe-Thompson Technologies Inc.

Pipe Thompson is the Microchip Rep for the Toronto Area. Along with providing technical information to Microchip customers, Pipe-Thompson can provide technical guidance and support.

4 Robert Speck Pkwy
Suite 1170

Mississauga, Ontario, Canada
L4Z 1S1

Phone: (905)281-8281
Fax: (905)281-8550
Email: pipethom@idirect.com

### Precision Design Services

Precision Design Services is an engineering consortium specializing in developing new products from idea through to a fully manufacturable, low cost, high quality product. Specific areas of expertise are: embedded microcontroller hardware and software design and very high speed receivers (in ECL—500 MBits/sec). PDS has extensive experience in the design and manufacturing processes of printed circuit boards from simple to double sided SMT and impedance controlled boards.

26006 View Point Drive East
Capistrano Beach, CA
92624-1224

Phone: (714)489-1064
Fax: (714)489-9299
Email: KASPER@EXO.COM

### Pragmatix

Pragmatix offers its customers a full service solution for their PICmicro® MCU based embedded systems needs. Our capabilities include product specification, hardware and software development, prototyping, volume production and systems integration. Also, Pragmatix offers unique parameterissable PICmicro® MCU cores for implementation in ASICs and FPGAs.

Bosuil 2
B2100 Deurne, Belgium

Phone: (+32) 3-3247733
Fax: (+32) 3-3247733
Email: preynen@innet.be

**http://www.geocities.com/SiliconValley/6766**

### PROBYTE Oy

Products for industrial use: temperature measuring devices, RF/Audio test equipment, PCM-test generators, data loggers. Microprocessor development tools: Protel CAD, Needham eprommers, Tech Tools emuators, Parallax PICmicro® MCUs and Stamp Tools, microEngineering PicBasic, Microchip Tools and PICmicro® MCUs.

Nirvankatu 31
Tampere, Finland
33820

Phone: Int+358-3-2661885
Fax: Int+358-3-2661886
Email: pri@sci.fi

**http://www.sci.fi/~pri**

### Rack and Stack Systems

Engineer (MSEE) with extensive experience in RF, microwave, semiconductors, GPS Time and Frequency. Programming in PICmicro® MCU assembler, HP Rocky Mountain Basic, and National Instruments LabView. All phases of system integration.

3425 Deerwood Drive
Ukiah, CA
95482-7541

Phone: (707)463.2380
Fax: (707)462-4333
Email: brooke@pacific.net

### Rochester MicroSystems, Inc.

Rochester MicroSystems, Inc. provides electronic design services to companies and research institutions. We provide: Electronic System Design, Electronic Circuit Design (analog and digital), Microprocessor and Interface Systems, Embedded Systems and Software, and Printed Circuit Board Layout Design. We have expertise in many application areas.

200 Buell Road, Suite 9
Rochester, NY
14624

Phone: (716)328-5850
Fax: (716)328-1144
Email: rmi@frontiernet.not

**http://www.frontiernet.net/~nmi/**

### Rocolec

Rocolec is an electronics design house, with experience in microcontroller/FPGA design, embedded software, printed circuit layout & prototyping for instrumentation & control.

3 Orchard Lane
Reepham, Norwich, Norfolk, UK
NR10 4NP

Phone: ++44-1603-879001
Fax: ++44-1603-879001
Email: info@rocolec.com

#### Solutions Cubed

Solutions Cubed provides embedded systems design, specializing in Microchip Technology microcontrollers. In addition to full design services and capabilities, Solutions Cubed provides a product line of Mini Mods. These miniature engineering modules are ideal for the electronic hobbiest and can be easily interfaced to Basic Stamps and their ilk.

256 E. First Street
Chico, CA
95928

Phone: (530) 891-8045
Fax: (530) 891-1643
Email: solcubed@solutions-cubed.com

**http://www.solutions-cubed.com**

#### SuperComputing Surfaces, Inc.

PICkle is an EDA tool for assembly programmers. Automates coding for hardware configuration. Supercedes functional libraries. Expert system with built in Wizards that generate peripheral handling code, math functions, precision delays, interrupts handlers and more. A true automatic code generator that optimizes register usage—Requires Windows.

8681 North Magnolia Avenue, Suite F
Santee, CA
92071

Phone: (619)562-5803
Fax: (619)562-3728
Email: info@pickleware.com

**http://pickleware.com**

#### Tato Computadores

ProPic programmer is a new programmer working under windows and can program almost any PICmicro® MCU.

R Santos Arcadio, 37
Sao Paulo/SP - Brazil
04707-110

Phone: (011)240-6474
Fax: (011)240-6474
Email: nogueira@mandic.com.br

**http://www.geocities.com/SiliconValley/Pines/6902/index.html**

### Telelink Communications

Our company specializes in Radio Telemetry/SCADA Systems, Data Logging over Radio Links, Short Haul Radio Modem Systems. A design and consultancy Service is also provided for specialized projccts where required.

Nugget Avenue
Bouldercombe, Queensland, Australia
4702

Phone: 61 79 340413, 61 418 7995551
Fax: 61 79 340413
Email: jack@networx.com.au

**http://www.networx.com.au/mall/tlink**

### Telesystems

The PicProg V3 is a low-cost production quality universal programmer. It supports a full range of the Microchip products including 12-, 14- and 16-bit PICmicro® MCUs, Serial and Parallel EPROM, EEPROM and Flash memory chips and Keeloq secure products. The PicProg supports AVR and AT89 MCUs by ATMEL, ACE MCUs by Fairchild, SX MCUs by Scenix and a wide range of E(E)PROM and Flash-memory chips by Intel and various other manufacturers. The PicProg operates through standard LPT port (printer transparent mode) or COM-port. It includes universal ZIF-panel for DIP packages (8 to 40 pins) and in-circuit programming connector.

The PicProg supports both a host and a stand-alone mode. It has a number of useful features including optimized programming algorithms and safe code protection enable for all grades of PICmicro® MCU devices.

The software package for PicProg includes both DOS command-line software (for batch programming) and Win32 software for Windows 9x/NT/2000. The PicProg is software/firmware expandable to future products/families.

Upgrades are available from our Web-site for free. The full package includes Programmer, Software, Manual, Parallel port Cable, Power supply and CD-ROM "Modern Microcontrollers" with various documentation and development tools.

Phone: +7-095-5301001
Fax: +7-095-5314840
Email: **ts@aha.ru**

**http://www.ts.aha.ru/english**

**Trinity Electronics Systems Ltd**

Trinity Electronics offers custom electronic design and manufacturing services. The company has completed over 600 design projects, over 300 PCB designs and dozens of PICmicro® MCU based designs. Products include:

Intrinsically safe designs
Industrial control systems
Industrial oven controls
Catalytic heater controls
Power measurement systems
Odometers for Heavy Equipment
Programmable Timers & Sequencers
DC motor speed controls
Professional audio and broadcast products
MP3 player (with speaker & amplifier)
Lighting controls
Special effects pyro control systems
Palm Pilot Accessories
RS-232 interfaces
LED & LCD displays
Heating and cooling controls
Automotive accessories

10708-181 Street
Edmonton, Alberta, Canada
T5S 1K8

Phone: (780) 489-3199
Fax: (780) 487-6397
Email: **sales@trinity-electronics.com**

**http://www.trinity-electronics.com**

**UltraWave Designs**

Design services in digital, analog, acoustic and RF for medical, military and commercial products. We can take your product from concept to production, or any stage in-between. Our skills exhibit more than 40 years of expertise.

Please visit our web site.

1129 State Street, Suite 3A
Santa Barbara, CA
93101

Phone: (800) 357-WAVE_or_(805) 564-3442__
Fax: (805) 683-9662
Email: info@ultrawave.com

**http://www.ultrawave.com**

APPENDIX F

# SIMMSTICK

**Figure F-1** Different SimmStick product PCBS

One of the most useful products to come along for developing microcontroller applications quickly (either for hobbyists or industrial applications) is the SiStudio/Dontronics "SimmStick" cards and bus. This system of cards will provide you with a basis for electrically wiring a microcontroller into an application. I have used SimmSticks for a number of applications and it has allowed me to create them generally in under an hour for very low cost.

The SimmStick was originally defined around the same time as the "PICStic" by Micromint became available. The PICStic, using a PIC16F84 could functionally simulate a BASIC Stamp (using microEngineering Lab's PicBasic Compiler) or be used as an application unique to itself by deleting or adding components to the various boards. The original "PICStic" was just designed for the PIC16C84; the SimmSticks were originally designed to provide prototype cards for the full range of PICmicro products. Now, along with the PICmicros® MCU, the SimmSticks can also support other microcontrollers as well.

The SimmStick can be loaded into a standard thirty pin memory SIMM socket and uses the following bus:

**TABLE F-1 SimmStick Bus Specification**

| PIN | LABEL | DESCRIPTION/COMMENTS |
|---|---|---|
| 1 | A1/Tx | RS-232 Transmit Line from the SimmStick |
| 2 | A2/Rx | RS-232 Receive Line to the SimmStick |
| 3 | A3 | Microcontroller General Purpose I/O |
| 4 | PWR | Unregulated Power In |
| 5 | CI | Line to Microcontroller's XTAL pins |
| 6 | CO | Line to Microcontroller's XTAL pins |
| 7 | +5V | Regulated +5 Volts |
| 8 | Reset | Microcontroller Negative Active Reset |
| 9 | Ground | System Ground |
| 10 | SCL | I2C Clock Line/Microcontroller General Purpose I/O |
| 11 | SDA | I2C Data Line/Microcontroller General Purpose I/O |
| 12 | SI | CMOS Logic Serial Line In/MCU General Purpose I/O |
| 13 | SO | CMOS Logic Serial Line Out/MCU General Purpose I/O |
| 14 | IO | Microcontroller General Purpose I/O and Timer In |
| 15 | D0 | Microcontroller General Purpose I/O |
| 16 | D1 | Microcontroller General Purpose I/O |
| 17 | D2 | Microcontroller General Purpose I/O |
| 18 | D3 | Microcontroller General Purpose I/O |
| 19 | D4 | Microcontroller General Purpose I/O |

**TABLE F-1 SimmStick Bus Specification (*Continued*)**

| PIN | LABEL | DESCRIPTION/COMMENTS |
|---|---|---|
| 19 | D4 | Microcontroller General Purpose I/O |
| 20 | D5 | Microcontroller General Purpose I/O |
| 21 | D6 | Microcontroller General Purpose I/O |
| 22 | D7 | Microcontroller General Purpose I/O |
| 23 | D8 | Microcontroller General Purpose I/O |
| 24 | D9 | Microcontroller General Purpose I/O |
| 25 | D10 | Microcontroller General Purpose I/O |
| 26 | D11 | Microcontroller General Purpose I/O |
| 27 | D12 | Microcontroller General Purpose I/O |
| 28 | D13 | Microcontroller General Purpose I/O |
| 29 | D14 | Microcontroller General Purpose I/O |
| 30 | D15 | Microcontroller General Purpose I/O |

I should discuss a few things about this bus. The first is that the number of pins available for digital I/O exceeds the number of pins on a PIC16F84 (or other eighteen-pin part). The 23 I/O pins were originally specified to provide I/O capabilities for a variety of microcontrollers. If the microcontroller that you are using does not have up to 23 I/O pins, then the extra pins can be used for other purposes (in the PICmicro® MCU Composite Video example, the output from the voltage divider is placed on the unused bus pins). For each different microcontroller and application, you will have to understand how the I/O lines are mapped out onto the SimmStick bus.

The bus specification is under review as I write this and may be changed slightly in the future. Please check the official SimmStick web site:

**http://www.simmstick.com**

or the Dontronics web site:

**http://www.dontronics.com**

to see if there have been any changes to this specification.

Serial data can be transmitted or received at TTL (CMOS) or RS-232 Levels. The RS-232 interface uses a Maxim MAX-232 chip that has vias and connections built into the SimmStick base card (like the "DT003"). These pins could also be used for digital signal I/O.

The SimmStick cards are capable of taking unregulated power, regulate it and distribute it throughout the application. Personally, I don't like to do this; instead I use a SimmStick backplane (either the DT001 or DT003) for power (and RS-232 connections) and eliminate the need for me to wire the backplane myself.

**Figure F-2** SimmStick backplane with built-in power supply and RS-232 interface

The microcontroller's clock lines are also available on the Bus. Either a Crystal or Ceramic Resonator can be mounted on the SimmStick board with optional capacitors using the built in vias and traces. The XTAL lines can also be distributed using the Bus (a clock is mounted on a single card and distributed to microcontroller SimmStick's on the bus). I would recommend that this line is cut to prevent the extra capacitances of the bus from affecting the operation of the PICmicro® MCU's clock. If the clock output is to be passed to other devices on the SimmStick bus, then I recommend buffering them on the SimmStick card to ensure the PICmicro® MCU's clock is not affected.

The SimmStick schematic for the eighteen pin PICmicro® MCUs is shown in Figure F-3 while its layout is shown in Figure F-4.

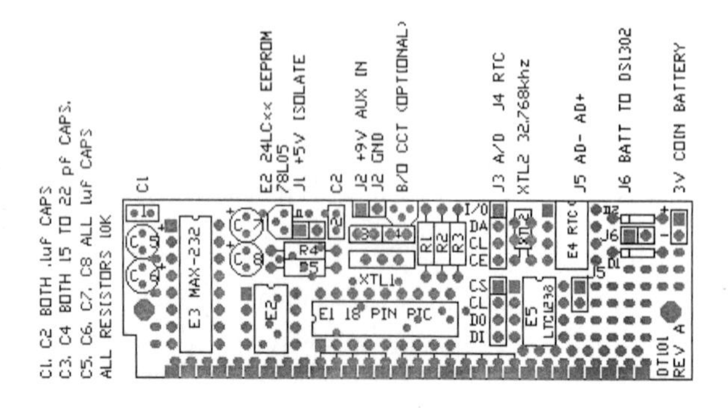


**Figure F-3** Board layout for the SimmStick Dt101 18-pin PICMicro® MCU board.

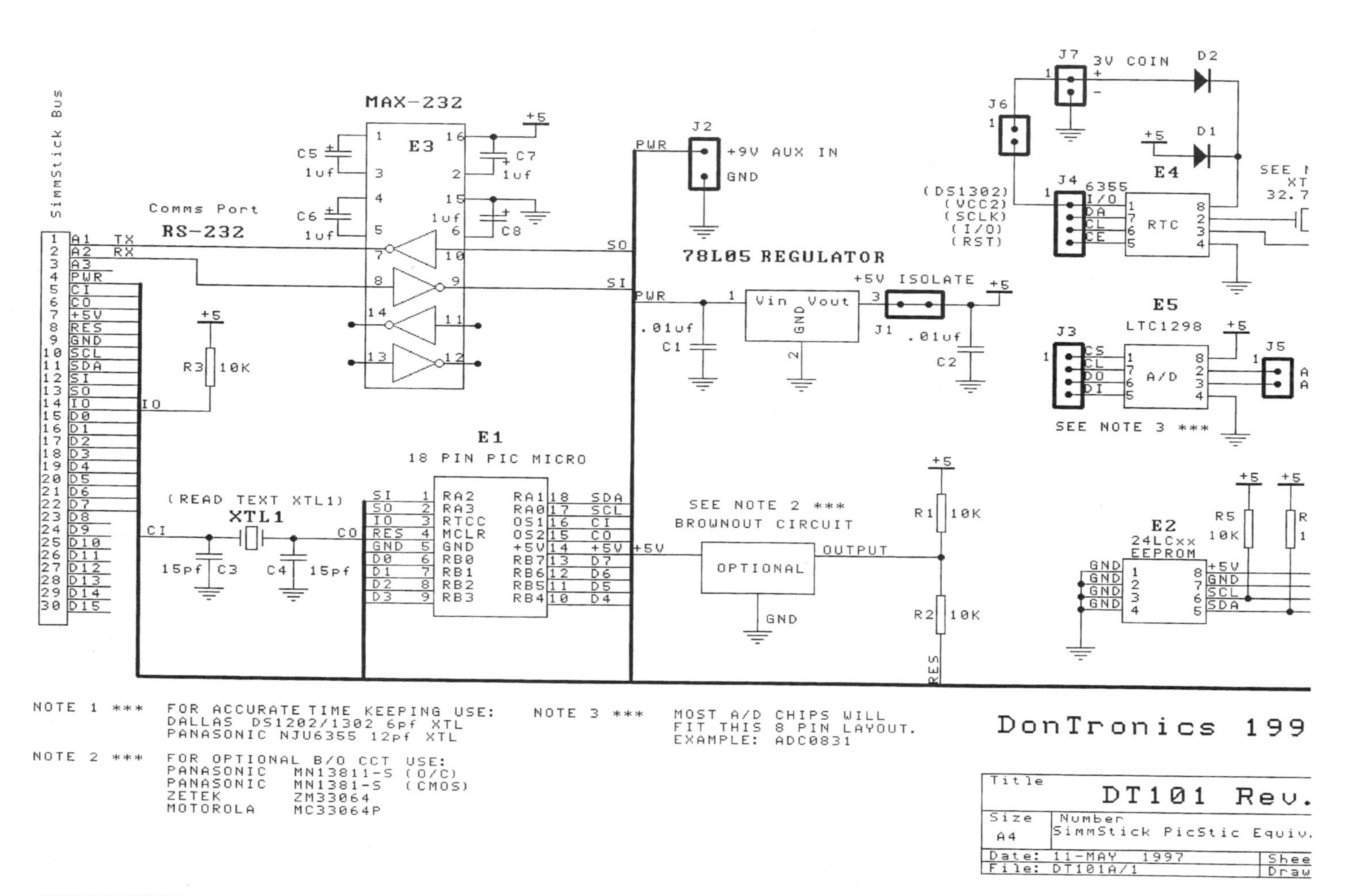


**Figure F-4** Schematic of SimmStick Dt101 18-pin PICMicro board.

**TABLE F-2 SimmStick Backplanes and Development Systems**

| PART NUMBER | FEATURES |
|---|---|
| DT001 | PICmicro® MCU Programmer/SimmStick BackPlane |
| DT003 | Backplane with integrated power and RS-232 |

**TABLE F-3 SimmStick Microcontroller Interface Boards**

| PART NUMBER | FEATURES |
|---|---|
| DT101 | 18 Pin PICmicro® MCU SimmStick board with full RS-232 and I2C/SPI EEPROM access and control |
| DT106 | 28/40 Pin PICmicro® MCU SimmStick board with built in LCD Interface and I2C/SPI EEPROM and RS-232/RS-485 |

**TABLE F-4 Interface SimmStick Boards**

| PART NUMBER | FEATURES |
|---|---|
| DT203 | 24 LED, 4/8 Switch Interface Board |
| DT204 | 4 Slot Expansion Board for SimmStick Bus |
| DT205 | 4 Relay Control Card with 4 LED Outputs |

**TABLE F-5 Wirz SimmStick Boards**

| PART NUMBER | FEATURES |
|---|---|
| Stamp Stick One | Parallax Inc, BASIC Stamp I (TM) |
| Stamp Stick 2SX | Parallax Inc, BASIC Stamp II (TM) |
| PICF876 Stick | PICmicro® MCU 16F876 |
| SX28 Stick | Scenix SX28AC50 |

Microcontroller reset is negative active only. This is not a problem for the PICmicro® MCU but for other microcontrollers (the Intel 8051 specifically), this can be an issue and will require the "Reset Inverter," which consists of an NPN bipolar or N-Channel MOSFET transistor inverter.

The current range of SimmStick products includes bus and development platforms, microcontroller interface boards, and a number of add on interface boards for the bus.

Wirz Electronics has developed a line of assembled, SimmStick compatible Microcontroller Boards. The boards utilize surface mount components leaving a large prototyping area for user circuitry.

The SimmStick can support five of the most popular microcontroller architectures available today (The PICmicro® MCU, Atmel's AVR and 20 Pin 8051, the Zilog Z8 (put in backwards into the Atmel 20 pin SimmStick) and Parallax's BS1 design). This gives you an interesting amount of capabilities because the devices can be interchanged in a single application to evaluate the performance of the different devices. This is why I like to keep the I/O peripherals on a separate SimmStick card.

What I have written here is really an introduction to the SimmStick. I highly recommend you looking at the Dontronics Website (http://www.dontronics.com) to see the options available for the SimmStick as well as go through the documentation to understand how the SimmStick can help you in your next application.

The official SimmStick web page:

**http://www.simmstick.com**

The SimmStick is available from:

**Dontronics**

P.O. Box 595
Tullamarine 3043, Australia
011 +(613) 9338-6286
Fax: 011 +(613) 9338-2935

email: don@dontronics.com

**http://www.dontronics.com**

The SimmStick is available in North America from:

**Wirz Electronics**

P.O. Box 457
Littleton, MA
01460-0457

email: sales@wirz.com

**http://www.wirz.com**

# APPENDIX G

# UMPS

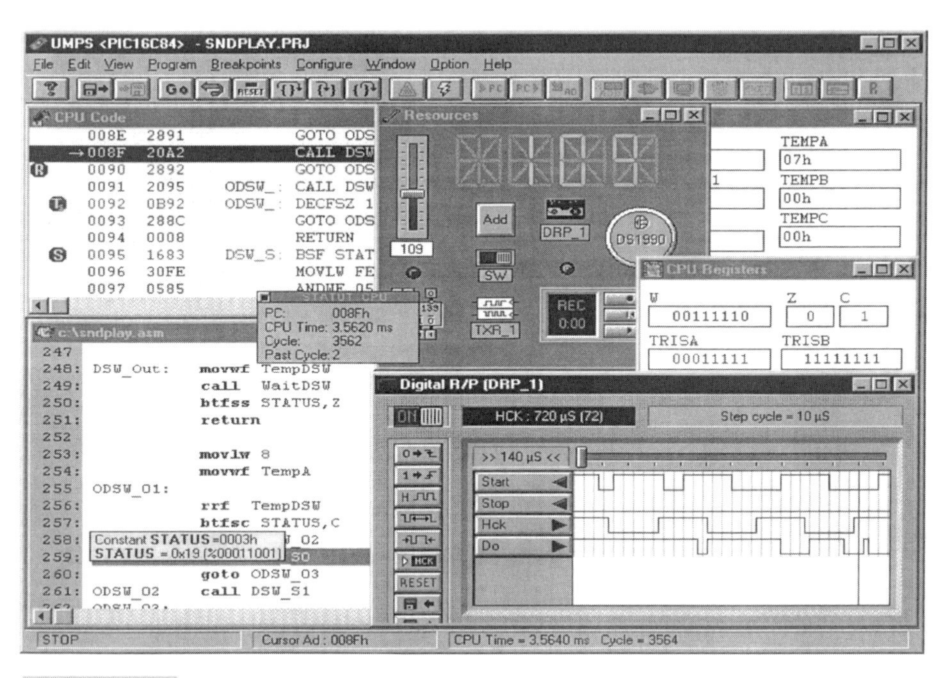


**Figure G-1** Sample UMPS application

I have been very lucky to discover a real treasure in the "UMPS" ("Universal Microprocessor Program Simulator") Integrated Development Environment ("IDE"). This program, which runs under Microsoft "Windows", is able to handle the development for a large number of different microcontrollers (basically everything presented in this book). Along with an editor and assembler, UMPS contains the ability to simulate devices with external hardware that really sets UMPS apart from other tools and eliminates the need for stimulus files. This capability of connecting UMPS to virtual devices gives you a real "what if" capability when you are designing your application.

UMPS runs under Microsoft Windows (3.11, 95, 98, NT and 2000) PC Operating Systems. Along with it's own internal assembler, UMPS can initiate the operation of the Microchip PIC Assembler ("MPASMWIN") as well as the Cosmic Assembler and "C" Compiler and use their symbolic output for the UMPS Simulator. More compilers will be added over time.

The most powerful and useful feature of UMPS is it's graphical "Resources" which can be "wired" to the device. At the time of writing, the following virtual hardware for the simulated microcontroller includes:

- LEDs
- Push Buttons
- Logic Functions (AND, OR, XOR, NOT)
- 7 Segment LEDs
- Square Wave Generator
- Digital Recorder/Player
- D/A Converter
- A/D Converter
- A/D "Slider"
- Serial Receivers/Transmitters
- I2C Memory (RAM & EEPROM)
- I2C Peripherals (LED Displays, Real Time Clocks, etc.)
- HD44780 Compatible LCDs
- Phase Locked Loops
- PWM Monitors
- PWM Generators
- Pull Up Resistors
- Serial A/D Converters
- 74LS138, 74LS139, 74LS373, 74LS374 TTL Chip Simulations
- CD4017 and CD4094 CMOS Chip Simulations
- Dallas Semiconductor DS1820 1-Wire Thermometer
- PWM Sounds
- PC Parallel Port

More simulations will be added as time goes on, but you should see that there really is a critical mass of different devices available that should make simulating many different applications possible. These simulated devices (which are known as "resources") are very full featured—I was able to simulate all of the "Experiments" and many aspects of the example applications using UMPs. When setting up the resources, you can move them around the resource window to arrange them as they make sense to you or how they will be laid out in the final application. You can also modify many of the I/O devices with different options (for example, the LCD display can be one or two rows and eight to twenty columns or along with the pull-up resistor, the I/O pins can be hard wired to Vcc or Ground). In the package, there is a simple tutorial showing how an application is developed under UMPS.

UMPS currently supports the following devices (with more being added all the time):

Microchip

- PIC12C5xx
- PIC16C5x
- PIC16C505
- PIC16C84, PIC16F84, PIC16F83, PIC16C554, PIC16C556, PIC16C558
- PIC16C671, PIC16C672
- PIC16C71, PIC16C710, PIC16C711

Dallas Semiconductor

- DS87C310, DS87C320

Intel

- 8031, 8032
- 8051

Atmel

- AT89C1051, AT89C2051, AT89S8252, AT89S53
- AT90C4434, AT90S8535
- ATMega603

Motorola

- 68HC705, 68HC705J1A, 68HC705P9, 68HC705B16, 68HC705B32
- 68HC11

SGS Thomson

- ST6210, ST6215, ST6220, ST6225
- ST6252, ST6253
- ST6260, ST6262, ST6263, ST6265

National Semiconductor

- COP820C

Samsung

- KS57C0408

Sunplus

The Assembler and Simulator Available in UMPS are very fast and I saw 50,000 instruction cycles per second on my 133 MHz Pentium PC. Compiler interfaces are also available.

The copy of "UMPS" on the CD-ROM included with this book is the latest demo version. This is a full version of UMPS, that will run for three months after being installed and there are a number of restrictions you should check in the "readme.txt" file in the "UMPS" directory of the CD-ROM.

Additional features included in this version of UMPs includes:

1. "Back Trace" as you single step, you can go back up to 16 instructions to see what happened.
2. A register can be incremented or decremented by selecting it and using the "+" or "−" keys.
3. In the UMPS assembler source, a binary word can be written in the following formats:
   - %010010110
   - 10010110B
   - B'10010110'
4. The ability to interface with hardware emulators.

To find out more about UMPS contact:

**Virtual Micro Design**
I.D.L.S
Technopole Izarbel
64210 Bidart
France

011-33(0)559.438.458
Fax: 011-33(0)559.438.401

email: p.techer@idls.izarbel.tm.fr

**http://idls.izarbel.tm.fr/entp/techer/index.htm**

UMPS is available in North America from:

**Wirz Electronics**
P.O. Box 457
Littleton, MA
01460-0457

email: sales@wirz.com

**http://www.wirz.com**

APPENDIX H

# GPASM/GPSIM LINUX PICmicro® MCU APPLICATION TOOLS

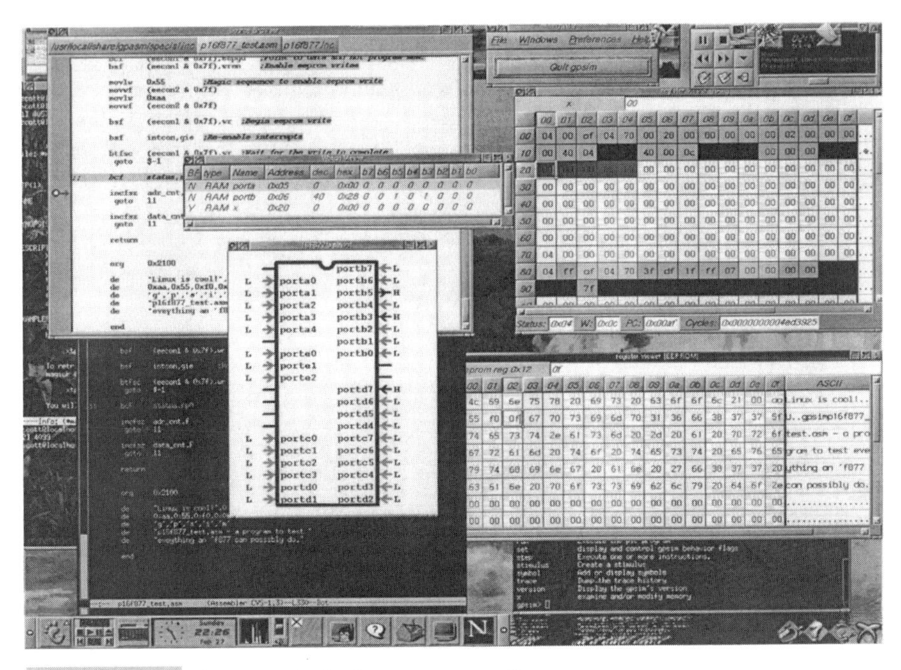


Figure H-1 Screen shot of gpsim operating

One of the fastest growing trends in the PC marketplace is "Linux" and the "Open Source" software concept. This movement has resulted in the availability of some remarkable free software, such as the "gpasm" and "gpsim" development tools that are available for the PICmicro. These "General Public License" ("GPL") tools are an MPASM compatible assembler and GUI simulator.

Scott Dattalo has provided a central repository for these tools and the versions that are included on the CD-ROM. The tools I have provided work very well and are being continually improved. If you are a Linux user, I think you will be very pleased at these GPL tools. The latest copies of gpasm and gpsim can be found on "SourceForge":

**https://sourceforge.net/project/?group_id=2298**

**https://sourceforge.net/project/?group_id=2341**

respectively. Additional information may be available on Scott's web page:

**http://www.dattalo.com**

The gpasm and gpsim tools themselves support a wide variety of PICmicro® MCUs. gpasm supports all the PICmicro® MCUs used in this book except for the PIC17Cxx devices. gpsim supports a number of the low-end, mid-range and PIC18Cxx PICmicros with the list being continually updated (along with the peripheral functions that are supported). James Bowman wrote the initial version of gpasm and the version on the CD-ROM that comes with this book is the result of his and several other people's contributions. The latest version of the gpasm assembler can be found in the "gpasm" subdirectory of the CD-ROM that comes with this book. This assembler is MPASM compatible, from the perspective of input and output. To "detar" the assembler, copy "gpasm-0.8.9.tar.gz" and enter:

```
tar -xvzf gpasm-0.8.9.tar.gz
```

This will create the subdirectory "gpasm-0.8.9" and will have all the archived files unloaded into it. Once this is done, run the Linux "make" utility to run to convert the assembler source code into an executable file. This is accomplished by entering the statements:

```
cd gpasm-0.8.9
make
su root
make install
```

For different versions of gpasm, the subdirectory change instruction ("cd") given above will change to reflect the version number.

gpasm is a command line tool and is invoked in the format:

```
gpasm [options] filename
```

where the options are:

gpasm will output a standard ".cod" file that can be used by MPLAB or gpsim. gpasm does not produce object files and there is no linking capability. gpasm uses the MPLAB ".inc" files for defining PICmicro® MCU register addresses.

gpsim is a full-featured PICmicro® MCU simulator created by Scott Dattalo that works with the low-end, mid-range and PIC18Cxx processors. A sample screen shot of its operation is shown in Figure H-1. gpsim is the fastest simulator available; when running straight simulation code (no stimulus or peripherals active), it can simulate the operation of a 20 MHz PICmicro® MCU in real time on a 400 MHz PII or PIII equipped PC.

The PICmicro® MCUs that gpsim can simulate include:

PIC12C508
PIC12C509
PIC16C84
PIC16CR83
PIC16F83
PIC16CR84
PIC16F84
PIC16F877—Not all Peripheral Functions are Supported
PIC16C71
PIC16C74—Not all Peripheral Functions are Supported
PIC16C61
PIC16C64—Not all Peripheral Functions are Supported
PIC16C65—Not all Peripheral Functions are Supported
PIC18C452—Not all Peripheral Functions are Supported

**TABLE H-1 GPASM Options**

| OPTION | DESCRIPTION |
|---|---|
| –a Format | Specify the output hex file format: inhx8m, inhx8s, inhx32 |
| –c | Source labels are case sensitive |
| –d Symbol = Value | Define a Symbol with a Value String |
| –h | Display the help information |
| –i | Specify a new include subdirectory |
| –l | List supported PICmicro® MCU processors |
| –m | Provide a Memory Dump |
| –o Filename | Specify an alternative hex output file name |
| –p PIC1xC\|Fxx | Specify the PICmicro the application is to be created for |
| –q | Quiet |
| –r | Specify the default radix: HEX, DEC, BIN |
| –v | output the gpasm version number |

The ADC of the PICmicros is supported along with the CCP modules and asynchronous features of the USART. The PSP and synchronous serial communications is not supported. Stimulus files are supported in gpsim, but its format is not compatible with MPLAB's. This is something that will probably be rectified in the future.

Included on the CD-ROM that comes with this book, in the "gpsim" subdirectory are the files:

gpsim-0.18.0.tar.gz
gpsim.ps
gpsim-0.18.0-0.i386.rpm

The ".rpm" file is a Linux "Redhat Package Manager" and allows you to add the package without having to "make" the source code. Use a similar "tar" statement as gpasm to "detar" gpsim and follow the instructions in the "readme.txt" to install the simulator.

To execute gpsim, use the format:

gpsim [options] hexfile [-c startup]

Where the options are shown in Table H-2:
The "base" gpsim operation consists of booting the "Control Window", from which the other windows, register viewer, object code browser, .asm source browser, Watch viewer and Breadboard viewer, can be initiated. In Figure H-2, the .asm source browser, register viewer, Watch viewer and Breadboard viewer are active. The "Bread board viewer" is used to monitor the operation of the I/O pins on the PICmicro® MCU.

gpsim has a number of features that are not available in MPLAB. gpsim has an execution trace buffer that is active during each debugging session. This functional allows you to go back and look at how the application executed. Trace buffers are an invaluable tool for debugging as they allow you to see what has happened. This eliminates the need to rerun the simulator repeatedly to try and find where the error is taking place. Normally they are available on emulators, but their inclusion here does make debugging a lot easier.

Another nice feature of gpsim is the ability to break on a register reading or writing a specific value to a register. This allows you to set more meaningful breakpoints to specific addresses, rather than repeatedly stop at the read/write instruction address and compare it for a value that indicates the problem area of the application has been reached.

**TABLE H-2 GPSIM Options**

| OPTION | DESCRIPTION |
|---|---|
| –c stpfile | Specify a Start Up Configuration |
| –h | Display the help information |
| –p PIC1xCIFxx | Specify the PICmicro® MCU to be Simulated |
| –s codfile | Specify an alternative .cod file |
| –v | output the gpasm version number |

An exciting feature that is being developed as this book was going to press is the LCD viewer and other peripheral viewers. In Figure H-2, the "LCD" window can be seen with sample characters on it. This feature gives gpsim the UMPS like capability of monitoring the execution of hardware during its simulation and adds a lot to understanding how the application works (and where are any potential problems). In the future, there will be additional devices simulated for gpsim to further help with developing and debugging applications.

I've been an admirer of the open source software movement for some time now. The range and quality of software that has become available (including the "Linux" operating system, various development tools and applications) is staggering and improves every day. The energy and work put into the code is amazing, especially since nobody is getting paid for it. gpasm and gpsim are examples of the open source concept at its very best.

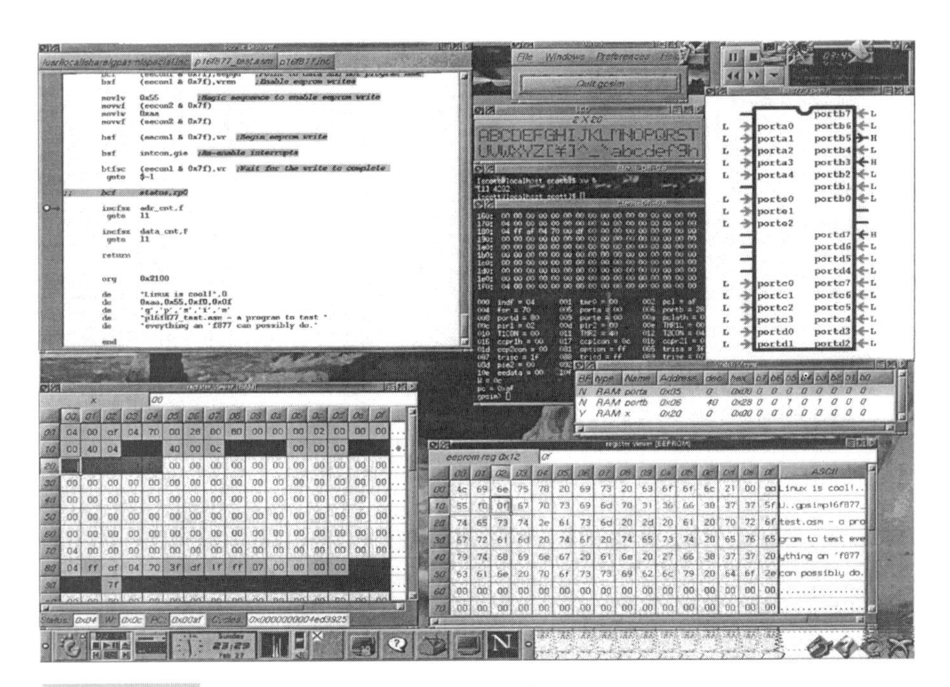


**Figure H-2** Other gpsim features operating

APPENDIX I

# PCB BOARDS

Along with the physical "El Cheapo" PCB, that is included with this book, I have included the "Gerber", "drill", "overlay" and "tool" files for the projects presented in this book which use a PCB. These files are located on the CD-ROM, in the "PCB" directory and can be accessed directly from the CD-ROM's HTML interface.

The following designs are included on the two PCB "panels". The "Board1" Panel 1 has:

- El Cheapo
- Yap-II
- +5 volt power supply
- Xmas lights application

The "Overlay" (or "silkscreen"), "top side" and "bottom side" copper designs are shown in Figure I-1, Figure I-2 and Figure I-3 respectively.

The "Board2" Panel has the following projects:

- Emu-II
- PIC17Cxx memory interface demo

The Overlay design is shown in Figure I-4, while the top side copper design can be found in Figure I-5 and the bottom side copper design is shown in Figure I-6.

These PCB's can be built at quick turn PCB manufacturers or assembled at home. If you assemble them at home, note that the boards are designed to be two sided, with "plated through" vias. If you do build the boards at home, I recommend that you solder wires

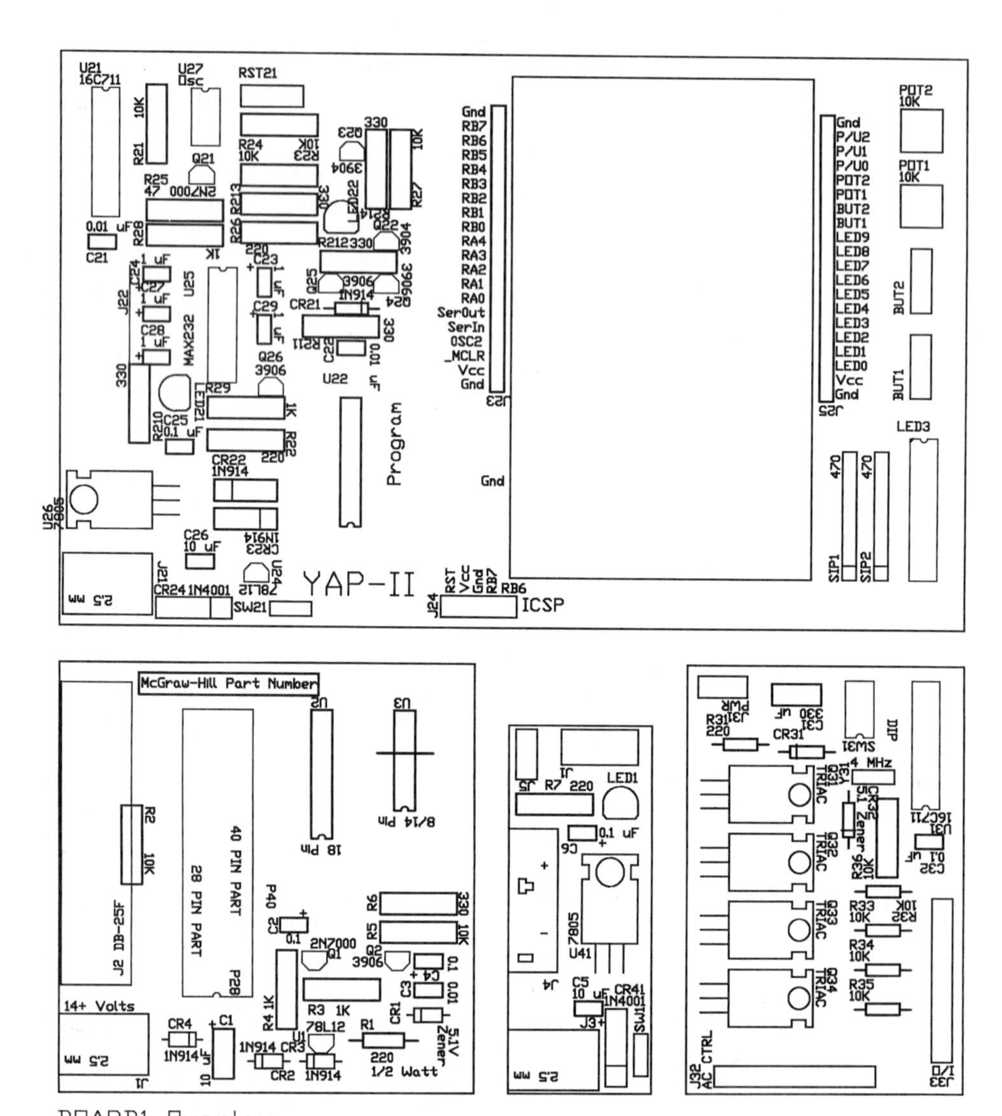

**Figure I-1**

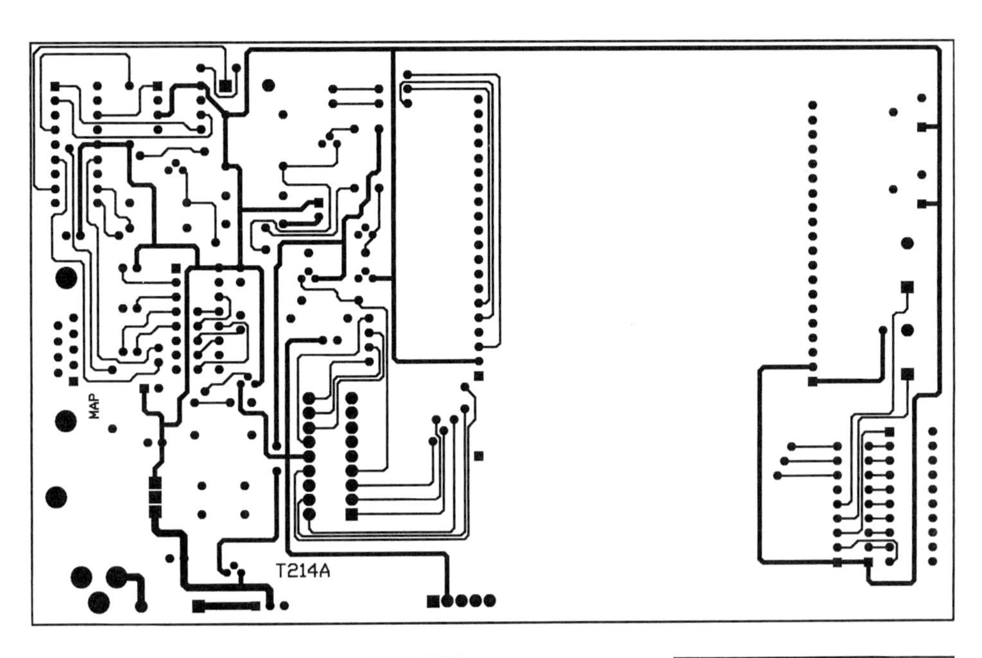


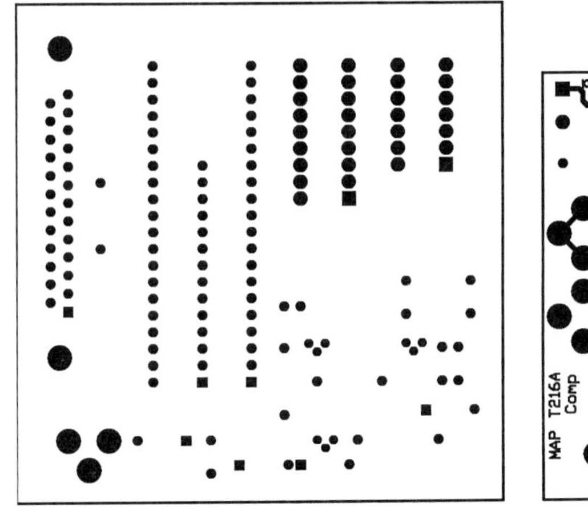

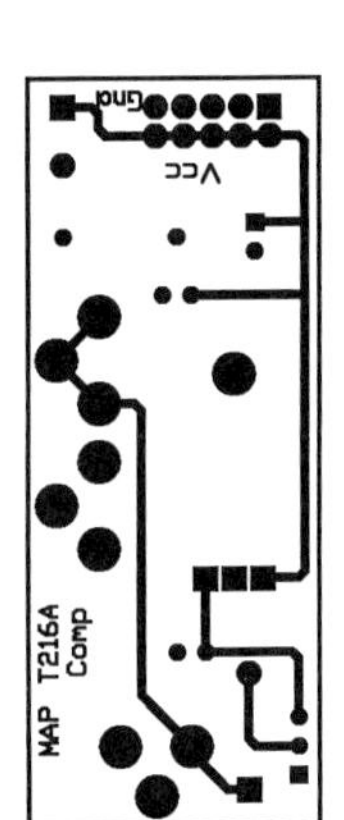


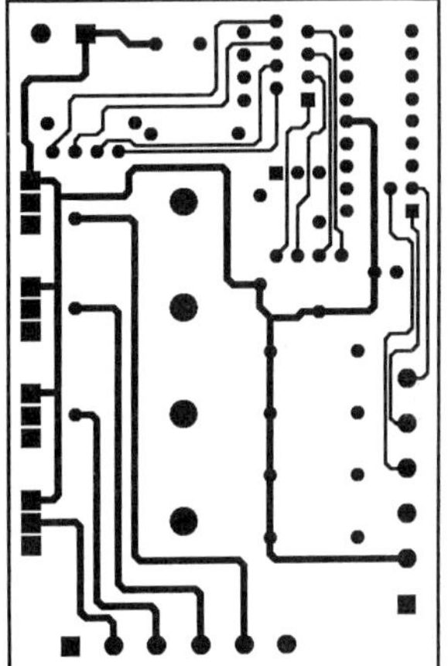

BOARD1 Top Layer

**Figure I-2**

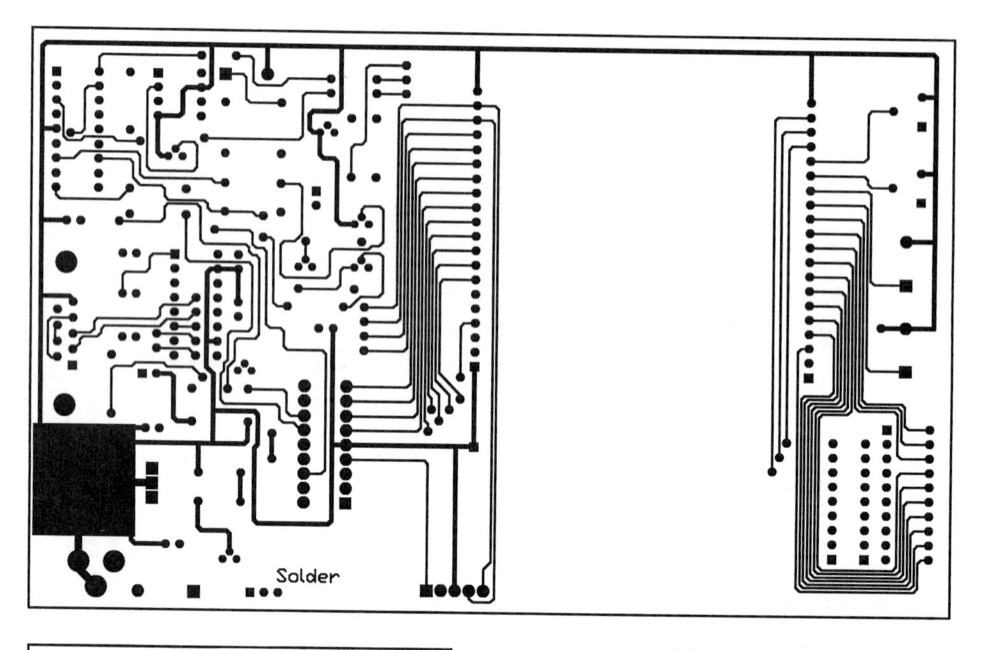


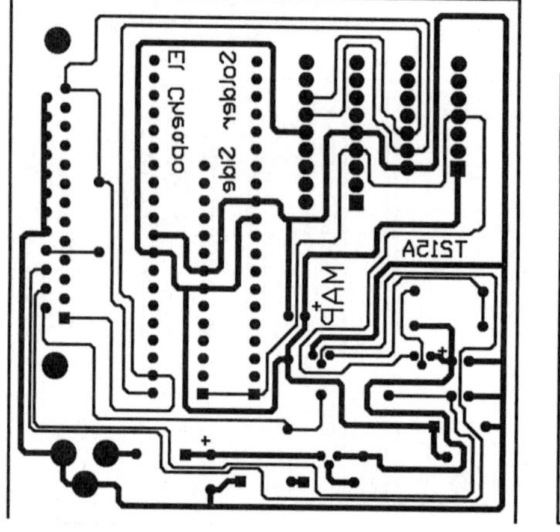

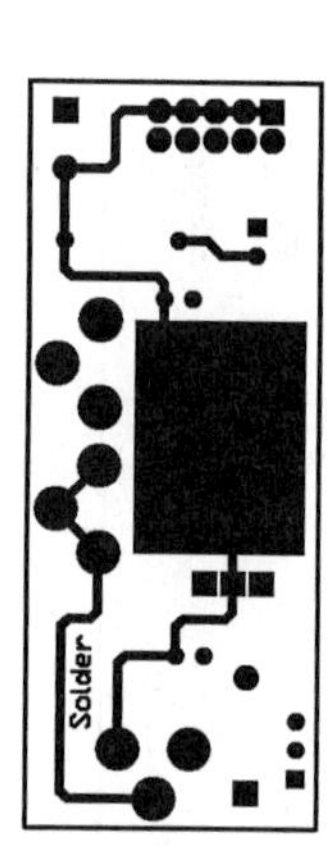


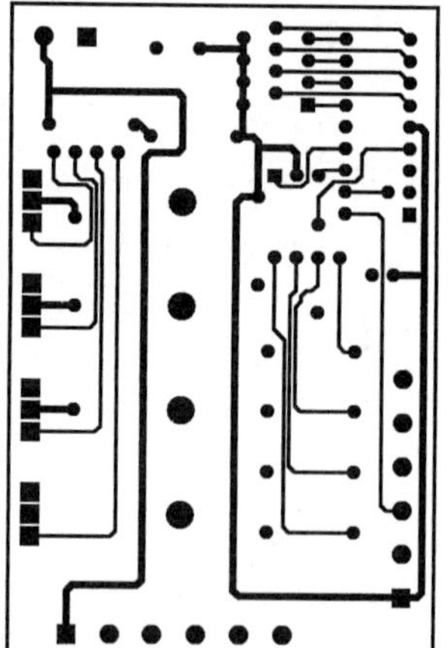

BOARD1 Bottom Layer

**Figure I-3**

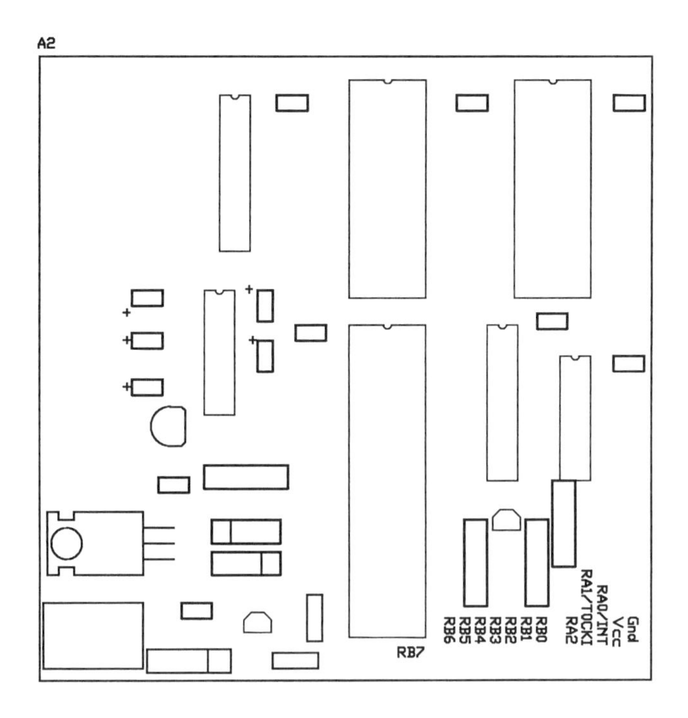


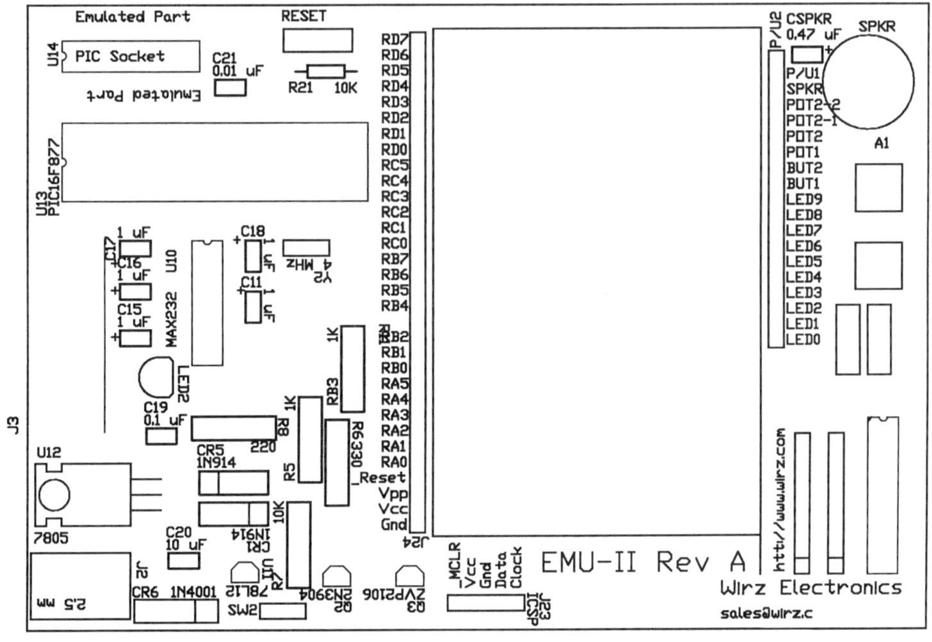


BOARD2 Overlay

**Figure I-4**

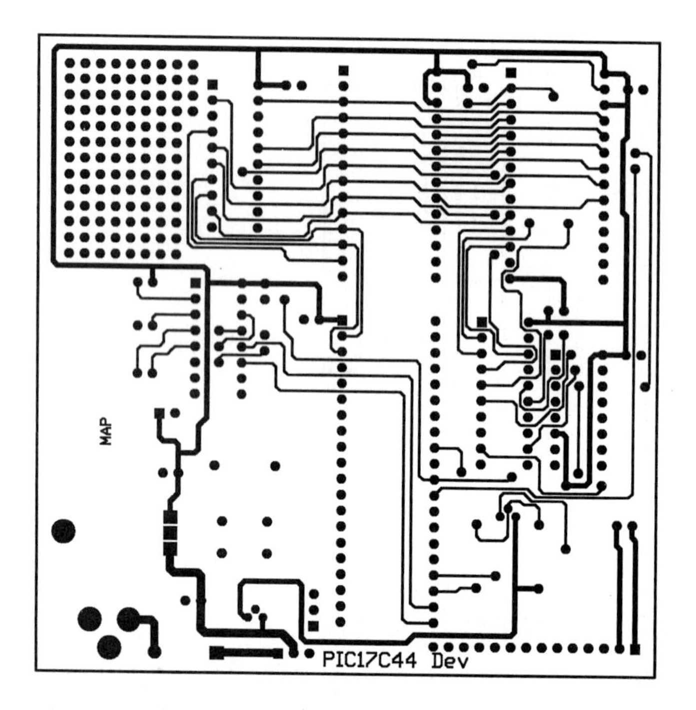


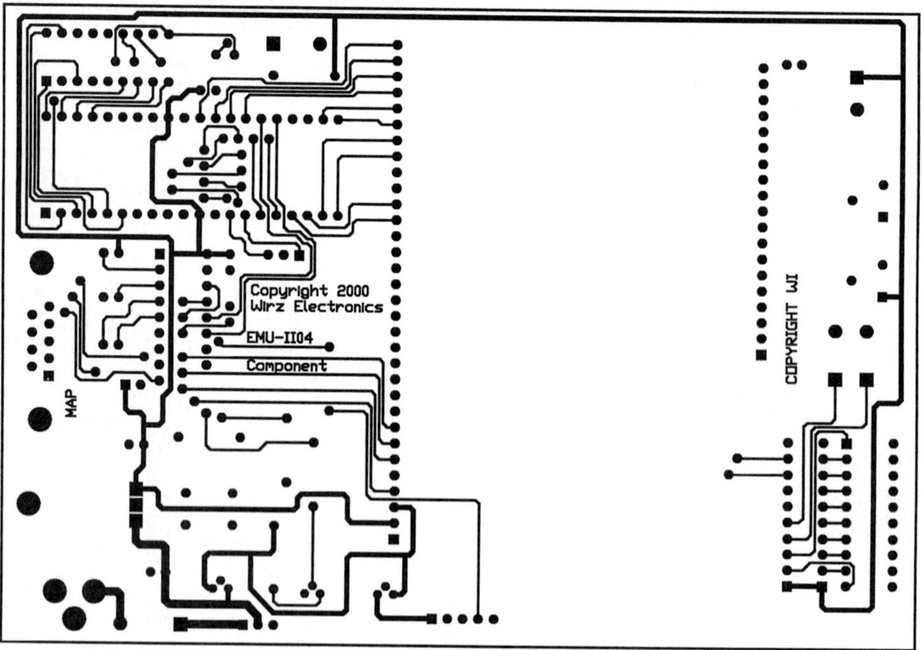


BOARD2 Top Layer

**Figure I-5**

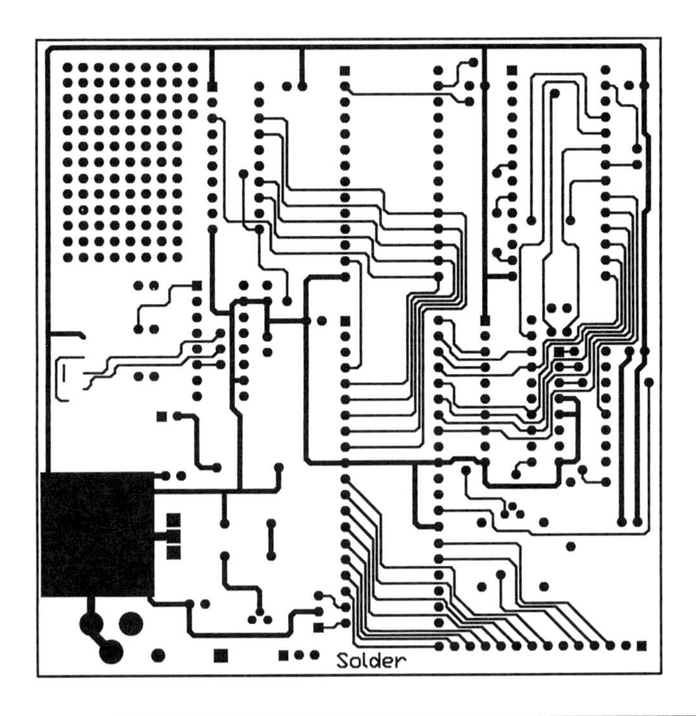


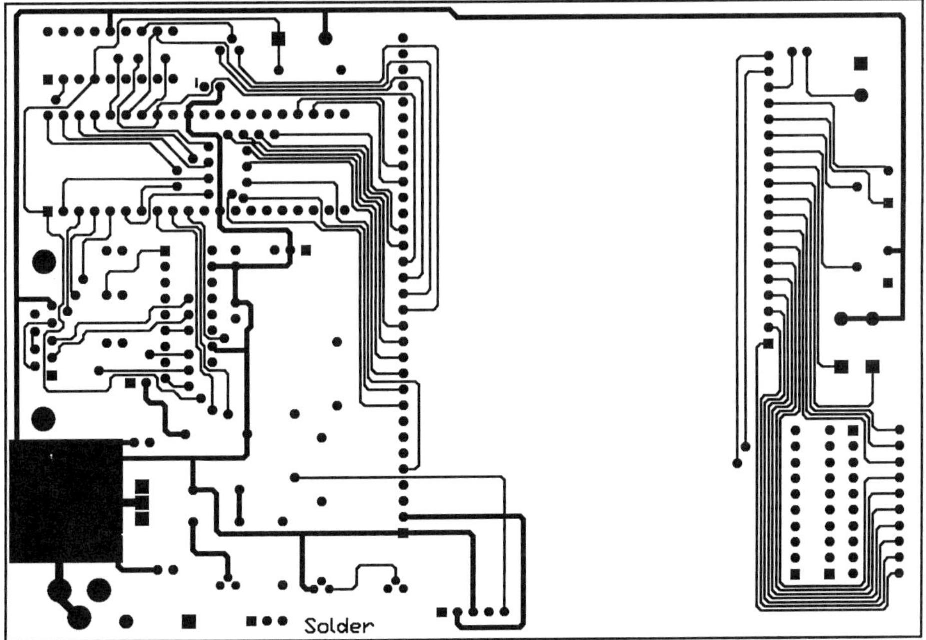


BOARD2 Bottom Layer

**Figure I-6**

through the topside/bottom side via connections and make sure the Pin-Through-Hole pins are soldered on *both* the top and bottom sides.

I am making these PCB designs available to you for your own personal use. By providing this information to you I am not giving you permission to sell the designs or redistribute them in any way. These designs are to be used by the purchaser of this book only. As well, these boards are not guaranteed in any way; while they are the same files I used for the projects in this book, I have not carried out any kind of signals analysis on the board or "rule check" to make sure that no signal parmeters are violated.

APPENDIX J

# CD-ROM

## CONTENTS AT A GLANCE



When I created the diskette that was shipped with the first edition of this book, I ended up scrambling twice to get the diskette together. The first time was to eliminate every unneeded byte and converting strings of blanks into tab sequences to save as much space as possible. This was repeated when I found out that just before the book was going into print that Microchip had "obsoleted" the 16C84 and I had to change a lot of the source to make sure it was compatible with the 16F84. If you have a copy of the first edition with a diskette, if you do a "dir" on the diskette, you will discover that there are "0" bytes left—I had to be really creative to get everything I wanted onto the diskette.

In later printings, I was able to include a CD-ROM with the book that included all the same files as well as Microchip PICmicro data sheets. In this book, I have eclipsed that with providing a much better interface for you, utilities for installing files on your PC's hard drive as well as a number of different tools you can sample to decide what is best for you.

I recommended at the start of the book that you install the CD-ROM's files on your PC's hardfile and get yourself familiarized with the HTML interface and what is available on the CD-ROM before you begin reading the book. By doing this, you will be able to easily find and browse the source code and Microchip data sheets as you go through the projects and applications presented in this book. In preparation to doing this, make sure you have downloaded the latest version of Adobes "Acrobat" and assigned ".asm" files to an editor (I recommend "WordPad" for Microsoft Windows systems). If you have any trouble doing this, there is additional information on the CD-ROM or through your systems "help" utility.

# Accessing the CD-ROM Files

When you are ready to start working through the experiments and applications I have presented, you will have to copy the files from the CD-ROM that comes with this book. The CD-ROM includes all the files required to create the experiments and projects I have presented. Along with these files, I have provided MS-DOS Microsoft Windows (95/98 NT/2000) and Linux tools for you to assemble ("build"), simulate and program PICmicro's. In terms of programming the "EL Cheapo" printed circuit board ("PCB") that is provided with this book, MS-DOS, Windows and Linux development tools are included on the CD-ROM as well.

The basic interface used on the CD-ROM is "hypertext markup language" ("HTML") pages, which you are probably familiar with this as the basic method of interfacing to the Internet. I have used this method because "browsers" are available free of charge for all PC and workstation systems, HTML can access web pages directly and HTML can issue MS-DOS and Windows commands directly by pointing and clicking. You will find loading applications to be very quick and easy from the CD-ROM's HTML interface, the first page of which is shown in Figure J-1.

The interface itself is based upon the hierarchical model shown in Figure J-2. Starting from the first page, you can jump to any other page on the CD-ROM on the Internet in one or two mouse clicks.

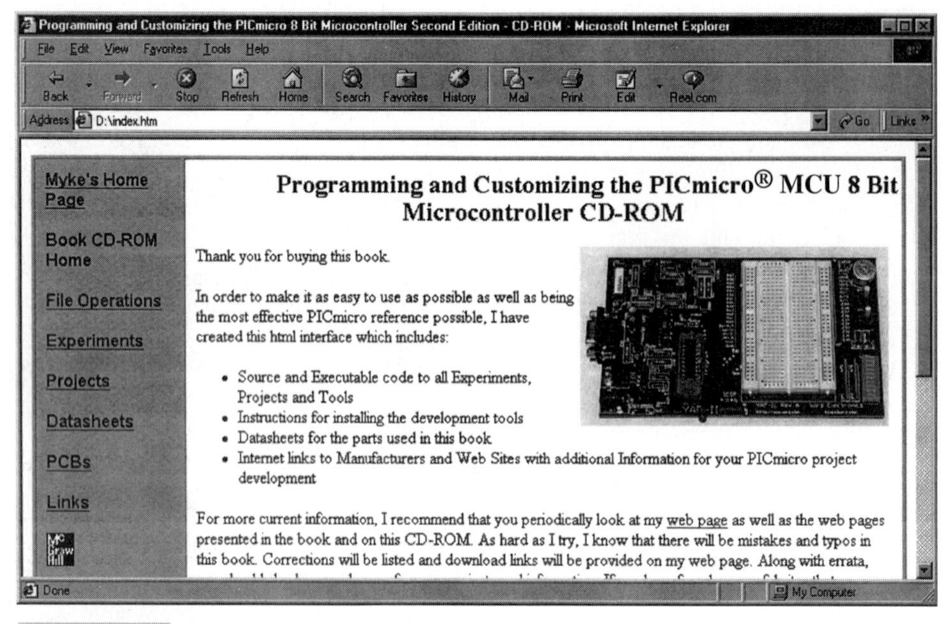


**Figure J-1** CD Rom HTML interface

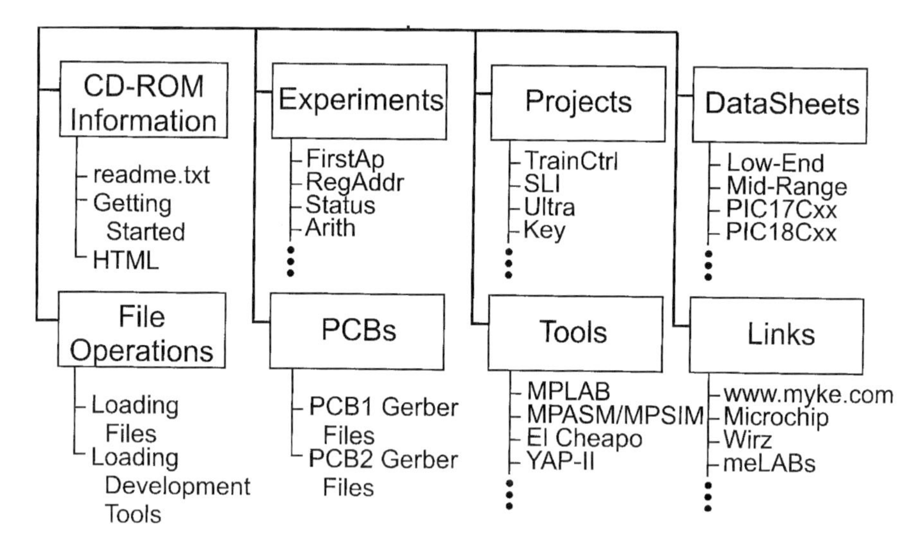


**Figure J-2** CD-ROM file architecture

I suggest that you spend some time going through the CD-ROM's HTML interface to familiarize yourself with what is available and how the source code and tools are loaded onto your PC's hard drive.

If you have Microsoft Windows 95/98/NT/2000 PC, the HTML interface should come up in a web browser immediately. If it does not, or if you are running Linux, or another operating system, then you should follow the steps listed below to access the HTML.

First you should initiate your system's web "browser" (see Figure J-3). The currently

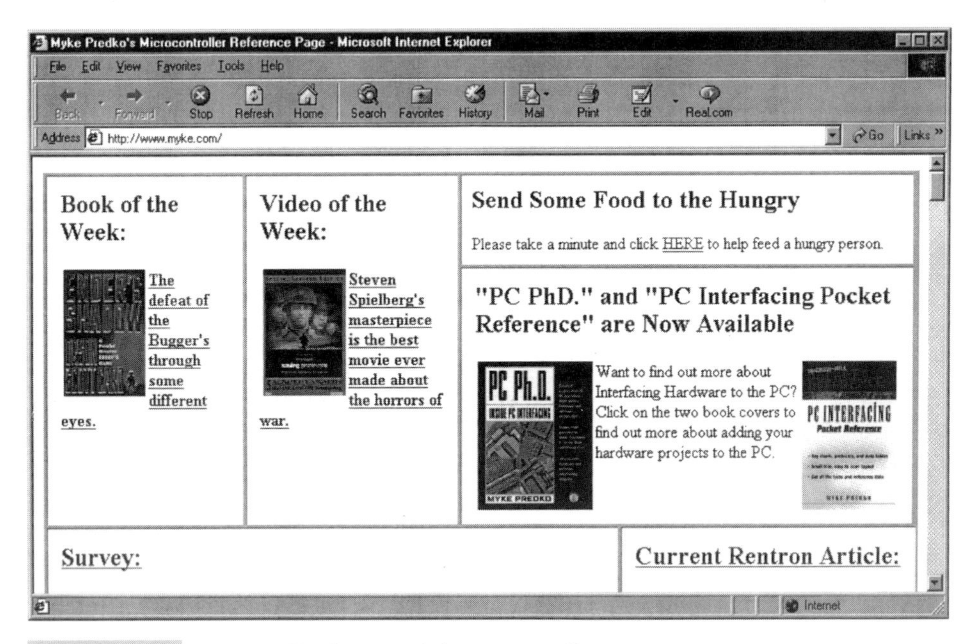


**Figure J-3** Internet Explorer web browser active

most popular browsers are Microsoft "Explorer" and Netscape "Navigator." To show how it can be done manually, I will be demonstrating how to load the HTML interface from "Internet Explorer" (Version 5.00).

This process is essentially the same for other web browsers and is simply finding the introductory HTML file on the CD-ROM.

Once you have your web browser operating, click on "File" and then "Open" (as shown in Figure J-4). From this, you will be given a file list similar to the one shown in Figure J-5 and "go back" until you are at "My Computer" and select the CD-ROM drive (Disk "D") on my PC. There should be a "McGraw-Hill" bitmap to indicate the CD-ROM has been read.

From here, select "index.htm", as shown in Figure J-6. Once you have clicked on the "index.htm" file you should get the CD-ROM's interface, shown in Figure J-1. Once you have this page available, you are ready to start working the CD-ROM. Files can be "down loaded" or executed with just a mouse click.

The tools and files on the CD-ROM are the latest available at the time of the *first* printing of this book. This is not to say that the CD-ROM in your copy of the book has not been re-mastered with later information after the first printing, but chances are, the CD-ROM in your copy of this book will have the tool set available in August of 2000. Please check my web site as well as the home pages of the tool providers on the CD-ROM for the latest versions of their code.

# Rentron Articles

In 1999, I started writing some articles for Bruce Reynolds' web site. These articles were a way for me to document ideas and circuits before they were published in this book. I have included links to these articles on the CD-ROM.

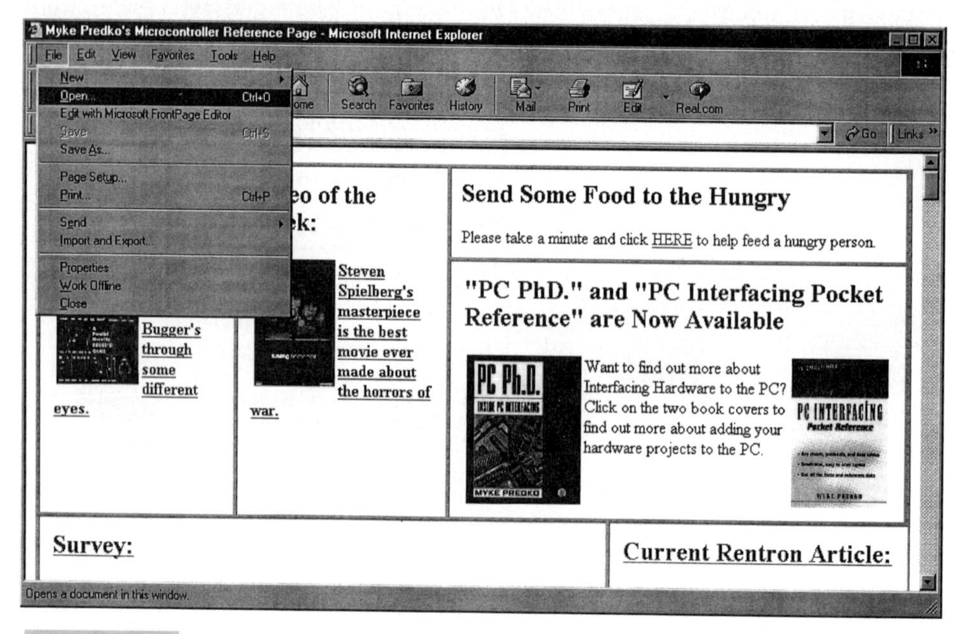


**Figure J-4** Internet Explorer file Open

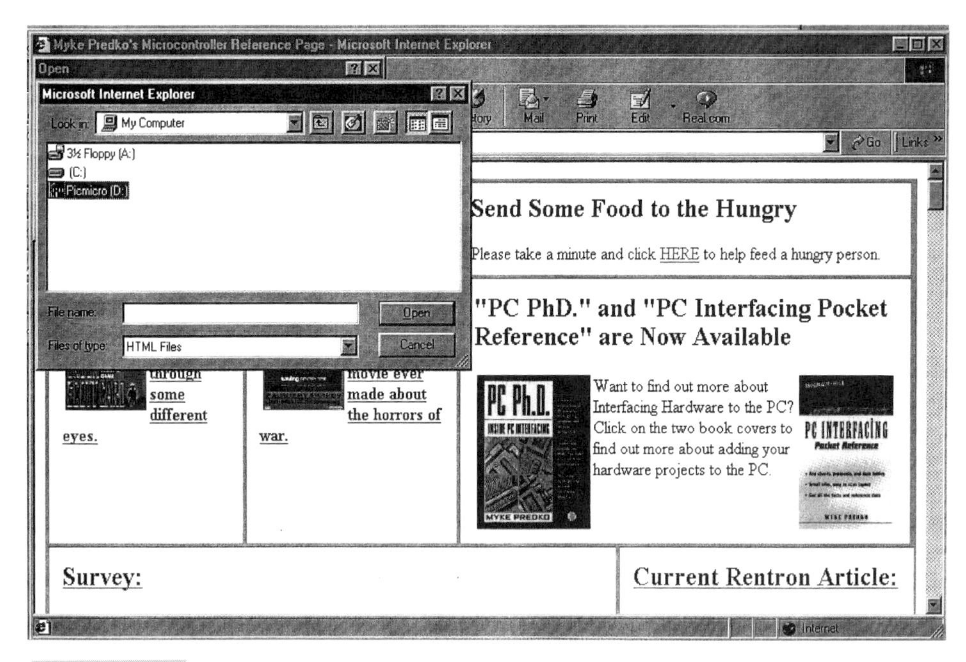


Figure J-5 Selecting the CD-ROM

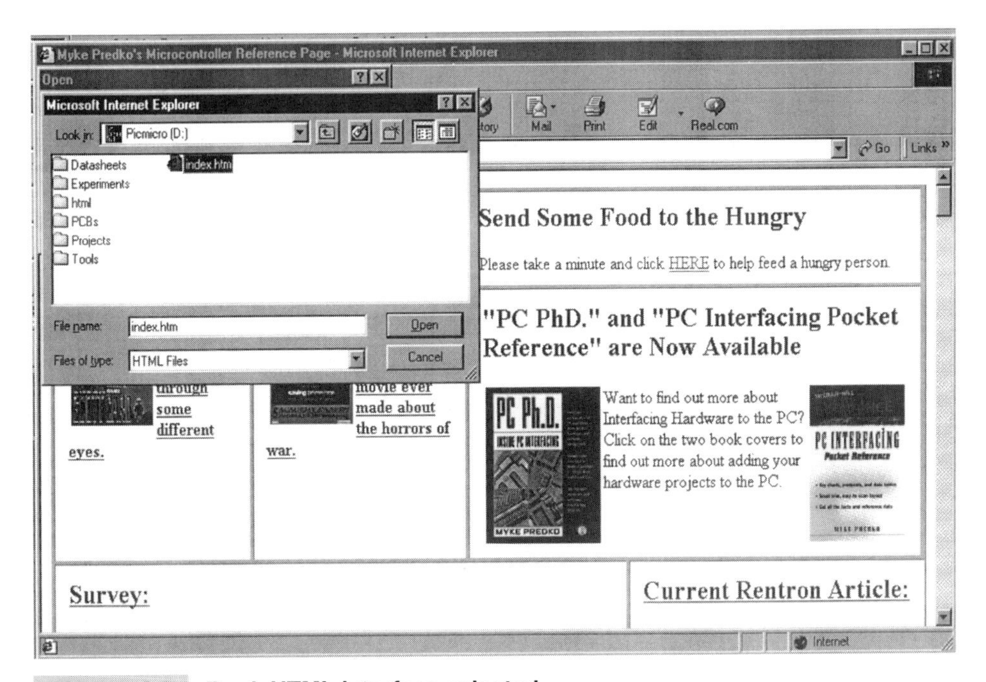


Figure J-6 Book HTML interface selected

From these articles, you will see the origin of the "El Cheapo", although not the first four iterations of it. Writing these articles has allowed me to try out ideas (like the "El Cheapo" and some of the train ideas) and refine them to the point of turning them into applications.

In many ways, the articles are not very polished, but they do have ideas which are not quite ready for prime time, but I did have a lot of fun writing them and finding out what people thought.

# Microchip Datasheets

When I was in university, I got some advice from a professor that has served me very well over the years. His advice was to only use electronic chips and products that are well documented and the company responds to queries from anyone, not just customers with multi-million dollar orders. This statement was made in the early 1980's when the Internet was still being called "ARPANET" (and was mostly used to link universities, colleges and military installations with an email and data file transfer capability) and faxes were still a few years away from general use.

Today, with the Internet and file formats like Adobe's "acrobat" providing good quality data sheets cheaply to potential customers very easily. As well, email and other Internet tools (such as list servers) responding to customers queries very easy and also distributes the information to a wider audience. Now, many companies are able to provide support at a level that was almost impossible twenty years ago.

Even though the Internet advances has resulted in a "leveling of the playing field", there are still some companies that continually "up the ante" and provide superior information and support to their customers. One of the best companies at supporting their customers is Microchip, which does go the extra mile in answering their customers and the people who develop their products have all the information, tools and parts needed to be successful.

On the CD-ROM, I have included the Microchip Adobe Acrobat ("PDF") data sheets of all the parts presented in this book along with data sheets on in-circuit programming, MPLAB and MPASM operation on the CD-ROM. These data sheets are excellent resources and will help to fill in the holes in the material I undoubtedly have in this book.

Microchip continues to update existing data sheets as well as introduce new products. In the three years since the first edition of this book has been available, over 120 new PICmicro part numbers have become available. I have no reason to believe a similar number of new devices will not be released in the next three years. For this reason, I suggest that the data sheets on the books CD-ROM be used as a starting place. Check Microchip's Internet web site or your Field Application Engineer ("FAE") for the latest data sheets and tools that you will help you with your applications.

# INDEX



**Boldface** numbers indicate illustrations; italic *t* indicates a table.